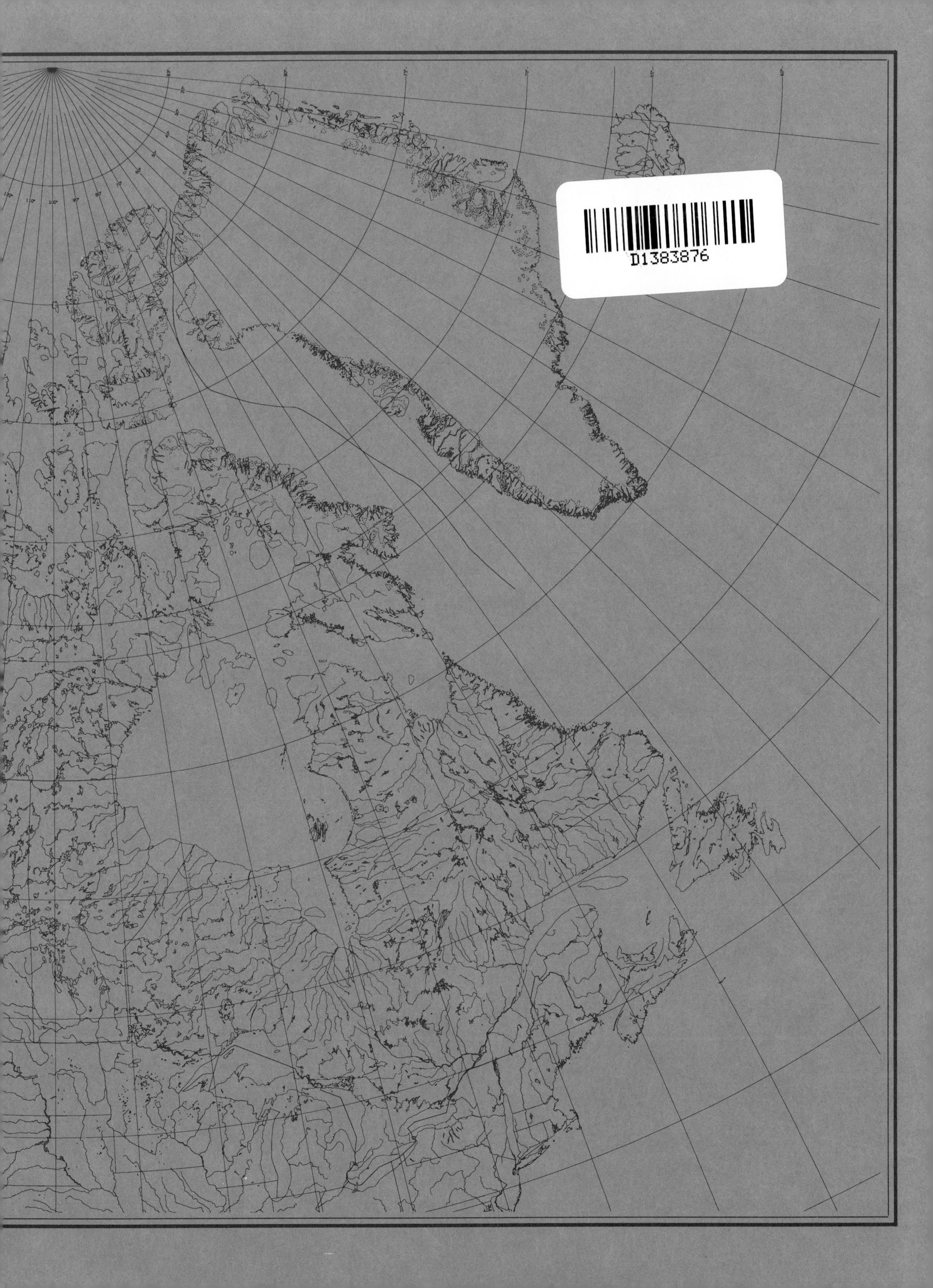

The Heritage of Engineering Geology; The First Hundred Years

William Otis Crosby, 1850–1925. The “Father of Engineering Geology in North America” and an active practitioner from 1883 to 1925.

Centennial Special Volume 3

The Heritage of Engineering Geology; The First Hundred Years

Edited by

George A. Kiersch
Professor Emeritus, Geological Sciences
Cornell University
Ithaca, New York 14853-1504
and
Kiersch Associates, Inc.
4750 N. Camino Luz
Tucson, Arizona 85718

1991

Acknowledgment

Publication of this volume, one of the synthesis volumes of *The Decade of North American Geology Project* series, has been made possible by members and friends of the Geological Society of America, corporations, and government agencies through contributions to the Decade of North American Geology fund of the Geological Society of America Foundation.

Following is a list of individuals, corporations, and government agencies giving and/or pledging more than $50,000 in support of the DNAG Project:

Amoco Production Company
ARCO Exploration Company
Chevron Corporation
Cities Service Oil and Gas Company
Diamond Shamrock Exploration Corporation
Exxon Production Research Company
Getty Oil Company
Gulf Oil Exploration and Production Company
Paul V. Hoovler
Kennecott Minerals Company
Kerr McGee Corporation
Marathon Oil Company
Maxus Energy Corporation
McMoRan Oil and Gas Company
Mobil Oil Corporation
Occidental Petroleum Corporation
Pennzoil Exploration and Production Company
Phillips Petroleum Company
Shell Oil Company
Caswell Silver
Standard Oil Production Company
Oryx Energy Company (formerly Sun Exploration and Production Company)
Superior Oil Company
Tenneco Oil Company
Texaco, Inc.
Union Oil Company of California
Union Pacific Corporation and its operating companies:
- Union Pacific Resources Company
- Union Pacific Railroad Company
- Upland Industries Corporation

U.S. Department of Energy

Published by the Geological Society of America, Inc.
3300 Penrose Place, P.O. Box 9140, Boulder, Colorado 80301

Printed in U.S.A.

Library of Congress Cataloging-in-Publication Data

The Heritage of engineering geology : the first hundred years / edited by George A. Kiersch.
p. cm.—(Centennial special volume ; 3)
"One of the synthesis volumes of the Decade of North American Geology Project series"—T.p. verso.
Includes bibliographical references and index.
ISBN 0-8137-5303-1
1. Engineering geology—History. I. Kiersch, George A., 1918– II. Geological Society of America. III. Decade of North American Geology Project. IV. Series.
TA705.H47 1991
624.1'51'09—dc20 90-25229
CIP

Cover Photo: New Croton (New York) masonry dam, completed in 1906. See chapter 1, Figure 21 for details. Photo by George A. Kiersch.

10 9 8 7 6 5 4 3 2

Contents

FAILURES, ERRORS OF JUDGEMENT, LITIGATION, AND THE GEOLOGIST'S RESPONSIBILITY

Preface

This volume is a contribution from the Engineering Geology Division of the Geological Society of America to the Decade of North American Geology (DNAG) Project as a part of the celebration of the Centennial of the Society. The able editorship of George A. Kiersch and the partial support of his editorial costs by the Engineering Geology Division are greatly appreciated.

In addition to four Centennial Special Volumes such as this, the DNAG Project includes a 29-volume set of syntheses that constitute *The Geology of North America,* six Centennial Field Guides that highlight 100 of the best geologic sites in the area of each of the six regional sections of the Society, 23 Continent/Ocean Transects, and seven wall maps at a scale of 1:500,000 that summarize the geology, tectonics, magnetic and gravity anomaly patterns, regional stress fields, thermal aspects, and seismicity of North America and its surroundings. Together, the books and maps of the DNAG Project are the first coordinated effort to integrate all available knowledge about the geology and geophysics of a crustal plate on a regional scale.

The products of the DNAG Project present the state of knowledge of the geology and geophysics of North America in the 1980s, and they point the way toward work to be done in the decades ahead.

Allison R. Palmer
Centennial Science Program Coordinator

Foreword

Preliminary planning for the development of a Geological Society of America (GSA) Centennial Special Volume on engineering geology was undertaken in 1980–81 by GSA Engineering Geology Division Chairman James W. Skehan with two volumes in mind: a history of engineering geology, and a sequel with state-of-the-art papers.

At the 1982 GSA Annual Meeting in Cincinnati, Ohio, a symposium entitled "The Role of Government Agencies in the Development of Engineering Geology" was held as the first step in producing the engineering volumes (chapter 3, this volume). By 1983 the original plans produced at the symposium had been modified, and Christopher C. Mathewson and Richard C. Jahns had agreed to serve as co-editors for one volume. With the death of Jahns, the Management Board of the Division asked me to join Mathewson as a co-editor for the volume.

By early 1984 an outline for the volume had been prepared. Mathewson and I met to make modifications to the outline and to agree on guidelines, and by July, selection of authors to be invited to contribute to the volume was complete. The outline was submitted to the GSA Decade of North American Geology Steering Committee and approved on November 12.

No invitations were sent to proposed authors until the summer of 1985, by which time some of the originally selected authors were no longer available. At the GSA Annual Meeting in Orlando, Florida, in 1985, with concern for the fast-approaching original publication date of 1987, the Division Management Board, chaired by David M. Cruden, reorganized the responsibility for the volume, appointed a Steering and Review Committee, and appointed me as chairman and editor-in-charge of the volume. By the following summer, three of the committee members were unable to continue their duties, and all responsibilities fell to me.

The three-year delay in getting this volume started was the major reason for many manuscripts not being submitted until 1988 and 1989. Because of time constraints, it was necessary to drop several scheduled topics, including: weak rock and soil, karst and weathering phenomena, radioactive waste disposal, urban geology, military geology, additional coverage on construction materials, selected exploration techniques, and computer technology.

This volume is not intended to be another textbook. Instead, it is a review of historical changes in engineering geology through time. The rationale for the subjects of history, geologic processes, natural materials, investigations for engineering works, and the geologists' responsibilities in litigation are related to the question, "How have the efforts of geologists for engineering works resulted in new technical knowledge and advances in the geosciences?"

Obviously, each area or subject does not offer equal possibilities for citing new knowledge or advances relevant to geologic processes, materials, and the reaction of man's works to the geologic environs; however, some examples from this volume include;

• The New Croton dam, New York, encountered foundation difficulties in 1895 when a deeply solutioned zone in limestone/marble was discovered below the groundwater table. This "first-known" occurrence of subsurface karst features and problems changed the evaluation of potential cavitation in calcareous rocks.

• The ancient channel of the Hudson River at the Storm King crossing of the Catskill Aqueduct was at least 500 feet below previously known levels. This changed geological thinking and expanded the potential for recognition of deep erosional features at many engineering sites in glaciated terrain.

• G. K. Gilbert, beginning in the 1870s, was the first geologist in North America to introduce quantitative and analytical solutions to geological questions. These included forecasting earthquakes and a remarkable insight into large-scale tectonic processes, such as the cycle of loading, subsidence, and rebound caused by a large body of water such as ancient Lake Bonneville.

• Warren J. Mead's studies on rock masses from 1925 through 1930 demonstrated the role of dilatency and the strain ellipsoid concepts in geological analysis and engineering design.

• Seismotectonic research in the 1960s and 1970s evaluated fault zone(s), including recency of last movement, for the design of nuclear power plants. This led to major advances in dating tectonic events for both science and the design of all engineering works by using young stratigraphy, associated minerals, and tectonic geomorphology to date movements of fault zones.

• The Pfeiffer litigation in California in 1960, regarding damage to a house and its foundation, became a precedent for landslide insurance and the payment of damages to the insured. Geological facts and events were central to the controversy over whether a landslide insurance policy on a dwelling includes repair of the foundation area as well as the aboveground building. The case clarified this crucial technical point; a judgment for repair of the dwelling included restoring the subsurface and the foundation to a safe condition, equivalent to that existing before the landslide.

• The career of William O. Crosby, "The Father of Engineering Geology in North America," and his projects related to engineering geology and research efforts are summarized.

ACKNOWLEDGMENTS

In addition to the efforts of the authors, completion of this volume has benefited from the assistance and support of many other individuals and organizations. They include: the GSA staff, particularly Allison R. (Pete) Palmer, Centennial Science Program Coordinator; the chairmen of the GSA Engineering Geology Division from 1986 through 1990, David M. Cruden, Christopher C. Mathewson, Ellis L. Krinitzsky, Thomas L. Holzer, and Jeffrey R. Keaton; and the reference staffs of the Cornell University and University of Arizona libraries. Colleagues and practitioners who kindly evaluated manuscripts and reviewed texts include M. G. Bonilla, D. M. Cruden, E. B. Eckel, R. H. Fakundiny, A. W. Hatheway, L. B. James, J. R. Keaton, E. L. Krinitzsky, and S. S. Philbrick. Most authors arranged for in-house reviews of their chapters. Kim Duffek of Kanoa Illustrations, Tucson, Arizona, prepared most of the line drawings for chapters 1, 2, 18, 22, 23, and 24, as well as revisions for other graphics. GSA and its Engineering Geology Division supplied funds for administration, editing, and other preproduction expenses.

George A. Kiersch
August 1990

Dedication

This volume is dedicated to

William Otis Crosby
1850–1925
The Father of Engineering Geology in North America
Practitioner 1893–1925

The Early Pioneers
1895 to 1930s
James F. Kemp, Heinrich Ries, Charles P. Berkey,
C. W. MacDonald, Warren J. Mead, Edwin C. Eckel, and Karl Terzaghi

Eminent Practitioners
1930s and 1940s
Sidney Paige, Thomas W. Fluhr, Frank A. Nickell,
Edward B. Burwell, Floyd Johnson, Robert H. Nesbitt,
Shailer S. Philbrick, Berlin C. Moneymaker, Roger Rhoades,
William H. Irwin, William F. Gardner, Chester Marliave,
Edwin B. Eckel, Arthur B. Cleaves, and Robert F. Legget

Geological Society of America
Centennial Special Volume 3
1991

Chapter 1

The heritage of engineering geology; Changes through time

George A. Kiersch
Geological Consultant, Kiersch Associates, Inc., 4750 Camino Luz, Tucson, Arizona 85718

"A very nice building, but I still think the ground is too soft on this side."
Peter Estin

Even in those days, "geological conditions" were recognized as critical to locating engineering works (permission Crowell-Collier Publishing Company, 1954 *in* Kiersch, 1955).

ANTIQUITY THROUGH MIDDLE AGES: PRIOR TO 1450 A.D.

Early engineering works and geologic craft/lore

The history of remarkable engineering construction feats is as old as man's records. Subsurface mining for copper ore on the Sinai Peninsula began at least 15,000 years ago (Stone Age), and tunneling (adit) was started about 3500 B.C. As civilization and commerce advanced and people congregated in cities, the problem of water-supply protection agianst the attacks of enemies became increasingly acute, and new methods, such as construction of aqueducts and reservoirs, had to be devised.

Use of "geologists" to assist in evaluating natural sites for engineering works and related legal implications has a long history if we include the lore of our forefathers regarding natural conditions and their meaning. In North America, early assistance and insight on geological reasoning for engineering purposes was fostered by a group of pioneers whose endeavors are described in this chapter; geological input for litigation and forensic purposes is discussed in Chapters 24 and 25 of this volume. However, any review of the early efforts in application of geology to engineering works in North America must recognize the fund of knowledge that had been acquired by earlier pioneers in Europe and Asia, and parts of Central and South America.

The numerous remnants and intact examples of remarkable construction feats built in past centuries represent a legacy to the early "engineer's" skills. It is not difficult to imagine a relation and interdependence between the "architect-engineer" and the "geologist," which began far back in ancient times. Obviously, even then, some individuals had an awareness of rock and soil conditions and offered counsel on excavations and the properties of natural materials for siting and construction of castles, canals, water tunnels, and aqueducts. Certainly, some broad concepts of geology have been used instinctively for thousands of years to guide the location and design of engineering works, although the formal designation for such practice—engineering geology—has only come about in this century. For instance, ancient cathedrals and castles are located on good natural sites; many, in fact, oc-

Kiersch, G. A., 1991, The heritage of engineering geology; Changes through time, *in* Kiersch, G. A., ed., The heritage of engineering geology; The first hundred years: Boulder, Colorado, Geological Society of America, Centennial Special Volume 3.

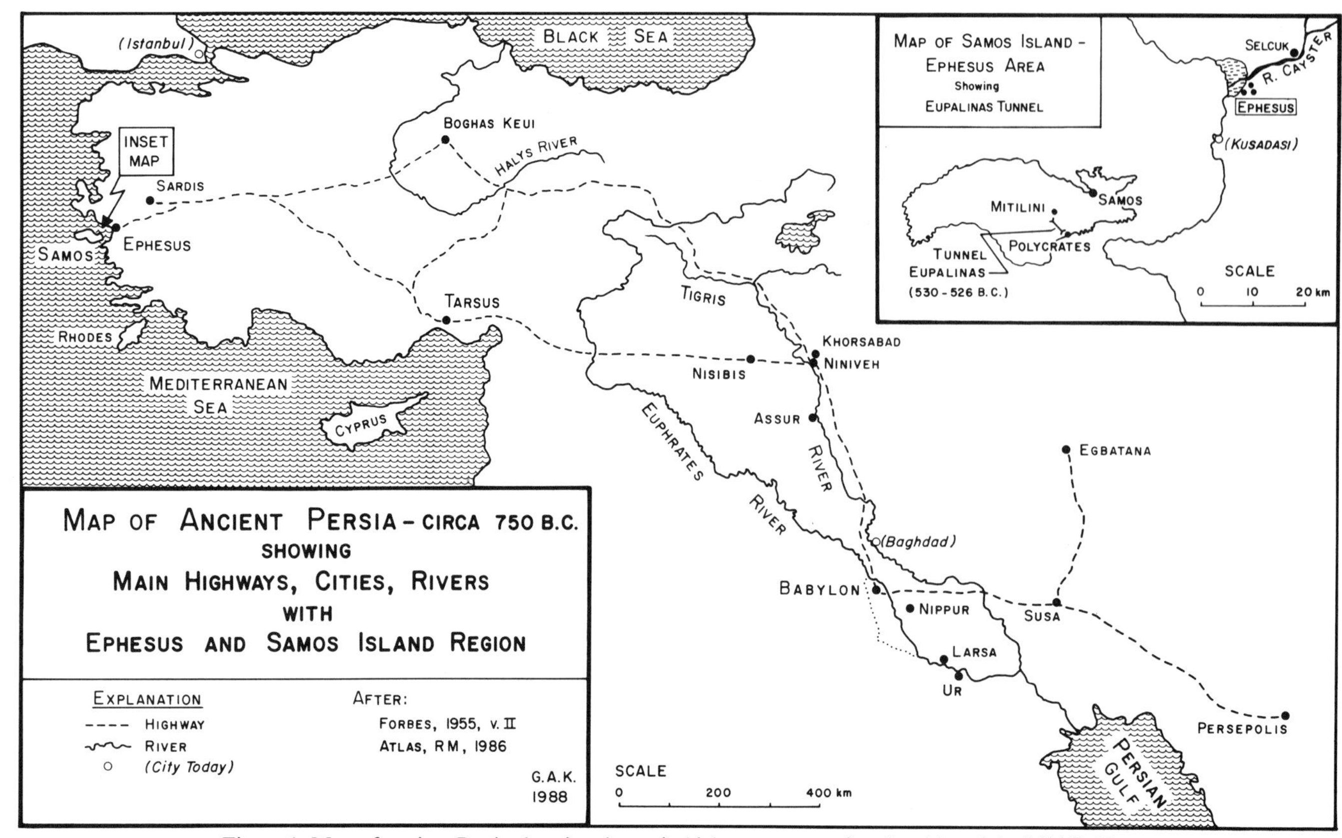

Figure 1. Map of ancient Persia showing the main highways connecting the cities of the Middle East. Inset map shows location of Ephesus (three sites) and the area of backfilling with subsequent sediment deposition by Cayster River. The Island of Samos and site of the Tunnel of Eupalinas near Polycrates is also shown.

cupy the most suitable locations for miles around. Yet even these "ideal sites" are today showing evidence of physical deterioration and/or movement within the rock foundation and require modern geotechnical stabilization or treatment (Marinos and Koukis, 1988). A few such cases from antiquity and early history will illustrate.

Prestige and religion were strong motives for large engineering works, such as the Egyptian pyramids, which were raised between 2700 and 2300 B.C. These monuments required not only the quarrying but the transport of well-shaped stone. Preceding the on-site building of the pyramids, workmen constructed the famous Pyramid Causeways, circa 3000 B.C., on which the large cut blocks of stone were transported from quarries to the Nile and then to the pyramid construction sites (Forbes, 1934, p. 129).

Highways. The wheel was invented around 3500 to 3200 B.C. Wheeled traffic and the use of bitumen for surfacing streets in Mesopotamia/Persia are reported from about 2,800 B.C. The earliest record of an attempt to pave a street with stones is near Molfettal (Bari), Italy, circa 2,600 B.C. The ancient Persian highway system established circa 750 B.C. (Fig. 1) linked Babylon with the palaces of Persian kings at Susa, Persepolis, and Egbatana and continued north and west through the Assyrian Empire (Assur) and Tarsus to Sardis and the harbor of Ephesus on the seacoast, about 2600 km distant.

Irrigation works. Canals and surface irrigation works were crudely developed circa 3500 B.C. in the ancient Near East, and by 3000 B.C. there were effective systems for perennial irrigation in Mesopotamia and basin irrigation in Egypt (Forbes, 1955, p. 72). Another early canal system was built by the Minoans on Crete circa 2300 B.C. After being destroyed by an earthquake, it was rebuilt around 1700 B.C., with parts remaining today. Also during this time (2200 B.C.), irrigation canals were built in China.

Mesopotamia, a land between the valleys of the Tigris and Euphrates Rivers, had many ancient towns located along the river banks (Fig. 1). Natural geologic processes associated with the heavy sediment load accompanying periodic flooding have built high natural levees along the rivers, particularly the Euphrates. Consequently, for long distances the river levels are above those of the surrounding farmlands and plains.

Beginning around 2000 B.C., canals were used to redistribute the Tigris and Euphrates waters. Although the Euphrates carries only 40 percent of the combined flow of the two rivers, it is the main source for water to irrigate the lands between them, for two geologic reasons: (1) the river bed is largely above the surrounding plain so that flow is more easily diverted, and (2) the peak flow, from snowmelt in Armenia, comes at the beginning of the summer when it is most needed, much later than that of the Tigris. The ancient Sumerians therefore built an intricate system of dikes and parallel and lateral canals to tap and divert the river water into reservoirs for use in the dry seasons.

Because of the great runoffs and the associated high sediment load, the systems of canals, weirs, and reservoirs required much maintenance. Thus, this system has provided engineers and geologists a large-scale laboratory with a 4,000-year history for the study of problems of sediment load, silting, and backfilling in both natural and man-made channels (Forbes, 1963, p. 16–19).

The first codification of irrigation laws was prepared by Chammurabi of Babylon in 1800 B.C., an outgrowth of the canal redistribution construction.

Canals. Circa 800 B.C., the Assyrians constructed large canals around the headwaters of the two rivers (Fig. 1) and transported the water southward into Mesopotamia for both irrigation and transport of seasonal supplies. Babylon, located 96 km south of Baghdad, Iraq, was once a city of about a million people.

Canals for shipping only were other early works of man; the first recorded was built by King Mernere in 2400 B.C. to avoid the dangers of a cataract on the Nile River at Elephantine (Forbes, 1955, p. 26). Later kings worked on the early Nile–Red Sea canal, and the first connection was built by Sesostris I in 1950 B.C. Yet, not until 1859 to 1869 was the modern Suez Canal built by French engineers (Ferdinand de Lesseps). Another early canal was attempted circa A.D. 60 by Nero across the Isthmus of Corinth, Greece. Later, circa A.D. 600, Periandros constructed "diolkos" (wooden rollers) and a paved roadway to portage ships over the Isthmus (Legget, 1962, p. 593).

Control of ground water. One of the greatest achievements in ground-water utilization dates back more than 2,500 years, to the construction of infiltration galleries or kanats (qanats in Persia), commonly kilometers in length. These subsurface collectors channeled the ground water from surrounding soft sediments and alluvial fan deposits and distributed the water to the villages and fields for irrigation (Fig. 2). The techniques of construction spread rapidly eastward to Afghanistan and China and westward to Egypt where one extensive system (circa 500 B.C.) irrigated 4,700 km^2 of land west of the Nile (Tolman, 1937).

Many kanats are still in operation. Perhaps the most famous and successful system, in the Turpan oasis in China, has thrived for 2,000 years. Turpan, located in the world's deepest dry depression (more than 3,900 km^2 that are as much as 155 m below sea level), has an extreme environment: one-half inch rainfall/year and temperatures of 104 to 120°F. Today this former stop on the silk route to central Asia is the center of production of China's finest long-staple cotton, and also has fields of melons and grapes. The city of Turpan and the surrounding agricultural district are supplied water by an ingenious system of underground aqueducts and galleries (pipe and covered canals) that collect water from the surface runoff and underground flow in alluvial deposits along the front of the Tien Shan Mountains. Water is

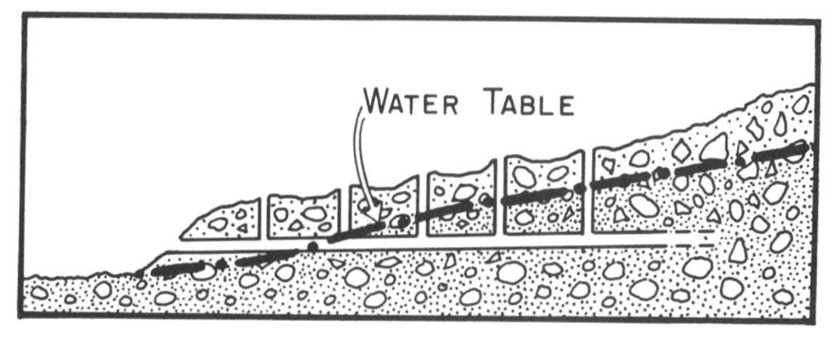

Figure 2. A hillside section of alluvial deposits illustrating an ancient water-collecting system (kanat). The use of shafts indicates the techniques is based on the older tradition of pitting.

distributed below ground to reduce evaporative loss from thousands of galleries, kanats, and wells fed by 972 tunnels traversing 2,700 km of desert. Most of the early kanats and canals, built centuries ago, are still in use. Since 1949 the Chinese government has built ten additional aqueducts (kanats, etc.) and 14 holding reservoirs to ensure an adequate supply of water for the Turpan district (Wren, 1983).

An exceptional understanding of ground-water origin and flow was demonstrated by the ancient Aztec Indians of Mexico when they conceived and built an amazingly ingenious project in the Valle de Mexico (region of Mexico City). At about 1,000 B.C. a large natural lake covered much of the area (Fig. 3). To improve the foundation conditions for support of buildings, temples, and pyramids, the Aztecs installed an effective system of drainages to dewater and stabilize the lake beds and building sites. The natural lake was fed by both freshwater springs from the nearby western volcanic mountains and by contaminated sulfurous waters from the other sectors (Sowers, 1981). Using a valid concept of underground water flow and source, the Aztecs proceded to construct a cutoff "curtain" of crude piling across part of the lake that supported a backfilled wall. This separated the fresh waters on the west from the mineralized waters flowing in from the north (Alcocer, 1935). Certainly this feat required a comprehension of the simple geological principles of ground-water origin and composition.

The early Romans demonstrated a similar capability with the construction of water aqueducts and attention to such environmental needs as subsurface drains to remove runoff. The Colosseum construction, A.D. 72 to 80, used a puzzolanic concrete, and the foundation contained a drainage system (Mocchegiani Corpano, 1984).

In North America, a prehistoric irrigation system was constructed with rude stone hoes and wooden sticks by the Hohokam Indians after their arrival in the Salt River Valley of central Arizona circa A.D. 100. They built 200 km of canals (9 m wide, 3 m deep) along the Salt River and an additional 100 km of

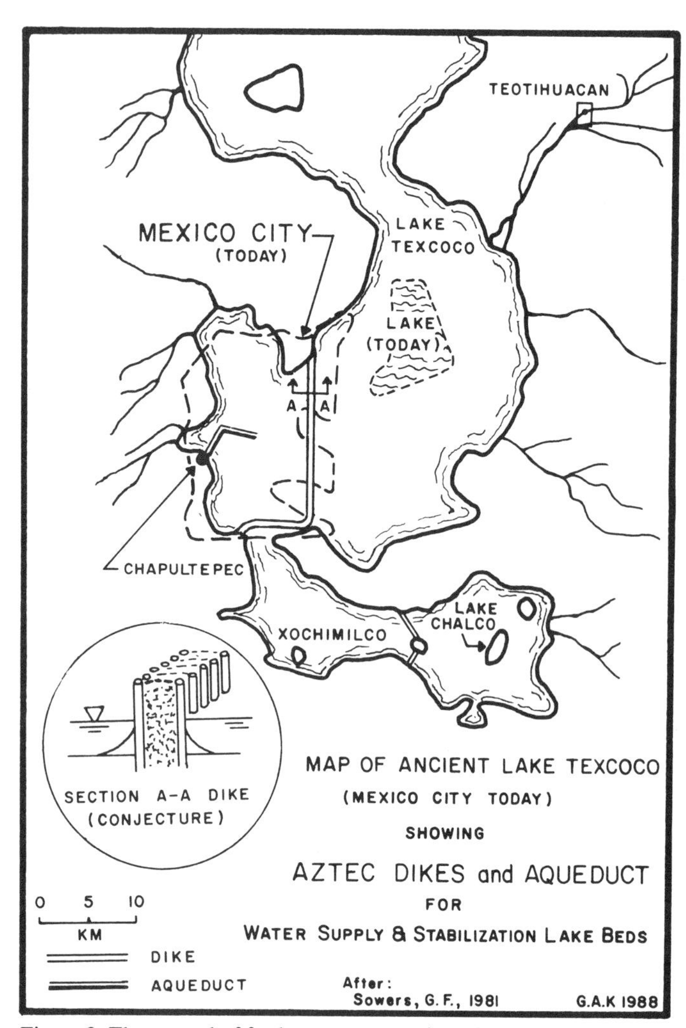

Figure 3. The control of fresh versus contaminated ground water in the Valle de Mexico by Aztec Indians, around 1000 B.C. Causeways and dikes were constructed to provide a fresh water supply, and the lake beds were stabilized by dewatering and drainages to improve foundation conditions for city.

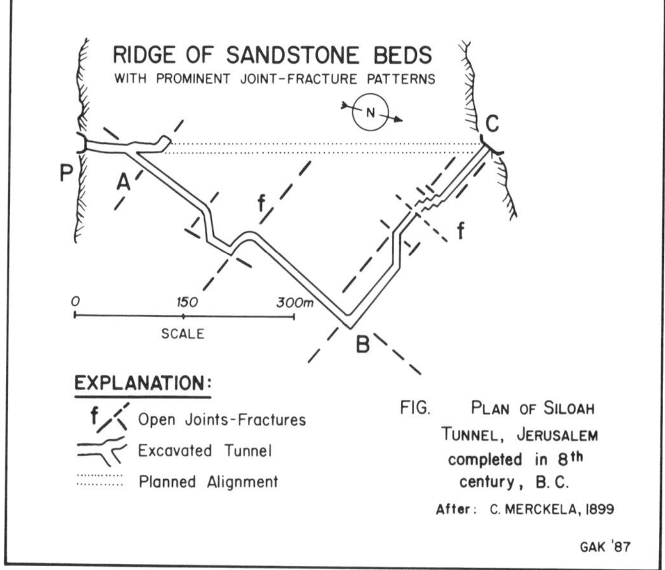

Figure 4. The Siloah tunnel constructed in eighth century Jerusalem. Tunnel direction was influenced by open fissures in the sandstone beds.

works along the Gila River before leaving circa A.D. 1400 (Schroeder, 1943). In South America the Incas and their predecessors were actively building diversion weirs and irrigation works from circa 100 B.C. to A.D. 600 (Forbes, 1955, p. 8) for irrigation and water supplies. Yet another kind of water-control works began in Europe during the period 109 to 31 B.C. when the Po Valley plains of Italy were drained by navigable canals from the Po River (Forbes, 1955, p. 45).

Tunnels for water transport. The tunnel constructed for the aqueduct of Jerusalem in the years 955 to 932 B.C. is probably the oldest known tunnel used for water transport, though an earlier tunnel to tap a spring is reported at nearby Gezer, circa 1900 B.C. (Forbes, 1964). The earliest tunnelers quickly learned that if the rock conditions were poor, the direction of the tunnel face would likely change. This was demonstrated in the eighth century B.C. along the 546-m tunnel of the Siloah Reservoir in Jerusalem (Fig. 4); though plotted to cross a system of open fissures in the host sandstone, the workmen instead followed the two directions of soft fissures, and the completed tunnel became a crooked V alignment between the portals (Merckel, 1899).

For protection and to serve trade, harbor facilities had to be developed at coastal locations, such as the Polycrates mole built circa A.D. 540 on the island of Samos (Fig. 1 inset map). Polycrates, ruler of Samos (550 to 522 B.C.) built the Polycrates wall (6.1 km) to surround the ancient city, the naval port, and the harbor (Heath, 1981). For a supply of fresh water, he directed Eupalinas of Megara, a great architect and hydraulic engineer and native of Samos, to construct the now-famous Tunnel of Eupalinas. Fresh spring water from the inland valley of Agiades near Mitilini was to be transported southward beneath Mount Castro to the coastal city of Polycrates (Fig. 5).

In designing and constructing the tunnel (530 to 526 B.C.), Eupalinus demonstrated an understanding of crude geologic principles. The Tunnel of Eupalinas is 1,045 m long and 1.75 m high. Although tunnels or adits were constructed much earlier (circa 3000 B.C.), this was the first known tunnel to be driven from both portals simultaneously. The two separate headings met underground with only minor horizontal (some 2 m) and vertical (0.75 m) offsets, according to Goodfield and Toulmin (1965). This amazing feat is one of the most remarkable engineering works surviving from antiquity (Van der Waerden, 1954). The undertaking and its success were due not only to Polycrates' rudimentary knowledge of geologic principles, but also to the ideal geologic conditions of the Mount Castro site. The tunnel location, design, and uniform grade were enhanced and aided or controlled, as was the mining, by the following favorable geologic conditions (Fig. 6): (1) the opening traverses flat-lying (to 4° dip), thin-bedded, partly cemented marly limestones with interbeds of sandy and/or silty limestones and fine-grained, calcareous/silty sandstones; (2) the host beds are 2.5 to 15 cm thick, medium soft, and air slack or partly deteriorate on exposure; and (3) rock is easily broken and/or excavated with crude hand equipment.

The uniform occurrence of soft beds enhanced the construction of an even grade, shape, and on-line tunnel. The tunnel alignment and azimuth were established by a geometric survey over the top of Mount Castro.

The builder reportedly applied Pythagorian geometrical principles to determine the relative elevations (Goodfield and Toulim, 1965). The alignment and differential "stair-step" sightings/leveling of backsights and foresights were made between reference posts to establish the elevation of the north portal relative to the south or coastal portal.

Figure 6. The typically marly limestones and thin-bedded, weakly cemented sandstone beds traversed by Eupalinas Tunnel. This quarried exposure near the north portal has a steeper dip than at the tunnel alignment. (Photo by G. A. Kiersch, 1983.)

Roman engineering works, land reclamation, and water wells. The magnificent roads of Mount Genevre Pass built in 75 B.C., and Great St. Bernard Pass, built about 57 B.C., or the short tunnels, such as the one at Furlo Pass in the Apennines, built in A.D. 78, and the first aqueduct, built in A.D. 134, all were constructed by Roman engineers during the period of expansion that culminated in the Roman Empire. Considerable thinking of a geological nature must have gone into the planning and construction of such engineering works, in order for them to have endured these many centuries (Clarke, 1910). History records the Romans as practical in outlook; other europeans did not approach the standard and scope of Roman works until the late eighteenth century (Anderson and Trigg, 1976, p. 2). For example, the design and construction of a normal Roman Highway (Fig. 7) included four layers of different materials, side drainage ditches, and a crown.

Figure 5. The Valley of Agiades near Mitilini, foreground, where an abundance of fresh water was available from springs in limestones around 530 B.C. The Tunnnel of Eupalinas (dotted line) was driven beneath Mount Castro (on skyline) simultaneously from portals on right of photo (north) and behind the slope on the left (south). (Photo by G. A. Kiersch, 1983.)

The reclaiming of low and/or water-logged lands began with primitive dikes constructed in the low countries of Europe between A.D. 600 and 800. From this, large-scale reclaiming of the marches located northeast of Bremen began in A.D. 1106 by Dutch colonists. By about 1250 A.D. the dikes built in north Holland province consisted of a clay embankment with a sand core, a forerunner of modern levee and earth-dam construction.

Although the forerunners of modern percussion methods for drilling water wells were developed independently in China and western Europe more than 1,500 years ago, flowing water wells were first discovered about A.D. 1100 in Flanders and, later in that century, in eastern England and northern Italy. The first wells were dug by Carthusian monks in A.D. 1126 near the village of Lillers (Davis and DeWiest, 1966). Soon thereafter, four wells were drilled several hundred feet deep and tapped pressurized water in a chalk formation in Gonnehem, Flanders (near Bethune); the well casing extended more than 3 m above ground level, and the flowing water had sufficient head to drive a water mill. The fractured chalk outcrop (infiltration) area was in the region of Artois, and these flowing wells became known as artesian wells (Brantly, 1961).

Impact of processes on works. Other important engineering works built in the Middle Ages demonstrate a lack of practical understanding in dealing with simple geological principles. The famous Tower of Pisa, Italy, completed in A.D. 1370, underwent uneven settling and began to lean during construction (Anderson and Trigg, 1976, p. 121; Frontispiece, this chapter). The Arno River Delta has backfilled the old Pisa sea front and channel during the intervening years, and the tower, built to observe the harbor and movement of ships, is now 13 km inland from the sea. Leghorn is the coastal port today.

Interestingly, between 2000 B.C. and A.D. 1000 the same geologic processes and progressive changes created a devastating environmental impact on the harbor and surroundings of the city of Ephesus (Uckun, 1989; Fig. 1, inset map). About 1000 B.C., the progressive silting and sediment backfilling of the ancient harbor at Ephesus by the River Cayster caused a complete ecological change throughout the immediate area. Swampy backwaters were formed along the inland channel of the ancient Aegean seacoast. Insects, particularly malaria-carrying mosquitoes, became a menace to the inhabitants, and the original city was moved to a nearby location. Unfortunately, the second location experienced similar problems, and about 300 B.C., in anticipation of a disease-free site, the city was shifted 2.5 km to a third location (the site of the ruins today). However, on-going geologic processes at the mouth of the River Cayster continued to backfill the harbor and channel unimpeded, and by A.D. 431, the living conditions were again unhealthy. By A.D. 500 the harbor was completely backfilled and cut off from the sea. When existence at the third site became unbearable, the inhabitants left, some moving to a fourth location inland at Selcuk. By A.D. 1000, Ephesus had lost its importance (Fig. 8), a victim of misunderstood geologic processes. Today the backfilled harbor and channel area of Ephesus are occupied by small farms (Fig. 9).

Figure 8. Photo of Ephesus ruins looking north across the former city. The harbor channel of River Cayster, now backfilled with sediments, is located immediately beyond the buildings in the upper part of the photo. This, the third location of Ephesus, was abandoned because of the diseases associated with swampy, backfilled lands that today are cultivated farms. (Photo by G. A. Kiersch, 1983.)

In spite of such experiences, engineers were slow to apply geological concepts to similar sites at other localities in Europe and the Middle East. Collapse of monuments, structures, and towers was a common occurrence into the 1800s, when the consequences of heavy loading on soils and weak rocks were still in the early stages of study. For example, at Ravenna, Italy, in the Po River Delta, a geologic setting similar to Pisa, four separate towers built in the 1700s all settled and lean prominently today (Fig. 10).

Some ancient thoughts about geology for engineering

As the earliest concepts about the Earth's history and its features began to emerge, ancient Greece was the scene of much geological thought, some relevant to engineering works. Herodotus (ca. 484 to 425 B.C.) observed that earthquakes cause large-scale fracturing (faults) across the landscape and thus shaped the Earth's surface; also , that the sediment volume carried by the Nile River caused growth of the great delta (Faul and Faul, 1983). Several writers of Grecian-Roman antiquity had an influence on the early concepts of practical geology. Among them, Pliny the Elder (ca. A.D. 23 to 79) described an eruption of Mount Vesuvius on August 24, A.D. 79, with the volcanic outpourings and ashfalls, the associated tremors and earthquakes, and the giant wave motions of the nearby sea (motions now recognized as a tsunami). The next day, while making observations on the beach near Capri, Pliny lost his life when he was overcome by the foul sulfurous fumes from a dense ashfall cloud.

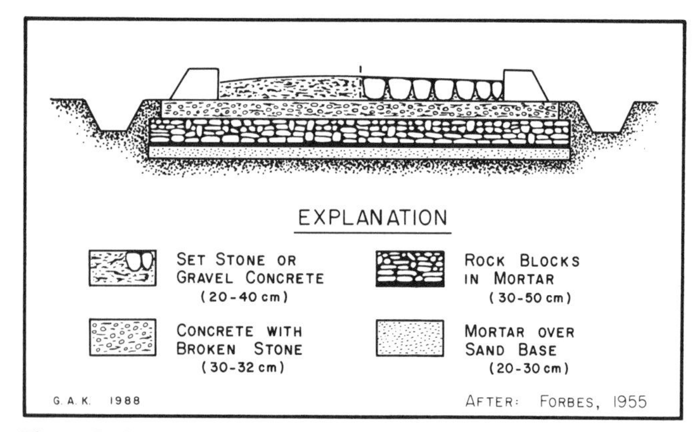

Figure 7. Cross section of normal Roman highway showing its four different materials and the development of a subgrade base course, side drainages, and a crown.

During this early time, Plato (427 to 347 B.C.) believed that ground water came from one large underground cavern fed by the ocean, but he still described the hydrologic cycle somewhat accurately (Davis and DeWiest, 1966). Aristotle (384 to 322 B.C.) modified the ground-water misconception by stating that some cavern water originated from rainfall. However, Marcus Vitruvius, circa 15 B.C., was the first to correctly grasp the hydrologic cycle. In contrast, Lucius Annaeus Seneca (4 B.C. to A.D. 65) differed with Aristotle's concept and denied that spring waters originated from rainwater. Seneca's position was widely held for more than 1,500 years and restricted advancement in scientific thinking about hydrogeology until the end of the Renaissance. This long status quo in geological thinking was due to several factors and was one reason for the lack of progress in developing scientific theory and relevant practical applications concurrently for earth science. For example, Plato and others insisted that philosophy and science be more or less separated from the interpretation of field observations, experimentation,

Figure 9. Photo of seacoast today at mouth of the former channel, River Cayster and harbor for Ephesus of A.D. 500 to 1,000. Site of Ephesus is several kilometers to the right of photo (see inset map, Fig. 1). The River Cayster's delta is today fertile and an area of small farms. (Photo by G. A. Kiersch, 1983.)

and practical applications. For centuries this created a wide gap between theory and practice and caused a lack of advancement in applied earth science. This sharp division in the teaching of science and engineering had a strong influence on the organization of European institutions of higher learning; the classic university offerings were separate from the Technical Universities, a distinction that in some countries continued into the twentieth century.

RENAISSANCE AND BAROQUE PERIODS: ca. A.D. 1450 to 1750

Geological concepts and the architect-engineer

As the Renaissance emerged, French, English, and Italian scientists came forth as the leaders who conceived new geological concepts, some of which became important applications for engineering works. A few of the principal contributors to our current understanding of geologic processes and phenomena important to planning and building engineering works are summarized from the History of Geology by Faul and Faul (1983) and others.

The first applied geologist, in the sense of a scientist, is widely held to be Leonardo da Vinci (1452 to 1519). Although a prominent hydraulic engineer, he is best known as a painter. For much of his career he was concerned with canals and construction of water projects (Clements, 1981), and many consider Leonardo da Vinci the first "applied geologist" for engineering works. His field observations of excavated rocks led him to reason that "the shells encased in the limestones of the Tuscany hills are former sea life and the rocks had been uplifted"—a milestone in early geological thinking. He was active in flood control design of canals and locks and, as Ingegnere Camerale of Milan (1498 to 1503), he improved and integrated the canal system of that city. He formulated a plan for control and improvement of the Arno River watershed with a series of flood-control dams, canals, and locks for navigation; costs were too great, however, and he returned to Milan. Engineer da Vinci studied wave action and the work of running water and concluded that valleys are cut by the rivers that occupy them; rocks transported by rivers gradually became rounded and smaller; and the material deposited by rivers in the sea with time forms various sedimentary rocks, which later may be uplifted to form mountains. This led to da Vinci's concept of geologic time, which is remarkably modern: he estimated that it took 200,000 years to accumulate the deposits of the Po River, and that this was not the whole of geologic time.

Perhaps most remarkable is da Vinci's concept of mountain "building"; rain and rivers constantly carry material to the sea, yet

Figure 10. Two of four prominent towers in Ravenna, Italy, are shown. The towers, built on deltaic deposits of the Po River, have settled and are today "leaning towers." (Photos by G. A. Kiersch, 1964.)

mountains are high and sea basins are low. He reasoned that the mountains were made lighter by the removal of material, and consequently, the "mountain" that is farthest from the center of gravity tends to move farther. In effect he stated the concepts of uplift and isostatic adjustment defined by Dutton in 1899!

Soon thereafter, Girolamo Cardano, a great mathematician and inventor, asserted, "All running water comes from rain and rain is caused by evaporation of the sea" (Cardano, 1550). The true source of river water was provided by two French scientists, Pierre Perrault (1608 to 1680) and Edmé Mariotte (1620 to 1684). Perrault measured rainfall in the Seine River basin for three years (1668 to 1670), estimated runoff, and concluded that rainfall could account for all stream water (Davis and DeWiest, 1966, p. 9). Mariotte (1717) observed the infiltration of rainwater in an underground cellar; he noted that the amount varied with the rainfall, as did the flow of springs at other sites, and concluded that the springs were fed by rainwater that filtered into the ground. Mariotte also verified Perrault's flow of the Seine River and concluded, "If one-third of the precipitation enters the ground there would be enough water to sustain the flow of wells and rivers in the basin" (Meinzer, 1942).

Engineering works

During the Renaissance period the first signs of a systematic geological approach appeared for planning such engineering works as irrigation projects, water supply, and infiltration galleries. A good example, the Rudolf gallery (1581 to 1593) is about 1 km long and still serves as an aqueduct for the King Gardens in Prague; the original drawings indicate the specialized professional skills of the designers and builders as related to the geologic conditions (Zaruba and Mencl, 1963). In a similar manner, V.L.B. Alberti in 1532 described the problems of foundations and methods for investigation of foundation soils in Paris. Unfortunately, during the Baroque period of the 1600s, less than adequate attention was given to natural phenomena and geologic conditions relevant to construction. For instance, in 1701, Andreas Schlüter began a reconstruction of the Mint Tower in Berlin to raise the height from 40 m to 100 m. The subsurface soil and rock conditions were not understood, and in spite of piles, the tower inclined with time and had to be dismantled. This continuing lack of understanding foundations was also evident for the leaning towers of Ravenna, Italy (described earlier), in the 1700s, and in America during the 1850s to 1880s when the Washington Monument was constructed.

By the early 1700s, several geologic processes had been recognized that were important to the construction of engineering works (Faul and Faul, 1983), such as the silting and backfill processes at many harbors that were opened in Roman times as described earlier (at Ephesus, A.D. 500, and Pisa, 1400 to 1500s), and retreat of the sea in the region of the Rhone Delta by the mid-1700s (Maillet, de Benoit, 1735).

Henri Gautier (1660 to 1737), the civil engineer to the King of France, provided several new and important geological ideas (some similar to Leonardo da Vinci's of the 1500s) that received little response in French geology. Among them: rainwater erodes the landscape, and the action of river erosion is correlated with the amount of sediment carried; as erosion proceeds, continents become lighter and the sea area heavier with sediments that harden to rock; equilibrium is reestablished when the sea is raised to form new mountains, while the eroded continent subsides into the sea (Gautier, 1721). Gautier spent his life designing bridges, canals, and locks in southern France and was intimately familiar with sedimentation and the action of rivers. Gautier was a pioneer in applying geological principles in the design and construction of engineering works; his findings predated many similar ones made by William Smith 75 years later in England.

INDUSTRIAL REVOLUTION TO THE NINETEENTH CENTURY: 1750 to 1900

Introduction

The concept that geologic conditions could influence the planning and construction of large-scale engineering works, such as roads, canals, tunnels, harbors and water supplies, was recognized during the period of the Industrial Revolution in eighteenth century Europe and nineteenth century North America.

The application of geology for engineering purposes played a small role in the early history and expansion of the United States up to the 1880s, as documented in a fine review by Radbruch-Hall (1987) that is freely utilized here. The early colonists from Europe settled along the eastern sea coast and the southeastern region. They established communities on natural harbors, or on lakes and rivers that were natural transportation routes; there was little or no benefit of geological knowledge in choosing the site. Many locations were in low, swampy terrain, and construction difficulties became common as the settlements expanded. Good examples are Washington, D.C., Boston, and later San Francisco.

Expansion of manufacturing and heavy industry brought a need for industrial sites; to keep abreast, buildings grew skyward in cities like New York and Boston and created the need for a better understanding of foundation conditions. However, America was expanding westward to new territory by the 1820s, and this expansion required construction of an improved network of roads and canals. Suddenly, in the middle of the century, the building of roads and canals was curtailed in favor of constructing a nationwide railroad network (1850s to 1870s). This rush to western lands and the Pacific region required both bold planning and unusual physical and human efforts, particularly in parts of the Rocky Mountain area and western territories, to complete links with the central states.

Any review of the historical milestones relevant to geology and engineering works in North America must recognize the accomplishment and early advances of the European investigators in the eighteenth and nineteenth centuries. Their experience and tested concepts became available to North American geologists and engineers when they undertook to serve the growing project demands of the 1800s. By the late 1890s, applied geological activities in North America were of a scope similar to those in Europe. A brief summary of some major milestones on both sides of Atlantic follows.

Some geological concepts in Europe that were relevant to engineering works

The first recorded recognition of extinct volcanoes was made in central France (Auvergne) by Guettard and Malesherves in 1751. Within a century, this became a classic geological area to support plutonism. Professor Abraham Gottlob Werner, of Freiburg Mining Academy, Saxony (1750 to 1817), the apostle of "neptunism," was a baby when the volcanoes of Auvergne were recognized.

The concept of plotting useful geological data on a map was proposed by Lister in 1684. Guettard (1751) made the first modern-type, rudimentary geological map in 1746, a map of England and France (Carte Mineralogique). A similar map of the Middle East was released in 1751 (Guettard, 1755), and one of both Switzerland and eastern North America in 1752; the maps were based on others' observations, as Guettard never set foot in the Middle East or America.

The success of his maps led Guettard and Antoine-Laurent Lavoiser, a chemist, in 1766, to undertake an overwhelming project to geologically map all 214 quadrangles of France for the Ministry of Mines. This constituted the first national field geological survey (1766 to 1777). After 11 years, they had finished 16 quadrangle sheets, with another dozen partially completed, when the Inspector General of Mines took over the project (Faul and Faul, 1983, p. 87). Rappaport (1969) has described these early maps in a geological atlas of the works of Guettard, Lavoisier, and Monnet.

Techniques of geological mapping developed rapidly in the late eighteenth century in France (Desmarest, 1804, 1806; Soulavie, 1780–1784; Cuvier and Brongniart, 1808–1811). Most geologic maps were directed toward specific problems, such as Desmarest's mapping (1763 to 1764) of the volcanic rocks of Auvergne (Volvic to the Mont-Doré), central France. In 1765, he presented a stunning set of field observations on volcanic mechanisms to the Academy of Science. Among his conclusions: that the prismatic and hexagonal pattern of columnar joints in the lava flows were the result of cooling and not desiccation; that ancient lavas from the volcano flowed downhill and followed the river valleys; that blocked rivers began cutting new valleys, and the ancient topography was shaped by running water; that old flows may cap flat-topped hills or can be found in valleys or on plains; and that the prismatic lavas stand on a bed of scoria and soil that overlay the old granite bedrock of the region, confirming the idea of "uniformitarianism" (Geikie, 1905).

Historically, earthquakes have been among the most feared geologic phenomena, and experiences in both Asia and Europe support this concern. Earthquakes have been one of China's great scourges for three millennia—a major event occurs every six years on the average. The most destructive quake ever recorded occurred in Kansu Province in A.D. 1556, killing 820,000 persons (Wilson, 1981). The Chinese have reliable records of all events during the past 2,750 years, and they are leaders in using the history of seismic cycles and events in hazard analysis and earthquake prediction.

The disastrous Lisbon earthquake of November 1, 1775, felt over much of Europe, brought forth an immediate and widespread rebirth of interest in geological phenomena. John Winthrop (1755) of Harvard College was the first American to attempt a physical explanation of the natural phenomenon of earthquakes after the Lisbon event and a small shock that year in Boston. John Michell (1724 to 1793), a perceptive English scientist, analyzed all available information on earthquakes and published a correct understanding of the geologic causes and associated events. He stated, "Earthquakes have their origin underground, are connected with volcanism and faulting, and the motion of the earth in earthquakes is partly tremulous and partly propagated by waves" (Michell, 1760). He estimated the velocity of waves in the Lisbon earthquake as more than 30 km per minute and proposed a method of determining the place of origin from the time of arrival at different geographic places (Michell, 1760, p. 569, p. 571–572, 574, 626). He further considered the compressibility and elasticity of the earth and observed that earthquake waves at sea (tsunamis) travel at velocities dependent on the depth of the water.

Many European scientists and engineers contributed to applied geological knowledge prior to early 1800s. Perhaps the three men most responsible for advancing the acceptance of the new science were Leonardo da Vinci of Italy (early 1500s), Henri Gautier of France, and William Smith of England, a civil engineer (1790s to 1827). Although Smith is credited with gaining the acceptance of geology by a large group of practicing engineers in the 1800s, he was not the first practitioner of geology for engineering works, as some writers have alluded. This honor is due Leonardo da Vinci (1452 to 1519), the earliest recorded "applied geologist" for engineered works (work discussed above).

During his construction of canals in England (1790s), Smith recognized the sequential succession of geologic strata and successfully applied this knowledge to canal location, excavation, and the elimination of gravity slides. His understanding of the strata led to work in the field of ground water, and Smith successfully drained the Prisley Bog and converted it into agricultural-land in 1801. Later, in 1810, he restored the flow of the hot springs at Bath, England, after they had gone dry (Adams, 1938). These feats attracted the attention of the engineering profession, and Smith was called on more and more often for counsel and opinions. His work led to the preparation of a colored geologic map of England in 1815; this was long after Guettard's (1746 to 1752) early maps that covered a large area in central France. Smith, an early "applied geologist," was a "founder of British stratigraphy" from his classifying and mapping of the soils of farmlands. He identified the residual fossils that had weathered from the underlying bedrock units and were characteristic of some soil types. Using them as distinctive marker, he identified proven productive soils and separated them from the weak or poorly productive soils on his early geologic maps (Bryan, 1939). Following Smith's work, G. B. Brocchi (1820) published a treatise on the foundation materials of Rome.

In spite of maps and improved geological concepts, canals and water-supply projects continued to experience engineering failures due to ignorance or misunderstanding of foundation conditions. An example is the Puentes Dam, Spain, in 1802. The 50-m-high dam was located across a deep, 17-m-wide inner canyon backfilled with gravel and sand. The 284-m-long gravity and masonry–type dam was anchored in sandstone abutments and built on a grillage supported on piles and protected by an upstream apron in the backfilled sector. The reservoir filled after 11 years, but the central foundation of unconsolidated deposits was weakened by inflow (probably piping) and a block 17 m wide and 33 m high failed (Anderson and Trigg, 1976, p. 20).

Environmental concerns. Throughout Europe by the 1860s, there was a growing demand for geological information on the soils and rock conditions beneath major cities. One response was a 300-page treatment by Professor Eduardo Suess, University of Vienna, on "The foundation materials and soils of Vienna" (*Der Boden der Stadt Wien*) in 1862; the milestone study contained a soils map of the famous city and described the shallow ground-water occurrence throughout the basin controlled by the Danube River.

Another early geologist to emphasize that the geologic environment had an affect on the health and development of inhabitants was a Frenchman, Bernard von Cotta (1866). His writings, from 1858 to the 1880s, described the effects of geologic structure on springs, vegetation, human activities, mining, industry, trade, and military action, as well as the human life span, cultural development, and land use. Another early French book on applied geology appeared in 1887 when Nivoit published a general text that included a short description of the basic principles affecting surface excavation, ground water, slope stability, and glacier action.

The inhabitants of Vienna experienced a serious cholera epidemic in the late 1300s. The city was again threatened in the 1860s, and community leaders feared an outbreak of sickness because the city's many hand-dug wells could be polluted by the near-surface release of sewage. Geologist Eduardo Suess was ac-

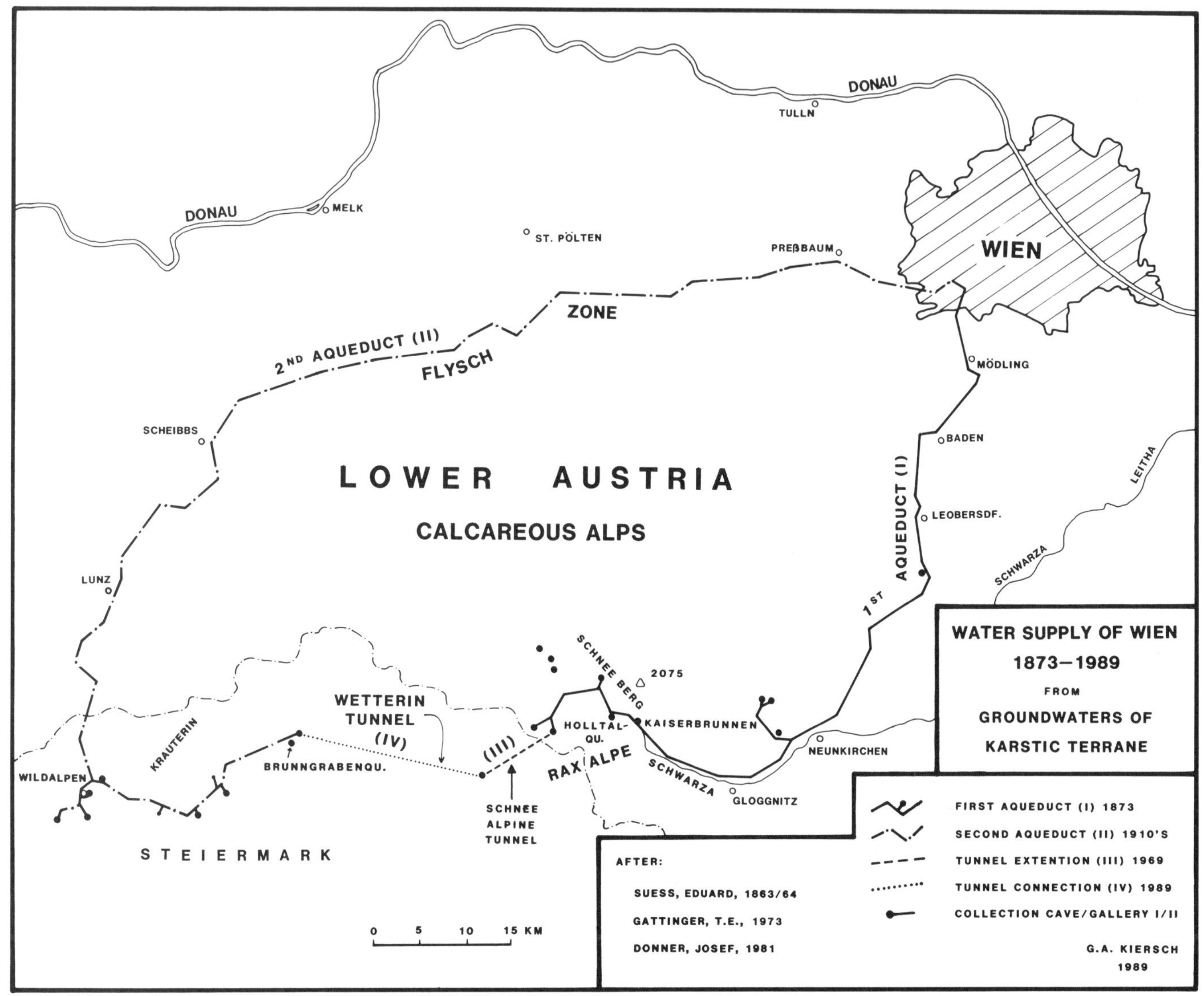

Figure 11. Plan map of the water supply of Vienna, derived from ground waters of the karstic terrane of the lower Austrian calcareous Alps, investigated by Eduardo Suess in 1863–1864. The first aqueduct (I) was completed and operating in 1873 from sources in Rax Alpe and Schneeberg regions; the second aqueduct (II) was built around 1910; extension III was built in 1969. The Wetterin tunnel (IV) that connects with aqueduct II is nearing completion in 1990.

tive in Vienna politics and advanced the concept of bringing in water from sources outside the area. The Burgermeister encouraged Suess to undertake a "feasibility" study immediately and recommend a project to supply the water needs of 1.8 million citizens.

Spring-fed water supply for Vienna. During 1863 to 1864, Suess investigated the occurrence, distribution, and runoff of surface and ground water in the karstic limestones in the Schneeberg and Rax Alpe regions, about 80 km southwest of Wien (Fig. 11), where the potential for a large-scale water supply existed.

The Schneeberg region consists mainly of Wettersteinkalk, a mid-Triassic mass of limestones and granular permeable dolomites as much as 1,000 m thick that rests on a basement of Ordovician to Permian, mildly metamorphosed schists. Solution action within the thick mass of carbonate rocks began in the Jurassic, but the main development of the interconnected network of openings developed during Tertiary time when a tropical climate prevailed; tectonic uplift accelerated the solution action. The amount of total dissolution accomplished by one large spring, flowing at 100 L/s throughout the past 1 m.y. would approach a cubic kilometer of the carbonate rocks, according to Kollmann (1983). Observations since 1873 indicate the openings have been widened due to ongoing solution action. The Schwarza River valley has downcut to near bedrock level, and today the dense schist basement acts as a downward barrier that aids collec-

Figure 12. Typical large-scale sinkhole of the upper surface, Schneeberg and Rax Alpe region seen in foreground. This sinkhole is part of the interconnected system of openings that collects runoff and channels it to springs and caverns near the floor of the Schwartza River valley where it is captured by aqueduct I. (Photo by G. A. Kiersch, 1964.)

tion of the ground water from the overlying karstic rocks. The catchment "basin" of the Schneeberg and Rax Alpe regions is more than 150 km^2 and includes about 50 important springs of 10 to more than 100 L/s flow. The mountain masses contain an interconnected network of solutioned openings that extend from the uppermost slopes, with large sinkholes (Fig. 12), to an intervening system of cavities, caves, and tubular openings that transmit the surface-water runoff to the Schwarza River valley below, as at Kaiserbrunnen (Fig. 13). Suess's initial studies (Suess, 1863 to 1864) forecast an adequate yearly water supply, with resources more than enough to meet the needs of Vienna's 1.8 million inhabitants (today only 1.4 million).

Suess acted quickly, following completion of his feasibility study, to convince the community leaders of the unusual potential for a water supply by personally leading excursions to the Schneeberg and Rax Alpe regions. The proposed water-supply system consisted of collection areas at the mouths of natural caves, aqueducts with gravity flow, arched aqueducts across valleys, oepn canals, wooden flumes, gate houses to control flow, harnessed flow for generating electricity, and storage reservoirs (Fig. 11). The city leaders and the Burgermeister, Herrn Cajetan Felder, au;thorized the nearly 100-km system of engineering works, and construction began in 1868 (Donner, 1981). The whole system, with several feeder aqueducts, opened on September 1, 1873. Water reached Vienna on October 24, 1873, only ten years after Suess began his conceptual geological studies—perhaps a record for from concept to completion. Farsightedly, the system was designed for 220,000 m^3/day capacity, although in 1873 the aqueduct carried only 138,000 m^3/day; other sources have been added since, and the system transmits at capacity today.

Spring-flows throughout the Schneeberg region are collected at the mouths of caves and tubular openings near the valley floor of the Schwarza River (Fig. 11), with the aid of adits and galleries driven into the carbonate cliffs to increase water flow. A series of weirs serves as a means of collecting the water. Since 1873, three major collecting and transporting systems have been added to supplement the original water aqueduct. System II was constructed in the 1910s so Vienna could tap the water resources of eastern Austria that drained east and northward to the Danube River (Fig. 11) and replace water lost along the leaky original aqueduct. In 1966 to 1969, the Schneealpen tunnel and gallery (system III) was constructed to tap the water just west of Hollental and the end of the first Schneeberg aqueduct (system I) (Fig. 11). This added 400 L/s or one-eighth of the total water demand (Gattinger, 1973). In 1985 to 1990, a connecting Wetterin tunnel

Figure 13. Rugged terrain of the eastern margin, Schwartza River valley. Sinkhole structure of Figure 12 is in the upper left margin, beyond the group of students from the Technische Hochschule, Wien, who are studying the geology and aqueduct features. (Photo by G. A. Kiersch, 1964.)

was driven from the Schneealpen tunnel to near Brunngrabenqu on the western drainage of the Salza River (Fig. 11). This provides a connection between the original systems I to III and system II aqueducts and allows for an interchange of water between the northwestern and southern slopes of the eastern Alps. Moreover, if slides or difficulties should occur within an unstable sector of flysch deposits along system II, the interconnected system will allow pumping to reroute all water through the southern connecting tunnels and via the original aqueduct (I) to Vienna. Though expensive, the current system can now bypass the potential slide hazards.

Suess, a pioneer in applied geology, investigated groundwater sources for many other cities in Europe and utilized his concept of ground-water containment resulting from fault blocks and solution action at depth that created an interconnected network of openings in the carbonate rocks.

Ground-water principles. Teodoro Ardemans, pioneer water-supply engineer of Spain (1664 to 1726), published extensively on practical problems of ground water and its supply that involved water chemistry, pollution, well construction, and water law. His two books in 1714 had an effect on the building codes of Madrid for almost 100 years, and two later books, in 1724, are important to the history of water engineering and hydrogeology (Davis, 1973).

By the mid-1800s, a strong scientifically oriented interest in water supplies and laws governing ground water had developed in Europe out of necessity. In 1856, Henri Darch, a well-known French engineer, stated the mathematical law governing the flow of ground water in a water-supply report for Dijon, France, that resembled a scientific monograph on hydraulic engineering (Davis and DeWiest, 1966). Seven years later, Jules Dupuit (1863) of France was the first scientist to develop a formula for the flow of water in a well. Adolph Thiem of Germany modified Dupuit's formula in 1870 to compute the hydraulic properties of an aquifer by pumping a well and observing the effect in other wells nearby (Thiem, 1906). During this period, the early studies of water chemistry were made by B. M. Lersch (1864) of Germany, while T. S. Hunt (1865) of Canada made some of the earliest geochemical interpretations (Davis and DeWiest, 1966).

One of the earliest treatises on the geology of ground water (Daubree, 1887) was published in France; and Lucas (1880) was probably the first person to use the term hydrogeology for the study of underground waters. Soon thereafter, in Austria, Forchheimer (1886) introduced the concept of three-dimensional equipotential surfaces of the hydraulic head that led to the construction and use of flow nets; he was the first scientist-engineer to apply higher mathematics to ground-water flow. Later, an American, C. S. Schlichter (1899), published concepts

Figure 14. Mt. Cenis tunnel in Alps, showing the boring machinery and conditions under which the tunnel was driven, from 1857 to 1872. The engineering techniques of drilling and blasting aided in driving the Hoosac tunnel in Massachusetts, completed in 1874. (From Kiersch, 1955, p. 15.)

on equipotential surfaces similar to those of Forchheimer (Davis and DeWiest, 1966).

Transportation routes via tunnels and canals. The growth of cities and need for better transportation routes signaled a new period for tunneling; the 1803 Napoleon-built tunnel, near Tronquori, France, was 6 km long. Soon thereafter followed roads and vehicular tunnels: the first tunnel attempted beneath the Thames River in London in 1808, by Threvethic, was a soft-ground (subaqueous) tunnel that collapsed (Kiersch, 1955, p. 14). The second attempt, started by Brunel in 1823, met with several setbacks, mainly due to the geologic conditions. Although 39 boreholes were drilled to explore the alignment, the tunnel was driven in part through young alluvial deposits instead of wholly within the underlying London Clay, as anticipated. Some of England's leading geologists warned that a water-bearing sand bed would be encountered at a depth of 15 m below the river bed. On this assumption, the tunnel was bored close beneath the channel of the river. Actually, this forecast proved wrong and was responsible for many of the construction problems and failures. By 1827, the tunnel had advanced to beneath the river, but hydrostatic pressure breached it. After many such setbacks, the Thames tunnel was completed in 1843 and still stands as the largest soft-ground tunnel in existence (White and White, 1953, p. 15), a milestone in geological experience and engineering skill for soft-rock tunneling beneath water. The disastrous and expensive history of the Thames tunnel undoubtedly influenced the decision to prepare a complete geological study and map that was critical to the planning and design of the Cofton tunnel in England (1838 to 1841)—the first recorded application of geology for such a purpose.

While the Thames tunnel was underway, construction of railway tunnels was beginning in England; Stephenson constructed the first railway tunnel on the Liverpool-to-Manchester line from 1826 to 1830.

The Alpine passes were the scenes of the first modern large-scale applications of geology in the construction of hard-rock tunnels. By 1857, when construction of the Mont Cenis tunnel (Fig. 14) began, modern views on alpine geology were being established by the pioneer work of De Saussure from 1740 to 1799, followed by a long line of prominent workers such as von der Linth, Suess, Wagner, Heim, Gertrand, Schardt, Lugeon, Argand, and others. The Mont Cenis (1857 to 1872) and such later tunnels as St. Gotthardt (1872 to 1882), Arlberg (1880 to 1883), Simplon (1898 to 1906), and Loetschberg (1906 to 1911) were preceded by preliminary geological studies and maps considered at that time to be exhaustive. The experience and knowledge gained in "forecasting" the conditions expected along these Alpine tunnels, combined with observations of the features exposed by construction, contributed many "firsts" to our knowledge of tunnel geology.

Some conditions were overcome, while others resulted in catastrophies and changes in tunnel alignment (White and White, 1953). For example, great quantities of ground-water inflow, rock temperatures to 106°F, and squeezing ground were first encountered in the St. Gotthardt tunnel (Fig. 15) when the alignment went beneath a deep ancestral valley backfilled with gravelly sediments. Boreholes drilled in 1944 from within the St. Gotthardt tunnel to investigate the excessive pressures and stresses revealed only 40 m of rock cover beneath the Andermatt valley. The Simplon tunnel encountered inflows of hot water with temperatures to 131°F when the bore was 2,150 m below the surface. The Loetschberg tunnel taught geologists additional lessons in forecasting tunnel conditions below a river valley. From the surface conditions (no borings along the alignment), geologists had forecast sound rock at 100 m below the Kander River valley, so the tunnel was constructed at a depth of 180 m below the valley floor. Without warning in 1908, water-bearing glacial debris was encountered in the heading beneath the river; the tunnel was flooded within 10 minutes, and 25 men lost their lives. Subsequent borings indicated the glacial debris extended to below 285 m, which necessitated changing the tunnel alignment across the valley. This "error" of judgment impressed on geologists that surface indications must be supplemented by subsurface borings when establishing tunnel grade, particularly beneath a valley of any type.

An interesting description of the geologic conditions encountered in the collapsed Unterstein tunnel (near Salzburg, Austria) is given by Wagner (1884); some of the typical field relations between the talc-schist rock slides and collapse of the railway tunnels are shown on Figure 16. Wagner (1884) presented an early "slip-circle" interpretation of landslide failure surfaces in rock.

In 1874, Rziha published a treatise on the influence of geologic factors in tunneling; his work still stands as an important early contribution to the application of geology to tunneling, along with the work on alpine tunnels by Wagner (1884) and later contributions of Fox (1907) and Stini (1950). A further milestone was recorded in 1874 when Ferdinand R. von Hochstetter delivered the annual vice-chancellor lecture at the Technische Hochschule, Vienna, on the subject of "Geology and Railway Building." Von Hochstetter is credited with an early use of the term engineering geology.

Perhaps the earliest attempt in Europe to describe the relevance of geologic features to engineering structures was published

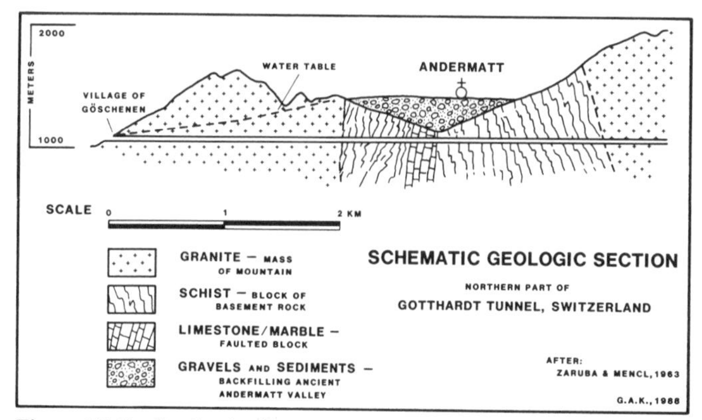

Figure 15. St. Gotthardt tunnel in Alps encountered high rock temperatures (106° F), a large ground-water inflow, and squeezing ground. The depth of gravels and sediments beneath Andermatt Valley was not well understood and contributed to the tunneling difficulties; ground conditions would have been less severe with a deeper grade.

by Penning in 1880; he recorded the effects of geology on a series of engineering works in England in a text on geology for engineers that followed the earlier approach of Thomassy (1860). The text presents only broad principles (by current standards), but is significant because Penning analyzed the importance of geological observation and reasoning in engineering practice. Wagner's (1884) later book on tunnel geology, like that of Rzhiha (1874), stressed the experience gained from the alpine tunnels; Wagner emphasized many of the factors mentioned by previous writers and additionally presented an interpretation of landslide phenomena in rock consistent with the current "slip-circle" explanation.

The Corinth Canal in Greece, undertaken by French capital and engineers in 1882, followed nearly the same alignment and used the same cuts prepared by the Romans under Nero: 35 shafts and a large cistern remained as evidence of the Roman efforts. Sprecht (1884) described the construction methods and the unique system for removing material from the uppermost cut of the canal excavation: along a tunnel located 47 m above sea level. The canal walls are cut on a 70° slope (maximum height, 78 m) with only two or three narrow berms (1-m wide) throughout (Fig. 17). The main rock consists of Pliocene marls overlain by Pleistocene conglomerates. These rocks are cut by numerous faults reflecting apparent horst and graben structures in this tectonically active region (Mariolakas and Stiros, 1987). The soft to medium-hard, brownish marls with some sandstone beds contributed to the ease of excavation and stability of cut slopes (inset, Fig. 17).

Summary. Interest in geology as applied to engineering works grew steadily throughout Europe in the late 1800s. By the early 1900s, geological counsel was commonly accepted for the planning of industrial expansion, as the progressively larger engineering structures usually meant a proportionately greater number of complex geological problems for practitioners.

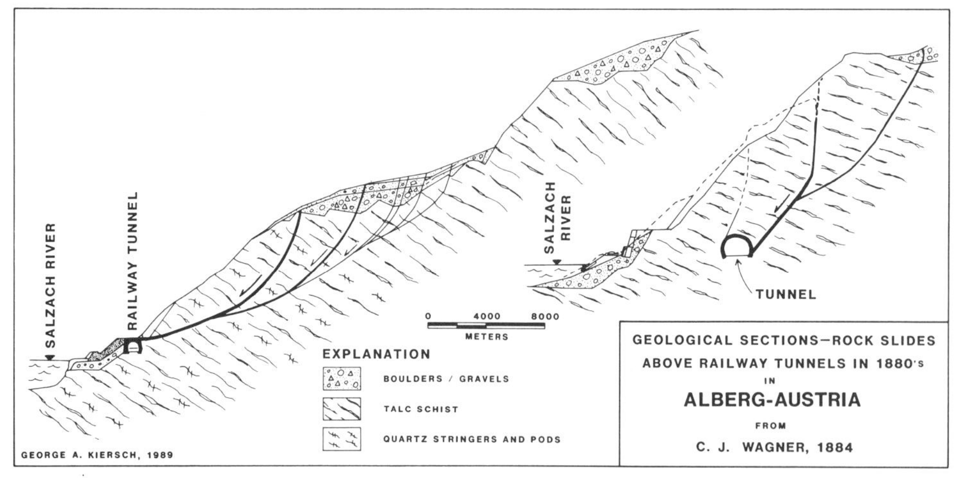

Figure 16. Geological sections of rock slides in the 1880s above railway tunnels at Alberg, Austria showing an early "slip-circle" interpretation of sliding phenomena in a rock mass.

Early applied geology in Mexico

During the pre-Hispanic period (200 B.C. to A.D. 1521), the Mayas and the Aztecs built the monumental pyramids of Teotihuacan, Chichen Itza, and El Tajin, demonstrating an early "geologic" understanding for problems related to construction on soft rocks or soils, as well as the management of water for reservoirs and irrigation (Fig. 3).

With the arrival of the Spaniards and the Conquest (1519 to 1521), most of the Aztec Empire was destroyed, along with their engineering skills and knowledge, and a new city was built on top of Tenochtitlan. The "new" structures, built by Spanish and French technicians, ignored the behavior and geologic properties of the "ground" in Mexico that were well-known to the Aztecs, such as the inherent problems related to settlement, weak bearing capacity, and slope stability. The Spanish were not equipped to solve these problems.

The early applications of geology to engineering works in Mexico go back to the eighteenth and first half of the nineteenth centuries. Engineers applied geological principles and know-how to building underground workings for metal mines, and dams for water reservoirs to serve mining enterprises and local irrigation. The majority were the efforts of private enterprise, including the dams and reservoirs of Saucillo (1730), Yuriria (1850), San Martin (1765), Torre Blanca (1850), Mal Paso (1865), La Mula (1868), and others (Fig. 18). More than 20 of the dams are still in operation, although badly clogged with sediments. By the end of the nineteenth century, the first intercontinental infrastructure system of railroads was built, by both government and private enterprise. A review of geology for engineering works in Mexico is given in Chapter 3 of this volume.

Some eighteenth and nineteenth century geological concepts in the United States relevant to engineering works

Historically there was no subject of "geology" in 1776, when the new nation was established. The term geology for earth science study was coined by Jean Andre DeLuc in 1778, but not until the nineteenth century did the word gain recognition and acceptance. However, terms for separate categories of geology were used much earlier: mineralogy in 1690, and lithology in 1716. Geologists began to emerge in America in the early part of the nineteenth century. By 1818, geology was growing as a descriptive science due to the work of Smith (1799) and Latrobe (1804), and the mapping activities of Maclure (Hazen, 1944) and others, who introduced a systematic classification of geologic materials in America. William Maclure made the first geologic map of America east of the Mississippi in 1809 and prepared an "Essay on Rocks" (Maclure, 1818). James F. and Samuel L. Dana (1818) prepared a colored map of the Boston area, while Gerard Troost (1826) described the geology of Philadelphia and Edward Hitchcock (1836) described the geology around Portland, Maine.

A strong group of state geological surveys was established in the period 1823 to 1837. The first survey was in North Carolina in 1823, followed by South Carolina in 1824, Massachusetts (1830), Tennessee (1831), Maryland (1834), New Jersey, Connecticut, and Virginia (1835), New York and Georgia (1836), and Ohio (1837); 14 state surveys in all were found between 1830 and 1837.

That geologic reasoning should be part of engineering practice was recognized in North America by some early geologists. Some of the early investigations by the state surveys were milestone contributions to applied geology for engineered works. Examples include: the historic evaluation by James Hall of the New York State Geological Survey (1839) of the engineer's unit-price classification for "solid rock" versus "weak shale" within a thick formational sequence of "slate rock and shale" to be excavated at the locks of Erie Canal in Lockport; and W. W. Mather and Charles Whittlesey of the Ohio Survey who analyzed the treacherous rotational slides and failures along the Lake Erie shoreline near Cleveland with "slip-circle" reasoning in the annual report of 1838. This report also described littoral-drift cur-

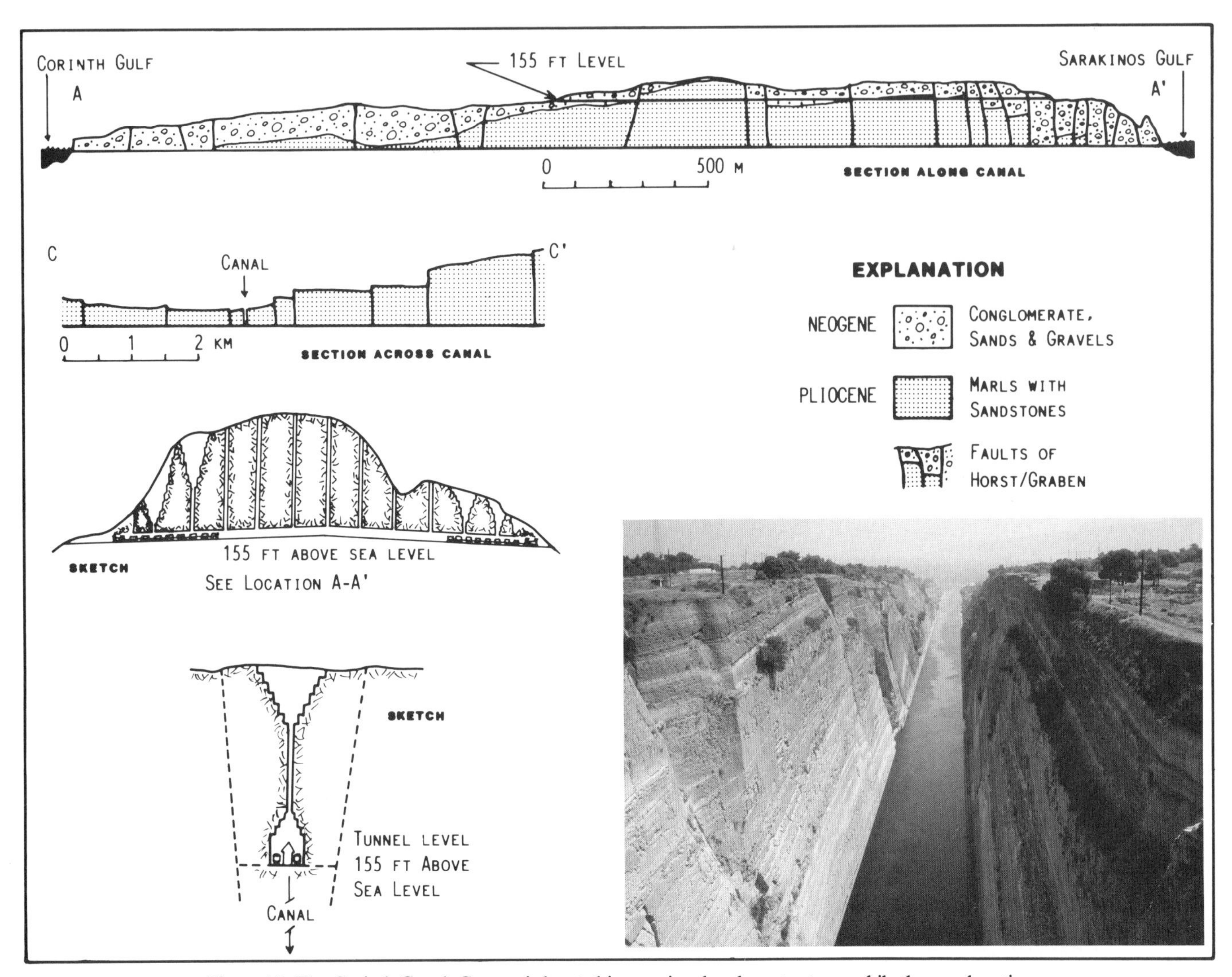

Figure 17. The Corinth Canal, Greece, is located in a regional graben structure, while the canal section discloses an apparent horst structure. Construction in the early 1880s excavated the uppermost part by means of a series of shafts to a tunnel level; mining progressed downward, as shown on the sketches, with waste removed through the tunnel. (After Specht, 1884; and section of Freyberg, 1973, *in* Mariolakas and Stiros, 1987. (Inset) The Corinth Canal, Greece, looking toward the eastern entrance. The soft to medium-hard light-brown, weakly bedded Neogene marly beds, with some sandstones, aided the excavation and stability of the deep, steep-sided (70°) open cuts in the 1880s. Much of the canal was built with pick and shovel work. (Photo G. A. Kiersch, 1983.)

rents and the use of groins (groynes) as a means of shoreline protection. The Ohio study, eight years before the French-published work of Collin (1846; see Skempton, 1946) on slope failure along canals of southern France (Legget and Karrow, 1983), exemplified the growing interest in applied geology throughout North America. During this period, Cozzens (1843) prepared a colored geologic map of Manhattan Island with eight map units that still make sense today. Cozzens' mapping and water-supply study was the first known interaction of geology and engineering that involved Manhattan Island and New York City, although Gale (1839, p. 186–87) had earlier described the glacial deposits of Manhattan Island and logged the shallow-water wells to explain the saline waters encountered.

About 1850 a French hydraulic engineer, Marie-Joseph Raymond Thomassy, took up residence in New Orleans and became fascinated with the physical characteristics of Louisiana and the dynamic power of the Mississippi River. His pioneering investigations along the river and throughout the basin were published as *Practical Geology of Louisiana* (1860). The work included an early description of alluvial deposits and the lower delta as the "key to Louisiana geology." This work is probably the earliest text on practical applied geology and engineering problems in North America (Thomassy, 1860; Skinner, 1979); it was followed by Ansted (1869).

The Appalachian Mountains were a barrier to westward expansion for colonists who settled on the Atlantic Coast. They did not start moving westward until the early part of the nineteenth century, and migration became a rush after gold was discovered in California in 1848. Soon railways were proposed to cross the continent, and serve the Pacific Coast and the vast area west of the Mississippi River. The railroad grant lands for construction, and for associated expeditions to explore territories, are briefly reviewed in Chapter 18 of this volume.

Routes of canals and roads. George Washington's dream of improving the Potomac River began in 1785 with construction of several short canals to skirt the waterfalls (Hahn, 1976) that characterized the "fall line" along the east side of the Appalachians. These waterfalls defined the upper limits of navigation and provided power for mills of all kinds; many early settlements were established along this zone (Merrill, 1964).

The earliest overland transportation project was the Philadelphia-Lancaster Turnpike in 1772. The Cumberland road, along the valley of the Potomac, was started in 1802 (Waggoner, 1958; Rose, 1976). Because the early roads and canals were built with a minimal understanding of geologic conditions and the terrain, advances frequently came during construction. One such geological revelation, while building the Cumberland road, was John L. McAdams' discovery in 1816 (Rose, 1976) that repeated passes of a heavy roller over a layer of broken stone produced grinding and powdered fines; if wetted, they became a firmly cemented mass of rock and paving (Shaler, 1895a). Basalt, limestone, and quartzite were the most suitable for this crushed-rock "macadam" road construction.

Canal fever began with the Erie Canal, across northern New York State, which was started in 1817 and completed in 1825; the initial system consisted of 83 locks, canals 12 m wide, and river portions that combined to stretch for almost 600 km, from the Hudson River to Lake Erie, forming the longest canal system

Figure 18. Map of Mexico showing the locations of six of the dams built between 1730 and 1868 in the central part of the country by private owners and miners.

of its day. In the subsequent decades, many improvements, including additional locks and a feeder canal (Oswego to Lake Ontario in 1828), were constructed; the overall system is still in use as the New York barge canal, for limited-size traffic. The Lehigh Canal to transport coal from the White Haven and Mauch Chunk anthracite regions to Easton and the Delaware River outlet canal to Chesapeake Bay was started in 1818 and opened in 1829.

The Chesapeake and Ohio Canal was an ambitious project planned from Georgetown in Washington, D.C., on the Potomac River to Pittsburgh, Pennsylvania, by way of the Younghiogheny and Monogahela Valleys; this route required a summit tunnel more than 6 km long. Work began on July 4, 1828; after 13 years of difficult construction, the canal reached Dam 6, 215 km from Washington, D.C. By 1850 it extended to Cumberland, Maryland. Most of the canal structures were built of stone or earth fill, and wherever possible the canal was excavated in soil or weak shale rock; embankments and slopes on the river side were riprapped with local stone. Some early applications of geology for engineering works are demonstrated among the 182 culverts, 11 aqueducts, 6 dams, and 74 locks (Davis, 1970) of this canal.

A rare geological investigation of a large engineering project, the proposed Kanawha Canal and Reservoir on the James River in the Allegheny Mountains, was undertaken by Toumey (1851). The investigation and report utilized geological cross sections and sketches and provided one of the earliest technical discussions on the disintegration of granite, the suitability of available rocks for building materials, and the sequence of rock types to be encountered in tunnels that were being proposed.

Early tunnels, subways, towers and monuments, and dams. There are numerous cases on record of geological guidance for engineering works during the nineteenth century. They include construction of a number of tunnels (earliest in 1821) and projects such as dams and reservoirs, subways, towers and monuments, and coastal installations (Kiersch, 1955, p. 14–17).

A great tunnel through Hoosac Mountain, Massachusetts, begun in 1851, progressed very slowly until engineering techniques of drilling and blasting, developed on the Mont Cenis tunnel (Fig. 14), were used. The Hoosac tunnel (7 km long) was finished in 1874 and stood as the longest tunnel in America until it was eclipsed by the Moffat tunnel, Colorado, in 1927.

Perhaps the most significant advances in the then little understood interactions between engineering design and geology were in soft-ground tunneling. In driving the Hudson River tunnel (1873 to 1904), White and White (1953) have documented the effective use of compressed air as a protective shield (first used in the Thames River tunnel in 1868); also, the use of clay to reduce leakage was successful for the first time on record in America. The river bed was blanketed to stop water inflow; to harden the silt and clay along the tunnel, flames from torches baked the clay in place, making a brick-like tunnel wall and back (White and White, 1953, p. 23).

The first subway in America started in Boston in 1890. New York completed its first section of subway in 1904. Actually, many cities in Europe had subways earlier, such as London (1863), Glasgow (1883), Budapest, Paris, and Berlin. To date, New York has built more than 375 km of subways. Most subways were built by the cut-and-cover method and resulted in developing the art of underpinning to extend building foundations to the rock line, or below subway grade.

A landmark early coastal project was Minot's Ledge Lighthouse, constructed off the coast of Cohasset, Massachusetts. The first structure (1847 to 1851) collapsed under high seas, and the second, present, tower was built in 1855 to 1860. The ingenious design and construction used steel piling, and the masonry building blocks were bolted and anchored to the granite foundation. The lowest course of masonry blocks had to be specially shaped to accommodate seams and irregularities of the foundation. Tower blocks locked and dovetailed so that the impact of each wave made the edifice stronger (Snow, 1954). Today the lighthouse is a "Landmark in American Civil Engineering" (Civil Engineer, 1988).

The common foundation problems inherent to unconsolidated sediments encountered along the eastern coastal plain are illustrated by the history of the Washington Monument. Initially considered in 1783, the first cornerstone was not laid until 1848, and construction had progressed to a height of only 47 m when stopped by the Civil War. After construction resumed in 1880, the engineers realized the original foundation, as designed, would not support the obelisk monument; settlement and tilting had occurred due to the soft underlying sediments. Consequently, the original square foundation area of 24 m on a side was enclosed by a large concrete buttress and spread footing of 38 m square (Radbruch-Hall, 1987), and the 169-m-high monument was completed in 1884 (Fig. 19). Not until the 1930s, when borings were made for the design of landscaping, was the foundation fully revealed to be of sand and clay, and sand with gravel and clay, overlying a soft blue clay bed that rests on the crystalline bedrock (Fig. 19). Because the site is within an ancestral channel of the Potomac River, any removal of earth from around the obelisk would likely endanger the foundation by allowing some squeezing of the soft clay (Reed and Obermeier, 1982). The landscaping project was reevaluated and abandoned (Hoover, 1934).

Although wells and cisterns sunk in the glacial deposits and the underlying bedrock beginning in 1664 provided an adequate water supply for New York City, growth by the 1800s forced the New York State legislature in 1834 to authorize the development of supplies outside Manhattan Island (Fluhr and Terenzio, 1984, p. 2–3). This led to the construction of the Old Croton Dam and Reservoir on Croton River (Westchester County) in 1837 to 1842 and construction of the first aqueduct system into Manhattan. Gale (1839, p. 181) describes the quarrying of granite for the aqueduct construction from a site near 10th Avenue and 48th Street.

The New Croton Dam of the New York Board of Water Supply began construction in 1892 and was involved with the earliest reported deeply weathered and solutioned cavernous limestone foundation in America; the condition was unsuspected by

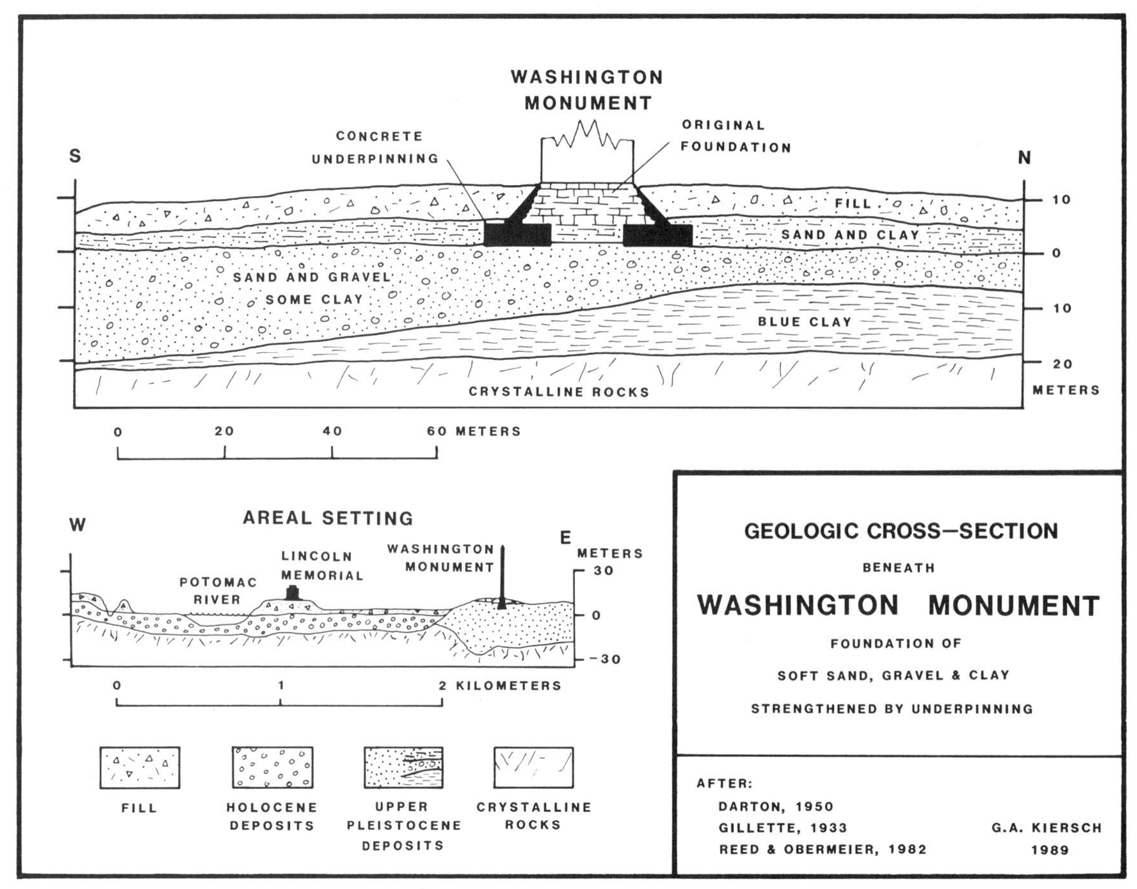

Figure 19. Section showing the soft sediments of the foundation beneath the Washington Monument, and the areal distribution of Holocene units beneath the Lincoln Memorial and Potomac River.

the consulting geologist, J. F. Kemp, until exposed in 1895 during construction (Gowen, 1900, p. 31). Investigation at that stage located a deeply weathered and solutioned zone along the contact of gneiss and marble units (Fig. 20). This unexpected occurrence required a redesign of the 356-m-long dam. Due to this delay for modification of the construction plans, and a partial reconstruction of the masonry dam (Gowen, 1900, p. 37), completion was not until January 17, 1906. The main gravity dam of rubble masonry was designed with an ungated side-channel spillway on the right abutment (Inset photo, Fig. 21); the water spills over the weir into a channel in rock.

The New Croton documentation of solutioned marble/limestone well below the ground-water table was a "first" for applied geologists in 1895 and changed the geological understanding of, and potential for, solution action and cavitation in calcareous rocks. Kemp's preconstruction report quoted W. M. Davis that "calcareous rocks were not solutioned below the groundwater table." Since the New Croton experience, exhaustive geological investigations at many U.S. sites have provided a clarification of, and working guidelines on, the "base levels of solution", ground-water flow below the water table, and the pattern of solutioning and openings formed (Rhoades and Sinicori, 1941; TVA, 1949).

Early highways. Although America expanded to new territory throughout the 1800s, an improved road system was not a major part of the transportation network until the last quarter of the century. The Civil War cut short early attempts at major road expansion, which were further curtailed in favor of a nationwide railroad network in the 1860s and 1870s. It was not until the invention of the bicycle in 1877 and the appearance of the automobile in the 1890s that the interest in improved roads was revived (Kiersch, 1955, p. 17–18). At first this need was met by constructing traffic-bound road surfaces, and later water-bound macadam-surface highways.

Reports on the influence of geology on roads appeared in European publications as early as 1888 (Huntting, 1945). The important geologic component of highway construction in both America and Europe was the immediate demand for serviceable crushed stone. As an outgrowth, geology made its initial contribution to highway construction with reports on the source of suitable road materials. The earliest such survey in the United States was on the road materials of Texas by Hill (1889), followed in the next decade by similar reports for North Carolina, Florida, Massachusetts, and New York. This initial developmental stage of geology as applied to highway construction extended from 1889 to about 1918. The growing acceptance of geologic data for

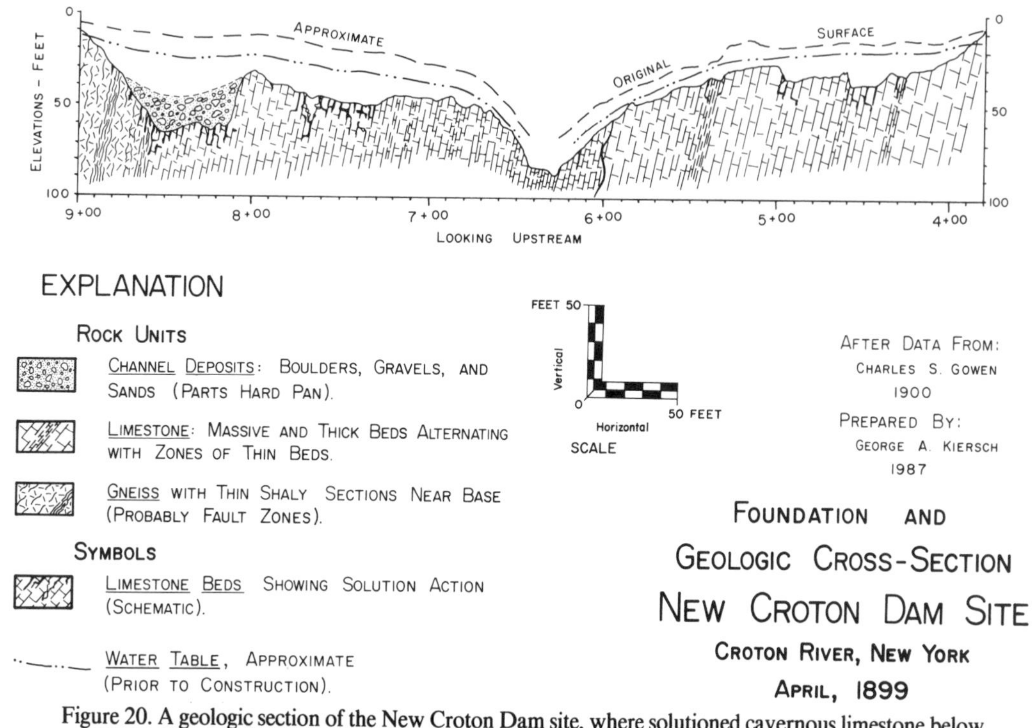

Figure 20. A geologic section of the New Croton Dam site, where solutioned cavernous limestone below water table was first encountered in America.

construction of roads throughout United States was reviewed by Shaler (1895a) and Merrill (1897).

The reports prepared by N. S. Shaler (1895b) on the road materials of Massachusetts, together with his responsible position as a member of the Massachusetts Highway Commission, support his designation as the first geologist for highways.

Early legal involvements. James Hall of the New York Geological Survey was requested to evaluate and classify a "slate rock and shale" sequence being excavated to enlarge the Erie Canal Locks at Lockport, New York, in 1839. The engineer's contract quoted a unit price for "solid rock" and a lower price for "slate rock and shale." The contractor claimed the total depth excavated, about 25 m, was in "solid rock" (Fig. 22).

Hall described the main section as a mass of shale with occasional layers of thin siliceous limestones. Furthermore, the soft "fresh shale" cleaved into blocks that decompose and crumble with time to form clay particles and small broken pieces. Although the shale was one of the softest rocks in the series, the engineer's contract contained some specifications that conflicted with the qualities of shale (contradistinction), and Hall was unable to state a clear classification of the shale as "not solid rock." Poor contract specifications concerning the classification of geologic materials and conditions of excavation are common even today, 150 years later. They continue to require the input of applied geoscientists (discussed in Chapter 24 and 25, this volume).

Another early case that claimed changes and damages to the environs was related to reclaiming the Cayuga marshes by constructing channels and drainages designed to improve a section of the Erie Canal. Originally authorized by the state of New York in 1825 (Emmons, 1838, p. 316), the early attempts were not very successful, and the New York State Legislature authorized further funding for engineering work in 1853. Specifically, the Erie Canal sector from Lake Onondaga westward, via the Seneca River past Jack's reef to Montezuma and through the Cayuga marshes and swamp lands, was to be improved by deepening and straightening the Seneca River, which would then provide effective drainage (Fig. 23). As work progressed, the landowners downstream filed claims for damage to their land; they contended that heavier flow of the spring runoff was caused by the river widening that cut across bars and enlarged the section of the channel from Montezuma and Jack's reef (Hayt, 1870, p. 48–57). Litigation was also filed by upstream landowners who claimed damages along the shores of Lake Onondaga and to the local salt works. This attempt to improve the Erie Canal constitutes a very early case of litigation in North America claiming a change to the environs caused by an operating engineering works. During this period, Lyell (1858) released an early text on the Earth's ongoing changes and their impact on man. Somewhat later, Williams (1886) published an extensive book on applied geology focused on America that aided legal considerations.

Early studies of earthquakes. During the period from 1811 to 1820, frequent earthquakes at localities in the central and eastern United States, along with volcanic activity in the western territories, aroused interest among American geologists. The New Madrid earthquake of 1811 to 1812, centered at New Madrid, Missouri, was probably the most serious earthquake in the history of the United States. The surface effects that occurred are a classic

Figure 21. New Croton masonry dam, faced with cut blocks of granite, looking northward along front from about station 4 + 50 to 9 + 00 (location Fig. 20). The unique, ungated, side-channel spillway was designed on the right abutment where an early stream existed (Sta. 8 + 00 to 9 + 00), as shown in the inset photograph. (Photos by G. A. Kiersch, 1988.)

example of all the changes an earthquake can effect to the terrain (Hodgson, 1964, p. 7–9; Fuller, 1912). The Charleston, South Carolina, earthquake of 1886 was investigated by geologists, and Dutton (1889) confirmed the suspected correlation between the strength of foundation materials and damage sustained.

Early consultants for engineering works

Professor William Otis Crosby, the "Father of Engineering Geology in America," began serving major engineering projects around Boston in 1893 where he was on the faculty of Massachusetts Institute of Technology. Although he was active in the field of mineral deposits and mining throughout America beginning in the 1880s, Crosby's activities after 1893 were mainly concerned with civil engineering works and his advice was sought in connection with many water-supply problems, dam and reservoir sites, and the alignments for tunnels and aqueducts (Appendix A).

James F. Kemp, a professor at Columbia University from 1895 to 1926, was appointed geological consultant to the Aqueduct Commission of New York City in 1895, the builder of the New Croton Dam and aqueduct, and forerunner of the New York City Board of Water Supply. Kemp's investigations of the already selected site and partly constructed Croton Dam showed problems with the deeply weathered and solutioned rock in the foundation (discussed above). Kemp was also consultant (1895 to 1898) on one of the earliest pressure tunnels in America, where the Croton aqueduct crosses beneath the Harlem River enroute to Manhattan. He located an unknown fault with a deeply weathered zone at the proposed tunnel level; this required a redesign of the alignment to minus 100 m elevation so that the pressure tunnel could be in satisfactory rock. This was a second early case where geological guidance changed the proposed design and location of an engineering works (Gowen, 1900).

Some geological concepts in Canada relevant to engineering works

Several early geological papers on Canadian projects had appeared by 1829 (Legget and Karrow, 1983). The Grenville Canal on the Ottawa River, built between 1819 and 1834, required the excavation of tough Precambrian bedrock by hand drilling and blasting. Similarly, construction of the 200-km-long Rideau Canal between 1826 and 1832 involved dams and locks founded on limestone and sandstone. Locating suitable quarry sites for aggregate and building stone required an understanding of geologic principles and concepts.

In 1847, a geological survey was established by the United Province of Canada (Quebec and Ontario) with William Logan as the director; this group became the Geological Survey of Canada in 1867. Logan had a serious appreciation of the importance of geology to engineering works and to natural hazards. One of Logan's early papers in1841 described the extensive landslides of April 1840, which formed a temporary dam on the Maskinonge River. The landslide mass involved what is now called the unusual and sensitive Leda Clay, responsible for many subsequent rapid slides and foundation difficulties throughout the St. Lawrence Valley region (see Chapter 2, this volume). Another early study concerned the location of the great Victoria Bridge across the St. Lawrence River at Montreal, completed in 1860, and other early surveys for routing the Canadian Pacific Railway (Chapter 3, this volume).

The construction of railways to link the widely scattered colonial settlements of modern Canada began with the Intercolonial Railway between Montreal, Halifax, and the Maritime Provinces (1864 to 1876). The chief engineer, Sanford Fleming, had an intuitive sense of the geologic conditions encountered, as demonstrated by his design of structures and protective works along the railway. For example, in 1870, test borings of the proposed locations for two multispan bridges across the Miramichi River in New Brunswick near Newcastle indicated one alignment did not have a satisfactory foundation. Consequently, engineer-geologist Fleming had additional borings drilled, and with improved understanding of the site, redesigned and built bridge piers that are still in use today (Legget and Karrow, 1983). Fleming's pioneering studies and exploration techniques in 1879 utilized in-place castings of test borings with insert rods in order to conduct penetration tests on the foundation strata 60 years before penetration tests became generally used in the 1930s.

The Canadian Pacific Railway from Montreal to the Pacific Coast was aided by geological studies in the 1880s through the encouragement of General Manager William Van Horne, himself an amateur geologist. The most serious geological problems encountered were landslides in the Thompson River Canyon sector of British Columbia (Leggett and Karrow, 1983).

Figure 22. The lock of the Erie Canal at Lockport, New York, where the section of "slate rock and shale" became a source of a contractor's dispute over unit prices for excavation in 1839. The uppermost rocks are exposed behind and left of the center building, and the lower units are exposed below the lock and bridge. (Photo by G. A. Kiersch, 1985.)

Academic training in geology for engineering works

The earliest lectures on geology for engineering works known in America were by Theodore B. Comstock in 1875 to 1876 at Cornell University in a practical *Structural and Economic Geology* course for architecture and engineering students. Comstock later pioneered geological activities in the western states that were a factor in establishing the Arizona School of Mines in 1891, and he served as the first president of the University of Arizona from 1893 to 1895. Near the end of the nineteenth century, several prominent geologists at American universities started to offer courses for engineering students on the applications of geology.

Professor William O. Crosby of the Massachusetts Institute of Technology offered the first continuing group of lectures in applied geology for engineered works in 1893 as part of the course in structural geology for students in civil and sanitary engineering and general studies. Professor R. S. Tarr beginning in 1894 offered a similar course at Cornell University called *Practical Geology/Dynamic Geology.* He was joined by Heinrich Ries in 1898 when Cornell offered coursework on geology to engineering students in three designated courses: *Mineralogy for Engineers, Economic Geology for Engineers,* and the aforementioned *Practical Geology/Dynamic Geology.* All were for one semester with some field trips.

Crosby's coursework was supported by a syllabus on the subject first published in 1893. He continued such a course for

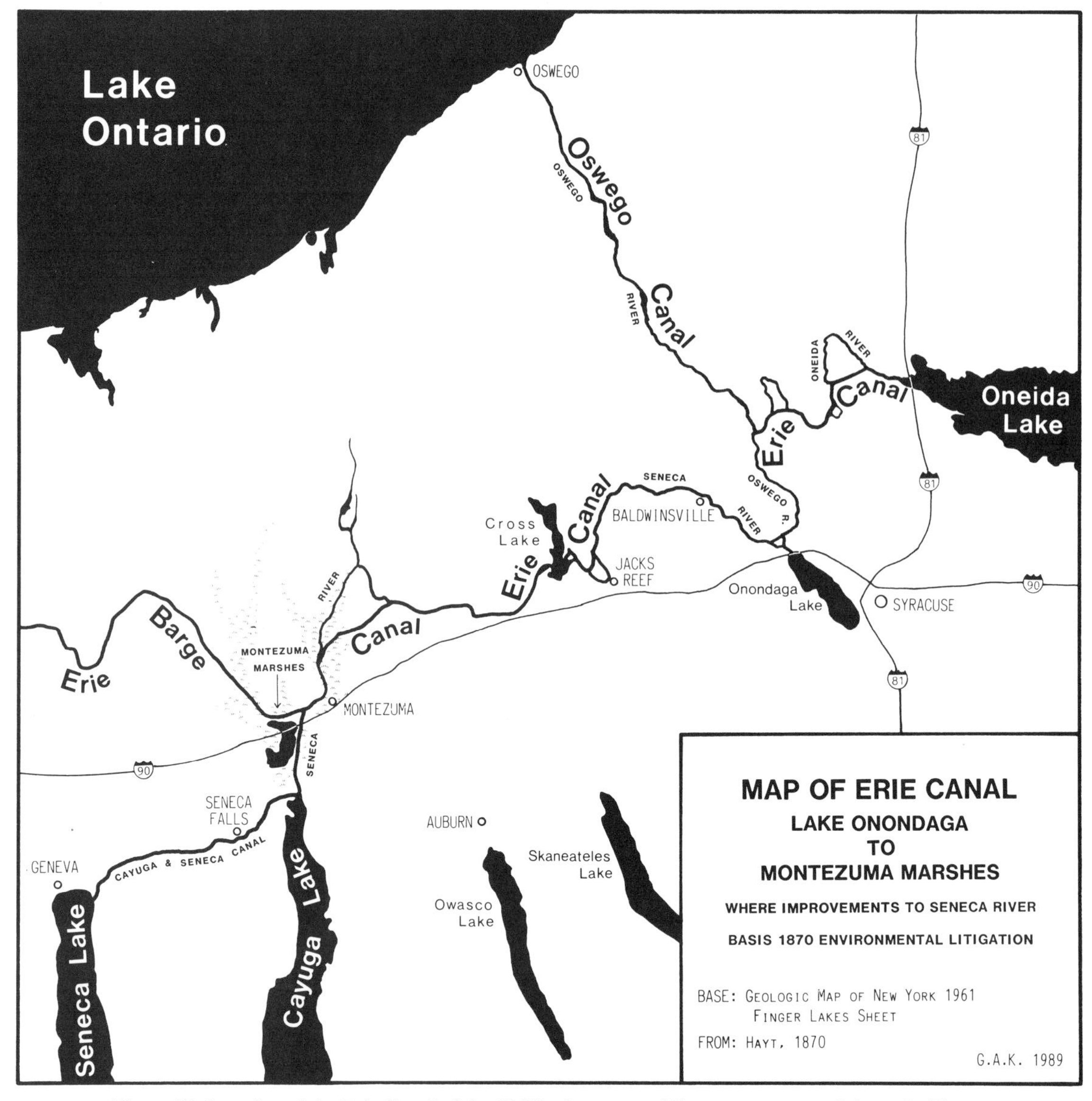

Figure 23. Location of the Erie Canal of the 1850's, from east of Syracuse to west of the main Finger Lakes in central New York. Engineering improvements, including the deepening and straightening of the channel of the reach of the Seneca River from Lake Onandoga westward past Jack's Reef, Montezuma, and the marshes at the head of Cayuga Lake, became the basis for legal claims of damage by landowners both upstream and downstream. This is a very early case of litigation claiming a change to environs by an operating engineering works. Adapted from S. T. Hayt, 1870.

civil engineering students until forced to resign because of deafness in 1907. He became professor emeritus and remained active as a full-time consulting engineering geologist until his death in 1925 (Shrock, 1972; Appendix-A).

On the West Coast, Professor John C. Branner of Stanford University was teaching some ideas and fundamentals of surficial geology to engineering students as presented in his forward-looking paper "Geology and its relation to topography" (Branner, 1898).

TWENTIETH CENTURY: 1900 TO 1940

Introduction

Although the unexpected reactions with time of many individual geologic processes and phenomena have frequently caused events of catastrophic proportions for engineered works, just such possibilities were partly responsible for the acceptance, rapid expansion, and practice of engineering geology during the first half of this century. Of equal importance, and sometimes more relevant to the growth of the profession was the influence of prominent engineers, such as J. R. Freeman, Lazurus White, Homer Hamlin, S. O. Harper, and Karl Terzaghi, and geologists such as William O. Crosby, John C. Branner, Charles P. Berkey, L. C. Glenn, Warren J. Mead, and Heinrich Ries. By the 1900s, engineers had begun to recognize the importance that geology might have on the success or failure of their works. Thus the early applied geologists were encouraged to explore and understand subsurface features by utilizing trenches, test pits and shafts, and various forms of drill holes to provide a three-dimensional subsurface interpretation for the engineer's design.

A dominant factor behind the accelerated interest in geology for engineering works in this century has been the increased governmental and state or province participation in irrigation, flood control, and hydroelectric projects, along with the broadening of soil-conservation measures in the 1930s.

Growth of geologic concepts relevant to engineering works

Boston Harbor and Dam. Between 1894 and 1903, John R. Freeman, an outstanding civil engineer, overcame strong political objections and technical controversy to gain acceptance for a plan to develop the Charles River basin and harbor of Boston (Cozort, 1981, p. 203–212). This early project for damming a river estuary system while preserving the shoals and natural channels of the harbor was largely dependent on how the harbor's geologic features would react to the changes induced by the proposed works. Based on an extensive two-year study and a team effort of eminent engineering and scientific specialists that included geologist William O. Crosby, the project was recommended by Chief Engineer J. R. Freeman in his report of June 24, 1903. Freeman and associates realized that "any change in the tidal prism would effect the preservation of the harbor, shoals and channels" (Freeman, 1903, p. 54–55) and furthermore "agreed a full knowledge of the geological conditions which formed and continued the estuary were required before deciding on a design." (p. 55). Professor Crosby had studied the currents and general features of the harbor during his 25 years at MIT, and was the applied geologist most capable of evaluating the facts on the origin and preservation of Boston Harbor. Crosby's data and interpretations were presented as Appendix No. 7 to Freeman's report (1903). His field investigation included borings in the Charles River basin and other coastal estuaries, mapping offshore islands, and plotting the bedrock contours to delineate areas of "bowlder clay and hard pan." Crosby concluded (p. 355): "the surging back and forth of tidal prism did more to shoal the harbor than to deepen it and the proposed damming would not cause serious changes; the harbor is essentially a drowned river valley, formed some 10,000 years ago; and since origin, the valley has slowly submerged due to a regional rate of 5 to 10 feet per thousand years, or overall 30 to 50 feet since origin."

Water-supply projects: New York area. Failures of the Colorado Dam near Austin, Texas, in 1900, made the engineering profession cognizant of the influence that geologic conditions have on the operation of a reservoir. Flood waters overtopped the dam, erosion removed support from the heel of the dam, and slip occurred along the thin clay and slate beds of the foundation, assisted by strong under-seepage and uplift pressure. The combined forces caused parts of the dam to move about 75 m downstream (Taylor, 1900).

The Metropolitan Board of Water Supply for the city of New York in 1905 appointed two consulting geologists, James F. Kemp and William O. Crosby, for the Catskill Aqueduct—an early substantial use of geology for engineering works in America. Kemp's work at New Croton Dam (1895 to 1898) resulted in his being appointed the first consulting geologist to the Board of Water Supply; he served until 1912, when the initial planning for dams and tunnels along the Catskill Aqueduct was completed. Kemp's activities for engineering works were mainly concerned with the New Croton Dam, the aqueduct, and the overall New York City Board of Water Supply (NYC BWS) system (Kemp, 1887, 1895, 1908, 1911, 1912a, b).

Investigations for planning the Ashokan Dam and Reservoir and the Catskill Aqueduct constituted important early American work on geology applied to dam, reservoir, and tunnel construction by Crosby and Kemp (see Fig. 35). In 1906, the first pressure tunnel in United States was undertaken as part of the Catskill Aqueduct at Rondout Valley, New York; the alignment traversed twelve highly folded and faulted rock formations of different physical character, all deeply covered by glacial deposits. The successful completion of Rondout tunnel was in part due to the contributions of geology; other tunnels along the aqueduct were at the Hudson River gorge and the 29-km pressure tunnel within the city of New York (Berkey and Sanborn, 1923; Sanborn, 1950).

The geological investigations by Crosby and Kemp during construction of the Catskill Aqueduct provided many contributions to understanding of geologic processes relevant to engineering works and these were tabulated by Kemp (1911). (1) Buried channels can occur that are not indicated by surface geology: Beneath the Rondout Creek channel, a buried channel extended to 3 m below sea level; the channel in granite at Storm King crossing unexpectedly extended 200 to 275 m below the Hudson River surface. (2) Relic stresses, which occurred in the Storm King Granite at depth, can cause spalling and/or rock bursts in shafts and tunnel openings.

The aqueduct consulting panel was joined by Charles P. Berkey in 1906, who was appointed assistant to J. F. Kemp. His duties included assembling reports on the field studies of consultants Crosby and Kemp and other supervisory investigators, such as tunnel engineer Sanborn (Sanborn, 1950). Berkey's timely geological efforts and distribution of the field reports on each part of the total aqueduct furthered the changing attitudes of civil engineers toward geology (Berkey, 1911).

An exceptional study and the best geological sections and interpretation of conditions to be encountered by a river crossing of the Catskill Aqueduct were prepared by J. Bernstein and

V. Zipzer about 1912 for the Harlem River crossing, according to White (1913). An exhaustive volume on *Geology of the work of the Board of Water Supply—City of New York* was prepared by Fluhr and Terenzio in 1973. A summary of the geologic conditions and inherent problems for each section of the aqueduct and the individual works of the system (1905 to 1978) was prepared by Fluhr and Terenzio (1984).

The Metropolitan Board also considered Long Island a potential source for new water supplies, and Crosby undertook the first geological investigation and evaluation of the water-bearing glacial deposits and underlying sedimentary rocks; this was part of a broader U.S. Geological Survey (USGS) study released in 1906 (Veatch and others, 1906) that determined the importance of selected glacial units as underground water supply for the western part of Long Island.

Crosby was engaged by the Board of Water Supply to make a second and more detailed geological study of Long Island in 1924 and 1925. This investigation provided the first description of the Mesozoic and Cenozoic bedrock units and updated the 20 map sheets and six geologic sections that Crosby had released earlier in 1911 as a report to the Board. Crosby's "Geology of Long Island" was being reviewed by the Board of Water Supply at the time of his death (1925), and Berkey did not arrange for its release. However, Crosby's pioneer work was widely used for the development of Long Island's water supplies (T. W. Fluhr, personal communication, 1983). His identification of bedrock units based on well borings, core records, cuttings, and other means confirmed that Triassic red sandstones occur beneath Long Island near Northport; this was the first recorded recognition of the southward trend of Triassic basin sediments into the New York region from Connecticut (Crosby, unpublished report to NYC BWS, 1925; T. W. Fluhr, personal communication, September 4, 1983).

After the Catskill aqueduct, Berkey served the Metropolitan Board of Water Supply on other projects until the mid-1920s (Fluhr, 1983). Then in the 1930s and 1940s his consulting activities shifted to Bureau of Reclamation projects, such as Boulder, Grand Coulee, Shasta, and Davis Dams, and some municipal and private projects. He became one of the foremost engineering geology consultants in America during this period. A review of his activities is given by Paige (1950), and his long service to the Geological Society of America and its Division of Engineering Geology are reviewed in Chapter 4, this volume. Berkey spent his entire career at Columbia University, where he taught the general courses in physical and historical geology.

Aqueducts for the Los Angeles area. During the early 1900s, major water-supply systems were also undertaken in California for the greater Los Angeles area. Construction of the first Owens River aqueduct by the Department of Water and Power, Los Angeles, began in 1907 and was completed in 1913. These works provided more "firsts" in engineering geology with respect to tunneling and the crossing of active fault zones, as did construction of the Mono Basin extension in 1934 to 1940. Los Angeles purchased 307,000 acres in Inyo and Mono Counties to protect water rights. The drama, intrigue, and legal maneuvers by the landowners to retain the water and by the builders (Los Angeles) to gain the water rights and construction right-of-way were depicted in the movie *Chinatown* in the 1970s. The Second Los Angeles aqueduct, constructed in 1965 to 1970, increased water delivered to city by 50 percent.

The first aqueduct of 375 km tapped the fresh waters of the Owens River (flowing into saline Owens Lake); the later Mono Basin extension (1940) extended the system 170 km northward for a total of 545 km (Fig. 24). The system's engineered works consist of more than 100 tunnels (120 km), many dams and powerhouses, and more than 400 km of confined or open canal flow. The tunnels and canals have required continued maintenance due to the numerous fault zones crossed and the highly varied rock conditions.

The Elizabeth tunnel, a part of the original 1913 aqueduct, carries water from the Fairmount Reservoir across a ridge and the San Andreas fault zone and discharges into a canyon for hydroelectric plants downstream. The horseshoe-shaped pressure tunnel, 8 km long, is mainly in granitic rock that varies from a hard to an altered and thoroughly crushed rock mass. The active San Andreas fault zone (about 1.5 km wide) is crossed orthogonally by the tunnel, which was an early "first" in applied geology. The underground exposure of the fault zone has been widely used since for various scientific investigations and extensive research on this active fault; to date no significant movement or damage to the tunnel has been reported (Wilson and Mayeda, 1966; Fig. 24).

The Mono Craters tunnel, driven later as an extension of the system (1934 to 1940), offered very different types of geological problems, and the following brief description is from Wilson and Mayeda (1966). The tunnel, 18 km long (3 m in diameter), pierces the volcanic necks underlying Mono Craters and some 20 inactive volcanic pumice cinder cones in the area between Mono Basin and Long Valley. Excavation required 5½ years from six headings; 67 percent of the tunnel is supported due to the wide variety of rocks penetrated: mainly rhyolite, tuff, volcanic ash, granite, metamorphics, sandstone, glacial deposits, lake beds, and alluvium. Serious geological problems included nearly every difficulty inherent to tunneling in one bore: large volumes of water under high pressure (some flows at 35,000 gpm); a high flow of carbon dioxide gas in a section of calcareous rocks (due to nearby volcanic activity); and squeezing and flowing ground at the fault contacts between metamorphic and granitic rocks where deeply weathered and material air-slaked on exposure.

An interesting sidelight on tunnel geology is related by Cloos (1947), a prominent German geologist. The long-standing riddle of the Rhine graben was settled when construction of a railway tunnel to approach Freiburg, near the Swiss border, exposed the contact between the Rhine Valley sandstone and ancient gneiss of the Black Forest.

Mapping major cities: Planning urban expansion. Around the turn of the century, the U.S. Geological Survey formulated an ambitious program to topographically and geologically map a group of major cities and release the maps as folios

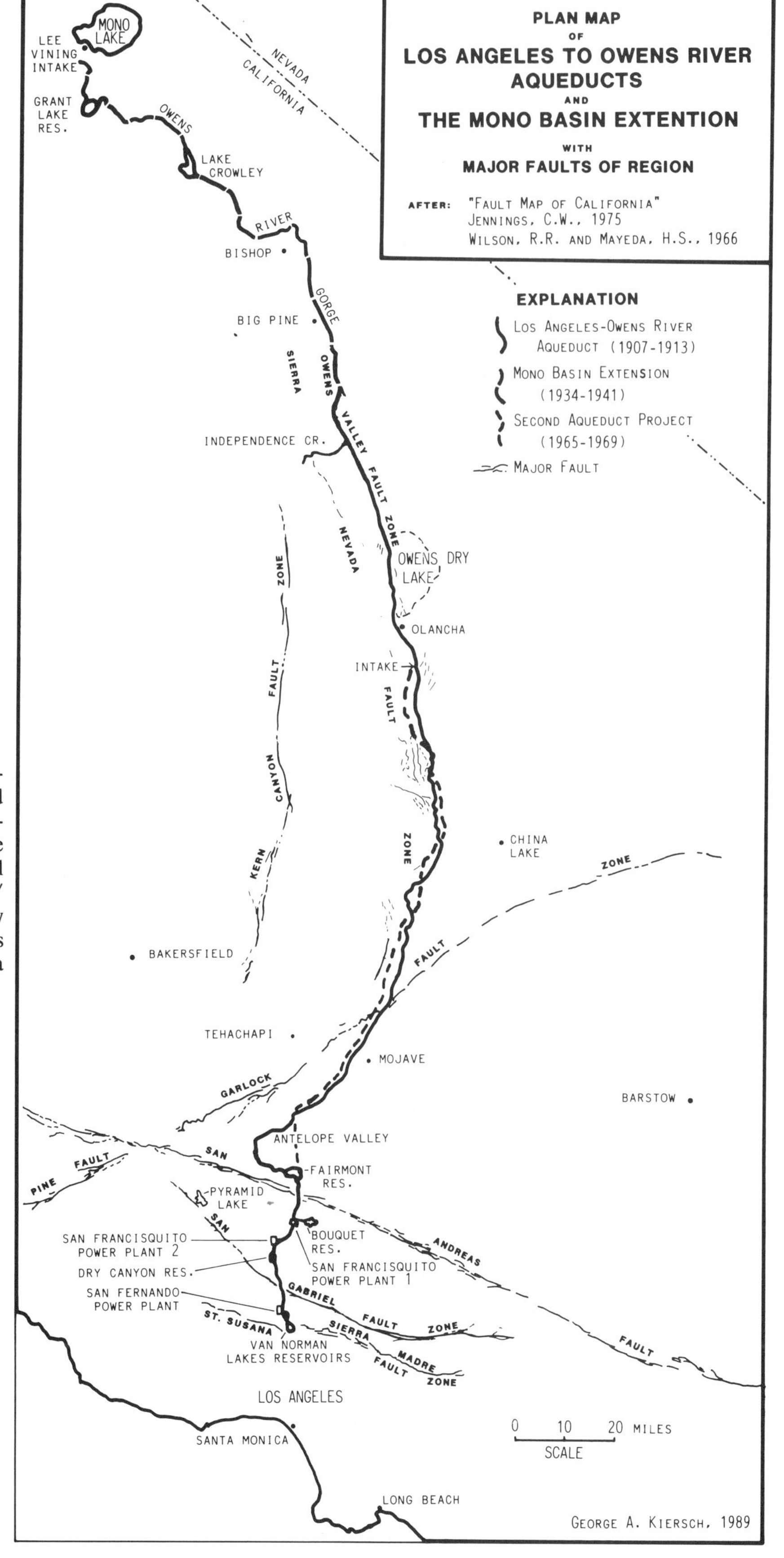

Figure 24. Map of the Los Angeles Aqueduct system; Lost Angeles to Owens River sector completed in 1913; Mono Basin extensions northward completed in 1941 with intake at Lee Vining; and the Second Aqueduct project that parallels the original system with an intake south of Owens Dry Lake/ Olancha, completed in 1969. This water-supply network of tunnels, canals, dams, and powerhouses crosses many active fault zones in the eastern Sierra Nevada and the Los Angeles region.

with supporting information on the surface and subsurface "environmental" features. This early approach to the support and evaluation of expected urban expansion and construction began with such cities as: Sacramento (Lindgren, 1894), Chattanooga (Hayes, 1894), Knoxville (Keith, 1895), Boise (Lindgren, 1898), Washington, D.C. (Darton and Keith, 1901), Chicago (Alden, 1902), San Francisco (Lawson, 1914), Minneapolis–St. Paul (Sardeson, 1916), and Detroit (Sherzer, 1917). Much of the eastern seaboard, now referred to as the Northeast Corridor, was largely covered by this program (Kaye, 1968).

Earlier, and strongly influenced by the German military effort of World War I, an engineering geology foundation map of the City of Danzig (Gdansk) was prepared in 1921 by H. Stremme and E. Moldenhauer. The difficult terrain of the city, built on alluvial flats of the Vistula valley, was skillfully analyzed, and the load-carrying foundation characteristics of four subsurface units were delineated and described, based on exploration and the records of geologic studies that began about 1910.

A geological guidebook for New York City and vicinity, prepared for the 16th International Geological Congress (Berkey and others, 1933), gives an excellent overview of the geologic features and glacial history of the region based on the early USGS folio studies. This is supported by a set of separate papers on engineering geology and impact of the major engineering works throughout region. Included are the City Water Supply network of aqueducts and tunnels, the many bridges and tunnels that connect traffic with Manhatten Island, the harbor facilities, and the highway system. The soft-rock tunnels built beneath the Hudson River are world-famous for the effective use of compressed air as a protective shield during construction. An early subsurface mapping project for greater New York City was supervised by T. W. Fluhr during the period 1936 to 1939. Data from 17,000 borings and other subsurface sources in the boroughs of Bronx, Brooklyns, Queens, and Richmond were correlated with a similar project for Manhattan Island, which totaled 27,000 borings and other data. The compilations were the basis for the bedrock/subsoil maps prepared for New York City (Murphy and Fluhr, 1944). The geologic environs of New York City and the underlying processes responsible for its distinctive geologic features have been reviewed by Schuberth (1968).

Western dams. The Reclamation Act of 1902 authorized the federal government to start a reclamation and irrigation program in the western United States under the Reclamation Service (an agency separated from the USGS Hydrologic Branch in 1907 and designated in 1923 as the Bureau of Reclamation). One of the earliest irrigation projects to be authorized was the development of the Salt River in Arizona, of which Roosevelt Dam (1906 to 1911) was the first project (Fig. 25). The principal investigation of the site was by drilling several holes in the channel section; the good-quality foundation of sandstone and quartzite required very little excavation (Fig. 26). Roosevelt Dam and most of the other early dam projects had little occasion to call on geologic counsel—the good-quality, natural sites available accommodated the moderate-height dams. Some of the dams and reservoir projects of the 1910s and 1920s that utilized geologic counsel are given in Appendix A.

Roosevelt Dam underwent rehabilitation and modification beginning in 1988. The dam height was raised, and the highway was shifted from the top of the dam to a new suspension bridge built upstream across an arm of the reservoir (see Chapter 2, this volume).

Earthquakes. World attention was directed to another phase of engineering geology by the great earthquake disaster in San Francisco, California, on April 18, 1906. Reports by the U.S. Geological Survey (Gilbert and others, 1907) and the Carnegie Institution (Lawson, 1908) on this cataclysm are classic in their scope and thoroughness. In the reports, all geologic phases were covered, including the effects of shock intensity on various rock and soil foundations. This event awakened the engineering profession to the natural hazards that since have commanded the attention of a large, science-oriented group of engineering geologists. Lessons learned at San Francisco and since have been applied throughout those parts of the world considered seismically active. Careful appraisal showed that shock intensity on the marshlands and in saturated, man-made fill was far greater than in the rocky hills and on natural, well-drained soils. This has been found true in evaluating every major shock of record throughout the world (Engle, 1952), and engineering structures are so modified in design and strengthened. Reid (1911) developed the concepts of elastic rebound and strike-slip movement along the San Andreas fault zone, which proved to be major milestones in understanding the causes of seismic events (Fig. 27). A photograph of a barn and manure pile torn apart by lateral movement illustrates the slip-displacement common to the San Andreas fault zone (Fig. 28).

The historic record of earthquakes in California dates back to 1769, but only since the early 1930s have earthquake studies become oriented to the safe and economical design of engineering structures (Richter, 1966). Although an early major earthquake affected a wide area around the San Francisco Bay area in 1868, little is known or on record. The movement, right-lateral along the now-designated Hayward fault, destroyed the village of Hayward. A geological study and report was prepared by George Davidson (U.S. Coast Survey) and others, but Lawson (1908) reported that authorities feared the release of data on the earthquake and the severity of damage would hurt the reputation of the San Francisco Bay area and so suppressed the report; no traces are known. This fault crosses beneath Memorial football stadium at UC-Berkeley, which was built in two separate halves so that the structure can shift laterally without major damage during an earthquake. The regional hazards of the San Francisco Bay area have been widely studied for several decades. The risks from natural hazards are summarized in Chapter 18 of this volume, along with graphics that show a correlation between the 1906 and 1989 earthquake damage.

Much of the progress in the field of engineering seismology is due to the efforts of scientists and engineers in California who founded the Seismological Society of America in 1907 to study

Figure 25. Roosevelt Dam, Salt River, Arizona, one of the earliest reclamation projects in western United States, was completed in 1911. This project underwent modification and rehabilitation in late 1980s, as described in Chapter 2. (Photo from Arizona Department of Transportation, 1985.)

earthquakes, and committees of the American Society of Civil Engineers that studied and reported on earthquake effects between 1907 and 1925. Historically the first sensitive seismograph stations in the San Francisco Bay area were installed in 1887 at UC-Berkeley and Lick Observatory. In 1921 the Carnegie Institution released a report recommending five topics for earthquake investigations in California, including: selected faults, surface displacements, isostasy, instrumental development, and a seismograph network in southern California (seven stations were installed by 1929).

The Kwanto earthquake of September 1, 1923, which devastated the cities of Tokyo and Yokohama, Japan, and destroyed more than 80 percent of the houses, was one of the most disastrous of modern times (Hodgson, 1964, p. 16–20). It generated further concern for earthquake studies in America. As at San Francisco, destruction was greatest where structures were built on alluvium and least where they were founded on bedrock. The floor of Tokyo Bay changed depth; this generated a destructive wave as high as 10 m, as well as widespread sand boils, craters, and ground-water fountains.

A second major earthquake in California, at Santa Barbara, July 1, 1925, came as if to dramatize the Congressional Act of January 1925, which authorized the Coast and Geodetic Survey to make investigations and reports in seismology. This milestone in engineering seismology was the beginning of our current methods of earthquake engineering studies and research in the United States. In the 1930s the Coast and Geodetic Survey began periodic measurements at selected localities along and across the San Andreas fault, such as from Monterey to Pacheco Pass, San Luis Obispo to Lost Hills, and San Francisco to Bakersfield (Richter, 1966). Since, and particularly following another major earthquake in 1933 at Long Beach, California, interest and accomplishments in connection with earthquake engineering studies have advanced at an ever-accelerating pace. For example, the U.S. Coast and Geodetic Survey obtained the first records of strong earthquake movements from the Long Beach earthquake (Neumann, 1952).

Landslides: Panama Canal and Europe. In 1902, under the auspices of the United States, construction of the Panama Canal was resumed and finally completed in 1914 (McCullough,

Figure 26. Roosevelt Dam, Salt River, Arizona, during construction in June 1907. View is of left abutment looking upstream. Note the good-quality foundation rock, the method of placing dimension stone blocks for the masonry dam, and other construction techniques of that day. (Photo courtesy U.S. Bureau of Reclamation/Glenn Lasson, and from Kiersch, 1955, p. 19.)

1977). Of unprecedented size, the project stands as an outstanding achievement in engineering. Noted for many firsts in the fields of engineering and medicine, the canal is also famous for landslides of such great magnitude that engineers were unsuccessful in stabilizing or mitigating them.

A preliminary investigation of the proposed canal route was made by the French geologists Bertrand and Zurcher in 1898, but apparently their warning that the Cucaracha Formation was an unsatisfactory material for deep cuts went unnoticed (Bertrand and Zurcher, 1906). Preliminary slides in 1906 and 1907 drew some geological attention, but until the famous Culebra slides started in 1910 (Fig. 29), geological counsel was not given serious consideration. At the request of President Taft, Hayes (1910) examined the Culebra slides and recommended that a geologist be attached to the canal staff. C. W. McDonald was assigned and thus became the first resident geologist in America on a major engineering project; later he was joined by Frank Tierney who subsequently became geologist for the Metropolitan Water Supply of Boston. Although the great slides were already past immediate control, McDonald recognized the character of the deformation within the soft, underlying Cucuracha beds and made a valuable contribution to engineering geology practice with his practical deductions (McDonald, 1915). The Panama Canal experience awakened the engineering profession to the major hazard of landslides and related problems common to all deep excavations; and furthermore, that landslide size is dependent on geologic conditions and design of the cut (see Chapter 7, this volume).

In Europe, additional construction on the extensive railway system was planned in the early 1900s. This focused attention on the need for an improved geological understanding of conditions along proposed routes. As a consequence, the Experimental Institute of State Railways of Sweden, founded by C. Segre in 1909, included a geological department. Disastrous landslides experienced on the Swedish railroads ultimately provoked the establishment of a special Geotechnical Commission in 1914, whose early studies laid the foundation for major advances in the emerging fields of soil mechanics and engineering geology. Moreover, the landslides experienced in Sweden, the soft-rock problems encountered in construction of Kiel Canal, Germany, and the

Figure 27. The San Andreas fault zone, strike-slip movement of 1906 in Marin County, California, is cited by Reid (1911) in his concept of elastic rebound. (From Heinrich Ries Collection, Cornell University.)

Figure 28. Lateral movement within the San Andreas fault zone occurred beneath the corner of the dairy barn building. Note that the manure pile originally at the "window" has been moved to the right in photograph and the corner of the wooden building has been twisted by action along the fault zone. (Source of photo: Heinrich Ries Collection, Cornell University.)

Figure 29. The Panama Canal, Culebra cut of Contractor's hill on February 7, 1913. View is looking north from west bank, south of Contractor's Hill. The typical, large-scale (2,900,000 yd^3) mass slides of Cucaracha Formation that delayed construction and caused damage to the heavy equipment are shown. (From U.S. Geological Survey, D. J. Varnes/R. L. Schuster.)

widespread sliding that occurred in the open cuts of the Panama Canal (all between 1910 and 1915) renewed interest in and laid the groundwork for advances in soil mechanics, and a better understanding of rock properties so critical to engineering geologists (Golder, 1957, p. 196).

Yet, despite the increasing number of satisfactory demonstrations for service and assistance to civil engineers, the construction profession was slow to make full use of applied geology. An outstanding example of this early attitude was dramatically demonstrated near Seattle, Washington, in 1918. The Cedar Reservoir was breached by piping, and the bank failed; destruction and flooding downstream resulted (see Chapter 23, this volume).

Attitudes changing among engineers. By the 1920s, the civil engineering profession realized the need for greater geological input and guidance for major works. Unfortunately, geological science did not respond immediately to the requests of the civil engineers for an improved knowledge of the physical properties of rocks and/or soft, unconsolidated sediments and soils. Consequently, civil engineers themselves began to provide input. A leader among the group of concerned engineers was Karl Terzaghi (Fig. 30), an early specialist in earth materials for construction. He worked for the U.S. Reclamation Service from 1912 to 1915, but returned to Europe to advance the combined fields of soil mechanics and geology for engineering works as professor at Roberts College, Constantinople (1915 to 1925), and the Technical Hochschule, Vienna (1929 to 1938). He returned to America as Professor of Foundation Engineering at Harvard from 1933 to 1963 and served as a prominent consultant for engineering works. Terzaghi based his soil mechanics techniques on sound geological knowledge (Terzaghi, 1955) and believed every soil-mechanics specialist should be half geologist; a combination he acknowledged later had not been followed by his successors, which was a major professional disappointment (Terzaghi, 1963).

During the early 1900s, Homer Hamlin first engaged in geology and then in engineering for several large California municipalities, and became one of the early engineer-geologists. Ultimately, his studies for municipalities on control of the Colorado River culminated in a 1920 proposal to put a dam in Boulder Canyon—a site later used for Hoover Dam (Nickell, 1942). Another early geologist involved with the investigation of dam sites along the Colorado River in the 1920s was Sidney Paige (Fig. 31), who became an eminent practitioner of engineering geology in the 1930s to 1950s (Paige, 1936, 1950).

Highways: Construction materials. The use of geologists to locate sources of adequate materials and provide guidance in planning routes for the developing nationwide highway system became a phase of applied geology in the early 1900s. By 1918, at least 49 papers on geology as applied to highway engineering had been published in America; this included the road-material sources of 24 states and reports on the relation of mineral composition to the engineering characteristics of the rock. In addition, a noteworthy contribution became available on the materials and manufacturing methods of both the portland-cement and paving-brick industries (Eckel, 1905). Probably the most extensive investigation was a statewide survey of aggregate sources for highway construction by W. C. Morse (1913) for the Ohio Department of Highways; potential aggregate sources, together with a brief description of the geology, are included in this two-volume report, which has served over the years as a basis for subsequent aggregate surveys.

Improvements in the automobile after World War I revived interest in good roads. The federal aid authorized by the Congressional Acts of 1912, 1916, and 1921, plus the first state gasoline taxes in 1919, provided funds for a new era in highway development. Concurrently, development of the second growth stage in geology applied to highway engineering began in 1918 when Ohio became the first state to employ a full-time staff geologist for highways. At the same time, standards of design and construction required the application of geology to more of the aspects of highway engineering. Engineering geologists were called on to apply geologic principles to all phases of highway construction. By 1924, more than 50,000 km of concrete rural roads had been built, and the mileage was increasing at 10,000 km/yr. Geological guidance was attempting to keep pace; for example, one of the first published reports of a concrete failure that could be traced to the aggregate used was published by Pearson and Loughlin (1923), although Stanton (1940) had observed the reaction even earlier. In the same year, Missouri employed staff highway geologists. By 1925, the design of highways changed, with emphasis on long-radius curves, passing-sight distance, and ruling grades, which caused an increase in cuts-and-fills and, quite naturally, increased the need for applied geology (Horner and McNeal, 1950).

The first example of a superhighway through mountainous country occurred in Italy. A superbly engineered road was constructed between the port of Genoa and the Po plain through the Apennine chain (Maddalena, 1935). Three railroads already occupied all the available space in some of the narrow valleys, and extreme difficulties were encountered in the rock formations that constituted the central core of the mountains. Much geologic assistance was required to execute the daring and unique plans to develop this highway. The Autobahns of Germany (1930s) are, likewise, milestones in current highway construction practice. From these and other pioneer superhighway projects, the present highway design and construction practices formulated in America since 1940s were adapted. D. G. Runner (1939) published a book on geology related to highway engineering.

Military geology. Reportedly, the first use of a geologist in a military operation was in 1813, when professor Karl August von Raumer assisted the Prussian general von Blücher; he analyzed the terrain of Silesia where Napoleon's army was later defeated at the battle of Katzback (Betz, 1984). Another early use of geology was by Rudolph von Bennigsen-Forder in 1843 who analyzed the terrain in Luxembourg for the French military (Marga, 1880–1882; Barré, 1897–1901).

Figure 30. Karl Terzaghi, an early specialist in earth materials, and leader in developing field of soil mechanics and foundation engineering over the period from 1908 to 1963. (Courtesy Ruth Terzaghi, photo taken 1957/and A. W. Hatheway.)

The Germans were responsible for developing the art of military geology by World War I, and went far in applying geology to terrain classification and construction of underground field fortifications and factories as a protective measure against heavy bombing (Kranz, 1913, 1927; Brooks, 1920). Although unprepared, the U.S. Army Expeditionary Force did organize a group of nine geologists to serve the field armies by 1918. They provided terrain analysis, water, and construction material supplies, and counsel on trench and tunnel excavations. The group included Kirk Bryan, E. C. Eckel, K. C. Heald, C. H. Lee, O. E. Meinzer, and Sidney Powers (Brooks, 1920). One effort in 1918 involved training students in *Military Mapping.* Professor W. J. Mead (Wisconsin) organized a U.S. Army training program for 1,500 students, which incorporated the principles of geology as relevant to the military and its engineering work (Shrock, 1977, p. 686).

Figure 31. Sidney Paige, an early engineering geologist, shown on the Colorado River expedition of 1921 examining damsites in the Grand Canyon and lower canyon regions. Paige was one of the pioneers to select the Boulder Canyon site (Hoover Dam) and was active in the 1930s and 1940s on U.S. Corps of Engineers projects. (From A. W. Hatheway; Photo taken by W. R. Chenoweth, chief of 1921 expedition.)

Generally, only meager attention was given to the military aspects of applying geologic data to planning and executing military operations in America during the 1920s and 1930s. However, Kurt von Bulow (1938) of the Germany army published a qualitative classification of common earth materials, which focused attention on this aspect of engineering geology. An excellent review of the influence of geology on military operations in northwest Europe during World Wars I and II was prepared in England by King (1951) and is reprinted by Betz (1975).

St. Francis Dam failure: Boulder Dam. On March 12, 1928, the dramatic and complete failure, within a very few minutes, of the St. Francis Dam near Saugus, California, proved a convincing disaster in the history of large engineering structures (described in Chapter 22, this volume). One of the many repercussions generated was a symposium to consider problems of dam and reservoir geology (AIME, 1929), with papers by C. P. Berkey (chairman), Kirk Bryan, O. E. Meinzer, Karl Terzaghi, G. H. Matthes, L. W. Fisher and R. D. Ohrenschall, C. K. Wentworth, L. C. Glenn, and H. T. Stearns. It directed the attention of engineers and geologists to the importance of adequate geological investigations and counsel in erecting dams and heavy construction works. The symposium was widely publicized, and constituted an important milestone in the growing efforts to bring before the engineering and geological professions at large the basic principles of dam and reservoir geology, and the importance of geology as an indispensable aid in civil engineering. Earlier that year, Heinrich Ries (1929) lectured before the Canadian Mining Institute on the subject of dam and reservoir geology.

Reverberation of the St. Francis disaster intensified the differences of opinion in early 1928 concerning construction of a proposed high dam at Boulder Canyon on the Colorado River. To appease all parties, Congress authorized the Secretary of the Interior on May 29, 1928, to appoint a board of five eminent engineers and geologists to examine the proposed site of the dam and to advise him as to matters affecting safety, economic and engineering feasibility, and adequacy of the proposed structure and incidental works (USBR, 1950, p. 11). The geologists were Charles P. Berkey and Warren J. Mead. The board's recommendations are now history, yet ironically, it required an engineering failure and the St. Francis catastrophe in America to gain recognition for the importance that geologic conditions may attain in heavy construction. Oddly, these events led to the final authorization and construction of Boulder (now Hoover) Dam, the world's highest dam (221 m), in the 1930s. The wide acceptance and utilization of geological input for planning and construction of Boulder Dam was strongly supported by the staff of the U.S. Bureau of Reclamation: S. O. Harper, Chief Engineer; W. R. Young, Assistant Chief; and J. L. Savage, Chief Design Engineer of the 1930s.

A close record of the geologic conditions and their significance was kept during construction of Hoover Dam. Much geological work went into the site selection and planning stages, and it is gratifying that the major engineering recommendations of the board members in 1928 to 1934 proved to be wise and economically sound. Many "firsts" in engineering geology were recorded at Boulder Dam; among them, the bureau engaged their first resident geologist, F. A. Nickell, for this project; and the 5 km of large-diameter (17 m) diversion tunnels were designed using no temporary roof or wall support. A complete history of geological investigations and construction problems is summarized in the Bureau of Reclamation Report on Boulder Dam (USBR, 1950).

Following close after Hoover Dam, and profiting from this experience, many large dams of all types were constructed throughout the world during the 1930s and 1940s. Many major dams were built by the U.S. Bureau of Reclamation (i.e., Grand Coulee, Shasta, Parker, Davis, and Friant), the U.S. Corps of Engineers (i.e., Muscle Shoals and Fort Peck), and the Tennessee Valley Authority (i.e., Norris, Kentucky, Douglas, Hiwassee, and Fontana), as part of government programs to conserve and utilize the natural resources. In most cases, geological investigations and counsel aided in their successful construction and operation. Several modest-size dams were constructed, however, with no geological input and became failures, such as the Lone Pine Dam, Arizona, built in 1934 to 1936 (discussed in Chapter 23, this volume).

Golden Gate Bridge controversy. Although engineers were aware of the need for geological input into the planning and design of major works by the 1930s, this acceptance and guidance could interject confusion and adverse effects into planning and construction if fundamental errors of judgment or misinterpretation of site conditions were made by a geologist. In addition, meddlesome and often adversarial approaches to construction problems, by an accuser or group who invests little money in research to support their concepts and accusations, may require owners to perform additional exploration, prepare arguments and reports to counter the criticism, and hold hearings to resolve them—a costly and time-consuming effort. This sometimes questionable use of scientific and engineering concepts and interpretations by intervenors became popular and was frequently overdone during the 1960s and 1970s relative to the construction and/or licensing of power plants in America.

An early case of intervenor opposition occurred in the exploration for and design of the foundation for the South Pier of Golden Gate Bridge, San Francisco, in 1931 to 1934 (Lutgens and others, 1934; Strauss, 1938; Schlocker, 1982). Opposition to the construction of the Golden Gate Bridge was supported by some San Francisco area corporate interests and citizen groups alike in the 1920s and early 1930s. This took the form of public challenges before construction began. One accusation that caused a delay was that the foundation conditions for the South Pier were unsafe and a redesign was required. The pier was located on and within a body of serpentine rock at a depth of 33 m below the channel surface (Fig. 32). The conclusions of consulting geologists Andrew C. Lawson (UC-Berkeley) and Allan E. Sedgwich (USC-Los Angeles) in 1932 were that the foundation as designed was safe and beyond question (Lutgens and others, 1934). The argument to stop the bridge was thus dropped, but the opposition took a different tack. They challenged the legality of the Bridge District in 1932 to sell bonds to finance construction. This attempt was based on plans for marketing bonds at a 5.25 percent rate of interest when district approval was only at a 5 percent rate. This maneuver to stop construction was solved when A. P. Giannini, Chairman of the Bank of America (originally Bank of Italy) pledged his bank's support—and quickly sold $3 million worth of district bonds at 5 percent. New bids were tendered on October 14, 1932, and construction began on January 5, 1933.

As construction progressed in 1933 to 1934, the safety of the South Pier was the opposition's sole remaining argument for delaying the bridge. By 1934, Bailey Willis, professor emeritus at Stanford University, began a concerted drive to discredit the consulting geologist's advice and interpretation of the geologic conditions of the pier foundation and surroundings (Fig. 33). Willis submitted his first report on April 7 to the Bridge District. Subsequently, Willis sent memo reports to Chief Engineer J. B. Strauss on August 22, 1934, and recommended suspension of construction until his points were clarified. Then, on October 19, 1934,

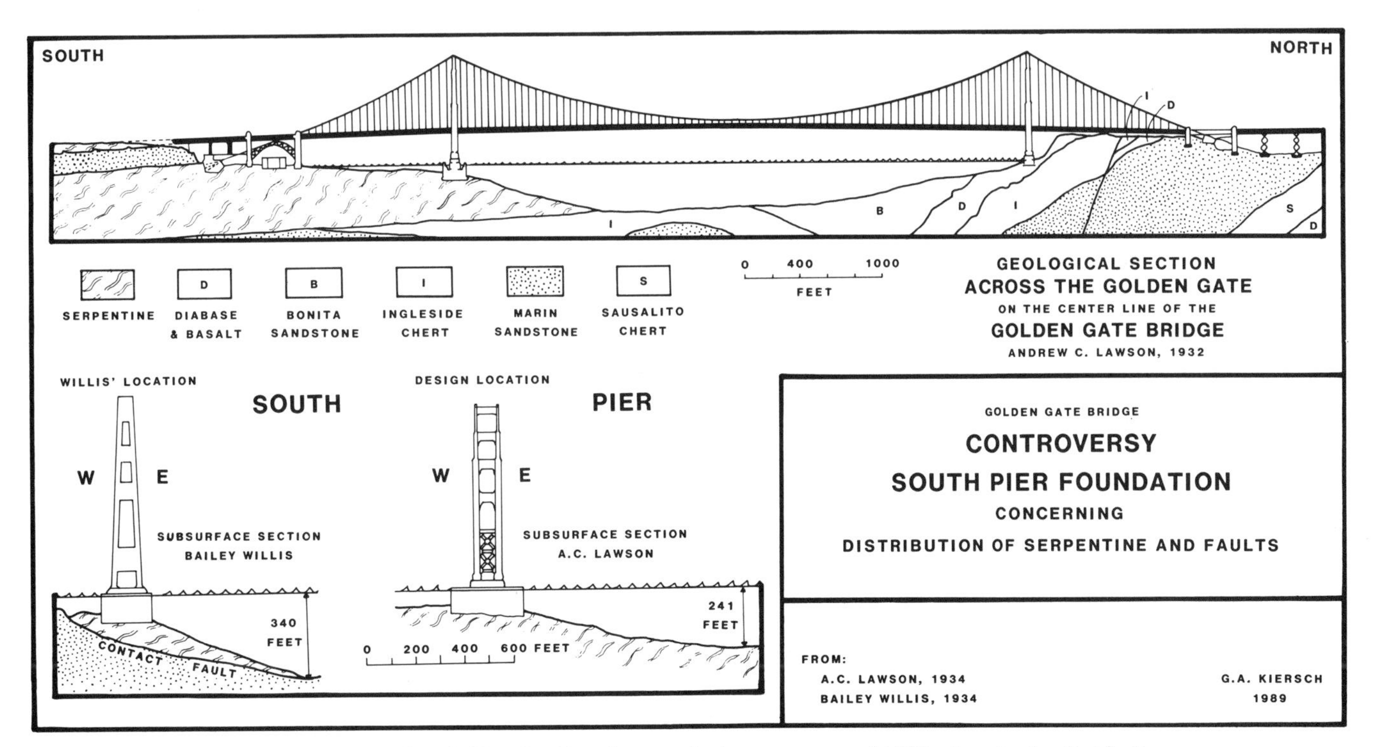

Figure 32. Golden Gate Bridge, San Francisco, south pier controversy of 1930s. A contention that faults occurred in a poor-quality serpentine foundation rock led geologist Bailey Willis to challenge findings of the consulting geologists A. C. Lawson and A. E. Sedgwich. The differences of interpretation are shown on subsurface sections of the south pier.

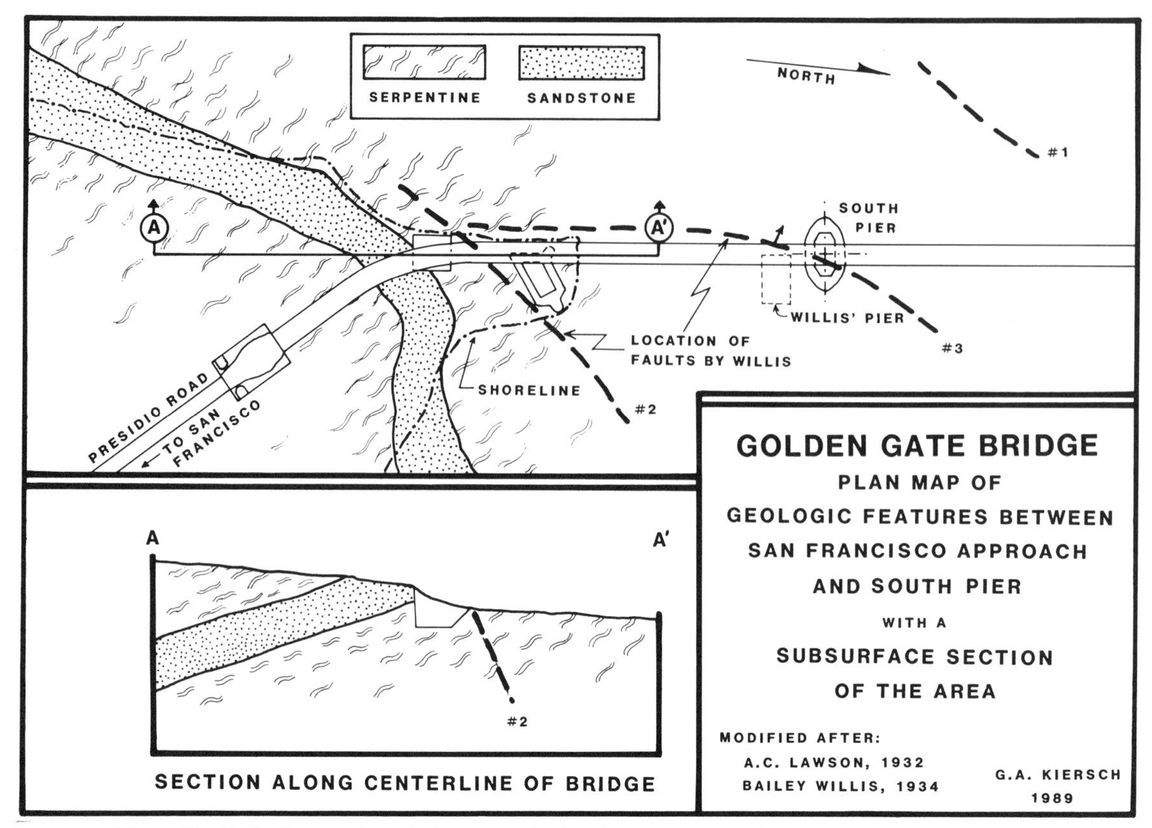

Figure 33. Geologic map of south pier area and subsurface section showing location of "faults" believed to occur and endanger the south pier foundation.

Willis (1934) published a two-page discussion of his arguments and technical points with diagrams and a model of the site. He described his feelings and fears about the serious stability hazards of serpentine foundation rock and the likelihood of large-scale sliding (Fig. 33), and requested the district stop construction and redesign the foundation. Willis's recommendations would have delayed the bridge and greatly increased the cost of constructing the South Pier by founding on a "sandstone mass" at a depth of some 250 feet beneath the "as-constructed" level (shown in Figs. 32 and 33).

After an extensive hearing and reconsideration in which each argument by Willis "has been carefully scrutinized and found erroneous as to fact or inference," the Building Committee concluded: a "sandstone mass" did not occur at depth nor did a fault plane beneath the pier site (Fig. 32), and furthermore, the serpentinized rock mass was a competent body, when confined, to carry the static load imposed by the bridge. The Building Committee on November 27, 1934, recommended that the directors disregard the arguments and recommendations of Professor Willis (Lutgens and others, 1934, p. 16). The long, sometimes bitter, and costly battle over geological arguments that the Golden Gate Bridge was unsafe was thereby closed. The bridge was open to vehicular traffic on May 27, 1937, the construction costs and bond were fully repaid on July 1, 1971, and no subsequent stability problems have been experienced in spite of several strong earthquakes.

Willis had claimed the serpentine rocks were treacherous materials, an idea then discredited (Fig. 34); that they would decompose, swell, and be an inadequate foundation; and that large-scale sliding would occur along the faults (as no. 3, Figure 33) and endanger the pier and channel. Exploration and deep cored borings supervised by Lawson and Sedgwich located neither the fault (no. 3) under the pier nor the "sandstone mass" at depth. Also, the channel depth shown by Willis (Fig. 32) was 30 m deeper than that actually found at the site.

Tunnels and aqueducts. As an outgrowth of industrial expansion during the 1920s, the demand for industrial water increased sharply. Many tunnels were constructed during the 1920s and 1930s to meet this need as well as the urge to improve transportation and communication. In America, large-scale tunnels were constructed for such purposes as aqueducts, railroads, and river diversion. Engineering predicaments developed for some of these projects; other, utilizing a more comprehensive application of geology, coped with similar circumstances with comparative ease.

The Moffat railroad tunnel was driven more than 10 km (1923 to 1927) through the Continental Divide in Colorado without benefit of an adequate preconstruction geological investigation (only a surface geological survey was performed). The tunnel encountered swelling ground and a wide fault zone, which caused delay in construction, expensive reinforcement, and has required periodic maintenance since (Lovering, 1928). The actual construction cost was approximately four times the original estimate based on the "forecast" that the tunnel would be driven entirely in solid rock. This "over-expenditure" became a financial burden for the railroad and the taxpayers of Denver during the following decades.

The Quabbin Aqueduct, central Massachusetts (1931 to 1935) accounted for another "first" in tunnel geology in America; it is the earliest reported case where a detailed "as-constructed" map of the tunnel was completed and the geology correlated with construction data and costs (Fahlquist, 1937, p. 736). The aqueduct was driven 40 km through rocks of many characteristics and properties without encountering any serious difficulties. Geological investigations, combined with the knowledge gained from the earlier tunnels in New York City, reduced construction uncertainties to a minimum.

The Hetch Hetchy Aqueduct of the city of San Francisco is another major project of this period. Its Coast Range tunnel, 46 km long, was the earliest (1927 to 1934) long-bore tunnel driven in the Pacific Coast section of the United States. This tunnel traversed some difficult sections, and the problems were instrumental in developing the use of gunite as a sublining to hold squeezing ground. The tunnel encountered active stresses in the host rock and required parts to be realigned. Sedimentary rocks were highly contorted, and the clayey matrix of some sandstones induced swelling and squeezing. Another hazard was methane and sulfuretted hydrogen gas (McAfee, 1934). Some of the faults traversed are known to be active.

A second major aqueduct constructed in the western United States was the Colorado River Aqueduct for the Metropolitan Water District of southern California. The San Jacinto tunnel, a section of the aqueduct, experienced serious setbacks in the early stages of construction (1934 to 1936) due to geologic conditions largely unknown to the contractor (Henderson, 1939). This tunnel experience further emphasized the seriousness and hazards of tunneling through a highly faulted mountain range where some faults are known to be active. Fault zones were encountered with enormous water inflows—7,500 gpm to 16,000 gpm at one location—which spotlighted the need for a comprehensive, preconstruction geological investigation to select the alignment and plan the construction. For example, control of inflow from a fault zone into a tunnel is possible by approaching the fault zone from the hanging-wall side so that the gradual inflow at the tunnel face has largely drained the interconnected fractures of the zone in advance of crossing the fault (such lessons learned at San Jacinto are described in Chapter 22, this volume).

After the successful completion of the Catskill Aqueduct in 1920, a geological reconnaissance to locate a second aqueduct, the Delaware Aqueduct (Fig. 35), was undertaken by the 1930s. Thomas W. Fluhr, geologist of the New York City Board of Water Supply, supervised the investigations for the planning and design of the new tunnel and related works. Geologists Horace Blank and Paul Bird joined this staff during the construction phase; subsequently, Blank joined the geology faculty of Texas A.&M. College and Bird was chief geologist for the New York State Department of Public Works. Active geological counsel contributed to the successful design and construction of the Delaware Aqueduct in 1937 to 1942, then the longest tunnel and underground aqueduct in the world. Fluhr served the board in many capacities over the intervening years, first as a staff member (1930 to 1963) on the construction of 225 km of aqueduct tunnels and four earth-fill dams, and subsequently as an active consultant to the board until his death in 1987. The principal structures and works of the water supply distribution system are shown on Figure 35. From 1933 to 1943, Fluhr was also asso-

Figure 34. Photograph of a large-scale test pit and exploration adits within the Franciscan Formation serpentine rocks on which the south pier of the Golden Gate Bridge was founded, at a depth of 33 m. Note size of the exploration exposure, side adits, and equipment used in the 1930s. The exceptionally large exposure reveals the physical characteristics of Franciscan serpentine rocks that change with depth. Exploration costs today usually prohibit this type of man-made deep shaft pit. (Source: Heinrich Ries Collection, Cornell University.)

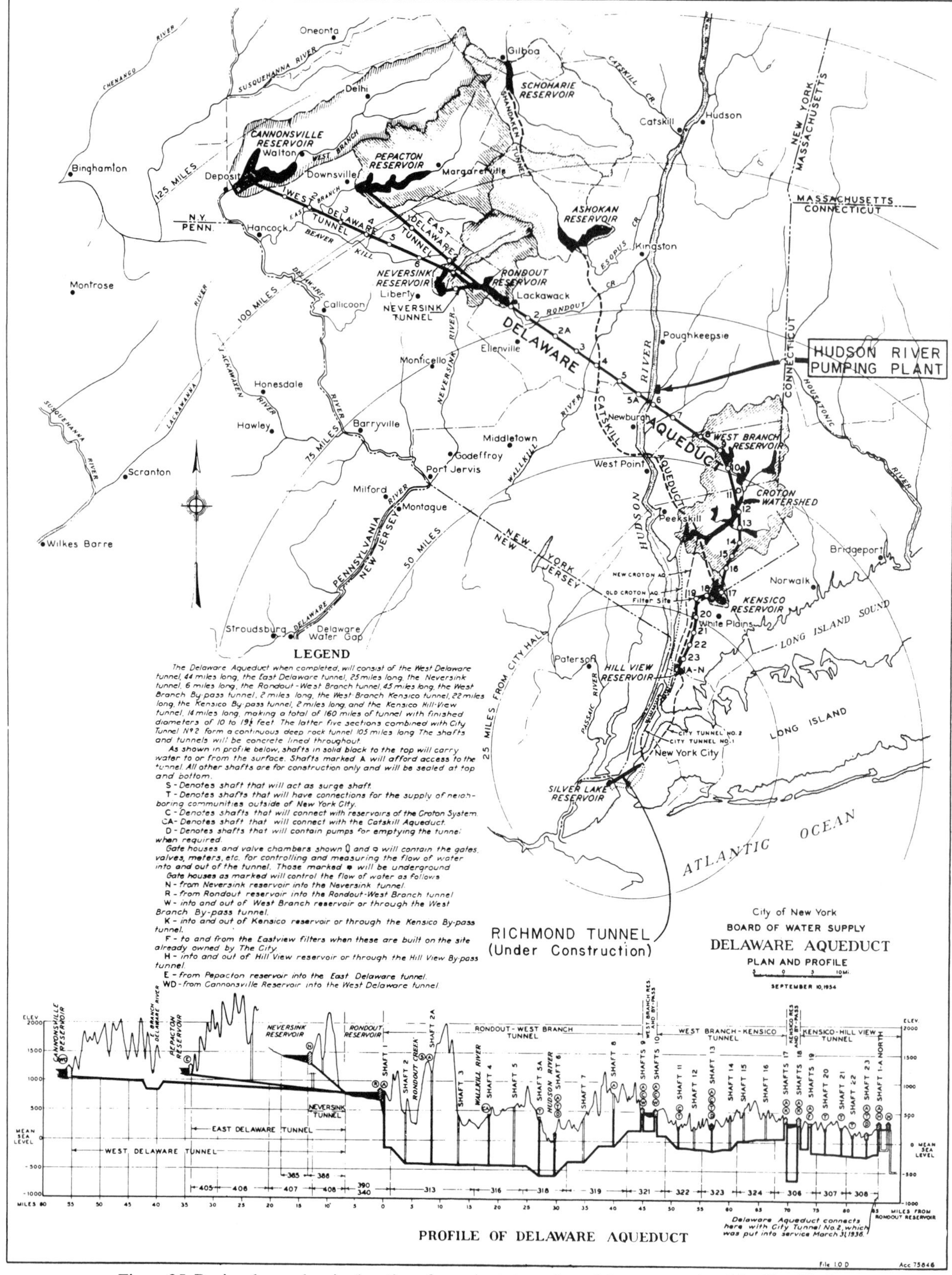

Figure 35. Regional map showing location of aqueducts, tunnels, and dams-reservoirs of the New York City Water Supply System of the 1970s along with a profile of Delaware Aqueduct system completed by 1950s. (From Fluhr and Terenzio, 1984.)

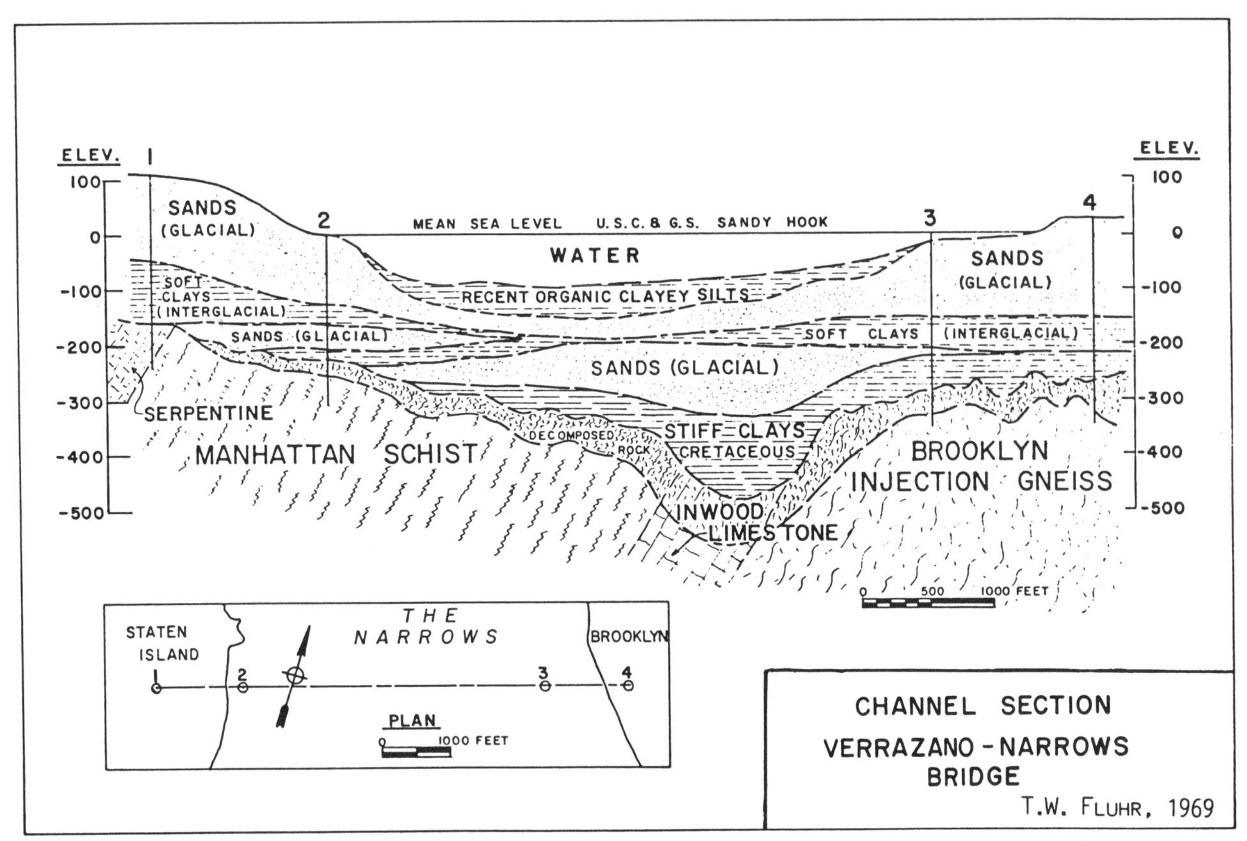

Figure 36. The channel section and sediments at the Verrazano Narrows, the entrance to New York Harbor. The location of Verrazano-Narrows Bridge, built in the 1960s between Staten Island and Brooklyn, is shown as are foundation rocks of both piers.

ciated with C. P. Berkey in geological work on several major underground projects around New York City: the Lincoln, Queens-Midtown, and Brooklyn-Battery subaqueous tunnels, Wards Island Interceptor Sewer Tunnel, and on foundations for the Triboro and Whitestone Bridges. Later, he was consultant on the Unionport Bridge, the 104th Street footbridge to Wards Island, Throgs Neck Bridge, and the Verrazano-Narrows Bridge (Fig. 36), as well as the East 63rd Street tunnel, of which the river section was constructed as a sunken tube (Fluhr, 1969). A full review of Fluhr's activities has been given by Fakundiny (1989).

Greater New York became the site of additional famous soft-ground tunnels during the 1920s to 1940s. The Holland vehicular tunnel, completed in 1927, was driven through the soft silt of the Hudson River, similar to that at the site of the earlier Pennsylvania Railroad tunnel (Fig. 37B) and in the channel beneath the George Washington Bridge (Fig. 37A). Soon thereafter, the Lincoln vehicular tunnel (1934 to 1937) was driven nearby through the same material. The most recent tunnel, the Brooklyn-Battery tunnel, was completed in 1949. This 2.8-km tunnel is a mixed-face tunnel that traverses both silt and bedrock—the most difficult to drive. The Brooklyn-Battery tunnel stands as one of the last great soft-rock tunnels constructed, as trench tunnels have become the most economical (White and White, 1953). A section for the East River tunnel from Manhattan to Queens via Welfare Island indicates that very different rock conditions occur in this area of the city (Fig. 38).

Research and training in geology for engineering works

The significant developments of past 100 years are reflected in the changing terminology, scope, and emphasis of engineering geology. Initially, the practice and academic training from the 1890s to the 1930s emphasized the application of geology for civil and mining engineering and architecture. This began with courses such as *Dynamic Geology and Economic Geology,* which included lessons on applications for engineered works. Then in 1905, the four branches of Applied or Economic Geology were defined by Johnson (1905, p. 244); one was the "Application of Geology to Various Uses of Mankind and Engineering Structures." Soon thereafter, Howe's (1909) treatment on mass wasting, *Landslides in San Juan Mountains, Colorado,* appeared with a section on causes and a classification. In Europe, Stini (1910) released a classification of slope phenomena that Sharpe (1938) later called "one of the most inclusive and satisfactory yet developed." Both treatises were some years before the Panama Canal studies of MacDonald in 1910 and 1911. The broad term *mass movement* and its connotation of essentially gravity movement of soil and rock was first introduced by Sharpe (1938).

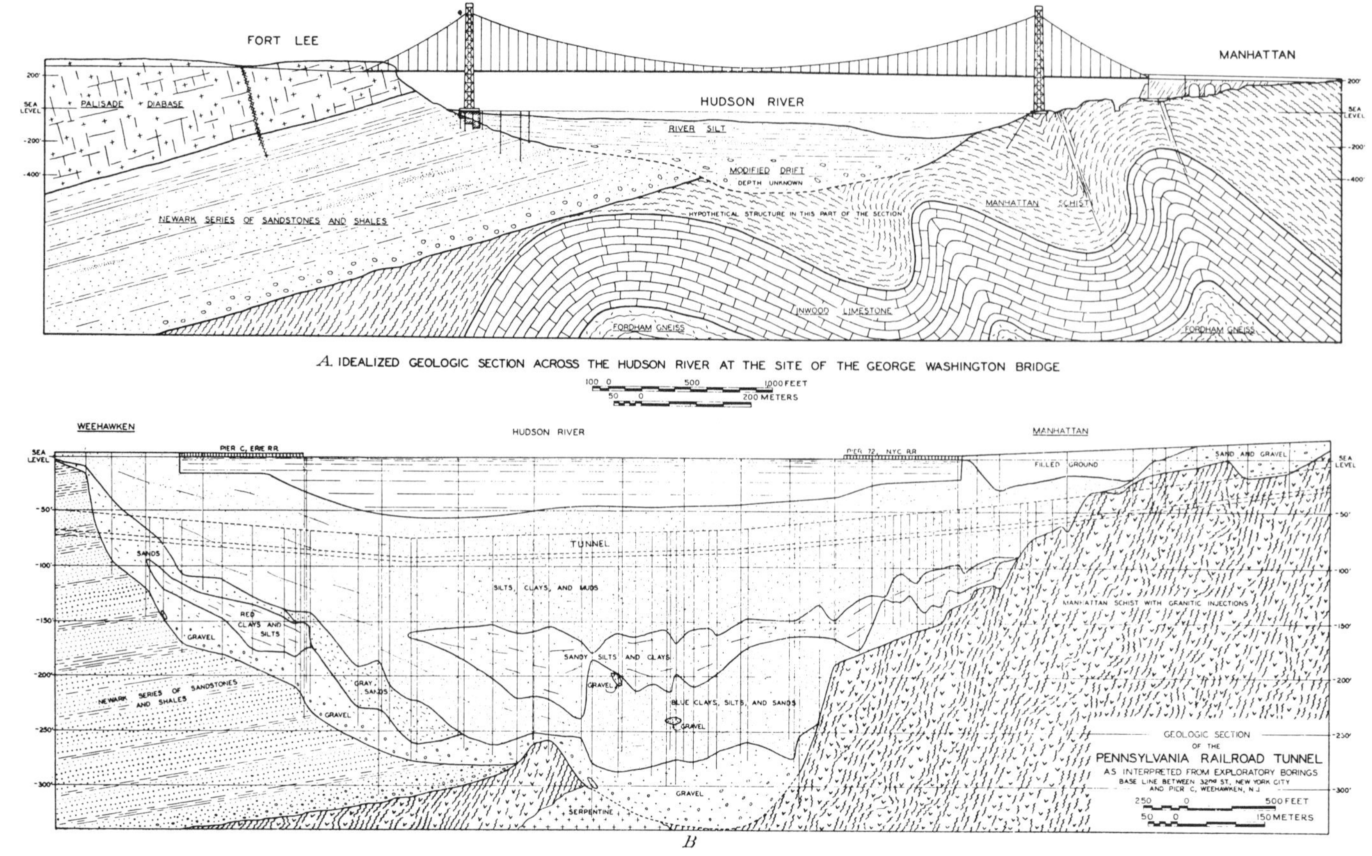

Figure 37. A. The Hudson channel at George Washington Bridge, showing the major rock units of New Jersey and Manhattan Island that underlie the channel sediments and are the foundation for bridge. B. The early Pennsylvania Railroad tunnel beneath the Hudson channel in soft sediments that required soft-ground tunneling techniques. (From Fluhr, 1969.)

During the early 1900s, Professor Heinrich Ries at Cornell organized his three separate courses in geology for engineering students into a strong one-semester course, *Geology for Engineers,* required of all civil engineering students (two lectures, two laboratories a week). This course continued as a requirement for civil engineering students until 1972, when Cornell shifted from a five- to a four-year curriculum and the course became an elective. Following Ries, the course was taught by applied geologists Alfred L. Anderson (1940 to 1963) and George A. Kiersch (1964 to 1978). The recognition and growing acceptance of geology courses offered to engineers and architects at several universities influenced Heinrich Ries to prepare the first textbook on the subject of geology for engineers, titled *Engineering Geology* (Ries and Watson, 1914). This text was republished in five editions (1936) and served two generations of civil engineers with instruction in the basics of geology. Other pioneer teachers included Warren J. Mead, University of Wisconsin (1916 to 1934) and Massachusetts Institute of Technology (1934 to 1954); T. T. Quirke, University of Minnesota (1915 to 1919) and Illinois (1919 to ca. 1951); C. K. Wentworth, University of Tennessee (1920s to 30s); and A. C. Lawson and G. D. Louderback, University of California–Berkeley (1920s to 1950s). The text by Ries-Watson was followed later with European texts by Fox (1935) and a second edition of Sorsbie (1938), first released in 1911.

In 1939, Robert F. Legget published the first edition of his widely used text, *Geology and Engineering,* a modification of Ries and Watson's earlier approach. This text has a strong emphasis on engineering case histories and less on the underlying geological processes responsible for some engineering problems.

G. K. Gilbert (USGS) was the first modern geologist in North America to relate the principles of mathematics, physics, and engineering to the solution of geological problems, and he seemed to solve puzzles in the manner of an engineer. For example, in 1909 Gilbert was one of the first to investigate forecasting of earthquakes. His studies of sediment transport in running water (1914), and debris flows as related to the mining debris of the Sierra Nevada (1917) together established engineering geology principles practiced today. Even more fundamental was his identification of the subsidence and rebound phenomena due to the loading and unloading of ancient Lake Bonneville, a concept critical to many engineered works (Yochelson, 1984).

The physical properties rock masses were being recognized in the earliest papers to address the need for an accurate rock classification in engineering contracts (Hall, 1839; and discussed earlier). In the early part of this century, additional important contributions on this subject were the discussion by G. J. Mitchell (1917) on "Rock Classification for Engineering" and supporting summaries by W. D. Smith (1919) and L. V. Pirsson (1919).

Professor Warren J. Mead, a pioneer in teaching applied geology to engineering students, was known worldwide for his research and geological consulting (Shrock, 1977, p. 692). As a specialist in physical and structural geology, his studies on rock properties and their relation to structural geology and failure of rocks (Mead, 1925, 1930) established several early principles relative to a rock mass and stress. These principles constituted the forefront of thinking on "rock mechanics" in the 1920s. He reviewed his findings on the suitability of rock masses for dam sites at the "International Conference on Large Dams" in 1936 (Mead, 1937a, b).

Kirk Bryan spent many years on field-oriented projects in the western states with the U.S. Geological Survey and was a contributor to the geology of reservoir and dam sites (Bryan, 1929) and the AIME Symposium (1929). In 1939, he reviewed the three distinct but broad categories requiring geological input for engineering works: the control of natural agents, processes and phenomena (dynamic geology to some); the placing of loads on Earth's crust (stability and durability of rock masses); and the utilization or control of ground-water circulation (permeability and flow of fluids).

Other major contributions to the literature for applied geologists included the geological treatises on sedimentation by Twenhofel (1932, 1939). Besides providing an interpretation and explanations for the origin of sedimentary rock materials, the volumes also covered the soft, unconsolidated, and soil-like deposits so common to engineering sites.

Besides the early contributors to the field of "engineering geology" already mentioned, additional geologists were prominent as investigators for sites, as consultants, and as contributors to the knowledge and growth of geology for engineering works between 1900 and 1940. This group includes: E. C. Eckel, who was first chief geologist of the Tennessee Valley Authority and served until his death in 1941, for work on the TVA and earlier projects; F. L. Ransome of California Institute of Technology for work on dams and the Los Angeles aqueducts; C. K. Wentworth of University of Tennessee for TVA and earlier projects; Sidney Paige of U.S. Corps of Engineers/Geological Survey; L. C. Glenn USGS, for work on dam sites of the Appalachian country; G. D. Louderback, University of California/Berkeley, for work on damsites; and Chester Marliave, private consultant on dams and tunnels in the western states.

The pioneering work of Karl Terzaghi in Europe on the mechanics of unconsolidated earth materials began about 1908; his soil-mechanics principles and techniques were first published in a textbook in German in 1925, *Physical Properties of Sediments and Soils.* Shortly thereafter, he joined with two European colleagues to publish an early text, in German, on Engineering Geology (Redlich and others, 1929), which received limited circulation in America. An exceptional review paper on the influence of geologic factors on the engineering properties of

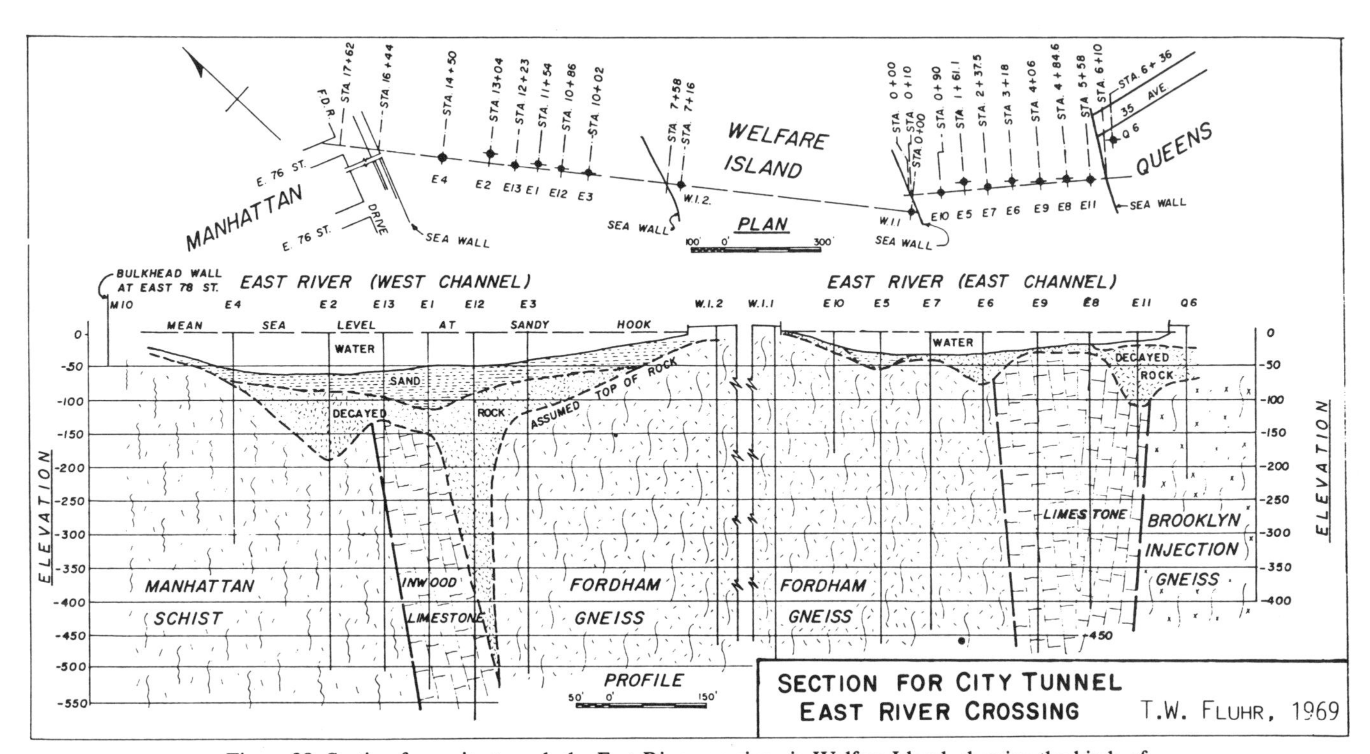

Figure 38. Section for a city tunnel, the East River crossing via Welfare Island, showing the kinds of rock conditions encountered in this area of the city by tunnels to Queens.

sediments, which reflects Terzaghi's correlation of geology and soil-testing analysis, was published on the occasion of the Fiftieth Anniversary of Economic Geology (Terzaghi, 1955).

A second prominent investigator in Europe during this period was Josef Stini, a professor of technical geology (1920s to 1940s) at the Technische Hochschule, Vienna. Stini was an exceptionally active researcher who published widely on applied geology and became a giant in his field throughout Europe. He founded the journal *Geologie und Bauwesen* about 1934, which was renamed *Felsmechanik und Ingenieurgeologie* (Rock Mechanics and Engineering Geology) in 1969, and published two early textbooks on technical geology (1919, 1922), followed by treatises on springs (1933) and tunneling geology (1950). He was among the first to introduce statistical joint measurements and originated the concept of the "Kluftkorper," or joint body, recognized circa 1905; he organized an early classification of slope phenomena (Stini, 1910); and he recognized slope creep movement (Talzuschub) and valley accretion (Stini, 1941; Fig. 39). During his 22 years at the Technische Hochschule, Stini was responsible for training a dedicated group of engineer-geological practitioners. Moreover, most engineering offices and construction firms of central Europe have long employed "construction geologists"—a reflection of his influence.

Another extensive and informative publication on a subfield of engineering geology during the first part of this century was the treatment on the *Geology of Dams* by Professor Maurice Lugeon, Universitad de Lausanne, in 1933. The volume includes case histories that review the exploration, design, and construction of dams and appurtenant works as practiced in Europe.

Hydrogeology and engineering. Many prominent geologists and engineers contributed to the modern principles of hydrology during the early twentieth century (Davis and DeWiest, 1966, p. 11–12). Advances in ground-water hydraulics continued at an accelerated pace from the 1890s, with general theory for base flow and ground-water discharge into streams, described by Edmond Maillet of France (1905). Mead (1919) subsequently defined "hydrogeology" as the study of the laws of the occurrence and movement of subterranean waters, and stressed that "ground water" is a geologic agent.

The most important single researcher in hydrogeology was O. E. Meinzer of the U.S. Geological Survey. His contributions between 1920 and 1940 covered many aspects of ground-water inventories and the importance of phreatophytes. His greatest impact was the orderly organization of the science of hydrogeology. Meinzer (1939, 1942) subdivided hydrology into two branches: surface hydrology and subterranean hydrology-hydrogeology, as Mead (1919) had earlier. Other geological contributors were: H. T. Stearns on the occurrence of ground water in volcanic rocks, based on investigations in Idaho, Oregon, Washington, and the Hawaiian Islands (1920s to 1960s); and the works of W. M. Davis and J. H. Bretz on the action of ground water in solutioning limestones and the occurrence of water in cavernous terrains.

Figure 39. A tributary canyon of the Colorado River in Utah. The narrow, slit-like part of the upper canyon wall demonstrates an area of valley accretion or slope movement (Talzuschub) as described by Stini (1941). The younger part of the canyon, near stream level, has not been similarly affected. (From U.S. Geological Survey photo files.)

A major advance in hydrogeologic theory was made in 1935 when C. V. Theis introduced an equation for nonsteady-state flow to a well, based on modification of an earlier formula published in 1928 in Germany by Herman Weber. Modern hydrogeochemical studies in America started with the works of F. W. Clarke published between 1910 and 1925—particularly chemical analyses of water with geochemical interpretations. Another early geochemist was Herman Stabler whose early regional studies of water chemistry in western America continued as principal sources of information until the 1950s.

OVERVIEW

Work of the federal construction agencies of the United States was delimited to spheres of responsibility by Congress in the 1930s. Probably the most important was the Flood Control Act of 1936, which designated the Department of the Army (Corps of Engineers) as the agency responsible for major works and flood control along navigable rivers and other waterways. The study and recommendation of works relating to improvement of watershed lands was assigned to the Department of Agriculture for such problems as reduction of flood-water and sediment damages, improving drainage of wet lands, providing better water supplies for irrigation and stock, and applying geology to problems of land use. Engineering geologists were employed by both the Soil Conservation Service and the Forestry Service to perform specialized phases of those programs (Griswold, 1950).

Throughout the 1930s, governmental sponsorship of conservation and public works programs accelerated applications of geology in engineering. Typical of these projects was the construction of a deep-water, inland channel connecting the city of Stockton, California, with the upper part of San Francisco Bay—the first such channel in the United States; similar channels have been constructed since at Houston, Texas; Wilmington, North Carolina; and other localities. Completed in 1935, the Stockton channel (Wood, 1932) is about 8 m deep and provides a waterway in or near the San Joaquin River for many kilometers. It necessitated extensive levee construction, as had been utilized for many years along such rivers as the Mississippi, Ohio, and Missouri.

With the advent of public works programs, knowledge gained from earlier small earth dams, levees, and foundations served as a basis for the development of soil mechanics—emerging since the early 1920s as a related field of engineering geology through the efforts of Karl Terzaghi in Europe and later Arthur Casagrande in America. These new soil techniques governed the design and construction of the many large earth-fill dams undertaken by private, state, and federal agencies beginning in the early 1930s. As aspects of soil mechanics rely on geological principles and the engineering geologist, this team effort introduced the geologist to mechanical and analytical concepts that have since been extended to other phases of the geosciences.

Closely paralleling the gradual endorsement of geology as an aid to the construction of dams, reservoirs, and tunnels has been a similar acceptance of geological advice in other branches of heavy construction. The historical impact of geology in the fields of irrigation and canals, foundations for power plants, buildings, bridges, and other types of engineering works demonstrates that geological principles and exploration techniques, and methods of evaluation formulated by experiences with dams and tunnels, were successfully applied to other phases of heavy construction and aided in solving unforeseen problems. Likewise, the understanding of geological problems and techniques germane to highway construction and maintenance was commonly utilized on railroads and, more recently, airfields. One tangible factor became evident—engineers were becoming more convinced that geological assistance in design and construction was important, feasible, and beneficial.

Engineering geology problems in the fields of ground water and petroleum, like other branches of engineering (discussed in Chapter 2, this volume), steadily increased in number and scope due to industrial growth. The encroachment of seawater on coastal areas, controlled operation of ground-water reservoirs, and in some instances, the subsidence of an area caused by the withdrawal of ground water and/or petroleum had become of increasing importance. For instance, construction of the sea-level canal across the Florida peninsula, started in 1935, was abandoned in 1936 mainly due to the fear of serious ground-water disturbances, such as its effect on artesian basins in the vicinity, and the encroachment of seawater (Paige, 1936). Subsidence of the Long Beach Harbor area, California (active since 1937), had been closely correlated with production in the Wilmington oil field. A decline of the pressure in the reservoir sands caused nearly 6 m of local subsidence (West, 1954). Many other areas in California, and elsewhere in America, began to undergo a similar subsidence by 1940. Solution or stabilization of this condition was one of the important problems confronting engineering geologists by the 1950s (discussed in Chapter 10, this volume).

While engineering geology had not attained anything close to maturity by the 1930s, the field of practice had grown in stature until it was regarded as a formal field of applied geological sciences, and furthermore was widely accepted as an essential aid to the area of civil engineering (Burwell, 1954).

Mead (1941), in his 50-year review of engineering geology, emphasized that any review of advancements in practice must entail the contributions by all divisions of geological sciences. Courses in geology had become a required part of the curricula for training civil engineers at most major engineering colleges in the country. However, Mead warned that such training did not qualify the student to handle any geologic situation and likened the course to instruction in first aid. Mead reviewed many facets of practice for engineering works active in 1939 and singled out foundation investigation and design aided by advances in soil mechanics, problems involving underground water, shoreline or coastal engineering problems, a growing consciousness of more effective methods of soil conservation and erosion control, and the value of geology to military operations as areas of growing importance to the field and professional practice.

APPENDIX A

William Otis Crosby, "The Father of Engineering Geology in North America"

The fifth professor of geology at Massachusetts Institute of Technology, William O. Crosby, was connected with the Institute for more than half a century. He first entered as a student in 1871 and died an emeritus professor 54 years later on the last day of 1925. In this long span of years as teacher, author, department head, and consulting geologist, he ably served his students, the Department of Geology, the Institute, and the public clientele that sought his professional advice. His students found him an enthusiastic and stimulating teacher who drew heavily on his diverse geological experiences. He was one of the very first geologists to work with engineers on excavation and construction projects and was a prominent pioneer in engineering geology. He left a record of hundreds of well-trained students; some 150 publications; and more than 100 confidential reports prepared for his clients.

This sketch briefly reviews four closely related aspects of the life of William Otis Crosby: (1) services to M.I.T. as teacher, department head, and collector-curator of specimens for teaching and research (1875 to 1907 to 1925); (2) authorship of some 150 published contributions to the literature of geology; (3) activities as a geological consultant on mineral deposits, and for major construction and engineering projects; and (4) his influence, closely coupled with that of his engineer classmate, John R. Freeman, in bringing the importance of geology to the attention of civil engineers (after Shrock, 1972, 1977).

William Otis Crosby was born in a log cabin on the northern bank of the Ohio River in Decatur, Ohio, on 14 January 1850. He was the first of six children born to Francis William Crosby (1823 to 1909) and Hannah Everett (Ballard) Crosby (1824 to 1909). His parent traced their ancestry to English forebears who emigrated to America in the 1630s. Francis William Crosby, Otis' father, served in the Union Army. His mother many years later managed Crosby House, a well-known and highly regarded hotel, in Washington, D.C., while his father, now Capt. Francis William Crosby, was in Europe most of ten years traveling and collecting geologic specimens for several museums.

Young Crosby's first serious interest in geology and the outdoors was aroused when at eighteen the Crosby family moved to Concord, North Carolina, where his father managed a gold mine. Here young Otis had his first opportunities to become acquainted with mineral deposits, mining, and the great outdoors. Later, young Crosby spent the summer of 1871 in Georgetown, Colorado, where his father operated a mill for a silver mine. Otis found time to visit many of the mines in the Georgetown district and to climb some of the higher peaks, and was soon recommended as a guide for visitors. When an M.I.T. party of 21 students, four professors, and President John D. Runkle appeared in Georgetown on 22 July 1871, young Crosby was recommended as their guide. President Runkle soon became impressed with Crosby's practical knowledge of mining methods, smelting practices, and regional geology, and asked him to build two of the Ballard furnaces for the new metallurgy laboratory at M.I.T. While there, he registered as a student to work toward a degree.

By the end of the first year he had decided to be a geologist. In his second year, he elected to study field science at a new summer school for science teachers organized by Harvard's famed Louis Agassiz on the Island of Penikese off the Massachusetts coast. Crosby was a keen observer, and the habits of collecting, recording, and classifying that he developed at Penikese served him for the next 50 years in his geological work in Massachusetts, other states, and many foreign countries. By the time Crosby reached his senior year, he had become deeply interested in the geology of the Boston area, and chose to write his bachelor's thesis (1876) on the "Geology of Eastern Massachusetts," the first geology thesis at M.I.T.

Crosby was appointed an assistant at the Museum of the Boston Society of Natural History in 1874 while still a student and soon became well-known for his lectures on the care of specimens, and how to identify and classify them for public view. He continued working at the museum until 1902, although his main activities were elsewhere.

Crosby was appointed to the M.I.T. staff on graduation in 1876 as an assistant in paleontology, later as instructor in geology and mineralogy, and in 1892 as assistant professor of structural and economic geology. This led to his offering the first coursework in geology for engineers (Dynamic Geology) in 1893, which continued until 1907 when he retired, due to his deafness, as director of the department and professor. Crosby taught about every aspect of geology during his M.I.T. service, and was indeed a broadly trained geological generalist, a factor in his acceptance as a special expert and consultant for engineering works.

Consulting Engineering Geologist

Among engineers, Crosby's advice on practical geological matters was highly valued. He was often called on in court cases to help solve legal controversies involving geological principles; and in his later years he gained an international reputation as an expert on water supply and construction of dams for water storage. His first work in the field started in 1895 when he was appointed consulting geologist to the Metropolitan Water Commission of Boston, and the Metropolitan Sewage Commission. This assignment led to investigation of the foundations for the Wachusett Dam, the North Dike, and conditions along the route of the proposed aqueduct tunnel, where he recognized deep preglacial gorges backfilled with gravels at a proposed dam site. He also investigated the Washington Street subway, Boston, in 1899.

Later he conducted geological work in preparation for construction of the dam across the tidal part of the Charles River and concluded the proposed dam would cause no serious changes—for excavation of the Boston subways and sewage tunnels, for the South Boston Dry Dock, and for the new M.I.T. buildings in Cambridge. Farther afield, Crosby was called on for advice on dam foundations and construction by many persons and organizations. His classmate, a well-known civil engineer of his day, John R. Freeman, commented:

> The Director of the U.S. Geological Survey remarked a few years ago that Crosby had become probably the best adviser to be found in the United States upon geological conditions at a dam site presenting difficult conditions, for tunnel alignments, and groundwater supplies in the period of 1893–1925 (Schrock, 1972, p. 40).

Crosby was the first to serve as consulting geologist for the United States Bureau of Reclamation Service (1921) in connection with several dam sites, including that for the Arrow Dam in Idaho, then the highest in the world, and was the first consulting geologist for the U.S. War Department (circa 1921) in regard to the foundations for the Muscle Shoals Dam across the Tennessee River. Earlier he was consulting geologist to the Board of Water Supply of New York City for six years (1906 to 1912), making geological investigations for the Ashokan and Kensico Dams, the deep Hudson River siphon, and the Catskill Aqueduct. His skillfully located exploration borings provided maximum information at minimum cost (Lane, 1930, p. 525). He undertook a study of the underground water of Long Island in the early 1900s and made an extensive investigation in 1924 to 1925, with 20 map sheets and six sections across the island. Certainly W. O. Crosby deservers to be called one of the earliest engineering geologists of North America, and indeed, the father of them all!

In the waning decades of the nineteenth century, far-sighted planners saw the need for greatly increased water and power supplies for the cities of the twentieth century, especially in eastern North America. Among the engineers involved with this planning was John Ripley Freeman, quoted above. He was one of the first engineers to realize the importance of geology in connection with the planning and design of major engineering works. Furthermore, Freeman was probably the greatest single factor in America to bring about the early acceptance for the need of services of geologists in the engineering profession (Shrock, 1972, p. 41). Professor Crosby served Freeman on many projects from the 1880s to 1925.

Some of Crosby's other major consulting projects in the 1920s were:

For Stone and Webster on a dam across the Mississippi River at Keokuk, Iowa at a narrow postglacial channel across a wide exposure of limestone bedrock; also on a Missouri River dam at Holter, Montana.

For F. S. Pearson Engineering Corporation on Medina Dam near San Antonio, Texas, and three large dams in Spain on the Ebro and Pallaresa Rivers (one was the highest dam in Europe).

For Great Western Power Company on the Big Bend and Big Meadows Dams on Feather River, California, sited in volcanic and granitic rocks.

For Mexican Northern Power Company on La Boquilla Dam, Conchos River, Mexico, the largest dam in Mexico at the time.

Hales Bar Dam, Tennessee, for control of serious leakage in the soluble limestone foundation; also, Rock Island Dam on the Caney Fork of the Cumberland River.

For Aluminum Company of America, North Carolina, on three dams, the Alcoa, Cheoa, and Fontana.

For Electric Bond and Share Company, North Carolina, four sites for dams on the Yadkin and Pedu Rivers.

Dams in West Virginia, Georgia, and on the St. Maurice River, Quebec, Canada. Also, consultant for water works and water-supply problems in Colorado Springs, Colorado; Little Rock, Arkansas; Birmingham, Alabama; Bloomington, Indiana; Williamsport, Pennsylvania; Providence, Rhode Island; and Westfield, Pittsfield, North Adams, and Lawrence, Massachusetts.

Crosby preferred field work and disliked preparing reports. One sidelight was that he insisted on wearing a rubber raincoat and carrying a gas mask (earlier, a bird in a cage) when underground, as protection against dangerous gases (recollection to writer by T. W. Fluhr, 1983).

REFERENCES CITED

Antiquity through Middle Ages

Alcocer, I., 1935, Ingeneria Hidraulica; Apuntes sobre La Antigna Mexico-Tenochititlan: Mexico, D.F.

Anderson, J. G., and Trigg, C. F., 1976, Case histories in engineering geology: London, Elek Science, 199 p.

Brantly, J. E., 1961, Percussion-drilling system, *in* Carter, D. V., ed., History of petroleum engineering: Tulsa, Oklahoma, American Petroleum Institute, p. 133–269.

Clarke, J., 1910, Physical science in the time of Nero: London.

Davis, S. N., and DeWiest, R.J.M., 1966, Hydrogeology: New York, John Wiley and Sons, 463 p.

Faul, H., and Faul, C., 1983, It began with a stone; History of geology from Stone Age to age of plate tectonics: New York, John Wiley and Sons, 230 p.

Forbes, R. J., 1934, Notes on the history of ancient roads and their construction: Amsterdam, Noord-Holland.

——, 1955, Studies in ancient technology, 1st ed.: Leiden, E. J. Brill, v. 1, p. 145–190; v. 2, p. 2–77.

——, 1963, Studies in ancient technology: Leiden, E. J. Brill, v. 7, p. 12–31.

——, 1964–1965, Studies in ancient technology, 2nd ed.: Leiden, E. J. Brice, v. 1 and 2 (above, 1955, 1st ed.).

Goodfield, J., and Toulmin, S., 1965, How was the tunnel of Eupalinas aligned?: IRIS, v. 56, no. 183, p. 46–55.

Heath, M., 1981, Samos-Pythagoras' Island: Athens, D. C. Davaris, p. 7.

Legget, R. F., 1962, Geology and engineering, 2nd ed.: New York, McGraw-Hill Brook Company, 857 p.

Marinos, P. G., and Koukis, G. C., eds., 1988, The engineering geology of ancient works, monuments, and historical sites, *in* Proceedings, International Symposium, Athens, Greece, September 1988: Rotterdam, A. A. Balkema, 3 volumes.

Merckela, C., 1899, Ingenieur technik im Alterthume: Berlin.

Mocchegiani, Corpano, C. M., 1984, The sewers; Ancient drains of ancient Rome, *in* Luciani, R., ed., Roma sotteranes (Underground Rome): Rome, Italy, Fratelli Palombi, p. 43–46 and 164–178.

Schroeder, A. H., 1943, Prehistoric canals in the Salt River Valley, Arizona: American Anthropologist, v. 8, and *in* Shertrone, H. C., 1945, ed., A unique prehistoric irrigation project: Smithsonian Institution Annual Report, p. 379–386.

Sowers, G. F., 1981, There were giants on the Earth in those days; 15th Terzaghi Lecture: American Society of Civil Engineers Journal of the Geotechnical Engineering Division, v. 107, no. GT4, p. 383–419.

Tolman, C. F., 1937, Ground water: New York, McGraw-Hill Book Company, 593 p.

Uckun, E., 1989, Ephesus; A brief history: Istanbul, Turkey, Keskin Color Ltd., Co., 48 p.

Van der Waerden, R. L., 1954, Science awakening; translated by Dresden, A.: Groningen, P. Noordhoff, p. 102–105.

Wren, C. S., 1983, Ancient labyrinth of aqueducts feed firey oasis in China: New York, New York Times, August 2, 1983, p. C–2.

Renaissance-Baroque Periods

Cardano, G., 1550, De Subtilitate: Nuremberg, Liber XXI.

Clements, T., 1981, Leonardo da Vinci as a geologist, *in* Rhodes, F.H.T., and Stone, R. O., eds., Language of the Earth: New York, Pergamon Press, p. 310–314.

Davis, S. N., and DeWiest, R.J.M., 1966, Hydrogeology: New York, John Wiley and Sons, p. 5–12.

Dutton, C. E., 1889, The Charleston earthquake of August 31, 1886: U.S. Geological Survey Annual Report 9, p. 203–528.

Faul, H., and Faul, C., 1983, It began with a stone; History of geology from Stone Age to age of plate tectonics: New York, John Wiley and Sons, 230 p.

Gautier, H., 1721, Nouvelles conjectures sur le globe de la terre: Paris.

Maillet, B. de, 1735, Description de l'Egypte: Paris, Genneau, Rollin.

Mariotte, E., 1717, Oeuvres de Mr. Mariotte, Van da Aa, P., ed.: Leiden, 2 volumes, 701 p.

Meinzer, O. E., ed., 1942, Hydrology: New York, McGraw-Hill Publishing Co., 712 p.

Zaruba, Q., and Mencl, V., 1963, Engineering geology; Developments in geotechnical engineering: Amsterdam, Elsevier Publishing Co., 483 p.

Industrial Revolution, Nineteenth century, 1750 to 1900s

Anderson, J.C.C., and Trigg, C. F., 1976, Case histories in engineering geology: London, 199 p.

Ansted, D. T., 1869, Earth's history or first lessons; A practical geology: Philadelphia, Pennsylvania, 244 p.

Branner, J. C., 1898, Geology and its relation to topography: Transactions of the American Society of Civil Engineers, v. 39, p. 53–95.

Brocchi, G. B., 1820, Map of Rome, *in* Thomas, R. G., 1989, Geology of Rome: Association of Engineering Geologists Bulletin, v. 26, no. 4, p. 415–426.

Bryan, K., 1939, Geology and the engineer: Harvard Alumni Bulletin, May 12, 3 p.

Civil Engineer, 1988, Landmarks in civil engineering: Civil Engineering, v. 58, no. 1, p. 46–48.

Collin, A., 1846, Landslides in clays, translated by Schriever, W. R., 1956: Toronto, University of Toronto Press.

Cozzens, I., 1843, A geological history of Manhattan or New York Island: New York, 114 p.

Cuvier, G., and Brongniart, A., 1808–1811, Essai sur la geographie mineralogique des environs de Paris: Philosophical Magazine, v. 35.

Dana, J. F., and Dana, S. L., 1818, Outlines of mineralogy of Boston and its vicinity, with a geologic map: American Academy of Arts Memoir 4, p. 129–223.

Daubree, A., 1887, Les eaux souterraines, aux epoques anciennes et a l'epoques actuelle: Paris, Dunod, 3 volumes.

Davis, W. E., 1970, Engineering geology of Raystown Dam and historical Chesapeake and Ohio canals, *in* Association of Engineering Geologists Annual Meeting Guidebook, Washington, D.C., Field Trip 7: Association of Engineering Geologists, 10 p.

Davis, S. N., 1973, Tedora Ardemans; Pioneer water-supply engineer of Spain: Water Resources Bulletin, v. 9, no. 5, p. 1028–1034.

Davis, S. N., and De Wiest, R.J.M., 1966, Hydrogeology: New York, John Wiley and Sons, p. 5–14.

De Saussure, C., 1779, Granite sheet structure: France, Journal des Mines, v. 7, p. 426.

Desmarest, N., 1804, Sur la constitution physique des couches de la colline de Montmarte: Paris, Memoires de l'Institut des Sciences, Lettres, Arts.

——, 1806, Memoire sur la determination de trois epoques de la nature par les produits des volcans: Paris, Memoires de l'Institut des Sciences, Lettres, Arts.

Donner, J., 1981, Eduardo Suess; Der Vater der I. Wiener Hochquellenleitung, *in* Suess, E., ed., Forscher und Politiker: Osterreichische Geologische Gesellschaft, Band 74–75, p. 41–51.

Dupuit, J., 1863, Etudes theoriques at pratiques sur le movement des eaux les eanaux decouverts et a travers les terrains permeables, 2nd ed.: Paris, Dunod, 304 p.

Dutton, C. E., 1889, The Charleston earthquake of August 31, 1886: U.S. Geological Survey Annual Report 9, p. 203–528.

Emmons, E., 1838, New York Geological Survey, 2nd Geological District 3rd Annual Report: New York Assembly, no. 200, p. 316.

Faul, H., and Faul, C., 1983, It began with a stone; History of geology from Stone Age to age of plate tectonics: New York, John Wiley and Sons, 254 p.

Fluhr, T. W., and Terenzio, V. G., 1984, Engineering geology of the New York City water supply system: New York Geological Survey Open-File Report 05.08-001, 183 p.

Forchheimer, P., 1886, Uber die Ergebigkeit von Vrunnen Anlaged und Sickerachlitzen: Zeitschrift des Architekten und Ingenieur, Vereins zu Hannover, v. 32, p. 539–564.

Fox, F., 1907, The Simplton Tunnel: London, Institution of Civil Engineers Proceedings, v. 168.

Fuller, M. L., 1912, The New Madrid earthquake: U.S. Geological Survey Bulletin 494, 119 p.

Gale, L. D., 1839, *in* Communication from the governor relative to the geological survey of New York State; Annual report: New York Assembly, February 27, 1839, no. 275.

Gattinger, T. E., 1973, Geologie und Baugeschichte des Schneealpenstollens der I. Wiener Hochquellenleitung (Steirmark-Niederosterreich): Wien, Geologische Bundesanstalt, Rasumofskygasse 23, p. 7–59.

Geike, A., 1905, The founders of geology, 2nd ed.: London.

Gowen, C. S., 1900, The foundation of the New Croton Dam: American Society of Civil Engineers Proceedings, v. 26, no. 1, p. 1–75.

Guettard, J. E., 1751, Memoire et carte mineralogique sur la nature et al situation des terrainsqui traversent at l'Anglettre: Paris, Royal Academy of Science Memoir, 1746, p. 363–392.

——, 1755, Memoire sur las granits de France: Paris, Royal Academy of Science Memoir, 1751, p. 164–210.

Hahn, T. F., 1976, George Washington's Canal at Great Falls, Virginia: Shepherdstown, West Virginia, American Canal and Transportation Center, 44 p.

Hall, J., 1839, Classification of excavation rock, Erie Canal lock at Lockport, *in* 3rd Annual Report of 4th Geological District, State of New York: New York Geological Survey, p. 297–339.

Hayt, S. T., 1870, Annual report of Erie Canal commissioners: State of New York in Assembly no. 4, January 4, 1870, p. 48–57.

Hazen, R. M., 1974, The founding of geology in America, 1771–1818: Geological Society of America Bulletin, v. 85, no. 12, p. 1827–1834.

Hill, R. T., 1889, Roads and materials for their construction in the Black Prairie regions of Texas: Texas University Bulletin, 39 p.

Hitchcock, E., 1836, Sketch of the geology of Portland, Maine, and its vicinity: Boston Journal of Natural History, v. 1, p. 306–347.

Hodgson, J. H., 1964, Earthquakes and earth structures: Englewood Cliffs, New Jersey, Prentice-Hall, Inc., 161 p.

Hoover, H., 1934, Improvement of the Washington Monument; Communication from the President of the United States: 72nd Congress, 2nd Session, U.S. House of Representatives Document 528, 54 p.

Huntting, M. T., 1945, Geology in highway engineering: Transactions of the American Society of Civil Engineers, v. 110, p. 271–344.

Institute of Civil Engineers, 1969, A century of soil mechanics: London, Institute of Civil Engineering.

Kiersch, G. A., 1955, Engineering geology: Golden, Colorado School of Mines Quarterly, v. 50, no. 3, 123 p.

Kollmann, W., 1983, Hydrogeologische untersuchungen in den nordlichen Gasausebergen: Graz, Austria, Berichte der Wasserwirtschaflichen Rahmenplanung, v. 66.

Latrobe, B. H., 1804, An account of the freestone quarries on the Potomac and Rappahannock Rivers: Transactions of the American Philosophical Society, v. 6.

Legget, R. F., and Karrow, P. F., 1983, Handbook of geology in civil engineering: New York, McGraw-Hill, Inc., variously paginated.

Logan, W. E., 1841, Landslide on Maskinonge River of April 4, 1840: Geological Society of London Proceedings, v. 3, p. 767–769.

Lucas, J., 1880, The hydrogeology of the lower greensands of Surrey and Hampshire: London, Institute of Civil Engineers Minutes of Proceedings, v. 61, p. 200–227.

Lyell, C., 1858, Principles of geology on the modern changes of the Earth and its inhabitants: New York, D. Appleton and Co., 834 p.

Maclure, W., 1809, Observations on the geology of the United States; Explanatory for a geological map: Transactions of the American Philosophical Society, v. 6, p. 411–420.

——, 1818, Essay on the formation of rocks: Philadelphia, Pennsylvania, Journal of the Academy of Natural Sciences, v. 2, p. 261–275, 285–309, and 327–344.

Mariolakas, I., and Stiros, S. C., 1987, Quaternary deformation of the Isthmus and Gulf of Corinth, Greece: Geology, v. 15, p. 225–228.

Mather, W. W., and Whittlesey, C., 1838, Geologic section and description of the lake front at Cleveland: Geological Survey of Ohio 1st Annual Report, 38 p.

Merrill, F.J.H., 1897, Road materials and road building in New York: New York State Museum Bulletin, v. 4, no. 17, 134 p.

Merrill, G. P., 1964, The first hundred years of American geology, revised 1924 ed.: New York, Hafner, 773 p.
Mitchell, J., 1760, Conjectures concerning the cause and observations upon the phenomena of earthquakes: London, Philosophical Transactions of the Royal Society, v. 51, part 2, p. 566–634.
Newman, J. R., 1981, The Lisbon earthquake, *in* Rhodes, F.H.T., and Stone, R. O., eds., Language of the Earth: New York, Pergamon Press, p. 59–63.
Nivoit, E., 1887, Geologie appliquee a l'art de l'ingenieur: Paris, Baudry, part 1.
Penning, W. H., 1880, Engineering geology: London, Balliere, Tindall, and Cox, 164 p., map; First published in 1879 as a series of papers in The Engineer.
Radbruch-Hall, D., 1987, The role of engineering; Geologic factors in the early settlement and expansions of the conterminous United States: Paris, International Association of Engineering Geology Bulletin of Engineering Geology, no. 35, p. 9–30.
Rappaport, R., 1969, The geological atlas of Guettard, Lavoisier, and Monnet, *in* Schneer, ed., Toward a history of geology: Cambridge, Massachusetts Institute of Technology Press, p. 272–287.
Reed, J. C., Jr., and Obermeier, S. F., 1982, The geology beneath Washington, D.C.; The foundation of a nation's capitol, *in* Legget, R. F., ed., Geology under cities: Geological Society of America Reviews in Engineering Geology, v. 5, p. 1–24.
Rhoades, R., and Sinicori, M. N., 1941, Pattern of groundwater flow and solution: Journal of Geology, v. 49, no. 8, p. 785–794.
Rose, A. C., 1976, Historical American roads: New York, Crown, 118 p.
Rziha, F., 1874, Lehrbuch der gesammten tunnelbaukenst, 2nd ed.: Berlin, Verlog von Ernst and Korn.
Schlichter, C. S., 1899, Theoretical investigations of the motion of ground waters: U.S. Geological Survey 19th Annual Report, part 2, p. 295–384.
Shaler, N. S., 1895a, Preliminary report on the geology of the common roads of the United States: U.S. Geological Survey Annual Report 15, p. 259–306.
—— , 1895b, The geology of the road-building stones of Massachusetts, with some consideration of similar materials from other parts of the United States: U.S. Geological Survey Annual Report 16, part 2, p. 277–341.
Shrock, R. R., 1977, History of the first hundred years of geology at Massachusetts Institute of Technology; Volume 1, Faculty: Cambridge, Massachusetts Institute of Technology, p. 271–300.
Skempton, A. W., 1946, Alexandre Collin (1808–1890); A pioneer in soil mechanics: Transactions of the Newcomen Society, v. 25, p. 91, also *in* 1969, A century of soil mechanics: London, Institute of Civil Engineers.
Skinner, H. C., 1979, Raymond Thomassy and the practical geology of Louisiana, *in* Schneer, C. J., ed., Two hundred years of geology in America; Proceedings of the New Hampshire Bicentennial Conference on the History of Geology: Hanover, New Hampshire, University Press of New England, p. 201–211.
Smith, T. P., 1799, Account of crystallized basalts found in Pennsyl;vania: American Philosophical Society Transactions, v. 4, p. 445–446.
Snow, E. R., 1954, Famous lighthouses of New England, 1st ed.: Boston, Massachusetts, Yankee Publishing Co., p. 251–291.
Soulavie, J.-L.G., 1780–1784, Historie naturelle de la France Meridionale: Paris, 8 volumes.
Specht, G. J., 1884, The Corinth Canal and proposed methods of blasting and removing the material: American Contract Journal, June 7, 1884, p. 279–281.
Stini, J., 1950, Tunnelbaugeologie: Wien, Julius Springer-Verlag, 366 p.
Suess, E., 1862, Der Boden der Stadt Wien: Wien, Wilhelm Braumuller, K. K. Hofbruchhandler.
—— , 1863–1864, Feasibility studies for a spring water supply for Vienna from Schneeberg region: Progress Reports to Vienna Burgermeisters (unpublished).
Thiem, A., 1906, Hydrologische methoden: Leipzig, Gebhardt, 56 p.
Thomassy, R., 1860, Geologie pratique de la Louisane (Practical geology of Louisiana): New Orleans, 263 p.
—— , 1863, Supplement a la geologie pratique de la Louisiane (le Petite-Anse): Society Geologie France, v. 20, p. 542–544, map.
Troost, G., 1826, Geologic survey of the environs of Philadelphia: Philadelphia, Pennsylvania.
Tuomey, M., 1851, Report on a geological examination of a portion of line and contemplated reservoirs of James River and Kanawha Canal, *in* Report to stockholders, James River and Kanawha Canal Company: Richmond, Virginia, p. 325–336.
TVA, 1949, Geology and foundation treatment; Tennessee Valley Authority Projects: Washington, D.C., U.S. Government Printing Office, Tennessee Valley Authority Technical Report 22, 548 p.
von Cotta, B., 1866, Die Geologie der Gegenwart (The geology of present day): Freiberg, Germany.
Wagner, C. J., 1884, Die beziehungen der geologie zu den ingenieurwissenschaften: Wien, Spielhagen, 88 p., 25 maps.
White, E., and White, M., 1953, Famous subways and tunnels: New York, Random House, 94 p.
Williams, S. G., 1886, Applied geology: New York, Appleton, 386 p.
Winthrop, J., 1755, A lecture on earthquakes: Boston, Massachusetts, Edes and Gill, 38 p.
Zaruba, Q., and Mencl, V., 1963, Engineering geology; Developments in geotechnical engineering: Amsterdam, Elsevier Publishing Co., 483 p.

Nineteenth Century, 1900 to 1940s

Adams, F. D., 1938, The birth and development of the geological sciences: Baltimore, Maryland, Williams and Wilkins Col, 506 p.
AIME, 1929, Geology and engineering for dams and reservoirs: American Institute of Mining and Metallurgy Technical Publication 215, 112 p.
Alden, W. C., 1902, Chicago, Illinois-Indiana folio: U.S. Geological Survey Geologic Atlas 81, 14 p., 12 maps.
Barré, O., 1897–1902, Cours de geographie; Croquis geographiques: Fontainebleau, Ecole d'Applica Artillery et Genie, 2 volumes.
Berkey, C. P., ed., 1911, Geology of the New York City Catskill Aqueduct: New York State Museum Bulletin 146; New York Education Department Bulletin 489, 283 p.
—— , 1942, The geologist in public works; Presidential address: Geological Society of America Bulletin, v. 53, p. 513–532.
Berkey, C. P., and Sanborn, J. F., 1923, Engineering geology of the Catskill Water Supply: Transactions of the American Society of Civil Engineers, v. 86, p. 1–91.
Berkey, C. P., and 6 others, 1933, New York City and vicinity, *in* 16th International Geological Congress Guidebook 9: Washington, D.C., U.S. Government Printing Office, 151 p.
Bertrand, M., and Zurcher, P., 1906, A geological study of the Isthmus of Panama, translated by Oaks, J. C.: Washington, D.C., Board of Consulting Engineers for Panama Canal Report, p. 146–163.
Betz, F., ed., 1975, Environmental geology: New York, Van Nostrand Reinhold Benchmark Series, p. 101–107.
—— , 1984, Military geology, *in* Finkl, C. W., Jr., ed., The encyclopedia of applied geology: New York, Van Nostrand Reinhold, p. 238–241.
Branner, J. C., 1898, Geology and its relation to topography: Transactions of the American Society of Civil Engineers, v. 39, p. 53–95.
Brooks, A. H., 1920, The use of geology on the western front: U.S. Geological Survey Professional Paper 128-D, p. 85–124.
Bryan, K., 1929, Problems involved in the geologic examination of dam sites: American Institute of Mining and Metallurgy Technical Publication 215, class 1, Mining Geology, no. 26, p. 10–18.
Burwell, E. B., 1954, The impact of geology on civil engineering: Virginia Polytechnic Institute Mineral Industries Journal, v. 1, no. 3, p. 1–4.
Cozort, D. A., 1981, Boston's Charles River basin; An engineering landmark: American Society of Engineers Journal of the Boston Society of Civil Engineers, v. 64, no. 4, 387 .
Cloos, H., 1947, Conversation with the Earth: Munchen, R. Piper and Co., 309 p.
Dale, T. N., 1907, Building stone and road metal; Recent work on New England granites: U.S. Geological Survey Bulletin 315-J, p. 356–359.

——, 1923, The commercial granites of New England: U.S. Geological Survey Bulletin 738, 488 p.

Darton, N. H., and Keith, A., 1901, Washington, D.C.-Maryland-Virginia folio: U.S. Geological Survey Geologic Atlas 70, 5 maps.

Davis, S.N ., and De Wiest, R.J.M., 1966, Hydrogeology: New York, John Wiley and Sons, p. 9–12.

Dutton, C. W., 1889, The Charleston earthquake of August 31, 1886: U.S. Geological Survey Annual Report 9, p. 203–528.

Eckel, E. C., 1905, Cements, limes, and plasters; Their materials, manufacture, and properties: New York, John Wiley and Sons, 172 p.; 3rd ed., 1928, 699 p.

Eckel, E. C., and staff, 1940, Engineering geology of the Tennessee River System: Tennessee Valley Authority Technical Monograph 47, 35 p.

Engle, H. M., 1952, Lessons from the San Francisco earthquake of April 18, 1906, *in* Earthquake and blast effects on structures: Earthquake Engineering Institute and University of California at Berkeley, p. 181–185.

Fahlquist, F. E., 1937, Geologic features, Quabbin Aqueduct: Transactions of the American Society of Civil Engineers, v. 102, p. 712–736.

Fakundiny, R. H., 1989, Memorial to Thomas W. Fluhr, 1898–1987: Geological Society of America Memorials, v. 19, p. 93–97.

Fluhr, T. W., 1969, Recent engineering data on the New York City group of formations: New York, Queens College Press Geological Bulletin 3, p. 1–2.

Fluhr, T. W., and Terenzio, V. G., 1973, Geology of the work of the Board of Water Supply, City of New York; Report of New York Board of Water Supply (unpublished): New York State Museum Library and New York City Public Library Open-File.

——, 1984, Engineering geology of the New York City water supply system: New York Geological Survey Open-File Report 05.08.001, 183 p.

Fox, C. S., 1935, A comprehensive treatise on engineering geology: London, Technical Press, Ltd., 392 p.

Freeman, J. R., 1903, Report of committee on Charles River Dam and the formation of Boston Harbor: Report of Chief Engineer Freeman, p. 38–109, Appendix 7; Report of W. O. Crosby, geologist, On geology of Charles River estuary and formation of Boston Harbor, p. 345–369, *in* Cozart, D. A., 1981, Boston's Charles River basin: American Society of Civil Engineers, Journal of Boston Society of Civil Engineers, v. 64, no. 4, p. 1–109.

Gilbert, G. K., 1909, Earthquake forecasts: Science, n.s., v. 29, no. 734, p. 121–136.

——, 1914, The transportation of debris by running water: U.S. Geological Survey Professional Paper 86, 263 p.

——, 1917, Hydraulic-mining debris in the Sierra Nevada: U.S. Geological Survey Professional Paper 105, 154 p.

Gilbert, G. K., Humphrey, R. L., Sewell, J. S., and Soule, F., 1907, The San Francisco earthquake and fire of April 18, 1906, and their effects on structures and structural materials: U.S. Geological Survey Bulletin 324, 170 p.

Glenn, L. C., 1915, Geology applied to dams and reservoirs: Proceedings of the Engineering Association of the South, v. 26, p. 99–113.

Golder, H. Q., 1957, Book review *of* 'Principles of engineering geology and geotechnics by Krynine, D. P., and Judd, W. R.': Geotechnique, v. 7, no. 4, p. 196.

Griswold, D. H., 1950, Applications of geology in soil conservation, *in* Van Tuyl and Kuhn, eds., Applied geology: Golden, Colorado School of Mines Quarterly, v. 45, no. 1B, p. 107–122.

Hall, J., 1839, Classification of excavation rock, Erie Canal Lock at Lockport, *in* 3rd Annual Report of 4th Geological District, State of New York: New York Geological Survey, p. 287–339.

Hayes, C. W., 1894, Chattanooga folio: U.S. Geological Survey Geologic Atlas Folio 6, 5 p., 4 maps.

——, 1895, Cleveland, Tennessee folio: U.S. Geological Survey Geologic Atlas Folio 20, 5 p.

——, 1910, Notes on geology and slides Culebra cut: Canal Record, v. 4, p. 115.

Henderson, L. H., 1939, Detailed geological mapping and fault studies of the San Jacinto tunnel line and vicinity: Journal of Geology, v. 47, no. 3, p. 314–324.

Hodgson, J. H., 1964, Earthquakes and earth structures: Englewood Cliffs, New Jersey, Prentice-Hall, Inc., 161 p.

Horner, S. E., and McNeal, J. D., 1950, Applications of geology to highway engineering, *in* Van Tuyl and Kuhn, eds., Applied geology: Golden, Colorado School of Mines Quarterly, v. 45, no. 1B, p. 154–191.

Howe, E., 1909, Landslides in San Juan Mountains, Colorado: U.S. Geological Survey Professional Paper 67, 58 p.

Irwin, W. H., 1938, Geology of the rock foundations of Grand Coulee Dam, Washington: Geological Society of America Bulletin, v. 49, p. 1627–1650.

Johnson, D. W., 1905, The scope of applied geology and its place in the technical school: Economic Geology, v. 1, no. 3, p. 243–256.

Kaye, C. A., 1968, Geology and our cities: Transactions of the New York Academy of Sciences, series 2, v. 30, no. 8, p. 1045–1051.

Keith, A., 1895, Knoxville folio: U.S. Geological Survey Geological Atlas 16, 6 p., 4 maps.

Kemp, J. F., 1887, The geology of Manhattan Island: Transactions of the New York Academy of Sciences, v. 7, p. 49–64.

——, 1895, The geological section of the East River, at Seventieth Street, New York: Transactions of the New York Academy of Sciences, v. 14, p. 273–276.

——, 1908, Buried channels beneath the Hudson and its tributaries: American Journal of Science, series 4, v. 26, no. 154, p. 301–323.

——, 1911, Geological problems presented by the Catskill Aqueduct of the City of New York: Canadian Mining Institution, Quaternary Bulletin, v. 16, p. 3–9.

——, 1912a, Geological problems presented by the Catskill Aqueduct of the City of New York: Canadian Institute of Mining and Metallurgy Journal, v. 14, p. 472–478.

——, 1912b, The Storm King crossing of the Hudson River: American Journal of Science, series 4, v. 24, no. 199, p. 1–11.

Kiersch, G. A., 1955, Engineering geology: Golden, Colorado School of Mines Quarterly, v. 50, no. 3, 123 p.

King, W.B.R., 1951, Influence of geology on military operations in northwest Europe: London, Advancements in Science, v. 8, p. 131–137.

Krantz, W., 1913, Militargeologie: Kreigstech feitschr, v. 1b, p. 464–471.

——, 1927, Die geologie in ingenieur-baufach: Stuttgart, F. Enke, 425 p.

Lane, A. C., 1930, William Otis Crosby (1850–1925): Proceedings of the American Academy of Arts and Sciences, v. 64, p. 518–526.

Lawson, A. C., 1908, The San Francisco earthquake of April 18, 1906: Carnegie Institution of Washington Publication 87, v. 1, p. 255–451.

——, 1914, San Francisco folio: U.S. Geological Survey Geologic Atlas 193, 24 p., 15 maps.

——, 1932, Geology of the proposed Golden Gate Bridge site: Report to Golden Gate Bridge and Highway District, private publication.

——, 1934, Report of data and remarks, submitted to the Building Committee, *in* Lutgens, H., Maxwell, T., and Kessling, F. V., eds., Investigation of criticism of foundation, Golden Gate Bridge, by Bailey Willis: Golden Gate Bridge and Highway District Report of Building Committee, November 23, 1934.

Legget, R. F., 1939, Geology and engineering: New York, McGraw-Hill Book Company, 650 p.; 2nd ed., 1962, 857 p.

——, 1973, Cities and geology: New York, McGraw-Hill Book Co., 578 p.

Lindgren, W., 1894, Sacramento, California, folio: U.S. Geological Survey Geologic Atlas 5, 3 p., 4 maps.

——, 1898, Boise, Idaho, folio: U.S. Geological Survey Geologic Atlas 45, 7 p., 4 maps.

Lovering, T. S., 1928, Moffat tunnel: Transactions of the American Institute of Mining and Metallurgical Engineers, v. 76, p. 337–346.

Lugeon, M., 1933, Barrages et geologie: Lausanne, Switzerland, Libraire de l'Universite, F. Kouge et Cie, 138 p.

Lutgens, H., Maxwell, T., and Kessling, F. V., 1934, Investigation of criticism of foundation, Golden Gate Bridge by Bailey Willis: Golden Gate Bridge and Highway District Report of the Building Committee, November 27, 1934.

McDonald, C. W., 1915, Some engineering problems of the Panama Canal in their relation to geology and topography: U.S. Bureau of Mines Bulletin 86.

Maddalena, L., 1935, Geologie appliquee aux autostrades modernes: Paris, Congres International des Mines, de la Metallurgie et de la Geologie Appliquee, v. 2, p. 569–573.

Maillet, E., 1905, Essais d'Hydraulique Souterraine und Fluviale: Paris, Librarie Scientifique A. Hermann, 216 p., 30 tables (Chapter 2, p. 34–63).

Marga, A., 1880–1882, Geographie militaire: Fontainbleau, France, Ecole d'Applica Artillery et Genie.

Mather, W. W., and Whittlesey, C., 1838, Geologic section and description of lake front at Cleveland, *in* 1st Annual Report on the geological survey: Geological Survey of Ohio, 38 p.

McAfee, L. T., 1934, How the Hetch Hetchy Aqueduct was planned and built: Engineering News Record, v. 113, no. 5, p. 134–141.

McCullough, D., 1977, The path between the seas; The creation of the Panama Canal 1870–1914: New York, Touchstone/Simon and Schuster, 653 p.

Mead, D. W., 1919, Hydrology: New York, McGraw-Hill Book Co., 626 p.

Mead, W. J., 1920a, Mechanics of geologic structures: Journal of Geology, v. 28, p. 505–523.

——, 1920b, Methods for making block diagrams: The Wisconsin Engineer, v. 25, p. 3–7.

——, 1921, Determination of attitude of concealed bedded formations by diamond drilling: Economic Geology, v. 16, p. 37–47.

——, 1925, The role of dilatancy: Journal of Geology, v. 33, p. 685–698.

——, 1930, Application of the strain ellipsoid (geological): American Association of Petroleum Geologists Bulletin, v. 14, p. 234–239.

——, 1937a, Geology of damsites in hardrock: Civil Engineering, v. 7, p. 331–334.

——, 1937b, Geology of damsites in shale and earth: Civil Engineering, v. 7, p. 392–395.

——, 1941, Engineering geology, *in* Berkey, C. P., ed., Geology, 1888–1938; 50th Anniversary volume: Geological Society of America, p. 573–578.

Meinzer, O. E., 1939, Discussions on subterranean waters; Definitions: Transactions of the American Geophysical Union, v. 4, p. 674–677.

——, ed., 1942, Hydrology: New York, McGraw-Hill Book Collection, 712 p.

Mitchell, G. J., 1917, The need for accurate rock classification in engineering contracts: Economic Geology, v. 12, p. 281.

Morse, W. C., 1913, Road construction materials in Ohio: Ohio Department of Highways, 2 volumes (unpublished).

Murphy, J. J., and Fluhr, T. W., 1944, The sub-soil and bedrock of the borough of Manhattan as related to foundations: The Municipal Engineering Journal, v. 30, p. 119–157.

Neumann, F., 1952, Some generalized concepts of earthquake motion, *in* Earthquake and blast effects on structures: Earthquake Engineering Research Institute and University of California at Berkeley, p. 8–19.

Nickell, F. N., 1942, Development and use of engineering geology: American Association of Petroleum Geologists, v. 26, p. 1797–1826.

Paige, S., 1936, Effect of sea-level canal on the groundwater level of Florida: Economic Geology, v. 31, no. 6, p. 537–570.

——, ed., 1950, Application of geology to engineering practice: Geological Society of America Berkey Volume, 327 p.

Pearson, J. C., and Loughlin, C. F., 1923, An interesting case of dangerous aggregate: American Concrete Institute Proceedings, v. 19, p. 142–155.

Pirsson, L. V., 1919, Rock classification for engineering: Economic Geology, v. 14, p. 264–266.

Redlich, K., Terzaghi, K., and Kampe, R., 1929, Ingenieurgeologie: Wien, Julius Springer, 708 p.

Reid, H. F., 1911, The elastic-rebound theory of earthquakes: University of California at Berkeley Department of Geology Bulletin, v. 6, p. 413.

Richter, R. C., 1966, California earthquake investigations; A review: Geological Society of America Engineering Geology Division, The Engineering Geologist Newsletter, v. 1, no. 3, p. 1–5.

Ries, H., 1929, Geology of dams and reservoirs: Montreal, Engineering Institute of Canada Engineering Journal, v. 12, no. 1, p. 3–7, no. 5, p. 329–332.

Ries, H., and Watson, T. L., 1914, Engineering geology: New York, John Wiley and Sons, Inc., 679 p. (5th ed., 1936, 750 p.)

Runner, D. G., 1939, Geology for civil engineers, as related to highway engineering: Chicago, Illinois, Gillette Publishing Co., 299 p.

Sanborn, J. F., 1950, Engineering geology in the design and construction of tunnels, *in* Paige, S., ed., Application of geology to engineering practice: Geological Society of America Berkey volume, p. 45–82.

Sardeson, F. W., 1916, Minneapolis–St. Paul, Minnesota, folio: U.S. Geological Survey Atlas 201, 14 p., 8 maps.

Schuberth, C. J., 1968, The geology of New York City and environs: Garden City, New York, National History Press, 278 p.

Sharpe, C.F.S., 1938, Landslides and related phenomena; A study of mass-movements of soil and rock: New York, Columbia University Press, 136 p. (Reprinted 1960, by Pageant Books, Inc., Patterson, New Jersey, 137 p.)

Sherzer, W. H., 1917, Detroit, Michigan, folio: U.S. Geological Survey Geologic Atlast 205, 22 p., 11 maps.

Shrock, R. R., 1972, The geologists Crosby of Boston; William Otis Crosby (1850–1925) and Irving Ballard Crosby (1891–1959): Cambridge, Massachusetts Institute of Technology, 96 p.

——, 1977, History of the first hundred years of geology at Massachusetts Institute of Technology; Volume 1, Faculty: Cambridge, Massachusetts Institute of Technology Press, William Otis Crosby, p. 271–300; Warren J. Mead, p. 679–696.

Smith, W.D., 1919, Rock classification for engineering: Economic Geology, v. 14, p. 180–183.

Sorsbie, R. F., 1938, Geology for engineers, 2nd ed.: London, G. Bell and Sons, 348 p. (1st ed., 1911, Chas. Griffin & Co.)

Stanton, T. E., 1940, Expansion of concrete through reaction of cement and aggregate: Proceedings of the American Society of Civil Engineers, v. 66, p. 1781–1811 (discussion by Stanton, v. 67, p. 1402–1418).

Stini, J., 1910, Die muren (Slope phenomena): Innsbruck, Wagner, 139 p.

——, 1919, Technische gesteinskunde: Wien, Waldheim-Eberle, 335 p. (2nd ed., 1929, 550 p.).

——, 1922, Technische geologie: Stuttgart, F. Encke, 789 p.

——, 1933, Die quellen: Vienna, Springer, 255 p.

——, 1941, Unsere Taler wachsen su (Valley accretion): Geologie Bauwes, v. 13, p. 71–79.

——, 1950, Tunnelbaugeologie: Wien, Julius Springer-Verlag, 366 p.

Strauss, J. B., 1938, The Golden Gate Bridge; Report of Chief Engineer to Board of Directors: Golden Gate Bridge and Highway District, September 1937, 246 p.

Stremme, H., and Moldenhauer, E., 1921, Engineering geologic foundation map of the City of Danzig, *in* Betz, F., ed., 1975, Environmental geology: New York, Van Nostrand Reinhold Benchmark Series, p. 336–340.

Taylor, T. U., 1900, The Austin Dam: U.S. Geological Survey Water-Supply Paper 40, 52 p.

Terzaghi, K., 1925, Erdbaumechanik auf bodenphysikalischer grundlage: Wien, Franz Deuticke, 399 p.

——, 1955, Influence of geological factors on the engineering properties of sediments, *in* Bateman, A. M., ed., Economic Geology 50th Anniversary Volume: Lancaster, Pennsylvania, Economic Geology Publishing Co., part 2, p. 557–618.

——, 1963, Karl Terzaghi's last writing on soils: Engineering News Record, v. 171, no. 21, p. 1–2.

Theis, C. V., 1935, The relation between the lowering of the piezometric surface and the rate and duration of discharge of a well using groundwater storage: Transactions of the American Geophysical Union, v. 16, p. 519–524.

Thiem, A., 1906, Hydrologische methoden: Leipzig, Gebhardt, 56 p.

TVA, 1949, Geology and foundation treatment; Tennessee Valley Authority Projects, *in* Tennessee Valley Authority Technical Report 22: Washington, D.C., U.S. Government Printing Office, 548 p.

Twenhofel, W. H., ed., 1932, Treatise on sedimentation: Baltimore, Maryland,

Williams and Wilkins, 626 p.
——, 1939, Principles of sedimentation: New York, McGraw-Hill Publishing Co., 707 p.
USBR, 1950, Boulder canyon project final reports; Part 3, Bulletin 1, Geological investigation: U.S. Bureau of Reclamation: Washington, D.C., U.S. Government Printing Office, 231 p.
Veatch, A. C., 1906, Fluctuations of the water level in wells with special reference to Long Island, New York: U.S. Geological Survey Water-Supply Paper 155, 83 p.
Veatch, A. C., Slichter, C. S., Bowman, I., Crosby, W. O., and Horton, R. E., 1906, Underground water resources of Long Island, New York: U.S. Geological Survey Professional Paper 44, 385 p.
Von Bulow, K., 1938, Wehrgeologie: Leipzig (Translated in 1943 by K. E. Lowe, ERO Translation T-23, Intelligence Branch, Office of Chief Engineers).
West, P. J., 1954, Foundation problems associated with the Terminal Island subsidence, California [abs.]: Geological Society of America Bulletin, v. 65, p. 1323–1324.
White, E., and White, M., 1953, Famous subways and tunnels: New York, Random House, 94 p.
White, L., 1913, The Catskill water supply of New York City; History, location and subsurface investigations, and construction: New York, John Wiley and Sons, 755 p.
Willis, B., 1934, Is the Golden Gate Bridge a $35,000,000 experiment?: The Argonaut, October 19, 1934, p. 2–4.
Wood, W. A., 1932, Connecting Stockton, California, with the sea: Engineers News Record, v. 109, no. 7, p. 183–185.
Yochelson, E. L., ed., 1984, The scientific ideas of G. K. Gilbert: Geological Society of America Special Paper 183, 148 p.

Manuscript Accepted by the Society July 24, 1990

ACKNOWLEDGMENTS

Preparation of the historical review on the heritage of engineering geology and its meaning has benefited from the assistance and input of many colleagues over the years. Besides the individuals and sources cited, I am particularly indebted to: William R. Brice, Cornell University, who alerted me to several early publications; the late Thomas W. Fluhr for his descriptions of New York water-supply projects and personnel of the 1920s to 1950s; Julius Schlocker, USGS, for copies of reports on the Golden Gate Bridge controversy; Shailer S. Philbrick, Cornell University, for background materials on early Corps of Engineers projects; the late Alois Kieslinger, Technische Hochschule, Wien, for publications on practice in Europe from the 1870s to the 1940s; and Walter Kollmann, Austrian Geological Survey, Wien, for assistance in assembling background on Eduardo Suess and the water aqueduct of the 1860s and 1870s. A. R. Palmer has exercised great patience and been most helpful with a review and comments that benefited and strengthened the preliminary mansuscript.

Three pioneers of applied geology: B. S. Butler, F. L. Ransome, and Waldemar Lindgren. The three are pictured while on an International Geological Congress field trip at Bisbee, Arizona, in 1933. Butler, an early student of Heinrich Ries, is well known for his work on major mining areas of Colorado. Ransome investigated the St. Francis Dam failure and was consultant for planning and construction of the Colorado Aqueduct. Lindgren, a former chief geologist for the U.S. Geological Survey and professor at the Massachusetts Institute of Technology, was an early advocate of engineering geology in the 1910s. (Photo courtesy of the M. N. Maxwell files.)

Printed in U.S.A.

Geological Society of America
Centennial Special Volume 3
1991

Chapter 2

Modern practice, training, and academic endeavors 1940s to 1980s

George A. Kiersch
Geological Consultant, Kiersch Associates, Inc., 4750 Camino Luz, Tucson, Arizona 85718

INTRODUCTION

World War II and the stress of war-time economics from 1940 to 1945 placed new demands on all phases of industry. These elements of change, combined with the effects of postwar economic expansion, were manifested in many geology-related problems. For example, (1) the need for increased supplies of industrial water in some coastal areas led to an excessive draw-down of the ground-water level and allowed sea-water encroachment; (2) construction of the Alaskan Highway in early 1940s for defense of territory led to an in-depth realization of the permafrost phenomenon and its impact on construction in arctic terrain; (3) the need for large underground storage and bomb-proof military facilities by the mid-1940s led to pioneering research in rock mechanics and the dynamic stress phenomenon of large-scale explosions (McCutchen, 1949; Kiersch, 1951); (4) the demands for terrain analysis to serve military actions resulted in several new aerial exploration and interpretive techniques that were later available for civil projects; and (5) more recently, geology was a major factor and was investigated extensively during the planning and construction of the Alaskan pipeline in the 1960s and 1970s (Péwe, Chapter 14, this volume).

The large reservoirs built in the 1920s and 1930s increased sediment-filling to the status of a serious geological problem; then, ironically, the cleansed reservoir water created leakage problems downstream in the very canals that were formerly self-sealed by the natural silt. Each difficulty ultimately provided an improved state of knowledge and progress in engineering geology.

By the 1950s, international interest was keen on a distinction between an artificial, underground nuclear explosion (test) signal and a natural seismic event; the "detectability of seismic signals" became the focus of major seismological research by the U.S. Air Force and experiments by the Terrestrial Sciences Laboratory (Haskell, 1957). During that decade, civilian activity greatly increased the demand for electric power to serve expanding industry; this required construction of new hydroelectric generating capacity and elaborate distribution systems. Among such projects were the combination transportation/shipping and hydrogeneration plants of the St. Lawrence Seaway project in the 1950s and 1960s, and the generating plants of the New York Power Authority.

Also during the 1950s, research projects at the Lawrence Radiation Laboratory, Livermore (Griggs and Teller, 1956), sponsored by the Atomic Energy Commission, investigated the peaceful use of nuclear explosions for construction and mining purposes (Bacigalupi, 1959; Teller and others, 1968). These included such projects as a major interstate highway cut in the Mojave Desert area (ENR, 1963); a new sea-level canal across the Panama Isthmus based on large-scale geological investigations of 1946 to 1948; the economic mining of the Athabaska oil and tar sands of Canada; the "manufacture" of geothermal steam or improved water-bearing characteristics of rocks; and inducing large-scale subsidence/compaction (Houser and Eckel, 1962). In the planning for the new sea-level canal, problems of unprecedented magnitude confronted geologists in the design of stable slopes for cuts over 600 ft deep in weak rock (Binger and Thompson, 1949). Typical of the problems was the threat of a major landslide at Contractor Hill along the canal in early 1954, when deep fractures caused by deep-seated deformation opened up across the top of the hill, posing a threat that cost millions of dollars to remedy (Thompson, personal communication, 1954).

During the post–World War II expansion, many countries undertook development of major drainage basins for multiple engineering projects. Notable among them were sections in India, Iraq, Africa, and Australia. The Snowy Mountain Project, Australia, ranks among the major efforts of its kind in the world (Park, 1953; Moye, 1955). The U.S. Bureau of Reclamation provided five senior advisors for the project, led by H. F. Bahmeier. A river was partially diverted to the opposite side of the Australian Alps for irrigation and electrical generation. The project included eight major dams, 16 power plants (mostly underground), 86 mi of tunnels, 490 mi of canals in mountainous terrain, and hundreds of miles of highways and railroads.

A typical international project given technical assistance was the long-range agricultural and industrial development of the Khuzestan region, Iran. Planning and technical advice for the

Kiersch, G. A., 1991, Modern practice, training, and academic endeavors, 1940s to 1980s, *in* Kiersch, G. A., ed., The heritage of engineering geology; The first hundred years: Boulder, Colorado, Geological Society of America, Centennial Special Volume 3.

Figure 1. Reza Shah Kabir Dam and power plant on the Karun River, Khuzestan, Iran, a key project in development of the Karun River power and irrigation potential. (Photo source: C. T. Main, Boston, 1986.)

Khuzestan Water and Power Authority, an Iranian government agency, were provided by the Development and Research Corporation of New York under a 1956 contract (Clapp, 1957; G. K. Clapp was a former chairman of Tennessee Valley Authority). Major projects included the Reza Shah Kabir Dam and powerhouse facilities on the Karun River, Khuzestan (Fig. 1), which was completed in the 1970s.

Since World War II, airport programs have created demands for larger sites with greater bearing capacities and enormously expanded construction-material needs. Creating similar demands have been the expansion of state and federal conservation measures, urban developments with necessity to reclaim marginal lands, and military planning both on land and underwater. In the 1960s, scientific demands required the active participation of engineering geologists for the needs of lunar and planetary exploration (Green, 1961).

The deterioration of highways during World War II, and the realization that roads are defense lines, caused highway construction to spurt, both in mileage and design standards. Turnpikes, toll roads, and freeways became the trend, and they required a greater use of geology for design and construction. The U.S. Congress in 1955 supported a far-sighted nationwide interstate highway system for construction over a 10-yr period. This activity engaged a large group of applied geologists at national and state agencies and with private firms.

Eight major dams failed around the world between 1959 and 1964. The two reservoir or dam failures in 1963, Vaiont (October) and Baldwin Hills (December), coming after the earlier failure at Malpasset (1958), initiated a period of reconsideration and evaluation of dam safety. This led to mandatory inspections of dams by the 1970s, with modifications or repairs, such as improving the stability of reservoir slopes to provide greater protection for the environs from the impact of a project. The need for increased electrical generating capacity, combined with forecasts for still larger future electrical needs, led to large-scale investigations, planning, and construction of nuclear power plants throughout the country, beginning in the mid-1960s. This trend continued until the late 1970s when the forecast for energy needs was cut back, in part by the pressure for greater environmental considerations, new federal policies, and the rising financial costs of construction. This initiated a nationwide cancellation of many nuclear generating plants under construction, and caused a temporary decline in the employment of engineering geologists. The Three Mile Island nuclear plant mishap of 1978 further contributed to a reduction in the construction of nuclear generating capacity. By this time, however, industrial pollution and burial of toxic waste were recognized as geologically related problems confronting the nation, and geologists joined the team effort to rectify them.

By the 1960s, concern within and outside of government was building for protection of the natural environment, specifically from the impact a proposed or already operating engineered works might have. An example is the 1969 leakage and blowout of an operating oil well in Santa Barbara channel, California. Nationwide attention to the spill led to many federal laws and regulations, beginning with the National Environment Policy Acts (NEPA) of 1969. The U.S. Environmental Protection Agency (USEPA) was established in 1970; the Water Quality Improvement Act of 1970 was passed that year, also. Other national, state, and local legislation followed in the 1970s and 1980s, such as the Resource Conservation and Recovery Act (RECRA) of 1976. These acts had become major guidelines for, and factors in, the practice of applied geosciences relevant to the environment by 1980 when Congress enacted the Superfund Act (CERCLA) with RECRA in 1984. The Low-Level Radioactive Waste Policy Act followed (Public Law 99-240) in 1985, and CERCLAIN in 1987. Projects viewed as having an impact on the environment included the disposal and deep burial (in rock) of nuclear wastes generated by both military and civil engineering works, the disposal of garbage and refuse in landfills, the occurrence and health hazards of trace elements in ground-water supplies, and the safety of flood plains and flood-hazard zoning of these lands. From this trend, two new areas of specialized practice have emerged: (1) the identification and mitigation of geologic hazards (Table 1), spurred by nuclear safety concerns and numerous large-scale engineering failures; and (2) burial-waste projects that focus on ground water as a contaminant carrier.

During the 1950s and 1960s, engineering geologists learned to extend the methods of traditional geological surface mapping and underground mine mapping to collect a wide variety of subsurface geologic data by using new techniques. This meant all available geologic data were being interpreted to try to reveal the physical characteristics responsible for the integrity, strength, deformability, and permeability of foundation materials. Furthermore, specific geological opinions were routinely expected and invited as a factor in project design requirements. By the 1980s, the capacity of digital computers had grown to unexpected proportions, and there emerged a general trend to emphasize the physical property measurements of earth materials. Unfortunately, new field observations and testing of representative samples are all too frequently minimal, and the computerized reevaluations are mainly recycled data only.

Expansion into many remote regions of the world in 1960s to 1980s, for mineral exploitation and related purposes, frequently involved the applied geologist with harsh climates and unique geological problems when constructing engineering works. Examples include oil production, for on- and offshore pipelines and drilling platforms in the Middle East, northern Canada–Baffin Island regions, and the North Sea; and roads, dams, and hydroelectric plants, and other facilities to serve new mines and industries in the Amazon, New Guinea, other Pacific Islands, and Africa. In yet another way, natural hazards such as volcanic eruptions became of concern to applied geologists, as demonstrated by Mount St. Helens, Washington, in 1980. This event focused attention on the special problems of ash falls, debris flows, and associated earthquakes relevant to engineering works in volcanic regions (Mullineaux, 1981). Of similar concern were the catastrophic debris flows, destruction, and collateral hazards associated with the eruption of Nevada de Ruiz, Columbia, in 1985 (Herd and others, 1986). Yet another, but quite different, volcanic event was the 1986 release of a deadly cloud of carbon dioxide–rich gas from Lake Nyos, Cameroon, a volcanic diatreme structure, in which over 1,700 people were killed (Tuttle, 1989).

Acceptance of the importance of geological principles and field-oriented judgment in planning, building, and operating engineering works was initially based on five critical factors to which the geologist can contribute: the cost, safety, and efficiency of design and construction, the welfare of the public, and the health aspects both of individuals and a region (Kiersch, 1955, p. 6). By the 1970s, the impact of operating engineering works on the environment had become a sixth critical factor, which included the legal ramifications of operating or modifying a major works. The repair and modernization of the nationwide infrastructure of highways, bridges, and dams, initiated in the 1980s, places a focus on several new and additional factors (Jansen, 1988), such as were experienced with the enlargement of reservoir capacity at Roosevelt Dam (Fig. 2) by raising the dam 77 ft.

Figure 2. Roosevelt Dam on the Salt River, Arizona, was the first major irrigation project in United States and has been operating since 1911. Beginning in 1988 the dam underwent rehabilitation and the capacity of the reservoir was increased by raising the dam. This required removing the highway from crest of dam and erecting the suspension bridge upstream across an arm of reservoir (completed 1990) as the highway replacement. (Photo courtesy of: Department of Transportation, State of Arizona/Walter Gray, 1988).

Common to all these geological advances have been the underlying needs of an expanding population for "social engineering projects," economic development of virgin lands, and the proportional decrease of acceptable natural sites for engineering purposes. From this seemingly disconnected and diversified growth, and to meet the day-to-day needs of engineering and applied science, has evolved the orderly grouping of the principles and techniques that constitute engineering geology.

Today, engineering geology is an interdisciplinary field of practice that is primarily concerned with the physical processes, phenomenology, and principles of the geosciences as they pertain to engineering works, applied sciences, and the needs of mankind. The practice of engineering geology is mainly associated with the construction efforts of engineers and scientists and comprises a complex and interdependent assemblage of subfields and specialties, as well as a small core of generalists. An example of this style of practice is the group of practitioners who joined with the diversified sciences and nongeological disciplines in the 1980s to serve the needs of environmentally concerned citizens. This led to support of most state geological surveys—and those of some counties and municipalities—with a variety of hazard-oriented map and inventory projects to benefit society at large. The other branches of applied geoscience—mining, petroleum, and ground water (hydrogeology)—are mainly occupied with the destructive efforts of engineers for exploitation of minerals and resources. The interrelations of professional practice in the applied geosciences are shown on Figure 3.

The sections that follow in this chapter include brief sketches of several subject areas that are not covered in the chapters on

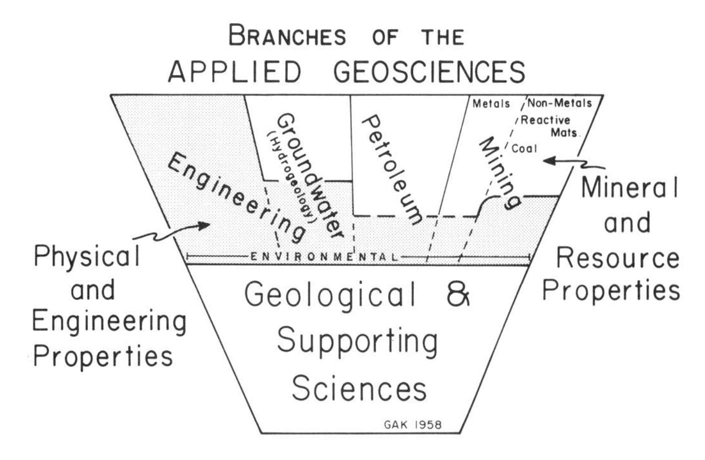

Figure 3. The interrelation of professional practice among the four main branches of applied geosciences. The physical/engineering properties of the natural materials and setting and the environment are characteristic of engineering geology activities. Although the other branches are primarily mineral/resources–oriented, varying proportions, up to one-half of the activities, are within the realm of engineering geology practice. (From Kiersch, 1958 and 1964.)

geological principles, processes, and events relevant to professional practice (Chapters 6 through 17), or the investigative techniques covered in Chapters 18 through 21. The responsibilities of the applied geologist for decisions and judgments rendered, which may lead to technical hearings or litigation, are reviewed in Chapters 22 through 25. The scope of modern engineering geology is summarized here, as are the advances in training and related academic endeavors in the field since 1940.

EFFECT OF WORLD WAR II

The advent of World War II (1939 to 1945) brought about the proliferation of applied geology on a scale hitherto unimagined. Of these new phases, application of geology to military operations, developed in Europe, was perhaps the most important advancement in engineering geology around midcentury (Kiersch, 1955, p. 29–30).

The Germans, who developed military geology during World War I, made many advances in the intervening years (von Bülow, 1938), which served to convince the allies that geologic insight and data are an important adjunct to military operations. A unique and very successful German Wehrmacht organization, the Forschungesstaffel (Research Detachment), was a special unit created in mid-1943 after a period of experimental operations in north Africa in 1942 (Betz, 1984a). The unit concentrated on preparation of terrain-evaluation maps, paying particular attention to trafficability for tanks. Maps were based on the combination method, a concept developed by a small group of German geographers. Field investigations were conducted by teams of physical geographers, plant ecologists, foresters, geologists, soil scientists, meteorologists, and other specialists, and photo interpretation and aerial reconnaissance were combined with ground reconnaissance and other available data to produce terrain evaluations. The terrain maps were ingenious presentations of geologically related data, and they aroused wide interest among military terrain specialists in other countries when they were studied after the war (Smith and Black, 1946; Wilson, 1948).

By 1940, military forces of many countries were beginning to use geologic data to plan their operations. For instance, South African geologists formed the nucleus of the 42nd South African Engineers Unit, whose work dealt largely with ground-water development and advice on a variety of military construction problems (Kent, 1952).

Acceptance of applied geology by the U.S. Armed Forces was slow, and did not occur until worldwide military operations confronted the allied forces with all types of terrain conditions (Erdman, 1943). At this point, the military realized geology could be important; in fact, it was almost indispensable in both the planning and field operations for the Army, Navy, and Air Force (Hunt, 1950; Russell, 1950; Whitmore, 1950). Experiences in World War II and military operations since in Korea, Vietnam, and other spots around the world indicate that the application of geology to military problems is successful in proportion to the degree to which the basic information (mapping) has been completed in advance (Betz, 1984a). Terrain data, including the subsurface, are basic in all terrain-intelligence interpretations, including the trafficability factor, sources of construction materials and water supply, and evaluation of sites for all installations, whether beachheads, airheads, supply routes, or underground installations. For example, samples of Normandy beach sands were collected in advance of the 1944 landings in France to forecast the trafficability characteristics of beaches. Throughout the Pacific region, advance reconnaissance teams with a terrain-specialist member reconnoitered future beachhead areas and surroundings to evaluate the geologic environs and select sites for airstrips, camps, and sometimes for amphibious landings.

As an outgrowth of building the highway to Alaska, permafrost and related geologic phenomena became important in Arctic operations of all kinds. Consequently, the Snow, Ice, and Permafrost Research Establishment was formed as part of the Corps of Engineers, along with other research organizations such as a group in the U.S. Geological Survey, to supply applied geologic data for construction activities of all types in the Arctic region (see Péwé, Chapter 14; Weeks and Brown, Chapter 17, this volume).

For the problems inherent to naval operations, many applications of submarine geology and sedimentation were developed, including the use of underwater sound, submarine mining, installation of underwater equipment, shore installations, and amphibious operations (Russell, 1950). The aerial-detection techniques for naval warfare have been adapted since to various types of geological exploration.

TABLE 1. THE COMMON NATURAL OR INDUCED HAZARDS

Rapid-Onset Event Short Duration/Catastrophic (Seconds to Days)	Scale	**Slow-Insidious Event** Long Duration (Weeks to Years)	Scale
NATURAL GEOLOGIC-HYDROLOGIC PHENOMENA/HAZARDS			
Slope Collapses/Slides**	A-S	**Many Slope Failures-Progressive****	A-S
. Rock-soil slides	A-S	. Creep	A-S
. Rockfalls	A	. Rock-soil slides	A-S
. Waves generated—lakes	A-S		
. Flooding	A-R	**Rising Sea Level**	R
. Earthquakes-locally	A	. Beach-shoreline erosion	R-A
Earthquakes	R	. Inundation river deltas-world, cities, lands, people	R-A
. Tsunamis (Seiche/R-A)	OW		
. Slides damming rivers, with	A-S	**Expansive Soils**	R-A
flooding on collapse	R-A		
		Most Subsidence Types**	R-A-S
Volcanic Eruptions/Outpourings	R-A	. Regional-areal features	
. Ashfalls	R	. Local forms	
. Flows and debris	R-A	. Man-induced, mines/withdrawal fluids	
. Damming rivers and lakes, causing	A-S		
flooding potential downstream	R-A	**Ground-water Contamination**	R-A
. Calderas—inflate/subside	A-S	. From natural events-sources	
. Earthquakes—modest scale	A		
		Volcanic Subsurface-Migration	A
Sinkhole Collapse-Subsidence**	A-S	. Magma reservoirs-rise and fall of surface; potential eruption	
Debris Flows	R-A-S		
		Soil Erosion	R-A
Gullying	A-S		
Snow Avalanches	R-A		
MAN-INDUCED GEOLOGIC-HYDROLOGIC "PHENOMENA"/HAZARDS			
Slope Collapse/Slides	A-S	**Slope-Creep**	A-S
. Slides-rock/soil	A-S		
. Waves—reservoirs/lakes,	A	**Ground-water Contamination**	A-S
flooding	R-A	. Chemicals/nuclear wastes locally buried	
NATURAL ATMOSPHERIC-HYDROLOGIC PHENOMENA/HAZARDS			
Floods-Most Types	R	**Climatic Changes**	
		. Jet stream shift	R
Cyclones-Anticyclones	R	. Blizzard and ice storm; hail	R-A
. Hurricanes and typhoons (Tropical cyclones)		. Volcanic eruptions	R-A
		Gases-atmospheric cooling; climatic variations	
Severe Thunderstorms	R	. Greenhouse effect; atmospheric warmup/melt; ice sheets, rise in sea level	R
. Tornadoes—violent (Most destructive disturbance)	R-A		
. Lightning	R-A	**Floods-Some Types**	R
Other Windstorm Types	R-A	**Droughts**	R
		Heat Waves	R

Explanation:
R = Regional implies events acting over 10s to 100s of square miles.
A = Areal means events acting over at least a 4 to 5 square mile area.
S = Site-specific means the one-half square mile surrounding an event of limited scale.
OW = Ocean-wide event.
** = Loss of life from slope failures averaged 600 per year worldwide from 1971 to 1974, (Varnes, 1981).
** = Damage losses from slides and subsidence alone totaled more than $1.5 billion per year in the United States during the past 50 years, irrespective of more precious loss of life; this is more than total loss per year from earthquakes, floods, hurricanes, and tornadoes (Kockelman, 1986).

ACTIVITIES RELATED TO MILITARY GEOLOGY

The success of applied geology during World War II actions resulted in the establishment of a Military Geology Branch within the U.S. Geological Survey in 1946; an earlier Military Geology Unit (MGU) started in 1942 (MGU, 1945). Functions of this unit were closely coordinated with the military branch, the Central Intelligence Agency, the U.S. Corps of Engineers, and the Army Map Service. Similarly, the U.S. Navy, through the Office of Naval Research and the Bureau of Ships, has financed research in oceanography, submarine geology, and sedimentation, which are now recognized as topics for planning the support of amphibious operations, as well as beach-erosion control.

World War II stimulated interest in the use of aerial photography, and since that time other types of imagery and sensors have provided invaluable data for basic and applied geoscientific research. Following that war, a vast amount of geoscientific data became available from programs sponsored by military establishments (Betz, 1984a). The Fukui earthquake of June 28, 1948, provided an unusual opportunity for geologists and engineers of the MGU to assess the importance of complex geologic conditions of sites for human habitation and acquire new data for design of earthquake-resistant structures (Collins and Foster, 1949).

Underground construction research

After 1945, realization of the destructive force of the atomic bomb and later the hydrogen bomb created concern in the minds of many that defense against their effects was nearly impossible. In response, the U.S. Corps of Engineers and government-sponsored research groups made tremendous strides in the design of protective construction to resist large-scale blasts, relying on knowledge of the geologic environs and properties of rock masses. Developments in destructive weapons dictated some underground locations for military command centers and storage facilities, for either tactical or operational purposes. Also, any relocation of vital plants needed to be highly selective and based on military importance, economic factors, and the inherent needs of the operation of the facility.

Geologic principles, and their application to the location, construction, and operation of subterranean installations designed to resist modest-scale subsurface explosions were reviewed by Kiersch (1949, 1951). The underground Explosion Test Program of U.S. Corps of Engineers (1947 to 1949) and research of Engineering Research Associates, Minneapolis (ERA, 1952–1953) established the principal geologic factors that impact on the design of a large-scale cavity. The Rand Corporation's underground construction symposium in 1959 recorded the state of knowledge regarding the design and construction of protective chambers (O'Sullivan, 1961).

Character of the overburden

A homogeneous medium, capable of dissipating energy equally in all directions from a subsurface blast is probably nonexistent in nature. Evaluation of rock reaction to induced stress is complicated by its heterogeneity; structural weaknesses result in unequal transmission of stress.

Igneous rocks can harbor a high residual stress. The physical characteristics of an ideal rock mass and its reaction to a subsurface blast have been investigated (Kiersch, 1951; O'Sullivan, 1961). The boundaries of the near-surface zone subject to a superimposed explosive stress vary locally, depending on the degree of rock inhomogeneity. At depth, the elastic limit of the rock is overcome by natural stresses; openings excavated below a critical depth will be located in the zone of instability where "shear" stresses are active and openings require supports. Depth to the zone of instability will vary with the specific rock type. For example, "shear" strength of an average homogeneous sandstone can be exceeded at a depth of 1,750 ft, while that of an average homogeneous plutonic rock occurs around 6,500 ft. Substantial residual stress or active stresses may also exist; when present they may render a site uneconomical for development.

Conditions of ideal cover for an underground installation may involve a combination of competent high-velocity rock units, separated by a low-velocity, weak rock unit. Formational boundaries partially dissipate the explosive energy in the overlying medium (Kiersch, 1951). Consequently, a rock mass may vary widely in resistance to explosive pressures, depending on its physical character, the magnitude and abundance of weakness planes, and their position with respect to the blast; these principles are reviewed by Duvall (p. 131–147), Judd (p. 255–294), and Smith (p. 313–335) in O'Sullivan (1961), and USCE (1961).

In the 1940s, an underground factory was located in an outcrop area of loess a short distance north of Linz, Austria (Fig. 4), where large areas of thick loess deposits are common to the Danube River valley. The underground facility was a modified room-and-pillar system (Kiersch, 1949, Fig. 3) with large, interconnected rooms up to 15 ft high. Entrance was off a village road along a small stream. The site successfully withstood aerial bombing attacks because the soft-to-spongy loess material dissipated the shock of conventional explosions.

Numerous advances of basic geological knowledge developed from underground construction during World War II. Two important cases involved the German Army. Construction of large submarine pens—massive, thick concrete structures—at Trondheim Bay, Norway, required that the bearing capacity of the beach sands and silt deposits be strengthened; an adequate foundation was prepared using an electro-osmotic "grouting" technique. The technique was successful because the natural mixture of sand and silt sizes at the Trondheim site was ideal for stabilization by electrogrouting (Casagrande, 1952). While attempts at some sites have not been successful, electro-osmosis is being exploited in some applied practical situations as a construction technique (Mitchell and Katti, 1981). The Trondheim structures have since been modified and today serve as modern office complexes.

The submarine pens of Narvik, Norway, north of Trondheim, were excavated in the granitic rock complex of the coastal

Figure 4. An underground factory site, World War II, located in loess deposits of the Danube River valley north of Linz, Austria. The entrance, lower right, is largely closed off. (Photo by G. A. Kiersch, 1963.)

area. Kieslinger recognized the importance of the joints and sheet structures, and the existence of an active residual stress field related to glacial rebound and tectonic history. The large-scale underground chambers were subsequently designed to compensate for the stresses and the sheet structures associated with unloading (Kieslinger, 1958, 1960; and personal communications, 1963).

UNDERGROUND INSTALLATIONS

Because underground installations offer a means of insuring both safety and economic feasibility through reduced costs and operational efficiency, the application of geologic principles for large-scale underground projects expanded significantly after World War II in both North America and Europe. Since the 1940s, several hundred underground rock installations have been built in Sweden, as have similar plants in Norway, France, Italy, and other European countries. Today there are over 20 different kinds of underground installations being built in hard-rock formations in Sweden and elsewhere, as shown on Figure 5 and described by Morfeldt (1983).

The trend to plan and construct underground is exemplified by the powerhouse at Kemano, British Columbia, Canada. This, the largest single underground excavation of its day, involved an exploratory tunnel to determine the physical characteristics of the site prior to the design. As constructed, the main arched chamber is up to 81 ft wide and 700 ft long with a height of 140 ft (Wise, 1952).

Two strategic United States military centers were built underground in the 1950s and 1960s: the Omaha, Nebraska, Command Control Center; and the NORAD Center in Cheyenne Mountain near Colorado Springs, Colorado. Such critical projects required particular attention to the geology, due to the site conditions or the size of underground chambers (Lane, 1971). The largest concentration of underground storage and business facilities today is beneath greater Kansas City, where they are located in former limestone quarries (Hasan and others, 1988).

Rapid transit systems underground have been installed in a number of cities throughout the United States since the 1960s; most have required large-scale geological investigations for their design, and further input during their construction. The Bay Area Rapid Transit (BART), built from 1966 to 1973, is underground beneath San Francisco Bay (cut-and-cover construction) and in San Francisco, and above ground throughout most of the East Bay region (Taylor and Conwell, 1981). The BART tube to Oakland beneath the bay became the only direct connection between the two cities after the Loma Prieta earthquake of October 17, 1989, damaged and closed the Bay Bridge (Fig. 6). This trans-bay tube was placed wholly on soft ground by designers to "cushion earthquake shocks."

Another region where design of engineered structures has been planned to successfully withstand the forces of earthquakes is Los Angeles (Heuer, 1977); construction of an underground metro system began there in late 1987 with the first phase scheduled for operation in 1993. Other cities with new subway systems include Washington, D.C., where construction of the underground downtown, and the above-ground system in the suburbs, began in the mid-1960s (Bawa, 1970; Bock, 1981). Atlanta, Georgia, built a system in the 1970s; Buffalo, New York, began construction of one in the late 1980s; and Boston extended the red line of its Metro beyond Cambridge station in the 1980s (Keville and Sutcliffe, 1983; see Chapter 24, this volume, and Meyer, 1983).

GROWTH OF ENGINEERING GEOLOGY

The post–World War II period witnessed the spread of applied geosciences and a substantial improvement of the status of activities in engineering geology. The greatly increased demand for the services of geologists to plan and participate in the construction of major engineering works, along with an even greater demand for aid in the discovery and exploitation of resources, changed the composition of the geoscientific community (Betz, 1984b, p. 241).

These changing demands required changes in both scientific thinking and professional practice for engineering geologists; the new focus was on physical processes and associated events, the reaction of natural environs to operating works, and the geologist's responsibilities. These are the main subjects covered in Chapters 6 through 25 of this volume. Because a number of topics and/or processes important to the field could not be included as separate chapters (see Preface and Chapter 1, this volume), brief discussions with graphics on several of the subjects are given below.

Federal and state agencies

Prior to WWII, a Beach Erosion Board was established within the Corps of Engineers in 1930 to minimize the effects of wave and current action along the seacoast and lake shores. However, not until the act was broadened (1935, 1945, and 1946) to include work at the entrances to rivers, and assistance in construction and maintenance of works designed to perpetuate public beaches, did the board concern itself with sedimentation in harbors, shore processes, and related engineering problems. This led to many new opportunities for a closer collaboration between geologists and engineers (Kiersch, 1955, p. 27–28). Today, sophisticated research, with lasers, prisms, sensors, and other measuring devices, is being carried out by the U.S. Corps of Engineers research facility at Duck, North Carolina, and other east-coast beaches to ascertain the movement of water and sand within the "near-shore" transition zone. This zone is emerging as a crucial factor in the erosion of beaches, a problem that threatens 70 percent of the nation's coastline (Dean, 1990; discussed by Leatherman in Chapter 8, this volume).

A separate Engineering Geology Branch was formed within the U.S. Geological Survey under Edwin B. Eckel in late 1944. The branch began functioning in 1946, and applied the principles of geology to engineering works on the largest possible scale

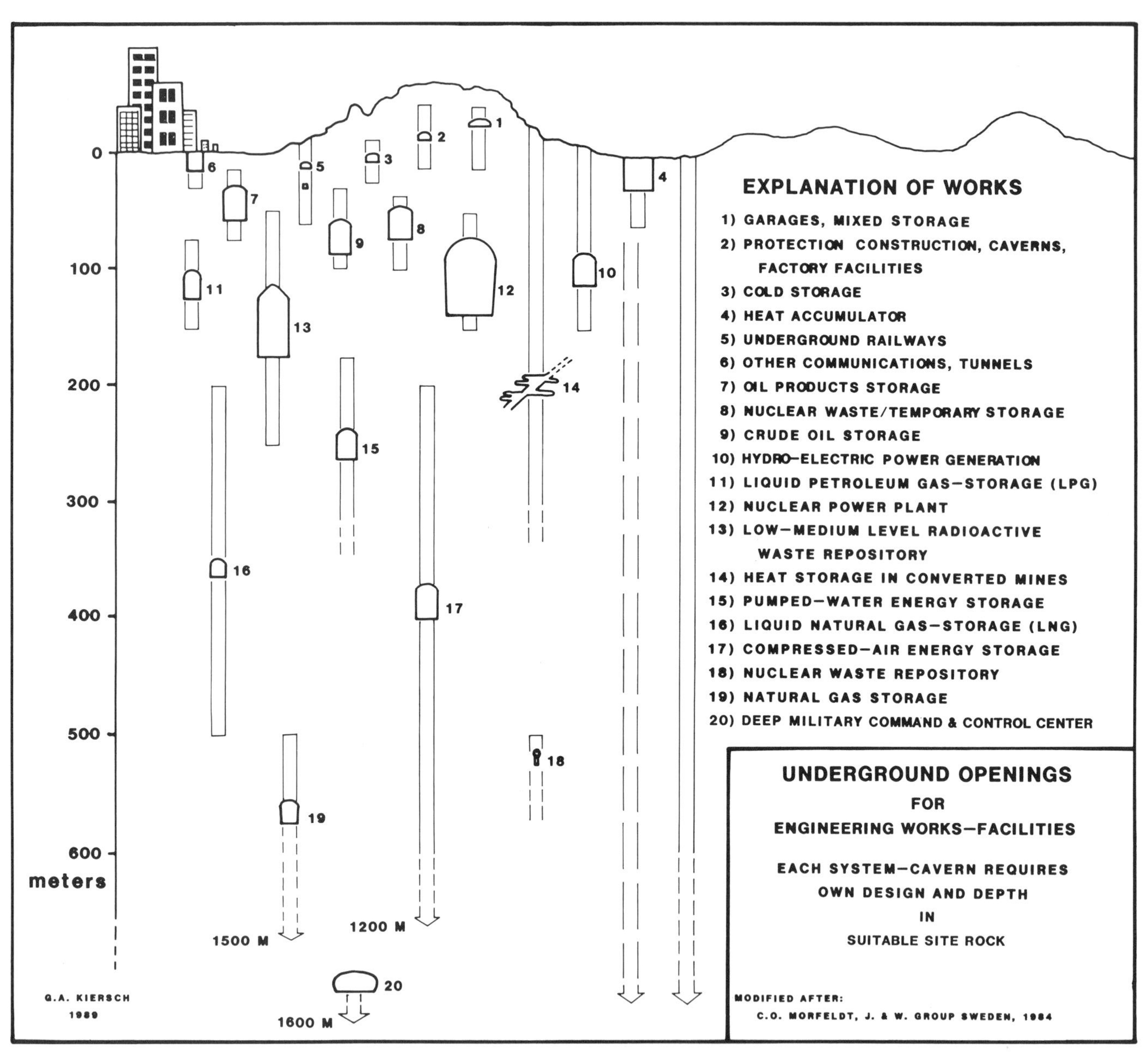

Figure 5. Twenty different systems of underground openings or caverns/chambers at varying depths for the location of engineering works and facilities. (Modified after C. O. Morfeldt, J. & W. Group, Sweden, 1984.)

throughout the country for over four decades (see Lee in Chapter 3, this volume). By the 1980s, activities of the branch had shifted to focus primarily on identifying natural hazards, risk evaluation, and mitigation techniques. Consistent with the growth of the field, a separate Engineering Geology Division of the Geological Society of America was established in 1947 (see Chapter 4, this volume).

In Canada, the applications of geology to engineering practice were being accelerated by the National Research Council of Canada and the Hydro-Electric Commisssion of Ontario, especially through the efforts of R. F. Legget, chairman of the Division of Building Research of the Council, and J. O. Gorman, chief geologist of the Commission. These agencies were following a pattern already established in other countries, such as South Africa, where the geological survey had for many years supplied personnel to act on engineering geology problems confronted by government departments, provincial administrations, and the South African Railway (Kent, 1952, p. 202). The growth of activities within the Geological Survey of Canada is described by Scott in Chapter 3 of this volume.

State agencies became more active in addressing engineering geology problems, particularly in connection with highways, water supply, and conservation after World War II, even though the earliest recognition was an engineering geology division of Illinois State Geological Survey under M. M. Leighton in 1927, with G. E. Ekblaw as branch chief. Typical of the trend was the planning and development of statewide water projects by the California Division of Water Resources under chief geologist E. C. Marliave in the 1950s and L. B. James in the 1960s and 1970s when over 100 geologists were engaged with the many phases of the California Water Plan. Concurrently, the California Division of Mines and Geology under Ian Campbell (1958 to 1969) initiated geologic hazard mapping during a period that saw burgeoning of concerns for the environmental and engineering applications of the geosciences (Oakschott, 1985, p. 332).

Employment of engineering geologists by the 1960s was largely in one of two categories: either on a large-scale basis related to regional/areal features, to provide background on geologic environs for planning engineering works; or on a more detailed scale, confined to small areas and considered site-specific geology (see Chapters 18 and 19, this volume). Site-specific studies include actual construction that may rely heavily on regional/areal geology (see Chapter 20, this volume).

Figure 6. San Francisco Bay Bridge showing the collapsed section caused by the Loma Prieta earthquake of October 17, 1989. (Photo courtesy of U.S. Geological Survey/H. Wilshire.)

Residual stresses in rock masses

The phenomena of rebound, relaxation, and associated uplift are reviewed in Chapter 13 by Nichols and Collins, and residual stress in rock elsewhere by Nichols and Varnes (1984). A rock mass eventually relieves itself of all internal stresses and, in so doing, goes through many deformations and possibly failures that frequently affect the stability of engineering works and/or site foundations. Sheet structures due to rebound were first observed by Niles (1871) in quarries of New England. Gilbert (1904) described the dome structures of Sierra Nevada; Dale (1907) described the "stressed granites" of Maine and Georgia; and Kemp observed "stressed rock" around New York City about 1910 (field notes). Both Gilbert and Dale attributed the structures to removal of the superincumbent load. Dale's description, based on the expansion of 4-ft holes bored in quarry rock, developed the concepts of sheet-like joints and natural breakage, as illustrated in Chapter 13. The phenomenon is well known at rock quarries in many areas in eastern United States. At Carrara, Italy, a few hours after a large block of marble was cut free and the rock thereby unloaded, rebound fractures appeared, followed by displacement and rock bursts (Kiersch, notes, 1964).

Holzhausen (1989) has published an extensive study of sheet structure, including morphology and boundary conditions, that relates the scattered observations of investigators over the past century to the mechanics of failure.

Highly stressed rock in New York City was identified by T. W. Fluhr, circa 1940, in the Queens anchorage of the East River Bridge; rebound fractures in the Ravenwood Granodiorite foundation caused the anchorage to crack, and the bridge truss was ripped apart. Fluhr observed similar features in the 63rd Street tunnel near Welfare Island that measured 1,400 psi, and again in the city tunnel No. 3 where 1,200 psi was measured (T. W. Fluhr, personal communication, 1983).

Serious and unexpected rebound activity was reported by Richey (personal communication, 1964) at a dam site in the Scottish Highlands in 1951. After the foundation for a large concrete dam was excavated several feet into the mid-Paleozoic rock mass, the exposed and unloaded rock reacted; fractures began to appear and within a few days, part of the foundation was described as a "popcorn" mass (Fig. 7).

Construction in late 1930s of the underground Festival Hall at Salzburg, Austria, unexpectedly encountered a strong pattern of rebound-relief structures formed parallel to the canyon walls of the nearby incised Salzburg River (Kieslinger, 1960, Fig. 5, personal communication, 1963). The fractures and relief structures extended below the foundation level, because the river flowed in a deeper, partially backfilled ancient channel; rebound joints occurred for 50 or more feet into the steep rock canyon wall. Although recognized after the early excavations, design changes were required to accommodate the 72,000 yds^3 of the underground chamber. Reportedly, some heads rolled because of the delays caused by the unforeseen and undetected rock defects of the site.

Small-scale structures caused by the postglacial unloading of confined rock are the folds and pop-ups common to the St. Lawrence Lowlands and the shore areas of Lake Ontario in New York State and Ontario. These features were extensively investigated in the 1970s, as possible evidence of present-day tectonic stresses, during the planning and design of several nuclear power plant projects in New York State (NYSEG/NYSP Authority and Niagara Mohawk Power Co.). The best known features, in the Alexandria Bay sector of the St. Lawrence Lowlands, are the Oak Point anticlinal feature, the Nelson "fold," the Ward "fold," and the Omar anticlinal feature and "fold/pop-up" near Omar, New York. All of the features were geologically mapped in 1977 (NYSEG, 1978, p. 2.5A to 17). Analysis of the Omar site data indicated that the pop-up is a ruptured fold uplifted 4 to 6 ft, with three distinct trends. Glacial striae are prominent on the polished Potsdam Sandstone surface of some structures, and part of the uplifted feature was formed prior to the last ice advance; as N55°E Woodfordian striae are in association with N20°E Fort Covington–age striae of 22,500 years ago. Similar features along the shores of Lake Ontario are also reported to have formed in part before the last ice advance (Kindle and Taylor, 1913; Stone and Webster, 1971). The "pop-ups" and low folds are similar in many ways to anticline (Fig. 8A) and fault (Fig. 8B) phenomena commonly encountered in the floors of quarries near Buffalo, New York, when beds of fine-grained limestone are rapidly unloaded over a wide area. The nontectonic pop-up features are attributed to residual stress; the compressive stresses are not high, and their potential for seismic activity is not a factor in the design of power plants in region.

Urban geology

Early efforts of the U.S. Geological Survey to map selected metropolitan areas (see Chapter 1, this volume) were revived on a more detailed scale in the 1950s to assist in planning for growth; two such areas were the San Francisco Bay region and the Los Angeles basin of California. John T. McGill, in his mapping of Los Angeles area, made some of first age classifications of the common landslide features so widespread in the region; furthermore, he specified the on-going risk state of each slide feature, based on its origin and recency of movement. He later designated the study of geology related to city areas as "urban geology" (McGill, 1964).

Kaye (1968) published a thought-provoking review on the study of geology in city areas, beginning with Boston and New York City in the nineteenth century (see Chapter 1, this volume). By the twentieth century, the towering buildings, subways, tunnels, bridges, and other structures were beginning to stress the ground beneath the cities in a way that was new to the engineer and geologist. The foundation of the city thus became another critical area for engineering activities, and by the late 1960s, geology was recognized as a full-fledged partner with engineering in city planning and building (Kaye, 1968, p. 1047).

Urban geology studies of worldwide societies for city and county planners were reviewed at the 23rd International Geological Congress in Czechoslovakia in a session on "Engineering Geology in Country Planning" (Zaruba, 1968). The policies, legislation, and techniques used in over 12 countries were discussed by well-known investigators of each nation.

Nearly 80 percent of our population lives in urban areas, and the need for understanding the geology beneath our cities has steadily grown in order to provide consistent and complete socioeconomic information on each geological problem. This can require action programs to reduce both human and financial losses due to the natural- and man-induced hazards common to urban regions (UGMP, 1973; and Legget, 1973).

Reservoirs

Reservoir perimeters. A need to broaden all dam design-site investigations and include the reservoir perimeter was dramatically demonstrated by the 1963 landslide on the valley wall of the Vaiont Reservoir in Italy (Chapter 22, this volume). A wave 410 ft high overtopped the dam and created widespread destruction downstream. This occurred even though earlier failures and warnings of such a danger had been experienced elsewhere: the rockfalls in 1905 and 1936 into the lake at Loen, Norway, caused waves up to 262 ft high (Slingerland and Voight, 1979); and the rock-soil slide at Pontessei Dam, Italy, in 1959 created a sudden wave at least 82 ft high (Walters, 1962). The common geology-related problems associated with reservoirs—excessive leakage and bank storage, slope instability throughout the banks and perimeter, silting and sedimentation in the basin, and induced earthquakes from the reservoir loading—are summarized by James and Kiersch (1988, p. 722–748).

Figure 7. Spillway damsite, Glen Shira project of the Scottish Highlands. Excavation in the mid-Paleozoic rock mass in 1951 caused parts of foundation rock to decrepitate on unloading due to long-time residual stresses. One area affected, to which engineer is pointing, is in the left foreground. (Photo from J. E. Richey, 1964, North of Scotland Hydro Electric Board.)

Slides on the perimeters of reservoirs may be due to a sudden release of large masses that generate reservoir waves, a slow sliding or creeping of mass into the reservoir with reduction of reservoir capacity (CDWR, 1970), or an excessive generation of silt and sediment from the banks or airborne sources that progressively backfills the basin (Dunne and Leopold, 1981). Geologically the instability of reservoir slopes may be due to occurrence of tectonic features or strongly fractured zones, nontectonic features like rebound joints and relief structures in brittle rocks, young surficial deposits and glacial debris, properties of the bedrock units, the recent geologic history, and/or the state of stress in rock masses. These principal parameters conducive to reservoir

Figure 8. A. A broad "anticlinal" uplift of limestone beds in the bottom of a quarry, after being unloaded, Buffalo area, New York. (Photo by G. A. Kiersch, April 1967.) B. A strong fracture with uplift that formed circa 1963 at the quarry of Figure 8A due to the unloading of the Lockport Dolomite beds. The uplift and strong fracturing tilted and damaged the crusher shown in the background. (Photo from Paul H. Bird, geologist New York State Department of Public Works, 1963.)

perimeter slides and instability are demonstrated by such well-known cases as: F. D. Roosevelt Lake, Columbia River (Fig. 9; and Jones and others, 1961; Lake Koocanusa, Kootenai River; Fort Randall Reservoir, Missouri River (Erskine, 1973); Kaunertal Reservoir, Austria; Bighorn Lake, Bighorn River; and Nurek Reservoir, USSR, which are reviewed elsewhere (Kiersch, 1988, p. 729–733).

Reservoir sedimentation. Sedimentation in reservoir basins is an economic liability. The sediments are composed primarily of clay, silt, sand, gravel, and rock-mineral fragments. Stream-borne sediment primarily comes from one or more of the following land sources: sheet erosion, stream-channel erosion, mass movements (CDWR, 1970), flood erosion, construction, mining and industrial wastes, and volcanic outpourings (Fry, 1950). Stream-channel erosion, the most important process, contributes 50 to 90 percent of sediment load (Robinson, 1973).

Dams retain the waterborne sediments of a river system, and the reservoir upstream from Hoover Dam, Lake Mead, is a benchmark case (Smith and others, 1960). The yearly sediment load carried by the Indus River, Pakistan, at Tarbela Dam demonstrates the importance of such investigations during the planning and design phases of a project. The suspended sediment load and bed load of gravel and cobbles at Tarbela will deplete the reservoir capacity of 6 million acre-feet within 40 to 60 years (Drisko, 1962).

A special source of sediment in reservoirs is provided by large-scale volcanic outpourings; the accompanying volumes of ejecta ash and fines may form an ash cloud that can overwhelm a lake or reservoir, as demonstrated by the Mount St. Helen's eruptions of 1980 (discussed in Chapter 11, this volume, and Foxworthy and Hill, 1982). Another outstanding case of volcanic sediment load with associated engineering problems occurred in Costa Rica from 1963 to 1965 with the 2-year activity of Irazu Volcano (Waldron, 1967).

The useful life of a reservoir is determined by the rapidity with which the yearly sediment accumulates and is not removed by natural actions or man-made devices. Many dams have been rendered ineffective or useless by the rapid "siltation" of the reservoir. The New Lake Austin on the Colorado River in Texas lost 95 percent of its capacity in 13 years; La Grange Reservoir on the Tuolumne River, California, lost 83 percent in 36 years; and the Habra Reservoir on the Habra River in Algeria lost 58 percent in 22 years. More dramatic is the history of the 120-ft Cat Creek Canyon arch dam near Hawthorne, Nevada. A half-hour flash flood in 1955 scoured the drainage and nearly filled the reservoir with sediment: a major sluicing project was required to regain the storage capacity.

Reservoir-induced earthquakes

Reservoir loading has been of concern to engineers and geoscientists ever since earthquakes (to magnitude 5) occurred in 1936 following the 80 percent filling of Lake Mead. Although few fully accepted proofs for such man-induced earthquakes are

Figure 9. Jackson Springs landslide of the Spokane arm of Franklin D. Roosevelt Lake, Washington. This earth slump of 14 million yds^3 occurred in 1969 during a period of extreme drawdown necessitated by excavation for a forebay dam preliminary to construction of the third powerhouse at Grand Coulee Dam (Schuster and Fleming, 1986). (Photo courtesy of U.S. Bureau of Reclamation and R. F. Schuster, U.S. Geological Survey, Denver.)

known, strong circumstantial evidence does exist at several major reservoirs in the world. However, there is a growing consensus among experts that reservoir-induced earthquakes are limited and normally should not be considered, except for very large and deep reservoirs with an appropriate geologic setting.

The milestone earthquake activity of Lake Mead and the Hoover Dam region was analyzed by Carder (1945, 1970); his findings correlated the release of seismic energy with the water load and movement of young faults within the regional fault blocks (Fig. 10). The concern for induced earthquakes was further strengthened by events in the 1960s at or near several large reservoirs (Table 2). The increase in seismic activity subsequent to reservoir impoundment, which inferred that loading caused or triggered the earthquakes, has been investigated at many reservoirs, as reported by Judd (1974), Milne (1976), and Rogers and Lee (1976). Unfortunately a reservoir-caused earthquake has no discrete signature to distinguish it from a naturally occurring seismic event.

Meade (1982) has made a critical assessment of the evidence for reservoir-induced events: less than 1 percent of world's reservoirs have been associated with macroseismicity (magnitude >3), and only 14 percent of the reservoirs deeper than 312 ft have had possible associated macroseismicity (Alexander and Mark, in Meade, 1982). Conservatism in seismic design to counter this vague threat is usually not justified because costs are high. Eight of the most widely reported cases of induced macroearthquakes in the world have been reviewed (Table 2), and only four sites are believed associated with reservoir-induced earthquakes. Two conclusive cases, Lake Mead and Nurek Reservoir, are briefly described, as they demonstrate differing geologic settings and evidence for a representative site (Meade, 1982).

Lake Mead, Nevada-Arizona. The world's largest reservoir until 1958, Lake Mead is a multivalley lake that extends 71 mi upstream from Hoover Dam. The postimpoundment seismicity is concentrated near the dam in Black Canyon (Fig. 10). Only one earthquake (intensity V) was reported in the region of the dam from 1852 to 1936; and no earthquakes were reported by local inhabitants in the 15-year period prior to reservoir impoundment (Carder, 1945).

Beginning with two earthquakes in September 1936 (intensity IV/V), recorded Lake Mead earthquakes numbered 12 through 1939 and totalled 30 by September 1965 (magnitude 3.5 to 5.0). The maximum seismicity occurred in May 1939 during a seasonal rise when the lake was lower than the 1938 maximum level. No events of importance have occurred since 1964 when Lake Mead became regulated by Glen Canyon Dam upstream and seasonal changes became much milder; Meade (1982) concluded that there is a positive correlation between the reservoir and the seismicity of Lake Mead. The geologic setting of Lake Mead includes young fault zones and pre-stressed rock units; a

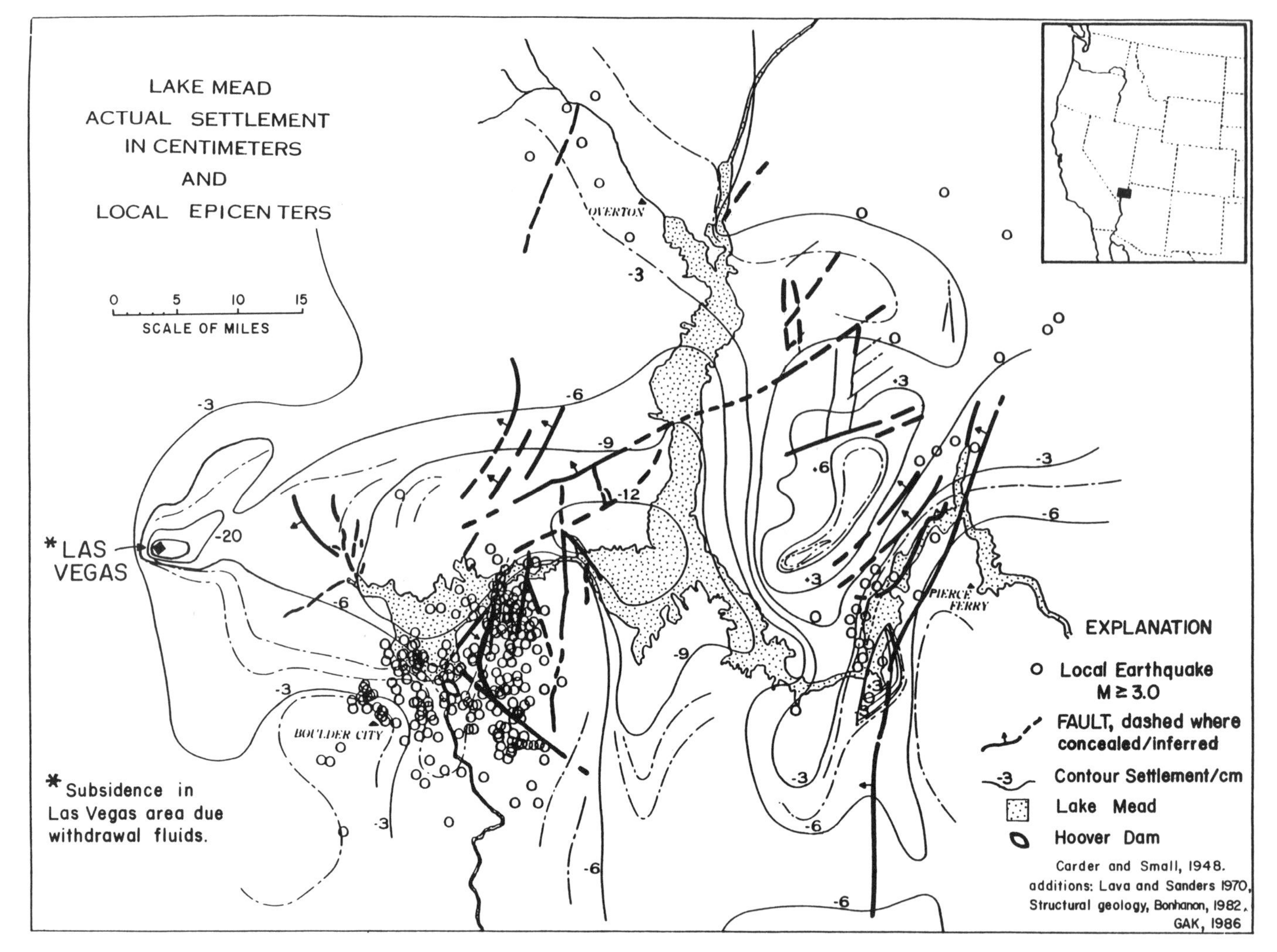

Figure 10. Areal map of Lake Mead, Nevada-Arizona, and the vicinity of Hoover Dam. The locations of local earthquake events (M ⩾3.0) for the period 1940 through 1947 are shown by symbols. Events since have followed the same pattern. The principal faults and crustal blocks that could be acted on by the reservoir loading are shown, and the relative settlement of the reservoir and marginal areas from 1940 to the 1960s is indicated by contours. (From Kiersch, 1988, Fig. 24-18.)

thick series of folded, faulted, tilted, and intruded sedimentary beds; and young volcanic rock units that are broken into regional blocks bounded by major fault zones. Some late Tertiary faults have experienced recurring movement due to tectonic forces. Carder (1945) concluded the loading by lake water caused areal subsidence (Fig. 10) that renewed movement along selected faults and caused seismic events.

Nurek Reservoir, USSR. This reservoir is formed by Nurek Dam, which is the highest dam in the world at 1,008 ft. The reservoir extends 43.5 mi upstream, with a maximum width of 3.1 mi (Fig. 11). The reservoir filling took place in several stages as construction progressed; first in late 1972 to 344 ft, second in late 1976 to 702 ft, and lastly in 1980 to a maximum depth of 984 ft. Nurek Reservoir is located in a region of high seismicity and was the subject of detailed investigations resulting from a high-sensitivity network that was installed in 1955. Nurek is the world's outstanding case of reservoir-induced seismicity, and seismic studies have been described by Soboleva and Mamadaliev (1976), and Simpson and Negmatullaveu (1981), with an assessment by Meade (1982).

The first seismic activity began in 1971 when the water level was at 164 ft; this increased to over 40 events early in the first quarter of 1972 and to 133 events in the last quarter when the reservoir was 344 ft deep. Initial seismicity occurred along two separate sectors of a fault through the reservoir; the portion in between appeared quiet. Yet, when the reservoir level was raised in the fall of 1972, seismicity occurred along the formerly quiet part of this fault.

The geologic structure and tectonic development of the area surrounding the Nurek Reservoir represent three different regions (Fig. 11). The northern part corresponds to the Gissar Range, while to the south, the Gissar valley is a depression, bounded by

TABLE 2. SELECTED CASES OF INDUCED MACROEARTHQUAKES ASSOCIATED WITH MAJOR RESERVOIRS OF THE WORLD*

Reservoir	Location	Depth (m)	Volume X 10^6, m^3	Largest Earthquake†
Hoover§ (Lake Mead)	U.S.A.	166	35,000	M_L = 5.0
Kariba	Zambia/Rhodesia	122	175,000	m_b = 5.8
Kremasta	Greece	120	4,750	M_s = 6.3
Koyna	India	103	2,780	M_s = 6.5
Kurobe	Japan	186	149	M_s = 4.9
Manic 3§	Canada	98	10,423	m_b = 4.3
Hsinfengkiang§	China	80	10,500	M_s = 6.1
Nurek§	U.S.S.R.	215	11,000	M_s = 4.5

*Meade, 1982.
†M_L = a local magnitude scale used; M_s = magnitude measured from amplitude of surface waves; m_b = magnitude measured from amplitude of body waves.
§Evidence supports reservoir-induced macroseismicity.

two major fault zones between the Gissar Range and the Tadjik depression (Fig. 11). The Tadjik depression is a region of recent deformation by a system of thrust and normal faults associated with the Illiak fault; all are critical to the seismic history and response of the Nurek Reservoir to water impoundment.

Subsurface geologic relations in the vicinity of the dam and reservoir site that influence the induced seismicity are described elsewhere (IGC, 1984; Kiersch, 1988, Fig. 24-9). The Ionakhsh fault and branches are seismotectonic with associated earthquakes up to intensity VII. Although a number of factors control whether and where induced seismicity occurs, water level alone appears to explain most of the temporal variations in induced seismicity at Nurek. Generally, the seismicity depends on the absolute water level, because virtually all major bursts of seismicity occurred when the water level was near or above a previous maximum. Once primed, the timing of activity is controlled by changes in the rate of filling; if high, seismicity remains low; decreases in filling rate cause increased seismicity. The largest earthquakes occurred after the rate of filling was decreased by more than 1.65 ft/day (Simpson and Negmatulaveu, 1981).

Dominant parameters. The dominant parameters that support or limit evidence for induced seismicity by reservoir loading, as summarized by Simpson and Negmatullaveu (1981) and Meade (1982), are:

- Induced seismicity appears to be a transient phenomenon. Activity decreases with time, and none of the largest events occur more than ten years after a reservoir is first filled. With time, the region reaches a new equilibrium and exerts no more effect on the seismic regimen than would a natural lake.
- A deep reservoir in a steep-walled, V-type canyon of reasonably young age with associated fault zones is the most likely setting for induced-seismic activity. The strongest activity occurs when water depth is over 300 ft, although some events occur when a reservoir is shallower.
- Weight of water alone is not a causative factor. The downward movement of reservoir water can change the stress regimen and decrease effective stress due to increase of hydrostatic pore pressures along a preexisting fault plane; friction values may decrease sufficiently to trigger a seismic event.
- Statistical techniques cannot properly evaluate evidence of induced seismicity. The response of seismic parameters may be different during rising and falling water levels, and the correlation of statistics cannot account for this.

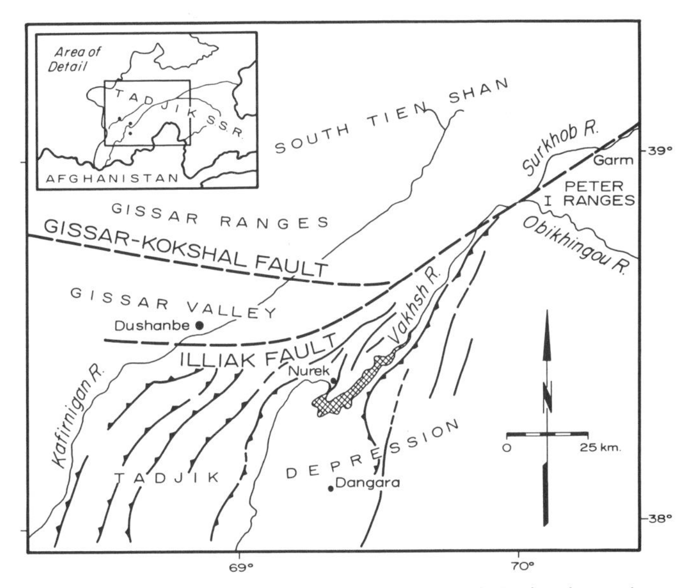

Figure 11. Regional structural setting of the Nurek Project in southern Tadjiskistan, U.S.S.R. The important Gissar-Kokshal and Illiak fault systems are responsible for the main crustal blocks and features of surrounding region. The widespread system of thrust and normal faults of the Tadjik depression associated with the Illiak fault is critical to the seismic history and response of the Nurek Reservoir to water impoundment. (From Kiersch, 1988, Fig. 24-8.)

• There is no unequivocal case in which a damaging earthquake (M = 6) has been triggered by reservoir impoundment.

Nuclear power plants

Commercial production of nuclear power in the United States began in 1959, at a time when geologic hazards were seemingly not as critical as in later years to the designers, constructors, and regulatory agencies. Geological investigations for plant siting and licensing are expensive and complex undertakings. Moreover, since the 1970s, siting work has required a more sophisticated appreciation of geologic hazards such as earthquake effects, subsidence, slope stability, and foundation integrity, as described by a group of experts in Hatheway and McClure (1979). The concerns for hazards were first formalized by the U.S. Atomic Energy Commission (AEC, 1971) in *Siting Criteria for Nuclear Power Plants* and after modification were released in the Code of Federal Regulations (1978). Based on historical seismicity, regional geology, and site-foundation conditions, the seismologist provides reasonable estimates of the safe-shutdown earthquake, using the parameters that govern the seismic response of the plant.

The required federal license to construct and operate nuclear power plants is awarded through a demanding process; permits are not granted without a high level of assurance as to suitability of the site. Consequently, applicants for permits and operating licenses must usually organize large teams of scientific and technical personnel to compile the Preliminary Safety Analysis Report (PSAR) required for each plant. Geologists are a key part of any team, which in the past decade has also included an environmental scientist; both the site-specific and regional assessment of the environs must withstand scrutiny to avoid costly delays in the design-and-construct sequence. Many owners initiated a regional geological study as the logical first step toward identifying candidate areas/sites (discussed Chapter 18, this volume). Furthermore, the licensing organization must not only assess the risk associated with safety-related geologic aspects of sites, but it must cope with intervenors raising real or imagined safety concerns.

Management techniques must be applied to coordinate data-collection efforts and the timely release of findings, as with the PSAR for review by the Nuclear Regulatory Commission (NRC); this complex document ultimately includes as amendments the responses to subsequent review questions involving geologic and other aspects. The Final Safety Analysis Report (FSAR) includes supplementary reports and an "as is" description and evaluation of the site and surroundings that includes the geologic findings revealed during construction.

Many of the key geological issues identified in siting and licensing during the 1960s and 1970s are analogous to problems confronting researchers in the geosciences today, such as evaluation of remote imagery, proof of subsurface stratigraphic continuity, evaluation of potential fault activity (Wallace, 1986), recency of fault movement that may involve datable minerals or marker paleosols (Shlemon, 1985), ground-water conditions affecting a site, and subsidence potential including evaluation of dissolution of the foundation material. A blend of both classical and engineering geology principles and practices is invariably required to resolve the critical issues.

The overriding goal of each owner is the construction of a safe power plant within budget. However, it became "impossible" to maintain construction schedules by the late 1970s and stay within allotted funds. This reality, along with a decreased demand for electricity, contributed to the termination of many nuclear plants in progress by 1980. Lessons learned indicate that effective management and judgment require efficient communication and coordination of scientific and technical data to achieve a timely and cost-efficient licensing of a plant. The conduct of the licensing process for a nuclear power plant is set forth in a flow chart by McClure and Hatheway (1979, Fig. 3); from 10 to 13 years have been required to satisfactorily resolve all issues and start up a new generating facility.

Exploration techniques

The pioneer investigation of Fisk (1944) on Lower Mississippi Valley alluvial deposits was followed soon after by the comprehensive study of subsurface exploration and sampling methods by Hvorslev (1949) sponsored by the American Society of Civil Engineers. This exhaustive study on sampling and exploration of soils and weakly consolidated sediments served as a guide for engineers and geologists for several decades.

Today, foundation investigations are trending toward in-situ determinations of soil and rock properties (Mitchell, 1986). Although new techniques for exploration and geological interpretation are evolving rapidly, there remains a need to improve drilling methods with respect to core recovery, cutting speed, and hole alignment. Perhaps future exploration will be accomplished by hydraulic cutting under downhole power, as has been demonstrated by the U.S. Bureau of Mines in the case of horizontal drilling (Kravits and others, 1987). There will always be a desire to replace boreholes with geophysical imaging. For the foreseeable future, the end of practicality in this arena may be the continued use of uncored boreholes, each logged by geophysical measurements, and each serving as a pathway or source for crosshole electromagnetic and seismic surveys.

An unusual boring was made at Grand Canyon Village, located below Yavapai Point on the South Rim of the canyon, where a long-standing problem has existed with its water supply. Since the 1930s, water has been uplifted by pumping from Indian Garden station within the canyon to the village 2,800 feet above. In 1985 the U.S. Park Service considered rebuilding the pipeline, which was exposed along the canyon wall, but a feasibility study recommended drilling a directional hole near the canyon face that would emerge near the Indian Springs pump station (Fig. 12); this boring would be the most efficient means of transporting water, have less impact on the environment, and be less expensive than the alternatives.

The 5,075 ft, high-angle 14¾-inch directional well bore was

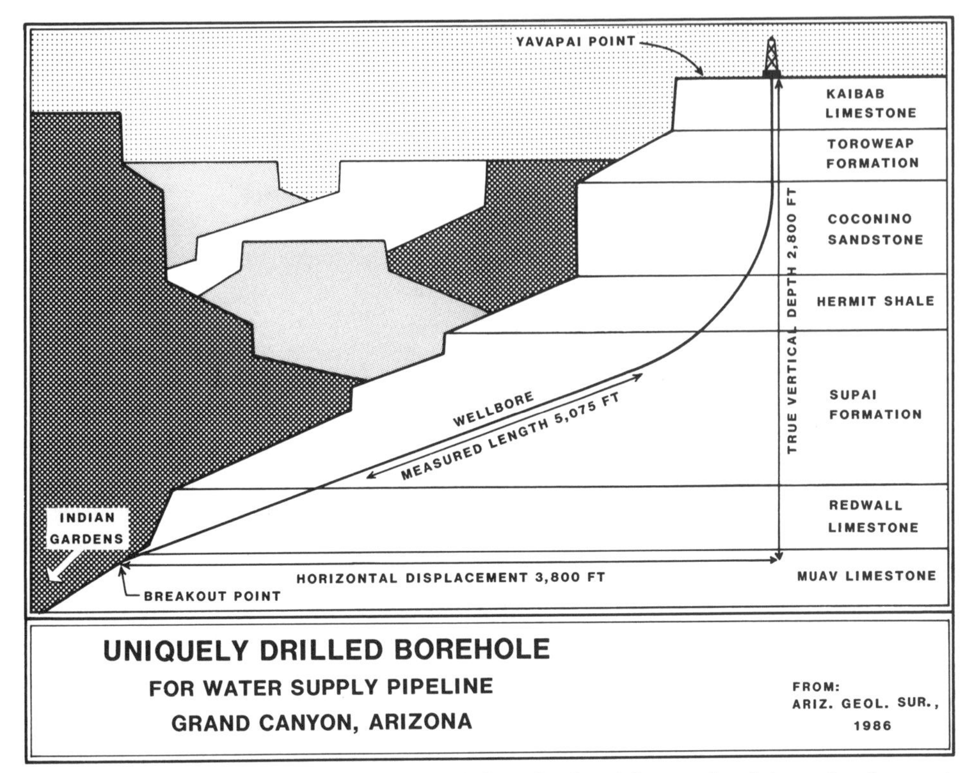

Figure 12. Uniquely drilled, large-diameter borehole at the Grand Canyon rim, Arizona, for placement of a water pipeline to supply Grand Canyon Village. The construction eliminated the over-the-cliff pipeline that was frequently damaged, and removed it as an environmental pollutant. (From Arizona Geological Survey.)

drilled in 49 days. The biggest challenge was boring at a 65° angle, using air to facilitate removal of cuttings and soapy water to cool the downhole equipment. Gyroscope and magnetic single-shot methods were used to guide the drilling, and the hole broke through the canyon wall within a few feet of the target. No structural features or faults were encountered throughout the total bore, which traversed six well-known rock formations (Fig. 12). The water line, installed in the cased hole, now delivers water from Indian Gardens to the village by pumping up to the South Rim through the well bore (AGS, 1986, p. 9).

Another ingenious boring explored the quality of a rock mass at Ripple Rocks, Seymour Narrows, British Columbia (Fig. 13), in 1955. The conventional diamond drill and wedge technique of boring no. 2 provided a sampling of the unexplored Vancouver series of volcanics within the rocks for the design of a tunnel from Maud Island (Fig. 13); it was impossible to drill from barges due to the strong currents. Raises were driven to position explosives, so the upper 40 ft could be demolished, thereby removing a famous navigational hazard (see Scott, Chapter 3, this volume). On logging hole no. 2 for contract-bid purposes, the writer was surprised at the good quality and high recovery of core. No serious difficulties were encountered along the tunnel and raises.

Constructed works

Geologic problems common to the maintenance and operation of major works such as dams, tunnels, reservoir slopes, bridge piers, nuclear plants, highways cuts, causeways, navigation locks, and pipelines are briefly described in other chapters of this volume. A prominent example in this category was the movement that occurred at the interface between the main embankment and underlying materials of San Luis Dam in central California on September 13, 1981 (Fig. 14).

The major slide occurred on the upstream, left abutment of the zoned earth dam, which had a crest length of 18,000 ft and a height of 385 ft. The slide scarp (Fig. 14) is more than 1,200 feet long and 35 feet high, with movement throughout a length of 1,700 ft and a disturbed zone 600 ft wide. Until stabilized, the slide mass moved 6 to 8 in per day. Failure occurred along the interface between the dam structure and underlying slope-wash material, which included plastic clay 1 to 40 ft thick, that overlay the rock foundation of bedded shale, sandstone, and conglomerate (Panoche Group). During construction by the U.S. Bureau of Reclamation (1963 to 1967), the clayey material was only removed in the cutoff trench of the dam, and only stripped (1 to 2 ft) over parts of the area upstream. A rapid drawdown of the reservoir, up to 2 ft per day, by the California Department of Water Resources, apparently contributed to foundation failure by increasing residual pore pressure within the slope wash. At failure, reservoir level was low and below most of the slide displacement. Another kind of problem associated with slope instability during reservoir drawdown is illustrated by the Jackson Springs landslide, which occurred along the shoreline of the F. D. Roosevelt Lake, Washington (Fig. 9).

Geological investigations to delineate the San Luis slide mass and features for the stabilization design included bore holes; undisturbed samples of slope-wash materials for testing; piezometers for pore pressures within embankment, slope wash, and foundation materials; and surveys to monitor slide movement (Howard, 1982, p. 17–19).

Karst terrain. Karst terrains can cause a spectrum of problems for construction; these have been recognized since the 1940s. This awareness has led to systematic pre-construction geological investigations of limestone regions to avoid loss of foundation support, surface subsidence and collapse, fluctuating or lowering ground-water surfaces, and/or differential settlement. All can be major problems for the engineering geologist at sites in such terrain (Foose and Humphreville, 1979, p. 353–354).

The Hershey Valley of Pennsylvania is a region of extensive karst conditions where a large number of catastrophic sinkholes have developed in response to lowering of the ground-water surface by pumping at mining operations (Foose, 1953, 1969). More spectacular and catastrophic was the response to lowering of the ground-water surface at mines within the Oberholzer Compartment near Carlstonville, Far West Rand, South Africa, in 1962; eight major "man-made" sink holes developed within this area of pinnacle-weathered dolomites concealed by a veneer of residual debris. One steep-walled, cyclindrical sinkhole (180 ft in diameter and 100 ft deep) opened with no warning under the crusher station; 29 lives were lost. Nearby, in 1964, an even larger sinkhole developed in the miner's village, claiming five lives and several homes (Foose, 1967).

The karst terrain of Florida and Alabama has been a region of sinkhole development due to man's activities in recent decades. For example, a major sink up to 520 feet across and 125 feet deep formed suddenly on May 22, 1967, and demolished two homes (Fig. 15). Such experiences led to establishing the Florida Sinkhole Research Institute at the University of Central Florida, Orlando, to investigate the engineering and environmental impacts of karst and sinkholes on the region.

Quick clays. The quick clays of marine origin in Scandinavia are similar to the Leda Clays in the St. Lawrence Valley,

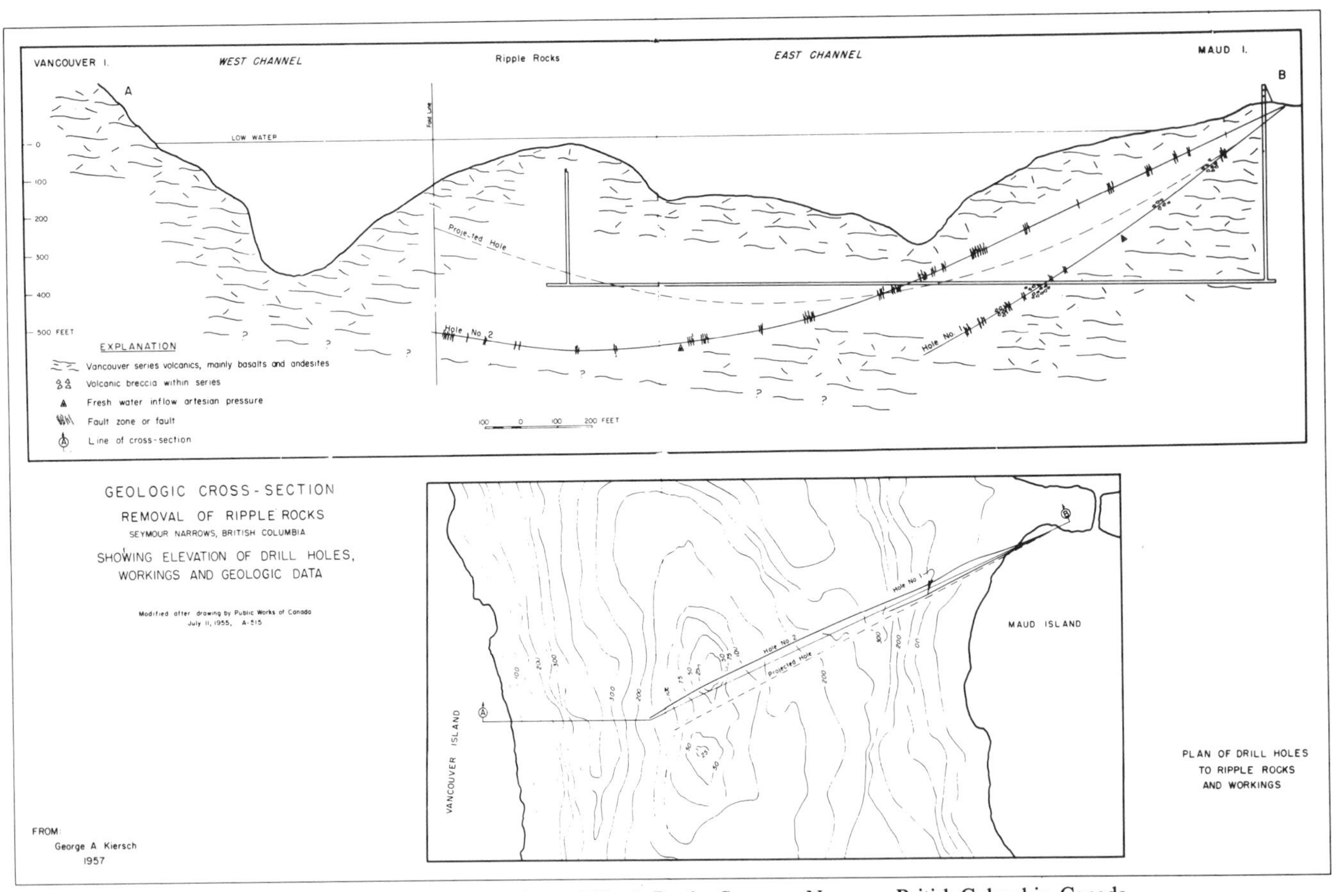

Figure 13. Plan and cross-section views of Ripple Rocks, Seymour Narrows, British Columbia, Canada. The locations and elevations of exploratory borings are shown as drilled in 1955 from the shore on Maud Island to within the main pinnacle of rocks. (From Kiersch, 1957, Fig. 2.)

Figure 14. Aerial view of San Luis Reservoir in Merced County, California, looking east, showing the left abutment of the dam and the bridge leading to the trash-rack structure. A 1,700-ft section of upstream face to the right of the bridge slipped toward the reservoir beginning on September 13, 1981, due to movement along the interface between the embankment and the foundation. (Photo by U.S. Bureau of Reclamation/courtesy, R. J. Farina.)

Canada (described by Scott, Chapter 3, this volume). These clays are very treacherous and can lose the greater part of their strength when mechanically disturbed, because they are in a weakened condition due to in-situ leaching of the natural salt content. The most sensitive clay can be liquified, and the dense liquid flows faster than water. An area north of Trondheim, Norway, that includes Nandal has been the scene of a great number of historic and prehistoric landslips (Fig. 16)—and even great landslides—involving sensitive clays. For example, the Vaerdalen slide of 1893 disturbed about 1 mi^2, and 72 million yds^3 of material were displaced (Holmsen, 1953). Similarly the Great Slide of Surte on the Gota River, Sweden, in 1950 occurred in less than three minutes, displaced 4 million yds^3 of material and destroyed 50 homes (Kerr, 1963, p. 135–138). In America, such events of 1955 (Crawford and Eden, 1963) have occurred in the Leda Clay throughout the St. Lawrence River valley and its tributaries of Canada (see Chapter 15, this volume).

SCOPE OF ENGINEERING GEOLOGY

The rapid increase in the number of practicing engineering geologists since the 1960s occurred in response to demands by several branches of engineering and construction, the applied sciences, land planners and investment counselors, legal firms, and society at large, including local governments. Consequently, in addition to a small core of generalists, a number of specialties developed within the field to serve the diverse demands. Major categories of practice are identified below.

An engineering geologist is a broadly trained geologist with an understanding of the engineering principles needed to acquire a database from which to evaluate the physical and engineering parameters of natural materials, and the engineering consequences of a geologic setting. Engineering geology includes any geoscience work relevant to the activities or well-being of mankind and concerned with the environs of engineering works. Prac-

Figure 15. Karst terrain near Bartow, Florida, showing the major sinkhole that formed suddenly on May 22, 1967, and demolished two homes. The sinkhole measured 520 by 125 ft and was 60 ft deep. (From U.S. Geological Survey, file no. PIO-70-6.)

tice mainly consists of geoscience-related work (up to 75 percent) that interacts with engineering and/or other disciplines. This approach is not a unique or a hybrid combination, but rather is common to all branches of applied geosciences and many professions today (Kiersch, 1955, p. 3):

> The engineering geologist bears a similar relation to the research or academic geologist as the industrial inventor to the research physicist or chemist. Although similarly trained and sympathetic—he is partly an artist dealing with specific problems and applications. The engineering geologist finds satisfaction in feeling himself an integral part of an overall constructive program, making lasting marks on the faces of both nature and society; a type of satisfaction which likely inspired the designers of the pyramids. The geologist supplies a point of view basically different from that of an engineer. The geologist's reasoning and judgment are based on observations and data collected by various investigative techniques as they concern the effect of natural forces or processes (effect to cause). Conversely, the engineer's design and logic are based on the principle of designing a structure to resist forces produced by a cause of calculated intensity (cause to effect). Each profession must consider the other's method of reasoning when analyzing a request or recommendation.

Professional activities are mainly concerned with physical processes and phenomenology, geologic events, and the deterministic characteristics or uncertainties related to the physical features and properties of rocks, soils, and fluids as relevant to the planning and design of engineering works (Fig. 3). This includes environmental or geologic reactions generated by the construction and operation of such works and any related impact on the health, safety, and welfare of mankind. The geological parameters and consequences must be recognized, interpreted, and adequately presented to concerned practitioners in numerous disciplines of applied science: construction-related engineering, hazard mitigation, environmental assessment, resource management, forensic activities, and other emerging specialties that interact with the geosciences.

Today, engineering geology practice comprises at least seven major components related to: (1) planning-construction-operation of engineering works; (2) impact of engineering works on the areal, urban, and regional environs; (3) hydrogeologic regimes relevant to hazardous wastes, including detection and monitoring of environmental contamination and health hazards, and restoration of uncontaminated ground water; (4) research on ways to mitigate or reduce natural- and man-induced hazards and their risk to mankind; (5) supply and quality of natural materials necessary for all phases of construction and operation of engineering works; (6) rehabilitation of existing engineering works and the nationwide infrastructure of highways, bridges, and tunnels; and (7) geologically-related litigation generated from all categories of practice.

A single practitioner cannot expect to develop expertise in all subfields of engineering geology today. Rather, normal practice will be restricted to some of the following major categories or typical duties:

- Investigation and evaluation of active and intermittent on-going regional or areal processes and historical events pertinent to evaluating selected local natural sites or materials (See Chapter 18, Fig. 1).
- Investigation, evaluation, and control of naturally occurring or man-induced geologic processes on a regional, areal, or site-related basis—including such hazards as earthquakes, floods, unstable slopes, volcanic eruptions, rising sea-level, subsidence, debris flows, expansive soils, and ground-water contamination that affect land use and/or impact public health, welfare, and safety (Table 1); or the geochemical interactions of gases, liquids, and solids, separately or in combination, on the environs or site areas.
- Investigation and evaluation of geologic conditions, areal or site, that affect engineering works, such as bridges, buildings, canals, dams, highways, open-pit mines, pipelines, power plants, towers, tunnels, and toxic waste-disposal/burial.
- Exploration, evaluation, and development of naturally occurring construction materials for project or resource use, such as concrete aggregate, rip-rap, quarry stone, and filter-grade materials, as well as the influence of geologic origin on the properties of such materials and the impact on the geologic environment of their exploitation.
- Control and development or mitigation of surface and ground-water resources affecting engineering works, whether in the design, construction, or operation phases of a project (see Chapter 18, Fig. 1).
- Investigation, evaluation, and testimony related to litigation claims or hearings concerned with issues such as the environmental changes, geologic reactions, and/or geologically related failures of engineering works, as well as those related to land evaluation, conditions of waste burial, investment counseling, and natural materials.

Figure 16. Slides and landslips that occurred in the sensitive clays of Nandal, Norway, in 1959. A. Shows the boundary surface of the main slide area with both intact and destroyed buildings founded on the underlying clays. B. Shows the typical pattern of landslips throughout a large area of displaced sensitive clays that underlie the farmlands of Nandal. (Photos by G. A. Kiersch, 1959.)

Rapid judgments and opinions are frequently requested of a generalist-practitioner that do not fall within the typical subfields or duties. Such decisions by an experienced generalist are based on a limited study and are likely influenced necessarily by the practitioner's knowledge of case-history data relevant to the site conditions and/or problem. By design, this aspect of engineering geology practice is not the outgrowth of systematic investigations and evaluations as given in Figure 1 of Chapter 18, but rather reflects the insight and experience level of the applied geoscientist.

Resolution of the underlying cause(s) of geological processes, hazards, and features is a primary responsibility of the applied scientist and engineering geologist. The subsequent solution phases, utilizing remedial and mitigation techniques, are normally a "team effort," involving both applied geologists and geological or geotechnical engineers (Kiersch, 1955, p. 4–5).

The engineering geologist is most effective as a member of a specialist team (Kiersch, 1964, p. 16; and Chapter 18, this volume). Plainly, the more mature and field-experienced the geologist, the more skillful and likely he or she is to solve the geological problems arising in engineering practice. Furthermore, some experience on construction projects and a practical bent are advisable. Without such attributes, an otherwise well-rounded geologist may not qualify to serve on an engineering project (Kiersch, 1955, p. 2).

Professionally, the engineering geologist must declare and express opinions without equivocation, stating the best evaluation or judgment based on the information available. Nothing is more humbling than a post-report excavation that reveals conditions different than those forecast. Consequently the engineering geologist must possess the perception necessary to separate geologic facts from inferences and hypotheses. Practice implies a professional risk and a liability for actions and geological errors relevant to engineering design and/or contract specifications. This practicality and responsibility is far removed from the pure scientist, whose pronouncements are for "knowledge" only.

TRAINING OF ENGINEERING GEOLOGISTS

Throughout the 1920s, 1930s, and 1940s, the main professional thrust was on the applications of geology to engineering practice, as reflected by the Berkey Volume papers (Paige, 1950). Professional responsibilities were concerned mainly with projects undertaken by federal agencies and the state highway systems. With the onset of World War II, applied geologists were requested for the first time to participate in such specialized tasks as terrain evaluation, military base and airfield construction, water supply and construction material development, and perfection of aerial detection techniques. Professionally, the tasks and responsibilities of the engineering geologist evolved ever more rapidly after World War II, and the need for improved training and more practitioner-oriented engineering geologists was apparent. Earlier, the principal centers for training of applied geologists were the mining schools of Colorado, Michigan, Missouri, and Minnesota, where options in mineral-oriented, geological engineering were well established.

By the 1950s, the field had taken on a distinct trend toward modern engineering geology, both in theory and practice, a trend that is reflected in the literature. Clear distinctions had developed between the "geology for engineers" training of the 1910s to the 1940s, the mineral-oriented geological engineering programs of the 1930s and 1940s, and the emerging, geologically oriented field of professional practice—engineering geology. Exposure of engineers to geologic principles alone did not constitute qualifications for the practice of engineering geology. However, the introductory background on the subject of geology provided to engineers, which Mead (1941) called instruction in first aid, served to promote geological input and geological awareness, whenever engineers planned, designed, and constructed engineering works.

Since 1950, two types of academic programs have been offered for training engineering geology practitioners; this has created some misunderstanding and conflicts. The geological engineering program described below requires a basic academic curriculum that includes engineering design courses and concepts, together with some geology courses. Curricula in geological engineering are reviewed by the Accredidation Board of Engineering and Technology, Washington, D.C., and 12 programs had been accredited by the 1980s (Appendix A). Accredition does not imply all curricula are the same; some are strongly engineering while others are more balanced between engineering and the geosciences.

The geoscience-oriented programs in engineering geology, also described below, are not subject to an accredidation and overview group. Yet there is wide acceptance among institutions of a "standard" curriculum that includes basic core courses in the sciences and engineering, and upper-level courses in the geosciences and related techniques or engineering specialties.

More importantly and regardless of the program, there is an inherent philosophical difference between the geologist, as a scientist, and the engineer. The geologist maintains the concept of multiple working hypotheses, and approaches a problem by observing the effect of a process or event to determine the causes and thereby recommend a solution. Reasoning is influenced by the recent past—a key to the near future. The engineer strives for the safest and most economical solution to a problem and approaches a design by specifying a cause and analyzing its effect.

A widely accepted definition of engineering geology, as the application of geosciences to the solution of engineering problems, has evolved. This implies that the problems of engineers and related discipines can be "solved" by a qualified geologist. Although the two professions overlap somewhat, both viewpoints are represented in the team approach discussed in Chapter 18 of this volume.

GeoScience-oriented programs

Beginning with the first program in 1948 at Washington University and continuing through the 1950s, a growing number of academic institutions began to offer a major in engineering geology as an undergraduate option in the geology and earth science departments (Kiersch and others, 1957). At least ten of the schools had graduate programs by 1955, and the number of schools offering undergraduate training in engineering geology to the geology major (Appendix B) had nearly doubled by the early 1960s.

Engineering geology programs and training are currently offered at more than 60 universities and colleges (Appendix C). The predominant training is in the geological sciences (geology, geophysics, seismology, hydrology), integrated with supporting elements of associated sciences and civil and/or environmental engineering. Additional interdisciplinary training in specific geological and/or geotechnical engineering subjects is desirable in subfields of anticipated practice. However, completion of designated coursework in geosciences and supporting disciplines or in engineering only does not qualify a geologist for responsible practice regardless of the experience level of the faculty; subsequent field experience and familiarity with specific aspects of engineering and construction projects are required. Ideally, academic training is supplemented by an extended period of field activities that includes geologic mapping and the study of varied geomorphic and physical features such as surficial deposits, and bedrock types, in diverse terrains and climates. Such an internship affords the optimum in field orientation to develop the maturity and judgment so essential to successful practice. A brief outline of subject matter for a typical advanced-level course of the 1960s is given in Appendix D.

The first modern texts on the geological principles important to engineering geology began to appear after World War II (Bendel, 1949; Paige, 1950; Trask, 1950; Keil, 1954; Kiersch, 1955; Krynine and Judd, 1957; Judd, 1964), but the adjective "engineering" was interjecting confusion about the actual field of practice. Renaming the subject "construction geology" (Stini, 1922; LeRoy and Leroy, 1977), "physico-geology," or "geotechnical geology" (Kiersch, 1964, 1986) has been proposed as more expressive and a way to clarify any misunderstanding. However, the title "engineering geology" has persisted even though the scope and meaning have shifted, because many practitioners feel the designation affords a close connection to the principal field of engineering and is clearer to employers. Realistically, this connection is not wholly reflective of the broad professional field today, as recognized in major countries throughout world (Reuter and others, 1980).

The field of engineering geology in the United States by the 1960s denoted a somewhat different technical background and professional practice than implied in many other countries (Kiersch, 1964). Geological science was being values in proportion to its quantitative performance in the applied sciences and engineering professions, because nearly 85 percent of professional geologists were involved to some degree with the applications of geology. Academic emphasis previously oriented toward the purely classical had undergone a reorientation in academia, and by the 1970s reflected major changes in the preparation for and practice in all branches of applied geology. Furthermore, the emphasis of professional practice had changed from a predominantly mineral-exploration science to one of diversification and acceptance by many professions. This expanded utilization was largely due to a stronger emphasis on the physical aspects of geoscience. An examination of engineering geology preparation in the United States since the 1960s illustrates the causes of change over the succeeding decades, and the societal needs, which influenced future academic training and professional practice.

Changes since the 1960s

Professional demands dictated that geology course content had to become strongly quantitative and less descriptive beginning in the 1960s. This shift required an interdisciplinary approach to the study of geologic events and processes, thereby providing insight into physical environmental studies. Facts and quantitative data, separated from inference, conjecture, and hypothesis, became the needs and demands not only of engineering, but of theoretical research as well.

Undergraduate curricula for training in applied geoscience began to change to include a mixture of the basic sciences, humanities, and engineering science, integrated with a few high-caliber geology courses (Kiersch, 1964). Consistent with the improved caliber of precollege and undergraduate training, geology courses thus became more rigorous and inclusive in content by combining the fundamentals of overlapping courses into one integrated subject. Typically, several two- and three-unit courses were integrated into one strong four-unit course in geoscience, by eliminating the artificial fences of some geological subjects. Teaching a strong integrated course often requires the cooperative efforts of two or more faculty specialists. Introductory courses and "Geology for Engineers" could also be offered at a more advanced level, because of the impact of the strengthened science curricula in secondary schools (Gatewood, 1963).

Today the curriculum in geological sciences for an applied geologist in many major universities is about equivalent to some geological engineering programs, with regard to the supporting subjects and basic science requirements. A major difference is the inclusion of design and related engineering courses in geological engineering, whereas a group of advanced-level geoscience courses are taken by the engineering geologist, strengthening the future practitioner's capability of unraveling complex geological issues. However, reasons for a strong distinction between the two kinds of programs are becoming fewer when compared to academic requirements of the pre-1970s. Another major requirement for the engineering geologist is substantial knowledge of, and a strong orientation toward, field observations and geological mapping. This ranges from the regional/areal reconnaissance stage through site-specific stages of completion and operation. In contrast, the geological engineer is more likely to deal with field data for design purposes, or be an administrator.

The environmental movement of the 1970s and 1980s, and the need to protect the health and safety of the public, have served to increase enrollment in engineering geology training programs. This has provided the impetus for enlargement of the applied geology faculty at several institutions and the establishment of engineering geology options and related programs. Most faculty for programs oriented toward engineering geology now include—at a minimum—an applied geomorphologist and a hydrogeologist, in addition to an engineering geologist, with access to geophysical, seismological, and other specialties through cooperating faculty.

Engineering geology is also widely recognized as the applied science of the earth in which "man" is a recognized element, and preparation is geologically oriented. Every discipline of the geological sciences is used to protect the health, safety, and welfare of the public. Moreover, engineering geologists are liable for their interpretations and actions, because errors of judgment kill people (discussed in Chapter 23).

Faculty specialists in engineering geology, often viewed as applied scientists, nevertheless teach basic undergraduate and graduate-level courses, offer their specialized courses, participate in scientific research, and direct graduate-level theses. The impact of this strong science input has caused a change in the meaning attached to engineering geology since the 1930s—from a "practical" to an "applied science" specialty. However, this has not been because of academia only. It also reflects the current state of knowledge, public needs, legal limitations, and national policies.

Offering separate courses and options in engineering geology theory and practice has been most successful when the instructors are qualified in a branch of geological science and have taught one or more required courses at both the undergraduate and graduate levels. This level of faculty participation has invariably been the measure of well-respected engineering geology academic programs in North America. Successful faculty sponsors of engineering geology programs have taught such required subjects as physical geology, structural geology, sedimentation, geomorphology, petrology-petrography, seismology, and field geology. In recent years this has included environmental scientists and specialists, as reflected in a text by Rahn (1986) and an earlier collection of 42 case-history papers on environmental issues by Tank (1973).

GeoEngineering-oriented programs

A geological engineer is trained as an engineer with a general knowledge of geological science. He or she often serves as an exploration or production specialist or administrator in any one of the four engineering fields allied with the geosciences. Geo-engineering programs provide a sound undergraduate training for advanced-level study in the geosciences and practice as an applied geologist in one of the four branches (Fig. 3). Many engineering geologists have been trained in this manner as undergraduates.

Geological engineering is the application of both engineering and geological principles to evaluation and design concepts as they relate to the Earth, its materials, structures, and forces. Besides being utilized in the mineral resources field (Fig. 3), some schools offer training for the planning and construction of engineering works (Neilson and others, 1960, p. 1051). By the 1960s, ten of the on-going geological engineering programs in the United States had established separate engineering geology or construction geology options, while the other 27 programs continued to be mineral oriented. Today, a few of the latter institutions offer an engineering geology option within their geological engineering programs, as do 12 institutions in Canada (Appendix A).

Geological engineering departments with an engineering geology option may have modest to strong ties with the geosciences through either joint departmental appointments or cooperating faculty. Reportedly, some students enroll in undergraduate geological engineering programs, because they can later qualify to register as professional engineers. Similarly, the engineering geologist is recognized as a professional, and practitioners can register and be licensed in many states.

Some of the major engineering schools offer a 4- or 5-year program for a Bachelor of Science degree, of which two or three years are liberal arts and basic sciences, and one or two years are the engineering sciences and allied courses. Professional-type courses are restricted to the graduate school. This diversified basic program for engineering and/or science students is ideally suited, with modifications, for conventional undergraduate training in geology; one difference is that selected geoscience courses are taken at the upper-division level in lieu of some engineering science courses. This approach is particularly suitable for the student who plans on further study and practice in engineering geology. The serious, mature student, with a sound background in the basic sciences and engineering, invariably grasps the fundamentals of undergraduate geology courses in a fraction of the time required by the poorly prepared student.

Graduate-level professional training in areas of the geosciences is expected if a student plans a career in engineering geology. Most graduate students concentrate on a specified field of the geosciences, with duration dependent on the individual's interests. Training to practice as both an engineer and a geological specialist is not encouraged, because professional demands render this difficult, except for routine design and construction activities.

Some engineering geologists are trained initially in one of the supporting disciplines, such as soil or rock mechanics. A means for evaluating subsequent academic and on-the-job training and experience, as well as the capabilities of these individuals, is available through professional registration. This is mandatory for both the principal engineering geologist and the geological engineer within a firm in many states. Acceptance for this is growing.

GEOLOGY FOR ENGINEERS

A geology for engineers course, which includes the elements of geological sciences, is often offered as a one- or two-semester course and constitutes a survey of the science (Kiersch and others, 1957). Geology for engineering students was taught at a small group of colleges throughout the United States by the 1910s (Chapter 1, this volume), and in Canada by 1925 (D. F. VanDine, personal communication, 1987). By the late 1950s, over 60 schools offered the course. Many schools such as Cornell initially called the course "Engineering Geology" but by 1960s this had been changed to the accepted name. A few schools, such as Texas A & M, adopted the title "Engineering Geology for Engineers," which implies a strong applications content. In either case, the limited training in no way qualifies the student for practice as an applied geologist, but rather provides the capability to recognize when a geologist is needed. Whenever possible, instruction has been by a senior member of the geology faculty with a record of practical experience; a young professor, regardless of teaching ability, is usually less effective with engineering students.

The text by Ries and Watson (1914) was widely used for the course into the 1940s; the fifth edition was published in 1936. T. T. Quirke (1947) of the University of Illinois published an early text and laboratory manual for the course. R. F. Legget's 1939 text became a leader after World War II, with revisions in 1962 and 1983. Other popular geology for engineers texts were by J. M. Trefethen (1949) and J. R. Schultz and A. B. Cleaves (1955). An excellent presentation of geological fundamentals combined with applications in engineering was published by E. C. Dapples (1959) of Northwestern University as a basic geology course for the science and engineering student. More recently, an early emphasis on the environmental aspects of geology for engineering projects was prepared by Flawn (1970), and a broad-based text on the elements of geology and applications for engineering was prepared by Mathewson (1981).

A second course, "Applications of Geology for Engineers," has been offered by a small group of schools (Kiersch and others, 1957, p. 17) since the 1940s. A group of topics on processes and exploration techniques appropriate for this course, suggested by Terzaghi (1961), is designed to further assist in recognizing potential geologic problems and when a geological specialist is needed.

Geology for Engineers, or *Technical Geology* of Stini in Austria has a distinguished history in Europe (described Chapter 1, this volume) and at other schools, such as the University of Grenoble, France (Jaeger, 1982, p. 66–67). Texts in English on geology for engineers were prepared by Blyth (1952, 1974) and Richey (1964), who was the chief geologist for Scottish Geological Survey and served on consulting boards for many hydro-projects in Scotland during the 1940s and 1950s. A Czechoslovakian edition of another early text was by Zaruba and Mencl (1956; translated and slightly revised, 1976). The growing European interest in geological training for engineers is exemplified by the publication of a recent British textbook (Bowen, 1984), a French text (Antoine and Fabre, 1980), and an Australian text (Beavis, 1985).

RESEARCH ENDEAVORS

Early directions

Planning and guidance for the very early research activities important to engineering geology were first addressed by such pioneers as G. K. Gilbert, C. E. Dutton, W. O. Crosby, and W. J. Mead (described in Chapter 1), and the group of practitioners that followed who organized the Division on Engineering Geology of the Geological Society of America in 1940s (in Chapter 4 of this volume). Brief summaries of other early research efforts are given below:

1. Based on "as-encountered" geological studies, 1-inch electrical strain-gauge rosettes were used to measure the residual stress in walls of the 13-mi Colorado-Big Thompson pressure tunnel (1940 to 1947). These studies resulted in the cost-saving recommendation that two-thirds of the proposed steel reinforcement could be eliminated (Rhoades and Irwin, 1947).

2. Paige and Rhoades (1948) summarized ongoing studies and needs in soil properties and testing, concrete aggregates and reactive materials, construction materials, sedimentation and silting of reservoirs, ground water, slope stability and mass wasting, clays and mineralogical properties, and measurement of residual stresses in rocks.

3. A farsighted early research project by Trask (1950) included 35 papers on the various aspects of sediments critical to engineering geology practice. This effort was funded by the Division of Geology and Geography, National Research Council; it reflected the concerns of the NRC for wider use of geological data by engineers.

4. Eckel (1951) reviewed some areas important to engineering practice: physical and mineral properties of rock and soil, a meaningful index for a shale unit, delineating units and features on a geologic map, forecasting slope stability, and techniques for acquiring data.

5. The first research in engineering geology to be sponsored by the Geological Society of America was by G. A. Kiersch in 1954 for "Study of rock weathering and significance to engineering and physical properties of rock and soil."

6. The U.S. Geological Survey, beginning in 1956, participated in research for protective construction to resist nuclear blasts at the Nevada Nuclear Test Site, with emphasis on the physical properties of site rocks and the hydrogeology of a rock mass (Eckel, 1968).

7. Kiersch (1958, 1961) reviewed deficiencies of 10 phenomena and materials critical to practice, as well as the applied research needs in 15 other categories, and the need for strengthening of instrumentation techniques, such as geophysical "mapping," tiltmeters, stress-strain rates on faults, and predicting volcanic eruptions.

8. Trask (1959) reported on sessions involving 35 engineering geologists at UC-Berkeley in 1957, mainly related to the California Water Plan and other projects in the western states. The group established an extensive list of geological research needs related to water-resources projects.

9. The National Science Foundation provided a one-year grant (1962-1963) to H. J. Pincus for studies on "Rock properties and testing" at the U.S. Bureau of Mines facilities in Minneapolis/St. Paul, Minnesota. The NSF also provided a senior faculty post-doctorate fellowship in 1963 and 1964 to G. A. Kiersch for studies in "Geomechanics and stresses in rock mass," at the Technische Hochschule, Vienna, Austria.

10. Judd (1964) reviewed the state of knowledge on stresses, with contributions by many experts, which resulted from an international symposium in 1963. This was the first major effort to coordinate and interpret the meaning of interdisciplinary data on stresses in rocks (described in Chapter 4, this volume).

Recent decades

Throughout the past two decades, research activities have been largely problem oriented. During the late 1960s and 1970s, research was strongly driven by nuclear construction projects. Examples include dating fault movements and forecasting the maximum earthquake effects on engineering works. By the 1980s, however, the emphasis had shifted to environmentally oriented problems, such as nuclear-waste disposal and burial, and contamination of the hydrologic regime by geochemical processes. Associated with this trend was a renewed awareness of the natural and man-induced geologic processes that cause local- to regional-sized disasters, such as earthquakes, floods, landslides and slope collapses, debris flows, volcanic eruptions, and groundwater contamination. These research activities included:

- "Dynamic problems in engineering geology," a symposium of invited speakers at the annual meeting of the Geological Society and Engineering Geology Division at Atlantic City in 1969. The state of knowledge surrounding five phenomena and techniques critical to practice were reviewed; these included reservoir-induced seismicity, seismotectonics and dating the recency of fault movements, and geophysical measurements for mapping purposes (Chapter 4, Appendix K).
- "Capable faulting and engineering works," a full-day symposium of invited speakers at the Seattle meeting of the Geological Society and Engineering Geology Division and cosponsoring groups in 1977. It provided a state-of-the-art summary on seismotectonics, age dating the recency of fault movement, and the relevance to engineering works.
- "Rock-Mechanics Research (Handin and others, 1981) . . .," reviews the needs, requirements, and priorities for rock-mechanics research in the United States as determined by seven panels of experts. Areas of research vital to the success of national projects were identified relative to resources recovery, construction, and earthquake-hazard mitigation; recommendations address the current state of knowledge and identify the deficient areas (Fig. 17).

Any attempt to define the research needs in engineering geology must address the diversified attitudes of the many-faceted practitioners. A 1983 list of research needs and ideas prepared by a workshop sponsored by the geotechnical engineering program of National Science Foundation (Sitmar, 1983) focused mostly on topics given by earlier investigators (Paige and Rhoades, 1948; Eckel, 1951; Trask, 1959; Kiersch, 1958, 1961; and Judd, 1964): slope processes, exploration-site characteristics, areal-regional features of materials, the hydrologic regime and environs, and seismotectonic studies–geomorphic processes.

The basic objectives or philosophy for research in engineering geology, according to a panel of the International Association of Engineering Geologists (Williams and others, 1985), should be: (1) based on a quantified need or geologic problem(s) related to practice that bears on the safety and welfare, the economic, or the environmental state of engineering works; (2) directed toward identifying and understanding the physical, chemical, biological, and social factors that act or interact to produce an impact on engineering works; and (3) to identify and understand the scientific principle(s) or concept(s) involved with the geologic phenomena and/or events encountered in engineering geologic practice.

A study panel of the Committee on the Solid-Earth Sciences, National Research Council, on "Geology, Land Use, and the Environment" made a critical assessment in 1988 of geology in the construction process and identified the research needs with long- and short-term priorities (Hatheway and others, 1990).

NATURAL HAZARDS

The impact of natural hazards (Table 1) on the health, safety, and welfare of mankind has been long recognized. These natural disasters have claimed more than 2.8 million lives worldwide in the past 20 years and adversely affected 820 million people with economic losses and hardships. A role for the science and engineering community to mitigate or reduce the impact of natural hazards on society was expressed by many countries and international organizations by 1986. In response, a U.S. Advisory Committee on the International Decade for Natural Hazards Reduction (IDNHR) prepared a review report on the need for a worldwide reduction of human and property losses due to catastrophic events (Housner, 1987). The United Nations Assembly on December 11, 1987, adopted a resolution stating that the United Nations should facilitate an "International Decade for Natural Disaster Reduction (IDNDR)" and designating the 1990s as the decade to foster international cooperation in natural-disaster reduction. The U.S. Advisory Committee on IDNDR subsequently prepared a second report, "Reducing disaster's toll" (Housner, 1989), in which the hazard-reduction process, needs and limitations of IDNDR, and a framework for the United States effort, are set forth. Any assessment of natural hazards and future threats, whether for warning, planning, policy making, or design, will require real-time, probabilistic predictions in order to mitigate the consequences (Cornell, 1988, p. A-82).

CRITICAL AREAS
REQUIRING
ROCK-MECHANICS RESEARCH

POROSITY , PERMEABILITY , FLUID FLOW
IN SITU MEASUREMENTS: GEOPHYSICAL & STORAGE COEFFICIENTS
PRIMARY-VS-SECONDARY OPENINGS
FLUID-FLOW MEASUREMENTS : TRACERS IN GROUNDWATER
CURRENT NEEDS :

MAPPING OF NATURAL/ARTIFICIAL FRACTURES
THREE-DIMENSIONAL PROBLEM
FRACTURES OBSERVED-VS-CONCEALED/EXTRAPOLATION
NUMEROUS GEOLOGICAL/GEOPHYSICAL TECHNIQUES
ACTIVE-PASSIVE
CURRENT NEEDS:

DETERMINATION OF IN-SITU STRESS
PREDICT ROCK RESPONSE TO CHANGING LOADS
UNDERSTAND PROCESSES--STRESSES
GRAVITATIONAL - TECTONIC - THERMAL - HYDRODYNAMIC - RESIDUAL - PALEOSTRESS
MEASURE BY ACTIVE OR PASSIVE MODE
CURRENT NEEDS:

ROCK FRAGMENTATION - DRILLING & EXCAVATION
ENERGY - AND MINERAL - RESOURCES/ROCK FRAGMENTATION
SUPPORT SCIENCE & ENGINEERING TECHNOLOGIES/
PHYSICAL PROCESSES
CURRENT NEEDS :

SCALING TEST DATA TO FIELD APPLICATIONS
REALISTIC DATA FOR ROCK-MASS CHARACTERISTICS
SCALE LOW COST LABORATORY DATA TO FIELD CONDITIONS
IN-SITU MODULI VALUES -- 0.2 TO 0.6 OF LABORATORY DATA
CURRENT NEEDS :

THERMO - PHYSICAL,- MECHANICAL,- CHEMICAL PROPERTIES
EFFECTS THERMAL PROPERTIES ON -- CONDUCTIVITY, DIFFUSIVITY, THERMAL INERTIA, RADIATIVE TRANSFER, EXPLOSION ON ROCK
CURRENT NEEDS :

NUMERICAL MODELING
STYLIZED REPRESENTATION ROCK MASS
FIVE FEATURES -- GEOMETRY, EQUATIONS MOTION, KINEMATICS, CONSTITUTIVE BEHAVIOR, SPECIAL FEATURES
CURRENT NEEDS :

FROM : "ROCK-MECHANICS RESEARCH REQUIREMENTS FOR RESOURCE RECOVERY, CONSTRUCTION AND EARTHQUAKE-HAZARD REDUCTION"
by
U.S. National Committee Rock Mechanics
1981

Figure 17. Critical areas requiring concerted interdisciplinary efforts in seven broad categories of rock-mechanics research; all are relevant to engineering geology practice. (From Handin and others, 1981.)

At least six natural hazards common to the United States and engineering geology practice should be investigated to improve confidence in our methods to predict and assess the ongoing behavior of hazards and thereby assist in reducing risk, loss of life, and property damage:

- Earthquakes and associated features (described below, and in Krinitzsky and Slemmons, 1990).
- Volcanic eruptions or outpourings, with such associated features as ash falls, pyroclastic flows, surges, and debris or mudflows, that dam rivers and lakes and cause flooding up and/or downstream, calderas with inflation or subsidence, modest earthquakes, and liquefaction of cohesionless materials resulting in slides (Chapter 11).
- Slope collapse or slides, both natural- and man-induced, that involve rock or soil; including rockfalls, some of which generate waves in lakes, and cause flooding and local earthquakes (Chapter 9).
- Debris flows and associated features (Chapter 7).
- Floods of most types (Chapter 6). Paleoflood hydrological studies can reconstruct histories to forecast magnitude and risks (Baker, 1987).
- Sinkhole collapse and subsidence, with associated features (described in this chapter).

Earthquakes

The hazard and risk assessment of earthquakes over recent decades has changed greatly due to rapid expansion and constant revision of the approach, methodology, scientific and engineering technology, and databases. These changes have influenced the more simplistic approaches and deterministic methods formerly used in estimating the appropriate design earthquake for engineering works. Presently, evaluations commonly use a "multi" approach that compares the results of several different methods with the quantitative data and may lead to an analysis by probabilistic methods (Slemmons, 1982; and Chapter 12, this volume). Moreover, there are other serious hazards and risks associated with an earthquake besides the physical destructiveness of the event. These include tsunamis and seiches, rockfalls that trigger destructive waves, and the damming of rivers by landslides, with initial flooding upstream and later destruction downstream when the temporary dam collapses.

In spite of great progress among scientists, engineers, and planners during the twentieth century, a major earthquake in any heavily populated area of the world could be a disaster, and even a catastrophe, as occurred at Agadir, Morocco (Fig. 18). A shallow-focus earthquake, less than 2 mi deep, with an epicenter near the city limits, completely destroyed Agadir on February 29, 1960. Damage was confined to a radius of 5 mi from the epicenter, leaving 12,000 dead and some 12,000 injured among the 33,000 inhabitants (Hodgson, 1964, p. 35–37). Engineering investigations concluded that the beam-to-column joint construction of SAADA Hotel in Agadir (Fig. 18) was inadequately reinforced and braced, so that the floors quickly pancaked (W. G. Kirkland, personal communication, May 1962).

Figure 18. SAADA Hotel, city of Agadir, Morocco, and the "Miami of North Africa," that was completely destroyed by a shallow-focus earthquake less than 2 mi deep. Engineering investigations concluded the building columns were inadequately constructed. (Photo courtesy of American Iron and Steel Institute/W. G. Kirkland, 1962.)

The varied effects of an earthquake, long recognized, may be due to such circumstances as focal depth, and complexity of areal-regional geologic conditions, as well as the man-made environs of construction sites, and population density (Fig. 19). By the early twenty-first century, major advances in techniques for gathering and interpreting seismic-event data, such as slip rates and absolute dates of paleoseismicity, and acceptance of the probabilistic approach will provide realistic evaluations and long-term earthquake predictions. Within the century, meaningful earthquake assessments will be available worldwide, and the "earthquake problem," both short- and long-term, will be largely solved (Allen, 1988, p. A-81).

PERSPECTIVES ON PROGRESS IN ENGINEERING GEOLOGY

The transitions and changes experienced in engineering geology since the 1940s were the basis for asking several promi-

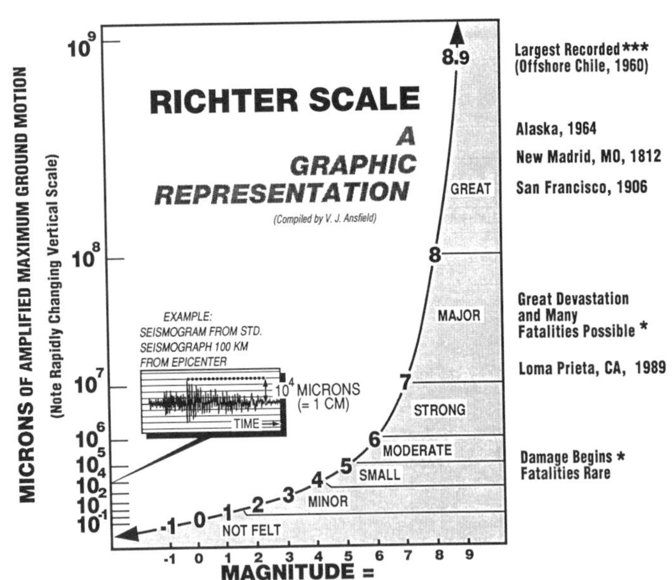

Figure 19. A graphic representation of the Richter scale correlated with the effects of earthquakes that may vary greatly due to the geologic environs and man-made impacts. (From A. J. Ansfield, personal communication, 1990.)

nent practitioners in 1988, "What have been the most profound changes and/or advancements during your career?" A few of their comments are listed below:

R. H. Nesbitt (formerly chief geologist, U.S. Corps of Engineers, Washington D.C.) cited continuing education available to practitioners after extended practice; the advances in drilling, sampling, and laboratory or in-situ testing, combined with borehole photography and geophysical surveys; improvements in excavation techniques and drainage for stabilization of soil and bedrock masses; pre- and post-construction instrumentation for predicting stresses and movements in soil or bedrock; and the improvement in theoretical knowledge for evaluating geologic conditions and developing a proper engineering design.

R. E. Gray (executive vice-president of AGI Consultants, Pittsburgh) cited the improved understanding of natural processes and rock properties that fostered the rock mechanics studies of the 1960s and 1970s, and led to advances in tunneling techniques; the wide use of pre-split blasting since the 1960s, which has reduced maintenance for rock slopes; studies on residual strength of a slope (Skempton, 1964) and on progressive slope failure (Bjerrum, 1966), which have provided major advances for landslide stabilization; and the recognition of stress-relief causes and procedures for measuring residual stress developed in the 1960s and 1970s.

N. R. Tilford (formerly chief geologist, Electric Bond and Share Company, EBASCO, New York) cited the sharp decline in the number of water resources projects; the emergence of more studies of geologic hazards, spurred by nuclear safety concerns and the failure of engineering works; and waste-buried projects as three growing areas of activity that are now of major importance in the profession.

W. V. Conn (consultant and formerly Division Geologist, U.S. Corps of Engineers, Atlanta) cited the major advance in stability of rock slopes and underground openings and chambers by use of resin-encapsulated rock bolts.

OUTLOOK

The geologically oriented specialist will be called upon to unravel and quantitatively evaluate a greatly increased number of complex geoscience problems that are related to the origin, geologic history, and changes with time of natural materials, processes, and phenomena. Besides the diverse group of problem areas discussed earlier, and at the end of Chapter 4 (this volume), engineering geologists will be expected to participate in solving such long-standing problems as (1) delineating near-surface magma migration, retarding or restricting volcanic flows, and providing eruption models and forecasts; (2) destressing large rock masses that retain high tectonic stress, prior to underground construction; (3) perfecting probabilistic analysis to realistically forecast long-term seismic hazards with meaningful short-term predictions to eliminate the "earthquake problem" (Allen, 1988); (4) utilizing the database of paleo-floods for a more factual understanding of the nature, magnitude, risk, and consequences of extraordinary flood phenomena (Baker, 1987); (5) refining geophysical techniques and measurements to more accurately monitor volcanoes, detect concealed sinkholes, "map" subsurface rock units/features, and provide a reliable, less costly replacement for some direct exploration methods; and (6) management of ground-water pollution with definition, monitoring contaminant source, modeling, and replacement of supplies.

Earth scientists have made tremendous progress during the past century toward understanding geologic processes and the causes of serious natural- and man-induced hazards (Table 1), and devising means for societal adjustment. Demographic trends and increased impact of hazards have created new challenges to reduce the risk potential in the twenty-first century, whether by mitigation techniques or a better realization of the causes. Mitigation of natural hazards requires the input of applied geoscientists for measurements, concepts, and providing predictions; for building design, construction control dams, avalanche barriers, and restraint techniques; and for land-use planning, evaluation of predictions, implementation of warnings, and evaluation of risk versus economic costs. In addition, the need is increasing for further research relevant to such man-made hazards as the disposal of nuclear wastes from civilian and military activities and the clean up as nuclear plants are decommissioned.

An alternative to deterministic geological interpretations as a measure of uncertainty is the probabilistic and/or statistical approach based on the use of logic trees (Coppersmith, 1988). This method of evaluation assesses the multiple, interrelated parameters and quantifies the uncertainties of such common geologic circumstances as the stability of slopes or open cuts, the conditions to be encountered along a tunnel, or the seismic potential of a fault zone.

Underlying the trend toward more complexity and diversity has been the ever-increasing demand that engineering geologists have stronger and more quantitative backgrounds. Practitioners in the twenty-first century will require even more diverse training in the geosciences, besides the ability to interrelate and contribute to the "team effort" so essential to an effective practice.

APPENDIX A. ACCREDITED ENGINEERING GEOLOGY OPTIONS IN GEOLOGICAL ENGINEERING PROGRAMS

Canada and the United States in the 1970s and 1980s
(given approximately in the order established)

CANADA (VanDine, 1984)	UNITED STATES (Hatheway, A. W., files, 1989; Kiersch and others, 1957, Appendix F)
University of British Columbia	University of California–Berkeley
Universite Laval	Colorado School of Mines
Queens University	Michigan Technological University
University of Saskatchewan	University of Missouri–Rolla
University of Toronto	South Dakota School of Mines
Ecole Polytechnique	University of Arizona
University of Manitoba	Princeton University
Universite de Quebec	University of Utah
University of Windsor	University of Idaho
Memorial University	University of Nevada
University of New Brunswick	Montana College of Mineral Sciences
University of Waterloo	University of Mississippi

APPENDIX B. ENGINEERING GEOLOGY PROGRAMS

Departments of Geology and Earth Sciences by the early 1960s
Undergraduate options and/or graduate programs

UNITED STATES
(from Kiersch and others, 1957, Appendices D and E, and author's files)

Washington University, 1948
Kansas State College, 1950
Illinois University, 1952
Syracuse University, 1955
Louisiana State University, 1950
University of Arizona, 1953
Notre Dame University, 1957

Late 1950s and early 1960s

Brigham Young University
University of Southern California
Cornell University
North Carolina State College
Princeton University
University of Utah
Texas A & M College
University of Washington
University of California–Los Angeles
University of Oregon
Virginia Polytechnic Institute

CANADA
(VanDine, 1984. Programs administered by Earth Science/Geological Science Departments in cooperation with Geotechnical Engineering.)

Universite Laval
Queens University

APPENDIX C. ACADEMIC PROGRAMS IN ENGINEERING GEOLOGY—1989

Canada and the United States*

School and training offered:
U = Undergraduate; M = Masters level; D = Doctorate

CANADA†

University of Alberta, U M D
University of Manitoba, U
Ecole Polytectinique de Montreal, U M D
McGill University, U M

UNITED STATES

Arizona
- Arizona State University, U M
- Northern Arizona University, U M

California
- California State University
  - San Diego State, U M
  - Fullerton, U
  - Long Beach, U
  - Fresno, U M
  - Los Angeles, U M
  - Sacramento, U
  - San Jose, U M
- University of California
  - Santa Cruz, U M D
  - Santa Barbara, U
- Stanford University, U

Colorado
- University of Colorado
  - Colorado Springs, U M
  - Denver, U

Florida
- University of South Florida, M

Hawaii
- University of Hawaii, M D

Indiana
- Purdue University, U M D

Iowa
- University of Iowa, M D

Illinois
- Olivet Nazarene University, U
- University of Illinois, M D

Kansas
- Emporia State University, U M
- Fort Hays State University, U M
- University of Kansaas, U

Kentucky
- Murray State University, U M

Louisiana
- Nicholls State University, U
- Northeast Louisiana University, M

Massachusetts
- Bridgewater State College, U

Missouri
- Southwest Missouri State University, U M
- University of Missouri
  - Kansaas City, M

Mississippi
- Mississippi State University, U

Montana
- Montana College of Mineral Science and Technology, U M

North Carolina
- University of North Carolina
  - Chapel Hill, U M D
- North Carolina State University
  - Raleigh, U

New Jersey
- Stockton State College, U

Nevada
- University of Nevada
  - Las Vegas, U

New York
- State University College
  - Geneseo, U
  - New Paltz, U
- State University of New York
  - Binghamton, U M D

Ohio
- Wright State University, U M
- Kent State University, U M D
- University of Akron, U M
- University of Dayton, U
- University of Cincinnati, U

Oregon
- Portland State University, M
- Southern Oregon State College, U

Pennsylvania
- Millersville University, U
- University of Pittsburgh, U M
- University of Pittsburgh
  - Bradford, U

Tennessee
- University of Tennessee
  - Knoxville, U

Texas
- Baylor University, U M
- Rice University, D
- Texas A & M University, U M D
- University of Texas
  - El Paso, M D
  - San Antonio, U M D
- Texas A & I University, U M
- Sul Ross State University, U M

Utah
- Utah State University, U M

Virginia
- Radford University, U

Washington
- Eastern Washington University, U
- Western Washington University, U M

West Virginia
- Marshall College, U

*Based on a survey of North American universities conducted in 1989 by the Association of Engineering Geologists and John W. Williams.
†Engineering geology options in geotechnical engineering are also administered under Earth Science/Geological Science Departments at University Laval, Queens University, Universite de Quebec, University of Windsor, and Memorial University (VanDine, 1984).

APPENDIX D. SUBJECT MATTER OF A COURSE IN ENGINEERING GEOLOGY IN THE 1960s
(from Kiersch, 1964, Appendix)

The brief outline of subject matter, given below, for an advanced-level course in engineering geology presupposes that the student has a background in the physical sciences, some engineering science and humanities, and a foundation in basic geology courses, including physical and historical geology, mineralogy, petrology/petrography, structural geology, geomorphology, sedimentation, and field geology.

PART 1. PRINCIPLES AND FUNDAMENTALS

A. Introduction.
   Scope; historical development; utilization.
B. Geologic phenomena of special importance.
   1. Properties of rock, mass vs. sample. In part, solid geophysics and geomechanics.
   2. Natural stresses and geomechanics. Surface excavations; underground openings, time factor, geologic versus human.
   3. Seismology.
   4. Weathering of rock; soil derivatives; properties of surficial materials.
   5. Sedimentation and silting.
   6. Erosion and mass wasting.
   7. Downslope gravity movements (landslides, etc.)
   8. Subsidence.
   9. Underground fluids (hydrogeology).
   10. Surface hydrology.
   11. Beach and coastal features.
   12. Frozen ground.
C. Geologic materials of special importance.
   Aggregate; cement and pozzolan; rock fill and earth fill; ice and snow; miscellaneous construction materials (e.g., dimension stone, riprap, sediment lining, stabilizer, coral, water for construction).

PART II. APPLICATIONS PLANNING/DESIGN, CONSTRUCTION, AND/OR MAINTAINENCE STAGES OF PROJECTS

A. Techniques and methods.
   1. Field exploration: direct and indirect.
      Drill holes; man-sized openings; aerial photos; borehole photos; TV observations; geophysical and geochemical methods; ground-water tests; surface and underground mapping; aerial sensing by infrared, radio wave, and radar.
   2. Field and laboratory work.
      Sampling of holes, openings, and cuts; soil mechanics–testing properties of surficial materials (see I-B-1, 4, 5); rock mechanics–testing properties of rock masses (se I-B-1, 2, 3, 7, 8); measuring inherent strain and analyzing active stresses on rock masses; time factor; petrography.
   3. Evaluation and interpretation of data (geostatistics).
      From direct and indirect exploration; field and laboratory tests; geologic models.
   4. Plans and specifications.
   5. Construction and maintenance (remedial).
      Grouting, drainage, dewatering, tracing ground-water flow and leakage, channel and shoreline stabilization, embankment and open-cut stabilization, excavation, and blasting.
B. Geology applied to engineering projects.
   1. Surface excavations and structures (exclusive of hydraulic works).
      Construction and mining of open cuts, quarrying, pipelines, buildings of all types, power plants, bridges and piers, highways, railroads, airports, launching pads, urban development, land reclamation.
   2. Hydraulic structures.
      Dams and reservoirs, surface and underground; canals and irrigation works; conservation works; stream-channel controls; beach and harbor control works; harnessing tidal energy.
   3. Underground excavations and structures.
      Tunnels, conduits, aqueducts, and siphons; mine workings and shafts; power plants; subterranean storage facilities and factories; drainage works; storage chambers for petroleum, gas liquids, and water; missile-launching installations.
   4. Construction materials. See Part I, C.
   5. Military.
      Intelligence-terrain factors; general military construction; ground, air, and naval operations; dynamic stresses.
   6. Planetary geology.
      Lunar and planetary operations; sites and resources for basing and protection; natural sources of heat energy.
   7. Regional economic development (partly overlaps items 1 to 4 above).
      Basis for interrelated development of natural resources, agriculture, and industry; land management; engineering and water-supply projects; city planning.
   8. Nuclear explosions for industrial uses.
      Surface excavation; induced subsidence and depression; underground fragmentation, reservoirs; earth- and rock-fill dams; underground storage of energy, geothermal heat; mineral exploitation, especially for oil in shale, tar sands; secondary recovery.
   9. Public health.
      Disposal of radioactive and industrial wastes; control of stream pollution.
   10. Urban geology.
      Largely background geology for engineering works; overlaps with B-7 above.
C. Legal aspects of geology in engineering practice
   1. Responsibility of project sponsor vs. contractor; changed conditions.
   2. Responsibility of insuree vs. insured.
   3. Mineral rights and claims; title ownership for construction materials, ground water, surface water, access roads, disposal areas; adequate rent or royalty payment.
   4. Accretion vs. avulsion of sedimentation and channels; riparian land rights lost or gained.
D. Failures of engineering works
   1. Failure due to a single geologic weakness or condition.
      Surface excavations and structures (B-1 above); hydraulic structures (B-2); underground excavations and structures (B-3); time factor.
   2. Failure due to a combinmation of adverse geologic conditions in natural environment or setting.
      Same subdivisions as in D-1.
   3. Failure due to man-made changes imposed on the geologic setting or site by operation of works.
      Same subdivisions as in D-1.
   4. Failure due to any combination of above circumstances.
      Same subdivisions as in D-1.

REFERENCES CITED

AEC, 1971, Seismic and geologic siting criteria for nuclear plants: Washington, D.C., Atomic Energy Commission (preliminary), 12 p.

AGS, 1986, An oil rig at the Grand Canyon: Arizona Geological Survey Fieldnotes, v. 16, no. 1, p. 9.

Allen, C. R., 1988, Earthquake hazard reduction in the past and next century: Geological Society of America Abstracts with Programs, v. 20, p. A81.

Antoine, P., and Fabre, D., 1980, Geologie au genie civil: Paris, France, Masson Publications, 284 p.

Bacigalupi, C. M., 1959, Large-scale excavations with nuclear explosives: Livermore, University of California Research Livermore Report 5457, 17 p.

Baker, V. R., 1987, Paleoflood hydrology and extraordinary flood events: Journal of Hydrology, v. 96, p. 79–99.

Bawa, K. S., 1970, Design and instrumentation of an underground station for Washington Metro system, *in* Brekke, T. L., and Jorstad, F. A., eds., Large permanent underground openings; Proceedings of an International Symposium, Oslo, Norway, 1969: Oslo, Universitetsforlaget, p. 31–42.

Beavis, F. C., 1985, Engineering geology: London, Blackwell Scientific Publications, 211 p.

Bendel, L., 1949, Ingenieu-geologie: Vienna, Springer-Verlag, 3 volumes.

Betz, F., ed., 1975, Environmental geology: New York, Van Nostrand Reinhold Benchmark Series, p. 101–107.

—— , 1984a, Military geology, *in* Finkl, C. W., ed., The encyclopedia of applied geology: New York, Van Nostrand Reinhold, p. 238–241.

—— , 1984b, Applied geology, *in* Finkl, C. W., ed., The encyclopedia of applied geology: New York, Van Nostrand Reinhold, p. 355–358.

Binger, W. V., and Thompson, T. F., 1949, Excavation slopes; Sea-level plan for the Panama Canal: American Society of Civil Engineers Transactions, v. 114, p. 734–738.

Bjerrum, L., 1966, Progressive failure in slopes of overconsolidated plastic clay and clay shale, *in* Terzaghi Lectures 1963–1972, Volume of 1974; Proceedings of the American Society of Civil Engineers: Journal of Soil Mechanics and Foundation Division, SM 5, p. 141–190.

Blyth, F.G.H., 1952, A geology for engineers: London, Edward Arnold and Co., 2nd ed., 336 p.; 6th ed, with M. H. de Freitas, 1974, 557 p.

Bock, C. G., 1981, The interrelationship of geologic conditions with construction methods and costs, Washington Metro: Bulletin of the Association of Engineering Geologists, v. 28, no. 2, p. 187–194.

Bowen, R., 1984, Geology in engineering: New York, Elsevier Applied Science Publishers, 392 p.

Carder, D. S., 1945, Seismic investigations in the Boulder Dam area, 1940–44, and influence of reservoir loading on earthquake activity: Seismological Society of America Bulletin, v. 35, p. 175–192.

—— , 1970, Reservoir loading and local earthquakes, *in* Adams, W. M., ed., Engineering seismology; The works of man: Geological Society of America Engineering Case Histories 8, p. 51–61.

Casagrande, L., 1952, Electro-osmotic stabilization of soils: Boston Society of Civil Engineers Journal, v. 39, p. 51–70 and 1949, Geotechnique, v. 1, p. 1959–1977.

CDWR, 1970, Middle Fork Eel River landslide investigations: Northern Distinct, California Department of Water Resources Memo-Report, 116 p.

Clapp, G. R., 1957, A "TVA" for the Khuzestan region, Iran: London, Huntington Aerosurveys, Ltd., Aerial Survey Review, no. 18, p. 2.

Code of Federal Regulations, 1978, Title 10, Energy; Part 100 (10 CFR 100), Reactor site criteria; Appendix A, Seismic and geologic siting criteria for nuclear power plants: Washington, D.C., U.S. Nuclear Regulatory Commission, 31 p.

Collins, J. J. and Foster, H. L., 1949, The Fukui earthquake, Hokuriku region, Japan, 28 June, 1948; Volume I, Geology; Volume II, Engineering: Far East Command, United States Army General Headquarters, Geological Surveys Branch, Office of Engineers, v. I, 81 p., v. II, 205 p.

Coppersmith, K. J., 1988, Addressing geological uncertainties in seismic hazard, *in* Final report of Diablo Canyon long-term seismic program: Diablo Canyon Power Plant, Pacific Gas and Electric Company Dockets 50-275 and 50-323.

Cornell, C. A., 1988, Probabilistic natural hazard assessment: Geological Society of America Abstracts with Programs, v. 20, p. A82.

Crawford, C. B. and Eden, W. J., 1963, Nicolet landslide of November, 1955, Quebec, Canada, *in* Trask, P. D. and Kiersch, G. A., eds., Geological Society of America Engineering Geology Case Histories Number 4, p. 45–50.

Dale, T. N., 1907, Building stone and road metal; Recent work; New England granites: U.S. Geological Survey Bulletin 315-J, p. 356–359.

Dapples, E. C., 1959, Basic geology for science and engineering: New York, John Wiley and Sons, 609 p.

Dean, C., 1990, "Nearshore" is plumbed for clues to explain how beach waves act: New York Times/The Environment, June 12, 1990, p. B-7.

Drisko, J. B., 1962, The sediment problem on the Indus at Tarbela: New York, Tippetts-Abbett-McCarthy-Stratton Memo-report, 22 p.

Dunne, T., and Leopold, L. B., 1981, Flood and sedimentation hazards in the Toutle and Cowlitz River System as a result of Mount St. Helens eruptions: Federal Emergency Management Agency, 92 p.

Eckel, E. B., 1951, Research needs in engineering geology; Presidental address: Denver, Proceedings of the Colorado Science Society, 11 p.

—— , ed., 1968, Nevada test site; Studies of geology and hydrology: Geological Society of America Memoir 110, 284 p.

ENR, 1963, A-blast asked for route excavation: Engineering News Record, v. 171, no. 25, p. 52.

ERA, 1952–1953, Underground explosion test program; Granite, limestone, and sandstone: Minneapolis, Minnesota, Engineering Research Associates, v. 1 and 2.

Erdman, C. E., 1943, Application of geology to the principles of war: Geological Society of America Bulletin, v. 54, p. 1169–1194.

Erskine, C. F., 1973, Landslides in the vicinity of the Fort Randall Reservoir, South Dakota: U.S. Geological Survey Professional Paper 675, 63 p.

Fisk, H. N., 1944, Geological investigations of the alluvial valley of the lower Mississippi River: Vicksburg, Mississippi, Mississippi River Commission, 78 p.

Flawn, P. T., 1970, Environmental geology; Conservation, land-use planning, and resource management: New York, Harper and Row, 298 p.

Foose, R. M., 1953, Groundwater behavior in the Hershey Valley, Pennsylvania: Geological Society of America Bulletin, v. 64, p. 623–645.

—— , 1967, Sinkhole formation by groundwater withdrawal, Far West Rand, South Africa: Science, v. 157, no. 3792, p. 1045–1048.

—— , 1969, Mine dewatering and recharge in carbonate rocks near Hershey, Pennsylvania, *in* Kiersch, G. A., and Cleaves, A. B., eds., Legal aspects of geology in engineering practice: Geological Society of America Engineering Geology Case Histories 7, p. 45–60.

Foose, R. M., and Humphreville, J. A., 1979, Engineering geological approach to foundations in karst terrain: Association of Engineering Geologists Bulletin, v. 16, no. 3, p. 353–381.

Foxworthy, B. L., and Hill, M., 1982, Volcanic eruptions of 1980 at Mount St. Helens: U.S. Geological Survey Professional Paper 1249, 122 p.

Fry, A. S, 1950, Sedimentation in reservoirs, *in* Trask, P. D., ed., Applied sedimentation: New York, John Wiley and Sons, p. 347–363.

Gatewood, C. E., 1963, Impact ahead: GeoTimes, v. 7, no. 5, p. 8–12.

Gilbert, G. K., 1904, Domes and dome structures of the High Sierra: Geological Society of America Bulletin, v. 15, p. 29–36.

Green, J., 1962, Geology of the lunar base: Los Angeles, California, North American Aviation Co., Space and Information Division Report SID 68-58, 127 p.

Griggs, D., and Teller, E., 1956, Deep underground test shots: Livermore, University of California Research Livermore Report 4659, 8 p.

Handin, J. R., Judd, W. R., and Pincus, H. J., 1981, Rock-mechanics research requirements for resource recovery, construction, and earthquake-hazard reduction: Washington, D.C., National Academy Press, 196 p.

Hasan, S. E., Moberly, R. L., and Caoile, J. A., 1988, Geology of greater Kansas

City, Missouri and Kansas, United States: Association of Engineering Geologists Bulletin, v. 25, no. 3, p. 281–341.

Haskell, N. R., 1957, An estimate of the maximum range of detectability of seismic signals: Bedford, Massachusetts, Terrestrial Science Laboratory Air Force Surveys in Geophysics 87, 42 p.

Hatheway, A. W., and McClure, C. R., eds., 1979, Geology in the siting of nuclear power plants: Geological Society of America Reviews in Engineering Geology, v. 4, 245 p.

Hatheway, A. W. (chmn), Erwin, J. W., Heim, G. E., Mathewson, C. C., and Steele, S. G., 1990, Geology, land use, and the environment, *in* Wylie, P., ed., Solid-earth Sciences Study: Washington, D.C., National Research Council, Board of Earth Sciences and Resources (in press).

Herd, D. G., and Columbia Camite de Estudios Vulcanologicos, 1986, The 1985 Ruiz volcano disaster: EOS Transactions of the American Geophysical Union, v. 67, no. 19, p. 458–460.

Heuer, R. E., 1977, Feasibility of tunneling for proposed regional core rapid transit line, Los Angeles, California: Report prepared for Southern California Rapid Transit District, March, 21 p.

Hodgson, J. H., 1964, Earthquakes and earth structures: Englewood Cliffs, New Jersey, Prentice Hall Inc., 161 p.

Holmsen, P., 1953, Landslips in Norwegian quick-clays: Geotechnique, v. 5, no. 3, p. 187–194.

Holzhausen, G. R., 1989, Origin of sheet structure; Morphology and boundary conditions: Engineering Geology, v. 27, no. 1–4 Special Issue, p. 225–278.

Houser, F. N., and Eckel, E. B., 1962, Induced subsidence by underground nuclear explosions: U.S. Geological Survey Professiona Paper 450-C, no. 66, p. 17–18.

Housner, G. W., chairman, 1987, Confronting natural disasters: Washington, D.C., National Academy Press, 60 p.

——, chairman, 1989, Reducing disaster's toll: Washington, D.C., National Academy Press, 40 p.

Howard, C., 1982, San Luis Dam slide: Association of Engineering Geologists Newsletter, v. 25, no. 1, p. 17–19.

Hunt, C. B., 1950, Military geology, *in* Paige, S., ed., Applications of geology to engineering practice: Geological Society of America Berkey Volume, p. 295–327.

Hvorslev, J., 1949, Subsurface exploration and sampling of soil for civil engineering purposes: Vicksburg, Mississippi, U.S. Army Corps of Engineers Waterways Experiment Station, 216 p.

IGC, 1984, Geological excursion 030, Tajik Soviet Socialist Republik: Moscow, International Geological Congress, p. 15–20.

Jaeger, C., 1982, Remarks *in* book review *of* 'Geologie Appliquee au Genie Civil': Engineering Geology, v. 19, no. 1, p. 66–67.

James, L. B., and Kiersch, G. A., 1988, Reservoirs, *in* Jansen, R. B., ed., Advanced dam engineering for design, construction, and rehabilitation: New York, Van Nostrand Reinhold, p. 722–748.

Jansen, R. B., ed., 1988, Advanced dam engineering for design, construction, and rehabilitation: New York, Van Nostrand Reinhold, 797 p.

Jones, F. O., Embody, D. R., and Peterson, W. L., 1961, Landslides along Columbia River Valley, northeastern Washington: U.S. Geological Survey Professional Paper 367, 93 p.

Judd, W. R., ed.,1964, State of stress in Earth's crust: Amsterdam, Elsevier Publishing Co., 733 p.

——, ed, 1974, Seismic effects of reservoir loading: Amsterdam, Engineering Geology, v. 8, no. 1–2, 212 p.

Kay, C. A., 1968, Geology and our cities: New York Academy of Sciences Transactions, series 2, v. 30, no. 8, p. 1045–1051.

Keil, K., 1954, Ingenieurgeologie und geotechnite: Halle, Seale, Wilhelm Knapp Verlag, 1132 p.

Kent, L. E., 1952, Geology as an essential aid in civil engineering: South African Institute of Civil Engineers Transactions, v. 2, no. 8, p. 201–215.

Kerr, P. F.,1963, Quick clay: Scientific American, v. 205, no. 5, p. 132–142.

Keville, F. M., and Sutcliffe, H., 1983, MBTA red line extension northwest; The influence of geology on alignment and grade, *in* Sutcliffe, H., and Wilson, J. W., eds., Proceedings of the 1983 Rapid Excavation and Tunneling Conference, Chicago: New York, Society of Mining Engineers and American Institute of Mining and Metallurgical Engineers, v. 1, p. 62–71.

Kiersch, G. A., 1949, Underground space for American industry: Mining Enginering, v. 1, no. 6, p. 20–25.

——, 1951, Engineering geology principles of subterranean installations: Economic Geology, v. 46, no. 2, p. 208–222.

——, 1955, Engineering geology: Golden, Colorado School of Mines Quarterly, v. 50, no. 3, 123 p.

——, 1957, Engineering Geology—an aid in estimating contract bids for heavy construction: Geological Society of America Bulletin, v. 68, p. 1831–1832.

——, 1958, Quantitative trends and research needs in engineering geology: Geological Society of America Bulletin, v. 69, p. 1597.

——, 1961, Some research needs in geoscience phenomena and techniques, with emphasis on application "Global System Engineering Air Force": Arctic Institute of North America report under contract, 97 p.

——, 1964, Academic trends in engineering geology: Journal of Geological Education, v. 12, no. 1, p. 14–21.

——, 1986, Response, engineering geology heritage: Association of Engineering Geologists Bulletin, v. 23, no. 1, p. 1–4.

——, 1988, Reservoirs; Reservoir problems other than leakage, *in* Jansen, R. B., ed., Advanced dam engineering: New York, Van Nostrand Reinhold, p. 729–750.

Kiersch, G. A., McGill, J. T., and Mann, J. F., 1957, Teaching aids and allied materials in engineering geology: Committee on Teaching Aids, Geological Society of America Division on Engineering Geology, 36 p.

Kieslinger, A., 1958, Restspannung und entspannung im gestein: Geologie und Bauwesen, v. 24, no. 2, p. 97–112.

——, 1960, Residual stress and relaxation in rocks: 21st International Geological Congess, Copenhagen, part 18, p. 270–276.

Kindle, E. M., and Taylor, F. B., 1913, Niagara folio: U.S. Geological Survey Geologic Atlas, Folio 190, 181 p.

Kockelman, W. J., 1986, Some techniques for reducing landslide hazards: Association of Engineering Geology Bulletin, v. 23, no. 1, p. 29–52.

Kravits, S. J., Sainato, A., and Finfinger, G. L., 1987, Accurate directional borehole drilling; a case study at the Navajo Dam, New Mexico: U.S. Bureau of Mines Report of Investigation 9102, 25 p.

Krinitzsky, E. L., and Slemmons, D. B., eds., 1990, Neotectonics in earthquake evaluation: Boulder, Colorado, Geological Society of America Reviews in Engineering Geology 8, 160 p.

Krynine, D. P., and Judd, W. R., 1957, Principles of engineering geology and geotectnics: New York, McGraw-Hill, 699 p.

Lane, K. S., ed., 1971, Underground rock chambers, *in* Proceedings of a Symposium, Phoenix, Arizona, January 13–14: American Society of Civil Engineers, 600 p.

Legget, R. F., 1939, Geology and engineering: New York, McGraw-Hill, 650 p.; 2nd ed., 1962, 857 p.; 3rd ed. with P. F. Karrow, 1983.

——, 1973, Cities and geology: New York, McGraw-Hill, 578 p.

LeRoy, L. W., and LeRoy, D. O., 1977, Subsurface geology; Petrology, mining, construction, 4th ed.: Golden, Colorado School of Mines, p. 5–6 and 594–618.

Mathewson, C. C., 1981, Engineering geology: Columbus, Ohio, Charles E. Merrill Publishing, 404 p.

McClure, C. R., and Hatheway, A. W., 1979, An overview of nuclear power plant siting and licensing, *in* Hatheway, A. W., and McClure, C. R., eds., Geology in the siting of nuclear plants: Geological Society of America Reviews in Engineering Geology, v. 4, p. 3–12.

McCutchen, W. R., 1949, The behavior of rocks and rock masses in relation to military geology: Golden, Colorado School of Mines Quarterly, v. 44, no. 1, 76 p.

McGill, J., 1964, The growing importance of urban geology: U.S. Geological Survey Circular 487, 4 p.

Mead, W. J., 1941, Engineering geology, *in* Berkey, C. P., ed., Geology, 1888–1938; 50th Anniversary volume: Geological Society of America, p. 573–578.

Meade, R. B., 1982, The evidence for reservoir-induced macroearthquakes, *in* State of the art for assessing earthquake hazards in the United States: Vicksburg, Mississippi, U.S. Army Corps of Engineers Waterways Experiment Station Report 19, paper S-73-1, 188 p.

Meyer, D. F., 1983, Metropolitan Boston Transit Authority red line project, Harvard Square to Porter Square tunnels, Cambridge, Massachusetts, *in* Sutcliffe, H., and Wilson, J. W., eds., Proceedings of Rapid Excavation and Tunneling Conference, volume 2: Society of Mining Engineers and American Institute of Mining and Metallurgical Engineers, p. 941–951.

MGU, 1945, The military geology unit: U.S. Geological Survey and U.S. Army Corps of Engineers, 22 p.

Milne, W. G., ed., 1976, Induced seismicity: Amsterdam, Engineering Geology, v. 10, no. 2–4, 385 p.

Mitchell, J. K., 1986, Practical problems from surprising soil behavior; 20th Karl Terzaghi lecture: American Society of Civil Engineers Journal of Geotechnical Engineers, v. 112, p. 259–289.

Mitchell, J. K., and Katti, R. K., 1981, Soil improvement, *in* State of the art report: Proceedings 10th International Conference on Soil Mechanics and Foundation Engineering, Stockholm, p. 261–317.

Morfeldt, C. O., 1983, The influence of engineering geological data on the design of underground structures, and on the selection of construction methods, *in* General report, Theme III, International Symposium on Engineering Geology and Underground Construction: Lisbon, International Associaton of Engineering Geology, p. III–1, to III–24.

Moye, D. G., 1955, Engineering geology for the Snowy Mountain scheme: Institute of Engineers of Australia Journal, v. 27, p. 281–299.

Mullineau, D. R., 1981, Hazards from volcanic eruptions, *in* Hays, W. W., ed., Facing geologic and hydrologic hazards; Earth science considerations: U.S. Geological Survey Professional Paper 1240-B, p. 86–101.

Neilson, J. M., Snelgrove, A. K., and Van Pelt, J. R., 1960, Curricula and professional aspects of geological engineering: Economic Geology, v. 55, p. 1048–1059.

Nichols, T. C., and Varnes, D. J., 1984, Residual stress, rock, *in* Finkl, C. W., ed., The encyclopedia of applied geology: New York, Van Nostrand Reinhold, p. 461–465.

Niles, W. H., 1871, Peculiar phenomena observed in quarrying: Proceedings of the Boston Society of Natural History, v. 14, January 4, 1871, p. 80–87; v. 16, p. 41–43 (1873).

NYSEG, 1978, Geology and seismology, New Haven nuclear plant site: New York State Gas and Electric Company Preliminary Safety Analysis Report, v. 4, 175 p.

Oakschott, G. B., 1985, Contributions of the state geological surveys; California as a case history, *in* Drake, E. T. and Jordan, W. M., eds., Geologists and ideas: Boulder, Colorado, Geological Society of American Centennial Special Volume 1, p. 323–335.

O'Sullivan, J. J., ed., 1961, Protective construction in a nuclear age: New York, The Macmillan Co., 2 vols., 836 p.

Paige, S., chairman, 1950, Applications of geology to engineering practice: Geological Society of American Berkey Volume, 327 p.

Paige, S., and Rhoades, R., 1948, Report of committee on research in engineering geology: Economic Geology, v. 43, p. 313–323.

Park, A. S., 1953, Snowy Mountain project: Compressed Air Magazine, v. 58, no. 7, p. 180–185.

Quirke, T. T., 1947, Geology for engineers and manual: Champaign, Illinois, Stipes Publishing Co., 182 p.

Rahn, P. H., 1986, Engineering geology; An environmental approach: Amsterdam, Elsevier Publishing Co., 590 p.

Reuter, F., Klengel, K. J., and Pasek, J., 1980, Ingenieurgeologie: Leipzig, VEB Deutscher Verlag fur Grundstoffindustrie, 451 p.

Rhoades, R., and Irwin, W. H., 1947, What the engineering geologist does: Engineering News Record, v. 139, no. 16, p. 90–94.

Richey, J. E., 1964, Elements of engineering geology: London, Pitman and Sons, Ltd., 153 p.

Ries, H., and Watson, T. L., 1914, Engineering geology: New York, John Wiley and Sons, Inc., 679 p; 5th ed., 1936, 750 p.

Robinson, A. R., 1973, Sediment: Our greatest pollutant, *in* Tank, R. W., ed., Focus on environmental geology: New York, Oxford University Press, p. 186–192.

Rogers, A. M., and Lee, W.H.K., 1976, Seismic study of earthquakes in the Lake Mead, Nevada-Arizona region: Seismological Society of America Bulletin, v. 66, no. 5, p. 1657–1681.

Russell, R. D., 1950, Applications of sedimentation to naval problems, *in* Trask, P. D., ed., Applied sedimentation: New York, John Wiley and Sons, p. 656–665.

Schultz, J. R., and Cleaves, A. B., 1955, Geology in engineering: New York, John Wiley and Sons, 559 p.

Schuster, R. L., and Fleming, R. W., 1986, Economic losses and fatalities due to landslides: Association of Engineering Geologists Bulletin, v. 23, no. 1, p. 11–28.

Shlemon, R. J., 1985, Application of soil-stratigraphy techniques in engineering geology: Association of Engineering Geologists Bulletin, v. 22, no. 2, p. 129–142.

Simpson, D. W., and Negmatullaveu, S. K., 1981, Induced seismicity at Nurek Reservoir, Tadjikistan, USSR: Seismological Society of America Bulletin, v. 71, no. 5, p. 1561–1586.

Sitmar, N., ed., 1983, Goals for basic research in engineering geology; Report of a workshop: Berkeley, University of California Department of Civil Engineering and National Science Foundation Geotechnical Engineering Branch, 25 p.

Skempton, A. W., 1964, Long-term stability of clay slopes: Geotechnique, v. 14, p. 77–102.

Slemmons, D. B., 1982, Lessons learned from earthquake hazard and risk assessments: Geological Society of America Abstracts with Programs, v. 14, p. 619.

Slingerland, R. L., and Voight, B., 1979, Occurrences, properties, and predictive models of landslide-generated water waves, *in* Voight, B., ed., Rockslides and avalanches: Amsterdam, Elsevier Publishing Co., p. 380–383.

Smith, T. R., and Black, L. D., 1946, German geography; War work and present: Geographic Review, v. 36, p. 398–408.

Smith, W. O., and others, 1960, Comprehensive survey of sedimentation in Lake Mead, 1948–49: U.S. Geological Survey Professional Paper 295, 248 p.

Soboleva, O. V., and Mamadaliev, U. A., 1976, The influence of Nurek Reservoir on local earthquake activity: Amsterdam, Engineering Geology, v. 10, p. 293–305.

Stini, J., 1922, Technische geologie: Stuttgart, F. Encke, 789 p.

Stone and Webster, 1971, Final analysis report, Fitzpatrick Nuclear Station, Scriba, New York: Power Authority of the State of New York.

Tank, R. W., ed., 1973, Focus on environmental geology: New York, Oxford Univresity Press, 474 p.; 2nd ed., 1976, 538 p.

Taylor, C. L., and Conwell, F. R., 1981, BART; Influence of geology on construction conditions and costs: Association of Engineering Geologists Bulletin, v. 28, no. 2, p. 195–205.

Teller, E., Talley, W. K., Higgins, G. H., and Johnson, G. W., 1968, The constructive uses of nuclear explosives: New York, McGraw-Hill, 313 p.

Terzaghi, K., 1961, Engineering geology on the job and in the classroom: Boston Society of Civil Engineers Journal, v. 38, no. 2, p. 109.

Trask, P. D., ed., 1950, Applied sedimentation: New York, John Wiley and Sons, 665 p.

——, 1959, Geological problems of dams in the United States, *in* Legraye, M., and Calembert, L., eds., Barrages et Bassins de Retenue: International Colloque l'Universite de Liege, v. 14, p. 29–42.

Trefethen, J. M., 1949, Geology for engineers: New York, Van Nostrand Co., Inc., 629 p.; 2nd ed., 1959, 632 p.

Tuttle, M. L., 1989, The 1986 gas release from Lake Nyos, Cameroon: Golden, Colorado, The Mines Magazine, v. 79, no. 1, p. 15–17.

UGMP, 1973, Urban geology master plan for California: California Division of Mines and Geology Bulletin 198, p. 51–74.

USCE, 1961, Design of underground installations in rock: U.S. Corps of Engineers EM1110-345-431, 68 p.

VanDine, D. F., 1984, Geological engineering applied to geotechnique; A review of the undergraduate university programs in Canada: Association of Engineering Geologists Bulletin, v. 31, no. 2, p. 171–178.

Varnes, D. J., 1981, Slope-stability problems of the Circum-Pacific region, *in* Halbouty, M. T., ed., Energy resources of the Pacific region: American Association of Petroleum Geologists Studies in Geology 12, p. 489–505.

Von Bülow, K., 1938, Wehrgeologie: Leipzig Quelle and Meyer, 170 p. (E.R.O. Translation T-23 by K. E. Lowe, Intelligence Branch, Office of the Chief of Engineers).

Waldron, H. H., 1967, Debris flow and erosion control problems caused by ash eruptions of Irazu Volcano, Costa Rica: U.S. Geological Survey Bulletin 1241-I, 137 p.

Wallace, R. E., chairman, 1986, Active tectonics; Studies in geophysics; National Research Council Panel on Active Tectonics: Washington, D.C., National Academy Press, 260 p.

Walters, R.C.S., 1962, Dam geology: London, Butterworth, 470 p.

Whitmore, F. C., 1950, Sedimentary materials in military geology, *in* Trask, P. D., ed., Applied sedimentation: New York, John Wiley and Sons, p. 635–665.

Williams, J. W., chairman, Mathewson, C. C., and Lee, F. T., 1985, Research activity and needs in engineering geology: U.S. Committee of International Association for Engineering Geology Research Panel, 31 p. (IAEG-files).

Wilson, L. S., 1948, Geographic training for the post-war world: A proposal: Geographic Review, v. 38, p. 575–589.

Wise, L. L., 1952, World's largest underground power plant: Engineering News Record, v. 149, no. 20, p. 31–36.

Zaruba, Q., ed., 1968, Engineering geology in country planning: Proceedings Session 12, 23rd International Geological Congress, 227 p.

Zaruba, Q., and Mencl, V., 1976, Engineering geology: Amsterdam, Elsevier Publishing Co., 493 p.

Manuscript Accepted by the Society September 10, 1990

ACKNOWLEDGMENTS

The author has gained insight on the scope and heritage of engineering geology from ideas and comments of colleagues and associates over five decades. These included the late P. Reiche, J. C. Haff, L. H. Henderson, C. P. Holdredge, T. F. Thompson, P. D. Trask, T. W. Fluhr, and E. B. Eckel, as well as Col. L. De Goes, L. B. James, S. S. Philbrick, V. Mencl, and Q. Zaruba. This summary has been strengthened by the reviews and comments on an early draft by Ellis L. Krinitzsky, Jeffrey R. Keaton, and Shailer S. Philbrick, and subsequent editorial review of Allison R. Palmer, for which I am very appreciative.

Spectacular karstic terrain of Laos and Thailand, showing the pinnacle remnants of Rat Buri limestones with their network of cavernous openings and sink-like structures. Oblique view of occurrence northwest of Vientiane, Laos, in the region of the Mekong River tributaries. (Photo by G. A. Kiersch, 1968.)

Printed in U.S.A.

Geological Society of America
Centennial Special Volume 3
1991

Chapter 3

Research efforts, governments of North America

Fitzhugh T. Lee
U.S. Geological Survey, Box 25046, Denver Federal Center, Denver, Colorado 80225
John S. Scott
Geological Survey of Canada, 580 Booth Street, Ottawa, Ontario K1A 0E9, Canada
Mariano Ruiz-Vazquez
School of Geology, University of Mexico, Mexico City, D.F.
Guillermo P. Salas *(Deceased)*
Sierra Gorda No. 12, 11010 Mexico, D.F.
Jorge I. Maycotte
Sociedad Geologicia Mexicana, Seccion de Ingenieria Geologica, Torres Bodet 176 (Cipres), 06400 Mexico, D.F.

ENGINEERING-GEOLOGICAL RESEARCH DEVELOPMENT IN THE UNITED STATES GOVERNMENT

Fitzhugh T. Lee

INTRODUCTION

This section presents a synopsis of the evolution, function, and distribution of engineering-geological activities in the United States government. Much of the background information was obtained from material supplied by each agency in response to a written request. Other information was obtained from colleagues and library documents. Even a casual inspection of this information reveals that federal-agency engineering-geology practice from its infancy in the 1930s to its present-day maturity has followed a course that mirrors the worldwide development of the discipline. This has come about naturally as the result of agency needs that have been driven by budget, mission, and research aims. The result has been a mélange of practical applications to various civil and mining projects supported by applied and theoretical research. Federal agencies that employ geotechnical staffs, but are predominantly regulatory (e.g., Nuclear Regulatory Commission) rather than being concerned with research or practice of engineering geology, were not included in this report.

Perhaps the highest tribute to the practice of engineering geology in government was stated long ago by Charles P. Berkey, himself a pioneer in the field: ". . .I claim a place of honor for these men who spend their lives in devising new ways of using their specialistic knowledge and experience and ingenuity for more effective public works and for the greater comfort and safety of men and women everywhere. . ." (Berkey, 1942).

U.S. GEOLOGICAL SURVEY

The U.S. Geological Survey (U.S.G.S.), Department of the Interior, has been involved in engineering geology for most of its 110-year life. In 1888, J. W. Powell, Director, began irrigation surveys, which were the first attempts at a national reclamation program and ultimately led to the establishment of the U.S. Bureau of Reclamation. From 1888 to the present, the U.S.G.S. has studied and classified the western hydrographic basins. Sites for canals and reservoirs have been identified, and construction materials have been noted. As the U.S.G.S. began to aid in building dams in the late nineteenth century, applied geology grew rapidly. Test drilling was done at some sties, but mainly to locate the depth to bedrock.

Most of the major dams built on the Colorado and other western rivers and their tributaries from 1900 to the present used sites first chosen by U.S.G.S. hydrologists, topographers, and geologists. Reports on many of these are found in U.S.G.S. Water Supply Papers and other publications. An excellent early textbook on the geology of reservoir and dam sites was written by Kirk Bryan (1929). O. E. Meinzer (1923a, b) contributed important textbooks on the occurrence of water, the measurement of hydrologic properties, and the influence of geologic structure on ground water. Geologists and hydrologists in the Water Resources Division have long been involved in engineering applications of hydrology through field and laboratory investigations. For example, a basic understanding was obtained of the fundamentals of land subsidence due to ground-water withdrawal that began in the 1920s in the San Joaquin Valley (Poland and Davis, 1969; Poland and others, 1975). In the last 20 to 30 years, U.S.G.S. scientists have joined in multidisciplinary studies of underground-weapons test effects, and of theoretical, experimen-

Lee, F. T., Scott, J. S., Ruiz-Vasquez, M., Salas, G. P., and Maycotte, J. I., 1991, Research efforts, governments of North America, *in* Kiersch, G. A., ed., The heritage of engineering geology; The first hundred years: Boulder, Colorado, Geological Society of America, Centennial Special Volume 3.

tal, and field investigations of radionuclide transmission in underground mined repositories. These studies have been particularly wide ranging at the Nevada Test Site where U.S.G.S. scientists have participated in geomechanical, geophysical, tectonic, hydrologic, and stratigraphic research that is directly applied to the siting of nuclear testing and waste-storage facilities.

Before the introduction of asphalt and concrete, the U.S.G.S. published reports in 1895 and 1896 on the geology of roads in the United States, and on road-building stones of Massachusetts. Maps of glacial deposits published between 1889 and 1915 were widely used by highway engineers in search of construction materials. In the late 1940s and early 1950s, the Engineering Geology Branch in the Geologic Division was heavily involved in highway geology, working directly with pioneers such as the Kansas Highway Department to demonstrate the usefulness of detailed geology in planning, building, and maintaining modern highways. This early work contributed greatly to the development of geologic staffs in many state highway departments. Highway geology studies led directly to research on landslides, particularly as they affect the highway engineer. U.S.G.S. personnel played key roles in preparing the widely known Highway Research Board report, "Landslides and engineering practice" (Eckel, 1958), and its successor, "Landslides: Analysis and control," published by the National Academy of Sciences (Schuster and Krizek, 1978, discussed by McClure and others, this volume). Landslide investigations have matured over the years; today they are the strongest single element of engineering geology research in the Geologic Division of the U.S.G.S.

The growth of engineering geology was a direct result of U.S.G.S. involvement in technical support of military operations in World War II. This successful application of practical information to specific military problems led to similar geologic guidance to help meet civilian construction needs. In November 1944, the Engineering Geology Branch (then Section) was established, and E. B. Eckel was appointed Chief of the Branch. Instead of focusing on specific construction sites, maps were produced and other data were obtained to provide background knowledge for any kind of future construction. Urban geologic mapping and studies of construction materials were emphasized.

In the 1950s and 1960s, many U.S.G.S. research efforts in engineering geology followed large national programs, including underground nuclear testing at the Nevada Test Site, effects of the 1964 Alaska earthquake, and the siting of nuclear-power reactors. At the same time, an increasingly large part of the program was devoted to such applied research as rock bursts in coal mines (Osterwald, 1961) and rock-mechanics studies of large tunnels (Robinson and Lee, 1964; Fig. 1). The U.S.G.S. has pioneered studies on the hazards presented by earthquakes, volcanoes, and landslides. As a result of geologic studies of Mount St. Helens, Crandell and Mullineaux (1978) predicted that this volcano would erupt in this century; it erupted on May 8, 1980. Recent geotechnical efforts have been concentrated on mining-induced subsidence and slope failure (Fig. 2), the nature of stresses in rock masses, and landslide process studies. These projects incorporate theoretical and laboratory studies as well as field measurements.

U.S. ARMY CORPS OF ENGINEERS

The Corps of Engineers, a component of the Department of the Army, has contributed to the development of engineering geology primarily through applications to large navigation, flood-control, and military-construction projects. The roots of the Corps' involvement in engineering geology started in 1911 when, with the Corps acting as the construction agency for the Panama Canal, C. W. MacDonald was hired as the first resident geologist on a major construction project (Fisher, 1981). MacDonald made substantial contributions to the understanding of landslide phenomena and remedial techniques (Fig. 3). Earlier, President Theodore Roosevelt had placed the construction of the canal under the direction of a Corps of Engineers officer, Colonel G. W. Goethals.

In 1931, the Corps of Engineers hired E. B. Burwell, Jr., as their first staff geologist. He organized and effectively teamed geologists with civil engineers as early as the Muskingum Conservation Project in the Zanesville, Ohio, District where 14 flood-control and conservation dams, levees, and other structures were completed between 1934 and 1937. The Missouri River Projects, begun in the mid-1930s, started with Fort Peck Dam and continued after World War II. Some of the world's largest dams were built along the Missouri River on some very weak rocks. The first full-face tunnel-boring machine was used on these projects to excavate many miles of openings in chalk and shale, and important advances were made in understanding the time-dependent behavior of weak rock.

In the early days of construction, the geologist's role was largely limited to water-supply development and location of construction materials. As the destructive power of weapons increased in the decades following World War II, the Corps' engineering geologists were increasingly involved in the design and construction of protective structures. There was Corps participation in Nuclear Weapons Effects Tests; Peaceful Nuclear Explosive Experiments; construction of the North American Air Defense (NORAD) underground facilities; the siting, design, and construction of the Intercontinental Ballistic Missile (ICBM) systems; and MX (missile) basing.

The district is the basic operating element of the Corps. In district offices, geologists do largely "site-specific" work on problems in planning, siting, and foundation design for major civil and military structures. Geologists in divisions and the Office of the Chief of Engineers serve as technical consultants and reviewers to the district. During the 1950s, the Corps of Engineers became the largest employer of engineering geologists in the world; in 1986, there were 231 geologists, geophysicists, and geographers on a geotechnical staff of 854.

The Corps has made significant contributions to several geotechnical areas, including stage grouting of foundation rock, in-

Figure 1. Emergency support measures for very heavy rock load or wet sections in the Harold D. Roberts Tunnel, Colorado. A. Steel-support failure caused by heavy squeezing ground necessitating use of rail spiling and temporary timber supports. B. Bulkhead installation; tunnel was filled with concrete to valves of pipes (foreground) before grouting to seal off water pressures of up to 70 kg/cm^2 at heading. Photographs by E. E. Wahlstrom.

Figure 2. Investigation of slope stability in open-pit coal mines near Sheridan, Wyoming.

situ measurement of rock properties, development of the NX-borehole camera, and methods of geophysical exploration. As more formal research efforts were initiated, the Corps established topical study facilities at the Waterways Experiment Station (Vicksburg, Mississippi), the Cold Regions Research and Engineering Laboratory (Hanover, New Hampshire), and the Coastal Engineering Research Center (Washington, D.C.).

U.S. BUREAU OF RECLAMATION

Formalized practice of engineering geology within the U.S. Bureau of Reclamation (U.S.B.R.) can be traced to the planning and design of Hoover Dam, Nevada, beginning in about 1930. A series of eminent geological consultants, including C. P. Berkey, W. J. Mead, F. L. Ransome, and Kirk Bryan, applied geological principles to engineering problems of unprecedented magnitude caused by the enormous size of that structure (Fig. 4). Berkey's influence was mainly responsible for the development of a large centralized geological group in the U.S.B.R. as the engineering geology program expanded. Prior to Hoover Dam, U.S.B.R. structures were relatively small, and attempts to incorporate site-specific geology in design were largely nonexistent. However, as the good dam sites were used, geological investigations became more important and more intense, with a corresponding growth of technical staff.

The main function of the U.S.B.R. is to provide water for the semiarid lands of the western United States. Irrigation is the primary objective, but most projects also involve collateral objectives, such as flood control and the development of power. To accomplish these objectives, the U.S.B.R. builds dams, tunnels, canals, pumping plants, powerhouses, and a variety of appurtenant structures for these major works. The demands of this engineering work require close integration of geological studies with the unique characteristics of each construction project. Typically, the geological staff participates in all phases of complex projects, from preliminary investigation and planning, to final design and construction, and to operation and maintenance. The emphasis is on quantitative solutions to geological problems.

The U.S.B.R. Engineering and Research Center in Denver, Colorado, is the focus for engineering geological studies. Field geologists are primarily responsible for gathering data, and the Division of Geology in Denver interprets and prepares the data for construction specifications. In 1987, there were 22 engineering geologists in the Denver office and another 144 at various field offices in the western United States. The Bureau's geotechnical staff also includes eleven ground-water geologists, four geophysicists, five petrographers, and ten seismologists.

Over the past 50 years, the Geotechnical Branch of the Division of Research and Laboratory Services in Denver has performed extensive laboratory and in-situ testing of soil and rock to determine their engineering properties. The staff of 31 professionals in the Geotechnical Branch includes three engineering geologists. Important geotechnical investigations on the physical properties of soil and rock have been conducted in this laboratory (e.g., Gibbs and Holland, 1960; Holtz and Gibbs, 1954; Balmer, 1953).

Geological field investigations are augmented by modern drilling methods, state-of-the-art geophysical equipment, and technical procedures developed internally and by others. The U.S.B.R. has been instrumental in developing a number of single- and multiple-purpose projects that utilize ground water as a sole source of water or in conjunction with surface water. Numerical and other modeling techniques determine well-field layout, long-term pumping influences, and other factors. The U.S.B.R. has undertaken field and laboratory research in ground-water control in slope stability, construction dewatering, salinity control, water salvage, and subsurface agricultural drainage.

TENNESSEE VALLEY AUTHORITY

The Tennessee Valley Authority (T.V.A.) was created by an act of Congress in May 1933 to accomplish certain specified activities in the drainage basin of the Tennessee River. The chief

Figure 3. View of the Culebra cut showing the Cucaracha slide, looking north, with dredges excavating slide material, February 8, 1914. Photograph by D. F. MacDonald.

Figure 4. Hoover Dam on the Colorado River near Las Vegas, Nevada. Photograph by U.S. Bureau of Reclamation.

Figure 5. Tennessee Valley Authority's Raccoon Mountain project, an underground power-plant excavation located near Chattanooga, Tennessee. Photograph by Tennessee Valley Authority.

activities were dam construction and river control for the basin; at the outset, Norris Dam was the only dam under consideration for immediate construction. The dam site and its surrounding area had already been examined by several geologists prior to the appointment of Major Edwin C. Eckel (father of Edwin B. Eckel) to the position of chief geologist in August 1933. Major Eckel organized a staff of engineering geologists to work on all phases of the Norris Dam project relating to geology. It is noteworthy that the T.V.A. was the first governmental agency to organize and maintain a staff of engineering geologists, as opposed to relying heavily on outside consultants. Since that early period, the number of full-time, professional engineering geologists employed by the T.V.A. has varied depending on the number of projects under construction at any given time. As many as 15 engineering geologists were on the staff in the late 1970s when the T.V.A. had a number of hydroelectric and nuclear power plants under construction (Fig. 5). There are currently nine geologists involved in engineering geology for the T.V.A.

During the first three decades of the T.V.A., their geologists followed a procedure for site investigations that included a study of the site and its surface geology, preliminary drilling to determine the general nature and condition of the foundation, and geologic mapping of as much as several square miles around the site. If, based on these findings, a decision was made to build on the site, a geologist was assigned to the project. His job was to log and interpret the drill core, map the foundation in detail as it was excavated, examine and describe the effects of any solution weathering, supervise grouting, locate sources of construction materials for the project, and note any effects construction activities might have on local water supplies.

During the early to mid-1960s, the engineering geology staff began utilizing instrumentation that would enable examination of foundation conditions in situ. Much of the Tennessee drainage basin is underlain by carbonate rocks with the usual weathering features that include an irregular bedrock-overburden interface, solution channels, and cavities. Some of the rock units also include very thin seams of clay that provide natural surfaces of weakness. In order to accurately determine the presence of these clay seams and to examine the effects of solution weathering in situ, a borehole TV camera was acquired and found to be a significant aid in foundation investigations. Borehole geophysical logging was added to the staff's inventory of exploration methods during the late 1960s and 1970s, and instrumentation included a small portable unit that measured variations in natural radiation (gamma) and electrical resistance associated with changes in lithology. The latest surface methods employed include electromagnetic and self-potential surveys.

Foundation investigations have long since been completed for the two nuclear-power plants currently under construction, as well as the older hydroelectric and coal-fired steam plants. Much of the present work is related to environmental engineering. This work includes projects such as determining the nature and extent of solution channels in limestone beneath an ash-disposal pond, and installing ground-water monitoring wells at the T.V.A.'s coal-fired steam plants. Current work includes dam-safety evaluations, and operation and maintenance investigations at existing power plants. The T.V.A.'s geological staff also provides geophysical expertise to other federal agencies, such as the Bureau of Reclamation and the Department of Energy.

As an outgrowth of T.V.A.'s involvement in a nuclear-power program, its engineering-geology staff maintains a network of 20 seismic stations in the region. This activity is coordinated with similar programs operated by the Georgia Institute of Technology, the University of Kentucky, the Virginia Polytechnic Institute and State University, and the Tennessee Earthquake Information Center at Memphis State University.

U.S. BUREAU OF MINES

The U.S. Bureau of Mines (U.S.B.M.) was formed in 1910 (formerly part of the U.S. Geological Survey) and charged with guiding research in mine safety, mining techniques, and metallurgy. The principal geotechnical expertise of the U.S.B.M. lies in rock mechanics, geophysics, studies of geology related to the design and safety aspects of subsurface and surface mines, and tunneling operations. Additional areas of interest include the disposal of mine wastes, the stability of spoil piles, and those complementary areas that contribute to the overall program of mining safety and economy.

In the 1940s, the U.S.B.M. began a program that was headquartered at the Applied Physics Laboratory in College Park, Maryland, to establish a more scientific approach to mining. The work was complicated by (1) lack of an adequate theory describing the stress distribution around a complex system of openings, (2) lack of standardized tests for determining the physical properties of mine rock, and (3) lack of procedures and instrumentation for making in-situ tests to validate designs. Through internal research, the U.S.B.M., over the next several decades, did cooperative studies with mining companies, provided effective public education, and played the lead role in changing U.S. mining practice from a largely trial and error procedure to a quantifiable and more predictable operation. Through the leadership of Leonard Obert, W. I. Duvall, and, later, V. E. Hooker, important advances were made in the understanding of underground rock behavior. Many such contributions were published in the U.S.B.M. Report of Investigations (RI) series. For example, Merrill and Morgan (1958) described a method for determining mine-roof strength, Obert (1940) measured pressures on rock pillars, and Obert and others (1946) described standardized tests for determining physical properties of mine rocks. A comprehensive textbook on the theory and practice of the design of underground openings in rock was written by Obert and Duvall (1967).

In the mid-1960s, the rock-mechanics personnel at College Park were transferred to the Denver Research Center. They investigated the use of subaudible noises for the prediction of rock bursts, in-situ rock stress, and anisotropic properties of rock.

The Spokane Mining Research Center in Washington State was opened in 1951. Early research at Spokane involved primarily regional problems, such as those of the deep lead and silver mines of the Coeur d'Alene Mining District in Idaho, and the underground phosphate mines of southern Idaho and Montana. Technical problems at these locations included ground support, mine waste rock disposal, and rock bursts produced by high stress levels (Fig. 6). In 1973, the facility's mission was expanded to include coal, as well as metal- and nonmetal-mining research. Projects included ground control in coal mines, rapid excavation of mine openings, improved productivity, and reclamation of coal-waste piles. Geotechnical studies at the Spokane Mining Research Center currently emphasize laboratory and field behavior of bedded rock and the geomechanics of several types of mined openings, as well as the slope stability of tailings dams and spoil piles.

The Twin Cities Research Center in Minneapolis, Minnesota, was opened in 1959, although its roots date back to 1915 when a metallurgical research station was authorized by Congress. Beginning in 1959, rock mechanics research focused on rock behavior, with emphasis on fragmentation by both conventional and novel techniques, including water-jet fragmentation, and systems employing lasers and microwaves. Fundamental studies of techniques to measure the geologic and hydraulic parameters of rock formations are being carried out to mitigate the problems of mine-water inflows. The staff of 160 people at the Twin Cities Center reflects the broad scope of a research program that includes mining engineering (which predominates), metallurgy, chemistry, geology, physics, geophysics, and mechanical engineering.

The Pittsburgh Research Center for many years has engaged in detailed surface and underground geotechnical studies of coal-mine problems. Much of the work has been concentrated on the effects of mining the Pittsburgh coal bed, consisting of roof stability and methane emission studies. For example, McCullough and Deul (1973) reported on the relation of clay seams and cleats in the coal to mine-gas and mine-support problems. More recently, geologists and engineers of the Roof Support Group have made subsidence studies for mines in Pennsylvania, Ohio, West Virginia, and Illinois, and have determined the influence of various geologic parameters on subsidence-prediction models.

SOIL CONSERVATION SERVICE

The U.S. Department of Agriculture's Soil Conservation Service (S.C.S.) was created in 1935 to develop and maintain a

Figure 6. Underground conditions caused by a rock burst in the Coeur d'Alene Mining District, Idaho. Photograph by U.S. Bureau of Mines.

national program of soil and water conservation. The agency had existed for two years as the Soil Erosion Service in the Department of the Interior. Hugh H. Bennett, the first chief of the S.C.S., was a graduate geologist who established erosion-control projects nationwide.

In the 1930s, the S.C.S. began studies to investigate conditions and processes of sedimentation on stream channels and valley lands in relation to accelerated erosion of tributary upland areas; reservoir sedimentation in representative sections of the country to determine the extent of damage and evaluate the effects of reservoir, climate, and watershed factors on the rate of sedimentation; the fundamental mechanics of entrainment and transport of sediment; and bed-load movement in natural streams by direct measurement. S.C.S. geologists C. B. Brown, L. C. Gottschalk, and G. M. Brune were pioneers in the field of sedimentation research and contributed important literature on the subject. In recent years, the S.C.S. has done channel-geomorphology studies to extend theories and correlations into applications of channel modifications for flood reduction and land drainage. They also have made predictions of channel responses to treatment.

Since 1950, S.C.S. geologists have been involved in planning, investigating, and designing of small- and intermediate-sized dams for retarding floodwater and storing water for other uses. Significant contributions from dam-site investigations include improved field-investigation techniques, and improved engineering and standards for dams. S.C.S. geologists have assisted in the construction of approximately 25,000 dams and have provided a greater understanding of the place of geology in soil- and water-conservation work. For example, the technique of using filtered floodwater to recharge irrigation aquifers by means of large-diameter injection wells was an S.C.S. innovation.

Basic soil classification and index tests were begun by the S.C.S. in the 1940s. Since then, geotechnical testing has been expanded to include tests ranging from basic to complex. Test results are used in analyses to design structures and assess the suitability of soil as a building or foundation material. Two significant contributions are in the identification and treatment of dispersive clay soils, and the development of filters for control of leakage and seepage from embankment dams.

Today, the S.C.S. operates under the National Conservation Act of 1977 and provides technical assistance to land users in almost 3,000 conservation districts and to various federal programs. Also, the S.C.S. aids 200 multicounty resource-conservation areas, leads the federal part of the National Cooperative Soil Survey, and runs the Great Plains Conservation Program. Through these programs, the S.C.S. has continuing involvement in landslides, grouting, subsidence, rock mechanics, mine reclamation, and flood-plain studies. The geotechnical staff includes 65 engineering geologists and is located in those geographic areas where maximum assistance can be provided to meet mandated conservation goals.

Figure 7. Road-building on national forests in southern California in the mid-1920s. Photograph by U.S. Forest Service.

U.S. FOREST SERVICE

The first use of engineering geologists in the Forest Service (Department of Agriculture) began in the late 1950s. Until that time, the work that would normally be accomplished by this discipline was done by other earth scientists such as hydrologists and soil scientists. The number of engineering geologists and geotechnical engineers in the Forest Service increased from less than five in 1960 to approximately 100 in 1986, most of whom are located in the western part of the United States. Because of the nature of the geology (extremely sensitive) and terrain (very steep) on much of the land administered by the Forest Service, engineering geologists are a critical element of the land-management team. The two major activities requiring these skills are timber harvesting and road building. The Forest Service has approximately 77 million hectares under its administration, most of which are covered with timber, and approximately 515,000 km of roads, many of which are unsurfaced single laned with steep grades and tight radii (Fig. 7).

Engineering geologists in the Forest Service are recognized worldwide as leaders in analyzing, predicting, and solving stability problems associated with land-disturbing activities. They have been innovators and pioneers in developing and using new methods, procedures, equipment, and materials applied to landslides, erosion, subsidence, and soil and rock strength. Some of the major geotechnical accomplishments include the first use of geotextiles (such as woven plastics) for retaining structures; development and use of prefabricated drain systems, many biotechnical vegetation methods for improving slope stability and erosion control, the unified rock classification system now used throughout the profession, and methods for improving marginal aggregates for road building.

The research arm of the Forest Service has undertaken original studies of land stability and erosion, such as an in-depth evaluation of the effects of tree roots on slope stability (Ziemer, 1981). As a result of their demonstrated expertise in analyzing and solving stability problems in steep, mountainous terrain, individual engineering geologists have been called on for assistance by other land-managing government agencies not only in the United States but by other countries such as Nepal and Peru.

Although the number of engineering geologists in the Forest Service has declined recently, their contributions continue to be invaluable to land managers. The Forest Service is presently involved in developing comprehensive forest land-management plans, and an important aspect of these plans is to minimize adverse environmental damage to the resources. Engineering geologists provide input to these plans and will be a part of the specialist groups involved with the implementation of the final plans.

The Role and Development of Engineering Geology in the Geological Survey of Canada*

John S. Scott

INTRODUCTION

Within democratic political systems of the western world, government organizations that serve the public are dynamic phenomena subject to the ever-changing economic, social, and political processes attendant upon society. Thus, in examining the role and development of a specific branch of applied science within one such government organization, it is useful to begin with a historical overview of both the organization itself and its primary mission.

In Canadian history, the Geological Survey of Canada occupies a unique place as the first science service in the country and as one of the government's oldest organizations. In fact, among geological surveys of the world, the Geological Survey of Canada ranks second only to the British Geological Survey in terms of length of service.

In 1841, the Legislature of the Province of Canada passed a resolution "that a sum not exceeding one thousand five hundred pounds sterling be granted to her Majesty to defray the probable expense in causing a Geological Survey of Province to be made." Thus, the Geological Survey of Canada was founded in the following year, 1842, by funding what was then envisaged by some as a short-term activity that would stimulate the mining industry in the colony and thereby contribute to enrichment of Provincial coffers. At that time no one could have imagined that the "Survey" would still be in progress nearly 150 years later.

For convenience the Survey's history can be assigned to three periods, designated as "Classic," "Growth and Decline," and "Modern." Each period contains distinct events of a scientific, economic, or administrative nature that bear directly on the conduct of engineering geological activities within the organization.

THE CLASSIC PERIOD

The Classic Period of the Geological Survey begins in 1842 with the appointment of William Edmond Logan as its first director. Logan was born in Montreal in 1798 of Scottish parentage and received his early education there and later at Edinburgh. Throughout his formative years he had developed a keen interest in chemistry, mineralogy, and geology. His association with family business interests in Wales provided an opportunity to pursue these interests, particularly geology, through study and detailed mapping of the coal measures of Wales. His work soon became recognized by the scientific community of Great Britain for its high standard of quality; this led directly to his election, in 1837, as a Fellow of the Geological Society. The wisdom of those scientists who recognized Logan's potential and recommended his appointment, and the equal wisdom of those who appointed him, was more than vindicated in the years that followed. During his 27 years as Director of the Geological Survey, Logan accumulated a truly remarkable record of achievements that are documented in his annual reports and in the publication in 1863 of his magnificent volume of 983 pages, *Geology of Canada*. His work was recognized internationally through the award of numerous medals and honors, including a knighthood bestowed upon him in 1856 by Queen Victoria. Yet for all of these achievements and honors, which reflect a man of scientific gift and genius throughout his entire career, Logan was a man who can best be described in his own words as one whose "whole connection with Geology is of a practical character."

In those early days the Survey was, in many ways, the Director, and it was the outstanding character of the man and the high standards of service to science and the public he established that shaped the Survey of his time. This tradition has persisted throughout the subsequent history of the Survey.

From 1842 until just after Confederation in 1867, the Geological Survey existed as a form of crown corporation that reported annually to the Secretary of State for the Provinces. While this arrangement provided the institution with a measure of autonomy and independence even greater than most departments of government existing then or now, it provided no assurance of continuity of funding from government sources. Funding was provided beyond the initial Act of Appropriation by a series of successive acts that varied in duration for up to 5 years. On more than one occasion the passage of these acts was due in no small measure to Logan's ability to convince the legislators by all fair means possible of the value of geology in the service of the nation.

In pre-Confederation days the Province of Canada consisted only of the southern parts of present-day Ontario and Quebec; an area relatively small in comparison with Canada's present area of 9.97 million km^2, but large by any standard and, then, devoid of any significant transportation system except that afforded by natural waterways. It was in this largely undeveloped land, without ready access or even rudimentary topographic maps, that Logan and his small staff undertook their pioneer work of Canada's first geological survey from their headquarters in Montreal.

In the conduct of their work, Logan and his assistants gave primary attention to mapping and systematic classification of the geology of Canada, including observations and interpretations that would lead to the discovery of useful mineral commodities. These included not only metalliferous ore but also fossil fuels and a wide variety of geologic materials of value for building and industrial uses. Annual reports of progress of the work of the Geological Survey and the summary volume of 1863 served both to foster the development of a mineral industry in Canada and to heighten awareness of the geological expertise resident within the staff members of the Survey.

*Geological Survey of Canada Contribution No. 19187.

Figure 8. Victoria Bridge, Montreal, circa 1863. Public Archives Canada/PA 126017.

The Victoria Bridge

Logan's geological advice was sought regarding the design of piers for the Victoria Bridge (Fig. 8) over the St. Lawrence River. This bridge, constructed between 1854 and 1859, was the earliest of the iron-girder bridges built in Canada. Its main span of 101 m was never surpassed in Canada for bridges of its type. Thus, in the earliest years of the Geological Survey, a pattern of service was established that, although placing primary emphasis on geological surveys in support of mineral development, included the wider application of geological knowledge to all forms of economic development—including engineering works. This initial pattern has persisted throughout the history of the Geological Survey and can be readily identified in the operational objectives of the present-day Geological Survey of Canada.

The Canadian Pacific Railway

Confederation of the provinces of Canada in 1867 brought with it significant changes both in administrative arrangements for the Survey and in a vastly expanded geographic area of operations. The provinces that joined in Confederation comprised those of Canada East and Canada West in the central region and Nova Scotia and New Brunswick in the Atlantic region. Within another four years the original four provinces had been joined by Manitoba and British Columbia. One of the prime conditions of the western provinces, particularly British Columbia, for entering confederation was the construction of a railway that would link the seaboard of British Columbia with the railway system of eastern Canada. At that time a rail line extended only as far west as the eastern end of Lake Nipigon, a point 2,500 miles east of the coast of British Columbia. This major commitment of the government to construct a transcontinental railway had a significant impact on the activities of the Geological Survey from the early 1870s until the mid-1880s.

Selection of potential routes for the railway was principally the task of engineers employed by the railway company. However, final selection of the route was made in consideration of both engineering aspects and the economic potential of the land through which the railway would pass. Accordingly, the Geological Survey was instructed to carry out regional surveys in large tracts of the Cordilleran and Prairie regions that would be of specific value in the planning and construction of the railroad (Zaslow, 1975).

As part of these surveys, the Geological Survey, in cooperation with the Canadian Pacific Railway Company, undertook in 1873 the purchase of a steam-driven diamond drill to, in the words of Selwyn, then Director of the Survey, "extend and hasten the exploration and survey in the North West Territory." In August of that year, six tons of equipment described by the engineer in charge (Waud, 1874, p. 12) as:

> Diamond Drill, a boiler with force pump and fittings, gearing for working the drill by horse power, 400 feet of 2½ in. diar., tubular drill rods, 150 feet of 3″ tubing (wrought iron), annular and hollow boring heads with diamonds and an independent steam pump with hose and other fittings

arrived at Winnipeg from the United States by Red River steamboat. Those who have had experience with drilling will find both interest in and sympathy with Mr. Waud's account of the difficulties he experienced in attempting to get his balky machine to perform and to install casing in holes that persisted in caving, let alone moving his equipment on carts drawn by oxen at a rate of about 1½ miles per hour. This was but one of many difficulties faced by field officers of the Survey as they contributed in a significant manner to early railway construction through their assistance in selection of routes and in the location of the various geologic commodities so necessary for such a construction venture.

By 1890, the Geological Survey, now with its headquarters in Ottawa, having moved from Montreal in 1881, had expanded to a staff of 52, including 34 professionals. Through an Act of Parliament of the same year the Geological Survey was established as a separate department of government; it thus regained, in an administrative sense, a stature that it enjoyed prior to Confederation. This administrative arrangement existed until 1906, which marks the end of the Classic Period of the Survey.

Forerunner of urban geology

Perhaps the clearest indication of the origins of engineering geological work in the Geological Survey can be found in the latter part of the Classic Period through the work of Dr. Henri

Ami, who served from 1882 to 1911 as a paleontologist. In 1900, Ami published in the Transactions of the Royal Society of Canada an excellent paper entitled "On the geology of the principal cities in Eastern Canada," which included synoptic descriptions of the geological formations underlying St. John, Quebec, Montreal, Ottawa and Toronto. Ami's motivitation for publishing this paper can be readily determined from his opening paragraph (vol. 6, sec. 4, p. 125) which reads as follows:

> The larger cities of our Dominion, as well as those of countries, are the centres of work and research in the pathways of science and economics. Whether as regards the question of boring for petroleum, gas, salt, fresh or mineral waters, numerous problems are involved, in which the underlying geological formations play an important part. What the drill has to penetrate in any of our larger centres of activity in Canada, before reaching the old Archaean or original crust of the earth in this portion of the North American continent covered by the areas under discussion, is a question not only of interest but also of economic value.

Two years later the Geological Survey was requested by the Department of Railways and Canals to undertake a geological examination of the materials encountered in borings for the abutments and piers of a major bridge, then under construction, across the St. Lawrence River at Quebec City. This task was assigned to Ami who published (Ami, 1903) carefully detailed descriptions with accompanying sketch logs for each of the boreholes plus a description of the geologic formations he had observed at the bridge site.

In acknowledging receipt of Dr. Ami's report, the chief engineer for the Department of Railways and Canals commented:

> This information you have been good enough to supply me with, is precisely what I required and I shall not be under the necessity of availing myself of your kind offer to supplement your report with further details.

and therein provided suitable testimonial to the care with which Dr. Ami had carried out his work.

GROWTH AND DECLINE

This period began in 1907 and continued until 1949. With the enactment in 1907 of legislation creating the Department of Mines, the Survey was reduced in administration stature to the rank of branch but continued to grow both in numbers of staff and stature at a scientific institution. In fact, the formation of a Department of Mines clearly defined the functions of the two component branches of the department—Mines, and Geological Survey. Toward the end of the nineteenth century, the distinction between the two became clouded through the formation of a separate Mines Branch within the Department of the Interior. From 1907 until the present the Geological Survey had remained within departments of government that bear mines and national mineral development as their primary mandate.

The Frank slide of 1903

Engineering geologists of today are commonly called upon to identify and to recommend remedial measures for geologic hazards that pose a threat or may already have claimed life or property. In April of 1903 a massive landslide (Fig. 9) destroyed much of the coal-mining town of Frank in the Crowsnest Pass area of southwest Alberta and claimed the lives of about 70 people. Officers of the Geological Survey were called on to investigate the cause of the slide and to report on the risk of further landslides. Initial reports were followed by a final report of the commission appointed to investigate the Turtle Mountain landslide (Daly and others, 1912). The commissioners concluded that coal-mining operations as well as natural causes such as oversteepened slopes, unfavorable orientation of joint planes, and seismic events contributed to the failure. They recommended limitations to future mining for coal and the use of heavy pillars in mining operations as well as relocation of the town away from the path of possible future landslides. A reexamination of the geology of the Frank slide (Cruden and Krahn, 1973) has revealed that bedding planes played a far more significant role in contributing to the landslide than was recognized by the original investigators, but this does not reduce the significance of the commission's findings regarding mining and safety.

Search for road-building materials

Throughout the first 30 years of this century, Canada continued to develop as a nation on political, social, and economic fronts. Attendant upon this development was the continuing need for improvement in transportation systems to link the provinces that extended across the breadth of the North American continent. Railroads continued to be the prime means for moving both industrial goods and people, and as the railroad network expanded, survey officers were continuously involved in geological explorations along the new routes. By 1913, however, the automobile was becoming something more than a mechanical curiosity, particularly in the more densely populated areas of southern Ontario and Quebec.

In 1913, the Geological Survey established a Road Materials Division, which carried out surveys, particularly in eastern Ontario and Quebec, for the location of both surficial and bedrock sources of road-paving material. Samples collected during these surveys were, until 1916, tested outside the facilities of the Mines Department. However, in 1916 the Mines Branch, a sister branch of the Geological Survey, established a Road Materials Laboratory (Clark, 1918). This laboratory took over the task of materials testing on behalf of the Geological Survey. By 1917, the responsibility for both surveys and testing of road materials had been transferred to the Mines Branch and was not subsequently resumed by the Geological Survey.

Problems with Leda Clay

An account of the drift on the Island of Montreal (Stansfield, 1914) constituted further pioneer work in Canadian urban

Figure 9. Rockslide on Turtle Mountain at Frank, Alberta. Geological Survey of Canada photo 201903.

geology. This work was resumed by others in the same area over 40 years later (Prest and Hode-Keyser, 1977). It is interesting that Stansfield was fully familiar with the notorious Leda Clay, which underlies much of the Ottawa–St. Lawrence Lowlands. He described the behavior of the Leda Clay at Montreal (Stansfield, 1914, p. 209) as:

> . . . strata being composed of a mixture of very fine sand and clay. The mixture is capable of retaining moisture and when wet will run under the slightest pressure. It is incapable of standing by itself and in excavations made within it. The work done in one day will be completely counteracted overnight by the oozing in of the quicksand. . . .

These characteristics of Leda Clay, a formation prevalent in the Ottawa area, were probably known by Survey geologists a few years earlier. Therefore, it is ironic that the site chosen for new headquarters in Ottawa for the Geological Survey, the Victoria Memorial Museum, was underlain by over 100 feet of Leda Clay. The government had not consulted the Survey on the geologic characteristics of the site chosen for the Victoria Memorial Museum, and the consequences of a poor choice of site and foundation design immediately became apparent upon completion of the building in 1911. According to Zaslow (1975, p. 268):

> As the unfortunate contractor learned to his sorrow, the site was underlain by about 140 feet of unconsolidated blue clay into which the building quickly began to settle. The heavier outer walls sank below the floors, which were held up by the inside walls. Cracks appeared in the pavement of the ground story and in the upper stories, and in the basement the tiled walls developed dangerous cracks and sheer planes. The weight of the tower caused the walls supporting it to come away from the floors and opened leaks in the roof that made it necessary to keep buckets under drip spots. Despite all efforts at patching, the crack widened and the tower began to develop a list; eventually in 1915 the tower had to be pulled down before it collapsed of its own accord.

Thanks to subsequent costly remedial work, the building still stands and is fully used for museum purposes.

River and canals

One of the earliest engineering projects in which Survey geologists had a direct, as opposed to peripheral, involvement was the geological investigation of the Fraser River Delta in British Columbia. This work began in 1919 with the purpose of assisting the Department of Public Works in determining the engineering methods necessary to improve the navigable parts of the river. Through study of the delta sediments and flow characteristics of the river, the principal investigator, W. A. Johnson, was able to identify those parts of the river that would be amenable to shore protection and river training and to point out localities where the stratigraphy of the sediments would pose particular difficulty in maintaining slope protection.

Late in 1926, the Department of Railways and Canals was involved in improving the St. Lawrence Waterway near Lachine, Quebec. The assistance of the Geological Survey was sought in interpreting the results of borings associated with the engineering investigations for lock construction. This service became the forerunner of a much closer involvement of Survey officers with the St. Lawrence Seaway Authority more than 25 years later.

Link between engineering and Pleistocene Geology

In 1930, W. H. Collins, then Director of the Geological Survey, gave recognition to the growing importance of the Pleistocene and Recent deposits as sources of construction materials, ground water, and as parent material of agricultural soils, all of which bore direct relation to the needs of an increasing population in the South. From the western provinces the Survey was also receiving requests to investigate water supplies for communities in the Interior Plains region. These events led to the formation of a new division within the Survey, under the direction of W. A. Johnson, given the somewhat cumbersome, yet functionally descriptive, title of "Pleistocene Geology, Water Supply, and Borings." This division and its successors in title were to become the administrative home of engineering geology within the Geological Survey.

Economic depression and World War II

The decade of the 1930s was marked by increasing economic depression, which brought about reduction in activity in all sectors of the economy. By 1935, severe overall reductions in government appropriations were widespread, but paradoxically, the Geological Survey received—from a government facing an election—a grant of $1 million with the intended purpose of creating employment. This sudden windfall immediately inflated the Survey's budget that year by a factor of ten, thereby generating a state of near chaos in the expanded field program that ensued. Apart from a massive collection of water-well data across the prairies, none of the activities of the "Million Dollar" year had much impact on the development of engineering geology in the Survey.

Late in 1936, the Survey was reduced to divisional status as the Bureau of Geology and Topography within the Mines and Geology Branch of the Department of Mines and Resources. This department was a strange amalgam of units from previously existing departments, which placed under one minister such divergent responsibilities as Mining and Geology and Indian Affairs and Immigration. However, also contained in the department was a Surveys and Engineering Branch with responsibilities for hydraulic structures and power development. In later years, this departmental association would have a significant impact on engineering geological work by the Survey.

With the advent of World War II and throughout the 5½ years of conflict that followed, the Survey again directed its efforts to the location of strategic minerals and other geologic commodities of vital necessity for the war effort. Although details are lacking in published reports, several officers were involved in the selection of sites for various military installations. Thus, it may be assumed that advice essentially of an engineering geological nature was being provided.

The Canol Project

In 1943, J. S. Stewart was assigned as liaison officer between the Department of Mines and Resources and the United States Army during construction of the Canol Project. This project, begun in May 1942 and completed in February 1944, was designed to deliver crude oil by pipeline from the oil field at Norman Wells on the Mackenzie River in the Northwest Territories across almost 965 km of mountainous and permanently frozen ground to Whitehorse in the Yukon Territory. The refinery products from Whitehorse were then to be piped along the Alaska Highway to serve the defense needs of that area. With the decline of the threat of war in Alaska, the need for the pipeline rapidly diminished; the project was abandoned in 1947.

It is doubtful if engineering geology played any significant role during the design or construction of the mammoth Canol Project. However, 30 years later, as the possibility of pipelines in the Mackenzie River valley again came into prominence, engineering geologists concerned with permafrost degradation problems were able to evaluate this phenomenon beneath roads constructed for the Canol Project (Isaacs, 1974).

Dam sites on the Columbia River

In the years immediately following World War II the return to civilian pursuits brought with it an upsurge in the construction of major engineering projects. One of these, of an international nature, was development of hydroelectric power and flood-control structures on the Columbia River drainage system. In response to a request from the Dominion Water and Power Bureau, a companion Branch to the Geological Survey in the Mines and Resources Department, Survey geologists were assigned to investigate dam sites along the Columbia and Kootenay Rivers. One of these geologists, Edward Hall, was loaned in 1947 to the Power and Water Bureau to assist in the evaluation of dam sites and interpretation of borings associated therewith. Hall's assignment, which included relocating from Ottawa to Revelstoke, British Columbia, continued for eight years.

By 1950, the period of decline in growth and activity of the Survey was past; in that year the status of branch was once again attained, with the formation of the Department of Mines and Technical Surveys.

THE MODERN SURVEY

Recognition of engineering geology

The new Department of Mines and Technical Surveys was established with the purpose of creating an integrated organization whose primary function was to provide technological assistance in the development of Canada's mineral resources. As a major branch within this department, the Geological Survey continued to be strongly identified with the departmental mission of promoting the discovery and development of mineral resources. However, for the first time, the Survey's function was also described as contributing geological information as an aid in the construction of such public works as dams, bridges, tunnels, and foundations. In order to better fulfill its now stated mission in engineering geology the Survey created a Division of Groundwater, Glacial, and Engineering Geology, which one year later, in 1951, was renamed as the Pleistocene and Engineering Geology Division. From that time until the present an engineering geology unit has existed within the Geological Survey under a variety of divisional labels but always closely associated with the Pleistocene or Quaternary geology unit.

Throughout the 1950s, Survey geologists contributed their expertise to such major engineering works as the construction of the St. Lawrence Seaway between Montreal and Lake Ontario. One geologist, E. B. Owen, was assigned to the St. Lawrence Seaway on a continuing basis for approximately five years ending

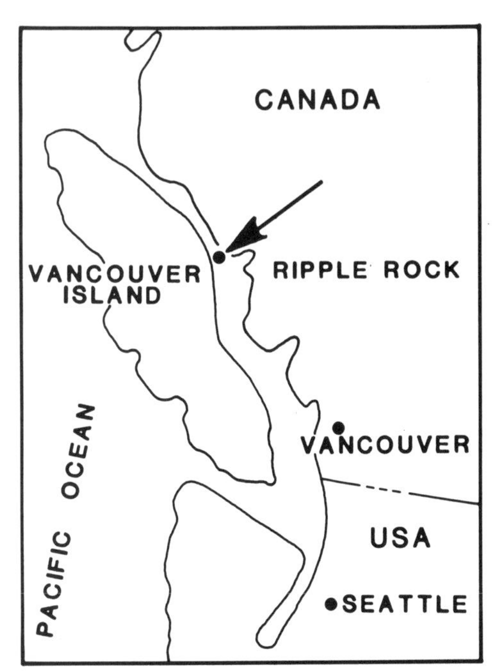

Figure 10. Locality map showing Seymour Narrows site in British Columbia, Canada. From Dolmage (1964).

in 1959. During this time he was responsible for logging much of the core for the foundations of locks and related hydraulic structures as well as locating suitable sources of construction material such as rip rap and armor stone.

Ripple Rocks

Ripple Rocks posed a serious hazard to navigation in Seymour Narrows, a major navigation channel situated 160 km north of Vancouver, British Columbia (Fig. 10). More than 30 ships were lost on the rocks, and the narrow channel, with tides at 12 to 14 knots, is extremely dangerous. Removal of Ripple Rocks, as described by Dolmage (1964), was accomplished by excavating a shaft to a depth of 174 m on the adjacent Maud Island and then tunneling 731 m beneath the channel, followed by rising 91 m into the main rock pinnacles (Fig. 11). The channel, rocks, and surrounding area are underlain by Mesozoic volcanics of mainly basalts and andesites, in part altered with clayey minerals (Fig. 12). Lateral tunnels 1.2 to 1.5 m in diameter were driven into the north and south parts of Ripple Rocks to "honeycomb" the upper reaches of rock mass. The system of lateral tunnels and coyote openings was loaded with 1,378 tons of explosives (nitramex 2 H) that were detonated practically instantaneously (slight delay at South Rock) to successfully "blow off" the upper 12 m (40 ft) of rocks and remove the projecting navigational hazard.

Credit for success of the operation must be accorded the firm of Dolmage and Mason of Vancouver who drew up the plans and supervised the work (Dolmage and others, 1958). However, credit is also due the late W. E. Cockfield, then officer-in-charge of the Vancouver office of the Geological Survey (a position previously held by Victor Dolmage). Cockfield was responsible for supervising the preliminary diamond drilling of the Ripple Rocks site and approaches, and it was his findings that established the feasibility of tunneling beneath Ripple Rocks. This unusual investigation included two inclined and curving cored borings drilled from Maud Island; one passed completely under the intervening channel and was wedged upward into the rocks with good core recovery for study. (Project described further in Chapter 2, this volume.)

Northern dam sites and Red River floodway

During the 1960s the Geological Survey was requested by other departments of the federal government to undertake preliminary geological investigations of potential dam sites within the Yukon and Northwest Territories. More than 90 sites were examined, although none has been subsequently developed for hydroelectric power generation.

In the spring of 1950, a major flood of the Red River caused such extensive property damage to the city of Winnipeg, Manitoba, that an investigation by the federal government was undertaken to determine the possibilities for alleviating the flood problem. This investigation resulted in a joint decision by the federal and Manitoba governments to proceed with construction of a diversion channel, the Red River floodway, around the city. This structure, a 48-km-long channel capable of carrying 821 m^3/sec of Red River floodwater, was excavated through soft Lake Agassiz clays and till that provided a confining layer for an artesian aquifer in the underlying Ordovician limestone.

During excavation of a test pit in 1961 to assess the stability of slopes in Lake Agassiz clays, substantial inflows of ground

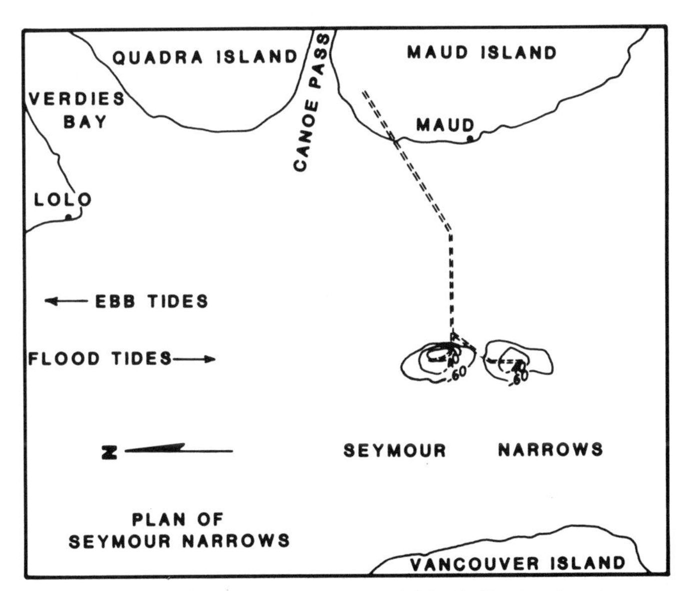

Figure 11. Plan of Seymour Narrows and Ripple Rocks, showing two rock pinnacles in center of channel 760 m wide. From Dolmage (1964).

Figure 12. Photo of volcanic rock series of site-area outcropping along shoreline of Maud Island and the location of shaft for tunnel driven beneath channel and into rocks, as shown on Figure 11. Note extremely dangerous whirlpool, and thus a navigational hazard, above rock pinnacles in the foreground as tides change. Photo by G. A. Kiersch, August 1955.

water occurred, which necessitated continuous dewatering to enable completion of the test pit. This event alerted the design engineers to the potential problems that artesian ground-water inflows could cause for both construction and for the extensive number of rural users of the aquifer as a source of water supply. Thus, a request was made by the Manitoba Water Control and Conservation Branch to the Geological Survey for assessing the magnitude of ground-water problems associated with construction of the floodway. Through analyses of pump test data, seismic profiling of the floodway centerline, and evaluation of water-well data in the region, it was possible to construct a model of the effect of construction on ground-water levels and to forecast those localities where artesian flows could be expected during construction (Hobson and others, 1964). This work was instrumental in significantly reducing costs that could have arisen due to adverse excavation conditions and to unwarranted claims for damage to industrial and domestic water wells.

Overconsolidated clay shales

In addition to providing engineering geological expertise to specific construction projects of concern to the federal government, the Geological Survey has endeavoured to undertake regional studies to provide information on the geologic factors that contribute to the engineering behavior of geologic materials. One such study (Scott and Brooker, 1968) concerned Upper Cretaceous shales in western Canada, which along with their stratigraphic equivalents throughout the interior plains of the United States, are prone to problems of stability for both slopes and foundations. Arising from this study was introduction of the term "overconsolidated clay shales," defined as a sedimentary deposit composed primarily of silt- and clay-sized particles dominated by members of the montmorillonite group of clay minerals. The deposits have been subjected to consolidation loads in excess of the present overburden, and disaggregation of the sediment can be effected by immersion in water. Following examination of numerous slope failures in overconsolidated clay shales throughout western Canada, it was concluded (Scott and Brooker, 1968) that the wedge method of stability analysis offered promise as an approach to stability prediction and that such factors as depositional environment, stress history, structure, stratigraphy, geomorphic processes, and ground water were all relevant to understanding slope behavior.

Analysis of the stability of slopes in overconsolidated shales required field measurement of pore pressures in addition to measurement of residual shear strength by use of direct shear apparatus. Piezometers of the open standpipe variety then available were inadequate for acquisition of pore-pressure measurements in the low-permeability medium provided by overconsolidated shales. In order to overcome this instrumental deficiency the University of Alberta and Geological Survey cooperated in the design and production of a highly sensitive piezometer using a diaphram-type electrical transducer as the sensing element. This instrument, described by Brooker and others (1968), was fitted with a strip chart recorder and was successfully used in the field study of slope failures.

Figure 13. Diamond drill equipped with drill-fluid chiller on location in Mackenzie River Valley, 1971. Geological Survey of Canada photo 203122–B.

Northern terrain and hydrocarbon development

In the late 1960s, the major oil discovery at Prudhoe Bay in Alaska accelerated petroleum exploration in the Canadian Arctic, particularly in the Mackenzie Delta and Beaufort Sea regions. With such exploration activity the possibility of oil and/or gas pipelines down the length of the Mackenzie Valley became real and, with this reality, the concern of government over the potential hazards that permafrost would create for such construction. In response to these concerns the Geological Survey, as well as industry, undertook a major program of geotechnical and terrain studies throughout the Mackenzie Valley to enable government to assess any application that might arise from industry for construction of pipelines.

In support of this work, diamond drilling equipment along with a specially designed drill-fluid chiller (Fig. 13) was air lifted to sites in the Mackenzie Valley to permit recovery of frozen soils with minimal thermal disturbance. Concern with northern terrain is a continuing preoccupation with units of the Geological Survey and extends to the offshore regions where both subsea permafrost and ice scour pose hazards to be contended with in exploration, well completion, and transportation of hydrocarbons. Applications of marine geophysical techniques, such as side-scan sonar for the detection of ice scours (Fig. 14) and high-resolution seismic profiling, have thus become standard techniques in assessing geologic hazards of the arctic offshore regions and in evaluating the geotechnical properties of sea-bottom sediments that may be dredged for construction of drilling platforms.

Recent activities

Recent engineering geological activities of the Geological Survey include studies of urban geology, site investigations for major airports, and major work currently in progress with Atomic Energy of Canada Limited for the geological disposal of high-level radioactive wastes. These and other studies of the engineering behavior of geological materials, as well as investigation of terrain hazards such as the major landslide at St. Jean Vianney, Quebec, that claimed over 30 lives in 1971 (Fig. 15) and studies of slope failures in mountainous regions (Eisbacher and Clague, 1984), exemplify the scope and diversity of engineering geological work carried out by the Geological Survey of Canada.

CONCLUSION

In conclusion, one might well ask, “What has been the role of the Geological Survey in the development of engineering geology?” On the basis of a historical overview of the Geological Survey as an institution, and on the work of some of its staff, it seems clear that Logan’s practical approach to the applications of geology of the mid-nineteenth century has persisted and that officers of the Survey, whether they be formally recognized as engineering geologists or they be geophysicists or other specialists, have contributed their geoscience expertise to the resolution of a wide variety of problems pertaining both to engineering projects and to Canadian terrain. Regardless of further institutional changes that will surely come, the Survey can be expected to continue its tradition of service and thereby further the growth of engineering geology in Canada.

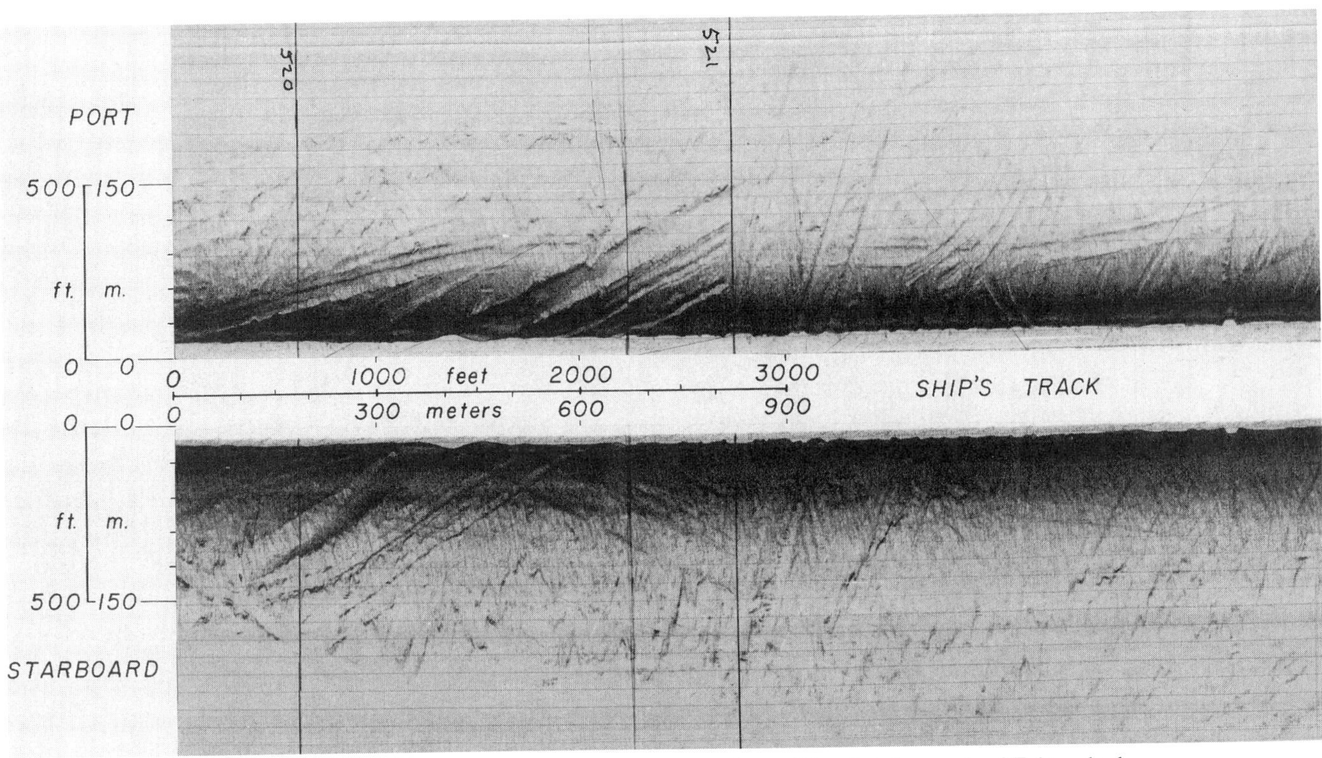

Figure 14. Sidescan sonagram (100 kH) of ice scours on floor of Beaufort Sea north of Tuktoyaktuk Peninsula, Northwest Territories. Depth of water approximately 35 m. Geological Survey of Canada photo 204416–N.

Figure 15. Massive flow slide in sensitive marine clays, St. Jean Vianney, Quebec, 1971. Geological Survey of Canada photo 201669.

HISTORICAL DEVELOPMENT OF ENGINEERING GEOLOGY IN MEXICO

Mariano Ruiz-Vazquez, Guillermo P. Salas, and Jorge I. Maycotte

EARLY HISTORY

During the prehispanic period (200 B.C. to A.D. 1521), the Mayas and the Aztecs built the monumental pyramids of Teotihuacan, Chichen Itza, and El Tajin, illustrating the first applications of geological concepts and principles to construction. Obviously they were aware of the problems that heavy constructions could cause when founded on soft rocks or soils, and how to manage water for reservoirs and crop irrigation.

With the arrival of the Spaniards came the Conquest (1519 to 1521) and the Colonial times (1540 to 1800) when most of the Aztec Empire was buried and their knowledge of engineering fundamentals was lost. A new city was built on top of Tenochtitlan. Later history is divided into other periods such as Independence (1810), the Imperial Interval (1867), and the Revolution (1910). At that time many structures were built by Spanish and French technicians who ignored the behavior and geologic properties of the Mexican ground; they faced problems related to settling, low bearing capacity, slope instability, and others, which they failed to recognize and did not know how to correct.

The mining industry became highly developed as a result of Spanish exploitation of gold and silver. Despite the fact that the mining processes were initially primitive and rustic, there were several mining districts in the country. By 1798, the technology involved became so developed that a treatise was written about mining in Mexico, and some years later (1825), the treatise was highly regarded by members of the mining profession, even those of Europe. This fame for the magnificent engineering in Mexico's mines was due to the mining engineers from Spain and they trained others at the Real Seminario de Mineria, established in 1792. These engineers applied their geological knowledge to underground mining works, for water dams and reservoirs built to further mining enterprises, and some local irrigation projects. Thus, the states of Aquascalientes, Chihuahua, Guanajuato, Hidalgo, Jalisco, Mexico, Michoacan, San Luis Potosi, and Zacatecas have today many early engineering works that are still in use. They include small mines, irrigation projects, and bridges and roads. These works have lasted for centuries, although many are now abandoned and subjected to weathering and pilfering, and are in ruins. However, the works and sites are relics of good engineering and sound geological concepts. Most of them represent the efforts of private enterprise, such as the dams and reservoirs at Saucillo (1730), Yuriria (1850), San Martin (1765), Torre Blanca (1850), Mal Paso (1865), and La Mula (1868); many others are distributed in some of the states previously mentioned. At least 20 of the old dams are still in operation, although badly clogged with sediments. (See Chapter 1 this volume.)

Most of the railroads that cross Mexico were built between 1873 and 1936. Six hundred km of road beds cross mountain ranges, through tunnels and along shear cuts in precipices; geological input and guidance was utilized for the design and construction. Hydraulic channels were developed between the years of 1925 and 1930 with the founding of Comisión Nacional de Irrigación. Of importance as an engineering feat with geological input was the design of La Boquilla dam, a masonry-gravity dam (74 m high with a reservoir of 30 million m^3 of water) completed in 1916, the largest such dam in the world.

MODERN ACTIVITY

Prior to 1930, several paved and unpaved narrow highways were built to connect Mexico City and other important cities. The most important highway was the one from Monterrey to Nuevo Laredo and Texas, which connects with U.S. highways. This engineering feat required a feasibility-planning geological study in order to cross precipitous mountains, deep canyons, and wide rivers.

In the 1930s, the Comisión Nacional de Irrigación constructed many large dams. The Presa Don Martin or Venustiano Carranza dam, constructed 1928 to 1932 in the states of Coahuila, was first one in the world to use gravity dam walls; it was constructed only after a thorough investigation of the geologic conditions relevant to the engineering design by Paul Waitz, a Swiss geologist and naturalized Mexican citizen. Calles dam in Aquascalientes, completed in 1931, was the first arch gravity dam constructed in Mexico (67 m high); it was followed later by the Endho dam in the state of Hidalgo.

During the 1930s and 1940s, geology was applied to construction of small earth and concrete dams, roads, and bridges by a few experienced professionals, many of whom had been associated with mining in Mexico. This group included geologists De la O. Carreño, Heins Lesser Johns, and Lorenzo Torres I., and mining engineers M. Alvarez Carvajal and Mario Veytia. During this period, the first soil mechanics laboratory for road-bed construction purposes was implemented in Mexico City by the Secretaría de Obras Públicas (S.O.P.) and a concrete-testing laboratory was organized by the Secretaría de Recursos Hidraúlicos (S.R.H.). The latter was the first technical institution devoted mainly to dam construction.

Dam site investigations and deep foundation excavations were evaluated by the early geologists to characterize the site for the civil engineer and construction organization, but no other geological records were made of the site conditions. In a few cases, the geologist visited a site to advise on a specific problem encountered during construction.

During the 1950s, the Comisíon Federal de Electricidad (C.F.E.) began to work in the field of applied geology for dam sites, coal exploration, and thermoelectric and geothermic plants. The practice of engineering geology and geotechnical engineering began on a large scale in 1954, with the help of Italian designers and technology, when the first large arch dams were designed for C.F.E. and construction began. To meet immediate demands,

Mexican engineers were required to learn how geological knowledge was applied to fulfill the requirements of the requisite feasibility studies, such as those involved with the arch dams at Apulco in the state of Puebla, Santa Rosa dam in the state of Jalisco, and El Novillo dam in the state of Sonora. For the early investigations, the studies of rock mechanics were performed by Italian technicians, but by the 1960s, Mexican engineers were capable of these duties following training in the United States and Europe.

When the construction of large arch and earth dams, along with large underground excavations, started in Mexico, several technical societies and college training programs were chartered in the capital. The Universidad Nacional Autónoma de Mexico (U.N.A.M.) began offering the Master's Degree in Engineering in 1957. Thereafter, the soil mechanics courses were strengthened and became very beneficial for civil engineers involved in the construction field. In 1958, the Director Assistant of Secretaría de Obras Públicas (S.O.P.) chartered the Instituto de Ingeniería and the Division of Estudios Superiores at the U.N.A.M., where hydraulic model studies and seismic-risk analysis were performed. Along with this research and academic support, S.O.P. built 20,137 km of roads and the largest bridges in Mexico at Coatzacoalcos and Alvarado in the state of Veracruz.

By 1961, soil mechanics was the leading subject in the broad category of geotechnical engineering in Mexico. To broaden the technical services available, C.F.E. founded the Laboratorio de Estudios Experimentales in 1964; rock and soil mechanics investigations were performed as well as concrete and material testing. By 1967, the largest equipment for testing rock samples in Mexico (1 m^2) was in operation at the University Laboratories in Mexico City.

With the construction of large dams throughout Mexico by the 1960s, the C.F.E. geologists were asked to deal with such special studies as detailed rock sampling of foundations, supervision of impervious grout curtains, and anchoring of rock masses for stability. This was the beginning of current-day engineering geology for the applied geologists of Mexico working with the diverse problems of slope stability, foundation treatment, and large-scale underground excavations. Such field investigations led the C.F.E. geologists to introduce the Lugeon field permeability tests for evaluating the rock units of dam sites, and the potential for leakage and/or bypass flow.

Within the last 25 years, application of geology and geophysics to engineering works has reached a high degree of development in Mexico. This has included increasing responsibility by Mexican geologists and geophysists in the planning and construction of the great hydroelectric dams on the Usumacinta River at Malpaso, Chicoasen, and Angostura in the state of Chiapas; at El Infiernillo and El Caraco in the state of Guerrero; on the Rio Bravo (Mexico–U.S.) at La Presa Falcon and La Amistad; and more recently the deep drainage system of Mexico City. Likewise, Mexican geologists have been involved with the construction and maintenance of the freeways; additional railroads; large bridges across the Papaloapan, Coatzacoalcos and Panuco Rivers; seaports; and modern airports.

Since the early 1970s, the technical growth of applied geology for engineering works has moved a big step forward; the most advanced geotechnology is practiced in Mexico by several foreign consultants and advisors and by the large body of Mexican professionals who have been trained in the United States and Europe.

Unfortunately, technical independence for Mexico was adversely affected in the early 1980s when the national budget and funds for construction activities were cut with the devaluation of the peso and the associated economic crisis. Similarly, imported technology was cut back, which interrupted most of the student scholarships in foreign countries.

Today, national engineering geology practice is mainly restricted to feasibility studies and an interrelation with reliable information on the rock mechanics and soil mechanics characteristic of a site and/or correlation with geomechanical investigative data. Such investigations can be carried out at the universities and federal research laboratories, as well as in private geotechnical engineering companies throughout the country; the latter sell their services abroad, especially to Central and South American countries. A few of the Mexican universities offer Master of Engineering (M.E.), Master of Science (M.S.), and Ph.D. training programs related to applied geoscience and engineering geology subjects.

Among the main engineering works built with the input and guidance of engineering geologists in Mexico are: the Latinoamerican Tower, Pemex Tower, Hotel Mexico Tower, the metro subway, and the deep sewage tunnel network. All these works were built in the metropolitan area, which is underlain by high-compressibility clays. Throughout the country, several large dams have been built such as: Chicoasen in Chiapas, the highest earth dam at 265 m with the largest underground facilities and power generation of 5,589 GWh; the Laguna Verde nuclear plant in the state of Veracruz. Likewise, there are open-pit mines and underground coal mining, and other copper and iron mining in the north of the country. The Tijuana water tunnel conduit, 843 m long, was excavated in migmatites utilizing a tunnel-boring machine of 3.66 m in diameter.

From 1910 to the middle of 1984, C.F.E. nad S.R.H. built approximately 550 dams and channels for irrigation, power generation, flood control, and water supply of all types.

REFERENCES CITED

Ami, H. M., 1900, On the geology of the principal cities in eastern Canada: Royal Society of Canada, Proceedings and Transactions, ser. 2, v. 6, p. 125–173.

—— , 1903, Notes of drillings obtained in six diamond-drill bore-holes in the bed of the St. Lawrence River at Victoria Cove, Sillery, 8 miles above Quebec City, Quebec: Geological Survey of Canada Summary Report for 1902, p. 326–336.

Balmer, G. G., 1953, Physical properties of some typical foundation rocks: Denver, Colorado, U.S. Bureau of Reclamation Concrete Laboratory Report SP–39, 15 p.

Berkey, C. P., 1942, The geologist in public works: Geological Society of America Bulletin, v. 53, p. 513–532.

Brooker, E. W., Scott, J. S., and Ali, P., 1968, A transducer piezometer for clay shales: Canadian Geotechnical Journal, v. 5, no. 4, p. 256–264.

Bryan, K., 1929, Geology of reservoir and dam sites, *in* Report on the Osyhee irrigation project, Oregon: U.S. Geological Survey Supply Paper 597–A, p. 1–72.

Clark, K. A., 1918, Road Materials Division, *in* Summary report of Mines Branch: Canada Department of Mines, p. 121–122.

Crandell, D. R., and Mullineaux, D. R., 1978, Potential hazards from future eruptions of Mount St. Helens volcano, Washington: U.S. Geological Survey Bulletin 1381–C, 26 p.

Cruden, D. M., and Krahn, J., 1973, A reexamination of the geology of the Frank Slide: Canadian Geotechnical Journal, v. 10, p. 581–591.

Daly, R. A., Miller, W. G., and Rice, G. S., 1912, Report of the commission appointed to investigate Turtle Mountain, Frank, Alberta: Geological Survey of Canada Memoir 27, 34 p.

Dolmage, V., 1964, Removal of Ripple Rocks, Seymour Narrows, British Columbia, Canada, *in* Kiersch, G. A., ed., Engineering Geology Case Histories, no. 5: Geological Society of America, p. 7–13.

Dolmage, V., Mason, E. E., and Stewart, J. W., 1958, Demolition of Ripple Rock: Canadian Mining and Metallurgical Institution Transactions, v. 61, p. 382–395.

Eckel, E. B., ed., 1958, Landslides and engineering practice: Highway Research Board Special Report 29, 232 p.

Eisbacher, G. E., and Clague, J. J., 1984, Destructive mass movements in high mountains; Hazard and management: Geological Survey of Canada Paper 84–16, 230 p.

Fisher, P. R., 1981, The role of the U.S. Army Corps of Engineers in the development of engineering geology: Geological Society of America Abstracts with Programs, v. 13, p. 452.

Gibbs, H. J., and Holland, W. Y., 1960, Petrographic and engineering properties of loess: Denver, Colorado, U.S. Bureau of Reclamation Engineering Monograph 28, November 1980, 37 p.

Hobson, G. D., Scott, J. S., and van Everdingen, R. O., 1964, Geotechnical investigation, Red River Floodway, Winnipeg, Manitoba: Geological Survey of Canada Paper 64–18, 43 p.

Holtz, W. G., and Gibbs, H. J., 1954, Engineering properties of expansive clays: Proceedings, American Association of Civil Engineers, v. 80, Sept. no. 516, October 1954, 28 p.

Isaacs, R. M., 1974, Geotechnical studies of permafrost, Fort Good Hope–Normal Wells region, Northwest Territories: Government of Canada Environmental Social Committee, Northern Pipelines Task Force on Northern Oil Development Report 74–16, 212 p.

McCulloch, C. M., and Deul, M., 1973, Geologic factors causing roof instability and methane emissions problems: U.S. Bureau of Mines Report on Investigations 7769, 25 p.

Meinzer, O. E., 1923a, The occurrence of ground water in the United States with a discussion of principles: U.S. Geological Survey Water-Supply Paper 489, 321 p.

—— , 1923b, Outline of ground water hydrology with definitions: U.S. Geological Survey Water-Supply Paper 494, 71 p.

Merrill, R. H., and Morgan, T. A., 1958, Method of determining the strength of a mine roof: U.S. Bureau of Mines Report of Investigations 5406, 22 p.

Obert, L., 1940, Measurement of pressures on rock pillars in underground mines, Part 2: U.S. Bureau of Mines Report of Investigations 3521, 11 p.

Obert, L., and Duvall, W. I., 1967, Rock mechanics and the design of structures in rock: New York, John Wiley and Sons, 650 p.

Obert, L., Windes, S. L., and Duvall, W. I., 1946, Standardized texts for determining the physical properties of mine rock: U.S. Bureau of Mines Report of Investigations 3891, 67 p.

Osterwald, F. W., 1961, Deformation and stress distribution around coal mine workings in Sunnywide No. 1 mine, Utah: U.S. Geological Survey Professional Paper 424–C, p. C349–C353.

Poland, J. F., and Davis, G. H., 1969, Land subsidence due to withdrawal of fluids, *in* Varnes, D. J., and Kiersch, G. A., eds., Reviews in Engineering Geology, v. 2: Geological Society of America, p. 187–269.

Poland, J. F., Lofgren, B. E., Ireland, R. L., and Pugh, R. G., 1975, Land subsidence in the San Joaquin Valley, California, as of 1972: U.S. Geological Survey Professional paper 437–H, p. H1–H78.

Prest, V. K., and Hode-Keyser, J., 1977, Geology and engineering characteristics of surficial deposits, Montreal Island and vicinity, Quebec: Geological Survey of Canada Paper 75–27, 29 p.

Robinson, C. S., and Lee, F. T., 1964, Geologic research on the Straight Creek Tunnel site, Colorado: Highway Research Board, Highway Research Record 57, p. 18–34.

Schuster, R. L., and Krizek, R. J., 1978, Landslides; Analysis and control: Transportation Research Board Special Report 176, 234 p.

Scott, J. S., and Brooker, E. W., 1968, Geological and engineering aspects of Upper Cretaceous shales in western Canada: Geological Survey of Canada Paper 66–37, 75 p.

Stansfield, J., 1914, The drift on the island of Montreal: Geological Survey of Canada Summary Report for 1913, p. 208–210.

Waud, W. B., 1874, Report of operations in Manitoba with the diamond-pointed steam drill: Geological Survey of Canada Report of Progress 1873–74, p. 12–16.

Zaslow, M., 1975, Reading the rocks; The story of the Geological Survey of Canada 1842–1972: Toronto, The Macmillan Company of Canada Limited, 599 p.

Ziemer, R. R., 1981, Roots and the stability of forested slopes; Proceedings of the Conference on Erosion and Sediment Transport in Pacific Rim Steeplands: Christchurch, New Zealand, International Association of Hydrological Sciences Publication 132, p. 343–361.

Manuscript Accepted by the Society December 23, 1988

ACKNOWLEDGMENT

The section on Historical development of Engineering Geology in Mexico is a compilation by editor George A. Kiersch of two unpublished papers: M. Ruiz-Vazquez and G. P. Salas on "Geology in Civil Engineering in Mexico" and J. I. Maycotte on "Historical Development of Engineering Geology in Mexico."

Printed in U.S.A.

Geological Society of America
Centennial Special Volume 3
1991

Chapter 4

History and heritage of Engineering Geology Division, Geological Society of America, 1940s to 1990

George A. Kiersch
Geological Consultant, Kiersch Associates, Inc., 4750 North Camino Luz, Tucson, Arizona 85718
Allen W. Hatheway
Department of Geological Engineering, University of Missouri, Rolla, Missouri 65401

INTRODUCTION

During the 1890s, the importance of an interrelation between geologic principles and guidance for construction of major engineering works was being clearly demonstrated through the efforts of Professor William O. Crosby (M.I.T.) and Professor James F. Kemp (Columbia). A half-century earlier, geologists in North America had begun to show this interdependence, as indicated by works of James Hall of the New York State Geological Survey in 1839 on rock cuts of the Erie Canal and William W. Mather of the Ohio Geological Survey in 1838 on rotational slides along the lake front at Cleveland. The contributions of these and other early workers are described in Chapter 1, as are other early geological studies for dams, tunnels, aqueducts, canals, and related works. However, the first organizations and groups formed to represent the early practitioners of applied geology only developed in the early 1900s.

HERITAGE: 1900s TO 1950s

The Economic Geology Publishing Company was formed to serve the interests of all applied/economic geologists in 1905, and the first issues of *Economic Geology* were released that year. This scientific journal soon developed a wide circulation, both domestic and foreign, and was the medium for all four branches of applied, or economic, geology, described by D. W. Johnson in 1906 as mining, petroleum, ground water, and "applications of geology to various uses of mankind and engineering structures." Waldemar Lindgren, chief geologist of the U.S. Geological Survey, in his 1913 textbook *Mineral Deposits,* defined these four branches and made reference to engineering geology practice. The main geological principles and their relevance to engineering works were put forth in the textbook by Ries and Watson (1914), *Engineering Geology.*

In 1913 the American Institute of Mining Engineers (AIME) established a Committee on Mining Geology, with Wal-

Frontispiece. Professor Charles P. Berkey, president of GSA in 1941 and first chairman of the Engineering Geology Division in 1947 and 1948.

Kiersch, G. A., and Hatheway, A. W., 1991, History and heritage of Engineering Geology Division, Geological Society of America, 1940s to 1990, *in* Kiersch, G. A., ed., The heritage of engineering geology; The first hundred years: Boulder, Colorado, Geological Society of America, Centennial Special Volume 3.

demar Lindgren as chairman, to represent the first specialists to separate from the original body of applied geologists. In 1919, a group of petroleum-oriented geologists from the Economic Geology group decided to organize the American Association of Petroleum Geologists (AAPG) (Bateman, 1955) and became the second separate group of applied geologists. However, they continued to publish most of their petroleum geology papers for several years in *Economic Geology* until the AAPG *Bulletin* was strongly established in the 1920s under editor R. C. Moore.

Society of Economic Geologists

The Society of Economic Geologists (SEG) was organized in 1920 and became an associated society of GSA. Annual meetings rotate between GSA and AIME. The purpose of the Society is to further the scientific field of economic geology and provide a forum for the presentation of papers and ideas on the four branches—mining, petroleum, ground water, and engineering applications. From the beginning and continuing through the mid-1950s, the SEG actively supported the discipline of engineering geology. Many strong engineering geology papers were published in *Economic Geology,* the bulletin of SEG, during this period, beginning with D. W. Johnson's paper in 1906 on "The Scope of Applied Geology." This long-standing interest is demonstrated by the nearly 50 papers on engineering geology subjects published in *Economic Geology* by 1960 (Appendix L). Good examples are the paper by S. Paige and R. Rhoades (1948), "Report of committee on research in engineering geology," and the three outstanding review papers in the Fiftieth Anniversary Volume that deal with aspects of engineering geology practice and are professional milestones: Legget (1955) on "Heritage"; Terzaghi (1955) on "Engineering Properties of Sediments"; and Ferris and Sayre (1955) on "Quantitative Approach to Ground-Water Investigations." Furthermore, the AAPG *Bulletin* published several strong papers during the 1920s through the 1940s, as did the AIME in *Mining Transactions* (e.g., the symposium on dams; AIME, 1929). This technical publication introduced into the literature the first truly engineering geology principles for dams. Soon thereafter, Ries and Watson (1914) revised their *Engineering Geology* text and included much of the symposium material; thus introducing modern geology for engineering works into the academic world.

Geological Society of America

Throughout his many years of service to the Society as secretary (1923 to 1940), Charles P. Berkey (chapter Frontispiece) made the relevance and importance of geology to planning and construction of engineering works a matter of first-hand familiarity to his GSA colleagues. For example, a brochure article in 1939 on engineering geology and his GSA presidential address on "The Geologist in Public Works" (Berkey, 1942) initiated further interest among geologists and engineers (Appendix I). Berkey's dedication to applied geology served as a guide and inspiration to many (Legget, 1973).

Henry Aldrich, who followed Berkey as the first permanent secretary of the Society in 1940, was a personal friend of many of the engineering geologists who ultimately founded GSA's Engineering Geology Division; these included Sidney Paige, one of the earliest to serve the U.S. Geological Survey on engineering works; Roger Rhoades, who served as the second chief geologist of the Tennessee Valley Authority; and Heinrich Ries, a pioneer applied geologist/teacher at Cornell University (Legget, 1973).

By 1946 the group of strong advocates of Engineering Geology and their associates, who had joined together informally in the late 1930s and during World War II, were eager to find an organization that would give identity to their geological specialization. Engineering geology had gained rapid acceptance among major engineering groups such as the Corps of Engineers, the Bureau of Reclamation, and the Tennessee Valley Authority, and with similar federal and state groups in the United States (Chapter 1, this volume). The Society of Economic Geologists, the American Institute of Mining Engineers, and the Geological Society of America had been recommended as suitable "homes"; each offered positive advantages as the logical organization for affiliation by the group of engineering geology practitioners.

In 1945 to 1946, proponents of a specialist organization among professionals were divided between affiliation with the SEG or GSA. Six of the eleven founders (Appendix A) were active SEG members and authors of numerous SEG *Bulletin* papers. Ultimately Berkey, with his long-time GSA affiliation, tipped the decision for the GSA, and the petition of 1947 (Appendix A) that founded the Engineering Geology Division soon followed. In overview and hindsight, the ability of GSA to assemble and publish papers and longer reports on engineering geology subjects has been a major benefit of affiliating with GSA—another is the opportunity for attendees at Section and Annual GSA meetings to keep up on changes and advances within the geosciences relevant to the practitioners' problems. These two major factors and strengths cannot be offered by specialty groups; attendance at GSA meetings provides the practitioner with the science-oriented information, while other groups may offer the more practical information. Many young engineering geologists do not seem to realize this tangible difference. GSA bylaws and statutes protect the Society as a nonprofit organization; thus the Society has not endorsed items such as building codes, registration, and professional practice mechanics, which are the main strengths and turf of other organizations.

FORMATIVE YEARS, 1946 TO 1955

Roger Rhoades, a leader among the founders of the Engineering Geology Division, consulted with Henry Aldrich at the 1946 annual meeting of the Geological Society of America in Chicago about forming a new division, and an informal session followed, attended by Sidney Paige, Arthur B. Cleaves, Parker Trask, Edward Burwell, William Irwin, and Shailer Philbrick. A separate section, the Cordilleran Section, had already been established; this precedent, and the personal attention of Secretary

Aldrich, strongly influenced acceptance of the petition (Legget, 1973). The minutes of the GSA Council meeting of December 1946 indicate the desire of engineering geologists to form a section within the Society, and the original petition of March 20, 1947, is given in Appendix A. The response of the GSA Council, granting the petition, is given in Appendix B; the bylaws were drawn up subsequently by Roger Rhoades, Arthur Cleaves, and George Woollard and approved by the Council at a meeting in April 1947. The Engineering Geology Division (EGD) was authorized as a topical group in contrast to a geographic section of the Society.

The Division became a reality, with Charles P. Berkey serving as the first chairman (Appendix E), during 1947 and 1948, Sidney Paige as vice-chairman, and Roger Rhoades, secretary. The first formal meeting occurred during the 1947 annual meeting of the Society in Ottawa, Canada; the program for this first meeting is given in Appendix C.

There have been regular sessions of the Division at every annual meeting of the Society since 1947, as well as programs and field trips at most of the section meetings. Today there are ten divisions of GSA whose bylaws and statutes all were patterned after those of the Engineering Geology Division.

Figure 1. Heinrich Ries, 1929 president of GSA and the third chairman of the Engineering Geology Division (1950).

Activities and accomplishments

The 1948 annual meeting was held in New York City and afforded Chairman Berkey an opportunity to assemble a strong program under William Irwin, S. S. Philbrick, and B. C. Moneymaker, in cooperation with T. A. Middlebrooks and Karl Terzaghi of ASCE; a stimulating field trip to major New York area projects was organized by Thomas F. Fluhr. After the enthusiastic New York meeting, the Division began to gather momentum with an initial group of sponsored activities that included the work of task committees on Teaching Aids (Cleaves, Wahlstrom, Happ); Bibliography of Engineering Geology (Eckel, Reiche, Carey); and Publications (Rhoades, Cleaves, Burwell).

The Society meeting in El Paso in 1949 and the Division's program and associated field trips attracted many western-states engineering geologists for the first time, including one of the authors (GAK). The single-session program emphasized on-going field projects (Appendix K), and the three task committees reported on progress and plans for the 1950s. Also, in 1946, President N. L. Bowen had appointed a working group of Sidney Paige (Chairman) and W. S. Mead, J. P. Buwalda, and B. C. Moneymaker to plan a special commemorative volume to honor the lifelong contributions to engineering geology by C. P. Berkey, long-time professor at Columbia University. Chairman Paige reported at the 1949 meeting that the major task of assembling this volume was completed and it would be released in 1950. The volume comprised twelve papers by specialists in various facets of engineering geology and was titled *The Application of Geology to Engineering Practice* (Appendix I). Few Society publications have enjoyed the sustained interest and demand of the Berkey volume, which had six printings between 1950 and 1967.

In 1950, Heinrich Ries, an eminent applied geologist of Cornell University and past president of GSA (1929), served as Division Chairman (Fig. 1). An enjoyable task for the Division during the formative years was publishing literature for the expanding field. The outstanding treatise on *Applied Sedimentation,* compiled by editor Parker D. Trask, was published in 1950 by John Wiley and Sons, with more than half of the 35 papers written by Division members. The two major publications of 1950, the Berkey and Sedimentation volumes (Appendix I; Trask, 1950), were the first of many conference volumes, symposia, texts, and special papers of the 1950s and 1960s that contributed to establishing the domain of engineering geology.

The need to bring about a better understanding of engineering geology and closer cooperation between geologists and engineers in the highway industry led to the organization of the Annual Highway Geology Symposium under the leadership of W. T. Parrott. The first meeting was in Richmond, Virginia, in February 1950 (see, McClure and others, this volume); several Division members, including S. E. Horner, J. D. McNeal, and Alice S. Allen, were involved with the early symposia. Perhaps the least formal among effective groups of geological specialists, the Symposium was still active in 1989, meeting annually in various parts of country, even though it has neither dues nor a membership list.

The expanded activities of 1951 under Chairman Roger Rhoades set in motion several new projects, including a single-task committee on the "Definition of the term Engineering Geology." This first definition (Appendix D)—a consensus of the professionals—has continued to express the principles and most widely accepted description of engineering geology, although a few voices over the past decades have argued that engineering geology is mainly an engineering discipline, or a hybrid discipline, or even a scientific field unto itself (discussed in Chapter 2, this volume).

In October of 1951, the Soil Mechanics and Foundation Division of the American Society of Civil Engineers (ASCE) voted to join with the Geological Society of America/Engineering Geology Division to form a Joint Committee on Engineering Geology. The initial Division representatives were Roger Rhoades and Edward B. Burwell; Karl Terzaghi of ASCE served as chairman (history described in McClure and others, this volume). This committee has sponsored many outstanding symposia, meetings, and other activities involving the Division since becoming activated in 1952.

In 1953, Professor Arthur B. Cleaves, an active practitioner since the 1930s and chief geologist of Pennsylvania Turnpike Authority, served as chairman. He had been compiling a census of all engineering geologists in North America since the 1940s and filed a list with the Secretary of 603 individuals associated with the field. Selected Reading List chairman S. C. Happ submitted a preliminary report and tabulation of references planned for publication in the GSA *Bulletin* (Appendix I). Task committees on "Products of Weathering of Bedrock" (T. H. Thornburn, chairman); "Influence of Geological Factors on Tunnel Construction" (A. B. Cleaves, chairman); "Nomenclature and Properties of Shales" (S. S. Philbrick, chairman); and "Properties of Rocks Exclusive of Shales" (V. Jones, chairman) presented reports on progress, including plans for future symposia on the four major areas critical to practice.

By 1955, the Division was realizing international recognition for its diversified programs and sponsored activities, enhanced by the efforts of individual members. The Committee on Landslide Investigation of the Highway Research Board under Edwin B. Eckel (chairman, 1951 to 1957) received strong support from Division members in the compilation of questionnaire data and other source materials for the publication "Landslides and Engineering Practice" (Eckel, 1958).

The Division chairman in 1955, Shailer S. Philbrick, reorganized the original Teaching Aids Committee and appointed three new members with experience in the training of engineering geologists in North America: John F. Mann (University of Southern California), John T. McGill (University of California, Los Angeles), and George A. Kiersch (University of Arizona), chairman. They were requested to prepare an evaluation and summary of North America teaching materials related to engineering geology. The Committee's efforts were published in 1957 as "Teaching Aids and Allied Materials in Engineering Geology" (Appendix I; Kiersch and others, 1957). The special publication included descriptions of coursework and curricula offered in engineering geology and geological engineering, and materials for presenting advanced-level coursework in "Applied Geology for Engineering." Another product was the Engineering Geology Reference List (Appendix I: Happ, 1955). Together with the teaching aids material, the two reports provided the earliest efforts by the profession to set forth guidelines for teaching purposes.

GROWTH YEARS, SINCE 1956

During the 1956 chairmanship of William H. Irwin, a publication panel on case histories was established to prepare a series of short histories on the geological and relevant engineering features of various types of construction projects. Parker Trask served as chairman, with members E. Dobrovolny, L. F. Grant, R. C. Marliave, L. E. Scott, and J. A. Trantina. The group acted with dispatch, and *Engineering Geology Case Histories No. 1,* consisting of nine separate papers, was published in May 1957. The informal style and offset printing allowed busy practitioners to present short accounts of their work for the benefit of students and the profession at large. Case Histories No. 2 and No. 3 followed in early 1958 and 1959. So far, twelve case histories have been published (Appendix I). The series has been very popular in both North America and many foreign areas, and most of the early numbers are now out of print.

At the 1958 meeting in St. Louis, Program Chairman Arthur B. Cleaves of Washington University arranged for a half-hour panel appearance on a local television station, along with incoming Chairman Robert F. Legget and Secretary George A. Kiersch. The public-affairs show covered geological activities of broad interest, with particular focus on the frequent slope failures of the St. Louis region associated with old Mississippi River terraces. The television appearance on geological endeavors was widely received by the Society as a new direction for the recognition of applied geological efforts and concerns for the environment. The technical program at the St. Louis meeting included a first-time review on research and related needs in the field of engineering geology.

The years 1959 and 1960 were action packed for the Division under Chairman Robert F. Legget and Chairman-elect Safford C. Happ. The California Association of Engineering Geology (CAEG), formed in 1957, had begun to extend membership to all engineering geologists of North America in 1958 (Appendix H). Subsequently, discussions were initiated between EGD and CAEG officers in 1960 to 1961 to explore the possibility of CAEG becoming the Cordilleran Section of the Division, but the proposal was dropped in 1961. CAEG felt the need for a special organization to deal with local problems such as matters of registration, building codes, and a distinction between engineering geologists and engineers. CAEG dropped California from its name in 1962 (as discussed in Chapter 5, this volume), and became the respected Association of Engineering Geologists.

During the 1959 term of Chairman Robert F. Legget, an informal *Newsletter* was established by Alice S. Allen, who had functioned as the assistant secretary of the Division since 1954. Alice Allen served as the resident liaison in Washington, D.C., for

Edwin B. Eckel, Chief of the Engineering Geology Branch, U.S. Geological Survey, located in Denver. As such, Alice Allen was aware of distinguished foreign visitors, symposia, conferences, and many other programs and meetings of interest to Division members, as well as research activities and related projects planned by major government agencies. The *Newsletter* included both broad professional news and brief reviews of new publications. Part of the news was submitted by four to six regional reporters and by individual members solicited by Allen during her editorship through 1966 (Appendix J).

At the 1959 annual meeting in Pittsburgh, the Division was approached by the ground-water contingent of its membership, under the leadership of George W. Maxey, Desert Research Institute, Reno, to withdraw from the Engineering Geology Division. At that time, most hydrogeologists were more interested in ground water as a resource than as a physical or adverse geologic factor relevant to engineering works. Although the EGD offered to change the Division title to include Hydrogeology, the hydrogeologists decided to petition the Council for separate status and were granted status as a new Division in November 1959.

An all-day field trip before the 1959 EGD meeting demonstrated the typical types of landslides, rock cuts, and subsidence over coal-mined areas of the Pittsburgh region. These were described in a guidebook assembled by editors Ackenheil and Philbrick. Program Chairmen Edwin B. Eckel and Robert F. Legget organized the presentation of a stimulating set of papers on "The Nevada Nuclear Test Sites" that was attended by over 400 people; some felt the symposium marked the "coming of age" of the Division (Legget, 1973). By the early 1960s, increasing numbers of the membership were involved with the siting, construction, and operation of nuclear power plant facilities and the concerns of nuclear industries.

A principal event of the year 1960 was the invitation by the American Institute of Mining Engineers for the Division to cosponsor a proposed "Society of Geological Engineers" (SGE) to be affiliated with the AIME. Professor John C. Maxwell of Princeton was chairman of the SGE group, and five representatives of the Division attended a conference with the SGE organizers at the Colorado School of Mines in Golden during the 1960 Denver meeting of the Society. Division Chairman George A. Kiersch expressed the feelings and consensus of EGD/GSA that "a further splintering of the applied geological profession by an SGE group was not in the best interests of the majority of practitioners." The proposal was debated further by members of several different organizations serving applied geologists during 1961 and 1962. Ultimately, the SGE group was formed within the AIME and functioned for a few years, but was disbanded around 1970 for lack of support outside of the immediate membership of AIME's Society of Mining.

During 1960 to 1961 the number of technical committees on special interests of Division members was expanded to include Legal Aspects, Coastal Engineering, Military Geology, Instrumentation, Engineering Seismology, Air-Photo Interpretation, Rock Mechanics, and River Engineering. These joined the long-standing committees on Dams and Reservoirs, Tunneling, Materials of Construction, and Publications. Likewise, the number of liaison activities was increased to seven in order to strengthen Division communications with the Cordilleran and Southeastern Sections of the Society, and technical organizations with overlapping interests, such as ASCE (Hydraulics), SEG, and AEG.

In 1962, the first Review Volume on major topics in engineering geology was published under the editorship of Thomas W. Fluhr and Robert F. Legget; eight volumes have been released to date (Appendix I).

An international conference on "State of Stress in Earth's Crust" was convened on June 13 to 14, 1963, in Santa Monica, California, under the leadership of William R. Judd. This first-ever symposium on stresses and related rock engineering/mechanics principles in North America was an outgrowth of the Division's Committee on Rock Mechanics, organized in 1960 with W. R. Judd as chairman. The papers and discussions from the stress conference were published (Judd, 1964) and subsequently received worldwide recognition among geoscientists and rock engineering specialists alike as a milestone of the engineering geology profession. The organization for the conference was largely composed of Division committee members, including Don U. Deere, Victor Dolmage, Nils Grosvenor, John W. Handin, Robert B. Merrill, Eugene C. Robertson, and Dart Wantland.

The Committee on Rock Mechanics of EGD/GSA became the National Academy of Science Committee (USNC) on Rock Mechanics in 1963. The Geological Society of America is an active supporting sponsor of the USNC/Rock Mechanics, and an appointed EGD representative is a member of the committee's management Board.

By 1965 the Division was in its seventh year of communicating with the membership through its simple, untitled, mimeograph-style typed newsletter. However, the members overwhelmingly requested a more formal news document. The Division reacted by establishing *The Engineering Geologist,* a quarterly newsletter; Alice S. Allen and Richard E. Goodman were coeditors, with a staff of six regional editors. Publication began in 1966, with a record four issues and 80 pages of information. The history of this regular Division release, which included a panel for the Chairman's yearly message, is tabulated as Appendix J.

At the Kansas City, Missouri, annual meeting in 1965, Chairman-elect Laurence B. James displayed the second "Berkey Gavel," fashioned from NX-rock core of Tejon Lookout Granite, along the Carley V. Porter tunnel of the California aqueduct, near the intersection of the Garlock and San Andreas fault zones.

Little is known about the first Berkey Gavel, other than that it was fashioned from a short segment of Manhattan Gneiss diamond drill core from New York City, provided by Thomas F. Fluhr and presented to the Division's first chairman, Charles Peter Berkey. In 1980 the gavel was recreated for a third time by Chairman Allen W. Hatheway (Fig. 2), and a limited edition of panel-gavels was made to be awarded each retiring Division

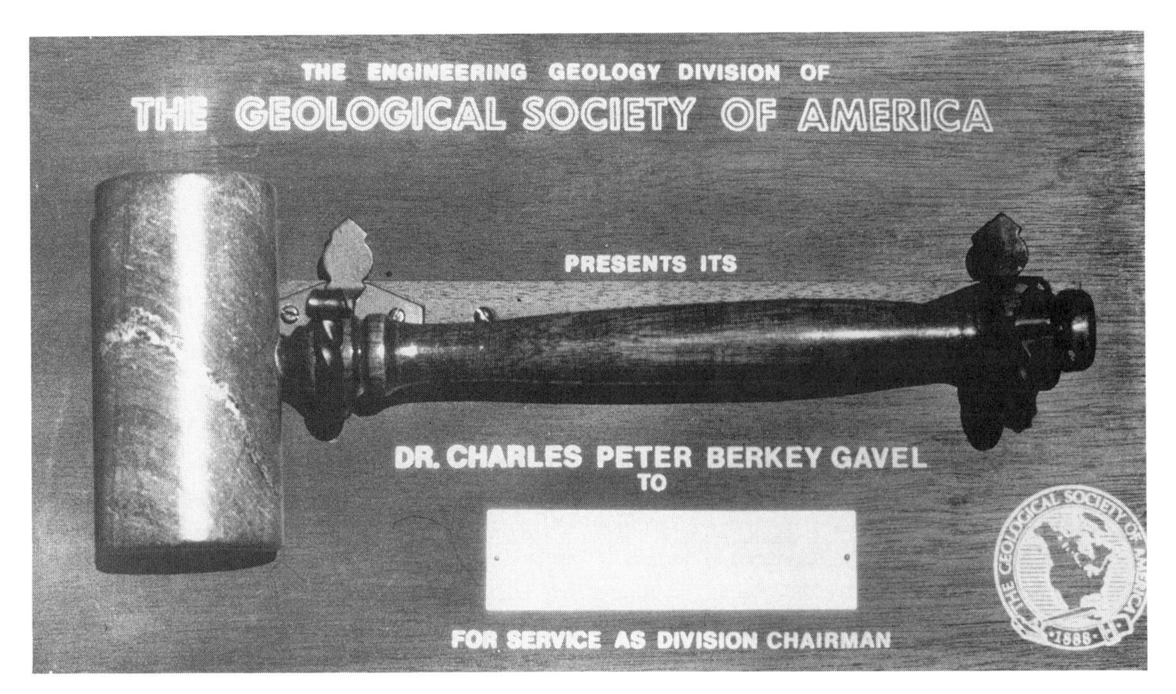

Figure 2. The Berkey Gavel.

chairman thereafter. This gavel employed NX-core of the famous Roxbury Conglomerate, chosen for its significance as being the bedrock of a portion of Boston in the area of practice of William O. Crosby, regarded as America's first full-time engineering geologist (see volume frontispiece).

The annual meeting of 1966 featured the record number of three strong Division symposia (Appendix K). They dealt with nuclear reactor siting, the Bay Area rapid transit system, and land subsidence; all very topical subjects for the coming decade of the 1970s. Past-Chairman Robert F. Legget (1959) served as president of GSA in 1966 (Fig. 3); his presidential address was entitled "Soil-Geology and Use."

A symposium on shale under the guidance of Shailer S. Philbrick was sponsored by the Joint Committee for Engineering Geology/GSA-ASCE in Denver, Colorado, during a Water Resources Conference in May, 1966; Division members contributed major efforts for field trips and the symposium session, which included L. B. Underwood, D. U. Deere, and C. S. Content.

The late 1960s were times of great activity for the profession. Dam and associated tunnel construction was booming, both at home and abroad. The scramble for siting of nuclear plants had begun, although the Atomic Energy Commission (Nuclear Regulatory Commission today) would not issue its standardized format until 1970 for Preliminary Safety Analysis Reports (PSAR) that were to occupy a substantial number of practicing engineering geologists in years ahead.

Chairman Donald H. McDonald began the regular *Newsletter* feature "Message from the Chairman" in 1968, a tradition that rightfully and needfully continues to this day. Also, the final task of the long-standing Teaching Aids Committee (1948 to 1962) was completed in 1962 with the publication of the "Directory of 109 black and white 16-mm Motion Pictures" by John T. McGill (Appendix I; discussed Chapter 2, this volume).

In 1968, Past-Chairman Edwin B. Eckel (Fig. 4) was appointed Science Editor of GSA. He later became executive secretary through 1974 and served with distinction. Afterward he prepared the first up-to-date history of the Society (Eckel, 1982).

W. Harold Stuart, long-time conscience and servant of the Division, served as its chairman in 1969. The Atlantic City annual meeting program included a research-oriented symposium, "Dynamic Problems in Engineering Geology," organized by George A. Kiersch, which reviewed typical problems awaiting in the 1970s, such as reservoir-induced seismicity and age dating of fault movement. This first-ever Division session on dynamic problems was well attended, and the lively discussions were indicative of the broad interest in research on problems within the profession.

Chairman Stuart had successfully carried his 1968 proposal for creation of the E. B. Burwell, Jr., Memorial Award (Fig. 5) to the Society Council, and the first award was made at the Atlantic City meetings to Lloyd B. "Spike" Underwood in 1969 (Appendix F), respected successor to Edward B. Burwell (chairman 1952) and to Robert H. Nesbitt (Chairman 1962) as the fourth chief geologist of the U.S. Army Corps of Engineers. His outstanding paper, the "Properties of Shale" (Underwood, 1967), continues to be a classic review on this major problem. W. Harold Stuart issued four separate chairman's messages during 1969 in *The Engineering Geologist*; they constituted a benchmark series of thoughtful remarks for the professional.

In its twenty-third year, 1970, the Division was led by

Chairman H. Garland Hershey. In accordance with past recommendations, the management board/EGD appointed a five-member Long-Range Planning Committee to consist of the current chairman, the three immediate past chairmen, and a fifth chosen by the other four. Matters of concern that affect the Division's future are referred to this committee, such as organizational changes, ways to further promotion of engineering geology, and revision of bylaws by the management board.

In 1970, the first geologically oriented results of the National Environmental Policy Act of 1969 were being felt in the form of required Environmental Impact Statements (EISs), and a number of county and state initiatives were in force calling for site-qualification reports by engineering geologists. California initiated its registration act for Professional Geologists and Certified Engineering Geologists (1969) through the efforts of CAEG/AEG, as the second state offering professional registration of geologists, after Arizona's pioneering effort (1921).

The Division's influence on GSA's older (1852) fellow organization, the American Society of Civil Engineers (ASCE), was at an all-time high, with considerable activity in the nine-person Joint Committee on Engineering Geology (JCEG). The JCEG was enjoying its peak acceptance as a quiet yet effective committee responsible for organizing sessions on engineering geology during the national or specialty meetings of both ASCE and GSA. Much of this cooperation was due to far-sighted efforts by EGD in recruiting influential ASCE members into GSA membership. Their recognized geological training, experience, and professional expertise enhanced the role and status of engineering geologists. Since the 1950s, a liaison had been in place with the Society of Petroleum Engineers (SPE), and the Society of Mining Engineers of the American Institute of Mining and Metallurgical Engineers (AIME). Both of these strong affiliates and early supporters of EGD activities are missing from Division membership today; since the 1970s, many of the engineering geologists in both petroleum and mining have become affiliated with the U.S. National Committee for Rock Mechanics (USNC/RM) and its reservoir engineering or open-pit activities.

The year 1970 also saw appointment of three EGD members (George O. Gates, Richard H. Jahns, and Donald R. Nichols) to the California Legislature Joint Committee on Seismic Safety, a powerful voice in what was later to bear fruit as the Alquist-Priolo Act (1974), creating active-fault study zones prior to issuance of permits for multifamily and public-service construction.

The Division was served by Howard C. Coombs as chairman in 1971, at a time when the interest of members was shifting away from the long-standing emphasis on dams and reservoirs. The foresight of the Joint Committee on Engineering Geology led to creation and sponsorship of a 1971 Conference at the annual ASCE meeting (Phoenix), on "Geologic Aspects of Subsurface Waste Disposal," a topic perhaps ten years ahead of its time.

By 1972, when Richard E. Gray served as chairman, the Division could claim 600 members. The U.S. National Committee on Tunneling Technology (USNCTT) of the National Research Council was formed that year, under the leadership of EGD member Don U. Deere. He had been affiliated with the Division's Committee on Tunneling and Underground Openings, which began in 1953, and the USNCTT efforts built on the interest and leadership of this group. EGD has maintained its representation in USNCTT ever since and provides a liaison representative to the committee. W. Harold Stuart completed another act of stewardship for the Division, producing a set of revised bylaws that reflected current practices in GSA and the Division.

Figure 3. Robert F. Legget, 1966 president of GSA and chairman of the Engineering Geology Division in 1959.

The Engineering Geology Division celebrated its Twenty-fifth Anniversary at the 1972 GSA annual meeting and was addressed by Robert F. Legget (chairman, 1959) at the Division's annual luncheon. Past-chairman Legget ended his Twenty-fifth Anniversary Address with the challenge for "every single member of the Division—to record the vital information that we know to be available in our special field" (Legget, 1973). Reacting to this challenge, the division chairman in 1973, Gordon W. Prescott, reactivated some of the dozen former technical panels that were

Figure 4. Edwin B. Eckel, executive secretary of GSA from 1970 to 1974, and chairman of the Engineering Geology Division in 1954.

established in 1960 to 1961 or earlier for reporting on the activities and new advances in the areas of construction materials, dams and reservoirs, engineering seismology, river engineering, and underground excavation.

Symposia and attendance at meetings reflected the general upswing in engineering geologic activity largely due to environmental concerns, nuclear plant siting and construction, and the growing awareness of risk from natural hazards. Division membership increased to 1,042 members in 1974, and to 1,339 members a year later.

The *Engineering Geologist,* or Division newsletter, came into a period of stability with the three-year term (1973 to 1975) of editor Mary E. Horne, which featured strong summaries on Division activities and engineering-geological activities within the geographic sections of GSA. The first back-to-back annual meeting with the Association of Engineering Geologists (AEG) was planned for Seattle in 1977, under the guidance of Richard W. Galster, chairman in 1978.

During the chairmanship of Paul L. Hilpman (1975), discussions took place concerning the viability of the Division, in the face of continued local section growth and expansion of the Association of Engineering Geologists (AEG). David J. Varnes, chairman-elect for 1977, spent considerable time tabulating a list of the specialty-categories among those who held dual membership in the Division and in AEG. This assessment formed the basis for an EGD Management Board decision to put the question of whether the Division should be deactivated to a vote of the membership; the motion to deactivate was defeated.

Since 1977, the Division has strengthened some of its activities and undertaken other new ventures to serve its diverse membership, with emphasis on the science-oriented aspects of professional practice dealing with the physical processes, phenomena, and environmentally related risks associated with engineering works. This emphasis is reflected in the subjects of the four Case Histories (Nos. 8, 9, 10, and 11) and Review Volumes III and IV published by the Division in the 1970s (Appendix I).

Many positive impacts on the profession were realized by the rush to site and permit nuclear power plants in the 1960s to 1970s. One of the most important was the development of techniques to assess "active" or "capable" faults, or those with the propensity and relatively high probability of producing earthquakes. The Joint Committee on Engineering Geology/GSA-ASCE-AEG, under the chairmanship of Robert L. Schuster, organized a symposium on "Capable Faulting Relevant to Engineering Works," held at the annual GSA meeting in Seattle, in 1977. This milestone symposium created major interest in engineering geology (several hundred attendees); the audience ranged from the more scientifically oriented members of GSA to experienced engineering practitioners. An additional stimulus was the large number of nuclear plants and facilities in the Pacific Northwest states. This joint activity (ASCE-GSA) was repeated at the annual meeting in 1980 with a symposium on "Geologic Hazards in Founding Offshore Structures."

In October 1977 an interdisciplinary Penrose Conference on landslides was held in Vail, Colorado. Convenors were Robert F. Fleming, GSA; Don C. Banks, chairman of the ASCE Committee on Rock Mechanics; and Robert L. Schuster, chairman of the Joint Committee on Engineering Geology (ASCE, GSA, AEG). The 70 participants focused on: (1) recognition and classification; (2) instrumentation, sampling, and testing; (3) analyses and design; and (4) remedial techniques.

In 1977 the Division, under the leadership of Chairman David J. Varnes, agreed to support a concerted effort to organize a U.S. National Group of the International Association of Engineering Geologists (USNC-IAEG). The proposed plan was to create the group under the aegis of the U.S. National Committee on Geology, a committee of the Board on Earth Sciences, National Academy Sciences/National Research Council. The United States was, at that time, the last major industrial nation without membership in the worldwide IAEG. The U.S. National Committee for the IAEG was ultimately formed in 1981 and has been guided largely by active EGD members since (see description of IAEG in Chapter 5, this volume). All three chairmen of the USNC/IAEG have been past chairmen of the Division. In 1987, the USNC/IAEG, now a subcommittee of U.S. National

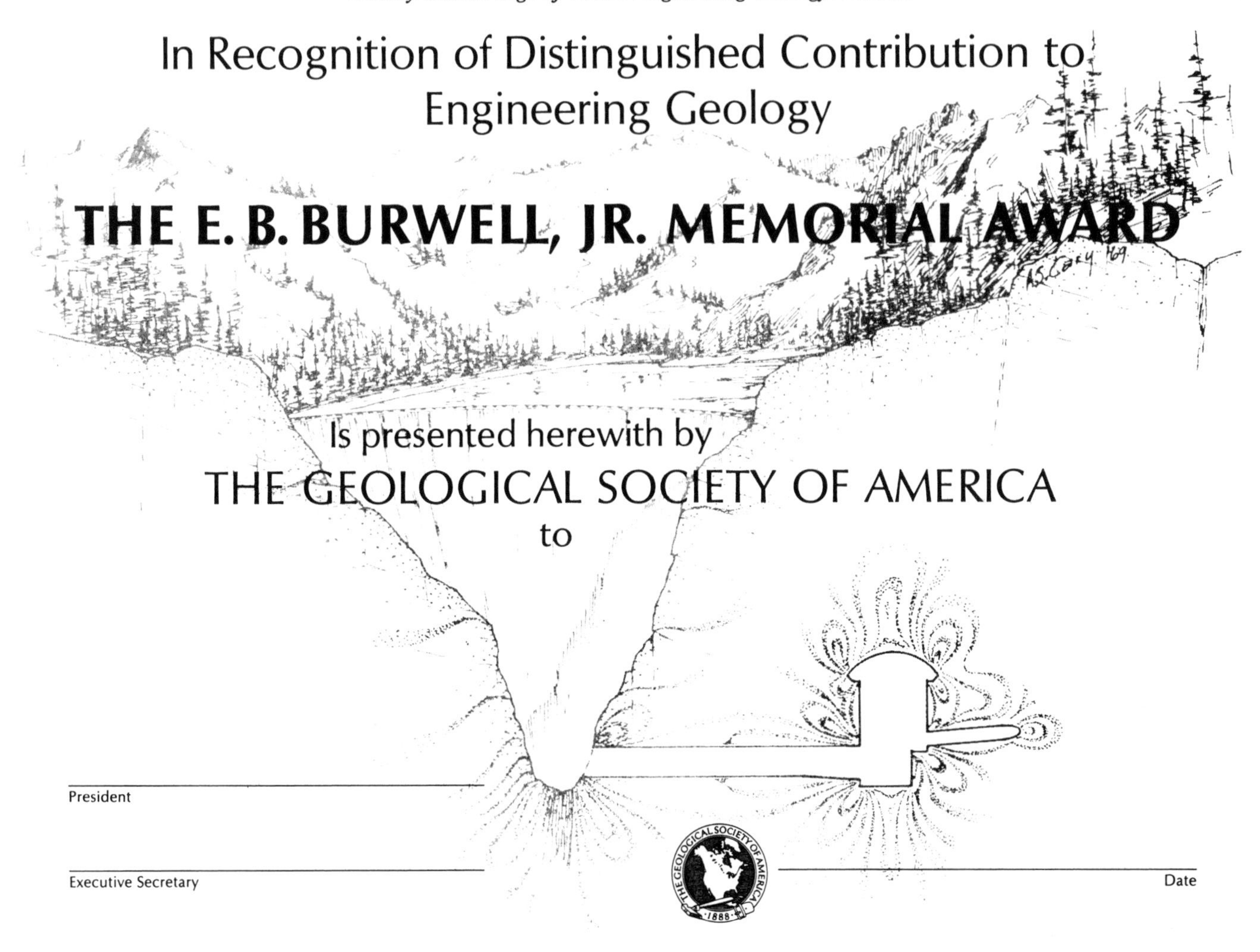

In Recognition of Distinguished Contribution to
Engineering Geology
THE E. B. BURWELL, JR. MEMORIAL AWARD
Is presented herewith by
THE GEOLOGICAL SOCIETY OF AMERICA
to
President
Executive Secretary
Date
THE GEOLOGICAL SOCIETY OF AMERICA
1888

Figure 5. The E. B. Burwell, Jr., Memorial Award plaque.

Committee on Geology, was granted full voting rights within the Committee, and its duties were broadened to include monitoring of ongoing research activities in the field as a means of keeping engineering geologists informed.

Intersociety cooperation was further enhanced in 1979 by Past Chairman James W. Skehan (1976), who developed a well-received joint EGD-NAGT (National Association of Geology Teachers)-AEG symposium on "Academic Training of Engineering Geologists," for the GSA annual meeting at San Diego.

The year 1980 was the "recognition year" for the truly threatening nature of uncontrolled hazardous waste disposal sites. Congress passed the Comprehensive Environmental Response, Compensation, and Liability Act (CERCLA:Superfund), and gave the task of developing its regulations to the Environmental Protection Agency (EPA). A massive yet diffuse attack on uncontrolled hazardous waste sites began, with multi-million-dollar technical contracts awarded to mainly environmental engineering consulting firms. Engineering geologists, however, were not in the vanguard of most of the early hiring, for the entire nature of site characterization and properties of waste was partly misunderstood. It was not until 1983 to 1984 that the need for guidance from experienced engineering geologists was recognized: historically and scientifically, much of the long-time practice in engineering geology involved the environment and its inherent risks.

With technological developments and geological applications for engineering developing quickly, the Division began a program in 1980 to develop data sheets, of the pocket, six-ring looseleaf-notebook variety, similar to those initiated by the American Geological Institute (R. W. Foose) that appeared in *Geotimes* between 1957 and 1964. The format was revised, and four new data sheets were published in *The Engineering Geologist* between 1982 and mid-1987.

In 1981, under Chairman John S. Scott, the Division decided on a Centennial publication for the Decade of North American Geology (DNAG), and the annual symposium was on the "Role of Government Agencies in the Development of Engineering Geology." In 1982 the EGD symposium at New Orleans continued this focus with the theme "Engineering Geologic Lessons Learned from History."

By the early 1980s, the Division also recognized that a

growing number of outstanding members deserved personal recognition. In 1982, two new awards that had been proposed earlier, the Distinguished Practice Award and the Meritorious Service Award, were established under Chairman Harry F. Ferguson. The guidelines for both awards were prepared by Past Chairman Richard E. Gray. The first recipient of the EGD Distinguished Practice Award was Neil B. Steuer in 1982, lately retired of the U.S. Nuclear Regulatory Commission, and a pioneer in the successful coordination of regional seismotectonic studies throughout the United States. Subsequent recipients of this award from 1983 through 1989 are given in Appendix G.

The 1984 Division symposium in Reno was a joint effort with the Quaternary Geology Division on "Debris Flows and Related Features" and attracted over 700 attendees. That year, the bylaws were amended so the terms of the Management Board representative coincided with the Division's officers; also, Robert H. Fickies completed a long and distinguished term as editor of the *Engineering Geologist.*

In October 1985 the Division Management Board approved creation of an AEG-GSA coordinating committee to include the EGD chairman-elect plus one member-at-large from the Division. The first Division technical short course was organized by Jeffrey R. Keaton, during the chairmanship of David M. Cruden (1986) and was presented as "Engineering Geology for Geologists" at the San Antonio GSA meeting in 1986. This was a revised version of an American Geological Institute course delivered at the 1978 AEG annual meeting.

In recognition of the Fortieth Anniversary of the Division in 1987 a committee chaired by David M. Cruden was appointed to raise funds and formulate guidelines for an annual grant to support research in engineering geology through the GSA Research Award Program, described below. At the 1988 Denver GSA meeting, the Division, in cooperation with the Association of Engineering Geologists, announced establishment of the Richard H. Jahns Distinguished Lectureship in Engineering Geology, in honor of the Division chairman in 1979 (Fig. 6) and past-president of Society in 1971.

Since the early 1980s, the Engineering Geology Division has undertaken and broadened its support of the most proven areas of EGD activity: symposia and publications on applied geoscience for engineering works. In addition, the Division has been a clearing house for studies of the specific processes of geological sciences relevant to the coordinated efforts of engineering geologists, geological engineers, geologists, and civil engineers concerned with the feasibility, design, construction, and operation of engineering works. To this end, the EGD technical sessions, as part of the GSA program and that of its associated societies, provide meeting places to hear from the leaders of the geoscience specialities each year. Invariably, symposia and research papers are presented on new advances in the state of knowledge, on processes, or on related developments in the geosciences. To a well-rounded engineering geologist, this research is important in blending the old ideas with new, up-to-date principles and knowledge when called on to understand increasingly complex geo-related problems associated with all kinds of engineering works, the environment, and/or the public's safety.

Figure 6. Richard H. Jahns, 1971 president of GSA and chairman of the Engineering Geology Division in 1979.

Publications

The EGD/GSA has devoted much effort to publications, as summarized in Appendix I. Division publications on the broad range of applied geosciences for engineering works offer far-ranging coverage through several series of publications, including *Special Volumes, Reports, Case Histories,* and *Reviews in Engineering Geology,* besides the opportunity to publish in the GSA *Bulletin* and *Geology,* or release short summaries or technical notes in *The Engineering Geologist.*

The *Special Volumes* were a new type of publication by GSA, as was the *Case History* series that provided a medium for short progress papers and immediate contributions for guidance to the practitioner; all have been self-supporting financially. The *Review Volumes* were started with the anticipation that they would prove to be a collection of annual reviews of major topics in engineering geology, with no repetition of a subject for at least ten years.

Awards by the Engineering Geology Division

Burwell Award. The Edward B. Burwell, Jr., Award, established by the Division in 1968, honors the memory of one of the founding members of the Division and the first chief geologist of the U.S. Army Corps of Engineers. It consists of an embossed Award Certificate shown in Figure 4. Award recipients are listed in Appendix F. This award is made to the author or authors of a published paper of distinction that advances knowledge concerning principles or practice of engineering geology, or of related fields of applied soil or rock mechanics where the role of geology is emphasized. The paper that receives the award must (1) deal with engineering geology or a closely related field, and (2) have been published no more than 5 years prior to its selection. There are no restrictions as to the publisher or publishing agency of the paper. The author or authors of the selected paper need not be a member(s) of the Engineering Geology Division or of the Geological Society of America and need not be a resident(s) or citizen(s) of the United States.

Distinguished Practice Award. The Division membership feels it is appropriate to recognize outstanding individuals for their continuing contributions to the technical and/or professional stature of engineering geology. Presentation of the Distinguished Practice Award recognizes such contributions. A nominee need not be a member of the Engineering Geology Division but must have made a major contribution to engineering geology in North America. The nomination consists of filing the nominee's resume and a brief statement of the technical and/or professional contributions that form the basis for the candidate's nomination. The selection of the awardee is made by the Committee on Distinguished Practice and Meritorious Service Awards. Approval of the selectee is the responsibility of the Division Management Board. Only one Distinguished Practice Award can be given each year, and the award is a citation and plaque. Recipients since 1982 are listed in Appendix G.

Meritorious Service Award. The Division's Meritorious Service Award was developed in 1982 to recognize outstanding efforts on behalf of deserving members. The Committee on Distinguished Practice and Meritorious Service Awards annually ranks the candidate nominees. The highest ranked nominee or, in unusual circumstances, the two highest ranked candidates receive the award. Approval of the recipient(s) is the responsibility of the Division Management Board. The Meritorious Service Award was preceded, in 1980, by the presentation of several Certificates of Appreciation to worthy members. A list of Meritorious Service Award recipients and the earlier Certificates of Appreciation are shown in Appendix G-1.

The Engineering Geology Division Fortieth Anniversary Research Fund

In recognition of the Engineering Geology Division's Fortieth Anniversary in 1987, the Management Board established an Anniversary Award Fund for research in engineering geology. This endeavor collected some $6,000 under the guidance of Past Chairman David M. Cruden. An annual grant of $500 for "research in engineering geology" will be made through the GSA Research Foundation Award Program.

To be eligible for the award, applicants must be members of the Engineering Geology Division or under the supervision of a member. The research proposed should promote the geosciences as applied to engineering works. Awardees will be selected by the GSA Committee on Research Grants, in coordination with the Division Management Board.

OUTLOOK: THE TWENTY-FIRST CENTURY

Today, engineering geologists and other applied geoscientists recognize that they are faced with a developed professional technology that is extremely diverse. There are few generalists among the engineering geologists of today; the basic geology or geoscience background of many practitioners can be as different as the interrelated geological and engineering principles involved. Experience, including an active career of diversified geo-related field work, data assimilation, analysis, and technical reading/research, has become the basis for excellence in engineering geology; a single structured course of university studies is not enough.

Although technical literature relevant to engineering geology was sparse before the late 1930s, as in related fields, this had changed by the 1950s, with the substantial increase in projects and geological practitioners (Appendices L and K). The problem of data assimilation has since developed for the profession; since the 1970s there has been more relevant information available than is possible to catalogue—let alone assimilate—by a single professional. Yet a familiarity with the new information is essential, as are the classic papers and studies of the earlier decades. Today, each practitioner must develop a strong reliance on the broad geoscience literature as well as that of the several specific subfields of practice in which it is possible to become proficient (described in Chapters 1 and 2, this volume).

Traditionally, the early engineering geologists had been academically trained as classical geologists, and they gained the relevance of engineering through project experience. This began to change as over two dozen university departments began to offer a specialty option in engineering geology or in geological engineering (for construction) during the 1950s and 1960s (Chapter 2, Appendices). With this group of newly trained personnel, relatively large numbers of young geologists are gaining professional employment with private engineering organizations and consulting firms, and the need for dissemination of technical information has burgeoned.

GSA President W. G. Ernst, in his retiring presidential address at San Antonio in 1986, reviewed the "Earth Sciences Status and Future: How Bad, How Good" and made several references to the growing importance of engineering geology (Ernst, 1987, p. 5). He cited numerous reasons for its steady expansion and described the enhanced image and potential for applied geoscience practitioners in the coming generations. The acceptance and growth of engineering geology, previously related to long-standing public awareness of the need for reliable site evaluations, is being broadened by increased governmental regulations for assessment of actions affecting the environment, including those related to demands for earth materials, safeguarding ground-water resources from contamination, and improving safeguards for human health. Requirements are increasing for geological input in land-use planning in light of rising concern about risks associated with natural and man-induced geologic hazards.

Robert F. Legget (1973) reminded the profession in his Twenty-Fifth Anniversary remarks that "The future waits, and the 21st century will be a very different world where the cities cover over twice the area of the 1970s." Within these "exploding environs" the challenges for the younger Division professionals and the next generation of new members will be even more diverse and scientifically provoking; demands on future leaders will open new and yet-to-be-conceived frontiers for applied geoscientists. The members of the Engineering Geology Division will have a leading role in carrying out the mandates and requests of mankind and thereby build for a better and safer twenty-first century.

For example, the 1988 GSA Division symposium (Appendix K) was on "Hazard Reduction in the 21st Century." Major advances in the next decades, discussed by speakers at that symposium, will include the widespread acceptance of probabilistic analysis to offer realistic long-term predictions of seismic hazards. This should lead to elimination of the "earthquake problem" within the twenty-first century (Allen, 1988), along with the control of earthquakes by lubrication of stressed faults and plate boundaries. Likewise, a reduction or alleviation of volcanic hazards, although difficult, may be accomplished by drilling into magma chambers and/or draining lakes in calderas to reduce the interaction of forces (Smith, 1985, p. 677).

APPENDIX A

Petition to the Council of The Geological Society of America:

We, the undersigned, desirous of creating a Section on Engineering Geology, and on behalf of a larger group known to have equal deep interests in such a Section, respectfully petition the Council to authorize the creation of such a Section in accordance with the appended By-laws which are believed to meet the requirements of Section VI of the By-laws of the Society.

We respectfully call attention to Article IX of the appended By-laws of the Section which indicate the procedure to put such a Section into immediate operation.

March 20, 1947

Charles P. Berkey, R. M. Leggette*, Chas. P. Theis*, H. Ries, John W Vanderwilt*, Irving B. Crosby, Sidney Paige*, Roger Rhoades*, George P. Woollard, T. T. Quirke, Olaf N. Rove*

*Also, member Society of Economic Geologists.

APPENDIX B

Engineering Section

Secretary Aldrich read a petition signed by 11 Fellows of the Society, asking the GSA Council "to authorize the creation of a Section on Engineering Geology, in accord with Section VI of the By-Laws." The purpose of the Section had been stated "to improve and promote the application of geology to engineering works, where the most effective realization of the engineering objective depends on geological interpretation of the natural conditions to which engineering structure or plan must be adapted." Vice-President Buddington suggested that the wording be modified to read, "the promotion of the science of geology as applied to engineering work."

Motion to grant the petition with this slight change in purpose was made by Councilor Anderson, seconded by Vice-President Gilluly and carried. Query was raised as to probable expenses of the Section, which would be cared for by the Society, and the Treasurer Hotchkiss commented that they would be expected to approximate the expenses of the Cordilleran Section. Motion carried.

APPENDIX C

Program, First Meeting—Engineering Geology Division
Ottawa, 1947
C. P. Berkey and Roger Rhoades, Cochairmen

1. 9:00—Business meeting and round-table discussion
2. 10:30—**William Irwin:** Application of field and laboratory geology to engineering works
3. 10:45—**Roger Rhoades:** Some fields of ignorance in Engineering Geology
4. 11:05—**Richard C. Mielenz, Kenneth T. Greene, and Elton J. Benton:** Chemical test for reactivity of concrete aggregates with cement alkalies
5. 11:20—**Robert H. Nesbitt:** Geology in concrete aggregate technology
6. 11:40—**Stafford C. Happ:** Geology of Kanopolis Dam, Kansas
7. 11:55—**Robert E. Barnett:** Effectiveness of geophysical explorations at Ft. Randall Dam, South Dakota
8. 12:15—**Shailer S. Philbrick:** Relationship of cyclothems to dam design
9. 12:35—**C. P. Berkey:** Geological contributions from the deep tunnels of New York
10. 12:55—**M. W. Bartley and R. F. Leggett:** Glacial geology at Steep Rock Lake, Ontario, and associated engineering problems

APPENDIX D

Report of Committee—Definition of
"Professional Engineering Geologist"

The Executive Committee of the Division on Engineering Geology GSA, during its informal meeting in November 1951, authorized the definition of the term Professional Engineering Geologist. Mr. E. B. Burwell, Jr., Chairman of the Division, appointed the undersigned by letter dated March 28, 1952, as a committee to prepare a proposed definition and submit it to members of the Executive Committee for comment.

This report summarizes the status of the definition and suggests a definition for consideration by Council members.

The term to be defined consists of three words. The first two words are adjectives which define a noun. The noun "Geologist" has a widely accepted, well established meaning which requires no comment. However, the two adjectives "Professional" and "Engineering" when coupled with the noun "Geologist" require careful examination in order to restrict their implications.

Let us consider the adjective "Engineering" in relation to the noun "Geologist". In the first place, an "Engineering Geologist" is not one who "engineers". Rather he is one who is associated with engineers in their undertakings, problems, plans and works. Furthermore, he is selective in the type of engineers with whom he associates. In the main they are civil engineers because it is the civil engineer whose endeavors in the field of construction have raised the demand for geological assistance in engineering construction. Occasionally the engineering geologist may be associated with architects, structural engineers, foundation engineers or others in the engineering profession whose work required geological assistance but does not include the search for and extraction of minerals unless they be materials of construction requiring only mechanical processing. Therefore the adjective "Engineering" in "Engineering Geologist" specifies the field in which this geologist operates. It also indicates that he is not in mining engineering because there the geologist would be identified as an "Economic Geologist" or "Mining Geologist". Likewise, not in petroleum engineering where the term for the associated geologist is "Petroleum Geologist". However, let us continue no more because further pursuit of the fields of engineering will shortly bring us to mechanical engineering where we might find the "Mechanical Geologist".

The adjective "Professional" implies several characteristics.

1. This geologist is not an amateur. He is a professional by training, experience, capability, and let us hope, remuneration.

2. He is not, however, a "Professional" in the sense that he has met State licensing requirements because State registration and licensing have not yet materialized. However, when such are required in the State in which he resides or practices he shall comply with the pertinent statutes and regulations.

3. One should remember that "Professional" was attached to "Engineer" to separate apprentice-trained, stationary, hoisting and locomotive engineers from the collegiate-trained civil, mechanical, etc., engineers.

With these ideas let us proceed to the definition:

Definition of Professional Engineering Geologist

1. A professional engineering geologist is a person who, by reason of his special knowledge of the geological sciences and the principles and methods of engineering analysis and design acquired by professional education or practical experience, is qualified to apply such special knowledge for the purpose of rendering professional services or accomplishing creative work such as consultation, investigation, planning, design or supervision of construction for the purpose of assuring that the geologic elements affecting the structures, works or projects are adequately treated by the responsible engineer.

2. A person is qualified as a professional engineering geologist by reason of his professional education if he holds a degree in geology from a college, university or institute of technology authorized, under the laws of the jurisdiction in which it is located, to grant academic degrees in sciences, and if he has acquired by practical experience or by professional engineering education the fundamental methods of engineering analysis.

3. A person licensed or registered to practice as a professional engineering geologist in any State, territory or possession of the United States or in the District of Columbia, is qualified as a professional engineering geologist.

Sidney Paige
Berlen C. Moneymaker
Shailer S. Philbrick, Chairman
October 6, 1952

APPENDIX E

Chairmen of the Division

1947 Charles P. Berkey, Columbia University, New York
1948 Charles P. Berkey, Columbia University, New York
1949 Sidney Paige, U.S. Geological Survey, Columbia University, New York
1950 Heinrich Ries, Cornell University, Ithaca, New York
1951 Roger W. Rhoades U.S. Bureau of Reclamation, Denver
1952 Edward B. Burwell, Jr., U.S. Army Corps of Engineers, Washington, D.C.
1953 Arthur B. Cleaves, Washington University, St. Louis
1954 Edwin B. Eckel, U.S. Geological Survey, Denver
1955 Shailer S. Philbrick, U.S. Army Corps of Engineers, Pittsburgh
1956 William H. Irwin, U.S. Bureau of Reclamation, Denver
1957 John R. Schultz, Harza Engineering Co., Chicago
1958 Ernest W. Dobrovolny, U.S. Geological Survey, Denver
1959 Robert F. Legget, National Research Council, Ottawa, Canada
1960 Stafford H. Happ, U.S. Atomic Energy Commission, Grand Junction, Colorado
1961 George A. Kiersch, Cornell University, Ithaca, New York
1962 Robert H. Nesbitt, U.S. Army Corps of Engineers, Washington, D.C.
1963 Thomas F. Thompson, Consultant, Burlingame, California
1964 John H. Melvin, Ohio State Geological Survey, Columbus
1965 George E. Ekblaw, Illinois State Geological Survey, Champaign
1966 Laurence B. James, California Dept. of Water Resources, Sacramento
1967 Elmer C. Marliave, Consultant, Sacramento, California, (until his death, 24 September 1967)
1967 Robert W. Karpinski, University of Illinois, Chicago
1968 Donald H. McDonald, H. H. Acres, Consultant, Niagara Falls, New York
1969 W. Harold Stuart, U.S. Army Corps of Engineers, Portland
1970 H. Garland Hershey, State Geologist of Iowa; Office of Water Resources, U.S. Department Interior, Washington, D.C.
1971 Howard A. Coombs, University of Washington, Seattle
1972 Richard E. Gray, General Analytics, Inc., Pittsburgh
1973 Gordon W. Prescott, U.S. Army Corps of Engineers, Washington, D.C.
1974 Howard J. Pincus, University of Wisconsin-Milwaukee
1975 Paul L. Hilpman, University of Missouri-Kansas City
1976 James W. Skehan, Boston College, Boston
1977 David J. Varnes, U.S. Geological Survey, Denver
1978 Richard W. Galster, U.S. Army Corps of Engineers, Seattle
1979 Richard H. Jahns, Stanford University, Palo Alto
1980 Allen W. Hatheway, Haley & Aldrich, Inc., Cambridge, Massachusetts
1981 John S. Scott, Geological Survey of Canada, Ottawa
1982 Harry F. Ferguson, USCE, Consultant, Pittsburgh
1983 Erhard M. Winkler, University of Notre Dame, South Bend
1984 Frank W. Wilson, Kansas Geological Survey, Lawrence
1985 Robert L. Schuster, U.S. Geological Survey, Denver
1986 David M. Cruden, University of Alberta, Edmonton
1987 Christopher C. Mathewson, Texas A&M University, College Station
1988 Ellis L. Krinitzsky, U.S. Army Engineers, Waterways Experiment Station, Vicksburg
1989 Thomas L. Holzer, U.S. Geological Survey, Menlo Park
1990 Jeffrey R. Keaton, Sergents, Hauskins, Beckwith, Salt Lake City

APPENDIX F

Recipients of the E. B. Burwell, Jr., Memorial Award
(Citations and responses given in the Division
Newsletter, or GSA *Bulletin* beginning in 1986)

1969–Lloyd B. Underwood (U.S. Corps of Engineers), Classification and Identification of Shales: Journal of Soil Mechanics and Foundations Division, America Society of Civil Engineers, SM6, p. 97–116, 1967.

1970–Glen R. Scott and David J. Varnes (of U.S. Geological Survey) General and Engineering Geology, United States Air Force Academy Site, Colorado: U.S. Geological Survey Professional Paper 551, 1967.

1971–Edwin B. Eckel (U.S. Geological Survey), The Alaska Earthquake, March 27, 1964, Lessons and Conclusions: U.S. Geological Survey Professional Paper 546, 1970.

1972–Richard J. Proctor (Metropolitan Water District of Southern California), Mapping Geologic Conditions in Tunnels: Association of Engineering Geologists Bulletin, v. 8, no. 1, 1970.

1973–Murray R. McComas and J. E. Hackett (Illinois Geological Survey), Geology for Planning in McHenry County: Illinois Geological Survey Circular 438, 1969.

1974–Robert F. Legget (National Research Council of Canada, ret.), Cities and Geology: McGraw-Hill, 1973.

1975–Erhard M. Winkler (Notre Dame University), Stone; Properties, Durability in Man's Environment: Springer-Verlag, 1973.

1976–David J. Varnes (U.S. Geological Survey), The Logic of Geological Maps with Reference to the Interpretation and Use for Engineering Purposes: U.S. Geological Survey Professional Paper 837, 1974.

1977–Richard E. Goodman (University of California, Berkeley), Methods of Geological Engineering in Discontinuous Rocks: West Publishing Co., 1976.

1978–Nicholas R. Barton (Norwegian Geotechnical Institute, Oslo), The Shear Strength of Rock and Rock Joints: International Journal of Rock Mechanics and Mining Science, v. 13, no. 9, 1976.

1979–Evert Hoek (Golder Associates, Vancouver) and John W. Bray (Imperial College of Science and Technology, London), Rock Slope Engineering: Institution of Mining and Metallurgy, London, 1977.

1980–Kerry E. Sieh (California Institute of Technology), Prehistoric Large Earthquakes Produced by Slip on the San Andreas Fault at Pallett Creek, California: Journal Geophysical Research, v. 83, no. 138, 1978.

1981–Allen W. Hatheway (University of Missouri-Rolla) and Cole R. McClure, Jr. (Bechtel Inc., San Francisco), Geology in the Siting of Nuclear Power Plants: Geological Society of America, Reviews in Engineering Geology, v. IV, 1978.

1982–Douglas R. Piteau (Piteau Associates, Vancouver) and F. Lionel Peckover (Vaudruil, Quebec), Landslides; Analysis and Control: Chapter 9 of Engineering of Rock Slopes, Special Report 176, Transportation Research Board, National Academy of Sciences, 1978.

1983–Peter W. Lipman and Donal R. Mullineaux (both of U.S. Geological Survey), The 1980 Eruptions of Mount St. Helens, Washington: U.S. Geological Survey Professional Paper 1250, 1981.

1984–Roy E. Hunt (Consultant), Geotechnical Engineering Investigations Manual: McGraw-Hill, 1984.

1985–Lawrence D. Dyke (Geological Survey of Canada), Frost heaving in bedrock in permafrost regions: Bulletin Association of Engineering Geologists, v. 21, no. 4, 1984.

1986–James F. Quinlan (U.S. National Park Service) and Ralph O. Ewers (Eastern Kentucky University), Groundwater flow in limestone terranes: strategy, rationale, and procedure for reliable, efficient monitoring of groundwater quality in karst areas, *in* Proceedings of the Fifth National Symposium on Aquifer Restoration and Ground Water Monitoring: Worthington, Ohio, National Water Well Association, 1985.

1987–Joseph I. Ziony (U.S. Geological Survey), Evaluating earthquake hazards in the Los Angeles region—An earth science perspective: U.S. Geological Survey, Professional Paper 1360, 1985.

1988–G. H. Eisbacher (University of Karlsruhe) and J. J. Clague (Geological Survey of Canada), Destructive mass movements in high mountains: hazard and management: Geological Survey of Canada, Paper 84-16, 1984.

1989–Robert B. Johnson and Jerome V. DeGraff, Engineering Geology: New York, John Wiley & Sons, 1988.

APPENDIX G

Distinguished Practice Award

1982 Neil B. Steuer, U.S. Nuclear Regulatory Commission
1983 Thomas W. Fluhr, New York City Water Supply Board
1984 Edwin B. Eckel, U.S. Geological Survey
1985 Shailer S. Philbrick, U.S. Army Corps of Engineers and Cornell University
1986 George A. Kiersch, USCE, Southern Pacific Corp., Cornell University, and Consultant
1987 David J. Varnes, U.S. Geological Survey
1988 Earl E. Brabb, U.S. Geological Survey
1989 William R. Judd, U.S. Bureau of Reclamation, Purdue University, and consultant

APPENDIX G-1

Meritorious Service Award
and
Certificate of Appreciation

1980 Donald R. Coates, George A. Kiersch, William J. Mallio, Richard W. Galster, W. Harold Stuart, David J. Varnes
1981 *No award given*
1982 Alice S. Allen
1983–84 *No awards given*
1985 Robert H. Fickes
1986 Ellis L. Krinitzsky
1987 *No award given*
1988 Richard E. Gray
1989 George A. Kiersch

APPENDIX H

Specialty Groups Formed
Involving Division Membership Since 1958

During the 1960s and 70s one new GSA Division and several National Committees and Groups were formed that involved Engineering Geology Division members.

Component of Division	Year	New Affiliation
California members (Reflects concerns, local problems, registration)	1958	California Association of Engineering Geologists, Later AEG in 1962
Hydrogeology	1959–60	Division on Hydrogeology in 1960(Third GSA Division)
Committee on Rock Mechanics of 1960–63	1963	U.S. National Committee/ Rock Mechanics, National Research Council, in 1963
Geological Engineers (part)	1963	Section Geological Engineers of AIME/SME in 1963 (Deactivated early 1970s)
Environmental Concerns and Policies–Practitioners		Committee on Environmen tal Policy of Geological Society, 1970s
Tunneling and Underground Openings Committee of 1960–72		U.S. National Committee on Tunneling Technology of National Research Council in 1972
Support and efforts of Division under guidance David J. Varnes (past-chairman, 1977)		The U.S. National Committee/International Association Engineering Geologists formed in 1979–81

APPENDIX I

Society Publications on Engineering Geology
Prior to Division

1939–Geology in engineering; Frontiers of geology: Charles P. Berkey, Geological Society of America Brochure, p. 31–34.

1942–The geologist in public works: Charles P. Berkey, Bulletin, Geological Society of America, v. 53, p. 513–532.

Publications of the Division
Special Publications

1950–Application of geology to engineering practice (Berkey Volume): Sidney Paige, Chairman, Geological Society of America, 327 p.

1955–Engineering geology reference list: Committee on Reference List, Stafford C. Happ, Chairman, in Bulletin Geological Society of America, v. 66, no. 8.

1957–Teaching aids and allied material in engineering geology: Committee on Teaching Aids, John F. Mann, John T. McGill, and G. A. Kiersch, Chairman, Special Report: Geological Society of America, 36 p.

1962–Films for teaching engineering geology: Committee on Teaching Aids, John T. McGill, editor, Distributed through Department Geological Sciences, Cornell University 1968 (describes 109 films and availability).

1991–Heritage of Engineering Geology—The First Hundred Years: George A. Kiersch, editor, (this volume). Decade of North American Geology, Centennial Special Volume, No. 3.

Case Histories in Engineering Geology

1957–Engineering Geology Case Histories No. 1: Parker D. Trask, Editor.

1958–Engineering Geology Case Histories No. 2: Parker D. Trask, Editor.

1959–Engineering Geology Case Histories No. 3: Symposium on Rock Mechanics; Parker D. Trask, Editor.

1964–Engineering Geology Case Histories No. 4: Parker D. Trask and George A. Kiersch, Editors.

1964–Engineering Geology Case Histories No. 5: George A. Kiersch, Editor.

1968–Engineering Geology Case Histories No. 6: George A. Kiersch, Editor.

1969–Engineering Geology Case Histories No. 7: Legal Aspects of Geology in Engineering Practice; George A. Kiersch and Arthur B. Cleaves, Editors.

1970–Engineering Geology Case Histories No. 8: Engineering Seismology; The Works of Man; William M. Adams, Editor.

1973–Engineering Geology Case Histories No. 9: Geological Factors in Rapid Excavation; Howard J. Pincus, Editor.

1974–Engineering Geology Case Histories No. 10: Geologic Mapping for Environmental Purposes; Harry F. Ferguson, Editor.

1978–Engineering Geology Case Histories No. 11: Decay and Preservation of Stone; Erhard M. Winkler, Editor.

1990–Engineering Geology Case Histories No. 12: Landslides of Southern California; James E. Slossan and Arthur Keene, Editors. (in prep.)

Reviews in Engineering Geology

1962–Reviews in Engineering Geology, v. I; General Thomas W. Fluhr, Robert F. Legget, Editors.

1969–Reviews in Engineering Geology, v. II; General, David J. Varnes, George A. Kiersch, Editors.

1977–Reviews in Engineering Geology, v. III, Landslides, Donald R. Coates, Editor.

1978–Reviews in Engineering Geology, v. IV, Geology in the Siting of Nuclear Power Plants; Allen W. Hatheway and Cole R. McClure, Jr., Editors.

1982–Reviews in Engineering Geology, v. V, Geology Beneath Cities of North America; Robert F. Legget, Editor.

1984–Reviews in Engineering Geology, v. VI, Man-Induced Land Subsidence; Thomas L. Holzer, Editor.

1987–Reviews in Engineering Geology, v. VII, Debris Flows/Avalanches; Process, Recognition, and Mitigation; John E. Costa and Gerald F. Wieczorek, Editors.

1990–Reviews in Engineering Geology, v. VIII, Neotectonics in Earthquake Evaluation; Ellis L. Krinitzsky and David B. Slemmons, editors (A Centennial Symposium of Division).

APPENDIX J

Editor—Newsletter and *The Engineering Geologist*
(Issues per year in parentheses)

1959–65 (2–3) Alice S. Allen, Editor and Assistant Secretary

1966 (4) Alice S. Allen and Richard E. Goodman

1967 (4) Richard E. Goodman

1968–71 (3–4) Lloyd B. Underwood

1972 (2) Richard H. Howe

1973–75 (2–3) Mary E. Horne

1976–78 (1–2) William E. Mallio

1979 (2) William E. Mallio and Allen W. Hatheway

1980–84 (3–4) Robert H. Fickies

1985 (2) Robert H. Fickies and Theodore C. Smith

1986–87 (1) Theodore C. Smith

1987–88 (2) Arthur Keene

1989 (1) Robert J. Larsen

APPENDIX K

PROGRAMS AND SYMPOSIA SPONSORED BY THE ENGINEERING GEOLOGY DIVISION AT ANNUAL MEETINGS OF GSA, 1947–1989

1947 Ottawa: General Session – Projects and Research with Round-Table Discussion, Charles P. Berkey and Roger Rhoades.

1948 New York: General Session – Projects of Northeastern States, William H. Irwin and Shailer S. Philbrick.

1949 El Paso: General Session – Dam Projects and Research with Discussion on Needs of Practice, Roger Rhoades and B. C. Moneymaker.

1950 Washington: Open-Forum – Problems in Engineering Geology, Edward B. Burwell, Jr.

1951 Detroit: General Session – Mainly Geological Problems of Highway Design and Construction, Parker Trask and Edward B. Burwell, Jr.

1952 Boston: General Session – Rebound, Uplift Pressures, Seismic Design, Roger Rhoades and Arthur B. Cleaves.

1953 Toronto: Toronto Subway, Edward B. Burwell, Jr.; General Session – Dams and Tunnels, Arthur B. Cleaves and Shailer S. Philbrick.

1954 Los Angeles: Two General Sessions – Project Geology, Logging and Mapping Problems; and Slope Stability, Thomas F. Thompson and Shailer S. Philbrick.

1955 New Orleans: Symposium – Products of Weathering of Bedrock and their Engineering Properties, Shailer S. Philbrick and Thomas F. Thornburn.

1956 Minneapolis: Two General Sessions – Foundation Problems of Bearpaw Shale, St. Peter Sandstone; and Pleistocene Beds, W. H. Irwin, Ernest Dobrovolny, A. B. Cleaves, and G. M. Schwartz.

1957 Atlantic City: Symposium – Geology of Three Major Dam Projects, John R. Schultz and Robert H. Nesbitt; General Session – Soils and Aggregate, Ernest Dobrovolny and E. B. Burwell, Jr.

1958 St. Louis: Symposium – Rock Mechanics, William R. Judd and Arthur B. Cleaves; General Session – Aggregate, Dam Sites, Quantitative Trends and Research Needs, Laurence B. James and Donald U. Deere.

1959 Pittsburgh: General Session – Typical Problems of the Appalachians, Paul H. Price and Robert H. Nesbitt; Symposium – Geological Investigations of Underground Nuclear Explosions at the Nevada Test Site, Edwin B. Eckel and Robert F. Legget.

1960 Denver: Symposium – Hydrogeologic Problems in Engineering Geology, David J. Varnes and Lawrence A. Warner; General Session – Rock Properties and Sites, Robert Carpenter and Christopher F. Erskine.

1961 Cincinnati: Symposium – Properties and Sites in Karstic Terrain, John H. Melvin and Robert W. Karpinski; General Session – Glacial Deposits and Projects, H. Garland Hershey and Carl W. A. Supp.

1962 Houston: Symposium – Young Sediments–Mississippi Drainage and Seashore, Jack R. Van Lopik and S. A. Lynch; General Session – Rock Properties, Stresses and Sites, Arthur B. Cleaves and William F. Tanner.

1963 New York: Symposium – Hydrogeology and Engineering Geology, N. M. Perlmutter and Jack B. Graham (co-sponsored Hydrology Division); General Session – Urban Geology and Related Projects, Thomas W. Fluhr and Robert H. Nesbitt.

1964 Miamia Beach: Symposium – Alaska Earthquake, Edwin B. Eckel and John H. Melvin; General Session – Engineering Geology and Hydrogeology, A. Nelson Sayre and Shailer S. Philbrick.

1965 Kansas City: Two General Sessions – Projects, Rock Mechanics, Exploration and Concrete Reactivity, Lloyd B. Underwood and Laurence B. James.

1966 San Francisco: Symposium – Problems of Siting Nuclear Reactors in California, Richard H. Jahns; Symposium – Geology of the Bay Area Rapid Transit System, Charles S. Content: Symposium – Land subsidence, Arthur M. Piper and William T. Ellis (co-sponsored Hydrogeology Division).

1967 New Orleans: Symposium – Significance of Pore Pressures in Problems of Engineering Geology, Laurence B. James and Ronald C. Hirschfeld (co-sponsored Joint Committee Engineering Geology GSA-ASCE).

1968 Mexico City: Two General Sessions – Projects of Mexico City, Dams, Tunnels, and Rock Mechanics Testing, Thomas F. Thompson and Leonardo Zeevaert.

1969 Atlantic City: Symposium – Dynamic Problems in Engineering Geology, George A. Kiersch and Ronald C. Hirschfeld (co-sponsored Joint Committee Engineering Geology GSA-ASCE).

1970 Milwaukee: Symposium – Geological Factors in Rapid Excavation, George Heim and James R. Swaisgood.

1971 Washington, D.C.: Two General Sessions – Environmental Aspects of Engineering Geology, Lynn Brown, James Knight, Larry Heflin, and Henry J. Coulter.

1972 Minneapolis: Symposium – Natural Construction Materials, Howard J. Pincus and Erhard M. Winkler; Symposium – Environmental Geology Mapping, Harry F. Ferguson and Richard E. Gray.

1973 Dallas: Symposium – Decay of Stone and Other Natural Construction Materials, Erhard M. Winkler and Ronald H. Gelnett; Symposium – Problems of Engineering Geology along the Mississippi River, Charles R. Kolb and Arthur B. Cleaves.

1974 Miami Beach: Symposium – Preservation of Stone, Erhard M. Winkler and Gerald A. Slater.

1975 Salt Lake City: Symposium – Nuclear Power Plant Siting, Allen W. Hatheway and Cole R. McClure, Jr.; General Session – Rock Properties, Seismicity, and Exploration, Vincent J. Murphy and Robert W. Fleming.

1976 Denver: Symposium – Expansive Soils, Christopher C. Mathewson; General Session – Hazards and Rock Properties, M. E. Milling and R. W. Fleming.

1977 Seattle: Symposium – Capable Faulting and Engineering Works (full-day), George F. Sowers and Lyman W. Heller (co-sponsored with ASCE and AEG).

1978 Toronto: Symposium – Geology and Space Beneath Cities, Richard W. Galster and Richard H. Jahns.

1979 San Diego: Symposium – Academic Training of Engineering Geologists, Richard H. Jahns (EGD), James V. O'Conner (NAGT), and Martin L. Stout (AEG).

1980 Atlanta: General Session – Engineering Geology in Environmental Planning, Theodore C. Smith. Symposium – Land Subsidence, Thomas L. Holzer (co-sponsored Hydrogeology Division honoring long-time EGD member Joseph F. Poland).

1981 Cincinnati: Symposium – Role of Government Agencies in Development of Engineering Geology, James W. Skehan and Richard H. Jahns; General Session – Rock Properties and Processes, J. W. Williams and T. R. West.

1982 New Orleans: Symposium – History of Engineering Geology, Christopher C. Mathewson and George A. Kiersch; General Session – Natural Risks, Susan M. DuBois and Jeffrey R. Keaton.

1983 Indianapolis: General Session – Geologic Disposal of Radioactive Wastes, Lokesh Chatuvedi; General Session – Processes and History, Robert L. Schuster and H. C. Skinner (co-sponsored History Division).

1984 Reno: Symposium – Mechanics of Debris Flow/Avalanche Generation and Mitigation in the Western U.S., Gerald Wieczorek and John Costa.

1985 Orlando: Symposium – Engineering Geology of Low-Energy Coastlines, Robert L. Schuster and Christopher C. Mathewson; General Session – Robert L. Schuster and David M. Cruden.

1986 San Antonio: Symposium – Engineering Geology in Public Policy (full-day session, Christopher C. Mathewson and John S. Scott.

1987 Phoenix: Symposium – Neotectonics in Earthquake Evaluation, Ellis L. Krinitzsky and David B. Slemmons; General Session – Rock Properties, John Rockaway and Thomas L. Holzer.

1988 Denver: Symposium – Hazard Reduction in the 21st Century, Thomas L. Holzer and F. Beach Leighton.

1989 St. Louis: Symposium (with the Hydrogeology Division) — Site characterizations for conditions of non-Darcian flow, John F. Harsh and Jeffrey R. Keaton.

APPENDIX L

Papers Concerned with
Engineering Geology
Published in *Economic Geology*
(arranged chronologically 1906–1960)

Johnson, D. W., 1906, The scope of applied geology, and its place in the technical school: v. 1, p. 243–256.
Howe, E., 1907, Isthmian geology and the Panama Canal: v. 2, p. 639–658.
Mitchell, G. J., 1917, The need of accurate rock classification in engineering contracts (letter): v. 12, p. 281.
Cleland, H. F., 1918, The geologist in war time-geology on the Western Front (letter); v. 13, p. 145–146.
Reinecke, L., 1918, Non-bituminous road materials: v. 13, p. 557–597.
Lindgren, W., 1919, Economic geology as a profession (editorial): v. 14, p. 79–86.
Smith, W. D., 1919, Rock classification for engineering (letter): v. 14, p. 180–183.
Pirsson, L. V., 1919, Rock classification for engineering (letter): v. 14, p. 264–266.
Bean, E. E., 1921, Economic geology and highway construction: v. 16, p. 215–221.
Patton, L. T., 1924, Geology and the location of dams in west Texas: v. 19, p. 756–761.
Patton, L. T., 1925, Geology and the location of dams on the Canadian River, Texas: v. 20, p. 464–469.
Ransome, F. L., 1928, Geology of the St. Francis Dam Site: v. 23, p. 553–563.
Ransome, F. L., 1933, What is rock (editorial): v. 28, p. 502–505.
Wilcox, S. W., and Schwartz, G. M., 1934, Reconnaissance of buried river gorges by the earth resistivity method: v. 29, p. 435–453.
Bradley, W. H., 1935, Geology of the Alcova Dam and reservoir sites, North Platte River, Natroma County, Wyoming: v. 30, p. 147–165.
Eckel, E. B., 1938, Geology of the Savage River, Maryland, dam and reservoir site: v. 33, p. 287–304.
Legget, R. F., 1942, An engineering study of glacial drift for an earth dam, near Fergus, Ontario: v. 37, p. 531–536.
Krynine, D. P., 1944, Some engineering aspects of river sand deposits: v. 39, p. 307–314.
Wagner, W.R.T., 1944, A landslide area in the Little Salmon River Canyon, Idaho: v. 39, p. 349–358.
Meilenz, R. C., and Okeson, C. J., 1946, Foundation displacements along the Malheur River syphon as affected by swelling shales: v. 41, p. 266–281.
Benson, W. N., 1946, Landslides and their relation to engineering in the Dunedia District, New Zealand: v. 41, p. 328–347.
Frink, J. W., 1946, The foundation of the Hales Bar Dam: v. 41, p. 576–597.
Paige, Sidney and Rhoades, Roger, 1948, Report of Committee on Research in Engineering Geology, Society of Economic Geologists: v. 43, p. 313–323.
Page, B. M., 1950, Geology of the Broadway tunnel, Berkeley Hills, California: v. 45, p. 142–166.
Lohmann, S. H., 1950, Report of Subcommittee on groundwater, Committee on Research, Society of Economic Geologists: v. 45, p. 70–71.
Kiersch, G. A., 1951, Engineering geology principles of subterranean installations: v. 46, p. 208–222.
Meilenz, R. C., Greene, K. T., and Schieltz, N. C., 1951, Natural pozzolans for concrete: v. 46, p. 311–328.
Laurence, R. A., 1951, Stabilization of some rockslides in Grainger County, Tennessee: v. 46, p. 329–336.
Wesley, R. H., 1952, Geophysical exploration in Michigan: v. 47, p. 57–63.
Crandall, D. R., 1952, Landslides and rapid flowage phenomena near Pierre, South Dakota: v. 47, p. 548–568.
Kiersch, G. A., and Hughes, P. W., 1952, Structural localization of groundwater in limestones–"Big Bend District," Texas-Mexico: v. 47, p. 794–806.
Legget, R. F., and Bartley, M. W., 1953, An engineering study of glacial deposits at Steep Rock Lake, Ontario, Canada: v. 48, p. 513–540.
Kiersch, G. A., and Treasler, R. C., 1955, Investigations, areal and engineering geology; Folsom Dam project, central California: v. 50, p. 271–310.
Wentworth, C. K., Mason, A. C., and Davis, D. A., 1955, Salt-water encroachment as induced by sea-level excavation on Angaur Island: v. 50, p. 669–680.
Winkler, E. M., 1955, Geology of important hydro-electric power stations in Austria: v. 50, p. 862–878.
Legget, R. F., 1955, Engineering geology; A fifty year review: Fiftieth Anniversary Volume (1905–1955), p. 534–556.
Terzaghi, K., 1955, Engineering properties of sediments: Fiftieth Anniversary Volume, p. 557–618.
Ferris, J. G., and Sayre, N. A., 1955, The quantitative approach to ground-water investigations: Fiftieth Anniversary Volume, p. 714–747.

Kiersch, G. A., 1958, Geologic causes for failure of Lone Pine Reservoir, east-central Arizona: v. 53, p. 854–866.

Perlmutter, N. M., Geraghty, J. J., and Upson, J. E., 1959, The relation between fresh and salty ground water in southern Nassau and southeastern Queens Counties, Long Island, New York: v. 54, p. 416–435.

Neilson, J. M., Snelgrove, A. K., and Van Pelt, J. R., 1960, Curricula and professional aspects of geological engineering: v. 55, p. 1048–1059.

REFERENCES CITED

Allen, C. R., 1988, Earthquake hazard reduction in the past and next century: Geological Society of America Abstracts with Programs, v. 20, p. A81.

AIME, 1929, Geology and engineering for dams and reservoirs: American Institute of Mining and Metallurgical Engineers Technical Publication 215, 112 p.

Bateman, A. M., 1955, Economic geology: Economic Geology 50th Anniversary Volume, p. 6.

Berkey, C. P., 1942, The geologist in public works: Geological Society of America Bulletin, v. 53, p. 513–532.

Eckel, E. B., ed., 1958, Landslides and engineering practice: National Research Council Highway Research Board Special Report 29, 232 p.

——, 1982, The Geological Society of America; Life history of a learned society: Geological Society of America Memoir 155, 167 p.

Ernst, W. G., 1987, United States earth sciences, status and future; How bad, how good?: Geological Society of America Bulletin, v. 99, p. 1–6.

Hall, J., 1839, Classification of "Slate rock and shale," Erie Canal Locks Construction, Lockport, New York: New York Geological Survey Annual Report, p. 287–339.

Johnson, D. W., 1906, The scope of applied geology, and its place in the technical school: Economic Geology, v. 1, p. 243–256.

Judd, W. R., ed., 1964, State of stress in the Earth's crust: Amsterdam, Elsevier Publishing Co., 733 p.

Kiersch, G. A., Mann, J. F., and McGill, J. T., 1957, Teaching aids and allied materials in engineering geology: Geological Society of America Engineering Geology Division Committee on Teaching Aids, 36 p.

Legget, R. F., 1973, 25th celebration issue: Geological Society of America Engineering Geologist, v. 8, no. 1, 9 p.

Lindgren, W., 1913, Mineral deposits, 1st ed.: New York, McGraw-Hill Brook Co., p. 1–2.

Mather, W. H., 1838, Geological section of Lake Erie shoreline at Cleveland: Columbus, Ohio Geological Survey Annual Report, p. 111–121.

Ries, H., and Watson, T. L., 1914, Engineering geology, 1st ed., 5th ed., 1936: New York, John Wiley and Sons, 679 p.

Smith, J. V., 1985, Protection of human race against natural hazards: Geology, v. 13, p. 675–678.

Trask, P. D., ed., 1950, Applied sedimentation: New York, John Wiley and Sons, 665 p.

Underwood, L. B., 1967, Classification and identification of shales: American Society of Civil Engineers Journal of Soil Mechanics and Foundations, v. SM6, p. 97–116.

MANUSCRIPT ACCEPTED BY THE SOCIETY OCTOBER 19, 1989

ACKNOWLEDGMENTS

Compilation of the Division history, events, and activities over the past half century has benefited from the assistance and direct input of many colleagues and the headquarters staff of the Society. Our appreciative thanks to Dorothy M. Palmer of the Society office for arranging the photographs of Division chairman and Society presidents, as well as copies of back issues of *The Engineering Geologist* and other file materials. In addition, Centennial coordinator/editor A. R. Palmer kindly provided a critique on and review of many historical items in an earlier text, which aided in shortening and improving the text.

The preliminary text received a review and beneficial comments on the early formative years of the Division from two past chairmen, Edwin B. Eckel and Shailer S. Philbrick, which has clarified and strengthened the text. The Twenty-fifth Anniversary Luncheon address of Past Chairman Robert F. Legget in 1972 provided documentation and background information on several early decisions of the Division founders and the group of early members. Similarly, Past Chairmen Richard E. Gray and Harry F. Ferguson have kindly provided background summaries on the Division Awards and related historical items.

Printed in U.S.A.

Geological Society of America
Centennial Special Volume 3
1991

Chapter 5

Professional practice and societal organizations

Cole R. McClure and Gail L. Sorrough
Bechtel Civil, Inc., Box 3965, San Francisco, California 94119
Richard E. Gray
GAI Consultants, Inc., 8 Black Walnut Drive, William Penn Estates, RD 3, Greensburg, Pennsylvania 15601
Richard W. Galster
18233 13th Avenue NW, Seattle, Washington 98177
David J. Varnes
U.S. Geological Survey, Box 94025, Denver Federal Center, Denver, Colorado 80225
George A. Kiersch
Geological consultant, Kiersch Associates, Inc., 4750 North Camino Luz, Tucson, Arizona 85718

GEOLOGY IN ENGINEERING-CONSTRUCTION COMPANIES

Cole R. McClure and Gail L. Sorrough

INTRODUCTION

Today, engineering geologists in private industry occupy key positions in the planning, design, and construction of many different kinds of engineering works. Since the beginning of this century, it typically has been the practice of engineering-construction companies to rely on outside consultants for projects requiring geological expertise. However, with the end of World War II and the rapid development of the early 1950s, engineering-construction companies in North America began to hire geologists as staff members.

A recent survey of the older major engineering-construction companies by Bechtel (1986) established that about half of the firms support engineering geology staffs in-house, while half rely solely on consultants, either individuals or specialty groups. Furthermore, many of the companies that retain engineering geologists in-house occasionally supplement their staff input with the services of outside consultants for a variety of reasons, including fulfilling contractual obligations, enhancing the work capabilities in a specific geographic area, or reinforcing expert opinions in controversial situations.

Today, some of the major engineering-construction companies that support their own in-house geoscience experts include Bechtel Civil, Inc.; EBASCO Services, Inc.; Fluor Engineers, Inc.; Harza Engineering Company; Morrison-Knudsen Engineering Company; United Engineers and Constructors; and Stone and Webster Engineering Corporation.

Geologists in engineering firms

Bechtel was one of the first engineering companies to hire staff geologists. In the early 1950s, they hired Ben Warner, Victor L. Wright, Robert J. Farina, and Charles P. Benziger to work on a project-by-product basis. However, lack of permanent job status and associated benefits, as well as the inability in those days to advance professionally within the company ranks, was not encouraging to the geologists or beneficial to the company, and consequently, many of these geologists moved on to other professional situations. Nevertheless, the need for geological expertise on projects continued to grow, and by 1960 a permanent in-house engineering geology group had been established. The attractive assurance of job security and benefits of a staff position brought stability to the group and facilitated its growth as a recognized team of experienced geologists. Initially the group was under the guidance of Charles S. Content, who had been a long-time senior geologist with the U.S. Bureau of Reclamation. Content retired in 1973, and management and expansion of the group became the responsibility of the senior author.

Morrison-Knudsen established an earth sciences division in 1957. Staffing, however, was minimal until 1967, when a reorganization placed an emphasis on staff development. Currently, the entire division numbers about 60 individuals with diverse professional capabilities and experiences, including engineering geologists. In addition, due to certain legal constraints delineated in a Morrison-Knudsen contract or because of the need for an outside opinion, the division occasionally engages consultants to work in conjunction with the in-house staff.

Stone and Webster Engineering Corporation relied on geological consultants from 1942 until 1966, when the first permanent staff geologist was hired for the Cabin Creek pumped

McClure, C. R., Sorrough, G. L., Gray, R. E., Galster, R. W., Varnes, D. J., and Kiersch, G. A., 1991, Professional practice and societal organizations, *in* Kiersch, G. A., ed., The heritage of engineering geology; The first hundred years: Boulder, Colorado, Geological Society of America, Centennial Special Volume 3.

storage project in Colorado. The staff has grown rapidly since then and now has some 50 geoscience specialists (S. C. Rossier, written communication, 1986).

EBASCO Services also relied on geological consultants until the late 1960s when a group of experienced geologists was assembled for the Keban dam project in Turkey. Subsequently these professionals became the nucleus of an EBASCO earth sciences group established at Greensboro, North Carolina, in 1973 under the direction of Norman R. Tilford.

United Engineers and Constructors followed a pattern similar to the other major firms: initially, they relied on consultants such as George A. Kiersch (1965–1977), or specialty firms, such as Weston-Geophysical, or on contract-staff members for power plant projects. Subsequently, in order to adequately serve increasing project demands for geological input, an in-house geotechnical group was established at Philadelphia in 1978.

Today, engineering geology groups in engineering-construction firms are composed of specialists from the many categories of the geosciences. Corporations have come to rely on the diverse expertise of an in-house applied geology group that can meet the changing demands of contemporary engineering projects with unique problems.

DEVELOPMENT OF ENGINEERING GEOLOGY GROUPS

The trend toward having engineering geologists in-house as part of the corporate structure was influenced by several factors in the post–World War II period. Most important was the significant increase in the construction of civil works of all types by both government agencies and private enterprises. This accelerated building provided opportunities for greater participation by engineering and construction firms. The increase in feasibility, design, and construction activities resulted in the need for more input and advice from applied geologists.

Furthermore, by the 1950s the need for input from experienced engineering geologists had been demonstrated dramatically by several failures of engineering works and costly remedial programs due to errors in judgment (see discussion in Chapters 22 and 23, this volume). For example, such early projects as Hales Bar Dam (1910s), St. Francis Dam (1920s), and Lone Pine Reservoir (1930s) were located, designed, and constructed without the benefit of geological guidance from competent engineering geologists. This exclusive engineering approach proved to be both unsound and risky; all too often a project resulted in severe economic losses, loss of life, and occasionally legal liabilities.

Private engineering-construction firms in North America were also influenced to employ full-time geological personnel by the major government agencies involved in engineering-construction activities. By 1948, the U.S. Army Corps of Engineers, the U.S. Bureau of Reclamation, the U.S. Geological Survey, and the Tennessee Valley Authority each had developed a division devoted exclusively to applied geology for engineering works (Paige and Rhoades, 1948, p. 319; and Chapter 3, this volume).

A post-war survey by the U.S. Department of Labor reveals that employment prospects for engineering geologists in private industry were expanding rapidly. For example, industries other than mining and petroleum, such as railroads, construction companies, and manufacturers of ceramic and abrasive products, had begun to employ geologists. This demand was opening up new areas for future employment of geologists, and particularly engineering geologists (U.S. Department of Labor, 1946).

An obvious advantage to having engineering geologists within the engineering company is the enhanced communication between the civil engineers and the applied geologists. This direct communication with experienced engineering geologists on a project allows for timely advice and input as the design or construction progresses, or as geologic conditions encountered change from those anticipated. Most importantly, project engineers are not required to rely solely on a report provided from an outside source describing the expected geologic conditions, for in such cases there is little or no opportunity for further discussions or questions once the report has been submitted and the contract completed.

Engineering geologists are important contributors to engineering projects in all phases of their development, as outlined on Figure 1, Chapter 18. They are involved in the feasibility and areal-siting studies, the site-specific investigations, the design and construction phases, and sometimes with the maintenance/operational phases. Their facility with geological information and their continued presence throughout the project to provide reviews and advice make important contributions in the design and construction stages, and often help to avoid schedule delays and costly changes in design.

This background source is the case at Morrison-Knudsen, where the presence of the engineering geology group has, in certain respects, compelled the company to focus on the importance of a particular specialty and use their expertise. This, in turn, has influenced project engineering decisions and cost estimates. Another advantage of having an engineering geology group in-house at Morrison-Knudsen is the continuity in engineering geological information that is provided and developed throughout the project. The same staff group can be involved from the inception and planning stages through the completion of a long-term project (Chapter 18, Fig. 1).

This continuity was especially advantageous to Morrison-Knudsen for the Itaipu hydroelectric project on the Parana River between Brazil and Paraguay. As the largest hydroelectric project in the world to date, Itaipu had technical complexities to match its size. Under the name of the consortium IECO-ELC for this project, the Morrison-Knudsen earth sciences personnel were involved in the prefeasibility studies, which began in 1970. Subsequently, when IECO-ELC became coordinators of project engineering to be performed by various Brazilian and Paraguayan companies (Lyra, 1982, p. 27), the same experienced geological staff was available for the role as coordinators of the enormous amount of geological data. By January 1982, 12 years after the first study, all site foundation treatment had been completed and

90 percent of the hollow gravity dam construction was completed (Caric, 1982, p. 44). The success of Itaipu, a significant international engineering achievement, was due, in part, to the Morrison-Knudsen earth sciences staff. The achievement is also a testament to the effectiveness of a unified team of engineering geology professionals. Today, a close working relationship is well established between the civil engineers and the earth sciences division at Morrison-Knudsen. There is a respected dependency on geologic information from engineering geologists, and this has been an important factor in many projects.

Functions and capabilities

Engineering geologists are a part of the same corporate structure as the staff engineers and other professionals with whom they work. Consequently, they are bound by the same legal constraints as the other design and construction staff members on a project. Generally, engineers feel this responsibility has a favorable influence on conservatism in design of the project works. Furthermore, they realize that the geologists are providing both the geological data for the project site and the attendant analysis and interpretation. This background is essential to developing the most appropriate design for the project without the additional conservatism that may be added by a consultant or consulting firms to protect themselves.

An excellent description of the engineering geologists' role and how it developed at Harza Engineering Company is documented by Franklin C. Rogers (1964). He notes that Harza geologists operate most frequently "as part of a team effort to which they contribute their special talents along with the foundation engineers, structural engineers or others who constitute the team. Usually they are outnumbered, so they tend to develop a powerful voice and dogged resistance to the persuasion of the engineers. The team operates on the assumption that each member is a specialist in his own right and that his views and opinions are to be respected as those of a specialist" (Rogers, 1964, p. 1–2).

Rogers also summarizes several examples of how the company "geologists and engineers demonstrated effective team action" on Harza projects. The Wanapum project on the Columbia River in Washington presented an especially complex situation. The erratic pattern of the basalt flows made it difficult to find an area of competent foundation rock suitable for the loads of a powerhouse and spillway. A team was assigned to study the complex rock sequence, and after 78,000 lineal ft of seismic survey and 26,000 lineal ft of drilling, followed by detailed identification and correlation of rock cores, a pattern emerged. Only one of the basalt flows in the sequence was judged suitable for the proposed powerhouse and spillway. The fact that the proposed engineering works could be, and have been, located in the desired reach of the Columbia River is largely the result of the Harza geologists' ingenuity and painstaking effort (Rogers, 1964, p. 2).

The Bechtel engineering geology group had expanded to some 60 professionals by the early 1970s. The expansion reflects both the increased demand for engineering geological expertise and an increase in the variety of specialists that compose the group. Rock mechanics, hydrogeologists, seismologists, remote sensing experts, and computer analysts are all necessary, in addition to experienced geologists and engineering geologists, in order for Bechtel to more efficiently fulfill its contractural obligations to worldwide clients. Over the years, Bechtel's engineering geology group has established its own rock mechanics and testing laboratory, acquired computer capability to enhance Landsat digital information, developed a computerized geographic information system, further developed specialty areas such as ground water, seismology, tectonics, and rock mechanics, and fostered the growth of a sizable geology library. Currently, the Bechtel engineering geology group has two separate functions: (1) provide geological support to company projects worldwide; and, (2) under its own management, engage in contracts that are specifically concerned with engineering geology problems.

RESEARCH AND NEW TECHNIQUES

Paige and Rhoades (1948) observed in their "Report of Committee on Research in Engineering Geology, Society of Economic Geologists" that engineering geology had not yet attained its full maturity. Although many changes have occurred in the engineering geology field in the intervening 40 years, this observation is still valid (described further in Chapters 1 and 2 of this volume).

Among other goals for engineering geology in 1948, Paige and Rhoades called for "development of concepts and techniques for a more quantitative solution of geologic problems; Geology in general has become steadily more quantitative through research, but the special needs of engineering geology are urgently pressing" (Paige and Rhoades, 1948, p. 315). These experienced practitioners also recognized that the pressure for more quantification in engineering geology came from civil engineers and designers. Most critical was the need to understand further the physical behavior of geologic materials under conditions imposed by engineered structures, and the engineers required this information in quantitative terms. Today, this requirement is largely met by the application of numerical modeling and computer analysis techniques to geologic data. Research in these and other specialty areas is constantly progressing and thereby affecting the practice of applied geology and the demands placed on engineering geologists. As a consequence, the in-house engineering geology groups in engineering-construction firms have expanded to accommodate these contemporary engineering demands. An interdisciplinary team of specialists has become essential to provide the necessary variety of reports, analyses, and opinions. Although the nucleus of an engineering geology group remains applied geologists, their expertise has been augmented by specialists in rock mechanics, ground water, geophysics, geochemistry, seismology, and remote sensing, as well as computer analysis and numerical modeling.

Computer analysis and numerical modeling

Computer analysis and numerical modeling have enhanced and further emphasized the quantitative aspects of engineering geology and thereby promoted closer cooperation between engineers and geologists. The empirical aspect of geological interpretations must now be supplemented with interpretations using any of a variety of numerical schemes to help account for the natural intricacies of the geologic environment. Some of the most widely used methods in geotechnical engineering analysis are the finite-element and finite-difference methods, numerical integration of governing equations, method of characteristics, the boundary-integral-equation method, and a combination of closed form and numerical schemes for solving governing equations (Desai and Christian, 1977, p. 2). Initially, the use of numerical models and computer programs for geological evaluations was not easy; many problems had to be solved before appropriate application of geological input was realized. The engineering geology group in the corporate environment provides an ideal set of circumstances to foster the research and develop the use of these models. Very simply, the corporate environment facilitates the accurate exchange of information between the geologists in the field, office, or laboratory, and the engineering and computer specialists.

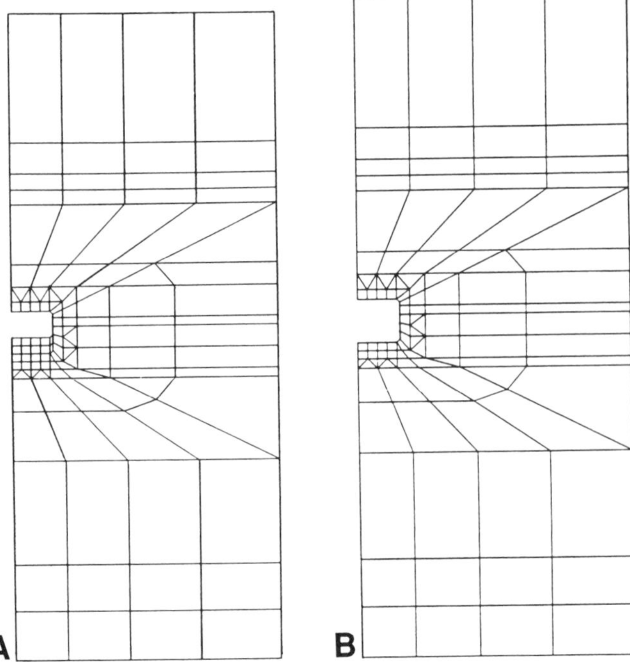

Figure 1. (A) a finite-element model used to predict closures, stresses, and deformed shapes as a function of time for an 8′ × 25′ opening. (B) the same finite-element model simulating 5′ of floor excavation 2 yr after the initial excavation. This model was generated for the Waste Isolation Pilot Plant in southeastern New Mexico, which is being developed by the U.S. Dept. of Energy to demonstrate the safe disposal of radioactive waste produced by U.S. defense programs. (Source: Bechtel Corporation.)

Rock mechanics

Initial attempts to apply computer techniques to rock mechanics problems proved especially frustrating for both the design-engineer and the engineering geologist. The designer needed relevant data as numbers for the computer model. The geologist supplied the requested data by extrapolating from various rock mechanics laboratory tests of limited core samples from the project site. In many cases, the resultant predictions made about the rock and the underground opening, tunnel, or slopes were quite different from the actual circumstances encountered in the field (Rose, 1975).

Today, engineering geologists who specialize in rock mechanics consider numerical simulation techniques, particularly methods that rely on finite-difference or finite-element analysis, powerful design and evaluation tools. Although geologic materials are heterogeneous with respect to their mechanical properties such as shear strength, Young's modulus, or Poisson's ratio, the pattern of variance throughout a rock mass is site-specific and highly dependent on such inherent geologic variables as spacing and orientation of joint sets and faults or fractures, the degree of weathering and chemical alteration, the state of stress, and the occurrence of certain minerals such as clays. Analyzing a rock mass as a homogeneous material with constant, nonvarying physical properties is generally inadequate.

Numerical methods allow the rock mechanics specialists to specify 1-, 2-, or 3-dimensional variations in rock properties using the "grid" concept. The spacing of grid points, called "nodes" is used to show the variation in properties on a scale that is appropriate for a given application. Commonly, the geologist will order rock-property testing on cores taken from the nodal points or request in-situ tests so that the data array input to the numerical model represents real values at the site being evaluated. If site-specific data are sparse, the experienced geologist can select reasonable ranges for interpolation and extrapolation of physical property values or use specialized statistical techniques for the handling of sparse data. The important factor is that the density of data selected should be appropriate to the project. As an enhancement feature, it is possible to hybridize grids, thereby increasing the density of nodal data wherever tight control is necessary, e.g., in the region of a proposed tunnel opening or a dam foundation.

Utilizing this geologically specified grid, the design engineer can work with the engineering geologist to superimpose proposed designs and evaluate the response of the rock mass (Fig. 1). This is an efficient technique for studying alternative design plans or for highlighting design flaws. Tabular and graphical output for the studied designs is readily available in hard copy, on tape, or by modem transfer to a remote office.

Ground water

Ground water is another area where modeling has come to play a significant role. Moreover, the increasing number of projects involved with ground water has had a corresponding effect on the development of this expertise within engineering geology groups. Hydrogeologists and modeling experts both play expanding roles as private industry becomes more involved in monitoring ground water and mitigating ground-water pollution problems. Coupled with the proliferation of personal computers, modeling, numerical analysis, and simulation are routinely applied to solving ground-water problems. Techniques for investigating flow in both saturated and unsaturated granular media, solving solute transport problems, modeling flow in fractured media, and developing parameter estimations and boundary integral methods continue to evolve. Consequently, hydrogeologists are required to stay abreast of research in order to provide the most appropriate solutions to project problems.

Seismology

In the early 1950s, earthquake hazard evaluation for dams was done largely by the engineering geologist and based mainly on the earthquake history of the region and site area. During the 1960s and 1970s the issue of siting critical engineering facilities, especially nuclear power plants, did much to solidify the autonomy of the geological profession within the engineering-construction business. The imposition of strict siting processes and regulatory guidelines necessary to meet the emerging complex environmental policies demanded the presence of professional seismologists in the engineering geology group.

Seismological theory and practice have experienced many changes in the last few decades. Significant advances in computer technology and the increase in quantity and quality of earthquake ground-motion recordings have allowed important gains to be made in the understanding of earthquake source mechanisms, spatial and temporal distribution, spectral content, and site effects.

Syntheses of the rapidly growing earthquake databases are now used in developing estimates of earthquake occurrence. This synthesis, combined with probabilistic assessment of the earthquake potential based on tectonics and geologic features, allows for more reliable estimates of the site-specific and time-specific earthquake hazard. This database has become essential for the design of most major engineered works. The appropriate application of such data requires cooperation and communication among seismologists, geologists, and engineers. Today, seismologists regularly contribute their expertise to the design teams for dams, bridges, nuclear power plants, and other large engineered structures.

More recently, seismologists in private companies have engaged in large-scale regional projects with private engineering companies and research organizations to quantify seismic hazards and seismic-hazard methodologies for large regions of the United States. Ultimately, this data can be used to provide the engineers and planners with an improved probabilistic basis and a more appropriate design of site-specific engineering works.

IMPACT ON ACTIVITIES USING CONTEMPORARY TECHNOLOGIES

Ground-water contamination

A recent ground-water contamination case in California is a typical example of the application of modeling and computer techniques to help analyze the data and find an appropriate means for mitigating the problem. Specific tasks of the project included: assessing the vertical and lateral extent of ground-water contamination; designing, constructing, and operating an interim ground-water treatment plant to prevent further degradation of ground-water resources; and preparing a ground-water model to generate data for use in the design of a final treatment plant.

Initially, the vertical and lateral extent of contamination was determined by collecting geochemical data from an extensive network of existing production wells and newly constructed monitoring wells. The investigation indicated that the active contaminant plume encompassed some 250 acres and had contaminated four separate aquifer units to a depth of approximately 200 ft. The interim extraction system consisted of three existing production wells, four newly constructed production wells, and three converted monitoring wells. Two ground-water models were selected to predict the long-term hydrogeologic and geochemical changes in the four aquifer units. One flow model designed to predict the hydrogeologic changes with respect to time in the four aquifer units is shown in Figure 2. Four two-dimensional transport models were designed to predict the geochemical changes, i.e., reduction in contaminant concentration with respect to time in each of the four aquifer units. Input to the design of these models included: aquifer characteristics showing transmissivity and storage coefficient determined from pumping tests in both existing and newly contracted wells; stratigraphic data (thickness, lateral and vertical extent of the aquifer units and their confining units) determined from geological logs of the wells; piezometric data determined from periodic water-level measurements in over 75 wells; and geochemical data from several rounds of sampling a monitoring network of 50 wells. Communication between the project hydrogeologist collecting and interpreting the field data and the in-house computer modeling specialist was essential in developing an appropriate plan to control and mitigate the contamination at the site.

Remote sensing for site analysis

Satellite remote sensing and digital image processing have also had a distinctive impact on engineering geology activities for private engineering firms and on developing a closer working relationship between engineering specialists and geologists. Remotely sensed satellite data (Landsat) are frequently used for regional and feasibility studies or as an aid to site analysis before

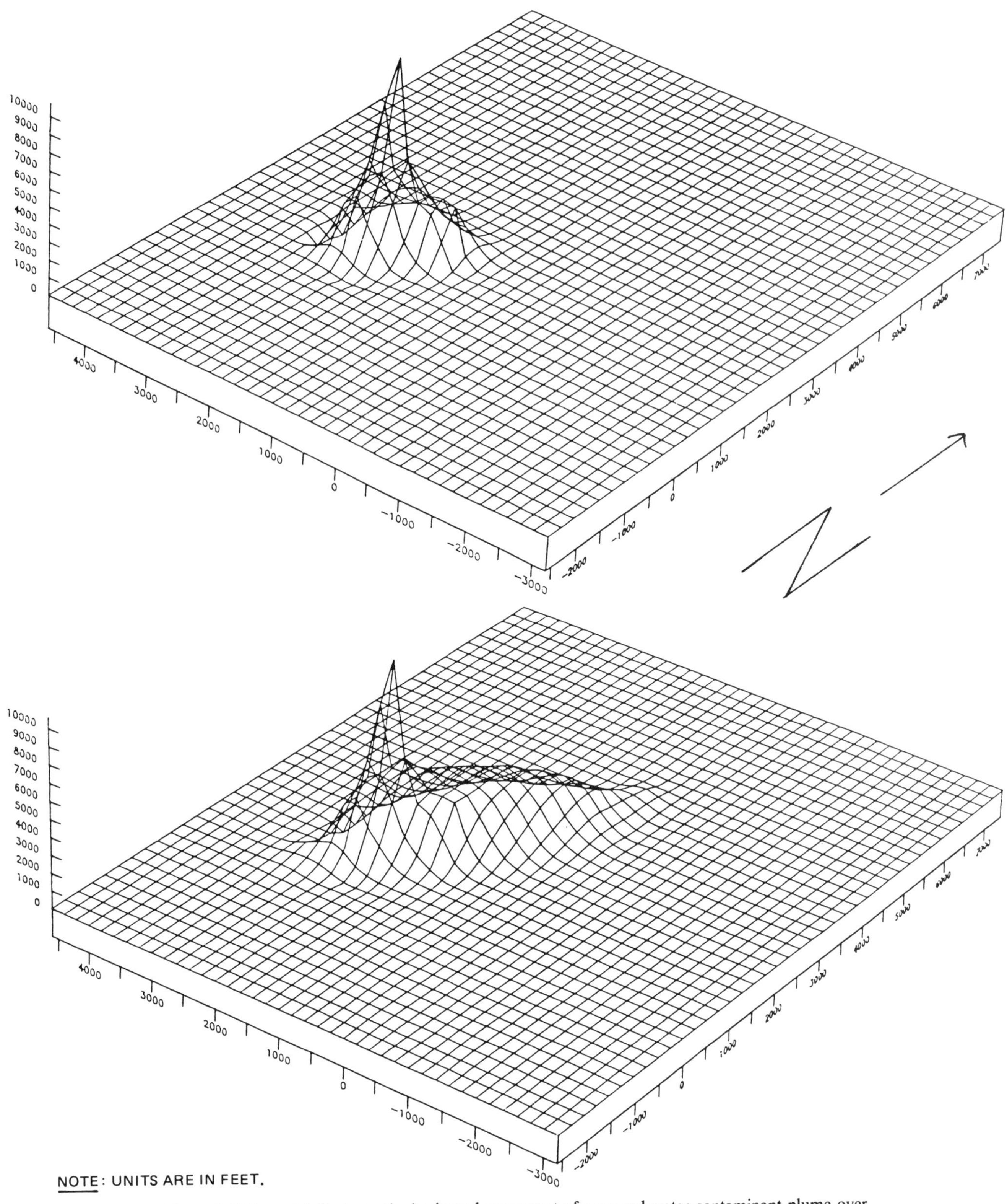

Figure 2. This model illustrates the horizontal movement of a ground-water contaminant plume over time, as predicted by numerical modeling. The horizontal axes represent areal extent, while the vertical axis represents the degree of contamination. The top view shows the extent of the plume in 1970. The peak in the graph is at the source of contamination. The bottom view shows the extent of the plume in the year 2000. Note that after 30 yr the plume is expected to spread, primarily to the northeast, while the peak of contamination remains at the source (Source: Bechtel Corporation.)

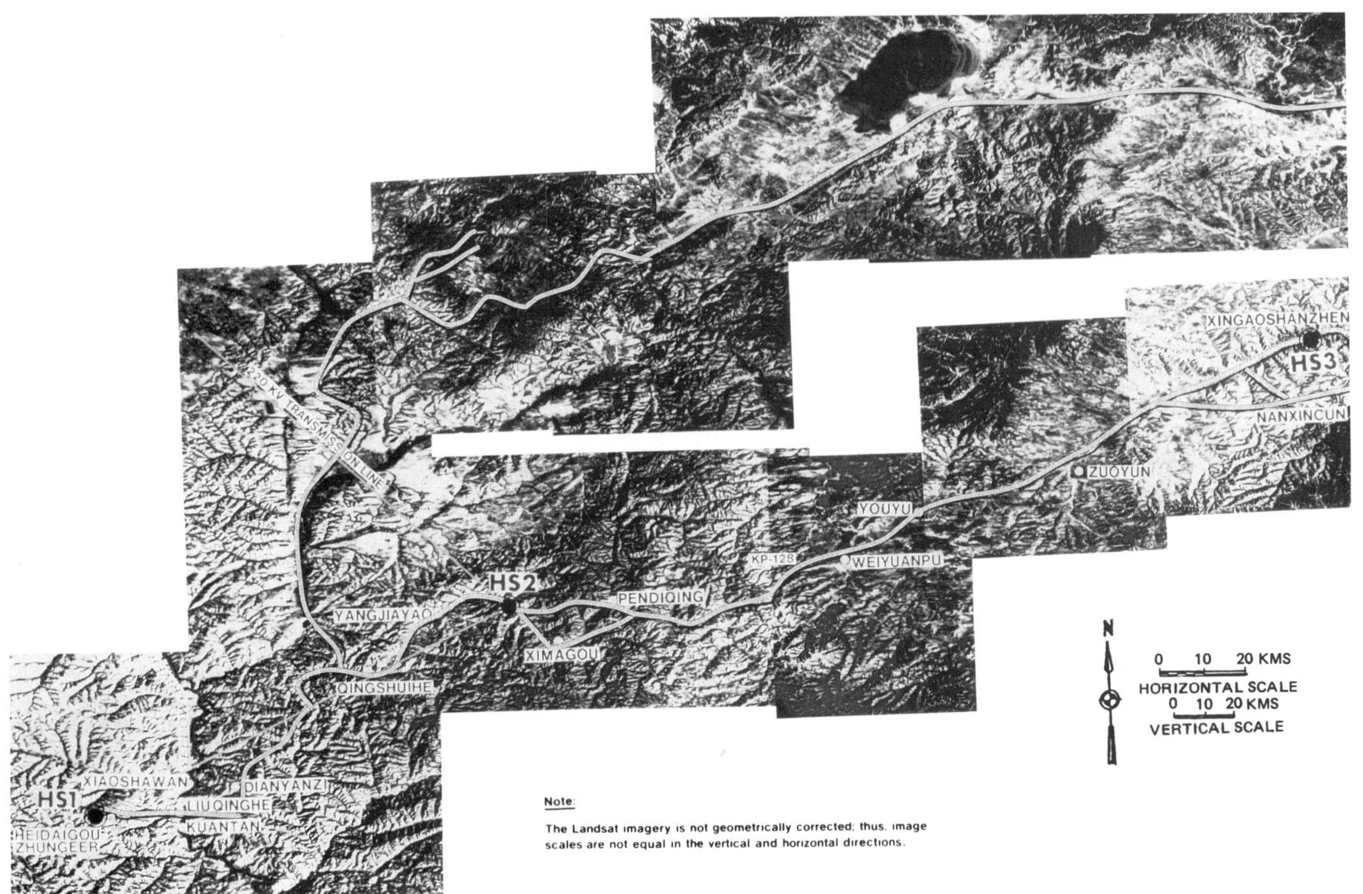

Figure 3. Composite Landsat image of a coal slurry pipeline route, People's Republic of China. (Source: Bechtel Corporation.)

and/or during project development (see Chapter 18, Fig. 1). The technique has proved especially valuable in remote locations where only a sketchy map base, or no map base at all, is available.

Bechtel recently made effective use of Landsat imagery for the selection of a coal slurry pipeline route in the People's Republic of China (Fig. 3). The pipeline project extended more than 700 km from the mine on the Yellow River in central China to a port city on the northeast coast. Current geologic and soils maps were not available for most of the pipeline route; therefore, ". . . Landsat MSS images were obtained and subscenes covering the alternative routes were identified, photographed, and assembled into a mosaic. Analyses of the mosaic and interactive digital enhancement of the subscenes on Bechtel's image processor provided the necessary geotechnical and cultural information to allow engineers to select the most suitable route" (Lees and others, 1985, p. 94).

Satellite imagery is now commonly used to facilitate the work of both engineers and geologists. Extensive global coverage and developments in computer enhancement capabilities of digital data will continue to increase the demand for satellite imagery for such tasks as identifying geologic, environmental, and cultural factors that could affect a project. These include: landslides, flood-prone areas, suitability of terrain, shallow-water bathymetry, or identifying lithologic and structural features that may have an impact on seismic design criteria. Furthermore, the distribution of towns, irrigation canals and reservoirs, identification of alternative dam sites, highway or pipeline routes, and potential construction aggregate sources are important for any preliminary site evaluation.

OUTLOOK

The use of new technologies and continued research in all areas of geology can help us provide more reliable and quantitative evaluations in engineering geology, which will contribute greatly to progress in the engineering-construction industry. The presence of an engineering geology group within the engineering-construction corporation provides an excellent means for communication and development of ideas among all the varied disciplines involved in design and construction.

JOINT COMMITTEE ON ENGINEERING GEOLOGY

Richard E. Gray

INTRODUCTION

The Joint Committee on Engineering Geology of the American Society of Civil Engineers (ASCE) and the Geological Society of America (GSA) have provided an active forum to discuss and investigate the common interests of applied geological science practitioners and civil engineering specialists since the early 1950s.

Committtee origin

In October 1951, the Executive Committee of the American Society of Civil Engineers' Geotechnical Engineering (GT) Division, then designated the Soil Mechanics and Foundations Division (SMFD), unanimously voted to join with the Geological Society of America and form a Joint Committee on Engineering Geology, with the following objectives: "To keep the membership of the two organizations informed on progress and developments in the field of engineering geology, particularly as such problems are related to soil mechanics and foundation conditions; to recommend the formation of specific task committees as may be required to keep the profession informed as stated above, through the compilation of special reports, papers, monographs, and symposia."

The concept of having a committee on engineering geology within the framework of the ASCE was developed in the Engineering Geology Branch of the chief engineer's office of the Bureau of Reclamation (USBR) in Denver, Colorado. The first formal document concerning a committee was written by bureau geologist William R. Judd, in July 1950, to the late R. F. Blanks, chief of the Laboratories Division (USBR) and a member of the Soil Mechanics and Foundation Division (SMFD) Executive Committee. This memorandum proposed: "The organization of a new committee under the Soil Mechanics and Foundations Division of the ASCE to encompass and deal with all problems pertinent to engineering geology."

In April 1951, the Executive Committee appointed an ad hoc committee to be responsible for the organization of the new committee. A. E. Cummings, chief research engineer of Raymond Concrete Pile Company, was selected as chairman. The other members of the group were R. F. Blanks; E. B. Burwell, chief geologist of the U.S. Army Corps of Engineers; S. Paige, Columbia University; and K. Terzaghi, Harvard University.

On October 23, 1951, the ad hoc committee concluded: "There is real need for better cooperation between engineers and geologists. The best way of developing this cooperation is by means of a joint committee composed of engineers interested in geology and geologists interested in engineering selected from the Soil Mechanics and Foundations Division of ASCE and from the Engineering Geology Division of GSA. . . .The work should be started by the appointment of an administrative committee consisting of three men from the ASCE and two men from the GSA, with an ASCE member as chairman. . . ."

Soon thereafter, R. F. Blanks became chairman of the SMFD Executive Committee and contacted K. Terzaghi to accept chairmanship of the group with the understanding that it would be Dr. Terzaghi's responsibility to set policies. In the interim, GSA named E. B. Burwell and R. Rhoades as their two representatives.

On March 3, 1952, the committee was approved by ASCE as a function of the SMFD. The first committee members were E. B. Burwell, H. O. Ireland of the University of Illinois, R. W. Spencer of the Southern California Edison Company, R. Rhoades of the Bureau of Reclamation (committee secretary), and K. Terzaghi (committee chairman). W. R. Judd was subsequently appointed assistant secretary to the committee and a nonvoting member.

EARLY CONTRIBUTIONS 1953–1959

The committee held its first meeting on January 24, 1953, in St. Louis. Eleven possible tasks previously suggested by the ad hoc committee were discussed for action. Four task committee activities (Table 1) were adopted: (1) nomenclature and engineering properties of shales; (2) nomenclature and engineering properties of rocks, exclusive of shales; (3) products of weathering of bedrock and their engineering properties; and (4) influence of geological factors on tunnel construction. R. Rhoades asked to be replaced, and W. R. Judd was appointed secretary.

The four task committees submitted reports to the joint committee prior to their discharge in 1959. Task committee 3 sponsored a symposium at the GSA meeting in November 1955 and again at the ASCE meeting in February 1957. Task committee 4 sponsored a symposium on "Geological Factors in Tunnel Construction" at the ASCE meeting in June 1955.

T. H. Thornburn replaced K. Terzaghi as chairman in October 1957, and the following year the committee's purpose was modified to read: "To keep the membership of the two organizations informed on the progress and developments in the field of engineering geology, particularly as applied to civil engineering problems, and still more particularly as such problems are related to soil mechanics and foundation conditions." In November 1958, the committee sponsored a session that led to publication of *Engineering geology case histories, number 3,* "Symposium on Rock Mechanics," under guidance of A. B. Cleaves and W. R. Judd (Trask, 1959). This symposium and further activity by W. R. Judd, chairman of the GSA-EGD Panel on Rock Mechanics, culminated in the establishment of the U.S. National Committee on Rock Mechanics within the National Research Council in 1968.

The final reports of the four task committees were released in 1959. Summaries are given in Appendix A, and the contributing members are listed in Table 1.

TABLE 1. ROSTER OF TASK COMMITTEE MEMBERS

TASK COMMITTEE 1–NOMENCLATURE AND ENGINEERING PROPERTIES OF SHALES

J. W. Early	R. H. Merrill
H. Forbes	S. S. Philbrick (chairman)
M. E. King	J. A. Trantina
K. S. Lane	V. Wolkodoff

TASK COMMITTEE 2–NOMENCLATURE AND ENGINEERING PROPERTIES OF ROCK, EXCLUSIVE OF SHALES

E. A. Abdun-Nur	L. A. Obert
N. E. Grosvenor	H. B. Stuart
J. Hardin	T. F. Thompson
V. Jones (chairman)	A. Van Valkenburg, Jr.
R. W. Lemke	D. O. Woolf
R. C. Mielenz (successor chairman)	

TASK COMMITTEE 3–PRODUCTS OF WEATHERING OF BEDROCKS AND THEIR ENGINEERING PROPERTIES

W. V. Conn	D. G. Moye
D. U. Deere	R. H. Nesbitt
P. P. Fox	F. R. Olmstead
J. P. Gould	P. Reiche
R. E. Grim	A. W. Root
J. W. Hilf	J. R. Schultz
K. B. Hirashima	G. F. Sowers
C. P. Holdredge	T. H. Thornburn (chairman)
G. A. Kiersch	

TASK COMMITTEE 4–INFLUENCE OF GEOLOGICAL FACTORS ON TUNNEL CONSTRUCTION

R. N. Carpenter	C. Rankin
A. B. Cleaves (chairman)	A. B. Reeves
T. W. Fluhr	E. B. Waggoner
A. N. Nicol	

CONTRIBUTIONS 1960–1988

A task committee on "Classification and Properties of Shales" was appointed in 1962 with S. S. Philbrick as chairman. The work of this committee culminated in a symposium on shale and a field trip at the ASCE Denver Water Resources Conference in May 1966.

R. C. Hirschfeld replaced T. H. Thornburn as chairman in 1967.

Continuing its practice of alternating sponsorship of sessions between ASCE and GSA meetings, the committee sponsored a session on "The Significance of Pore Pressures in Engineering Geology" (Table 2A lists the papers presented) at the GSA meeting in November 1967. In 1969, the committee sponsored a session on "Legal Aspects of Construction Engineering Geology" (Table 2B) at the ASCE meeting. A second endeavor in 1969 was a session "Dynamic Problems in Engineering Geology" (Table 2C) organized by G. A. Kiersch for the GSA meeting.

R. E. Gray replaced R. C. Hirschfeld as chairman in 1970. A session on "Geologic Aspects of Subsurface Waste Disposal" was held in January 1971 at the ASCE meeting. The committee also sponsored a paper on "Engineering Geology Education" (Kenney and Spencer, 1971).

TABLE 2A

PAPERS ON PORE PRESSURES
NEW ORLEANS, NOVEMBER 1967

CHAIRMEN: L. B. JAMES AND R. C. HIRSCHFELD

Principles of effective stress
R. C. Hirschfeld

Pore pressures in dams and foundations
K. V. Taylor

Effect of pore pressures on the stability of slopes
D. U. Deere and F. D. Patton

Role of pore pressures in subsidence caused by ground water withdrawal
J. F. Poland

Pore pressure effects in soils during earthquake
H. B. Seed

TABLE 2B

PAPERS ON LEGAL ASPECTS OF
CONSTRUCTION ENGINEERING GEOLOGY
NEW ORLEANS, FEBRUARY 1969

CHAIRMAN: E. WAGGONER

The contractor, the law, and engineering geology
Alfred M. Petrofsky, Jacobs Associates, San Francisco

Owner, designer, and the geologist
Arthur R. Reitter, Tudor Engineering Company, San Francisco

The changed conditions clause in construction contracts–an appraisal from the standpoint of the owner and his counsel
B. Palmer King, Bureau of Reclamation

Legal aspects of construction engineering geology
Abraham E. Margolin, Attorney, Kansas City, Missouri

In November 1972 at the GSA meeting, the committee sponsored a session on "Environmental Geology Mapping;" the papers were published in *Engineering geology case histories, number 10,* as "Geologic Mapping for Environmental Purposes" (Ferguson, 1974).

In January 1974, two committee-sponsored sessions were held during the ASCE meeting: "Recent Developments in the Classification and Index Properties of 'Shale'" and "The Geotechnical Environment of Valleys."

R. L. Schuster replaced R. E. Gray as chairman in 1974. In that year, both the management board of GSA's Engineering Geology Division and the Association of Engineering Geologists (AEG), requested expansion of the committee to include representatives of AEG. The executive committee of the Geotechnical Engineering Division invited AEG to appoint two representatives

TABLE 2C

PAPERS ON DYNAMIC PROBLEMS IN ENGINEERING GEOLOGY
ATLANTIC CITY, NEW JERSEY, NOVEMBER 1969

Some engineering geology aspects of surface faulting and related tectonic movements
M. G. Bonilla

Geologic investigation of landslides along the Middle Fork Eel River, California
Michael E. Huffman, R. G. Scott, P. J. Lorens, and Lawrence B. James

Seismic events related to newly built reservoirs
William I. Gardner

The in-situ measurement of compressional and shear wave velocities for use in foundation design
Richard J. Holt and V. J. Murphy

American Falls–preservation and enhancement study
Thomas A. Wilkinson and J. D. Guertin

Horizontal configuration and the rate of erosion of Niagara Falls
Shailer S. Philbrick

TABLE 3. JOINT COMMITTEE ON ENGINEERING GEOLOGY

GSA MEMBERS 1951–1988	TERM
E. B. Burwell	1951–1955
R. Rhoades	1951–1953
E. B. Eckel	1953–1959
P. D. Trask	1957–1961
A. B. Cleaves	1958–1963
R. C. Mielenz	1958–1961
S. S. Philbrick	1961–1966
L. B. James	1963–1969
J. M. Kellberg	1963–1964
G. A. Kiersch	1967–1970
H. F. Ferguson	1969–1978
P. L. Hilpman	1970–1979
J. B. Ivey	1978–1984
C. R. McClure, Jr.	1979–1985
L. Chaturvedi	1984–1987
R. T. Pack	1985–1988
JOINT-COMMITTEE CHAIRMAN – ASCE MEMBERS	
K. Terzaghi	1951–1957
T. H. Thornburn	1957–1967
R. C. Hirschfeld	1967–1970
R. E. Gray	1970–1974
R. L. Schuster	1974–1978
T. Howard	1978–1985
N. Sitar	1985–

to the committee, and the GSA Council ratified the expansion and change in committee membership. By 1975 the initial joint committee became the ASCE-GSA-AEG Committee on Engineering Geology.

The committee, with three Geotechnical Engineering Division committees, co-sponsored an ASCE specialty conference on Rock Engineering for Foundations and Slopes in August 1976 (ASCE, 1977). This was the first ASCE specialty conference to feature problems in rock.

At the 1977 GSA Annual Meeting, the committee organized a symposium on "Identification of Capable Faults."

T. R. Howard replaced R. L. Schuster as chairman in 1978.

The committee co-sponsored the symposium, "Review of the Causes of the Failure of Teton Dam," in 1978.

At the GSA meeting in 1980 the committee sponsored a symposium, "Geologic Hazards in Founding Offshore Structures," which was organized by R. L. Schuster.

N. Sitar replaced T. R. Howard as chairman in 1985.

A session on "Landslide Dams" was sponsored by the committee at the April 1986 meeting of ASCE (Schuster, 1986).

Due to a shift in interests, the Association of Engineering Geologists formally withdrew from membership in the joint committee in April 1986. Currently, committee membership represents and functions as the original joint committee of ASCE and GSA. A 1½-day symposium on "Remote Sensing" was sponsored by the joint committee at the 1988 annual meeting of AEG.

Table 3 lists the GSA representatives and the chairmen of the joint committee from 1951 to 1988.

HIGHWAY GEOLOGY SYMPOSIUM

Richard E. Gray

BACKGROUND

Born of the need to establish a better understanding and closer cooperation between geologists and civil engineers in the highway industry, a symposium was organized on "Geology as Applied to Highway Engineering" and sponsored by the Virginia Department of Highways April 14, 1950, in Richmond. W. T. Parrott, geologist of the Virginia Department of Highways, was a driving force behind the symposium, which was attended by representatives of highway departments of Georgia, South and North Carolina, Virginia, Kentucky, West Virginia, Maryland, and Pennsylvania, along with the members of federal and state surveys, U.S. Army Corps of Engineers, Department of Agriculture, Bureau of Roads, and several regional academic institutions.

Papers at the first symposium reflected the concerns of both geologists and engineers and included presentations by: Jasper L. Stuckey, state geologist, North Carolina; A. Stinnett, USGS (ground water in unconsolidated sediments); R. A. Laurence, USGS regional geologist (landslides and rockfalls); L. W. Currier, USGS (geologic materials); and R. W. Moore, U.S. Bureau of Roads (geophysical methods for subsurface exploration).

A second symposium was also held in Richmond in 1951; a third at the Virginia Military Institute, Lexington, in 1952. The fourth symposium, held in 1953, began the move to nearby states, with the meeting in Charleston, West Virginia (HGS, 1950–52).

To date, 38 consecutive annual meetings have been held in 24 different states. Prior to 1962 all meetings were held east of the Mississippi River, with Virginia, Ohio, West Virginia, Maryland, North Carolina, Pennsylvania, Georgia, Florida, and Tennessee serving as the host states. The symposium moved west for the first time in 1962 and met in Phoenix, Arizona. Since then, it has alternated, for the most part, back and forth between eastern and western locations, with all symposia enjoying strong support by the host state highway department and academic institutions. The symposium was renamed the "Symposium on Highway Geology" in 1960 to reflect the growing acceptance of engineering geology and its numerous sub-fields.

FUNCTIONS

Unlike most technical groups and organizations that meet on a regular basis, the highway geology symposium has no central headquarters, no annual dues, and no formal membership requirements. The governing body of the symposium is a steering committee made up of about 20 engineering geologists and geotechnical engineers from state and federal agencies and academic institutions, as well as technical services companies and consulting firms throughout the country. Steering committee members are elected for three-year terms; their elections and reelections are largely determined by their interests, participation in, and contributions to, the annual symposium.

The symposia are generally set up for 2½ days, with 1½ days for technical papers and a full day for the field trip that usually occurs on the second day. In most cases, the trips traverse 150 to 200 miles, provide for six to eight scheduled stops, and require about eight hours. Occasional cultural stops are scheduled around geologic and geotechnical points of interest.

At the technical sessions, case histories and state-of-the-art papers are the norm. The papers presented at the technical sessions are published in the annual proceedings. While some are out of print, copies of most of the last 18 proceedings may be obtained from the current treasurer of the symposium, David Bingham, of the North Carolina Department of Transportation in Raleigh, North Carolina.

An excellent review of highway geology principles and problems of practice was published by S. E. Horner and J. D. McNeal (1950), geologists of the State Highway Commission of Kansas.

TRANSPORTATION RESEARCH BOARD

Richard E. Gray

SUMMARY

The Transportation Research Board (TRB) and its forerunner, the Highway Research Board (HRB), have made significant contributions to the broad field of engineering geology, and its impact on the route selection, design, and construction of transportation systems. The board sponsors technical sessions at its annual meetings, and the papers are grouped in appropriate publications. The annual meeting is held each January in Washington, D.C.

The best-known publications of the TRB and earlier HRB that relate to engineering geology are on landslides:

Eckel, E. B., ed., 1958, Landslides and engineering practice: Highway Research Board Special Report 29; National Academy of Science–National Research Council Publication 544, 282 p.

Schuster, R. L., and Krizek, R., eds., 1978, Landslides; Analysis and control: Washington, D.C., National Academy of Sciences Transportation Research Board Special Report 176, 228 p.

The following members of the Engineering Geology Division/GSA made significant contributions to these reports: Highway Research Board Special Report 29—Edwin B. Eckel, David J. Varnes, Arthur B. Cleaves, John D. McNeal, and Shailer S. Philbrick; Transportation Research Board Special Report 176—Robert L. Schuster, David J. Varnes, George F. Sowers, and David L. Royster.

ASSOCIATION OF ENGINEERING GEOLOGISTS

Richard W. Galster

INTRODUCTION

During the decade that followed establishment of the Engineering Geology Division (EGD) within the Geological Society of America (GSA) in 1947, significant changes were occurring in the field of engineering geology. The great post-war demographic changes that brought an influx of population to the American west coast was perhaps most dynamically felt in California where the requirements for additional housing, transportation networks, greater and more prudent use of ground water, conservative use of surface water, and flood control for urbanizing valleys became rapidly apparent. The development of relatively inexpensive hydropower in the Pacific Northwest was proceeding rapidly, together with plans for irrigating many arid areas of the American West and flood control for larger urbanizing areas. The small cadre of staff engineering geologists that served the infant field on the major civil projects of the 1930s (Grand Coulee, Boulder, Bonneville, and TVA), had swollen to a much larger group working in various federal and state agencies on civil projects and ground-water problems. In addition, an increasing number were establishing private practices to service the expanding housing industry, especially in California where the urbanization of hillsides was creating significant problems and a rash of grading ordinances, both of which required applied geological solutions.

Thus, while the post-war mania for civil projects was proceeding throughout North America, it was the number of large-scale water projects underway in California by both the federal and state governments, plus the population influx and demand for additional housing and other improvements, that resulted in a concentration of engineering geologists in California during the 1950s. Although many were working under various titles that did not always convey their geological applications, they were communicating geology and geological recommendations and constraints to engineers, builders, and public officials. Unlike the engineers with whom they worked, they had no recognized professional status in terms of legal state registration and no ready means of frequent communication of technical data and sharing of professional experience within a region. The meetings of other societies were not frequent enough to provide such exchanges in a timely manner in this rapidly expanding field. Out-of-state travel to meetings was restricted. Moreover, they were mainly scientific societies and thus not committed to items relating to professional practice or legal registration. Many of the practicing engineering geologists of the 1950s were not members of GSA and consequently did not attend the annual Cordilleran or national technical sessions and field trips. No major texts existed that related to the actual practice of engineering geology until the 1950 publication of the Berkey volume, *Applications of geology to engineering practice.* Earlier texts like Ries and Watson (1914–39) and Legget (1939) were "geology for engineers" (reviewed in Chapters 1 and 2, this volume). In 1957 the first case histories and review volumes in engineering geology were being prepared by the EGD/GSA but had not yet been published.

TABLE 4. SECTIONS OF THE ASSOCIATION OF ENGINEERING GEOLOGISTS (AEG, 1988)

Section (former name)	Date Established
Southern California (Los Angeles)	May 1958
Sacramento	Sept. 1958
San Francisco	Sept. 1958
Washington State	June 1963
Rocky Mountain (Denver)	April 1964
Texas (Ft. Worth-Dallas)	June 1964
Baltimore-Washington-Harrisburg (Washington, D.C.-Baltimore)	Jan. 1965
Oregon (Portland)	Apr. 1965
New York-Philadelphia	Jan. 1966
Kansas City-Omaha (Kansas City)	Jan. 1967
United Kingdom (London)	Feb. 1967
At Large	Oct. 1967
St. Louis	May 1969
Montreal	Aug. 1969
North Central	Jan. 1971
New England	Apr. 1971
Southeastern	Apr. 1971
South Africa	Oct. 1971
Allegheny-Ohio	Nov. 1975
Carolina	Feb. 1976
Lower Mississippi Valley	June 1980
Utah	Feb. 1981
Oklahoma	Sept. 1981
Alaska	Oct. 1984

HISTORY

In March 1957 a conference on "Geological Engineering Problems of Water in California" was held at the University of California, Berkeley. The conference was called by Parker Trask of that institution and included leading engineering geologists in California. Among the participants was John T. McGill of the U.S. Geological Survey Engineering Geology Branch, whose comments on geological requirements for a new building code sent a shock wave through the engineering geologists in Sacramento.

After a number of informal discussions among several of the actors in the drama, a group of ten geologists met in Sacramento, California, on the evening of June 3, 1957, and began to formulate an organization to be called the California Association of Engineering Geologists (CAEG). These included Claire P. Holdridge and Bruce M. Hall of the U.S. Army Corps of Engineers, William I. Gardner and Charles E. Hall of the Bureau of Reclamation, Lawrence B. James and H. D. Woods of the California Division of Water Resources, Theodore L. Sommers and Robert W. Reynolds of the California Division of Highways, and John C. Manning and Elmer C. Marliave, private consultants. A month later they were joined by Joseph F. Poland and George F. Worts of the U.S. Geological Survey to make up the founding group (Galster, 1982). Establishing requirements for membership and state registration, and defining engineering geology were early priorities for the group. By late 1957, action by the City of Los Angeles in establishment of a list of qualified engineering geologists to render reports required for city construction permits, stimulated the founding group into an early expansion of membership. The necessity was recognized early of establishing rather autonomous sections in areas of concentrated membership in order to properly review qualifications of applicants and to organize frequent technology transfer at local levels. Prior to the fledgling association's first general (annual) meeting in Sacramento in October 1958, autonomous sections had been established in Los Angeles, Sacramento, and San Francisco, and the Los Angeles section had already sponsored three field trips.

During the formative years of 1958 to 1963, CAEG was occupied with establishing rules of ethical behavior for members and working toward obtaining registration of engineering geologists in California. Other major activities included planning for a technical periodical to provide a forum for its members. Beginning in 1958, membership was extended to engineering geologists outside California. The response was surprising, and late in 1962 the word "California" was eliminated from the Association name by vote of the predominately California membership. Subsequently the first section outside California was established in Washington State in June 1963, followed in 1964 by sections in Denver (now Rocky Mountain) and Fort Worth–Dallas (now Texas). The growth and maturity of the 3,000-member association can be tracked by the establishment of sections listed on Table 4, and by the home sections of the past presidents and the coast-to-coast mobility of annual meeting locations (AEG, 1988). An additional commitment to "overseas" membership was generated in 1983 by establishment of a correspondent program in areas where insufficient membership does not warrant section status. Correspondents are now located in Hawaii, British Columbia, New Zealand, Australia, Indonesia, the Philippines, Hong Kong, Brazil, and Saudi Arabia. Administration of the association since 1973 has been through an executive director: Floyd T. Johnston, 1973 to 1983; Noel M. Ravneburg, 1982 to 1986; Patricia S. Osiecki, 1986 to 1987; and Edwin A. Blackey, 1988 to present (AEG, 1988). The association is managed by a board of directors consisting of a representative from each section and an executive council consisting of the four association officers and the immediate past president.

PUBLICATIONS

The association has published a series of field trip guidebooks for their annual meetings along with meeting programs and abstracts. The first edition of its prime technical periodical was issued in January 1964 under the name *Engineering Geology; The Bulletin of the Association of Engineering Geologists* as a semi-annual publication in a diminutive 6″ × 9″ format. The *Bulletin* has undergone several changes in style and format over the years; the "Engineering Geology" part of the name was dropped in 1968, publication was increased to quarterly in 1972, and the present 8½″ × 11″ format was adopted in 1981. The content of the *Bulletin* covers the entire spectrum of engineering geology, from case histories and methods utilized solving geologically related engineering problems to mapping techniques and professional and educational matters. Of special note is the "Geology of Cities" series, which appears from time to time in the *Bulletin* and features the geology of major cities of the world and its relation to urban development and construction. The association normally has one or more symposia at its annual meetings. Many have been published in the *Bulletin,* beginning with the first issue—a symposium on grouting (Goldschmidt, 1964)—held at the 1963 annual meeting in San Francisco.

Subsequent published symposia include such subjects as reservoir leakage (Stuart, 1969), instrumentation (Galster, 1972), rock support systems (James, 1972), expansive clays (Lee, 1974), engineering geophysics in South Africa (Lee, 1978), land use in coal mining areas (Dunrud and Morris, 1978), engineering geology in karst terrain (Foose, 1979), rapid transit construction costs (Proctor, 1981), hazardous waste disposal (Stirewalt and Tilford, 1981), radioactive waste disposal (Tilford and Saleem, 1981), education and registration (Williams, 1984), engineering geologic mapping (Keaton, 1984), hydrogeology (Bean, 1985), and computer applications (Tarkoy, 1986). The Southern California section (formerly Los Angeles section) has published two major volumes on engineering geology, mainly relating to California but applicable worldwide (Lung and Proctor, 1966; Moran and others, 1973). In 1983 the association published the proceedings of the 24th U.S. symposium on rock mechanics (Hoskins, 1983). The newsletter, the association's longest-running publication, was initially published in 1958 and began quarterly publication in 1972. It contains important technical notes as well as news of association activities and the profession. An annual directory of association members contains information about the association—its history, by-laws, policies, awards, and committees (AEG 1988).

CONTRIBUTIONS

The association has had a long-time commitment to future engineering geologists. In 1968 the Marliave Scholarship Fund was created to assist worthy and needy students in completing their education with low-interest loans and grants. In 1969 the first student chapter of the association was chartered at the University of Missouri at Rolla, the forerunner of 18 such chapters in the United States and Canada (AEG, 1988). Students are encouraged to prepare technical student papers, and the winning entries are published annually in the *Bulletin.*

In 1968, efforts to legally register geologists in California succeeded with the passage of A.B. 600 by the California Assembly. The law provided for registration of all geologists and the specialty certification of engineering geologists. The California law has served as a model for registration laws in several other states, and the association has been at the forefront in working for such laws in the various state legislatures. The association has been a consistently action-oriented organization in other areas of professional practice, including building codes as applied to seismic activity, ground stability and foundation conditions, industrial safety, and problems of professional practice. The latter culminated in publication of *Guidelines for professional practice* (Brown and Proctor, 1981). Early commitment to a strong code of ethics brought on some legal problems in the early 1980s and a replacement of the original code by the present Principles of Ethical Behavior (AEG, 1988). The association's early commitment to technical policy statements on dam safety (in 1979) and disposal of high-level radioactive wastes (in 1980) are typical of its action-oriented attitude (AEG, 1988). Association presidents and technical committee chairmen have frequently served as spokesmen for the profession on matters ranging from seismic safety to public policy. The association has maintained close ties to the GSA Engineering Geology Division (EGD) throughout its history. Four presidents of AEG have also served as EGD chairmen: E. B. Eckel (1965), F. W. Wilson (1973), R. W. Galster (1983), and A. W. Hatheway (1985). Similarly the association and its autonomous sections have maintained close liaisons with numerous engineering councils, national geological committees, and other geological and engineering organizations (see lists, AEG, 1988). The association spans an important area of interest between strictly technical and scientific societies and engineering-oriented societies, and provides representation for both categories in the diverse field of engineering geology.

INTERNATIONAL ASSOCIATION OF ENGINEERING GEOLOGISTS IN NORTH AMERICA

David J. Varnes and George A. Kiersch

BACKGROUND

History of North American participation in the International Association of Engineering Geology (IAEG) goes back to the time of the association's conception. At the International Geological Congress (IGC) held in New Delhi in 1964, the idea of forming an international organization in engineering geology was put forward by A. Shadmon of Israel, largely because the IGC lacked a section on engineering geology or on materials used in construction. An initial interested group of nine individuals, including George Bain of the United States, proposed first to form a commission with the International Union of Geological Sciences (IUGS) to explore the possible overlap with existing organizations. When it became apparent that this procedure would entail delays, perhaps even four years to the next IGC, a now augmented body promptly convened and established the IAEG on December 21, 1964, to be administered by a provisional committee of 14 with A. Shadmon, president, and M. Arnould of France, secretary. This group was enlarged by the addition of other prominent engineering geologists, including IAEG vice-president Arthur M. Hull of the U.S., then president of the U.S.-based Association of Engineering Geologists (AEG), but acting individually. The group met in Paris in 1967 and adopted statutes to be presented at the first general assembly of IAEG in 1968. In 1967, the IAEG applied for and was accepted as an affiliate of the International Union of Geological Sciences (IUGS). At the first general assembly, Professor Quido Zaruba was elected president, and Lloyd S. Cluff of the U.S., then AEG president but, like Hull, acting individually, was elected vice-president for North America. During the period 1969 to 1970, the IAEG grew in size and activities; its *Bulletin* was established, with an editorial board that included A. M. Hull, who served through 1972. Commissions were appointed on engineering geologic mapping and on landslides and other mass movements, which included Dorothy Radbruch-Hall and Eugene Kojan of the U.S., respectively, and a working group on earthquakes from an engineering point of view was established, with Lloyd Cluff as chairman.

From 1967 to 1975, there were repeated efforts on the part of IAEG to have AEG, or a part of it, become the U.S. National Group of IAEG. Yet, despite the individual efforts of AEG presidents Hull, Cluff, and Frank W. Wilson, who also served on the IAEG executive committee from 1972 to 1976, this action was consistently resisted by the executive council of AEG. Part of the reason was that AEG, in this period, was itself establishing sections in Canada, the United Kingdom, and South Africa; part lay in the general disinterest of U.S. members of AEG in international affairs; part in problems relating to distribution of the IAEG *Bulletin* and dues; and part, the organizational policies of GSA's Engineering Geology Division. In any event, efforts to form a U.S. IAEG group decreased and became dormant.

Meanwhile, there was growing activity by Canadian engineering geologists in the work of IAEG. P. M. Crepeau was elected to the IAEG Executive Committee for the 1968–1972 term. Subsequently, in 1972, Canada hosted the 24th International Geological Congress in Montreal, and this involved much preparation by IAEG members, both in Canada and abroad. Commission meetings were held during the Congress: the General Assembly elected John S. Scott to be IAEG vice-president for North America, and Owen L. White was appointed to the editorial board. He later succeeded Scott as vice-president for the term 1978–1982.

ORGANIZATION IN CANADA

In 1974, the Engineering Geology Division of the Canadian Geotechnical Society (CGS) was established as the Canadian National Group of IAEG (CNG/IAEG). One of its roles was to represent CGS as the Canadian National Group (CNG) of IAEG. Today the CGS is an independent learned society, representing some 1,200 geotechnical and engineering geology members throughout Canada. The society is affiliated with the International Society of Soil Mechanics and Foundation Engineering (ISSMFE), the Associate Committee on Geotechnical Research of the National Research Council of Canada (ACGR), and through the EGD it represents Canada on the Council of IAEG.

Through the Engineering Geology Division of CGS, members of the Canadian National Group of IAEG play prominent parts in the organization of, and contributions to, the activities of the CGS, such as the annual Canadian geotechnical conferences, the *Canadian Geotechnical Journal,* cross-Canada lectures and international conferences, and hosting engineering geological delegations from foreign countries (reviewed by Scott in Chapter 3, this volume). The Engineering Geology Division of CGS (and thus CNG/IAEG) represents the CGS on the Canadian Geoscience Council. Furthermore, EGD sponsors the Thomas Roy Award that is annually presented to the best Canadian engineering geologist in the past three years.

During the late 1970s, the CNG/IAEG experienced remarkable growth, attaining a high of 877 members in 1979—making it the largest national group within IAEG. Since then, the membership has declined and currently is about 300 of the total 1,200 members of the CGS. The 35th annual conference of the CGS was held at Montreal in 1982 in parallel with the 25th annual meeting of the Association of Engineering Geologists. Conference Chairman Luc Boyer stated "this brings together the two societies, for a rapprochement between their members, as well as between AEG and IAEG."

ORGANIZATION IN THE UNITED STATES

During the 25th International Geological Congress (IGC) in 1976 in Sydney, Australia, Professor M. Arnould, then president of IAEG, asked V. E. McKelvey, director of the U.S. Geological Survey and U.S. Representative to the IUGS, about the possibility of having the Engineering Geology Division of the Geological Society of America (EGD/GSA) become the U.S. National Group of IAEG. The request was referred to David J. Varnes, then chairman-elect of EGD/GSA, and discussions were held with M. Arnould and later with the EGD/GSA management board and with GSA officials.

Although the suggestion was attractive, it presented several difficulties. First, the Geological Society of America is not confined to the United States, but comprises and represents all of North America; therefore this proposal conflicted with both Canadian and Mexican organizations. Furthermore, at this time there was active consideration being given to the possibility that the Engineering Geology Division of GSA would decrease the scope of its activities in view of the growing importance of the Association of Engineering Geologists outside North America. Finally, Charles L. Drake, incoming president of GSA, advised that moves to form a National Group of IAEG could be properly undertaken only through the U.S. Committee on Geology, which operates under the joint aegis of the U.S. National Academy of Sciences–National Research Council (NAS–NRC) and the U.S. Department of Interior–Geological Survey (USGS) to effect appropriate participation of the geological community in international relations. The USNC/Geology represents the U.S. in the International Union of Geological Sciences (IUGS), to which the IAEG is affiliated, and other international geological activities, including the International Geological Congress (IGC).

The objective, as outlined by Linn Hoover, secretary USNC/Geology, was to organize the U.S. National Group of IAEG under a U.S. Committee for IAEG, which would, itself, be a subcommittee of the U.S. Committee on Geology that functions under the Board on Earth Sciences at the National Academy of Sciences. This procedure would require approval by both the USNC/Geology and the Governing Board of the Council of the National Academy of Sciences. Successive action by these two bodies, which meet infrequently, took time. More critically, the approvals required the demonstrated endorsement by the two active groups in the United States that would be represented on the committee: the Engineering Geology Division of the Geological Society of America and the Association of Engineering Geologists. Such broad support was furnished in 1979 by Richard H. Jahns, chairman of EGD/GSA and Richard Proctor, president of AEG.

The request to form a U.S. committee for IAEG was approved by the USNC/Geology and by the Governing Board of the National Academy of Sciences in 1979. Nominees for the eight elected members to serve on the USC/IAEG were obtained during the following year from both AEG and EGD/GSA. From the beginning it was agreed that the U.S. committee would represent the varied sub-fields and categories concerned with the applications of geological sciences to engineering works. Membership includes: (1) the current chairman of the Engineering Geology Division of GSA, a strong science- and research-oriented group; (2) the current president of the Association of Engineering Geologists, a practitioner's group with more of an engineering orientation; and (3) the chairman of USNC/Geology as ex-officio voting members. Of the eight elected committee members, two are, by agreement, from the U.S. Geological Survey (a financial supporter of USNC/Geology); and of the remaining six, at least two have academic-research orientation, two are from industry or consulting practices, and there must be some representation from other federal or state agencies. Having both the EGD/GSA and AEG directly involved has been beneficial to the subsequent successful operations of the USC/IAEG. The list of eight proposed members of the initial USC/IAEG was approved by the USNC/Geology on May 30, 1980, and by the National Academy of Sciences in July (Table 5, committee membership 1980–88). Meanwhile, in July 1980, Linn Hoover formally presented a request to form the USC/IAEG to the IAEG Council in Paris, which was meeting on the occasion of the 26th International Geological Congress. The request "was confirmed unanimously and with acclamations." Thus, the United States became the 40th National Group and the last of the major countries to join IAEG.

Since 1978, the U.S. membership in IAEG has grown from 40 individual members to a National Group of about 250 members in 1988. The USC/IAEG, through Frank Press, presi-

TABLE 5. MEMBERSHIP OF U.S. NATIONAL COMMITTEE/IAEG 1980–1988

MEMBER	TERM	EX-OFFICIO MEMBERS USNC/GEOLOGY	TERM
Johns, Richard H.	1980-81	Drake, Charles	1980-81
Wilson, Frank M.	1980-81	Curtis, Doris S.M.	1981-82
Deere, Don U.	1980-82	Maxwell, John C.	1982-83
Proctor, Richard	1980-82	Gould, Howard R.	1983-85
Varnes, David J.*	1980-87	Haun, John G.	1985-87
Kiersch, George A.†	1980-87	Skinner, Bryan J.	1987-89
		EGD/GSA	
Ivey, John B.	1981-84	Hatheway, Allen W.	1980-81
Tilford, Norman R.	1981-84	Scott, John C.	1981-82
Williams, John W.	1982-85	Ferguson, Harry F.	1982-83
Sydnor, Robert H.	1982-85	Winkler, Erhard M.	1983-84
Radbruch-Hall, Dorothy	1983-85	Wilson, Frank W.	1984-85
Pincus, Howard J.§	1983-89	Schuster, Robert L.	1985-86
		Cruden, David C.	1986-87
		Mathewson, C. C.	1987-88
		AEG	
Long, Joseph S.	1984-87	Ivey, John B.	1980-81
Rockaway, John W.	1984-87	Depman, Albert J.	1981-82
Benziger, Charles P.	1985-88	Paris, William C.	1982-83
Bonilla, M. G.	1985-88	Galster, Richard W.	1983-84
Underwood, Lloyd B.	1985-88	Valentine, Robert M.	1984-85
Cabrera, John G.	1986-89	Hatheway, Allen W.	1985-86
Fleming, Robert W.**	1986-89	Tilford, Norman R.	1986-87

*Chairman, 1980-83, and Secretary-Treasurer, 1980-87.
†Chairman, 1983-87.
§Chairman, 1987-89.
**Secretary-Treasurer, 1987-89.

dent, National Academy of Sciences, made an unsuccessful bid for the 5th International IAEG Congress in 1986 to be held in San Francisco. During the formative years of the USNG/IAEG (1980–83), David J. Varnes served as chairman of the U.S. Committee for IAEG, collected dues, and with secretarial help furnished by the U.S. Geological Survey, handled the mailing of IAEG *Bulletin* and newsletters. The fact that the USC/IAEG is an arm of the U.S. Committee on Geology, which derives support from the USGS, made such activity official business, and this has been essential to the committee's operations.

George A. Kiersch succeeded David J. Varnes as chairman for the period 1983 to 1987, but Varnes continued as secretary-treasurer until 1987. Concurrently for the period 1982 to 1986, David J. Varnes served as IAEG vice-president for North America, succeeding Owen L. White of Canada.

Chairman Kiersch broadened the emphasis of the U.S. Committee in 1984 with the establishment of four separate panel groups to further the planning and ongoing activities: Research Relevant to Engineering Geology (J. W. Williams, chairman); Public Relations (J. W. Rockaway, chairman); participation in IAEG 5th International Congress, 1986 (J. S. Long/J. G. Cabrera, co-chairmen); and Council Meeting in Washington, D.C., 1985 (H. J. Pincus, chairman).

Beginning in 1986, the U.S. Committee initiated planning for technical sessions and field trips on engineering geology during the 1989 International Geological Congress (IGC), Washington, D.C. Some 15 special symposia dealing with varied aspects of engineering geology received the support of the USC/IAEG, within the broad section on "Surface and Near-Surface Processes" and "Applied GeoSciences: The Environment, Hazards, and Engineering Works."

Several extended field trips pertaining to engineering geology were scheduled, including: engineering geology of western United States urban centers, engineering geology of major dams on the Columbia River, geology and engineering geology of the New York metropolitan area, landslides in central California, and geology of San Francisco and vicinity. Short trips will study geology and engineering problems of the Washington, D.C., area and shoreline erosion in the upper Chesapeake Bay area.

The USNC/Geology revised their constitution during 1986–87, and modifications adopted at the April 7, 1987, meeting in Washington, D.C., had the following impact on the six subcommittees as USC/IAEG: Chairmen of the standing subcommittees became voting ex-officio members of USNC/Geology. This placed the engineering geology community through IAEG in contact with the principal research and action groups of geological sciences, the Board on Earth Sciences, USNC/Geology, and U.S. Geodynamics Committee, and indirectly with the International Lithosphere Program.

At this same meeting, the USNC/Geology also accepted the request to amend the statutes, item III-G, "Functions of U.S. Committee/IAEG" in response to past requests from the community at large, and agreed to the following broader activities:

To coordinate a U.S. Program in Engineering Geology that includes: (1) advisory functions and an awareness of inter-related professional and society sponsored activities of the Engineering Geology Division of Geological Society of America, the Association of Engineering Geologists, American Society of Civil Engineers, and others as appropriate; (2) monitor and advise in the categories of research, training and education, and practice in Engineering Geology . . . the application of geological sciences for engineering works; and (3) support the further study, publication and dissemination of important scientific and engineering-oriented findings relevant to the field. Coordination efforts will exclude activities of the National Research Council, federal or state agencies, and academic or private groups; however, the Program will keep the profession-at-large aware of such activities.

The request to broaden the USC/IAEG activities was an outgrowth of several earlier discussions on this matter, such as one at the annual meeting of U.S. Committee/IAEG in San Antonio, Texas, on November 11, 1986, and another during a general workshop on "Goals in Engineering Geology" sponsored by the National Science Foundation and the University of California–Berkeley, January 21–23, 1983.

The International Council meeting in Buenos Aires on October 19, 1986, elected George A. Kiersch vice-president for North America to serve a four-year term, 1986–1990.

Contributions

As mentioned earlier, U.S. members of IAEG participated in the formation and subsequent productivity of working groups and commissions. One of the earliest and most continuously active of these is the Commission on Engineering Geological Mapping (D. H. Radbruch-Hall, secretary), which produced the comprehensive and well-illustrated *Engineering geological maps; A guide to their preparation,* published by Unesco in 1976, and later recommendations on map symbols and descriptions and classification of rocks and soils. D. J. Varnes and the Commission on Landslides and Other Mass Movements on Slopes prepared Landslide hazard zonation; A review of principles and practice, published by Unesco in 1984.

In October 1985, the U.S. Committee/IAEG co-sponsored with AEG an international symposium on "Management of Chemical Waste Sites," which was held at Winston-Salem, North Carolina, and organized by meeting chairman Norman R. Tilford and technical chairman Zubair A. Saleem. This first such international symposium in the United States was sponsored by two major engineering geologic organizations of the world when IAEG and AEG joined in a common enterprise. Preceeding the symposium, the international council of IAEG held its annual meeting at the National Academy of Sciences headquarters, Washington, D.C., on October 6–7. On the morning of October 8, the U.S. Committee hosted a field trip to the U.S. Geological Survey national headquarters, Reston, Virginia; this was followed in the afternoon by a geologically guided bus tour and travel to Winston-Salem, North Carolina, for the international symposium and technical meetings. The council sessions and field tour were arranged by panel chairman Howard J. Pincus. The two-day symposium was well attended; 54 papers and 21 posters by both foreign and U.S. authors provoked stimulating discussions; the sessions were a superior performance by the participants and organizers.

The 5th IAEG International Congress, Buenos Aires, Argentina, October 20–25, 1986, accepted 13 technical papers from members of the U.S. National Group of IAEG. Travel restrictions limited participation to ten papers, which were of high quality and very well received.

During the annual meeting of USC/IAEG in Phoenix, Arizona, on November 11, 1987 (Geological Society of America meetings), the new committee officers assumed their duties: Howard J. Pincus, chairman; and Robert W. Fleming, secretary-treasurer. Because of the restructuring and modification of the USNC/Geology constitution during 1986–87, the past chairman and secretary were asked to serve an extra year, until elections could be held and certified.

The U.S. Committee/IAEG will become an active participant in the forthcoming "International Decade for Natural Disaster Reduction" (IDNDR) of the 1990s. Members of the U.S. Committee and National Group are serving as liaison representatives to the U.S. Advisory Committee on IDNDR, and others are representatives within Coordinating Committee 1 on "The Environment—'Hazards'—and Geophysics" of the International Commission on Lithosphere and the U.S. Geodynamics Committee.

ORGANIZATION IN MEXICO

The Mexican National Group of IAEG, originally formed as a section on engineering geology within the Geological Society of Mexico, was approved by the IAEG Council in Madrid in 1978. The group did not thrive, however, and the initial members instead continued their respective activities within other organizations, such as the Geological Society of Mexico and the International Society of Rock Mechanics. One member, Jorge Maycotte, prepared a summary in 1984 on the "Historical Development of Engineering Geology in Mexico" for the editors of the proposed IAEG comprehensive book on engineering geology. Although this effort ultimately changed direction, Maycotte's efforts have been summarized in Chapter 1, this volume.

Though not formally organized, the applied geologists of Mexico have been very active since 1972 with technical symposia and/or meetings at least once each year. By 1983 this interest had broadened, and since then, three or four workshops and/or national meetings have been held each year in association with the Sociedad Mexicana de Mecánica de Rocas, or Geological Society. In March 1986, the IAEG National Group of Mexico was re-activated when about 30 members affiliated as the National Group within the Sociedad de Mecánica de Rocas; membership in 1987 was some 50 members when L. Espinosa Graham was elected president of the National Group/IAEG for 1987–1988.

The Second Binational U.S.–Mexico Symposium was held in Tucson, Arizona, in June 1987. The first was in Mexico City, February 24–26, 1983.) The Mexican Group of IAEG was well represented at the Tucson sessions. In recent years the *Bulletin* and symposia papers of group members have been mainly focused on underground excavations or slope stability analysis and stabilization techniques. A representative of the National Group of Mexico was invited to attend the 1987 IAEG Council meetings in Beijing, China.

APPENDIX A. SUMMARY OF REPORTS BY TASK COMMITTEES

Task committee-1; Nomenclature and engineering properties of shales

The work of the committee was largely limited to a review of published data available in the files of the Bureau of Reclamation and the U.S. Army Corps of Engineers. It was soon evident that the greatest difficulty lay in the physical identification of the various types of materials classified as shale by different workers. Shale to one investigator was siltstone, mudstone, indurated clay, or immature shale to another. Furthermore, the great range in physical properties of the shales emphasized the necessity for an exact description/classification of the many lithologic types. Evaluation of the problems of classification among the committee members quickly established that shales should be compared not only on the basis of field and hand specimen characteristics, but on mineralogical characteristics, cement, and grain size as well, since these might dominate any rational classification.

The task committee report revealed the inadequacy of present classification procedures and the lack of fundamental information with respect to the physical properties of shales and associated argillaceous rocks. On this basis, the Joint Committee on Engineering Geology concluded that a definitive evaluation of shales was not possible with the data available and discharged the task committee.

This study, however, did stimulate further investigation of shales and their engineering properties by others and was a factor in the preparation of a fine treatise by L. B. Underwood in 1965, "Classification, nomenclature and identification of shales," presented at a joint committee symopsium (Underwood, 1967).

Task committee-2; Nomenclature and engineering properties of rocks exclusive of shales

This committee undertook a program that included (1) classification of rocks for engineering purposes, (2) geological distribution of rocks, (3) engineering implications of rock properties, (4) techniques of exploration and testing, and (5) a bibliography on rock properties. Following the initial task committee meeting in 1955 that discussed an approach to the five problem areas, activities were limited mainly to items 1 and 5. A classification of rocks for engineering purposes envisioned a scheme whereby rocks are described in terms of the characteristics and properties of concern in engineering. Any such scheme should be based on the qualities/features observable in both hand specimens and the rock mass, rather than the conventional geologic origin or geologic history.

The task committee, therefore, evaluated a system in which the basis of rock classification consists of texture, structure, and composition. Progress was made on subdividing rocks into two classes: particulate and nonparticulate or amorphous. The particulate textures comprised two groups: granular without interlocking, and granular with interlocked grain boundaries. Although no general acceptance of this scheme by the task committee was ever realized, it is believed that the general approach possesses considerable merit as a basis for more fundamental evaluation of rocks for engineering purposes. Progress was also made on item 5 in the development of a bibliography on stress–strain relationships in rock.

Task committee-3; Products of weathering of bedrock and their engineering properties

The task committee initially prepared and distributed to interested parties a set of forms on the products of weathered bedrock and their engineering properties. Response was minimal.

At the GSA Annual Meeting in 1955, the task committee sponsored the following symposium: "Weathered Rock and Engineering Problems." Abstracts are published in the GSA Bulletin, v. 66, no. 12, part 2, 1955.

G. A. Kiersch, "Rock weathering in engineering geology—problems and need for research."
W. V. Conn and G. F. Sowers, "Engineering properties of bedrock weathering products in the southeastern Piedmont."
C. P. Holdredge, "Residual soils of the West Slope of the Sierra Nevada and their engineering properties."
D. U. Deere and T. H. Thornburn, "Soil mechanics properties of weathered volcanic rocks in Hawaii."
T. H. Thornburn and D. U. Deere, "Engineering problems associated with weathered limestone."

A second symposium on weathering was organized by the task committee for the ASCE convention in 1957:

G. F. Sowers, "Soil and foundation conditions in the Piedmont and Blue Ridge regions."
T. H. Thornburn, "Physical properties of some residual sandstone soils."
D. U. Deere, "Physical characteristics of some tropical residual soils."

A fourth paper, "Utilization of weathered formation materials for Trinity Dam embankment" by J. W. Hilf, was presented at the ASCE convention in 1957.

Task committee-4; Influence of geological factors on tunnel construction

This task committee organized a symposium at the ASCE convention in 1955. The following papers were presented under the general title, "Geological Factors in Tunnel Construction":

C. R. Rankin, "Heavy ground loads in tunnel construction in faulted areas."
A. B. Reeves, "Geology and tunnel design."
H. L. Scharon and A. B. Cleaves, "Geophysical investigations for the Lehigh Tunnel."
A. B. Cleaves, "Tools and techniques."

The papers by Reeves, Scharon and Cleaves, and Cleaves were published in the *Journal of Soil Mechanics and Foundations Division, Proceedings, ASCE,* v. 84, no. SM2, May 1958.

REFERENCES

AEG, 1988, Association of Engineering Geologists directory: Lawrence, Kansas, Allen Press, 234 p.
ASCE, 1977, Rock engineering for foundations and slopes, v. 1–2; Proceedings Geotechnical Engineering Division Specialty Conference, University of Colorado, Boulder, August 1976: New York, American Society of Civil Engineers.
Bean, R. T., ed., 1985, Hydrogeology symposium: Bulletin of the Association of Engineering Geologists, v. 22, p. 229–286.
Bechtel Civil, Inc., 1986, Survey of engineering-construction companies concerning engineering geologists: Bechtel Geology Group, unpublished, 25 p.
Brown, G. A., and Proctor, R. J., 1981, Professonal practice guidelines: Lawrence, Kansas, Allen Press, 145 p.
Caric, D. M., 1982, The Itaipu Hollow gravity dam: International Water Power and Dam Construction, v. 34, no. 5, p. 30–44.
Desai, C. S., and Christian, J. T., eds., 1977, Numerical methods in geotechnical engineering: New York, McGraw-Hill, 783 p.
Dunrud, C. R., and Morris, R. H., eds., 1978, Seminar on coal mining activities and concepts of multiple land use: Bulletin of the Association of Engineering Geologists, v. 15, p. 145–251.
Ferguson, H. F., ed., 1974, Geologic mapping for environmental purposes: Geological Society of America Engineering Geology Case Histories, no. 10, 40 p.
Foose, R. M., ed., 1979, Engineering geology karst terrain: Bulletin of the Association of Engineering Geologists, v. 16, p. 353–447.
Galster, R. W., ed., 1972, Instrumentation, practical applications, and results: Bulletin of the Association of Engineering Geologists, v. 9, p. 139–212.
——, 1982, A history of the Association of Engineering Geologists: Bulletin of the Association of Engineering Geologists, v. 19, p. 207–249.
Goldschmidt, A. R., ed., 1964, Untitled grouting symposium in Engineering Geology: Bulletin of the Association of Engineering Geologists, v. 1, p. 1–67.
HGS, 1950–52, Proceedings of the First, Second, and Third Symposia; Geology as Applied to Highway Engineering: Richmond, Virginia, Department of Highways, 5 p., 82 p., 88 p., respectively.
Horner, S. E., and McNeal, J. D., 1950, Application of geology to highway engineering, *in* Van Tuyl, F. M., and Kuhn, T. H., eds., Applied Geology: Colorado School of Mines Quarterly, v. 45, no. 1B, p. 151–191.
Hoskins, E., 1983, Proceedings, 24th U.S. Symposium on Rock Mechanics: Texas A&M University, Association of Engineering Geologists, and U.S. National Committee on Rock Mechanics, National Research Council, 856 p.
James, L. B., ed., 1972, Rock support systems—Underground and open excavations: Bulletin of the Association of Engineering Geologists, v. 9, p. 213–326.
Keaton, J. R., ed., 1984, Engineering Geology Mapping Symposium: Bulletin of the Association of Engineering Geologists, v. 21, p. 253–364.
Kenney, C., and Spencer, R., 1971, Engineering geology education: Geotimes, v. 6, no. 10, p. 24–25.
Lee, F. T., ed., 1974, Active clays in engineering and construction practice: Bulletin of the Association of Enginering Geologists, v. 11, p. 259–419.
——, 1978, Engineering geophysics in South Africa: Bulletin of the Association of Engineering Geologists, v. 15, p. 1–144.
Lees, R. D., Lettis, W. R., and McClure, C. R., 1985, Applications of remote sensing techniques to engineering and construction services, *in* Proceedings, International Symposium on Remote Sensing of Environment, Fourth Thematic Conference, "Remote Sensing for Exploration Geology," San Francisco, 1985: Ann Arbor, Environmental Institute of Michigan, v. 1, p. 93–97.
Legget, R. F., 1939, Geology and engineering: New York, McGraw-Hill Book Company, Inc., 650 p.
Lung, R., and Proctor, R. J., eds., 1966, Engineering geology in Southern Califor-

nia: Glendale, Special publication of the Los Angeles Section, Association of Engineering Geologists, 389 p. with maps.

Lyra, F. H., 1982, Planning and construction sequence at Itaipu: International Water Power and Dam Construction, v. 34, no. 5, p. 27–30.

Moran, D. E., Slossen, J. E., Stone, R. O., and Yelverton, C. A., eds., 1973, Geology, seismicity, and environmental impact: Association of Engineering Geologists Special Publication, 445 p., with map.

Paige, S., and Rhoades, R., 1948, Report of Committee on Research in Engineering Geology, Society of Economic Geologists: Economic Geology, v. 43, no. 4, p. 313–323.

Proctor, R. J., ed., 1981, Rapid transit construction costs related to local geology: Bulletin of the Association of Engineering Geologists, v. 18, p. 133–220.

Ries, H., and Watson, T. L., 1914, Engineering geology: New York, John Wiley and Sons, 679 p.

Rogers, F. C., 1964, Geologists; growing role in dam design: Bulletin of the Association of Engineering Geologists, v. 1, no. 2, p. 1–5.

Rose, D., 1975, Civil engineering for the engineering geologists, part 2: Association of Engineering Geologists Annual Meeting, 49 p.

Schuster, R. L., ed., 1986, Landslide dams; Processes, risk, and mitigation—Proceedings of session by Geotechnical Engineering Division at Seattle, 1986: New York, American Society of Civil Engineers, 172 p.

Stirewalt, G. L., and Tilford, N. R., eds., 1981, Geologic and hydrologic factors in the disposal of hazardous wastes in the southeastern United States: Bulletin of the Association of Engineering Geologists, v. 18, p. 225–276.

Stuart, W. H., ed., 1969, Reservoir leakage and groundwater control: Bulletin of the Association of Engineering Geologists, v. 6, p. 1–94.

Tarkoy, P. J., ed., 1986, Computer applications in engineering geology: Bulletin of the Association of Engineering Geologist, v. 23, p. 219–286.

Tilford, N. R., and Saleem, Z. A., eds., 1981, Geological disposal of high-level radioactive wastes: Bulletin of the Association of Engineering Geologists, v. 23, p. 349–443.

Trask, P. D., ed., 1959, Symposium on Rock Mechanics: Geological Society of America Engineering Geology Case Histories, no. 3, 76 p.

Underwood, L. B., 1967, Classification and identification of shales: Journal of Soil Mechlanics and Foundation Division, American Society of Civil Engineers, v. 93, no. SM6, p. 97–116.

U.S. Department of Labor 1946, National roster of scientific and specialized personnel; Geology as a profession: Washington, D.C., U.S. Department of Labor Vocational Booklet no. 1, 19 p.

Williams, J. W., ed., 1984, Education-registration Workshop Proceedings: Bulletin of the Association of Engineering Geologists, v. 21, p. 157–178.

Manuscript Accepted by the Society December 14, 1988

Golden Gate Bridge. Engineers and geologists inspecting the conditions of serpentine foundation rock at the bottom of the excavation for the south pier, 108 ft below the channel surface, circa 1934. (Photo courtesy of the Heinrich Ries Collection, Cornell University.)

Printed in U.S.A.

Geological Society of America
Centennial Special Volume 3
1991

Chapter 6

Surface water and flooding

Perry H. Rahn
Department of Geology and Geological Engineering, South Dakota School of Mines and Technology, Rapid City, South Dakota 57701

INTRODUCTION

Historical perspective

Rivers provide numerous benefits to man as avenues of commerce and sources of water supply. Flood plains have rich agricultural soils, and some of the earth's most heavily populated areas occur along rivers.

Yet the riverine environment can be hazardous. Ancient civilizations struggled against floods while trying to earn a living from the fertile land adjacent to the Nile, Tigris, Euphrates, Indus, and Yellow Rivers. The worst geologic disaster known occurred in 1887 when about 800,000 Chinese lost their lives from the Yellow River flood (Costa and Baker, 1981). Another 100,000 Chinese died when the Yangtze River flooded in 1911. The great flood of 1927 extensively damaged the lower Mississippi River Valley and a flood-control plan was quickly adopted. The Arno River ripped through Florence, Italy, in 1966 and damaged one of the foremost art centers of the world.

Legget (1973, p. 66) has noted that in Central Europe the builders of medieval towns generally avoided the flood plains. They were rightly afraid of floods, and recognized the difficulty of construction on the wet ground adjacent to rivers. The early sectors of the older cities were usually located on the higher ground provided by river terraces. This wise practice was not based consciously on geological training, but utilized the same craft-lore demonstrated by the builders of the Pyramids and other early works (Chapter 1, this volume). Unfortunately, since the late 1800s, buildings on flood plains has steadily increased throughout the world with the growth of major cities to satisfy pressures for urban living.

In the United States, flood-related deaths continue, and annual flood damage is increasing (Rahn, 1986). In one four-year interval (1969–1973), floods produced the following disasters:

- August 1969, Hurricane Camille caused flooding of the James River and other streams in Virginia that left 152 people dead or missing.
- February 1972, the failure of a coal-waste dam sent a flood wave down Buffalo Creek Valley, West Virginia, leaving 125 dead or missing.
- June 1972, floods in the Black Hills left 238 dead and 8 missing.
- June 1972, Hurricane Agnes produced floods from Virginia to New York that killed 105 people and caused $3.1 billion damage (1972 dollars).
- Spring of 1973, floods on the Mississippi River produced a record 88 days of flood stage at Vicksburg, Mississippi. The flood inundated more than 49,000 km^2 of land and damaged more than 30,000 houses.

This review identifies the reasons for the increased flood losses throughout the U.S. and means for reducing such losses. Insight from flooding and/or construction of engineered works to control rivers has provided some advances in geological knowledge relative to the impact on natural environs.

The traditional approach

The U.S. government first became involved in water resource projects in 1888 when Congress appointed John Wesley Powell to establish the Irrigation Survey, which became the Reclamation Service in 1902 and the Bureau of Reclamation in 1923. Since then the USBR has built many multipurpose dams in the West. The Mississippi and Ohio River floods of 1927, 1936, and 1937 and the floods of 1936 and 1938 on the Connecticut and other rivers in New England led directly to the Flood Control Acts of 1936 and 1938. The subsequent construction of many dams and reservoirs was an attempt to mitigate the hazard. Floods and droughts in the 1930s provided the ideal stimulus for the maintenance of public and political interest in water and land. Expansion of the work of the Bureau of Reclamation and the Department of Army (Corps of Engineers), and the creation of the Tennessee Valley Authority (1933) and the Soil Conservation Service (1935), were derived from a combination of the Great Depression, political upheaval, population shifts, and natural disasters (Kiersch, Chapter 1, this volume). This set of circumstances promoted large expenditures for water development and the era of big dam building.

Over the past century, the occurrence of any disastrous flood

Rahn, P. H., 1991, Surface water and flooding, *in* Kiersch, G. A., ed., The heritage of engineering geology; The first hundred years: Boulder, Colorado, Geological Society of America, Centennial Special Volume 3.

has invariably resulted in the demand for dam construction. For example, when a June 16, 1965, flood hit Denver, Colorado, the Corps of Engineers was pressured and responded with the $85 million Chatfield Dam on the South Platte River.

Since Congress designated the Corps of Engineers responsible for flood-control planning along navigable rivers and waterways in 1936, the Corps has grown immensely. Currently (1986) the Corps is designing or building 124 dams at a cost estimated at more than $15 billion. Besides flood control, the Corps' dams generally include multipurpose benefits such as hydropower, recreation, water storage, or redevelopment of land around the reservoir. Yet, despite federal expenditures of about $15 billion on engineered control works since 1936, the annual flood losses have increased.

Figure 1. Oblique aerial photograph of a flood plain caused by the meandering of Rapid Creek, South Dakota.

FLUVIAL PROCESSES AND HYDROLOGIC ANALYSIS

Geomorphology

Flood plains. Floods are natural geomorphological phenomena and are most hazardous where they inundate flood plains. This review concentrates on floods in the common classical form of a stream and flood plain. Braided channels and alluvial fan morphology and processes are described elsewhere by Collinson (1986) and Ritter (1986).

A flood plain, the relatively flat constructional landform bordering a stream, is inundated to some degree when the stream overflows its banks (Fig. 1). Flood plains are not caused by floods per se, but originate by the lateral migration of meanders (Fenneman, 1906).

A flood plain may include a diversity of fluvial features such as: a river channel, oxbow lakes in various stages of being filled with sediment and swamp muds, meander scrolls formed by channel migration, sloughs formed as meander depressions or abandoned anastomosing channels, natural levees, and backswamp areas formed by slack water ponded between natural levees and the valley wall (Wolman and Leopold, 1957; Costa and Baker, 1981). Valley lowlands, which might be called "flood plain" and traditionally mapped as alluvium, may include landslide deposits, colluvium, and low-angle alluvial fans deposited from tributary canyons. Slope wash from the valley margins may accumulate as debris on a flood plain and form a gently inclined surface that slopes toward the main valley (Leopold and Miller, 1954).

Bankfull stage and flood-plain inundation. The frequencies of bankfull stage in rivers of diverse size and physiographic settings are similar. The recurrence interval is about 1.5 yr, although there is some variability (Williams, 1978). Aggrading stream reaches experience more frequent overbank flooding (Richards, 1982). Geological features also affect the overbank flooding frequency; Costa and Baker (1981) conclude the 50-yr flood for a 260 km^2 basin in Texas may be several times that of basins in Wisconsin where the presence of permeable glacial surficial deposits allows for rapid infiltration of runoff. A return period of 1.5 yr for bankfull discharge for any river is an oversimplification. Nevertheless, the recognition of this approximate recurrence interval conveys the general frequency of floods.

Likewise there does not seem to be a simple relationship between recurrence interval and the amount of flood plain inundation. Sections of the lower Mississippi River, before man-made channelization, overtopped the banks and inundated portions of the flood plain as much as five times a year. Yet upstream in Wisconsin, the 1952 Mississippi River flood (recurrence interval 40 yr) inundated only 56 percent of the local flood plain (Chorley and others, 1984). The amount and frequency of flood plain inundation is not only a matter of discharge frequency, but also is related to the temporal adjustments in the channel capacity. Some changes may be induced by man.

Methods utilized to study flood plain inundation include multispectral composites of Landsat imagery taken before, during, and after flooding, such as the Mississippi River Valley in 1973 (Deutsch and Ruggles, 1974). Rango and Salomonson (1974) demonstrate a correlation between the hydrologic maps of flood-prone areas as delineated on U.S. Geological Survey hydrologic-hydraulic maps with the flood plains outlined by remote sensing techniques. The use of Holocene stratigraphy, soils, and vegetation as an aid in the determination of flood-frequency events is described by Baker (1977), Costa (1978), and Costa and Baker (1981). The age and elevation of Holocene slackwater alluvium has been used to reconstruct paleoflood history of rivers in Texas (Kochel and others, 1982), Colorado (Costa, 1983), Arizona (Ely and Baker, 1985), and Utah (O'Conner and others, 1986).

Terraces. A stream terrace is an abandoned flood plain that results from a lowered base level. Terraces are generally safe from floods.

Terraces may be paired or unpaired (Fig. 2). The continuity

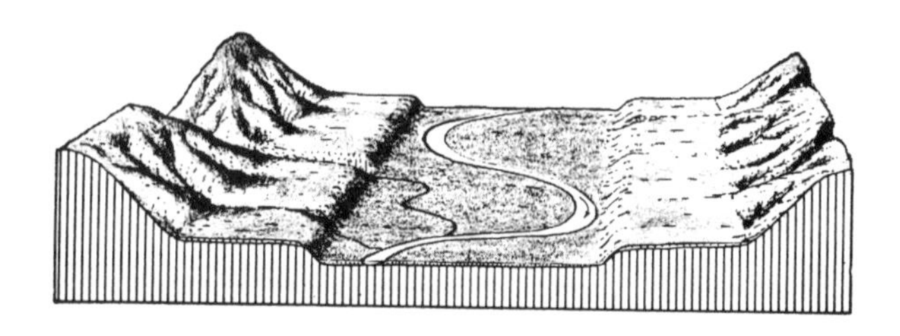

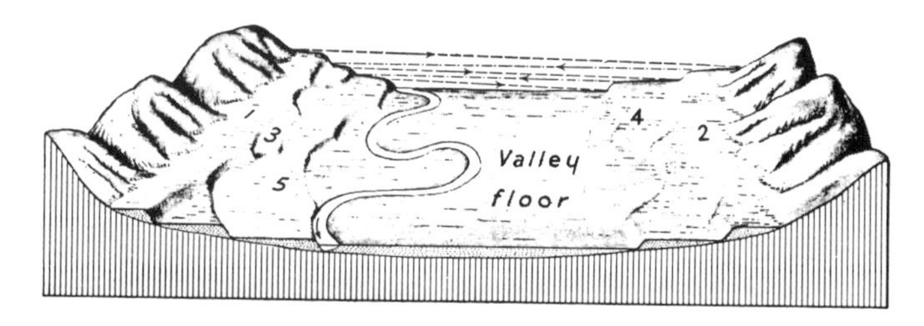

Figure 2. Sketches showing paired (above) and nonpaired (below) terraces (from Longwell and others, 1948).

of a terrace along a valley, as paired on both sides, may be correlated to a graded stream condition. "In contradistinction, a stream downcutting progressively will leave along the valley sides isolated flat spurs; these may not be paired, and they are of irregular height because they never were portions of a single continuous surface" (Leopold and others, 1964, p. 465). Unpaired terraces imply continued downcutting accompanied by lateral erosion (Thornbury, 1948). The stream shifts back and forth from one side to the other over the valley, and the valley floor is gradually lowered.

Unpaired terraces are of special interest as flood plain hazards because some very low terraces may be inundated during floods. Jahns (1947) used alluvial stratigraphy to evaluate the inundation of low terraces along the Connecticut River during the 1936 catastrophic flood when low terraces were inundated that had not been affected for several hundred years.

The distinction between terraces that may be flooded and those above flood levels may be complicated by changes in aggradation or degradation. Such changes can be caused by natural processes or by man. Cyclic episodes of aggradation or degradation may be due to overgrazing or climatic change (Leopold and Miller, 1954), or to sensitive interbasin factors that control runoff or sediment production (Womack and Schumm, 1977; and Patton and Boison, 1986). A low terrace may be practically undiscernable from the flood plain. Most rivers have complex terrace sequences that can be both paired or unpaired. For example, the terraces along the lower Mississippi River were formed due to degradation accompanying sea-level lowering during glacial maxima, while the terraces along the upper Mississippi River were formed due to degradation that accompanied reduced sediment loads during interglacial time. The active flood plain width of the lower Mississippi River is 16 km, lying within a terraced valley floor varying between 40 to 200 km wide (Fig. 3). The study of morphology of the lower Mississippi River and its relationship to the effect of climate and sea level during the Quaternary was pioneered by Fisk (1944, 1947, 1951).

A number of basic concepts or principles of geomorphology are the result of past geological investigations concerned with floods and flood-control works. Unfortunately this knowledge and understanding of surficial features and the effects of young alluvial sediments throughout a river valley are sometimes overlooked by the planners, design engineers, or builders of the extensive levee and dam systems to control riverine development (Kammerer and Baldwin, 1963).

Hydrologic analysis

Flood magnitude. Stream discharge is regularly measured by the U.S. Geological Survey at some 10,000 sites in the U.S. Stream discharge is usually determined at a gaging station by use of a rating curve (plot of stage versus discharge). Larger floods commonly fall beyond the limits of the rating curve, and consequently their discharge cannot be accurately determined. However, the Manning equation can be used to estimate flood discharge (Dalrymple and Benson, 1967) using slope and cross-sections as measured in the field with a roughness factor from standard manuals (Barnes, 1967). The Manning equation relates discharge to other parameters as:

$$Q = \frac{A}{n} R^{2/3} S^{1/2}$$

where: Q = discharge, m^3/s; A = cross-sectional area, m^2; n = roughness; R = hydraulic radius (m); and S = slope of channel.

Tice (1968) has determined the relationship between flood discharge and size of drainage area for selected eastern U.S. rivers. Peak discharge also depends on the interrelationships of the geologic setting, physiography, and precipitation (Riggs, 1973; Costa and Baker, 1981). Engineering techniques to determine peak discharge for selected regions in the U.S. are given by the U.S. Bureau of Reclamation (1973).

Flood discharge, and total stream runoff from a basin, increases with man's encroachment and increased use of the land. For example, the progressive conversion of a forested watershed to rangeland, cropland, and ultimately to an urban area results in large increases in runoff. In general, peak discharges and sediment loads are increased due to urbanization, while lag times are reduced (Leopold, 1968; Hollis, 1975).

Statistical analysis of flood frequency. Statistical methods for analyzing flood frequency have been discussed by Hoyt and Langbein (1955), Chow (1964), Reich (1973), Hjelmfelt and Cassidy (1975), Wanielista (1978), Linsley and Franzini (1979), and Cobb (1983). Analyses of flood frequency are attained by utilizing an appropriate probability density function such as log-normal Gumbel or log-Pearson (Overton and Meadows, 1976). Typically an array of yearly flood peaks will contain many low

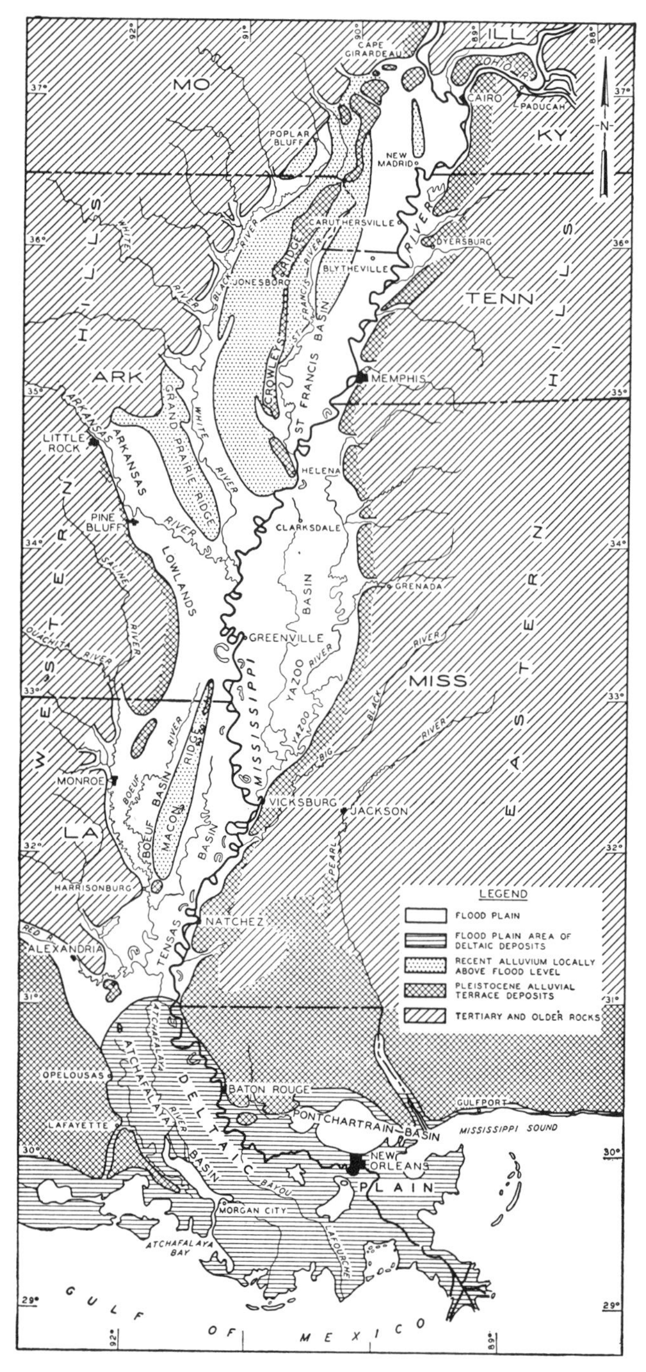

Figure 3A. Map of lower Mississippi River showing extent of flood plain (from Turnbull and others, 1950).

values but only a few high values, so the distribution is log-normal.

The traditional U.S. Geological Survey procedure for flood-frequency analysis is described by Dalrymple (1960). Data are assembled showing the highest flood for each year in a period of N years of record. The data are ranked (m), and the recurrence interval (T) is calculated for each flood as:

$$T = \frac{m}{1+N}$$

The recurrence interval is determined for each flood, and plotted with the flood frequency curve as the x-axis and discharge as the y-axis (Fig. 4). The data are plotted on Gumbel paper, which contains a logarythmic x-axis (modified in the interval 1.01 to 10 yr) and a y-axis of either an arithmetic or logarithmic scale, whichever provides the nearest to a straight-line plot. The mean annual flood discharge for any river typically has a recurrence interval of 2.3 yr.

To standardize flood data analyses among federal agencies, the Water Resources Council (WRC) chose one standard method in 1967. The council recommended that all agencies try to fit the log-Pearson Type III probability distribution to annual flood peak series, using a log-probability plot (Fig. 5). If skewness is zero, the data should plot as a straight line. Regional skew values for the U.S. are available (Benson, 1968, 1972; Hardison, 1974). The Water Resources Council (1967) did recommend another distribution or technique for unique conditions. Reich (1972) determined that Gumbel's method worked satisfactorily for small watersheds in Pennsylvania.

The U.S. Federal Emergency Management Administration (FEMA) uses the 100-yr flood as the base flood elevation in establishing regulatory requirements. The 100-yr flood refers to the flood discharge (or stage) with a statistical occurrence probability of 1 percent (P = 0.01) in any given year. The probability that this flood will not occur in a given year is (1-P) = 0.99 or 99 percent.

From statistics (Mason and others, 1983), it can be shown that the probability of occurrence, P(x), of some event having a binomial probability distribution, can be obtained from statistical tables or the formula:

$$P(x) = \frac{n!}{x!\,(n-x)!} P^{x} (1-P)^{n-x}$$

where n = number of trials; x = number of observed successful events; and P = probability of success of each trial. For example, consider a flood with a recurrence interval of 100 yr (P = 0.01). What is the probability that no floods of 100-yr discharge (or greater) will occur during 100 yr? Although it may seem there is a 100 percent chance of occurrence of a 100-yr flood in 100 yr, this is not true. No hydrologic phenomenon has a 100 percent chance of occurring, although it may approach 100 percent. In this case, where P = 0.01, n = 100 years, and x = 0 events:

$$P(x) = \frac{100!}{(0!)\,(100-0)!} (0.01)^{0} (1-0.01)^{100-0} = 0.366 = 36.6 \text{ percent}$$

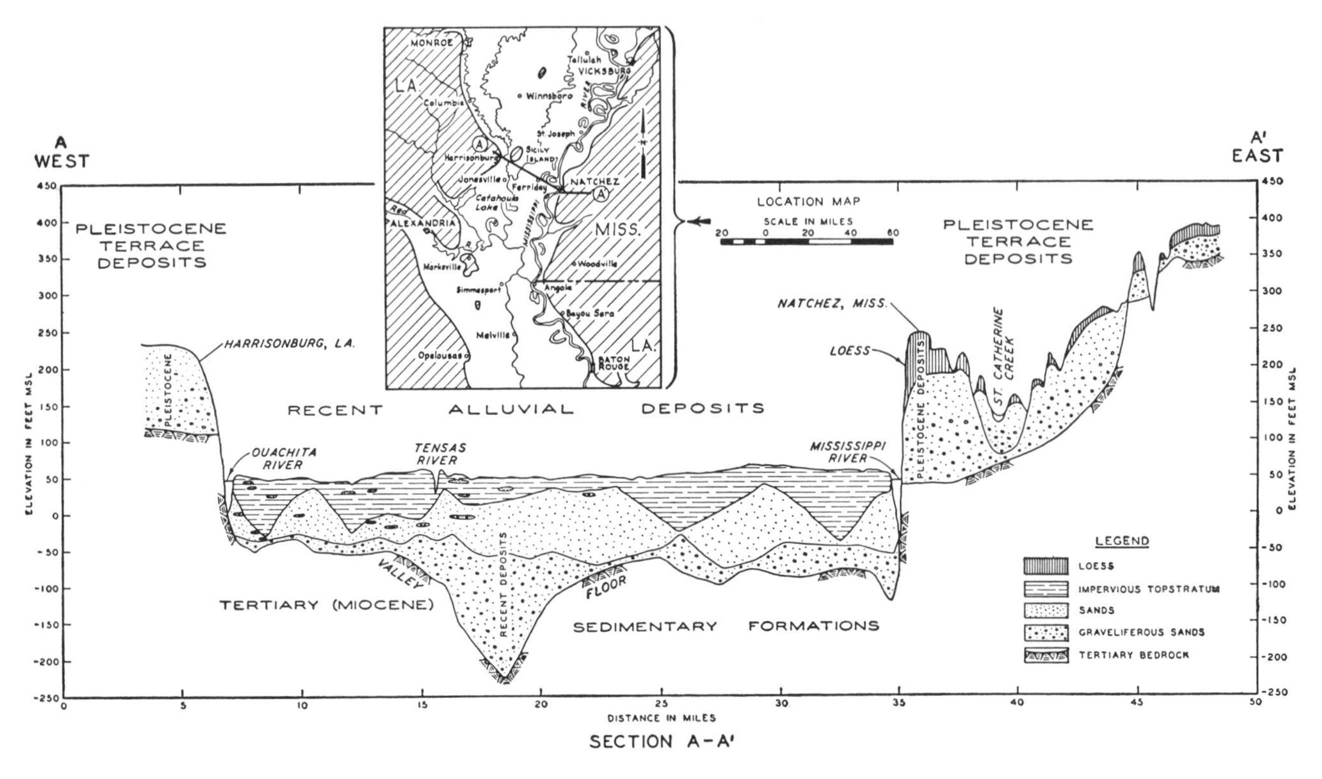

Figure 3B. Geological cross-section near Natchez, Mississippi (from Turnbull and others, 1950).

Since the probability of *not* occurring is 36.6 percent, the probability that one or more 100-yr floods *will* occur during 100 yr is 63.4 percent. Probability tables for other recurrence intervals and durations are given by Linsley and Franzini (1979) and Costa and Baker (1981).

By similar analysis, during 100 yr the chance of occurrence of the 500-yr flood is 18.1 percent. Thus, of those areas protected only for the 100-yr flood, during any 100-yr interval about 18 percent of these "protected" areas are likely to be inundated by a 500-yr flood. Furthermore, the damage to these areas would be severe because most inhabitants are under the impression that they are safe from all floods.

A major problem in flood-frequency analysis is the assumption that the available flood data are random events from the same population of all floods. The presence of an extremely large number in a set of flood discharge data presents a dilemma to the hydrologist. For example, Figure 5 shows the probability plot of annual flood peak discharge for Rapid Creek at Rapid City, South Dakota. The log-Pearson III curve includes the 1972 peak discharge value. It could be argued that the 1972 event be treated as a "statistical outlier" with the 100-yr flood discharge determined from a pre-existing curve based only on the remainder of the data. Overton and Meadows (1976) conclude in the case of one large outlier that the value should be "censored" and removed from the data set.

An example of the magnitude of error in extrapolation beyond the range of existing data may be found in the record of

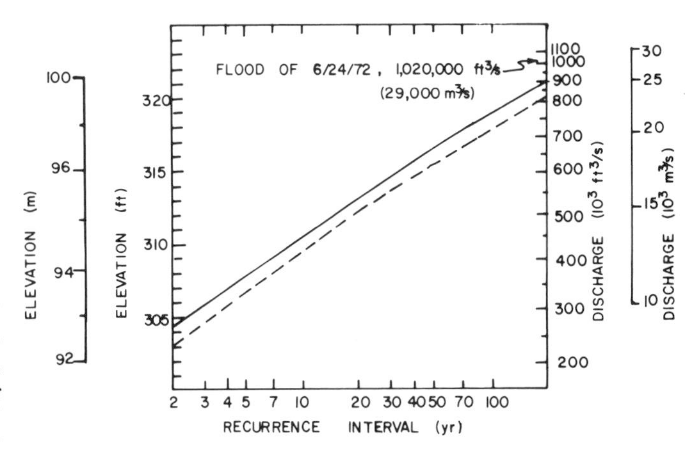

Figure 4. Recurrence interval plot for floods on the Susquehanna River at Harrisburg, Pennsylvania (from Page and Shaw, 1973). The solid line represents natural discharge; the dashed line represents regulated discharge (1972 conditions).

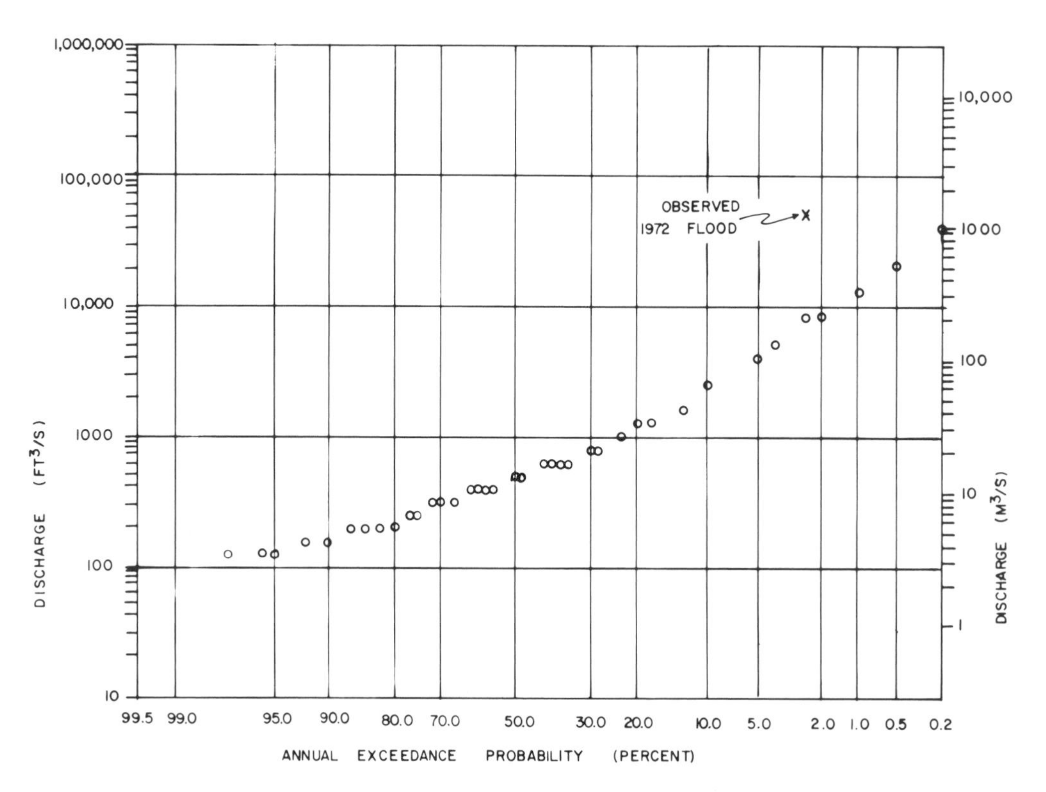

Figure 5. Log-probability plot of annual floods (1946–1983) for Rapid Creek at Rapid City, South Dakota (from Benson and others, 1985). The plotted data indicate the observed peaks (for greater than 5 percent probability).

maximum 24-hr rainfall at Hartford, Connecticut (Hershfield and Wilson, 1957). Based on the 50-yr record through 1954, the 100-yr value was 15 cm, but a 1955 hurricane produced 31 cm in 24 hrs. Consequently, the estimated 100-yr precipitation was revised to 22 cm, a 40 percent increase. Another approach is provided by the flood stages of the Connecticut River at Hartford over several centuries. From the historical record, there is little basis for a stage greater than 9 m, the height of the 1854 flood. Yet floods in 1936 and 1938 produced stages of 12 and 10 m, respectively. Using the pre-existing curve, and treating the 1936 and 1938 events as outliers, the Connecticut River thereby produced (supposedly) a 400-yr flood in 1936 and a 200-yr flood in 1938, which seems highly unlikely. More realistically, when floods of such low probability occur frequently, the curve should be revised. The omission of large events (censored outliers), will cause a bias toward underprediction, particularly for large return periods. A critical factor in forecasting floods (and a reason for not censoring outliers) is the spread of agricultural lands and urbanization, which causes increased floods within many stream basins.

It has been said that "Nature has neither a memory nor a conscience." In other words the occurrences of rare hydrologic events can be considered independent random trials. Thus, the occurrence of a flood confers no immunity against its immediate recurrence. If the 100-yr flood occurs in one year, its chance of occurrence the next year remains the same. Every year is a new spin of the roulette wheel. For example, Kansas City, Missouri, experienced two separate and distinct rainfall events in 24 hours in 1965, and each exceeded the 100-yr rainfall frequency (Hjelmfelt and Cassidy, 1975).

CATASTROPHIC FLOODS

Catastrophic floods are classified as either flash floods or inundation floods. Flash floods are typically caused by local convective thunderstorms (cloud bursts), and are most common on small watersheds in steep topography. Inundation floods are typically caused by widespread cyclonic precipitation and are most common on large watersheds in low topography. However, as in any classification of natural phenomena, floods occur that do not fit this simple classification.

The following discussion includes examples of both kinds of floods. Subsequent sections contain information on attempts to control floods.

Flash floods

Cases of flash floods caused by intense precipitation include the 1903 Heppner, Oregon, cloud burst flood; the 1969 Hurricane Camille, Virginia, floods; the 1972 Rapid City, South Dakota, flood; and the 1976 Big Thompson Canyon, Colorado, flood.

The Heppner flood of June 14, 1903, was the result of a cloud burst which fell on a 52 km^2 drainage basin. An unknown amount fell during a half hour, causing a peak discharge of 1,020 m^3/s on Willow Creek. The sudden flood wave through the steep Willow Creek valley (gradient of 5.3 m/km) destroyed much of the town, with a loss of some 200 lives (Murphy, 1904).

Hurricane Camille (1969), in its passage over the Blue Ridge Mountains of Virginia, produced flash floods in many streams. Some of these streams crested 6 m or more above normal flood stage in a matter of 4 to 8 hours, taking more than 150 lives and causing an estimated $112 million in property damage.

The June 9, 1972, flood in the Black Hills was one of the largest natural disasters in the U.S. At least 238 people died, mostly in Rapid City where Rapid Creek reached a peak discharge of 1,430 m^3/s (Larimer, 1973; Schwartz and others, 1975). Geologically, there is nearly complete agreement between the area inundated along Rapid Creek and the distribution of Quaternary alluvial deposits shown on published geologic maps (Rahn, 1974). Figure 6 shows part of the flooded area on three vertical aerial photographs of Rapid City: the 1952 photo shows the preflood condition, the 1972 photo shows the 1972 flood destruction, and the 1982 photo shows the modern setting with a floodway established along Rapid Creek.

The Big Thompson Canyon flood resulted from intense thunderstorms on July 31, 1976. The Big Thompson River, with a drainage basin of 155 km^2, had a peak discharge of 884 m^3/s (Maddox and others, 1977). At least 139 people were killed, and $36 million in property damage occurred. Both the Rapid City flood of 1972 and the Big Thompson flood of 1976 have much in common. They illustrate the threat of cloud burst–generated torrential floods adjacent to mountainous regions. Both floods occurred near dams which provided no protection because the main precipitation occurred downstream from the dams. Furthermore, the storms were short lived (about 12 hr), confined to a fairly local area, and provided little time for warning. Extremely fast velocities can occur, as in the Big Thompson flood. Streamside campers were evacuated to high ground, even though no rain had fallen at their localities; minutes later a flood wave in excess of 6 m swept down the canyon.

If timely precipitation data are available, analytical techniques can alert the public to an imminent flood hazard (Cooper, 1978). Discharge prediction for large inundation-type floods (which typically take a half-day or longer to crest) can be made and warnings given if adequate rainfall and streamflow data are made available to hydrologists. In contrast, flash floods develop suddenly and crests occur within very short periods of time, ranging from several minutes to a few hours. Therefore it is not feasible to collect and analyze hydrologic data in time to predict flash floods, even with computer processing. The National Weather Service of the National Oceanic and Atmospheric Administration issues general warnings using an analysis of synoptic weather information available from radar, satellite, and telemetry of rainfall. The best a local forecaster can do is alert the public to the potential for a flash flood. The use of an unmanned water-level detecting device, which is placed on a stream above a community, can warn of an imminent flash flood, but so far has only limited application.

Inundation floods

Inundation floods typically occur in large river valleys and last for a much longer time than flash floods. Deaths are generally not as numerous, but damage due to prolonged inundation is excessive. Two typical inundation floods are described and evaluated.

1972 Susquehanna River. Hurricane Agnes brought torrential rain to the mid-Atlantic states in late June and early July of 1972. Total precipitation at several locations from New York to Virginia was in excess of 38 cm during June 19–23 (Bailey and others, 1975) with isohyets showing maximum values near Harrisburg, Pennsylvania (Engman and others, 1974). The widespread flooding caused Agnes to be called the most destructive storm in U.S. history, claiming 117 lives and causing damage estimated at $3.1 billion in 12 states. Damage was particularly great in Pennsylvania, New York, Maryland, and Virginia. Damage in Pennsylvania reached $2.1 billion, and included destruction of houses and other structures, flooding of public water and sewage facilities, inundation of industrial and public utility plants, and loss of crops (DeAngelis and Hodge, 1972). Fortunately, owing to the timely public warning, the total death toll was comparatively low considering the severity of the storm.

The outstanding aspect of the 1972 floods in the mid-Atlantic states was their great areal extent. Peak discharges (per km^2 of drainage area) for some stations approached the highest ever observed in the U.S. At an instrumented experimental watershed on Mahantango Creek north of Harrisburg, Engman and others (1974) were able to show discharges of up to approximately 11 $m^3/sec/km^2$, in excess of the 100-yr estimates for peak discharge. The Susquehanna River at Harrisburg (drainage area 62,400 km^2) crested at 9.9 m (4.9 m above flood stage) on June 24, representing a discharge of 29,000 m^3/s (Fig. 4). The previous highest stage dating from 1786 was 8.9 m set in 1936, representing a discharge of 21,000 m^3/s. Bailey and others (1975) estimate the Susquehanna, Schuylkill, and James rivers experienced peak flows having recurrence intervals in excess of 100 yr throughout most of their lengths.

In terms of dollar damage, Hurricane Agnes was the largest single natural disaster in the U.S. This happened despite the careful investigations and elaborate dam and levee construction programs in the mid-Atlantic States since passage of the federal Flood Control Act of 1936. For example, levee works constructed to withstand the "standard project flood" were overtopped by 2 m at some places, requiring tens of thousands of people to be evacuated.

A flood frequency using annual peak discharge at Harris-

1952

1972

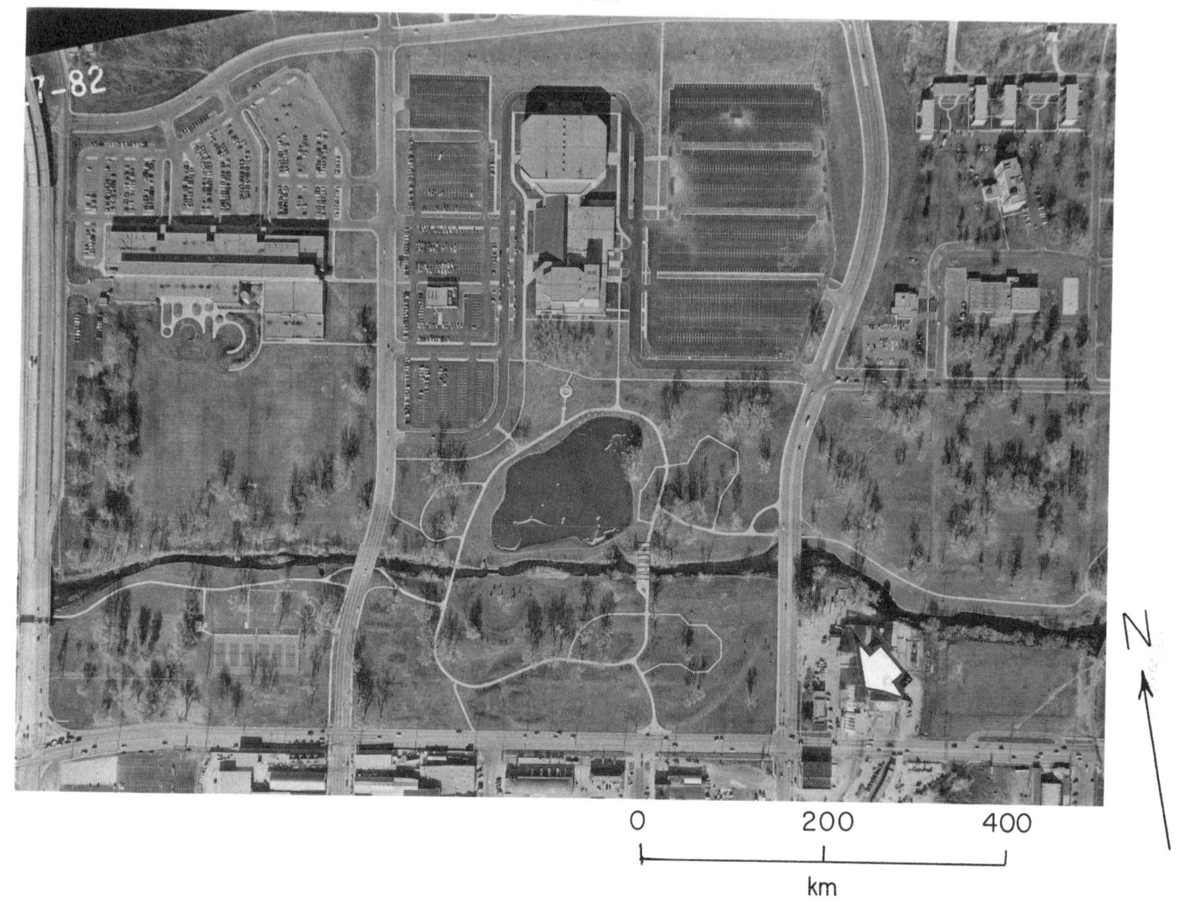

Figure 6 (this and previous page). Vertical aerial photographs of downtown area of Rapid City, South Dakota, showing change in urban landscape associated with the June 9, 1972, flood. 1952: Pre-flood condition with houses along Rapid Creek (U.S. Soil Conservation Service photo). Arrow locates 5th St. Bridge across Rapid Creek. 1972: Two days after catastrophic flood (South Dakota Remote Sensing Institute photo). The 5th St. Bridge and houses on the flood plain were destroyed. 1982: The floodway (photo by Horizons, Inc.). Houses north of Omaha Street have been removed and the floodway now consists of open park, a lake, tennis courts, parking lots, and a bicycle path. Except for the massive Trisco flour mill at 5th and Omaha streets (arrow), the floodway has no buildings. The new Central High School and Civil Center (northern part of this photo) were built in an area supposedly above the 100-yr flood elevation.

burg was made by Page and Shaw (1973) using the log-Pearson Type III method. Based on this curve, the June 24 discharge (29,000 m^3/s) had a recurrence interval exceeding 100 yr (Fig. 4). The discharge of the Susquehanna is controlled to some degree by dams on the tributaries. The 1972 relationship of natural-flow and regulated-flow attenuation is shown on Figure 4. This curve demonstrates that for a 100-yr flood the engineering structures effectively reduce a discharge of some 22,000 m^3/s to about 20,000 m^3/s.

About 10 percent of Harrisburg is situated on the flood plain (Schneider and Goddard, 1974) as shown by the flooding in 1972 (Page and Shaw, 1973). For example, Three Mile Island, where a nuclear power plant was about to be constructed, was almost completely inundated.

At Wilkes-Barre, Pennsylvania, the Susquehanna River crested at 12.4 m and the central square downtown turned into a lake 4 m deep. Many old homes were destroyed (Fig. 7). The alluvial plain of the Susquehanna River throughout the anthracite-rich Wyoming Valley is evident from a study of the topographic map (Fig. 8A). The geologic map by Hollowell (1971) shows Pleistocene glacial kame terrace sediments, lake sediments, and glacial outwash overlain by Holocene alluvium and alluvial fan deposits within the alluvial plain of Wyoming Valley (Fig. 8B). Terraces along the Susquehanna River are re-

Figure 7. Oblique aerial photograph of Wilkes-Barre, Pennsylvania, during 1972 flood (from Bailey and others, 1975).

lated to aggrading conditions accompanying glacial maxima, followed by degradation during interglacial stages of the Quaternary period (Peltier, 1949). Interestingly, the areas inundated by the 1972 flood were mostly confined to the areas of alluvium or alluvial fan deposits throughout most of the Wyoming Valley.

1973 Mississippi River. The Mississippi River, the largest river in the United States, drains 40 percent of the area in the 48 contiguous states. The lower reach of the Mississippi River has been building up its bed since the waning of the last ice age some 30,000 years ago, when sea level was over 100 m lower than at present (Fig. 3B). As it aggrades, the levees are built up higher than the flood plain. Presently the gradient of the river is so low that the channel bottom of the Mississippi River is only 30 m above sea level as far upstream as 760 km above the Gulf of Mexico. Below Baton Rouge, the river assumes the form of a tidal estuary some 320 km long (Matthes, 1951).

The Mississippi River has a flood history extending back to 1543, when the Spanish explorer, De Soto, was stopped by floods. LaSalle, exploring the Mississippi for the French in 1684, found vast portions of the lower valley inundated. New Orleans, founded in 1717, has been subject to repeated inundations; the city lies partly below sea level but is protected from the Mississippi by extensive levees of varying cross section with a top

Figure 8. Maps of Wilkes-Barre, Pennsylvania, area. A. Portion of 1:24,000 scale topographic map (Wilkes-Barre West quadrangle). The topography of the valley has been affected by coal mining in that culm banks more than 30 m high and subsidence pits are abundant. The area inundated by the June 1972 flood (from Flippo and Lenfest, 1973) is shown by diagonal lines. B. Geologic map (from Hollowell, 1971). Quaternary alluvium (Q_{al}) and outwash (Q_o) are shown.

A

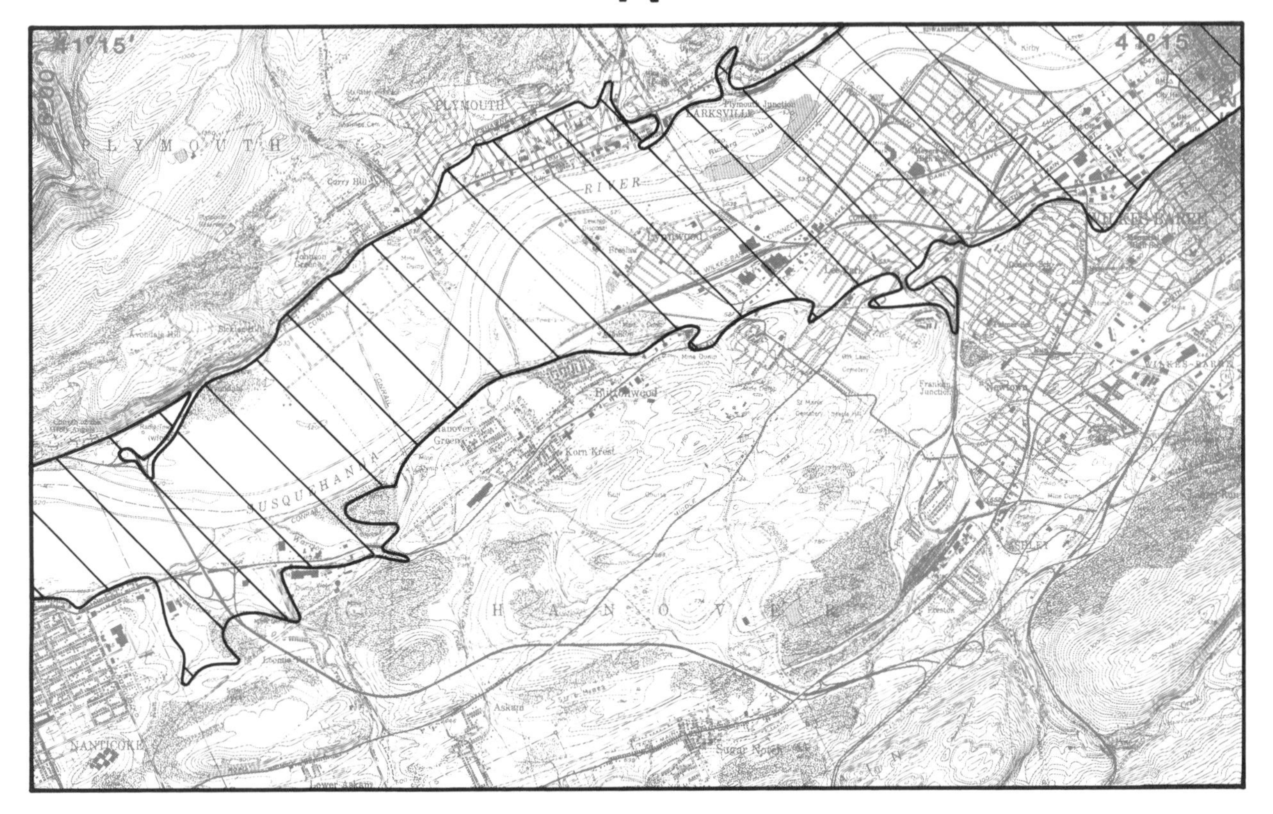
41°15'
PLYMOUTH
RIVER
SUSQUEHANNA
Buttonwood
Hanover Green
Korn Krest
NANTICOKE
Sugar Notch
Askam

B

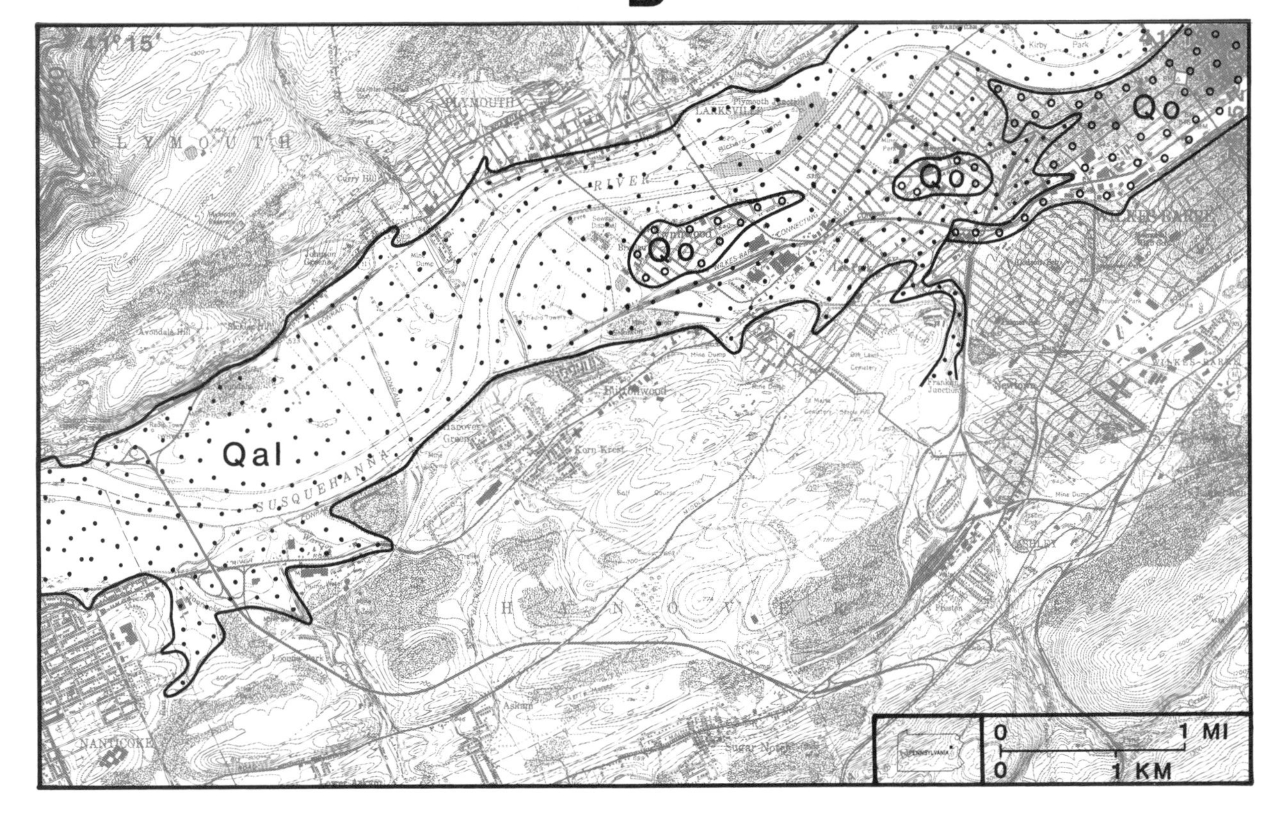
41°15'
PLYMOUTH
RIVER
SUSQUEHANNA
Qal
Qo
Qo
Qo
Buttonwood
Korn Krest
NANTICOKE
Askam
0
1 MI
0
1 KM

elevation of 7 m above sea level. As a consequence of levee construction and the confinement of flood waters between the levees, flood stages have risen with time. In the course of a century, the high-water mark mounted 2 m at New Orleans. High-water stages for upstream communities have been lowered, however, by artificial cutoffs that eliminated some of the largest meanders and shortened the river about 270 km. The highest land in New Orleans is on the natural levees along the banks of the river. The land slopes away from the levees, decreasing about 4 m over a distance of 2 km into a saucer-shaped depression in which the city is built. Storm runoff from intense rains collects in the basins and is pumped into the river. The high-water table is controlled by dewatering wells, and 19 major pumping stations have a capacity of 1,100 m^3/s (Ports, 1985).

The Mississippi River has a long history of man's effort to control floods. The first artificial levees were built in 1717 after New Orleans flooded. In 1879 the Mississippi River Commission was established as a response to public outcry for greater flood-control structures. Although the Commission spent $10 million on levees, the floods of 1882 caused breaks at 284 places. Levees were built higher and stronger until by 1926 more than 2,900 km of levees were in place along the Mississippi, standing an average 6 m high. The commission felt the area was safe, but the major floods of 1927 breached levees at 225 places. Nearly 50,000 km^2 of land was flooded, 183 people died, and $500 million of damage was done.

The federal government, through the U.S. Army Corps of Engineers, built five major dams upstream on the Missouri River to reduce the flood hazard and continued to reinforce levees downstream. Potential flood waters at New Orleans were relieved by means of diversions constructed through the Atchafalaya Floodway and the Bonnet Carre Spillway. Locally, diversion is also accomplished along the Mississippi River by allowing water to pour through levee gates into adjacent flood plain areas. Today the flooding problem at New Orleans is primarily related to hurricanes. Hurricane Betsy (1965) and Hurricane Camille (1969) prompted the federal government (Corps of Engineers) to build a huge hurricane protection system. A series of floodwalls and levees protecting against flood waters up to 5 m has been constructed to encircle New Orleans. The $690 million hurricane protection project should be completed by the year 2006 (Ports, 1985).

Prolonged rain in March and April of 1973 raised the Mississippi River above flood stage for as long as 97 consecutive days in some places. More than 50,000 km^2 of land were flooded, 50,000 people evacuated, and damage exceeded $400 million. Measured flood stages set records along the main branch of the Mississippi, even though water was diverted through spillways into uninhabited areas in many places. The devastation of the 1973 floods, which occurred despite extensive flood-control efforts by engineering structures, raised the question of what level of flood control is adequate or desirable. Higher levees and more dams could be built to protect against floods of low probability, such as the 500-yr flood. However, the costs would be astronomical, and may exceed the existing property damage expected from such events. According to Chin and others (1975, p. 1):

> The Mississippi basin flood in the spring of 1973 was exceptional in its duration, magnitude, and areal extent. Flooding began along portions of the Upper Mississippi River in early March as a result of much heavier than normal precipitation, and by April 3 the main stem was well above flood stage along its entire course below Cairo, Illinois. All major tributaries experienced continuous flooding with parts of 10 states affected to some extent—Minnesota, Wisconsin, Iowa, Illinois, Missouri, Tennessee, Kentucky, Arkansas, Mississippi, and Louisiana. More than 12 million acres (30 million ha) of land were inundated during the worst of the flooding, and 28 deaths were attributed to the floods.

Flood stages were the highest ever observed at St. Louis, where the river reached its maximum crest stage of 13.2 m on April 28, exceeding the historical crest stage of 12.6 m on June 27, 1844. Interestingly, the peak stage on April 28th did not represent the greatest peak discharge of record; the 1973 peak discharge of 24,100 m^3/s was actually exceeded during the floods of 1785, 1844, 1883, 1892, 1903, 1909, and 1927. Construction of the extensive modern levee–floodwall system caused an increase in stage for a given discharge at St. Louis (Belt, 1975).

The Mississippi River is bound by some 2,600 km of earthen levees. Cutoff and straightening on the river has been engineered between Cairo, Illinois, and the mouth, a river length of 1,400 km (Happ, 1950). In this reach, 530 km of river channel were shortened by 190 km with the construction of meander cutoffs during 1933–1936. Such cutoffs seek to halt a migrating meander channel, but are successful only to the extent that the cutoffs can be stabilized. The extensive levee systems on the Mississippi require maintenance due to undercutting and overtopping. During high stages, piping under a levee through a permeable alluvial deposit can produce sand boils on the lower parts of the adjacent flood plain, eventually leading to failure (Noble, 1976). Constant vigilance and repair is essential to prevent catastrophe.

The Missouri River is the largest tributary of the Mississippi River. More large engineering works have been built on this river than any in the world. During April, 1952, Omaha, Sioux City, Pierre, Bismark, and other Missouri River towns were flooded by a combination of melting snow, frozen ground, ice jams, and rain. Since 1950 the Pick–Sloan plan for water utilization and stream control has been carried out by the Corps of Engineers and the Bureau of Reclamation in a construction program unprecedented in magnitude, initially estimated to cost $10 billion. By 1986 more than 100 reservoirs on the tributaries of the Missouri had been built in conjunction with six major dams on the Missouri. Yet today the region remains subject to floods. For example, heavy rain during a three-week period in June 1984 caused extensive flooding near Sioux City, Iowa, and record discharges were recorded in the James, Vermillion, and Little Sioux rivers. The Missouri River at Sioux City (even with controlled releases from dams just upstream on the Missouri) reached 2,900 m^3/s, its second-highest discharge on record. A discharge of 3,200 m^3/s

was recorded at Omaha, but flooding did not exceed the devastation that occurred on April 14, 1952, when the Missouri crested with a discharge of 12,500 m^3/s.

In spite of all the engineered works on the Missouri–Mississippi river system, there is apprehension for the inhabitants of the lower valley. According to the National Science Foundation (1980, p. 221), "The lower Mississippian River has long presented the nation with its greatest flood problem. The implications of any major failure of the extensive control system now in place are staggering."

RELEVANCE OF ENGINEERING WORKS

Engineering structures

Dams. The geological investigations required for the planning, design, construction, and operation of dams for flood control and other purposes have provided many new technical gains and advances in geological knowledge, particularly concerned with: (1) the characteristics, physical properties, and structural features of rock masses, (2) structural patterns and seismotectonics, and (3) surficial deposits. Over the past century, some of the worst floods resulting in high loss of life have been due to breaching or other failures of dams; many were due to overtopping, others occurred because of inadequate engineering design or construction, and some failed due to the adverse geological conditions of the site/foundation (Kiersch, chapter 1, and James and Kiersch, chapter 22, this volume).

Engineers and planners have traditionally suggested the construction of dams and levees as a solution to many flood problems. Optimistically, the anticipated engineering solution to flood problems is that flood-control dams help to control floods. Realistically, dams cannot be viewed as a final solution to flooding because the useful life of any dam may be limited by its durability or the sediment accumulation that fills the reservoir. As a related issue, a dam can cause unusual downstream effects; for example, the loss of sediment load can increase shoreline erosion (Mathewson and Minter, 1981) or cause excessive scour around bridge piers (Schumm, 1971; Williams and Wolman, 1984).

The analysis of the benefits and costs of dams and other flood-reducing structures is beyond the scope of this chapter. Chandler (1984) found that after 50 years of operation the benefits of the Tennessee Valley Authority (TVA) dams have not come close to balancing the cost. Floods still devastate the lower Mississippi Valley despite the TVA dams. The rising annual flood damage in the U.S. indicates that government programs of dam construction have not eliminated the flood problem.

Levees. A common method of local flood control is to build a dike along the river's edge to help contain the river in its channel. People living in flood plains in some places in the U.S. have artificially increased levee heights so that today many streams have earth or concrete levees more than 10 m high. Some 3,200 km of artificial levees exist in the lower Mississippi River Valley. While these structures serve in a limited way to protect part of a flood plain from small floods, they promote a false sense of security because they get overtopped or undermined by larger floods. Numerous levees were built along the Susquehanna River at Wilkes-Barre, but did not contain the 1972 flood (Fig. 8A).

Construction of a levee on one side of a river adds to the magnitude of flooding on the other side of the river or upstream (Waananen and others, 1977). For example, levee construction in the St. Louis, Missouri, area was promoted by the Corps of Engineers with the idea that flood water would be quickly carried out of the area. This only compounded the problem for residents of Charles County downstream. Surprisingly, the 1973 Mississippi River flood broke stage records between Burlington, Iowa, and Cape Girardeau, Missouri, yet the discharge was less than several other floods, including the 1927 flood. Records show that post-levee floods have much higher stages than pre-levee floods. With no levees, the 1973 flood would probably have crested 3.3 m below the stage that was observed. If the 1927 flood runoff were to happen today, the flood crest would be higher than in 1927.

The construction of a levee is a green light for urbanization to proceed behind it. Consequently, if a levee fails, the damage realized exceeds damage to the region had the levee never been built. For example, in February, 1986, a levee gave way on the Mokelumne River at Thornton, California, in the San Joaquin Valley, forcing more than 1,300 residents to evacuate. Similarly, in Yuba County, California, 28,000 people were forced to flee on February 20, 1986, after a levee on the Yuba River washed out; the breach came only 1 hour after the levee had been inspected. About 72 km^2 of low-lying lands were inundated; the collapse was so sudden that residents of Linda and Olivehurst had only 15 minutes to evacuate before the towns were inundated.

Channelization. Channelization may locally reduce the amount and duration of flooding, but the increased velocity in the channelized reach commonly produces other long-range problems. The adverse effects of channel modifications have been described by Happ (1950), Keller (1976), Ritter (1986), Morisawa (1985) and others, Scour, for example, may be initiated on channelized reaches (Daniels, 1960; Petts, 1984). Emerson (1971) shows that increased flooding and scour leading to bridge failure occurred below a channelized reach of the Blackstone River, Missouri. Channelization may also trigger gullying and tributary network expansion (Ruhe, 1970). Man-made meander cut-offs can cause channel scour, which endangers bridge foundations (Bray and Kellerhalls, 1979). Sixteen artificial cutoffs on the Mississippi River between 1932 and 1942 decreased the length of the river by 244 km (Stevens and others, 1975). The cutoffs generally resulted in increased velocity of flood waves, which caused degradation of the channel; where sandy banks occur, bank-caving and widening, aggradation, and braided conditions occurred locally. Revetments placed piecemeal along a channel may cause extensive erosion where unprotected (Palmer, 1976). Costa and Jarrett (1981) show that stream channelization efforts in mountainous areas of Colorado have been ineffective because debris-flow deposits block the channel, causing subsequent flow in new directions.

Channels and other engineered structures are common in the Los Angeles area, where they are built on alluvial fan complexes. The Los Angeles basin contains 17 cities and 12 million people. Concrete-lined channels and levees faced with concrete, asphalt, or rock have been built along the Santa Ana River and other streams. A network of flood-control dams is located in headwater areas. Retarding basins have been constructed along the stream channels to infiltrate excess water and recharge the ground water. These combinations of engineering structures functioned as anticipated during floods in the winter of 1969, although a peak discharge of 2,850 m^3/s caused considerable damage to the levees. However, if the flood of 1862 (estimated at 12,000 m^3/s) occurred today, the levee system would be overwhelmed, and much of western Orange County, including Disneyland, would be subject to inundation (Ruhle, J. L., 1976, personal communication).

Channelization may be channel improvement to a hydraulic engineer, but it can be ecological disaster (Ruhe, 1970; Reuss, 1973). The Corps of Engineers and Soil Conservation Service have channelized more than 55,000 km of streams in the U.S. The effort provides limited protection but is devastating to the fluvial environment (Keller, 1976).

Alternatives to engineering structures: flood plain management

There is a simple and effective alternative to engineering structures such as dams, levees, and channel improvements. And that is to restrict construction on flood plains. The basic idea of flood plain management is for a governing agency to zone a flood plain in such a way as to prohibit residences and/or other types of structures. Ideally, the land can be used for agriculture, recreation/parks, railways and highways, or limited commercial enterprises that would be little affected by floods.

It is instructive to follow the development of the flood plain management program in Rapid City, South Dakota, following the devastating June 9, 1972, flood. Prior to the flood, there was complete lack of concern about flood hazards. Following the flood, with a loss of 238 lives and great economic hardship, the city, through a $48 million federal grant, purchased all homes and most commercial establishments on the flood plain. All remaining homes on the flood plain were removed, and the land was converted into recreational areas including a golf course, soccer and baseball fields, tennis courts, a 12-km-long bike path, and a low-maintenance parkway (Fig. 6). Thus, future floods will be able to pass relatively unobstructed through the town and will present little hazard to people or residences (Rahn, 1984).

A review of the Pearl River flood of April 1979 at Jackson, Mississippi, provides a valuable lesson about flood management. The Pearl River has a relatively flat gradient, averaging about 0.2 m/km through south-central Mississippi. Jackson, the state's capital, borders the west bank. In 1961 the Ross Barnett Dam was constructed on the Pearl River upstream from Jackson. Levee construction and channel modification were done by the Corps of Engineers between 1964 and 1968 to accommodate a standard project flood discharge of 5,900 m^3/s, estimated to reach a stage of 13.1 m. Unfortunately, the Ross Barnett Reservoir was nearly full when the April 1979 flood crest arrived, and available storage was negligible. A discharge of 3,700 m^3/s of sediment-free water over the spillway compounded the flood problems downstream. The Pearl River rose to an unprecedented flood stage of 13.2 m at Jackson. The high stage was partially due to the new buildings located on the flood plain, which impeded water flow. Flood damage to Jackson was estimated at $500 million, and provided many lessons for public policy both locally and nationally. The National Science Foundation (1980, p. 159), reported an overconfidence in structural flood control: "The boundless optimism of public officials during the 1960s concerning the 'elimination of the flood threat' for Jackson through levee construction was obviously misplaced." In addition, the flood-control structures encouraged construction on the flood plain: "Concomitant to the misplaced trust in the dams and levees has been a virtual abdication of public responsibility on both sides of the Pearl for floodplain management." The National Science Foundation report (1980, p. 5) described the overconfidence in dams and levees as well as the failure to restrict flood plain encroachment as key factors responsible for flood problems:

> Floods are the most consistently destructive natural hazards in the United States, and the upward trend in damages continues in spite of the Nation's investment in structural and nonstructural measures.... Potential benefits from structural flood control measures are often lost through subsequent, unwise development in supposedly protected areas, hence, it should be legally required that no federally funded structural or nonstructural flood damage reduction measures shall be carried out unless there are accompanying floodplain restrictions.

The recent coming-of-age alternatives to flood protection other than structural measures can be traced to geographer Gilbert White. During a period of several decades, White (1942, 1972) and others amassed information on the effects of flooding and the use of alternatives to reduce flood damage. The studies included evaluation of the major engineering works and failures used since 1936 to reduce riverine flood hazards in the U.S. According to Kates (1985), as hydraulic engineering projects, "they worked well, but as social engineering project efforts, they had a perverse effect. While they reduced the frequency and magnitude of flooding, they also encouraged the development of flood plains. Thus, there were fewer floods but greater damages." The technological fix only accelerated flood plain development, setting the scene for a potential catastrophe. Flood protection by engineered structures is feasible—within limits. The term "flood control" tends to imply complete control; but every dike, flood wall, or dam has limitations.

Federal Emergency Management Agency

Local governing bodies usually do not have sufficiently restrictive measures to prevent encroachment of flood plains.

Boulder, Colorado, exemplifies a community built to a large degree on the flood plain of a large river. Despite dozens of topographic surveys and numerous floods, skepticism of the flood hazard in Boulder remains; past efforts to create a serious flood plain management district have achieved little success. The flood plain is densely occupied today, and the serious potential for a catastrophe remains (Leveson, 1980).

In the face of mounting flood losses and escalating costs to taxpayers, the U.S. Congress created the National Flood Insurance Program (NFIP) in order to reduce future damage and provide insurance protection for property owners. The NFIP was established with the passage of the National Flood Insurance Act of 1968. The insurance is administered by the Federal Insurance Administration of the Federal Emergency Management Agency (FEMA). The program is based on an agreement between local communities and the federal government to the effect that if a community will implement programs to reduce future flood risks, the federal government will make flood insurance available.

FEMA works with the states and local communities to identify flood hazard areas, and publishes a flood-hazard boundary map. If a community joins the program and complies, the federal government makes flood insurance available. Flood hazard areas are determined on the basis of river stage records and discharge, as well as hydraulic and meteorological analysis. FEMA adopted the 100-yr flood as the design level of risk for the Federal Flood Insurance Program, a compromise between the rare catastrophic floods and the more frequent floods. Guidelines for the preparation of a flood insurance study involve many site-specific field investigations and analyses as outlined by FEMA (1985). Discharges must be adjusted for urbanization using techniques described by Sauer and others (1983). Flood elevations are normally determined by the "step-backwater" computer program (Corps of Engineers, 1984; FEMA, 1985). Where historic flood information is available, high-level marks can be used to determine the statistical probability of different stage elevations.

FEMA (1985, p. 5) defines a floodway as a critical part of the flood plain which must be kept free of any encroachment that could increase the flood stage: "Normally, the floodway will include the stream channel and that portion of the adjacent land areas required to pass the 100-year flood discharge without cumulatively increasing the water-surface elevation at any point more than one foot above that of the pre-floodway condition." Methods to determine the floodway are explained in the flood insurance study guidelines. Limited development is permitted on the area between the floodway and the 100-yr flood plain boundary, termed the floodway fringe: "The floodway fringe area encompasses the portion of the flood plain that could be completely obstructed without increasing the water-surface elevation of the 100-yr flood by more than 1.0 foot at any point."

Testimony in congressional hearings showed the range of perceived harmful effects of FEMA flood plain regulations (FEMA, 1981). A common complaint is that there would be a reduction in property value when flood plain regulations prohibit certain development; this condition was thought to reduce estate tax assessments. However, there still may not be a tax base loss in totality for the community because there is usually other land outside the flood hazard area, but still within the tax jurisdiction, that is equally suited for the development. In that case there is no net tax loss, just a shift to a more appropriate location. Attempts by landowners to use legal methods to try to defeat the enactment of flood plain regulations have not found favor in court; by upholding flood plain regulations, the courts couch their decision in terms of public harms that are being prevented.

CONCLUSION

Since 1936, the national approach to flood problems generally has been for the federal government to assume the major obligation to protect developed areas by building dams and other engineering structures. In addition, federal agencies have operated with other governmental groups in providing relief and rehabilitation assistance for flood disasters. Under original federal policies, flood plain property owners bore only a portion of the flood costs of relief and rehabilitation. The general public, by bearing all or a major part of the cost of flood protection works, in effect subsidizes flood plain occupants in their use of the flood plain. Consequently, 90 percent of the taxpayers who do not live in flood-prone areas are contributing to the welfare of the 10 percent who do.

The more recent governmental FEMA program involving the regulation of flood plain development is a step to reduce flood hazards and more fairly distribute the burden of costs and responsibility. However, because the government uses the 100-yr flood as criteria for flood hazards, too much emphasis may go into determining whether a particular piece of property lies one foot above the elevation established as the level of the 100-yr flood. If above, the property may be developed into a shopping center while the property below may not be developed because of zoning regulations or lack of financing if flood insurance is not available. Consequently, the engineer–geologist is under considerable pressure to come up with an accurate estimate of the 100-yr flood level.

There is merit to recognition of the entire flood plain as a hazardous area, rather than only the 100-yr floodway (Fig. 9). In many areas the flood plain is quite well defined and represents only a small percentage of the total land surface within a region (Fig. 10). In places where the 100-yr floodway almost reaches to the edge of the flood plain it seems prudent to recognize the complete hazard area, rather than try to protect only the 100-yr flood zone.

An increasing percentage of the annual national flood loss results from the very large floods of low probability such as occurred on the Susquehanna River in 1972. Because FEMA regulates flood plains on the basis of the 100-yr flood, lowlands outside the 100-yr floodway frequently are inundated by rare large floods. The binomial probability distribution formula shows, for example, that there is an 18 percent probability of the

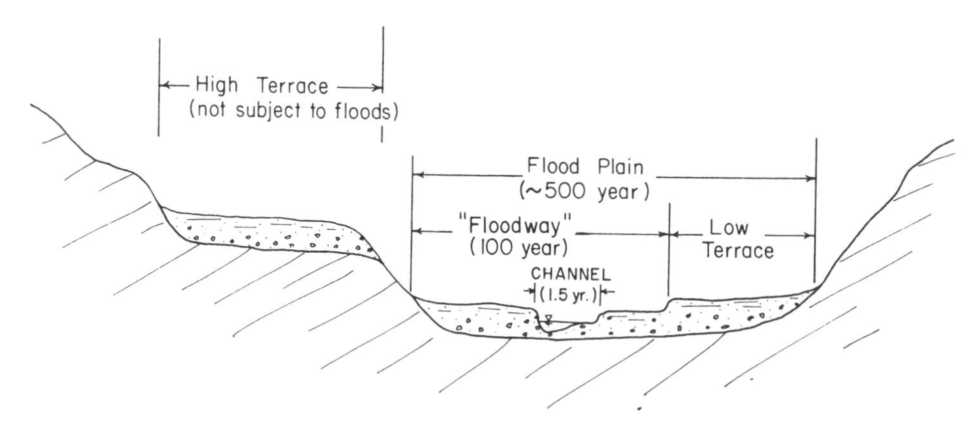

Figure 9. Sketch illustrating relationship of flood inundation frequency to flood plain and terraces.

Figure 10. Oblique aerial photograph of the Kentucky River during the 1963 flood at Hazard, Kentucky. There is a clear division between the flood plain (which was inundated by this flood) and the adjacent hills.

500-yr flood occurring in 100 yr. Thousands of communities in the U.S. have protection from the 100-yr flood but not for the 500-yr flood; hence, future catastrophic flood disasters will occur. Engineering structures, as levees, are built to protect the most flood-prone areas; the existence of any residual risks then tends to be forgotten (Kirby, 1978). Floods become rare but catastrophic events (Yevjevich, 1972).

The shortcomings of the use of engineering structures to mitigate floods have been discussed by the Geological Society of America's Committee on Geology and Public Policy (Moss and others, 1978, p. 6):

> Long and hard experience has shown that structural engineering works are rarely, by themselves, complete solutions to the elimination of flood hazards. . . .An unfortunate error is often made by the public in equating some structural measures, such as dams or levees, with "flood control." The problem so often overlooked by the general public is that there is always a small probability that a flood will occur that is larger than the flood for which the structure was designed. Such a flood would not be controlled. Indeed, overreliance on structures alone may foster increased floodplain development because of a false sense of security. Damage from an excessive flood event would then be greater, and indeed has been greater in some places, than it would have been without the initial degree of protection. "Flood control" is probably a misnomer when applied to engineering works. Dams do not give flood control, but only a "specific amount of flood protection." (Leopold and Maddock, 1954)

Although the FEMA program is not perfect, the federal policy has been changed from mitigation by subsidized dam construction for the purpose of saving the foolish few who live on flood plains, to a policy of amelioration by the discouragement of urbanization in areas of high flood potential. Nevertheless there are shortcomings, and improvements are needed: (1) Presently there is no mechanism to account for flood hazards to urban areas already built on flood plains. (2) The reliance on the 100-yr flood as a criterion for federal policies seems to be shortsighted, especially urbanized flood hazard areas just barely above the 100-yr flood stage. (3) There is constant pressure at local levels to permit encroachment on flood plains. Frequently, people have a strong resistance to any government interference in their private lives. The FEMA program may be disregarded since participation is at the discretion of local governments.

Geologists can provide important input to prevent future flood disasters by the careful mapping of valley deposits and landforms. All large-scale geologic maps published by federal or state agencies should delineate specific types of alluvial deposits. Too frequently such deposits are mapped simply as Quaternary alluvium. However, for purposes of flood-hazard identification, geologists should provide recognition of the varied morphological features common to stream valley deposits, because they represent a wide range of features and flood-hazard potentials. Special geologic-hazard maps would show the various fluvial features as related to flood-inundation potential. The U.S. Geological Survey currently publishes 1:24,000 scale maps that delineate areas prone to flooding (for example, Waananen and others, 1977).

FUTURE RESEARCH

An important area for future geologic research on surface runoff and floods is to determine the condition of grade of a stream, and Holocene and/or man-made influences on the ever-changing pattern of aggradation or degradation. Valley landforms can change in the course of 100 years, and updated geologic maps are needed. The impact of changing hydrologic regime, brought about in part by man (especially agriculture and urbanization), can have a marked effect on flood frequency.

A better understanding of geologic controls for estimating flood frequency for ungaged watersheds is needed, as well as better documentation, first-hand description of floods, and improved forecasting techniques for flash floods.

A geologic map or report serves a purpose only if it is utilized. If a local agency approves a subdivision in a flood-prone terrain, geologists of the region should alert the public to the hazards. Geologists should participate in local forums where land-use decisions are made and provide guidance for the planners and administrators, as well as for those involved in the technical aspects.

REFERENCES CITED

Bailey, J. F., Patterson, J. L., and Paulus, J.L.H., 1975, Hurricane Agnes rainfall and floods, June–July 1972: U.S. Geological Survey Professional Paper 924, 403 p.

Baker, V. R., 1977, Stream-channel response to floods with examples from central Texas: Geological Society of America Bulletin, v. 88, p. 1057–1071.

Barnes, H. H., Jr., 1967, Roughness characteristics of natural channels: U.S. Geological Survey Water-Supply Paper 1849, 213 p.

Belt, C. B., Jr., 1975, The 1973 flood and man's constriction of the Mississippi River: Science, v. 189, p. 681–684.

Benson, M. A., 1968, Uniform flood-frequency estimating methods for federal agencies: Water Resources Research, v. 4, p. 891–908.

——, 1972, Thoughts on the design of the design flood, *in* Schulz, E. F., Koelzer, V. A., and Mahmood, K., eds., Floods and droughts: Proceedings of the Second International Symposium in Hydrology, Water Resources Publication, Fort Collins, Colorado, p. 27–33.

Benson, R. D., Hoffman, E. B., and Wipf, V. J., 1985, Analyses of flood-flow frequency for selected gaging stations in South Dakota: U.S. Geological Survey Water-Resources Investigation Report 85-4217, 202 p.

Bray, D. I., and Kellerhalls, R., 1979, Some Canadian examples of the response of rivers to man-made changes, *in* Rhodes, D. D., and Williams, G. P., eds., Adjustments of the fluvial system: Dubuque, Iowa, Kendall Hunt, p. 351–372.

Chandler, W. U., 1984, The myth of the TVA: Cambridge, Massachusetts, Ballinger Publishing Co., 240 p.

Chin, E. H., Skelton, J., and Guy, H. P., 1975, The 1973 Mississippi River Basin flood; Compilation and analyses of meteorologic, streamflow, and sediment data: U.S. Geological Survey Professional Paper 937, 137 p.

Chorley, R. J., Schumm, S. A., and Sugden, D. E., 1984, Geomorphology: London, Methuen, 605 p.

Chow, V. T., 1964, Handbook of applied hydrology: New York, McGraw-Hill.

Cobb, E. D., 1983, Evaluation of streams in selected communities for the application of limited-detail study methods for flood-insurance studies: U.S. Geological Survey Water-Resources Investigation Report 85-4098, 54 p.

Collinson, J. D., 1986, Alluvial sediments, *in* Reading, H. G., ed., Sedimentary environments and facies: Oxford, Blackwell Scientific Publications, p. 20–62.

Cooper, A. J., 1978, Streamflow forecasting, *in* Geophysical predictions; Studies in geophysics: Washington, D.C., National Research Council, National Academy of Sciences, p. 193–201.

Costa, J. E., 1978, Holocene stratigraphy in flood-frequency analysis: Water Resources Research, v. 14, p. 626–632.

—— , 1983, Paleohydraulic reconstruction of the flash-flood peaks from boulder deposits in the Colorado Front Range: Geological Society of America Bulletin, v. 94, p. 986–1004.

Costa, J. E., and Baker, V. R., 1981, Surficial geology, building with the Earth: New York, John Wiley and Sons, 498 p.

Costa, J. E., and Jarrett, R. D., 1981, Debris flows in small mountain stream channels of Colorado and their hydrological implications: Association of Engineering Geologists Bulletin, v. 18, p. 309–322.

Dalrymple, T., 1960, Flood-frequency analysis: U.S. Geological Survey Water-Supply Paper 1543-A, 80 p.

Dalrymple, T., and Benson, M. A., 1967, Measurement of peak discharge by the slope-area method: U.S. Geological Survey Techniques of Water-Resources Investigation, Book 3, Ch. A2, p. 1–12.

Daniels, R. B., 1960, Entrenchment of the Willow Drainage Ditch, Harrison County, Iowa: American Journal of Science, v. 258, p. 161–176.

DeAngelis, R. M., and Hodge, W. T., 1972, Preliminary climatic data report, Hurricane Agnes, June 14–23, 1972: Boulder, Colorado, National Oceanic and Atmospheric Administration Technical Memorandum EDS-NCC-1, 62 p.

Deutsch, M., and Ruggles, F. H., Jr., 1974, Optical processing and projected applications of the ERTS-1 imagery covering the 1973 Mississippi River Valley floods: American Water Resources Association, Water Resources Bulletin, v. 10, p. 1023–1039.

Ely, L. L., and Baker, V. R., 1985, Reconstructing paleoflood hydrology with slackwater deposits; Verde River, Arizona: Physical Geography, v. 6, no. 2, p. 103–126.

Emerson, J. W., 1971, Channelization; A case study: Science, v. 173, p. 325–326.

Engman, E. T., Parmele, L. H., and Gburek, W. J., 1974, Hydrologic impact of tropical storm Agnes: Journal of Hydrology, v. 22, p. 179–193.

Federal Emergency Management Agency, 1981, Evaluation of the economic, social, and environmental effects of floodplain regulation: Washington, D.C., Federal Emergency Management Agency, FIA-8, 168 p.

—— , 1985, Flood insurance study, guidelines and specifications for study contractors: Washington, D.C., Federal Emergency Management Agency, FEMA 37.

Fenneman, N. M., 1906, Floodplains produced without floods: American Geographic Society Bulletin, v. 38, p. 89–91.

Fisk, N. H., 1944, Geological investigations of the alluvial valley of the lower Mississippi River: Vicksburg, Mississippi, Mississippi River Commission, 78 p.

—— , 1947, Fine-grained alluvial deposits and their effects on Mississippi River activity: Vicksburg, Mississippi, Waterways Experiment Station, 2 vols., 82 p.

—— , 1951, Mississippi River Valley geology relative to river regime: American Society of Civil Engineers Transactions, v. 117, p. 667–682.

Flippo, H. N., Jr., and Lenfest, L. W., Jr., 1973, Flood of June 1972 in Wilkes-Barre area, Pennsylvania: U.S. Geological Survey Hydrologic Investigations Atlas, HA-523.

Happ, S. C., 1950, Stream-channel control, *in* Trask, P. D., Applied sedimentation: New York, John Wiley and Sons, p. 319–335.

Hardison, C. H., 1974, Generalized skew coefficients of annual floods in the United States and their application: Water Resources Research, v. 10, p. 745–752.

Hershfield, D. M., and Wilson, W. T., 1957, Generalizing of rainfall-intensity-frequency data: Extrait des Comptes Rendus et Rapports-Assemblee generale de Toronto, tome I, p. 499–506.

Hjelmfelt, A. T., Jr., and Cassidy, J. J., 1975, Hydrology for engineers and planners: Ames, Iowa, Iowa State University Press, 210 p.

Hollis, G. E., 1975, The effect of urbanization on floods of different recurrence interval: Water Resources Research, v. 11, p. 431–35.

Hollowell, J. R., 1971, Hydrology of the Pleistocene sediments in the Wyoming Valley, Luzerne County, Pennsylvania: Pennsylvania Geologic Survey, Fourth Series, Water Resource Report 28, 77 p.

Hoyt, W. G., and Langbein, W. B., 1955, Floods: Princeton, New Jersey, Princeton University Press, 469 p.

Jahns, R. H., 1947, Geologic features of the Connecticut Valley, Massachusetts, as related to recent floods: U.S. Geological Survey Water-Supply Paper 996, 158 p.

Kammerer, J. C., and Baldwin, H. L., 1963, Water problems in the Springfield–Holyoke area, Massachusetts: U.S. Geological Survey Water-Supply Paper 1670, 68 p.

Kates, R. W., 1985, Success, strain, and surprise: Issues in science and technology, National Academy of Science, v. 2, p. 46–58.

Keller, E. A., 1976, Channelization; Environmental, geomorphic, and engineering aspects, *in* Coates, D. R., ed., Geomorphology and engineering: Stroudsburg, Pennsylvania, Dowden, Hutchison, and Ross, Inc., p. 115–140.

Kirby, W., 1978, Prediction of streamflow hazards, *in* Geophysical predictions; Studies in geophysics: Washington, D.C., National Research Council, National Academy of Sciences, p. 202–215.

Kochel, R. C., Baker, V. R., and Patton, P. C., 1982, Paleohydrology of southwestern Texas: Water Resources Research, v. 18, p. 1165–1183.

Larimer, O. T., 1973, Flood of June 9–10, 1972, at Rapid City, South Dakota: U.S. Geological Survey Hydrologic Atlas, HA-511.

Legget, R. F., 1973, Cities and geology: New York, McGraw-Hill Book Co., 624 p.

Leopold, L. B., 1968, Hydrology for urban land planning; A guidebook on the hydrologic effects of urban land use: U.S. Geological Survey Circular 554, 18 p.

Leopold, L. B., and Maddock, T., Jr., 1954, The flood control controversy: New York, Ronald Press Co., 278 p.

Leopold, L. B., and Miller, J. P., 1954, A postglacial chronology for some alluvial valleys in Wyoming: U.S. Geological Survey Water-Supply Paper 1261, 90 p.

Leopold, L. B., Wolman, M. G., and Miller, J. P., 1964, Fluvial processes in geomorphology: San Francisco, W. H. Freeman, 522 p.

Leveson, D., 1980, Geology and the urban environment: New York, Oxford Press, 386 p.

Linsley, R. K., and Franzini, J. B., 1979, Water-resources engineering: New York, McGraw-Hill Book Co., 716 p.

Longwell, C. R., Knopf, A., and Flint, R. F., 1948, Physical geology: New York, John Wiley and Sons, 602 p.

Maddox, R. A., Caracena, F., Hoxit, L. R., and Chappell, C. F., 1977, Meteorological aspects of the Big Thompson flash flood of 31 July 1976: National Oceanic and Atmospheric Administration Technical Report ERL 388-APCL 41, 83 p.

Mason, R. D., Lind, D. A., and Marchal, W. G., 1983, Statistics; An introduction: San Diego, Harcourt Brace Jovanovich, Publisher, 626 p.

Mathewson, C. C., and Minter, L. L., 1981, Impact of water resource development on the hydrology and sedimentology of the Brazos River, Texas, with implications on shoreline erosion: Association of Engineering Geologists Bulletin, v. 18, p. 39–53.

Matthes, G. H., 1951, Paradoxes of the Mississippi: Scientific American, v. 8, p. 19–23.

Morisawa, M., 1985, Rivers, form, and process: New York, Longman, Inc., 222 p.

Moss, J. H., Baker, V. R., Doehring, D. O., Patton, P. C., and Wolman, M. G., 1978, Floods and people; A geological perspective: Boulder, Colorado, Geological Society of America, Report of the Committee on Geology and Public

Policy, 7 p.
Murphy, E. C., 1904, Destructive floods in the United States in 1903: U.S. Geological Survey Water-Supply Paper 96, 81 p.
National Science Foundation, 1980, A report on flood hazard mitigation: Washington, D.C., National Science Foundation, 253 p.
Noble, C. C., 1976, The Mississippi River flood of 1973, *in* Coates, D. R., ed., Geomorphology and engineering: Stroudsburg, Pennsylvania, Dowden, Hutchinson and Ross, p. 79–98.
O'Connor, J. E., Webb, R. H., and Baker, V. R., 1986, Paleohydrology of pool-and-riffle pattern development; Boulder Creek, Utah: Geological Society of America Bulletin, v. 97, p. 410–420.
Overton, D. E., and Meadows, M. E., 1976, Stormwater modeling: New York, Academic Press, 358 p.
Page, L. V., and Shaw, L. C., 1973, Floods of June 1972 in the Harrisburg area, Pennsylvania: U.S. Geological Survey Hydrologic Atlas, HA-530.
Palmer, L., 1976, River management criteria for Oregon and Washington, *in* Coates, D. R., ed., Geomorphology and engineering: Stroudsburg, Pennsylvania, Dowden, Hutchinson and Ross, p. 329–346.
Patton, P. C., and Boison, P. J., 1986, Processes and rates of formation of Holocene alluvial terraces in Harris Wash, Escalante River Basin, south-central Utah: Geological Society of America Bulletin, v. 97, p. 369–378.
Peltier, L. C., 1949, Pleistocene terraces of the Susquehanna River, Pennsylvania: Pennsylvania Geologic Survey, Fourth Series, Bulletin G23, 158 p.
Petts, G. E., 1984, Impounded rivers, perspectives for ecological management: New York, John Wiley and Sons, 326 p.
Ports, M. A., 1985, Computer keeps New Orleans' head above water: Civil Engineering, v. 55, no. 8, p. 56–58.
Rahn, P. H., 1974, Lessons learned from the June 9, 1972, flood in Rapid City, South Dakota: Association of Engineering Geologists Bulletin, v. 12, no. 2, p. 83–97.
—— , 1984, Flood-plain management program in Rapid City, South Dakota: Geological Society of America Bulletin, v. 95, p. 838–843.
—— , 1986, Engineering geology, an environmental approach: New York, Elsevier Science Publishing Co., 589 p.
Rango, A., and Salomonson, V. V., 1974, Regional flood mapping from space: Water Resources Research, v. 10, p. 473–484.
Reich, B. M., 1972, Log-Pearson Type III and Gumbel Analysis of floods, *in* Schulz, E. F., Koelzer, V. A., and Mahmoud, K., eds., Floods and droughts: Proceedings of the Second International Symposium in Hydrology, Water Resources Publication, Fort Collins, Colorado, p. 291–303.
—— , 1973, How frequently will floods occur?: Water Resources Bulletin, v. 9, no. 1, p. 187–188.
Reuss, H. S., 1973, Stream channelization; Hearings between the Conservation and Natural Resources Subcommittee of the House Committee on Government Operations: U.S. Government Printing Office, part 5, p. 2780–3288, and part 6, p. 3789–3811.
Richards, K., 1982, Rivers, form, and process in alluvial channels: London, Methuen, 358 p.
Riggs, H. C., 1973, Regional analyses of streamflow characteristics: U.S. Geological Survey Techniques of Water-Resources Investigations, chapter B3, book 4, 15 p.
Ritter, D. F., 1986, Process geomorphology: Dubuque, Iowa, William C. Brown, 579 p.
Ruhe, R. V., 1970, Stream regimen and man's manipulation, *in* Coates, D. R., ed., Environmental geomorphology: State University of New York Binghampton, Publications in Geomorphology, p. 9–23.
Sauer, V. B., Thomas, W. O., Jr., Sticker, V. A., and Wilson, K. V., 1983, Flood characteristics of urban watersheds in the United States; Techniques for estimating magnitude and frequency of urban floods: U.S. Geological Survey, Water-Supply Paper 2207, 63 p.
Schneider, W. J., and Goddard, J. E., 1974, Extent and development of urban flood plains: U.S. Geological Survey Circular 601-J, 14 p.
Schumm, S. A., 1971, Fluvial geomorphology, channel adjustments, and river metamorphosis, *in* Shen, H. W., ed., Fluvial geomorphology in river mechanics: Fort Collins, Colorado, Water Resources Publications, ch. 4 and 5.
Schwartz, F. K., Hughes, L. A., Hansen, E. M., Petersen, M. S., and Kelly, D. B., 1975, The Black Hills–Rapid City flood of June 9–10, 1972; A description of the storm and flood: U.S. Geological Survey Professional Paper 877, 47 p.
Stevens, M. A., Simmons, D. B., and Schumm, S. A., 1975, Man-induced changes of middle Mississippi River: American Society of Civil Engineers Journal of Water Ways, v. 101, WW2, p. 119–133.
Thornbury, W. D., 1948, Principles of geomorphology: New York, John Wiley and Sons, 618 p.
Tice, R. H., 1968, Magnitude and frequency of floods in the United States; Part 1-B, North Atlantic Slope basins, New York to York River: U.S. Geological Survey Water-Supply Paper 1672, 858 p.
Turnbull, W. J., Krinitzsky, E. L., and Johnson, S. J., 1950, Sedimentary geology of the alluvial valley of the lower Mississippi River and its influence on foundation problems, *in* Trask, P. D., ed., Applied sedimentation: New York, John Wiley and Sons, p. 210–226.
U.S. Army Corps of Engineers, 1984, Generalized computer program HEC-2, water-surface profiles, users manual: Vicksburg, Mississippi, U.S. Army Corps of Engineers, Hydrologic Engineering Center.
U.S. Bureau of Reclamation, 1973, Design of small dams: U.S. Bureau of Reclamation, 816 p.
Waananen, A. O., Limeros, J. T., Kockelman, W. J., Spangle, W. E., and Blair, M. L., 1977, Flood-prone areas and land-use planning; Selected examples from the San Francisco Bay region, California: U.S. Geological Survey Professional Paper 942, 75 p.
Wanielista, M. P., 1978, Stormwater management; Quantity and quality: Ann Arbor, Michigan, Ann Arbor Science Publishers, Inc., 383 p.
Water Resources Council, 1967, A uniform technique for determining flood-flow frequencies: Washington, D.C., Bulletin 15, 15 p.
White, G. F., 1942, Human adjustments to floods; A geographical approach to the flood problem in the United States: Chicago University Department of Geography Research Paper 57, 236 p.
—— , 1972, Prospering with uncertainty, *in* Schulz, E. F., Koelzer, V. A., and Mahmood, K., eds., Floods and droughts: Proceedings of the Second International Symposium in Hydrology, Water Resources Publications, Fort Collins, Colorado, p. 9–15.
Williams, G. P., 1978, Bank-full discharge of rivers: Water Resources Research, v. 14, p. 1141–58.
Williams, G. P., and Wolman, M. G., 1984, Downstream effects of dams on alluvial rivers: U.S. Geological Survey Professional Paper 1286, 83 p.
Wolman, M. G., and Leopold, L. B., 1957, River flood plains, some observations on their formation: U.S. Geological Survey Professional Paper 282-C, p. 87–107.
Womack, W. R., and Schumm, S. A., 1977, Terraces of Douglas Creek, northwestern Colorado, an example of episodic erosion: Geology, v. 5, p. 72–76.
Yevjevich, V., 1972, Analysis of risks and uncertainties in flood control, *in* Schulz, E. F., Koelzer, V. A., and Mahmood, K., eds., Floods and droughts: Proceedings of the Second International Symposium in Hydrology, Water Resources Publications, Fort Collins, Colorado, p. 363–374.

Manuscript Accepted by the Society April 24, 1987

ACKNOWLEDGMENT

George A. Kiersch, David M. Cruden, Jack A. Redden, and Arden D. Davis kindly reviewed this paper and made many useful suggestions for improvement.

Printed in U.S.A.

Geological Society of America
Centennial Special Volume 3
1991

Chapter 7

Erosion, sedimentation, and fluvial systems

Lawson M. Smith
Geotechnical Laboratory, U.S. Army Engineer Waterways Experiment Station, Box 631, Vicksburg, Mississippi 39180
David M. Patrick
Department of Geology, University of Southern Mississippi, Box 5044, Hattiesburg, Mississippi 39406

INTRODUCTION

The development of knowledge in erosion and sedimentation parallels the growth of the geological sciences. In his *Illustration of the Huttonian Theory of the Earth,* Playfair (1802) provides lucid descriptions of erosional processes, illustrating their significance in the evolution of landscapes. Sir Charles Lyell (1830) described in uniformitarian terms the nature and importance of erosion and sediment transport. The power of rain to erode surface materials was discussed by Greenwood in 1857. Reports of the exploration of the American Southwest by the U.S. Geological Survey in the latter half of the nineteenth century are replete with the consideration of the impact of erosion and sedimentation in shaping the landscape. Most notable among these reports are those of G. K. Gilbert, whose keen observations and analytical powers allowed him to develop the basis for many of today's important concepts in fluvial geomorphology (Gilbert, 1880).

As the geological sciences moved into the twentieth century, Gilbert continued to provide theoretical bases for the comprehension of erosional and sedimentary processes. His classic discussion "The transport of debris by running water" was the result of years of flume studies and field observations (Gilbert, 1914). Gilbert's contributions in this paper include not only a detailed discussion of processes but one of the first analytical statements regarding the impact of man on a fluvial system. Twenhofel's (1932) famous *Treatise on Sedimentation* advanced our fledgling knowledge of sedimentary processes. From the field of soil conservation, Bennett (1939) synthesized existing knowledge of the impact of agricultural practices on erosional processes and sedimentation. Bennett's description of the historical significance of man in soil erosion, the importance of natural factors, and the utility of agricultural and soil engineering practices in reducing erosion and sedimentation was monumental. They became a guidebook for soil conservation practice for the next 30 years.

A number of significant papers appeared in the early 1940s that expanded our knowledge of erosion, transport, and depositional processes. Among these important contributions were Bagnold's (1941) discussion of the *Physics of Blown Sand,* Zingg's (1940) examination of the importance of slope degree and length in soil erosion, and Einstein's (1942) formulae for sediment transport. Happ and others (1940) described the impact of man on accelerated upland erosion and stream and valley sedimentation. Ellison's (1945) studies of raindrop erosion complemented earlier studies of the soil erosion process, providing data for the subsequent development of soil erosion equations (Wischmier and Smith, 1965).

From the field of hydraulic engineering and hydrology, Robert Horton's contributions to our initial understanding of erosional processes and sedimentation have been paramount. Based on his earlier work on runoff plat experiments (Horton, 1940), he described numerically the erosional evolution of landscapes and drainage systems (Horton, 1945). His paper provided the springboard for the development of numerical (mechanical) analysis of erosional processes and landforms and, as discussed later, is a major contribution to the development of the concept of the "fluvial system."

Beginning in 1948, the first of the "Federal Interagency Sedimentation Conferences" was held. Drawing scientists and engineers from a wide spectrum of disciplines interested in erosion and sedimentation, these conferences have been held periodically to discuss advancements in the knowledge of erosion, transport, and deposition of sediment (USBR, 1948; USDA, 1963; and Water Resource Council, 1976). The most recent interagency conference was held in Tucson in 1986. The hundreds of detailed papers contained within the proceedings of these conferences provide a comprehensive review and discussion of the evolution of knowledge in erosion and sedimentation.

During the 1940s, our knowledge and practice of the principles of applied sedimentology were significantly advanced by Fisk (1944, 1951), whose monumental studies of the geology of the Mississippi Alluvial Valley for the Corps of Engineers laid the framework for all subsequent investigations in this region. These early applications of the geosciences to engineering works in alluvial systems were further developed by later researchers, including Turnbull and others (1950), Krinitzsky (1965), Saucier

Smith, L. M., and Patrick, D. M., 1991, Erosion, sedimentation, and fluvial systems, *in* Kiersch, G. A., ed., The heritage of engineering geology; The first hundred years: Boulder, Colorado, Geological Society of America, Centennial Special Volume 3.

(1974), and Winkley (1977). The applied aspects of erosion and sedimentation were brought to the forefront by the 1950 publication of P. D. Trask, *Applied Sedimentation,* in which he reviewed a number of interrelated research activities and projects that focus on application of the geosciences to engineering works.

Several publications, primarily proceedings of conferences, issued in the last 20 years, contain valuable papers on erosion and sedimentation. Among the more noteworthy are Carson's (1971) text on the mechanics of erosion, Shen's (1972) proceedings of a conference on sedimentation, the American Society of Civil Engineers' manual on *Sedimentation Engineering* (Vanoni, 1975), *Soil Erosion: Prediction and Control* (Soil Conservation Society of America, 1976), and two separate texts entitled *Erosion and Sediment Yield* (Laronne and Moseley, 1982; Hadley and Walling, 1984).

DEVELOPMENT OF THE CONCEPT OF FLUVIAL SYSTEMS

Definition of fluvial system

Like all environmental systems, river (fluvial) systems are characterized by a number of important qualities, such as: (1) the limits of the fluvial system are environmental; (2) the elements of the system interact; (3) the fluvial system is controlled by previous actions; (4) a single element usually dominates the fluvial system; (5) the system evolves through time; (6) energy and matter flow through the system; and (7) the dynamics of the system are influenced by thresholds. The concept and characterization of fluvial systems as reviewed in this paper are defined and described by Schumm (1977).

The limits of a fluvial system are the drainage divide of its basin and the mouth of the river. Elements of the fluvial system, such as the tributary channels and the main channel, interact; that is, a change in the tributary network will have an impact on the character of the main channel, and vice versa. Fluvial systems are controlled by the dominant element of climate, which through precipitation and temperature, determines the amount of water flowing through the system. A river system is also controlled by previous actions, such as a long-term response to the processing of large amounts of glacial meltwater, a condition that still influences some rivers today. River systems evolve over time geomorphically as they adjust their physical character to the influence of major internal and external parameters. Fluvial systems transfer energy through raindrop impact on hillslopes to the exertion of velocity-induced shear stresses on the streambanks. Mass, primarily in the form of water and sediment, is transported from the farthest drainage divide to the channel mouth through the expenditure of energy in the system. Thresholds such as critical discharge levels or channel slopes influence system dynamics by changing the importance of certain processes, such as channel or streambank erosion.

Subsystems exist within an overall fluvial system and may be considered as important and separate elements of the system. The major subsystems of a fluvial system are the main channel and its flood plain, the tributary network, and hillslopes. These subsystems are characterized by energy and mass transfer through them and interaction between them. The hillslope subsystem transfers water and sediment overland during precipitation through the generation of sheet flow and eventual erosional development of rills and gullies. Water and sediment in the gullies then enter the tributary drainage network and are transported downstream, modifying the tributary channels progressively downstream. Upon entering the main channel, the collective influence of the many tributaries on the main channel is expressed by the rate and type of sediment production from the tributary and hillslope subsystems and by hydrologic smoothing of the processing of precipitation and consequent streamflow to the main channel. The resulting mainstream reflects an adjustment to the tributary and hillslope subsystems, in terms of process and form.

Factors controlling fluvial systems

Fluvial systems are controlled by a number of external and internal factors. Major external factors influencing the evolution and character of fluvial systems are time, geologic setting, initial relief, climate, and of course, man. Factors within the fluvial system, which are related to the external factors, are vegetation, local relief of the basin, hydrology, drainage network morphology, and hillslope morphology. The external variables, geology and initial relief, are established at the onset of river-basin evolution. The influence of climate changes as the river system evolves through time, or as the climate changes.

Time is a useful yardstick by which to estimate the evolution of a river system. In terms of fluvial systems, time is usually considered as having four scales: cyclic, graded, steady, and instantaneous. These timescales are discussed in the next paragraphs, as they relate to equilibrium conditions in fluvial systems.

As products of the external factors, the internal factors of vegetation, local relief of the basin (maximum elevation minus minimum elevation), hydrology (water and sediment discharge), drainage network morphology (tributary network subsystem), and hillslope morphology (hillslope subsystem) evolve through time and respond to changes in the external factors (primarily climate). The significance of cyclic, graded, and steady time scales on the role of the external and internal factors of fluvial system evolution is illustrated in Table 1 (Schumm and Lichty, 1965). The role of climate as a dynamic variable in influencing the internal variables of a fluvial system is illustrated on Figure 1 and discussed further in Schumm (1969). The roles of time and climate are described by Wolman and Gerson (1978).

Systems analysis

Several salient concepts of "systems analysis" are especially appropriate to the consideration of the mechanics of fluvial systems. These concepts include equifinality, feedback, relaxation

TABLE 1. SIGNIFICANCE OF DRAINAGE BASIN VARIABLES DURING VARIOUS TIMESCALES*

Drainage Basin Variables	Status of Variables During Designated Time Spans		
	Cyclic	Graded	Steady
1. Time	Independent	Not relevant	Not relevant
2. Initial relief	Independent	Not relevant	Not relevant
3. Geology (lithology, structure)	Independent	Independent	Independent
4. Climate	Independent	Independent	Independent
5. Vegetation (type, density)	Dependent	Independent	Independent
6. Relief or volume of system above base level	Dependent	Independent	Independent
7. Hydrology (runoff and sediment yield per unit area within system	Dependent	Independent	Independent
8. Drainage network morphology	Dependent	Dependent	Independent
9. Hillslope morphology	Dependent	Dependent	Independent
10. Hydrology (discharge of water and sediment from system)	Dependent	Dependent	Dependent

*Source: Schumm and Lichty, 1965, Table 1, p. 112.

time, thresholds, and equilibrium conditions. A number of equilibrium conditions exist in fluvial systems at any instant. These various equilibrium states have been described by Schumm (1977) and will be reviewed here in terms of time scales.

Equifinality. This term refers to the development of similar geomorphic features from different processes, an important concept in the analysis of river systems. Several reaches of a river may appear similar, even though these reaches have been profoundly influenced by different factors and have substantially different histories. In the complex evolution of rivers, various processes acting on different materials result in some features of the river system, which on casual observance, would appear to have a similar origin and history.

Feedback. This reaction occurs in river systems through the impact of the output of the tributary system on the main channel, which in turn may become locally regulated by a tributary influence and change the direction of the evolution of the main channel. An example is the growth of large alluvial fans in a river valley and at the confluence of a major tributary. These large bodies of sediment deflect the channel of the main river and provide a substantial increase in local sediment available for transport. Since the main river cannot transport the sediment brought to it by the tributary, positive feedback occurs and the channel of the river is changed.

Relaxation time. The time period required for a fluvial system to readjust its system operation to a new state of equilibrium following a change in equilibrium is the relaxation time. Since there are various time scales for different equilibrium states, the relaxation time between each equilibrium state is also variable. Many of the world's rivers have been profoundly affected by

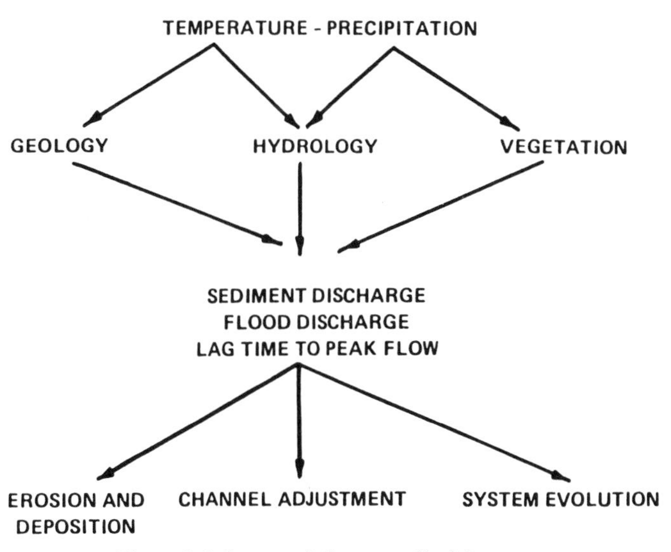

Figure 1. Influence of climate on fluvial systems.

large-scale influences, such as glaciation, that still substantially control their character. That is, the relaxation time required for these rivers to adjust to a new state of equilibrium, as a stream not presently being influenced by glaciation, has not been reached since the cessation of direct influence of glaciation, over 10,000 yr ago.

Thresholds. Change in equilibrium states of fluvial systems usually involve a change in the influence of an external or internal factor or the passage of a threshold. Examples of thresholds include the form of a river channel; for instance, the width-to-depth ratio as it adjusts to increased or decreased sediment transport. When sediment transport decreases due to a reduction of sediment availability, the width-to-depth ratio of the channel may decrease due to scouring of the channel bed. As the streambanks increase in height, a critical (threshold) bank height is reached, which may initiate mass failure of the streambank, and concomitant increase in channel width, width-to-depth ratio, and locally, the amount of sediment available for transport. In terms of system mechanics, a threshold was passed interrupting an equilibrium state, and positive feedback in the form of increased sediment production from streambank erosion returned the system to a new equilibrium state after a given relaxation time (Schumm, 1973).

Equilibrium states. The concept of equilibrium states in fluvial systems has been recognized since the time of Leonardo da Vinci, who described the apparent natural adjustment of rivers and valleys. Over 100 years ago, G. K. Gilbert (1877) outlined the basic tenets of the concept of "dynamic equilibrium" later developed by Hack (1960). Most fluvial geomorphologists today recognize four separate equilibrium states that may be seen over various time scales in a fluvial system. These equilibrium states are: (1) decay, (2) steady state, (3) dynamic, and (4) dynamic metastable (Fig. 2). Decay equilibrium occurs over the longest time (cyclic period, reflecting the entire history of the evolution of the drainage basin, as much as 10,000 to 100,000 yr). Decay equilibrium conditions are the product of the long-term erosional development of the drainage basin, and exist when the rate of change of form decays through time from relatively fast to slow change (Chorley and others, 1984).

During the long-term decay (erosional development) of a fluvial system, there are smaller scale fluctuations that collectively result in the overall decay of the system. Consequently, the fluvial system is dynamic about a mean trend, and the condition is termed "dynamic equilibrium." An example of dynamic equilibrium of a river would be the response of the river to the introduction of sediment and meltwater during periodic melting of the late Pleistocene glaciers. Dynamic equilibrium is equivalent to the graded condition in terms of time scales.

Within the dynamic equilibrium time period or condition, thresholds are surpassed that cause interruptions in the stability of the fluvial system. The dynamic equilibrium condition is then interrupted until a new stability condition is reached. Collectively, these multiple stability conditions within the longer period of dynamic equilibrium are termed "dynamic metastable equilibria" (Chorley and Kennedy, 1971).

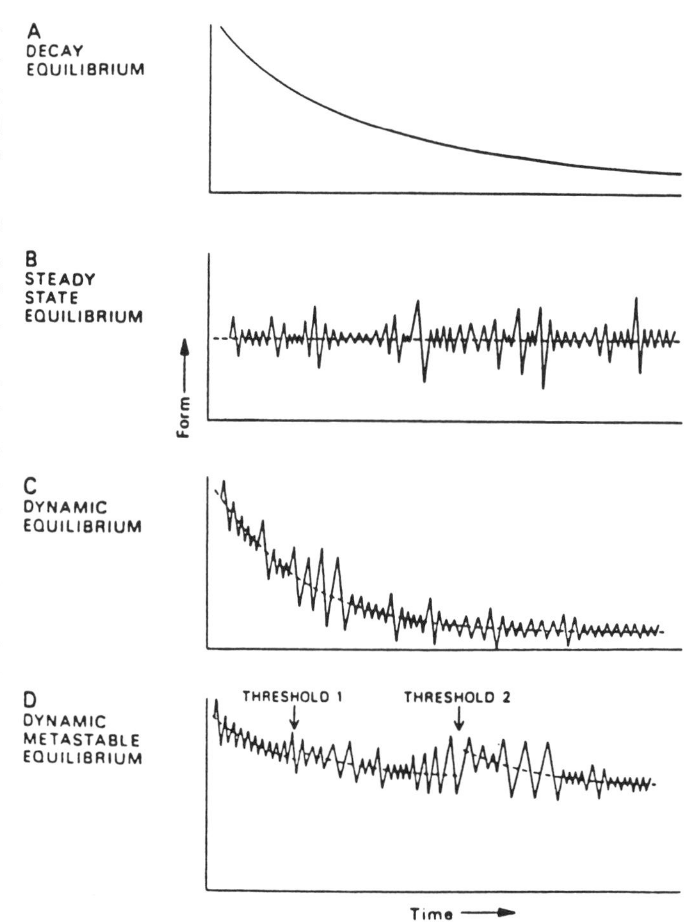

Figure 2. Equilibrium conditions of fluvial systems (after Chorely and others, 1984).

At any time, a river is in steady-state equilibrium when active processes control present form. Steady-state, or instantaneous, equilibrium occurs in the river bed as the depth and velocity of streamflow controls the amount of sediment transport at a given location. Appropriate time scales for steady-state equilibrium may be minutes to hours.

Man's interaction with fluvial systems

Previous discussions of the mechanics of fluvial systems mentioned that man acts as an external variable in influencing the evolution of fluvial systems. On many of the major rivers of the United States and elsewhere, man's modification of the fluvial system has been substantial. In many cases, the effects of these modifications have been equivalent to several thousands of years of adjustment to a major change in climate or change in geologic environment. On many of the principal navigational streams of the United States, the present channel has only a minimal resemblance to the former natural river of 100 yr ago. For example, of the 3.5 million miles of rivers and streams in the United States, erosion has been occurring on over half a million miles of banklines along these streams (USCE, 1978). Man modifies a fluvial system by changing the character of the internal variables of the system, such as vegetation, hydrology, and drainage network morphology. The result of man's modification of the fluvial system is feedback in the form of system changes in process and form (Rahn, 1986).

An appropriate way to illustrate the impact of man on a fluvial system is to examine the relation between water and sediment discharge in a river and certain characteristics of the river, such as the following parameters: W, stream width; D, stream depth; s, stream slope, Mw, meander wavelength; S, sinuousity; W/D, width-depth ratio; Qw, water discharge; and Qs, sediment discharge.

Schumm (1977, p. 133–137) demonstrated empirical relationships between these parameters as given in the proportionalities below:

$$Qw \propto \frac{W \times D \times Mw}{s} \quad (1)$$

$$Qs \propto \frac{W \times s \times Mw}{D \times S} \quad (2)$$

Under these relationships, an increase in water in the stream, through a diversion of water into the system, would result in an increase in the width, depth, and meander wavelength, and a decrease in channel slope (equation 1). Diversion of water from the river system would cause an opposite effect. When the sediment discharge of a stream is increased through disturbance of the natural vegetative cover (agriculture or forestry), the channel width, slope, and meander wavelength increase while the channel depth and sinuousity decrease (equation 2). Trapping of channel sediment behind reservoirs would result in a decrease in channel width, slope, and meander wavelength, as well as an increase in channel depth and sinuousity. These general relationships are useful in understanding not only the impact of man's works on a fluvial system but also the complex adjustment of fluvial systems to changes in hydrology. Other examples of these relationships are developed by Santos-Cayade and Simons (1972).

THE IMPACT OF ENGINEERING WORKS ON EROSION AND SEDIMENTATION

Land use

Overview. The impact of engineering works, particularly those pertaining to river engineering, on erosion and sedimentation must be examined and evaluated in terms of related activities that are also known to impact erosion and sedimentation balances. For example, in searching for causes and solutions of fluvial problems associated with dams or navigational structures, the possible effects of land use cannot be excluded. Furthermore, the identification of causes becomes complex due to the fact that an integrated system is involved; the subsets of the system cannot be examined in isolation from other subsets. An overview of erosion and sedimentation problems has been described in Laronne and Mosley (1982).

Urbanization and agricultural land use. The effects of urbanization and agricultural land use (including forestry) on erosion and sedimentation are well known and documented in Coates (1973), while the effects of highway construction are described by Parizek (1970). Overally, these practices usually result in higher water discharges within the drainage basin, and overextensive agriculture and forestry usage generally cause higher sediment discharges and yields. The effects of these increased discharges of water and sediment have significant effects on tributary and master streams, which include: (1) increased flood hydrograph peaks, (2) excessive sedimentation and clogging of portions of the basin, and (3) channel degradation and bank erosion in other parts of the basin. With respect to increased flood hydrograph peaks and engineering design, many of the dams designated unsafe after inspections initiated by the National Dam Safety Act of 1978 were unsafe due to a lack of spillway capacity; this was based on a required capacity calculated on more modern techniques and a more complete rainfall-runoff data base than those available to dam designers. Quite possibly a portion of the required capacity was due to urbanization and increased agricultural land use since these dams were originally designed. The Eel River Basin in northern California and the Yazoo Basin Uplands of northern Mississippi are examples of river basins in which the erosion and sedimentation effects of overextensive forestry and agriculture are well documented. In the Eel River Basin, a combination of clear-cutting, road building, and agricultural practices acting together in an environmental of steep slopes, highly erodable materials, and tectonic activity has produced the highest sediment yield in North America as well as channel degradation and aggradation and bank erosion (Smith and Patrick, 1979). In the Yazoo Basin Uplands, deforestation and overextensive agricultural practices early in this century produced excessively high sediment discharges that clogged countless basin tributaries and contributed to flooding. However, conservation measures such as sediment traps and reforestation during the 1930s and 1940s have significantly reduced sediment yields in this basin (Happ and others, 1940; Whitten and Patrick, 1981; Grissinger and Murphey, 1982). Even so, channelization, also a conservation measure, is still affecting the streams in this basin.

Bridges. The relationships between fluvial erosion and sedimentation and highway and railroad bridges are important in terms of the integrity of the structure as well as conditions on the stream. Erosion around bridge piers during flood events is well documented and has caused the failure of numerous bridges and loss of life (Shen, 1971). The effects of such erosion will be increased if the channel is also undergoing degradation by the upstream migration of a knickpoint. The impact of the bridge on the stream, under certain conditions, may also be important. The position of the bridge in terms of the geometry of the stream may result in current being directed toward nearby banks, resulting in bank erosion.

Impact. The overall impact of land use on the environmental quality of fluvial systems in terms of erosion and sedimentation is often severe, which may, in turn, cause further adverse impact on engineering works pertaining to safety and maintenance. Also, the impact of land use on fluvial systems may be difficult to discern and distinguish from that of a given engineering work, thereby complicating the process of identifying causes and effects of erosion and sedimentation.

Dams and reservoirs

Overview. River engineering projects such as dams and reservoirs, and navigational and channel improvement structures, are intended to mitigate adverse fluvial conditions or to enhance the utilization of water resources. The application of geological sciences to these engineering activities requires an interdisciplinary geological approach, specifics of which are dependent on the type or function of the project. The geological information must contribute to both the overall geotechnical and hydraulic investigations of the site or reach of river involved. Geological input into geotechnical investigations includes foundation and materials investigations; input into hydraulic studies involves investigations of the effects of the project on the geomorphology of the river system. Traditionally, the emphasis of engineering geology has been on foundation conditions and natural materials; however, during the last decade, there has been increased interest in and need for fluvial geomorphological data and interpretation of related geological data. The increased interest in related geoscience aspects has resulted from the realization that although many river engineering projects satisfy the design purposes, they may also cause, or at least trigger, other conditions—some possibly adverse. The presence of a dam on a stream effectively changes the hydrologic regimen of the stream, which in turn may affect the hydraulic geometry and fluvial geomorphology upstream and downstream of the dam. Furthermore, wildlife habitats both upstream and downstream of the dam may be adversely affected (Mahmood, 1973).

Reservoir sedimentation. The reservoir behind the dam raises the base level of the stream, which in turn, causes deposition of sediment, initially in the most distant upstream parts of the reservoir; with time, deposition extends farther downstream in the reservoir (Simons, 1979). This sedimentation in the reservoir will decrease storage capacity and may affect water quality, recreational facilities, and wildlife habitat in the reservoir area. Reservoir sedimentation may also cause aggradation (sedimentation and channel build-up) in nonflooded, upstream tributary channels flowing into the reservoir. Reservoir sedimentation and channel aggradation problems are concerns that must be addressed and evaluated in selecting reservoir sites. The determination of sediment yield within the basin and sediment discharge in the stream will provide comparative data for site selection. After construction and reservoir filling, sedimentation must be managed by the conduction of sediment surveys within the basin; slope instability throughout banks and perimeter is a major contributor (Kiersch, 1988). Furthermore, the impoundment of water in the reservoir may result in the raising of water tables within the reservoir area, depending on the geologic setting.

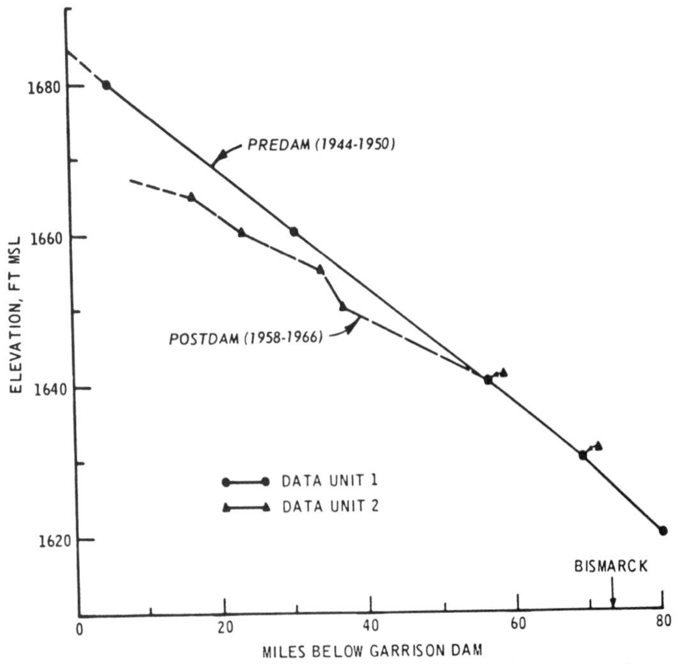

Figure 3. Longitudinal profile on the Missouri River downstream from Garrison Dam, North Dakota (USA) showing channel erosion (degradation) due to clear-water release (after Patrick and others, 1982).

Erosion and sedimentation downstream of dams. Damming a river and controlling the discharge through the dam result in two conditions that may adversely affect the river regimen downstream. The initial condition is the clear water released, or the sediment-free nature of the outflow water produced by blocking the stream bed load in transport. Subsequently, the controlled release of impounded water can satisfy power, irrigation, or flood-control requirements.

The clear-water release produces channel-bed degradation with deepening immediately downstream of the dam, which decreases the downstream channel slope (Fig. 3). This degradation may be accompanied by bank erosion. Less commonly, aggradation may occur immediately downstream of certain dams when flows have been so reduced that there is insufficient energy to transport incoming sediment delivered by downstream tributaries carrying high sediment loads (Gessler, 1971). Also, the sediment-free water may no longer seal unlined irrigation canals, as occurs in the western states, resulting in excessive leakage (Kiersch, 1955, p. 113).

A change in stream planform may occur along the degraded reach or farther downstream (Fig. 4). Whether meandering or braided conditions occur depends on the channel slopes and mean bankfull discharge. Thus, at constant bankfull discharge and with a decrease in channel slope, a braided stream will become meandering; the same change will result from a constant channel slope with a decrease in bankfull discharge (Leopold and Wolman, 1957). Typically, however, reaches downstream of dams become braided due, directly or indirectly, to the degradation process itself. Consequently, to replenish its sediment load, the stream will erode either the channel bed or the banks, depending on their relative erodability. Although bed erosion is probably more common, the erosion of banks will result in divided flow and braided conditions. Also, reaches downstream from areas experiencing bed erosion will receive influxes of sediment derived from this erosion, which may also produce braided conditions in these downstream reaches (Rahn, 1977; Whitten and Patrick, 1981).

Channel degradation below the dam produces a lowered base level, which will affect the regimen of tributary streams downstream of the dam (Simons, 1979). The lowering of base level may cause these streams to degrade their channels some distance upstream by a process called headcutting (Fig. 5).

Erosion in unlined emergency spillways. Most large dams have emergency spillways by which reservoir levels may be safely lowered without overtopping the dam during periods of high water. Often these spillways are not a part of the dam structure, but rather are excavated through a portion of an abutment and are therefore founded in a variety of soil and/or rock types. Some spillways are lined with concrete; many of either the lined or unlined bypasses have never experienced a design flow. Generally, small dams have unlined emergency spillways. In recent years, both the Corps of Engineers and the Soil Conservation Service have experienced failure or near failure of unlined spillways that have, in certain cases, threatened the dam itself (Cameron and others, 1986). The privately owned Delta, Melville, Abraham, and Deseret Irrigation Company (DMAD) dam in west-central Utah catastrophically lost impoundment due to the failure of its emergency spillway by erosion in July of 1983. Figure 6 shows the emergency spillway at DMAD shortly before it failed. Interestingly, these failures and near failures have occurred in spillways founded in both soil and apparently durable rock, and at flows significantly lower than their design capacities. Ongoing research has suggested that the flow through the spillway, and downstream into the channel, produces knickpoints due to the inhomogeneity of the soil or rock mass along the flowpath. These knickpoints may migrate upstream toward the abutment in a manner similar to the type of headcutting commonly associated with channelization. If unrestrained, the headcut could lead to failure of the spillway and the loss of the impoundment. Obviously, such a failure could produce catastrophic conditions downstream as well as significant erosion and sedimentation in downstream areas. Flume studies suggest that the upstream migration of knickpoints is controlled by rather sensitive flow (discharge) thresholds, which in turn, control the location of back eddies between the water flowing over the knickpoint and the face of the knickpoint (Cameron and others, 1988a and b; May, 1988).

Impact. Overall, the most serious downstream effect of dams is the occurrence of bank erosion, which may accompany both degradation and braiding. Although post-dam erosion may or may not prove appreciably greater than expected, it may occur in a fashion and at locations quite different than occurred prior to closure of the stream by the dam (Rahn, 1977).

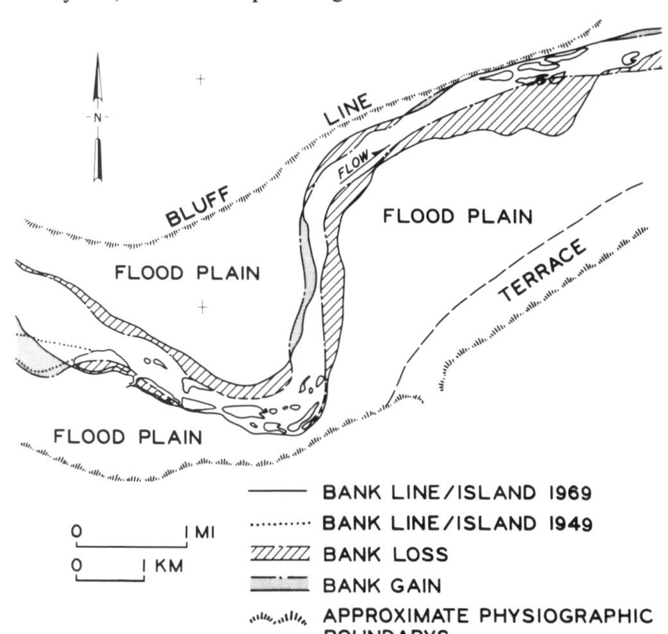

Figure 4. Historical planform changes identified from topographic maps of the Missouri River downstream from Garrison Dam, North Dakota (USA) showing erosion and accretion (after Patrick and others, 1982). Note bank loss on both inside and outside of meander bends.

Navigational and channel improvement structures

Overview. Included in this section are river engineering projects designed to improve navigation and decrease flooding; these include channel realignment (channelization), river training structures (dikes), locks, and levees. As indicated earlier, traditional geological input to such projects has mainly addressed foundation and material aspects of the design; however, there is an ever-increasing tendency for engineering geologists to become more involved in the fluvial geomorphology of the river system and for designs to be more compatible with the natural system (i.e., work with the river rather than against it).

Channelization. Channelization is the method of improving flood control or navigation that has been used extensively on large and small streams (see Rahn, this volume). In general, the method involves various procedures, from clearing trees, brush, debris, and snags from along a stream course to the construction of a new channel. In extreme cases, the new channel will be developed primarily by cutting off meander loops and straightening and widening the old channel. Channelization results in improved hydraulic efficiency by decreasing channel roughness and increasing channel slope. These new conditions result in im-

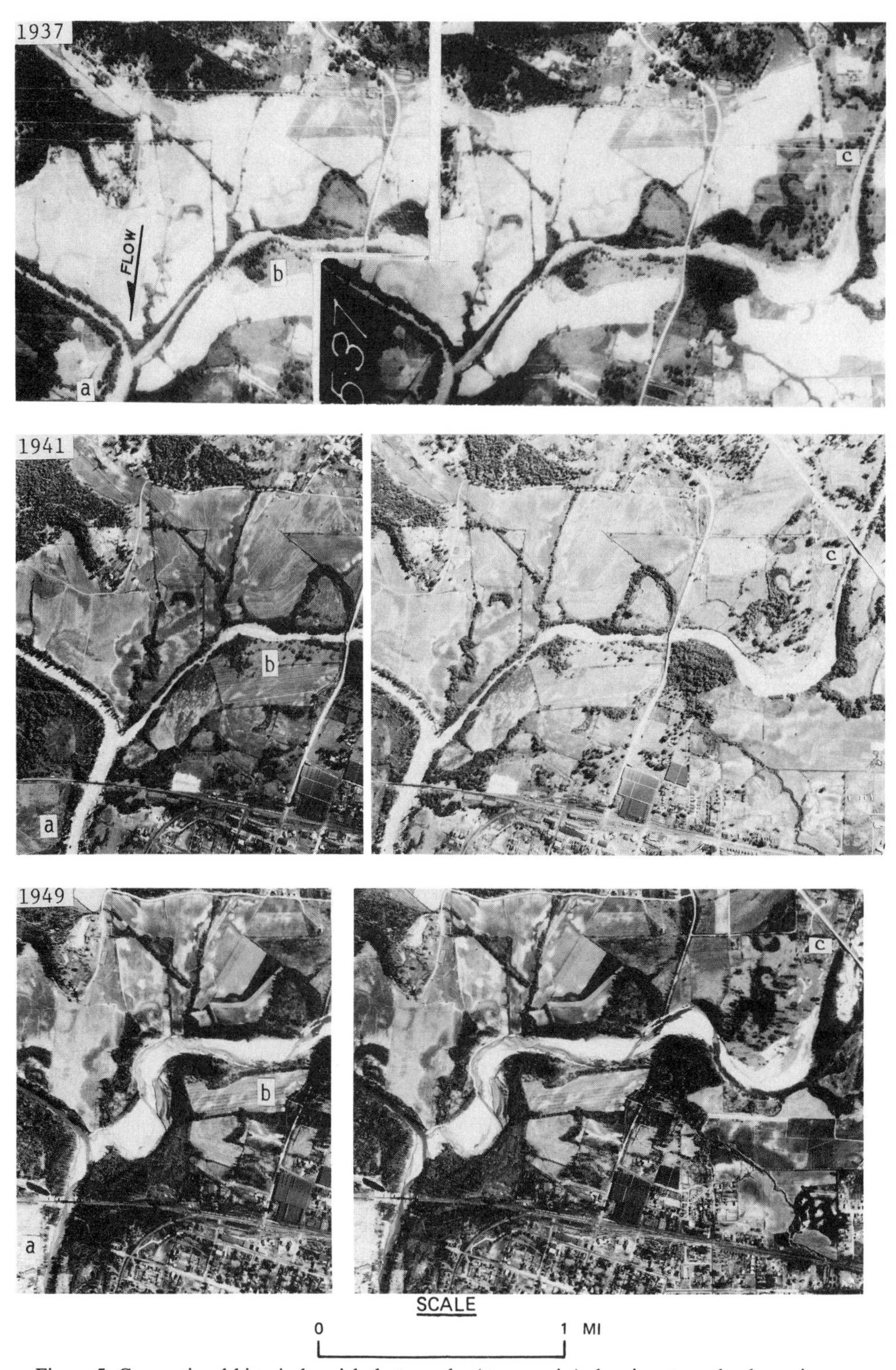

Figure 5. Conventional historical aerial photography (stereo pairs) showing streambank erosion produced by channel degradation and headcutting near the confluence of the Yalobusha River (a) and Batupan Bogue (b) at Grenada, Mississippi (USA). The upstream extent of significant channel degradation and streambank erosion can be seen on Batupan Bogue (c). The confluency is approximately 2 km downstream of Grenada Dam (on the Yalobusha River), which has exhibited downstream degradation due to clear-water release. The Yalobusha River has also been channelized downstream of the confluence with Batupan Bogue (after Whitten and Patrick, 1981, p. 159).

proved drainage of flood waters and decreased flooding through the channelized reach; in cases of navigable streams, navigation is improved by decreasing channel sinuousity. The siting of these cutoffs or new channels and the design of bank slopes require geological investigations of the soils and sediments in the channel and on the flood plain, and of the fluvial geomorphology of the stream system.

Several adverse conditions, which may or may not be predictable, may result from channelization. Such conditions may produce excessive erosion and sedimentation. The improved hydraulic efficiency of the channelized reach results in decreased transit times and increased discharges downstream of the channelized reach.

Probably the most serious adverse condition resulting from channelization is channel degradation along and upstream of the channelized reach and on tributary streams that enter the channelized stream along or above the channelized reach. Channel degradation occurs because of increased velocities produced by the increased slope. The scouring action may proceed considerable distances upstream and will result in the establishment of a new, lower longitudinal profile. The upstream movement of degradation—headcutting—may be temporarily slowed by the occurrence of erosion-resistant soils or rocks in the stream channel (Grissinger and others, 1982). When these erosion-resistant materials are exposed in the channel as waterfalls or rapids, they are termed knickpoints (Fig. 7). The increased slope along the channelized reach decreases base level for all upstream tributaries, permitting headcutting to extend throughout the basin (Fig. 8). Sediments derived from upstream headcutting may be delivered to unchannelized downstream areas where aggradation may occur.

Streambank erosion affects riparian landowners and is perhaps the most serious problem derived from extreme channelization. The extent to which bank erosion occurs is a function of current velocity and the properties of soils making up the banks. The movement of the headcut and lowering of the channel results in the banks becoming relatively higher and steeper, which causes bank undercutting, slope instability, and failure (Whitten and Patrick, 1981).

Generally, many of the fluvial instability problems that currently confront applied geologists and engineers have been caused by inadequately designed channelization projects undertaken many years ago by individual landowners, local interests, or government agencies. Current practice minimizes extreme channelization and limits channel modification to the clearance of vegetation and debris from stream banks. Extreme channelization, with ensuing channel degradation, aggradation, and bank failure, can have serious adverse effects on wildlife habitats by destroying nesting or spawning areas, in addition to introducing more sediment into the water.

Headcutting can be halted or at least retarded by the construction of grade-control structures along the channel (Barnes, 1971). A grade-control structure, or weir, can be constructed of reinforced concrete or sheet piling. It protects the stream bed and banks, dissipates stream energy with baffles, and retards headcutting. These structures should be located on erosion resistant materials whenever possible to maximize their effectiveness.

Figure 6. The emergency spillway at DMAD Dam in west-central Utah shortly before spillway failure by erosion and catastrophic loss of impoundment in July, 1983 (after Cameron and others, 1986).

Dikes. Jetties or dikes are masonry, steel, or timber structures that are anchored into the bank and extend into the channel; they divert current action away from the banks. The installation of dikes on one or both sides of the stream is intended to maintain the thalweg in a particular part of the channel and thereby deepen the channel and improve its navigability. This occurs because of the decreased effective channel width and the sedimentation occurring behind the dikes. Another effective use of dikes is to protect streambanks against erosion (USCE, 1981b).

Locks. Locks and the dams associated with them are navigational structures designed to increase water elevations in a stream to improve the navigational capacity. The effects of the lock and dam on fluvial erosion and sedimentation are unclear. The Ohio River, at one time or another, has had over 50 locks and dams constructed between Pittsburgh, Pennsylvania, and Cairo, Illinois. This reach of the river has experienced serious streambank erosion at a number of locations. The severity of the erosion has resulted in litigation and rather detailed studies of the causes and effects of erosion, paticularly in the confines of the Louisville District, U.S. Army Corps of Engineers. These studies have shown that the locks and dams have had no causal relationship with nearby streambank erosion (USCE, 1981a).

Levees. Flood protection can often be realized by the construction of levees along the stream. Generally, the levee will be constructed of locally available materials, which are frequently granular. Much useful information has been derived from investigations and observations pertaining to the levees along the Mississippi River. Extensive levee building began on the Mississippi River during the last century, and the levee system has been expanded and the levees increased in height on an almost contin-

Figure 7. Photograph of a 2-m-high waterfall (knickpoint) produced by channelization-induced degradation on a tributary to the Middle Fork of Tillatoba Creek, northwestern Mississippi (USA; after Whitten and Patrick, 1981, p. 61).

uous basis since the work was initiated (Winkley, 1971). The effect of the levees in terms of river stage (level) and discharge over the years has been an increase in stage (level) for a given discharge. Stated another way, serious floods are now occurring on the Mississippi at lower discharges (Belt, 1975).

The effects of levees on erosion and sedimentation are also important and are, in part, related to discharges. The levee system also prevents overbank sedimentation (vertical accretion), which in turn, results in increased channel sedimentation and/or increased sedimentation at the mouth or delta of the river. In the lower Atchafalaya Basin, excessive sedimentation and the formation of lacustrine deltas over the last 50 years or so have significantly reduced the areas of Grand and Six-Mile Lakes and adjacent wetlands (Smith and others, 1986).

Impact. As discussed, certain engineering works have a significant impact on erosion and sedimentation in the fluvial environment. These same works can have equally significant impact downstream on mixed and marine environments, including estuaries, bays, deltas, and beaches (see Leatherman, this volume). A problem affecting the lower Atchafalaya Basin and the coastal environments of Louisiana is land loss. This problem may be caused by subsidence, sea-level rise, and/or coastal erosion; however, it may also be due to the constraining effects of levees and other structures that control and have decreased the flow in the distributary channels, such as Old River, in the lower reaches of the Mississippi River system (Nummedal, 1983).

OUTLOOK

A number of unsolved problems complicate our ability to understand the relationships between erosion and sedimentation, and operating engineering works. The more important ones are addressed here along with some suggested solutions.

Interdisciplinary studies

There is no question that erosion, sedimentation, and fluvial behavior are highly interdisciplinary in nature. Soil scientists, geologists, physical geographers, geotechnical and hydraulic engineers, as well as other scientists and engineers have made important contributions to our understanding of these phenomena. As the effects of the processes impact the general population, there is a need for planners, politicians, and other nonscientists to understand the generalities of these problems. All too often, however, scientific and engineering studies have not been sufficiently interdisciplinary, and consequently, the concepts of fluvial geomorphology were not incorporated by either the hydraulic engineers and/or by many applied geologists.

Geological applications

Most geologists, particularly in Federal and state agencies, are involved in the more traditional approach to geotechnical and engineering applications, such as foundation, ground-water, and materials investigations. Frequently these same geologists have not become fully aware of, or adequately trained in, the concepts of process geomorphology and sedimentation (Patrick and others, 1982). Thus, the fluvial problems of the agencies have been mainly the domain of the hydraulic engineers.

Need for early investigations

All too frequently, the potential erosion and sedimentation problems of a proposed engineering work have not received serious attention during the early geological investigation phases of a project (Kiersch, Ch. 18, Fig. 1, this volume). Generally, such investigations have been conducted only when geology-related

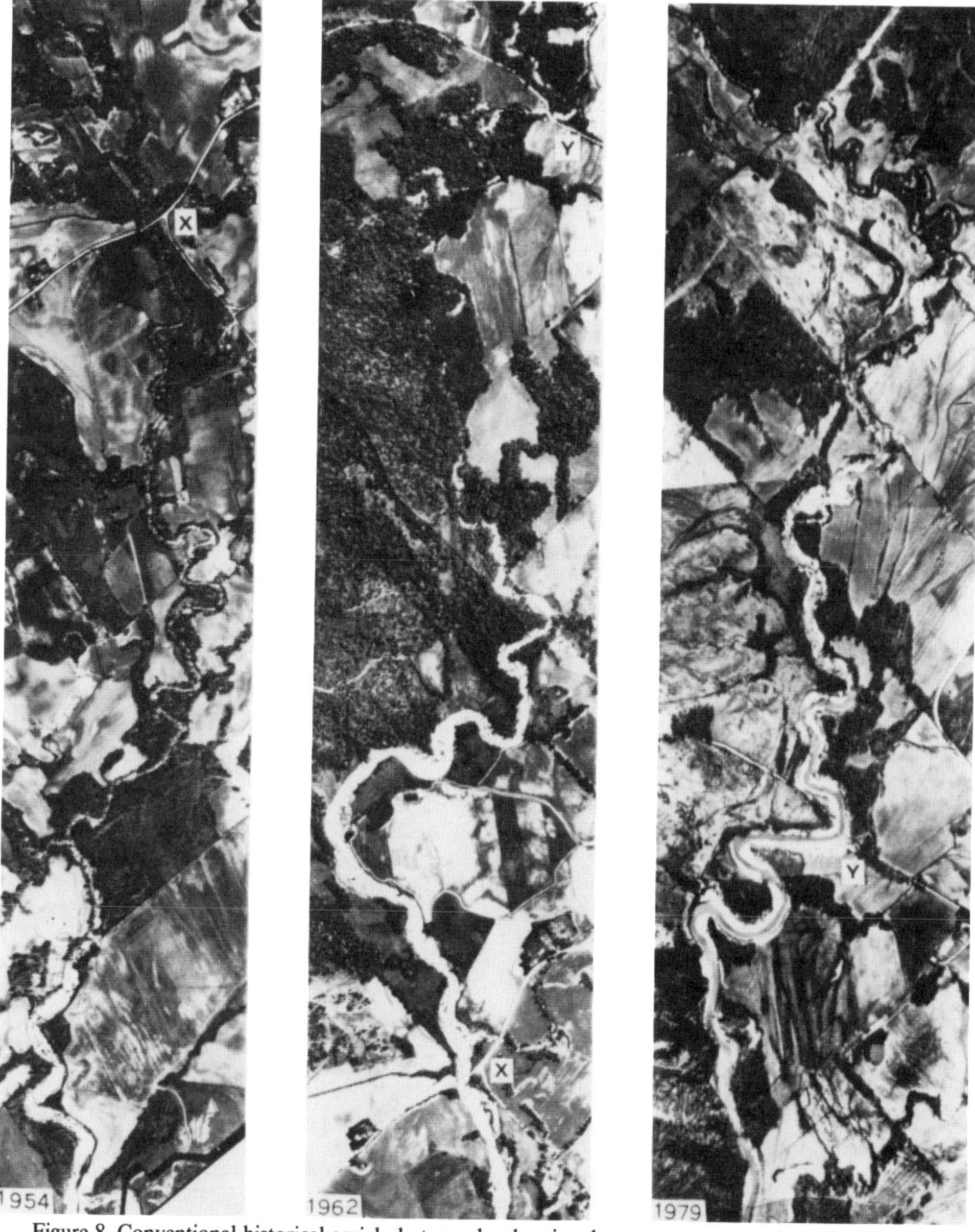

Figure 8. Conventional historical aerial photography showing the upstream progression of knickpoints (headcutting) that occurred between 1954 and 1979 along the South Fork of Tillatoba Creek in northwestern Mississippi (USA). Note that points "x" and "y" represent the same respective locations in the successive scenes; the geomorphic changes at these points (point "x" between 1954 and 1962, and point "y" between 1962 and 1979) are distinctive and profound (after Whitten and Patrick, 1981, p. 67).

problems developed during operation of the project. In such a situation, the time gap between project completion and the first evidence of a geologic reaction and an erosion/sedimentation problem may be a number of years. Consequently, when geological studies are finally initiated, a portion of the base-line data critical to understanding the origin and causes of the erosional and/or sedimentation reactions may be unobtainable.

Litigation

There is a common tendency of the population living in the riparian areas to seek redress by means of litigation against the action agencies—such as the Corps of Engineers, Bureau of Reclamation, Tennessee Valley Authority, or Soil Conservation Service—for real or perceived problems caused by the operation of engineering works. Either before or after the litigation, such agencies are unlikely to admit any responsibility even though there may be only sketchy evidence, one way or the other. When litigation occurs, the affected agency must develop field evidence hastily to support its legal position. Furthermore, even though the engineering works improved overall conditions, such as flooding on a river, this benefit may not matter when the riparian landowners are suing over streambank erosion.

Partial solutions

The geologic problems discussed in this chapter can lead to approaches by the action agencies to improve our understanding of the relationships between erosion and sedimentation, and operating engineering works. Some circumstances that can contribute to the mitigation of adverse impacts are: (1) Process geomorphology must be understood and applied by both geologists and engineers. (2) River sites and reaches must be considered as subsets of an interactive system. (3) Sites of engineering works on rivers should be periodically monitored, prior to and after construction, in terms of items (1) and (2), thereby developing base-line data to identify trends (imagery is particularly helpful). (4) River structures must be designed to work with the river rather than against it.

The future

Unquestionably, a continued river modification policy, as well as over-extensive land use, will result in continued erosion and sedimentation problems of all kinds. However, evidence to date suggests that in the absence of continued modification, and after a number of years, a degree of stability will occur.

REFERENCES CITED

Bagnold, R. A., 1941, Physics of blown sand and desert dunes: London, Methuen, 265 p.

Barnes, R. C., Jr., 1971, Erosion control structures, *in* Shen, H. W., ed., River mechanics, v. II: Ft. Collins, Colorado, H. W. Shen, p. 28-1 to 28-6.

Belt, C. B., Jr., 1975, The 1973 flood and man's constriction of the Mississippi River: Science, v. 189, p. 681–684.

Bennett, H. H., 1939, Soil Conservation: New York, McGraw-Hill Book Company, Incorporated, 993 p.

Cameron, C. P., Cato, K. D., McAneny, C. C., and May, J. H., 1986, Geotechnical aspects of rock erosion in emergency spillway channels: Vicksburg, Mississippi, U.S. Army Engineer Waterways Experiment Station, Technical Report REMR-GT-3, 101 p.

Cameron, C. P., Cato, K., Patrick, D. M., and May, J. H., 1988a, Geotechnical aspects of rock erosion in emergency spillway channels—Report 2, Analysis of field and laboratory data: Vicksburg, Mississippi, U.S. Army Engineer Waterways Experiment Station, Technical Report REMR-GT-3, 49 p.

Cameron, C. P., Patrick, D. M., Bartholomew, D. O., Hatheway, A. W., and May, J. H., 1988b, Geotechnical aspects of rock erosion in emergency spillway channels—Report 3, Remediation: Vicksburg, Mississippi, U.S. Army Engineer Waterways Experiment Station, Technical Report REMR-GT-3, 63 p.

Carson, M. A., 1971, The mechanics of erosion: London, Pion, Ltd., 174 p.

Chorley, R. J., and Kennedy, B. A., 1971, Physical geography; A systems approach: London, Prentice-Hall, 370 p.

Chorley, R. J., Schumm, S. A., and Sugden, D. E., 1984, Geomorphology: London, Methuen, 605 p.

Coates, D. R., ed., 1973, Environmental geomorphology and landscape conservation: Stroudsburg, Pennsylvania, Dowden, Hutchinson and Ross, Incorporated, Benchmark Papers in Geology, 485 p.

Einstein, H. A., 1942, Formulas for the transportation of bed load: American Society of Civil Engineers Transactions, v. 107, p. 561–573.

Ellison, W. D., 1945, Studies of raindrop erosion: Agricultural Engineering, v. 25, p. 131–136, 181–182.

Fisk, H. N., 1944, Geological investigation of alluvial valley of the lower Mississippi River: Vicksburg, Mississippi, Mississippi River Commission, 78 p.

——, 1951, Mississippi River geology in relation to river regime: Transactions of the American Society of Civil Engineers, v. 117, p. 667–682.

Gessler, J., 1971, Aggradation and degradation, *in* Shen, H. W., ed., River mechanics, v. I: Ft. Collins, Colorado, H. W. Shen, p. 8-1 to 8-24.

Gilbert, G. K., 1877, Report on the geology of the Henry Mountains, Montana: U.S. Geographical and Geological Survey of the Rocky Mountain Region, p. 18–98.

——, 1880, Report on the geology of the Henry Mountains: Washington, D.C., U.S. Government Printing Office, p. 115–117, 123–124, 140–142.

——, 1914, The transportation of debris by running water: U.S. Geological Survey Professional Paper 86, 263 p.

Grissinger, E. H., and Murphey, J. B., 1982, Present problems of stream channel instability in the bluff area of northern Mississippi: Journal of the Mississippi Academy of Sciences, v. 27, p. 117–128.

Grissinger, E. H., Murphey, J. B., and Little, W. C., 1982, Late Quaternary valley-fill deposits in north-central Mississippi: Southeastern Geology, v. 23, no. 3, p. 147–162.

Greenwood, G., 1857, Rain and rivers: London, Longmans, Green and Company, Limited, p. 53–54, 75–74, 102–104, 173–176.

Hack, J. T., 1960, Interpretation of erosional topography in humid temperate regions: American Journal of Science, v. 258, no. 80, p. 97.

Hadley, R. F., and Walling, D. E., eds., 1984, Erosion and sediment yield; Some methods of measurement and modeling: Cambridge University press, 218 p.

Happ, S. C., Rittenhouse, G., and Dobson, G. C., 1940, Some principles of accelerated stream and valley sedimentation: U.S. Department of Agriculture Technical Bulletin 695, 139 p.

Horton, R. E., 1940, An approach toward a physical interpretation of infiltration capacity: Soil Science Society of America Proceeding, v. 5, p. 399–417.

——, 1945, Erosional development of streams and their drainage basins; hydrophysical approach to quantitative morphology: Geological Society of America Bulletin, v. 56, p. 275–370.

Kiersch, G. A., 1955, Enginerring geology; historical development, scope, and utilization: Golden, Colorado School of Mines Quaterly, v. 50, no. 3, 123 p.

——, 1988, Reservoirs, *in* Jansen, R. B., ed., Advanced dam engineering for design, construction, and rehabilitation: New York, Van Nostrand Reinhold Company, Ch. 24 p. 729–750.

Krinitzsky, E. L., 1965, Geological influences on bank erosion along meanders of Lower Mississippi River: Vicksburg, Mississippi, U.S. Army Engineer, Waterways Experiment Station, Investigation Report 12-15, 47 p.

Laronne, J. B., and Moseley, M. P., 1982, eds., Erosion and sediment yield: Stroudsburg, Pennsylvania, Hutchinson Ross Publishing Company, Benchmark Papers in Geology, 375 p.

Leopold, L. B., and Wolman, M. G., 1957, River channel patterns; braided, meandering, and straight: U.S. Geological Survey Professional Paper 282-B, p. 39–85.

Lyell, C., 1830, Principles of Geology: Dorsetshire, p. 174, 188–193.

Mahmood, K., 1973, Fish facilities at river development projects, *in* Shen, H. W., ed., Environmental impact on rivers (River mechanics, III): Ft. Collins, Colorado, H. W. Shen, p. 18-1 to 18-54.

May, J. H., 1988, Geologic and hydrodynamic controls on knickpoint erosion [Ph.D. thesis]: College Station, Texas A&M University, 198 p.

Nummedal, D., 1983, Rates and frequencies of sea-level changes, a review with an approach to predict future sea levels in Louisiana: Gulf Coast Association of Geological Societies Transactions, v. 33, p. 361–366.

Parizek, R. R., 1970, Impact of highways on the hydrogeologic environment, *in* Coates, D. R., ed., Environmental geomorphology: Binghamton, State University of New York, p. 151–199.

Patrick, D. M., Smith, L. M., and Whitten, C. B., 1982, Methods for studying accelerated fluvial change, *in* Hey, R. D., Bathurst, J. C., and Throne, C. R., eds., Gravel-bed rivers: London, John Wiley, p. 783–815.

Playfair, J., 1802, Illustrations of the Huttonian theory of the earth: London, Cadell and Davis, p. 102–105, 350–352.

Rahn, P. H., 1977, Erosion below main stem dams on the Missouri River: Association of Engineering Geologists Bulletin, v. 14, no. 3, p. 157–181.

—— , 1986, Engineering geology, an environmental approach: New York, Elsevier, p. 241–297.

Santos-Cayade, J., and Simons, D. B., 1972, River response, *in* Shen, H. W., ed., Environmental impact on rivers (River mechanics III): Ft. Collins, Colorado, H. W. Shen, p. 1-1 to 1-25.

Saucier, R. T., 1974, Quaternary geology of the Lower Mississippi Valley: Little Rock, Arkansas Archaeological Survey Research Series no. 6, 26 p.

Schumm, S. A., 1969, Geomorphic implications of climatic changes, *in* Chorley, R. J., ed., Water, earth, and man: London, Eyre Methuen, Limited, p. 525–534.

—— , 1973, Geomorphic thresholds and complex response of drainage systems, *in* Morisawa, M., ed., Fluvial geomorphology: Binghamton, State University of New York Publications on Geomorphology, p. 299–310.

—— , 1977, The fluvial system: New York, John Wiley and Sons, 338 p.

Schumm, S. A., and Lichty, R. W., 1965, Time, space, and causality in geomorphology: American Journal of Science, v. 263, p. 110–119.

Shen, H. W., ed., 1971, Scour near piers, *in* Shen, H. W., ed., River mechanics, v. II: Ft. Collins, Colorado, H. W. Shen, p. 23-1 to 23-25.

—— , ed., 1972, Sedimentation: Symposium to Honor Professor H. A. Einstein, Colorado State University, Fort Collins, 683 p.

Simons, D. B., 1979, River and canal morphology, *in* Shen, H. W., ed., Modeling of rivers: New York, John Wiley and Sons, p. 5-1 to 5-81.

Smith, L. M., and Patrick, D. M., 1979, Engineering geology and geomorphology of streambank erosion—Report 1, Eel River Basin, California: Vicksburg, Mississippi, U.S. Army Engineer, Waterways Experiment Station Technical Report GL-79-7, 80 p.

Smith, L. M., Dunbar, J. B., and Britsch, L. D., 1986, Geomorphological investigation of the Atchafalaya Basin, Area West, Atchafalaya Delta, and Terrebonne Marsh, v. 1: Vicksburg, Mississippi, U.S. Army Engineer Waterways Experiment Station Technical Report GL-86-3, 85 p.

SCS, 1976, Soil erosion; Prediction and control: Proceedings of the National Conference of Soil Erosion, Ankeny, Iowa, Soil Conservation Society of America, 393 p.

Trask, P. D., ed., 1950, Applied sedimentation: New York, John Wiley and Sons, 707 p.

Turnbull, W. J., Krinitzsky, E. L., and Johnson, S. J., 1950, Sedimentary geology of the alluvial valley of the lower Mississippi River and its influence on foundation problems, *in* Trask, P. D., ed., Applied sedimentation: New York, John Wiley and Sons, p. 210–226.

Twenhofel, W. H., 1932, Principles of sedimentation: New York, McGraw-Hill Book Company, Incorporated, 673 p.

USBR, 1948, Proceedings of the Federal Inter-Agency Sedimentation Conference: Washington, D.C., U.S. Department of the Interior, Bureau of Reclamation, 314 p.

USCE, 1978, Interim report to Congress, section 32 program; Streambank erosion, control, evaluation, and demonstration act of 1974: Washington, D.C., U.S. Army Corps of Engineers, Office of the Chief of Engineers, 137 p.

—— , 1981a, Final report to Congress, section 32 program; Streambank erosion control, evaluation, and demonstration act of 1974; Appendix D, Ohio River demonstration projects: Washington, D.C., U.S. Army Corps of Engineers, Office of the Chief, 351 p.

—— , 1981b, Final report to Congress, section 32 program; Streambank erosion control, evaluation, and demonstration act of 1974; Appendix E, Missouri River demonstration projects, v. 1: Washington, D.C., U.S. Army Corps of Engineers, Office of the Chief, 324 p.

USDA, 1963, Proceedings of the Federal Inter-Agency Sedimentation Conference: Washington, D.C., U.S. Department of Agriculture, Agriculture Research Service Miscellaneous Publication no. 970, 933 p.

Vanoni, V. A., ed., 1975, Sedimentation engineering: New York, American Society of Civil Engineers, 745 p.

Water Resources Council, 1976, Proceedings of the Third Federal Inter-Agency Sedimentation Conference: Denver, Colorado, Sedimentation Committee, 954 p.

Whitten, C. B., and Patrick, D. M., 1981, Engineering geology and geomorphology of streambank erosion; Report 2, Yazoo Basin Uplands, Mississippi: Vicksburg, Mississippi, U.S. Army Engineer Waterways Experiment Station Technical Report GL-79-7, 178 p.

Winkley, B. R., 1971, Practical aspects of river regulation and control, *in* Shen, H. W., ed., River mechanics, v. I: Ft. Collins, Colorado, H. W. Shen, p. 19-1 to 19-79.

—— , 1977, Man-made cutoffs on the lower Mississippi River; Conception, construction, and river response: Vicksburg, Mississippi, U.S. Army Engineer District, Potamology Investigations Report 300-2, 209 p.

Wischmier, W. H., and Smith, D. D., 1965, Predicting rainfall-erosion losses from cropland east of the Rocky Mountains: Washington, D.C., U.S. Department of Agriculture Agricultural Handbook no. 282, 47 p.

Wolman, M. G., and Gerson, R., 1978, Relative scales of time and effectiveness of climate in watershed geomorphology: Earth Surface Process, v. 3, p. 189–208.

Zingg, A. W., 1940, Degree of and length of land slope as it affects soil loss in runoff: Agricultural Engineering, v. 21, p. 59–64.

MANUSCRIPT ACCEPTED BY THE SOCIETY NOVEMBER 6, 1987

Printed in U.S.A.

Geological Society of America
Centennial Special Volume 3
1991

Chapter 8

Coasts and beaches

Stephen P. Leatherman
Laboratory for Coastal Research, University of Maryland, College Park, Maryland 20742

INTRODUCTION

Coasts, often sedimentary in nature, serve as the dynamic interface between land and sea. While rocky shores exist along much of New England and the West Coast, the preponderance of United States coastal urbanization has occurred along sedimentary coasts. Indeed, much of the outer shoreline along the U.S. East and Gulf coasts is characterized by barrier systems.

The study of sedimentary coasts is a multidiscipline effort involving geologists, physical geographers, and coastal engineers, including hydraulic engineers and fluid hydrodynamicists. These specialities can all be considered under the general field of coastal geomorphology wherein the morphological development of the coast, acting under the influence of winds, waves, currents, and sea-level changes, is the subject of these physical science investigations. Coastal engineering, while primarily a branch of civil engineering, leans heavily upon coastal and geological sciences. Their charge is to address both the natural and human-induced changes in the coastal zone, design structural and nonstructural devices and procedures to intercede against such changes, and evaluate the impacts of proposed solutions on these problem areas.

Because numerous factors govern the development and evolution of coastal areas, solutions devised for one area will often fail if blindly applied to another. This stems from the wide-ranging morphologies and energy conditions found along the coast. Coastal engineering, therefore, is site specific, and project success requires careful collection and evaluation of all pertinent physical data from the geosciences.

HISTORICAL DEVELOPMENT

The coastal engineering literature is replete with coastal defense or harbor failures due to a lack of understanding of coastal processes. The importance of applied coastal research is clearly demonstrated by harbors built in Dublin, Ireland, during the nineteenth century. Two unsuccessful attempts were made at maintaining desirable channel depths due to high rates of longshore transport, with sediment moving straight into the harbor. The third attempt succeeded because of the advantageous configuration of the harbor entrance and consideration of tidal current circulation patterns. It was unfortunate that the basic principles of coastal morphology and littoral drift technology were not considered; a "glass of cold research ice water" would have permitted a more thorough and carefully designed plan from the beginning (Bruun, 1954a). In retrospect, planning and design engineers of the time cannot be blamed for all their mistakes, as applied geoscience/coastal geomorphology was in its infancy.

The field of coastal geomorphology was initiated primarily by physical geographers; the giants of the time (1910 to 1930) were Douglas W. Johnson and William Morris Davis. Davis introduced the concept of time-mandated evolution of coastal morphologies. For Davis (1912), geologic structure, process, and stage (youthful, mature, and old age) were the principal components of his time classification of landform features. While the "geographical cycle of coastal development" was a step forward, it eventually stymied some thinking because coastal evolution can be interrupted or greatly altered by certain "catastrophic" events so that "ultimate grade" may never be realized.

Johnson (1919) had the most profound effect on the thinking of the time, and his classic book *Shore Processes and Shoreline Development* was the first comprehensive treatment of coastal morphologies. It also served as an early guide to coastal engineers in the United States and abroad. Of particular note was Johnson's influence on the then new specialist to the field, M. P. O'Brien.

In the mid 1920s, the National Research Council appointed a committee to report on the status of coastal erosion in the United States. The concensus was that there was little expertise to predict coastal erosion and no real move to correct this deficiency. In response to this report, governors' delegates from the coastal states formed in 1926 the American Shore and Beach Preservation Association (ASBPA). At the same time an advisory "Board on Sand Movement and Beach Erosion" was appointed by the Chief of U.S. Army Corps of Engineers. These newly created organizations asked D. W. Johnson to compile a list of studies needing attention; there had been too much armchair theorizing and the immediate need was reliable field data. The

Leatherman, S. P., 1991, Coasts and beaches, *in* Kiersch, G. A., ed., The heritage of engineering geology; The first hundred years: Boulder, Colorado, Geological Society of America, Centennial Special Volume 3.

Figure 1. Hurricane destruction of Galveston, Texas in 1900 (photo courtesy of Rosenberg Library).

board asked M. P. O'Brien to initiate beach studies along the New Jersey coast in 1929.

The early interaction between Johnson and O'Brien (ultimately "Dean" of U.S. coastal engineering) was significant. O'Brien, a civil engineer who worked on jet pumps among other things for General Electric, had little knowledge of shoreline morphology and sedimentology. Johnson, the master in coastal geomorphology, set the research agenda for the process-oriented studies. Unfortunately, there were no geologists or physical geographers coming up the ranks to continue and foster this interaction with the coastal engineers, except for notables like Parker Trask and W. C. Krumbein. The latter worker was influential in applying quantitative and statistical techniques to geology-based shoreline studies. Geographers seemed to abandon the coastal arena, and few geologists were involved in shoreline research. Consequently, there was a minimum of applied coastal geological studies before the early 1970s; O'Brien (personal communication, 1986) feels that work of the board was greatly affected by their absence.

In 1930, Congress established the Beach Erosion Board in response to the loss of beaches at early resort areas, particularly along the New Jersey coast (Quinn, 1977). Engineers were already building groins, jetties, and seawalls in an attempt to stem the tide of erosion and keep navigational channels between the sandy coasts open. In 1963, the board was reorganized and its mission broadened to become the Coastal Engineering Research Center (CERC).

Coastal disasters perhaps played the greatest role in promoting public awareness and scientific investigation of shoreline processes. The hurricane of 1900 swept across Galveston Island, devastating the most improtant city in Texas at that time and killing more than 6,000 people (Fig. 1). This catastrophe still stands as the worst natural disaster in North America in terms of loss of human life. A 1926 hurricane levelled Miami Beach, Florida, and essentially presented developers with a clean slate to lay out the modern-day resort area. In August 1933 a major hurricane struck the Maryland coast, causing considerable damage at Ocean City and cutting a new inlet at the southern terminus of the town. The Great New England Hurricane of 1938 was not so kind in terms of loss of property and lives. The south shore

Figure 2. Southampton Beach, Long Island, New York, was leveled by overwash and cut to ribbons by inlets during the New England hurricane of 1938 (photo courtesy of State of New York Parks and Beaches).

of Long Island, New York, was already urbanized, and people were totally unaware of the approaching 4 to 5 m storm surge. The barrier island communities at Westhampton Beach, Quoque, and Southampton Beach were cut to ribbons by 12 incipient inlets, and most of the barrier was flattened by overwash surges (Fig. 2). The Rhode Island barriers suffered a similar fate, and streetcars were bobbing in storm-flooding waters in downtown Providence, Rhode Island. A poststorm field investigation by Beach Erosion Board members (Wilby and others, 1939) concluded that a better scientific understanding of coastal meteorology as well as related geological processes was critically needed.

World War II served as a catalyst to accelerate coastal process investigations. Not only was technological capability expanded with the invention of radar (capable of tracking coastal storms), but there were tremendous improvements likewise in aircraft and cameras; stable platforms became available to host metric-quality aerial cameras for shoreline mapping studies. Furthermore, major advancements were made in understanding wave/beach interactions in preparation for the D-Day Normandy invasion (1944) and the amphibious landings in the Pacific theater, as well as the physical properties of beach materials for construction of military works (Hunt, 1950).

O'Brien, with the assistance of Joe Johnson, both mechanical engineering professors at the University of California, Berkeley, set up the World War II waves project, whose principal objective was to determine the beach and wave characteristics that would hinder landing craft on approaching enemy-held shores. O'Brien sent the field party to the U.S. northwest coast where John Issacs and Willard Bascom used Dukws (amphibious six-wheeled trucks) and rode through 6-m high, plunging breakers to measure subaqueous beach profiles. As Bascom (1980) recounts, these surf rides were wilder than any rollercoaster could ever provide—so dangerous that the Coast Guardsmen served notice they were working at their own risk. Indeed, O'Brien moved the emphasis of coastal processes research from theorizing to the collection of hard field data.

The University of California at Berkeley became the first U.S. center for coastal studies. This group included M. P. O'Brien, Joe Johnson, Willard Bascom (geologist), John Issacs, and eventually Robert Wiegel and Parker Trask (geologist). Here, H. A. Einstein shared an office with Johnson and thus interacted with the coastal research group.

Coastal engineers dominated applied shoreline research from the early 1930s until the late 1960s with Berkeley serving as the center of expertise. Later, the wave dynamics group at the Massachusetts Institute of Technology became prominent, principally through the efforts of A. T. Ippen, P. S. Eagleson, and D.R.F. Harleman. The University of Florida initiated coastal process and engineering investigations in 1954 under the leadership of Per Bruun.

By the mid-1960s the Coastal Studies Institute at Louisiana State University was the premier center for coastal geomorphological studies based on Office of Naval Research (ONR) Geography Program funding. The three foremost professors were R. J. Russell (now deceased), J. P. Morgan (now retired), and Jim Coleman, the present director. Other centers of importance in applied coastal processes and morphology were developed at oceanographic institutions—Francis Shepard, Walter Munk, and Doug Inman at Scripps and K. O. Emery and John Zeigler at Woods Hole. Throughout the 1970s, the Coastal Research Center at the University of South Carolina under Miles Hayes' direction was a leading center for beach process research, and Hayes' students now head many of the coastal programs along the East and Gulf coasts. Chris Kraft's coastal geomorphology program at the University of Delaware has yielded top students, and hence scientific papers, regarding three-dimensional stratigraphy of coastal areas, particularly barrier islands.

Today there are four principal centers of expertise in coastal engineering in the United States: U.S. Army Coastal Engineering Research Center, Waterways Experiment Station; University of Florida; University of Delaware; and University of California–Berkeley. In addition, there are a number of scattered groups specializing in coastal geomorphology, with the largest assemblage still concentrated at Louisiana State University. Coastal geomorphologists are historically affiliated with geological sciences in the United States, but most frequently with geography departments abroad.

COASTAL PROCESSES AND CONTRIBUTIONS

Introduction

In tracing the development of applied coastal and geological sciences and their role in engineering decision making, one must never forget that without fundamental research, applied research of any importance is impossible. In this review, purely geological studies (e.g., sand grain roundness) that have no engineering implications are not considered. Rather, emphasis is placed on avenues of research that further the understanding of basic scientific principles and relationships and likewise offer practical guidance in the protection and use of the coastal zone.

The streams of research approach provides perspective and shows the advance of concepts through time. A broad listing of principal areas of investigation in coastal processes relevant to engineering geology are: coastal configuration, depth of closure, equilibrium beach profiles, sediment budgets, inlet morphodynamics, barrier island dynamics, and shoreline change measurements.

Coastal configuration

Beach planform was first considered by de la Beche (1832), and Lewis (1938) later elaborated on this work. Their contention was that beaches turn in a direction perpendicular to dominant (storm) wave approach. If the storm waves approach at an oblique angle, then longshore sediment transport will be directed so as to cause the coast to realign itself normal to the wave approach direction. Schou (1945), working on pocket beaches in northern Europe, emphasized the role of resultant winds and the direction of maximum fetch in controlling coastal orientation. His formulation of wind-resultant relationships was most appropriate for the intricate and sheltered beaches around Denmark for which they were developed (King, 1972).

Davies (1959) suggested that dominance of long swells along open coasts helped to explain beach orientation. The amount of refraction that long swells undergo as they pass through shallow water is a major determinant of beach planform configuration. Therefore, offshore relief (nearshore bathymetry) must be a consideration. Bascom (1954) pointed out yet another effect of wave refraction on beach planform in controlling berm height along the shore. Constructive (swell) waves can build up berms, and beach berms are lowest where wave height is low due to spreading of wave energy by refraction. Wave refraction analysis has since been used to determine relative vulnerability along the shore and predict changing island planform morphologies in the vicinity of tidal inlets (Goldsmith and others, 1975).

Yasso (1964) showed that some recurved spits and headland-bound pocket beaches fit logarithmic spiral and half-heart shapes, respectively. Based on his wave tank tests of beach shape caused by constant swell direction, Silvester (1960) concluded that certain geometric forms of beaches may be an indication of equilibrium planform. He further deduced that this natural log-spiral form could be utilized for stabilizing stretches of curved or straight coasts. Artificial headlands, created along the shore by constructing breakwaters, would segment the longshore transport system and cause the shoreline between the hard structures to assume a stable crenulate shape. This design has been successfully

Figure 3. The "Five Sisters" of Winthrop have served to stabilize the shore, but have also resulted in beach tombolos accreting outward in the lee of these offshore structures (photo courtesy of D. FitzGerald).

employed in the U.S. at Winthrop Beach, Massachusetts (Fig. 3) as well as worldwide (Singapore and Japan).

Geometric forms have been recognized in the outline of many beaches worldwide. The planform configuration of certain beaches and spits suggests the net direction of littoral drift, an important consideration in any shoreline plan and project. As mentioned earlier, harbor and port construction must take into consideration coastal processes in order to avoid costly mistakes, as witnessed in the three attempts to establish a harbor at Dublin, Ireland (Bruun, 1954a). Coastal engineers also draw from these studies to design structures and to simulate natural coastal configurations. Finally, recognition of equilibrium planforms can be used to assess future changes in coastal configuration and to assist in stabilizing beaches subject to erosion.

Depth of closure

One of the most critical parameters in designing shore-control engineering structures is defining the depth, and hence distance, offshore of active sediment movement by waves. Breakwater location as well as length of groins and jetties must take this into account for a successful design (Fig. 4).

Depth of closure or wave base has been the subject of discussion and investigation during the last century by coastal geomorphologists and marine geologists. Cornaglia (1887) first considered this subject. He theorized that a "neutral line" existed on the subaqueous beach profile, whence material moves both landward and seaward by wave action. For the Mediterranean Sea, 10 m was cited as the critical depth, although it is not clear how he arrived at this determination, which is in general agreement with the spatially averaged value often cited today. The term wave base was actually introduced by Gulliver (1899), and Fenneman (1902) defined it as "the depth to which wave action ceases to stir the sediments." Through the publication of his classic textbook, Johnson (1919) played a major role in shaping the thinking of researchers for many generations. Unfortunately, he conceived that wave base extended to a depth of as much as 180 m, hence the entire continental shelf was considered wave graded. As Dietz (1963) pointed out, Johnson's lack of reliable field data—apart from two profiles off little known or studied Madagascar—resulted in a confused and largely fallacious theory. In essence, his field "facts" were simple empiricisms since there was no true theoretical or conceptual foundation of understanding. Nevertheless, Johnsonian concepts became the firmly entrenched classical view of coastal processes.

Later, Inman (1949) speculated that depth of closure was related to sediment changes along the offshore profile based on SCUBA sampling of the southern California coast. Sediment grading is still considered the best field indication of the transition between highly active and inactive zones. Dietz and Menard (1951) suggested use of "surf base" for the active zone, shoreward of which there is a 95 percent bottom loss of wave energy. Dietz (1963) showed how wave base was directly related to the profile of equilibrium (see following section). In addition to sediment characteristics along the subaqueous profile, he implied that the seaward limit to appreciable sediment transport could be revealed by nearshore bathymetry—extending to the zone where water-depth contours are parallel to a relatively straight shoreline. Dietz also persuasively argued that the classical "wave base" was too vague and confusing; hence the adoption of profile of closure.

Subsequently, Hallermeier (1981) proved to be another prime mover in this area when he proposed a tripartite partition of the beach profile: (1) littoral zone of constant sand agitation; (2) shoal zone where extreme waves can carry some littoral-zone sand into its landward section; and (3) the offshore zone, where large sand bodies (shoals) are essentially relict. The zone limits and boundaries can be calculated by using the local (significant) wave height and period coupled with sand characteristics. Hallermeier (1983) also provided some practical applications of sand transport limits in shore-control structure designs. For instance, breakwaters intended to provide shelter from wave action for the nearshore to beach sector should be located outside the littoral zone as defined for the site-specific area. Unfortunately, some

Figure 4. Groins emplaced along the Rockaway shoreline, South Shore of Long Island, New York, were poorly designed in their placement and perhaps spacing along the shore and length offshore (1979, photo by S. P. Leatherman).

breakwaters serve as artificial headlands and block the littoral drift; this has resulted in serious downdrift beach erosion (as case of Santa Monica breakwater, California; see Fig. 7). Knowledge of sand transport limits, as well as a delineation of the profile of equilibrium, are paramount considerations in an effective and functional design of coastal structures, such as shore-normal groins and jetties, shore-parallel breakwaters, and offshore oil and waste-water pipelines through the surf zone.

Equilibrium beach profile

Profiles of equilibrium and closure depth, although intimately related subjects, differ because closure depth is considered a position or limiting point along a profile of change. Indeed, Dietz (1963) provided a valid and illuminating comparison between river profiles of equilibrium and their marine equivalent. Such a profile is achieved when transporting power and sedimentation are in balance, as reflected by a dynamic equilibrium between erosion and deposition. For rivers, sea-level position provides the ultimate base level, whereas wave base serves essentially the same role in the marine environment. Clearly, the Johnsonian (1919) view is incorrect, as the continental shelf surface is drowned and totally out of equilibrium with coastal currents. While wave-induced bed motion falls off asymptotically with depth, many other motions (especially tidal) pervade the entire ocean so there is no abrupt and specific depth at which sediment transport suddenly abates. Observations from deep submersible vehicles show sand transport at great depths (1,000s of meters); the polishing action of moving sand grains on the bottom—as photographed by the *Alvin* of the *Titanic* wreckage—a recent, graphic testimony to this process.

Fenneman (1902), a noted physical geographer, is generally credited as the first to explain shoreface slope and shape as an equilibrium response to wave regime. Johnson (1919) proposed the adoption of the Davisian geographic cycle of youth, maturity, and old age in a subaqueous sense to explain shoreface and offshore slope development. Unfortunately, these vague and largely incorrect ideas persisted until Dietz's (1963) classic paper.

The equilibrium profile concept was first addressed quantitatively by Bruun (1954b), in which an empirical equation is given between water depth, h, and distance, x, from the shore: $h = Ax^{2/3}$, where A is a function of sediment (and possibly wave) characteristics. He also suggested that "an equilibrium beach profile is a statistical average profile which maintains its form apart from small fluctuations including seasonal (winter-summer) fluctuations." Dean (1977) utilized a set of 502 beach profiles along the U.S. Atlantic and Gulf coasts to define the equilibrium shape, and demonstrated a goodness-of-fit analysis that closely followed the empirical determination made by Bruun (1954b). This concept was later employed in an analysis of sea-level rise as a cause of shore erosion.

For a given equilibrium beach profile, a rise in sea level would hypothetically result in (1) a shoreward displacement of the beach profile as the upper beach is eroded, (2) offshore movement of the eroded material for deposition on the nearshore bottom, and (3) a rise of the nearshore bottom due to this deposition equal to the rise in sea level, maintaining a constant water depth (Bruun, 1962). The "Bruun Rule" has become one of the principal approaches to the modeling of shoreline change in response to sea-level rise, and a critical consideration in the forecast of an accelerated rise rate due to the greenhouse effect. The Bruun concept represents the balance between eroded and deposited quantities of sediments in an on/offshore direction without consideration of longshore transport. Practically, the Bruun Rule is difficult to confirm or quantify without precise bathymetric surveys and an integration of complex nearshore geologic profiles over a long period of time. Definition of the active profile boundaries also necessitates selection of the depth of closure. Furthermore, attempting an on/offshore sediment balance between a relatively confined zone of erosion (the narrow beach/dune zone) and a broad zone (like the shoreface/inner shelf of thinly spread sediments) is very difficult.

Hands (1976) was the first to provide convincing evidence that the Bruun Rule was valid, although Schwartz (1965) had conducted small-scale wave-tank tests to verify the hypothesis. Hands used Lake Michigan as a "full-scale" natural laboratory to document effects of rising water levels on shore position, obtaining good agreement between predicted and measured shore and nearshore changes. Hallermeier (1981), following Hand's work, showed dependency of profile closure on local wave conditions, thus establishing the seaward termination of the equilibrium profile.

Everts (1978) suggested that a geometric seaward limit exists, defined at the juncture between the upper curved shoreface profile and lower plane ramp profile. This consideration is critical in evaluating disequilibrium since the placement of groins and seawalls are cited as causing shoreface steepening.

Rising water levels establish the potential for shore erosion, but energy is required to adjust the profile to the new equilibrium. This disequilibrium could arise in response to long-term (decades) sea-level rise, or almost instantaneously (days) as the super-elevated surge waters of a major storm. Dean (1983) derived a numerical model to forecast hurricane-induced beach erosion based upon the need to maintain an equilibrium profile, adjusted to a particular water level, which is in response to equal wave energy dissipation per unit of wave run-up.

Sediment budgets

Beaches are in a delicate balance between sand supply and natural forces of erosion. Human intervention in the coastal zone has a long and interesting history. Alexander the Great built a land bridge to invade the island city of Tyre in 332 B.C. This littoral barrier created a large, artificial tombolo at Tyre, a major feature that persists today along the Lebanese coast (Komar, 1976).

Only during the last half-century has intervention with ongoing coastal processes become a worldwide phenomenon of major significance. Dam construction, for example, has resulted in sediment trapping, preventing nourishment of many coasts and beaches. The High Aswan Dam on the Nile River is such an example of interference; the deltaic and downdrift beaches are currently undergoing dramatic and arguably catastrophic erosion. Other practices, such as mining of beach and dune sands, often predate our understanding of littoral sediment budgets and are authorized by historical precedents.

The sediment-budget approach is the tried and tested method of quantifying sources and sinks for a given control volume. However, straight-forward application is hampered by data requirements for annualized hard numbers on such transport pathways as littoral drift, inlet losses, and offshore leakage. A sediment budget assists coastal scientists and engineers to (1) identify relevant processes, (2) estimate volume rates required for design purposes, and (3) single out significant processes for special consideration (USCE, 1984). Application of sediment budgets allows some prediction of what effects man-induced alterations may have on nearshore processes. For example, if a river to the sea is dammed, how much sand will be lost to the beach? Or, will the protection of a cliff being eroded by a seawall cause a major sand depletion of the adjoining beach? Engineering-wise, what is the net littoral drift of sand along a specific beach, and what effect will placement of jetties have on the updrift and downdrift sections? By conducting basic geological science studies, the applied coastal geomorphologists have contributed substantially to the resolution of such questions through definition of the sources and sinks for a sediment budget.

The principal sources of sand to beaches include: rivers, sea-cliff erosion, long-shore transport into an area, and possibly onshore transport, as well as biogenous and hydrogenous deposition. Stream transport of large quantities of sand directly to the ocean is the major source of nearshore sediments for some coastal areas. In southern California, Inman and Frautschy (1966) divided the coast into four discrete sedimentation cells, with rivers being the principal sediment source and submarine canyons (Shepard, 1963) the chief littoral sinks for each cell. By contrast, rivers are an unimportant source of sand for most Atlantic and Gulf Coast beaches, although the Mississippi River is an important exception. The wide, flat coastal plains and shallow estuaries effectively separate the Piedmont rivers from the ocean shore.

Cliffs composed of unconsolidated sediments are a major source of material for the adjacent beaches, particularly during a marine transgression—a rise in sea level with respect to the land surface that results in shore retreat. Eroding cliffs are important not only in maintaining the adjacent sandy beaches, but also in the continuance of the downdrift, accreting sand spits. Kuenen (1950) provided worldwide estimates of the relative contribution of cliff erosion, while Zeigler and others (1959) conducted investigations in a restricted area and defined the rate of cliff retreat along outer Cape Cod, Massachusetts. While there have been several engineering schemes proposed to stabilize the outer Cape shoreline (USCE, 1979), the ongoing quantities of sediment eroded from the cliff face currently make such an undertaking uneconomical.

Onshore transport of sediments eroded from the shoreface and inner shelf is difficult to quantify. However, several experiments conducted by the U.S. Army Corps of Engineers have shown that sandy material dumped in water greater than 6 m deep will not significantly nourish the beach by onshore transport (USCE, 1984).

Longshore sediment transport is responsible for the largest amount of sand movement along a beach. Therefore, defining the net and gross rates of littoral drift is of paramount importance in realistically assessing the sediment budget. Calculating the quantity of sand impoundment on littoral barriers, such as jetties, or computing the longshore sediment transport empirically from statistical wave data are the two major means of estimating sediment rates (Komar, 1976). Natural and artificial tracers can also provide pertinent information, such as the source of beach sediments. By determining the relative proportion of augite to quartz, Trask (1952) demonstrated that a significant proportion of sand filling Santa Barbara harbor travelled from Morro Bay, 160 km to the north. Recently, fourier analysis of grain shape has proved to be a powerful method of delineating source contributions (Ehrlich and Weinberg, 1970).

The fluorescent coating of indigenous quartz grains, colored glass, brick fragments, oolitic grains, and artificially made aluminum pebbles have also been used successfully as tracers (Yasso, 1966; Ingle, 1966). Likewise, the Coastal Engineering Research Center (CERC) in 1966 initiated a radioisotope sand tracing (RIST) program for in situ observations and rapid data collection in the littoral zone (Duane, 1976).

Inlet morphodynamics

Perhaps the most important features of a sandy coastline in terms of littoral processes are estuaries and inlets. Estuaries, such as the Chesapeake and San Francisco Bays, occur at mouths of large rivers. The hydraulic connection to the sea is large, allowing propagation of tidal waves up the estuary. Inlets are small with respect to the interior basin, and tidal currents originate because of hydraulic head differences between the ocean and bay. Along the U.S. barrier island coastlines, which stretch almost continuously from New York to Texas, tidal inlets provide the watercourses between the back-barrier environments and the ocean. Since Colonial days, these hydraulic connections between the open ocean and the calm bay waters have been treacherous to navigate in their natural condition (e.g., Oregon Inlet, North Carolina). For some coastal states, such inlets provide the only access to ports and commercial centers. Inlets are capable of changing their cross-sectional dimensions (depth/width) quickly (during a single storm), migrating rapidly (tens to hundreds of meters per year) along the shore, and even closing completely. To control or alleviate this potential, the coastal specialist requires much basic geological information to understand dynamic watercourses and evaluate possible problems.

Critical to any commercial shipping is maintaining a navigable depth. Whether a particular tidal inlet will require corrective measures to limit the changes due to the interaction of waves, currents, and winds depends on the size and number of ships using the entrance. Corrective control work usually involves dredging a channel, constructing jetties, or a combination of both. O'Brien (1931) was the first investigator to recognize the relationship between tidal prism (volume carried from low to high tide) and cross-sectional inlet area. More recently, Bruun and Gerritsen (1957), Bruun and Lackey (1962), O'Brien (1969), and Jarrett (1976) have refined these principles and provided an improved understanding of inlet hydraulics.

The tidal entrance area expands as the tidal prism increases along sandy coastlines. Inlets tightly confined and controlled by jetties will become deeper, and thereby can cause local jetty and bridge instability. This principle governs the evolution of the barrier island chain along the Louisiana deltaic coast. The substrate is subsiding rapidly, where affected by withdrawal of subsurface gas and fluids, causing the tidal prisms of enclosed bays to become much larger with time. As the tidal inlets are expanding, increasing quantities of sand are emplaced in the flood and ebb-tidal deltas. Consequently, the ongoing set of geologic reactions causes sand starving of the adjacent barrier islands.

Breaching of barrier islands that create new inlets or the stabilization of existing inlets results in impoundment of sand in shoals, thereby reducing sediment supply to adjacent beaches. For example, Dean and Walton (1975) noted that tidal inlets represent the largest sink of beach sand and showed that the most critically eroding areas in the state of Florida are adjacent to tidal inlets. Moreover, mechanical bypassing of sand at inlets would be most beneficial to the local environs as demonstrated by many studies (see section on impact of engineering works).

Inlet opening and long-term persistence, either naturally or by shore-control measures, can exert a significant impact on the adjacent lagoons along a barrier island coast. For instance, Shinnecock Bay along the south shore of Long Island, New York, provided suitable anchorage for hundreds of ocean-going sail ships in Colonial times. A direct and permanent inlet to this small bay was provided by jetty construction in the early 1950s; subsequently a continual influx of littoral materials derived from adjacent beaches has entered the inlet and shoaled the bay. Today, Shinnecock Bay is very shallow, and pleasure boats must carefully navigate the shoaled waters.

Lucke (1934) outlined the progressive stages of lagoonal infilling behind barrier islands when inlets remain stationary and open for hundreds of years. While Lucke's model was based only on an interpretation of maps from the New Jersey coast, the salient points of his geomorphic reasoning have been only recently appreciated by coastal specialists.

Recently, coastal geologists, notably Hayes (1975) and his students, have directed their investigations toward understanding tidal inlets morphodynamically. Before the inspired work of O'Brien (1931, 1969), most engineers had viewed inlets as essentially a problem in open-channel flow (Henderson, 1966), wherein the rigid sides and bottom permitted the application of standard fluid-flow mechanics. Hayes (1975) noted that all inlets are not alike and described the morphologic detail of mesotidal tidal inlets (tidal range of 2 to 4 m) along the Massachusetts coast. Furthermore, wave refraction around the large ebb-tidal deltas cause a local reversal in longshore sediment transport, and this accounts for the paradoxical buildup of sand on the downdrift jetty. Such processes operating in a natural setting result in the formation of bulbous updrift ends of mesotidal barriers in the Georgia bight (Hayes, 1979).

As microtidal inlets (<2 m tidal range) are more dynamic than their mesotidal counterparts, a distinction is important to the practicing coastal engineer. Inlets along the Outer Banks of North Carolina have historically opened and closed throughout the entire shoreline reach. Characteristically, microtidal inlets have been unstable, as documented by the large-scale inlet migration at Fire Island, New York. Moreover, inlet hydraulics is the principal process that promotes barrier island migration through construction of their flood-tidal deltas, and tampering with microtidal inlets through stabilization will eventually interfere with this natural migration process (Leatherman, 1985a).

Another important function served by existing inlets is draining off and drawing down the storm-induced head of water (surge) driven by offshore winds onto the bayside of the barrier island. Early settlers on the Outer Banks of North Carolina often fled to the bayside to escape the hurricane-generated surf, only to drown in the storm surge as the hurricane moved northward and the winds turned offshore, driving the super-elevated waters over the low-lying barrier flats. In fact, most inlets open as outlets during hurricane conditions.

Some mesotidal inlets, with their characteristic ebb-tidal deltas, are associated with a coastal plain stream or small river system, which promotes development of extensive seaward shoals. However, the most important geological factors in ebb-delta growth relevant to shore control are low wave energy and large tidal prism, the means and rate of natural inlet sediment bypassing, and flood versus ebb dominance of channels.

A widely held concept among coastal geomorphologists is that mesotidal-type inlets occupy Pleistocene drainage channels. Specifically, as sea level rose during the Holocene, the existing Pleistocene watercourses were converted to the present-day tidal inlets, and the higher interfluves become the loci for barrier island genesis (Halsey, 1979). As such inlets have remained open and relatively stable during historic times, they are believed to have behaved similarly in geologic time.

The impact of human intervention on a microtidal inlet can result in extensive ebb shoaling. For instance, Ocean City, Maryland, was stabilized by jetties in 1934 and 1935, and today the ebb delta extends several kilometers offshore and contains over 6 million m^3 of beach sand.

Barrier island dynamics

Engineers commonly treated barrier beaches like any other shoreline in their design of shore-control structures. Geological specialists have argued, however, that these abundant East and Gulf Coast features are inherently migratory. Interruption of landward island movement could ultimately lead to in-place drowning, depending upon the degree of interference and time frame for landward migration (Leatherman, 1983a).

Initially, most coastal engineers viewed overwash and inlet dynamics as destructive to the barriers, as these forces severely stress the barrier environment. Yet the plants involved recover rapidly, and new dunes can form within a decade. The long-held impression that barriers can wash away, coupled with severe beach erosion and damage to urbanized islands, led coastal engineers to initiate barrier-island stabilization utilizing rigid groins, jetties, and seawalls or nonrigid beach nourishment and dune building shore-control measures.

More recently, coastal geomorphologists and geological specialists have delineated the natural migratory behavior of barrier islands and offered new concepts on their stability. Many existing barrier islands are known to have originated 7,000 to 8,000 years ago and have migrated intermittently to their present positions with Holocene sea-level rise. Studies by Dillon (1970) and Pierce (1969) showed that the subsurface barrier stratigraphy is the result of inlet, overwash, and eolian deposition, and that these processes control the sediment budget of a landward-migrating barrier island. Although there have been earlier studies of barrier stratigraphy (e.g., Fisk, 1959; LeBlanc and Hodgson, 1959) funded by oil companies, none of the investigators focused on the interrelationship of barrier dynamics and shore-control measures so critical to engineering geology practices.

Research along the Delaware coast has dispelled any further doubt among coastal scientists about barrier migration; numerous cores and borings were used to trace the Holocene marine transgression and barrier migration during the last 7,000 years (Kraft, 1971). The long-term effect of this "rolling-over process" (Dillon, 1970) is continued movement upward and landward in time and space. The antecedent topography and sediment type/erodibility are important considerations in coastal engineering analyses and are discussed by Belknap and Kraft (1985).

Although most coastal specialists accepted this geomorphic cycle and concept by the mid-1970s, many contend such processes occur in time frames that are too long to be considered in the design of shore-control structures. Most coastal engineering projects have life expectancies of 50 to 100 years, while barrier island dynamics involve geologic time (thousands of years). However, this is not necessarily true; some coastal barriers, such as Nauset Spit, Massachusetts, can experience significant "rollover" in as little as three decades (Leatherman, 1983a). Elsewhere, some barriers, such as western Fire Island, New York, indicate little change for over a thousand years, as confirmed by radiocarbon dating of retrieved peat buried in the barrier sediments. Obviously, in planning shore-control measures, each barrier must be evaluated on a site-specific basis using both the historic and geological processes data.

The differences in shoreline processes and forms were considered to be a result of wave and tidal energy by Price (1953), a theory since refined by Hayes (1979). Mesotidal or "drumstick" barriers function differently than their microtidal counterparts, which tend to be more dynamic and subject to overwash and inlet breaching. Mesotidal inlets with greater tide ranges are more adjustable to high tides during storm surge conditions than microtidal barriers; the implications for urbanization should be clear. The U.S. East and Gulf Coasts are primarily microtidal except for parts of New England and the South Carolina and Georgia coasts. Microtidal barrier islands unfortunately have hosted the major urbanization, such as Atlantic City, New Jersey, Miami Beach, Florida, and Galveston Island, Texas.

Principles of barrier morphodynamics can be applied to ascertain the varying responses to potentially disruptive alterations involved in offshore oil and gas exploration. The importance of critical barrier width in island migration was first described by Leatherman (1979), based on historical shoreline analysis of Assateague Island, Maryland. Specifically, barriers narrow to some critical width before island migration (rollover) takes place by overwash and inlet processes. For example, in the 200- to 300-m-width range, sufficient sand was transported across Assateague so that bayside deposition equalled seaside erosion, and true barrier migration occurred. This principle has since been applied throughout the Louisiana coastal region to locate pipeline canals and to implement barrier-restoration projects; a specified barrier island width should be maintained to prevent sediment loss by overwash and encourage dune growth.

A nationwide assessment of barrier island dynamics determined that inlets play the major role in landward barrier migration (Leatherman, 1985b). Washover and dune-building processes are largely responsible for vertical island accretion. Locally, such as at South Padre Island, Texas, eolian processes can serve as the major transport mechanism. In this semiarid region of strong onshore winds, hurricane washovers become pathways of landward movement of the fine beach sand until barrier dune recovery. The large dunes that are created in this manner (Mathewson and others, 1975) migrate across the barrier surface to the adjacent bay waters of Laguna Madre, a sea-to-bayside transport of sand.

Shoreline change measurements

Changes occurring in the configuration of shorelines have always been of special interest to applied geoscientists and coastal geomorphologists, as reviewed by Shepard and Wanless (1971) and El-Ashry (1977). Study of early explorer charts can provide much information about the past history and evolution of a dynamic coast. For example, comparison of Champlain's 1605 map of Nauset Harbor with modern maps indicates the loss of an anchoring glacial island and a major reorientation of the spit (Leatherman, 1982). Such maps provide a gross representation of past conditions, rather than precise data for calculation of historical shoreline movement rates.

Techniques of map making that can be used to obtain quantitative shoreline movement data have changed considerably over the past 150 years. The earliest maps of sufficient accuracy were derived from careful ground measurements. The U.S. Coast and Geodetic Survey relied upon survey teams with alidade and planetable to obtain ground data to construct shoreline maps (Shalowitz, 1964). Although triangulation techniques and new instruments improved the speed and accuracy of map making, it was time consuming. In recent decades, modern photogrammetric techniques permit cost-effective and accurate map making of coastlines and shore features from aerial photography.

Accurate data on historical shoreline changes are prerequisite to understanding the extent (magnitude) of present beach erosion, as well as for forecasting future changes relative to estimating sediment transport rates, particularly littoral drift, and predicting the effects of a proposed shore-control engineering structure. In some states, historical shoreline changes are used to establish building-setback lines.

The beach serves as a recreational area and provides a buffer between the high-energy surf and shorefront buildings. Therefore, much attention has been placed on understanding beach dynamics and determining both the short-term changes (winter-summer and storm induced) and long-term trends. Historical shoreline change maps have been completed for much of the U.S. coastline, but this information has large differences in accuracy, a fact too frequently overlooked by land planners, applied geoscientists, and engineers.

A number of techniques are currently employed to acquire information on historical shoreline changes (Leatherman, 1983b). The simplest approach utilizes enlarged air photos with shoreline changes (waterline, dune-line, first line of stable vegetation, etc.) traced onto a base map. Although easy to execute and inexpensive, the quality is suspect, but adequate if only generalized trends are desired.

Stafford (1971) developed a procedure for using historical aerial photographs to quantify coastal erosion along the Outer Banks of North Carolina. His studies demonstrated that mean high water was the best indicator of the land-water contact, and all subsequent coastal mapping projects have used this interface as the shoreline marker for comparative purposes. Compensation must be made for seasonal and storm-induced beach changes in order to distinguish between long-term natural fluctuations and short-term artificial manipulations (e.g., beach nourishment).

A recent technique for shoreline mapping uses the zoom transfer scope (ZTS), considered by many coastal specialists to represent state-of-the-art mapping. While the ZTS is very efficient in removing scale differences between photographs, error introduced by tilt cannot be fully eliminated. The method most often employed by professional photogrammetrists to construct maps is the use of stereoplotters. This instrument can correct for lens and atmospheric distortions as well as irregularities in airplane flight (tilt, tip, and yaw). Although very accurate, this method requires sophisticated and expensive equipment, is labor intensive, and best utilized for constructing topographic maps.

A new automated technique of shoreline mapping (Metric Mapping) employs computer routines to emulate photogrammetric techniques (Leatherman, 1983b). The high-quality geometric-image maps produced meet or exceed National Map Accuracy Standards. While previous mapping procedures have relied upon aerial photographs as sources, this computerized technique allows compilation of shoreline data from all accurate information, and maps are drawn by a computer-driven plotter.

The earliest United States maps that can be used for quantitative shoreline changes are the National Ocean Survey (NOS) "T" sheets (formerly U.S. Coast Survey, subsequently U.S. Coast and Geodetic Survey). NOS "T" sheets in some coastal areas date from the 1830s, and are commonly available for most East and Gulf Coast areas from the 1860s. Incorporation of shoreline data from both air photos and NOS "T" sheets extends the record to 150 years in some cases; the longer and more complete the historical record, the more reliable the assessment of long-term trends.

The first national appraisal of shore erosion problems was the National Shoreline Study conducted by the U.S. Army Corps of Engineers (USCE, 1971). Of the 134,400 km of United States ocean and Great Lakes shorelines, significant erosion occurs along 32,800 km or 25 percent of the total. Excluding Alaska, it can be shown that 43 percent of the shoreline is undergoing "significant" erosion. A large portion of the shoreline is categorized by the U.S. Army Corps of Engineers (USCE, 1971) as noncritical, meaning that the erosion problem is amenable to land-use controls and other management techniques rather than relying upon shore-control measures to halt erosion.

Historical aerial photographs were used by May and others (1983) to define shoreline changes. The national average (unweighted) erosion rate is 0.4 m/yr; throughout the Atlantic coast the average rate of recession is 0.8 m/yr, while the Gulf coast is 1.8 m/yr, and the Pacific coast is stable on average. Even though average rates are virtually meaningless due to the complexity of coastal landforms—ranging from "non-erodible" crystalline rock to ephermeral sand spits—the preponderance of shore retreat indicates the magnitude of a national geologic process and hazard. In addition, historical tide gauge records (Hicks and others, 1983) indicate a rise in relative sea levels along the U.S. Atlantic and Gulf Coasts, which correlates with the persistent historical trend in shore retreat. The tectonically active Pacific and Alaskan coasts have shown varying trends of shore change. For example, tide-gauge records document a rapidly falling relative sea level for parts of the Alaskan coast.

The National Ocean Service of NOAA has also prepared shoreline change maps for portions of the U.S. continental coasts. The Delaware, Maryland, Virginia, North–South Carolina region, and parts of the California coasts have been mapped, and the data indicate a general pattern of pervasive shore retreat except for local anomalies (D. Carroll, 1985, personal communication). Shoreline mapping on a state by state basis is being continued by the University of Maryland team utilizing metric mapping (Leatherman, 1983b), and to date the south shore of Long Island, New York, Massachusetts, and New Jersey have been completed.

IMPACT OF ENGINEERING WORKS

This section summarizes several important cases of shore-control structures that have met with varying degrees of success. Other such cases are given by Krumbein (1950), Mason (1950), and Rosenbaum (1976). A basic premise or dictum of engineers is, "Do something concrete." This tendency is clearly reflected in the omnipresent Victorian-age vertical seawalls along the English coast. This stabilizing measure is often achieved at the expense of the above-water beach; today 25 percent of British beaches are in this condition, and the legacy continues. In defense of coastal engineers, such major structures save lives and valuable buildings, as graphically illustrated by the Galveston, Texas, sea wall.

A significant problem of shoreline structures is that severe erosion downdrift of such works prompts construction of additional structures in an attempt to stop that erosion (Rosenbaum, 1976). The history of shoreline works in Milwaukee, Wisconsin, along Lake Michigan typifies the way in which shoreline structures propagate downdrift. A harbor breakwater, built in stages from 1881 to 1929, had caused severe enough downdrift erosion that in 1916 the city of Milwaukee had to construct a second breakwater (Fig. 5). Over time the breakwater was extended downdrift, but it was found that the point of greatest erosion on the shore kept pace. Eventually, groins had to be emplaced even further downdrift to arrest the accelerated beach and bluff erosion. These third-generation structures caused additional erosion downdrift, and so on along the shore.

Seawalls, bulkheads, revetments

The city of Galveston is located on the northeast end of a long, low microtidal barrier island. Due to its location and deep-water anchorage along the bayside, Galveston became the most prosperous and important city in Texas by the late 1800s. Unfortunately, the primary sand dunes along the Gulf front were leveled to permit access to the beach and to serve as fill elsewhere. On September 8, 1900, Galveston was demolished by a major hurricane with an estimated 4.6-m storm tide. Six thousand people were killed, and the buildings were reduced to rubble (Fig. 1). Just before the devastating hurricane, a storm protection plan had been developed, but financial difficulties and public apathy left the plans on the drafting board. To protect the city, a 4.9-m-high seawall with a curved frontal face was constructed in 1904 in conjunction with increasing the land elevation through placement

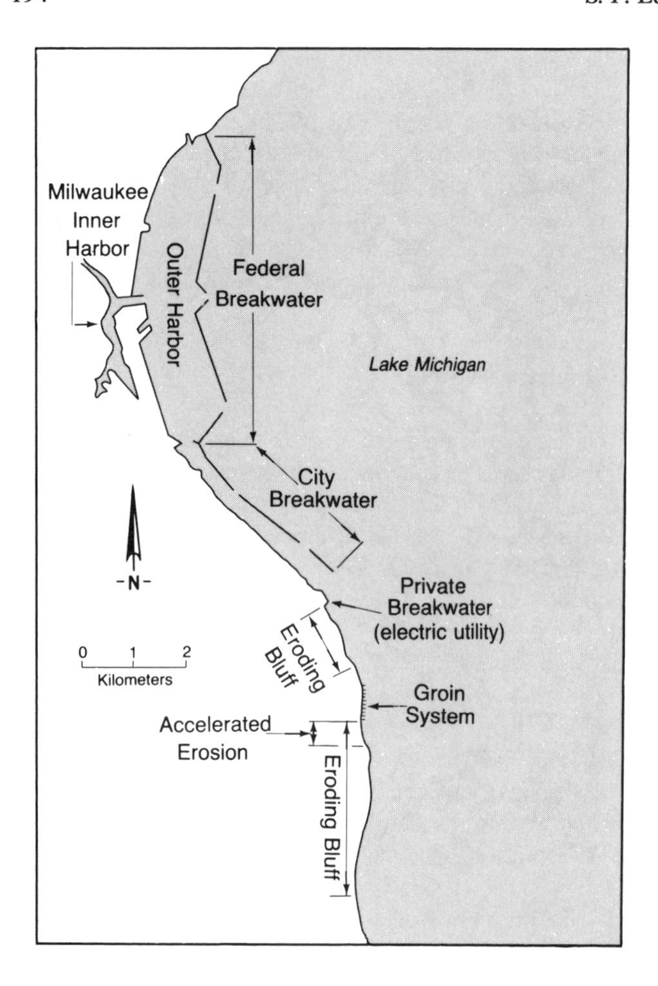

Figure 5. Lake Michigan shoreline structures at Milwaukee, Wisconsin. Federal breakwater was built in stages from 1881 to 1929; city breakwater was constructed in stages from 1916 to 1930; private breakwater emplaced in 1920; and groins constructed in 1933 to 1934 (from Rosenbaum, 1976).

Figure 6. Galveston seawall still stands today, but the recreational beach has been lost (1980, photo by S. P. Leatherman).

of dredged material; later, extensions were added to protect larger portions of the island (Davis, 1961).

In 1915 the seawall was tested by a second hurricane with similar storm surge, but only 12 deaths were recorded, and there was less property damage than caused by the 1900 hurricane. The wall protected the landward-flanking buildings, but the large waves scoured sand from the foot of the seawall. By 1934 the shoreline generally coincided with the toe of riprap along the seawall face. While the seawall has functioned well to protect the people, the recreational beach was lost (Fig. 6). Since that time, additional riprap and groins were emplaced to protect the seawall toe and prevent failure by undermining during a severe storm. While the seawall seems secure today, there has been no major accumulation of sand to produce a recreational beach.

By contrast, there is still a small beach in front of the O'Shaughnessay seawall, constructed in 1929 along Ocean Beach, San Francisco, California. This high-energy, open-coast beach probably receives sediment from onshore movement of sand from the crescent-shaped offshore bar, formed in response to tidal currents ebbing from San Francisco Bay.

Breakwaters

The Santa Barbara, California, breakwater, constructed in 1927 to 1928 to provide safe anchorage for recreational boats, is considered a classic study in the effects of engineering structures on adjacent beaches. The breakwater interrupted littoral drift, moving generally southward along the coast, and caused considerable deposition on its updrift side and resulting erosion of the beaches downdrift (Johnson, 1957; Wiegel, 1959). The tombolo that built from the updrift beach to meet the offshore breakwater then formed a spit of sand that threatened to seal the harbor entrance.

To prevent closure of the harbor, sand has been continuously dredged from the harbor and dumped on the downdrift beaches. This bypassing alleviated the harbor sedimentation problem as well as benefitted the severely eroding beaches downdrift, where houses were toppling into the surf for reasons not understood at that time. This example, and that of detached breakwater construction in 1933 at Santa Monica (Fig. 7), have provided guidance and served to emphasize that serious beach erosion can be caused by interruption of the natural littoral drift. Offshore breakwaters elsewhere, such as the detached segments at Winthrop Beach in Boston Harbor, Massachusetts, seem to have functioned reasonably well. In fact, the Santa Monica breakwater did not work correctly until it was subsequently damaged in a 1950s storm to allow some wave energy to pass through and prevent the building of a tombolo (Wiegel, 1964).

Figure 7. Accretion of beach sand behind the Santa Monica breakwater since construction in 1933 has back-filled a large part of the protected harbor. Design of the breakwater did not consider the importance of sand carried by longshore current and the severe erosion of downdrift beaches that occurred. Aerial view of Santa Monica Bay, California, June 22, 1947 by Fairchild Aerial Surveys, Incorporated (photo from Kiersch, 1955, Fig. 9).

Figure 8. Ocean City Inlet jetties have caused a major offset along the Maryland coast, resulting in updrift accretion as a sedimentary wedge of beach sand at Ocean City and rapid beach erosion and island migration of northern Assateague Island (photo by S. P. Leatherman, 1975).

Jetties

Jetties constructed at the entrances to tidal inlets also block longshore sediment transport by direct entrapment and diversion offshore into a deep-water depositional sink. One such jettied inlet, Lake Worth Inlet near Palm Beach, Florida, has had minimal impacts on the adjacent shores because of mechanical sand bypassing. Although this project was well designed in its day (1930s), much has been learned since about the impact of jetties on the environs. Unfortunately these principles have not always been applied in designing coastal engineering projects.

Ocean City Inlet in Maryland was formed by bayside breaching as a severe hurricane in August 1933 moved northward. Jetties were constructed by the Corps of Engineers in 1934 and 1935 to stabilize the channel and keep it navigable for recreational boats. As the updrift shoreline accreted, Ocean City fishing pier had to be lengthened twice, and the zone behind the north jetty was filled to capacity within a decade. Since this time the sand, transported along the shore at a net rate of 114,000 m^3/yr (Dean and others, 1978), has formed a huge, ebb-tidal delta of 6,000,000 m^3 (Fig. 8).

The jetties have completely blocked the sediment moving southward along the coast so that northern Assateague Island has been sand starved, rapidly retreating landward with an average erosion rate of 11 m/yr (Leatherman, 1979). The island narrowed to the critical width to sustain barrier rollover, which occurs in a punctuated fashion during winter northeasters. At the historical rate of landward barrier migration, the northern end of the island, which has already migrated more than its width into the adjacent bay since inlet stabilization, will collide with the mainland by the year 2000. This will result in a 3-km wide breach in the barrier island continuity along the Maryland shore. Although the problem has now been recognized by the National Park Service, which administers Assateague Island National Seashore, current budgetary constraints will probably dictate no action. Furthermore, the Corps of Engineers and Ocean City are not expected to install a sand-bypassing system, which is critically needed.

Some jetties have been designed on insufficient information or on erroneous analysis of the baseline data. An example is a weir jetty at the east pass of Choctawhatchee Bay along the panhandle of Florida; this early design allowed sediment to overtop a low section of the updrift jetty. The bypassed sand could be efficiently dredged from the protected waters as needed and pumped across the inlet to the downdrift beaches. Unfortunately, the weir was placed in the wrong jetty at Choctawhatchee Bay; the design engineers did not realize that the net, long-term direction of longshore sediment transport was actually in the opposite direction. Apparently, only the short-term wave data had been used to predict littoral drift, and anomalous and insufficient baseline geological information resulted in an incorrect design and failure of the engineering works.

Groins

The most common and important engineering structure to trap littoral sediments, groins have received mixed acceptance relevant to their effectiveness and efficiency as shore-protective structures. The Westhampton Beach groin field along the south shore of Long Island, New York, is such a case. The 15 groins emplaced in the 1960s along Westhampton Beach have retained a considerable amount of sand fill, yet there has been rapid erosion just downdrift of the last groin. Furthermore, the sand depletion downdrift has been so severe that the integrity of Moriches Inlet was threatened. In 1980 a winter northeaster opened an inlet adjacent to this stabilized inlet. Closure of this new breach required a massive influx of sand that was pumped from the adjacent flood tidal delta and trucked from the mainland; the total cost to the Corps of Engineers of this minor repair was $11 million in emergency funds. In addition, a number of beach houses have been lost during storm conditions, and the 3-km section downdrift of the groins to Moriches Inlet is overwashed annually and remains vulnerable to future breaching. Concerned inhabitants point to the problems created by the Westhampton Beach groin field. On the other hand, coastal specialists and engineers consider the shortcomings to be an implementation problem and not the design. Actually, only half of the planned groins were built, and then no sand fill was placed as required in the original plans. This case highlights the kind of politics involved with shoreline erosion and providing adequate protection.

Beach nourishment

The impacts of rigid coastal engineering structures on adjacent and downdrift beaches, as well as aesthetic considerations, have been cited by many as a reason to opt for soft approaches to shore erosion, namely beach nourishment and dune building. The first possible case of beach nourishment occurred during the reign of Cleopatra (30 to 40 B.C.). She reportedly insisted that sand be brought by vessel from Egypt for placement on the Turkish shore so that she would not have to step on foreign soil. More serious needs for beach fill have arisen along many shores, and numerous engineering projects have been undertaken in recent decades.

An early beach restoration program was conducted in 1951 to 1952 at Harrison County, Mississippi (USCE, 1984). Offshore material was placed along some 42 km (26 mi) of Mississippi Sound shoreline to build a beach with a width of 61 m (200 ft). This project is generally considered to have performed well, having weathered some heavy storm activity. Hurricane Camille in 1969 resulted in significant sand losses, and the area was renourished in 1972. This man-made beach has provided both protection to the upland during hurricanes and has also served as an important recreational area. The longevity of this project can perhaps be attributed to its sheltered location.

One of the largest beach nourishment projects ever undertaken was completed in 1980 by the U.S. Army Corps of Engineers at Miami Beach, Florida. More than 10 million m^3 of sand were pumped from several kilometers offshore for placement along a 17-km section of beach, at a total cost of $65 million. The resulting beach was more than 61 m wider than before project initiation, when the waves lapped against the seawalls of many famed hotels (Fig. 9). International tourism had dropped dramatically with the loss of a recreational beach, and the Miami Beach mayor proclaimed the area to be a waterfront slum. With the high value per mile of the grand hotels, the benefit-cost ratio of the beach fill as storm protection and for recreational pursuits was overwhelmingly favorable.

The nourished beaches have withstood moderate hurricane activity (e.g., Hurricane David in 1979), and the successful restoration project has met the needs of the participating coastal cities (Miami Beach, Surfside, and Bal Harbour). However, how many more such areas (e.g., Atlantic City, New Jersey, and Ocean City, Maryland) will require continued nourishment in the future? Consequently, what burden should the federal and/or state taxpayer bear in financing these expensive, site-specific coastal restoration projects?

OUTLOOK AND RESEARCH

Although it is often difficult to forecast the future, the direction of applied geosciences for coastal engineering works seems fairly certain. A major lesson learned from a century of geomorphic research and coastal engineering experience is that rigid engineering structures are expensive to construct and maintain and often have detrimental effects. Therefore, there will be greater reliance on soft engineering approaches worldwide (i.e., beach nourishment), particularly along the U.S. East and Gulf Coast barrier shorelines. This will require continued advancements in understanding physical processes and resulting landforms.

Major engineering works (e.g., Galveston seawall) will still be maintained since structural failure would be catastrophic. Today Galveston is a thriving major city with high-rise centers. The same is true for other rigid engineering structures (e.g., repair is presently underway at Sea Bright, New Jersey, to fortify old coastal structures such as rubble seawalls and groins). This illustrates the cost of the struggle against the sea. In fact, it could be argued that once a major, rigid engineering device has been emplaced, a commitment has also been made to hold the line in perpetuity. It seems that no such engineering structures have ever been abandoned or removed along the U.S. coast, although some have clearly failed (e.g., groin field at Cape May is now separated from the eroded beach). By the same token, it is doubtful that any new, major engineering structures (especially seawalls) will be constructed in the future. Indeed, several states (e.g., North Carolina and Maine) have disallowed the further use of rigid shoreline engineering devices.

Figure 9. Aerial view of Miami Beach, Florida, during beach fill by the onshore pumping of offshore coralline sands (photo courtesy of U.S. Army Corps of Engineers, 1979).

Beach nourishment, and dune building where practical, will be the principal approaches to engineering beaches in the future. The cost, however, will become much higher with increased shore erosion caused by accelerated sea level rise due to the "greenhouse effect." Many of the recreational beaches in Florida are already nourished periodically (with annual expenditures of more than $15 million), and this practice is gaining in popularity elsewhere. Atlantic City, New Jersey is nourished at least every decade, and a $30 million beach replenishment project is scheduled to begin at Ocean City, Maryland, in 1988. The Corps of Engineers' stabilization plans for the south shore of Long Island, New York, will probably never be realized as originally conceived in the 1960s. Public support for further groin construction is not forthcoming, and the usefulness of sand dunes as storm surge dikes is limited. Any actual work to be conducted in this long-authorized project will probably only entail beach fill.

The concept of building sand dunes as dikes along the U.S. outer barrier island coast to serve as a "Maginot-type" line has been discredited by applied coastal geomorphologists, and the idea has been abandoned by most coastal engineers. Elsewhere, millions of Dutch people live and work below sea level, and more than half of the country would be beneath the sea without dikes. It has been said that God created the Earth, but the Dutch made Holland! Clearly the Dutch are the dike-building masters, and their struggle with the sea spans ten centuries.

A major exception to the reliance on soft engineering approaches involves the construction of storm-surge barriers across major rivers and estuaries. The British have recently completed a movable barrier across the River Thames. When open, the gates allow minimal interruption of the flow and easy passage of ships, but they can be quickly closed to prevent a storm surge from propagating upstream to flood London. Significant loss of life and property occurred around the North Sea in the 1953 storm, and a relative rise in sea level of nearly a third of a meter has taken place in the past century, compounding the problem.

The Dutch followed suit, enabling the Haringvliet estuary to be closed off during storms by massive sluice gates, thus completing the protection of Holland from flooding. This is the world's largest coastal engineering project, with a $2 billion cost for the surge barriers alone. It also was prompted by the 1953 storm, which killed 1,800 and flooded about 8 percent of the country.

Research opportunities for applied coastal geomorphologists

range from three-dimensional geomorphic analyses to process studies. Research must be undertaken to determine the sedimentary dynamics of the nearshore zone. This is an important area of study since the subaerial beach represents only about 10 percent of the active profile. Also, accelerated sea-level rise due to the "greenhouse effect" means that increased rates of beach erosion can be anticipated. This should focus research attention on the equilibrium profile and result in much reflection on the applicability of the Bruun Rule in different geological environs.

Accurate, long-term shoreline change data are required for quantitative sediment analyses, which are prerequisite to the design of any coastal engineering works. With the data available, all coastal states, following the lead of North Carolina and Florida, will probably adopt building setback restrictions. Likewise, FEMA through the Federal Insurance Administration is also expected to require that shore erosion data be coupled with flood levels as basis for establishment of actuarial rates.

In addition to understanding the physical processes and morphologies in natural settings, applied coastal geomorphologists must also concern themselves with the developed beaches. The role of coastal engineering structures in promoting beach erosion must be more thoroughly studied, especially in areas of wave reflection from seawalls and offshore losses due to long groins. Minimal work has been done, apart from the preliminary studies by Herbich (1969) and Everts (1979), and beach erosion relevant to engineering works is a significant area of future research. Despite the considerable advancements made to date, much remains to be done, and future progress will be fostered by a closer interaction of applied coastal geoscientists and engineers.

REFERENCES CITED

Bascom, W., 1954, The control of stream outlets by wave refraction: Journal of Geology, v. 62, p. 600–605.

—— , 1980, Waves and beaches: Garden City, New York, Anchor Books, 366 p.

Belknap, D. F., and Kraft, J. C., 1985, Influence of antecedent geology on stratigraphic potential and evolution of Delaware's barrier systems: Marine Geology, v. 63, p. 235–262.

Bruun, P., 1954a, Coast stability: Copenhagen, Danish Technical Press, 400 p.

—— , 1954b, Coastal erosion and development of beach profiles: U.S. Beach Erosion Board Technical Memorandum 44, 88 p.

—— , 1962, Sea level rise as a cause of shore erosion: American Society of Civil Engineers Journal of Waterways and Harbors Division, p. 117–130.

Bruun, P., and Gerritsen, F., 1957, Natural bypassing of sand at coastal inlets: American Society of Civil Engineers Journal of Waterways and Harbors Division, v. 85, p. 75–107.

Bruun, P., and Lackey, J. B., 1962, Engineering aspect of sediment transport, *in* Varnes, D. J., and Kiersch, G. A., eds., Reviews in engineering geology; Geological Society of America, v. 2, p. 39–103.

Cornaglia, P., 1887, Sul regime della spiagge e sulla regulazione dei porti: Turin, p. 187.

Davies, J. L., 1959, Wave refraction and the evolution of shoreline curves: Geographic studies, v. 5, p. 1–14.

Davis, A. B., Jr., 1961, The Galveston sea wall: Shore and Beach, v. 29, p. 6–13.

Davis, W. M., 1912, Die beschreibende Erklarung der landformen: Berlin, Leipzig, 565 p.

Dean, R. G., 1977, Equilibrium beach profiles, United States Atlantic and Gulf coasts: Newark, University of Delaware, Oceanographic Engineering Technical Report 12, 45 p.

—— , 1983, Shoreline erosion due to extreme storms and sea-level rise: Gainesville, University of Florida, Department of Coastal and Oceanographic Engineering, 58 p.

Dean, R. G., and Walton, T. L., Jr., 1975, Sediment transport processes in the vicinity of inlets with special reference to sand trapping: New York, Academic Press, Estuarine Research, v. 2, p. 129–149.

Dean, R. G., Perlin, M., and Dally, B., 1978, A coastal engineering study of shoaling at Ocean City Inlet: Newark, University of Delaware, Department of Civil Engineering, 135 p.

de la Beche, H. T., 1832, A geological manual: London, 535 p.

Dietz, R. S., 1963, Wave base, marine profile of equilibrium, and wave-built terraces; A critical appraisal: Geological Society of America Bulletin, v. 74, p. 971–990.

Dietz, R. S., and Menard, H., 1951, Origin of abrupt change in slope at the continental shelf margin: American Association of Petroleum Geologists Bulletin, v. 35, p. 1994–2016.

Dillon, W. P., 1970, Submergence effects on a Rhode Island barrier and lagoon and inferences on migration of barriers: Journal of Geology, v. 78, p. 94–106.

Duane, D. B., 1976, Sedimentation and coastal engineering; Beaches and harbors, *in* Stanley, D. J., and Swift, D.J.P., eds., Marine sediment transport and environmental management: New York, John Wiley, p. 493–517.

Ehrlich, R., and Weinberg, B., 1970, An exact method for characterization of grain shape: Journal of Sedimentary Petrology, v. 40, p. 205–212.

El-Ashry, M. T., ed., 1977, Air photography and coastal problems: Stroudsburg, Pennsylvania, Benchmark Papers in Geology, v. 38, 425 p.

Everts, C. H., 1978, Geometry of profiles across inner continental shelves of the Atlantic and Gulf coasts of the United States: U.S. Army Corps of Engineers Coastal Engineering Research Center Technical Report 78-4, 92 p.

—— , 1979, Beach behavior in the vicinity of groins; Two New Jersey field examples, *in* Proceedings of Coastal Structures: American Society of Civil Engineers, p. 853–867.

Fenneman, N. M., 1902, Development of the profile of equilibrium of the subaqueous shore terrace: Journal of Geology, v. 10, p. 1–32.

Fisk, H. N., 1959, Padre Island and the Laguna Madre flats, coastal south Texas, *in* Proceedings of the 2nd Coastal Geography Conference: Baton Rouge, Louisiana State University, p. 103–152.

Goldsmith, V., Byrne, R., Sallenger, A., and Drucker, D., 1975, The influence of waves on the origin and development of the offset coastal inlets of the southern Delmarva peninsula, Virginia, *in* Cronin, L. E., ed., Estuarine research: New York, Academic Press, p. 183–200.

Gulliver, F., 1899, Shoreline topography: American Academy of Arts and Sciences, v. 34, p. 151–258.

Hallermeier, R. J., 1981, A profile zonation for seasonal sand beaches from wave climate: Coastal Engineering, v. 4, p. 253–277.

—— , 1983, Sand transport limits in coastal structure designs, *in* Proceedings of Coastal Structures 83: American Society of Civil Engineers, p. 703–716.

Halsey, S. D., 1979, Nexus; New model of barrier island development, *in* Leatherman, S. P., ed., Barrier islands: New York, Academic Press, p. 185–210.

Hands, E. B., 1976, Observations of barred coastal profiles under the influence of rising water levels, eastern Lake Michigan, 1967–1971: U.S. Army Corps of Engineers Coastal Engineering Research Center Technical Report 76-1, 116 p.

Hayes, M. O., 1975, Morphology of sand accumulation in estuaries: Proceedings, 2nd International Estuarine Research Federation Conference, Myrtle Beach, South Carolina, p. 3–22.

—— , 1979, Barrier island morphology as a function of tidal and wave regime, *in* Leatherman, S. P., ed., Barrier islands: New York, Academic Press, p. 1–27.

Henderson, F. M., 1966, Open channel flow: New York, MacMillan Company,

522 p.
Herbich, J. B., 1969, Beach scour at seawalls and natural barriers: College Station, Texas A & M University, Coastal and Ocean Engineering Division Report, 23 p.
Hicks, S. D., Debaugh, H. A., Jr., and Hickman, L. E., Jr., 1983, Sea level variations for the United States, 1855–1980: Rockville, Maryland, National Oceanic and Atmospheric Administration, 170 p.
Hunt, C. B., 1950, Military geology, *in* Paige, S., chairman, Applications of geology in engineering practice: Geological Society of America, Berkey Volume, p. 295–327.
Ingle, J. C., 1966, The movement of beach sand, *in* Developments in sedimentology 5: Amsterdam, Elsevier, p. 86–100.
Inman, D. L., 1949, Sorting of sediments in light of fluid mechanics: Journal of Sedimentary Petrology, v. 19, p. 51–60.
Inman, D. L., and Frautschy, J. D., 1966, Littoral processes and the development of shoreline, *in* Proceedings of Coastal Engineering Specialty Conference: American Society of Civil Engineers, p. 511–536.
Jarrett, J. T., 1976, Tidal prism-inlet area relationships: Vicksburg, Mississippi, U.S. Army Waterways Experiment Station General Inlet Tidal Investigations Report 3, 55 p.
Johnson, D. W., 1919, Shore processes and shoreline development: New York, John Wiley, 584 p.
Johnson, J. W., 1957, The littoral drift problem at shoreline harbors: American Society of Civil Engineers, Journal of Waterways and Harbors Division, p. 1–37.
Kiersch, G., 1955, Engineering geology: Quarterly Journal of the Colorado School of Mines, v. 50, no. 3, p. 81.
King, C.A.M., 1972, Beaches and coasts: New York, St. Martins Press, 570 p.
Komar, P. D., 1976, Beach processes and sedimentation: Englewood Cliffs, New Jersey, Prentice-Hall, 429 p.
Kraft, J. C., 1971, Sedimentary facies patterns and geologic history of a Holocene marine transgression: Geological Society of America Bulletin, v. 82, p. 2131–2158.
Krumbein, W. C., 1950, Geological aspects of beach engineering, *in* Paige, S., chairman, Applications of geology in engineering practice: Geological Society of America, Berkey Volume, p. 195–223.
Kuenen, P. H., 1950, Marine geology: New York, John Wiley, 568 p.
Leatherman, S. P., 1979, Migration of Assateague Island, Maryland, by inlet and overwash processes: Geology, v. 7, p. 104–107.
—— , 1982, Barrier island handbook: College Park, University of Maryland, 109 p.
—— , 1983a, Barrier dynamics and landward migration with Holocene sea-level rise: Nature, v. 301, p 415–418.
—— , 1983b, Shoreline mapping; A comparison of techniques: Shore and Beach, v. 51, p. 28–33.
—— , 1985a, Geomorphic and stratigraphic analysis of Fire Island, New York: Marine Geology, v. 63, p. 173–195.
—— , 1985b, Barrier island migration; An annotated bibliography: Monticello, Illinois, Vance Bibliographies, 54 p.
Le Blanc, R. J., and Hodgson, W. D., 1959, Origin and development of the Texas shoreline: Gulf Coast Association of Geological Societies Transactions, v. 9, p. 197–220.
Lewis, W. V., 1938, Evolution of shoreline curves: Geologists' Association of London Proceedings, v. 49, p 107–127.
Lucke, J. B., 1934, A study of Barnegat Inlet: Shore and Beach, v. 2, p. 45–94.
May, S. K., Dolan, R., and Hayden, B. P., 1983, Erosion of United States shorelines: EOS Transactions of the American Geophysical Union, v. 64, p. 551–552.
Mason, M. A., 1950, Geology in shore-control problems, *in* Trask, P. D., ed., Applied sedimentation: New York, John Wiley, p. 276–290.
Mathewson, C., Clary, J., and Stinson, J., 1975, Dynamic physical processes on a south Texas barrier island; Impact on construction and maintenance: International Electrical and Electronics Engineers Ocean 75, p. 327–330.
O'Brien, M. P., 1931, Estuary tidal prisms related to entrance areas: Civil Engineering, v. 1, p. 738–739.
—— , 1969, Equilibrium flow areas of inlets on sandy coasts: American Society of Civil Engineers Journal of Waterways and Harbors Division, p. 43–52.
Pierce, J. W., 1969, Sediment budget along a barrier island chain: Sedimentary Geology, v. 3, p. 5–16.
Price, W. A., 1953, Shorelines and coasts of the Gulf of Mexico: College Station, Texas A & M University Contribution in Oceanography and Meteorology, v. 1, p. 191–217.
Quinn, M. L., 1977, The history of the Beach Erosion Board: U.S. Army Corps of Engineers, 1930–1963, U.S. Army Coastal Engineering Research Center Miscellaneous Report 77-9, 181 p.
Rosenbaum, J. G., 1976, Shoreline structures as a cause of shoreline erosion; A review (2nd edition), *in* Tank, R., ed., Focus on environmental geology: New York, Oxford University Press, p. 166–179.
Schou, A., 1945, Det marine forland: Folia Geographica Danica, v. 4, 236 p.
Schwartz, M. L., 1965, Laboratory study of sea level rise as a cause of shore erosion: Journal of Geology, v. 73, p. 528–534.
Shalowitz, A. L., 1964, Shore and sea boundaries: Washington, D.C., U.S. Department of Commerce Publication 10-1, v. 2, 749 p.
Shepard, F. P., 1963, Submarine geology: New York, Harper and Row, 557 p.
Shepard, F. P., and Wanless, H. R., 1971, Our changing coastlines: New York, McGraw-Hill, 579 p.
Silvester, R., 1960, Stabilization of sedimentary coastlines: Nature, v. 188, p. 467–469.
Stafford, D. B., 1971, An aerial photographic technique for beach erosion surveys in North Carolina: U.S. Army Coastal Engineering Research Center Technical Memorandum 36, 115 p.
Trask, P. D., 1952, Source of beach sand at Santa Barbara, California, as indicated by mineral grain studies: Beach Erosion Board Technical Memorandum 28, 24 p.
USCE, 1971, National shoreline study: Washington, D.C., U.S. Army Corps of Engineers, 59 p.
—— , 1979, Cape Cod easterly shore beach erosion study: Waltham, Massachusetts, U.S. Army Corps of Engineers, New England Division, 3 volumes.
—— , 1984, Shore protection manual: Washington, D.C., U.S. Army Corps of Engineers, Coastal Engineering Research Center, 3 volumes.
Wiegel, R. L., 1959, Sand bypassing at Santa Barbara, California: American Society of Civil Engineers Journal of Waterways and Harbors Division, p. 1–30.
—— , 1964, Oceanographical engineering: Englewood Cliffs, New Jersey, Prentice-Hall, 532 p.
Wilby, F. B., and 6 others, 1939, Inspection of beaches in path of the hurricane of September 21, 1938: Shore and Beach, v. 7, p. 43–47.
Yasso, W. E., 1964, Plan geometry of headland bay beaches: Office of Naval Research, Geography Branch Technical Report 7, 2 p.
—— , 1966, Formulation and use of fluorescent tracer coatings in sediment transport studies: Sedimentology, v. 6, p. 287–301.
Ziegler, J. M., Hayes, C. R., and Tuttle, S. D., 1959, Beach changes during storms on outer Cape Cod, Massachusetts: Journal of Geology, v. 67, p. 318–336.

MANUSCRIPT ACCEPTED BY THE SOCIETY JUNE 3, 1987

ACKNOWLEDGMENTS

This chapter benefitted greatly from lengthy discussions the author had with three of the nation's premier coastal engineers—Per Bruun, Joe Johnson, and M. P. O'Brien. The manuscript was critically reviewed by Per Bruun, Duncan FitzGerald, John Housley, Evelyn Maurmeyer, Tom Moslow, Malcolm Thomas, and Robert Wiegel. D. M. Cruden reviewed an early version, and George A. Kiersch undertook the task of revising and editing the final manuscript.

Printed in U.S.A.

Geological Society of America
Centennial Special Volume 3
1991

Chapter 9

Slope movements

Robert W. Fleming and David J. Varnes
U.S. Geological Survey, Box 25046, MS 966, Denver Federal Center, Denver, Colorado 80225

INTRODUCTION

Hillslopes are a fundamental unit of a landscape, comprising that reach of ground between a drainage divide and a valley floor, and thus much effort has been expended in their study. Early research by Davis (1899) and Penck (1924) was directed toward developing unified theories of slope formation and evolution. Subsequently, emphasis has shifted toward morphometric description of slopes (Strahler, 1956) and study of hillslope processes (Schumm, 1956). Currently, geomorphologists are making impressive advances in understanding hillslope forms and processes (Carson and Kirkby, 1972; Scheidegger, 1970, 1975; Huggett, 1985). While unified theories of hillslope formation and evolution are apparently many years away, the next generation of models can be based on carefully obtained measurements of hillslope processes.

Slope movements of several different types are among the principal processes by which hillslopes evolve. Slope movements are downward and outward movements of slope-forming materials composed of natural rock, soils, artificial fills, or combinations of these materials. This definition is identical to the definition of landslide used by Eckel (1958a). The terms landslide and mass wasting are sometimes used synonymously for slope movement. Slope movements, however, include some processes that involve little or no true sliding, such as falls and flows, and do not include some processes contained in mass wasting such as subsidence (Sharpe, 1938).

Historical developments

Progress in understanding and control of slope movements has been the result of a truly interdisciplinary effort involving geological scientists, engineers, physicists, and hydrologists. Most of the major practitioners in applied geology in the 19th and 20th centuries have contributed significantly to our understanding of slope movement types and processes. Voight (1978, 1979) dedicated his monumental two-volume work on rock slides and avalanches to Heim and Stini. Heim's (1882) work in the disastrous landslide at Elm in Switzerland has been widely cited in the literature of engineering geology. The work of Stini, perhaps not so well known to North American geologists, produced 333 publications, including several on slope movements, and one of the first classifications of landslides (Müller, 1979). Other notable advances in our understanding of landslides during the early 20th century came from the work of Howe (1909) on landslides in the San Juan Mountains of Colorado and Sharpe's (1938) descriptions and classification of slope-movement types. In addition, several publications reported details of major landslide events at Turtle Mountain, Alberta, landslides associated with construction of the Panama Canal, and the landslide failure of Fort Peck Dam, Montana.

The development of soil mechanics by Terzaghi, beginning with publication of *Erdbaumechanik* in 1925, provided the tools for analytical analyses of slope movements. His discovery of the principle of effective stress is one of the major achievements of this century (Skempton, 1960), and Terzaghi's impact on understanding the stability of soil and rock slopes has been enormous. His major papers on landslides (Terzaghi, 1950, 1962) provided an understanding of the mechanisms of landslide formation and movement. Equally important, his interdisciplinary approach to analysis of slope movements has served as a role model for a generation of applied geologists and engineers.

Beginning about 30 years ago, all the pieces were in place to achieve rapid improvements in analysis and control of slope movements. Criteria for recognition and classification of unstable slopes were well developed (Eckel, 1958b), several methods of analysis had been formulated and tested (Taylor, 1937; Bishop, 1955; Janbu, 1954), quantitative geomorphology was emerging to provide a basis for process studies (Schumm and Chorley, 1964), and the economic significance of different types of slope movements was becoming recognized (Smith, 1958). In recent years, sufficient progress has been achieved that the basic methods of study of slope movements have been synthesized in comprehensive general works (Zaruba and Mencl, 1969, 1982; Schuster and Krizek, 1978; Sidle and others, 1985; Abrahams, 1986). The growing level of worldwide interest is evidenced by the International Symposium on Landslides held in Japan in 1972 and 1977, in India in 1980, in Canada in 1984, and to be

Fleming, R. W., and Varnes, D. J., 1991, Slope movements, *in* Kiersch, G. A., ed., The heritage of engineering geology; The first hundred years: Boulder, Colorado, Geological Society of America, Centennial Special Volume 3.

TYPE OF MOVEMENT			TYPE OF MATERIAL		
			BEDROCK	ENGINEERING SOILS	
				Predominantly coarse	Predominantly fine
FALLS			Rock fall	Debris fall	Earth fall
TOPPLES			Rock topple	Debris topple	Earth topple
SLIDES	ROTATIONAL	FEW UNITS	Rock slump	Debris slump	Earth slump
	TRANSLATIONAL	FEW UNITS	Rock block slide	Debris block slide	Earth block slide
		MANY UNITS	Rock slide	Debris slide	Earth slide
LATERAL SPREADS			Rock spread	Debris spread	Earth spread
FLOWS			Rock flow (deep creep)	Debris flow	Earth flow
				(soil creep)	
COMPLEX	Combination of two or more principal types of movement				

Figure 1. Abbreviated classification of slope movements (after Varnes, 1978).

continued in Switzerland in 1988. Works on special topics include those on analysis (Chowdhury, 1978), hazard mapping (Varnes, 1984), instrumentation (Broms, 1975; Bhandari, 1984; Fukuoka, 1980; Hanna, 1985), construction and remedial methods (Winterkorn and Fang, 1975); and stabilization (Veder, 1981).

TYPES OF SLOPE MOVEMENTS

There are numerous classifications of slope movements in the published literature. The classification already mentioned by Stini (1910) pertains to debris flows. One of the best known and most enduring general classifications is by Sharpe (1938). He places landslides as gradational elements between transport of earth material by ice and water and emphasizes the gradational nature of many geomorphic processes. Some classifications have been developed to emphasize distinctions between certain types of slope movements such as mudflows and landslides (Campbell and others, 1985) or to emphasize the causes of slope movements (Chowdhury, 1980).

Probably the most widely used classification of slope movements is by Varnes (1958, 1978). The newer version (Varnes, 1978) contains terms as two-or-more-word descriptors that name the materials involved and the predominant type of movement as shown in Figure 1. Types of movement are divided into falls, topples, slides, spreads, and flows. Types of materials are bedrock and two types of engineering soils—predominantly coarse grained and predominantly fine grained. The suggested name for a particular slope movement is based on the type of movement, such as slide, modified by the type of material such as rock to produce "rock slide."

Falls

A fall is the movement of a mass that has detached from a steep slope or cliff and descends mostly through the air by free fall, rolling, or bouncing. Movements are rapid to extremely rapid. The travel distance of one or a few rocks in a rock-fall event is of practical concern to human activity in the vicinity of steep slopes. Ritchie (1963) provides empirical guidelines to predict rock-fall runout. Piteau and Peckover (1978) describe a computer program to predict the travel distance of many simulated falls for variable amounts of friction and coefficients of restitution in the potential travel path.

Rock falls are the most common form of slope movement triggered by earthquakes (Keefer, 1984), and they can be extremely destructive. For example, thousands were killed by rock fall in an earthquake in Pakistan in 1974 (Keefer, 1984; Hewitt, 1976). Ackenheil (1954) described a rock-fall event near Pittsburgh, Pennsylvania, that crushed a bus killing 22 people.

Specific rock-fall events are difficult to predict by means of static analysis, but they may be amenable to prediction by instrumentation (Tilling, 1976). In a general way, the frequency of rock-fall events can be estimated by means of such dating methods as lichenometry, dendrochronology, rock and mineral weathering, etc. Methods of analysis and prediction are described elsewhere in this volume. Some control methods for rock fall are described by Piteau and Peckover (1978).

A complex form of slope movement (Fig. 1), which has caused great destruction, has been variously termed rock-fall avalanche, debris avalanche, and sturzstrom (Hsü, 1975). The failures are characterized by tremendous kinetic energy and typically are a combination of rock fall on a steep-slope segment followed by flow of the debris over long distances on low slopes. Some of the avalanches are saturated with water, which could partially account for their great mobility, but others are apparently dry (Howard, 1973; Hsü, 1975). The mechanics of movement of these very large forms of slope movement have been studied extensively (Heim, 1882; Eisbacher, 1979; Crandell and Fahnestock, 1965; Plafker and others, 1971; Chowdhury, 1980; Melosh, 1983; Shreve, 1968; Cruden, 1985; Lee and Duncan, 1975; and

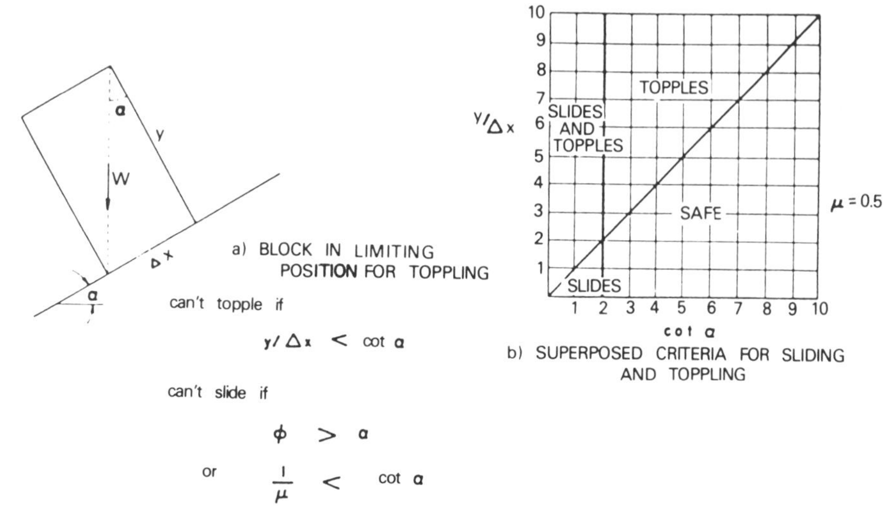

Figure 2. Overturning and sliding criteria for a single block on an inclined plane: (a) block in limiting orientation for toppling, (b) superposed criteria for sliding and toppling (after Goodman and Bray, 1976; and Hoek and Bray, 1974).

several papers in Voight, 1978, 1979). Larger volume events tend to travel farther than smaller events (Hsü, 1975), and empirical relations relating volume and travel distance have been derived from published accounts (Li, 1983).

At present, control methods for rock-fall avalanches are not feasible because of their uncertain locations, large volumes, and high velocities. However, case histories of events such as Elm, Switzerland; Turtle Mountain, Canada; and Mayunmarca, Peru (Heim, 1882; McConnell and Brock, 1904; and Kojan and Hutchinson, 1978), suggest that small movements may precede catastrophic failure, and warnings of impending failure may be possible for some events.

Topples

A topple consists of forward rotation of a mass about a pivot point on a hillslope. The toppling may culminate in abrupt falling or sliding, but the form of movement is tilting without collapse (Varnes, 1978). Until recently, topples were not recognized as a distinct form of slope failure. Topples may involve a single fragment with a volume of a few cubic meters to as much as 1 billion m^3 (deFreitas and Watters, 1973). Caine (1982) described topples in Tasmania comprising up to 1 million m^3 of rock that moved as a single event and resulted in cliff retreat at a rate of 0.2 mm/yr during the last 100,000 years.

Simple criteria to test for stability against sliding and toppling of a single block on an inclined plane are presented in Figure 2 (after Goodman and Bray, 1976; and Hoek and Bray, 1974). Goodman and Bray (1976) extended the criteria in Figure 2 to include physical interactions among several blocks. Hocking (1978) and Evans and others (1981) also have proposed methods for analysis of toppling failure.

Data on rates of movement and control measures for topples are sparse. A fascinating description of a topple at Chaco Canyon National Monument, New Mexico, by Schumm and Chorley (1964) recounts a long history of very slow movement. The topple involved a block of massive sandstone about 50 m long by 30 m high and 10 m thick resting on a shale unit. The site is adjacent to the ruins of cliff dwellings, and Indians long ago had attempted to halt the toppling movement by wedging trees beneath the rock and constructing a terrace to prevent erosion of the underlying shale. The trees were dated by radiocarbon at about 1,000 A.D. National Park Service personnel measured the toppling displacement during the last five years before the rock fell in 1941.

The record of displacement clearly showed an increasing rate of displacement with time (Schumm and Chorley, 1964; Varnes, 1983). Projections of the data backward in time from the abrupt collapse in 1941 indicated that movement began about 2,500 years ago (Schumm and Chorley, 1964). The data from the topple are apparently amenable to analysis and prediction of the time to failure. Thus, topples that culminate in abrupt failure may be predictable from stability considerations (Goodman and Bray, 1976) and on the basis of measurements in the field.

Slides

True slides of soil or rock involve downslope displacement along one or several surfaces of failure (Varnes, 1978). Slides are subdivided on the basis of whether the movement is rotational or translational, and whether the slide debris is composed of a few units or many units (Fig. 1). Slides with a strong backward rotational component to their movement have a curved, concave-upward-shaped failure surface and are called slumps. Translational slides that involve little or no rotation of the slide materials have relatively planar failure surfaces and move downslope on a path roughly parallel to the failure surface.

The distinction between slumps and translational slides is useful in analyzing potential for movement, planning a testing program or field investigation, and designing corrective measures. The curved failure surfaces associated with slumps are characteristic of materials such as clay that possess a significant amount of cohesion and are relatively homogeneous. In addition, the lower parts of slumps, if lying on a shear surface that dips into the slope, offer some passive resistance that is usually absent in translational slides. Many of the early studies of landslides were directed toward analysis of slumps (Collin, 1846; Fellenius, 1927; Taylor, 1937).

Movement of a translational slide is commonly controlled by a strong discontinuity, such as the contact between soil-like materials and bedrock or along soft zones or fracture surfaces in rock. The failure surface may follow the discontinuity in part, and follow stronger, intact materials in part.

Lateral spreads

Lateral spreads are forms of movements characterized by large components of distributed, extensional lateral displacement. Examples of several types of large-scale spreads in rock are described in Varnes (1978). The processes of spreading and the implications to engineering works in rock are not well documented; however, movement rates are apparently very slow. Most reported cases of rock lateral spreads are from Europe (for example, Záruba and Mencl, 1969).

Lateral spreads in engineering soils, as contrasted with rock, are common, rapid, and destructive. Some failures of fine-grained, sensitive soils such as quick clays are lateral spreads. Very sensitive soils undergo a dramatic loss of strength if they are remolded or disturbed. A well-documented example of failure of a slope in quick clay at Rissa, Norway, was summarized by Gregersen (1981). A small fill from an excavation, which was placed along the shore of Lake Botnen, triggered a series of complex failures, many of which were lateral spreads. The event was captured in an excellent motion picture prepared by the Norwegian Geotechnical Institute.

Characteristics of the slope failures and physical properties of the sensitive clays of Canada and Scandinavia are summarized by Viberg (1984). Detailed descriptions of individual spreads are given by Carson (1979) for an event at Ste. Madeleine de Rigaud in Quebec, and by Hansen (1965) for spreads triggered by the 1964 Alaskan earthquake.

Other materials that commonly produce lateral spreads are loose, granular soils. The mechanism of spreading is through liquefaction of the loose soils. The liquefaction can occur spontaneously, presumably caused by changes in pore-water pressures (Varnes, 1978), or in response to vibration such as produced by strong earthquakes. The saturated granular deposits that are initially loose become more densely packed during shaking and, because the grains attempt to consolidate more rapidly than excess pore water can be dissipated, the grains and any exterior loads become virtually supported by the fluid, and a condition of liquefaction exists. Lateral spreads induced by liquefaction typically occur on low slopes (as low as 0.3° [Youd, 1975; Keefer, 1984]). The ground surface of the spread is rafted laterally in coherent fashion on a layer of liquefied material at depth, and the surface of the spread is broken into grabens and horsts characteristic of extensional deformation.

Flows

Flows of soil and rock are characterized by shear strains that are distributed throughout the mass of material, rather than concentrated on one or more surfaces. Flow in rock includes deformations that are distributed among many large, small, or microfractures. Flow of rock, sometimes called rock creep or sackung, is being recognized in many parts of the world in areas of high relief. Movement rates are apparently very slow, and the engineering significance of the deformations is only beginning to be appreciated (Varnes, 1978).

Flows in engineering soil (unconsolidated geologic materials) probably constitute the most destructive form of slope movement in North America. The flows may be dry or wet and can occur in materials of virtually any distribution of grain sizes and vary in volume more than ten orders of magnitude (Fig. 3). Similarly, the rates of movement of flows can vary from extremely slow to extremely rapid.

Flow of sensitive clays, liquefaction of granular deposits, and flow of rock-fall avalanches have been mentioned in connection with lateral spreads and falls. In addition to these special cases of complex slope movement that include significant amounts of flow, other movements termed debris flows, mud flows, and earth flows are widespread and common over much of North America. The more common flow types are transitional with sliding movement. As the amount of distributed shearing deformation increases within the sliding body, the type of movement becomes more and more like a flow. The amount of distributed shear necessary for the feature to be termed a flow rather than a slide is apparently variable. Keefer and Johnson (1983), in a field study of earth flows in California, found that only a small percentage of the total displacement was due to distributed shearing; the remainder consisted of sliding on a basal slip surface. Crandell and Varnes (1961) measured a large amount of shear in narrow zones on the flanks at the surface of the Slumgullion earth flow in Colorado, and they found lesser amounts of relative deformation at the ground surface in the central region. Thus, most of the earth flow was transported as a plastic plug with shearing concentrated on the flanks and presumably along the base of the earth flow. At the other extreme, flow-type slope movements also

Figure 3. View southward toward Mount Shasta Volcano, California. The scattered mounds and ridges in the foreground are part of a debris-flow deposit that originated on the slopes of ancestral Mount Shasta between 300,000 and 360,000 years ago. The deposit covers an area of at least 450 km^2 and has an estimated volume of about 25 km^3, making it the largest known landslide of Quaternary age on Earth (Crandell and others, 1984; photograph by D. R. Crandell).

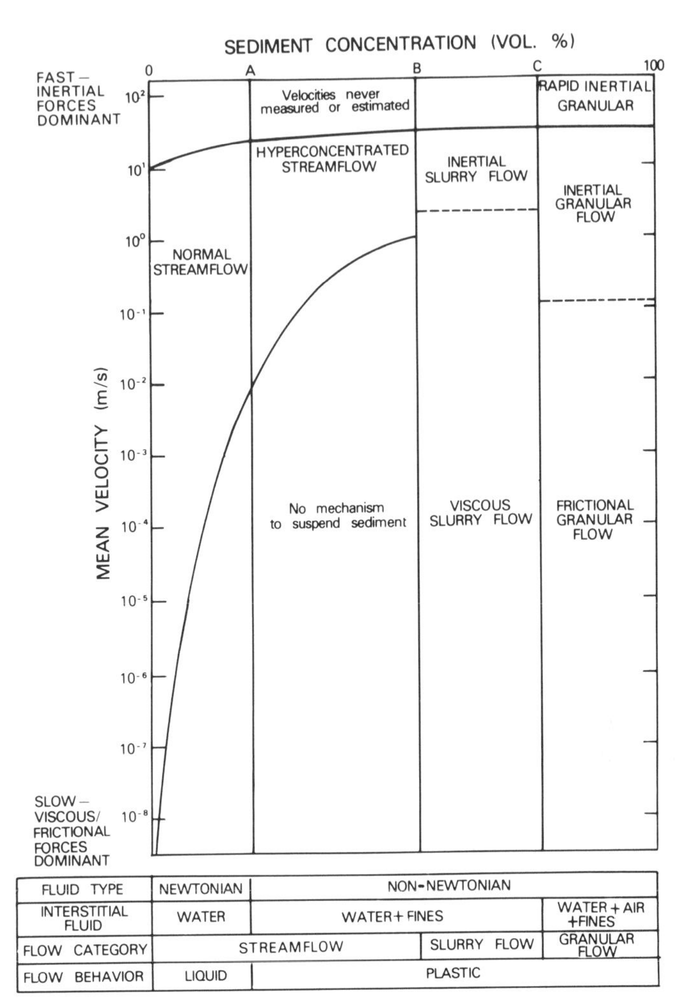

Figure 4. Rheological classification of sediment-water flows illustrating position of mud and debris flows (slurry flow) in a continuum between stream flow and granular flow. Vertical boundaries are rheological thresholds that are functions of sediment concentration and texture. The boundary between viscous slurry flow and inertial slurry flow divides laminar-flowing debris and mud flows from debris and mud avalanches (from Pierson and Costa, 1987).

have transitional boundaries with transport of sediment by flowing water—that is, fluvial processes. Pierson and Costa (1986) have proposed a classification (Fig. 4) for flows that emphasizes the transitional nature of the boundaries between fluvial flow, slope-movement flow such as mud and debris flow, and true sliding.

For the three common types of flow—the very fluid debris flows and mud flows and the stiffer, slow-moving earth flows—there are systematic differences in composition that produce a difference in ease of mobilization and flow rate. The principal difference between mud flow and debris flow is that mud flows lack a significant fraction of coarse-grained material. Johnson (1984) and Rodine (1974) have shown that the characteristics of the flows are determined by the characteristics of the slurry in the matrix of the flow. Thus, the extremes in behavior of flows of engineering soils are represented by rapidly flowing mud and debris flows at one extreme and slowly flowing earth flows at the other.

Materials in earth flow are typically more clay rich and, thus, cohesive. Clay-rich materials undergo a reduction in strength with increasing water content; however, the reduction is gradual. A clay-rich material with a liquid limit of 50 has about the same strength at a water content of 50 percent as it does at a water content of 55 percent. In order to make a clay-rich deposit (excepting quick clays) very fluid, it is necessary for the deposit to incorporate a great deal of water, and water can normally be incorporated only by remolding it into the deposit. Because it is difficult to remold natural materials to increase the water content, flows in clay-rich deposits tend to move slowly and the materials are typically stiff.

Materials in mud and debris flows tend to have a smaller clay-mineral fraction than those in earth flows. Deposits with a small clay fraction undergo a drastic change in fluidity with a small change in water content. Consistency of some materials changes from plastic to very fluid with a change of water content of as little as 2 or 3 percent. Only a small clay fraction of 2 to about 8 percent is apparently necessary to produce a slurry capable of transporting the coarser particles in a mud or debris flow (Rodine, 1974; Johnson, 1984). Very fluid flows may travel at a rate of many meters per second, and an earth flow may travel at only a few meters per year or less.

TABLE 1. TYPES OF NORTH AMERICAN SLOPE MOVEMENTS RELATED IN TERMS OF MATERIALS OR TRIGGERING AGENTS

Landslide Type	Locations	References	Remarks
Landslides in colluvium	Appalachian Mtns. and Plateau, Ohio River Valley	Deere and Patton, 1971; Hamel and Flint, 1972; Gray and Gardner, 1977; Fleming and others, 1981	
Loess Failures	Mississippi and Missouri River valleys	Lutton, 1971; Handy, 1973	Commonly transform from slides to flows
Earthquake-induced landslides	Areas that may experience large or great earthquakes	Keefer, 1984; Tinsley and others, 1985; Wilson and Keefer, 1985	Includes all types of slope movements
Quick or sensitive clay landslides	St. Lawrence Valley, coastal Alaska	Kenney and Drury, 1973	Involves both sliding and flowing, triggered by disturbances or may occur without warning
Submarine landslides	Widespread on continental shelf and in deltaic deposits	Prior and Coleman, 1982; Sangrey and Garrison, 1977	Involves both sliding and flowing
Overconsolidated clays--clay shales	Panama Canal, Missouri River valley, Cretaceous-Tertiary shales in western U.S. and Canada	Skempton, 1964, 1970; Bjerrum, 1967; Lutton and others, 1979	
Volcanic landslides	Western North America	Crandell, 1971; Voight and others, 1983	Includes lahars that are flows or avalanches. Slope movements of all types associated with volcanoes
Rock-fall avalanches	Areas of high relief in western North America	Voight, 1978; Cruden, 1985; Melosh, 1983; Cruden and Hungr, 1986	Includes flowing, falling, and/or sliding
Tectonic landslides	Heart Mountain and other areas of gravitational movement	Prostka, 1978	Slope movements so large that they overlap with tectonic-caused features
Sackung(en)	Mountainous areas of western North America	Radbruch-Hall, 1978	Features are at least partly flow
Soft clays	Widespread, may occur wherever normally consolidated clays occur	Kenney, 1968	

Control of flows involving engineering soils includes defensive structures to retain or divert flows (Mears, 1977; Hollingsworth and Kovacs, 1981), grading regulations to improve construction practices in materials susceptible to flowing, and avoiding construction in areas susceptible to flows. All three types of control methods have been used successfully in California to reduce damages from flows (Kockelman, 1986).

Slope movements as related to materials or triggering agents

The foregoing classification of slope movements, based on type of movement and whether the material is rock or soil, is purely descriptive. It does not attempt to predict the most likely type of failure that could result from certain genetic kinds of geologic materials or triggering agents. Yet it is well known that special characteristics of materials of different geologic origins and histories may be crucial to the analysis of stability and control of movement. Some of the special cases of slope-movement types that have impacted engineering works in North America are listed in Table 1, with selected references that are entries to the published literature for each case.

EVALUATION OF HAZARD

Analytical evaluation of hazard

Engineering geologists and geotechnical engineers have long been interested in methods to analyze slope movements, for analysis affords the essential key to understanding the effects of different factors that tend to improve or diminish slope stability. Initially, analytical methods for evaluation of stability were applied to static cases for true landslides (Taylor, 1937; Terzaghi, 1925). More recently, methods have been derived to analyze virtually every type of slope movement with the additional influences of transient loading, such as earthquakes and reservoirs.

There are two fundamentally different approaches to the mechanics of slope analysis. The more common approach uses limit-equilibrium methods that estimate forces and/or moments that promote or resist movement for a body of assumed or known shape. The output of a limit-equilibrium analysis is usually a factor of safety, which may be variously expressed in terms of, for example, strength parameters (Morgenstern and Sangrey, 1978) or forces (Taylor, 1948). Loosely interpreted, the factor of safety represents the calculated margin of stability (F.S. = 1 at equilibrium) for a given set of conditions including geometry, strength, density, water conditions, and perhaps other special circumstances such as effects of tree roots (Wu and Swanston, 1980; Riestenberg and Sovonick-Dunford, 1983) or earthquakes. If the location of the potential or actual failure surface is not known, it is necessary to conduct iterations of the analysis to locate the surface that results in a minimum value of factor of safety.

Because failure of a hillslope occurs when the factor of safety is equal to or less than one, a rigorous analysis of a previously failed slope is an excellent way to evaluate strength parameters and water conditions at the time of failure. Descriptions of the different limit-equilibrium methods are contained in Morgenstern and Sangrey (1978), Veder (1981), Fang (1975), Chowdhury (1978), Fredlund (1984), and Pariseau and Voight (1979).

A second approach to analytical slope analysis is by means of numerical methods such as finite-element techniques. The numerical methods generally result in computations of forces and displacements at many points on the surface and within a hillslope (Duncan, 1972). The most commonly used method is the finite-element method (Morgenstern and Sangrey, 1978). A cross section of a slope or embankment is idealized by dividing it into an array of triangular elements. Each element contains nodal points at the tips of the triangular elements that are shared with adjacent elements. Computation of forces and displacements at the nodal points requires that the stress field must satisfy equilibrium at every nodal point and that deformations must be compatible among all the elements. Conditions on the boundaries of the hillslope and a force-deformation relation for the materials in the hillslope must be initially specified. The calculations solve force-displacement equations for each nodal point. Early finite-element models used elastic force-deformation relations, but improvements in computer codes and computers now allow different types of rheology and simulation of construction sequences in the analyses (Chowdhury, 1978).

The finite-element method is the foundation of the deformation approach to slope stability that has been adopted by many European investigators (Záruba and Mencl, 1982). Mencl (1977) reported several situations where finite-element methods provide a different perspective to slope analysis than do limit-equilibrium methods. These include conditions where equilibrium is not developed simultaneously over the entire failure surface (progressive failure), slopes undergoing creep, and slopes that are statically very complicated. In the deformational approach, analyses of slopes for very complex situations use both limit-equilibrium and deformational methods, and assessments of stability are judged from results of several trials. In particular, the numerical methods provide a rational technique for selection of sites for instrumentation, appropriate laboratory-test procedures, and evaluation of deformations associated with construction (Chowdhury, 1978).

The deterministic and numerical methods of slope analysis result in an estimate of the factor of safety and amount of displacement of a slope. However, the methods do not take into account the fact that many of the input parameters in the calculations are random variables (Chowdhury, 1984). Effects of variations in the random variables can be evaluated using probabilistic assessments of stability. In practice, a probability distribution is assigned to each random variable, and factors of safety are calculated for different values of the variables (Vanmarcke, 1977). Generally, the variables are assigned a normal distribution. Likelihood of failure can be expressed in terms of a probability, and slope design can be evaluated in terms of increased cost to create greater reliability. Chowdhury (1978) predicted that the probabilistic approaches will be increasingly applied to special stability problems where uncertainties exist in the values of pore pressures, strength parameters, failure geometries, and other less-obvious variables such as initial state of stress, anisotropy, and deformation history.

Areal evaluation of hazard

In many instances, decreasing the risk from various forms of slope movement can be achieved only by avoiding high-hazard areas. The concept of avoidance has been formally developed by land-use planners (McHarg, 1969), wherein selection of the most suitable area for a particular land-use activity is based on minimizing impacts from several cultural and physical constraints including slope movements.

Engineering geologists have been providing information that can be applied to damage reduction through preparation of maps that contain estimates of the relative (or absolute) likelihood of a slope failure. Varnes (1984) identified three principles that have guided the areal evaluation of slope-movement hazards: (a) the past and present are the keys to the future, (b) the main conditions that cause landsliding can be identified, and (c) degrees of hazard can be estimated. The first principle, a twist on the Principle of Uniformitarianism, predicts that slope failures in the future will most likely be in physical situations that have experienced past and present slope failures. The second and third principles follow from the first in that study of specific examples of slope failures leads to a recognition of the factors that caused or permitted the failure. If the relative contribution of each causative or resisting factor can be identified, the degree of hazard can be estimated. Examples of maps based on these principles range from maps of landslide deposits (Lessing and others, 1976) to those showing units of equal values of factors of safety, in which the factors of safety were computed by limit-equilibrium methods for the specific physical conditions at many different points in an area (Ward, 1976). Many areal evaluations are achieved by superimposing maps of different physical attributes such as steepness of slope, hydrologic conditions, genesis of hillslope deposits, strength characteristics, and interpreted seismic response (Varnes, 1984).

The areal studies of slope movements can be divided into two classes (Brabb, 1984). The most direct and simple class is represented by inventory maps that show areas that have failed previously. The inventories may show only locations, or they may show actual landslide outlines and indicate other relevant information such as type of failure. All the maps of this class basically show facts about hillslopes. The other class of map is interpretative, and the maps are termed susceptibility maps. These maps depict areas likely to experience slope failures in the future by combining geologic units, steepness of slopes, and perhaps other factors.

Some susceptibility maps of ground-failure hazards illustrate relative slope stability on a gradational, qualitative scale (Radbruch and Crowther, 1973), while others may result in map units that specify recommended land uses (Hoexter and others, 1978), and still others may include ground-failure hazards as part of more comprehensive engineering geologic maps or maps of several types of geologic hazards (Legget, 1973). Several examples of complex slope-movement maps, particularly from Europe, are described by Varnes (1984). Some of these, such as the ZERMOS mapping in France, are connected to a national program of hazard reduction (Humbert, 1977). Others were completed as research projects in complex areas of slope movement (Kienholz, 1978, 1985).

Because areal assessments of hazards of slope movement have been actively pursued for only about 20 years, there have been few tests of their accuracy. In one test, in Contra Costa County, California, most of the landslides that damaged manmade structures during the period 1950 to 1971 occurred in areas where abundant landsliding has taken place in the past (Nilsen and Turner, 1975). Thus, their locations were relatively predictable. A major storm in the San Francisco Bay area that triggered more than 13,000 slope-movement events produced many landslides in unexpected locations (Brabb, 1985). About 80 percent of these slope failures occurred in areas not previously mapped as landslide deposits. In general, deep-seated landslide movement produces recognizable deposits that endure for perhaps many hundreds of years. Shallow landslides of surficial deposits that mobilize into debris flows may not be recognizable in a few years after a movement episode. In the San Francisco Bay area, where both types of failure occur, the hazards of deep-seated landsliding were well known, but hazards of shallow-soil landslides that occur during intense storms were poorly known (Brabb, 1985).

A test of areal hazard evaluation of a different type was provided by the May 18, 1980, eruption of Mount St. Helens volcano in southwestern Washington. A map by Crandell and Mullineaux (1978) showed areas that could be affected by volcanic mud and debris flows during an eruption. The depictions of various hazard areas were based on careful study of the distribution of deposits from past eruptive events, and the 1980 flows followed predictions to a remarkable degree (Miller and others, 1981).

Predictions of slope movements

A complete and accurate prediction of a slope-movement event should specify the location, type, magnitude, and time. The current stage of the art in slope-movement predictions is better than for most natural hazards, and significant reductions in damages are attainable (Brabb, 1984). The analytical methods of slope analysis, including limit-equilibrium and finite-element methods, result in one kind of prediction of slope performance. The areal methods of slope evaluation, including inventory and susceptibility mapping, are predictive tools of another kind. Both these approaches, when combined with other specialized techniques, can provide a reasonable appraisal of slope performance.

Indications of the location of a potential slope-movement event can be obtained from detailed study of failures in similar materials and in similar physical situations. For example, habitats of the source areas of debris flows in Marin County, California, caused by a severe storm in 1982, showed a strong preference for concave slopes (Ellen and Wieczorek, 1988). Further, if a typical debris flow traveled more than about a hundred meters on a planar or convex slope, it thinned by spreading laterally and stopped. Debris flows that travel long distances apparently must occupy a channel (Johnson, 1970). In addition to topographic constraints on the locations of debris flows, there were material constraints. Only specific types of geologic materials were mobilized into debris flows during the storm. In this sense, locations of potential future events are predictable in Marin County to levels that are adequate for general land-use planning. Similarly, locations of other types of slope movements may be predictable by combining various types of information about susceptible materials and physical habitats of slope movements from past events.

The magnitude of a slope-movement event includes size or volume and rate of movement, including travel distance. General estimates of magnitude of slope movements come from the accumulated body of knowledge and experience derived from many thousands of past events. Predictions of magnitude of a specific potential failure must be based on what can be learned from examination of nearby localities supplemented by careful monitoring for precursors to movement. Direct field observations and measurements should be evaluated in the context of what is known about types of failures associated with different geologic materials as well as geometries and potential movement rates.

The fourth element of prediction—time—is the most difficult. Moreover, the levels of precision that may be attainable will vary among the different types of slope movements and potential

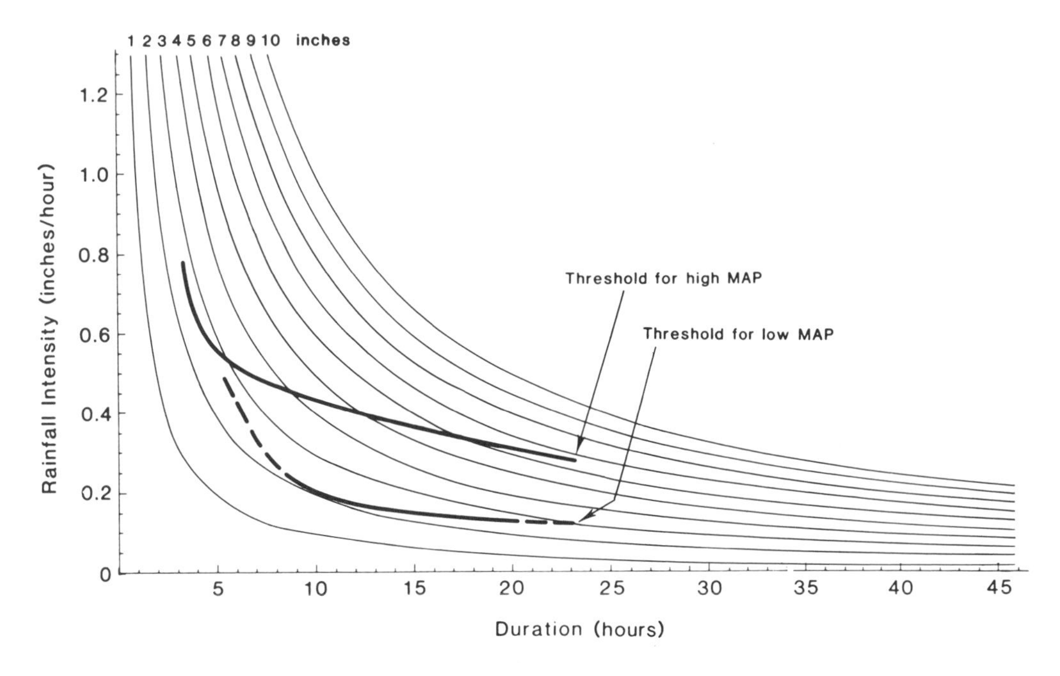

Figure 5. Storm characteristics and thresholds for abundant debris flows in the San Francisco Bay region for areas of high and low mean annual precipitation (map). The threshold line is drawn where precipitation intensity and duration values for a storm in 1982, which resulted in abundant debris-flow events throughout most of the region, exceeded values for major storms that did not produce abundant events. Set of gray lines represents total precipitation from a storm (after Cannon and Ellen, 1985).

triggering causes, and the precision necessary will depend on the risks involved. Prediction of time of abrupt failure of a slope in an operating open-pit mine needs to be much more precise than for evaluation of hazard of a planned-construction activity. In the broadest sense, times of failure may be predictable from recurrence intervals in much the same way that volcanic eruptions and earthquakes are. The methods that may be used to establish recurrence are any of the several Quaternary dating techniques (Colman and Pierce, 1977) including lichenometry, dendrochronology, historical records, soil development, weathering profiles, etc. Similarly, recurrence intervals for certain types of slope movements that are keyed to specific triggering events are the same as the recurrence intervals for the triggering events. These include earth lateral spreads caused by earthquake-induced liquefaction (Youd and Perkins, 1978) and debris flows triggered by severe storms.

Debris flows appear to respond to a narrow threshold of rainfall-triggering conditions. Prediction of the rainfall event that equals or exceeds known threshold conditions essentially predicts the time of debris flows. For the Santa Monica Mountains in southern California, Campbell (1975) found that after 25 cm of seasonal antecedent rainfall, a storm producing rainfall at an intensity of 0.6 cm/hr will trigger significant numbers of debris flows. A similar threshold (Fig. 5) for debris flows in the San Francisco Bay region was determined by Cannon and Ellen (1985). There, Cannon and Ellen found that the level of the threshold was sensitive also to the amount of rainfall that an area normally receives. Areas of low mean-annual precipitation have a much lower threshold for triggering abundant debris flows than areas of high mean-annual precipitation. Thus, predictions of time of failure can vary from crude estimates of recurrence intervals of years, decades, or centuries to predictions that may not be location-specific, but may be reasonably accurate in terms of the nature of triggering activity.

In many situations, it is desirable to predict the time of abrupt failure of a hillslope as precisely as possible. Kennedy and others (1971) describe the successful prediction of the collapse of portions of the open-pit Chuquicamata Mine in Chile. The prediction, which was made about a month before a rapid failure and was based on extrapolation of displacement data, permitted modification of mine-transportation systems and avoided injuries and damages to equipment.

A method to forecast time of rapid slope failure from displacement data was developed and used successfully in Japan by Saito (1969, 1980). Varnes (1983) generalized the Saito equation and derived a number of linear relations among strain (or displacement), time, and time to failure that permit analysis of displacement data to forecast time of failure. The method of data analysis contains a test for goodness of fit of the data to a predictable failure time. Measurements of tertiary (accelerating) creep in a wide variety of natural and man-made materials reveal that the time to failure can be predicted from the expression:

$$\frac{d\epsilon}{dt} = \frac{C}{(t_f - t)^n}$$

where t is the time of observation, t_f is the time of failure, $d\epsilon/dt$ is the strain rate, and C and n are constants.

Several methods are now available to estimate magnitude, type, location, and time of a slope-movement event. The estimates contain varying degrees of precision depending in part on the type of slope movement anticipated, but they provide an excellent framework for slope-movement hazard evaluation.

INSTRUMENTATION

The purposes of slope-movement instrumentation are for analysis of stability, to determine areal extent and rates of movement, to monitor performance of corrective measures, and to provide warnings. A large number of instruments have been designed specifically for the study of slope movements, and many others have been developed for other purposes and adapted to slope-movement investigations. In general, measurements of greatest sensitivity and accuracy are required in the earliest stages of movement (Terzaghi, 1950). Precise surveys and carefully conducted measurements of subsurface phenomena are required, for instance, in monitoring the initial performance of dams and embankments. During active movement, less accurate measurements are adequate for most purposes. More comprehensive discussions of slope-movement instrumentation are contained in Wilson and Mikkelsen (1978); Broms (1975); Bhandari (1984); Pilot (1984); and Hanna (1985).

Measurement of surface movements

Surface movements are commonly monitored using different surveying techniques; the most accurate systems are capable of accuracy of one part in 300,000 for distance and about one millimeter vertical (Wilson and Mikkelsen, 1978). As a rule, points should be established on stable as well as moving ground as an internal check of accuracy.

The interpretation of survey data on a single point may be ambiguous, and measurements should be made on several points. If points are initially set along lines normal and parallel to the movement direction, errors are more readily detected and interpretation is simplified.

Additional information about surface slope movements at specific points on a landslide can be obtained if four points are placed on the ground at the corners of a quadrilateral. Measurement of the perimeter and diagonals of the quadrilateral with a tape or tape extensiometer is analogous to the braced quadrilateral method of surveying in which one of the measurements is redundant. Thus, the accuracy of the measurements can be checked, and data can be used to compute a finite strain of the ground surface. Computation of strains have been shown to be useful by Ter-Stepanian (1980) and Záruba and Mencl (1982).

Nearly all the ground-based techniques suffer from the requirement that the most critical locations for measurement points be known at the beginning of an investigation. On the other hand, aerial photogrammetric methods are not limited to a few preselected points. Analytical plotters, utilizing stereo pairs of sequential photography, can detect movement in the subcentimeter range from 1:2,000-scale photography under optimal conditions (Fraser and Gruendig, 1985). Once primary control for the photography has been established, virtually every other identifiable point of the photographs, such as rocks, bushes, or cultural features, can be used to monitor displacement. Additionally, the plotters can utilize oblique, hand-held photography, provided a few parameters are measured in the field (Dueholm and Garde, 1986). More widespread use of photogrammetric methods would be useful to clarify the possibilities and limitations of the technique. In particular, measurement of deformation at many points on the surface of a landslide could reveal differences in stability of different landslide elements.

Measurements of surface deformations in one dimension can be obtained for extremely small movements with laser distance-measuring devices or, for short distance, with an invar rod and dial gage. Recording extensiometers, which record the change in length of a wire, cable, or rod, are commonly used where a closely spaced time record of movement is desired. Such devices commonly consist of a tensioned wire connected to an electrical potentiometer or mechanical device. Deformations can be recorded digitally or plotted on an analog chart.

Measurement of subsurface movements

Displacements can be measured in a borehole with a sensitivity of up to about 1:10,000 (Wilson and Mikkelsen, 1978) using an inclinometer, a device that measures tilt of a specially constructed, grooved casing. The grooves in the casing serve to maintain alignment of the tiltmeter as it is lowered or raised in the boring. Readings of tilt are made at selected intervals in the casing. Later readings are taken at the same depth points as initial readings, and differences in tilt are integrated to produce a displacement profile. Displacement profiles from two sets of readings taken at 90° to one another can be added vectorially to obtain a resultant displacement profile. Because movement is referenced to the bottom of the boring, the end of the casing must extend through the material that may be involved in movement. Wilson and Mikkelsen (1978) recommend that the position of the top of the casing be checked by precise survey methods to confirm the upward-integrated displacement and, thus, that the deepest failure surfaces have been penetrated. The inclinometers are most useful in locating the failure surface(s), in monitoring displacement during the very early stages of movement, and in checking the effectiveness of remedial measures when displacements are small.

For actively moving slopes where large displacements would disrupt an inclinometer casing in a short time, other methods of locating the failure surface are more economical.

Toms and Bartlett (1962) described a method where metal rods of increasing length are lowered down a 2.5-cm diameter plastic casing. The length of rod just able to pass a given point in the casing is a measure of curvature of the casing at that point. Also, in installations of this type, a weight attached to a thin cable is left at the bottom of the casing. If the casing is sheared by slope movement, the weight can be pulled up to the bottom of the sheared zone, and the top of the zone can be plumbed from the ground surface (Wilson and Mikkelsen, 1978).

The approximate depth to the failure surface and the amount of displacement can be measured using an array of weights attached to 8-mm cables. The weights are lowered to different depths in the boring, and the boring is backfilled. Movement of the landslide pulls cables down the boring if the weights are below the failure surface; the lengths of cable sticking out of the boring do not change if the weights are above the failure surface(s). Cables can be connected to a potentiometer to produce a record of displacement versus time; such installations are useful in the interior of very large landslides where no stable reference exists for monitoring rate of movement.

Several other direct and indirect methods of locating the failure surface have been described by Hutchinson (1983). Among these is a technique for estimating depth to the failure surface from grabens, which is described by Jakobson (1952) for the landslide at Surte, Sweden, and by Hansen (1965) for landslides triggered by the Alaskan earthquake. Carter and Bentley (1985) and Cruden (1986) have shown that the shape of the failure surface can be predicted from survey measurements at the ground surface. Assuming that a slope failure occurs as a rigid body with no change in volume, the path of a point on the surface of a landslide will be parallel to the failure surface at depth. If the position of the failure surface is known where it intersects the head or toe of the landslide or from drilling, it is possible to estimate the position of the failure surface for other points by following the paths of movement of points on the landslide surface.

Subsurface water

In addition to locating potential or actual failure surfaces and monitoring displacement, instrumentation is necessary for understanding the behavior of water in hillslopes. Fluid pressures are a key aspect of stability analysis. On simple slopes amenable to analysis by infinite-slope techniques, the factor of safety against sliding is larger by about a factor of two if the water level is deeper than the potential failure surface compared to a level coincident with the ground surface.

Unfortunately, subsurface water conditions are usually complex, and proper instruments and techniques for fluid-pressure monitoring are not generally used. Thus, slope analysis based on inaccurate water-pressure data may be seriously deficient (Patton, 1983).

The two principal types of devices for measuring fluid pressures are open tubes and piezometers. Open tubes connected to porous tips and sealed at the depth of interest have been used for many years to determine water pressures (Casagrande, 1949). The water level in the tube, which can be measured or recorded by several methods, is intended to reflect the fluid pressure in the sealed interval. Open tubes work well in simple situations where the permeability is relatively large. In complex situations where permeabilities are low, the open tubes may have an unacceptably long response time or may provide a meaningless average of water levels in different pressured zones. The great value of open tubes is in their simplicity and reliability.

Piezometers comprise a group of pressure-measuring devices that are in physical contact with the pore fluid. The principle of operation may be based on deformation of a diaphragm with strain gages attached or with a piezometric crystal. The advantages of piezometers over open tubes is that a very small volume of water will deform the diaphragm, and measurement of pore pressure is less affected by the installation and presence of the instrument.

Patton (1983) provides some guidelines for design and installation of piezometers for slope-stability problems. Piezometers should be installed in narrow depth zones that are well sealed to prevent leakage. The depth zones should be located to provide measurements of pressures at the positions of interest with respect to stability analysis, including just above and just below the potential failure surface. Sufficient piezometers should be installed to provide redundant data. The piezometers themselves should not affect the values measured, and the system should be designed so that it can be calibrated and tested after installation. Some of the systems that satisfy these requirements are described by Londe (1982) and Patton (1983).

RELEVANCE TO ENGINEERING WORKS

Slope-movement hazards are a significant part of many major construction projects, including transportation routes, energy development, water storage and transmission, and buildings. Considerable impetus has been given to studies of slope movements by a growing awareness of the costs of damage, and a conviction that many of the damages can be avoided or reduced. The famous engineering geologist, C. P. Berkey, in a letter to C.F.S. Sharpe (1938, p. 5), wrote:

> I am convinced that the question of landslides is a matter of much larger importance than is usually assumed. Recent experience leads to the belief that it is of special significance in connection with many practical problems, particularly those connected with engineering projects.

There are many examples of the relevance of slope-movement problems to engineering works. One of the best known is the construction and operation of the Panama Canal. Landslides in overconsolidated clays severely disrupted construction of the canal, delaying completion of the canal by nearly two years. The original French estimate for complete excavation of the canal was about 18 million m^3. A total of 57 million m^3 of landslide debris alone was removed from the canal between the start of construction and 1940 (MacDonald, 1942; see James and Kiersch, Chapter 22, this volume).

In addition, landslide problems have plagued construction of dams, powerhouses, and reservoirs, including several along the Missouri and South Saskatchewan rivers (Knight, 1963; Matheson and Thomson, 1973; Peterson, 1956), Revelstoke Dam (Piteau and Peckover, 1978; Patton, 1983), and the Lower Baker Powerhouse (Peck, 1967). Landslides have damaged offshore oil facilities, interfered with construction and operation of highway systems, and contributed to the failure of reservoirs (Davies and others, 1972).

In response to increasing damages from slope-movement events, the National Research Council of the United States has established a Committee on Ground Failure Hazards. The committee (National Research Council, 1985) has proposed a national program to reduce losses from damaging landslides.

The first attempt to assemble costs of landslide damage for the U.S., as well as some costs in Canada, was by Smith (1958). The estimate of damages, based largely on responses by highway and railroad organizations to a questionnaire, was in the range of hundreds of millions of dollars annually. Subsequently, Krohn and Slosson (1976) estimated that annual costs of damages to privately owned structures and their sites is about $400 million (1971 dollars). This estimate did not include damages to public property, railroads, and agricultural and forest lands. Fleming and Taylor (1980), in a survey of costs of damages for smaller geographical areas, found that public costs of landslide damage are typically larger than private costs. Using the above information, Schuster and Fleming (1986) concluded that the direct and indirect loss to public and private property in the U.S. is about $1.5 billion annually. As far as we know, no comprehensive estimates of damages have been completed for Canada. However, the many technical reports of specific landslide events in Canada indicate that the damages are significant there also.

Where costs of damage from slope movements have been assembled for small areas such as King County, Washington (Tubbs, 1974); the San Francisco Bay area, California (Taylor and Brabb, 1972); San Diego County, California (Shearer and others, 1983); and the Pittsburgh, Pennsylvania, area (Briggs and others, 1975), costs are larger than anticipated. Also, costs of damages from slope movements are not routinely assembled by any private or governmental organization, and incomplete recordkeeping generally results in lower reported costs than those actually incurred (Fleming and Taylor, 1980).

Portions of the western U.S. have been experiencing, in recent years, a cycle of wetter-than-average climate. In California, significant landslide damages occurred about once every four years (Nilsen and others, 1976) during the 30-year period following World War II. However, since 1978, there have been significant slope-movement problems nearly every year. In 1980, high-intensity rainfall in southern California produced slope-movement damages estimated at $500 million (Slosson and Krohn, 1982). In 1982, a major storm triggered thousands of debris flows and a few larger landslides in northern California. The losses included about 30 fatalities and damage or destruction to 6,500 homes and 1,000 businesses. Estimated damage costs exceeded $280 million (Smith and Hart, 1982) and, within six months of the disaster, lawsuits totaling $298 million had been filed against city and county agencies in the San Francisco Bay region.

Figure 6. The 1983 Thistle, Utah, landslide (right) and its temporary impoundment, Thistle Lake (left). The lake was drained late in 1983 by a bedrock tunnel constructed through the mountain at center. (U.S. Forest Service photograph).

Two individual landslides occurred in 1983 that have competed for the most costly single landslide in the history of North America. One of these, the Big Rock Mesa landslide on the Malibu coast of southern California, began moving slowly in the summer of 1983. A few houses were damaged or destroyed by the movement, but continued movement threatened more than 300 additional homes. The aggregate of legal claims against Los Angeles County for this landslide exceeded $500 million as of July 1984 (Association of Engineering Geologists, 1984). Reportedly, attempts to stop the movement have been at least temporarily successful, and the costs of damages apparently will not reach their full potential.

The Thistle, Utah, landslide was the most noteworthy of numerous landslides and debris flows that occurred throughout Utah in 1983 and 1984 (Fig. 6). In April 1983, an old landslide was reactivated, and movement rates of up to 2 m/hr produced a blockage of Spanish Fork Canyon. The landslide occurred in about the worst place imaginable to produce large monetary damages. It severed the transcontinental tracks of the Denver and Rio Grande Western Railroad, blocked U.S. Highways 6, 50 and 89, dammed Spanish Fork, and formed a large lake. The lake flooded the small town of Thistle, causing complete loss of several homes and businesses. The railroad was relocated through tunnels in the canyon wall opposite the landslide, and the highway was relocated to the east of the landslide. Ultimately, the lake was drained by a low-level drainage tunnel.

The University of Utah Bureau of Economic and Business Research and others (1984) evaluated both direct and indirect costs of the Thistle landslide. Direct costs, which included emergency measures to prevent failure of the landslide dam and costs of relocations of the transportation routes, were about $200 million. Indirect costs were not totaled but may have exceeded the direct costs. For example, during the period that transportation was disrupted through the canyon, more than 2,500 jobs were lost in the mining industry, and the railroad lost $81 million in revenue.

In addition to producing enormous costs of damages to property, slope movements in the U.S. are responsible for an average of more than 25 fatalities per year (Krohn and Slosson, 1976). In general, fatalities from slope failures in North America have been much fewer than in other parts of the world. Eisbacher and Clague (1984) listed 17 slope-movement events in Europe between the 13th and 20th centuries that produced more than 100 fatalities each. Debris avalanches in 1962 and 1970 from Mt. Huascaran in Peru killed more than 20,000 people. Fatalities from the debris avalanche from the slopes of Nevada del Ruiz in 1985 in Colombia killed at least 20,000 people (Herd, 1986). In North America, the major slope-movement events in this century in terms of fatalities include debris avalanches triggered by Hurricane Camille in 1969 in central Virginia (Williams and Guy, 1973) and the disastrous slope movements in Puerto Rico during 1985 that included one landslide that killed at least 129 persons (Jibson, 1986).

Technology is available to greatly reduce losses from different types of slope movements by avoiding or preventing them. With respect to hillside developments, Leighton (1976) estimated that a reduction in damaging slope failures of 95 to 99 percent is attainable through use of three levels of investigation: (1) regional-areal, (2) tract or community, and (3) site, with more detailed investigations progressively applied to the smaller areas. Terzaghi (1950, p. 110) commented that most slope-movement types are predictable:

> It has often been stated that certain slides occurred without warning. Yet no slide can take place unless the ratio between the average shearing resistance of the ground and the average shearing stresses on the potential surface of sliding has previously decreased from an initial value greater than one to unity at the instant of the slide. . . . Hence, if a landslide comes as a surprise to the eyewitnesses, it would be more accurate to say that the observers failed to detect the phenomena which preceded the slide.

These comments about precursors to slope movements provide encouragement for the continuing study of hillslope processes and development of techniques and instruments with the ultimate aim to predict the most likely sites for failures of various types, movement rates, and the times and magnitudes of the failures.

The best example of success of a landslide-hazard-reduction program is that enacted by the City of Los Angeles, California. In the years immediately following World War II, a great deal of hillside development was undertaken in Los Angeles with little or no regulation. Severe storms during the winter of 1951–1952 produced erosion, sedimentation, and landslide damages estimated at $7.5 million (Jahns, 1969). The mayor appointed an ad hoc committee, which included private citizens, representatives of the construction industry, and professional people in private industry and city government. The committee prepared a grading ordinance, which was adopted by the city. After the grading regulations were modified and strengthened several times during the 1950s, a severe storm in 1962 produced extensive damages. Groups were organized to review and revise the grading regulations and, in 1963, a new ordinance was adopted. The new ordinance provided a more restrictive grading code and required greater geologic and engineering participation in design, construction, final inspection, and certification of a grading project (Jahns, 1969). Thus, construction on hillslopes in Los Angeles occurred during three periods of time, each with different sets of grading regulations. Prior to 1952, there was virtually no regulation. During the period 1952–1963, there was an evolving code with limited constraints on grading. Since 1963, there has been a stringent grading code in effect (Slosson, 1969).

The winters of 1968–1969 and 1977–1978 provided stern tests of the effectiveness of the different regulations. The Department of Building and Safety of the City of Los Angeles conducted a thorough analysis of the damage caused during 1969 according to when a particular property was developed. Results are presented in Table 2. Comparing the data for damage to sites constructed during each of the three time periods with different grading codes, the sites constructed prior to 1952 sustained 18 times as much total damage as did those developed after 1963. Less than 3 percent of the total damage occurred to sites developed after the stronger grading regulations. Similarly, the near-record rainstorms of 1977–1978 produced major damages, 93 percent of which were to sites developed before the grading regulations of 1963 (Fleming and others, 1979).

OUTLOOK

The impressive advances in slope-movement technology during the past 30 years will certainly continue in methods of analysis, evaluation of hazards, instrumentation, and predictions. As classification of slope movements becomes more precise, clearer recognition of the different types of slope movements also will lead to a better understanding of the processes controlling the movements.

Methods of analysis are needed to refine empirical estimates of rate of movement and travel distance, particularly of earth flows and debris flows. Research is underway to provide improvements in analytical methods (Iverson, 1985, 1986; Cannon,

TABLE 2. DAMAGE ASSOCIATED WITH DESTRUCTIVE STORMS OF 1969 IN HILLSIDE AREAS OF LOS ANGELES
(Adapted from Slosson, 1969)

	Sites Developed Prior to 1952	Sites Developed 1952-1962	Sites Developed 1963-1969
Number of sites constructed	10,000	27,000	11,000
Total damage	$3,300,000	$2,767,000	$182,400*
Average damage per site	$330	$100	$7**
Percentage of sites damaged	10.4	1.3	0.15

*More than $100,000 of the $182,400 in damages occurred to sites that were currently being graded. Even the best of grading projects are susceptible to damage during construction.

**If the total damage value is used, the average damage value per site is about $17. The value of $7 per site was obtained by deducting the damages to sites under construction.

1986; Savage and Smith, 1986), but progress in analytical evaluations has been hampered by a lack of high-quality field and laboratory data. Improved laboratory testing methods are needed to investigate the properties of flows as well as strength characteristics of soil and rock masses. Field measurements of subsurface water flow and pressures and internal deformations are needed to improve methods of analysis and predictions.

For certain types of slope-movement problems, predictions of hazards are best made as areal evaluations and the results presented in map form. Experimentation with mapping different types of slope movements should continue, and the accuracy of previously published maps should be assessed as failure events put them to the test.

While clearly the technology to cope with slope-movement problems will improve, it is less certain that the improvements will result in a proportionate reduction of damages. During the past 30 years, in spite of much improved technology, damages from slope movements have continued to increase. In part, this may be the result of more complete reporting of damage, but largely it appears to be the result of increasing use of less favorable sites for housing, highways, and other construction and the failure of communities to impose reasonable constraints on development. Long-term reductions in damages require the continuing education, support, and mutual cooperation of all levels of government, the general public, and those supplying the technology.

REFERENCES CITED

Abrahams, A. D., ed., 1986, Hillslope processes; 16th Annual Geomorphology Symposium, State University of New York at Buffalo, September 28–29, 1985: Winchester, Massachusetts, Allen and Unwin, Inc., 416 p.

Ackenheil, A. C., 1954, A soil mechanics and engineering geology analysis of landslides in the area of Pittsburgh, Pennsylvania [Ph.D. thesis]: Pittsburgh, Pennsylvania, University of Pittsburgh, 121 p. (Ann Arbor, Michigan, University Microfilms, Publication no. 9957, 1962.)

Association of Engineering Geologists, 1984, Southern California; Section news: Newsletter, v. 27, no. 3, p. 34.

Bhandari, R. K., 1984, Simple and economical instrumentation and warning systems for landslides and other mass movements, *in* Proceedings, IV International Symposium on Landslides, Toronto, September, v. 1: Downsview, Ontario, Canadian Geotectonical Society, p. 251–273.

Bishop, A. W., 1955, The use of the slip circle in the stability analysis of slopes: Geotechnique, v. 5, no. 1, p. 7–17.

Bjerrum, L., 1967, Progressive failure in slopes of overconsolidated plastic clays and clay shales, *in* Proceedings, American Society of Civil Engineers: Soil Mechanics and Foundation Engineering Division Journal, v. 93, no. SM5, p. 1–49.

Brabb, E. E., 1984, Innovative approaches to landslide hazard and risk mapping, *in* Proceedings, IV International Symposium on Landslides, Toronto, Canada, September, v. 1: Downsview, Ontario, Canadian Geotechnical Society, p. 307–323.

——, 1985, On the line; Losing by a landslide: Boulder, University of Colorado, Institute of Behavioral Sciences, Natural Hazards Observer, v. 10, no. 2, p. 6.

Briggs, R. P., Pomeroy, J. S., and Davies, W. E., 1975, Landsliding in Allegheny County, Pennsylvania: U.S. Geological Survey Circular 728, 18 p.

Broms, B. B., 1975, Landslides, *in* Winterkorn, H. F., and Fang, H., eds., Foundation engineering handbook: New York, Van Nostrand Reinhold Co., p. 373–401.

Caine, N., 1982, Topping failures from alpine cliffs on Ben Lomond, Tasmania: Earth Surface Processes and Landforms, v. 7, p. 133–152.

Campbell, R. H., 1975, Soil slips, debris flows, and rainstorms in the Santa Monica Mountains and vicinity, Southern California: U.S. Geological Survey Professional Paper 851, 51 p.

Campbell, R. H., and 7 others, 1985, Landslide classification for identification of mud flows and other landslide hazards, *in* Campbell, R. H., ed., Feasibility of a nationwide program for the identification and delineation of hazards from mud flows and other landslide hazards: U.S. Geological Survey Open-File Report 85-276A, 24 p.

Cannon, S. H., 1986, The lag rate and the travel-distance potential of debris flows: Geological Society of America Abstracts with Programs, v. 18, p. 93.

Cannon, S. H., and Ellen, S., 1985, Rainfall conditions for abundant debris avalanches; San Francisco Bay region, California: California Geology, v. 38, no. 12, p. 267–272.

Carson, M. A., 1979, Le glissment de Rigaud (Québec) du 3 Mai 1978; Une interpretation du mode de rupture d'aprés la morphologie de la cicatrice: Geographie Physique et Quaternaire, v. 33, no. 1, p. 63–92.

Carson, M. A., and Kirkby, M. J., 1972, Hillslope form and process: London, Cambridge University Press, 475 p.

Carter, M., and Bentley, S. P., 1985, The geometry of slip surfaces beneath landslides; Predictions from surface measurements: Canadian Geotechnical Journal, v. 22, p. 234–238.

Casagrande, A., 1949, Soil mechanics in the design and construction of the Logan Airport: Boston Society of Civil Engineers Journal, v. 36, no. 6. Reprinted in Contributions to Soil Mechanics 1941–1953, Boston Society of Civil Engineers, Boston, Massachusetts, 1953, p. 176–205.

Chowdhury, R. N., 1978, Slope analysis, *in* Developments in geotechnical engineering 22: Amsterdam, Elsevier Scientific Publishing Co., 423 p.

—— , 1980, Landslides as natural hazards; Mechanisms and uncertainties: Geotechnical Engineering, v. 11, no. 2, p. 135–180.

—— , 1984, Risk estimation for failure progression along a slip surface, *in* Proceedings, IV International Symposium on Landslides, Toronto, September, v. 2: Downsview, Ontario, Canadian Geotechnical Society, p. 281–386.

Collin, A., 1846, Landslides in clays, *translated by* W. R. Schriever: Toronto, University of Toronto Press, 1956, 181 p.

Colman, S. M., and Pierce, K. L. (compilers), 1977, Summary table of Quaternary dating methods: U.S. Geological Survey Miscellaneous Field Studies Map MF-904.

Crandell, D. R., 1971, Postglacial lahars from Mount Rainier Volcano, Washington: U.S. Geological Survey Professional Paper 677, 75 p.

Crandell, D. R., and Fahnestock, R. K., 1965, Rockfalls and avalanches from Little Tahoma Peak on Mount Rainier, Washington: U.S. Geological Survey Bulletin 1221-A, 30 p.

Crandell, D. R., and Mullineaux, D. R., 1978, Potential hazards from future eruptions of Mount St. Helens Volcano, Washington: U.S. Geological Survey Bulletin 1383-C, 26 p.

Crandell, D. R., and Varnes, D. J., 1961, Movement of the Slumgullion earthflow near Lake City, Colorado, Article 57 *in* Geological Survey Research 1961; Short papers in the geologic and hydrologic sciences: U.S. Geological Survey Professional Paper 424-B, p. B136–B139.

Crandell, D. R., Miller, C. D., Glicken, H. X., Christiansen, R. L., and Newhall, C. G., 1984, Catastrophic debris avalanche from ancestral Mount Shasta Volcano, California: Geology, v. 12, p. 143–146.

Cruden, D. M., 1985, Rock slope movements in the Canadian cordillera: Canadian Geotechnical Journal, v. 22, p. 528–540.

—— , 1986, Discussion *on* The geometry of slip surfaces beneath landslides; Predictions from surface measurements: Canadian Geotechnical Journal, v. 23, p. 94.

Cruden, D. M., and Hungr, O., 1986, The debris of the Frank slide and theories of rockslide–avalanche mobility: Canadian Journal of Earth Sciences, v. 23, p. 425–432.

Davies, W. E., Bailey, J. F., and Kelly, D. B., 1972, West Virginia's Buffalo Creek flood; A study of the hydrology and engineering geology: U.S. Geological Survey Circular 667, 32 p.

Davis, W. M., 1899, The geographical cycle: Royal Geographical Society of London Geographical Journal, v. 14, p. 481–504.

Deere, D. U., and Patton, F. D., 1971, Slope stability in residual soils, *in* Proceedings, Panamerican Soil Mechanics and Foundation Engineering Conference, 4th, San Juan, Puerto Rico: New York, American Society of Civil Engineers, p. 87–170.

deFreitas, M. H., and Watters, R. J., 1973, Some field examples of toppling failure: Geotechnique, v. 23, no. 4, p. 495–514.

Dueholm, K. S., and Garde, A. A., 1986, Geologic photogrammetry using standard color slides: Reprinted from Research Activities for 1985, Institute of Surveying and Photogrammetry, Technical University of Denmark, 10 p.

Duncan, J. M., 1972, Finite element analysis of stresses and movements in dams, excavations, and slopes; State of the art, *in* Desai, C. S., ed., Application of the finite element method in geotechnical engineering; A Symposium: Vicksburg, Mississippi, U.S. Army Engineer Waterways Experiment Station, p. 267–326.

Eckel, E. B., 1958a, Introduction, *in* Eckel, E. B., ed., Landslides and engineering practice: National Academy of Sciences, National Research Council Publication 544, Highway Research Board Special Report 29, p. 1–5.

—— , ed., 1958b, Landslides and engineering practice: National Academy of Sciences, National Research Council Publication 544, Highway Research Board Special Report 29, 232 p.

Eisbacher, G. H., 1979, Cliff collapse and rock avalanches (sturzstroms) in the MacKenzie Mountains, northwestern Canada: Canadian Geotechnical Journal, v. 16, p. 309–334.

Eisbacher, G. H., and Clague, J. J., 1984, Destructive mass movements in high mountains; Hazard and management: Ottawa, Geological Survey of Canada Paper 84-16, 230 p.

Ellen, S. D., and Wieczorek, G. F., eds., 1988, Landslides, floods, and marine effects of th storm of January 3–5, 1982, in the San Francisco Bay region, California: U.S. Geological Survey Professional Paper 1434 (in press).

Evans, R., Valliappan, S., McGuckin, D., and Raja Sekar, H. L., 1981, Stability analysis of a rock slope against toppling failure, *in* Proceedings, International Symposium on Weak Rock, Tokyo, p. 665–670.

Fang, H., 1975, Stability of earth slopes, *in* Winterkorn, H. F., and Fang, H., eds., Foundation engineering handbook: New York, Van Nostrand Reinhold Co., p. 354–372.

Fellenius, W., 1927, Erdstatische Berechnungen (Calculation of stability of slopes): Berlin, W. Ernst und Sohn (revised edition, 1939), 40 p.

Fleming, R. W., and Taylor, F. A., 1980, Estimating the costs of landslide damage in the United States: U.S. Geological Survey Circular 832, 21 p.

Fleming, R. W., Varnes, D. J., and Schuster, R. L., 1979, Landslide hazards and their reduction: U.S. Geological Survey Yearbook, Fiscal Year 1978, p. 13–21.

Fleming, R. W., Johnson, A. M., and Hough, J. E., 1981, Engineering geology of the Cincinnati area (Geological Society of America guidebook, field trip 18), *in* Roberts, T. G., ed., Cincinnati '81 field trip guidebooks; Geomorphology, hydrogeology, geoarcheology, engineering geology: American Geological Institute, v. 3, p. 543–570.

Fraser, C. S., and Gruendig, L., 1985, The analysis of photogrammetric deformation measurements on Turtle Mountain: Photogrammetric Engineering and Remote Sensing, v. 51, no. 2, p. 207–216.

Fredlund, D. G., 1984, Analytical methods for slope stability analysis, *in* Proceedings, IV International Symposium on Landslides, Toronto, September, v. 1: Downsview, Ontario, Canadian Geotechnical Society, p. 229–250.

Fukuoka, M., 1980, Instrumentation; Its role in landslide prediction and control, *in* Proceedings, International Symposium on Landslides, New Delhi, India, April, v. 2, p. 139–153.

Goodman, R. E., and Bray, J. W., 1976, Toppling of rock slopes, *in* Proceedings, Special Conference on Rock Engineering for Foundations and Slopes, Boulder, Colorado: American Society of Civil Engineers, v. 2, p. 201–234.

Gray, R. E., and Gardner, G. D., 1977, Processes of colluvial slope development at McMechan, West Virginia: International Association of Engineering Geology Bulletin, no. 16, p. 29–32.

Gregersen, O., 1981, The quick clay landslide in Rissa, Norway, *in* Proceedings, Tenth International Conference of Soil Mechanics and Foundation Engineering, Stockholm, v. 3, Session 11, p. 421–426.

Hamel, J. V., and Flint, N. K., 1972, Failure of a colluvial slope, *in* Proceedings, American Society of Civil Engineers: Soil Mechanics and Foundations Division Journal, v. 98, no. SM2, p. 167–180.

Handy, R. L., 1973, Collapsible loess in Iowa, *in* Proceedings: Soil Science Society of America, v. 37, no. 2, p. 281–284.

Hanna, T. H., 1985, Field instrumentation in geotechnical engineering: Germany, Clausthal-Zellerfeld, Trans Tech Publications, Series on Rock and Soil Mechanics, v. 10, 843 p.

Hansen, W. R., 1965, Effects of the earthquake of March 27, 1964, at Anchorage, Alaska: U.S. Geological Survey Professional Paper 542-A, 68 p.

Heim, A., 1882, Der Bergsturz von Elm: Zeitschrift der Deutschen Geologischen Gesellschaft, v. 34, p. 74–115.

Herd, D. G., 1986, 1985 eruption of Nevado del Ruiz, Columbia; Overview: EOS American Geophysical Union Transactions, v. 67, no. 16, p. 402–403.

Hewitt, K., 1976, Earthquake hazards in the mountains: Natural History, v. 85, no. 5, p. 30–37.

Hocking, G., 1978, Analysis of toppling-sliding mechanisms for rock slopes, *in* Proceedings, U.S. Rock Mechanics Symposium, 19th, Stateline, Nevada, p. 288–295.

Hoek, E., and Bray, J. W., 1974, Rock slope engineering: London, Institution of Mining and Metallurgy, 309 p.

Hoexter, D. F., Holzhausen, G., and Soto, A. E., 1978, A method of evaluating the relative stability of ground for hillside development: Engineering Geology, v. 12, no. 4, p. 319–336.

Hollingsworth, R., and Kovacs, G. S., 1981, Soil slumps and debris flows; Prediction and protection: Association of Engineering Geologists Bulletin, v. 18, no. 1, p. 17–28.

Howard, K. A., 1973, Avalanche mode of motion; Implications from lunar examples: Science, v. 180, no. 4090, p. 1052–1055.

Howe, E., 1909, Landslides in the San Juan Mountains, Colorado: U.S. Geological Survey Professional Paper 67, 58 p.

Hsü, K. J., 1975, Catastrophic debris streams (sturzstroms) generated by rockfalls: Geological Society of America Bulletin, v. 86, p. 129–140.

Huggett, R. J., 1985, Earth surface systems: Berlin, Springer-Verlag, 270 p.

Humbert, M., 1977, Risk-mapping of areas exposed to movements of soil and subsoil; French "ZERMOS" maps [in French with English summary]: International Association of Engineering Geology Bulletin, v. 16, p. 80–82.

Hutchinson, J. N., 1983, Methods of locating slip surfaces in landslides: Association of Engineering Geologists Bulletin, v. 20, no. 3, p. 235–252.

Iverson, R. M., 1985, A constitutive equation for mass-movement behavior: Journal of Geology, v. 93, p. 143–160.

——, 1986, Unsteady, nonuniform landslide motion; 2. Linearized theory and the kinematics of transient response: Journal of Geology, v. 94, p. 349–364.

Jahns, R. H., 1969, Seventeen years of response by the City of Los Angeles to geologic hazards, *in* Olsen, R. A., and Wallace, M. M., eds., Geologic hazards and public problems; U.S. Office of Emergency Preparedness, Region 7, Conference Proceedings, San Francisco, Calif.: Washington, D.C., U.S. Government Printing Office, p. 283–296.

Jakobson, B., 1952, The landslide at Surte on the Gota River, *in* Proceedings: Royal Swedish Geotechnical Institute, v. 5, 121 p.

Janbu, N., 1954, Stability analysis of slopes with dimensionless parameters: Cambridge, Massachusetts, Harvard University, Harvard Soil Mechanics Series, no. 46, 80 p.

Jibson, R. W., 1986, Evaluation of landslide hazards resulting from the 5–8 October 1985, storm in Puerto Rico: U.S. Geological Survey Open-File Report 86-26, 40 p.

Johnson, A. M., 1970, Physical processes in geology: San Francisco, Freeman, Cooper and Company, 577 p.

——, with contributions by Rodine, J. D., 1984, Debris flow, *in* Brunsden, D., and Prior, D. B., eds., Slope instability: New York, John Wiley and Sons, p. 257–361.

Keefer, D. K., 1984, Landslides caused by earthquakes: Geological Society of America Bulletin, v. 95, p. 406–421.

Keefer, D. K., and Johnson, A. M., 1983, Earth flows; Morphology, mobilization, and movement: U.S. Geological Survey Professional Paper 1264, 56 p.

Kennedy, B. A., Niermeyer, K. E., Fahm, B. A., and Bratt, J. A., 1971, A case study of slope stability at the Chuquicamata Mine, Chile: American Institute of Mining and Metallurgical Engineers, Society of Mining Engineers Transactions, v. 250, p. 55–61.

Kenney, T. C., 1968, A review of recent research on strength and consolidation of soft sensitive clays: Canadian Geotechnical Journal, v. 2, p. 97–119.

Kenney, T. C., and Drury, P., 1973, Case record of the slope failure that initiated the retrogressive quick-clay landside at Ullensaker, Norway: Geotechnique, v. 23, no. 1, p. 33–47.

Kienholz, H., 1978, Geomorphology and natural hazards of Grindelwald, Switzerland, scale 1:10,000: Arctic and Alpine Research, v. 10, no. 2, p. 169–184.

——, 1985, Assessment of slope stability in the Nepalese Middle Mountains, *in* Proceedings, IVth International Conference and Field Workshop on Landslides, Tokyo, August, p. 5–10.

Knight, D. K., 1963, Oahe dam; Geology, embankment, and cut slopes: American Society of Civil Engineers, Soil Mechanics Foundation Division Journal, v. 89, no. SM2, p. 99–125.

Kockelman, W. J., 1986, Some techniques for reducing landslide hazards: Association of Engineering Geologists Bulletin, v. 33, no. 1, p. 29–52.

Kojan, E., and Hutchinson, J. N., 1978, Mayunmarca rockslide and debris flow, Peru, *in* Voight, B., ed., Rockslides and avalanches; 1. Natural phenomena: Amsterdam, Elsevier Scientific Publishing Company, Developments in Geotechnical Engineering 14A, p. 315–361.

Krohn, J. P., and Slosson, J. E., 1976, Landslide potential in the United States: California Geology, v. 29, no. 10, p. 224–231.

Lee, K. L., and Duncan, J. M., 1975, Landslide of April 25, 1974, on the Mantaro River, Peru: Washington, D.C., National Academy of Sciences, National Research Council, Committee on Natural Disasters, Commission on Sociotechnical Systems, 71 p.

Legget, R. F., 1973, Cities and geology: New York, McGraw-Hill Book Co., 624 p.

Leighton, F. B., 1976, Urban landslides; Targets for land-use planning in California, *in* Coates, D. R., ed., Urban geomorphology: Geological Society of America Special Paper 174, p. 37–60.

Lessing, P., Kulander, B. R., Wilson, B. D., and others, 1976, West Virginia landslides and slide-prone areas: West Virginia Geological and Economic Survey, Environmental Geology Bulletin 15, 64 p., 28 maps, scale 1:24,000.

Li, T., 1983, A mathematical model for predicting the extent of a major rockfall: Zeitschrift fur Geomorphologie N. F., v. 27, no. 24, p. 473–482.

Londe, P., 1982, Concepts and instruments for improved monitoring, *in* Proceedings: American Society of Civil Engineers; Geotechnical Engineering Division Journal, v. 108, no. GT6, p. 820–834.

Lutton, R. J., 1971, A mechanism for progressive rock mass failure as revealed by loess slumps: International Journal of Rock Mechanics and Mining Science, v. 8, no. 2, p. 143–151.

Lutton, R. J., Banks, D. C., and Strohm, W. E., Jr., 1979, Slides in Gaillard Cut, Panama Canal Zone, *in* Voight, B., ed., Rockslides and avalanches; 2. Engineering sites: Amsterdam, Elsevier Scientific Publishing Company, Developments in Geotechnical Engineering 14B, p. 151–224.

MacDonald, D. F., 1942, Panama Canal slides; The third locks project: Balboa Heights, Canal Zone, The Panama Canal Company, Department of Operation and Maintenance, Special Engineering Division, 73 p. (Reprinted 1947.)

Matheson, D. S., and Thomson, S., 1973, Geological implications of valley rebound: Canadian Journal of Earth Sciences, v. 10, no. 6, p. 961–978.

McConnell, R. G., and Brock, R. W., 1904, Part 8 of Report on the great landslide at Frank, Alberta: Department of the Interior (Canada), Annual Report for 1903, 17 p.

McHarg, I. L., 1969, Design with nature: New York, Natural History Press, 197 p.

Mears, A. I., 1977, Debris-flow hazard analyses and mitigation; an example from

Glenwood Springs, Colorado: Colorado Geological Survey Information Series 8, 45 p., map scale 1:4,800.

Melosh, H. J., 1983, Acoustic fluidization: American Scientist, v. 71, p. 158–165.

Mencl, V., 1977, Modern methods used in the study of mass movements: International Association of Engineering Geology Bulletin, no. 16, p. 185–197.

Miller, C. D., Mullineaux, D. R., and Crandell, D. R., 1981, Hazards assessments at Mount St. Helens, *in* Lipman, P. W., and Mullineaux, D. R., eds., The 1980 eruptions of Mount St. Helens, Washington: U.S. Geological Survey Professional Paper 1250, p. 789–799.

Morgenstern, N. R., and Sangrey, D. A., 1978, Methods of stability analysis, *in* Schuster, R. L., and Krizek, R. J., eds., Landslides; Analysis and control: Washington, D.C., National Academy of Sciences, Transportation Research Board Special Report 176, p. 155–171.

Müller, L., 1979, Josef Stini; Contributions to engineering geology and slope movement investigations, *in* Voight, B., ed., Rockslides and avalanches; 2. Engineering sites: Amsterdam, Elsevier Scientific Publishing Co., Developments in Geotechnical Engineering 14B, p. 94–109.

National Research Council, Committee on Ground Failure Hazards, 1985, Recommendations for reducing losses from landsliding in the United States: Washington, D.C., National Academy of Sciences, 41 p.

Nilsen, T. H., and Turner, B. L., 1975, Influence of rainfall and ancient landslide deposits on recent landslides (1950–71) in urban areas of Contra Costa County, California: U.S. Geological Survey Bulletin 1388, 18 p.

Nilsen, T. H., Taylor, F. A., and Dean, R. M., 1976, Natural conditions that control landsliding in the San Francisco Bay region; An analysis based on data from the 1968–69 and 1972–73 rainy seasons: U.S. Geological Survey Bulletin 1424, 35 p.

Pariseau, W. G., and Voight, B., 1979, Rockslides and avalanches; Basic principles and perspectives in the realm of civil and mining operations, *in* Voight, B., ed., Rockslides and avalanches; 2. Engineering sites: Amsterdam, Elsevier Scientific Publishing Co., Developments in Geotechnical Engineering 14B, p. 1–92.

Patton, F. D., 1983, The role of instrumentation in the analysis of the stability of rock slopes: International Symposium on Field Measurements in Geomechanics, Zurich, September 1983, p. 719–748.

Peck, R. B., 1967, Stability of natural slopes: American Society of Civil Engineers, Journal of the Soil Mechanics and Foundations Division, v. 93, no. SM4, p. 403–417.

Penck, W., 1924, Die Morphologische Analyse; Stuttgart (Translated as "Morphological Analysis of Landforms"): London, MacMillan, 429 p., 1953.

Peterson, R., 1956, Rebound in the Bearpaw shale, western Canada: Geological Society of America Bulletin, v. 69, p. 1113–1124.

Pierson, T. C., and Costa, J. E., 1987, A rheologic classification of subaerial sediment-water flow, *in* Costa, J. E., and Wieczorek, G. F., eds., Debris flows/avalanches; Process, recognition, and mitigation: Geological Society of America, Reviews in Engineering Geology, v. 7, p. 1–12.

Pilot, G., 1984, Instrumentation and warning systems for research and complex slope stability problems, *in* Proceedings, IV International Symposium on Landslides, Toronto, September, v. 1: Downsview, Ontario, Canadian Geotechnical Society, p. 275–305.

Piteau, D. R., and Peckover, F. L., 1978, Engineering of rock slopes, *in* Schuster, R. L., and Krizek, R. J., eds., Landslides; Analysis and control: Washington, D.C., National Academy of Sciences, Transportation Research Board Special Report 176, p. 192–228.

Plafker, G., Ericksen, G. E., and Concha, J. F., 1971, Geological aspects of the May 31, 1970, Peru earthquake: Seismological Society of America Bulletin, v. 61, no. 3, p. 543–578.

Prior, D. B., and Coleman, J. M., 1982, Active slides and flows in underconsolidated marine sediments on the slopes of the Mississippi Delta, *in* Saxov, S., and Nieuwenhuis, J. K., eds., Marine slides and other mass movements: New York, Plenum Publishing Corporation, p. 21–49.

Prostka, H. J., 1978, Heart Mountain fault and Absaroka volcanism, Wyoming and Montana, U.S.A., *in* Voight, B., ed., Rockslides and avalanches; 1. Natural phenomena: Amsterdam, Elsevier Scientific Publishing Company, Developments in Geotechnical Engineering 14A, p. 423–437.

Radbruch, D. H., and Crowther, K. C., 1973, Map showing areas of estimated relative amounts of landslides in California: U.S. Geological Survey Miscellaneous Geologic Investigations Map I–747, scale 1:1,000,000.

Radbruch-Hall, D. H., 1978, Gravitational creep of rock masses on slopes, *in* Voight, B., ed., Rockslides and avalanches; 1. Natural phenomena: Amsterdam, Elsevier Scientific Publishing Company, Developments in Geotechnical Engineering 14A, p. 607–657.

Riestenberg, M. M., and Sovonick-Dunford, S., 1983, The role of woody vegetation in stabilizing slopes in the Cincinnati area, Ohio: Geological Society of America Bulletin, v. 94, no. 4, p. 506–518.

Ritchie, A. M., 1963, Evaluation of rockfall and its control: National Academy of Sciences, National Research Council Publication 1114, Highway Research Board, Highway Research Record 17, p. 13–28.

Rodine, J. D., 1974, Analysis of the mobilization of debris flows [Ph.D. thesis]: Stanford, California, Stanford University, 226 p.

Saito, M., 1969, Forecasting time of slope failure by tertiary creep, *in* Proceedings, International Conference on Soil Mechanics and Foundation Engineering, 7th, Mexico City, Mexico, v. 2, p. 677–683.

——, 1980, Evidential study on forecasting occurrence of slope failure: Tokyo, OYO Corporation, Report No. RP-4116, Technical Note 38, 17 p.

Sangrey, D. A., and Garrison, L. E., 1977, Submarine landslides: U.S. Geological Survey Yearbook Fiscal Year 1977, p. 53–63.

Savage, W. Z., and Smith, W. K., 1986, A model for the plastic flow of landslides: U.S. Geological Survey Professional Paper 1385, 32 p.

Scheidegger, A. E., 1970, Theoretical geomorphology (2nd rev. ed.): Berlin and New York, Springer-Verlag, 435 p., originally published 1961.

——, 1975, Physical aspects of natural catastrophes: Amsterdam, Elsevier Scientific Publishing Co., 289 p.

Schumm, S. A., 1956, Evolution of drainage systems and slopes in badlands at Perth Amboy, New Jersey: Geological Society of America Bulletin, v. 67, no. 5, p. 597–646.

Schumm, S. A., and Chorley, R. J., 1964, The fall of threatening rock: American Journal of Science, v. 262, no. 9, p. 1041–1054.

Schuster, R. L., and Fleming, R. W., 1986, Economic losses and fatalities due to landslides: Association of Engineering Geologists Bulletin, v. 23, no. 1, p. 11–28.

Schuster, R. L., and Krizek, R. J., eds., 1978, Landslides; Analysis and control: National Academy of Sciences, National Research Council, Transportation Research Board Special Report 176, 234 p.

Sharpe, C.F.S., 1938, Landslides and related phenomena; A study of mass movements of soil and rock: New York, Columbia University Press, 137 p.

Shearer, C. F., Taylor, F. A., and Fleming, R. W., 1983, Distribution and costs of landslides in San Diego County, California, during the rainfall years of 1978–79 and 1979–80: U.S. Geological Survey Open–File Report 83–582, 15 p.

Shreve, R. L., 1968, The Blackhawk landslide: Geological Society of America Special Paper 108, 48 p.

Sidle, R. C., Pearce, A. J., and O'Loughlin, C. L., 1985, Hillslope stability and land use: Washington, D.C., American Geophysical Union, Water Resources Monograph 11, 140 p.

Skempton, A. W., 1960, Significance of Terzaghi's concept of effective stress, *in* Terzaghi, K., From theory to practice in soil mechanics: New York, John Wiley and Sons, p. 42–53.

——, 1964, Long-term stability of clay slopes: Geotechnique, v. 14, no. 2, p. 75–102.

——, 1970, First-time slides in over-consolidated clays: Geotechnique, v. 20, no. 3, p. 320–324.

Slosson, J. E., 1969, The role of engineering geology in urban planning, *in* The Governor's Conference on Environmental Geology: Colorado Geological Survey Special Publication 1, p. 8–15.

Slosson, J. E., and Krohn, J. P., 1982, Southern California landslides of 1978 and 1980, *in* Proceedings of a symposium; Storms, floods, and debris flows in

Southern California and Arizona 1978 and 1980, Pasadena, California, September 1980, p. 291–304, Available from U.S. Department of Commerce, National Technical Information Service, Springfield, Virginia, (Rep. CSS-CND-019).

Smith, R., 1958, Economic and legal aspects, *in* Eckel, E. B., ed., Landslides and engineering practice: Washington, D.C., National Academy of Sciences, National Research Council Publication 544, Highway Research Board Special Report 29, p. 6–19.

Smith, T. C., and Hart, E. W., 1982, Landslides and related storm damage; January 1982, San Francisco Bay region: California Geology, v. 35, no. 7, p. 139–152.

Stini, J., 1910, Die Muren; Versuch einer Monographie mit besonderer Berücksichtigung der Verhältnisse in den Tiroler Alpen: Innsbruck, Wagner, 139 p.

Strahler, A. N., 1956, Quantitative slope analysis: Geological Society of America Bulletin, v. 67, no. 5, p. 571–596.

Taylor, D. W., 1937, Stability of earth slopes: Boston Society of Civil Engineers Journal, v. 24, no. 3, p. 197–246.

——, 1948, Fundamentals of soil mechanics: New York, John Wiley and Sons, 700 p.

Taylor, F. A., and Brabb, E. E., 1972, Maps showing distribution and cost by counties of structurally damaging landslides in the San Francisco Bay region, California, winter of 1968–69: U.S. Geological Survey Miscellaneous Field Studies Map MF-327, scale 1:500,000.

Ter-Stepanian, G., 1980, Measuring displacements of wooded landslides with trilateral signs, *in* Proceedings, International Symposium on Landslides, 3rd, New Delhi, India, v. 1, p. 355–359.

Terzaghi, K., 1925, Erdbaumechanik auf bodenphysikalischer Grundlage: Vienna, Deuticke, 399 p.

——, 1950, Mechanism of landslides, *in* Paige, S., ed., Application of geology to engineering practice: Geological Society of America, Engineering Geology, Berkey Volume, p. 83–123.

——, 1962, Stability of steep slopes on hard unweathered rock: Geotechnique, v. 12, p. 251–270.

Tilling, R. I., 1976, Rockfall activity in pit craters, Kilauea Volcano, Hawaii, *in* Proceedings, Andean and Antarctic Volcanology Problems Symposium, Santiago, Chile, September 1974: International Association of Volcanology and Chemistry of the Earth's Interior, p. 518–528.

Tinsley, J. C., Youd, T. L., Perkins, D. M., and Chen, A.T.F., 1985, Evaluating liquefaction potential, *in* Ziony, J. I., ed., Evaluating earthquake hazards in the Los Angeles region; An earth-science perspective: U.S. Geological Survey Professional Paper 1360, p. 263–315.

Toms, A. H., and Bartlett, D. L., 1962, Applications of soil mechanics in the design of stabilizing works for embankments, cuttings, and track formations, *in* Proceedings, Institution of Civil Engineers, London, v. 21, p. 705–711. Reprinted in National Academy of Sciences, National Research Council, Transportation Research Board, Landslide Instrumentation, Transportation Research Record 482, 1974, 51 p.

Tubbs, D. W., 1974, Landslides and associated damage during early 1972 in part of west-central King County, Washington: U.S. Geological Survey Miscellaneous Investigations Map I-852-B, scale 1:48,000.

University of Utah Bureau of Economic and Business Research; Utah Department of Community and Economic Development; and Utah Office of Planning and Budget, 1984, Flooding and landslides in Utah; An economic impact analysis: Salt Lake City, Utah, University of Utah Bureau of Economic and Business Research, 123 p.

Vanmarcke, E., 1977, Reliability of earth slopes, *in* American Society of Civil Engineers, November, v. 103: Geotechnical Engineering Division Journal, no. GT11, p. 1247–1265.

Varnes, D. J., 1958, Landslide types and processes, *in* Eckel, E. B., ed., Landslides and engineering practices: National Academy of Sciences, National Research Council Publication 544, Highway Research Board Special Report 29, p. 20–47.

——, 1974, The logic of geological maps, with reference to their interpretation and use for engineering purposes: U.S. Geological Survey Professional Paper 837, 48 p.

——, 1978, Slope movement types and processes, *in* Schuster, R. L., and Krizek, R. J., eds., Landslides; Analysis and control: National Academy of Sciences, National Research Council, Transportation Research Board Special Report 176, p. 11–33.

——, 1983, Time-deformation relations in creep to failure of earth materials, *in* Proceedings, Southeast Asian Geotechnical Conference, 7th, Hong Kong, November 1982, v. 2, p. 107–130.

——, 1984, Landslide hazard zonation; A review of principles and practice: United Nations Educational, Scientific, and Cultural Organization (UNESCO), 7 place de Fontenoy, 75700 Paris, France, 63 p.

Veder, C., 1981, Landslides and their stabilization, with contributions by F. Hilbert: New York, Vienna, Springer-Verlag, 247 p. (Translated by Erika Jahn.)

Viberg, L., 1984, Landslide risk mapping in soft clays in Scandinavia and Canada, *in* Proceedings, IV International Symposium on Landslides, Toronto, September, v. 1: Downsview, Ontario, Canadian Geotechnical Society, p. 325–348.

Voight, B., ed., 1978, Rockslides and avalanches; 1. Natural phenomena: Amsterdam, Elsevier Scientific Publishing Co., Developments in Geotechnical Engineering 14A, 833 p.

——, ed., 1979, Rockslides and avalanches; 2. Engineering sites: Amsterdam, Elsevier Scientific Publishing Co., Developments in Geotechnical Engineering 14B, 850 p.

Voight, B., Janda, R. J., Glicken, H., and Douglass, P. M., 1983, Nature and mechanics of the Mount St Helens rockslide-avalanche of 18 May 1980: Geotechnique, v. 33, p. 243–273.

Ward, T. J., 1976, Factor of safety approach to landslide potential delineation [Ph.D. thesis]: Fort Collins, Colorado State University, 119 p.

Williams, G. P., and Guy, H. P., 1973, Erosional and depositional aspects of Hurricane Camille in Virginia, 1969: U.S. Geological Survey Professional Paper 804, 80 p.

Wilson, R. C., and Keefer, D. K., 1985, Predicting areal limits of earthquake-induced landsliding, *in* Ziony, J. I., ed., Evaluating earthquake hazards in the Los Angeles region; An earth-science perspective: U.S. Geological Survey Professional Paper 1360, p. 263–315.

Wilson, S. D., and Mikkelsen, P. E., 1978, Field instrumentation, *in* Schuster, R. L., and Krizek, R. J., eds., Landslides; Analysis and control: National Academy of Sciences, National Research Council, Transportation Research Board Special Report 176, p. 112–138.

Winterkorn, H. F., and Fang, H., eds., 1975, Foundation engineering handbook: New York, Van Nostrand Reinhold Co., 751 p.

Wu, T. H., and Swanston, D. N., 1980, Risk of landslides in shallow soils and its relation to clearcutting in southeastern Alaska: Forest Science, v. 26, no. 3, p. 495–510.

Youd, T. L., 1975, Liquefaction, flow, and associated ground failure, *in* Proceedings, U.S. National Conference on Earthquake Engineering, 1st, Ann Arbor, Michigan, 1975: Earthquake Engineering Research Institute, p. 146–155.

Youd, T. L., and Perkins, D. M., 1978, Mapping liquefaction-induced ground failure potential, *in* Proceedings, American Society of Civil Engineers: Geotechnical Engineering Division Journal, v. 104, no. GT4, p. 433–446.

Záruba, Q., and Mencl, V., 1969, Landslides and their control: Amsterdam, Elsevier Publishing Co., and Prague, Academia, Developments in Geotechnical Engineering, v. 2, 214 p.

Záruba, Q., and Mencl, V., 1982, Landslides and their control (2nd edition): Amsterdam, Elsevier Scientific Publishing Co., Developments in Geotechnical Engineering 31, 324 p.

Manuscript Accepted by the Society April 20, 1987

Printed in U.S.A.

Geological Society of America
Centennial Special Volume 3
1991

Chapter 10

Nontectonic subsidence

Thomas L. Holzer
U.S. Geological Survey, 345 Middlefield Road, Menlo Park, California 94025

INTRODUCTION

Geologists have investigated many different types of subsidence (Table 1) in North America during the past 100 years. Their principal contribution has been a better understanding of subsidence processes associated with the sudden formation of sinkholes, volcanic activity (Schuster and Mullineaux, this volume), tectonism (Bonilla, this volume), and sediment compaction induced by withdrawal or natural expulsion of underground fluids. Although major advancements in the understanding of other types of subsidence processes have been made primarily by engineers and soil scientists, geologists have outlined the geologic framework within which these subsidence processes are active. Allen (1969) provides an overview of the geologic processes that contribute to subsidence and the geologic setting of subsidence.

This chapter traces the evolution during the past 100 years of the conceptual understanding of land subsidence in North America caused by compaction of unconsolidated sediment induced primarily by the withdrawal of underground fluids. It reviews the ways these concepts have been applied, both to development of the theory of fluid flow through porous media and to gain insight into natural geologic processes. Four important case histories are examined, and the chapter concludes with discussions of the outlook for future investigations of subsidence in North America and a summary of research needs. Terminology used in the chapter follows Poland and others (1972).

Subsidence associated with man-induced compaction is one of man's major inadvertent engineering feats. At least 34 areas in Mexico and the United States have subsided (Fig. 1); an aggregate area of about 22,000 km^2, approximately equal to the area of New Jersey, has been lowered more than 30 cm. The largest areas affected by subsidence—the San Joaquin Valley, California, and the Houston, Texas, area—are 13,500 km^2 (Poland and others, 1975) and 12,200 km^2 (Gabrysch and Bonnet, 1975) respectively, and include areas that have subsided less than 30 cm. Maximum observed subsidence is 9 m in the San Joaquin Valley.

The principal effect of subsidence on engineering works is to submerge them or to increase their susceptibility to flooding. In addition, differential subsidence can affect the operation of canals or dams and can damage structures outright if it is very localized. Well casings also may collapse when adjacent formations compact. Costs from subsidence in North America vary greatly from area to area, but locally are high. Aggregate costs to mitigate subsidence in the Wilmington oil field at Long Beach, California, were \$150 million (Mayuga, 1970); annual costs of subsidence from 1969 to 1974 in Houston, Texas, were \$31.7 million (Jones, 1977). Flooding was the principal concern in both areas. Subsidence has caused few casulaties because it usually develops slowly. The greatest loss of life was five people who were killed in 1963 by flooding associated with the subsidence-induced failure of the Baldwin Hills Reservoir (see Chapter 22 this volume) at the Inglewood oil field in Los Angeles, California (Kresse, 1969). The potential for loss of life, however, may be increased in low-lying coastal areas subject to storm surges where subsidence has aggravated an already serious flood hazard. For example, more than 300,000 people near Houston, Texas, live in an area that is subject to flooding from hurricane storm surges and has undergone subsidence (Fig. 2).

HISTORICAL DEVELOPMENT AND CONTRIBUTIONS

This section reviews the historical development of the understanding of land subsidence processes and the ways in which studies of land subsidence have contributed to both the development of concepts important to the withdrawal of underground fluids from porous media and our understanding of geologic processes. Investigations of land subsidence contributed to the recognition of the significance of the compressibility of aquifer and petroleum-reservoir systems, to the application of the principle of effective stress to geologic problems, and to the understanding of the maintenance of pore pressures in geopressured systems.

Land subsidence associated with withdrawal of underground fluids from unconsolidated sediment is caused by sediment compaction induced by increases of effective stress. Stress increases are caused in turn by decreases of pore pressure induced by the fluid withdrawal. At effective stresses typically encoun-

Holzer, T. L., 1991, Nontectonic subsidence, *in* Kiersch, G. A., ed., The heritage of engineering geology; The first hundred years: Boulder, Colorado, Geological Society of America, Centennial Special Volume 3.

TABLE 1. TYPES OF LAND SUBSIDENCE

	Natural	Man Induced
Compaction	Sediment loading* Liquefaction	Ground water pumping* Hydrocarbon withdrawal* Geothermal development* Geopressured development* Surface wetting
Collapse into Voids	Karst Piping	Karst Piping Mining Nuclear explosions
Biological Oxidation		Drainage of organic soils
Tectonic	Volcanic Seismic Aseismic	

*Discussed in this chapter.

Figure 1. Map of areas of subsidence caused by withdrawal of underground fluids in North America. Subsidence was caused by withdrawal of ground water unless otherwise indicated.

tered in aquifers, clay is more compressible than sand. Thus, most of the compaction usually occurs in clay. The model of compaction of clay as fluids drain slowly into the pumped sands of the aquifer is known as the aquitard-drainage model (Helm, 1984). As depth and effective stress increase, the compressibility of sand decreases at a smaller rate than that of clay. Laboratory investigations indicate that compressibilities of sand and clay may become comparable at depths as shallow as 1,000 m (Roberts, 1969), the depth of shallow petroleum reservoirs. Thus, compression of sand may be significant where subsidence is caused by withdrawal of fluids at depths below 1,000 m.

The earliest description of the concept of the aquitard-drainage model is by Pratt and Johnson (1926), who investigated a 1-m subsidence from 1917 to 1925 at the Goose Creek oil field near Houston, Texas. The subsidence was the first documented example of this type in North America. They proposed that the subsidence was caused by compaction of clays as pore water drained into the oil-producing sands.

> . . . it should be noted that the coastal plain beds at this locality consist largely of unconsolidated clays or gumbos, and that the oil sands occur in discontinuous lenses. Under such circumstances the removal of oil and water from the sands is not followed by a free flow of water into the pore spaces from distant localities, because the compact nature of the clays, which completely surround and isolate the sand lenses, prevents such ready flow of water. The pore spaces are therefore occupied by water draining in more slowly from the adjacent clays; and it is a well-known fact that the draining of clays causes them to become more compact. This, in turn, would permit subsidence of the overlying layer. (Pratt and Johnson, 1926, p. 590)

Pratt and Johnson also recognized that compaction of the oil-bearing sands and actual removal of sand during oil production probably contributed to the subsidence, but they suggested that both processes were not as significant as compaction of clay. Minor (1925) earlier had proposed that the compaction was confined entirely to the sands and was due to rearrangement of individual sand grains caused by flow of fluids through the sand to the wells.

Snider (1927) challenged Pratt and Johnson's (1926) aquitard-drainage mechanism at Goose Creek and proposed that the subsidence was caused by collapse of free pockets of oil and gas in the sands. His objection was based on the assertion that clays retain water very tenaciously. Geologists who were concurrently developing the concept of elastic storage in confined aquifers (for example, Meinzer, 1928) seized upon Snider's proposal that the compaction was seated in the sand. These hydrogeologists were primarily investigating short-term effects and had not yet addressed aquitard compressibility.

Subsidence at Goose Creek remained the best-known example of subsidence caused by withdrawal of underground fluids from unconsolidated sediment until 1932–1933, when releveling in the Santa Clara Valley in northern California revealed that parts of the valley had subsided 1.25 m since 1912 (Rappleye, 1933). Gayol (1929) had presented evidence for subsidence in Mexico City, but his work does not appear to have been widely known. Tolman and Poland (1940) attributed subsidence in the Santa Clara Valley to compaction of the clays interbedded in the aquifer system. Their principal evidence was a correlation between subsidence and clay-bed thickness They drew an analogy between subsidence and a laboratory experiment of clay compaction (Terzaghi, 1927) and restated the concept of the aquitard-drainage model.

> If the water-level in the outer cylinder is lowered, the pressure in the sand layers is reduced in direct proportion to the lowering of the water-level and this pressure-difference allows the water in the clay to drain into the sand-layers and the clay-compaction is measured by the volume of water driven out of the clay bed into the sand. (Tolman and Poland, 1940, p. 33)

Harris and Harlow (1947) were the first to use the aquitard-drainage model concept to predict subsidence when they applied Terzaghi's (1943) theory of consolidation to the subsidence at the Wilmington oil field in California (Fig. 1). Harris and Harlow (1947) proposed that the withdrawal of oil and water had increased the effective stress in both the oil-bearing sands and the confining shales. On the basis of compressibilities determined from consolidation tests on shale cores, they predicted the ultimate compaction beneath many points in the Wilmington field. They also evaluated the time required for pore pressure to dissipate in the shales but dismissed it as being too short to be of significance to alter their predictions. Although others who investigated subsidence at Wilmington have proposed that the compaction occurred in the oil-bearing sands (Gilluly and Grant, 1949), the methodology employed by Harris and Harlow (1947) has remained unchallenged from a conceptual perspective.

Figure 2. Damage at Brownwood subdivision near Houston, Texas, caused by flooding during Hurricane Alicia in August 1983. Area has subsided about 2.5 m since 1915 because of ground-water withdrawal.

Carrillo (1948) was the first to apply Terzaghi's (1943) diffusion equation to the dissipation of pore pressures in an aquitard. He attempted to model the 1939–1947 subsidence in part of Mexico City. Carrillo assumed very simple conditions, including a constant rate of head decline and an aquifer with four identical aquitards interbedded in a hydraulically connected sand. Despite the simplicity of the model, Carrillo obtained good general agreement between observed and calculated history and magnitude of subsidence from 1939 to 1947. He also recognized a problem that would plague investigations for two more decades—determination of representative values of permeability and compressibility.

Gibbs (1960) and Miller (1961) adapted Harris and Harlow's (1947) approach to predict the ultimate subsidence at sites in the San Joaquin Valley, California, by using aquitard properties determined from consolidation tests on cores. They also estimated post-1959 residual compaction (caused by pore-pressure dissipation after 1959) by using a solution to Terzaghi's (1943) diffusion equation.

These early analyses were based on methodology already in broad use by geotechnical engineers to predict settlements of structures built on compressible soils; however, geotechnical engineers and geologists used different terminology to characterize correlative engineering properties of soils and aquitards (Jorgensen, 1980). Domenico and Mifflin (1965) derived the relation between these correlative properties and demonstrated the similarity between the ground-water flow equation (Jacob, 1940) and the diffusion equation for transient pore pressures in a clay layer (Terzaghi, 1943). They showed that, if the compressibility of water is ignored, the specific storage of an aquitard, S_s (the volume of water released from storage per unit volume of aquitard per unit head decline), was related to the coefficient of compressibility, a_v (the change of void ratio per unit change of effective stress), according to the following equation:

$$S_s = a_v G_w/(1+e)$$

where G_w = specific weight of water and e = void ratio.

Domenico and Mifflin (1965) quantified the effect on subsidence from incomplete pore-pressure dissipation within the aquitard and pointed out that aquitards become overconsolidated if water levels partially recover after they have declined. They recognized that during periods of ongoing water-level decline or partial water-level recovery, aquitards typically behave as if they were normally consolidated or overconsolidated, respectively. Green (1964) had documented that specific storage of aquitards in Santa Clara Valley, California, was smaller during periods of water-level decline that followed periods of partial water-level recovery than it had been during the original decline.

The application of geotechnical concepts to aquitard compaction was extended by Holzer (1981) to include Casagrande's (1936) concept of natural preconsolidation. Casagrande had showed that the compaction of soils under load was affected by previous loads. If the current load were smaller than the highest of the previous loads, the compaction of the soil would be smaller than when the load exceeded the previous load. He suggested that the load at which the compaction behavior of the soil changes be called the preconsolidation load. Analysis of the historical relation between subsidence and water-level decline in six ground-water basins in the U.S. indicated that most of the aquifer systems behaved as if they had been naturally preconsolidated by amounts equivalent to water-level declines of 16 to 63 m.

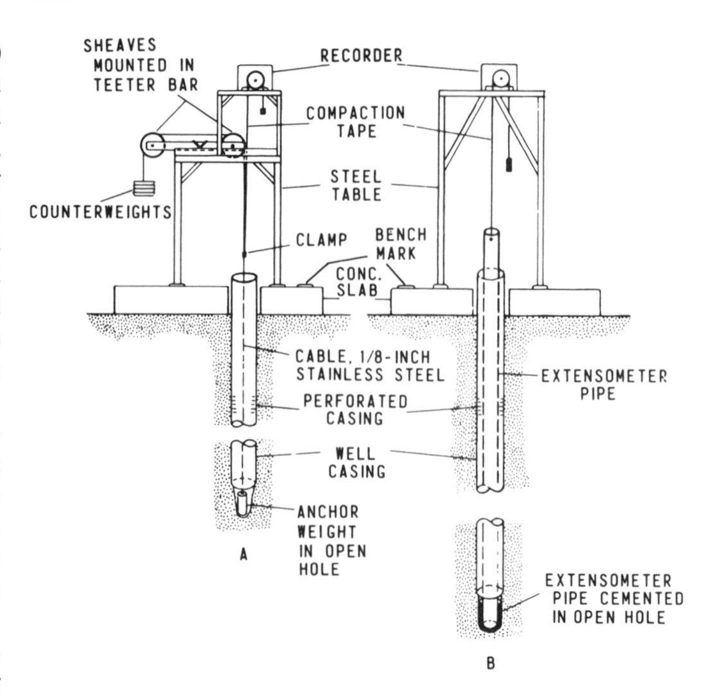

Figure 3. Schematic drawings of cable (A) and pipe (B) vertical extensometers used to measure compaction of aquifer systems (from Poland and Yamamoto, 1984).

Concurrent with the theoretical and laboratory developments in the 1950s and 1960s, extensive field investigations in California in the Santa Clara Valley (Green, 1964) and the San Joaquin Valley (Johnson and others, 1968; Bull, 1975) confirmed the general validity of the aquitard-drainage model concept under field conditions. The vertical distribution of compaction was measured directly with vertical extensometers (Fig. 3) that simultaneously measure aquifer water levels and compaction of the aquifer system above the base of the extensometer. The measurements confirmed that compaction was occurring primarily in the aquitards (Lofgren, 1968). A need still remained, however, for techniques to determine representative aquitard properties. The breakthrough was made by Riley (1969) with his analysis of data from recording borehole extensometers that had been installed in central California beginning in 1955. Following Lofgren's (1968) suggestion that the change of effective stress equalled the change of water level, Riley (1969) pointed out that a plot of water-level change versus compaction (Fig. 4) was

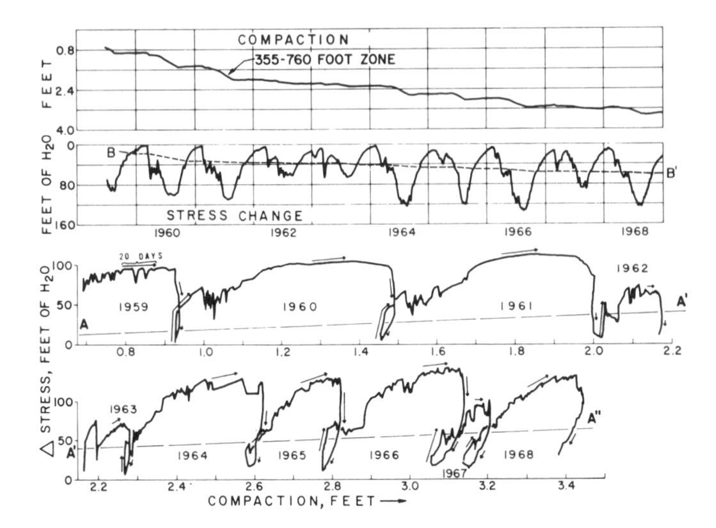

Figure 4. Extensometer record showing history of compaction and stress change and the relation between stress change and compaction near Pixley, California. Line A-A′ is the boundary between elastic and permanent compaction. Water-level changes above and below A-A′ cause permanent and elastic compaction, respectively (from Riley, 1969).

analogous to a stress-strain diagram for the aquifer system. He identified two types of deformation in these diagrams: elastic and permanent (Fig. 4). For a site near Pixley, California, Riley estimated field-based values of elastic and plastic specific storage and vertical permeability of the aquitard.

Riley's insight provided the incentive to develop more sophisticated predictive models. The most widely used model was developed by Helm (1975). He used numerical methods to solve Terzaghi's diffusion equation for the dissipation of pore pressures in an aquitard, and designed a model that could be easily calibrated by field measurements from borehole extensometers. Since its development, the aquitard-drainage model has been successfully applied in North America to subsidence problems in Arizona, California, and Texas (Helm, 1984).

Horizontal displacements

Although most formulations of the aquitard-drainage model have been one dimensional and consider only vertical deformation, damaging horizontal strains and ground failure demonstrate that horizontal displacement can be associated with land subsidence. Horizontal displacements have been measured in North America in a water-well field at Lake Texcoco, Mexico (Marsal, 1969), and in three oil fields in southern California: Inglewood (Castle and Yerkes, 1976), Buena Vista (Whitten, 1961), and Wilmington (Grant, 1954; Berbower, 1959). The maximum measured displacement is 3.4 m at the Wilmington oil field (Yerkes and Castle, 1969). The centripetal pattern of the displacements with respect to the subsidence bowl and the location of zones of extension and compression at the margin and center of the subsidence bowl, respectively, have suggested to many investigators that the horizontal displacements are caused by differential compaction that bends the overburden. A model for such deformation, developed by Lee and Shen (1969), produced reasonable agreement between observed and calculated values at Inglewood and Wilmington (Castle and Yerkes, 1976; Lee, 1980). In addition, horizontal displacements that are localized near ground failures have been measured in several areas of ground-water withdrawal (Holzer and Pampeyan, 1981; Holzer, 1986). The ground failures occur along zones of differential subsidence caused by localized differential compaction. Jachens and Holzer (1979, 1982) concluded that the mechanism of failure was similar to that documented in oil fields in that the horizontal strains were produced by bending of the overburden. Although numerical methods have facilitated solution of three-dimensional consolidation problems (Safai and Pinder, 1980; Bear and Corapcioglu, 1981), they have not yet been widely applied to field problems.

Ground failure

Ground failure, ranging from tension cracks to shear failure, is widely associated with land subsidence caused by withdrawal of underground fluids (Fig. 5). The first failure was observed sometime between 1918 and 1926 at the Goose Creek oil field, Texas, when three fault scarps formed around the margin of the subsidence bowl (Fig. 6). The faulting remained unique until additional faulting was recognized in about 1942 in the Houston, Texas, subsidence area (Lockwood, 1954). Ground failure in oil fields and ground-water basins has been summarized by Yerkes and Castle (1969) and Holzer (1984), respectively. In general, ground failure is rare during the early stage of subsidence. It has been the major cause of damage in some noncoastal subsidence areas, however, where loss of surface elevation may have little practical impact if regional tilting is small. For example, failure of the Baldwin Hills Reservoir, California, in 1963, which was caused by leakage along a ground failure that offset the reservoir lining, claimed five lives and damaged more than $15 million worth of property; a catastrophic but less damaging reservoir failure also occurred in the 1970s in the Lerma Valley, west of Mexico City (for further discussion of both, see James and Kiersch, Chapter 22 this volume). In the Houston, Texas, area, more than 86 faults having an aggregate length of 240 km have damaged hundreds of engineering works. Although most ground failures grow by slow, aseismic creeping motion, seismicity is associated with a few failures (Yerkes and Castle, 1976). More than one mechanism of failure appears to apply. Most tension cracks (Fig. 7) appear to be caused by tensile strains generated by differential compaction (Jachens and Holzer, 1982). Cracks in a few areas, however, form polygonal patterns and are more readily explained by desiccation associated with water-table declines (Neal and others, 1968). Proposed mechanisms of faulting include regional stress changes from differential compaction (Castle and Youd, 1972), localized differential compaction caused by

Figure 5. Map of areas of ground failure caused by withdrawal of ground water or petroleum in North America. Failure was caused by withdrawal of ground water unless otherwise indicated.

differential water-level declines across pre-existing faults (Reid, 1973), and shear failure caused by compaction of sediment with pre-existing zones of weakness (O'Neil and Van Siclen, 1984).

Contributions of subsidence studies

Investigations of land subsidence have fostered (1) recognition of the significance of the compressibility of the porous framework of aquifers and petroleum reservoirs and (2) application of geotechnical concepts to geologic processes. Beginning with Meinzer's (1928) classic treatise on elastic compressibility of artesian aquifers, investigations of aquifer-system and petroleum-reservoir compressibility have repeatedly applied insights gained from land subsidence (Poland, 1961). Examples of applications of geotechnical concepts to geologic processes include maintenance of pore pressures in geopressured systems and the role of effective stress in overthrust faulting.

Artesian aquifer systems originally were thought to behave only as conduits for ground water as it flowed from areas of recharge to wells or other areas of discharge. Confining beds were regarded as essentially the impermeable boundaries of the conduit. The concept of aquifer and aquitard storage was not recognized. Meinzer and Hard (1925, p. 90–91) proposed that aquifers are not rigid but have volumetric elasticity: ". . .the [Dakota] sandstone has a volume elasticity, so that, as the buoying force of artesian pressure within the sandstone was relieved, the sandstone underwent a certain amount of compression. . . ." When Meinzer (1928) published his classic paper on the compressibility of aquifers, one of the direct lines of evidence that he cited for such compressibility was the subsidence at the Goose Creek oil field in Texas.

Jacob (1940) also took note of subsidence in oil fields and the implied compressibility when he derived the fundamental differential equation governing the flow of water in an artesian

Figure 6. Fault scarp that formed on northern margin of Goose Creek oil field between 1918 and 1926 near Houston, Texas, during oil production. The land to the left (south) has dropped about 0.41 m (from Pratt and Johnson, 1926).

Figure 7. Approximately 1-year-old tension crack in south-central Arizona that has been enlarged by erosion to form an earth fissure. Fissure is about 1 m wide.

aquifer. His estimate of specific storage included consideration of the compressibility of the aquitards. Although he assumed for theoretical convenience that drainage was instantaneous, he clearly stated the concept of the aquitard-drainage model. "In the actual case, however, a third term is required to account for the water derived from storage in the adjacent and included clay-beds. This term would in general be a function of the rate and perhaps also the magnitude of the decline of pressure, of the thickness and distribution of the intercalated clay-beds, and of the permeability and modulus of compression of the clay" (Jacob, 1940, p. 577). Jacob also recognized the importance of the aquitards when he suggested that they were probably the "chief source of the water derived from storage within an artesian aquifer." Hantush (1960), a student of Jacob, later took the compressibility of the aquitard more rigorously into account in his formulation of the theory of leaky aquifers.

Early theoretical treatments of fluid flow in petroleum reservoirs ignored the compressibility of the reservoir itself, although fluid compressibility was recognized (Muskat, 1937). Land subsidence at oil fields, however, prompted Carpenter and Spencer (1940) to conduct an extensive series of laboratory measurements of the compressibility of typical oil-bearing sandstones. They concluded that compaction of sandstones caused by declines of fluid pressure was too small to account for observed subsidence. They inferred that compaction was occurring in the shales adjacent to the reservoir. Development of the concept of specific storage in hydrogeology, however, did not go unnoticed by petroleum engineers. Muskat (1949, p. 541) took special notice of Jacob's (1940) work, and noted that field-determined values of total reservoir system compressibility included a component attributable to compressibility of the reservoir. Hall (1953) pointed out that neglect of reservoir compressibility in the extreme case of low-porosity rock could cause erroneous estimates of hydrocarbon reserves in newly discovered reservoirs that were undersaturated with gas. Errors were potentially as high as a factor of two. Today, rock compressibility is routinely considered as a component of the reservoir system total compressibility (Earlougher, 1977).

Theoretical investigations of the build-up and maintenance of anomalously high fluid pressures in thick sedimentary sequences usually are formulated in terms of the aquitard-drainage model. For example, Bredehoeft and Hanshaw (1968) used this model to estimate the permeability that was required to maintain excess pore pressure in thick sedimentary deposits undergoing rapid burial. Their application of both concepts from geotechnical engineering and a mathematical solution to Terzaghi's (1943) diffusion equation mirrored the methodology applied by investigators of land subsidence (for example, Domenico and Mifflin, 1965). Their application was refined by Sharp and Domenico (1976) to include the transport of thermal energy; this work received the Geological Society of America's Meinzer Award in 1979.

Application of the principle of effective stress to tectonic deformation was the basis for one of the major advances in structural geology and tectonophysics in the past 100 years. Its application to explain overthrust faulting and induced earthquakes demonstrated the important role played by fluid pressure in the movement of crustal blocks. Although geologists properly associate the application to thrusting with Hubbert and Rubey (1959), Meinzer (1928, p. 264) clearly stated the principle in his

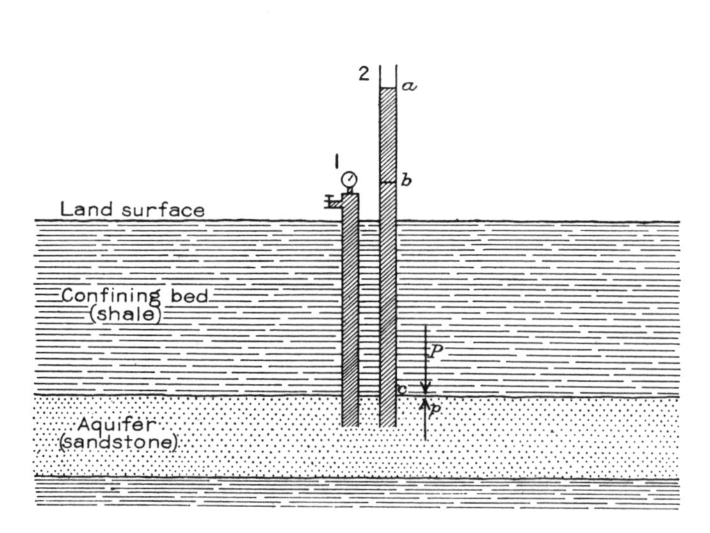

Figure 8. Meinzer's (1928, Fig. 1) idealized cross section outlining the principle of effective stress as applied to an artesian aquifer. The downward stress P is due to the weight of the confining bed and the pressure p is the pore pressure in the aquifer. Only the difference, P–p, is borne by the aquifer.

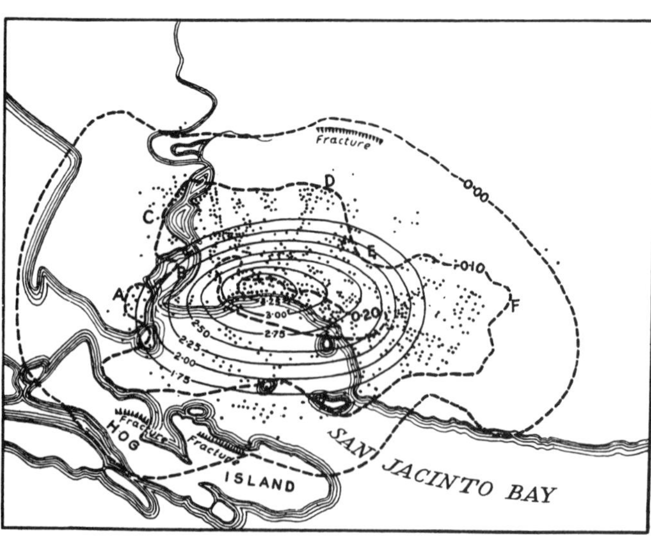

Figure 9. Contours of equal subsidence (in feet) from 1917 to 1925 (light solid lines) and 1924 to 1925 (heavy broken lines) at the Goose Creek oil field, Texas. Dots represent wells. Fractures are fault scarps that formed contemporaneously with subsidence (see Fig. 6). Maximum dimension of subsidence area is about 4.8 km (from Pratt and Johnson, 1926).

paper on the Dakota Sandstone (Fig. 8): "As the specific gravity of the shale that forms the confining bed is much greater than that of water, the downward pressure P is greater than the upward pressure p. Only the difference, P–p, is borne by the sandstone that forms the aquifer." The concept, of course, is intrinsic to the aquitard-drainage model and was in widespread use by geologists who were investigating land subsidence. Although Hubbert and Rubey (1959) provide an unusually long list of acknowledgments in their introduction and it is likely that they were aware of Meinzer's work, they do not mention Meinzer (1928) or the aquitard-drainage model. Rubey (Hubbert and Rubey, 1959) acknowledges his personal indebtedness to Meinzer, and Hubbert (1940), in a classic treatise on ground-water flow, acknowledges both Meinzer and Jacob.

IMPACT OF SUBSIDENCE ON ENGINEERING WORKS

Damage from land subsidence caused by withdrawal of underground fluids has been costly in both urban and rural areas of many parts of North America; a complete review of its impact on engineering works is beyond the scope of this chapter. However, the following selected examples are representative of the effects of subsidence on engineering works and society.

Goose Creek oil field, Texas

Subsidence of 1 m that accompanied development of the Goose Creek oil field on the north shore of Galveston Bay in Texas (Fig. 9) was the first occurrence of subsidence caused by sediment compaction described in North America. Approximately 12 km^2 of land subsided from 1917 to 1925 (Pratt and Johnson, 1926). The primary impact of the subsidence was the submergence of low-lying parts of the field. Although the area was sparsely settled and only the engineering works associated with oil development were at risk, the subsidence has the unique distinction of having had its cause decided in a court of law. Submerged land in Texas belongs to the state, and only the state can grant oil and gas leases for such land. Consequently, after part of the field subsided below sea level, Texas sought title to the oil. The court ruled that the subsidence was an act of man and not the result of natural processes. Thus, rights were retained by the leasees. Fortunately for geologists, the litigation provided the incentive to determine the cause of the subsidence at the Goose Creek oil field.

Wilmington oil field, California

Long Beach, California, the location of one of the major ports on the West Coast, sits partially atop one of California's most prolific oil fields, the Wilmington oil field. Before development, approximately half of the 55-km^2 field consisted of tidal or submerged lands. Land subsidence was recognized early in the development of the field and threatened to cause inundation of most of the port facilities, a U.S. Navy shipyard, and other commercial and public facilities (Fig. 10). Before subsidence was stopped in the 1960s, approximately 64 km of waterfront had been affected and more than 50 km^2 of land had subsided more

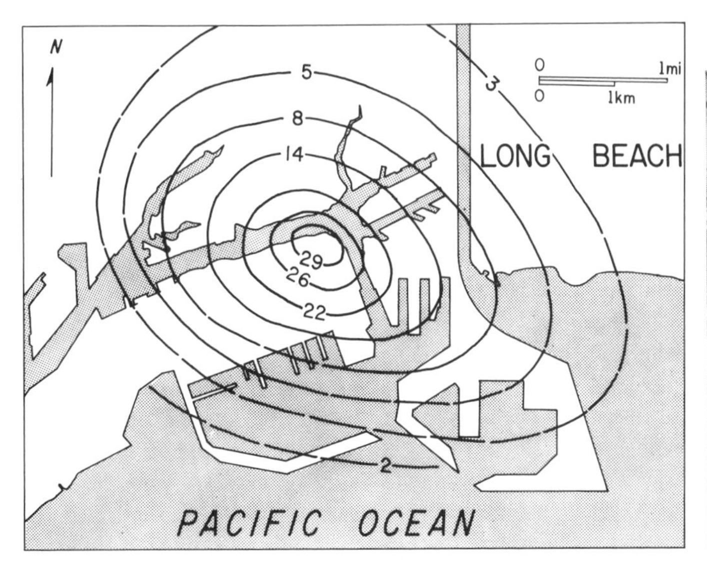

Figure 10. Contours of equal subsidence (in feet) from 1926 to 1967 at the Wilmington oil field, Long Beach, California (after Mayuga, 1970).

Figure 11. Rising building in Mexico City, Mexico. Building is built on piles that are seated on firm strata in the compacting interval in the aquifer system. Land surface has subsided beneath and around building.

than 0.6 m (Mayuga, 1965). If remedial measures had not been taken, approximately 13 km^2 of urban real estate would have have subsided below sea level. Maximum subsidence from 1926 to 1967 was 8.8 m (Mayuga, 1970).

The threat of submergence prompted the first serious effort in North America to predict subsidence from fluid withdrawal. The method developed at Wilmington, using Terzaghi's (1943) theory of consolidation and taking cores (Harris and Harlow, 1947), has been widely used, in North America and around the world, to estimate subsidence. The field investigations conducted at Wilmington also provided one of the most thoroughly documented case histories of land subsidence. Investigations of rates of pore-pressure dissipation, the relative compressibilities of sands and clays, subsurface distribution of vertical deformation, and surface horizontal deformation are just a few examples of the types of studies that were undertaken.

Subsidence at Wilmington was mitigated by repressuring the field by water flooding. The success of the repressuring in arresting the subsidence provided the first example in the world of subsidence control by fluid injection. The subsidence also prompted California to pass the first subsidence control act in the U.S. The California Subsidence Act of 1958 (California Public Resources Code Sections 3315 to 3347) authorizes the State Oil and Gas Supervisor to require unitization of an oil field for the purpose of repressuring oil and gas reservoirs if exploitation threatens coastal areas with submergence.

Mexico City, Mexico

Subsidence of Mexico City probably began before 1891 when the first precise leveling surveys were conducted. Maximum subsidence is unknown but at least 8.5 m has been measured (Figueroa Vega, 1977). Rates of subsidence in the central part of the city, which has the longest record of relevelings, have ranged from 0.045 to 0.460 m/yr. Rates have been decreasing since about 1951. Approximately 225 km^2 have subsided.

The most serious problem caused by the subsidence in Mexico City derives from its unusually large magnitude. The subsidence is not uniform and has adversely affected the gradient of sewers, increasing the potential for flooding by sewage. Until the recent construction of a sewer tunnel, sewage had to be pumped to the main discharge channel, which was above the general elevation of the city. A second problem is with structures built on piles seated on firm layers within the compacting aquifer system. This type of foundation is necessary because the subsoils are very compressible and unable to support the weight of heavy buildings. As the ground beneath structures built on piles subsides in response to water-level declines, the structures appear to rise above the land surface (Fig. 11). This change disrupts utility service and requires maintenance of adjacent sidewalks and access structures.

Mexico City also provides a rare constructive example of land subsidence (Herrera and others, 1977). Land subsidence was intentionally induced in Lake Texcoco, a dry lake bed, to create four 8-m-deep depressions for water impoundment. Special wells were drilled to cause the designed subsidence.

San Joaquin Valley, California

Approximately 13,500 km^2 of the San Joaquin Valley, half of the valley, have subsided due to withdrawal of ground water

Figure 12. Approximate location of maximum subsidence in the San Joaquin Valley, California. Location is about 15 km southwest of Mendota. Subsidence of 9 m occurred from 1925 to 1977 due to withdrawal of groundwater. Signs on telephone pole indicate former elevations of land surface. (Photograph by Richard L. Ireland).

(Poland and others, 1975). It is the largest subsidence area in the world. Maximum subsidence from 1926 to 1977 locally equalled 9 m (Fig. 12). The subsidence affected the operation and construction of major water projects in California and prompted one of the world's most thorough investigations to understand the mechanics of subsidence caused by withdrawal of ground water. The program, under the leadership of Joseph F. Poland, included detailed investigations of the relation between subsidence and geologic (Miller and others, 1971), hydrologic (Bull and Poland, 1975), and geotechnical (Johnson and others, 1968) parameters. In addition, vertical extensometers were developed to measure compaction in sediments undergoing changes of water level (Lofgren, 1961; Riley, 1986), and predictive techniques were developed that are the basis for current state-of-the-art methods (Helm, 1975).

Subsidence in the valley affected two major water projects: the Federal Central Valley Project and the California State Water Project. These projects built more than 1,000 km of canals at average gradients of 0.005 percent. These low gradients made canal operation very vulnerable to tilts caused by land subsidence. Subsidence was not recognized in the early phase of the Central Valley Project, and expensive rehabilitation of canals affected by subsidence was required. Prokopovich and Marriott (1983) estimated $16.3 million was spent on this rehabilitation. Subsidence was a major consideration during the planning stage of canals for the State Water Project (see this volume). During the peak period of planning and construction, 128 engineering geologists were employed to address geologic problems relevant to the many engineering works, including subsidence, along the canal route (James, 1974). Some of the applied geologists investigated alternative routes to avoid or minimize subsidence. A route along the margin of the subsidence area was selected, together with a program to pre-consolidate the sediments along the alignment by flooding and hydrocompaction (James and Kiersoh, this volume).

OUTLOOK

The future

All of the occurrences of land subsidence in North America caused by withdrawal of underground fluids were unforeseen when withdrawals began. As subsidence was recognized in one area after another, knowledge of the phenomenon increased. Today a basic understanding of subsidence processes has emerged. This does not imply, however, that North America will not have major subsidence problems in the future.

First, determination of representative field parameters is difficult. Although generalizations on the basis of geologic age and composition of deposits offer some guidance (Poland and Davis, 1969), predictions of subsidence in virgin basins from which fluids have not yet been withdrawn generally are inaccurate. To date, accurate predictions have been restricted to areas with established subsidence. Preconsolidation also complicates prediction. Areas may be essentially stable until water-level declines are sufficiently large to cause the preconsolidation stress to be exceeded, at which point subsidence begins (Fig. 13).

Second, institutional and legal barriers may inhibit efforts to curtail pumping of underground fluids. Particularly in the U.S., tort recovery theories, which establish rights to claims for damage, commonly are in conflict with legal doctrines that establish rights to resource recovery (Amandes, 1984). Thus, even though fluid withdrawal may cause damaging subsidence, legal recourse to stop pumping may not be available.

Third, experience with subsidence in North America is primarily restricted to withdrawal of ground water and shallow petroleum and gas. other situations involving removal of underground fluids have yet to be significantly exploited. In particular, withdrawal of fluids in conjunction with the extraction of deeper energy resources looms as a new frontier in the future as the U.S.'

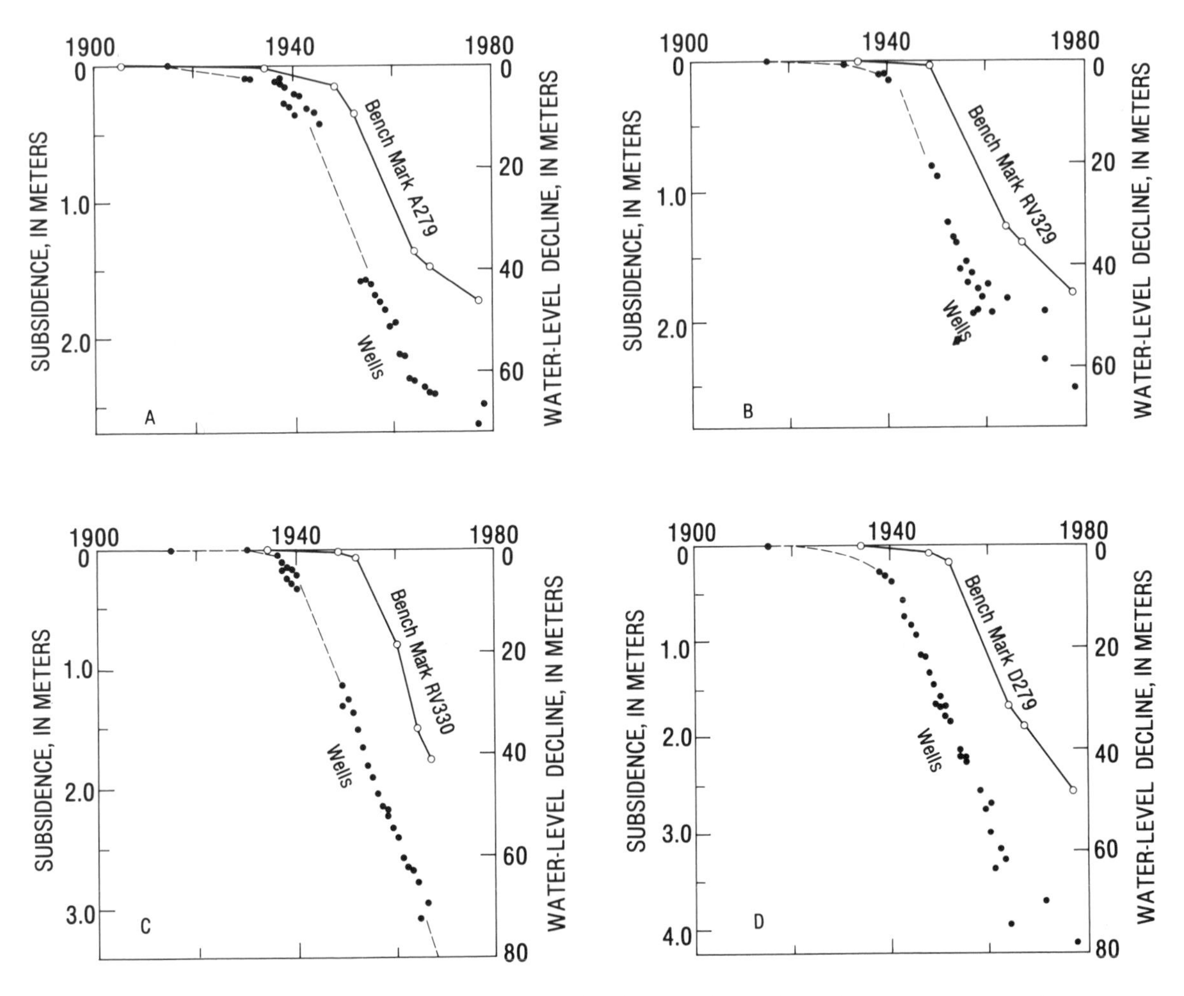

Figure 13. Subsidence (open circles) and water-level decline (solid circles) data at four locations in south-central Arizona that illustrate preconsolidation effects. Major subsidence did not begin until water levels declined about 31 m (from Holzer, 1981).

dependence on foreign oil increases. For example, the vast, untapped, geopressured reservoirs of the Gulf Coast appear to be capable of generating several meters of subsidence if not carefully managed when fluid pressures are lowered (Papadopulous and others, 1975). Modest geothermal development in North America already has caused subsidence. Part of The Geysers in Northern California has subsided more than 0.13 m (Lofgren, 1978). Although the impact from geothermal subsidence on engineering works in North America has been trivial, the potential impact of geothermal development is great. Maximum subsidence associated with with the withdrawal of geothermal fluids at Wairakei, New Zealand, was 4.5 m from 1964 to 1975 (Stillwell and others, 1976). Finally, in situ gasification of coal may induce subsidence if it is not carefully conducted.

Fourth, subsidence associated with natural compaction may become significant to engineering works as the duration of man's permanent settlement of North America increases. Geodetic leveling surveys across the Mississippi River delta reveal natural rates of subsidence of more than 8 mm/yr above the thick deposits of the Mississippi River delta (Fig. 14). Already more than 45 percent of the urbanized area of New Orleans, Louisiana, is below sea level, some parts by as much as 2 m (Kolb and Saucier, 1982). Coupled with subsidence caused by drainage of organic soils and fluid withdrawal, the greater New Orleans area could become a prominent subsidence case history.

A major future contribution of the aquitard-drainage model may be in the field of economic geology. Geologists (for example, Jackson and Beales, 1967; Oliver, 1986) are just beginning to recognize the potential importance to ore deposition and petroleum accumulation of fluids expelled from sedimentary basins by tectonic and sedimentary compaction. As new insights are gained into the history of formation of these economic resources, the whole approach to mineral exploration and petroleum prospecting may be altered.

Research needs

Although the ability to predict subsidence has improved tremendously since the 1917–1925 subsidence at the Goose Creek oil field, conceptualization of field problems remains simple. Compaction usually is idealized as one dimensional. In addition, compaction seated within the aquitard is evaluated separately from fluid flow in the ground-water or petroleum reservoir. More comprehensive field-deformation data, particularly surficial horizontal deformation and subsurface deformation, would permit more sophisticated analyses with refined models of subsidence processes. In addition, new methods are needed to determine representative field values of compressibility, preconsolidation, and permeability. Finally, as field data improve, the physics of compaction of geologic sediment should be reevaluated. Present approaches emphasize hydrodynamic compaction and ignore secondary phenomena. Investigations in geotechnical engineering suggest that this conceptual understanding may be simplistic (Mitchell, 1986) and that nonhydrodynamic deformation may be a significant component of total deformation. Considering the major impact that concepts developed in applied geological science and geotechnical engineering have had on the understanding of subsidence processes, advances in these disciplines should continue to improve our knowledge of the physics of land subsidence.

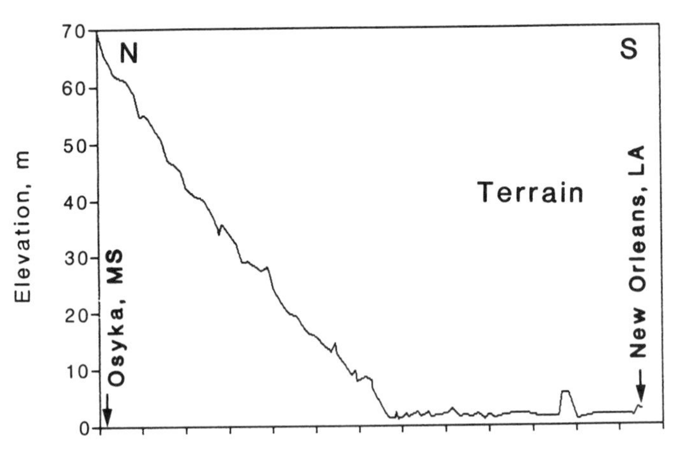

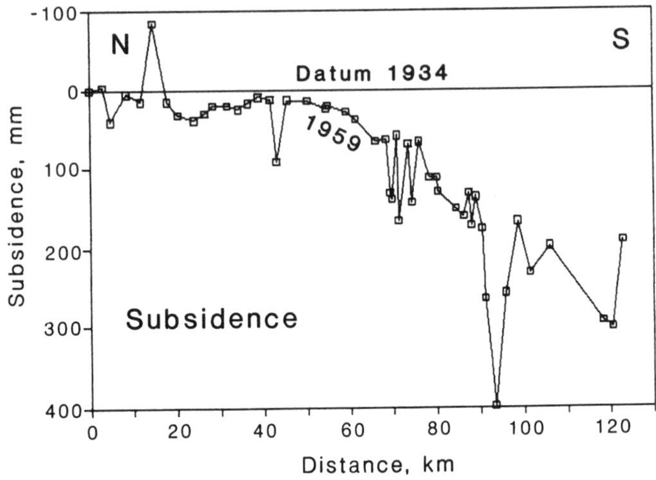

Figure 14. Terrain and subsidence profiles across Mississippi River delta from Osyka, Mississippi, to New Orleans, Louisiana. Profiles illustrate natural subsidence of the delta. Data from National Geodetic Survey.

REFERENCES CITED

Allen, A. S., 1969, Geologic settings of subsidence, *in* Varnes, D. J., and Kiersch, G. A., eds., Reviews in engineering geology, v. II: Geological Society of America, p. 305–342.

Amandes, C. B., 1984, Controlling land surface subsidence; A proposal for a market-based regulatory scheme: University of California at Los Angeles Law Review, v. 31, no. 6, p. 1208–1246.

Bear, J., and Corapcioglu, M. Y., 1981, Mathematic model for regional land subsidence due to pumping; 2. Integrated aquifer subsidence equations for vertical and horizontal displacements: Water Resources Research, v. 17, no. 4, p. 947–958.

Berbower, R. F., 1959, Subsidence problems in the Long Beach Harbor: American Society of Civil Engineers Waterways and Harbors Division Journal, v. 85, no. WW2, p. 43–80.

Bredehoeft, J. D., and Hanshaw, B. B., 1968, On the maintenance of anomalous fluid pressures; I. Thick sedimentary sequences: Geological Society of America Bulletin, v. 79, no. 9, p. 1097–1106.

Bull, W. B., 1975, Land subsidence due to ground-water withdrawal in the Los Banos–Kettleman City area, California; Part 2, Subsidence and compaction of deposits: U.S. Geological Survey Professional Paper 437–F, 90 p.

Bull, W. B., and Poland, J. F., 1975, Land subsidence due to ground-water withdrawal in the Los Banos–Kettleman City area, California; Part 3, Interrelations of water-level change, change in aquifer-system thickness, and subsidence: U.S. Geological Survey Professional Paper 437-G, 62 p.

Carpenter, C. B., and Spencer, G. B., 1940, Measurements of compressibility of consolidated oil-bearing sandstones: U.S. Bureau of Mines Report of Investigations 3540, 20 p.

Carrillo, N., 1948, Influence of artesian wells in the sinking of Mexico City, *in* Proceedings, International Conference on Soil Mechanics and Foundation Engineering, 2nd, Rotterdam, v. 7, p. 156–159.

Casagrande, A., 1936, The determination of the pre-consolidation load and its practical significance, *in* Proceedings, International Conference on Soil Mechanics and Foundation Engineering, 1st, Cambridge, v. 3, p. 60–64.

Castle, R. O., and Yerkes, R. F., 1976, Recent surface movements in the Baldwin Hills, Los Angeles County, California: U.S. Geological Survey Professional Paper 882, 125 p.

Castle, R. O., and Youd, T. L., 1972, Comment on "The Houston fault problem": Association of Engineering Geologists Bulletin, v. 9, no. 1, p. 57–68.

Domenico, P. A., and Mifflin, M. D., 1965, Water from low-permeability sediments and land subsidence: Water Resources Research, v. 1, no. 4, p. 563–576.

Earlougher, R. C., Jr., 1977, Advances in well test analysis: Society of Petroleum Engineers Monograph 5, 264 p.

Figueroa Vega, G. E., 1977, Subsidence of the City of Mexico; A historical review, *in* Proceedings, International Land Subsidence Symposium, 2nd, Anaheim, California: International Association of Hydrological Sciences Publication 121, p. 35–38.

Gabrysch, R. K., and Bonnet, C. W., 1975, Land-surface subsidence in the Houston–Galveston region, Texas: Texas Water Development Board Report 188, 19 p.

Gayol, R., 1929, Breves apuntes relativos a las obras de Saneamiento y Desague de la Capital de la Republica de las que, del mismo genero, necesita con grande urgencia: Mexico, D. F., Revistas Mexicana de Ingenieria y Arquitectura, v. 7.

Gibbs, H. J., 1960, A laboratory testing study of land subsidence, *in* Proceedings, Pan American Conference on Soil Mechanics and Foundation Engineering, 1st, Mexico City, 1959, v. 1, p. 13–36.

Gilluly, J., and Grant, U. S., 1949, Subsidence in the Long Beach Harbor area, California: Geological Society of America Bulletin, v. 60, p. 461–530.

Grant, U. S., 1954, Subsidence of Wilmington oil field, California, *in* Jahns, R. H., ed., Geology of Southern California: California Division of Mines Bulletin 170, Chapter X, p. 19–24.

Green, J. H., 1964, The effect of artesian-pressure decline on confined aquifer systems and its relation to land subsidence: U.S. Geological Survey Water–Supply Paper, 1779–T, 11 p.

Hall, H. N., 1953, Compressibility of reservoir rocks: American Institute of Mining, Metallurgical, and Petroleum Engineers Petroleum Transactions, v. 198, p. 309–311.

Hantush, M. S., 1960, Modification of the theory of leaky aquifers: Journal of Geophysical Research, v. 65, no. 11, p. 3713–3725.

Harris, F. R., and Harlow, E. H., 1947, Subsidence of the Terminal Island–Long Beach area, California: American Society of Civil Engineers Proceedings, v. 73, p. 1197–1218.

Helm, D. C., 1975, One-dimensional simulation of aquifer system compaction near Pixley, California; 1. Constant parameters: Water Resources Research, v. 11, no. 3, p. 465–478.

——, 1984, Field-based computational techniques for predicting subsidence due to fluid withdrawal, *in* Holzer, T. L., ed., Man-induced land subsidence: Geological Society of America Reviews in Engineering Geology, v. VI, p. 1–22.

Herrera, I., Iimas, J. A., Graue, R., and Hanel, J. J., 1977, Development of artificial reservoirs by inducing land subsidence, *in* Proceedings, International Land Subsidence Symposium, 2nd, Anaheim, California: International Association of Hydrological Sciences Publication 121, p. 39–45.

Holzer, T. L., 1981, Preconsolidation stress of aquifer systems in areas of induced land subsidence: Water Resources Research, v. 17, no. 3, p. 693–704.

——, 1984, Ground failure induced by ground-water withdrawal from unconsolidated sediment, *in* Holzer, T. L., ed., Man-induced land subsidence: Geological Society of America Reviews in Engineering Geology, v. VI, p. 67–105.

——, 1986, Ground failure caused by ground-water withdrawal from unconsolidated sediment; United States, *in* Proceedings, International Land Subsidence Symposium, 3rd, Venice, 1984: International Association of Hydrological Sciences Publication 151, p. 747–756.

Holzer, T. L., and Pampeyan, E. H., 1981, Earth fissures and localized differential subsidence: Water Resources Research, v. 17, no. 1, p. 223–227.

Hubbert, M. K., 1940, The theory of ground-water motion: Journal of Geology, v. 48, no. 8, pt. 1, p. 785–944.

Hubbert, M. K., and Rubey, W. W., 1959, Mechanics of fluid-filled porous solids and its application to overthrust faulting: Geological Society of America Bulletin, v. 70, no. 2, p. 115–166.

Jachens, R. C., and Holzer, T. L., 1979, Geophysical investigations of ground failure related to ground-water withdrawal: Ground Water, v. 17, no. 6, p. 574–585.

——, 1982, Differential compaction mechanism for earth fissures near Casa Grande, Arizona: Geological Society of America Bulletin, v. 93, no. 10, p. 998–1012.

Jackson, S. A., and Beales, F. W., 1967, An aspect of sedimentary basin evolution; The concentration of Mississippi Valley-type ore during late stages of diagenesis: Canadian Petroleum Geology Bulletin, v. 15, p. 383–433.

Jacob, C. E., 1940, On the flow of water in an elastic artesian aquifer: American Geophysical Union Transactions, v. 21, pt. 2, p. 574–586.

James, L. B., 1974, The role of engineering geology in building the California State Water Project, *in* Calembert, L., ed., La Geologie de l'ingenieur: Liege, Societe Geologique de Beligique Centenaire Colloque, p. 237–258.

Johnson, A. I., Moston, R. P., Morris, D. A., 1968, Physical and hydrologic properties of water-bearing deposits in subsiding areas in central California: U.S. Geological Survey Professional Paper 497–A, 71 p.

Jones, L. L., 1977, External costs of surface subsidence; Upper Galveston Bay, Texas, *in* Proceedings, International Land Subsidence Symposium, 2nd Anaheim, California: International Association of Hydrological Sciences Publication 121, p. 617–627.

Jorgensen, D. G., 1980, Relationship between basic soils; Engineering equations and basic ground-water equations: U.S. Geological Survey Water–Supply Paper 2064, 40 p.

Kolb, C. R., and Saucier, R. T., 1982, Engineering geology of New Orleans, *in* Legget, R. F., ed., Geology under cities: Geological Society of America Reviews in Engineering Geology, v. V, p. 75–93.

Kresse, F. C., 1969, Baldwin Hills Reservoir failure of 1963, *in* Lung, R., and Proctor, R., ed., Engineering geology in Southern California: Association of Engineering Geologists (Los Angeles Section) Special Publication, p. 93–103.

Lee, K. L., 1980, Subsidence earthquake at California oil field, *in* Saxena, S. K., ed., Evaluation and prediction of subsidence: New York, American Society of Civil Engineers, p. 549–564.

Lee, K. L., and Shen, C. K., 1969, Horizontal movements related to subsidence: American Society of Civil Engineers Soil Mechanics Division Journal, v. 94, no. SM6, p. 140–166.

Lockwood, M. G., 1954, Ground subsides in Houston area: Civil Engineering, v. 24, no. 6, p. 48–50.

Lofgren, B. E., 1961, Measurement of compaction of aquifer systems in areas of land subsidence, *in* Geological Survey Research, 1961: U.S. Geological Survey Professional Paper 424–B, p. 49–52.

——, 1968, Analysis of stresses causing land subsidence, *in* Geological Survey Research, 1968: U.S. Geological Survey Professional Paper 600-B, p. 219–225.

——, 1978, Monitoring crustal deformation in The Geysers–Clear Lake geothermal area, California: U.S. Geological Survey Open–File Report 78-597, 19 p.

Marsal, R. J., 1969, Development of a lake by pumping induced consolidation of soft clays, *in* Contribution of Texcoco Project to International Conference on Soil Mechanics and Foundation Engineering, 7th, Mexico City, p. 229–256.

Mayuga, M. N., 1965, How subsidence affects the City of Long Beach, *in* Proceedings, Geologic Hazards Conference, 2nd, Los Angeles, May 26–27, 1965: State of California Resources Agency, p. 122–129.

——, 1970, Geology and development of California's giant; Wilmington oil field, *in* Halbouty, M. T., ed., Geology of giant petroleum fields: American Association of Petroleum Geologists Memoir 14, p. 158–184.

Meinzer, O. E., 1928, Compressibility and elasticity of artesian aquifers: Economic Geology, v. 23, p. 263–291.

Meinzer, O. E., and Hard, H. A., 1925, The artesian-water supply of the Dakota Sandstone in North Dakota with special reference to the Edgeley quadrangle: U.S. Geological Survey Water–Supply Paper 520-E, p. 73–96.

Miller, R. E., 1961, Compaction of an aquifer system computed from consolidation tests and decline in artesian head, *in* Geological Research, 1961: U.S. Geological Survey Professional Paper 424–B, p. 54–58.

Miller, R. E., Green, J. H., and Davis, G. H., 1971, Geology of the compacting deposits in the Los Banos–Kettleman City subsidence area, California: U.S. Geological Professional Paper 497–E, 46 p.

Minor, H. E., 1925, Goose Creek oil field, Harris County, Texas: American Association of Petroleum Geologists Bulletin, v. 9, p. 286–297.

Mitchell, J. K., 1986, Practical problems from surprising soil behavior: American Society of Civil Engineers, Journal of the Geotechnical Engineering Division, v. 112, no. GT3, p. 259–289.

Muskat, M., 1937, The flow of homogeneous fluids through porous media: New York, McGraw-Hill Book Company, 763 p.

——, 1949, Physical principles of oil production: New York, McGraw-Hill Book Company, 922 p.

Neal, J. T., Langer, A. M., and Kerr, P. F., 1968, Giant desiccation polygons of

Great Basin playas: Geological Society of America Bulletin, v. 79, no. 1, p. 69–90.
Oliver, J., 1986, Fluids expelled tectonically from orogenic belts; Their role in hydrocarbon migration and other geologic phenomena: Geology, v. 14, no. 2, p. 99–102.
O'Neil, M. W., and Van Siclen, D. C., 1984, Activation of Gulf Coast faults by depressuring of aquifers and an engineering approach to siting structures along their traces: Association of Engineering Geologists Bulletin, v. 21, no. 1, p. 73–87.
Papadopulos, S. S., Wallace, R. H., Jr., Wesselman, J. B., and Taylor, R. E., 1975, Assessment of onshore geopressured-geothermal resources in the northern Gulf of Mexico Basin, *in* White, D. E., and Williams, D. L., eds., Assessment of geothermal resources of the United States; 1975: U.S. Geological Survey Circular 726, p. 125–146.
Poland, J. F., 1961, The coefficient of storage in a region of major subsidence caused by compaction of an aquifer system, *in* Geological Research, 1961: U.S. Geological Survey Professional Paper 424–B, p. 52–54.
Poland, J. F., and Davis, G. H., 1969, Land subsidence due to withdrawal of fluids, *in* Varnes, D. J., and Kiersch, G. A., eds., Reviews in Engineering Geology, v. II: Geological Society of America, p. 187–268.
Poland, J. F., and Yamamoto, S., 1984, Field measurement of deformation, *in* Poland, J. F., ed., Guidebook to studies of land subsidence due to groundwater withdrawal: United Nations Educational, Scientific, and Cultural Organization, Studies and Reports in Hydrology, p. 17–35.
Poland, J. F., Lofgren, B. E., Ireland, R. L., and Pugh, R. G., 1975, Land subsidence in the San Joaquin Valley, California, as of 1972: U.S. Geological Survey Professional Paper 437–H, 78 p.
Poland, J. F., Lofgren, B. E., and Riley, F. S., 1972, Glossary of selected terms useful in studies of the mechanics of aquifer systems and land subsidence due to fluid withdrawal: U.S. Geological Survey Water-Supply Paper 2025, 9 p.
Pratt, W. E., and Johnson, D. W., 1926, Local subsidence of the Goose Creek oil field: Journal of Geology, v. 34, no. 7, p. 557–590.
Prokopovich, N. P., and Marriott, M. J., 1983, Cost of subsidence to the Central Valley Project, California: Association of Engineering Geologists Bulletin, v. 20, no. 3, p. 325–332.
Rappleye, H. S., 1933, Recent areal subsidence found in releveling: Engineering News Record, v. 110, p. 848.
Reid, W. M., 1973, Active faults in Houston, Texas [Ph.D. thesis]: Austin, University of Texas, 122 p.
Riley, F. S., 1969, Analysis of borehole extensometer data from central California, *in* Proceedings, International Land Subsidence Symposium, 1st, Tokyo, v. 2: International Association of Hydrological Sciences Publication 89, p. 423–431.
——, 1986, Developments in borehole extensometry, *in* Proceedings, International Land Subsidence Symposium, 3rd, Venice, 1984: International Association of Hydrological Sciences Publication 151, p. 169–186.
Roberts, J. E., 1969, Sand compression as a factor in oil field subsidence, *in* Proceedings, International Land Subsidence Symposium, 1st, Tokyo, v. 2: International Association of Hydrological Sciences Publication 89, p. 368–375.
Safai, N. M., and Pinder, G. F., 1980, Vertical and horizontal land deformation due to fluid withdrawal: International Journal for Numerical and Analytical Methods in Geomechanics, v. 4, p. 131–142.
Sharp, J. M., Jr., and Domenico, P. A., 1976, Energy transport in thick sequences of compacting sediment: Geological Society of America Bulletin, v. 87, no. 3, p. 390–400.
Snider, L. C., 1927, A suggested explanation for the surface subsidence in the Goose Creek oil and gas field: American Association of Petroleum Geologists Bulletin, v. 11, p. 729–745.
Stillwell, W. B., Hall, W. K., and Tawhai, J., 1976, Ground movement in New Zealand geothermal fields, *in* Proceedings, Second United Nations Symposium on the Development and Use of Geothermal Resources, San Francisco, May 1975, v. 2: Washington, D.C., U.S. Government Printing Office (Berkeley, University of California Lawrence Laboratory), p. 1427–1434.
Terzaghi, C., 1927, Principles of final soil classification: Massachusetts Institute of Technology, v. 63, p. 41–43.
——, 1943, Theoretical soil mechanics: New York, John Wiley and Sons, 510 p.
Tolman, C. F., and Poland, J. F., 1940, Ground-water, salt-water infiltration, and ground-surface recession in Santa Clara Valley, Santa Clara County, California: American Geophysical Union Transactions, v. 21, pt. 1, p. 23–34.
Whitten, C. A., 1961, Measurement of small movements in the earth's crust: Academy of Science Fennicae Annales, series A, v. 3, Geologica–Geographica, Suomalainen Tiedeakatemia, no. 61, p. 315–320.
Yerkes, R. F., and Castle, R. O., 1969, Surface deformation associated with oil and gas field operations in the United States, *in* Proceedings, International Land Subsidence Symposium, 1st, Tokyo, v. 1: International Association of Hydrological Sciences Publication 88, p. 55–66.
——, 1976, Seismicity and faulting attributable to fluid extraction: Engineering Geology, v. 10, p. 151–167.

Manuscript Accepted by the Society April 15, 1987

ACKNOWLEDGMENTS

I thank Francis S. Riley and Robert F. Yerkes for their suggestions and reviews of the manuscript. John D. Bredehoeft and William E. Brigham helped me trace the history of investigations of reservoir compressibility in petroleum engineering. I also thank Joseph F. Poland, who originally encouraged my interest in land subsidence and who has been an inspiration for my subsidence-research career.

Printed in U.S.A.

Geological Society of America
Centennial Special Volume 3
1991

Chapter 11

Volcanic activity

Robert L. Schuster and Donal R. Mullineaux
U.S. Geological Survey, Box 25046, MS 966, Denver Federal Center, Denver, Colorado 80225

INTRODUCTION

This chapter discusses some of the ways volcanic activity can affect structures, developments, facilities, and other engineering works. It is vitally important to recognize the possible effects of volcanic activity on these works of man. The large variety of volcano-related hazards now recognized records a marked recent growth in the understanding of these phenomena. The increasing application of systematic observation and analysis to current volcano-related events and to old deposits of previous events throughout the world has greatly increased our understanding of volcanic phenomena; the threat to life and property from these hazards can now be significantly reduced or avoided by careful forethought by geological scientists, engineers, and planners. Public demand, however, may prevent the withdrawal from use of lands that are affected only once in hundreds or thousands of years, and a low degree of risk in such places may be judged to be economically and socially acceptable.

Volcanic eruptions are widely known and feared, yet few of their specific effects are familiar to most people. Volcanoes affect people in many positive ways; for example, erupted materials commonly produce highly fertile soils. However, the immediate effects of volcanoes on people and their engineering works are frequently negative. Although in many countries volcanic activity is a constant and severe threat, it represents only an occasional danger in the United States. That danger is serious enough, however, to warrant preparatory planning in Hawaii, Alaska, and other western states.

Nonexplosive eruptions that produce lava flows are familiar volcanic phenomena; although they present little danger to people's lives, they can severely damage property. Explosive eruptions, in contrast, can be extremely dangerous to both people and property; several recent events have emphasized this danger. The 1980 eruptions of Mount St. Helens in southwestern Washington spectacularly devastated an area of about 600 km^2, killed nearly 60 people, and spread ash that was thin but highly disruptive to transportation and communications into several states. Violent eruptions of El Chichón volcano in Mexico in 1982 spotlighted the occurrence and effects of pyroclastic flows, and of possible effects of volcanic aerosols on climate (Duffield and others, 1984). In 1985, small-scale eruptions of Nevado del Ruiz in Colombia caused mudflows that killed more than 24,000 people (Katsui and others, 1986); that tragedy illustrates the threat from mudflows even at distances of tens of kilometers from an erupting volcano. In 1986, small amounts of ash erupted by Augustine volcano in Alaska more than once closed the airport at Anchorage, 300 km away.

Effects of eruptions on engineering works range from merely troublesome to devastating. The greatest variety and severity of effects occur close to a source vent, but eruptions can significantly affect a complex, modern society at distances of hundred of kilometers from the source. Transportation can be severely curtailed by reduced visibility during the fall of ash, by mudflow deposits on roads and runways, and by damage to engines from ingested ash and clogged filters. Utilities such as sewers and water supply systems are susceptible to clogging by air-fall ash and mudflow deposits, and their machinery can be damaged by ash. Communications systems can also be severely hampered if wet ash short-circuits electrical equipment or physically disrupts communication or power lines, or if ash suspended in the air interferes with broadcast signals.

Huge volcanic eruptions, which ejected as much as 1,000 times the volume of material erupted by Mount St. Helens in 1980, have occurred in the geologic past. The effects of similarly large eruptions in the future would be far greater than any eruption that humans are known to have experienced. Because the effects of these huge eruptions would be so massive, and because the probability of their occurrence is so low for any time period for which planning is likely to be undertaken, such eruptions are beyond the scope of this chapter and are not discussed herein.

The term "hazard" in this chapter refers to a hazardous volcanic event or condition, or the products of volcanism; these hazards vary widely in probability and severity from place to place. The term "risk" refers to the likelihood of loss to people, such as the loss of life, homes, or productive land.

DESCRIPTIONS AND EFFECTS OF VOLCANIC HAZARDS

Several volcanic processes directly endanger people and property by the materials ejected during eruptions; their general

Schuster, R. L., and Mullineaux, D. R., 1991, Volcanic activity, *in* Kiersch, G. A., ed., The heritage of engineering geology; The first hundred years: Boulder, Colorado, Geological Society of America, Centennial Special Volume 3.

Figure 1. Aerial view of Mount St. Helens lava dome, 1983. The dome has been built by many small-volume eruptions of viscous lava that initially piled up directly over the vent and later formed small lobes that built up over the vent and spilled out over the flanks of the growing dome. (Photograph by Lyn Topinka, U.S. Geological Survey.)

character and effects are discussed in many publications (e.g., Macdonald, 1972; Crandell and others, 1984; Blong, 1984; and references therein). These include not only nonexplosive eruptions that produce lava flows, but also explosive eruptions that can eject great volumes of volcanic ash and coarser fragments into the air, and can produce hot masses of flowing fragments and gas—pyroclastic flows and surges—and laterally directed blasts of debris and gas. Landslides, ranging in size from huge debris avalanches to small, shallow slips, commonly result from volcanic activity, and most volcanic events can cause volcanic mudflows, debris flows, and floods. All kinds of eruptions produce volcanic gases.

Several other kinds of events associated with volcanic eruptions, especially earthquakes and tsunamis, can endanger people and property. Surface subsidence and ground fractures as a result of volcanic activity can result in a significant hazard locally, as on Hawaiian volcanoes. Subsidence and fractures generally don't endanger lives, but can damage engineering works.

Lava flows

Lava flows and volcanic domes originate from nonviolent extrusions of lava. Highly viscous lava generally piles up over a source vent to form either a dome (Fig. 1) or a short, thick flow. Lava that is more fluid streams away from the source vent until it cools and solidifies, where it forms a hard and brittle but fractured rock. Generally, the fronts of lava flows advance less rapidly than an average person can walk. However, lava can flow at much greater speeds. Most fast-moving lava is confined within well-established channels, and the flow fronts move less rapidly than lava in the channels. Occasionally, flow fronts also advance rapidly; in 1950, a lava flow from Mauna Loa advanced 25 km to the sea at about 9 km/hr (Macdonald, 1972), and extremely fluid lava from a lava-lake breakout on the African volcano Niragongo killed at least 60 people as it advanced at an average velocity estimated to have been 30 km/hr (Tazieff, 1977).

Lava flows generally extend downslope in tongue-like lobes (Fig. 2) or broad sheets. Most reach no more than 10 to 20 km from the source vents, although highly fluid flows have extended to distances of more than 50 km. Such long flows require eruption of large volumes of lava at rapid rates. Some ancient lava flows in the western U.S. have covered tens of thousands of square kilometers, but such flows are uncommon and have not occurred during historic time. The paths of lava flows are controlled by the shape of the ground surface; the flows move downslope to lower terrain, and locally can move diagonally across slopes or up over obstacles.

Lava flows commonly crush, partly inundate, or bury structures and other immovable objects they encounter, and cover land and fertile soil; combustible structures, materials, and crops that are close to flow margins frequently are burned.

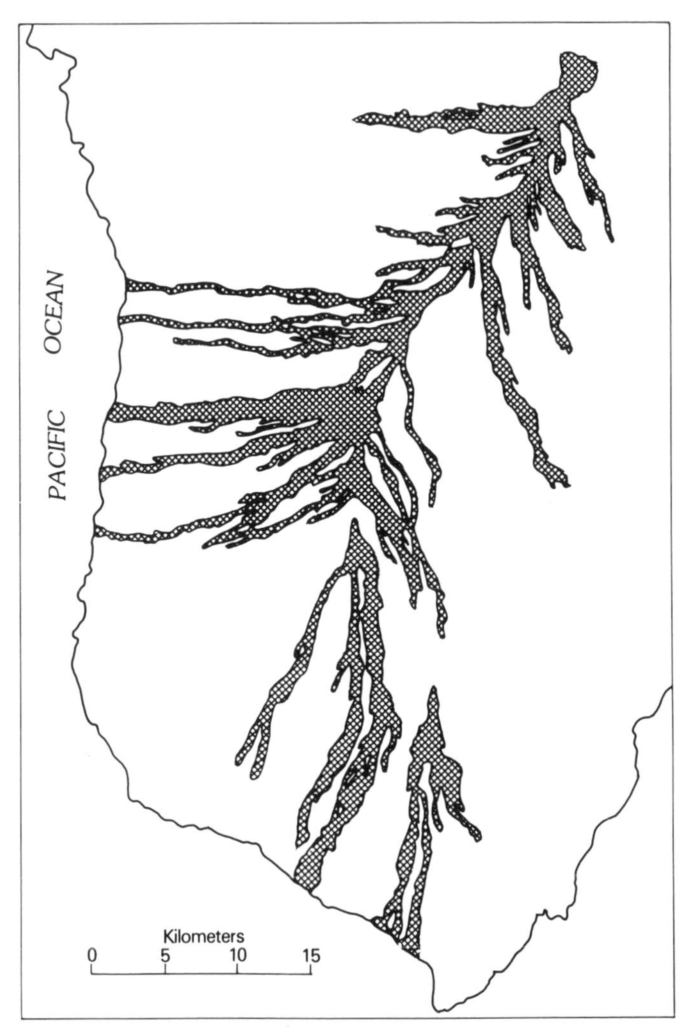

Figure 2. Map showing pattern of historical linear lava flows along southwest rift of Mauna Loa, Hawaii Island (after Lipman, 1980).

Figure 3. Vertical eruption column rising over Mount St. Helens on May 18, 1980, and beginning to drift downwind (eastward) toward the right. (Photograph by Austin Post, U.S. Geological Survey.)

Tephra

Tephra (coarse and fine fragmental volcanic ejecta) is produced chiefly by explosive eruption of molten and solid fragments that are ejected above the volcanic vent in near-vertical columns (Fig. 3). Most large fragments follow arcuate, ballistic trajectories and fall back close to the vent, whereas smaller ash particles fall less rapidly and are carried greater distances downwind. Ash typically forms a continuous layer on the ground surface, which decreases in thickness and particle size with increasing distance from the volcano. Tephra eruptions vary greatly in volume; some eruptions deposit only a trace even near the volcano, and others deposit several tens of centimeters of ash at distances of hundreds of kilometers downwind.

Much of the volcanic ash from a voluminous eruption is carried away along a narrow track by high-speed (jet-stream) winds in the upper troposphere. Lesser amounts can be dispersed much more widely by lower-speed winds blowing in different directions at other altitudes. Although ash from the largest eruption of Mount St. Helens in 1980 thins within 100 km to only a few centimeters (Fig. 4), measurable amounts were deposited as much as several hundred kilometers downwind from the volcano. Because volcanoes are present in most of the western U.S., virtually no locality in that region is free of long-term risk from ash fall.

Volcanic ash can cover broad areas; the severity of its effects depends on many factors, such as thickness, rate of fall, and particle size. Thick ash deposits can load rooftops sufficiently to cause collapse, and can cover agricultural land, killing crops and burying fertile soil. Ash can impede traffic by forming thick deposits on roads, reducing visibility, and damaging engines. Water, sewer, and other utility operations may be hampered as ash clogs pipes and filters. Near the volcano, copious ash or large, hot rock fragments can cause fires. People can be endangered by the fall of large fragments, by breathing ash-laden air, by collapse of roofs, and by fires. Wilcox (1959) noted that a mantle of ash more than a few centimeters thick can also drastically increase runoff, which can affect water supply, transportation, and agriculture for some time after an ash fall. Even a light fall of ash can severely decrease visibility, contaminate forage and water, and result in abrasion to mechanical equipment. Although ash falls cause a wide variety of problems, they seldom threaten human lives.

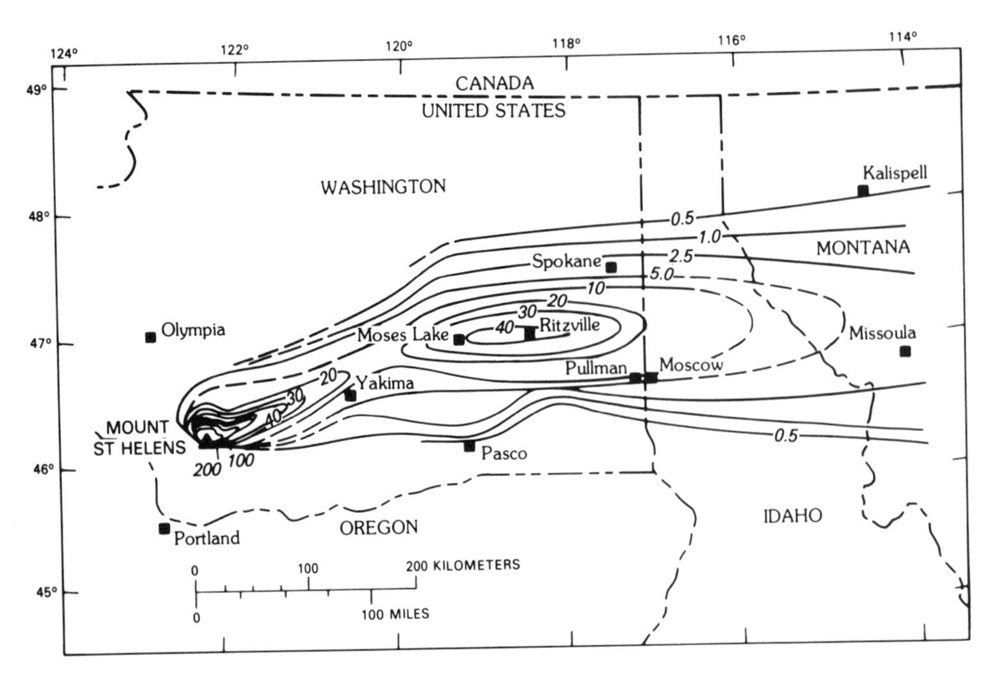

Figure 4. Isopach map of ash erupted from Mount St. Helens on May 18, 1980. Lines represent accumulated thickness in millimeters (after Sarna-Wojcicki and others, 1981).

Large volcanic-ash eruptions can clearly affect local weather and even climate for several years. Although the effects on climate of the eruption of Mount St. Helens in 1980 were virtually unmeasurable, the eruption of El Chichón volcano in 1982 had a much greater effect, mainly because that eruption ejected a large amount of sulfur as an aerosol (Toon, 1982).

Pyroclastic flows and surges

Pyroclastic flows and surges are denser-than-air masses of hot, dry rock debris that move like a fluid and owe their mobility to hot air and other gases mixed with the debris (Fig. 5). These phenomena are also frequently called hot avalanches, nuées ardentes, and glowing avalanches. They may be gradational with lateral or directed blasts. Pyroclastic flows and surges can move at high velocity, gaining speed from eruptive force or gravity; they commonly attain velocities of 40 km/hr, and can reach speeds estimated to be as high as 150 to 200 km/hr.

Pyroclastic flows commonly include a coarse basal mass of hot rock debris that hugs the ground surface; pyroclastic surges consist of less-dense concentrations of generally finer material. Both are usually accompanied by clouds of hot dust. Both can originate from outward ejection of hot rock fragments, from fall-back of vertically erupted material onto the volcano's flanks, or from collapse and flow of the side of a growing volcanic dome.

Because of their mobility, pyroclastic flows can rapidly extend beyond the base of a source volcano. Although most pyroclastic flows do not extend more than about 15 km from source vents, some large ones have moved downvalley as much as 60 km (Crandell and others, 1984). Pyroclastic surges generally reach no more than about 10 km from source vents. The distribution of pyroclastic flows is moderately controlled by topography; they generally are channeled into valleys or basins. Surges more commonly sweep outward from their vents with only minimal regard to the underlying topography.

Pyroclastic flows and surges are especially lethal and destructive because of their great velocities and high temperatures. The danger from a pyroclastic flow results from (1) the basal flow that can bury and incinerate people and objects in its path, and (2) the cloud of hot ash and gases that may accompany the basal flow. Such clouds and pyroclastic surges can cause asphyxiation and burning of lungs and skin by hot dust or steam and other gases. In addition, rock fragments carried in them can cause serious injury or damage by impact. Hot ash clouds accompanying pyroclastic flows can sweep laterally over adjacent terrain, burning materials not buried by the main mass of the flow.

Directed or lateral blasts

Directed or lateral blasts are violent outward ejections of fragments and gases; they are similar to pyroclastic flows and surges and may be gradational with those events. Blasts commonly result from sudden releases of pressure from bodies of magma or steam-pressured rock, as can occur with the removal of a large landslide block from a volcano. The outwardly directed ejection of debris and gas sweeps over the ground surface, initially almost

Figure 5. Pyroclastic flow descending the north flank of Mount St. Helens, August 7, 1980. The coarse, basal parts of the pyroclastic flow underlie the leading edges of the flow, and clouds of fine ash rise above the basal units (Hoblitt, 1986).

uncontrolled by topography. Lateral blasts are characterized by high velocities, as much as several hundred kilometers per hour, and can be either cold or hot. Their deposits commonly are unsorted mixtures of blocks, ash, and even fragments of soil and trees. Some blast deposits, such as those from Mount St. Helens, contain stratigraphic units that are well sorted.

Most lateral blasts extend less than 10 km outward from source vents, but some have reached much greater distances. The lateral blasts of Mount St. Helens in 1980 and of Bezymianny on the Kamchatka Peninsula in the eastern U.S.S.R. in 1956 reached about 30 km from their sources (Gorshkov, 1959; Hoblitt and others, 1981; Moore and Sisson, 1981).

Lateral blasts are extremely dangerous because of their high velocities and entrained coarse debris. They knock down or dislocate most structures in their paths, and they impact and abrade whatever is not removed.

Volcanic gases

Volcanic gases are emitted during all eruptions, and can be given off at other times even when no molten or solid rock material is being ejected. Volcanic gases consist chiefly of water vapor, carbon dioxide, carbon monoxide, and various compounds of sulfur, chlorine, and nitrogen; some volcanoes also emit small but potentially dangerous amounts of fluorine.

Gases commonly are concentrated and strong near a vent, but disperse rapidly and become diluted downwind. Distribution of the gases is mostly controlled by wind, and dilute gas odors have been reported many tens of kilometers downwind from vents.

Volcanic gases can be dangerous to health or life as well as to property (Wilcox, 1959, p. 442–444). Gases are potentially injurious to people, mainly because of the effects of acid and ammonia compounds on eyes and respiratory systems, but they may also have other adverse effects. Carbon dioxide gas, because it is heavier than air, can collect in local basins and suffocate unwary animals or people. Gases commonly harm plants, and gases adhering to ash can poison animals that eat ash-covered plants. Gases also can corrode metal, fabric, and other materials. Cumulative effects of even dilute volcanic gases over a long period may cause substantial property damage.

Landslides

Slope movements described as landslides here are restricted to falling (earth and rock falls), sliding (earth and rock slides), and avalanching (debris and rock avalanches) movements related directly to volcanic activity. Mudflows and debris flows of volcanic origin (lahars) are discussed in a separate section. Falls, slides, and avalanches generally occur on steeper slopes than do mudflows and debris flows, and the appearance of a mudflow or debris flow is more obviously that of a body behaving as a fluid.

Landslides vary widely in size, from small rock falls and slides that might be dangerous only to an individual caught in them, to massive debris avalanches involving more than a cubic kilometer of material. The largest and most catastrophic landslides caused by eruptive activity have been high-velocity, gravity-driven movements resulting from collapse of the flanks of volcanic cones due directly to eruptions or to related seismic activity. Landslides caused by volcanic eruptions usually begin as rock falls or rock slides that break up into high-velocity rock or debris avalanches on the steep slopes of the volcano (Schuster and Crandell, 1984). The avalanche masses originally may be dry, or they may contain considerable meteoric water or water derived from intruded molten rock. On snow-covered volcanoes, the mass may gain water from melting snow. If the mass is fluid-charged upon reaching the lower slopes of the volcano, it can change downstream successively into a debris flow, a mudflow, and a mud flood or common flood as the rock particles settle out.

Most volcanic landslides are limited to the slopes of their respective volcanoes. Some large debris avalanches, however, extend far beyond, reaching distances up to 60 km from their sources (Schuster and Crandell, 1984). Locally, high-velocity slides and avalanches sweep up over ridges or hills, but most landslides are strongly controlled by topography and are channeled into valleys or basins.

Small to large landslides can batter, sweep away, or bury even large engineered structures. The effects can also extend far beyond the limits of a slide if the mass smashes into a water body, generating a wave (tsunami) that can be dangerous along the shoreline of the entire water body.

Two well-studied historic volcanic landslides provide case histories of large landslides and their effects. In 1888, a low-temperature phreatic explosion blasted away a part of the north flank of Bandai, an 1,818-m-high volcano in northern Japan, releasing a debris avalanche with an estimated volume of 1.2 km^3 (Sekiya and Kikuchi, 1889). The avalanche rushed 5 km down the mountain slopes into streams at the base, forming lahars that traveled down stream valleys, destroying villages and farmlands, and killing more than 400 people (Macdonald, 1972, p. 176; Nakamura, 1978, 1981). This example demonstrates an important secondary danger of volcanic landslides; they are commonly transformed into lahars that may continue downvalley many kilometers from the volcano.

The largest landslide in recorded history is the 1980 volcanic debris avalanche (Fig. 6) caused by multiple collapses of the northern sector of the cone of Mount St. Helens, which had been weakened by earthquakes, hydrothermal activity, and magmatic intrusion. These collapses resulted in a 2.3-km^3 rock slide, which quickly degenerated into an enormous avalanche of hot and cold rock debris that swept some 8 km northward and 22 km westward (Fig. 7), burying the upper part of the North Fork Toutle River Valley system to an average depth of 45 m and a maximum depth of 195 m; the estimated volume of this debris avalanche is 2.8 km^3 (Voight and others, 1983).

Figure 6. Debris-avalanche deposit in the upper valley of the North Fork Toutle River, southwestern Washington. View east from the distal margin of the avalanche toward the devastated cone of Mount St. Helens. (Photograph by R. M. Krimmel, U.S. Geological Survey.)

Lahars

Lahars are mudflows and debris flows of volcanic material; they commonly originate from eruptions of hot material onto snow or ice, eruptions through crater lakes, or saturation and mobilization of loose volcanic material during heavy rains. They vary in texture from fine-grained mudflows to extremely coarse debris flows, and can include soil particles and rock hydrothermally altered to clay, as well as volcanic ash. Mixed with this fine material may be blocks of volcanic rock as large as several meters in diameter, volcanic bombs, and even the trunks of large trees.

Because of the mobility of lahars and the steep slopes on which they usually originate, they often flow at high velocities and to great distances; both of these characteristics increase the hazard to humans and their developments. Lahars move down the sides of volcanoes, following stream courses, with speeds dependent on steepness of slope, freedom from obstructions, and viscosity of the flowing mass (Macdonald, 1972, p. 170). Lahar velocities of 30 to 50 km/hr are common, and speeds of 70 to 180 km/hr have been estimated.

The distance traveled by a lahar depends largely on the steepness and nature of the terrain over which it flows, though loss of water plays an important part in slowing and stopping the flow. Lahars originating on the steep slopes of a volcanic cone often continue only short distances after they reach the flatter slopes at the base of the cone, but some of large volume continue for great distances (Macdonald, 1972, p. 170). Lahars that have traveled 10 to 15 km are common, and distances of 80 to 120 km have been recorded. The Toutle River lahars resulting from the 1980 eruption of Mount St. Helens extended about 120 km down the Toutle and Cowlitz Rivers and into the Columbia River (Fig. 7).

In addition to their impressive lengths, some lahars cover large areas and have huge volumes. For example, the 23-km-long Pleistocene Nirasaki mudflow or debris avalanche from Yatsuga volcano, Japan, had an estimated volume of about 9 km^3 and covered 120 km^2 (Mason and Foster, 1956). The Osceola mudflow, which originated on Mount Rainier, Washington, about 5,000 years ago, had a volume of about 2 km^3 and covered over 300 km^2 (Crandell, 1971).

Lahars have been among the most destructive of the processes related to volcanic activity. Major loss of life caused by lahars in historical times has mostly been in the Circum-Pacific region, and especially in Central and South America (>24,000 killed), Japan (>11,650 killed), and Indonesia (>9,300 killed) (Neall, 1976; Katsui and others, 1986). Lahars are capable of removing most manmade objects from their paths because of their high specific gravities (relative to flood waters) and velocities, which enable them to carry objects of tremendous size and weight in a buoyant manner.

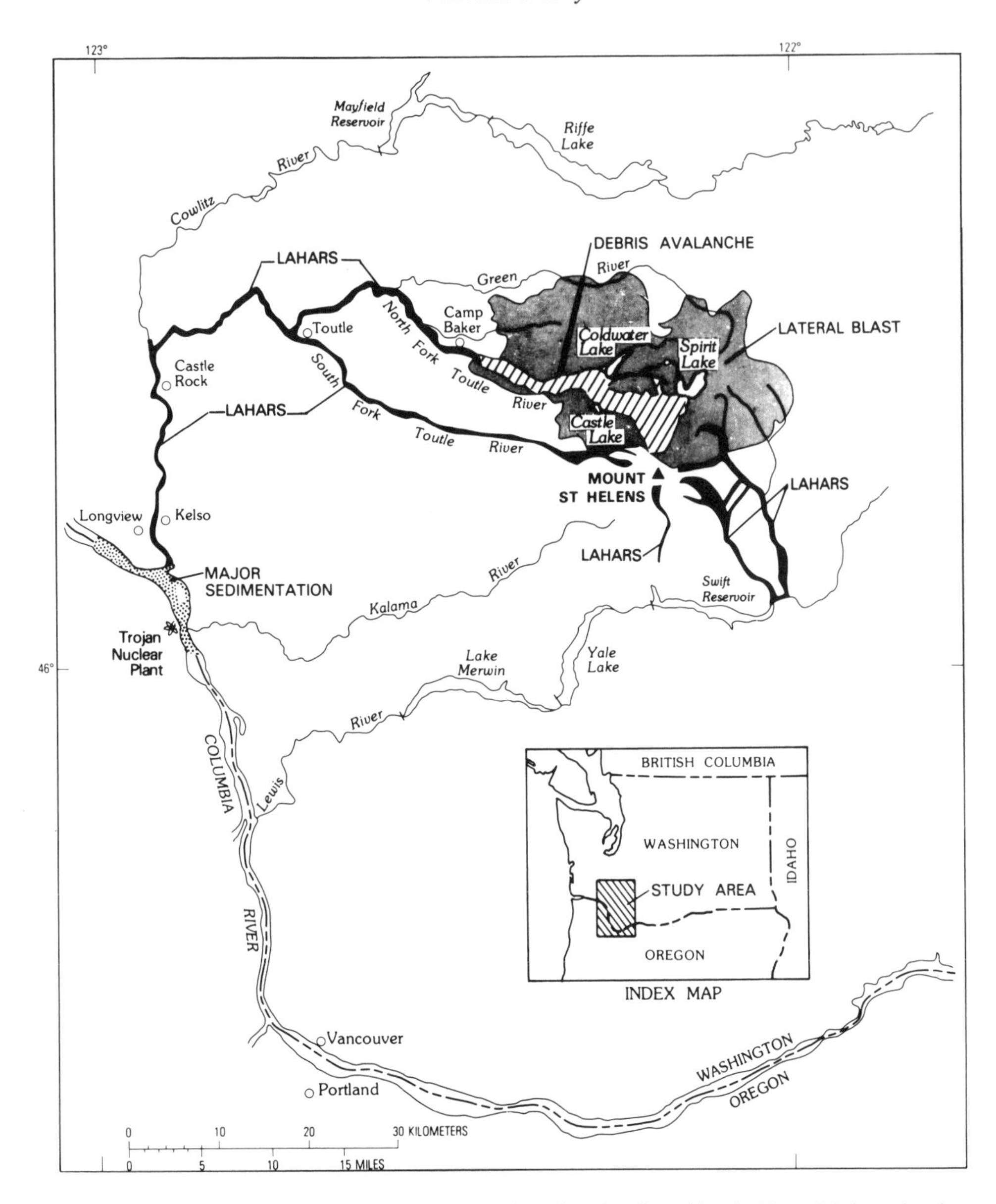

Figure 7. Location map of area near Mount St. Helens directly affected by the blast, debris avalanche, and lahars, from the May 18, 1980, eruption.

Water floods

There are many documented cases of floods related to volcanism. As might be expected, sediment-laden water floods of volcanic origin are difficult to distinguish from very fluid lahars. Obviously, an individual water flood may grade into a lahar, or vice versa. For example, melting of snow on the slopes of a volcano may initiate a water flood. As the flood moves down the slopes of the volcano, however, it may acquire enough sediment to be transformed into a mudflow or debris flow. It may move as a lahar for many kilometers as it leaves the volcano and travels down a river valley. At some point, however, if the gradient becomes so flat that much of the sediment is deposited, a lahar may be transformed into a flood that continues downstream. Such a water flood derived from a lahar may cause damage from inundation and sedimentation for many miles beyond the terminus of the lahar. This was the case for the Toutle-Cowlitz River mudflows from the May 1980 eruptions of Mount St. Helens in which flood waters continued many kilometers down the Columbia River beyond the limit of the true mudflow.

The causal mechanisms of water floods related to volcanic activity are much the same as those that cause lahars; the resulting product depends primarily on the degree of slope and the availability of material to become sediment load. Any one of the following mechanisms can provide the large amounts of water required for floods: (1) melting of snow and ice on the slopes of a volcano due to an eruption; (2) release of natural impoundments ("breakout" floods) by breaching of natural dams that were formed by volcanic deposits or by landslides related to volcanic activity; (3) ejection of water from a crater lake due to a volcanic eruption; (4) sudden draining of a crater lake by a noneruptive mechanism (commonly by collapse of a crater wall); (5) release of ponded water beneath or behind glaciers due to an eruption (the resulting floods are known as "jökulhlaups", the name given them in Iceland); and (6) heavy rains on products of eruptions, resulting in increased runoff.

Earthquakes

Earthquakes are frequently caused by volcanic activity. Most seismic activity associated with volcanism is local and not strong. Most such earthquakes are caused by subsurface movement of magma; others are caused by adjustment to differential load within the volcanic pile, and can occur either within or underneath the volcano itself. The various subsurface stresses break the rock, causing small to moderately large earthquakes; if the displacement extends to the ground surface, it may be visible as a fault.

A few earthquakes associated with volcanoes are large enough to cause significant damage. Fractures and ground displacement at the surface generally are limited to areas near the source of the earthquakes, but ground shaking is widespread and frequently triggers landslides. Severe shaking can extend hundreds of kilometers from the source of an earthquake.

Ground shaking is an especially pervasive, as well as the most extensive, effect of earthquakes. Shaking can damage virtually any kind of structure, and can be intensified in soft, saturated sediments. Shaking also commonly causes settlement in soft materials. Landslides caused by shaking can displace entire structures, and ground fractures can rupture structures, roadways, and utility lines. Fracture of water and sewer lines has frequently caused severe secondary damage.

Tsunamis

Tsunamis, also called seismic sea waves or tidal waves, are long, low, rapidly moving water waves, usually in the ocean. Large tsunamis generally are caused by abrupt crustal movements along the floors or margins of water bodies, causing displacement of large masses of water. Nevertheless, submarine eruptions have caused tsunamis, as have the entrance into water bodies of pyroclastic flows and volcanic landslides and lahars.

Tsunamis travel outward from their points of origin, reaching distances of tens of thousands of kilometers at velocities of hundreds of kilometers per hour. As the waves enter shallower water, they become shorter and higher, and reportedly have swept onshore to heights as great as 30 to 40 m above sea level (Blong, 1984; Crandell and others, 1984). The heights and distances inland reached by tsunamis are strongly affected by local offshore and onshore topography and the direction of wave approach, and thus are difficult to predict. However, tsunamis commonly affect no more than a narrow band of coastline only a few kilometers wide.

Tsunamis can move onshore as turbulent waves that damage or destroy virtually everything in their paths. People can be battered or drowned, buildings moved off their foundations, trees uprooted, and boats carried inland. As the waves recede, they can carry people and property out to sea. Tsunamis can also rise quietly and inundate nearshore areas. The salt water can kill crops, poison soil, corrode metal, and damage objects and structures in many other ways.

Ground subsidence and fracture

Ground subsidence and fractures at volcanoes are commonly caused by withdrawal of magma, landsliding, and collapse of lava tubes. In Hawaii, entire islands are settling as the ocean floor subsides because of the load placed on it by the volcanoes (Apple and Macdonald, 1966). Elsewhere, subsidence and fracturing are limited to areas on or immediately adjacent to volcanoes, and do not extend far beyond them.

Subsidence and fracturing can dislocate and fail structures and other facilities, and subsidence increases susceptibility to inundation by water or lava flows. They generally do not pose a serious threat to people's lives unless the people are caught in structures that are damaged.

EFFECTS OF VOLCANIC ACTIVITY ON ENGINEERING WORKS

Factors relating to potential damage

Potentially damaging volcanic hazards include two general groups: those that travel along the ground surface, such as lava flows, pyroclastic flows and surges, lahars, and floods, and whose courses and distributions are determined chiefly by pre-existing topography; and those that travel through the air, such as large ballistic fragments and volcanic ash, which are independent of topography, and whose distribution is determined by the heights attained by the airborne materials and the direction and velocity of winds.

This section discusses effects of several types of volcano-related hazards on structures, developments, and facilities. In most cases these hazards do not act independently; they may grade into each other, and accompany or initiate others. The cumulative effects of volcanic gases are not discussed in this report; they may be significant over a long period of time, but generally do not endanger engineering works. Nor are the effects of volcano-related earthquakes, tsunamis, and ground fractures discussed; they are similar to effects of these same events unrelated to volcanoes, which have received extensive treatment in many other reports.

Effects of lava flows

Lava flows have destroyed vast amounts of property and innumerable manmade structures. There are many well-

Figure 8. Tower of the church of San Juan Parangaricutiro, Mexico, protruding from the rubbly surface of a lava flow from Parícutin volcano, July 1945. (Photograph by W. F. Foshag, U.S. Geological Survey.)

documented cases of such destruction in Italy, Iceland, and Mexico, and on the island of Hawaii, chiefly by burial, partial inundation, and crushing.

Since the modern cone of Mount Vesuvius, Italy, began to grow in A.D. 172, flows repeatedly have invaded villages on the lower slopes of the mountain (Bolt and others, 1975, p. 82). One of the most destructive of these lava flows was that of 1906; highly fluid lava flowed down the mountain and spread out on the lower slopes with a thickness of as much 7 m, burying houses to the second story and pouring in through windows and doors. Some houses were knocked down, but, because of the relatively small thrust exerted by fluid lava, others were only filled with lava without destroying the weak masonry walls (Bolt and others, 1975, p. 82).

During the 1669 eruption of Mount Etna, Italy, a large lava flow moved southward until it was temporarily stopped by the wall of the port city of Catania. After resisting the force of the lava for several days, the wall gave way and the lava destroyed part of the city, knocking down weak buildings and burying more resistant ones (Bolt and others, 1975, p. 83–84). The 1928 eruption of Etna produced a 12-m-thick flow that buried the town of Mascali. This lava was more viscous than that of Vesuvius in 1906, and exerted greater thrust on structures it encountered, so that the masonry walls of many buildings were toppled and crushed beneath the lava (Bolt and others, 1975, p. 83).

The largest historic lava eruption was the 60-km-long 1783 Lakagigar flow in Iceland, which covered an area of over 565 km^2 and had a total volume of about 13 km^3, overwhelming 14 farms and damaging 30 others (Einarsson, 1974). In 1973, lava from the eruption of Eldfell volcano on Iceland flowed into the town of Vestmannaeyjar, destroying about 200 buildings. Vestmannaeyjar harbor and important factories were saved by pumping water on the flow, thus chilling it and reducing its viscosity (Einarsson, 1974).

In the nine-year eruptive history (1943–52) of Parícutin volcano in Mexico, lava flows with a volume of 700 million m^3 covered an area of 24.8 km^2 (Fries, 1953). In 1944, flows from Parícutin destroyed all of the village of Parícutin and most of the town of San Juan Parangaricutiro, where lava partly buried the old masonry church (Fig. 8). Small amounts of lava dribbled in through windows and doors, but the walls were not crushed (Bolt and others, 1975, p. 85). This provides another example of the small amount of force that fluid lava may exert on structural walls.

Lava from the 1955 eruption of Kilauea volcano in Hawaii destroyed 17 houses and buried 10 km of public roads. Another flow in 1960 destroyed a schoolhouse, a commercial fishpond, several beach residences, and most of the village of Kapoho (Bolt and others, 1975, p. 85). Smaller flows in 1969 buried roads in Hawaii Volcanoes National Park (Fig. 9). Future lava flows from Kilauea can be expected to have even greater effects on engineering works because of increasing residential development in the vicinity of the volcano.

Effects of tephra

This section deals with the effects of tephra that has been ejected at high angles into the air, and falls back freely to the surface. Large fragments also are commonly ejected during eruption of pyroclastic flows and volcanic blasts.

Coarse fragments and projectiles. Many large fragments of rock or still-molten lava are thrown through the air on ballistic trajectories. Most of those projectiles fall close to their source vents, and impact and penetration from them can damage nearby structures. However, in some eruptions many projectiles can be hurled as far as 5 to 10 km, and a few may be propelled much farther. As an extreme example, a bomb weighing about 10 kg is reported to have been thrown about 33 km from the vent of Hekla volcano in Iceland in 1947 (Thorarinsson, 1954). Because large fragments are also ejected in conjunction with lateral blasts and pyroclastic flows, it frequently is unclear just what kind of event produced the ballistic fragments.

Figure 9. Lava flow from Kilauea volcano spilling over rim of Aloi Crater onto park highway, Hawaii Volcanoes National Park, April 1970. The lava was scraped off the road surface by a highway maintenance crew. (Photograph by D. A. Swanson, U.S. Geological Survey.)

Figure 10. Houses in town of Vestmannaeyjar partially buried by basaltic tephra from the 1973 eruption of Eldfell volcano, island of Heimaey, Iceland (U.S. Geological Survey, 1975).

One instance of damage to buildings from large volcanic projectiles occurred in the 1783 eruption of Asama volcano in Japan, in which lumps of red-hot pumice measuring more than 50 cm across penetrated the roofs of houses 11 km from the volcano (Aramaki, 1956). In their study of the 1902 eruption of Soufrière volcano on the island of St. Vincent, Anderson and Flett (1903) noted that large stones frequently perforated wooden or iron roofs of buildings. Stones more than 0.3 m in diameter that fell in the town of Georgetown, 8 km from the crater, commonly penetrated roofs. However, projectiles "almost the size of a hen's egg" did little damage to buildings in the town of Kingstown (Anderson and Flett, 1903).

Other examples of roof damage from ballistic fragments occurred in the 1957 eruption of Manam volcano, Papua New Guinea (Palfreyman and Cooke, 1976); the 1975 eruption of Ruapehu volcano, New Zealand (Nairn and others, 1979); and the 1976 eruption of Augustine volcano, Alaska, (Kienle and Swanson, 1980).

Numerous experiments have been conducted and observations made to determine the effects on common building materials of being struck by projectiles with different impact energies. Blong (1981; 1984, p. 201–206) has extrapolated the results of several of these studies to provide an interrelationship between volcanic projectile size, density, and impact energy, and the potential damage to a range of building materials.

Volcanic ash—roof loads. Roof failures can also be caused by loads of thick volcanic ash. Failure is generally by collapse, rather than by penetration. An early detailed description of the effects of heavy ash fall on a human development is that of the 79 A.D. eruption of Vesuvius. The 1906 eruption of Vesuvius also had a devastating effect; nearly all the deaths in that eruption resulted from the collapse of flat roofs under the weight of accumulated tephra. Most of these failures were in the towns of Ottaviano and San Giuseppe, 5 and 6 km from the mountain. A 10-cm depth of this ash was usually sufficient to cause the collapse of weak, flat roofs; steeply inclined roofs were used in reconstruction (Perret, 1950). The principal loss of life was in San Giuseppe, where the people crowded into the local church just before the roof collapsed due to a load of ash (Hobbs, 1906). Many other weak house roofs covered with heavy tiles also collapsed.

Similar cases of roof collapse due to ash loads ranging in thickness from about 10 cm to 3 m have resulted from eruptions of Tambora volcano (1815) in Indonesia (Quarterly Journal of Science and the Arts, 1816); Soufrière volcano (1812 and 1902) on St. Vincent (Anderson and Flett, 1903); Mount Katmai (1912) in Alaska (Wilcox, 1959); Volcan Fuego (1971) in Guatemala (Bonis and Salazar, 1973); and Eldfell volcano (1973) in Iceland (Einarsson, 1974) (Fig. 10).

Detailed data are sparse regarding the effects of fine-grained tephra loads on buildings. However, there is an obvious parallel between the effects on buildings of tephra loads and those of snow loads, for which there is an abundance of data (Peter and others, 1963; Taylor, 1979). For design purposes, snow can be assumed to have a density of about 0.2 to 0.3 g/cm^3. Volcanic-ash densities commonly range from about 0.4 to 0.7 g/cm^3 in the newly fallen state (Einarsson, 1974; Blong, 1981); in a compacted and saturated state, this value can increase to about 1.5 g/cm^3. Thus roofs can carry about 20 percent as thick a tephra mantle as they can a snow cover.

Volcanic ash—other effects. The May 18, 1980, eruption of Mount St. Helens provided a well-documented example of the effects of small amounts of ash on a modern society. It ejected an estimated 490 million tons of tephra (Sarna-Wojcicki and others, 1981), nearly all consisting of ash that fell in a broad band across eastern Washington, northern Idaho, and western Montana (Fig. 4). This ash, with an estimated minimum uncompacted volume of 1.1 km^3, greatly affected most engineering works and operations. Depths of as much as 5 cm were recorded at 50 km from the volcano, but average depths were much less (Fig. 4). This thin cover of ash caused little structural damage to buildings, but cleanup costs were large (Dillman and Roberts, 1982).

The ash fall paralyzed transportation for several days in parts of eastern Washington and northern Idaho by blanketing highways, resulting in reduced visibility, damage to unprotected engines, and, in some cases, blocking of the roads by drifted ash. Within a few hours, 2,900 km of state highways were closed in eastern Washington because of the accumulation and drifting of ash (Fig. 11). Interstate 90, which crosses Washington from Seattle to Spokane, was closed for a week.

Air transportation in the area was temporarily paralyzed by the ash cloud itself, due to poor visibility and the probability of pitting of windshields and damage to engines. Ash accumulation on airport runways, taxiways, and aprons curtailed aircraft operations for a much longer time. For example, the Spokane International Airport was closed for 3 days; during this period 576 commercial flights were canceled (Schuster, 1983).

Almost all municipalities in areas of eastern Washington that received more than about 2 cm of ash experienced severe problems due to clogging of storm sewers, and in a few cases their sanitary sewage systems became plugged and the plants were inoperative until cleaned (the plant at the city of Moses Lake was plugged for five weeks).

Figure 11. Removal of ash from the May 18, 1980, eruption of Mount St. Helens—Interstate 90, eastern Washington. The Washington State Highway Department estimated that its maintenance crews removed 540,000 tons of ash from state highways as a result of the May and June 1980 eruptions of Mount St. Helens. (Photograph by Washington State Highway Department.)

Effects of pyroclastic flows and surges

Studies of the effects of pyroclastic flows and surges on engineering structures have been few; effects range from minor to complete destruction of developments in their paths. Of particular interest are the pyroclastic flows and surges, frequently described as nuées ardentes or glowing clouds, that accompanied the 1902 Caribbean eruptions of Soufrière volcano on the island of St. Vincent, and Mount Pelée on the island of Martinique.

Soufrière erupted in May 1902. A glowing cloud of gas and ash swept down the volcano, killing more than 1,500 people, although many escaped by taking refuge in cellars or stone buildings that survived the force of the hot cloud (Anderson and Flett, 1903). The next day, Mount Pelée erupted more violently than Soufrière. A glowing cloud with a velocity of about 165 km/hr struck the city of St. Pierre (8 to 10 km from the crater) with tremendous force, killing some 30,000 people. The effects of this hot cloud have been summarized by Macdonald (1972, p. 145) as follows: "Masonry walls 3 ft thick were knocked over and torn apart, big trees were uprooted, 6-inch cannon were torn from their mounts, and a 3-ton statue was carried 40 ft from its base." Macdonald noted that the temperature of the hot cloud that swept over the city was between 700° and 1,000°C.

Anderson and Flett (1903) observed that the violence of the glowing cloud at St. Pierre was terrific, as shown by toppled masonry walls, twisted ironwork of the building verandas, the ruined cathedral, and uprooted and dismembered trees. In the north end of the town, some 8 km from the crater, all houses were leveled, except where they were protected by a sea cliff. The ash-laden cloud had twisted the iron stanchions of the bridge crossing the Rio Roxelane where it passes through the town, and had planed off the upper parts of houses that stood in the shelter of the steep northern bank of the river. At the southern edge of the town (about 10 km from the crater), at the outer margin of the hot cloud, destruction was less severe because both the force and the temperature were lower there; walls were left standing in many houses.

Destructive pyroclastic flows and surges have occurred in this century at many other volcanoes. Pyroclastic flows during the 1951 eruption of Mount Lamington in New Guinea devastated a 230-km^2 area, taking nearly 3,000 lives (Taylor, 1958). In December 1951, a pyroclastic flow swept down the side of Hibok-Hibok volcano in the Philippines, killing 500 people and destroying the outskirts of the port town of Mambajao (Macdonald and Alcaraz, 1956). Besides destroying masonry buildings in the village of Francisco Leon (Fig. 12), 5 km south of the volcano, the pyroclastic flows from the April 1982 eruption of El Chichón volcano in southern Mexico dammed the Rio Magdalena, creating a 5-km-long lake of hot water (Silva and others, 1982).

Effects of volcanic blasts

Because volcanic blasts often are directed along or close to the ground surface, they are particularly devastating to engineered structures and developments. The rock debris they carry at high velocity damages, destroys, or buries structures and developments, and kills "virtually all life by impact, abrasion, burial, and heat" (Crandell and others, 1984, p. 16).

Noteworthy examples of directed blasts occurred in the 1888 eruption of Bandai volcano in Japan, the 1956 eruption of Bezymianny volcano on the Kamchatka peninsula, U.S.S.R., and the 1980 eruption of Mount St. Helens.

Bandai volcano (Japan), 1888. The eruption of Bandai volcano is an example of a phreatic or steam-blast eruption, partly in the form of a lateral blast, unaccompanied by lava flows or pumice ejection. The eruption was caused by explosive expansion of ground water superheated by hot rock beneath the volcano's surface (Sekiya and Kikuchi, 1889). Some 15 to 20 explosions occurred within a minute or two, blasting away part of the volcanic cone. Several of these explosions were inclined; the last was almost horizontal. The force of the blasts, which reached estimated velocities of about 150 km/hr, and the rock fragments propelled by them, did tremendous damage on the lower north slopes of the mountain (Sekiya and Kikuchi, 1889).

On the slopes most exposed to the blasts, houses were leveled and trees with diameters of more than a meter were knocked down. The villages of Shibutani and Shirakijo, 4 km from the crater, were badly damaged by the blasts, which funneled down a valley on the lower slopes of the mountain.

Bezymianny volcano (U.S.S.R.), 1956. On March 30, 1956, a tremendous lateral explosion shot eastward from Bezymianny at an angle of 30° to 40° from the horizontal, destroying the entire mountain top (Gorshkov, 1959). Because the area hit by the blast was uninhabited, there were no casualties and little economic loss. However, at a distance of 25 km from the volcano, trees 0.3 m in diameter were snapped off; 30 km away, the bark of living trees was scorched, and deadwood was set afire.

Figure 12. Remains of a building in the village of Francisco Leon, 5 km south of El Chichón volcano in southeastern Mexico. The building and village were destroyed by pyroclastic flows from the April 1982 eruption of El Chichón. The bent steel reinforcing bars provide an indication of the force and direction of the pyroclastic flow. (Photograph by R. I. Tilling, U.S. Geological Survey.)

Mount St. Helens (U.S.), 1980. Engineered works in the area immediately north of Mount St. Helens were suddenly and totally destroyed by a lateral blast and debris avalanche on May 18, 1980. The blast included a shock wave, thermal effects, and massive amounts of hot, airborne pyroclastic debris moving at more than 350 km/hr (Kieffer, 1981).

The blast affected a 600-km^2 area that extends generally northward from the mountain through a broad arc from west–northwest to east–northeast (Fig. 7). Within this arc, destruction was essentially total within a radial distance of about 13 km from the crater of Mount St. Helens; almost all vegetation and much soil were removed from slopes that face the volcano, and remaining tree stumps were shredded. Within this zone, temperatures of the pyroclastic cloud constituting the blast were more than 350°C (Moore and Sisson, 1981).

Beyond the zone of total devastation, in an area extending to about 20 km east-northeast and 25 km north and northwest of the crater, old-growth timber was blown down, and lesser vegetation was killed by the blast, buried by entrained debris, or both. In this zone, falling trees extensively damaged recreational and logging roads.

Surrounding this blown-down zone was a marginal area about 2 to 3 km wide within which vegetation was killed by the heat of the blast, but remained standing; within this zone, damage to engineered works was minimal.

Effects of landslides

Although most landslides caused by volcanic eruptions have been restricted to the sides of the volcanic cones, and thus directly affect very few structures or facilities, there have been some notable exceptions. Prehistoric rock and debris avalanches on Mount Rainier, Washington, have traveled much farther, as have historic avalanches on Papandajan and Merapi volcanoes in Indonesia; Asama, Unzen, Bandai, and Tokachi-dake volcanoes in Japan; Sheveluch volcano in the U.S.S.R.; and Mageik volcano (Alaska) and Mount St. Helens (Washington) in the U.S. This section mainly describes the effects of the 1980 debris avalanche from Mount St. Helens, which is well documented, and although it killed fewer than 10 people was one of the world's most destructive volcanic landslides.

The collapse of the upper northern portion of Mount St. Helens in 1980 caused a debris avalanche (Figs. 6 and 7) that buried the remaining northern slope of the mountain, the southern margin of Spirit Lake, and the valley floor of the North Fork Toutle River north of the volcano. The following engineered works and facilities were affected by the debris avalanche (Schuster, 1983):

(1) Public and private buildings and facilities on the south shore of Spirit Lake, which were buried by as much as 60 m of debris-avalanche material.

(2) State Highway 504, which was buried from its terminus at timberline on Mount St. Helens to the lower end of the debris avalanche in the North Fork Toutle River valley, a roadway length of some 32 km. The debris avalanche destroyed two major bridges on this stretch of the highway. In addition, many kilometers of private logging roads and five private and two U.S. Forest Service bridges were destroyed by the avalanche.

No major structures or facilities, such as dams and reservoirs, powerplants, or large buildings, were in the area devastated by the debris avalanche. Had the avalanche gone to the south, however, it probably would have inundated Swift Reservoir (Fig. 7) and caused flooding of the lower valley of the Lewis River, and might have affected the Trojan Nuclear Plant on the Columbia River downstream from the mouth of the Lewis River.

The destruction of engineered works by the debris avalanche was not due to inadequate design. The only way that this destruction could have been avoided would have been by land-use planning that prevented development in the devastated area. However, nearly all the development occurred before Crandell and Mullineaux (1978) identified and publicized the potential hazards from Mount St. Helens. The lessons learned here should not be forgotten; restrictions on future development in this area, and in similar areas near other potentially hazardous volcanoes, should be considered carefully in regard to the effects of future eruptions.

Effects of lahars

Mudflows and debris flows (lahars) historically have been among the most destructive of the processes related to volcanic activity. In addition to loss of life, they have caused extensive damage to buildings, transportation and communication systems, bridges, farmland, and forests. Highly destructive lahars in the last few years have been caused by the 1980 eruption of Mount St. Helens and the 1985 eruption of Nevado del Ruiz in Colombia.

Mount St. Helens (U.S.), 1980. Probably the best-documented instance of destruction caused by lahars is that of the debris flow/mudflow complexes that originated from the Mount St. Helens eruption of May 18, 1980. The case history of these lahars provides considerable information that is of value in planning and siting future developments and engineered facilities.

As a result of the 1980 eruption, mud and associated debris flowed down several of the valleys that radiated from Mount St. Helens (Fig. 7). The largest and most destructive lahars occurred in the valleys of the North Fork and South Fork Toutle River and in the Cowlitz River downstream from its confluence with the Toutle. These lahars and associated debris (mainly logs) passed down the Toutle and Cowlitz Rivers into the Columbia River, damaging roads and highways, railways, bridges (Fig. 13), housing, vehicles (Fig. 14), water and sewage treatment facilities, and agricultural and forest lands (Schuster, 1983). In addition, lahars in tributaries of the Lewis River on the east and southeast sides of the mountain destroyed forest roads and bridges, and flowed into Swift Reservoir (Janda and others, 1981).

Nevado del Ruiz (Colombia), 1985. On November 13, 1985, an eruption of Colombia's Nevada del Ruiz volcano caused lahars that killed more than 24,000 people and injured another 10,000 in valleys beyond the volcano. The eruption was not large, but it was accompanied by one or more small pyroclastic flows or surges that caused rapid melting of ice and snow on the upper reaches of the mountain, triggering disastrous lahars. About 21,000 perished as a lahar flowed out of a canyon of the Lagunilla River to demolish much of the city of Armero, 50 km from the mountain; about 75 percent of the population of Armero died in the disaster (Katsui and others, 1986). At least 5,150 homes, 50 schools, 2 hospitals, and 200 km of road were destroyed or damaged by the lahars (Herd, 1986).

Figure 13. Remains of private log-haul bridge destroyed by May 18, 1980, Mount St. Helens lahar along the North Fork Toutle River. Note smear from this mudflow on tree trunks in middle distance.

Figure 14. Logging-company employee bus heavily damaged and partially buried by lahar on the North Fork Toutle River, 1980. Bus was unoccupied when struck by the flow.

Effects of water floods

The effects of water floods related to volcanic processes are essentially the same as those of the more common floods resulting from heavy rains, rapid melting of heavy snowpacks, etc. A flood of any of those origins could wreak havoc on structures, facilities, and developments in its path.

This section will deal with floods and potential flooding related to the 1980 eruption of Mount St. Helens, because this case history provides considerable information on (1) the effects of floods, and (2) mitigative measures for the prevention of post-eruptive floods from lakes created by volcanic events.

Beaulieu and Peterson (1981) have discussed several ways in which floods were caused by the Mount St. Helens eruption and the possible causes of future flooding due to activity of that volcano. The following paragraphs summarize the modes of flooding presented by Beaulieu and Peterson, who discussed (1) the actual flooding in the lower Cowlitz River valley and on the lower Columbia River due to the large lahar that entered the Cowlitz River from the Toutle River, (2) the potential flood hazard in the lower Cowlitz and Toutle River valleys due to clogging of the pre-eruptive river channels by sediment from lahars, and (3) the flood hazard due to damming of the North Fork Toutle River and its tributaries by a massive debris avalanche.

Flooding in the lower Cowlitz River valley and on the lower Columbia River. The huge volcanic mudflow that passed from the Toutle River to the Cowlitz River (Fig. 7) as a result of the 1980 eruption of Mount St. Helens caused severe flooding in the Cowlitz River valley and minor flooding along the larger Columbia River. It is generally considered that the mudflow graded into a true water flood where the Cowlitz River entered the Columbia River valley (Fig. 7).

The Toutle River mudflows (Fig. 7) deposited as much as 5 m of sediment in the channels of the Toutle and lower Cowlitz Rivers (Fig. 15; Lombard and others, 1981). This sediment reduced the channel flow capacity to about one tenth of pre-eruption capacity. Thus a potential existed for unusually high flood levels during succeeding natural runoffs. To counter the flood threat, the U.S. Army Corps of Engineers by October 1981 excavated 14 million m^3 from the lower 19 km of the Toutle River and 43 million m^3 from the lower Cowlitz River channel. The total amount removed from the Toutle, Cowlitz, and Columbia River channels during the period May 1980 to October 1981 was about 75 million m^3 (Schuster, 1983).

At Rainier, Oregon, the level of the Columbia River was raised 1.4 to 2.0 m on May 18–19 (Kienle, 1980). Although flood damage in the Columbia River valley was minor, 34 million m^3 of sediment were deposited in the river 105 to 125 km upstream from its mouth (Bechly, 1980). By October 1981, the Corps of Engineers had dredged from the Columbia River about 18 million m^3 of sediment resulting from the mudflow and flood.

The Trojan Nuclear Plant (Fig. 7) is probably the most important engineering structure along the segment of the Columbia River affected by sedimentation from the flood. This plant, with a rated electrical production of 1.13 million kw, is on the shore of the river 8 km upstream from the mouth of the Cowlitz River. It is cooled by water pumped from the Columbia River immediately adjacent to the plant. Soundings conducted by the Corps of Engineers a few days after the 1980 eruptions showed that as much as 12 m of sediment had been deposited in the channel adjacent to the plant, resulting in a minimum bottom depth of 11 m. Because the intake structure was located at a depth of only 3 m, this posed no serious threat to the cooling system.

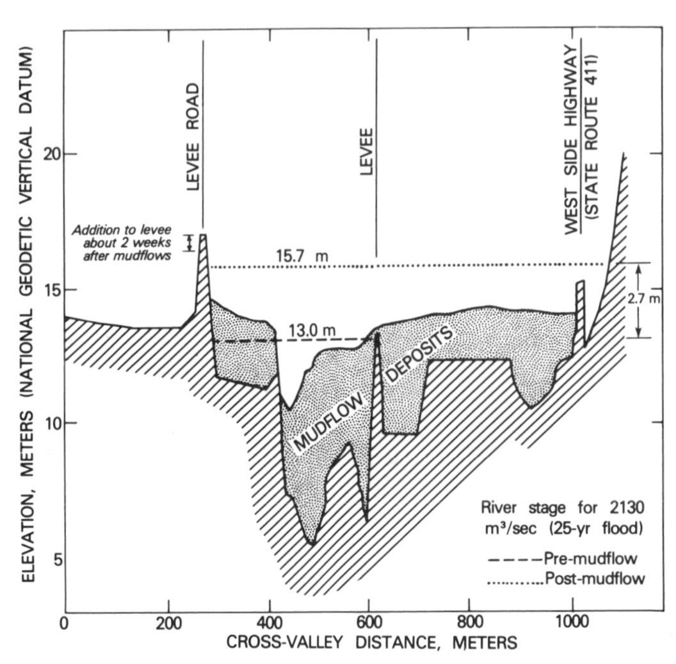

Figure 15. Channel and valley conditions of the Cowlitz River near Castle Rock, Washington, prior to and after the lahars of May 18, 1980 (Lombard and others, 1981).

Flood hazard due to damming of North Fork Toutle River and its tributaries. The Mount St. Helens debris avalanche (Figs. 6 and 7) formed several new lakes by damming tributaries and the main stem of the North Fork Toutle River, into which Spirit Lake drained before the 1980 eruption. These impoundments created a set of conditions that, without installation of control structures, could eventually have led to overtopping and possible breaching of the individual debris dams, and possibly severe downstream flooding. To assure stability of their blockages, the levels of the three largest impoundments, Spirit Lake (259 million m^3), Coldwater Lake (83 million m^3), and Castle Lake (24 million m^3) (Fig. 7), are now controlled by outlet structures constructed by the U.S. Army Corps of Engineers.

MITIGATION OF VOLCANIC HAZARDS

Volcanic hazards can be mitigated by avoidance, by physical control measures, and by minimizing their effects.

Avoidance

Hazards can be avoided on a long-term basis by land-use zoning that prevents building of structures in dangerous areas. On a short-term basis, people and some property—though generally not engineering works—can be evacuated from threatened areas when an eruption is imminent or even underway. Methods of assessing volcanic hazards and showing their locations on maps for these uses are described by Crandell and others (1984).

Predictions of future eruptions can help people prepare for and respond effectively to these natural hazards. Long-range forecasts based on past eruptions can include hazard assessments stating the probable frequency of hazardous events as well as areas they could affect. These forecasts can be accompanied by hazard-zonation maps that show the relative magnitudes of certain hazards in different areas. For example, Crandell (1976) has published a preliminary map of volcanic hazards in the State of Washington, and similar maps are being prepared for Oregon and California. Methods of assessing volcanic hazards and showing their locations on maps for these uses are described by Crandell and others (1984). Such information can be used for making decisions regarding both long-term land use and short-term response to eruptions that either are threatening or are in progress.

Short-range forecasts are based chiefly on geophysical monitoring that measures the effects, especially earthquakes and deformation, of magma rising into the volcano. Such forecasts should provide a general warning of impending eruptions, but specific and reliable warnings of time and magnitude are not yet possible.

Control and diversion

Eruptions cannot be prevented or controlled at this time. However, in some cases it may be possible to control, or at least to divert, some of the hazards resulting from eruptions. Control and diversion efforts require favorable topographic conditions and sources of barrier materials, and may raise legal and social problems.

Lava flows have been diverted, at least temporarily, using explosives to break down lava-flow channels and disrupt the flowing lava, by constructing barriers and diversion channels, and by using water to cool and solidify the lava. Generally, the larger and more rapidly moving the lava flows are, the more difficult it is to control or divert them. If the lava flow continues to be fed by a voluminous eruption, it most likely will overwhelm any control or diversion attempt.

The path taken by volcanic ash is dictated by regional wind patterns; thus, the ash cannot be controlled or diverted. The products of lateral blasts spread widely at high velocities and with great momentum, and generally cannot be diverted effectively.

It is difficult, if not impossible, to physically control or divert most landslides resulting from volcanic eruptions, because they commonly are so large, because volcano slopes are steep, and because their locations are difficult to predict. Conversely, many lahars can be controlled because they are smaller and follow valleys. The most commonly used structures to control lahars are check dams built across valleys to form debris-retention basins. These are commonly used in Indonesia and Japan to control lahars by providing reservoir volume. Check dams are designed to sustain "full reservoir" fluid pressures plus dynamic loading due to arrival of the flow.

Unlike lahars, some water floods caused by volcanic activity may not occur until considerably after an eruption, thus allowing time for construction of control measures. This is the case for breakout floods due to breaching of natural dams formed by volcanic deposits. Mitigative measures commonly used to reduce the hazard from such natural dams include artificial breaching of the blockage before it impounds much water and prevention of overtopping by construction of a spillway or other outlet works. Drainage of crater lakes has proven successful at some volcanoes in eliminating much or all of the flood hazard due to potential release of lake water by future volcanic activity.

Minimizing effects

Little can be done to minimize effects of lava flows, which virtually destroy anything that is overridden. However, some structures that are approached but not overridden by lava flows can be protected from fire.

Ash falls can cover large areas, but are particularly amenable to measures that will minimize their effects. Structures in high-hazard zones can be designed to withstand loads of ash and impacts from falling rock fragments. In addition, air-tight construction and filtering systems can reduce the infiltration of dust-sized particles into structures and mechanized equipment. Ash can be cleaned off roadways and roofs more readily than most other volcanic materials, although it is easily reworked by wind and tends to recollect on cleared areas long after eruptions have ended.

Protection from pyroclastic flows and surges and from lateral blasts would require structures that are capable of resisting high-speed impacts by rock fragments, dislocation and damage by high-speed flow, and heat. Each of these volcanic processes endangers people's lives and engineering works throughout its extent. Because the effects decrease in severity outward from the source, the chance of survival of people by means of adequate shelter and strongly constructed engineering works increases with increasing distance from the source of the eruption.

CONCLUSIONS

It is well known that volcanic events can damage or destroy structures and developments that are vital to social and economic well-being. Although volcanic activity cannot be prevented, important engineering works can be designed and constructed to prevent or lessen damage. Other mitigative measures affecting the volcanic process itself, such as deflection of lava flows or prevention of flooding by construction of dikes or embankments, may be at least partially successful.

The best mitigative measure, however, and probably the only possible one to reduce the hazard of most kinds of volcanic activity, is avoidance. Structures and developments ideally should be located in areas of minimal volcanic hazard. In the past within the U.S., volcanic hazards have generally been disregarded in land-use planning; for example, this disregard has characterized land development along the entire Cascade Range, with its many dormant but potentially dangerous volcanoes (Crandell and others, 1979). We hope that some of the examples cited here will help to guide planning of future developments and facilities in the vicinity of volcanoes, and will demonstrate that a conscious choice should be made between (1) acceptance of the possibility of increased costs due to changes in location or design of engineered developments and facilities, and (2) accepting the possible economic and social costs of an eruption.

REFERENCES CITED

Anderson, T., and Flett, J. S., 1903, Report on the eruptions of the Soufrière, in St. Vincent, in 1902, and on a visit to Montagne Pelée, in Martinique, Part 1: Royal Society of London Philosophical Transactions, Series A, v. 200, p. 353–553.

Apple, R. A., and Macdonald, G. A., 1966, The rise of sea level in contemporary times at Honaunau, Kona, Hawaii: Pacific Science, v. 20, no. 1, p. 125–136.

Aramaki, S., 1956, The 1783 activity of Asama Volcano, Part 1: Japanese Journal of Geology and Geography, v. 27, no. 2-4, p. 189–229.

Beaulieu, J. D., and Peterson, N. V., 1981, Seismic and volcanic hazard evaluation of the Mount St. Helens area relative to the Trojan nuclear site; Highlights of a recent study: Oregon Geology, v. 43, no. 12, p. 159–169.

Bechly, J. F., 1980, Mt. Saint Helens eruption; Restoration of Columbia and Cowlitz River channels: College Station, Texas A & M University, Dredging Seminar, paper presented November 6, 52 p.

Blong, R. J., 1981, Some effects of tephra falls on buildings, *in* Self, S., and Sparks, R.S.J., eds., Tephra studies: Proceedings of the NATO Advanced Study Institute, Tephra Studies as a Tool in Quaternary Research, Laugarvtan and Reykjavik, Iceland, June 18–29, 1980, p. 405–420.

—— , 1984, Volcanic hazards: Sydney, Academic Press, 424 p.

Bolt, B. A., Horn, W. L., Macdonald, G. A., and Scott, R. F., 1975, Geological hazards: New York, Springer-Verlag, 328 p.

Bonis, S., and Salazar, O., 1973, The 1971 and 1973 eruptions of Volcan Fuego, Guatemala, and some socio-economic considerations for the volcanologist: Bulletin Volcanologique, v. 37, p. 394–400.

Crandell, D. R., 1971, Postglacial lahars from Mount Rainier Volcano, Washington: U.S. Geological Survey Professional Paper 677, 75 p.

—— , 1976, Preliminary assessment of potential hazards from future volcanic eruptions in Washington: U.S. Geological Survey Miscellaneous Field Studies Map MF-774, scale 1:1,000,000.

Crandell, D. R., and Mullineaux, D. R., 1978, Potential hazards from future eruptions of Mount St. Helens Volcano, Washington: U.S. Geological Survey Bulletin 1383-C, 26 p.

Crandell, D. R., Mullineaux, D. R., and Miller, C. D., 1979, Volcanic-hazards studies in the Cascade Range of the western United States, *in* Sheets, P. D., and Grayson, D. K., eds., Volcanic activity and human ecology: New York, Academic Press, p. 195–219.

Crandell, D. R., Booth, B., Kusumadinata, K., Shimozuru, R., Walker, G.P.L., and Westercamp, D., 1984, Sourcebook for volcanic-hazards zonation: Paris, UNESCO, 97 p.

Dillman, J. J., and Roberts, M. L., 1982, The impact of the May 18 Mount St. Helens ashfall; Eastern Washington residents report on housing-related damage and cleanup, *in* Keller, S. A. C., ed., Mount St. Helens; One year later: Cheney, Washington, Eastern Washington University Press, p. 191–198.

Duffield, W. A., Tilling, R. I., and Canul, R., 1984, Geology of El Chichón Volcano, Chiapas, Mexico: Journal of Volcanology and Geothermal Research, v. 20, p. 117–132.

Einarsson, T., 1974, The Heimaey eruption in words and pictures: Reykjavik, Iceland, Heimskringla, 56 p.

Fries, C., Jr., 1953, Volumes and weights of pyroclastic material, lava, and water erupted by Parícutin Volcano, Michoacan, Mexico: EOS American Geophysical Union Transactions, v. 34, no. 4, p. 603–616.

Gorshkov, G. S., 1959, Gigantic eruption of the volcano Bezymianny: Bulletin Volcanologique, v. 20, p. 77–112.

Herd, D. G., 1986, The 1985 Ruiz Volcano disaster: EOS American Geophysical Union Transactions, v. 67, no. 19, p. 457–460.

Hoblitt, R. P., 1986, Observations of the eruptions of July 22 and August 7, 1980, at Mount St. Helens, Washington: U.S. Geological Survey Professional Paper 1335, 44 p.

Hoblitt, R. P., Miller, C. D., and Vallance, J. W., 1981, Origin and stratigraphy of the deposit produced by the May 18 directed blast, *in* Lipman, P. W., and Mullineaux, D. R., eds., The 1980 eruptions of Mount St. Helens, Washington: U.S. Geological Survey Professional Paper 1250, p. 401–419.

Hobbs, W. H., 1906, The grand eruption of Vesuvius in 1906: Journal of Geology, v. 14, p. 636–655.

Janda, R. J., Scott, K. M., Nolan, K. M., and Martinson, H. A., 1981, Lahar movement, effects, and deposits, *in* Lipman, P. W., and Mullineaux, D. R., eds., The 1980 eruptions of Mount St. Helens, Washington: U.S. Geological Survey Professional Paper 1250, p. 461–478.

Katsui, Y., Takahashi, T., Egashira, S., Kawachi, S., and Watanabe, H., 1986, The 1985 eruption of Nevado del Ruiz Volcano, Colombia, and associated mudflow disaster: Report of Natural Disaster Scientific Research, no. B-60-7, Japan, 102 p. (in Japanese with English abstract).

Kieffer, S. W., 1981, Fluid dynamics of the May 18 blast at Mount St. Helens, *in* Lipman, P. W., and Mullineaux, D. R., eds., The 1980 eruptions of Mount St. Helens, Washington: U.S. Geological Survey Professional Paper 1250, p. 379–400.

Kienle, C. F., 1980, Evaluation of eruptive activity at Mt. St. Helens, Washington, March through June, 1980: Oregon, Portland General Electric Company, Portland, Oregon, 183 p. (prepared by Foundation Sciences, Inc., Portland, Oregon).

Kienle, J., and Swanson, S. E., 1980, Volcanic hazards from future eruptions of Augustine Volcano, Alaska: Geophysical Institute, University of Alaska, Report to Bureau of Land Management, National Oceanic and Atmospheric Administration, and Alaska Department of Natural Resources, 122 p.

Lipman, P. W., 1980, Rates of volcanic activity along the southwest rift zone of Mauna Loa Volcano, Hawaii: Bulletin Volcanologique, v. 43-4, p. 703–725.

Lombard, R. E., Miles, M. B., Nelson, L. M., Kresh, D. L., and Carpenter, P. J., 1981, The impact of mudflows on the lower Toutle and Cowlitz rivers, *in* Lipman, P. W., and Mullineaux, D. R., eds., The 1980 eruptions of Mount St. Helens, Washington: U.S. Geological Survey Professional Paper 1250, p. 693–699.

Macdonald, G. A., 1972, Volcanoes: Englewood Cliffs, New Jersey, Prentice-Hall, Inc., 510 p.

Macdonald, G. A., and Alcarez, A., 1956, Nuées ardentes of the 1948–1953 eruption of Hibok-Hibok: Bulletin Volcanologique, v. 18, p. 169–178.

Mason, A. C., and Foster, H. L., 1956, Extruded mudflow hills of Nirasaki,

Japan: Journal of Geology, v. 64, p. 74–83.

Moore, J. G., and Sisson, T. W., 1981, Deposits and effects of the May 18 pyroclastic surge, *in* Lipman, P. W., and Mullineaux, D. R., eds., The 1980 eruptions of Mount St. Helens, Washington: U.S. Geological Survey Professional Paper 1250, p. 421–438.

Nairn, I. A., Wood, C. P., and Hewson, C.A.Y., 1979, Phreatic eruptions of Ruapehu—April 1975: New Zealand Journal of Geology and Geophysics, v. 22, no. 2, p. 155–173.

Nakamura, Y., 1978, Geology and petrology of Bandai and Nekoma volcanoes: Sendai, Japan, Science Reports of the Tohoku University, Faculty of Science, Tohoku University, series 3, v. 14, no. 1, p. 67–119.

——, 1981, Mode of volcanic dry avalanche from the 1888 explosive eruption of Bandai Volcano, Japan, *in* Arc Volcanism, 1981 IAVCEI Symposium, Tokyo and Hakone, August 28–September 9, 1981, Abstracts: Volcanological Society of Japan and International Association of Volcanology and Chemistry of the Earth's Interior, p. 254.

Neall, V. E., 1976, Lahars as major geological hazards: International Association of Engineering Geology Bulletin 14, p. 233–240.

Palfreyman, W. D., and Cooke, R.J.S., 1976, Eruptive history of Manam Volcano, Papua New Guinea, *in* Johnson, R. W., ed., Volcanism in Australasia: New York, Elsevier, p. 117–131.

Perret, F. A., 1950, Volcanological observations: Carnegie Institution of Washington Publication 549, 162 p.

Peter, G. W., Dalgleish, W. A., and Schriever, W. R., 1963, Variations of snow loads on roofs: Division of Building Research, National Research Council of Canada, Research Paper No. 189, 11 p.

Quarterly Journal of Science and the Arts, 1816, Miscellaneous observations on the volcanic eruptions at the islands of Java and Sumbawa, with a particular account of the mud volcano at Grobogar: v. 1, p. 245–258.

Sarna-Wojcicki, A. M., Shipley, S., Waitt, R. B., Jr., Dzurisin, D., and Wood, S. H., 1981, Areal distribution, thickness, mass, volume, and grain size of air-fall ash from the six major eruptions of 1980, *in* Lipman, P. W., and Mullineaux, D. R., eds., The 1980 eruptions of Mount St. Helens, Washington: U.S. Geological Survey Professional Paper 1250, p. 577–600.

Schuster, R. L., 1983, Engineering aspects of the 1980 Mount St. Helens eruptions: Association of Engineering Geologists Bulletin, v. 20, no. 2, p. 125–143.

Schuster, R. L., and Crandell, D. R., 1984, Catastrophic debris avalanches from volcanoes: Proceedings of IVth International Symposium on Landslides, Toronto, v. 1, p. 567–572.

Sekiya, S., and Kikuchi, Y., 1889, The eruption of Bandai-san: Japan Imperial University, College of Science Journal, v. 3, pt. 2, p. 91–172.

Silva, L., Cocheme, J. J., Canul, R., Duffield, W. A., and Tilling, R. I., 1982, The March-April, 1982, eruptions of Chichónal Volcano, Chiapas, Mexico; Preliminary observations: EOS American Geophysical Union Transactions, v. 63, no. 45, p. 1126.

Taylor, D. A., 1979, A survey of snow loads on the roofs of arena-type buildings in Canada: Canadian Journal of Civil Engineering, v. 6, p. 85–96.

Taylor, G. A., 1958, The 1951 eruption of Mount Lamington, Papua: Australia Bureau of Mineral Resources Geology and Geophysics Bulletin 38, 117 p.

Tazieff, H., 1977, An exceptional eruption; Mt. Niragongo, June 10th, 1977: Bulletin Volcanologique, v. 40, fasc. 3, p. 189–200.

Thorarinsson, S., 1954, The tephra-fall from Hekla on March 29th, 1947, *in* Einarsson, T., Kjartansson, G., and Thorarinsson, S., eds., The eruption of Hekla, 1947–1948: Visindafelag Islendinga Rit, v. 2, no. 3, 68 p.

Toon, O. B., 1982, An overview of the effects of volcanic eruptions on the climate: EOS American Geophysical Union Transactions, v. 63, no. 45, p. 901.

U.S. Geological Survey, 1975, Man against volcano; The eruption of Heimaey, Vestmann Islands, Iceland: U.S. Geological Survey Publication INF–75–22, 19 p.

Voight, B., Janda, R. J., Glicken, H., and Douglass, P. M., 1983, Nature and mechanics of the Mount St. Helens rock-slide avalanche of 18 May 1980: Geotechnique, v. 33, no. 3, p. 243–273.

Wilcox, R. E., 1959, Some effects of recent volcanic ash falls, with especial reference to Alaska: U.S. Geological Survey Bulletin 1028–N, p. 409–476.

Manuscript Accepted by the Society April 20, 1987

Printed in U.S.A.

Geological Society of America
Centennial Special Volume 3
1991

Chapter 12

Faulting and seismic activity

Manuel G. Bonilla
U.S. Geological Survey, 345 Middlefield Road, MS 977, Menlo Park, California 94025

INTRODUCTION AND HISTORY

This chapter traces some of the ideas and concepts leading to the current understanding of the process of faulting and earthquake generation, gives examples of engineering geology investigations contributing to that understanding, describes some engineering projects that have been strongly influenced by the process, and suggests needed research. Each of these topics is discussed in sequence.

The understanding of faulting and earthquakes and of the significance of these to engineering has developed over several centuries. John Michell in 1761 was probably the first to publish a cross section of a clearly recognizable fault (Adams, 1938, Fig. 66). Michell did not attribute earthquakes to faulting, but proposed the important idea that seismic vibrations were the result of the propagation of elastic waves in the earth (Adams, 1938). Charles Lyell (1830) emphasized the uplift and depression of land that accompanies earthquakes. He did not attribute earthquakes to faulting, but a contemporary of his evidently did, for the following statement appeared in a review of Lyell's book (Scrop, 1830, p. 463):

> The sudden fracture of solid strata by any disruptive force must necessarily produce a violent vibratory jar to a considerable distance along the continuation of these strata. Such vibrations would be propagated in undulations, which may be expected, when influencing a mass of rocks several thousand feet at least in thickness, to produce on the surface exactly the wave-like motion, the opening and shutting of crevices, the tumbling down of cliffs and walls, and other characteristic phenomena of earthquakes.

This idea was apparently disregarded, and coseismic faulting, some of which reached the ground surface, was generally considered to be the result rather than the cause of earthquakes until the time of G. K. Gilbert. Gilbert was the first to clearly state that faulting is the cause of earthquakes. Furthermore, Gilbert used the "modern" concepts of stick-slip, elastic strain accumulation, and seismic gaps to make an earthquake forecast for the Salt Lake City, Utah, area in 1883 (Gilbert, 1884). However, these ideas and others related to the study of faults were overlooked for many decades (Wallace, 1980).

The definitive paper relating earthquakes to faulting was that of Reid (1910) who proposed that the 1906 California earthquake resulted from slow accumulation of elastic strain followed by faulting and elastic rebound. More recently, Tocher (1958) demonstrated that surface faulting usually accompanied earthquakes of magnitude 6.5 or greater in northern California and Nevada; he was the first to relate earthquake magnitude to surface rupture length and to the product of length and displacement. The relation of rupture length to earthquake magnitude was applied by Benioff (1965) and by Albee and Smith (1967) to the problem of estimating earthquake size in connection with the Malibu reactor site, described in this chapter. Systematic compilation of data relating surface-fault dimensions to earthquake magnitude have been made by various investigators, including Bonilla (1967, 1970) for events in North America, and for events throughout the world by Bonilla and Buchanan (1970), Slemmons (1977, 1982), and Bonilla and others (1984).

Using earlier theoretical work, Aki (1966) introduced a new measure of earthquake size, the seismic moment, which is related to the area of the fault surface, the average fault displacement, and the shear modulus of the rock. By this method, the average coseismic subsurface displacement on a fault can be estimated from seismograms. Brune (1968) showed that the rate at which seismic moment is released can be translated into slip rate on the fault. The converse, translation of fault slip rate into rate of production of earthquakes in terms of seismic moment, has led to the use of geologic data on slip rate to estimate earthquake rate in a given time interval (Anderson, 1979; Molnar, 1979; Wesnousky and others, 1984). Seismic moment can be converted to the more commonly used earthquake magnitude over a wide range of magnitudes (Hanks and Kanamori, 1979).

Geologic studies of the recent history of faults, especially using exploratory trenches, have led to the identification of prehistoric episodes of faulting and, by inference, of prehistoric seismic events. The use of exploratory trenches for such studies became common in the 1970s (e.g., Malde, 1971; Clark and others, 1972; Bonilla, 1973; Sieh, 1978) and allowed inferences to be made regarding the seismic history of particular faults or regions extending back thousands of years (Sieh, 1981).

The "characteristic earthquake" or "maximum earthquake"

Bonilla, M. G., 1991, Faulting and seismic activity, *in* Kiersch, G. A., ed., The heritage of engineering geology; The first hundred years: Boulder, Colorado, Geological Society of America, Centennial Special Volume 3.

model of fault behavior, if it is proved to be substantially correct, will have an important effect on estimates of expected earthquakes and fault dimensions. This model suggests that the earthquakes and the dimensions of faulting on a given fault or fault segment are of nearly equal size in each succeeding event. This idea was implicit in papers by Allen (1968), Wallace (1970), and Matsuda (1975), and was explicitly developed in papers by Swan and others (1980), Wesnousky and others (1983), and especially Schwartz and Coppersmith (1984).

The significance of faulting to engineering works was incisively discussed by Louderback (1937, 1942, 1950), and data on faulting were compiled for engineering purposes by Bonilla and colleagues and by Slemmons in papers cited above. Other papers have discussed the importance of faulting, including design considerations, for particular types of structures. Among such papers are Sherard and others (1974) on dams; and Newmark and Hall (1974, 1975), Hall and Newmark (1977), Taylor and Cluff (1977), and McCaffrey and O'Rourke (1983) on pipelines.

CONTRIBUTIONS OF INVESTIGATIONS FOR ENGINEERING WORKS TO KNOWLEDGE OF FAULTING AND SEISMICITY

Geological and geotechnical investigations for various kinds of engineering works have contributed directly to the sum of knowledge about faulting and seismic activity and, perhaps more importantly, have stimulated topical research. Site-specific investigations have provided concentrated information on local and regional geology and faults. Topical research has shed light on the process of faulting and earthquake generation, its effects on engineering works, and on ways to investigate the process. Research initiated in response to problems on one project is commonly found to be of general application. A few examples follow.

Sheffield and Lower San Fernando dams

Many earth dams have been damaged by earthquakes, but only two are discussed here. Study of the failures was important in advancing the art of analyzing the stability of slopes during earthquakes. The Sheffield Dam near Santa Barbara, California, constructed in 1917, was 8 m high and contained a pool 5 m deep at the time of the 1925 Santa Barbara earthquake. The earthquake produced a maximum acceleration estimated to be 0.15 g at the dam, under which the embankment failed completely (Seed and others, 1969, 1978); fortunately little damage and no loss of life resulted. Failure was by liquefaction, apparently in the silty sand and sandy silt just below the dam embankment.

The Lower San Fernando Dam (sometimes called the Lower Van Norman Dam) northwest of Los Angeles, California, constructed by hydraulic fill methods in 1916, was enlarged to a final height of 43 m in 1930. The older part of the dam has a clay core and outer zones of silty sand. The water depth was 27 m at the time of the February 9, 1971, earthquake, which produced a maximum acceleration of about 0.6 g at the dam (Scott, 1973; Seed, 1979). Liquefaction in the upstream sand zone led to a major slide that involved the clay core and took out the crest along nearly half the length of the dam, leaving a rim of cracked fill only 1.4 m above water level (Fig. 1). Because of this dangerous condition, 80,000 people living downstream were evacuated (Ross, 1975; Seed, 1979). Analyses of these and other case histories by Seed and his colleagues showed that the conventional method of pseudostatic analysis of slopes under seismic loading is inadequate when applied to saturated cohesionless soils, and dynamic analyses are needed (Seed and others, 1975; Seed, 1979). As a result of the near catastrophe in San Fernando, the State of California instituted a reevaluation of existing embankment dams throughout the State (Babbitt and others, 1983).

Figure 1. Remains of crest of Lower San Fernando Dam, damaged by a landslide following the magnitude 6.5 earthquake of February 9, 1971. Men on left are looking down at reservoir water, which is only 1.4 m below them. One outlet tower is visible in the right background; the other tower was knocked down by the landslide and was submerged in the reservoir. Photograph by T. L. Youd, U.S. Geological Survey, February 9, 1971.

Bodega Head and Malibu nuclear reactor sites

The Bodega Head and Malibu sites were the first nuclear sites in the United States for which active faults were an important consideration. Investigation of these sites in the 1960s stimulated research on fault behavior and led to concepts regarding site evaluation that have been applied to the siting of reactors, dams, pipelines, waste-disposal facilities, and other engineering works.

The Bodega Head reactor site, 82 km north-northwest of San Francisco, California, is about 0.3 km from the poorly defined western edge of the San Andreas fault zone and about 2.1 km west of the 1906 fault trace (Fig. 2). Investigation and site preparation began in 1958 and terminated in 1964 (Novick, 1969). Because of the proximity of the site to the San Andreas fault, site excavations were minutely examined as they progressed. During excavation of the shaft for the reactor, a fault (Fig. 3) was found in the unconsolidated sediments overlying the quartz diorite bedrock. Mapping of the shaft fault showed that (1) where exposed in the sediments, it died out horizontally in at least one direction; (2) in some places it died out downward in the sediments, but in other places it joined a fault zone in the bedrock; (3) it was not visible in certain sedimentary layers but was clearly visible above and below those layers; and (4) it could not be traced to the ground surface (Schlocker and Bonilla, 1963, 1964). The discovery of the fault and its characteristics raised several debated questions. Was the 'fault' merely the edge of a landslide in the unconsolidated sediments whose flank happened to coincide with an underlying fault in bedrock? Moreover, inasmuch as the fault was not visible in certain sedimentary layers, and strike-slip faults such as the shaft fault commonly have discontinuous traces (Schlocker and Bonilla, 1963), was the last displacement on the fault older than the apparently overlying beds or had it just become unrecognizable or died out upward? Although located outside the San Andreas fault zone, was the shaft fault likely to have displacement when the next movement occurred on the San Andreas fault? To answer some of these questions, a thorough review was made of the surface ruptures accompanying the 1906 earthquake (Schlocker and Bonilla, 1964).

The study showed that even though the field investigation of the 1906 faulting was mostly limited to the main trace, several ruptures were reported at a distance from the main trace, including some ruptures several kilometers away, well outside the San Andreas fault zone itself. This finding had an important bearing on the evaluation of the fault at the Bodega Head site and raised questions about the existing Atomic Energy Commission regulations, which stated that a nuclear reactor should not be placed "closer than one-fourth mile from the surface location of a known active earthquake fault" (U.S. Atomic Energy Commission, 1962, p. 3510). In performing research sponsored by the Atomic Energy Commission on this problem, Bonilla compiled data on subsidiary faulting and, at the same time, on the relation of earthquake magnitude to surface rupture length and displacement for historic events in the U.S. (Bonilla, 1967, 1970). This research

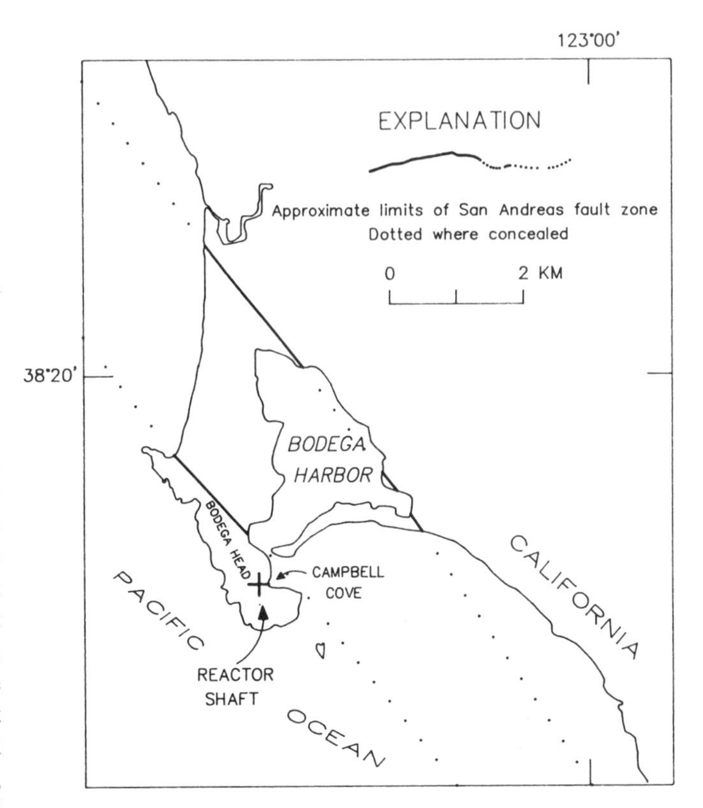

Figure 2. Map showing relation of Bodega Head nuclear reactor site to San Andreas fault zone. The 1906 surface rupture on the San Andreas fault was near the northeastern edge of the fault zone. Modified from Schlocker and Bonilla, 1963, Plate 1.

Figure 3. Shaft fault as seen in vertical section during excavation of reactor shaft at Bodega Head site. In Quaternary beds, the maximum observed vertical separation on the fault was 0.36 m; a strike-slip component was recognizable but was of unknown amount. In the granitic bedrock the observed strike-slip component was 7.3 m and the fault was 0.6 to 3.0 m wide. Photograph by Julius Schlocker, U.S. Geological Survey, 1963.

was later extended to include historic surface ruptures in other parts of the world (Bonilla and Buchanan, 1970; Slemmons, 1977, 1982; Bonilla and others, 1984). The phenomenon of fault traces being unrecognizable or obscure in certain layers led, much later, to research that is currently in progress on the conditions under which this might occur and how it may affect the interpretation of exploratory trenches across faults (Bonilla, 1985).

The Malibu nuclear reactor site, on the coast 45 km west of Los Angeles, California, lies within the Malibu coast zone of deformation and just south of the Malibu coast thrust fault, a major east-west fault with principal movement between late Miocene and late Pleistocene. Investigations related to the project showed that a fault of unknown age traversed the proposed location of the reactor containment vessel and that several faults having displacement in the late Pleistocene existed in the Malibu coast zone outside the plant site (Yerkes and Wentworth, 1965). In addition to these local conditions, the regional setting was an important factor in evaluation of this site. The site lies in an east-west belt of moderate seismicity that contains the Malibu coast zone. This zone forms the northern border of a structural block whose eastern border is inferred to be 30 km to the east, along the right-lateral, northwest-trending Newport-Inglewood zone of faults and folds, which has been active in the Holocene (Fig. 4). The inference was made that the structural block is currently moving relatively northward, with right-slip on its eastern border and thrusting on its northern border. This combination of local and regional evidence led to the conclusion that the east-west structural zone containing the nuclear reactor site is tectonically active (Yerkes and Wentworth, 1965; Marblehead Land Company, 1966). The value of such regional analysis is recognized in the U.S. nuclear plant siting criteria, which, in defining a capable fault, includes those faults that exhibit "... a structural relationship to a capable fault ... such that movement on one could be reasonably expected to be accompanied by movement on the other" (U.S. Nuclear Regulatory Commission, 1977, p. 413).

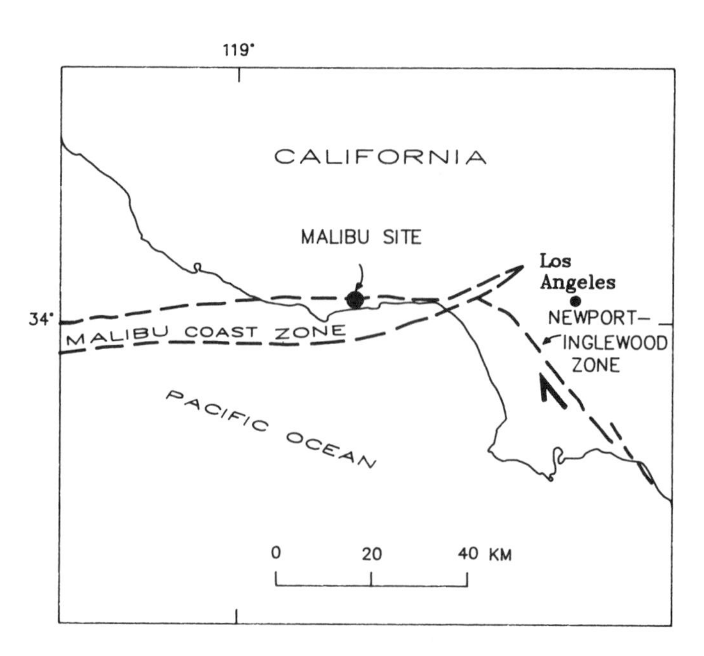

Figure 4. Sketch map showing relation of Malibu Coast zone of deformation to the Newport-Inglewood zone of faults and folds, and location of Malibu nuclear reactor site. Modified from Yerkes and Wentworth (1965, Fig. 44).

The few existing data relating fault length to earthquake magnitude were used to some extent in estimating the size of potential earthquakes near the Malibu site (Benioff, 1965; Albee and Smith, 1967). The need for more complete data led to compilations of fault length versus magnitude during the study of historic faulting by Bonilla (1967) that was started as a result of the problems related to the Bodega Head investigation.

Investigations directly or indirectly related to the Bodega Head and Malibu nuclear reactor sites had effects on the U.S. reactor site criteria in addition to those already mentioned. The term "known active earthquake fault" in the criteria during the early 1960s was ambiguous and taken by some to mean a fault with historic surface rupture and by others a fault with rupture in the past 10,000 years. At Malibu, faults whose most recent displacement was between 10,000 and about 180,000 years ago were nevertheless considered capable of surface rupture for purposes of reactor design (Atomic Safety and Licensing Board, 1966; U.S. Atomic Energy Commission, 1967). This finding strongly influenced the wording in the new site criteria, which were in preparation, and led to the inclusion of "Movement at or near the ground surface at least once within the past 35,000 years or of a recurring nature within the past 500,000 years" as part of the definition of capable fault (U.S. Nuclear Regulatory Commission, 1977, p. 413). The necessity to consider evidence at a distance along the strikes of the San Andreas and Malibu coast zones to evaluate the risk of faulting at the Bodega Head and Malibu sites is also reflected in the wording of the criteria. Furthermore, the requirement to determine the capability of faults within 8 km of a reactor site and the inclusion of monoclinal flexure in the criteria are a direct outcome of information uncovered in the review of historic surface faulting in the U.S. that resulted from the Bodega Head investigations. Much of the wording and many of the concepts in the U.S. Nuclear Regulatory Commission (1977) criteria have since appeared, in modified form, in guidelines of organizations that are concerned with dams, with disposal of hazardous waste, or with liquefied natural gas facilities.

Trans-Alaska Pipeline System

The Trans-Alaska oil pipeline, which cost more than $8 billion (Godfrey, 1978), was the most costly engineering project ever undertaken by private industry. Solution of the associated environmental problems contributed to knowledge in many fields, including permafrost, seismic hazard evaluation, and the engineering of pipelines that must cross permafrost, areas of high

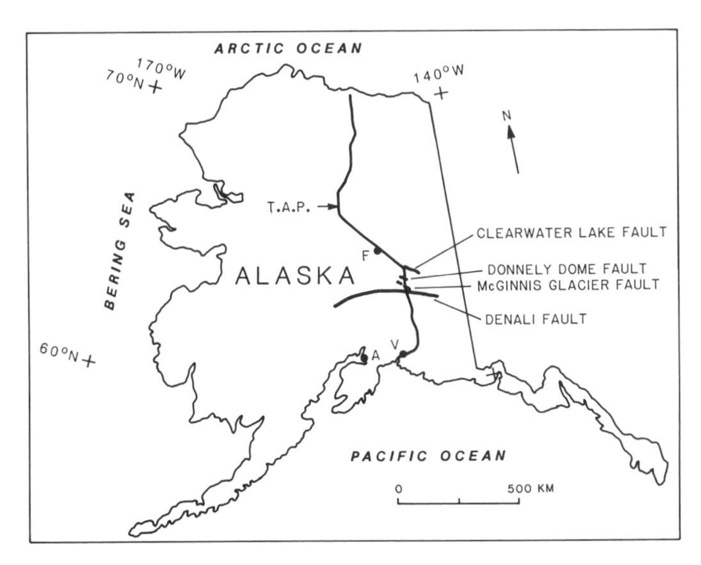

Figure 5. Outline map of Alaska showing Trans-Alaska pipeline (T.A.P.) and the four faults identified as active and of concern to the pipeline. F, Fairbanks; A, Anchorage; V, Valdez. Faults are from Brogan and others (1975, Fig. 2); location of pipeline from Brew (1974, Fig. 1).

seismicity, and active faults. The 1.2-m-diameter steel pipeline, planned and built in the period 1968 through 1977 (Roscow, 1977), crosses Alaska from north to south, a distance of 1,287 km (Fig. 5).

In the early stages of planning, not much was known about the seismicity or the activity of faults in the vast area to be crossed by the pipeline. On the basis of available data, five seismic zones having earthquake potential ranging from magnitude 5.5 to 8.5 were outlined, the required fault investigations were described, and the stipulation was made that no storage tank or pump station was to be located in an active fault zone (U.S. Federal Task Force on Alaskan Oil Development, 1972).

Investigations to identify active faults covered about one-third of the state of Alaska. These investigations included extensive use of helicopters and fixed-wing aircraft, and study of aerial photographs, satellite imagery, radar imagery, and low-sun angle photography. About 8,000 lineaments were identified; of these, four that crossed the pipeline were faults considered to have the potential for surface rupture (Cluff and others, 1974; Brogan and others, 1975). The fault investigations just described were done for the Alyeska Pipeline Service Company, but other investigations on a smaller scale were also done by the U.S. Geological Survey and by environmental groups. These studies greatly enhanced understanding of the location, dimensions, and Quaternary activity of faults in a very large part of the state of Alaska.

The potential effects of the design earthquakes stipulated in the environmental impact report (U.S. Federal Task Force on Alaskan Oil Development, 1972) had to be translated into the design of the pipeline. The existing data on earthquake ground motion were critically reviewed and presented in the form of plots of peak acceleration, peak velocity, peak displacement, and duration as a function of earthquake magnitude and distance to the slipped fault (Page and others, 1972). Duration of shaking was expressed in terms of a new parameter—the interval between the first and last acceleration peak equal to or greater than 0.05 g—that could be readily scaled from existing accelerograms and would approximate the time over which the ground accelerations exceed the given level. This parameter was subsequently termed bracketed duration (Bolt, 1974). From these data, estimates of near-fault peak ground motions and durations were drawn for magnitudes 5.5 through 8.5 (Page and others, 1972). These ground-motion estimates, together with estimates of surface fault displacements, were then used by engineers as a basis for design of the pipeline to resist both shaking and faulting (Newmark and Hall, 1974, 1975). The estimates of surface fault-displacement (Cluff and others, 1974) were based partly on data (Bonilla and Buchanan, 1970) that were compiled because of the problems associated with the Bodega Head and Malibu nuclear reactor sites.

The Trans-Alaska pipeline project contributed to geoscientific knowledge in several ways, only a few of which are described above. Investigation of relevant geological problems resulted in (1) an improved understanding of seismic zonation and Quaternary faulting in Alaska, (2) translation of expected earthquake magnitudes into data useful in design, and (3) design of pipelines to resist both earthquake shaking and surface faulting. The near-fault ground motion estimates in the report by Page and others (1972) incorporated an extensive new suite of strong-motion recordings from the 1971 San Fernando, California, earthquake (magnitude 6.6); the estimates were highly controversial because they indicated more intense shaking would occur close to the fault than had been typically assumed by seismic engineers for the design of important structures (see for example, Housner, 1970, p. 79–80). These ground-motion estimates affected not only the design of the Trans-Alaska pipeline system but also the design of other critical facilities such as nuclear reactors (see Diablo Canyon Nuclear Power Station, described below) and dams (e.g., the Los Angeles Dam, Wesson and others, 1974). These studies for critical structures also raised concern about the earthquake resistance of more ordinary structures (Page and others, 1975). Some effects of the potential earthquakes and the presence of active faults on the design and construction of the pipeline are discussed in a following section.

Little Cojo Bay Liquefied Natural Gas Terminal

In connection with the planned Little Cojo Bay Liquefied Natural Gas Terminal in Santa Barbara County, California, extensive local and regional geological and geophysical studies were made, and topical research was initiated. During the investigation, faults considered capable of surface rupture were identified at the site (Slosson and Associates, 1981). The necessity to consider faulting led to additional research on three different topics:

(1) centrifuge and numerical modeling of the propagation of faults through soils to better interpret the geologic history of displacements of faults and to assist in design of structures in fault zones (Roth and others, 1981a, 1982), (2) laboratory testing and analysis of the process of fault rupture through alluvium to better predict the width and inclination of rupture zones associated with dip-slip faults (Cole and Lade, 1984; Lade and others, 1984), and (3) numerical and laboratory modeling of bedding-plane faulting accompanying flexural slip (Roth and others, 1981b). The results will no doubt be applied to other projects, if due consideration is given to the differences between laboratory and field conditions.

Nuclear waste repository sites

Investigations for repository sites for nuclear wastes include special aspects that set them apart from typical engineering works. Most projects have a design lifetime and therefore a period of concern for safety of 50 to 100 years, whereas for geologic disposal of nuclear wastes the period of concern is 10,000 years or more (U.S. Department of Energy, 1985). To anticipate the effects of geologic processes so far into the future requires a profound knowledge of the rates and changes in rates of those processes in the vicinity of the site. Consequently these studies produce a wealth of geoscience information on various topics, including current and past seismicity, tectonics, and stress fields. For example, the reader is referred to the studies for the potential high-level nuclear waste repository site at Yucca Mountain, Nevada (U.S. Geological Survey, 1984; Stock and others, 1985; Frizzell and Zoback, 1987).

IMPACT OF FAULTING AND SEISMICITY ON ENGINEERING WORKS

The impact of faulting and seismicity on engineering works is of two general kinds: direct damage, and increased costs incurred to avoid damage from future events. Among the increased costs are design changes, delays, and complete abandonment of some projects. The emphasis in this section is on projects that were strongly affected by anticipated future faulting and earthquakes. Costs are given in dollars if readily available; however, the costs are not directly comparable between projects because they are not adjusted for inflation.

Structures of many kinds have been directly damaged, to various degrees, by faulting. The damaged structures include embankment dams, tunnels, bridges, roads, railroads, nursing homes, commercial buildings, apartments, houses, canals, storm drains, water wells, and water, gas, and sewer lines. Fault damage to structures is described in many reports (Lawson and others, 1908; Louderback, 1937; Ambraseys, 1960; Duke, 1960; California Department of Water Resources, 1967; Subcommittee on Water and Sewerage Systems, 1973; Niccum and others, 1976; Hradilek, 1977; Youd and others, 1978; Sylvester, 1979; Gordon and Lewis, 1980; McCaffrey and O'Rourke, 1983; Pampeyan, 1986). Damage to the Baldwin Hills reservoir made possible by displacement on a fault is discussed in Holzer (this volume) and James and Kiersch (chapter 22, this volume). Innumerable buildings and a few dams, including the Sheffield and Lower San Fernando dams described above, have been severely damaged by earthquake vibrations. Other projects have been affected by postulated future events; some of these projects are described below.

Bodega Head nuclear reactor site

This reactor site was abandoned, after an expenditure of $4 million for planning and construction (Novick, 1969), because of the shaft fault described above and the possibility of a large earthquake on the nearby San Andreas fault during the 50-year design life of the plant. The proximity of the shaft fault to the San Andreas, its similar sense of displacement, and the uncertain age of its last displacement meant that the possibility of faulting under the reactor had to be considered. The applicant proposed a design that would accommodate 0.9 m of either horizontal or vertical shearing by faulting. The proposed design was approved in principle by the Advisory Committee on Reactor Safeguards, but the Division of Reactor Licensing of the Atomic Energy Commission and its consultants held that, although the design might protect the reactor from faulting, the piping and other connections leading out of the reactor probably could not withstand both the faulting and the shaking from the expected magnitude 8+ earthquake on the San Andreas fault. Furthermore they stated that "experimental verification and experience background on the proposed novel construction method are lacking," and concluded that "Bodega Head is not a suitable location for the proposed nuclear power plant at the present state of our knowledge" (U.S. Atomic Energy Commission, 1964, p. 13, 14). On October 30, 1964, three days after this statement was released, the utility announced that it had abandoned plans for a nuclear power plant at the site (Novick, 1969).

Malibu nuclear reactor site

After public hearings and other reviews of the problems relating to faulting and seismicity, some of which are discussed above, the Atomic Energy Commission ordered that the proposed Malibu reactor would have to be designed to accommodate fault displacement. Both the amount of differential ground displacement and the adequacy of any proposed design criteria to accommodate the displacement were required to be determined (U.S. Atomic Energy Commission, 1967). Six years later, the applicant withdrew the construction permit application (letter from Los Angeles Department of Water and Power to U.S. Atomic Energy Commission dated May 30, 1973), and the project was abandoned. No estimate of the cost of the project to the applicant is at hand but it is surely greater than the $4 million cost of the Bodega Head project.

Diablo Canyon Nuclear Power Station

The discovery of an offshore fault while the Diablo Canyon Nuclear Power Station was under construction caused many delays and greatly increased the cost of the plant. Construction of the plant, which is on the central California coast 20 km south-southwest of San Luis Obispo, started in 1968. A report on petroleum provinces of the United States, published in 1971, contained a map showing an unnamed fault lying a short distance offshore from the plant site (Hoskins and Griffiths, 1971). A small-scale cross section in their report indicates that the contact between lower and upper Pliocene sedimentary units has been affected by faulting, but the fault was not discussed in the text.

In 1973 the U.S. Atomic Energy Commission supported offshore work by the U.S. Geological Survey to investigate faulting and other possible hazards. The investigation included subbottom acoustic reflection profiling, bathymetry, and recording of the magnetic field over a track length of 1,200 km that covered an area some 13 km wide by 124 km long (Wagner, 1974). This study confirmed the presence of the fault reported by Hoskins and Griffiths and named it the Hosgri fault after them. The investigation determined that the fault has strike-slip as well as dip-slip displacement, that it possibly cuts late Quaternary sediments in some places, and that some earthquakes have occurred along its length (Wagner, 1974).

Various other studies and analyses related to the seismic potential of offshore faults were subsequently performed by the applicant and other groups (Earth Sciences Associates, 1974). After review of these and other studies, the U.S. Geological Survey concluded in 1975 that a magnitude 7.5 earthquake could occur on the Hosgri fault about 5 km from the nuclear plant (U.S. Nuclear Regulatory Commission, 1976, Appendix C) and that the ground-motion values derived for the Trans-Alaska pipeline system (Page and others, 1972) should be used as a starting point for the seismic design analysis (U.S. Nuclear Regulatory Commission, 1976, Appendix C, p. C-16). The suggested magnitude-7.5 earthquake and associated ground-motion parameters were much larger than those considered in the design and construction of the plant; consequently, extensive reanalyses of the plant were required, and many parts of the plant had to be modified (Lawroski, 1978; Piper, 1981). Furthermore, a license condition that was added to the operating license for the plant requires that the design earthquake and resulting ground motions be reevaluated by about 1988, using new information and interpretations (Brand, 1985). The original cost estimate for Diablo Canyon Nuclear Power Station was $350 million, but by 1979 its cost was $1.4 billion (San Jose Mercury News, 1984). An unknown, perhaps substantial, fraction of the increase in cost is attributable to the postconstruction discovery of the Hosgri fault, and more costs were subsequently incurred when flaws in the seismic design were found in 1982 (San Jose Mercury News, 1984). In considering the significance of the increase in cost given above, one should keep in mind that many nuclear plants, including some in the eastern part of the United States where earthquakes are not a major problem, have cost six to nine times more than originally projected (Dallaire, 1981; Cook, 1985).

Trans-Alaska Pipeline System

Permafrost presented the most critical geologically related problems to the pipeline (Lachenbruch, 1970; Roscow, 1977), but earthquakes and faulting were also important—economically and technically—in several ways. The pipeline system had to accommodate potential effects of the design earthquakes and surface faulting. Accelerographs were installed at 11 places along the pipeline to automatically detect, evaluate, and transmit strong-motion data to a central terminal at Valdez (Roscow, 1977; Péwé and Reger, 1983). Because of extreme temperature variations and other considerations, the elevated segments of the pipeline normally rest on shoes that allow the pipe to move sideways and upward, except at certain anchor points. This construction mode, which applies to about half of the pipeline, can accommodate some movement resulting from earthquakes and faulting as well as from thermal effects.

Fault design parameters were given by Cluff and others (1974) for four faults considered to be active (Fig. 5). The four faults and the corresponding design parameters, given in terms of fault shift (the combination of fault slip and fault distortion or drag across the fault zone) in the strike-slip and dip-slip directions, respectively, are: Denali fault, 6.1 m and 1.5 m; McGinnis Glacier fault, 2.4 m and 1.8 m; Donnelly Dome fault, 0.9 m and 3.0 m; and Clearwater Lake fault, 2.1 m and 3.0 m. Suggested special designs for fault crossings (Newmark and Hall, 1974; Hall and Newmark, 1977) included placing the pipe on beams at ground level on which it can slide and, in bedrock, placing the pipe in a shallow gravel-filled trench whose sides slope at 45° or less so that the pipe can move up and out of the trench. In the above-ground mode of installation the pipe can resist about 1.2 to 1.5 m of vertical displacement without undue strain, especially if it is free to accommodate the displacement over large distances (Hall and Newmark, 1977). Because the exact location of active fault traces within the active fault zones was not known and the pipeline crossed the zones at oblique angles, design modifications had to be considered for long segments of the pipeline.

The crossing of the Denali fault, on which the largest fault displacements were anticipated, is illustrated in Figure 6. The part of the pipeline in the fault zone rests on beams placed on a gravel berm, in contrast to adjacent parts that rest on either steel or concrete beams elevated above ground level. At the Denali fault the beams are 14 m long and about 18 m apart over a pipeline distance of nearly 600 m (Péwé and Reger, 1983, p. 93–94). Between the pipe and the beams are special shoes that permit the pipe to move horizontally and vertically with respect to the ground.

Special designs were used at two other fault crossings. Nearly 1,700 m of the pipeline lies in the McGinnis Glacier fault zone. The design analysis used the fault parameters given above. Design changes from the standard elevated configuration included substitution of friction supports for some anchor supports and addition of more bumper stops than usual. If the design faulting occurs, some of the supports are expected to behave plastically. For the 2,600-m crossing of the Donnely Dome fault, analysis using the fault parameters given above indicated that in the elevated configuration the pipe would accommodate the de-

Figure 6. Special design of the Trans-Alaska pipeline to accommodate surface faulting at the Denali fault. The part of the pipeline in the background, beyond the active fault zone, is elevated and has the normal above-ground design. The part of the pipeline in the fault zone (foreground) is at ground level and rests on steel or concrete beams that allow the pipe to move horizontally and vertically. The design fault displacements are 6.1 m strike shift and 1.5 m dip shift. The pipe is 1.22 m (48 in) in diameter. Photograph by O. J. Ferrians, Jr., U.S. Geological Survey, September 19, 1977.

sign fault displacements, although some supports may behave plastically.

For various nonseismic reasons the pipeline is buried rather than above ground in the 6-km segment where it obliquely crosses the Clearwater Lake fault identified in the early studies. Review of the design showed that the buried pipe could not safely resist the design fault displacement at the oblique angle at which the fault was crossed. Options at this point included the following: (1) make no changes but conform to some very severe requirements, including seismic monitoring of the fault, making provisions for rapid shutdown, and a site-specific contingency plan for oil-spill control; (2) reorient the pipeline so that the angle of intersection with the fault would be closer to 90°; or (3) conduct a detailed investigation to determine whether an active fault actually crosses the pipeline and, if so, exactly where (Williams, 1982, Appendix B). Option 3 was chosen, and a program that included geologic mapping of Quaternary deposits and land forms, logging of trenches, radiocarbon dating, and gravity and magnetic surveys was performed by the consulting firm that had done the original studies. The conclusion from the detailed studies was that the Clearwater Lake fault identified in the earlier studies did not represent an active fault and that no design for fault displacement was needed (Williams, 1982, Appendix B).

General Electric Test Reactor

The General Electric Test Reactor, located near Pleasanton, about 50 km east of San Francisco, California, was licensed in 1959 and was the first commercial reactor to be licensed in the U.S. Before being shut down, the facility was an important producer of medical radioisotopes. In 1977, during review of the geology and seismology related to renewal of the operating license, a new map was released by the U.S. Geological Survey. This map (Herd, 1977) shows the trace of the Verona fault about 60 m from the reactor, whereas an earlier map (Hall, 1958) shows the fault about 900 m from the reactor. After limited trenching and other studies were done, the Nuclear Regulatory Commission issued an order to put the reactor in a cold shutdown condition and show cause why the shutdown should not continue. Intensive investigations were carried out by the licensee, the Nuclear Regulatory Commission, and the U.S. Geological Survey. Among the stipulations agreed to by all directly involved parties was that the plant is located within a zone of tectonic faulting, the Verona fault zone (Atomic Safety and Licensing Board, 1982, p. 14). Two aspects of the studies relating to the plant were unusual. Probability analyses were used in evaluating the possibility of faulting beneath the reactor. These analyses were accepted with some reluctance by the staff of the Geosciences Branch of the Nuclear Regulatory Commission and by only two of the three members of the Atomic Safety and Licensing Board (1982). The other unusual feature was the licensee's conclusion that should the design fault displacement (reverse oblique slip of 1 m) occur, it would be deflected around the reactor foundation. This analysis was accepted by the staff of the Geosciences Branch and, with reservations on the part of one member, by the Atomic Safety and Licensing Board (1982). To my knowledge, this is the first time that probability analyses and a fault-deflection proposal have been applied and accepted with regard to possible faulting under a nuclear reactor. The Atomic Safety and Licensing Board (1982) approved the restart of the reactor provided that specified changes were made, but the reactor has not been restarted. During the shutdown period of 1977 to 1982, when the possibility of faulting at the plant site was being investigated and hotly debated, the General Electric Company lost its medical isotope business to Canadian firms (Meehan, 1984, p. 127). Additional discussions of the impact of geological issues on the nuclear industry are given in James and Kiersch (Chapter 23, this volume).

Auburn Dam

The Auburn Dam, in the Sierran foothills 50 km northeast of Sacramento, California, was planned to be the world's longest doubly curved thin-arch concrete dam, having a length of more than 1,200 m and a height of more than 200 m. The dam site is in an area of historically low seismicity, and although located within a regional fault zone, no active faults were known in the area. The foundation of the dam was being excavated in 1975 when a magnitude 5.7 (M_L) earthquake accompanied by minor surface faulting occurred 68 km away, also in the Sierran foothills. This unexpected event led to detailed regional and local studies of the potential for faulting and earthquakes near the dam site (Packer and others, 1978). Catastrophic failure of the dam could result in great loss of life and property—one estimate placing the potential deaths at 260,000 (Rose, 1978)—and therefore the problem was reviewed by several agencies and consulting boards. Estimates of possible fault displacements at the site ranged up to 1 m. The State of California recommended "... a design that would permit the structure to withstand a fault displacement of three-quarters of a foot..." (letter from H. D. Johnson, secretary for Resources, State of California, to C. D. Andrus, Secretary of the Interior, March 5, 1979). In a news release on July 30, 1979, the Secretary of the Interior announced that the Bureau of Reclamation would seek to develop such a design. Estimates of the cost of the project before it was put in abeyance range from $200 million (Wallace, 1986, p. 13) to $300 million (San Francisco Chronicle, 1979).

Other dams

Besides the Auburn Dam and the Sheffield and Lower San Fernando dams discussed above, which were affected by actual or potential earthquakes or faulting, several dams, all in California, have been modified because of the possibility of faulting in the foundations. During the planning stage for Morris Dam, east-northeast of Los Angeles (Fig. 7), a fault was discovered in the foundation bedrock at the dam site, and the design of the 75-m-high concrete dam was modified to include a special joint that would allow fault displacement. When the sediments overlying the bedrock fault were excavated during construction, they were found to be unfaulted, leading to the conclusion that the possibility of future faulting was slight; however, the special joint was retained when the dam was built in 1935 (Louderback, 1950; Legget, 1962, p. 515). The earth-fill, 37-m-high Coyote Dam, built across the Calaveras fault 30 km southeast of San Francisco in 1936, is designed to accommodate 6.1 m of horizontal and 1 m of vertical fault displacement (Louderback, 1937, 1950; Sherard and others, 1974). An earth dam built in 1891 across the San Andreas fault near Palmdale (Fig. 7) was completely reconstructed in 1969 because of the possibility of faulting. The new Palmdale Dam, of zoned-embankment type, was designed with the expectation that as much as 6.1 m of horizontal and 1 m of vertical fault displacement could occur (Sherard and others, 1974). The earth- and rock-fill Cedar Springs Dam (Fig. 7), 66 m high, was built over faults, considered active, that cross the site longitudinally. Vertical displacement in alluvium on one of the faults was as much as 1.5 m, but the amount of horizontal fault displacement, if any, could not be determined. Changes in the original design because of the faults included reduction in height, shifting of the axis, and major changes in the zonation of the dam (Sherard and others, 1974). The earth-fill, 47-m-high Los Angeles Dam (Fig. 7) was built in 1976 to replace the severely damaged Lower San Fernando Dam, described above. After exhaustive study, it was concluded that the dam site was within a zone of faults capable of future vertical displacements of 1 to 2.7 m at the time of a postulated local earthquake of about magnitude 7.5 (Yerkes and others, 1974; Wesson and others, 1974). Ground motions from the postulated earthquake and the estimates of potential surface faulting were incorporated in the design of the dam (Civil Engineering, 1978; Federal Disaster Assistance Administration, 1975, p. I-D-14, I-D-15). The design values for the earthquake ground motions and faulting were partly based on research on those topics stimulated by the investigations for the Trans-Alaska pipeline system and the Bodega Head nuclear reactor site, described above.

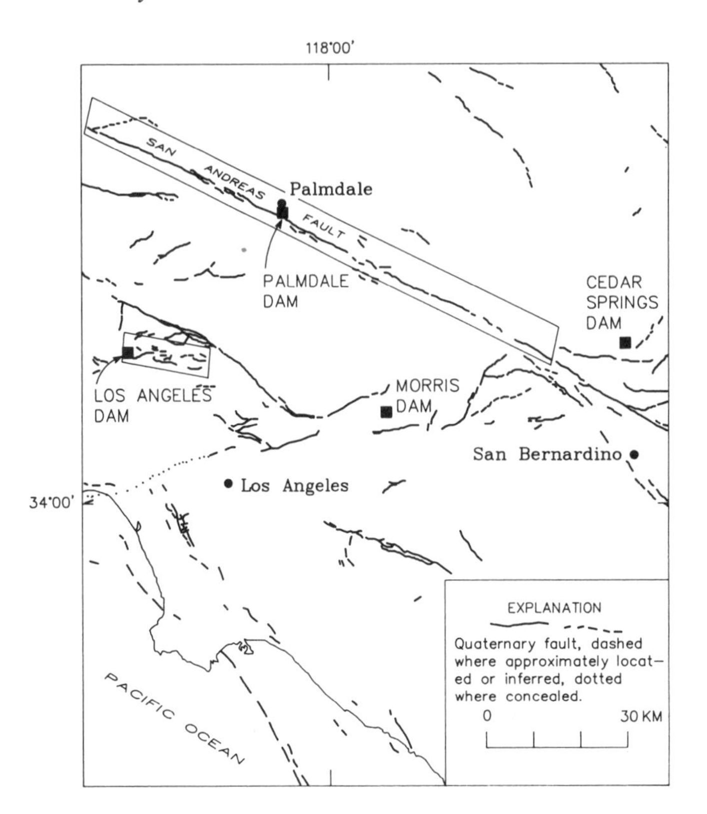

Figure 7. Map of part of southern California showing Quaternary faults and locations of four dams that have been designed to accommodate fault displacement. The reservoir behind Cedar Springs Dam is called Silverwood Lake. Fault traces from Jennings (1975). Coseismic surface faulting has occurred in historical time within the areas outlined by the two parallelograms.

San Francisco sewer outfall

A sewer outfall extending into the ocean off San Francisco, California, had to cross the San Andreas fault (Gilbert and others, 1981; Murphy and Eisenberg, 1985). A design consideration was the possibility of as much as 5 or 6 m of horizontal and about 1 m of vertical fault displacement during a magnitude 8+ earthquake. Under the assumption that such a large seismic event probably will not occur during the 75-year design life of the outfall, the design philosophy was to accommodate smaller fault displacements but minimize damage and provide for easy repair should the larger displacements occur. For a distance of 366 m centered over the fault zone, special flexible joints were provided between the individual sections of the reinforced concrete pipe, which has an inside diameter of 3.7 m. Provision was made for access to the pipe and for diverting the sewage should the pipe be broken by faulting.

RESEARCH NEEDS AND OUTLOOK

An American Society of Civil Engineers committee on the siting of nuclear facilities included among their conclusions the statement, "... definition of geologic and seismic characteristics of potential sites remains the area in greatest need of additional effort to reduce current conservatisms . . ." (Kruger, 1979, p. 498). This statement could be applied to other types of projects as well. The severe impact of actual or potential faulting and earthquakes on a wide variety of projects is apparent from the descriptions given herein. Clearly, the appraisals of the risks from faulting and earthquakes must be accurate to avoid costly overconservatism or underconservatism, and research leading to a better understanding of the process should continue. The following summary of some specific research needs is based on case histories summarized above, recommendations of a National Science Foundation workshop (Sitar and others, 1983), recommendations of a committee of the National Research Council (Wallace, 1986), and my experience.

New and improved techniques are needed for dating Quaternary materials and events. Accurate techniques are necessary to improve our knowledge of prehistoric tectonic events and their attendant earthquakes. Better dating techniques are also required for improvement in the evolving field of tectonic geomorphology, which can contribute much to the analysis of recent and current deformation. Probability analyses can be expected to play an increasing role in decisions regarding future risk from faulting and earthquakes; however, improvement is clearly needed in presentation to nonspecialists of the value, methods, assumptions, sensitivity to the assumptions, and limitations of probabilistic analyses. Particularly, such analyses should only be made with a thorough, qualified knowledge of the geological and geophysical environment.

The 'characteristic earthquake' model needs to be tested. It suggests that a given fault or fault segment will produce faulting and earthquakes of nearly the same size in successive events. Of great significance is the question of whether, at a particular site, several displacements of similar size can be followed by a much greater displacement.

Further research is needed on the interaction between an engineered structure and a propagating fault. In the 1972 Managua, Nicaraugua, earthquake, strike-slip surface faulting of small displacement (about 17 cm, according to Niccum and others, 1976) was apparently deflected by an underground vault that was stronger than the slightly-to-strongly indurated sand and gravel in which it was embedded (Niccum and others, 1976). A laboratory-model study of a structure in a strike-slip fault produced a similar result (Duncan and Lefebvre, 1973). An analysis of the General Electric Test Reactor, described above, predicted that reverse oblique slip of 1 m will bypass the reactor. Additional research on this type of interaction is needed and should include the following: (1) the physical properties of the material surrounding or below the structure and the three-dimensional variation of those properties; (2) the type, amount of displacement, and rupture velocity of the faulting; and (3) the inertia and physical properties of the engineered structure. Better understanding of fault-structure interaction would be applicable to both existing and future projects.

Only a few inconclusive data are available regarding the absorption or dispersal of fault ruptures as they pass upward through unconsolidated materials or weak rock (Bonilla, 1970, p. 58–59). The phenomenon is important in at least two ways. One way is in determining the time of most recent displacement on a fault, such as the shaft fault at the Bodega Head reactor site. The second way is determining whether a fault in bedrock will propagate upward through thick, unconsolidated deposits in such a manner as to damage a structure near the ground surface. An analysis of the problem in connection with nuclear reactors was done by finite-element methods (Scott and Schoustra, 1974), and other laboratory and numerical studies (Roth and others, 1981a, 1981b, 1982; Cole and Lade, 1984; Lade and others, 1984) are also applicable. More empirical, laboratory, and theoretical studies are needed.

None of the dams and other structures discussed above that have been designed to accommodate fault displacement have yet been tested by actual faulting. This fact emphasizes the need for research such as that outlined in the preceding two paragraphs.

Better ways are needed to distinguish marine and nonmarine terraces that have been relatively uplifted gradually (e.g., by folding or change in sea level) from those that have been uplifted suddenly by coseismic fault displacement. Sudden uplift can be assumed to be accompanied by earthquakes and tsunamis; furthermore, rapid uplift could have serious consequences such as shoaling of harbors or lifting the cooling-water intake of a nuclear reactor above water level.

The history of research in the field of engineering geology suggests what the future will bring. The historic pattern has been that a practical problem results in limited research that provides an immediate but commonly incomplete basis for a decision on the problem; frequently the initial research is followed by further research that provides better solutions to similar practical problems and at the same time contributes to the geological sciences. The field of engineering geology in the future, as in the past, will include research that both solves practical problems and contributes to an improved scientific understanding of geological and geophysical processes. Sponsorship and performance of the problem-oriented research will be done by both the public and private sectors of society.

REFERENCES CITED

Adams, F. D., 1938, The birth and development of the geological sciences: New York, Dover (1954 reprint), 506 p.

Aki, K., 1966, Generation and propagation of *G* waves from the Niigata earthquake of June 16, 1964; Part 2, Estimation of earthquake moment, released energy, and stress-strain drop from the *G* wave spectrum: Tokyo University, Earthquake Research Institute Bulletin, v. 44, p. 73–88.

Albee, A. L., and Smith, J. L., 1967, Geologic criteria for nuclear power plant locations: Society of Mining Engineers Transactions, v. 238 (December), p. 430–434.

Allen, C. R., 1968, The tectonic environments of seismically active and inactive areas along the San Andreas fault system, *in* Dickinson, W. R., and Grantz, A., eds., Proceedings of Conference on Geologic Problems of San Andreas fault system: Stanford University Publications in the Geological Sciences, v. 11, p. 70–82.

Ambraseys, N. N., 1960, On the seismic behavior of earth dams: World Conference Earthquake Engineering, 2nd, Tokyo and Kyoto, Japan, 1960, Proceedings, v. 1, p. 331–358.

Anderson, J. G., 1979, Estimating the seismicity from geological structure for seismic-risk studies: Seismological Society of America Bulletin, v. 69, no. 1, p. 135–158.

Atomic Safety and Licensing Board, 1966, Initial decision in the matter of Department of Water and Power of the City of Los Angeles, Malibu Nuclear Plant Unit no. 1: U.S. Atomic Energy Commission Docket 50-214, 48 p.

——, 1982, Initial decision in the matter of General Electric Company (Vallecitos Nuclear Center–General Electric Test Reactor), Operating License no. TR-1:216 p., 2 appendixes.

Babbitt, D. H., Bennett, W. J., and Hart, R. D., 1983, California's seismic reevaluation of embankment dams, *in* Howard, T. R., ed., Seismic design of embankments and caverns: New York, American Society of Civil Engineers, p. 96–112.

Benioff, H., 1965, Testimony in the matter of Department of Water and Power, City of Los Angeles, Malibu Nuclear Plant, Unit no. 1: [mimeo.] Reports to U.S. Atomic Energy Commission, Docket no. 50-214, July 1, 1965, 6 p.

Bolt, B. A., 1974, Duration of strong ground motion: World Conference on Earthquake Engineering, 5th, Rome, Italy, June 1973, Proceedings, v. 1, p. 1304–1313.

Bonilla, M. G., 1967, Historic surface faulting in continental United States and adjacent parts of Mexico: U.S. Geological Survey Open-File Report, 36 p.; also U.S. Atomic Energy Commission Report TID-24124, 36 p.

——, 1970, Surface faulting and related effects, *in* Wiegel, R. L., ed., Earthquake engineering: Englewood Cliffs, New Jersey, Prentice-Hall, p. 47–74.

——, 1973, Trench exposures across surface fault ruptures associated with the San Fernando earthquake, *in* Geological and geophysical studies, v. III of San Fernando, California, earthquake of February 9, 1971: U.S. Department of Commerce, National Oceanographic and Atmospheric Administration, p. 173–182.

——, 1985, Surface faulting studies, *in* National Earthquake Hazards Reduction Program, Summaries of technical reports volume XX: U.S. Geological Survey Open-File Report 85-464, p. 71–72.

Bonilla, M. G., and Buchanan, J. M., 1970, Interim report on worldwide historic surface faulting: U.S. Geological Survey Open-File Report, 32 p.

Bonilla, M. G., Mark, R. K., and Lienkaemper, J. J., 1984, Statistical relations among earthquake magnitude, surface rupture length, and surface fault displacement: Seismological Society of America Bulletin, v. 74, no. 6, p. 2379–2411.

Brand, D. A., 1985, Letter to H. R. Denton, Director, Office of Nuclear Reactor Regulation, U.S. Nuclear Regulatory Commission, transmitting Long Term Seismic Program, Program Plan, Pacific Gas and Electric Company, Diablo Canyon Power Plant, U.S. Nuclear Regulatory Commission Docket nos. 50-275 and 50-323, 199 p (on file at Public Documents Room, U.S. Nuclear Regulatory Commission, 1717 H Street NW, Washington, D.C.).

Brew, D. A., 1974, Environmental impact analysis: The example of the proposed Trans-Alaska pipeline: U.S. Geological Survey Circular 695, 16 p.

Brogan, G. E., Cluff, L. S., Korringa, M. K., and Slemmons, D. B., 1975, Active faults of Alaska: Tectonophysics, v. 29, p. 73–85.

Brune, J. N., 1968, Seismic moment, seismicity, and rate of slip along major fault zones: Journal of Geophysical Research, v. 73, p. 777–784.

California Department of Water Resources, 1967, Earthquake damage to hydraulic structures in California: California Department of Water Resources Bulletin 116-3, 200 p.

Civil Engineering, 1978, Los Angeles Reservoir is safe from earthquakes: v. 48, no. 6, p. 88–89.

Clark, M. M., Grantz, A., and Rubin, M., 1972, Holocene activity of the Coyote Creek fault as recorded in sediments of Lake Cahuilla, *in* The Borrego Mountain earthquake of April 19, 1968: U.S. Geological Survey Professional Paper 787, p. 112–130.

Cluff, L. S., Slemmons, D. B., Brogan, G. E., and Korringa, M. K., 1974, Summary report, basis for pipeline design for active-fault crossings for the Trans-Alaska Pipeline System: Woodward-Lundgren and Associates report prepared for Trans-Alaska Pipeline System, Houston, Texas, Appendix A-3.1111, 115 p.

Cole, D. A., Jr., and Lade, P. V., 1984, Influence zones in alluvium over dip-slip faults: American Society of Civil Engineers, Journal of Geotechnical Engineering, v. 110, no. 5, p. 599–615.

Cook, J., 1985, Nuclear follies: Forbes, v. 135, no. 3 (Feb. 11), p. 82–100.

Dallaire, G., 1981, Are America's utilities sorry they went nuclear?: Civil Engineering, v. 51, no. 1, p. 37–41.

Duke, C. M., 1960, Foundations and earth structures in earthquakes: World Conference on Earthquake Engineering, 2nd, Tokyo and Kyoto, Japan, 1960, Proceedings, v. 1, p. 435–455.

Duncan, J. M., and Lefebvre, G., 1973, Earth pressures on structures due to fault movement: American Society of Civil Engineers, Soil Mechanics and Foundations Division Journal, v. 99, no. SM12, p. 1153–1163.

Earth Sciences Associates, 1974, Geology of the southern Coast Ranges and the adjoining offshore continental margin of California, with special reference to the geology in the vicinity of the San Luis Range and Estero Bay, *in* Pacific Gas and Electric Company, Final Safety Analysis Report, Units 1 and 2, Diablo Canyon site: U.S. Atomic Energy Commission Docket nos. 50-275, 50-323, v. 3 [with amendments through 1975], 194 p.

Federal Disaster Assistance Administration, 1975, Final federal environmental impact statement, Los Angeles Dam and Reservoir Project: U.S. Department of Housing and Urban Development, Federal Disaster Assistance Administration, Region Nine, 830 p.

Frizzell, V. A., Jr., and Zoback, M. L., 1987, Stress orientation determined from fault slip data in Hampel Wash area, Nevada, and its relation to contemporary regional stress field: Tectonics, v. 6, no. 2, p. 89–98.

Gilbert, G. K., 1884, A theory of the earthquakes of the Great Basin with a practical application: American Journal Science, 3rd ser., v. 27, p. 49–53.

Gilbert, O. H., Jr., Eisenberg, Y., and Treadwell, D. D., 1981, Seismic design of the San Fransicso ocean outfall, *in* Prakesh, S., ed., International Conference on Recent Advances in Geotechnical Earthquake Engineering and Soil Dynamics, St. Louis, 1981: Rolla, University of Missouri, Proceedings, v. 3, p. 1133–1138.

Godfrey, K. A., Jr., 1978, Trans Alaska pipeline: Civil Engineering, v. 48, no. 6, p. 59–69.

Gordon, F. R., and Lewis, J. D., 1980, The Meckering and Calingiri earthquakes, October 1968 and March 1970: Geological Survey of Western Australia Bulletin 126, 229 p.

Hall, C. A., Jr., 1958, Geology and paleontology of the Pleasanton area, Alameda and Contra Costa counties, California: Berkeley, University of California, Publications in Geological Science, v. 34, no. 1, p. 1–89.

Hall, W. J., and Newmark, N. M., 1977, Seismic design criteria for pipelines and facilities: American Society of Civil Engineers, Technical Council on Lifeline Earthquake Engineering Specialty Conference, Los Angeles, 1977, Proceed-

ings, p. 18–34.
Hanks, T. C., and Kanamori, H., 1979, A moment-magnitude scale: Journal of Geophysical Research, v. 84, no. B5, p. 2348–2350.
Herd, D. G., 1977, Geologic map of the Las Positas, Greenville, and Verona faults, eastern Alameda County, California: U.S. Geological Survey Open-File Report 77-689, 25 p., map scale 1:24,000.
Hoskins, E. G., and Griffiths, J. R., 1971, Hydrocarbon potential of northern and central California offshore, *in* Cram, I. H., ed., Future petroleum provinces of the United States; Their geology and potential: American Association of Petroleum Geologists Memoir 15, v. 1, p. 212–228.
Housner, G. W., 1970, Strong ground motion, *in* Wiegel, R. L., ed., Earthquake engineering: Englewood Cliffs, New Jersey, Prentice-Hall, p. 75–91.
Hradilek, P. J., 1977, Behavior of underground box conduit in the San Fernando earthquake: American Society of Civil Engineers, Technical Council on Lifeline Earthquake Engineering Specialty Conference, Los Angeles, 1977, Proceedings, p. 308–319.
Jennings, C. W., 1975, Fault map of California: California Division of Mines and Geology, [California] Geologic Data Map Series, Map no. 1, scale 1:750,000.
Kruger, P., chairman, 1979, Nuclear facilities siting; Report of Task Committee on Nuclear Effects: American Society of Civil Engineers, Journal of the Environmental Engineering Division, v. 105, no. EE3, p. 443–502.
Lachenbruch, A. S., 1970, Some estimates of the thermal effects of a heated pipeline in permafrost: U.S. Geological Survey Circular 632, 23 p.
Lade, P. V., Cole, D. A., Jr., and Cummings, D., 1984, Multiple failure surfaces over dip-slip faults: American Society of Civil Engineers, Journal of Geotechnical Engineering, v. 110, no. 5, p. 616–627.
Lawroski, S., 1978, Letter report from Chairman, Advisory Committee on Reactor Safeguards to J. M. Hendrie, Chairman, U.S. Nuclear Regulatory Commission, on Diablo Canyon Nuclear Power Station Units 1 and 2, 7 p. (on file at Public Documents Room, U.S. Nuclear Regulatory Commission, 1717 H Street NW, Washington, D.C.).
Lawson, A. C., and others, 1908, The California earthquake of April 18, 1906, Report of the State Earthquake Investigation Commission: Carnegie Institution of Washington Publication 87, v. 1, 451 p.
Legget, R. F., 1962, Geology and engineering (2nd edition): New York, McGraw-Hill, 884 p.
Louderback, G. D., 1937, Characteristics of active faults in the central Coast Ranges of California with application to the safety of dams: Seismological Society of America Bulletin, v. 27, no. 1, p. 1–27.
——, 1942, Faults and earthquakes: Seismological Society of America Bulletin, v. 32, no. 4, p. 305–330.
——, 1950, Faults and engineering geology, *in* Paige, S., chairman, Application of geology to engineering practice: Geological Society America, Berkey Volume, p. 125–150.
Lyell, C., 1830, Principles of geology, being an attempt to explain the former changes of the earth's surface, by reference to causes now in operation: London, Murray, 2 vols., 841 p.
Malde, H. E., 1971, Geologic investigation of faulting near the National Reactor Testing Station, Idaho: U.S. Geological Survey Open-File Report, 167 p.
Marblehead Land Company, 1966, Brief of Intervenor, Marblehead Land Company, in support of findings of fact and conclusions of law: U.S. Atomic Energy Commission Docket no. 50-214, 152 p.
Matsuda, T., 1975, Magnitude and recurrence interval of earthquakes from a fault: Zisin, ser. 2, v. 28, p. 269–283 (in Japanese with English abs.).
McCaffrey, M. A., and O'Rourke, T. D., 1983, Surface faulting and its effect on buried pipelines: Ithaca, New York, Cornell University, School of Civil and Environmental Engineering, Geotechnical Engineering Report 83-10, 181 p.
Meehan, R. L., 1984, The atom and the fault: Cambridge, Massachusetts, MIT Press, 161 p.
Molnar, P., 1979, Earthquake recurrence intervals and plate tectonics: Seismological Society of America Bulletin, v. 69, no. 1, p. 115–133.
Murphy, G. J., and Eisenberg, Y., 1985, San Francisco outfall; The champ?: Civil Engineering, v. 55, no. 12, p. 58–61.
Newmark, N. M., and Hall, W. J., 1974, Seismic design spectra for Trans-Alaska pipeline: World Conference on Earthquake Engineering, 5th, Rome, Italy, June 1983, Proceedings, v. 1, p. 554–557.
——, 1975, Pipeline design to resist large fault displacements, *in* U.S. National Conference on Earthquake Engineering, Ann Arbor, Michigan, 1975, Proceedings: Oakland, California, Earthquake Engineering Research Institute, p. 416–425.
Niccum, M. R., Cluff, L. S., Chamorro, F., and Wyllie, L., 1976, Banco Central de Nicaragua; A case history of a high-rise building that survived surface fault rupture, *in* Humphrey, C. B., ed., Engineering Geology and Soils Engineering Symposium, no. 14, Boise, Idaho, 1976: Boise, Idaho, Idaho Transportation Department, Division of Highways, p. 133–144.
Novick, S., 1969, The careless atom: Boston, Houghton-Mifflin, 225 p.
Packer, D. R., Cluff, L. S., Schwartz, D. P., Swan, F. H., III, and Idriss, I. M., 1978, Auburn Dam; A case history of earthquake evaluation for a critical facility: International Conference on Microzonation, 2nd, San Francisco, California, 1978, Proceedings, v. 1, p. 457–470.
Page, R. A., Boore, D. M., Joyner, W. B., and Coulter, H. W., 1972, Ground motion values for use in the seismic design of the Trans-Alaska Pipeline System: U.S. Geological Survey Circular 672, 23 p.
Page, R. A., Blume, J. A., and Joyner, W. B., 1975, Earthquake shaking and damage to buildings: Science, v. 189, p. 601–608.
Pampeyan, E. H., 1986, Effects of the 1906 earthquake on the Bald Hill outlet system, San Mateo County, California: Association of Engineering Geologists Bulletin, v. 23, no. 2, p. 197–208.
Péwé, T. L., and Reger, R. D., 1983, Delta River area, Alaska Range, *in* Péwé, T. L., and Reger, R. D., eds., Richardson and Glenn highways, Alaska, guidebook to permafrost and Quaternary geology: Alaska Division of Geological and Geophysical Surveys Guidebook 1, p. 47–136.
Piper, C. F., 1981, Letter from news bureau representative, Pacific Gas and Electric Company: San Jose Mercury News, October 31, 1981, p. 11B.
Reid, H. F., 1910, The mechanics of the earthquake, *in* The California earthquake of April 18, 1906, Report of the State Earthquake Investigation Commission: Carnegie Institution of Washington Publication 87, v. 2, 192 p.
Roscow, J. P., 1977, 800 miles to Valdez; The building of the Alaska pipeline: Englewood Cliffs, New Jersey, Prentice-Hall, 227 p.
Rose, D., 1978, Risk of catastrophic failure of major dams: American Society of Civil Engineers, Journal of the Hydraulics Division, v. 104, no. HY9, p. 1349–1351.
Ross, F. I., 1975, Disaster response; Assistance to individuals, *in* Oakeshott, G. B., ed., San Fernando earthquake of 9 February 1971: California Division of Mines and Geology Bulletin 196, p. 437–441.
Roth, W. H., Scott, R. F., and Austin, I., 1981a, Centrifuge modeling of fault propagation through alluvial soils: Geophysical Research Letters, v. 8, no. 6, p. 561–564.
Roth, W. H., Sweet, J., and Goodman, R. E., 1981b, Numerical and physical modeling of flexural slip phenomena and potential fault movement: Rock Mechanics, Supplement 12, p. 27–46.
Roth, W. H., Kalsi, G., Papastamatiou, O., and Cundall, P. A., 1982, Numerical modelling of fault propagation in soils: International Conference on Numerical Methods in Geomechanics, 4th, Edmonton, Canada, May 31–June 4, 1982, Proceedings, v. 2, p. 487–494.
San Francisco Chronicle, 1979, New report on safety of Auburn Dam: June 21, 1979, p. 5.
San Jose Mercury News, 1984, Diablo Canyon's stormy history: April 29, 1984, p. 4A.
Schlocker, J., and Bonilla, M. G., 1963, Engineering geology of the proposed nuclear power plant site on Bodega Head, Sonoma County, California: U.S. Geological Survey Trace Elements Investigations Report TEI-844, 37 p., 4 plates, 13 figures.
——, 1964, Engineering geology of the proposed nuclear power plant on Bodega Head, Sonoma County, California: U.S. Geological Survey Report for U.S.

Atomic Energy Commission, 31 p., 1 fig.
Schwartz, D. P., and Coppersmith, K. J., 1984, Fault behavior and characteristic earthquakes; Examples from the Wasatch and San Andreas fault zones: Journal of Geophysical Research, v. 89, no. B7, p. 5681–5698.
Scott, R. F., 1973, The calculation of horizontal accelerations from seismoscope records: Seismological Society of America Bulletin, v. 63, no. 5, p. 1637–1661.
Scott, R. F., and Schoustra, J. J., 1974, Nuclear power plant siting in deep alluvium: American Society of Civil Engineers, Journal of the Geotechnical Engineering Division, v. 100, no. GT4, p. 449–459.
Scrop, G. P., 1830, Review *of* 'Principles of geology, being an attempt to explain the former changes of the earth's surface, by a reference to causes now in operation,' by Charles Lyell, 1830: The Quarterly Review, v. 43, no. 86, p. 411–469.
Seed, H. B., 1979, Considerations in the earthquake-resistant design of earth and rockfill dams: Geotechnique, v. 29, no. 3, p. 215–263.
Seed, H. B., Lee, K. L., and Idriss, I. M., 1969, Analysis of Sheffield Dam failure: American Society of Civil Engineers, Journal of the Soil Mechanics and Foundations Division, v. 95, no. SM6, p. 1453–1490.
Seed, H. B., Idriss, I. M., Lee, K. L., and Makdisi, F. I., 1975, Dynamic analysis of the slide in the lower San Fernando Dam during the earthquake of Feb. 9, 1971: American Society of Civil Engineers, Journal of the Geotechnical Engineering Division, v. 101, no. GT9, p. 889–911.
Seed, H. B., Makdisi, F. I., and De Alba, P., 1978, Performance of earth dams during earthquakes: American Society of Civil Engineers, Journal of Geotechnical Engineering Division, v. 104, no. GT7, p. 967–994.
Sherard, J. L., Cluff, L. S., and Allen, C. R., 1974, Potentially active faults in dam foundations: Geotechnique, v. 24, no. 3, p. 367–428.
Sitar, N., Brekke, T. L., and Jahns, R. H., 1983, Goals for basic research in engineering geology, Report on a workshop held in St. Helena, California, Jan. 21–22, 1983: Berkeley, University of California, Department of Civil Engineering, 28 p.
Sieh, K. E., 1978, Prehistoric large earthquakes produced by slip on the San Andreas fault at Pallett Creek, California: Journal of Geophysical Research, v. 83, no. B8, p. 3907–3939.
—— , 1981, A review of geological evidence for recurrence times of large earthquakes, *in* Simpson, D. W., and Richards, P. G., eds., Earthquake prediction; An international review: American Geophysical Union, Maurice Ewing Series, v. 4, p. 181–207.
Slemmons, D. B., 1977, Faults and earthquake magnitude: U.S. Army Engineer Waterways Experiment Station, Vicksburg, Mississippi, Miscellaneous Paper S-73-1, Report 6, 129 p. and appendix.
—— , 1982, Determination of design earthquake magnitudes for microzonation: International Earthquake Microzonation Conference, 3rd, Seattle, Washington, 1982, Proceedings, v. 1, p. 119–130.
Slosson and Associates, 1981, Summary of observations during final geotechnical investigations monitored at the proposed Cojo Bay LNG terminal site near Point Conception, Santa Barbara County, California: Van Nuys, California, Slosson and Associates, 85 p., 4 appendixes.
Stock, J. M., Healy, J. H., Hickman, S. H., and Zoback, M. D., 1985, Hydraulic fracturing stress measurements at Yucca Mountain, Nevada, and relationship to the regional stress field: Journal of Geophysical Research, v. 90, no. B10, p. 8691–8706.
Subcommittee on Water and Sewerage Systems, 1973, Earthquake damage to water and sewerage facilities, *in* Utilities, transportation, and sociological aspects, vol. II *of* San Fernando, California, earthquake of February 9, 1971: U.S. Department of Commerce, National Oceanic and Atmospheric Administration, p. 75–193.
Swan, F. H., III, Schwartz, D. P., and Cluff, L. S., 1980, Recurrence of moderate-to-large magnitude earthquakes produced by surface faulting on the Wasatch fault zone, Utah: Seismological Society of America Bulletin, v. 70, no. 5, p. 1431–1462.
Sylvester, A. G., 1979, Earthquake damage in Imperial Valley, California, May 18, 1940, as reported by T. A. Clark: Seismological Society of America Bulletin, v. 69, no. 2, p. 547–568.
Taylor, C. L., and Cluff, L. S., 1977, Fault displacement and ground deformation associated with surface faulting: American Society of Civil Engineers Technical Council on Lifeline Earthquake Engineering Specialty Conference, Los Angeles, 1977, Proceedings, p. 338–353.
Tocher, D., 1958, Earthquake energy and ground breakage: Seismological Society of America Bulletin, v. 48, no. 2, p. 147–153.
U.S. Atomic Energy Commission, 1962, Reactor site criteria: Code of Federal Regulations, Title 10, Part 100, Chapter 1, Section 100.10.
—— , 1964, Public announcement dated October 27, 1964, including letter from Advisory Committee on Reactor Safeguards dated October 20, 1964 (4 p), and Summary analysis for Docket no. 50-205 by the Division of Reactor Licensing dated October 26, 1964, 14 p.
—— , 1967, Decision in the matter of Department of Water and Power of the City of Los Angeles Malibu Nuclear Plant Unit no. 1, Docket no. 50-214, 17 p.
U.S. Department of Energy, 1985, General guidelines for the recommendation of sites for nuclear waste repositories: Code of Federal Regulations, Title 10, Part 960, Subpart C, Postclosure Guidelines.
[U.S.] Federal Task Force on Alaskan Oil Development, 1972, Introduction and summary, v. 1 *of* Final environmental impact statement, proposed Trans-Alaska pipeline: U.S. Department of Interior Interagency Report, 386 p.; available from the National Technical Information Service, U.S. Department of Commerce, Springfield, VA 22151, NTIS PB-206921-1.
U.S. Geological Survey, compiler, 1984, A summary of geologic studies through January 1, 1983, of a potential high-level radioactive waste repository site at Yucca Mountain, southern Nye County, Nevada: U.S. Geological Survey Open-File Report 84-792, 103 p.
U.S. Nuclear Regulatory Commission, 1976, Supplement no. 4 to the Safety Evaluation Report by the Office of Nuclear Reactor Regulation, U.S. Nuclear Regulatory Commission, in the matter of Pacific Gas and Electric Company Nuclear Power Station Units 1 and 2, Docket nos. 50-275 and 50-323, 50 p.
—— , 1977, Seismic and geologic siting criteria for nuclear power plants: Code of Federal Regulations, Title 10, Part 100, Appendix A.
Wagner, H. C., 1974, Marine geology between Cape San Martin and Point Sal, south-central California offshore: U.S. Geological Survey Open-File Report 74-252, 17 p.
Wallace, R. E., 1970, Earthquake recurrence intervals on the San Andreas fault: Geological Society of America Bulletin, v. 81, p. 2875–2890.
—— , 1980, Gilbert's studies of faults, scarps, and earthquakes, *in* Yochelson, E. L., ed., The scientific ideas of G. K. Gilbert: Geological Society of America Special Paper 183, p. 35–44.
—— , chairman, 1986, Active tectonics: Washington, D.C., National Academy Press, 266 p.
Wesnousky, S. G., Scholz, C. H., Shimazaki, K., and Matsuda, T., 1983, Earthquake frequency distribution and the mechanics of faulting: Journal of Geophysical Research, v. 88, no. B11, p. 9331–9340.
—— , 1984, Integration of geological and seismological data for the analysis of seismic hazard; A case study of Japan: Seismological Society of America Bulletin, v. 74, no. 2, p. 687–708.
Wesson, R. L., Page, R. A., Boore, D. M., and Yerkes, R. F., 1974, Expectable earthquakes and their ground motions in the Van Norman Reservoirs area, *in* The Van Norman Reservoirs area, northern San Fernando valley, California: U.S. Geological Survey Circular 691-B, 9 p.
Williams, J. R., 1982, Design review, Trans-Alaska oil pipeline, 1974–1976: U.S. Geological Survey Open-File Report 82-225, 29 p., 3 appendixes.
Yerkes, R. F., and Wentworth, C. M., 1965, Structure, Quaternary history, and general geology of the Corral Canyon area, Los Angeles County, California: U.S. Geological Survey report prepared for U.S. Atomic Energy Commission, 215 p., 4 appendixes.
Yerkes, R. F., Bonilla, M. G., Youd, T. L., and Sims, J. D., 1974, Geologic

environment of the Van Norman Reservoirs area, northern San Fernando Valley, California: U.S. Geological Survey Circular 691-A, 35 p.

Youd, T. L., Yerkes, R. F., and Clark, M. M., 1978, San Fernando faulting damage and its effect on land use, *in* American Society of Civil Engineers Specialty Conference on Earthquake Engineering and Soil Dynamics, Pasadena, Calif., June 19–21, 1978, Proceedings, v. 2, p. 1111–1125.

MANUSCRIPT ACCEPTED BY THE SOCIETY JUNE 3, 1987

ACKNOWLEDGMENTS

Several individuals have helped in the preparation of this chapter. Discussions with Julius Schlocker, C. M. Wentworth, and R. F. Yerkes gave information about the Bodega Head and Malibu nuclear reactor site investigations and about the development of the U.S. Nuclear Regulatory Commission reactor site criteria. Discussions with J. R. Williams, R. A. Page, O. J. Ferrians, Jr., and W. B. Joyner provided insights and information regarding the planning and construction of the Trans-Alaska pipeline system.

Santa Helena Escarpment and dip slope of Mesa de Anguilla, composed of the thick Georgetown and Edwards Limestones. The Big Bend depression and the confluence of Terlingua Creek and the Rio Grande are in the foreground. The canyon was initially a proposed site for a high dam, but reservoir capacity was minimal due to the tilted surface of the Mesa block and the natural topographic features upstream at Lajitas (in distance) where any water that is impounded would bypass the uplifted Mesa. (Photo courtesy of the International Boundary and Water Commission, 1951.)

Printed in U.S.A.

Geological Society of America
Centennial Special Volume 3
1991

Chapter 13

Rebound, relaxation, and uplift

Thomas C. Nichols, Jr., and Donley S. Collins
U.S. Geological Survey, Box 25046, MS 966, Denver Federal Center, Denver, Colorado 80225

INTRODUCTION

Definition and impact

Rebound is defined as the expansive recovery of surficial crustal material, either instantaneous, or time dependent, or both, and is initiated by the removal or relaxation of superincumbent loads (Nichols, 1980). The displacements caused by rebound allow elastic and inelastic relaxation of the crustal masses to occur. The outward and upward movements associated with rebound are uplift displacements related to rebound processes. Rebound of geological materials is attributed to stress relief, but the process is poorly understood and the basis for predicting time-dependent rebound has not been clearly established. Not only are changes of stress important to the rebound process, but so are fabric, material properties, and anisotropy of the geologic materials, as well as external environmental factors such as moisture and temperature (Nichols, 1980). The problem of rebound is one with which design and construction engineers must deal, whenever the equilibrium of geologic materials is disturbed, especially in large excavations both surface and underground. In areas where rebound deformations can significantly affect engineering structures, it is desirable to understand rebound behavior, and to determine practical guidelines for prediction of the short- and long-term consequences of rebound.

Background

The phenomenon of rebound undoubtedly has been observed by ancient as well as modern quarry operators, civil engineers, and applied geologists. In the United States and England, recorded accounts of rebound and actual measurements made to study the phenomenon appear in the mid-19th century (Nichols and Varnes, 1984) describing spontaneous and explosive expansion in a wide range of quarried rock types (sandstones, granites, vein rock). "Pop ups" (Fig. 1), buckles, and sheared slabs (Fig. 2) often are found on quarry floors. Many accounts reveal the sometimes violent nature of rebound deformations and their apparent relation to quarrying practices and the geologic environs.

In southeastern Connecticut, excessive time-dependent rebound displacements have occurred in lower Paleozoic metamorphic rocks as the result of an engineered road excavation in 1970 (Block and others, 1979). Since 1970, as much as 30 cm of fault-like displacements toward the southeast have occurred (Walter, 1982). Block and others (1979) conclude that the displacements are related to an active fault zone. Walter (1982) discusses an alternative hypothesis: the displacements are caused by the unloading of rocks that are under high-level regional stresses.

Natural rebound phenomena have been observed and noted by many geological scientists. Záruba (1958) examined numerous accounts by European geologists that discussed the naturally occurring bulging of valley bottoms, that is, the anticlinal rises of strata beneath valley bottoms. Origins of these anticlines were thought to be "squeezing out" or plastic flow of weak clayey rocks from beneath rocks on adjacent valley slopes, that is, diapiric flow toward the site which experienced unloading. In addition, swelling of clayey rocks was thought to be a major factor. Based on his evaluation of these reports and his field investigations, Záruba believed that plastic flow and swelling of clays occur upon removal of loads by erosion. He stated that some of the bulging may be related to abnormally high moisture conditions of the Pleistocene. Záruba also discussed the engineering excavation and design problems related to valley bulging.

Ferguson (1967) and Hofmann (1966) observed valley-bottom bulging, folding, buckling, and faulting in interbedded rocks containing sandstones, siltstones, shales, and limestones in the eastern U.S. that seem to have resulted from unloading by valley erosion. In addition, Ferguson observed tensile valley-wall fractures also thought to be caused by valley unloading.

Sheeting or exfoliation fractures—commonly observed parallel fracture systems often bounding large massive rock bodies (Fig. 3)—are thought to be rebound features. These fractures are always more numerous at or near rock surfaces, and diminish slowly inward. Because they represent the apparent relaxation and deterioration of rock masses caused by weathering and erosional rather than tectonic processes, geologists have long speculated as to their exact cause. Gilbert (1904), in his studies of dome

Nichols, T. C., Jr., and Collins, D. S., 1991, Rebound, relaxation, and uplift, *in* Kiersch, G. A., ed., The heritage of engineering geology; The first hundred years: Boulder, Colorado, Geological Society of America, Centennial Special Volume 3.

Figure 1. Pop up in Barré Granite quarry, Vermont. Arrows show separation between pop up and granite floor. (Photograph by T. C. Nichols, Jr.)

Figure 2. A brittle failure in granite at Mount Airy, North Carolina, produced by quarrying the adjacent constraining rock (Nichols and Able, 1975).

structures of the Sierra Nevada, evaluated the hypotheses that sheeting was due either to thermal effects of the sun's radiation or to the physical–chemical effects of weathering. Upon examining the evidence, he favored a third causative process, that of dilation upon removal of superjacent materials. Dale (1923), in his study of New England granites and building stone, summarized the current ideas regarding granites and their structure. He accepted Gilbert's hypothesis of sheeting origin for the horizontal fractures that so facilitate the quarrying of otherwise massive granite. Jahns (1943), in his study of sheet structure in granite, similarly concludes that ". . .possible causes of the structure indicates that relief, through removal of superincumbent load, of a primary confining pressure is chiefly responsible for the large-scale exfoliation involved. Insolation, the progressive hydration and formation of chemical alteration products in certain susceptible minerals, and the mechanical action of fire, frost, and vegetation are possible minor contributory causes." More recently, Kieslinger (1960) concluded that the removal of confining loads by erosion allows the relief of definitely oriented near-surface residual stress fields (introduced into rock masses by tectonic or other past loading) followed by the slow relaxation and volume increase of rocks that results in the formation of sheeting fractures. Twidale (1973), on the other hand, feels that several possible explanations of sheeting may be valid in particular areas, but the best general explanation is that involving lateral compression, either relict or modern, where the sheet fractures develop parallel to the direction of maximum compression. After the removal of superincumbent loads, arched fractures are developed.

Contemporary research and understanding

From these and other observations, it is apparent that rebound deformations are ubiquitous in their occurrence and of significant frequency and magnitude to be of concern to modern-day engineering practices. There are numerous accounts of rock and soil deformations that occur as a result of engineering needs and natural excavation processes. Engineering and construction practitioners are concerned especially with long-term deformations that may cause severe damage to foundations and other engineered works. Many studies have been made attempting to determine the causes of rebounding rock masses and resulting failures. These studies have included detailed measurements of rock and soil rebound deformations, stress determinations, laboratory and in-situ determinations of physical and geotechnical properties, investigations of the effects of natural excavation processes (primarily erosion and mass wasting), the effects of engineering excavations, and theoretical analyses of rock rebound response to unloading. Several recent investigations (Matheson and Thomson, 1973; Carlsson and Olsson, 1982; Neuzil and Pollock, 1983; Nichols and others, 1986) describe natural and excavation-induced rebound that caused a deterioration or change or rock fabric and pore-water conditions; such changes can significantly affect the design, engineering practice, and operation of major works.

REBOUND: THE RESULT OF NATURAL AND ANTHROPOMORPHIC UNLOADING PROCESSES

Rebound deformations in rock masses primarily involve dilation, either elastic or inelastic or a combination, resulting from change of the state of stress due either to natural or man-made excavations. Rebound displacements due to the viscous mantle response are not explicitly considered here.

Figure 3. Sheeting fractures in quartz monzanite on Shuteye Peak, Sierra Nevada Mountains, California. (Photograph by N. King Huber.)

Elastic and inelastic deformation and relaxation

As overburden loads are removed from underlying rock masses, either by natural processes or by engineering excavations, the initial deformation elastic rebound is followed by much slower, time-dependent inelastic rebound. The elastic displacements are the nearly instantaneous, nondestructive, recoverable displacements, whereas, the inelastic displacements are the time-dependent, often destructive and nonrecoverable displacements. The nonrecoverable displacements cause significant relaxation of rock fabric that allows penetration of external agents, which causes further relaxation. Rebound, caused by the slow removal of overburden through natural processes, consists of large components of both elastic and inelastic displacements, whereas rapid removal of overburden through natural or engineering excavation will result primarily in elastic displacements that mask slower inelastic displacements within the adjacent unexcavated rock mass. The nearly instantaneous elastic rebound displacements depend only on rock stiffness and internally stored strain energy that is immediately relieved by the removal of confining rock loads. The time-dependent rebound displacements rely on internally stored strain energy and other sources of external and internal energy associated with long-term geologic processes and conditions that allow rocks to deform, relax, and deteriorate. An attempt is being made here to identify some of the more obvious processes and describe their effects on rebounding rock masses.

Removal of rock overburden loads by natural excavation

Rebound of rock masses caused by natural excavation processes and removal of overburden loads, primarily by slope movements, fluvial erosion, and possibly glacial retreat, is usually more complete than that caused by engineering practice. This is because the natural processes generally are much slower, allowing time-dependent chemical and mechanical fabric disturbances to become more complete. Thus, stored strain energy in rock masses surrounding natural excavations is likely to be less than in those surrounding artificial excavations.

Erosion. One of the most notable studies demonstrating erosional rebound in the north-central U.S. and south-central Canada is that of Matheson and Thomson (1973). The authors demonstrated the nearly ubiquitous occurrence of valley rebound where major rivers have incised valleys into flat-lying sedimentary rocks having a low modulus of elasticity. Based on borehole logs, geologic cross sections were constructed for the valleys at major dam and bridge sites along the Missouri, north and south Saskatchewan, Red Deer, and Pembina rivers, and Long Creek. Additional data were obtained from field and airphoto observations along the Wapiti, Little Smokey, Athabaska, and Peace rivers in Canada, as well as along the above-mentioned rivers. The cross sections clearly demonstrate the existence of anticlinal structures, which commonly follow the valleys and raised valley rims. The measured amount of valley upwarping appears to relate to the modulus of elasticity (E) of the rocks being incised, and for the sites examined, ranged from 3 to 10 percent of the valley depth. The examined sites were predominantly located in a soft Upper Cretaceous sedimentary terrain where E was about 50,000 psi (344.8 MPa) or less.

Matheson and Thomson's (1973) investigation is significant because they determined not only the magnitude of vertical rebound displacements that occurred in the valley anticlines and rims, but also that such displacements consisted of large time-dependent and elastic components. The time-dependent components imply that natural rebound may take place over long periods of time. Also, Matheson and Thomson observed several bedrock gouge zones in bridge abutment excavations below the valley-bottom level, and attributed the gouge development to interbed flexural slip resulting from rebound of the valley floor. To test this hypothesis, they conducted a numerical-theoretical simulation of the flexural slip, which modeled the thin layers of low-elastic-modulus beds near the valley bottom and demonstrated the feasibility of this type of origin. The development of such gouge zones near valley bottoms, resulting from rebound in soft sedimentary beds, is of considerable concern to engineers in the design of stable slopes adjacent to bridges and dams.

Other documentation of natural rebound occurrences has been more qualitative in nature and without quantitative verification. Documentation often includes observations of rebound features but not observations of associated rebound displacements. Observed features associated with rebound deformation generally include valley anticlines and faults, raised valley rims, and, especially, extensional fractures paralleling rock surfaces (sheeting fractures). In some cases, estimates of rebound and related rebound behavior have been based on models of geologic sections theoretically subjected to unloading by removal of the overburden.

For example, Neuzil and Pollock (1983) theoretically demonstrated the effect of rebound on pore pressures in hydraulically "tight" rocks. They showed that by eroding 500 m of overburden above a "tight" rock in 5×10^6 yrs, deficient or negative pore pressures may be developed to significant depths below the eroded surface. Borehole measurements made in South Dakota demonstrated that the pore pressures in the Upper Cretaceous Pierre Shale at 99 m were significantly deficient as a result of the rebound effect. Similarly, Koppula and Morgenstern (1984) theoretically demonstrated the occurrence of deficient pore pressures that resulted from eroding a soil mass as a function of depth and time.

Observed natural slope instability in saturated clay rocks may be related to the rebound equilibration of near-surface pore pressures. Such near-surface equilibration of pore pressures resulting from the excavation of rebounding saturated clay rock has been demonstrated by Vaughan and Wahlbancke (1973). The authors suggested that pore-pressure equilibration is a dominant factor in controlling delayed failure of excavated slopes.

In addition, geotechnical and geological measurements have been used to illustrate natural rebound in rock masses. Collins and Nichols (1987) used measurements of eroded geologic overburden to verify in-situ determinations of overconsolidation caused by rebound of the Pierre Shale. Also, Nichols and others (1986) demonstrated the feasibility that horizontal fracture zones found in the Pierre Shale, at depths of 72 and 87 m, are a result of erosional rebound.

Deglaciation. Rebound due to glacial unloading is widely reported, but generally, the isostatic response of the mantle is considered to be the primary causative mechanism. However, Strazer and others (1974), in their foundation studies for the construction of Interstate 5 through Seattle, Washington, alluded to the overconsolidation and rebound of the proglacial Lawton Clay Member caused by the advance and recession of the Vashion ice mass, which is estimated to have been about 915 m thick. Their laboratory tests confirm that the clay has become heavily overconsolidated with the removal of the ice mass. In addition, the clay fabric includes vertical extension fractures that were caused by rebound deformations as the ice overburden was removed.

Although evidence is yet lacking, glacial unloading of the heavily overconsolidated shales in the glaciated areas of the North American midcontinent probably caused similar rebound features within these shales. Glacial unloading of rocks will cause a certain amount of elastic expansion as well as inelastic deformation. In addition to possibly imparting stored strain energy, as glaciers override rock masses, they pluck away loose rock fragments created by sheeting and other near-surface fracture sets. Frictional resistance of the overriding mass and increased rock pore pressures, caused by the weight of the ice, enhance further surface fracturing and rock fabric relaxation. Therefore, upon removal of the ice mass, resulting rebound deformations are inelastic as well as elastic.

Removal of rock loads by engineering excavations

Removal of rock overburden by engineering excavation and the resulting rebound has been extensively observed and recorded. Many surface and underground engineering projects have been disrupted by rebound deformations that resulted from rapid exposure in excavations. Often, the disruptive deformations take place soon after excavation, but they may also continue long after the engineered facilities are completed. Severe rebound response can require significant changes in design and construction procedures; such changes are often costly and time consuming. The deleterious effects can be particularly expensive and dangerous because problems often go unnoticed until severe damage occurs to the works, sometimes catastrophically.

Surface excavations. Commonly, rebound deformations affecting construction practice occur in large surface excavations such as for quarries, dams, highways, building and bridge foundations, and canals. Often these excavations are located in eroding valleys where natural rebound processes are taking place more quickly than on the adjacent upland surfaces, and therefore, are susceptible to very active rebound instabilities.

Quarry operations. Rebound deformations commonly occur as a result of quarry operations. Although usually not dangerous, quarry rebound can affect the quality of the rock to the extent that the commercial feasibility of the quarry is influenced. Consequently, the quarry may be abandoned or a better quarrying procedure developed. For example, in the H. E. Fletcher granite quarry near Chelmsford, Massachusetts, rockbursts occur frequently, originating in areas where quarrying activity has locally concentrated the stress field. The quarrying procedure consists of excavating large granite blocks, typically 25 m high, along the quarry walls (Holzhausen, 1977). The blocks are freed by cutting channels into the walls containing usually two or more thick sheets separated by subhorizontal sheet fractures. As the channels are cut into the high vertical walls, rockbursts commonly occur at the ends of the channels, accompanied by sudden unstable propagation of one or more steeply dipping fractures. According to Holzhausen (1977), these fractures run approximately parallel to the quarry walls and may run tens of meters before stopping. Stress data from Hooker and Johnson (1969) and Holzhausen (1977) indicate that regional compressive stresses are on the order of 300 bars (30 MPa) and, in areas of stress concentration, they may be 500 bars (50 MPa) and greater. Because of the geometry and orientation of the blocks with respect to the stress field, Holzhausen assumed that conditions approximating plane stress prevail and that the blocks are being compressed uniaxially. He observed that the cut channels, normal to the direction of loading, are therefore stress concentrators that cause the rockbursts. Based on these observations, Holzhausen conducted a series of laboratory uniaxial compression tests on granite specimens with flaws cut into each specimen. Speciments that contained notches cut inward from the edge or internal slits cut inthe middle, normal to the direction of loading, developed throughgoing fractures parallel to the direction of loading. In all cases, the fractures intersected the tips of the cut flaw and propagated through zones of theoretically determined maximum shear stress, making the applied axial load considerably less than the compressive strength. The similarity of the notch failures in laboratory tests compared to the failure of the channel cuts in the Fletcher quarries led Holzhausen to conclude that the channel cuts are the stress concentrators causing the damaging rockbursts during quarrying.

Dam construction. Construction of the Oahe Dam in the Pierre Shale of South Dakota by the U.S. Army Corps of Engineers is typical of large dams built on overconsolidated clay shales, which are highly susceptible to rebound deformations. During the excavation for the Oahe Dam Stilling Basin, rebound deformations occurred in the shale that were in excess of the predicted rebound deformations (Underwood and others, 1964). The excavation, which removed between 30 and 60 m of shale overburden (approximately 10,300,000 m^3), was accomplished between April 1952 and March 1955. On December 13, 1954, with about 98 percent of the excavation completed and the landward half of the Stilling Basin floor bladed to level grade, the contractor stopped work for the winter holidays. After the contractor resumed work on January 13, 1955, newly discovered rebound displacements were measured. The average rebound of the excavated floor was .09 m with a maximum differential movement of 0.34 m along existing fault planes. The differential rebound displacements along the fault planes were much in excess of the predicted theoretical elastic rebound displacements that agreed with the earlier measurements made on December 1, 1954.

Because of the large differential rebound deformations, the Corps of Engineers found it necessary to conduct further investigations to determine the desirability of changing the design of the dam. The investigation consisted of making horizontal- and vertical-movement measurements on numerous steel surface pins installed at depths from 0.5 to 1.0 m and from 3 to 9 m, and on a steel tube marker set at 11.6 m below the bottom of the excavation. The measurements were collected from January 1955 until early spring of 1956. Vertical movements at and near the surface to a depth of 3.0 m varied considerably, ranging from 7.6 to 25.4 cm of upward movement in the first 12 months. During the same time period, vertical movements at depths from 3.0 to 12 m ranged from 13 to 38 mm and varied less than the shallower measurements. All these measurements decreased with time. The primary cause of the abrupt displacement along the fault planes was believed to be the excavation-induced concentration of lateral stress in the 9 to 12 m of shale below the bottom of the basin. Continued movements that occurred at diminished rates were attributed to plastic deformations in the shale. Near the upper surface, the plastic deformation was thought to be a result of swelling caused by infiltration of surface water.

Because of the large differential rebound and subsequent upward rebound deformation, the Corps of Engineers decided that the original foundation of the Stilling Basin structures should be modified to avoid possible severe damage that could result from continued rebound. The two criteria used for the redesign were: (1) within certain cost limits, the shale in the upper 6 to 12 m of the foundation had to be restrained to reduce further rebound; and (2) the individual structures had to be designed to accommodate any differential rebound that still occurred. The significant design changes for the Stilling Basin structures included the addition of many inclined anchors to depths of 8 and 12 m intended to retard the expansion of the top layers of shale, to "sew" the faults and prevent sliding, thereby eliminating the need for a key structure to stabilize the foundation. The redesigned structure was a compact, integral, very strong block that could resist structural damage from continuous foundaton movement.

Since the completion and filling of the Oahe Dam in 1961, rebound displacements have been controlled and there has been no apparent major structural damage to the Stilling Basin structures.

The recognition of and compensating for any relaxation structures that occur throughout a dam foundation or the abutments can be critical to the success of a major project. This interdependence is described in Chapter 22 (this volume), for the Vaiont Reservoir, Italy.

Highway construction. Unpredicted rebound deformations are often damaging to highway construction. During and after the construction of Interstate 5 through Seattle, Washington, slope instability resulting from horizontal rebound deformations became a major problem (Strazer and others, 1974). As slopes were excavated, movement along deep-seated, nearly horizontal slip surfaces occurred within the Lawton Clay, a proglacial, varved lake deposit. Six areas of major instability were encountered during construction. Failures occurred even though cut slopes and conventionally designed and constructed retaining walls, intended to provide lateral earth support along the uphill side of the freeway, were considered to have adequate safety factors based on laboratory peak-shear strength tests. The lateral rebound deformations causing the slides were apparently effected by high locked-in lateral stresses that resulted from the overconsolidation of the Lawton Clay.

According to Strazer and others (1974), clays of the Lawton Clay Member have an overconsolidation ratio of 15, caused by the removal of 915 m of glacial ice. Based on K_0 (coefficient of earth pressure) tests performed by Sherif (1968) on the clay, there is an estimated horizontal stress double that of the vertical stress "locked in" the clay through the overconsolidation process. Strazer and others (1974) hypothesized that upon release of lateral confinement by excavation, the excessive horizontal-stress condition caused lateral deformation (rebound) resulting in nearly horizontal failures. Nichols and others (1986) also suggested that the possible cause of some nearly horizontal fracture zones in the Pierre Shale of South Dakota was the overconsolidation process and erosional relief.

Based on laboratory shear tests, Palladino (1971) further suggested that slippage occurred along preexisting slip surfaces where decreased residual shear strength had been developed, and therefore, residual shear strengths should be used for design purposes.

Subsurface excavations. Subsurface excavations for engineered works in urban areas generally are relatively shallow and usually within several hundred meters of the surface. Tunnels for transportation systems, water transport, pipelines, and utilities are usually less than 20 m in diameter. Larger, but less common, underground facilities may include parking lots, recreation and storage facilities, air raid shelters, and caverns for military purposes.

The deeper and larger underground excavations, generally away from urban areas, are mines, power-plant installations, caverns and storage repositories for hydrocarbon fuels and waste, or protective construction. For instance, some of the gold mines in the sedimentary and volcanic rocks of South Africa extend to nearly 3,100 m below the surface (McGarr, 1971). In the granitic rocks of Sweden, there are shallow, underground, arched-ceiling caverns, used for hydrocarbon fuel storage, that are at least 30 m high and 20 m wide. The lengths of these caverns are usually longer than 20 m but are theoretically unlimited (Jansson, 1976). These excavations approach the practical limits of most underground openings.

Depending on the rock conditions, excavation-induced rebound displacements in the very deep mines of South Africa, caused by excessively high rock pressures up to 100 MPa (Gay, 1979), stress the rock beyond its elastic limit. Resulting rock failures occur in a continuous nonviolent manner or occasionally as violent, damaging, seismic events (McGarr, 1971) that cause loss of life and affect mine production.

Within U.S. coal mines as deep as several hundred meters, there has been loss of life and production due to rock squeeze that is manifested as bumps, heaves, bucklings, floor synclines, and floor anticlines, all of which are caused by rebound (Fig. 4) (Osterwald and others, 1987).

For the large caverns excavated in Sweden at shallow depths of several hundred meters, dangerous rebound displacements are caused by rock discontinuities rather than by high-magnitude stress fields. In Sweden, where there is an abundance of relatively unfractured crystalline rock, large caverns are easily designed. With proper evaluation of the rock quality, these caverns are generally very safe to excavate and use. However, where the rock quality deteriorates, design is more difficult and the cavern sizes must be smaller; construction and use, therefore, usually are more hazardous.

Because design and construction of underground facilities and mines are commonplace throughout the world, elastic rebound phenomena are well understood and almost always consid-

Figure 4. Floor anticline in thin-bedded siltstone seen in Sunnyside No. 1 Mine, Utah. The crest of the anticline, marked by irregular longitudinal fracture, trends obliquely to the direction of the mine opening. (Photograph by J. C. Witt.)

ered for design of an underground opening. Much more problematic are the unexpected large magnitude rebound displacements due to unexpected rock loading and time-dependent rebound displacements, rock failures, and general degradation of rockmass properties that contribute to unsafe conditions. These phenomena are not well understood and generally not considered until rock response problems occur.

The following selected case histories of underground excavation describe some of these problems; their solutions illustrate what has and/or can be done.

Tunnel excavations–highway construction. The Thorold Tunnel, constructed in Middle Silurian–age rock and overlain by 9 m of Pleistocene-age overburden, is a four-lane, twin tube reinforced concrete structure that passes under the Welland Canal (Bowen and others, 1976). It is located on Ontario 58, south of St. Catherines, Ontario. The tunnel has east and west service buildings that provide ventilation and electrical services as well as dams to retain the Welland Canal waters. The Paleozoic rocks of the area are known to have anomolous rebound displacements that have affected engineered facilities. Measured horizontal-stress levels for depths less than 30 m are commonly more than 1,200 psi (8.2 MPa) and as much as 3,200 psi (22.1 MPa) (Lo and others, 1979). Time-dependent displacements (rock squeeze) have caused significant structural damage to underground facilities similar to the Thorold Tunnel during the past 70 years.

Shortly after the completion of the Thorold Tunnel in 1968, an increasing amount of horizontal fracturing occurred in the walls of that tunnel near the west service building. The fracturing was a result of rock squeeze—loading against the structure created by a time-dependent inelastic rebound of the host rock (Bowen and others, 1976). During 1972, strain measurements in the roof struts gave a maximum concrete stress value of 2,500 psi (17.2 MPa) and a predicted value of 3,275 psi (22.6 MPa) for later in the year. By December 1972, it was determined that the existing stress levels and their rate of increase were becoming unacceptable, requiring that the west service building structure be relieved. In-situ measurements also showed that the west service building was subjected to an excessive north–south loading at the tunnel-roof level. Possible loading sources considered were internal structural restraint, shrinkage, hydrostatic load, swelling of bentonite (used in waterproofing between the tunnel concrete and host-rock wall), and freezing and thermal effects. However, based on in-situ measurements, it was concluded that only displacement of rock outside the service structure could provide the loading required to produce the observed stresses. To relieve this stress, a single relief slot was cut into the rock on the south side of the west service building, thus isolating the structure from the rock. The location of the slot was chosen where cracking and measured in–situ stresses were most pronounced. The slot was filled with a clay–bentonite mixture that was impervious and could easily compress under pressure due to inward rebound displacements. The results of the corrective action begun in 1973 and completed in 1974, were indicated by a slight closing of the horizontal cracks in the tunnel wall as well as by a reduction of stresses that appeared to be at acceptible limits for the tunnel structure (Bowen and others, 1976).

Tunnel excavations–dam construction. The Oahe Dam is located about 8 km northwest of Pierre, South Dakota. During construction, 12.9 km of soft rock tunnels were excavated with tunneling machines. Construction of six-outlet work tunnels was begun in December 1954 and completed in April 1957; the excavation of seven power tunnels was begun in February 1959 and completed in May 1961 (Underwood, 1965). The tunnels were constructed mainly in the DeGrey Member of the Upper Cretaceous Pierre Shale Formation (Johns and others, 1963). Strength values for the shale, derived from several hundred unconfined compression tests, vary widely, from 5 to 182 tons/ft^2 (0.48 to 17.4 MPa). Values of Young's modulus (E) for the shale also vary widely, from 20,000 psi (137.9 MPa) to 140,000 psi (965.5 MPa). Based on these tests on previous deep excavations and rebound experiences with the Pierre Shale, an E of 100,000 psi (689.7 MPa) and a compressive strength of 9 tons/ft^2 (0.86 MPa) were adopted for design purposes (Johns and others, 1963; Knight, 1963).

During Atterberg testing, shale samples containing original moisture were slow to absorb additional moisture. Samples that were oven or air dried and subsequently moistened, readily absorbed water, resulting in a thorough breakup of the colloidal materials (Knight, 1963). During tunnel excavation, a similar phenomenon was observed in the shale as it was exposed to air. The shale decomposed and became jointed, faulted, and developed horizontal parting planes along bentonite seams. To avoid this problem a bituminous sealer was applied within 2 hr after the shale was exposed. This procedure was followed within 48 hr by either placing a mortar liner or protective concrete slab over the

Figure 5. Cable-suspended crossframe supporting shale in tunnel crown at Oahe Dam site, South Dakota.

bituminous sealer. In some places, moisture condensation was allowed to form on the shale surface at a maintained humidity, which provided reliable surface protection up to 100 days, before the emplacement of a concrete tunnel lining (Johns and others, 1963).

Overbreak and fallout occurred at intersections with fault surfaces, closely spaced bentonite beds, and joints above the tunnel springline and crown. This slowed, and in some cases, stopped machine tunneling. To prevent fallout, holes were drilled from the ground surface through the tunnel roof; heavy steel cables lowered through these holes and fastened to roof-bearing plates were stressed by hydraulic jacks and then tied off at the ground surface (Fig. 5). Sometimes this method failed; a resulting void, due to fallout, had the potential of growing to the surface. In order to prevent the void from reaching the surface, concrete and sanded grout were poured through holes bored from the surface into the void (Underwood, 1965).

The construction of the Oahe tunnels proves that with careful planning, laboratory tests of samples, accurate geologic maps, and knowledge of past construction in soft or weak rock, tunneling can be accomplished through ground such as the Pierre Shale, desepite the difficulty of numerous rebound and slabbing problems, concentric cracks, and "squeezing ground". Since the tunnels came into full operation in 1965, there have been very few problems.

Tunnel excavations–mining and power use. Broch and Nilsen (1979) discussed stress-related problems found in an exploration tunnel excavated in 1976 adjacent to an open-pit iron mine at Orfjellin Rana in the central part of the Caledonian Mountain range, Norway. The host rocks are Cambrian and Silurian in age and contain intensely folded marble and mica schists. The schistosity of the rock strikes approximately parallel to the long axis of the pit. Three-dimensional stress measurements were conducted in 1977 at two different locations in the 20-m^2 exploration tunnel. The first measurement was taken in a marble having pronounced foliation that dips about 30° northwest. The second measurement was located in an intensely folded mica schist with an average schistosity dip of 30° north. The overburden at both measurement sites was about 250 m thick. Measurements at both sites showed that the major principal stress (compressive) was nearly horizontal with a nearly east–west direction. The calculated vertical overburden stresses were found to be similar to the measured vertical stresses of 687.5 T/M^2 (6.1 MPa). The stress ratios, σ_3 (vertical), σ_2, and σ_1, for the sites were approximately 1:2:2.5 (site 1) and 1:2:4 (site 2). The stress variations were probably due to differences in stiffness caused by rock type and foliation direction. Brock and Nilsen (1979) noted that for the Ortfjell area, stress ratios indicate that the rock masses are subjected to high tectonic stresses. Orientation of the flat-lying major principal stress is approximately parallel to the strike of rock and the long axis of the pit.

These stresses produced rockburst problems when the tunnel excavation reached 400 m from the tunnel entrance under an overburden of about 150 m. The rockburst intensity varied from less violent loosening of blocks of rock and flakes to more violent, noisy rockbursts. Variations of the rockmass properties are believed to be the reason for the intensity variations observed in the rockbursts. Moreover, in the Ortfjell area, rock bursting was not observed in the nearby granite despite the presence of high tectonic stresses. Apparently, the major principal stress is oriented along the direction of the rock schistosity and, coupled with an increase of overburden, contributed to rockbursts and spalling in the tunnel. Various patterns of rockbolting were employed to stabilize this failure. For instance, where the tunnel paralleled the major principal stress, which in turn paralleled the schistosity of the rocks, rockbursting activity was found to be relatively high, and more extensive rockbolting was necessary.

Carlsson and Olsson (1982) presented a similar study that was conducted in tunnels excavated under 5 to 15 m of gneissic granite cover and located in the Forsmark area of southern and central Sweden. In-situ triaxial-stress measurements in the superficial rock mass gave values greater than 20 MPa for the principal stresses in the horizontal plane. These stresses measurements show a clear correspondence of the highest compressive stress directions with rock foliation, and vertical- and horizontal-joint sets. The three-dimensional stress tensor suggests that the horizontal fractures are subjected to the smallest closure pressure, which is the vertical stress. As a result of the stresses in the rock mass, the horizontal joint set has a noticeably higher aperture value than the vertical sets. Subsequent rockbursting observed within the tunnels is believed to be a propagation of existing small fractures, not breakage in the rock matrix. Carlsson and Olsson (1982) suggest that the stresses observed in the Forsmark area are uniaxial in the horizontal plane.

EFFECTS OF GEOLOGIC CONDITIONS

Rock stresses

Rock stresses are the primary source of stored strain energy that causes rebound deformations. Rock stresses are introduced into rocks through gravity, thermal, and tectonic loading. In addition, residual stresses (Voight, 1966) of large magnitude can be imparted to and stored in rock masses. However, very near the surface, these residual stresses are often relieved by natural processes. Because the stress field profoundly influences the long-term integrity of a rock mass in all rock types, it is a significant factor in construction practice. Knowledge of the stress field and rock displacements caused by its relief are essential criteria for design. Near-surface stress fields in rock masses are usually quite unpredictable because they are dramatically affected by the character and geometry of the rock mass and by the external influences. In many rock masses, thermal stresses and the effect of topography on gravity stresses can significantly affect the stress field at shallow depths (Swolfs and Savage, 1984, 1985). In addition, residual stress can superpose anomalous effects on both shallow and deep stress fields (White and others, 1973; Gay, 1979; Nichols, 1986). Because of these variations, stress measurements are necessary to evaluate rock-mass stresses.

Terrain

Geologic terrains significantly affect rebound deformations. The properties of different rock types—especially strength and stiffness—affect stress magnitudes, rate of relaxation, and associated time-dependent rebound displacements. The softer, weaker sedimentary rocks generally are less highly stressed than crystalline shield rocks (Nichols, 1980). Also, weathering and erosion rates of the soft sedimentary shelf rocks are much faster than those of the more resistant shield rocks. Stored strain energy and external energy sources drive inelastic time-dependent displacements at a faster rate in the soft rocks than in the hard rocks. Some of the more dense sedimentary rocks such as crystalline limestones and dolomites, similar to crystalline shield rocks, contain high levels of stored strain energy very near the surface (White and others, 1973). Rapid excavation in these rocks gives rise to time-dependent deformations and to brittle failure as well as to low-magnitude earthquakes (Pomeroy and others, 1975).

Rebounding rock masses in all types of terrain are ultimately being driven to brittle or ductile failure through elastic and inelastic displacements. The failure process seems to depend on the brittle or ductile strength of the rock mass, the magnitude and orientation of the stress field with respect to excvation geometry, the rate of removal of constraining loads, and the rate at which external energy sources can be introduced into the rock mass.

Lithology

Lithology is a very important factor in the rebound process, because ultimately, it is responsible for the mechanical behavior of a rock mass. Especially important are the mineral constituents, grain orientation and size, intergranular bonding, and porosity of the rocks (Mitchell, 1976); all play important roles in the elastic and inelastic rebound process. The strengths of the mineral grains, bonding cements, grain size, and amount of porosity and pore fluids determine the compressibility and brittle strength of rocks (Farmer and Attewell, 1973). The mineral alignments control strength anisotropy. The long-term or ductile strength of rocks is controlled by the rate of stable micro-crack growth and interactions that are in turn controlled by the rate of granular or intergranular inelastic displacements (Costin and Mecholsky, 1983; Dey and Wang, 1981). Thus, the long-term strengths of the grains and intergranular bonds are important. Porosity and the contained pore fluids enhance long-term failures. In addition, the intergranular and intragranular slip mechanisms of many rock fabrics are the mechanisms that enable the near-surface rocks to store large amounts of strain energy (Carter, 1969).

Chemistry

Chemical composition of the rebounding rock, including pore water chemistry, and reactions with external elements are very important to the physical changes of the rock (Russell and Parker, 1979), especially during the later stages of rebound. Where groundwater is free to circulate in permeable rocks, oxidation and ion exchange are taking place during the early stages of rebound. In hydraulically tight rocks, however, little chemical activity is present during the initial elastic rebound response. For both of these hydraulic conditions, as the inelastic displacments of time-dependent rock rebound increase, meteoric water, air, and changing thermal gradients permeate the rock mass from the surface downward. Thus, oxidation, ion exchange activity, and solutioning become very active, further weakening (Rehbinder and Lichtman, 1957) and disintegrating rock fabric bonds. In some cases, the tight rock will act as a semipermeable membrane, creating pore-pressure gradients that will affect the mechanical behavior of the rock (Hanshaw, 1972). This also greatly enhances the inelastic displacements and deterioration of the rebounding rock mass until the internal energy reaches zero.

Hydrology

The hydrology of a rebounding rock mass is of primary importance to the rebound process because it provides transport for chemical agents that react with the mineral constituents of the rock mass. Without exchange of internal fluids, chemical reactions in most cases would be retarded. Hydrology also provides a flushing action through aquifers to expel connate waters and maintain pore pressures that react mechanically with rock pressures in such a way as to enhance rock failures (Handin and others, 1963).

The mechanical behavior of rocks is very much dependent on pore pressures, which in many rocks are unknown. For instance, in hydraulically tight rocks, pore pressures are difficult to measure; therefore, the mechanical rock behavior is difficult to determine (Neuzil and Pollock, 1983). In addition, pore-water flow caused by the removal of overburden loads from above saturated rocks with low permeabilities may be a very important factor controlling the rate of rebound displacements and failure. Vaughan and Wahlbancke (1973) demonstrated that near-surface equilibration of pore pressures, caused by artificial excavation, occurred at a rate consistent with Terzaghi's (1950) consolidation theory. They implied that the ground-water flow induced by overburden removal and resulting porosity increase is a reversal of the elastic consolidation theory. Also, they observed that time-dependent failure of slopes is a function of the reduced pore pressures caused by equilibration rather than an intrinsic reduction of rock strength. Unfortunately, changing pore pressures in deforming rock specimens are next to impossible to determine in the laboratory with the present technology. An understanding of hydrologic influences on the rebound process in tight rocks is difficult to assess. Similarly, it is difficult to assess the effect of pore fluid on rebound behavior of rocks with low porosity and low permeability such as relatively unfractured intrusive rocks. In rocks less tight, knowledge of the rate of flow or hydraulic conductivity is necessary to understand mechanical rebound processes.

Rate of erosion

The rate of erosion influences the rate at which rebound processes of rock deterioration occur. The rate of overburden removal controls the rate of rock mass exposure and subsequent mechanical and chemical weathering. In similar geologic terrains, the zones of mechanical fracturing and chemical weathering appear to be thicker and more stable at locations of low erosion rates compared to locations of higher erosion rates (Nichols and others, 1986). This implies that mechanical and chemical activity are progressing at faster rates in the more actively eroding areas.

OUTLOOK

Where we are headed

Civil and mining engineers as well as applied geological scientists working in the surface and underground excavation and construction industry have been making significant progress toward understanding rock rebound. As the capability for in-situ measurement becomes more sophisticated, the amount and accuracy of observations will increase, and the data base will become more adequate for design requirements. In addition, more analytical and numerical models are being developed that will give a better understanding of the physical behavior of rock masses during the rebound process. Excellent laboratory and field data are being obtained describing the physical and chemical processes active within rebounding rock masses and their constituents. Most of the data being generated are assimilated by engineers and geological scientists and used either for model verification, design, or more commonly, as a basis for exploration programs to gather future geoscience and design data.

Needed research

Continued research is needed in the development of in-situ measuring techniques and instrumentation. Some of the poorly understood aspects of rock rebound displacements are the in-situ mechanical and elastic rock properties and their variations as a function of time, and overburden removal and applied thermal graidents. In-situ conditions of pore pressure and the state of stress, especially in hydraulically tight rocks, are often misunderstood because of inadequate measuring techniques or devices. Though analytical and numerical models are available for some of these areas of inquiry, more reliable field data is necessary to validate the models. More research is needed to understand localized near-surface stress fields such as those related to rock structures, thermal gradients, anomalous gravity loads, and geologic history. Strain energy generated by these stress fields can cause destructive displacements during shallow excavations. With the continuing sophistication and advances of scientific and engineering technologies, our modeling efforts and data collection should be significantly expanded, thereby allowing new and existing rebound theories to be developed, refined, and used. Not only will these be useful to the construction and engineering industry, but also to geological scientists studying orogenic processes taking place in the earth's crust. Most of the measurements and data obtained for the study of rebound processes also have been and will be used to investigate crustal rock deformations and their origin. For instance, stress measurements and modeling techniques will be especially useful for determining far field stresses generated by oceanic plates (Savage and Swolfs, 1986). Measurements of pore pressures needed for the interpretation of fault mechanics (Byerlee and Brace, 1972) and microcrack studies (Dey and Wang, 1981) will be necessary to understand the earth physics and environment of rock failure in crustal rocks.

REFERENCES CITED

Block, J. W., Clement, R. C., Lew, L. R., and deBoer, J., 1979, Recent thrust faulting in southeastern Connecticut: Geology, v. 9, p. 79–82.

Bowen, C.F.P., Hewson, F. I., MacDonald, D. H., and Tanner, R. G., 1976, Rock squeeze at Thorold Tunnel: Canadian Geotechnical Journal, v. 13, p. 111–126.

Broch, E., and Nilsen, B., 1979, Comparison of calculated, measured and observed stresses at the Ortfjell open pit (Norway), *in* Proceedings of the 4th Congress of the International Society for Rock Mechanics: Rotterdam, A. A. Balkeman, v. 2, p. 49–56.

Byerlee, J. D., and Brace, W. F., 1972, Fault stability and pore pressure: Seismological Society of America Bulletin, v. 62, no. 2, p. 657–660.

Carlsson, A., and Olsson, T., 1982, Rock bursting phenomena in a superficial rock mass in southern central Sweden: Rock Mechanics, v. 15, no. 2, p. 99–110.

Carter, N. L., 1969, Flow of rock-forming crystals and aggregates, *in* Reicker, R. E., ed., Rock mechanics seminar: Bedford, Massachusetts, Air Force Cambridge Research Laboratory, v. 2, p. 509–594.

Collins, D. S., and Nichols, T. C., Jr., 1987, Comparing geotechnical to geologic estimates for past overburden in the Pierre–Hays, South Dakota area; An argument for in-situ pressure meter determination: Mountain Geologist, v. 24, no. 2, p. 51–54.

Costin, L. S., and Mecholsky, J. J., 1983, Time dependent crack growth and failure in brittle rock, *in* Mathewson, C. C., ed., Rock Mechanics; Theory-Experiment-Practice: College Station, Texas, Proceedings of the 24th U.S. Symposium on rock mechanics, p. 385–394.

Dale, T. N., 1923, Commercial granites of New England: U.S. Geological Survey Bulletin 738, 488 p.

Dey, T. N., and Wang, Chi-Yuen, 1981, Some mechanisms of microcrack growth and interaction in compressive rock failure [abs.]: International Journal of Rock Mechanics and Mining Sciences and Geomechanics, v. 18, p. 199–209.

Farmer, I. W., and Attewell, P. B., 1973, The effect of particle strength on the compression of crushed aggregate: Rock Mechanics, v. 5, no. 4, p. 237–248.

Ferguson, H. F., 1967, Valley stress release in the Allegheny Plateau: Engineering Geology, v. 4, p. 63–71.

Gay N. C., 1979, The state of stress in a large dyke on E.R.R.M., Boksburg, South Africa [abs.]: International Journal of Rock Mechanics and Mining Sciences and Geomechanics, v. 16, no. 3, p. 179–185.

Gilbert, G. K., 1904, Domes and dome structure of the High Sierra: Geological Society of America Bulletin, v. 15, p. 29–36.

Handin, J., Hager, R. V., Friedman, M., and Feather, J. N., 1963, Experimental deformation of sedimentary rocks under confining pressure-pore pressure tests: American Association of Petroleum Geologists Bulletin, v. 47, no. 5, p. 717–755.

Hanshaw, B. B., 1972, Natural-membrane phenomena and subsurface waste displacement, *in* Underground waste management and environmental implications: American Association of Petroleum Geologists Memoir 18, p. 308–317.

Hofmann, H J., 1966, Deformational structures near Cincinnati, Ohio: Geological Society of America Bulletin, v. 77, p. 533–548.

Holzhausen, G., 1977, Axial and subaxial fracturing of Chelmsford granite in uniaxial compression tests, *in* Wang, Fun-Den, and Clark, G. B., eds., Energy resources and excavation technology: Proceedings, 18th U.S. Symposium on Rock Mechanics, Boulder, Colorado, Johnson Publishing Company, p. 3B7-1 to 3B7-6.

Hooker, V. E., and Johnson, C. F., 1969, Near-surface horizontal stresses including the effects of rock anisotropy: U.S. Bureau of Mines Report of Investigations 7224, 29 p.

Jahns, R. H., 1943, Sheet structure in granites; Its origin and use as a measure of glacial erosion in New England: Journal of Geology, v. 51, p. 71–98.

Jansson, G., 1976, Rock excavation for underground oil storage plants, *in* Swedish underground construction mission to the United States of America, October 4–15, 1976: 15 p.

Johns, E. A., Burnett, R. G., and Craig, C. L., 1963, Oahe Dam; Influence of shale on power structures design: Journal of Soil Mechanics and Foundations Division, Proceedings of the American Society of Civil Engineers, v. 89, no. SM1, p. 95–113.

Kieslinger, A., 1960, Residual stress and relaxation in rocks: 21st International Geological Congress Report, Copenhagen, Denmark, Pt. 18, p. 270–276.

Knight, D. K., 1963, Oahe Dam; Geology, embankment, and cut slopes: Journal of the Soil Mechanics and Foundations Division, Proceedings of the American Society of Civil Engineers, v. 89, no. SM2, p. 99–125.

Koppula, S. D., and Morgenstern, N. R., 1984, Deficient pore pressures in an eroding soil mass: Canadian Geotechnical Journal, v. 21, no. 2, p. 277–288.

Lo, K. Y., Lukaic, B., Yuen, C.M.K., and Hori, M., 1979, In-situ stresses in a rock overhang at the Ontario Power Generating Station, Niagara Falls *in* Proceedings of the 4th Congress of the International Society for Rock Mechanics, v. 2: Rotterdam, A. A. Balkeman, p. 343–352.

Matheson, D. S., and Thomson, J., 1973, Geological implications of valley rebound: Canadian Journal of Earth Sciences, v. 10, p. 961–978.

McGarr, A., 1971, Violent deformation of rock near deep-level, tabular excavations; Seismic events: Bulletin of the Seismological Society of America, v. 61, no. 5, p. 1453–1466.

Mitchell, J. K., 1976, Fundamentals of soil engineering: New York, John Wiley and Sons, 422 p.

Neuzil, C. E., and Pollock, D. W., 1983, Erosional unloading and fluid pressures in hydraulically "tight" rocks: Journal of Geology, v. 91, no. 2, p. 179–193.

Nichols, T. C., Jr., 1980, Rebound, its nature and effect on engineering works: Quarterly Journal of Engineering Geology, v. 13, p. 133–152.

——, 1986, A study of rock stress and engineering geology in quarries of the Barre Granite of Vermont: U.S. Geological Survey Bulletin 1593, 72 p.

Nichols, T. C., Jr., and Abel, J. R., Jr., 1975, Mobilized residual energy; A factor in rock deformation: Association of Engineering Geologists Bulletin, v. 12, no. 3, p. 213–225.

Nichols, T. C., Jr., and Varnes, D. J., 1984, Residual stress, rocks, *in* Finkl, C. W., Jr., ed., The encyclopedia of applied geology: New York, Van Nostrand Reinhold Company, v. 13, p. 461–465.

Nichols, T. C., Jr., Collins, D. S., and Davidson, R. R., 1986, In-situ and laboratory geotechnical tests of the Pierre Shale near Hayes, South Dakota; A characterization of engineering behavior: Canadian Geotechnical Journal, v. 23, p. 1818–194.

Osterwald, F. W., Dunrud, C. R., and Collins, D. S., 1987, Coal mine bumps as related to geologic features in the northern part of the Sunnyside District, Carbon County, Utah: U.S. Geological Survey Professional Paper (in press).

Palladino, D. J., 1971, Slope failures in an over-consolidated clay, Seattle, Washington [Ph.D. thesis]: Urbana-Champaign, University of Illinois, 200 p.

Pomeroy, P. W., Simpson, D. W., and Sbar, M. L., 1975, The Wappingers Falls, New York, earthquake of June 7, 1974, and its aftershocks: Geological Society of America Abstracts with Programs, v. 7, no. 7, p. 1331.

Rehbinder, P. A., and Lichtman, V., 1957, Effects of surface active media on strain and rupture in solids, *in* Proceedings, 2nd International Congress

Surface Activity: New York, Academic Press, v. 3, p. 563–580.
Russell, D. J., and Parker, A., 1979, Geotechnical, mineralogical, and chemical interrelationships in weathering profiles of an overconsolidated clay: Quarterly Journal of Engineering Geology, v. 12, p. 107–116.
Savage, W. Z., and Swolfs, H. S., 1986, Tectonic and gravitational stresses in long symmetric ridges and valleys: Journal of Geophysical Research, v. 91, no. B3, p. 3677–3685.
Sherif, M. Z., 1968, Cylinder pile design on the basis of new soil test procedures: University of Washington, Soil Engineering Research Series N. 2.
Strazer, R. J., Bestwick, L. K., and Wilson, S. D., 1974, Design considerations for deep retained excavations in overconsolidated Seattle clays, *in* Active clays in engineering and construction practices: Association of Engineering Geology Bulletin, v. 11, p. 379–397.
Swolfs, H. S., and Savage, W. Z., 1984, Site characterization studies of a volcanic cap, *in* Dowding, C. H., and Singh, M. M., eds., Rock mechanics in productivity and protection: Proceedings, 25th Symposium on Rock Mechanics, New York, Society of Mining Engineers of the American Institute of Mining, Metallurgical and Petroleum Engineers, p. 370–380.
—— , 1985, Topography, stresses, and stability at Yucca Mountain, Nevada, *in* Ashworth, E., ed., Research and engineering applications in rock masses: Proceedings, 26th U.S. Symposium on Rock Mechanics, Boston, A. A. Balkema, v. 2, p. 1121–1129.
Terzaghi, K., 1950, Mechanisms of landslides, *in* Paige, S., ed., Applications of geology to engineering practice: Geological Society of America, Berkey Volume, p. 83–123.
Twidale, C. R., 1973, On the origin of sheet jointing: Rock Mechanics, v. 5, p. 163–187.
Underwood, L. B., 1965, Machine tunneling of Missouri River Dams: Journal of the Construction Division, Proceedings, American Society of Civil Engineers, v. 91, no. CO 1, p. 1–27.
Underwood, L. B., Thorfinnson, S. T., and Black, W. T., 1964, Rebound in redesign of Oahe Dam hydraulic structures: Journal of Soil Mechanics and Foundations Division, Proceedings of the American Society of Civil Engineers, v. 90, SM2, p. 65–86.
Vaughan, P. R., and Wahlbancke, H. J., 1973, Pore pressure changes and the delayed failure of cutting slopes in overconsolidated clay: Geotechnique, v. 23, no. 4, p. 531–539.
Voight, B., 1966, Restspannungen im Gestein, *in* Proceedings, 1st Congress, International Society of Rock Mechanics, Lisbon: Laboratory National Civil Engineering, v. 2, p. 45–50.
Walter, K. L., 1982, Analysis of in-situ stress release features in southeastern Connecticut [M.S. thesis:] Amherst, University of Massachusetts, 50 p.
White, O. L., Karrow, P. F., and Macdonald, J. R., 1973, Residual stress relief phenomena in southern Ontario, *in* Proceedings, 9th Canadian Rock Mechanics Symposium, Montreal: December 1973, p. 323–348.
Záruba, Q., 1958, Bulged valleys and their importance for foundation of dams: Sixth International Congress on Large Dams, Comptes Rendus, Transactions, v. 4, p. 509–515.

Manuscript Accepted by the Society April 21, 1987

Mourne Mountains near Dublin, Ireland, showing two stages of rebound-relaxation structures in the ancient rock mass. (Photo courtesy of J. E. Richey, 1964.)

Printed in U.S.A.

Geological Society of America
Centennial Special Volume 3
1991

Chapter 14

Permafrost

Troy L. Péwé
Department of Geology, Arizona State University, Tempe, Arizona 85287

INTRODUCTION

The long, cold winters and short, cool summers in the polar regions result in the formation of a layer of frozen ground that does not completely thaw during the year. This perennially frozen ground, known as permafrost, affects many human activities in the Arctic, as well as in the Subarctic and at high altitudes, and causes problems that are not experienced elsewhere.

Permafrost is a naturally occurring material that has a temperature below 0°C continuously for two or more years (Muller, 1943, p. 3). This layer of frozen ground is designated exclusively on the basis of temperature. Part or all of its moisture may be unfrozen, depending upon the chemical composition of the water or depression of the freezing point by capillary forces. For example, permafrost with saline soil moisture, such as that found under the ocean immediately off the arctic shores, might be colder than 0°C for several years but would contain no ice and thus would not be firmly cemented. Most permafrost is consolidated by ice; permafrost with no water, and thus no ice, is termed dry permafrost. The upper surface of permafrost is called the permafrost table. In permafrost areas, the surficial layer of ground that freezes in the winter (seasonally frozen ground) and thaws in summer is called the active layer. The thickness of the active layer under most circumstances depends mainly on the moisture content; it varies from 10 to 20 cm in thickness in wet organic sediments to 2 to 3 m in well-drained gravels. Permafrost is a widespread phenomenon in the northern part of the Northern Hemisphere, underlying an estimated 20 percent of the land surface of the world (Fig. 1).

Although the existance of permafrost was known to the inhabitants of Siberia for centuries, not until 1836 did scientists of the Western world take seriously isolated reports of thick frozen ground existing under northern forest and grasslands. Then, Alexander Theodor von Middendorff measured temperatures to depths of approximately 107 m in permafrost in the Shargin shaft, an unsuccessful well dug for the governor of the Russian-Alaskan Trading Company at Yakutsk, and estimated that the permafrost was 214 m thick (Péwé, 1974).

For the last 100 years, scientists and engineers in the Soviet Union have been pioneers actively studying permafrost and applying results to development of the northern country (Melnikov, 1984). Similarly, prospectors and explorers have been aware of permafrost in the northern part of North America for many years.

In the early part of the twentieth century, a few railroads were built in permafrost terrain in North America, mainly the Alaska Railroad and the railroad to Fort Churchill on the south shore of Hudson Bay, Canada. Although permafrost was and is the predominant and most serious cause of engineering problems that affect the northern part of these railroads, no systematic studies of the perennially frozen ground were made prior to and during construction. In the 1950s, detailed engineering geology studies were made by scientists of the U.S. Geological Survey in critical areas of permafrost trouble along the Alaska Railroad (Péwé, 1949; Wahrhaftig and Black, 1958, Péwé and Paige, 1963; and Fuglestad, 1986).

Prior to World War II, the industry in North America most concerned with problems created by permafrost was placer gold mining in the interior of Alaska and in the Yukon Territory of Canada. It was a small operation until the 1920s; then large-scale dredging for gold was undertaken in the perennially frozen deposits. The U.S. Smelting, Refining and Mining Co. (Boswell, 1979), operating widely throughout central and western Alaska, studied the distribution and engineering characteristics of permafrost in detail in its operations. This organization kept meticulous records and drawings of its experiences with permafrost and how they successfully solved engineering geology problems with frozen ground long before the word *permafrost* was invented and the subject became a national concern (Boswell, 1979; Péwé, 1975b, footnote 1). For example, they were the first company in North America to attempt to delineate the distribution of permafrost using geophysical techniques (Joesting, 1941). Unfortunately, almost all of their records concerning permafrost, collected over 40 years, are proprietary or destroyed (Péwé, 1975a).

During World War II, permafrost became a national and international engineering geology concern in North America with the construction of the Alaska Highway in Canada and Alaska, Army air bases in Alaska and Canada (Barnes, 1946; Wilson, 1948), and the oil pipeline from Norman Wells south to

Péwé, T. L., 1991, Permafrost, in Kiersch, G. A., ed., The heritage of engineering geology; The first hundred years: Boulder, Colorado, Geological Society of America, Centennial Special Volume 3.

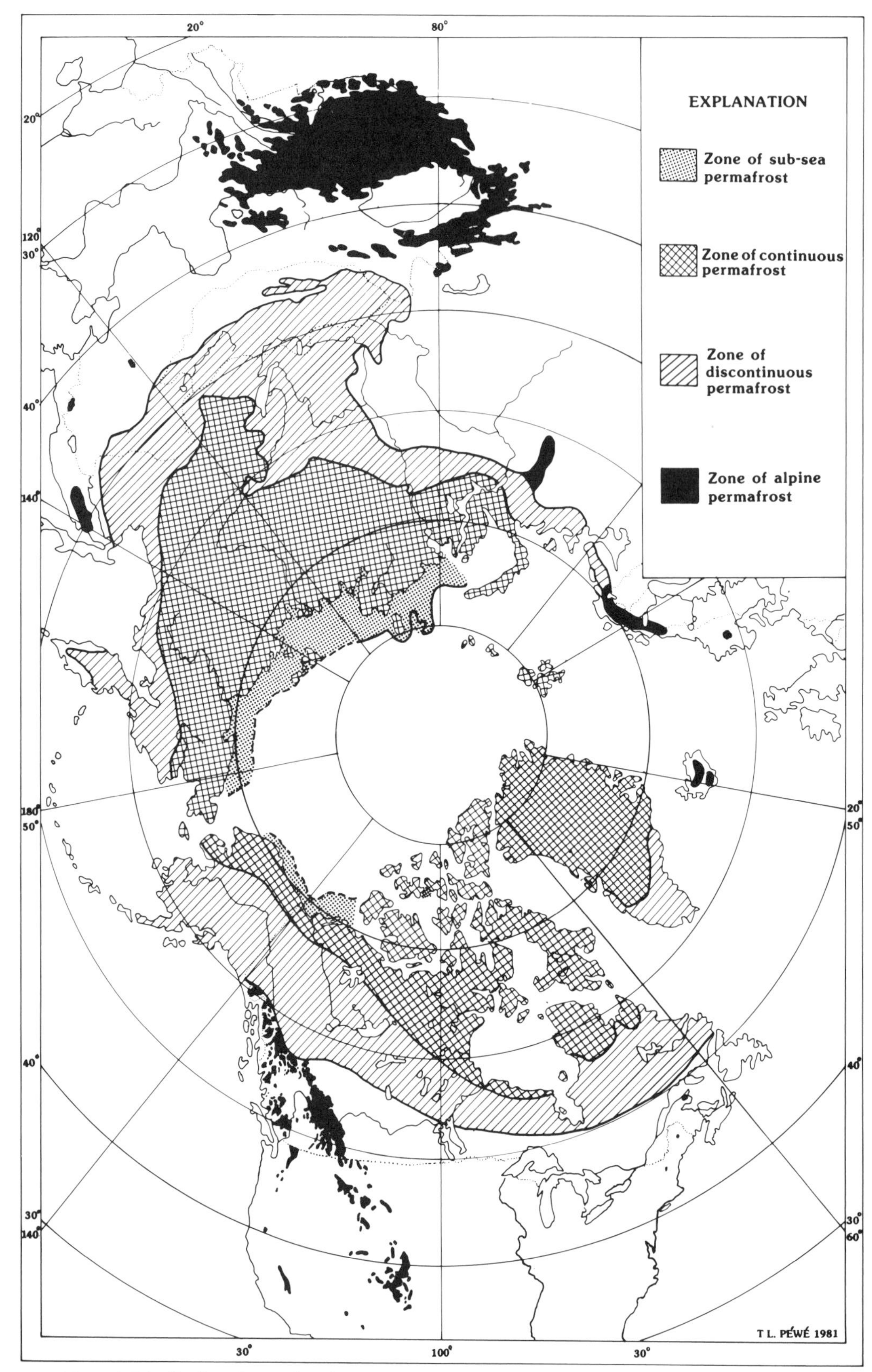

Figure 1. Distribution of permafrost in the Northern Hemisphere. Isolated areas of alpine permafrost not shown on the map exist in the high mountains and outside the map area in Mexico, Hawaii, Japan, and Europe. (From Péwé, 1983.)

the Alaska Highway near Whitehorse in Canada (Hemstock, 1949 a, b). Millions of dollars were lost in trying to overcome engineering problems created by thawing of ice-rich permafrost (Péwé, 1948). The U.S. Smelting, Refining and Mining Co. provided the U.S. Army with data concerning their experiences in engineering on permafrost. Dr. S. W. Muller of the U.S. Geological Survey compiled for the U.S. Army Corps of Engineers the now classic first book in English on permafrost and related engineering problems (1943), which was to remain the textbook on permafrost in North America for the next 20 to 30 years. The word *permafrost,* coined by Muller in 1943, is now in international use.

Immediately after the end of World War II, widespread systematic studies of permafrost and seasonal frost were undertaken by many scientists and engineers of North America, mainly those associated with the U.S. Geological Survey, the Geological Survey of Canada, the U.S. and Canadian military forces, the U.S. National Academy of Sciences–National Research Council, and the National Research Council of Canada (U.S. Army Corps of Engineers, 1946, 1947, 1949, 1951, 1954, 1956; Black, 1950; Péwé, 1948; Hardy and D'Appolonia, 1946; Highway Research Board, 1948; Johnson, 1952). Such studies and many others were the vanguard of those that now permit more intelligent planning and construction in permafrost areas of the world.

With continued expansion of the utilization of areas in the high latitudes and high altitudes, the understanding of permafrost and related engineering problems has indeed become of international concern (Fig. 1). This has given rise to important and constructive international conferences to discuss and publish advances and problems of the subject: United States, 1963; Yakutsk, USSR, 1973; Canada, 1978; Alaska, 1983; and Norway, 1988 (National Academy of Sciences–National Research Council, 1966, 1973, 1978, 1983a, 1984; National Research Council of Canada, 1978; French, 1982). Major textbooks on various aspects of permafrost are now available in North America (for instance, French, 1976; Washburn, 1980; Johnston, 1981). A milestone of the last decade in the science and engineering of permafrost research and applications was the establishment of the International Permafrost Association in 1983. After 10 years of preparation, the four founding countries—Canada, China, the United States, and the USSR—established the association at the Fourth International Permafrost Conference at Fairbanks, Alaska. The association now has Adhering National Bodies from 15 different countries and is concerned with coordination for the Fifth International Conference on Permafrost to be held in Norway in 1988.

Permafrost profoundly affects human activities in the Arctic and Subarctic and requires that conventional engineering construction techniques and design be modified at additional costs (Corte, 1969). Agriculture, mining, water supply, sewage disposal, and construction of all types are seriously affected by subsidence of the ground surface caused by thawing of the perennially frozen ground; furthermore, additional problems are brought on by the associated soil flowage and frost action. Unless planners, engineers, and builders have a thorough understanding of the geological, thermal, and mechanical problems unique to permafrost, impassable roads and railroads, unusable airstrips, and destroyed or abandoned buildings and pipelines may result.

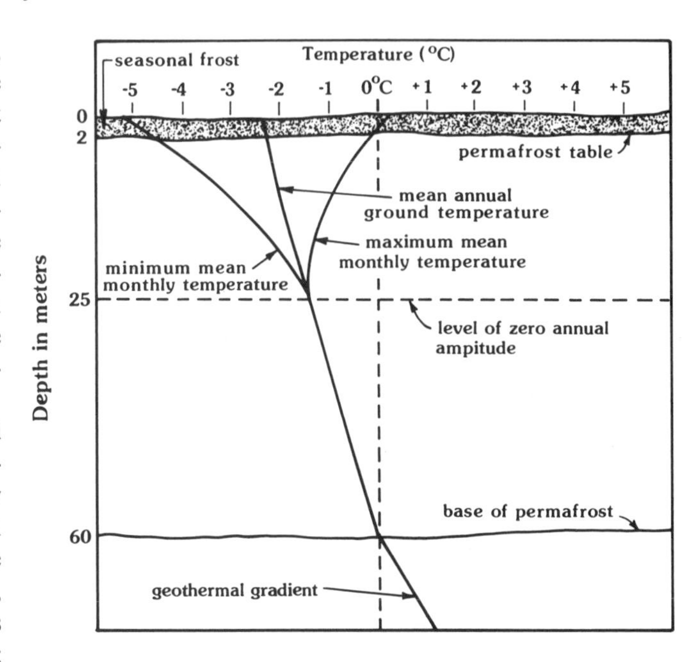

Figure 2. Hypothetical example of a temperature profile and thickness of permafrost in central Alaska. (From Péwé, 1975b.)

ORIGIN AND THERMAL REGIME OF PERMAFROST

In areas where the mean annual air temperature drops below 0°C, some of the ground frozen in the winter will not be completely thawed in the summer; therefore, a layer of permafrost will form and continue to grow downward in small increments from the seasonally frozen ground. The permafrost layer will become thicker each winter; the thickness is controlled by the thermal balance achieved between the heat flowing upward from the Earth's interior and that flowing outward into the atmosphere—a balance that depends upon the mean annual air temperature and geothermal gradient. The average geothermal gradient is about 1°C increase in the temperature of the Earth for every 30 to 60 m of depth. Eventually the thickening permafrost layer reaches an equilibrium depth at which over several years the same amount of geothermal heat reaching the permafrost is lost into the atmosphere. A state of equilibrium takes thousands of years to be reached where permafrost is hundreds of meters thick.

An example of the change of temperature of frozen ground with depth and the upper and lower limit of permafrost is illustrated in Figure 2. The annual fluctuation of air temperature from winter to summer is reflected in a subdued manner in the upper few meters of the ground. This fluctuation diminishes rapidly with depth; it is only a few degrees at 8 m and is barely detectable at 15 m. The level at which the fluctuations are hardly detectable (10 to 15 m) is termed the level of zero amplitude. Below this depth the temperature increases steadily under the influence of geothermal heat. The temperature of permafrost at the depth of minimum annual seasonal change varies from near 0°C at the southern limit of permafrost to −10°C in northern Alaska and −13°C in northeastern Siberia. In the continuous zone, temperature of the permafrost is colder than −5°C. Permafrost is the result of present climate; however, many temperature profiles show that permafrost is not in equilibrium with the present climate at the sites of measurement, and in such areas, much of the permafrost is a product of a colder past climate.

CHARACTERISTICS OF PERMAFROST

Distribution and thickness

Permafrost is essentially a phenomenon of the polar regions. It occurs in half of the Soviet Union and Canada, in 85 percent of Alaska (Ferrians and others, 1969) (Fig. 1), 20 percent of China, and probably all of Antarctica. In the Northern Hemisphere, permafrost is more widespread and extends to greater depths in the northern than in the southern regions. It is 740 m thick in northern Alaska, 1,600 m thick in northern Siberia, and thins progressively toward the south. Permafrost is generally differentiated into two broad zones on land in the polar areas of the Northern Hemisphere: the continuous and the discontinuous. In the continuous zones (Fig. 1), permafrost is nearly everywhere except under the large lakes and rivers that do not freeze to the bottom. The discontinuous zone includes numerous permafrost-free areas that increase progressively in size and number from the north to the south.

The thickness and areal distribution of permafrost are directly affected by natural surface features such as snow and vegetation cover, topography, and bodies of water, in addition to the Earth's interior heat and the temperature of the atmosphere. The most conspicuous change in thickness of permafrost is related to climate. If the mean annual air temperature is the same in two areas, the permafrost will be thicker where the conductivity of the ground is higher and the geothermal gradient is less. Lachenbruch and others (1982) report an interesting example from northern Alaska. The mean annual air temperatures at Cape Simpson and Prudhoe Bay are similar, but permafrost thickness is 305 m at Cape Simpson and about 740 m at Prudhoe Bay because rocks at Prudhoe Bay are more siliceous and therefore have a higher conductivity and a lower geothermal gradient than do the rocks at Cape Simpson (Fig. 3). Lachenbruch (1968) has graphically illustrated that bodies of water—lakes, rivers, and the sea—have a profound effect on the distribution of permafrost (Fig. 4). Inasmuch as south-facing slopes of hills receive more incoming solar energy per unit area than other slopes, they are warmer; permafrost is generally absent on these in the discontinuous zone (Fig. 5) and is thinner in the continuous zone (Péwé, 1982).

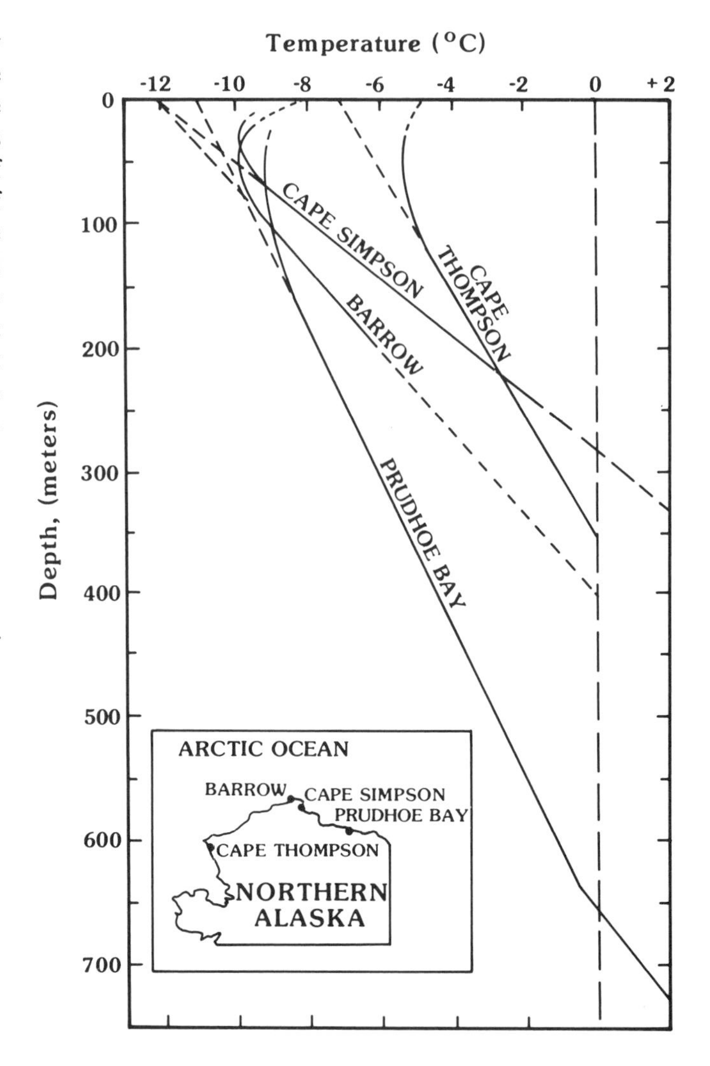

Figure 3. Generalized profiles of measured temperature on the Alaskan Arctic Coast (solid lines). Dashed lines represent extrapolations. (From Lachenbruch and others, 1982.)

The main role of vegetation in permafrost areas is to shield perennially frozen ground from solar energy. Snow cover also influences heat flow between the ground and the atmosphere and therefore affects the distribution of permafrost. Thus, permafrost is not present in areas of the world where great snow thicknesses persist throughout most of of the winter.

Alpine Permafrost. In addition to the widespread perennially frozen ground in the polar areas of the earth, permafrost also exists at high altitudes in the lower latitudes and has been termed alpine permafrost (Fig. 1). Although information about permafrost in the polar areas has been systematically accumulating for many years, data about frozen ground in high plateaus and mountains are sparse (Péwé, 1983).

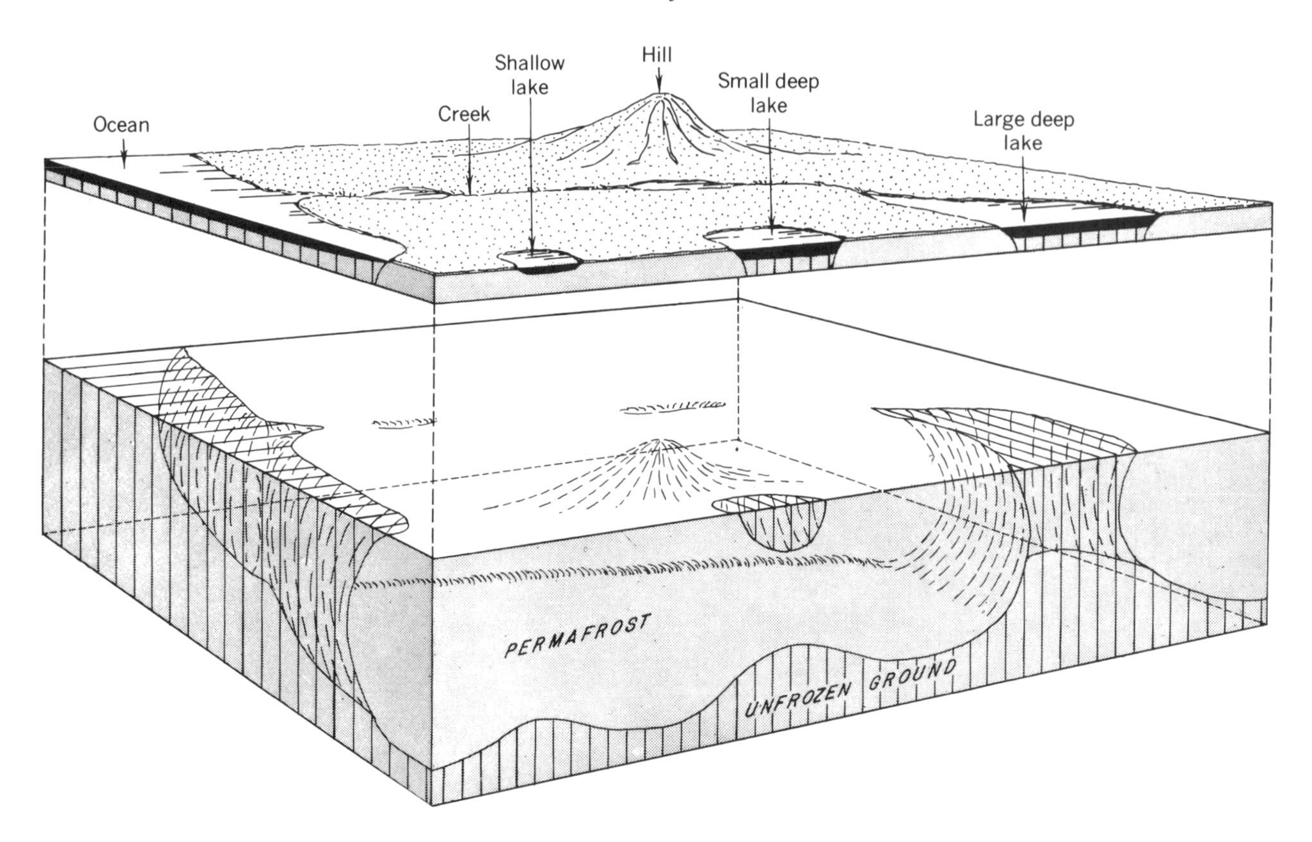

Figure 4. Cutaway block diagram with surface lifted, showing the effect of surface features on the distribution of permafrost in the continuous permafrost zone. (After Lachenbruch, 1968, p. 837).

In the contiguous United States, alpine permafrost is almost entirely limited to the high mountains of the western United States. About 100,000 km^2 is known to exist (Péwé, 1983). Permafrost occurs as low as 2,500-m elevation in the northern states and about 3,500-m elevation in Arizona. The largest area of alpine permafrost in the world is in western China (Fig. 1), where 1,500,000 km^2 of permafrost is known (Péwé, 1981; Tong, 1981; Péwé, 1986).

The best evidence for the existence of perennially frozen ground in alpine areas is temperature measurements taken below the active layer that indicate temperatures of 0°C or colder for two or more years. Such temperature measurements are relatively rare. Active ice-cemented (lobate) rock glaciers (Haeberli, 1985), ice wedges, active cryoplanation terraces (Reger and Péwé, 1976), pingos, ice lenses, and pore ice are all evidence of the presence of perennially frozen ground in alpine areas. Moreover, the distribution of alpine permafrost is affected by the interaction of solar radiation and elevation, snowfall, orientation and slope of land, and vegetation; the thickness, in addition, is affected by the thermal conductivity of the rock (Péwé, 1983; King, 1984).

Subsea Permafrost. One of the most active and exciting areas of current permafrost research concerns the distribution and properties of permafrost under the Arctic Ocean—on the continental shelf—termed subsea or offshore permafrost. The occurrence is unique and has no real analog on land (Sellmann and Hopkins, 1984). Because of the great hydrocarbon resources on the arctic continental shelves, investigations into subsea permafrost have progressed rapidly in the last 25 years (generally much faster than permafrost investigations on land).

Knowledge of the distribution, type, and water or ice content of subsea permafrost is critical for planning petroleum exploration, locating production structures, burying pipelines, and driving tunnels beneath the sea bed. Furthermore, the temperature of the sea bed must be known in order to predict potential sites for accumulation of gas hydrates or areas in which ground water or artesian pressures are likely. In addition, knowledge of the distribution of the subsea permafrost permits a thorough interpretation of the regional geological history and the position of ancient sea levels.

The following scenario suggests that the origin, distribution, and characteristics of subsea permafrost have a rather simple explanation. At the height of the glacial epochs, especially about 20,000 years ago, most of the continental shelf in the Arctic Ocean was exposed to the polar climates for thousands of years;

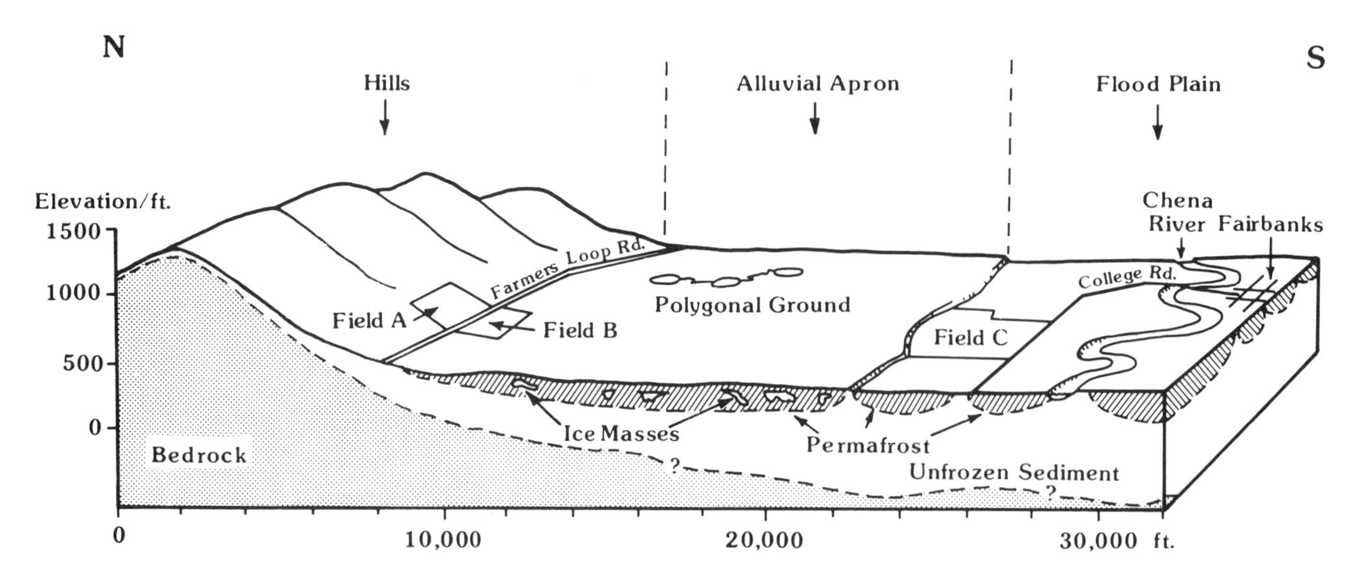

Figure 5. Character and distribution of permafrost in Fairbanks, Alaska area. (From Péwé, 1954.)

inner parts were exposed longer than the outer parts. The climate caused cold permafrost to form to depths of more than 700 m (Hopkins and others, 1977; Hunter and others, 1976). Subsequently, within the last 10,000 years, the level of the Arctic Ocean rose and the sea advanced over a frozen landscape to produce a degrading relict subsea permafrost. The perennially frozen ground is no longer exposed to a cold atmosphere, and the saline water causes a reduction in strength and consequent melting of the ice-rich permafrost bonded with fresh-water ice. The temperature of subsea permafrost, near –1°C, is no longer as cold, and therefore is sensitive to the warming from internal geothermal heat and encroachment activities of man (Lachenbruch and others, 1982; Sellmann and Hopkins, 1984).

Ground ice

The ice content of permafrost is probably the most important feature relevant to human life in the North. Ice in perennially frozen ground exists in various sizes and shapes, with definite distribution characteristics grouped into five main types: pore ice, segretated or Taber ice, foliated or ice-wedge ice, pingo ice, and buried ice.

When investigators consider the subject of ground ice, the question is generally raised as to how much ice actually exists in the ground. Because such information would be interesting and valuable from a historical standpoint as well as essential in solving engineering problems posed by permafrost (Lachenbruch, 1970), the question is being considered. Estimates of the volume of worldwide ground ice range from 0.2 to 0.5 million km^5, less than 1 percent of the total volume of the earth (Shumskiy and Vtyurin, 1966; Shumskiy and others, 1964, p. 433).

On the basis of an examination of ice in the ground in many bore holes near Barrow, Alaska, and the extrapolation of this bore-hole information to the rest of the coastal plain, Brown (1967) estimated that 10 percent by volume of the upper 3.5 m of permafrost of the coastal plain of Alaska is composed of ice wedges (foliated ground ice). Taber ice is the most extensive type of ground ice—in places representing 75 percent of the ground by volume. Brown (1967) calculated that the pore and Taber ice content in the depth between 0.5 and 3.5 m (surface to 0.5 m is seasonally thawed) is 61 percent; between 3.5 and 8.5 m, 41 percent by volume. The total amount of pingo ice is less than 0.1 percent of the permafrost. The total amount of perennial ice in the permafrost of the Arctic Coastal Plain of Alaska is estimated to be 1500 km^3, and below 8.5 m, most of that is present as pore ice.

Ice wedges

The most conspicuous and controversial type of ground ice in permafrost is the large ice wedges or masses characterized by parallel or subparallel foliation structures. Most foliated ice masses occur as wedge-shaped, vertical, or inclined sheets or dikes 1 cm to 3 m wide and 1 to 10 m high where seen in transverse cross section (Fig. 6). Some masses seen on the faces of frozen cliffs may appear as horizontal bodies a few centimeters to 3 m in thickness and 0.5 to 15 m in length. The true shape of these ice wedges can be seen only in three dimensions. Ice wedges are parts of a polygonal network of ice enclosing cells of frozen ground 3 to 30 m or more in diameter. They reflect a polygonal microrelief pattern on the surface of the ground that is a characteristic feature of permafrost terrain (Fig. 7).

Figure 6. Large mass of underground ice (foliated ice wedge), in perennially frozen silt underlying larch forest at Mamontova Gora, Yakutia, Siberia, USSR. The right limit of Aldan River is 310 km upstream from its mouth. (Photograph no. 3430 by T. L. Péwé, July 22, 1973.)

The origin of ice wedges is now generally accepted as being explained by the thermal contraction theory (Leffingwell, 1915, 1919). During the cold winter, polygonal thermal contraction cracks about 1 to 2 cm wide and 2 to 3 m deep form in the frozen ground. In early spring, water from the melting snow runs down these tension cracks and freezes and, with accumulating hoarfrost, produces a vertical vein of ice that penetrates permafrost. When the permafrost warms and reexpands during the following summer, horizontal compression results in the upturning of the frozen sediment by plastic deformation. During the next winter, renewed thermal tension reopens the vertical ice-cemented crack, which may be a zone of weakness. Another increment of ice is added in the spring when melt water enters and freezes. Over the centuries the vertical wedge-shaped mass of ice is produced (Lachenbruch, 1962).

Ice wedges require a more rigorous climate to grow than does permafrost. The mean annual air temperature alone does not control the formation of ice wedges (Péwé, 1966, 1975b); rather, the ground cracks when the temperature of the upper part of the permafrost is perhaps colder than −15°C and a winter "cold snap" occurs, rapidly cooling the permafrost further and causing it to crack. Regions of the world where ice wedges are actively growing over widespread areas have a mean annual air temperature of about −6 to −8°C or colder (northern Alaska, for example). Ice wedges occasionally form in restricted areas (Péwé, 1966) or local cold spots or during colder periods of a few years duration, in regions with a general mean annual temperature listed as slightly warmer than −6°C.

IMPACT ON ENGINEERED WORKS

Development of the polar regions demands understanding and ability to cope with problems of the environment dictated by permafrost. The most dramatic, widespread, and economically important examples of the influence of permafrost on life in the North deal with construction and maintenance of roads, rail-

Figure 7. Petroleum camp at Prudhoe Bay, Alaska, built over raised-edge ice wedge polygons. The roads are on a 1.5-m gravel fill, and the buildings are constructed on piles frozen into permafrost. (Photograph no. 3088 by T. L. Péwé, July 8, 1970.)

roads, airfields, bridges, buildings, dams, sewers, oil and gas pipelines, and communication lines (Fig. 8).

Principles of land use

A thorough study of frozen ground should be included in the planning of any project in the North. Except in cases of very thin permafrost where thawing is a possibility, it is generally best to disturb frozen ground as little as possible to maintain a stable foundation for engineering structures. Construction techniques that preserve permafrost are referred to as the passive method; those that destroy permafrost are termed the active method (Muller, 1943).

Muller (1943, p. 85, 86) stated, "Once the frozen ground problems are understood and correctly evaluated, a successful solution is, for the most part, a matter of common sense, whereby the frost forces are utilized to play the hand of the engineer and not against it." With few exceptions, all building, railroad, and highway construction in permafrost areas has involved installation of a pad or gravel fill before construction (Fig. 9). Dry gravel is generally a better conductor of heat than the underlying silt or vegetation mat. In the cooler parts of the Arctic, it is possible to install a layer of gravel thick enough (about 2 m) to contain seasonal freezing and thawing of the ground. In these instances, there is no thawing of the underlying permafrost, and in some cases the permafrost table moves upward into the pad. However, the deep active layer in subarctic areas precludes containment of seasonal freezing and thawing in a gravel fill. Alternate procedures are necessary, including palcement of insulating materials under fill.

Four fundamental types of permafrost-related, land-use problems are: (1) thawing of ice-rich permafrost with subsequent surface subsidence under unheated structures such as roads, airfields, agricultural fields, and parks; (2) ground subsidence under heated structures; (3) frost action, generally intensified by poor drainage caused by permafrost; and (4) freezing of buried sewer, water, and oil lines.

Ground subsidence

The most ubiquitous and unique geologic hazard of Arctic and Subarctic regions results from the thawing of ice-rich permafrost. This thawing promotes a loss of bearing strength, high moisture content, and subsidence of the ground surface. Melting of large ground-ice masses produces dramatic differential settlement and can result from man's disturbance of the thermal equilibrium of the ground or from climatic change.

Permafrost has adversely affected agricultural development in many parts of the Subarctic by influencing water supply, soil drainage, the stability of roads and buildings, and especially the topography of cultivated land (Péwé, 1954). The destructive effects of permafrost on cultivated fields result from the thawing of large ground-ice masses. Care must be used in selecting areas for cultivation, because thawing permafrost may force abandonment or modification to pasturage only a few years after clearing (Péwé, 1954). Although fields containing thermokarst mounds and pits can be utilized with repeated grading, excess time, money, and soil are expended in the struggle (Péwé, 1982).

Railroads. Serious construction and maintenance problems created by thawing of ice-rich permafrost affect railroads in the Soviet Union, Canada, China, and Alaska. The northern part of the Alaska Railroad has been plagued with construction and maintenance problems dealing with relatively warm permafrost in the discontinuous zone (Péwé, 1949; Péwé and Paige, 1963; Wahrhaftig and Black, 1958; Péwé and Bell, 1975). Differential settlement of the roadbed is extreme when the sensitive permafrost thaws. For example, in the 1960s, the annual maintenance cost of the railroad grade in a 50-km section west of Fairbanks averaged about $200,000. Because of permafrost thawing, an exceptionally wide roadbed was constructed on a thick layer of gravel over frozen ice-rich silt, but it has not prevented the track from settling up to 0.6 m per year, even after 50 years of construction (Péwé, 1982).

The most spectacular engineering geology permafrost problem dealing with railroads in the United States is in the Nenana River Gorge (Healy Canyon) of the Alaska Range in central Alaska (Wahrhaftig and Black, 1958; Fuglestad, 1986). For 10 miles the river flows through a two-story canyon. The outer canyon is a glacial valley with flaring walls that rise 770 m above the 0.7- to 1.2-km-wide canyon floor. At the beginning of the canyon and again on its last 8 km, the river flows in a 150-m-wide postglacial inner gorge that has walls 60 to 90 m high (Fuglestad, 1986). In the gorge are deposits of a glacial lake clay that are perennially frozen and ice rich. Numerous unstable bedrock slopes, and the thawing of frozen lake clays along with the downcutting and eroding of the river, create landslide conditions (Fig. 10) that are exacerbated by railroad construction and maintenance that greatly disturbs the thermal equilibrium of the permafrost.

Railway service has been delayed for several days at a time as huge slides up to 1 km wide and 610 m long occur. The track is continually moved downward and laterally toward the river. Slumping has been as much as 1.2 m/day. The most critical area is the Moody landslide area (Fig. 11).

Figure 8. Differential subsidence of roadbed of Copper River and Northwestern Railroad near Strelna, 120 km northeast of Valdez, Alaska. Thermal equilibrium of the fine-grained sediments underlying the roadbed was disrupted during the construction, and the permafrost started to thaw. Maintenance and use of the railroad was discontinued in 1938. Subsidence, as well as lateral displacement, has continued, however. (Photograph by L. A. Yehle, 1960.)

Roads, highways, and airfields. Such engineered works as roads, highways, and airfields are especially susceptible to deterioration and eventual destruction by thawing of ice-rich permafrost. Tremendous strides have been made in the last 20 years in utilizing detailed permafrost mapping to aid design prior to construction (Esch, 1984). Numerous techniques are being developed to prevent heat flow from the warm atmosphere in the summer into the cold, underlying ground (Fig. 9). These efforts have met with considerable success; however, it has not been economically feasible as yet to insulate great distances, or great areas, by artificial means. One test in central Alaska (Péwé and Reger, 1983) along the Richardson Highway demonstrated that if prethawed and preconsolidated peat were placed beneath the

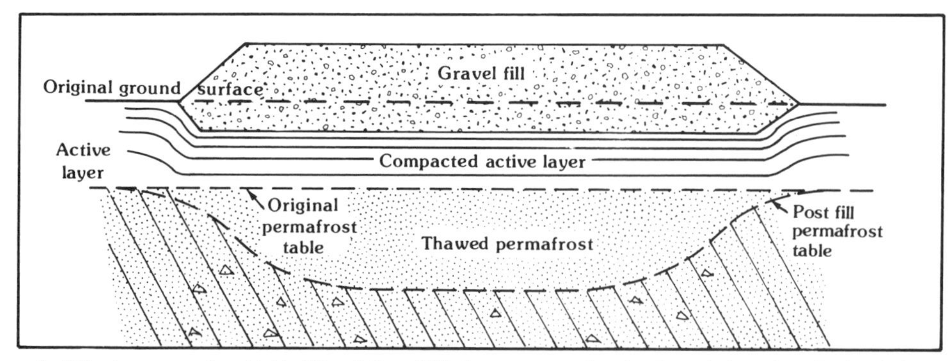

A. Effect on permafrost table if insulating of fill plus compacted active layer is less than the insulating effects of original active layer.

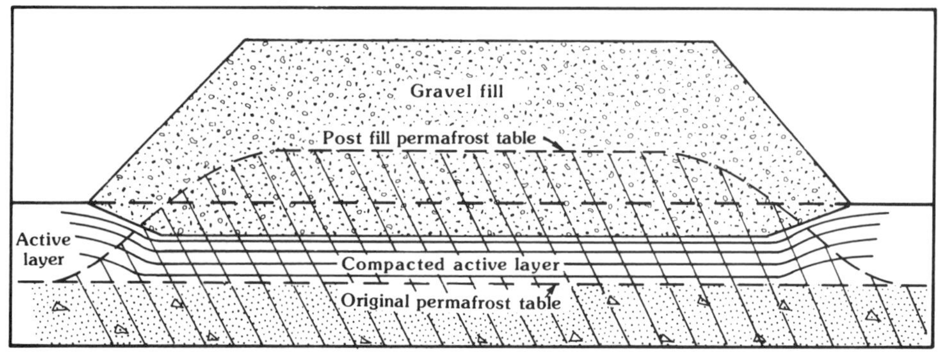

B. Effect on permafrost table if insulating effect of fill and active layer is greater than the insulating effect of the original active layer.

Figure 9. Diagram showing the effect of different thicknesses of gravel fill upon the thermal regime of the ground and the modification of the level of the permafrost table under the fill. (From Ferrians and others, 1969.)

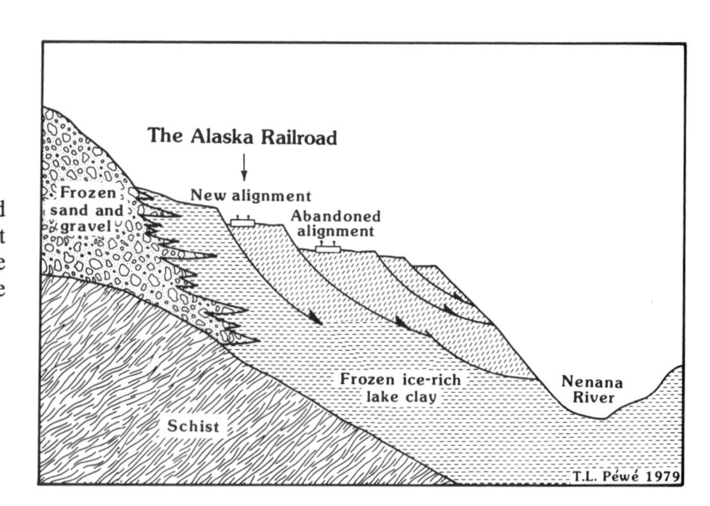

Figure 10. Diagrammatic sketch of landslides along the Alaska Railroad in perennially frozen lake clay. After the vegetation is removed, the heat that is absorbed from the channeled surface-drainage water thaws the lake clay. As the clay thaws, large blocks of sediment slip toward the Nenana River canyon, and the railroad track must be realigned.

Figure 11. View of the most critical section of the Moody landslide area, on the Alaska Railroad. The railroad leaves the west canyon wall at Mile 353.0 on a long curve to the right and traverses thick lake-clay deposits before it reaches the schist bedrock in the immediate foreground, formerly the east wall of the ancient river gorge. In addition to lateral erosion of the Nanana River, much activity within the slide area is attributed to ground water that originates as surface water on the upper slopes. The railroad has tried to divert this water into flumes or vertical standpipes drilled next to the track on the inside ditch line. To prevent the track from sliding downhill, in 1967, timber pilings were driven on the outside shoulder of the track and tied back with cable to a row of pilings ("deadmen") driven on the inside shoulder. Further sliding occurred, and additional "deadmen" were added in 1974. Many "deadmen" driven in 1967 have rotated downhill and are now beneath the track. (Caption from Fuglestad, 1986. Photograph no. 4663, by T. L. Péwé, June 22, 1981.)

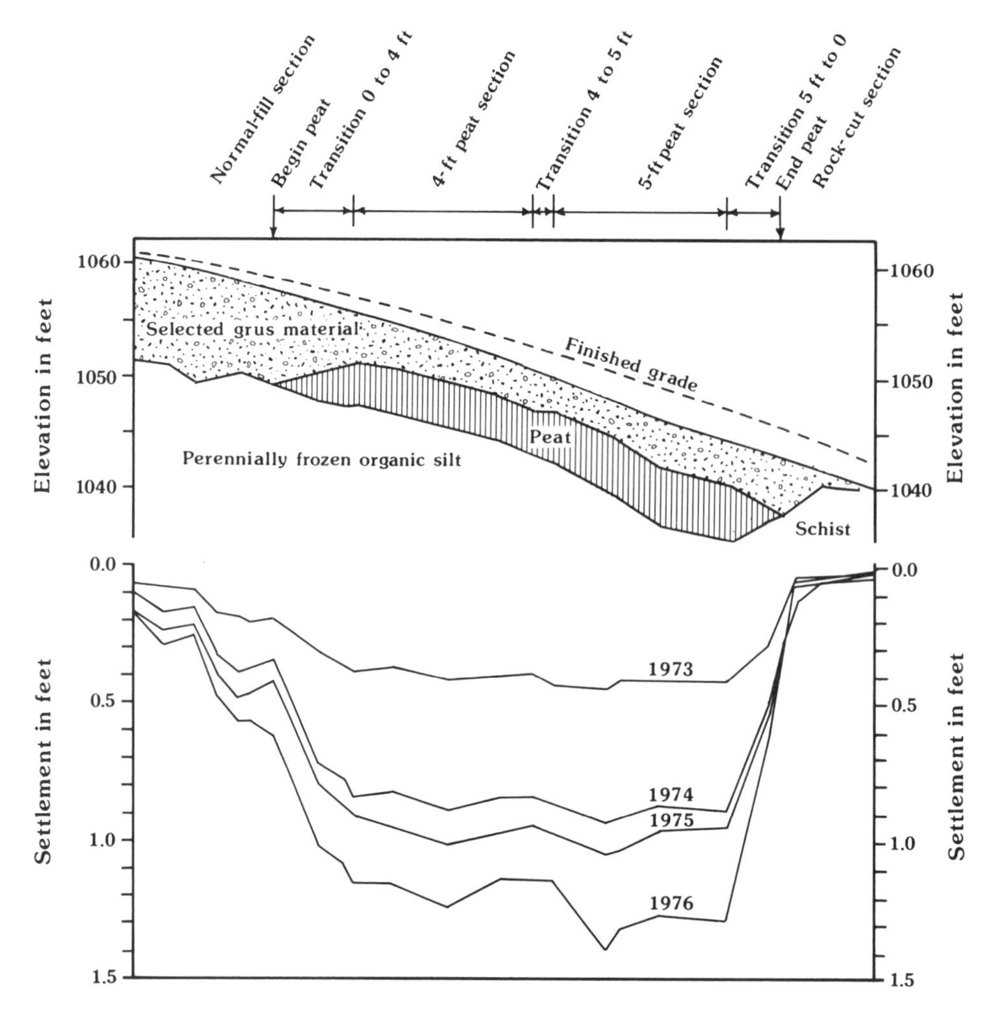

Figure 12. Comparison of longitudinal (centerline) profile of peat test section and amounts of settlement after four years (1973 to 1977), Mile 300.7, Richardson Highway, Alaska (Péwé and Reger, 1983).

road fill, it would delay the thawing of permafrost for some time and require a thinner thickness of road fill (Fig. 12). The test section freezes much sooner and the underlying permafrost chills much colder than other sections.

An example of high-cost road maintenance in ice-rich permafrost environment occurred at the east entrance of the University of Alaska at Fairbanks. In 1962, a road was rerouted across ice-rich peat and organic silt at the east entrance. This 0.25-km rerouting cost $170,000. Shortly after the subgrade was paved, differential settlement occurred because of thawing of the underlying ground ice and ice-rich peat and subsequent compacting of the peat. Releveling of the road shoulders and similar annual maintenance, including leveling with asphalt, has been necessary since 1974.

Another example at the University of Alaska–Fairbanks has been called the Farmers Loop "sinkhole" because of the tremendous amount of thaw-subsidence that has occurred during the past few years. An accumulated pavement thickness of up to 1.8 m has been formed from patches in limited areas (Fig. 13). To further the ongoing investigation, the road was painted white in 1983 (Fig. 13) to promote reflection of solar radiation. Thermal heat tubes were also installed at the west edge of the road in an attempt to freeze the ground beneath the highway and prevent further thawing and subsidence.

Destruction of highways from ice-rich permafrost continues to be a serious problem in the polar areas. The best defense against this natural hazard is preconstruction geological investigations. Subsequently a relocation of roads to nonfrozen areas, or areas of sand and gravel containing scant ground ice, may be advisable. If these alternatives are not practical or economical, the placement of insulation between the gravel overlay and the subgrade may reduce the problem, but special maintenance may be necessary (Fig. 9).

Buildings. Heated buildings introduce more heat into the

Figure 13. Farmers Loop Road at the University of Alaska, Fairbanks. Thawing of underlying ice-rich permafrost caused the tilting of telephone poles and light standards in the distance. The area in the foreground was painted white to reflect solar radiation; it has still subsided more than 2 m in 10 years. Locally, as much as 1.8 m of asphalt has been added to keep the surface level. In 1982, thermal heat pipes (left side of the road) were installed to freeze peat and sediments under the road. (Photograph no. PK 25443 by T. L. Péwé, May 15, 1983).

ground and cause more thawing of ice-rich permafrost in a shorter time than a highway, railroad, cleared field, or other area where the natural surface has been disturbed. Instead of gradual subsidence over a large area, there may be local differential lowering of the ground surface 12 to 25 cm or more a year (Fig. 14).

Such distortion and destruction of structures are generally spectacular and emphasize the hazards of thawing permafrost and its effect on people's pocketbooks. Most people are only vaguely aware of the maintenance or replacement costs of deformed highways and airfields. However, the cost and importance of the ice-rich permafrost hazard are dramatically realized when one observes tilted homes, abandoned cabins, and ill-placed split-level homes (Fig. 15).

It is disheartening to notice the deformation and destruction of homes in certain areas of the world because of improper construction on ice-rich permafrost. Even though detailed information about local permafrost has been available from local and federal agencies since the 1940s—and earlier from some mining companies—buildings continue to be constructed in a manner that will result in deformation, increased maintenance costs, and perhaps abandonment. Obviously, the ideal solution is to construct in permafrost-free areas or on permafrost with a relatively low ice content.

If construction must proceed in areas of ice-rich permafrost, special engineering techniques can be employed to preserve the permafrost intact so that building heat does not enter the ground. Placement on piles is one common solution; another is to provide openings under the building for cold-air circulation and building-heat dispersion (Fig. 16).

Sufficient scientific and engineering expertise is now available for the trouble-free construction of engineered works on ice-rich permafrost. Unfortunately, frequency and avoidance costs of permafrost problems are higher in areas of "warm" sensitive permafrost, such as Fairbanks, Alaska, than in areas where permafrost is colder and less sensitive.

Effects on buried pipelines

General statement. The extraction and transportation of commercial large-scale oil and natural gas in the Arctic have introduced a new set of permafrost-related geological, engineering, and environmental problems. Although natural gas and oil

Figure 14. Bert and Mary's Roadhouse at Mile 275.7 of the Richardson Highway, central Alaska. This large log building was seriously deformed and eventually destroyed by ground subsidence caused by thawing of ice-rich permafrost. Heat from the furnace, located in the front of the building, caused that part of the structure to sink most rapidly. Built in 1949, the building showed little deformation until 1952. By the fall of 1953, the building had subsided 0.7 m below the level of the unheated porch; by 1962, it had subsided 2 m. The building was razed in 1964. See Péwé (1982) for a 10-year series of sequential photographs. (Photograph no. 2072 by T. L. Péwé, May 29, 1962.)

have been extracted in the Arctic for many years, production and transportation have been relatively limited. Small, temporary pipelines for crude oil and refined products were built in northwestern Canada (Norman Wells) and Alaska (along the Alaska Highway) during World War II and shortly thereafter.

The most favored means of transporting petroleum products is by large-diameter, above- or below-ground pipeline. Natural gas may assume the temperature of its surroundings, or it may even be chilled and thus cause little thawing of the permafrost; nevertheless, the burial of a pipeline, with the stripping of surface vegetation, causes a disturbance of the thermal equilibrium of the frozen ground. Oil pipelines can cause a major disruption of the thermal equilibrium, because the flowing oil is warmer than permafrost. The basic problems involved with pipeline construction in permafrost regions are the thawing of permafrost if pipe is buried, and frost action of the supporting piles if built above ground.

A few short- to long-distance natural-gas pipelines exist in permafrost regions of world. These pioneer lines are mostly above ground, even though the cold gas would not significantly thaw the permafrost. More importantly, the difficulty of excavation in permafrost and the resulting disturbance of the thermal equilibrium have favored above-ground construction. Such pipelines are exposed and subject to expansion and contraction, weathering, and damage by humans and natural elements, especially wind. Also, the above-ground line serves as a barrier for humans and animals and must be locally buried or elevated.

Major natural-gas pipelines in the north are known only in the Soviet Union. A 72-cm-diameter unshielded line from Messoyakha to Norilsk for 300 km is about 1 m or more above ground and supported either by logs lying on the ground or bents supported by piles of either wood, concrete, or steel (Fig. 17). Locally the line is about 3 to 4 m above the ground on longer piles to create artificial underpasses. The first line was built during 1967 and 1968, and the new line, incorporating more advanced designs, during 1971 through 1973. It is reported that wind caused destruction of a part of the older lines.

A gas pipeline about 30 cm in diameter is buried 1 m in

Figure 15. Tilted, modern log cabin with full, heated concrete basement built on ice-rich permafrost (Pewe, 1954). The location is 3 mi north of Fairbanks near the junction of McGrath and Farmers Loop Roads. Heat from the 3- to 4-year-old cabin is thawing the permafrost; the greatest subsidence (about 1 m) is near the furnace and fireplace. The foundation is severely cracked, and distortion of the house has broken windows and drain pipes and jammed doors and windows. Recent repairs have entailed breaking up the basement slab, leveling the house with jacks, back-filling with gravel, and repouring the basement slab and wall foundation. Because thawing is not complete, additional settling is expected. (Photograph no. 4001 by T. L. Péwé, July 14, 1977.)

Figure 16. Modern construction over permafrost at Yakutsk, Siberia, USSR. Apartment houses and other buildings are being built on concrete piles frozen into the underlying ground. Air space under the buildings prevents heat from the structures from going into and causing thawing of the underlying perennially frozen ground. (Photograph no. PK 17180, by T. L. Péwé, July 17, 1973.)

Figure 17. Above-ground 72-cm-diameter (28.3 in) gas pipeline near Norilsk, Siberia, built in the period 1971 to 1973. (Photograph by L. A. Viereck, U.S. Forest Service, 1974.)

loess at a point 4 km west of Yakutsk, Siberia. Where this line crosses low terraces of the Lena River near Yakutsk, it divides into two smaller lines, and they are elevated on wood piles to form an underpass. These piles, and those on the Norilsk line, are subject to extreme frost heaving and differential jacking that deform the pipelines and destroy the continuity of supports (Péwé, 1976, fig. 13).

Crude oil transported in pipelines is warmer than the temperature of permafrost. The consequence—that initial heat of the oil plus frictional heating of the pipe will thaw the surrounding perennially frozen ground is the basic problem of oil pipeline construction in permafrost regions. The rate and amount of thaw are dictated by the temperature of the oil and permafrost, if other parameters are stable.

Trans-Alaska Pipeline System. The long-anticipated, vast petroleum potential of northern Alaska was in part realized with the discovery of oil at Prudhoe Bay in 1968. The 9.6 billion barrels of proven reserves suggested that one-third of the oil reserves of the United States in the 1970s were in northern Alaska. The warm crude oil was transported from the North Slope to an ice-free port at Valdez (Fig. 18) by construction of a 1.2-m-diameter pipeline that was 1,285 km long. The Trans-Alaska Pipeline System (TAPS), a remarkable construction

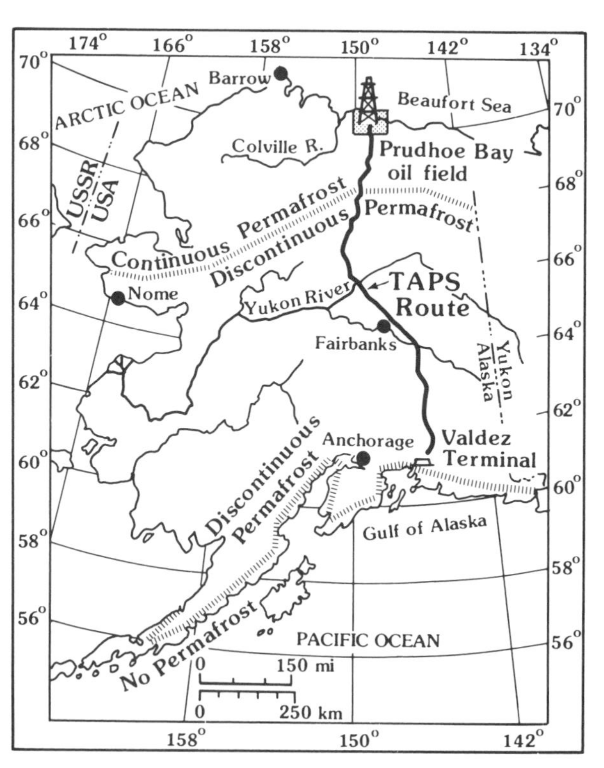

Figure 18. Index map of Alaska indicating the route of the Trans-Alaska Pipeline system (TAPS). (From Péwé, 1982.)

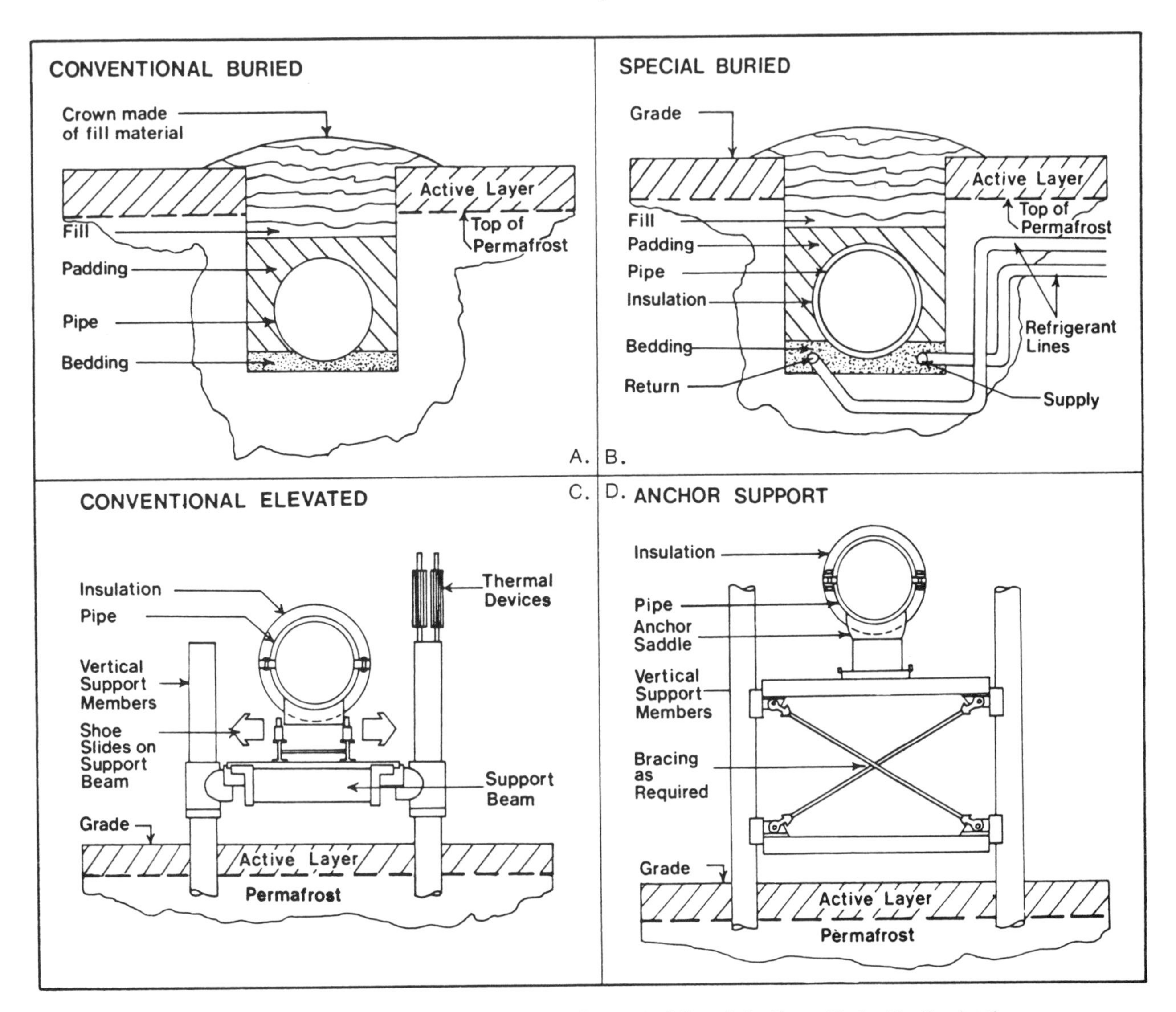

Figure 19. Different construction modes used for the building of the Trans-Alaska Pipeline by the Alayeska Pipeline Service Company. A, conventional buried. B, special buried. C, conventional elevated. D, Anchor support. (Diagram courtesy Aleyeska Pipeline Service Company.)

achievement in a permafrost environment, was completed in 1977. The pipeline is designed to transport as many as 2 million barrels per day. The total cost of constructing TAPS was about \$8 billion. Of this, some \$1 billion was spent to investigate and understand the characteristics of perennially and seasonally frozen ground, demonstrating the impact frozen ground had on the cost of construction.

The pipeline was originally designed for burial in permafrost along most of the route. With the oil temperature (at full production) estimated at 70 to 80°C, such an installation would have thawed the adjacent permafrost. Thawing of the widespread ice-rich permafrost by a warm-oil line can cause liquefaction, loss of bearing strength, and soil flow. Differential settlement of the line can occur, and mudflows of thawed soil may form on slopes. The greatest differential settlement could occur in areas of ice wedges, where troughs could form and deflect surface water into the trenches, causing erosion and more thawing.

The pipeline was built in three construction modes, depending on the environment, terrain, and permafrost conditions (Fig. 19). Oil is pumped through the pipe at temperatures up to 63°C, depending on production rates and heat generated by pumping and friction with the pipe. Consequently, the potential effects of the heat on the specific frozen ground along the route determined the mode of pipeline installation. For example, in areas where the ice content is very low or absent or where no permafrost exists (some 658 km), the pipe is buried in the conventional manner, as it would be in most areas of the world (Fig. 19A). In contrast, seven short sections of the pipeline (total

11.2 km) were buried and then frozen into the ground (Fig. 19B). These sections can provide crossings for caribou and other animals and are located in both ice-poor and ice-rich permafrost environments. The temperature of the permafrost in which the pipe is buried is maintained by pumping refrigerated brine through pipes emplaced beneath the pipeline.

About half the pipeline (615 km) is elevated above ground with anchor support because of the presence of ice-rich permafrost (Fig. 19D). Although an above-ground pipeline successfully discharges its heat into the air and does not directly affect the underlying permafrost, other problems caused by permafrost and associated phenomena must be considered. For example, to eliminate frost heaving of the 120,000 vertical-support members (VSM) (Fig. 19C, D), each VSM is frozen firmly into the permafrost using a special thermal device as shown on Figure 19C. To compensate for expansion of the above-ground pipe caused by the warm oil, and contraction caused by extremely cold air temperatures in winter, the line is built in a flexible zigzag configuration (Fig. 20), which converts expansion of the pipe into lateral movement and likewise accommodates pipe motions induced by earthquakes (Fig. 19C). As the line expands or contracts, the pipe slides across the beam; the pipe is anchored on the crossbeams at the end of each zigzag configuration (every 240 to 540 m).

Utility lines. Water, steam, gas, sewer, and other utility lines are commonly buried in the permafrost areas. To prevent freezing, such lines are usually placed in underground boxes, called utilidors, from 0.3 m to more than 2 m wide (Fig. 21). Construction of buried utilidors in ice-rich permafrost generally creates problems similar to those of a heated building, triggering thawing of permafrost and subsidence of the utilidor. This causes pipe breakage and the eventual destruction of the system (Fig. 21).

Dams

In permafrost areas, the design and construction of dams are more complex because of the thawing effect of the impounded water on the perennially frozen ground underlying the reservoir and structures (Johnston, 1981, Sayles, 1987). Thawing may cause differential movement and seepage. Most dams are earth-fill structures in permafrost areas, and two types may be constructed: impervious or semipervious. The former, designed to maintain the foundation and embankment in a frozen condition, is best in the continuous permafrost zone. The latter allows for thawing under the structure.

Only a few small structures have been built on permafrost in North America, and they are, for the most part, in the discontinuous zone (Rice and Simoni, 1966). Several large and small earth-fill dams have been built in the Soviet Union in both the discontinuous and continuous zones (Fig. 1) (Johnston, 1981).

The largest dam built on permafrost (in 1968) is reported by Péwé (1973) to be on the Vilyuy River in Siberia at the Vilyuyskaya hydroelectric power station near the village of Chernysheviskiy. The dam is an earth-fill structure 75 m high and 600 m long and contains about 5,000,000 m^3 of material. It is built on jointed diabase that contains ice in the joints. The underlying bedrock has been thawing at a rate of 4 to 8 m per year; it is grouted as the ice melts.

Figure 20. View (to the south) of the crossing of the Richardson Highway by the Trans-Alaska Pipeline at Mile 243.5. Vertical thermal piles equipped with heat-radiating fins keep the ground around the pipe on either side of the highway frozen. In the background, the elevated pipeline crosses ice-rich permafrost in till. (Photograph no. 4652 by T. L. Péwé, June 29, 1981).

GROUND WATER AND PERMAFROST

Ground-water flow systems in permafrost regions are greatly modified. The frozen ground acts as an impermeable layer that (1) restricts recharge, discharge, and movement of groundwater, (2) also acts as a confining layer, and (3) limits the amount of liquid water that can be stored in unconsolidated sediments and bedrock (Williams, 1970). Ground water may occur above, within, and below permafrost.

Research investigations on ground water in permafrost regions in North America essentially began immediately after World War II, even though some observations were reported much earlier (Tyrell, 1903). Perhaps the first systematic studies were in connection with sanitation engineering by Alter (1950 a, b) in Alaska. Both the U.S. Geological Survey and the Canadian Geological Survey began detailed studies, some of which continue on a reduced scale (Cederstrom, 1952; Hopkins and others, 1955; Cederstrom and Péwé, 1961; Williams, 1965, 1970; Brandon, 1965; and many others).

The earlier International Permafrost Conferences contained

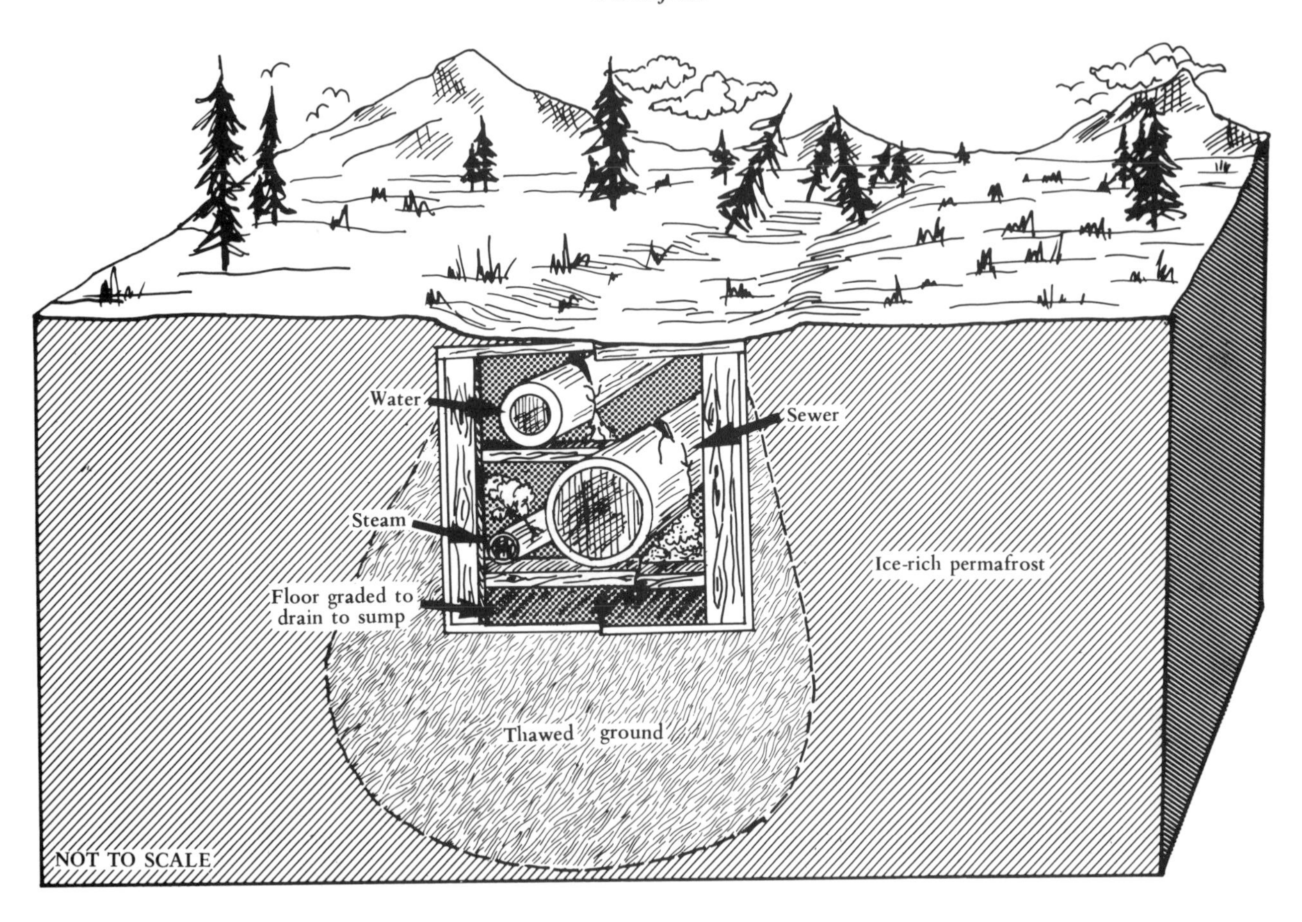

Figure 21. Diagrammatic sketch (exaggerated) of a utilidor buried in ice-rich permafrost. Heat from the utilidor has thawed the ground, and allowed the irregular subsidence of the utilidor that resulted in broken pipes. This is not an uncommon occurrence in permafrost areas, especially where the ice content is not uniformly distributed. (From Péwé, 1982).

entire sections devoted to the subject of the science and engineering of ground-water problems (Brandon, 1966; Williams and Waller, 1966; Williams and Van Everdingen, 1973). Ground-water basin studies were expounded in regard to investigations for pipeline projects and petroleum production installations. Hydrologic basin model studies are continuing in Alaska and Canada (Williams and Van Everdingen, 1973). The most detailed studies concerned with ground-water movement and the origin of pingos (Fig. 22) are those by MacKay (1966, 1968, 1973, 1979, 1986).

OUTLOOK FOR PERMAFROST RESEARCH: AN ASSESSMENT OF FUTURE NEEDS

Preliminary statement. After World War II, permafrost research was well established, but in the 1970s a new stimulus appeared: the discovery of petroleum in northern Alaska. In the 1970s, major advances occurred in the understanding of permafrost (National Academy of Sciences–National Research Council, 1983b), especially the design and structure of the Trans-Alaska oil pipeline across about 1,285 km of permafrost terrain. At that time the National Academy of Sciences–National Research Council spearheaded special studies to outline trends and needs for permafrost research: (1) basic research (1974), (2) research associated with the Trans-Alaska Pipeline (1975), and (3) problems and priorities in offshore permafrost (1976). With the economic impact of permafrost finally becoming better known, it is anticipated that in the next decade scientists and engineers will be called on to solve problems related to onshore and offshore oil and gas explorations in polar areas, transportation across permafrost terrain, and use of alpine permafrost areas for recreation and military facilities. The challenges of sea ice and related phenomena to offshore exploration are discussed in Weeks and Brown (this volume).

Recommendations for permafrost research in the immediate future. The National Academy of Sciences–National

Figure 22. Large ice-cored closed system pingo in loess, 75 km east of Yakutsk, Siberia, USSR. (Photograph no. 3448 by T. L. Péwé, July 27, 1973.)

Research Council (1983b) recognized a serious need to improve our ability to detect and determine the configuration of ground-ice masses by drilling, geophysics, and applied sciences because of the serious engineering- and environment-related problems caused by the degradation of ice-rich permafrost. Furthermore, it recognized the need for a much improved understanding and knowledge of the origin of all types of ground ice.

The report concluded that to advanced engineering technology, it is necessary to: (1) carefully monitor prototype existing facilities constructed in permafrost, and (2) develop improved methods to predict heat and mass transport within permafrost and across its boundaries with adjacent media. Furthermore, research must be continued and expanded to determine the physical, chemical, mechanical, thermal, and electrical properties of permafrost, especially saline permafrost.

Subsea permafrost investigations also remain high on the list of subjects that need increased knowledge. To attain this, both industry and government should continue to collect and analyze data concerning distribution, thickness, and properties of permafrost and also continue refinement of theoretical models of subsea permafrost, critical to predict the depth and thickness of ice-bonded permafrost.

There is now a unique opportunity for scientists and engineers to work together, collecting new data and analyzing and developing a better understanding of processes that affect permafrost and its long-term behavior. There is a clear relationship between the protection of permafrost environments and the long-term performance of manmade structures, which must be considered by designers (National Academy of Sciences–National Research Council, 1983b).

REFERENCES CITED

Alter, A. J., 1950a, Arctic sanitary engineering: Washington, D.C., Federal Housing Administration, 106 p.

—— , 1950b, Water supply in Alaska: American Water Works Association Journal, v. 42, p. 519–532.

Barnes, L. C., 1946, Permafrost; A challenge to engineers: Military Engineering, v. 38, p. 9–11.

Black, R. F., 1950, Permafrost, *in* Trask, P. D., ed., Applied sedimentation: New York, John Wiley and Sons, p. 247–275.

Boswell, J. C., 1979, History of Alaskan operations of U.S. Smelting, Refining, and Mining Company: Fairbanks, University of Alaska, 126 p.

Brandon, L. V., 1965, Groundwater hydrology and water supply in the District of MacKenzie, Yukon Territory, and adjoining parts of British Columbia: Canadian Geological Survey Paper 64-39, 102 p.

—— , 1966, Evidences of groundwater flow in permafrost regions, *in* International Conference on Permafrost, Lafayette, Indiana, 1963, Proceedings: Washington, D.C., National Academy of Sciences, National Research Council Publication 1287, p. 176–177.

Brown, J., 1967, An estimation of the volume of ground ice, coastal plain, northern Alaska: U.S. Army Materiel Command Cold Regions Research and Engineering Laboratory Technical Note, 22 p.

Cederstrom, D. J., 1952, Summary of ground-water development in Alaska, 1950: U.S. Geological Survey Circular 169, 37 p.

Cederstrom, D. J., and Péwé, T. L., 1961, Ground-water data, Fairbanks area, Alaska: Alaska Department of Health and Welfare Hydrological Data Report 9, 28 p.

Corte, A. E., 1969, Geocryology and engineering, *in* Varnes, D. J., and Kiersch, G. A., eds., Reviews in engineering geology II: Boulder, Geological Society of America, p. 119–185.

Esch, D. C., 1984, Design and performance of road and railway embankments on permafrost, *in* Permafrost, Fourth International Conference, Final Proceedings, Fairbanks, Alaska: Washington, D.C., National Academy of Science, p. 25–30.

Ferrians, O. J., Jr., Kachodoorian, R., and Green, G. W., 1969, Permafrost and related engineering problems in Alaska: U.S. Geological Survey Professional Paper 678, 37 p.

French, A. M, 1976, The periglacial environment: London, Longman, 309 p.

—— , ed., 1982, Proceedings of the Fourth Canadian Permafrost Conference; The Roger J. E. Brown Memorial Volume: Ottawa, National Research Council of Canada, 591 p.

Fuglestad, T. C., 1986, The Alaska Railroad between Anchorage and Fairbanks, guidebook to permafrost and engineering problems: The Alaska Division of Geological and Geophysical Surveys, Guidebook no. 6, 82 p.

Haeberli, W., 1985, Creep of mountain permafrost; Internal structure and flow of alpine rock glaciers: Zurich, Mittgeilugen der Versuchsanstalt fur Wasserbau Hydrologie und Glaziologie, no. 77, 142 p.

Hardy, R. M., and D'Appolonia, E., 1946, Permanently frozen ground and foundation design: Engineering Journal of Canada, v. 29, p. 1–11.

Hemstock, R. A., 1949a, Engineering in permafrost in Canada's Mackenzie Valley: National Research Council of Canada Technical Memoir 13, 3 p.

—— , 1949b, Permafrost at Norman Wells; N.W.T.: Calgary, Canada, Imperial Oil Limited, 100 p. (pub. in 1953).

Highway Research Board, 1948, Bibliography on frost action in soils: Washington, D.C., Highway Research Board Bibliography no. 3, 47 p.

Hopkins, D. M., Karlstrom, T.N.V., and others, 1955, Permafrost and groundwater in Alaska: U.S. Geological Survey Professional Paper 214-F, p. 113–146.

Hopkins, D. M., and 13 others, 1977, Earth Science studies, *in* Beaufort Sea synthesis report: National Oceanic and Atmospheric Administration Arctic Special Bulletin 15: Fairbanks, University of Alaska, p. 43–72.

Hunter, J.A.M., Judge, A. S., MacAualy, H. A., Good, R. L., Gagne, R. M., and Burns, R. A., 1976, Permafrost and frozen sub-sea bottom materials in the southern Beaufort Sea: Canada, Department of the Environment, Beaufort Sea Project report 22, 174 p.

Joesting, H. R., 1941, Magnetometer and direct-current resistivity studies in Alaska: American Institute of Mining and Metallurgical Engineers Transactions, v. 164, p. 66–87.

Johnson, A. W., 1952, Frost action in roads and on fields; Review of the literature 1765–1951: Highway Research Board Special Report 1, 287 p.

Johnston, G. H., ed., 1981, Permafrost; Engineering design and construction: New York, Wiley and Sons, 340 p.

King, von L., 1984, Permafrost in Scandinavia: Heidelberg, Heidelberger Geographische Arbeiten no. 76, 174 p.

Lachenbruch, A. H., 1962, Mechanics of thermal contraction cracks and ice-wedge polygons in permafrost: Geological Society of America Special Paper 70, 69 p.

—— , 1968, Permafrost, *in* Fairbridge, R. W., ed., The encyclopedia of geomorphology: New York, Reinhold Publishing Corporation, p. 833–839.

—— , 1970, Some estimates of the thermal effects of a heated pipeline in permafrost: U.S. Geological Survey Circular 632, 23 p.

Lachenbruch, A. N., Sass, J. H., Marshall, B. V., and Moses, T. H., Jr., 1982, Permafrost, heatflow, and the geothermal regime at Prudhoe Bay, Alaska: Journal of Geophysical Research, v. 84 no. 11, p. 9301–9316.

Leffingwell, E. de K., 1915, Ground-ice wedges, the dominant form of ground-ice on the north coast of Alaska: Journal of Geology, v. 23, p. 635–654.

—— , 1919, The Canning River region, northern Alaska: U.S. Geological Survey Professional Paper 109, 251 p.

Mackay, J. R., 1966, Pingos in Canada, *in* International Conference on Permafrost, Lafayette, Indiana, 1963, Proceedings: Washington, D.C., National Academy of Sciences, National Research Council Publication 1287, p. 71–76.

—— , 1968, Discussion of the theory of pingo formation by water expulsion in a region affected by subsidence: Journal of Glaciology, v. 7, p. 346–351.

—— , 1973, The growth of pingos, western Arctic coast, Canada: Canadian Journal Earth Science, v. 10, p. 979–1004.

—— , 1979, Pingos of the Tuktoyaktuk Peninsula area, Northwest Territories: Geographic Physique et Quaternaire, v. 33, p. 3–61.

—— , 1986, Growth of Ibyuk Pingo, western Arctic coast, Canada, and some implications for environmental reconstructions: Quaternary Research, v. 26, no. 1, p. 68–80.

Melnikov, P. I., 1984, Major trends in the development of Soviet permafrost research, *in* Permafrost, Fourth International Conference, Final Proceedings, Fairbanks, Alaska: Washington, D.C., National Academy of Sciences, p. 163–166.

Muller, S. W., 1943, Permafrost or permanently frozen ground and related engineering problems: U.S. Army, Office, Chief of Engineers, Strategic Engineering Study Special Report 62, 231 p.; 1945, second printing with corrections; reprinted in 1947, Ann Arbor, Mich., J. W. Edwards, Inc.

National Academy of Sciences–National Research Council, 1966, International Conference on Permafrost, 1963, Proceedings, Lafayette, Indiana: Washington, D.C., National Research Council Publication 1287, 563 p.

—— , 1973, Permafrost, the North American contribution to the Second International Conference, Yakutsk, Siberia, USSR: Washington, D.C., National Research Council, 783 p.

—— , 1974, Priorities for basic research on permafrost: Washington, D.C., National Research Council, 54 p.

—— , 1975, Opportunities for permafrost-related research associated with the Trans-Alaska Pipeline System: Washington, D.C., National Research Council, 37 p.

—— , 1976, Problems and priorities in offshore permafrost research: Washington, D.C., National Research Council, 43 p.

—— , 1978, Permafrost; the USSR contribution to the Second International Conference, Yakutsk, Siberia, USSR: Washington, D.C., National Research Council, 866 p.

—— , 1983a, Permafrost; Fourth International Conference, Proceedings, Fairbanks, Alaska: Washington, D.C., National Research Council, 1524 p.

——, 1983b, Permafrost research; An assessment of future needs: Washington, D.C., National Research Council, 103 p.

——, 1984, Permafrost; Fourth International Conference, Final Proceedings, Fairbanks, Alaska: Washington, D.C., National Research Council, 413 p.

National Research Council of Canada, 1978, Proceedings; Third International Conference on Permafrost, Edmonton: Ottawa, National Research Council of Canada, v. 1, 947 p.; v. 21, 255 p.

Péwé, T. L., 1948, Terrain and permafrost of the Galena Air Base, Galena, Alaska: U.S. Geological Survey Permafrost Program Progress Report 7, 52 p.

——, 1949, Preliminary report of permafrost investigations in the Dunbar area, Alaska: U.S. Geological Survey Circular 42, 3 p.

——, 1954, Effect of permafrost on cultivated fields, Fairbanks area, Alaska: U.S. Geological Survey Bulletin 989-F, p. 315–351.

——, 1966, Ice wedges in Alaska; Classification, distribution, and climatic significance, *in* International Conference on Permafrost, Lafayette, Indiana, 1963, Proeedings: Washington, D.C., National Academy of Sciences, National Research Council Publication 1287, p. 76–81.

——, 1973, Permafrost Conference in Siberia: Geotimes, December, p. 23–26.

——, 1974, Permafrost: Encyclopaedia Britannica, v. 14, p. 89–95.

——, 1975a, Quaternary stratigraphic nomenclature in unglaciated central Alaska: U.S. Geological Survey Professional Paper 862, 32 p.

——, 1975b, Quaternary Geology of Alaska: U.S. Geological Survey Professional Paper 835, 145 p.

——, 1976, Permafrost: 1976 Yearbook of Science and Technology; New York, McGraw-Hill Book Co., p. 32–47.

——, 1981, Tibetan science updated: Geotimes, v. 26, no. 1, p. 16–20.

——, 1982, Geological hazards of the Fairbanks area, Alaska: Division of Geological and Geophysical Survey Special Report 15, 109 p.

——, 1983, Alpine permafrost in the contiguous United States; A review: Arctic and Alpine Research, v. 15, no. 2, p. 145–156.

——, 1986, China expands research in frozen ground; A report of the Third Chinese Conference on Permafrost: Zeitschrift für Gletscherkunde und Glazialgeologie, v. 22, p. 1–7.

Péwé, T. L., and Bell, J. W., 1975, Map showing distribution of permafrost in the Fairbanks D2 NW Quadrangle, Alaska: U.S. Geological Survey Map FM 688 A, scale 1:24,000.

Péwé, T. L., and Paige, R. A., 1963, Frost heaving of piles with an example from the Fairbanks area, Alaska: U.S. Geological Survey Bulletin 1111-I, p. 333–407.

Péwé, T. L., and Reger, R. D., 1983, Richardson and Glenn Highways, Alaska; Guidebook to permafrost and Quaternary geology: Alaska Division of Geological and Geological Survey, Guidebook no. 1, 263 p.

Reger, R. D., and Péwé, T. L., 1976, Cryoplanation terraces; Indicators of a permafrost environment: Quaternary Research, v. 6, p. 99–109.

Rice, E. F., and Simoni, O. W, 1966, The Hess Creek dam, *in* International Conference on Permafrost, Lafayette, Indiana, 1963, Proceedings: Washington, D.C., National Academy of Sciences, Natural Research Council Publication no. 1287, p. 436–439.

Sayles, F. H., 1987, Embankment dams on permafrost: U.S. Army Corps of Engineers, Cold Regions Research and Engineering Laboratory Special Report 87-11, 109 p.

Sellmann, P. V., and Hopkins, D. M., 1984, Subsea permafrost distribution on the Alaskan shelf, *in* Permafrost; Fourth International Conference, Final Proceedings, Fairbanks, Alaska: Washington, D.C., National Academy of Sciences, p. 75–82.

Shumskiy, P. A., and Vtyurin, B. I., 1966, Underground ice, *in* Permafrost International Conference, Lafayette, Indiana, 1963, Proceedings: Washington, D.C., National Academy of Science–National Research Council Publication 1284, p. 108–113.

Shumskiy, P. A., Krenke, A. N., and Zotikov, I. A., 1964, Ice and its changes in solid earth and interface phenomena; V. 2, Research and geophysics: Cambridge, Massachusetts Institute of Technology Press, p. 425–460.

Tong, Boling, 1981, Some features of permafrost on the Qinghai-Xizang Plateau and factors influencing them, *in* Liu Dongsheng, ed., Geological and ecological studies of Qinghai-Xizang Plateau; V. 2, Environment and ecology of Qinghai-Xizang Plateau: Beijing, Science Press, p. 1795–1801.

Tyrell, J. B., 1903, A peculiar artesian well in the Klondike: Engineering Mining Journal, v. 75, p. 188.

U.S. Army Corps of Engineers, 1946, Airfield pavement design; Frost conditions, *in* An interim engineering manual for War Department construction: U.S. Army Corps Engineers, pt. 12, ch. 4, 10 p.

——, 1947, Report on frost investigation 1944–1945: U.S. Army Corps Engineers, 66 p.

——, 1949, Addendum No. 1, 1945–1947, report on frost investigation 1944–45: U.S. Army Corps Engineers, 50 p.

——, 1951, *et seq.* Bibliography on snow, ice and permafrost, v. 1–16, U.S. Army SIPRE; v. 7–20, U.S. Army CRREL; Bibliography on snow, ice, and frozen ground, with abstracts, v. 21–22; Bibliography on cold regions science and technology, v. 23–40.

——, 1954, Arctic and subarctic construction building foundations, *in* Engineering manual for War Department construction: U.S. Army Corps Engineers, pt. 15, ch. 4, 14 p.

——, 1956, Engineering problems and construction in permafrost regions, *in* The Dynamic North: U.S. Office Naval Operations [Polar Projects] Bk. 2 (OPNAY P03-17), 53 p.

Wahrhaftig, C., and Black, R. F., 1958, Engineering geology along part of the Alaska Railroad, *in* Wahrhaftig, C., and Black, R. F., eds., Quaternary and engineering geology in the central part of the Alaska Range: U.S. Geological Survey Professional Paper 293, p. 69–119.

Washburn, A. L., 1980, Geocryology: New York, John Wiley and Sons, 406 p.

Williams, J. R., 1996, Ground water in permafrost regions; An annotated bibliography: U.S. Geological Survey Water-Supply Paper 1792, 294 p.

——, 1970, Ground water in the permafrost regions of Alaska: U.S. Geological Survey Professional Paper 696, 83 p.

Williams, J. R., and Van Everdingen, R. O., 1973, Ground water investigations in permafrost regions of North America; A review, *in* Permafrost, The North American contribution to the Second International Conference, Yakutsk, Siberia, U.S.S.R.: Washington, D.C., National Academy of Science, p. 435–446.

Williams, J. R., and Waller, R. M., 1966, Groundwater occurrence in permafrost regions of Alaska, *in* International Conference on Permafrost, Lafayette, Indiana, 1963, Proceedings: Washington, D.C., National Academy of Sciences, National Research Council Publication 1287, p. 159–164.

Wilson, W. K., Jr., 1948, The problem of permafrost: Military Engineering, v. 40, p. 162–164.

Manuscript Accepted by the Society November 3, 1987

ACKNOWLEDGMENTS

The author is deeply indebted for excellent suggestions, revisions, and additions offered by George A. Kiersch, consulting geologist, Tuscon, Arizona, and Oscar J. Ferrians, Jr. of the U.S. Geological Survey and Chairman, Committee on Permafrost, U.S. National Academy of Sciences.

Printed in U.S.A.

Geological Society of America
Centennial Special Volume 3
1991

Chapter 15

Glacial deposits

Donald R. Coates
Department of Geological Sciences, State University of New York at Binghamton, Binghamton, New York 13901

INTRODUCTION

Today, glaciers cover 10 percent of the Earth's land area, whereas 30 percent was covered during the maximum glaciation of the Quaternary Period. Although 20 percent of the Earth's surface has a direct glacial heritage, another 12 percent was impacted by glacially related climates and sediments. Moreover, wind-blown silt, or loess, and long-travelled outwash have also affected many terrains, as in China, eastern Europe, and the Missouri–Mississippi Valley region. The glacial deposits that cover 10 million km^2 in North America have profoundly influenced the terrain and must be accounted for when making land-use decisions, whether for agriculture, urbanization, or planning and construction.

The applied geological scientist must consider and understand the vast differences that occur in the many different types of glacial deposits in order to successfully utilize the natural materials and/or sites for man's use and his engineering works. Each of the different glacial sediments, whether till, ice contact, or outwash, will exhibit different physical characteristics when subjected to man-induced changes. These different properties become manifest when the deposits are exposed in excavations, bearing loads are emplaced, or the material is utilized for construction purposes. Furthermore, there are many dissimilarities among glacigenic sediments even within broad categories. For example, lodgement till has entirely different values of consolidation, permeability, and density than either melt-out till, or flowtill. Many failures of engineering works have occurred when these quality differences have not been recognized in the field (see Kiersch, chapters 1 and 22, this volume).

Glacial deposits create their own landforms and landscapes, and proper identification of these terranes can prove important to the applied geologist. One thrust of this chapter is the close relationship between glacigenic deposits and landforms that constitutes a system, which when understood, can be of great assistance for the evaluation and management of the geological environment. A major objective of this review on glacial deposits is the importance of understanding glacial processes as they affect the origin of glacial materials and the landscapes formed. The mapping and testing of glacial landforms and deposits to ensure the integrity of the Trans-Alaskan Pipeline Project is a good example of the use of glacial knowledge in geoengineering affairs (Krieg and Reger, 1976).

History of glacial terminology

The term "drift" was first applied to unconsolidated sediments that seemed anomalous and could not be explained by everyday processes in England (Fairbridge, 1968, p. 291). Its usage stemmed from the belief that such materials originated from the biblical Noah's Flood. A term with somewhat similar meaning, "diluvium," largely replaced the use of drift by the late eighteenth century. However, by the late 1830s, the term "glacial drift" was used to explain the deposits formed by the actions of drifting icebergs and polar ice (Lyell, 1839). Surprisingly, it was not until the mid-1800s that Agassiz's theory of glacial ice emplacement became widely accepted (Goldthwait, 1975, p. 2).

Geikie (1863, p. 185) was the first to use the term "till," which he defined as ". . . .a stiff clay full of stones varying in size up to boulders produced by abrasion carried on by the ice sheet as it moved over the land." Some investigators continue to use the term "glacial till," but this is redundant because till is both a sedimentologic and a genetic term. By 1877 a further distinction had been made of till dividing it into two types; lodgement till formed under the ice, whereas ablation till moved down through the ice as the glacier melted. For the first time, two till types were distinguished by their process of emplacement and location.

By the twentieth century a further subdivision and classification of glacial drift had been made with the recognition of non-stratified and stratified deposits (Chamberlin, 1894). Stratified deposits formed within the context of the glacier were referred to as "ice contact," whereas materials washed considerably beyond the ice margin were called "outwash." When the processes of origin and sedimentation could be recognized further, the additional distinctions of glaciofluvial (or fluvioglacial) and glaciolacustrine were used. More recently the terms glaciomarine and glacioeolian have been used when describing sediments derived from or adjacent to the ice. Moreover, glacial sediments make up many different types of landforms, which are closely linked to the mode of deposition of the glacial debris. The somewhat tradi-

Coates, D. R., 1991, Glacial deposits, *in* Kiersch, G. A., ed., The heritage of engineering geology; The first hundred years: Boulder, Colorado, Geological Society of America, Centennial Special Volume 3.

TABLE 1. CLASSIFICATION OF GLACIAL DEPOSITS*

Dominant Materials		Conjectured Situation		Shape or Morphology
Lodgement till subglacial, basal, nonbedded, compact	*Till* till dominated, poorly sorted	*Ground moraine* under ice and off retreating edge	I	*Till plain* or rolling hills washboard moraine †minor moraines
Ablation till superglacial, loose lens, bedded, contorted		*Streamlined* sliding melting base	II	*Drumlin* grooved till crag-tail
		End moraine at standing or advancing ice edge	III	*Lobate/looped moraine* push/thrust boulder belt lateral/interlobate moraine †kame moraine
		Disintegration stagnant, decaying marginal area, buried ice masses	V	*Controlled/uncontrolled disintegration* dead ice knobs/rings disintegration ridge inverted lake
Glaciofluvial coarse cobble to silt, channeled or cross-bedded	*Wash* well- sorted	*Ice contact* dipping, deformed, irregular beds in ice pit, channel, or tunnel	IV	*Esker* †crevasse filling chain (of kames) *Kame* †field/kame and kettle kame moraine moulin kame kame terrace/plain
Glaciolacustrine fines: fine sand to colloid clay, laminated		*Proglacial* at grade, uniform beds extending away from ice	V	*Outwash plain/fan* valley train kettled/pitted outwash †collapsed outwash *Lacustrine*/marine delta strand/raised beach glacial *varve*

*From Goldthwait (1975).

†Other forms often attributed to stagnant ice disintegration.

tional classification of glacial deposits and their associated landforms is given in Table 1.

A resurgence of research of modern glaciers in the 1950s has produced a new array of terminology refinements and an increasing understanding of the processes that contribute to the formation of glacial deposits. For example, Dreimanis (1969) schematically showed the relationship of different till types with their mode of transport. Moreover, new terms came into use, such as "flowtill," and melt-out till. These ideas became some of the new standards by which nonstratified drift was judged (Fig. 1). Earlier, Rodgers (1950, p. 298–299) had argued that two types of classification schemes are needed for sedimentary rock: (1) a descriptive one that conveys the physical character of the sediment; and (2) a genetic one that indicates ". . . .what is known, inferred, or believed to be the origin of a given rock." Some sediments may have the appearance of a till; Flint and others (1960) named these non-sorted terrigenous sediments that contain a wide range of particle sizes, regardless of origin, "diamicton."

In the late 1970s and in the 1980s the continuing work of Boulton (1978), Eyles (1979, 1983), and Lawson (1979, 1981) has added a new dimension to the understanding of glacial deposits, their origin, and the precision with which they can be classified. Lawson documented the importance of knowing the total

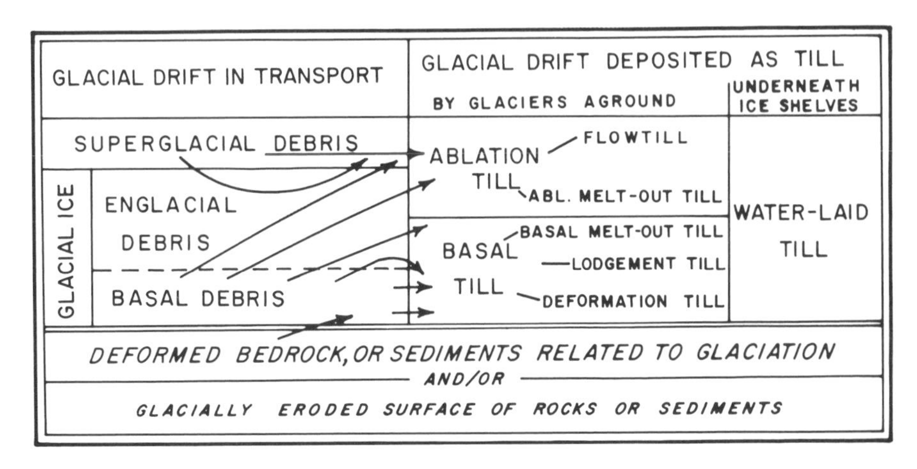

Figure 1. Genetic classification of tills, and their relationship to glacial drift in transport. Modified after Dreimanis (1969).

glacial environs, such as distinguishing between primary and secondary depositional processes and identifying the origin and characteristics of a deposit, as the critical factors needed to establish the identification and genetic classification scheme for glacial deposits (Lawson, 1981, p. 15; Table 2).

For a broader understanding of glacial processes, including syntheses of deposits and landforms, the reader is referred to the following sources: Flint (1971), Price (1973), Embleton and King (1975), Goldthwait (1975), Sugden and John (1976), Davidson-Arnott and others (1982), Coates (1982), and Drewry (1986).

GLACIAL PROCESSES AND DEPOSITS

Till

Till is the most variable of glacial sediment relevant to the processes of emplacement and inherent texture, composition, and structure (Fig. 1). However, in spite of the normally vast differences, till can attain an unusual degree of homogenization throughout many hundreds of kilometers. Under special conditions of entrainment, transportation, and deposition, till sediment may be continually deposited, transported, and redeposited by the various glacial cycles (Dreimanis, 1976).

Till is usually defined as a sediment that has been transported and deposited by glacial ice with little or no sorting by water or other processes; however, some authors prefer to elaborate and describe till as ". . . sediment deposited directly by the glacier ice that has not subsequently undergone disaggregation and resedimentation" (Lawson, 1981, p. 2). In addition, a special type is called "flowtill" by some and "sediment flow" by others (Table 2). For increased precision in nomenclature and application, till and "till-like" diamictons are separated into primary and secondary sediment groups. The primary glacigenic sediments are unique to ice deposition, whereas secondary deposits through resedimentation may occur in other nonglacigenic environments. The primary processes release the glacial debris directly and create deposits that are inherently related to the ice and its mechanics, while the secondary processes mobilize, rework, transport, and redeposit the debris at a site removed from its original position. The secondary deposits develop nonglacial properties and should be considered sedimentary materials formed by other geomorphic processes. "It is imperative to recognize the fundamental distinctions between the origins of the properties of sediments deposited in the glacial environment, particularly as they will react to engineering works. Otherwise, it is perilous to attempt the interpretation of glacial stratigraphic sequences in terms of their former depositional environment" (Lawson, 1981, p. 15).

Numerous investigations and studies on the many different facets of tills and diamictons are discussed by Goldthwait (1971), and Legget (1976). Specialists agree that careful study of many different properties is required to identify the origins of diamictons and to differentiate primary from secondary glacigenic-related deposits. These include texture, internal structure, bed contacts, bed forms, sediment geometry, stratigraphic associations, deformational features, clast fabric, and others. For example, some of the other subaerial processes cited as creating diamictons can be confused with those of glacial origin, including mudflows, flash-flood alluvium, fan alluvium-colluvium, volcanic breccia, fault breccia, collapse breccia, slumped debris, and other landslide deposits. Even some subaqueous processes may produce deposits that resemble till, such as those formed by turbidity currents and submarine slides.

The debate over possible glacigenic deposits also extends into Quaternary diamictons and stratified beds. For example, disagreements have raged for 70 years as to whether the Lu Shan Mountains of eastern China were glaciated during the Quater-

TABLE 2. COMPARISON OF PROPERTIES OF MELT-OUT TILL AND SEDIMENT FLOW DEPOSITS
(Lawson, 1981)

Deposit	Texture* Type 1) mean (ϕ) 2) σ (ϕ)	Internal Organization: General	Internal Organization: Structure	Internal Organization: Pebble Fabric	Surface Forms	Contacts and Basal Surface Features	Penecontemporaneous Deformation	Geometry and Maximum Dimensions
Melt-out till	Gravel- 1) 1 to 6 sand- 2) 1.8 to 3.5 silt; silty sand; sandy silt	Clasts randomly dispersed in fine-grained matrix.	Massive; may preserve individual or sets of ice-debris strata as subparallel laminae and lenses.	Strong; unimodal parallel to local ice flow; low angle of dip	Similar to ice surface; may be deformed or faulted.	Upper sharp, may be transitional; sub-ice probably sharp.	Possible internally and in overlying material during ice melt	Sheet to discontinuous sheet; several km² in area, 1 to 6 m thick.
Sediment flow deposits (Increasing water content ↓)	Gravel-sand-silt sandy silt, silty sand 1) 2 to 3 2) 3 to 4	Clasts dispersed in fine-grained matrix	Massive	Absent to very weak; vertical clasts	Generally planar; also arcuate ridges, secondary rills and desiccation cracks	Nonerosional conformable contacts; contacts sharp; load structures	Possible subflow and marginal deformation during and after deposition	Lobe: 50 x 20 m, 2.5 m thick
	Gravel-sand-silt, sandy silt, 1) 2 to 3 2) 3 to 4	In plug zone, clasts dispersed in fine-grained matrix	Massive; intraformational blocks common.	Absent to very weak; vertical clasts	Arcuate ridges; flow lineations, marginal folds, mud volcanoes, braided and distributary rills.	Nonerosional, conformable contacts; contacts indistinct to sharp; load structures.	Possible subflow and marginal deformation during and after deposition.	Lobe: 30 x 20 m 1.5 m thick; sheet of coalesced deposits.
		In shear zone, gravel zone at base, upper part may show decreased silt, clay and gravel content; overall, clasts in fine-grained matrix.	Massive; deposit may appear layered where shear and plug zones are distinct in texture.	Absent to weak; bimodal or multimodal; vertical clasts.				
	Gravelly sand to sandy silt 1) -2.5 to 2.5 2) 3.5 to 2	Matrix to clast dominated; lack of fine-grained matrix possible; basal gravels.	Massive; intraformational blocks occasionally.	Moderate, multimodal to bimodal, parallel and transverse to flow direction.	Irregular to planar; singular rill development, mud volcanoes.	Nonerosional, conformable contacts; contacts indistinct to sharp.	Generally absent; possible subflow deformation on liquefied sediment.	Thin lobe or fan wedge: 30 x 65 m, 3.5 m thick, less often sheet of coalesced deposits.
	Sand, silty sand, sandy silt 1) >3.5 2) <2.5	Matrix-dominated except at base where granules possible.	Massive to graded (distribution, coarse-tail).	Absent	Smooth, planar; mud volcanoes possible	Contacts conformable; indistinct	Absent.	Thin sheet: 20 x 30 m, 0.3 m thick. Fills surface lows of irregular size and shape.

* ϕ = mean grain size, σ = standard deviation.

nary Period. Those who support the glacial origin cite the occurrence of diamictons with a lack of stratification, the varied occurrence of grain sizes, large boulders, and some bedrock striations (Li, 1947, 1975). Conversely, the nonglacial origin is championed by such workers as Si (1981), Huang (1982), and Zhou (1983) who cite the lack of stratification and consider the different grain sizes as due to mudflows. Instead of stratified ice contact and outwash deposits, they are attributed to deposition from snowmelt and not glacial meltwaters.

Till possesses a vast array of different physical properties because of the many different types (Tables 1 and 2). These properties are dependent upon such factors as the mode of deposition, place of deposition, lithologic character of the parent bedrock, and its provenance, length of transport, position within the glacier, and even postdepositional modification. The usual features attributed to till are: (1) lack of sorting, (2) lack of stratification, (3) mixture of many sediment sizes from clay to boulders, (4) striations on clasts, (5) facets on clasts, (6) geometric fabric of clasts, (7) greater compactness than adjacent sediments, (8) angularity of clasts, (9) bimodal or polymodal grain-size distribution, (10) presence of nearly horizontal joints, and (11) deformational structures. It is unusual for a single till to contain all these features, but when many of these characteristics occur in a diamicton the chances of its having a glacigenic heritage are greatly increased. Some workers believe there are still other features that are distinctly glacigenic. Folk (1975, 1977) and Orr and Folk (1983) cite the presence of etch marks and chattermarks on garnet grains as evidence of glacial processes.

Although applied geologists are commonly interested in whether a diamicton is tight and compact or loose and less coherent, an understanding of the habitat of the specific till and its characteristics is usually critical for planning and predicting their reaction to engineering works. The following overview and discussion of tills has used names that occur most frequently in publications useful for the engineering-oriented practitioner.

Lodgement till. The name implies that rock particles are "lodged" at the base of the glacier; this occurs with "warm" or temperate ice. However, "cold" ice is frozen to the ground, the rock-sediment surface, at the pressure melting point. Lodgement till occurs most frequently upglacier from the terminus but may form along shear bands near the margin. Consequently, two or more till units with significantly different characteristics can be deposited within the same time frame and stratigraphic sequence. The lodgement till occurs where the tractive force imposed by the moving ice cannot maintain in motion the debris or the debris-rich ice against the frictional resistance of the ground surface. Thus the frictional coefficient between the clasts and the basal material is greater than that between the ice and the bed; clast motion is thereby retarded relative to the confining ice. This plastering process involves sediment of all sizes, and there can be accompanying deformation, crushing, and particle abrasion. The resulting till is generally very compact and may take on properties of fissility. The total thickness of lodgement till units at a single locality has a wide range, but thicknesses greater than 70 m occur in the Binghamton, New York–Great Bend, Pennsylvania area (Fig. 2).

Figure 2. Lodgement till at Great Bend, Pennsylvania. Photograph by the author.

Ablation till. A general term applied to diamictons that form as the result of ice wastage at or near the surface of the glacier. When the ice melts and loses volume, the entrained debris is released in a variety of fashions; it may move to a new location by subsiding, slumping, flowing, and falling. The resulting diamicton is usually loose, noncompact, and nonfissile. Ablation till is usually divided into two principal groups of diamicton, based on whether: (1) the debris primarily follows a vertical path to its final resting place, or (2) the material moved laterally and was aided by other geomorphic processes, as discussed by Lawson (1981) and shown in Table 2.

Melt-out till forms when ice surrounding and below the rock fragments melts, thereby releasing the debris to move by gravity downward (Table 3). The original location of the debris can be subglacial or englacial in addition to the surface materials. The melting horizon moves downward or upward into the ice mass and the debris released under confining conditions inhibits mixing and deformation. The release of sediment causes readjustment of grain contacts and increases particle packing. The finer grained material moves into pore spaces between larger grains. The extent of preservation of features is a function of the volume, distribution, and rate of ice wastage. The normal end result is production of a till that is less dense and consolidated with imperfect fabric when compared with lodgement till.

Flowtill is sediment that flows downslope and consists of sediment-water slurries derived from the ice surface under the force of gravity. This type of diamicton generally forms from meltwater-mudflow mixtures during the retreat phase. The ablation of the ice core destabilizes and undermines the sediments, thereby causing their secondary transport away from the source.

TABLE 3. CHARACTERISTICS OF MELT-OUT AND LODGEMENT TILLS
(Lawson, 1981)

Deposit	Texture* Type 1) mean (ϕ) 2) σ (ϕ)	Internal Organization			Surface Forms	Contacts and Basal Surface Features	Penecontemporaneous Deformation	Geometry and Maximum Dimensions	Miscellaneous Properties
		General	Structure	Pebble Fabric					
				Buried Ice Melt					
Melt-out till	Gravel- 1) 1 to 6 sand- 2) 1.8 to 3.5 silt; silty sand; sandy silt.	Clasts randomly dispersed in matrix.	Massive; may preserve individual or sets of ice strata.	Strong; unimodal parallel to local ice flow; low angle of dip; S_1>0.75.	Similar to ice surface; may be deformed.	Upper sharp, may be transitional; sub-ice probably sharp.	Possible; observable if structured sediments present.	Sheet to discontinuous sheet; km² to m² in area, m thick.	Internal contacts of strata are diffuse; loose.
				Lodgement at Glacier Sole					
Lodgement* till	Gravel- sand- silt; silty sand	Clasts randomly dispersed to clustered in matrix	Massive; shear foliation, other "tectonic" features	Strong; unimodal(?) pattern; orientation influenced by ice flow and substrate; low angle of dip.	Similar to base of ice.	Image of substrate.	Possible subglacial.	Discontinuous pockets or sheets of variable thickness and extent.	Usually dense, compact.

*From Boulton (1970, 1971), Lavrushin (1970a, 1970b), Mickelson (1973), and Boulton and Dent (1974).

Such material rolls, slides, and falls to the slope base where it is further mixed with other debris. The continued influx of meltwater and thawing of underneath ice develops excess pore pressure and seepage pressure within the mass. These changes further reduce the resistance to flow and failure. Consequently, flowtill is characterized by multiple mechanisms of grain support and transport with combinations of pebble fabric, internal sedimentary structures, bedforms, geometry, internal grain size variations, and glacigenic settings. These features distinguish flowtill from melt-out till and lodgement till, as shown in Table 3.

Ice contact

Glacigenic ice contact deposits and outwash are the two basic groups of stratified drift. Although there may be many physical differences, the principal distinction is location of deposition (Shepps, 1968). Whereas the ice-contact materials are formed in the presence of the glacier, outwash is debris washed considerably beyond the ice margin onto a proglacial setting. Furthermore, the majority of ice-contact sediments develop from stagnant or "dead," non-active ice, in which the meltwater is released within the ice mass.

The intimate association of the glacial debris with the meltwater component and the flushing action of the water cause ice-contact materials to be stratified. Although most of these deposits are glaciofluvial in origin, glaciolacustrine beds may occur associated with the ponding of waters in the ice-contact environs. Furthermore, because transport distances are usually short, ice-contact deposits invariably show large variations in texture, composition, and strata geometry. Of course, glaciofluvial facies are more variable and coarser grained than the glaciolacustrine facies, which may range from clay to fine sands; deltaic-type structures may occur.

Identification and delineation of ice-contact deposits versus outwash materials can be critical to assessing aquifer conditions, construction material potential, foundation characteristics, and other aspects to be considered in siting and for operation of engineering works. When compared to outwash, ice-contact deposits contain the following features (Price, 1973): (1) rapid change vertically; (2) greater textural differences laterally; (3) larger textural ranges, from clay to boulders; (4) gravel-cobble sizes show less rounding; (5) clasts contain more evidence of striations; (6) sediments may show deformation features, caused by subsidence or collapse when underlying ice core melts; (7) dip of strata and angle of repose may be greater than originally deposited; (8) greater compositional range, with retention of softer lithologies; and (9) individual strata may be poorly sorted and contain several grain sizes.

Outwash

Outwash deposits, or outwash, are formed from sediment that was within the glacier and was thereafter transported some distance away from the ice margin by meltwater. Consequently, emplacement is much farther away from the glacier than the ice-contact deposits, and outwash deposits are usually well sorted and of uniform texture (Fig. 3). Furthermore, the strata have greater continuity both vertically and horizontally; each stratum has relatively few textural sizes. The meltwater streams that form these glaciofluvial deposits often have a braided-stream type of environment, so that individual beds can have all the features of nonglacial streams. Thus the strata may possess such primary bed forms as superimposed linguoid bars, longitudinal bars, bar-edge sand-wedge bedding, channel cross-lamination and festoons, lateral accretion, and overbank sediments (Miall, 1983). Although the total sediment package of outwash may contain multistory units, when viewed in its totality, outwash is much more consistent, uniform, and predictable than ice contact. Therefore, such qualities enhance its utility and usage for many different engineering and societal needs.

Figure 3. Outwash deposits in a valley-train terrace near Chemung, New York. This sediment formed from glaciofluvial meltwaters in a high flow regime. Photograph by the author.

Loess

The term "loess" has been applied to a variety of different sediments, and for many years debate raged over its genesis (Lugn, 1962). However, most modern workers assign a wind origin for its entrainment and deposition. There is also agreement that to qualify as loess the material should be mostly homogenous, nonstratified, and nonindurated, with a preponderance of silt and subordinant amounts of other textures, such as fine sand and clay (Pecsi, 1968). Thus, loess has the double characteristic of possessing both a well-defined lithology and genesis.

Loess may have layers that range in thickness from centimeters to more than 5 m. It may be very crudely stratified, with alternating layers of paleosols, solifluction material, and even admixtures of sand or gravel. Because all known loess is Quaternary in age, its provenance has often been linked to glaciated areas, or regions affected by glacially related environments. Loess is only moderately sorted, with loosely coherent grains from 0.01 to 0.05 mm in diameter for the dominant grain size, which generally represents at least 50 percent of the sediment. However, appreciable clay-size particles below 0.005 mm, or sand over 0.25 mm may be present. The color is usually buff to yellowish but may be other colors such as gray.

The principal provenances of loess are from desert environments, such as the Gobi Desert of China. Equally important source areas for loess are in proglacial environments along the margin of the great ice sheets. The Laurentide ice sheets of North America gave rise to the extensive loess deposits of the Missouri–Mississippi Valley region, which has a total coverage of 1.6×10^6 km^2. Washington and Oregon have loess areas of more than 200,000 km^2. Similarly, large parts of Rumania, Hungary, and eastern Europe received the wind-blown silt from the margin of the Fenno-Scandinavian ice sheets that now covers 1.8×10^6 km^2 (Flint, 1975).

The geological and engineering properties of loess are dependent upon its ancestral setting and postdepositional history. Although loess is unconsolidated when above the water table, it has the ability to stand with cliffs 30 m in height. Desiccation and weathering phenomena can produce extensive joint patterns in loess, thereby giving them a secondary porosity and permeability for the movement of fluids. DuMontelle and others (1971) describe some of the inherent engineering and land-use problems associated with loess terranes. In referring to the Peoria loess with a thickness of 9.1 m along the bluffs that border the Mississippi Valley of Illinois they state:

> The loess deposits exert the main control on construction conditions along the bluff. The deposits of thick loess have high dry strength and assume a near vertical angle of repose as long as the materials are relatively dry. . . .Construction of ponds, excessive watering of lawns, and use of septic tanks can cause these materials to become unstable. Surficial creep of the soils and failure of foundation and retaining walls is likely to occur in such areas. (Dumontelle and others, 1971, p. 207–208)

GLACIAL LANDFORMS

Some assemblages of glacial materials can form diagnostic and specific landforms—such as eskers—that are unique to the glacigenic environment. However, in other cases glacial deposits may be nondescript and have surface forms that are similar to those created by other geomorphic processes. Thus, confusion can occur because of the principle of equifinality—a landform with a particular geometry that can be developed by more than a single process. For example, it is possible for a landform at one locality to consist of stratified sediments, while at another locality a similar landform is composed of till. Thus, extreme caution is urged when attempting to link some terrains with a prediction of the character of the underlying materials without thoroughly examining the substrata first.

Table 1 presents the traditional view for relating glacial deposits with their commonly derived landforms. Fookes and others (1978) describe glacial landforms and discuss them in terms of the types of glacial deposits and inherent engineering properties. They use the "landsystem" approach, as does Eyles (1983), with chapters on the Subglacial Landsystem, the Supraglacial Landsystem, and the Glaciated Valley Landsystem. In each system the characteristics of the glacial sediments are related to the specific landform or landscape.

Till-dominated landforms

The dominant constituent of most moraines, drumlins, and till sheets is till. However, exceptions can occur, such as the "kame moraine" or the "heads-of-outwash end moraine." These latter features may contain considerable, and even predominant, deposits of stratified drift.

Moraines. These accumulations of glacial sediment usually have a particular surface expression. The term "ground moraine" has been prevalent in the older literature and refers to a variety of sediments and topographies. Some of the terrain has no diagnostic features but may possess a somewhat hummocky surface. Moraines can be classified in several different manners, including:

1. Dynamics. Whether active or inactive. Active glaciers may transport surface debris as geometric units, both along the side as bands or in the main body of the ice.

2. Position. A terminal moraine represents the farthest advance of the glacier. End moraines form at the ice margin, and if the margin retreats to a new position a recessional moraine may develop. These hummocky belts of hills have wide ranges in geometry, size, and perfection of form. Lateral moraines can form at the sides of active glaciers, and if two ice bodies join, an interlobate moraine may form.

3. Process of formation. The forces that produce the moraine can create push moraines, thrust moraines, ablation moraines, and others.

4. Geometry. Moraines can be grouped, whether they are linear or nonlinear. Fluted moraines contain elongated ridges parallel to ice motion, whereas rogen (ribbed) moraines contain linear features transverse to direction of glacier movement.

5. Environmental setting. Describes moraines in terms of relationships with surrounding terrain, such as cross-valley moraines, choker moraines, nunatak collar moraines, and DeGeer moraines.

Drumlins. These landforms are usually streamlined hills elongated in the direction of ice motion; thousands of drumlins may occur in some large fields. Drumlins rarely exceed a height of 60 m and have length-width ratios that may range from 1.5:1 to more than 4:1; the up-ice slope is two or more times steeper than the lee slope. They are composed predominately of till but occasionally have some stratified horizons and beds of sand and gravel.

Nonstratified drift is also the primary constituent of crag and tail landforms and till shadow hills (Coates, 1966a) common throughout the central part of the glaciated Appalachian Plateaus. The upland hills that rise a few hundred meters above the valley bottoms are asymmetric, with north-facing slopes twice as steep as south-facing slopes. The underlying till is 8 to 10 times thicker on the southern slopes and in places reaches thicknesses of 80 m.

The till plains of the central United States, and also those of northern Europe, can be nearly flat and featureless for hundreds of square kilometers. The terrain is commonly underlain by a series of till sheets of different ages with paleosol separations. Average thickness of the total till substrate is about 40 m.

Ice-contact landforms

These stagnant ice features result from ice disintegration and give rise to a large variety of landforms that contain granular sediment of all textures and degree of stratification. A common element that the features share is an origin by topographic inversion. The sediment is deposited by glaciofluvial processes into a depression, hole, stream channel, or opening that is lower than the surroundings and the adjacent ice. Upon complete melting of the glacier, the sediment formerly positioned at a lower topographic elevation remains as a remnant that is now higher than the adjacent terrain. Such eminences have a large range in size and geometry.

Kames. This family of landforms includes a large array of hills and hummocky terrain. There can be single and isolated kames to clusters of kames with interruptions by kettles (or kettle holes). Such hybrids of terrain can be referred to as a kame complex, kame-and-kettle, or knob-and-kettle topography. Kame moraines form at the distal or marginal stagnation zones of ice sheets where belts of kamey terrain form the dominant topography. Here the underlying composition of deposits is a mixture of various tills and stratified materials. Kame terraces form by glaciofluvial action in streams at the contact of the valley walls and the lateral side of a stagnating ice mass. With deglaciation, what was formerly a series of beds in stream channels now remains as a terrace-like sequence of somewhat flat-topped terrain (Fig. 4B). Kame deltas are still another form in this grouping.

Eskers. Eskers are the most distinctive and unique landforms created in the glacigenic environment. Their stratified drift was deposited in former stream channels that could have occurred in superglacial, englacial, or subglacial channelways. Although they may possess wide texture variations, bedding relationships, and sizes, their common feature is the sinuous elongated ridge that rises above the adjacent topography. Most eskers have heights less than 40 m, and widths less than 500 m. However, their length can vary from a few tens of meters to tens of kilometers. Their orientation is roughly parallel to the principal vector of ice motion, in spite of the fact that they form in an ice-disintegrating environment. Eskers may have branching ridges and can also have a beaded appearance. Beaded eskers often show inheritance from a history that produced both glaciofluvial and glaciolacustrine sediments. Some authors would prefer to classify eskers as one of the varieties of crevasse or channel-fill deposits, but such sediments generally contain excessive amounts of till with only minimal transportation by running water.

Outwash landforms

Outwash deposits can take the topographic form of plains, fans, and aprons. These features are mostly composed of sand and gravel, with smaller amounts of clay and silt or occasional clasts larger than gravel. The outwash plains (sandurs) contain very gentle slopes, such as those of southern Long Island, with gradients of only 1 or 2 percent. When the outwash zone represents a position where glacial recession was uneven, leaving behind temporary stranded ice masses, the outwash may ultimately contain kettle holes (areas of ice melt-out). Such features are then called "pitted outwash plains." Although the sediment in these outwash landforms is usually well bedded, well sorted, and continuous, interruptions can occur to provide for more erratic sedimentation. The events that produce abnormal stratigraphy are "jökulhlaups," the Icelandic term for the rather sudden discharge of lakes impounded within the glacier. These breakouts may be somewhat normal occurrences for certain glaciers, but are generally not usual for the majority of glaciers.

When outwash is confined within a valley and unable to spread out laterally, the resulting landform is called a valley train. The sedimentation that results from a glacial episode produces constructive landforms. Upon deglaciation the renewed fluvial processes may incise such terrain, thereby creating a terrace (Fig. 3). In such circumstances it may be important to determine whether this new feature is a kame terrace or a valley-train terrace. Careful study of the geologic setting and characteristics of the deposits can determine the origin and inherent properties of the feature.

In some special instances, glacial geologists have used a systems approach when uniting the characteristics of ice-contact landforms with those of outwash landforms. When these sediments are in a valley setting, Koteff (1982) prefers to classify this series of features into a morphological sequence model. Shear zones at the live-ice/stagnant-ice interface are suggested as the major sediment source. Debris from this zone is transported at and beyond the margin to form a series of different sediments and landforms. During recession of the ice front, it is possible for a single major south-sloping valley to have a series of these sequences, each representing a chronologic group of features developed during a stillstand of the ice margin. These features are prominent throughout southern New England and south-flowing drainages in south-central New York. However, this view of these deposits has been challenged by Gustavson and Boothroyd (1987) who believe such features originate by subglacial and englacial meltwater deposition.

GEOLOGICAL/ENGINEERING PROPERTIES OF GLACIAL DEPOSITS

All too often, engineers are only interested in the appearance or behavior of earth materials, with diamicton-type sediments

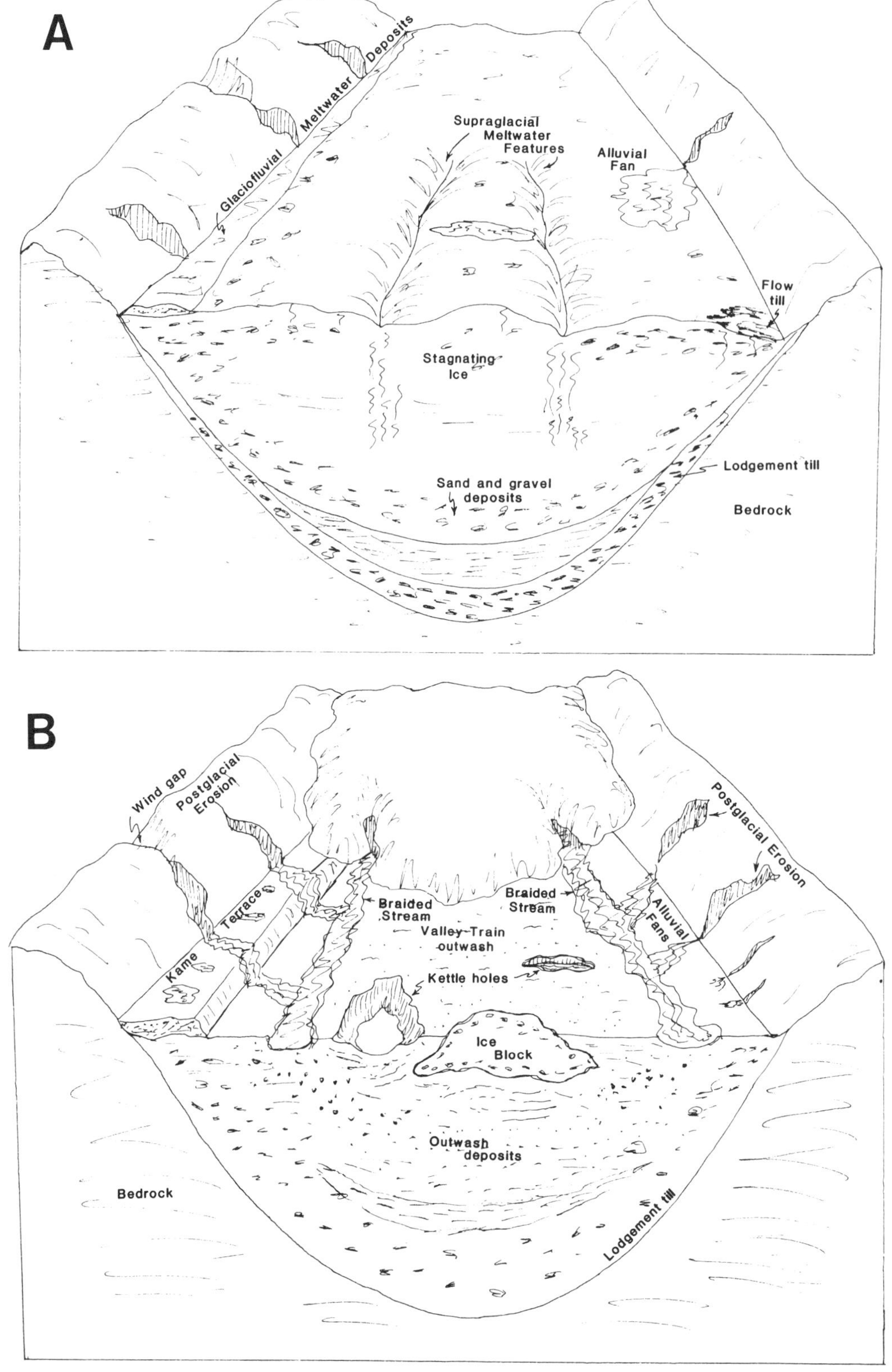

Figure 4. A. Initial phase of deglaciation of a stagnant glacier. B. Recession of the ice margin has left a series of ice-contact and outwash deposits and landforms. Braided streams form the valley-train outwash. Kettle holes and an ice block create a pitted surface on the outwash. A kame terrace is formed along the valley side and is undergoing some dissection from postglacial stream erosion. Postglacial alluvial fans are present, and lodgement till has formed the sediment above bedrock.

TABLE 4. SUMMARY OF ENGINEERING PROPERTIES AND BEHAVIOR FOR THREE CATEGORIES OF GLACIAL DEPOSITS
(Fookes and others, 1978)

Land System and Category of Glacial Materials	Bearing Capacity	Settlement	Slope Stability	Excavation	Use as Fill Material	Construction Materials
Lodgement Till	Usually good, but liable to contain soft patches. Silt lenses liable to frost heaves.	Usually small. Long-term settlement. Differential settlement may be anticipated.	Generally stable at quite steep slopes. Water-bearing sand/silt layers can cause instability.	Lodgement tills can be very tough and excavation hindered by high plasticity and boulders.	Good impermeable fill, but sensitive to moisture changes. May be wet of optimum.	Generally unstable due to variability.
Ablation Till	Generally good but may be variable.	Slightly higher, but more rapid settlement than lodgement tills.	As lodgement tills but with lower cohesion. When low fines content stand near angle of rest.	Excavation may be hindered by boulders, and high silt content can hamper excavation in wet conditions.	Good fill material. Can be used as impermeable fill. Silt content causes sensitivity to moisture-content changes.	Unsuitable without much screening and washing.
Fluvio-glacial Cohesionless Deposits	Generally good but lenses of till or openwork gravel can occur, and relative density is variable (especially in kames).	Settlement largely during construction, but differential settlement due to long-term consolidation of clay or till or lacustrine clay lenses.	Materials generally all stand at angle of rest. Locally, clay/silt lenses may be unstable.	Usually easy. Face shovel preferred to mix sand and gravel. Sometimes hindered by wet silt/clay layers or by boulders.	Good granular, generally free-draining fill. Some selective digging may be required, particularly in kames.	May provide good sources of coarse and fine aggregate, but selective digging and washing often required.

forming one group and granular materials another group. This approach does not provide sufficient allowance for the many variations that occur in the heterogeneities of glacial deposits. Furthermore, a restrictive engineering assessment of earth materials, without appreciation and understanding of their origin as given in engineering classifications of glacial materials (Legget, 1976; Geo Abstracts 1978; Eyles, 1983), limits the scope and accuracy with which predictions can be made for land-use purposes. To assist such endeavors, the New York Geological Survey is currently producing Quaternary maps of the state that are specially designed to assist engineers, resource managers, and land-use planners (Muller, 1977; Cadwell, 1986). Fookes and others (1978) and Boulton (1976) discuss the different attributes of glacial deposits and their utility for engineering purposes (Table 4). Fiksdal (1982) used glacial landforms in Washington State to show their importance to hazard determination and engineering applications (Table 5).

Depositional properties

The mode of till deposition, whether as melt-out till, flowtill, or lodgement till, determines its geologic/engineering properties. The following factors affect the utility of till for engineering purposes: weight of the ice; water content during and after emplacement of debris; composition, size, and arrangement of clasts; texture of matrix; and desiccation and other postdepositional effects. For example, the initial grain-size distribution of the sediment will define the values obtained for angle of internal friction, cohesion, compressibility, and the consistency limits. Consequently, knowledge of both the depositional history of the sediment and its postdepositional changes is necessary for a meaningful understanding and prediction of the engineering behavior of the materials (Milligan, 1976; Sladen and Wrigley, 1983).

Water plays an important role in the emplacement of all glacial deposits. Pore pressures exist throughout firn ice as well as in the basal zone where slurries may develop. The ground condition under the ice, whether frozen or not, will be a factor in determining ground-water pressures. If effective drainage is inhibited, only a minor degree of sediment consolidation can take place because the weight of the ice overburden is buoyed; the materials are only weakly consolidated by the low confining pressure and thus have low strength (Boulton, 1976). The state of consolidation in melt-out till and flowtill is not related to ice loading, but instead to the sediment overburden, water table fluctuations, and freeze-thaw episodes. Some fluid flows may dry out rapidly, leading to partial saturation and producing complex patterns of consolidation. The melt cycle causes high local pore-water pressure and sediment instability.

Lodgement till is overconsolidated with very high preconsolidation loading. Flowtill may be slightly overconsolidated but not due to preconsolidation pressure. Melt-out tills are of a normal consolidation mode (Table 6).

TABLE 5. LANDFORMS AND ASSOCIATED HAZARDS, ENGINEERING CHARACTERISTICS, AND CONSTRUCTION MATERIALS
(Fiksdal, 1982)

Landforms	Total Land Area (%)	Hazards*									Engineering Characteristics			Construction Materials			
		Landslides	Ponding	Gully Erosion	Runoff	Settlement	Creep	Flooding	Groundwater Contamination	Slope	Foundation Stability	Infiltration	Slope Stability	Sand	Gravel	Topsoil	Fill
Drumlin	25			sli	sli		sli				good	poor	good				good
Ground Moraine	5		mod								good	poor	good				good
Constructional Slope	5	sli		sli	sli		sli				good	poor	good				mod
Outwash Surface	5								mod		good	good	good	good	good		good
Outwash Valley Train	10		mod						mod		good	var	good	good	good		good
Kame Terrace	5		sli						mod		good	good	good	mod	mod		good
Terrace Undifferentiated	2										good	good	good	mod	mod		good
Collapse Slope	2	sli		sli			sli				good	var	mod	mod	mod		good
Outwash Delta	1									sev	good	good	poor	good	good		good
Valley Train	20		mod			mod		sev	mod		var	mod		mod	mod	good	good
Erosional Slope	10	sev		sev	mod		sev			sev	mod		poor		mod		good
Mudflow Surface	5		mod								good	poor	good				mod
Modified Slope	1		mod				mod				good	poor	mod				poor
Kettles	1		sli									poor					
Esker	1										good	good	good	mod	mod		good
Modified Area	3		sli			mod			sli		var	var	var				poor

*Abbreviations: sli = slight; mod = moderate; sev = severe; var = variable

No symbol indicates non-applicability, no hazard, or no resource.

TABLE 6. DEPENDENCE OF SOME GEOTECHNICAL PROPERTIES ON THE GLACIAL PROCESSES
(Boulton, 1978)

		Geotechnical Effects					
Processes	**Geological Effects**	Friction	Cohesion	Consolidation Rate	Compressibility	Consolidation State	Jointing
Erosion and Incorporation			•	•	•		
Transport		•		•	•		
Deposition							
Lodgement Till	clast aggregation	•		•			
	shearing						•
	ice loading					overconsolidated	
	unloading						•
Flow Till	loss of fines	•	•	•	•		
	gain of fines	•	•	•	•		
	presence of stratified beds			•	•		
	drying out					overconsolidated	•
Melt-out Till						normally consolidated	
Post-depositional changes	water table changes						
	freezing						•
	washing out	•	•	•	•		

Fractures in till

The postdepositional history of till can greatly influence its physical behavior and in many tills can lead to significant fracturing. Terzaghi and Peck (1967) have shown that joints in overconsolidated sediments can considerably reduce their in situ strength below values measured for intact material in the laboratory. Such joints can originate from a variety of processes, including: pressure release during and after ice unloading, shrinkage during desiccation, contraction during freezing, and failure associated with shearing forces.

Some tills may show as many as three different types of joints. A set of closely spaced horizontal or subhorizontal fractures in lodgement till can occur as a result of elastic rebound. A second set of subhorizontal shear joints may possess slickensides and cut the till into a series of shear lenses with spacings that range from several decimeters to millimeters. A third set of vertical joints may appear as conjugate shear pairs with orientations that reflect their development during ice movement. Flowtills may contain vertical joints in their surface horizons caused by desiccation, and occasionally may also have imperfect horizontal fracturing. Melt-out tills are generally not jointed.

Fractured tills in the interior plains of Canada were first described by Horberg (1952). Here, a series of fractures exists in the vicinity of thrust planes in till near Lethbridge, Alberta. More recently, fractures and joints have been mentioned by several authors—especially in the hydrogeology literature—because of their importance in changing the hydraulic properties of a till that would otherwise be a low-permeability medium (Grisak and others, 1976). Jointing must also be an important consideration in slope stability design and for engineering structures (Rowe, 1972). Jointing characteristics and incidence are often related to the clay component of tills; jointing is most common in clay-loam textures.

In eastern North Dakota, numerous excavations in till, some as deep as 21 m, were required for the installation of ICBM missiles. Fractures in the till were observed in 70 percent of the excavations, with vertical fractures the most obvious. However, horizontal fractures were also present (Grisak and others, 1976). Most of the vertical and near-vertical fractures ended at the Cretaceous bedrock contact, but in some instances the till shared the same single-fracture system. There is also a correlation of fracture orientations of bedrock and till at the Whiteshell Nuclear Research Establishment, which suggests the tension fracturing resulted from crustal rebound following glacial unloading. Thin till on crests of upland ridges in the Finger Lakes Region of New York has joint systems that mirror the underlying Devonian bedrock.

Till strength

The engineering quality of most tills ranges from compact and nonplastic to weak, plastic clayey tills. The Atterberg limits serve as a good quantitative indicator of the role of grain size in influencing the engineering properties and potential uses of till. When clay content is high, till develops high cohesion with a relatively small angle of internal friction. Time-dependent deformation is usually not a problem except for the plastic tills or water-lain clay tills. Crawford and Burn (1962) and Dejong and Morgenstern (1973) point out that most tills are dense and overconsolidated. They regard them as good foundation sediments for spread and raft footings and for support of end-bearing piles.

The modulus of elasticity of tills can vary from 145 to 670 kg/cm^2 and their undrained compressive strength from 1.4 to 29 kg/cm^2 (Geo Abstracts, 1978). Such large differences are due to such factors as the size of the loaded area, the effect of the time/rate of loading, and the variability of the deposit. Because of possible gross anisotropy of tills within the same sediment area, it may be necessary to sample more than one position to determine whether settlement problems under loading will occur.

The majority of tills show poor sorting compared with most other sediments. The relatively high cohesion of the clayey tills also imparts low permeability and poor drainage. This can lead to difficulties due to increased sensitivity to moisture-content changes; it can also cause problems with compaction control, and lead to natural moisture content exceeding the optimum when artificially emplaced into engineering structures. Silty and sandy tills are easy to compact to a dense stage, provided the soil has the proper water content at the time of compaction. A water content of 2 to 3 percent above optimum is usually too soft for compaction. Degrees of compaction higher than 1.02 can only be attained with substantial increases in effort. A change in water content of only 2 percent may change the remolded, undrained strength by 50 percent in clayey tills, whereas relatively little change occurs in sandy tills. Thus, the compressibility of till is relatively independent of its initial density, except for tills with unusually large amounts of plastic fines. The compaction of tills to higher density does not appreciably change their compressibility and shear-strength character. If permeability is not a controlling factor in the design or use of engineering works, it is possible to use till that contains a small amount of compaction, because the sediment can then be compacted further with minimum energy. The strength envelope of many tills lies between 35° and 40° for levels of normal effective stresses that are larger than 5 kg/cm^2. Numerous authors provide additional information on till strength, loading, and construction uses; these include Anderson (1983), Loiselle and Hurtubise (1976), and McGown and others (1978).

Stability of till excavations

The slope stability of tills is dependent on the original character of the till, the geologic setting, and inherent moisture conditions of the site. In particular, these special factors include the bulk undrained and drained shear strength, consolidation or swelling properties, permeability and pore water pressures, the degree of sediment homogeneity, and its texture and fabric (Eden, 1976).

Lodgement tills have low rates of permeability and swelling, so that undrained conditions apply during slope excavation and drained conditions govern the long-term behavior. The stress relief of unloading produces water suction in cut slopes that dissipates with time, making stability more critical with regard to long-term pore pressures. In compacted fills, the end-of-construction pore pressures are generally high. In high embankments, consolidation under self weight will tend to produce a density increase, with permeability reduction at depth. When water percolation exceeds evaporation, a perched water table may develop.

Clayey tills fail at lower slopes than others (Milligan, 1976). For example, some dense tills in British Columbia have stability slopes shallower than 20° because of thin zones of highly plastic, slickensided clay materials. Well-developed sand tills of low plasticity have relatively high drained strength and are of low sensitivity and brittleness. Excavations in these tills have operational drainage strength and produce test results similar to intact specimens, unless there are fissures with a critical orientation. Therefore, relatively steep slopes of 1:1 can be adopted, but with more plastic tills the stable slope is about 2.5:1 (Cocksedge, 1983). The low brittleness of sandy tills produces failures that are usually shallow, with movement by short-distance slips. Slope failures in tills of greater cohesion, and other fine-grained glacial sediments, are deep seated and rotational in type (Fig. 5). The majority of lodgement tills have a reduced plastic index and, therefore, reduced brittleness.

IMPACT ON ENGINEERING WORKS

Glacial clays/slope failures

Clays of glacial origin are notorious for damage they have produced and for causing great difficulties for engineering geologists involved in construction. (Figs. 5, 6). They may be of fluvial, marine, or estuarine environments, but they all originate from sediment washed out from the glacier. For this reason they may possess properties quite different from clays of nonglacial origin (Legget, 1942), such as having densities greater than normal and incorporating "glacial flour" type of fines. The terms "quick clay" and "sensitive clay" are widely used to describe their behavior (Crawford, 1961). Furthermore, Crawford (1961) has shown that many of the clays have extraordinarily high salt content that can be the key to destabilizing these materials (Table 7). The ill-famed Leda Clay of the St. Lawrence Valley has a strain of about 1 to 2 percent. Its sensitivity is strongly related to the liquidity index because an increase in pore pressure also increases the distance between the grains and thus reduces cohesion. In addition, the solid portion of the clay is less than one-third the total volume, so that high compressibility can lead to high rates of settlement. The Leda Clay also has a tendency to shrink when dry, which can cause another range of problems when near buildings. Failures in the Leda have been caused by rapid surface erosion, construction or other types of vibration, electric currents, and buildup of excess moisture. The latter has produced a number of catastrophic landslides throughout the St. Lawrence region (Table 8). At St. Thuribe, Canada, on May 7, 1898, nearly 3×10^6 m^3 of earth flowed through a 60-m opening in less than

Figure 5. Aerial photograph of landslide features and stream dissection of fine-grained glacial sediments along the Swift Current Creek, south-central Saskatchewan Province, Canada. The silts and clays have undergone failure creating entire arrays of imbricate slumps common to such glacial deposits. Courtesy Richard Parizek.

Figure 6. Piping in the glaciolacustrine sediments near Chenango Forks, New York. Diameter of the opening is 55 cm. Note mud slurry being discharged through the conduit, which can be visually traced more than 6 m into the hillside. At this locality numerous pipes have aided in hillside degradation and landsliding (Coates, 1977).

four hours and left a scar 500 m by 900 m, with a 15-m depression. The Nicolet slide in Quebec occurred November 12, 1955; it killed three people and caused $5 million in property damage (Crawford and Eden, 1963). At St. Jean Vianney, Canada, a new housing development was built on a terrace underlain by 30 m of Leda Clay. Although a few cracks appeared on some nearby blacktop streets on April 23, 1971, the real tragedy did not happen until 11 days later:

> The earth simply dissolved to a depth of nearly 100 feet; in the canyon thus formed, a river of clay—sometimes as deep as 60 feet—flowed at a rate of 16 miles per hour toward the Saguenay River, two miles away. At its widest, the canyon was a half-mile across, and it extended for approximately one mile (Blank, 1971).

The few hours of destruction claimed 31 lives, and 38 homes disappeared into the liquefied material. As in all these cases, exceptionally heavy rains had overloaded the sediments, thus triggering the landslide.

Glacial clays can be instrumental in other types of geologic hazards. The Alaskan earthquake of March 27, 1964, caused widespread damage and devastated large parts of Anchorage, Alaska. In the Turnagain Heights section, 280 ha were liquefied by the shaking. Developers of this part of the city failed to consider the sensitive clays underlying the area. More than 100 people were killed and 750 homes destroyed. Here the glacio-estuarine Bootlegger Cove Clay underwent liquefaction, causing massive flowage and slab sliding (Hansen, 1965).

A large grain elevator at Transcona, Manitoba, was constructed in 1911 to 1913. The bin structure was 30.6 m high and 23.1 by 58.5 m in width and length. On October 18, 1913, when 31,500 m^3 of grain had been stored, a vertical settlement of 30 cm occurred within an hour. This was followed by a tilting motion, and within 24 hours, the building rested at a 26°53′ angle to the vertical. The west side was 7.2 m below ground, and the east side rose 1.5 m. Inspection for the cause of the damage discovered the structure was emplaced on glacial Lake Agassiz clay that had a bearing capacity of only 31,300 kg/m^2. This compared to the 30,300 kg/m^2 bearing pressure at failure (Baracos, 1957; Legget and Karrow, 1983).

The opposite situation occurred on Long Island, New York. It had been the consensus that the Lake Flushing varved silt and clay deposits had not been loaded by the ice sheet and were therefore considered poor supporting materials (Parsons, 1976). Therefore, prior to the 1950s the practice had been to construct buildings on piles that supported the foundations on the sturdier and deeper tills or bedrock. However, after the calculated amount of settlement did not occur, a testing program showed the preconsolidation pressures of the silt and clay beds were much higher than predicted. Further geologic mapping of the area showed that the glacier had indeed readvanced over the strata, thereby improving their strength. Subsequently, the New York City Housing Authority was able to employ foundation rafts instead of piles, saving millions of dollars in construction costs (Parsons, 1976).

Buildings and foundations

Preliminary investigations in planning the construction of the St. Lawrence Seaway Project indicated that outcrops and test pit exposures of the superglacial till were loose and sandy, and were expected to overlay the dolomitic bedrock. Confirmation of a thick bed of tough, dense, consolidated and partly cemented basal till underlying the superglacial till came as a geologic surprise. The basal till was so tough and firmly bonded that even the least expensive methods of excavation required extensive drilling and blasting. Yet when wetted by the frequent rains, additional problems occurred because of the lack of stability of the material (Cleaves, 1963). The basal till had a liquid limit of 16.2 percent, a plastic index of 4.9 percent, and 79 percent of the material had a sand-to-clay matrix. Thus, it had high saturation, low porosity and permeability, and high density. In May 1958, claims by contractors for cost overruns were made for more than $27 million, although they subsequently settled for about $5 million. The claims contended the material was different than as described in the bid specifications. Actually, the failure to recognize the occurrence of basal till at construction sites throughout the St. Lawrence–Great Lakes regions was a result of inadequate preconstruction geological exploration and evaluation, according to Cleaves (1963, p. 56).

In the early 1970s, elastic utility companies were anticipating construction of additional nuclear power plants in upstate New York—in the vicinity of the St. Lawrence Lowland. Earlier studies in California had documented that earthquakes can cause deformation features to form in fine-grained unconsolidated sed-

TABLE 7. PROPERTIES OF SOIL AND SOFT SEDIMENTS AT NICOLET LANDSLIDE, QUEBEC PROVINCE, CANADA
(Crawford and Eden, 1963)

Boring	Sample	Depth (ft)	Geodetic Elevation (ft)	w/c (%)	LL (%)	PL (%)	Clay size (%)	Activity	Salt Content* (g/l)	Sensitivity† (St)	Average Shear Strength§ (tons/ft²)	Preconsolidation Load (tons/ft²)
1	68-1	8.0	20.0	47.8	50.1	25.1	54	0.46	8.3	8.8	0.44	
1	68-2	9.8	18.2	52.7	44.4	24.4	50	0.49	10.2	13.8	0.65	2.0
1	68-3	11.4	16.6	52.9	45.9	22.4	47	0.50	6.6	12.1	0.45	1.3
1	68-4	12.9	15.1	52.6	51.0	23.1	47	0.59	11.0	16.3	0.62	
1	68-5	14.6	13.4	54.0	51.0	25.0	50	0.52	7.8	13.8	0.36	1.4
2	68-6	12.0	38.0	56.6	50.1	24.3	75	0.34	0.2	9.7	0.30	1.2
2	68-7	14.0	36.0	62.9	60.3	25.2	71	0.49	0.8	9.1	0.23	1.2
3	68-8	Clay remolded by landslide		57.7	44.5	23.3	69	0.31	0.9	79.0		

*Grams of salt per liter of extracted pore water.

$$^{\dagger}St = \frac{\text{unconfined compression shear strength}}{\text{Laboratory vane shear strength (remolded)}}$$

§By unconfined compression test.

iments. Therefore, a geological investigation was made throughout the St. Lawrence Lowland to evaluate similar-appearing glacial structures and features that might have been induced by seismic activity (Coates, 1975). The wealth of glacial sediments throughout the region provided numerous possibilities for various types of soft-sediment deformation. Whereas the California sediments were not glacigenic, those in New York generally had the overprint produced by different types of glacier-related processes. Throughout the St. Lawrence area a full range of deformed features was noted that included faulting, folding, and décollement-like structures common to a glacial environment impacted by seismic events. However, such features could also have been due to the processes of ice push, ice shove, ice thrust, basal overriding, static loading and ice pressing, melt-out collapse, slumping, and even frozen ground pressures.

Kaye (1976) provides another side to the principle of equifinality in citing a project in the Boston area where foundation engineers failed to identify the glacial features. The foundation design by engineers and architects for several buildings in the Beacon Hill area assumed the hills were drumlins and that the subsurface sediments would be a clay-rich glacial till. The till was believed to be very compact and capable of withstanding great loading stresses. Furthermore, the till was predicted to be largely impermeable and incapable of presenting any serious water-flow or seepage problems. Unfortunately these were not the conditions encountered when the foundations for several buildings were excavated. Because of the erroneous misconceptions and lack of preconstruction investigation, several of the buildings developed severe construction problems, causing time delays and cost overruns. For example, the Boston Common garage was designed to have a 13.5-m-deep excavation into a dry clay till. Instead, the site encountered large ground-water flows that required costly changes and new construction methods, thereby delaying the project several months.

The Leverett Saltonstall State Office building was designed initially with a pile foundation that would penetrate into the tough and durable till. However, the piles never reached the accepted resistance stage; rather, they failed because the subsurface granular materials were not supportive. Instead of the hills being drumlins composed of lodgement till, the terrain was an end-moraine deposit that consisted of a series of imbricate slices of granular materials, ranging from clay to boulders. The stratified morainal sediments had entirely different physical properties than the predicted till. Although the landforms appeared similar, the subsurface deposits had been created by entirely different glacigenic processes and contained vastly different sediments. Kaye (1976) concluded the Boston cases demonstrated that: (1) the erosive power of ice sheets is variable, (2) ice sheets do not necessarily obliterate all preexisting deposits, (3) morainal deposits can be redistributed by meltwater, (4) surface drift should not be assumed as the last ice sheet, (5) lack of deep weathering does not always indicate age of drift, and (6) englacial transport of debris can cause confusion in interpreting its history.

Dam-reservoir construction

Fluhr (1964) and Legget (1982) have demonstrated the importance of knowing the geologic setting, origin, and characteris-

TABLE 8. EARTHFLOWS AND LANDSLIDES IN LEDA CLAY
(modified from Crawford, 1961)

Number	Date	Name	River	Acres	Yd³	Victims
1	Dec. 7, 1955	Hawkesbury		12	500,000	
2	Nov. 12, 1955	Nicolet	Nicolet	0	250,000	3
3	Aug. 3 and 6, 1951	Rimouski	Rimouski	25	1,000,000	
4	May 18, 1945	St. Louis	Yamaska	7		
5	Sept. 9, 1938	Ste. Genevieve	Batiscan	15		
6	July 24, 1935	St. Vallier	Des Meres	15		
7	Fall, 1924	Matteau Farm	St. Maurice	10		
8	Apr. 6, 1908	Notre-Dame de la Salette	Lievre	0		33
9	Oct. 11, 1903	Poupore	Lievre	100		
10	May 7, 1898	St. Thuribe	Blanche	80	3,500,000	1
11	Sept. 21, 1895	St. Luc de Vicenne	Champlain	5		5
12	Apr. 27, 1894	St. Alban	Ste. Anne	1,000	25,000,000	4
13	Apr. 4, 1840	Maskinonge	Maskinonge	84		
14	----	Green Creek	Ottawa	30	2,000,000	
15	----	Azatika Creek	Ottawa	10,000		

tics of site-related materials for successful dam construction. Six major dams were constructed in the Catskill Mountains for the New York City water supply between the years 1907 and 1963 (Fluhr, 1964). The dams were designed with a central impermeable core of glacial till, with semipermeable soil shoulders. This design was selected because the Catskill region has a great assortment of glacial materials for such projects. For example, the Downsville Dam has impervious till above the bedrock floor, and additional till was used for the dam core. A nearby kame contained suitable glaciofluvial deposits for use as the semipervious shell. At the Cannonsville Dam, a large hill of glacial materials formed a natural topographic site for extension of the dam across the West Branch Delaware River. The hill contained till overlain by varved clays, with another till unit above, and was capped by glaciofluvial sand and gravel. A till deposit on the opposite valley side was used for core and embankment material. Again a nearby kame supplied the necessary sediment for the shell.

The Portage Mountain Dam on the Peace River, British Columbia, is 200 m wide and 1,250 m long. It is an earth-fill dam composed of glacial materials containing 1.63×10^6 m^3 of impervious sediment. Nearby moraines contained the appropriate material, helping to make the dam cost effective (Legget, 1982). Philbrick (1976) provided a case history of the Kinzua Dam, Pennsylvania, and showed the necessity of making a complete areal investigation of terrain conditions and possible sources of construction materials. The early predesign investigations located an ideal dam site relative to foundation level and sources of embankment fill and aggregate.

Unfortunately, several critical factors relevant to glacial deposits were not considered in construction of the Lake Lee Dam near Lee, Massachusetts. The dam broke on March 28, 1968; the resulting flood killed two people, and caused damages of $10 million. Failure was attributed to piping of the glacial sediments, which produced collapse after underground seepage became channeled under the structure through the destabilized piped conduits.

A similar catastrophe was averted in the Garrison Dam, North Dakota, by construction of relief wells to prevent piping of the glacial sediment and the accompanying development of sand boils (Arnold, 1964).

The Binghamton region of New York has 20 small dams to the north of the urban area to protect it from flood damages. The dams were built under provisions of Public Law 566, the Watershed Protection and Flood Prevention Act. The successful bidder for one of the dams—Old Forge Construction Company, based in the Adirondack Mountains—began work in April 1974. Unfortunately, workers encountered numerous delays because the character of the substrate materials was more formidable than had been anticipated. Subsequently the company requested additional money from the owners, but funds were denied. The contract documents had provided all the information necessary to predict the character of the materials, but Old Forge had not bothered to evaluate all the bid documents. Instead they had assumed the materials at the site would be the same as those that occur in the Adirondack valleys, where the glacial sediments are sandy, loose, and easily removable. In contrast, the till in southern New York is exceptionally tough, compact, and uncommonly difficult to work when wet. The southern New York firms were aware of the difficult till, and that is why their cost estimates were higher than Old Forge.

For many years the U.S. Army Corps of Engineers had contemplated construction of a dam on the Cowanesque River of north-central Pennsylvania. Intensive field investigations commenced in 1965 and included mapping the surficial features and

Figure 7. New roads and curbs at the relocated site for the village of Nelson, Pennsylvania. The ponding occurs at pressure release positions of piped glaciolacustrine sediment from the hillslope to the left of the photograph. Subsidence and failures are also present along the roadway. Courtesy of John Judski.

deposits of the area to be impacted by the reservoir pool and adjacent lands. Special attention was given to those glacial sediments that might produce instabilities or landslides, contributing to the failure of a structure (Coates, 1966b, 1977). The east-flowing Cowanesque River has a drainage of 764 km^2. During glaciation, ice imponded its outlet, causing a series of glaciolacustrine and deltaic sediments to form in the proglacial lakes. The special types of information that were needed included: (1) analysis of the sedimentary materials for the dam, (2) character of materials and their wetted stability at the lateral sides of the reservoir, and (3) nature of deposits throughout alignment of rerouted roads.

In 1978, contracts were given to several construction firms to begin the process of building new roads and relocating the village of Nelson, Pennsylvania, which lay in the valley to be flooded by the reservoir. The engineering contractor was not provided a copy of the geological report, which had delineated the areas of potential problems throughout the site for the village relocation, particularly the area of notorious glaciolacustrine sediments. Unfortunately, specifications for the job made no reference to such potential difficulties nor the type of material to be encountered; all materials were described as "normal soils." The successful contractor was not alerted to the potential difficulties ahead.

Soon after work commenced, unexpected problems began to occur. The problems included excessive water flooding the construction sites; failure of slopes; slurrying of sediment; local ponding of water; and subsidence of many structures such as roadways, curbs, drains, and buildings (Fig. 7). These conditions were caused by the piping within glacial materials that released abnormal amounts of sediment and water, accompanied by mud boils and ground heaving. Figure 8 shows the attempt to grade hillsides and the resulting piped terrane when the bank was excavated; Figure 7 shows the excess water created through the piped

Figure 8. This roadcut excavation near the new location for the village of Nelson, Pennsylvania, has destabilized the adjacent hillslope. Piping has occurred in these glaciolacustrine sediments of the Cowanesque Valley. Courtesy of John Judski.

terrane. Because of the disregard for the known geologic conditions and materials, an entire new drainage system had to be installed, and some 15,000 tons of rock, sand, and gravel were imported to stabilize the area. This oversight resulted in a cost overrun of more than $844,000.

Other considerations

Throughout North America, more than 10 million km^2 have been glaciated and covered by glacial deposits. The need for an understanding of the history and characteristics of these glacial deposits is vital whenever engineering works are to be built and operated. Land-use evaluations prior to construction of foundations, buildings, dams, roads, water-supply systems, and landfills must include an understanding of the geological environment and the character of the near-surface materials.

Water content in till. Eden (1976) has demonstrated the importance of water content in till relevant to its construction suitability. Many problems were encountered in the cut-and-fill operation when constructing the alignment for the Quebec North Shore and Labrador Railway. The route traversed a ground moraine with drumlinized ridges for about 240 km. In the cut sections the equipment became mired in sediment that liquefied, and nearly equal difficulties occurred in fill areas. The unstable till had a 14.3 percent water content, so that about 4 percent of water content needed to be extruded to reach optimum dry density of 10.2 percent. However, because of the impermeable nature of the till, the attempts at densification were initially unsuccessful due to high pore-water pressure and the unstable behavior of the sediment. Dikes were necessary in the construction of the Churchill Falls Power Project in 1969 to 1970. Unfortunately, till was the only available nearby borrow material, but it proved to be unstable. It could not be sufficiently compacted to the design density and permeability because the water content exceeded specifications. Drier material had to be transported from great distances, thus increasing the project costs (Eden, 1976).

Orientation of features. The Susquehanna section of the glaciated Appalachian Plateau has several unique features caused by glacial processes; knowledge of such factors can be critical in land-use planning for engineering works (Coates, 1966b, 1972). For example, the till-shadow effect of the upland hills has great influence on the positioning of housing developments, roads, and other facilities. Such sites are also the only successful locations for solid-waste disposal in landfills. Furthermore, south-flowing streams have more regularized streamflow than north-flowing streams. Their flood peaks are more subdued, and their baseflow is 5 to 10 times greater than that of similar size drainages that flow north. Simply analyzed, during deglaciation, south-flowing meltwaters flushed out the fine materials, leaving as lag mainly sand and gravel. However, the north-flowing waters became ponded against a gradually receding ice margin, and thereby deposited more fine-grained sediments than their south-flowing counterparts. Consequently, the design of any flood-control measures can be different for a north versus south-flowing stream. The availability of water is greater for south-oriented drainages, and any related sewage facilities would require less elaborate design because of greater sustained flows.

Disposal siting. Many liquid and solid wastes are emplaced into the substrate for disposal and incarceration. In glaciated terrane, this invariably means that glacial sediments will be the host material. The design of leach fields from septic systems relies on the permeability of the glacial soils. Special sand filters are often required when such soils fail to pass percolation tests. In southern New York, many of the older waste-disposal landfills were sited in highly permeable glacial deposits; this has resulted in unacceptable levels of leachate moving offsite. The Landstrom landfill near Spencer, New York, is located in valley-fill sediments of the Cayuga trough with a mixture of Valley Heads moraine and granular glacial sediments. The Colesville landfill near Windsor, New York, is located in a valley-train terrace where leachate has moved offsite, impacting nearby farms and houses. The Conklin landfill south of Binghamton is also situated on some sloping terrain of the Susquehanna River where the substrate glacial materials are granular with moderately high hydraulic conductivity. Within the past 20 years, the leachate has moved offsite and impacted 30 downgradient homeowner's wells. Major lawsuits are presently in the courts on all three of these landfills.

The Chemung County landfill is located on a till-shadow hill east of Elmira, New York. The site has been in operation since 1972, and by 1986 the permitted area had nearly been filled to capacity. A geological investigation (Empire Soils, 1986) concluded that outwash occurred at some depth below the landfill, and indicated the presence of other glacial materials with relatively high hydraulic conductivities. Because of these findings, the New York State Department of Environmental Conservation requested Chemung County to abandon the site for disposal of solid waste. Fortunately, because of the \$1.3 million expenditure that would be needed to relocate for a new landfill operation, county officials requested permission to make a detailed hydrogeologic investigation of the present site to determine the feasibility for expansion to contiguous properties. Two new studies were completed in 1986 (Coates and Timofeeff, 1986a, 1986b) to determine: (1) whether outwash occurs under the site, (2) the hydraulic conductivities for the different types of glacial deposits, (3) a three-dimensional analysis of the sediment package under the disposal trenches, (4) whether site materials could be recompacted as suitable material for the lining, (5) could on-site material be used as cover, (6) a dewatering system for a confined upgradient aquifer, and (7) the data base for design of trenches, drain systems, holding ponds, and off-site drainage.

Four different kinds of glacial deposits occur at the Chemung County landfill: lodgement till, melt-out ablation till, glaciolacustrine silt-clay, and ice-contact deposits. The glacial deposits were mapped, and a series of 49 exploratory wells and borings drilled, some to depths of 52 m. From these data, various maps were prepared that included depth to bedrock, potentiometric contour maps, isopach contour maps of the four different deposits, and depth to water. Pumping tests established cones of influence and the hydrologic properties of the sediments; laboratory analyses calculated texture ratios, permeability, and recompaction properties. The hydraulic conductivity of the ice contact deposits was 10^{-5} cm/sec, but for the other sediments this ranged between 10^{-6} and 10^{-8} cm/sec. The geological investigations documented that the landfill site was suitable for expansion because of the presence of a thick buffer of glacial sediments and of materials with low hydraulic conductivities; the materials could be used both as lining for the landfill and as cover material; also, if leachate did move off-site, it would take more than 2,000 years to reach any aquifer in the Chemung Valley. Thus reevaluation and detailed analysis of the site materials proved to be wise and beneficial to Chemung County.

CONCLUSIONS AND OUTLOOK

Glacial deposits make up the most complex and variable group of surficial sediments formed by any of the geomorphic processes. Because of the great diversity of these materials, it is very easy, and common, to misinterpret them. Therefore, it is important when use or application of such deposits is anticipated that preconstruction geoscience studies be multidisciplinary—including such specialists as engineering geologists, glacial geologists, and hydrogeologists—to minimize the effect of possible damage failures, and cost overruns.

Glacial deposits are broadly placed in two groups—nonstratified and stratified. Till is a diamicton that is nonsorted, nonstratified, and formed directly from glacial processes. Stratified sediments are either ice contact or outwash. From a genetic viewpoint the difference is one of position, wherein ice-contact materials form from meltwaters on, within, or under the ice, and outwash is carried beyond the ice margin by fluvial processes. The geotechnical properties of these three different kinds of material can show vast variations. Because till is not granular, its porosity, permeability, bearing strength, and other engineering qualities are greatly different than the stratified sediments.

Some glacial landforms are distinctive and possess the type of glacial deposits that can be predicted; drumlins are usually composed of till, whereas kames and eskers contain stratified sediments. Unless extreme care is taken, however, landforms can be confused. In the Beacon Hill area of Boston, Massachusetts, failure to recognize such differences led to construction damage and extra costs.

Many successful engineering projects can be cited, such as the construction of the Trans-Alaska Pipeline, where the accurate knowledge of glacial deposits was critical (Péwé, this volume). However, lessons can also be learned from those case histories where inadequate or incorrect attention was given to the glacial substrate, such as in the real-estate development of the Turnagain Heights section of Anchorage, Alaska, devastated by the 1964 Alaskan earthquake. Similarly, the urban planning in many Canadian communities in the St. Lawrence Valley placed buildings on the glacial Leda Clay, which subsequently triggered damaging landslides. Likewise, preconceived ideas about underlying glacial deposits have led to inadequate geological investigations and testing at many construction sites, such as Flushing, New York, and the St. Lawrence Seaway Project. Similarly, cost overruns have also been incurred by engineering firms working on dams and highways. Frequently, faulty information regarding the character of the glacial deposits is provided by the engineer-designers to construction crews.

In overview, land-use planners and environmental managers must work in close liaison with geoscientists and engineers when engineering works are undertaken in glaciated terrane. This interdependence will become even more important in the future because, increasingly, the more suitable lands have been used. Those that remain are in less desirable locations, or contain potentially hazardous conditions (see Kiersch, Chs. 1, 2, and 18, this volume). Consequently, increased levels of geological investigation and geotechnical testing and analysis will be required, which will lead to an improvement in the ability to predict how glacial deposits will respond to the construction and operation of engineered works.

One subfield where an improved knowledge of glacial deposits is becoming increasingly crucial is waste disposal. New guidelines are constantly being drafted, upgraded, and amended that require highly sophisticated geoscience and engineering studies at sites proposed for waste disposal and landfilling. In these situations, glacially oriented geologists must work closely with hydrogeologists to determine the pertinent characteristics and properties of materials involved. Ten years ago the geological and engineering costs to obtain a state-approved landfill permit in New York totalled a few hundred thousand dollars; these costs have now escalated to several million dollars because of the requirements for providing a multitude of new and different tests of the glacial deposits.

Engineers are called upon to use glacial materials most frequently in the form of sand and gravel deposits for construction. As these materials become more scarce, it is necessary to locate new sources as close as possible to the project. The skill to accomplish this must come from applied geoscientists who understand the origin and occurrence of glacial deposits. In a similar manner, till can be used for embankment and fill purposes, but some varieties are more suitable. Excavations in glacial materials must invariably assess the hydrogeologic conditions to be intercepted. Fortunately, new techniques are constantly being developed by geoscientists that enhance and aid the efforts of the applied geologist for engineering works in glaciated terrane. Yet paramount is a broad background in glacial geology and the field characteristics of glacigenic deposits.

REFERENCES CITED

Anderson, W. F., 1983, Foundation engineering in glaciated terrain, *in* Eyles, N., ed., Glacial geology: Oxford, Pergamon Press, p. 275–301.

Arnold, A. B., 1964, Relief well on the Garrison Dam and Snake River embankment, North Dakota, *in* Kiersch, G. A., ed., Engineering case histories no. 5: Boulder, Colorado, Geological Society of America, p. 45–52.

Baracos, A., 1957, The foundation failure of the Transcona elevator: Engineering Journal, v. 40, p. 973.

Blank, J. P., 1971, The town that disappeared: Reader's Digest, v. 99, no. 12, p. 86–90.

Boulton, G. S., 1970, On the deposition of subglacial and melt-out tills at the margins of certain Svalbard Glaciers: Journal of Glaciology, v. 9, p. 231–245.

——, 1971, Till genesis and fabric in Svalbard, Spitsbergen, *in* Goldthwait, R. P., ed., Till; A symposium: Columbus, Ohio State University Press, p. 41–72.

——, 1976, The development of geotechnical properties in glacial tills, *in* Leggett, R. F., ed., Glacial till: Royal Society of Canada Special Publication 12, p. 292–303.

——, 1978, The genesis of glacial tills; A framework for geotechnical interpretation, *in* The engineering behavior of glacial materials: Geo Abstracts, p. 52–58.

Boulton, G. S., and Dent, D. L., 1974, The nature and rates of post-depositional changes in recently deposited till from southeast Iceland: Geografiska Annalar, v. 56, Series A., p. 121–134.

Cadwell, D. H., 1986, Quaternary geology of New York, Finger Lakes sheet: New York State Museum and Science Service Map and Chart Series no. 40, scale 1:250,000.

Chamberlin, T. C., 1894, Proposed genetic classification of Pleistocene glacial formations: Journal of Geology, v. 2, p. 517–538.

Cleaves, A. B., 1963, Engineering geology characteristics of basal till, St. Lawrence Seaway Project, *in* Trask, P. D., and Kiersch, G. A., eds., Engineering geology case histories no. 4: Boulder, Colorado, Geological Society of America, p. 51–57.

Coates, D. R., 1966a, Glaciated Appalachian Plateau; Till shadows on hills: Science, v. 152, p. 1617–1619.

—— , 1966b, Report on the geomorphology of the Cowanesque Basin, Pennsylvania: Baltimore District, U.S. Army Corps of Engineers Cowanesque Reservoir Study, 27 p.

—— , 1972, Hydrogeomorphology of Susquehanna and Delaware Basins, *in* Morisawa, M., ed., Quantitative geomorphology: Binghamton, State University of New York, p. 273–306.

—— , ed., 1975, Quaternary deformed sediments of the St. Lawrence Lowland as an index of seismicity: New York State Atomic and Space Development Authority, 264 p.

—— , 1977, Landslide perspectives, *in* Coates, D. R., ed., Landslides: Boulder, Colorado, Geological Society of America Reviews in Engineering Geology, v. 3, p. 3–28.

—— , ed., 1982, Glacial geomorphology: London, George Allen and Unwin, 398 p.

Coates, D. R., and Timofeef, N. P., 1986a, Hydrogeologic conditions of the Chemung County landfill expansion site: Chemung County Solid Waste Disposal District, 103 p.

—— , 1986b, Hydrogeologic assessment for expansion of the Chemung County landfill, part 2: Chemung County Solid Waste Disposal District, 132 p.

Cocksedge, J. E., 1983, Road construction in glaciated terrain, *in* Eyles, N., ed., Glacial geology: Oxford, Pergamon Press, p. 302–317.

Crawford, C. B., 1961, Engineering studies of Leda Clay, *in* Leggett, R. F., ed., Soils in Canada: Royal Society of Canada Special Publication 3, p. 100–217.

Crawford, C. B., and Burn, K. N., 1962, Settlement studies on the Mt. Sinai Hospital, Toronto: Engineering Journal, v. 45, p. 31–37.

Crawford, C. B., and Eden, W. J., 1963, Nicolet landslide of November 1955, Quebec, Canada, *in* Trask, P. D., and Kiersch, G. A., eds., Engineering geology case histories no. 4: Boulder, Colorado, Geological Society of America, p. 45–50.

Davidson-Arnott, R., Nickling, W., and Fahey, B., eds., 1982, Research in glacial, glacio-fluvial, and glacio-lacustrine systems: Norwich, Geo Books, 318 p.

DeJong, J., and Morgenstern, N. R., 1973, Heave and settlement of two tall building foundations in Edmonton, Alberta: Canadian Geotechnical Journal, v. 10, p. 261–281.

Dreimanis, A., 1969, Selection of genetically significant parameters for investigation of tills: Zesz, Nauk. Univ. im. A., Mickiewicza W. Posnanium, Geografica, v. 8, p. 15–29.

—— , 1976, Tills; Their origin and properties, *in* Legget, R. F., ed., Glacial till: Royal Society of Canada Special Publication 12, p. 11–49.

Drewry, D. J., 1986, Glacial geologic processes: Baltimore, Maryland, Edward Arnold, 288 p.

DuMontelle, P. B., Jacobs, A. M., and Bergstrom, R. E., 1971, Environmental terrane studies in the East St. Louis area, Illinois, *in* Coates, D. R., ed., Environmental geomorphology: Binghamton, State University of New York, p. 201–212.

Eden, W. J., 1957, The Hawkesbury landslide, *in* Proceedings, Tenth Canadian Soil Mechanics Conference: National Research Council, Associate Committee on Soil Mechanics, Technical Memorandum 46, p. 14–22.

—— , 1976, Construction difficulties with loose glacial tills on Labrador Plateau, *in* Legget, R. F., ed., Glacial till: Royal Society of Canada Special Publication 12, p. 391–400.

Embleton, C., and King, C.A.M., 1975, Glacial geomorphology: London, Edward Arnold, 573 p.

Empire Soils, 1986, Chemung County landfill hydrogeologic investigation: Groton, New York, Thomsen Associates, various pagination and appendices.

Eyles, N., 1979, Facies of supraglacial sedimentation on Icelandic and Alpine temperate glaciers: Canadian Journal of Science, v. 16, p. 1341–1361.

—— , 1983, Glacial geology; A landsystem approach, *in* Eyles, N., ed., Glacial geology: Oxford, Pergamon Press, p. 1–18.

—— , 1983, Glacial geology: Oxford, Pergamon Press, 409 p.

Fairbridge, R. W., ed., 1968, The encyclopedia of geomorphology: New York, Reinhold Book Corporation, 1295 p.

Fiksdal, A. J., 1982, Landforms for planning use in part of Pierce County, Washington, *in* Craig, R. G., and Craft, J. L., eds., Applied geomorphology: London, George Allen and Unwin, p. 44–54.

Flint, R. F., 1971, Glacial and Quaternary geology: New York, John Wiley and Sons, 892 p.

Flint, R. F., Sanders, J. E., and Rodgers, J., 1960, Diamictite; A substitute term for symmictite: Geological Society of America Bulletin, v. 71, p. 1809–1810.

Fluhr, T. W., 1964, Earth dams in glacial terrain, Catskill Mountain region, New York, *in* Kiersch, G. A., ed., Engineering geology case histories no. 5: Boulder, Colorado, Geological Society of America, p. 15–29.

Folk, R. L., 1975, Glacial deposits identification by chattermark trails in detrital garnets: Geology, v. 3, p. 473–475.

—— , 1977, Glacial deposits identification by chattermark trails in detrital garnets; Reply: Geology, v. 5, p. 249.

Fookes, P. G., Gordon, D. L., and Higginbottom, I. E., 1978, Glacial landforms; Their deposits and engineering characteristics, *in* The engineering behaviour of glacial materials: Midland Soil Mechanics and Foundation Engineering Society, Geo Abstracts, p. 18–51.

Geikie, A., 1863, On the phenomenon of the glacial drift of Scotland: Geological Society of Glasgow Transactions, v. 1, pt. 2, 190 p.

Geo Abstracts, 1978, The engineering behaviour of glacial materials: Norwich, England, Geo Abstracts, 240 p.

Goldthwait, R. P., ed., 1971, Till: Columbus, Ohio State University Press, 402 p.

—— , ed., 1975, Glacial deposits: Stroudsburg, Pennsylvania, Dowden, Hutchinson, and Ross, 464 p.

Grisak, G. E., Cherry, J. A., Vonhof, J. A., and Blumele, J. P., 1976, Hydrogeologic and hydrochemical properties of fractured till in the Interior Plains region, *in* Legget, R. F., ed., Glacial till: Royal Society of Canada Special Publication 12, p. 304–335.

Gustavson, T. C., and Boothroyd, J. C., 1987, A depositional model for outwash, sediment sources, and hydrologic characteristics, Malaspina Glacier, Alaska; A modern analog of the southeastern margin of the Laurentide Ice Sheet: Geological Society of America Bulletin, v. 99, p. 187–200.

Hansen, W. R., 1965, Effects of the earthquake of March 27, 1964, at Anchorage, Alaska: U.S. Geological Survey Professional Paper 542-A, 68 p.

Horberg, L., 1952, Pleistocene drift sheets in the Lethbridge region, Alberta, Canada: Journal of Geology, v. 60, p. 303–329.

Huang, P., 1982, Quaternary climatic evolution of China and glacial evidences problems in Lu Shan: Glaciation and Frozen Earth, v. 4, no. 3, p. 96–102.

Kaye, C. A., 1976, Beacon Hill end moraines, Boston; New explanation of an important urban feature, *in* Coates, D. R., ed., Urban geomorphology: Geological Society of America Special Paper 174, p. 7–20.

Koteff, C., 1982, The morphological sequence concept and the deglaciation of southern New England, *in* Coates, D. R., ed., Glacial geomorphology: London, George Allen and Unwin, p. 121–144.

Krieg, R. A., and Reger, R. D., 1976, Preconstruction terrain evaluation for the Trans-Alaska Pipeline Project, *in* Coates, D. R., ed., Geomorphology and engineering: Stroudsburg, Pennsylvania, Dowden, Hutchinson, and Ross, p. 55–76.

Lavrushin, Y. A., 1970a, Reflection of the dynamics of glacier movement in the structure of a ground moraine: Litologiya i Poleznye Iskopaemye, v. 1,

p. 115–120. (in Russian)

—— , 1970b, Recognition of facies and subfacies in ground moraine of continental glaciation: Litologiya i Poleznye, v. 6, p. 38–49. (in Russian)

Lawson, D. E., 1979, Sedimentological analysis of the western terminus region of the Matanuska Glacier, Alaska: Cold Regions Research and Engineering Laboratory Report 79-9, 112 p.

—— , 1981, Sedimentological characteristics and classification of depositional processes and deposits in the glacial environment: Cold Regions Research and Engineering Laboratory Report 81-27, 16 p.

Legget, R. F., 1942, An engineering study of glacial drift for an earth dam near Fergus, Ontario: Economic Geology, v. 37, p. 531–556.

—— , ed., 1976, Glacial till: Royal Society of Canada Special Publication 12, 412 p.

—— , 1982, Glacial landforms and civil engineering, *in* Coates, D. R., ed., Glacial geomorphology: London, George Allen and Unwin, p. 351–374.

Legget, R. F., and Karrow, P. F., 1983, Handbook of geology in civil engineering: New York, New York, McGraw-Hill Book Company, various pagination.

Li, S., 1947, Lu Shan in glacial stages: Central Geological Institute of China Special Edition 2, p. 73–84.

—— , 1975, Quaternary glaciation of China: China, Science Press, p. 312–388.

Loiselle, A. A., and Hurtubise, J. E., 1976, Properties and behaviour of till as construction material, *in* Legget, R. F., ed., Glacial till: Royal Society of Canada Special Publication 12, p. 346–363.

Lugn, A. L., 1962, The origin and sources of loess: Lincoln, University of Nebraska Studies, new series, v. 26, 105 p.

Lyell, C., 1839, Nouveaux elements de geologie: Paris, Pitois-Levraulat and Cie, 648 p.

McGown, A., Anderson, W. F., and Radwan, A. M., 1978, Geotechnical properties of tills in west-central Scotland, *in* The engineering behaviour of glacial materials: Norwich, Geo Abstracts, p. 81–91.

Miall, A. D., 1983, Glaciofluvial transport and deposition, *in* Eyles, N., ed., Glacial geology: Oxford, Pergamon Press, p. 168–183.

Mickelson, D. M., 1973, Nature and rate of till deposition in a stagnating ice mass, Burroughs Glacier, Alaska: Arctic and Alpine Research, v. 5, p. 17–27.

Milligan, V., 1976, Geotechnical aspects of glacial tills, *in* Legget, R. F., ed., Glacial till: Royal Society of Canada Special Publication 12, p. 269–291.

Muller, E. H., 1977, Quaternary geology of New York, Niagara Sheet: New York State Museum and Science Service Map and Chart Series 28, scale 1:250,000.

Orr, E. D., and Folk, R. L., 1983, New scents on the chattermark trail; Weathering enhances obscure microfractures: Journal of Sedimentary Petrology, v. 53, no. 1, p. 121–129.

Parsons, J. D., 1976, New York glacial lake formation of varved silt and clay: American Society of Civil Engineers, Journal of the Geotechnical Engineering Division, v. 102, no. GT6, p. 605–638.

Pecsi, M., 1968, Loess, *in* Fairbridge, R. W., ed., The encyclopedia of geomorphology: New York, Reinhold Book Corporation, p. 674–679.

Philbrick, S. S., 1976, Kinzua Dam and the glacial foreland, *in* Coates, D. R., ed., Geomorphology and engineering: Stroudsburg, Pennsylvania, Dowden, Hutchinson, and Ross, p. 175–197.

Price, R. J., 1973, Glacial and fluvioglacial landforms: Edinburgh, Oliver Boyd, 242 p.

Rodgers, J., 1950, The nomenclature and classification of sedimentary rocks: American Journal of Science, v. 248, p. 297–311.

Rowe, P. W., 1972, The relevance of soil fabric to site investigation practice: Geotechnique, v. 22, p. 195–300.

Shepps, V. C., 1968, Glacial deposits, *in* Fairbridge, R. W., ed., The encyclopedia of geomorphology: New York, Reinhold Book Corporation, p. 430–431.

Si, Y., 1981, Was there really Quaternary glaciation in Lu Shan?: Dialectic Natural Bulletin, no. 2, p. 68–74.

Sladen, J. A., and Wrigley, W., 1983, Geotechnical properties of lodgement till; A review, *in* Eyles, N., ed., Glacial geology: Oxford, Pergamon Press, p. 184–212.

Sugden, D. E., and John, B. S., 1976, Glaciers and landscape; A geomorphological approach: New York, John Wiley and Sons, 376 p.

Terzaghi, K., and Peck, R. B., 1967, Soil mechanics in engineering practice: New York, John Wiley and Sons, p. 729.

Zhou, T., 1983, Palaeogeography: Beijing, China, Beijing Teachers' University Press, p. 141–290.

Manuscript Accepted by the Society November 6, 1987

Geological Society of America
Centennial Special Volume 3
1991

Chapter 16

Aggregates

Katharine Mather
Consultant, 213 Mount Salus Road, Clinton, Mississippi 39056
Bryant Mather
Chief, Structures Laboratory, U.S. Army Waterways Experiment Station, Vicksburg, Mississippi 39180-0631

INTRODUCTION

The geologist concerned with aggregate should be aware that under the term *aggregate* there is a very broad range of geological materials. In the foreword to Dolar-Mantuani (1983) we wrote:

> Every kind of rock and every mineral species that occurs on this planet as a solid particle, grain, or mass—except ice—is potentially subject to evaluation for use as concrete aggregate. Therefore there is no kind of rock and no mineral species that is not of potential interest to the petrographer working on concrete aggregates. However those substances that can significantly affect the performance of concrete for better or worse when they occur as aggregate constituents are more important and, among these, those that occur most frequently are of greater interest.

Any bit of solid material used as an inclusion in a matrix or binder in a composite construction material is considered to be aggregate if it remains solid at the temperature of use and does not dissolve in the environment of use, so long as it is used as particulate material. As noted in the quotation, ice is not evaluated as aggregate for concrete, but ice could be used as aggregate in an appropriately low-temperature environment. Rock salt—halite (NaCl)—which would dissolve in many environments, is for that reason, among others, rarely used as an aggregate, yet rock salt has been quite successfully used as fine and coarse aggregate with portland cement, and with a saturated NaCl solution used as mixing water, to make concrete to plug a tunnel in a salt mine (Polatty and others, 1961). Interestingly, crushed rock salt is being considered as the backfill material in connection with underground disposal of radioactive waste, and work on salt-saturated grout for use in this same work is under way (Wakeley and Burkes, 1986).

Historical background

It is impossible to set a definite date when "applied geologists" began contributing to the search for and selection of natural aggregates for use in engineering works. People who did this sort of thing in ancient times did not consider themselves "geologists," as it was long before the term had been coined, yet they had an insight and skill with earth materials (see Kiersch, Chapter 1, this volume). Even today, there are people who do not consider themselves geologists who have assimilated from education and experience sufficient awareness of the relations of location of deposit, nature of material, use to which it is to be put, probable processing and transportation costs, and requirements for mechanical and other properties to do a very creditable job of seeking out and selecting aggregate deposits. In the United States the early emergence of engineering geology, as related to the search for aggregates, came with the need for road building. Kiersch (1955), p. 17–18 wrote:

> Generally, America was expanding to new territory before 1900 instead of concentrating on building up one small area as in sections of Europe. As part of this territorial expansion, some improved roads were constructed during the early 1800s. However, before the middle of the century, road building was curtailed in favor of constructing a nationwide railroad network. It was not until the invention of the bicycle in 1877 and the appearance of the automobile in the 1890s that the interest in improved roads was revised. At first this need was met by constructing traffic-bound roads and later water-bound macadam surface highways. Reports on the influence of geology on roads appear in English publications as early as 1888 (Huntting, 1945). The important geologic aspect of this highway construction, both in America and Europe, was the immediate demand for serviceable crushed stone. As an outgrowth, geology made its initial contribution to highway construction with reports on the source of suitable road materials. The earliest survey in the United States was on the road materials of Texas by Hill (1889), followed by similar reports published for North Carolina in 1892 and 1893 and Florida in 1893.
>
> The reports prepared by N. S. Shaler (1895) on the road materials of Massachusetts, together with his responsible position as a member of the Massachusetts Highway Commission, support his designation as the first highway engineering geologist. This initial development stage of geology as applied to highway construction actually extended from 1889 to about 1918, and was almost entirely concerned with the location and description of suitable construction materials.
>
> In keeping with the trend of an increased demand for surfaced roads after 1900, a growing number of reports on highway materials appeared in the geologic literature between 1900 and 1918. Some were made by

Mather, K., and Mather, B., 1991, Aggregates, *in* Kiersch, G. A., ed., The heritage of engineering geology; The first hundred years: Boulder, Colorado, Geological Society of America, Centennial Special Volume 3.

the Federal Survey, but most of the studies were conducted by the state geological surveys, i.e., W. B. Clark, H. F. Reid, and G. B. Shattuck (Maryland), J. A. Holmes and J. H. Pratt (North Carolina), and W. O. Hotchkiss (Wisconsin). The latter work was, however, a cooperative endeavor by the state geological surveys and highway commissions. L. W. Page, who directed the U.S. Office of Public Roads from 1905 to 1919, was, at one time, geologist for the Massachusetts Highway Commission. H. A. Buehler, well-known geologist, was an ex-officio member of the State Highway Commission of Missouri for many years.

By 1918 at least 49 papers on geology as applied to highway engineering had been published. These discussed the road-material sources of 24 states and included several reports on the relationship of mineral composition to the engineering characteristics of the rock. Probably the most extensive investigation of this group was a statewide survey of aggregate sources for highway construction by W. C. Morse for the Ohio Department of Highways in 1913 (Marshall and Maxey, 1950). Potential aggregate sources were tested and results, together with a brief description of the geology and potential aggregate sources, are included in a two-volume report entitled "Road Construction Materials in Ohio," an unpublished report on file in the Highway Department. This report has, over the years, been of considerable value to highway engineers and likewise has served as a basis for subsequent aggregate surveys.

In addition, a noteworthy contribution to highway construction in this country became available in the early 1900s on the materials and manufacturing methods of both the portland-cement and paving-brick industries (Eckel, 1905)."

GEOLOGICAL PROPERTIES OF AGGREGATES

Introduction

The literature on aggregate materials, aggregate material properties, requirements for aggregate, and behavior of aggregates in service is extensive and readily available in the context of particular classes of use. Hydraulic-cement concrete, asphaltic concrete, road basecourse, railroad ballast, filter media, and rock-fill for dams are among the classes. The ASTM (1948) symposium provides as broad a coverage as any we know of, including what is still the best review of aggregate distribution (Woods, 1948). The more recent ASTM state-of-the-art reports, such as ASTM (1978), are narrower in scope, dealing only with concrete aggregates, but they cover such topics as petrographic examination (R. C. Mielenz); grading (W. H. Price); shape, surface texture, surface area, and coatings (M. A. Ozol); weight, density, absorption, and surface moisture (W. G. Mullen); porosity (W. L. Dolch); abrasion resistance, strength, toughness, and related properties (R. C. Meininger); thermal properties (H. K. Cook); chemical reactions other than carbonate reactions (S. Diamond); chemical reactions of carbonate aggregates in cement paste (H. N. Walker); and soundness and deleterious substances (L. Dolar-Mantuani).

In the American Concrete Institute, Committee 221 was organized in 1936 to deal with normal-weight aggregates, and Committee 213 was organized in 1946 to deal with lightweight aggregates. Each committee has published a guide (Lamond, 1984; Holm, 1979). Popovics (1979) discussed aggregates in nine of the 15 chapters in his book. He treated aggregate properties as physical, chemical, and geometric. In our foreword to Dolar-Mantuani (1983), we commented that it was, so far as we knew, the first book in English on aggregates since that of Knight and Knight (1948) and the first in North America. We further remarked:

> As persons educated as geologists who have spent their professional careers in research, development, testing, and evaluation related to concrete and concrete aggregates, we are aware of the need to bring together in one volume a wide variety of summarized scientific and technological information on concrete aggregates. Now that such a compilation exists it can, should, and will be studied and referred to by scientists, engineers, and technicians who need to learn how to do these things and by other scientists, engineers, executives, administrators, and lawyers who need to understand the significance of what is done by scientists and engineers who evaluate concrete aggretates.

Paige and Rhoades (1948) found it gratifying that applications of geology and petrography to concrete and concrete aggregates had become greatly expanded, especially through the adoption of petrographic methods; Fears (1949) provided an annotated bibliography on mineral aggregates for the Highway Research Board; Blanks (1950) reviewed progress largely through research at the U.S. Bureau of Reclamation as it related to concrete aggregate; Lewis (1950) summarized research on concrete aggregates from the standpoint of road building; Landgren and Sweet (1952) reported on durability of Wyoming aggregates; Happ (1955) included 49 references on aggregates, of which seven were categorized as "outstanding"; Melville (1955) reported on aggregate reactions in concrete in Virginia; and Mather, K. (1958) presented an introduction to potentially deleterious reactions of aggregate constituents with alkalies in concrete.

The engineering geologist who practices in the area of aggregates must learn to differentiate between the requirements for successful performance of the aggregates in the particular application being considered and in the anticipated environment of service; such an evaluation will include how the specific needs are met, or not met, by the various materials available from the candidate sources. Although rock and mineral names are very useful clues, by themselves they are never an adequate basis for acceptance or rejection. In some projects, requirements will be imposed that reject the natural materials available because they are believed to be unsatisfactory. The applied geologist who believes that local materials would prove acceptable should bring this argument to the relevant decision-making authorities; in some cases this may significantly reduce construction costs. In other cases, even though a material could comply with the project requirements, it is recognized that the overall actions of the engineering works would be subjected to a greater risk of unsatisfactory performance than is reasonable. Similarly, by bringing these relationships to the attention of the appropriate authorities early, the geologist may avoid serious likelihood of any distress in service of materials, often at little or no increase in cost.

Deleterious behavior potential

As previously noted, almost every type of rock can be counted on to be present in some collection of rock particles being considered for use as or being used as aggregate. Therefore one must be aware of which rock types are capable of deleterious behavior in the intended service. K. Mather (1953) reviewed efforts going back to 1905 to correlate petrographic information on rocks with their performance as aggregates. Lovegrove (1905) determined that loss to aggregate samples in the Deval abrasion testing machine and resistance to fracture by impact were related to texture, mineral composition, and the extent of alteration of the rocks. Dense, unaltered igneous rocks with micrograntic, micrographic, and ophitic textures showed least loss by attrition. Lord (1911) reported that igneous rocks of deep-seated origin and with granular textures were harder but less tough than their fine-grained, near-surface equivalents. Tomlinson (1915) described a method developed under the direction of M. O. Withey and A. M. Winchell for use in the survey of concrete aggregates of Wisconsin being carried out by Withey. The method involved separating sand samples by sieve fractions and then separating the fractions into four density classes using heavy liquids, then examining each class of each size using a microscope and computing weighted average composition by rock types and mineral species. This is the forerunner of the petrographic procedure for examining aggregate for concrete (Mather and Mather, 1950) now covered by ASTM C 295 as given in the appropriate volume of any Annual Book of ASTM Standards. The use of these techniques was discussed by Mielenz (1946, 1954).

The earliest known investigation of a failure of concrete ascribed to properties of the constituent aggregate particles was that reported by Pearson and Loughlin (1923). The concrete was cast stone manufactured using an altered anorthosite as aggregate. The concrete was examined, the quarry was visited, and thin sections of aggregate and concrete were examined. It was found that much of the feldspar in the aggregate particles had altered to the zeolite, laumontite, which loses water and disintegrates when exposed to dry air. Loughlin (1927, 1928) discussed qualifications of stone for use as aggregate and the usefulness of geology, especially petrology, in studying limestones for use as aggregate. In the 1928 work, he reported, for the first time, the potential for deleterious behavior of aggregate particles consisting of limestone with swelling clay ("montmorillonite" = smectite).

Reactive aggregates

That particles of rocks and minerals are unstable in specified environments is well known. Ice melts at temperatures of about 0°C, halite dissolves in water unless the water is already a saturated solution of sodium chloride, calcite is soluble in mineral acids, and native iron rusts; many other well-known examples could be cited. The deleterious behavior of the unstable zeolite mineral laumontite (Pearson and Loughlin, 1923), and of limestone with smectite (Loughlin, 1928) when used as aggregate has already been noted. Stanton (1940) reported failure of concrete by expansion and cracking in California that had been observed since at least 1923 from Monterey to Los Angeles and inland near the San Joaquin River. In a subsequent paper (Stanton and others, 1942) additional data were presented tying the expansion to a chemical reaction associated with water-soluble alkalies (reported as Na_2O or K_2O on chemical analysis) in the cement and "siliceous magnesian limestone" from the Monterey group. This was the beginning of the study of alkali-aggregate reaction (AAR). The classical AAR involves thermodynamically metastable forms of silica such as opal, chalcedony, strained quartz, tridymite, cristobalite, artificial glass, acid volcanic glass, and less well-defined constituents of such metamorphic rocks as phyllites and graywackes. These sorts of materials can react in the highly alkaline medium present in the pore fluid of concrete made using cement having more than 0.6 percent Na_2O equivalent ($\%Na_2O + 0.658 \times \% K_2O$) by chemical analysis. Under these conditions the pH of the pore fluid may be as high as 13. A tremendous amount of literature has been written, and its volume seems to be increasing as affected concrete structures keep turning up in places where none were previously known, such as England, South Africa, and Japan. The chapters by Diamond and Walker in ASTM (1978) recount the story up to that date. Chapter 7 of Dolar-Mantuani provides an excellent summary of this topic. Relevant test methods were reviewed by Mielenz and Witte (1948).

Of the two different reactions, alkali-silica and alkali-carbonate rock, that involve aggregates, the most important is the alkali-silica reaction. This involves the breakdown of silica or siliceous material to an alkali silica gel that has the capacity to imbibe water, swell, and exert hydrostatic pressure on its surroundings, so as to disrupt concrete. Typically, after swelling and inducing cracks, the gel ultimately takes in sufficient water to become quite fluid and then migrates through the cracks, fills cracks and voids, and exudes on the exterior of specimens that have been kept moist. Often the reactive constituent of an aggregate is given its rock name when only one of the constituents of the rock is reactive. Andesite, rhyolite, dacite, and basalt are associated with reactivity because of the glass they contain; chert and flint because of the opal or chalcedony they contain; and granite, sandstone, and quartzite because of the thermodynamically metastable quartz they contain.

If a nonreactive aggregate of equal quality and equal or lower cost is available for use, the prudent step to avoid damage to concrete by alkali-aggregate reaction is to avoid the use of reactive aggregate. However, when there are reasons that make use of a reactive aggregate desirable or mandatory, the customary precaution taken to avoid distress is to require low-alkali cement, i.e., cement in which, on chemical analysis, the value for $\% Na_2O + 0.658 \times \% K_2O \leqslant 0.60\%$. This will prevent damage in most cases. An exception, described by Hadley (1968), involved environmental concentration of alkali hydroxides by upward leaching, evaporation, and near-surface precipitation in concrete slabs on grade.

Another procedure that should permit the concurrent use of reactive aggregate and high-alkali cement is to use a sufficient quantity of an effective pozzolan or of ground granulated iron blast-furnace slag. These options are mentioned in the ASTM Specifications for Concrete Aggregates (C 33), and a test for

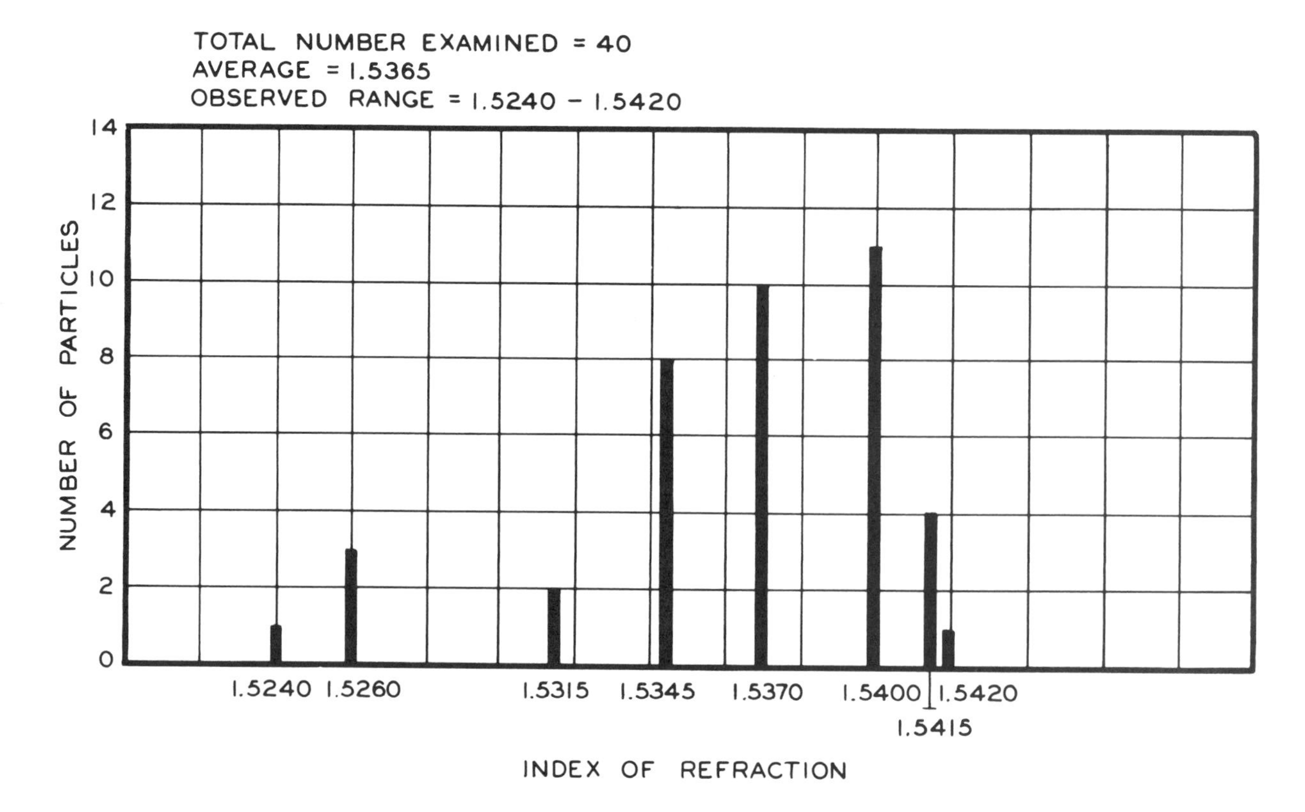

Figure 1. Distribution of chalcedonic chert coarse aggregate particles selected from concrete from Tuscaloosa Lock, with respect to index of refraction (B. Mather, 1952).

effectiveness of such materials is available (ASTM: C 441). Specifications for pozzolan and for ground granulated iron blast-furnace slag have been published (ASTM: C 618, C 989). Appendix 3 to C 989 discusses specifically the effectiveness of such ground slag in preventing excessive expansion of concrete due to the alkali-aggregate reaction.

The Tuscaloosa Lock and Dam on the Warrior River at Tuscaloosa, Alabama, was the first Corps of Engineers structure to be extensively studied and found to have widespread cracking due to alkali-silica reaction. The results of this study (B. Mather, 1952) suggested that the reactive material was chalcedony (Fig. 1) included in a portion of the chert particles making up the coarse aggregate. Figures 2 through 4 show the distribution of reaction product in the concrete as seen on sawed surfaced of core samples. Figures 5 and 6 illustrate the degree to which the reaction product is capable of taking up moisture from the air and swelling and exuding.

The subject of reactive aggregates was exhaustively reviewed at the 7th International Conference on Alkali-Aggregate Reaction held in August, 1986 under the sponsorship of the National Research Council of Canada (Grattan-Bellew, 1987). Proceedings of earlier conferences have been published: Idorn (1974), Asgeirsson (1975), Poole (1976), B. Mather (1978), Oberholster (1981), and Idorn and Rostam (1983).

A thorough review of research on alkali-carbonate rock reactions through 1964 is given by Newlon (1964). B. Mather (1974) has summarized the state of the art in this area; four types of alkali-carbonate rock reaction occur that may be recognized in concrete. It is possible that future work will show that one or more of these four types is merely a different manifestation of the same reaction shown by different rocks under a variety of circumstances. The four types of reactions are discussed below.

Nondolomitic carbonate rocks. Some rocks that contain little or no dolomite may be reactive (K. Mather and others, 1963; Buck, 1965). The reaction is characterized by reaction rims visible along the borders of cross sections of aggregate particles; etching these cross-sectional surfaces with dilute hydrochloric acid reveals that the rims are "negative" rims—that is, the reaction rim zone dissolves more rapidly than the interior of the particle. The evidence to date indicates that the reaction is not harmful to concrete and may even be beneficial.

Dolomite or highly dolomitic carbonate rocks. The reaction of dolomite or highly dolomitic aggregate particles in concrete has been reported (Tynes and others, 1966). The reaction was characterized by visible reaction rims on cross sections of the aggregate particles. When these cross-sectional areas of aggregate particles were etched with acid, the rimmed area dissolved at the same rate as the non-rimmed area. No evidence has been reported that this reaction was damaging to concrete.

Impure dolomitic rocks. The rocks of this group have a

Figure 2. Honeycombed concrete from core 5-1, Tuscaloosa Lock, depth 4 to 4.3 m (×0.8). Direction of placement toward bottom of photograph. Arrows indicate gel pockets; the large voids in the lower section are gel-lined (B. Mather, 1952).

characteristic texture and composition. The texture is such that larger crystals of dolomite are scattered in and surrounded by a fine-grained matrix of calcite and clay. The rock consists of substantial amounts of dolomite and calcite in the carbonate portion, with significant amounts of acid-insoluble residue consisting largely of clay. Two reactions have been reported with rocks of this sort: dedolomitization and rim-silicification.

Dedolomitization reaction. Dedolomitization is believed to have produced harmful expansion of concrete (Hadley, 1961). Magnesium hydroxide, brucite [$Mg(OH)_2$], is formed by this reaction; its presence in concrete that has expanded and that contains carbonate aggregate of the indicated texture and composition is strong evidence that this reaction has taken place.

Rim-silicification reaction. Rim-silicification reaction is not definitely known to be damaging to concrete, although there are some data that suggest that a retardation in the rate of strength development in concrete is associated with its occurrence. The reaction is characterized by enrichment of silica in the borders of reacted particles (Bisque and Lemish, 1958). This is seen as a positive or raised border at the edge of cross sections of reacted

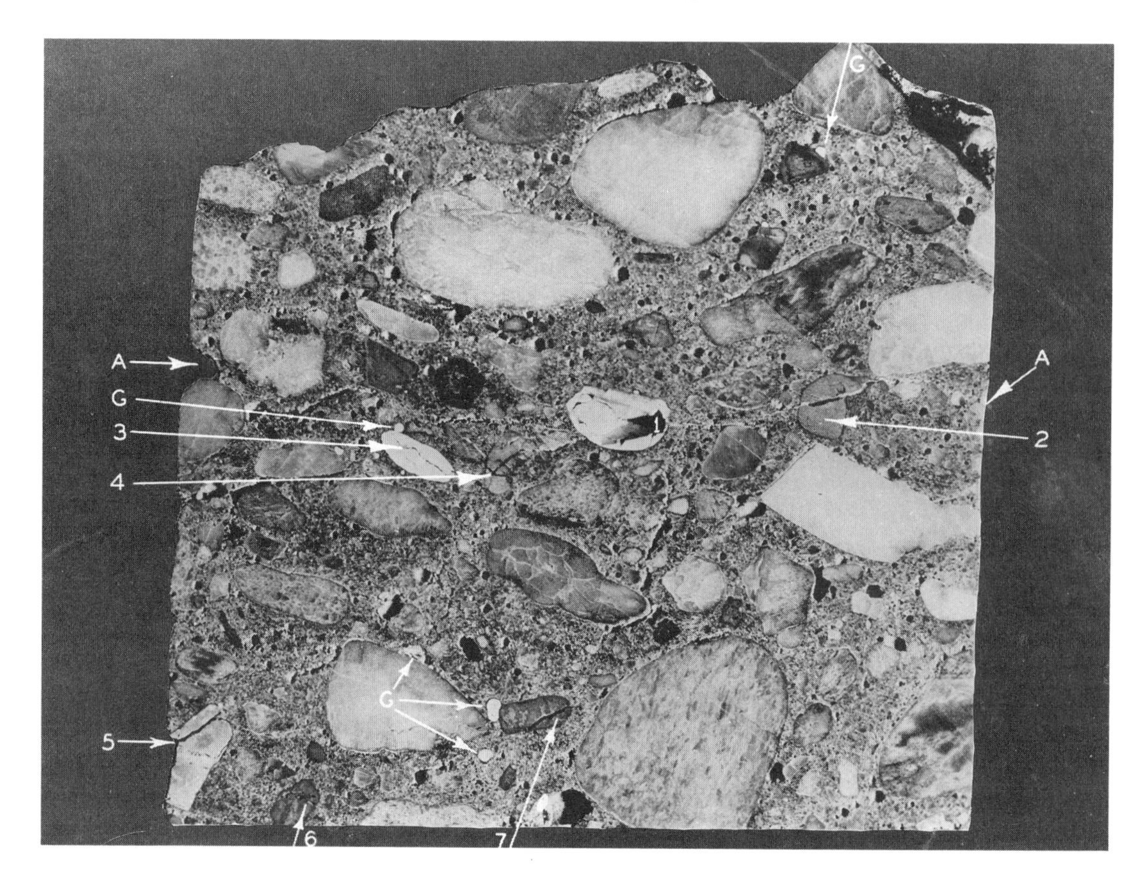

Figure 3. Sawed surface of core cut longitudinally, Tuscaloosa Lock. The direction of placement is toward the top of photograph. A-A indicates the ends of a crack system that can be traced across the core, passing through two chert pebbles (1 and 2). The white chert pebble, 1, has an outer zone very firmly bonded to the matrix and the center of the pebble separated by cracks from the outer zone. The chert pebbles indicated by arrows (2 through 7) show the wide cracks common in chert particles in this concrete. In pebble 2 the gel filling of parts of the crack system can be seen at the left and right. G indicates gel pockets (B. Mather, 1952).

particles after they have been etched in dilute hydrochloric acid. Reaction rims may be visible before the concrete surfaces are etched. Fortunately, carbonate rocks that contain dolomite, calcite, and insoluble material in the proportions that cause either the dedolomitization or rim-silicification reactions are relatively rare in nature as major constituents of the whole product of an aggregate source.

The following criteria for recognition of potentially harmfully reactive carbonate rocks indicate those dolomitic carbonate rocks capable of producing the dedolomitization or rim-silicification reaction. Since the reactions generated by some very dolomitic or by some nondolomitic carbonate rocks are not known to be harmful to concrete, no attempt is made to provide guides for recognition of these rocks.

1. When petrographic examinations of quarried carbonate rock or of natural gravels containing carbonate-rock particles are made according to ASTM: C 295, the data obtained concerning texture, calcite-dolomite ratio, the amount and nature of the acid-insoluble residue, or some combination of these parameters will be adequate to recognize potentially reactive rock. Rocks associated with observed expansive dedolomitization have been found to be characterized by fine grain size (generally 50 μm or less), with the dolomite largely present as small, nearly euhedral crystals generally scattered in a finer grained matrix in which the calcite is disseminated. The tendency to expand, other things being equal, appears to increase with increasing clay content from about 5 to 25 percent by weight of the rock and also appears to increase as the calcite-dolomite ratio of the carbonate portion approaches 1:1.

2. Samples of rock recognized as potentially reactive by petrographic examination should be tested for length change during storage in alkali solution in accordance with ASTM:

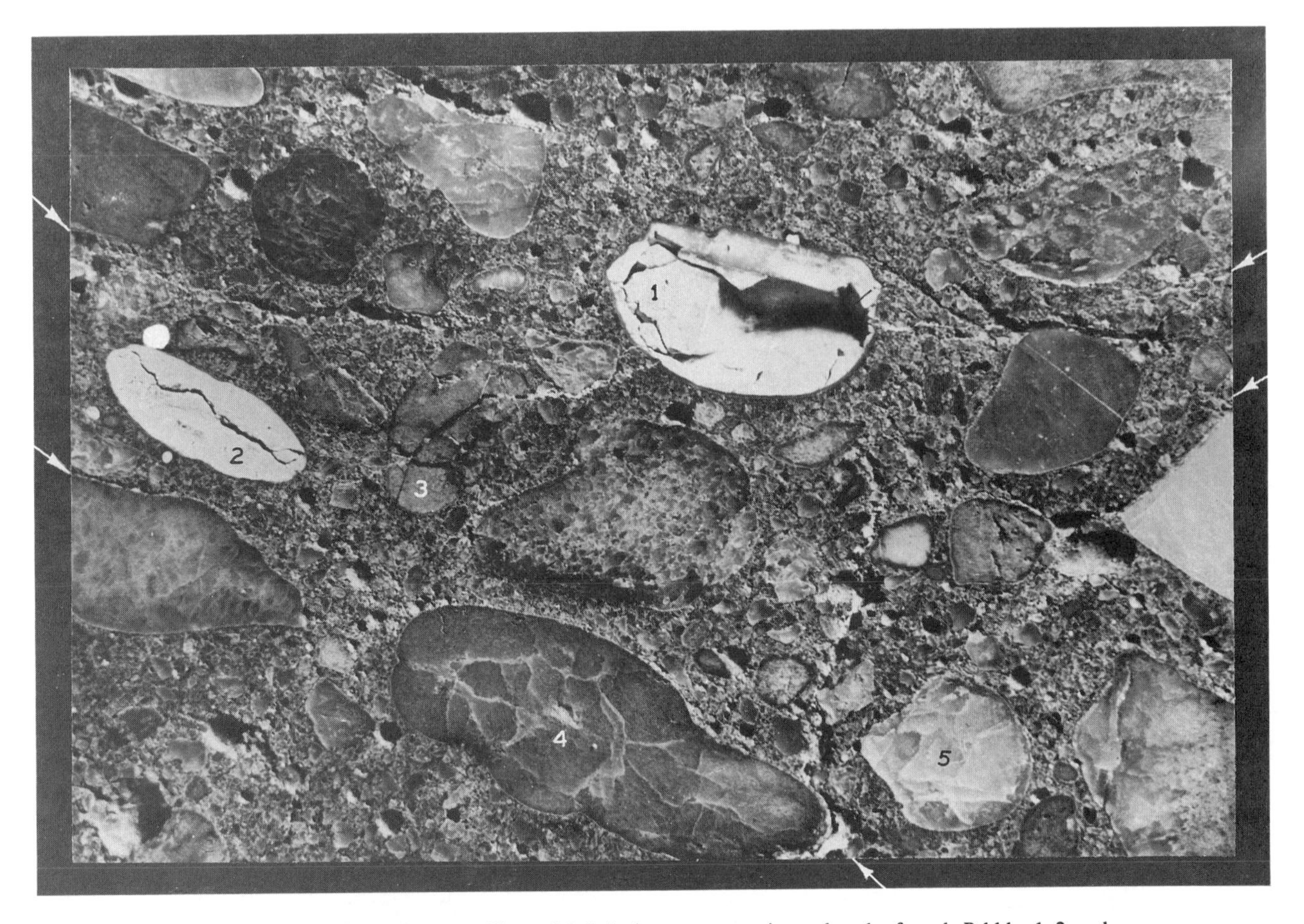

Figure 4. Part of area shown on Figure 3 (×3.4). Arrows at margin mark ends of crack. Pebbles 1, 2, and 3 are chert with well-developed wide cracks; those in pebble 3 are locally gel-filled. Pebbles 4 and 5 are cracked, but the cracks are narrower than those of pebbles 1, 2, and 3 (B. Mather, 1952).

C 586. Rock characterized by expansion of 0.1 percent or more by or during 84 days of test by ASTM: C 586 should be classified as potentially reactive.

3. If adequate reliable data are available to demonstrate that concrete structures containing the same aggregate have exhibited deleterious reactions, the aggregate should be classified as potentially reactive on the basis of its service record.

The application of engineering judgment will be required in making the final decision as to which rocks are to be classified as innocuous and which are to be classified as potentially reactive. Once a rock has been classified as potentially reactive, one of the following actions should be taken.

1. Use procedures such as selective quarrying to avoid use as aggregate of rock classified as potentially reactive.

2. If it is not feasible to avoid the use of rock classified as potentially reactive, then specify the use of low-alkali cement, the minimum aggregate size that is economically feasible, and dilution so that the amount of potentially reactive rock does not exceed 20 percent of the coarse or fine aggregate or 15 percent of the total if reactive material is present in both.

3. If it is not practical to enforce the first two conditions, then the aggregate source that contains potentially reactive rock should not be indicated as a source from which acceptable aggregate may be produced.

CONCLUSION

Historically it has been customary to describe materials used as aggregate as *inert* and, in fact, to use the term *inert aggregate* when differentiating the relatively less reactive from the more reactive constituent, for example, the cement paste in concrete. Research and service records over the past 50 years, however, clearly indicate that no aggregate particle is completely inert. All

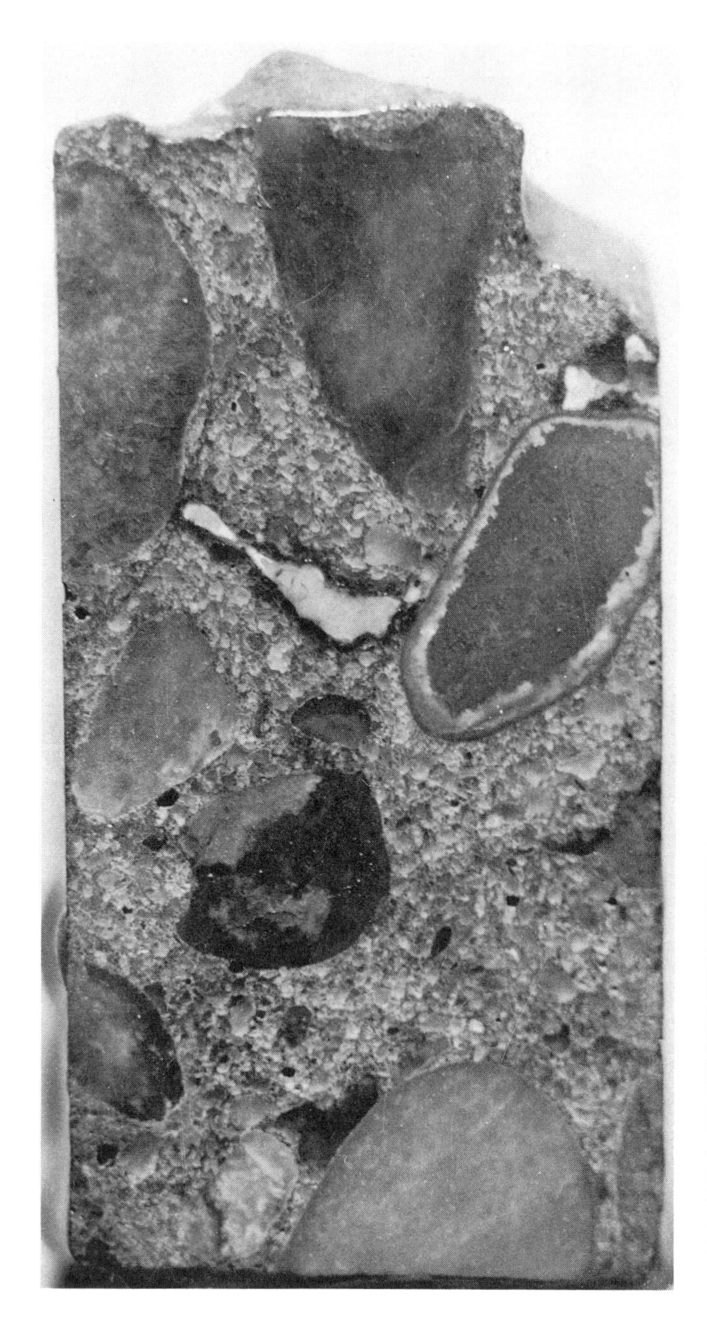

Figure 5. A thin-section blank impregnated with resin, ground, and then exposed to warm humid air for about 36 hours; magnification 4×, photographed about August 23, 1949. Two large gel pockets (upper right and upper center) had taken up moisture and swelled. The chalcedonic pebble (lower left center) had taken up moisture, and a wet spot appeared on a sand grain at the lower left (B. Mather, 1952).

Figure 6. Same specimen as shown in Figure 5 after storage over water at room temperature for period August 24 to October 18, 1949. The gel pockets shown in Figure 5 have enlarged, and many more have appeared. Cracks that were not obvious before are now traced in reaction product (B. Mather, 1952).

aggregate particles are reactive—they merely differ in kind and degree of reactivity.

All aggregate particles have linear coefficients of thermal expansion and thus will undergo volume change on temperature change that will probably be different from that of the cement-paste matrix; this will create stress at the aggregate-paste interface. All aggregate particles have a modulus of elasticity and hence will have a strain response to stress that will probably be different from the cement-paste matrix. Furthermore, this will affect the degree to which the aggregate restrains the tendency of the matrix to change in volume as it changes in moisture content. The chemical activity of certain aggregate particles has received the most attention of researchers and engineers, since it has resulted in the greatest economic consequences, but the physical activity of aggregate particles should also be appreciated and considered relevant to the operation of engineering works.

In general however, most aggregates that are selected according to well established practices and found to comply with appropriate specifications, after testing of representative samples, perform in concrete with completely satisfactory results.

REFERENCES CITED

Asgeirsson, H., ed., 1975, Symposium 16 on Alkali-Aggregate Reaction; Preventative Measures: Reykjavik, Rannsoknastfnun Byggingaridnadarins, 270 p.

ASTM, 1948, Symposium on mineral aggregates: Philadelphia, Pennsylvania, American Society for Testing and Materials Special Technical Publication 83, 233 p.

—— , 1978, Significance of tests and properties of concrete and concrete-making materials: Philadelphia, Pennsylvania, American Society for Testing and Materials, Special Technical Publication 169B, 4th edition, p. 539–764.

Bisque, R. E., and Lemish, J., 1958, Chemical characteristics of some carbonate aggregate as related to durability of concrete: Washington, D.C., Highway Research Board Bulletin 196, p. 29–45.

Blanks, R. F., 1950, Modern concepts applied to concrete aggregate (with discussion): New York, American Society of Civil Engineers Transactions, v. 115, p. 403–437.

Buck, A. D., 1965, Investigation of a reaction involving nondolomitic limestone aggregate in concrete: Vicksburg, Mississippi, U.S. Army Engineer Waterways Experiment Station Miscellaneous Paper 6-724, 40 p.

Dolar-Mantuani, L., 1983, Handbook of concrete aggregates; A petrographic and technological evaluation: Park Ridge, New Jersey, Noyes Publications, 345 p.

Eckel, E. C., 1905, Cements, limes, and plasters: New York, John Wiley and Sons, 712 p.

Fears, F. K., 1949, Bibliography on mineral aggregates (annotated): Washington, D.C. Highway Research Board Bibliography 6, 89 p.

Grattan-Bellew, P. E., ed., 1987, Concrete Alkali-Aggregate Reactions-Proceedings of the 7th International Conference: Park Ridge, New Jersey, Noyes Publications, 509 p.

Hadley, D. W., 1961, Alkali reactivity of carbonate rocks; Expansion and dedolomitization: Washington, D.C. Highway Research Board Proceedings, v. 40, p. 462–474, 664.

—— , 1968, Field and laboratory studies on the reactivity of sand-gravel aggregates: Journal of the Portland Cement Association Research and Development Laboratories, v. 10, no. 1, p. 1–17.

Happ, S. C., 1955, Engineering geology reference list: Geological Society of America Bulletin, v. 66, p. 993–1030.

Hill, R. T., 1889, Roads and materials for their construction in the Black Prairie regions of Texas: Austin, University of Texas Bulletin, 39 p.

Holm, T. A., chairman, 1979, Guide for use of structural lightweight aggregate concrete: Detroit, Michigan, American Concrete Institute Manual of Concrete Practice 213R-79, 30 p.

Huntting, M. T., 1945, Geology in highway engineering (with discussions): American Society of Civil Engineers Transactions, v. 110, p. 271–344.

Idorn, G. M., ed., 1974, Seminar on Alkali-Silica Reaction: Karlstrup, Denmark, Aalborg Portland Cement Company, 25 p.

Idorn, G. M. and Rostam, S., eds., 1983, 6th International Conference on Alkalis in Concrete; Research and Practice: Copenhagen, Dansk Betonforening, 532 p.

Kiersch, G. A., 1955, Engineering geology; Scope, development, utilization: Golden, Colorado School of Mines Quarterly, v. 50, no. 3, 122 p.

Knight, B. H., and Knight, R. G., 1948, Road aggregates; Their uses and testing: London, Edward Arnold and Company, 304 p.

Lamond, J. F., chairman, 1984, Guide for use of normal weight aggregates in concrete: Detroit, Michigan, American Concrete Institute Manual of Concrete Practice 221R-84, 26 p.

Landgren, R., and Sweet, H. S., 1952, Investigation of durability of Wyoming aggregates: Highway Research Board Proceedings, v. 31, p. 202–217.

Lewis, D. W., 1950, Research on concrete aggregates, *in* Proceedings, 36th Annual Road School: West Lafayette, Indiana, Purdue University Engineering Bulletin, v. 34, no. 3, p. 70–86 (Extension Service no. 71).

Lord, E.C.E., 1911, Examination and classification of rocks for road building: U.S. Department of Agriculture, Office of Public Roads Bulletin no. 37, 28 p.

Loughlin, G. F., 1927, Qualification of different kinds of natural stone for concrete aggregate: American Concrete Institute Proceedings, v. 23, p. 319.

—— , 1928, Usefulness of petrology in the selection of limestone: Rock Products, v. 31, p. 50.

Lovegrove, E. J., 1905, Attrition Tests of Road-making Stones, with Petrological Reports by John S. Flett, Petrographer, H. M. Geological Survey, and J. Allen Howe, Curator and Librarian of the Museum of Practical Geology: London, The Surveyor and Municipal and County Engineer, v. 28, pt. 1, p. 568–572; pt. 2, p. 601–605; pt. 3, p. 632–635; pt. 4, p. 652–655; pt. 5, p. 684–687; pt. 6, 716–720; pt. 7, p. 744–748; pt. 8, p. 768–776.

Marshall, H. E., and Maxey, J. S., 1950, The role of the geologist in constructing Ohio's highways: Columbus, Ohio, Engineering Experiment Station News, v. 12, no. 2, p. 12–14.

Mather, B., 1952, Cracking of concrete in the Tuscaloosa Lock, Alabama: Highway Research Board Proceedings, v. 31, p. 218–233.

—— , 1974, Developments in specification and control, *in* Cement-aggregate reactions: Washington, D.C., National Academy of Science–National Research Council, Transportation Research Board Transportation Research Record 525, p. 38–42.

Mather, B., ed., 1978, Proceedings of the Fourth International Conference on Effects of Alkalies in Cement and Concrete: West Lafayette, Indiana, Purdue University School of Civil Engineering, Publication CE-MAT-1-78, 376 p.

Mather, K., 1953, Applications of light microscopy in concrete research, *in* Symposium on light microscopy: American Society for Testing and Materials Special Technical Publication 143, p. 51–70.

—— , 1958, Cement-aggregate reaction; What is the problem? *in* Trask, P., ed., Engineering geology case histories no. 2: Geological Society of America, p. 17.

Mather, K., and Mather, B., 1950, Method of petrographic examination of aggregates for concrete (with discussion): American Society for Testing and Materials Proceedings, v. 50, p. 1288–1313.

Mather, K., Luke, W. I., and Mather, B., 1963, Aggregate investigations; Milford Dam, Kansas; Examination of cores from concrete structures: Vicksburg, Mississippi, U.S. Army Engineer Waterways Experiment Station Technical Report 6-629, 81 p.

Melville, P. L., 1955, Concrete aggregate reaction in Virginia, *in* Proceedings, 6th

Annual Symposium on Geology as Applied to Highway Engineering: Baltimore, Maryland, State Roads Commission, p. 32–35.
Mielenz, R. C., 1946, Petrographic examination of concrete aggregates: Geological Society of America Bulletin, v. 57, p. 309–318.
—— , 1954, Petrographic examination of concrete aggregate: American Society for Testing and Materials Proceedings, v. 54, p. 1188–1218.
Mielenz, R. C., and Witte, L. P., 1948, Tests used by the Bureau of Reclamation for identifying reactive concrete aggregates: American Society for Testing and Materials Proceedings, v. 48, p. 1071–1107.
Newlon, H. H., ed., 1964, Symposium on alkali-carbonate rock reactions: Washington, D.C., Highway Research Record No 45, 244 p.
Oberholster, R. E., ed., 1981, Proceedings of the Fifth International Conference on Alkali-Aggregate Reaction in Concrete, Cape Town: Pretoria, National Building Research Institute, 548 p.
Paige, S., and Rhoades, R., 1948, Report of the committee on research in engineering geology: Economic Geology, v. 43, no. 4, p. 313–323.
Pearson, J. C., and Loughlin, G. F., 1923, An interesting case of dangerous aggregate: American Concrete Institute Proceedings, v. 19, p. 142.
Polatty, J. M., Goode, T. B., Bendinelli, R. A., and Houston, B. J., 1961, Project Cowboy; Drilling and grouting support: Vicksburg, Mississippi, U.S. Army Engineer Waterways Experiment Station Miscellaneous Paper 6-419, 45 p.
Poole, A. B., ed., 1976, The Effect of Alkalies on the Properties of Concrete, Proceedings of a Symposium held in London: Wexham Springs, U.K., Cement and Concrete Association, 374 p.
Popovics, S., 1979, Concrete-making materials: New York, McGraw-Hill, 370 p.
Shaler, N. S., 1895, The geology of the road-building stones of Massachusetts, with some consideration of similar materials from other parts of the United States: U.S. Geological Survey Annual Report 16, pt. 2, p. 277–341.
Stanton, T. E., 1940, Expansion of concrete through reaction of cement and aggregate (with discussion): American Society of Civil Engineers Proceedings, v. 66, p. 1781–1811.
Stanton, T. E., Porter, O. J., Meder, L. C., and Nicol, A., 1942, California experience with the expansion of concrete through reaction between cement and aggregate (with discussion): American Concrete Institute Proceedings, v. 38, p. 209.
Tomlinson, C. W., 1915, Method of making mineralogical analyses of sand: American Institute of Mining and Metallurgical Engineers Bulletin 101, p. 947.
Tynes, W. O., Luke, W. I., and Houston, B. J., 1966, Results of laboratory tests and examinations of concrete cores, Carlyle Reservoir spillway, Carlyle, Illinois: Vicksburg, Mississippi, U.S. Army Engineer Waterways Experiment Station Miscellaneous Paper 6-802, 30 p.
Wakeley, L. D., and Burkes, J. P., 1986, Distribution of chloride in a salt-saturated grout in contact with halite rock: Cement and Concrete Research, v. 16, no. 3, p. 267–274.
Woods, K. B., 1948, Distribution of mineral aggregates: American Society for Testing and Materials Special Technical Publication 83, p. 4–19.

MANUSCRIPT ACCEPTED BY THE SOCIETY NOVEMBER 3, 1987

Early "granite quarry" of New England in the late 1800s by James Hope (1819–1892). (Photo courtesy of the Museum of Fine Arts, Boston; Y. and M. Karolik collection.)

Geological Society of America
Centennial Special Volume 3
1991

Chapter 17

Snow and ice

Wilford F. Weeks
Geophysical Institute, University of Alaska at Fairbanks, Fairbanks, Alaska 99775-0800
Robert L. Brown
Department of Civil Engineering and Engineering Mechanics, Montana State University, Bozeman, Montana 59717

INTRODUCTION

Glaciology, the study of snow and ice, encapsulates many aspects of the more conventional world of geology, including the study of igneous (lake, river, and sea ice), sedimentary (snow packs), and metamorphic (glaciers, ice sheets, and shelves) ice masses. However, glaciology avoids some of geology's inherent problems because snow and ice bodies always occur on or near the surface of the earth and are, geologically speaking, quite thin: snow, lake, river, and sea ice are less than a few tens of meters; ice shelves and glaciers are less than a few hundred meters; and ice sheets are less than 4,000 m. Consequently, their investigation by either direct sampling (coring) or indirect geophysical methods is comparatively straightforward. They are also basically monomineralic; ice I(h) and gas, plus in the case of sea ice, a liquid phase with minor amounts of a few simple solid salts. This is clearly a simplification when one considers the mineralogical complexities of most rock masses. Furthermore, color presents few problems, basic white on white.

In considering the differences between the behavior of natural ice masses and the more typical materials considered by engineering geologists, one must remember that ice on the Earth's surface invariably exists at or near its melting temperature, as contrasted with surficial rock masses that occur at temperatures far below melting and are, relatively speaking, the real "frozen" bodies. From a research point of view, the "high" temperatures of natural ice masses are an advantage in that igneous, sedimentary, and metamorphic processes occur rapidly, resulting in changes that are invariably measurable in a few years and more commonly in a few hours or days; geologically near-lightning speeds. Also, the near-melting temperatures of snow and ice can be tolerated by humans willing to endure minor discomfort; a fact that greatly simplifies experimental procedures and field observations. From an engineering point of view, just the opposite is true. Near-melting temperatures result in nonlinear, time-dependent property behavior that is difficult to analyze, and the resulting undesirable material characteristics are then coupled with environmental conditions that impede standard construction procedures.

HISTORY

Engineering interest in snow and ice results from the fact that it covers a significant portion of the earth's surface. More than 50 percent of the land surface is covered by snow during some portion of the year. In a similar sense, sea ice covers 10 percent of the surface of the sea, and if the regions visited by icebergs are included, 23 percent of the world ocean must be considered. In fact the world's most ephemeral surface features are its seasonal snow and sea-ice covers. As man carries out highly varied sets of operations on these same surfaces, difficulties with snow and ice are inevitable.

As mankind's difficulties with snow and ice are hardly of recent origin, one might surmise that glaciological processes would be reasonably well understood, so that the results would be useful in modeling the behavior of more complex rock systems. Although there are cases where this is true, the study of snow and ice in general, and of engineering glaciology in particular, is still in its infancy. Only since the end of World War II, with advances in air transportation and field equipment, have scientific studies of the world's larger ice masses become relatively convenient. Even today the remoteness of many sites of glaciological interest causes field operations to be both expensive and infrequent. Also, glaciology, as contrasted with glacial geology, is a subject that is not commonly included in earth science curricula. As a result, practicing glaciologists approach their subject with widely varying backgrounds; a fact that in our opinion has significantly contributed to the diversified development of the field.

Some important applied problems and projects that have produced or been associated with significant developments in engineering glaciology are as follows: difficulties with river ice crossings, leading to the solution of the problem of a plate on an elastic foundation (Hertz, 1884); problems associated with ship operations on the Soviet Northern Sea Route, starting studies of ice-structure interactions (Makarov, 1901) and ultimately leading to improved icebreaker designs (Kashteljan and others, 1968); avalanche problems in the Alps, leading to the development of snow mechanics (Bader and others, 1939); bearing capacity problems for floating ice sheets, resulting in a theory for the strength of

Weeks, W. F., and Brown, R. L., 1991, Snow and ice, *in* Kiersch, G. A., The heritage of engineering geology; The first hundred years: Boulder, Colorado, Geological Society of America, Centennial Special Volume 3.

sea ice (Assur, 1958; Anderson and Weeks, 1958) and procedures for dealing with vertical variations in ice-plate properties (Assur, 1967); studies of snow and ice foundations carried out at Camp Century, making possible the construction of the DEW line stations in central Greenland (Tobiasson, 1979); the discovery of offshore oil and gas in the North American Arctic, leading to extensive studies of ice forces on fixed offshore structures and of ice gouging (Weeks and Weller, 1984); continuing difficulties with river ice jams and associated flooding, resulting in major improvements in understanding these phenomena and in associated amelioration procedures (Ashton, 1986); mining activities near the margins of both valley glaciers and continental ice sheets, leading to interesting applications of glacier flow theory (Untersteiner and Nye, 1968; Colbeck, 1974); problems with ice accumulation on propellers, power lines, and ships, resulting in varied studies of these complex phenomena (Colbeck and Ackley, 1983; Itagaki, 1977); and continuing difficulties with drifting snow, resulting in impressively realistic simulations of the process (Anno, 1986).

It is impossible to cover the range of subjects that should be treated in any consideration of engineering glaciology. In the following, we will first briefly review the structure and properties of snow, of glacier ice, and of lake, river, and sea ice. Following this, four examples of glaciological problems of current engineering interest will be discussed: avalanches, foundations, forces on offshore structures, and ice-induced gouges in the sea floor. Selected references are given throughout as a window to the surprisingly large and somewhat elusive literature on engineering aspects of snow and ice.

STRUCTURE AND PROPERTIES

Snow

Structure. From an engineering viewpoint, snow is a difficult material to describe quantitatively as its properties are extremely variable. When initially deposited on the ground, snow crystals, although invariably of the classic hexagonal internal structure (ice Ih; Hobbs, 1974), can have a wide variety of external shapes, ranging from plates to columns to prisms to needles, depending on the exact environmental conditions (temperature, degree of supersaturation) under which the crystals grew. In fact, various combinations of these shapes can occur if the crystals experience substantial environmental changes during their formation. The time scale for these processes, once precipitation starts, is on the order of the few minutes required for the snow crystals to reach the ground.

In addition to the formation processes that take place in the atmosphere, snow can form at the snow cover surface during the evening due to the condensation of vapor when the air adjacent to the surface becomes supersaturated. The large, feathery crystals that result are usually referred to as surface hoar, have exceptionally poor structural integrity, and are a major source of avalanches. As much as 5 cm of new snow can be added to an existing snow cover by this process in an evening.

Even after snow crystals are incorporated into a snow cover, conditions are invariably sufficient to cause continual changes in the structural nature of the snow as it attempts to achieve a thermodynamically stable configuration. Commonly the larger grains grow at the expense of the smaller grains, and intergranular bonds develop, resulting in a stronger, more stable material. In the case of dry snow ($T < 0°C$), this process is very slow since the transfer of mass from the small ice grains to the larger ones occurs via vapor diffusion across the pores. At low temperatures, substantial property changes via this mechanism alone can take years. However, the process is accelerated somewhat by pressure sintering due to the overburden produced by subsequent snowfalls. This process, which dominates when temperature gradients in the snow are less than 10°C/m, is often referred to as equitemperature or equilibrium metamorphism; however, both terms are misleading because constant temperatures and equilibrium are conditions that rarely, if ever, exist in natural snowpacks.

When large temperature gradients are applied to the snow cover, a process often called temperature gradient metamorphism becomes dominant, resulting in a snow type referred to as a kinetic form. These conditions frequently occur during the early winter when a cold spell couples with a thin snow cover and near-freezing ground temperatures. Striking changes in snow fabric can result in a few days, with major associated changes in material properties such as strength, ductility, and conductivity. The most advanced form of this kinetic process is "depth hoar:" large-faceted crystals with very poorly developed intergranular bonding. Often the strength of such snow drops to below 10 percent of its original value within a week's time. Depth hoar is believed to play a central role in avalanche initiation, and as a result, its formation has been investigated extensively since the 1930s (Bader and others, 1939).

Properties. Figure 1 shows a surface section of a snow sample illustrating its granular structure. The darker areas represent air space, which in newly formed snow can constitute over 90 percent of the total volume. For seasonal alpine snow the porosity typically varies between 50 and 85 percent. This open structure allows a variety of deformation mechanisms to function when the material is subjected to loads, which include: pressure sintering (neck growth), intergranular glide, grain rotation, grain deformation, and bond fracture. Because many of these mechanisms are both irreversible and nonlinear, the mechanical behavior of the snow can be quite complicated. In addition, since snow is characteristically within 30°C of its melting temperature, it tends to exhibit viscous behavior, with time-dependent properties.

Since about 1950, considerable effort has been directed toward determining the mechanical properties of snow. Detailed reviews of these results have been provided by Mellor (1974) and Salm (1982). In this work it has been fairly common to treat snow in terms of its engineering properties such as strength, yield stress, elastic modulus, and Poisson's ratio. In particular, the properties analyzed have tended to be characterized by stress-strain equations that model snow as a linear viscoelastic material (i.e., strain is taken to be a time-dependent variable linearly re-

Figure 1. Surface section of alpine snow. The grid shown is 1 mm.

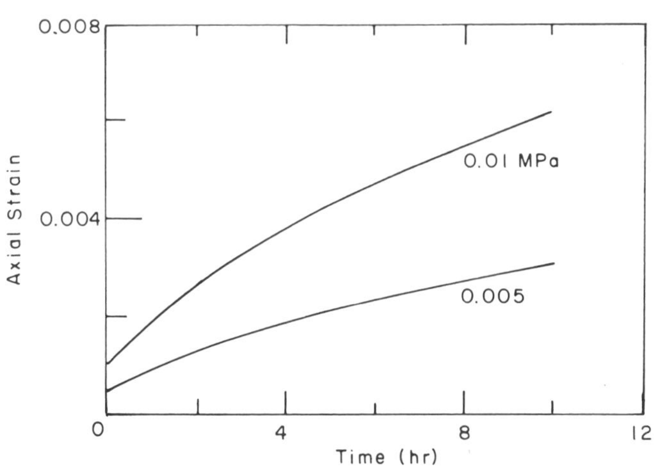

Figure 2. Typical constant stress creep curves for alpine snow having a grain size of 0.5 mm and a density of 300 kg/m^3.

lated to stress). Although these methods of describing snow may be suitable for solving certain types of problems, they fail to accurately reflect many aspects of snow behavior.

One significant shortcoming of past studies is that the effects of variations in the microstructure or texture of the snow on its bulk properties have not been considered. Here, we define microstructure to include such variables as grain size and shape, pore size and shape, and intergranular bonding. All of these factors are now known to have dramatic effects on the mechanical properties of snow. Another shortcoming of past work is the failure to adequately reflect the fact that snow does not respond linearly to high stress and cannot be generally represented as a linear material.

To date, the following conclusions about the properties of snow seem valid:

(i) At low stresses ($<$0.01 MPa), snow can be approximately represented as a linear viscoelastic material. Figure 2 shows the creep curves for snow for two constant stress levels. As can be seen from the figure, the strain rate $\dot{\epsilon}$ eventually approaches a constant value that is proportional to the stress. At these low stresses, doubling $\dot{\epsilon}$ requires that the applied stress be approximately doubled. When stresses exceed 0.1 MPa, the stress-strain rate relationship is increasingly nonlinear, and it becomes important to use nonlinear viscoelastic models in representing snow behavior.

(ii) The fracture strength for any given snow type is highly variable. If deformed at a high rate, snow will fracture in a brittle manner at relatively low stresses. However, if allowed to go through a deformation process before being subjected to high strain rates, snow may behave in a highly ductile manner with increased strength. To date, a simple fracture criterion that describes the strength of snow for a wide variety of loading conditions has not been developed.

(iii) Under steady stress or flow conditions, it may be possible to adequately characterize snow as a compressible fluid ($\dot{\epsilon}$ is assumed to be directly related to stress). Although this is as simple an approximation as one can realistically apply to snow, it has been used successfully in dealing with a variety of engineering problems.

(iv) Given a particular snow type, snow properties can be quite adequately specified in terms of snow density; properties such as strength, elastic modulus, and viscosity all increase systematically with density. However, as snow microstructure also strongly affects such properties, density alone is not adequate for precise material property predictions. Unfortunately the measurement and statistical evaluation of the appropriate microstructural parameters (e.g., mean bond size, mean number of bonds per grain, grain size, grain shape) is both difficult and time consuming. In an attempt to make such information more readily available, computer-based image analysis systems have been used with increasing frequency during the last 15 years to determine these parameters from surface sections such as shown on Figure 1.

(v) Because snow normally occurs at near-melting temperatures, temperature significantly affects snow properties, particularly above –10°C.

Although new formulations are being developed that include microstructural parameters in snow property descriptions

(Gubler, 1978; Hansen and Brown, 1986), it will be some time before progress is made to the point where such constitutive equations can be readily applied. In the meantime, the advances and concepts developed during 1960 to 1980 (Salm, 1982) can adequately serve the on-going needs of the engineering profession, if used with insight and care.

Glacier ice

Structure. A glacier is any large ice mass, formed by the precipitation of snow from the atmosphere, that is permanent on the human time scale and possesses a gravity-induced movement. Glaciers may be as small as a square kilometer and as large as the Antarctic Ice Sheet, which has an area in excess of 10^7 km^2. Of all the naturally occurring ice masses, glaciers have received the most attention. The reasons for this are several: glaciers are fascinating entities in their own right; they cover 10 percent of the earth's land surface at present and have covered roughly three times this area during the geologically recent past; their occurrence extensively modifies the surface of the land; the release or storage of water substance from or in glaciers is the major factor influencing the rise and fall of global sea level; glaciers are involved both directly and interactively in the process of climatic change; and because glaciers form from snow and trap both atmospheric gases and dust, they offer a unique record of climatic history during the last 100 to 500 thousand years.

Most glaciers that are not significantly out of balance can be subdivided into an accumulation area and an ablation area. Although during the winter the entire surface of a glacier may be covered with snow, as summer progresses the snow melts off the glacier ice at progressively higher elevations until the snowline reaches some maximum elevation late in the summer. This line, which approximately marks the boundary between the accumulation area above and the ablation area below, is called the firn line, and the wet snow that survives the summer above this line is called firn. In the high-altitude portions of polar glaciers where no melting occurs, the term firn is not used. As the firn and the cold, dry snow located at higher elevations on many glaciers are covered with subsequent winter snowfalls, they gradually densify. The warmer summer temperatures, the overburden, and when present, summer snowmelt all combine to accelerate the densification process. This densification proceeds until at a density of roughly 830 kg/m^3, the pores within the ice mass become disconnected and the ice mass ceases to be permeable to air. This change, by definition, marks the transition from snow (permeable to air) to ice (impermeable to air). The distance from the snow-ice transition to the surface of the glacier varies from 30 to 100 m and changes with climatic conditions.

Studies in the dry-snow facies of Greenland and Antarctica show that during this densification process the average grain size systematically increases from a fraction of a mm^2 to between 2 to 4 mm^2. This increase varies as an exponential function of the mean temperature of the site and linearly with the time since deposition (Gow, 1971). The initial crystal orientations in the upper parts of ice sheets and glaciers are usually random, although exceptions have been reported. However, at depths exceeding roughly 0.3h, where h is the total ice thickness, preferred orientations begin to develop. These fabrics are of a variety of different types (Patterson, 1981) and at some sites can be quite striking. For instance, at Byrd Station in West Antarctica, an orientation (multimaximum or girdle) begins to develop at depths between 0.2 and 0.3h and continues until roughly 0.6h, where a strong single maximum fabric occurs with all the c-axes of the ice crystals oriented close to vertical. This fabric persists to 0.83h, below which it again changes back to a multimaximum pattern (Gow and Williamson, 1976). As might be expected, these fabric changes are associated with pronounced changes in the rheological properties of the ice; under identical conditions, strongly oriented ice with the easy-glide planes oriented in the plane of shear will deform four times more readily than similar ice with a random crystal orientation (Russell-Head and Budd, 1979).

Properties. As most engineering works involving major construction on glaciers occur in the near-surface portions of the dry snow facies of large ice sheets, the surficial material encountered is usually high-density snow and rarely fine- to medium-grained ice with a random crystal orientation. Maintenance of structures at locations where ice occurs at the glacier surface (i.e., where significant surface melt commonly occurs) is sufficiently difficult that such sites are rarely selected. Other engineering problems in coarse-grained, strongly oriented ice usually are associated with tunneling in the near-melting ice near glacier margins where tunnel closure rates are of concern. There are, of course, exceptions: for instance, during the early 1900s, 8.8 km of the Copper River and Northwestern Railway was laid across the Allen Glacier in Alaska with the ballast resting directly upon the ice. Some disadvantages of this route are reviewed in Tarr and Martin (1914). A particular increase in maintenance headaches occurred in 1912 when the "stagnant" glacier resumed its advance. As the general approach to such engineering problems during this pioneer period can be summarized in the statement "give me enough snoose (snuff) and dynamite and I'll build you a road to Hell," it is doubtful that the early builders and engineers were overly concerned with the fine points of glacier dynamics and the rheological properties of ice (Janson, 1975).

The general mechanical properties of glacier ice can be described as follows (Weeks and Mellor, 1983):

(i) Under moderate hydrostatic pressures and low temperatures, ice compresses elastically with a bulk modulus of about 9 GPa.

(ii) Under deviatoric stress, ice deforms as a nonlinear viscoelastic solid, changing its fabric and structure during the process. Under constant stress a complete creep curve usually results, showing an initial elastic response, then a deceleration followed by acceleration. Figure 3 shows the effect of temperature on the initial portions of such curves. Below −10°C, this effect can be described by an Arrhenius relation with an activation energy of about 70 kJ/mole. However, once the load is released, a significant delayed elastic strain is recovered. This is in

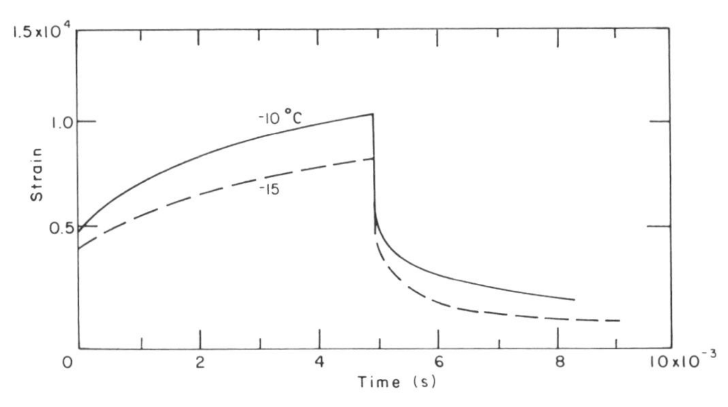

Figure 3. Constant stress creep curves for glacier ice obtained at two different temperatures. At times of less than 5×10^3s, the stress was 0.25 Mpa, at times greater than this value the stress was zero.

marked contrast to snow, which displays very little strain recovery. Under constant strain rate, a complete stress/strain curve shows stress rising to a maximum before falling and tending asymptotically to a limit. Figure 4 shows typical stress response curves for polycrystalline ice deformed at strain rates of 1×10^{-7} and 2.5×10^{-8} s^{-1}. The yield drop occurs because during initial deformation the density of mobile dislocations grows slowly until a sufficiently large number are generated to accommodate the flow at a lower stress.

(iii) High sensitivity to strain rate and temperature causes ice to display a broad range of rheological properties; elastic behavior dominates at high strain rates and low temperatures, and ductile behavior dominates at low strain rates and near-melting temperatures. Commonly, both elasticity and nonlinear viscosity make significant contributions to deformation and rupture. In the elastic range the compressive strength of ice is both temperature and strain-rate dependent. In contrast, the tensile strength is not strongly dependent on either of these factors. These temperature-induced variations are shown on Figure 5.

(iv) Under multiaxial stress states, the compressive bulk stress has little effect on the deviatoric stress/strain rate relation when the deviatoric stress is low and temperature is well below the melting point. However, at high strain rates, moderate confining pressure suppresses the formation of internal microcracks and increases the deformation resistance. If bulk compressive stresses are large enough to cause ice to approach the temperature of the ice-water phase transition, a decrease in the deformation resistance occurs that is essentially independent of the magnitude of the deviatoric stress.

Additional important factors affecting the strength and deformation resistance of ice are:

Porosity. Strength and deformation resistance decrease as porosity increases. Over the range of naturally occurring porosities (n) encountered in the near-surface layers of ice sheets ($0 \leqslant n \leqslant 60$ percent), large property variations result (i.e., compressive strengths of 0.5 to 11 MPa and elastic moduli of 0.1 to 9 GPa). Under such conditions there also appears to be a change of slope in the physical property-versus-porosity curves when porosity is approximately 40 percent (Anderson and Benson, 1963; Robertson and Bentley, 1975). This porosity corresponds to the density at which snow grains are presumed to settle into a random, close-packed arrangement.

Grain size. Fracture strength decreases with increasing grain size, at least up to grain sizes (d) of 5 mm, according to the relation $\sigma \propto d^{-1/2}$. In the ductile range, however, grain size does not appear to have a major effect on strength and deformation resistance. It should be mentioned that in temperate glacier ice, grain shapes are very complex and have been little studied (Rigsby, 1968).

Crystal orientation. Deformation occurs most readily when ice crystals are preferentially oriented so that their easy glide (basal) planes are parallel to the plane of resolved stress. Although it is necessary to consider orientation effects in problems of glacier flow and dynamics; such considerations have, to date, not been included in most engineering plans or design of works.

Useful reviews of the properties of glacier ice are given in Patterson (1981), Weeks and Mellor (1983), and Mellor (1986).

Lake, river, and sea ice

Structure. The overall geometry of lake, river, and sea ice bodies is, of course, very different from that of glacier ice masses. There are two main types of lake, river, and sea ice: congelation and frazil. Congelation ice can, in turn, be divided into three different subtypes. Congelation refers to ice that grows via the loss of heat through the ice sheet to the atmosphere. The crystals have a generally columnar shape, with the axis of the columns oriented parallel to the direction of heat flow. Strong preferred crystal orientations invariably develop in such ice; the exact nature of the orientations, as well as the grain sizes and shapes, depend on the nature of the initial ice skim. If the initial skim develops under clear, calm conditions without external seeding, large tabular c-axis vertical crystals commonly form and "pinch out" any c-axis horizontal crystals that might initially occur. Grain boundaries are very irregular, and it can be difficult to locate individual boundaries because crystal orientations are nearly identical. Ultimately the ice sheet that develops behaves like a giant single crystal with its c-axis vertical. If, on the other hand, the initial ice skim is nucleated externally, a fine-grained skim with random c-axis orientations results. In this case, during subsequent growth the c-axis horizontal crystals "pinch out" crystals with a c-axis vertical orientation. The resulting ice sheet shows well-defined grain boundaries, and crystals have simple polyhedral shapes when viewed in horizontal thin sections. Again there is a general increase in average grain size as the ice thickens, with the elimination of smaller, less favorably oriented grains. Although the sequence of events leading to these two very different fabrics is now well documented (Gow, 1986), the physical mechanisms in-

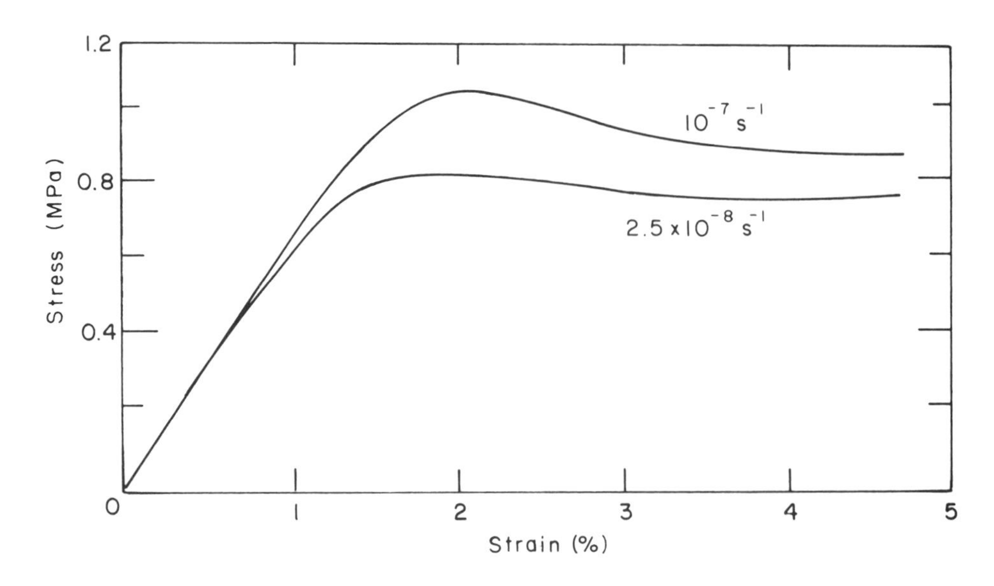

Figure 4. Response of glacier ice to two different constant strain rates.

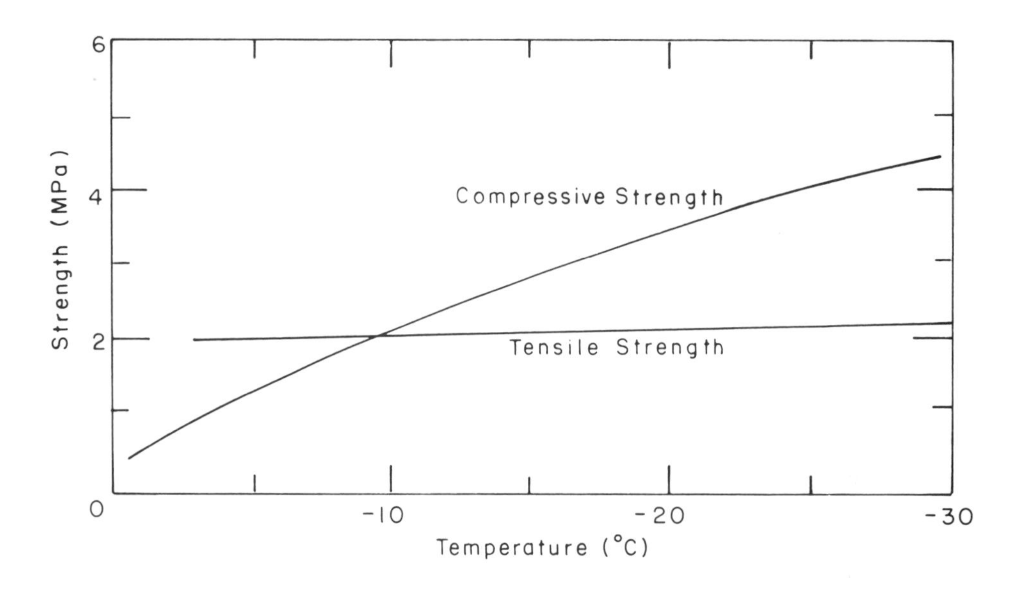

Figure 5. Brittle fracture strength of glacier ice (grain size 1 to 2 mm) as a function of temperature.

volved are not well understood. In sea ice, the c-axis horizontal structure always develops, and this proceeds with either c-axes randomly oriented in the horizontal plane or directionally aligned in the horizontal plane. Nearly constant c-axis alignment directions occur over large areas of near-shore sea ice along both the Alaskan and Siberian coasts. Both observations and theory suggest that the alignment directions are related to the direction of the current at the growing ice-water interface (Weeks and Gow, 1978). The presumed cause of these alignments is associated with small differences in the solute distribution at the growing ice-water interface.

The engineering importance of these differences is that both c-axis vertical and c-axis horizontal (random) ice sheets are transversely isotropic; they show property variations in the vertical direction, but properties are independent of direction in the horizontal plane. On the other hand, c-axis horizontal (aligned) ice sheets are orthotropic, showing property variations in three directions that are orthogonal to each other.

Frazil ice, the other principal lake, river, and sea ice type, results from the development of discs, plates, and spicules of ice in the water column itself. Frazil formation has long been considered an important process in the formation of river ice covers where it is associated with turbulent mixing in the fluid. However, the process is now known to be of importance in certain sea-ice areas (particularly in the Antarctic) where the fine-grained, equiaxed, c-axis random ice that results from frazil generation can make up a significant portion of thick, multiyear ice floes (Weeks and Ackley, 1986). There is no agreement on the exact reasons for the apparent differences in the amount of frazil present in the Antarctic versus the Arctic seas. One interesting and plausible explanation advanced suggests that much of the Antarctic frazil results from the supercooling of sea water by adiabatic decompression as it flows out and up from beneath the extensive ice shelves that surround the Antarctic continent (Lewis and Perkins, 1986). As expected by its structure, frazil ice samples are isotropic with the properties independent of direction. In-situ sheets of frazil ice are, however, transversely isotropic as a result of the vertical gradient in ice temperature.

Two other aspects of floating ice that exert significant structural influences on property variations are included gas and salt and the microstructure of the ice sheet:

Included gas. The presence of included gas in most river and lake ice specimens is obvious upon casual observation because the ice matrix is transparent. The bubbles are strikingly arrayed in patterns that change with growth velocity due to changes in past meteorological conditions and can be used as markers from lake to lake. Excellent examples of these inclusion patterns are given by Swinzow (1966) and Gow and Langston (1975, 1977). Observations of gas bubbles in sea ice are more difficult because sea ice is less transparent than fresh-water ice and the effects of gas inclusions are overshadowed by variations in the volume of liquid brine included in the ice. Furthermore, until recently (Cox and Weeks, 1983), there was no rapid, convenient field procedure for determining the gas content of sea ice. Limited data suggest that the gas contents of lake and sea ice are similar, in the range of 0.5 to 1 percent with maximum values up to 4 percent. Generally, the gas content of multiyear sea ice is appreciably higher than that of first-year ice as the result of brine drainage. Values of 10 percent are common, and extreme values reach 20 percent (Cox and others, 1984). The higher values are usually found in the upper portions of the ice.

Substructure. During the formation of sea ice, growth conditions are invariably in a range in which a nonplanar solid-liquid interface is stable. The resulting cellular substructure results in ice crystals that are composed of plates of pure ice separated by films or pockets of included brine or gas. This structure is shown strikingly in thin sections of sea ice (Fig. 6) with the detailed spacings between the inclusion arrays (measured parallel to the c-axis) changing as a function of the growth velocity. These changes in void patterns result in subtle changes in ice properties (Weeks and Assur, 1963). In addition to the millimeter-scale substructure associated with brine entrapment, there are larger brine drainage channels (tubular treelike features with dendritic arms) through which brine drains from the ice sheet. Most channel diameters range in size between 0.1 and 1 cm and are spaced so that there is roughly 1 channel per 180 cm^2. Although these large flaws clearly have an effect on the gross properties of sea ice sheets, to date the subject has not been studied.

Composition. The amount of salt entrapped in sea ice is a function of the salinity of the sea water and the ice growth rate; higher growth rates result in more entrapment. Once entrapment has occurred, a number of processes ensue that all lead to the drainage of brine from the ice. A detailed description of these mechanisms was presented by Weeks and Ackley (1986). As a result, sea-ice salinity profiles have systematic shapes that change regularly during the ice growth season. Several numerically simulated profiles for first-year sea ice of different thicknesses are shown on Figure 7 (Cox and Weeks, 1988). Given a sea-ice sheet with known salinity and temperature profiles, the brine volume profile is uniquely determined because at each temperature the composition of brine in equilibrium with the ice is specified by the phase diagram for sea water at below-freezing temperatures (Assur, 1958). As most first-year sea ice has salinities in the range of 4 to 12‰ and temperatures between −2° and −30°, brine volumes can be expected to vary between 30 and 300‰ (3 and 30 percent). For any given salinity, brine volumes (and changes in brine volumes) are always greatest at near-melting temperatures (at locations near the bottom of the ice sheet). Figure 8 shows the brine volume profiles that would result from a combination of the salinity profiles of Figure 7 with linear temperature profiles in which the ice surface temperature was calculated from climatological averages for the Arctic Basin. The kinks in the brine volume curves in the upper left hand corner of the diagram are the result of eutecticlike changes resulting from the crystallization of the solid salt $NaCl \cdot 2H_2O$. Because lake and river ice freeze with a planar ice-water interface, there is no solute entrapment and the above compositional complications do not apply.

More detailed descriptions of the structural and compositional aspects of lake, river, and sea ice can be found in papers by Ashton (1986) and Weeks and Ackley (1986).

Properties. As the ice phase in sea, lake, and river ice is identical to the ice in glacier ice, the differences in physical and mechanical properties between these ice classes are largely the result of changes in grain structure and in bubble and brine pocket shapes and distributions. Comprehensive discussions of the properties of lake and river ice are presented by Michel (1978) and Ashton (1986) and of sea ice by Weeks and Mellor (1983), Mellor (1986), Weeks and Assur (1968, 1972), Weeks and Cox (1984), and Weeks and Ackley (1986).

Important factors that influence the mechanical properties of lake, river, and sea ice are: structure, load orientation relative to substructure, gas and brine volume, temperature, and strain or stress rate.

A few points of interest to engineering works are:

(i) The mechanical properties of sea ice approach those of bubble-free lake or river ice as the void volumes (porosities) of

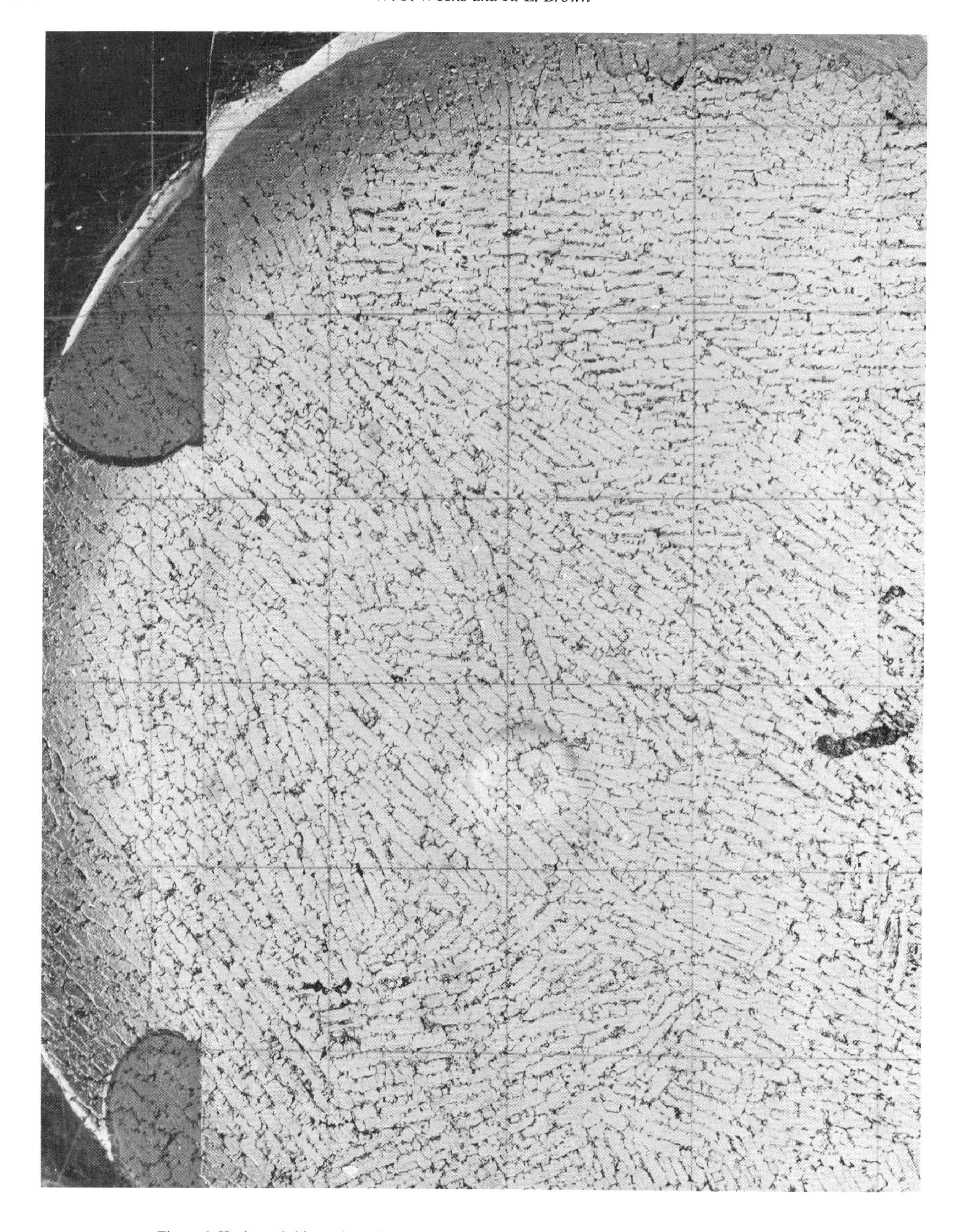

Figure 6. Horizontal thin section of sea ice showing the substructure delineated by the brine inclusions.

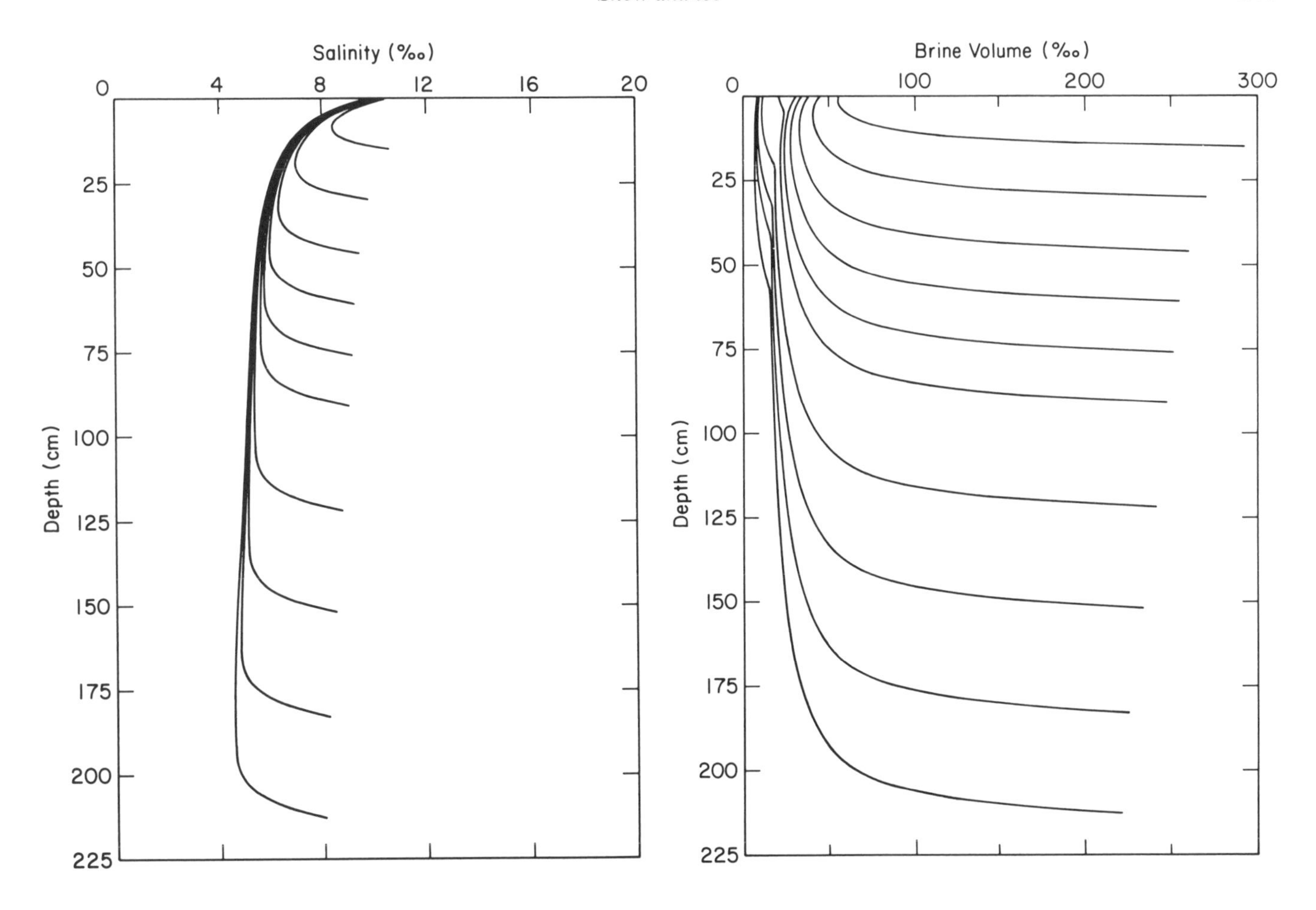

Figure 7. Numerically simulated salinity profiles for first-year sea ice of several thickness (Cox and Weeks, 1988).

Figure 8. Brine volume profiles for first-year sea ice of several different thicknesses based on the salinity profiles of Figure 7 and linear temperature profiles (Cox and Weeks, 1988).

gas and brine in the sea ice approach zero. This means that lake, river, and even glacier ice properties can be used as a limit in studies of sea-ice property variations.

(ii) Ice properties are strongly affected by changes in void volume. For lake and river ice, the void volume is the result of gas inclusions; for first-year sea ice, it is primarily the result of brine inclusions; and for multiyear sea ice, both gas and brine effects can be significant. Because the substructure of sea ice is well defined by its growth conditions, simple geometric models have been useful in analyzing strength and modulus changes caused by void volume changes. Figure 9 shows tensile strength profiles for first-year sea ice calculated by such a model from the brine volume profiles of Figure 8. The vertical variation in the strength of the ice is striking. Similar approaches could be useful in lake and river ice studies if quantitative relations could be established between growth conditions and bubble structures.

(iii) Sea, lake, and river ice are both stiffer and stronger when loaded parallel to the long axis of the columnar crystals.

(iv) There is a pronounced increase in the compressive strength of sea, lake, and river ice with increasing strain rate. On the other hand, the tensile strength is comparatively unaffected by such changes.

(v) There is a general decrease in strength as sample size increases.

(vi) In the strain rate range of 10^{-6} to 10^{-3} s^{-1}, confinement has little effect on the strength of either granular snow ice or frazil sea ice samples. Similar results are obtained for columnar ice when the crystals are confined in a direction normal to the axis of elongation. However, when columnar grains are constrained in the plane of the ice sheet, strength increases by a factor varying from 2 to 5 (Timco and Frederking, 1986).

(vii) The three-dimensional yield surface for columnar ice changes in both size and shape with changes in loading rate and temperature. Figure 10 shows yield surfaces for both granular and columnar fresh-water and sea ice as suggested by Timco and Frederking (1984). The development of improved general con-

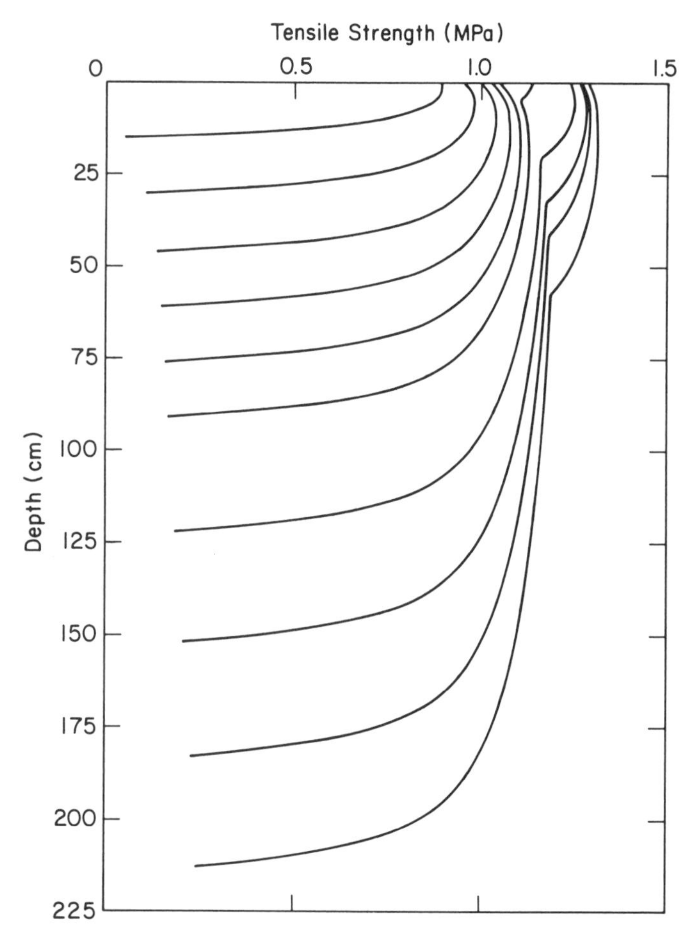

Figure 9. Tensile strength profiles for sea ice of varied thicknesses calculated by a geometric model from the brine volume profiles shown in Figure 8 (Cox and Weeks, 1988).

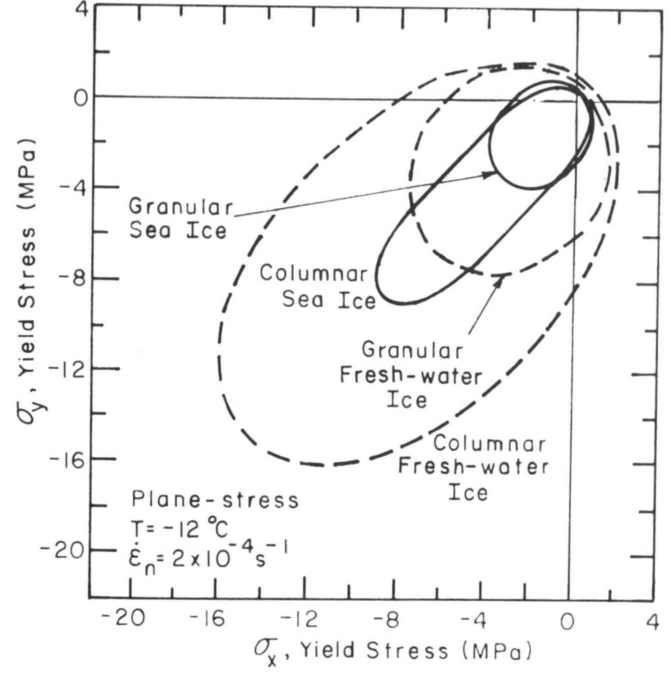

Figure 10. Yield envelopes for plane-stress conditions in the plane of the ice cover for granular/discontinuous columnar sea ice, columnar sea ice, and granular and columnar fresh-water ice at T = –12°C and $\dot{\epsilon}$ = 2 × 10^{-4} s^{-1}. The figure shows the pronounced influence of the brine in the ice and the grain structure on the failure envelope of the ice (Timco and Frederking, 1984).

stitutive relations for sea, lake, and river ice continues to be an area of active research interest (Brown and others, 1986; Sunder, 1986).

EXAMPLES OF ENGINEERING PROBLEMS

Although problems in engineering works and glaciology are extremely varied, the glaciologist is usually asked to provide an engineering design team with quantitative estimates of snow and ice patterns, forces, and behavior. The following four examples are chosen to provide the reader with an appreciation of a few of the complexities of such endeavors. Note that these activities fall into two general classes. In the first case, the site and the glaciological setting are presumed to be both known and relatively static. The question then becomes, how will this stationary environment interact with different engineering options? The design of structural foundations for snow and ice terrains is such an example. The other three cases cited concern the forces associated with specified events plus estimates of the probabilities of the occurrence of the different events. The design of offshore structures for ice-covered seas is a classic problem of this latter type.

Avalanches

The snow avalanche is a complex phenomenon that, although studied for many years, is still not well enough understood that accurate predictions of avalanche occurrence and magnitude can be made. Weaknesses within a snow cover can develop rapidly as the result of new precipitation, changes in air temperature, or both. For instance, a four- or five-day cold spell can completely destroy the strength at the bottom of a shallow (<1 m) snow cover as the result of the formation of depth hoar. Ensuing snowfalls can then load the snow cover beyond its reduced load-carrying capacity, resulting in an avalanche. Other factors leading to avalanches include additional loadings produced by snowfall, snow drifting, wind, earthquakes, skiers, or explosives.

The majority of avalanches are not large, having fracture lines less than 0.5 m deep and encompassing areas of less than 3,000 m^2. Even so, these avalanches are dangerous; they can bury skiers and destroy property. The forces generated are truly amazing (Lang and Brown, 1980). Some exceptionally large avalanches with fracture lines in excess of 20 m depth do occasionally occur in the mountain ranges of Alaska, the Andes, and the Himalayas. These avalanches entrain tremendous masses of snow, mud, and rock, travel large distances, and have the energy to destroy large areas of forest.

Avalanches are in many ways analogous to landslides in that they commonly result from internal weaknesses in the snow or rock mass. Once started, they also have similar flow characteristics so the sliding mass can normally be modelled as a non-Newtonian fluid or a granular material. The mechanics of the interactions between the sliding or flowing materials and the underlying surfaces is also quite similar, although the snow avalanche has the potential for higher velocities. In some cases, speeds in excess of 200 km/hr are achieved. Part of the explanation for this difference is that snow avalanches may more readily become airborne. Also the apparent coefficient of friction in snow avalanches can decrease to very small values due to turbulent particle motion near the base of the avalanche. Although this phenomenon can also occur in landslides (Fleming and Varnes, this volume), the friction reduction is not believed to be as significant. Finally, mudslides, such as those which occurred during the Mount St. Helens eruption (Schuster and Mullineaux, this volume), also bear some resemblance to the snow avalanche. In fact, computer programs used to model snow avalanches have been successfully used to model both mudflows (Lang and Dent, 1987) and landslides (Trunk and others, 1987).

As the snow avalanche has considerable destructive potential, modelling the avalanche flow and its impact with obstacles such as buildings is of practical engineering importance. During the past 30 years, essentially two methods presented below for modelling avalanche dynamics have evolved. However, no model developed to date can correctly model all aspects of even a dry snow avalanche. Normally such an avalanche has three parts, as shown on Figure 11. The upper part is a powder cloud containing suspended snow particles, which commonly obscures the other two parts. The second part is light, flowing snow with a density of 5 to 30 kg/m^3, which rides on top of the dense lower part (60 to 200 kg/m^3) and constitutes most of the avalanche mass. In the case of a wet snow avalanche, the avalanche is composed almost totally of the dense, lower part, which may have a density approaching 400 kg/m^3. To date, modelling has primarily been restricted to the dense lower portion of the avalanche.

The most widely used avalanche model is that of Voellmy (1955), because it is both easy to develop and understand and has been reasonably successful. The model is based on the theory of the steady open channel flow of a fluid. The slope is normally broken into three parts: (1) the starting zone where the avalanche is initiated and quickly accelerates to a terminal velocity, (2) the avalanche track with constant slope and avalanche speed, and

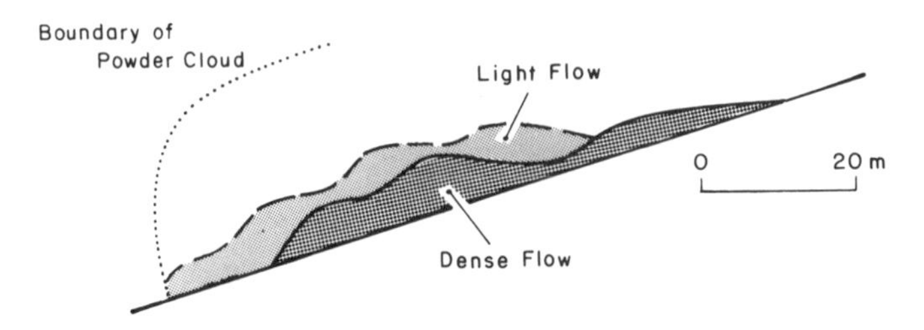

Figure 11. Schematics of an avalanche. Normally the powder cloud obscures the denser portions.

(3) the runout zone where the avalanche decelerates as the slope decreases.

In the middle section, the terminal velocity can be calculated by the direct application of equilibrium principles from:

$$v = [\xi h(\sin\theta - u\cos\theta)]^{1/2}$$

where θ is the slope angle, u is a coefficient of friction, h is the flow depth, and ξ is a term containing the fluid (snow) viscosity and the friction coefficient. Voellmy calculated the runout distance S by using a simple energy balance equation to obtain:

$$S = v^2/[2g(u\cos\theta - \tan\theta) + v^2g/\xi h_m)$$

where h_m is the mean deposition depth of the avalanche debris.

Although Voellmy's theory appears simple, it has proven difficult to use reliably since considerable judgement is needed in estimating ξ, u, h, h_m, and the point separating the avalanche track and the runout zone. Consequently, several extended versions of the theory have been formulated (Salm, 1966; Perla and others, 1980) involving more reasonable estimates of the drag forces on the avalanche and the effects of variable avalanche mass. All of these models suffer from the number of unverified assumptions required and from lumping the avalanche into one mass.

Lang and Brown (1980) have used a different approach that avoids some of the problems of Voellmy's method; they performed a direct numerical solution of the Navier-Stokes equations to model the flowing mass of the avalanche. The leading and upper surface was taken to be stress free, whereas the lower surface was prescribed with a friction coefficient. The flowing mass was assumed to be representable by a Newtonian fluid. Figure 12 shows the calculated leading edge of the avalanche before and during impact on a wall placed normal to the slope. Since the computer solution allows the calculation of normal and shear stress distributions over the wall during impact, the net forces applied to the wall can be calculated. For the calculations shown in Figure 13, a 1-m-deep flow with a density of 200 kg/m^3 would produce an average load of 200,000 N for each meter of structure width. In this case the avalanche speed was about 30 m/s. The forces are clearly large enough to inflict heavy

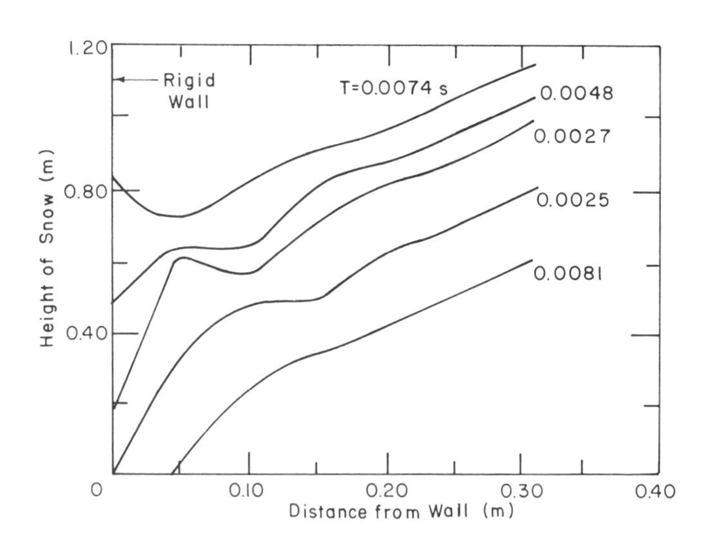

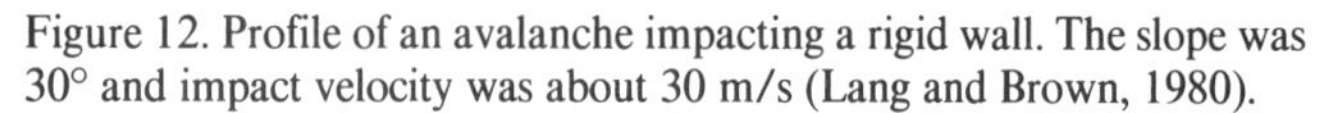
Figure 12. Profile of an avalanche impacting a rigid wall. The slope was 30° and impact velocity was about 30 m/s (Lang and Brown, 1980).

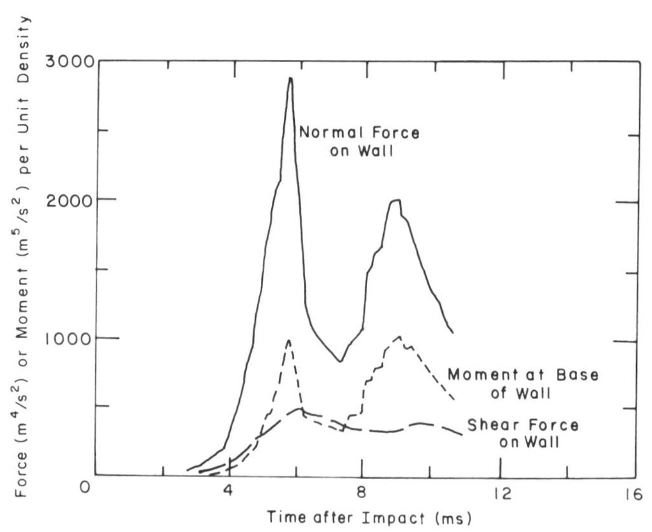

Figure 13. Loads applied to a wall placed normal to the slope (Lang and Brown, 1980).

damage on conventionally designed buildings, a result that is certainly in agreement with observations.

Later, Dent and Lang (1980) improved this model by including the effects of turbulence and allowing the snow to lock up once the velocities drop below a critical value. This model has proven to give highly useful results and has since been utilized to model both mudflows and landslides.

While these later methods have provided more information than the Voellmy formulation, they also have inherent problems and are more complicated, requiring judgement in picking viscosity and friction coefficients. However, it has now been demonstrated that with appropriate use of either method, reasonable estimates of both the speed and destructive force of avalanches can be obtained.

Foundations

Any platform or building foundation placed on snow or ice will settle, since snow and ice creep under sustained loads. Here we will consider the evaluation of foundation settlement into a snow cover because, as mentioned earlier, most major construction on large glaciers has occurred at sites where the foundation material is dense snow. Examples include early warning radar stations and the research facilities at Camp Century, Byrd Station, and the South Pole. Of course, snow accumulation rates and drifting around a building may be appreciably larger than foundation settlement in regions where there is a net yearly accumulation of snow. At any rate, an evaluation of settlement is needed to determine the significance of this process. If such creep is allowed to continue for an appreciable time, structural loading ensues that in extreme cases can result in the failure of structural members (Fig. 14).

Foundation settlement is a slow, steady, long-term process requiring on the order of years for substantial settlement into firn to develop. Observations have shown that the penetration of a foundation depends on snow density, temperature, load intensity, footing size, rigidity, and shape. Surprisingly, the amount of analytical work on such foundation settlement problems has been very limited. Kerr (1962) treated the problem by representing the snow cover as an assemblage of springs and dashpots, much as one builds a viscoelastic model. This allowed him to model the sinkage rate of the structure into the snow cover, provided correct values were given to the spring and dashpot constants. However, it does not allow one to study the detailed response of the snow under the footings.

Rather than using Kerr's approach, the method of Dandekar and Brown (1986), who considered a flexible footing that produces a uniform pressure over the snow cover, will be discussed here. The snow is modelled as a compressible non-Newtonian fluid, and assuming symmetry, only one-half of the footing need be considered. The viscosity of the fluid is taken to be strongly density dependent, so that the resistance increases significantly as the material under the foundation densifies. The settlement is then calculated by the finite difference method using input parameters that were chosen to be consistent with foundation tests run at Camp Century (Reed, 1966). Figure 15 compares the calculated settlement values with the experimental results. Figure 16 compares the density changes calculated under a flexible footing with the profile induced by gravity alone (Dandekar and Brown, 1986). In these curves the snow was assumed to have an initial density of 400 kg/m^3, and test temperatures were approximately –20°C for the two-year test period. The observed settlements of 0.1 m would have been substantially larger if the temperatures had been higher.

a

b

Figure 14. (a) The failure of structural members in the base of the radar station DYE-3 caused by snow settlement. (b) Air photograph of DY3-3 taken after the 64-m sideways move necessitated by snow accumulation and structural failures resulting from snow creep. A sense of the scale of the structure can be gained by noting the crane in the left (Tobiasson, 1979).

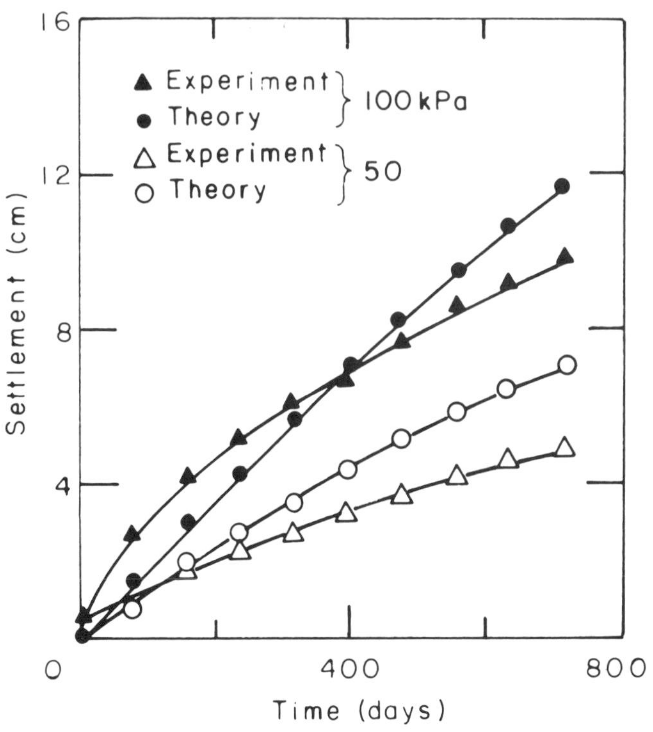

Figure 15. Settlement of a foundation into snow cover for a 0.5 × 0.5 m test pad (Reed, 1966).

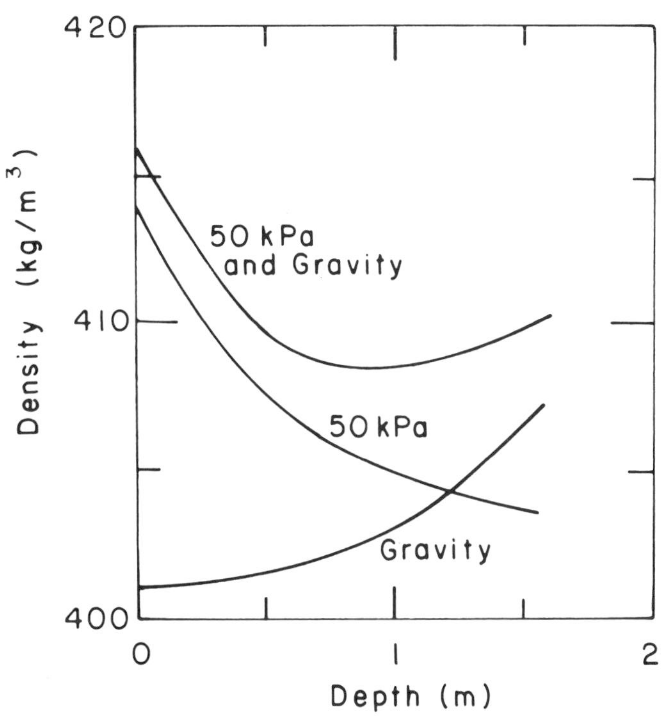

Figure 16. Calculated density changes under a flexible footing. The three curves compare the effects of the foundation load, gravity, and combined gravity and foundation (Dandekar and Brown, 1986).

Although the calculated settlements adequately predict the observed values for time periods of less than 500 days, at longer times the calculated settlements begin to exceed the measured values. This was expected, since the coefficients used to represent the material behavior were taken to be constant. However, the coefficients are not constant, because snow is a thermomechanically active material and is invariably in a state of change. In the absence of any large temperature gradients, the snow will slowly strengthen as the ice grains increase their intergranular bonding by diffusion of water vapor from convex surfaces to the bonds. Consequently, this bonding increases the material stiffness, with the net effect of decreasing the settlement rate. Had this effect been incorporated into the settlement calculation, a closer correlation with Reed's measurements would have resulted. As observed, foundation settlement is a slow, steady process, requiring years to become significant if the overburden pressures are small. Similar solutions have also been obtained for rigid foundations. In this case, settlements were of the same magnitude, although somewhat larger.

Forces on offshore structures

This is a complex subject with a long history that still receives considerable attention in the engineering literature. The initial interest in this class of problems was associated with the design of bridge piers that could withstand the ice forces occurring during the spring-ice debacle. More recently, activity on this subject has been spurred by interest in the design of "permanent" structures for the exploration and production of offshore oil resources from beneath the continental shelves bordering the Arctic Ocean and the seas off eastern Canada. Ice conditions in these regions are both highly variable and formidable: in the high Arctic, one is concerned with impacts from a variety of thick sea-ice features and ice islands, while off the coast of eastern Canada the primary concern is impact from icebergs. The discussion herein will focus on the former problem. Useful reviews of this general subject can be found in Ashton (1986), Coon and others (1981), Untersteiner (1986, 1988), and in the Third IAHR State-of-the-Art Report on Ice Forces (1986). A view of more current activities can be obtained by examining papers in recent IAHR (1986), OMAE (1986), OTC (1986), POAC (1985), and Polartech '86 (1986) symposia.

In order to obtain realistic estimates of the forces that a specified ice mass will exert on a given type of offshore structure, a variety of approaches has been utilized. As mentioned, a number of small-scale tests have been utilized to obtain estimates of the mechanical properties of sea ice. These values can then be utilized to calculate the forces necessary to cause the ice to fail via different mechanisms. For instance, ice can fail by bending as it rides up the surface of a structure; by crushing if in-plane stresses exceed the crushing load; by buckling if in-plane stresses reach the buckling load; and by splitting and spalling. For large structures, failures may occur at different locations and over different spatial domains so that the ice rarely fails uniformly along the complete ice-structure contact. In fact, several types of ice failures may occur simultaneously. Moreover, experimental observations indicate that the strengths of ice samples decrease as the sample sizes increase. This effect is presumed to be related to the distribution of different types of defects in natural ice samples; however, there is as yet no agreement on scaling for this phenomenon. Considering all this, what are the peak localized ice stresses and the overall ice forces exerted on a structure?

To assist in bridging the gaps between the properties of small specimens, the results of physical modeling experiments, and the behavior of thick real ice sheets and features, in-situ tests have been performed on thick ice. Such tests are time consuming, few in number, and very expensive. For instance, a series of full-scale compressive tests recently performed by EXXON reportedly cost $4 million. As might have been expected, full-scale tests give low strength values. Admittedly the problem could be approached more directly by building a near-full scale test structure designed to obtain ice force measurements. Not only is this option very expensive, but even if it were exercised it would be difficult to guarantee that the test structure would actually encouter thick multiyear ice capable of producing forces approaching the design condition. Because this type of data does not exist, attempts have been made to measure the deformation in gravel-fill islands used for exploratory drilling platforms as ice masses fail against their berms. Analysis of such data has been difficult because these failures invariably occur on the edges of the large, partially grounded piles of ice rubble that develop on the island flanks. Recently, attempts have been made to circumvent this coupling problem by placing arrays of stress gauges on the ice surrounding caisson-retained drilling structures in the Beaufort Sea (Fig. 17). A difficulty has been that these systems of gauges have been hard-wired to one or more central recorders. Unfortunately, a significant ice movement/force event invariably resulted in the failure of the hard-wiring just at the time the data became interesting. One particularly interesting program has measured the stresses and strains in thick old ice floes as they fractured against the sides of Hans Island, a steep-sided rock island located in Kennedy Channel between Ellesmere Island and Greenland. Although only a small amount of the data from the Hans Island tests has been released, indications are the observed ice forces are significantly lower than anticipated from a plastic-limit analysis (Ralston, 1980). This is reasonable in that plastic-limit analysis would be expected to provide an upper limit. However, until these differences can be understood and predicted, the basis for design of offshore structures will remain less than satisfactory. The general opinion, at present, is that the low observed values are the combined result of scale effects coupled with mixed mode failures; easy to imagine but not so easy to unravel.

In addition, most ice-force observations have been made at sites where ice types and ice thicknesses do not show the variations that would be expected at exposed sites in deeper water near the edges of the continental shelf in the Beaufort or Chukchi Seas. The appropriate coupling of ice thickness, ice type, ice velocity, and offshore structure necessary to provide realistic ice-force time series remains an elusive goal.

Papers of particular interest in this general area include Croasdale (1980), Maes and Jordaan (1986), Kreider and Vivatrat (1983), Kry (1980), Sanderson (1984), and Wheeler (1981).

Ice-induced gouges in the sea floor

This brief survey of a few aspects of engineering glaciology concludes with a problem where ice meets rock (or, at least, surficial sea-floor sediments) with consequences most unattractive to offshore petroleum production. We refer specifically to the deep gouges that are produced when the keels of grounded ice features are pushed across the surfaces of the shelves of ice-covered seas. An artist's depiction of this process is shown on Figure 18. Along the coasts of the eastern Canadian Arctic,

Figure 17. A view of the caisson retained offshore drilling structure "Kadluk," the man-made spray-ice island that was utilized as a protective structure, and the surrounding ice conditions. Beaufort Sea, March 1984 (Johnson and others, 1985).

gouges are primarily produced by icebergs, while along the Beaufort Sea coast they are primarily produced by pressure ridges and ice-island fragments. The following discussion focuses on this latter region where gouges more than 3 m deep are common and gouges in excess of 8 m have been reported. Clearly this phenomenon must be considered in the design of offshore pipeline systems with the immediate question being, "How deep must a pipeline be buried along a specific route to reduce the chances of the line being ruptured by the keel of a grounded ice feature to some acceptable level?"

To answer this question it is essential to know the characteristics of existing gouges (Barnes and others, 1984), which can be summarized as follows:

(i) Gouges occur at water depths of less than 80 m, although gouges at depths greater than the maximum keel depths of contemporary pressure ridges (~50 m) are believed to be relicts from periods when sea level was lower.

(ii) Within the gouged zone the frequency of observed gouges is highly variable, ranging from near zero to in excess of 200 gouges/km. The highest frequencies occur on the seaward sides of shoals in water 20 to 30 m deep.

(iii) The depths of both existing and newly formed gouges are described well by a negative exponential distribution (Fig. 19), with the character of the exponential falloff varying with water depth (Weeks and others, 1984). As expected, there are fewer deep gouges in shallow water as the large ice masses required to produce such gouges have grounded farther out to sea. For example, in water 5 m deep a 1-m gouge has an exceedance probability of 10^{-4} (1 gouge in 10,000 will, on the average, be expected to have a depth equal to or greater than 1 m). In water 30 m deep, a 3.4-m gouge has the same probability of occurrence.

(iv) Gouge orientations generally parallel the isobaths.

(v) The distribution of distances between gouges is well described by a negative exponential distribution, which suggests that the spatial occurrence of gouges can, within limits, be modeled as a Poisson process.

(vi) The exact ages of observed gouge sets are unknown. Current opinions on this matter expressed in the literature are highly varied and range from a few tens of years (i.e., since the last major storm) to 6,000 years (when the sea initially covered the area). Recent simulation studies (Weeks and others, 1985) that consider the infilling of newly formed gouges tend to support the "few tens of years" hypothesis.

To date, three attempts have been made to estimate the

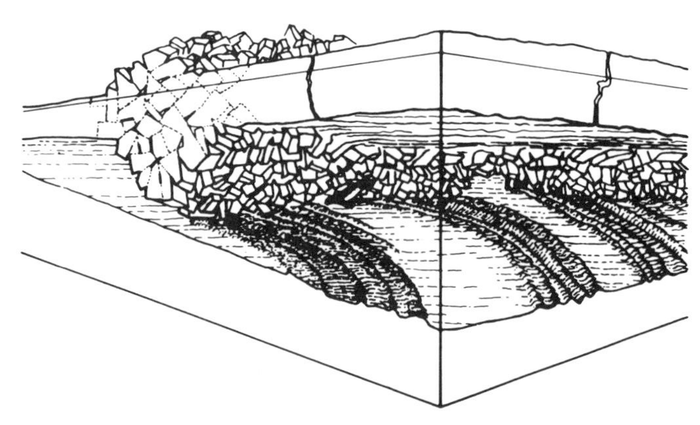

Figure 18. Schematic drawing of the ice-gouging phenomenon (Reimnitz and others, 1973).

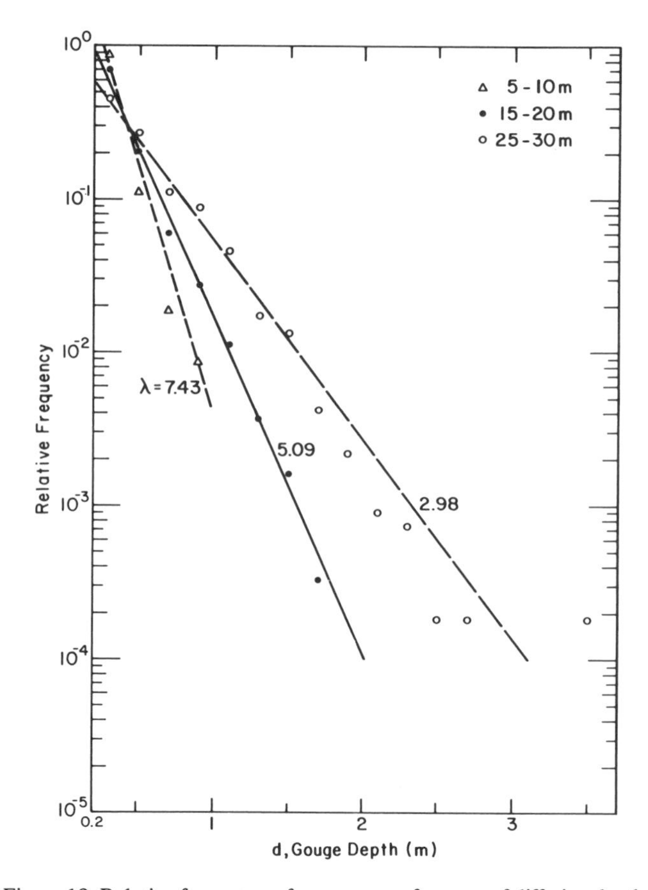

Figure 19. Relative frequency of occurrence of gouges of differing depths for three different water depth intervals along the Alaskan Beaufort Sea coast. The symbol λ represents the free parameter in the negative exponential probability density function (Weeks and others, 1984).

depths of gouges with specified recurrence intervals. In two of these, information on pressure ridge keels or sails and pack-ice drift rates has been combined to calculate gouging rates (Pilkington and Marcellus, 1981; Wadhams, 1983). In the third study, the limited data on observed rates and depths of newly formed gouges were used (Weeks and others, 1984). The estimated depth of the 100-year gouge that would be expected along a 76-km pipeline in water 25 m deep varies by over a factor of 2: 4.7 and 7.0 m versus 3.1 m. Deep burial of offshore pipelines is both extremely expensive and time consuming. Furthermore, pipeline burial depths of 7 m are at the margins of existing offshore trenching technology. It also should be remembered that although deep burial avoids the gouge problem, it could create another ice problem in that the hot pipeline is placed nearer to the top of the thermally unstable offshore permafrost. Melting of the interstitial ice in this material would result in soil compaction, with the associated possibility of a pipeline failure.

Research currently under way includes expanded replicate mapping of the sea floor using sidescan sonar and precision fathometry (Rearic, 1986); improved simulations of processes leading to gouge infilling; and better characterization of sea-floor sediments. It is also possible to estimate the maximum gouge depth conceivable in a given soil type from calculations of the forces necessary to cause the cutting edge of an ice keel to fail. The difficulty, however, lies in selecting the values that should be assumed for the strength of a pressure ridge keel (a jumble of ice blocks in which the interblock bonding is presumed to be highly variable). Experimental determinations of this quantity would be far from simple. At present, the installation costs resulting from pipeline burial can be an important factor in determining whether or not a given offshore oil or gas field is economic.

CONCLUSION

We have attempted to provide the reader with a glimpse of the varied engineering activities that are concerned with snow and ice in its many forms. However, we have not discussed the use of snow and ice as a construction material. The reason for this is that in general such applications have not been overly successful. This is particularly true if a structure is expected to have an appreciable lifetime. Perhaps the most successful of such applications is the use of flooding to rapidly increase the thickness of ice roads and drilling pads; of disaggregation, heat, and recompaction to increase the bearing strength of snow surfaces for runways and parking lots, and of spray icing to build a barrier around offshore drilling structures (see Fig. 17).

We hope to have convinced the reader that although snow and ice always appears wearing a white hat, its behavior is extremely complex and creates a variety of challenging engineering problems that require sophisticated geological and geophysical input to achieve adequate solutions. Although engineering glaciology will always remain a small speciality as compared with the more standard geological and engineering disciplines, we expect to see steadily increasing activity in this area as the population and the complexity of human activities in the polar regions increase.

REFERENCES CITED

Anderson, D. L., and Benson, C. S., 1963, The densification and diagenesis of snow, *in* Kingery, W. D., ed., Ice and snow; Properties, processes, and applications: Cambridge, Massachusetts Institute of Technology Press, p. 391–411.

Anderson, D. L., and Weeks, W. F., 1958, A theoretical analysis of sea-ice strength: Transactions American Geophysical Union, v. 39, no. 4, p. 632–640.

Anno, Y., 1986, Snow deflector built at the edge of a road cut: Cold Regions Science and Technology, v. 12, p. 121–129.

Ashton, G. D., ed., 1986, River and lake ice engineering: Littleton, Colorado, Water Resources Publications, 485 p.

Assur, A., 1958, Composition of sea ice and its tensile strength, *in* Arctic sea ice: Washington, D.C., National Academy of Sciences, National Research Council Publication 598, p. 106–138.

—— , 1967, Flexural and other properties of sea ice sheets, *in* Oura, H., ed., Physics of snow and ice; Proceedings, International Conference on Low Temperature Science 1966: Hokkaido University, Institute of Low Temperature Sciences, v. 1, part 1, p. 557–567.

Bader, H., Haefeli, R., Bucher, E., Neher, J., Eckel, O., and Thams, Chr., 1939, Der Schnee und Seine Metamorphose: Bern, Beitrage zur Geologie der Schweiz Geotechnische Serie, Hydrologie Lieferung 3 (translated 1954 as Snow and its metamorphism: Snow, Ice, and Permafrost Research Establishment Translation 14, 313 p.).

Barnes, P. W., Rearic, D. M., and Reimnitz, E., 1984, Ice gouging characteristics and processses, *in* Barnes, P. W., Schell, D., and Reimnitz, E., eds., The Alaskan Beaufort Sea; Ecosystems and environments: New York, Academic Press, p. 185–212.

Brown, R. L., Richter-Menge, J. A., and Cox, G.F.N., 1986, An evaluation of the rheological properties of columnar ridge sea ice, *in* Murthy, T.K.S., Connor, J. J., and Brebbia, C. A., eds., Ice technology: New York, Springer-Verlag, p. 55–66.

Colbeck, S. C., 1974, A study of glacier flow for an open-pit mine; An exercise in applied glaciology: Journal of Glaciology, v. 13, no. 69, p. 401–414.

Colbeck, S. C., and Ackley, S. F., 1983, Mechanisms for ice bonding in wet snow accretions on power lines: Cold Regions Research and Engineering Laboratory Special Publication 83-17, p. 25–30.

Coon, M. D., Brown, C. B., Cox, G.F.N., Ralston, T. D., Shapiro, L., and Weeks, W. F., 1981, Research in sea ice mechanics: Washington, D.C., National Research Council Marine Board, National Academy Press, 80 p.

Cox, G.F.N., and Weeks, W. F., 1983, Equations for determining the gas and brine volumes in sea ice samples: Journal of Glaciology, v. 29, no. 102, p. 306–316.

—— , 1988, Numerical simulations of the profile properties of undeformed first-year sea ice during the growth season: Journal of Geophysical Research, v. 93, p. 12449–12460.

Cox, G.F.N., Richter-Menge, J. A., Weeks, W. F., Mellor, M., and Bosworth, H., 1984, Mechanical properties of multiyear sea ice; Phase I, Test results: U.S. Army Cold Regions Research and Engineering Laboratory Report 84-9, 105 p. (See also Phase II, Test results: CRREL Report 85-16, 81 p.)

Croasdale, K., 1980, Ice forces on fixed, rigid structures, *in* Carstens, T., ed., Working group on ice forces on structures: U.S. Army Cold Regions Research and Engineering Laboratory Special Report 80-26, p. 34–106.

Dandekar, B., and Brown, R. L., 1986, A numerical evaluation of footing settlement into snowcover: Cold Regions Science and Technology, v. 12, p. 131–138.

Dent, J., and Lang, T. E., 1980, Modeling of snow flow: Journal of Glaciology, v. 26, no. 94, p. 131–140.

Gow, A. J., 1971, Depth-time-temperature relationships of ice crystal growth in polar glaciers: U.S. Army Cold Regions Research and Engineering Laboratory Research Report 300, 18 p.

—— , 1986, Orientation textures in ice sheets of quietly frozen lakes: Journal of Crystal Growth, v. 74, p. 247–258.

Gow, A. J., and Langston, D., 1975, Flexural strength of lake ice in relation to its growth structure and thermal history: U.S. Army Cold Regions Research and Engineering Laboratory Research Report 349, 28 p.

—— , 1977, Growth history of lake ice in relation to its stratigraphic, crystalline, and mechanical structure: U.S. Army Cold Regions Research and Engineering Laboratory Report 77-1, 24 p.

Gow, A. J., and Williamson, T., 1976, Rheological implications of the internal structure and crystal fabrics of the West Antarctic ice sheet as revealed by deep core drilling by Byrd Station: Geological Society of America Bulletin, v. 87, p. 1665–1677.

Gubler, H., 1978, Determination of the mean number of bonds per snow grain and of the dependence of tensile strength of snow on sterological parameters: Journal of Glaciology, v. 20, no. 83, p. 329–342.

Hansen, A. C., and Brown, R. L., 1986, The granular structure of snow; An internal state variable approach: Journal of Glaciology, v. 32, no. 112, p. 434–438.

Hertz, H., 1884, Uber das Gleichgewicht schwimmender elastischer Platten: Weidmann's Annalen der Physik und Chemie, v. 22, p. 449–455.

Hobbs, P. V., 1974, Ice physics: Oxford, Clarendon Press, 837 p.

IAHR, 1986, IAHR Symposium on Ice 1986, Proceedings, 2 volumes: Iowa City, Iowa, Institute of Hydraulic Research, 1108 p.

Itagaki, K., 1977, Icing on ships and stationary structures under maritime conditions; A preliminary literature survey of Japanese sources: Cold Regions and Engineering Laboratory Special Publication 77-27, 22 p.

Janson, L. E., 1975, The Copper Spike: Anchorage, Alaska Northwest Publishing Company, 175 p.

Johnson, J., Cox, G.F.N., and Tucker, W. B., 1985, Kadluk ice stress measurement program: Proceedings, 8th International Conference on Port and Engineering under Arctic Conditions v. 1, p. 88–100.

Kashteljan, V. I., Poznjak, I. I., and Ryvlin, A. Ia., 1968, Ice resistance to the motion of a ship: Leningrad, Sudostroenie (in Russian 238 p.; translated 1969 by Marine Computer Application Corporation. 339 p.).

Kerr, A., 1962, Continuity in foundation models and related problems: U.S. Army Cold Regions Research and Engineering Laboratory Research Report 62-81.

Kreider, J. R., and Vivatrat, V., 1983, Ice force prediction using a limited driving force approach: Journal of Energy Resources Technology, v. 105, p. 17–25.

Kry, P. R., 1980, Ice forces on wide structures: Canadian Geotechnical Journal, v. 17, no. 1, p. 97–113.

Lang, T. E., and Brown, R. L., 1980, Snow-avalanche impact on structures: Journal of Glaciology, v. 25, no. 93, p. 445–456.

Lang, T. E., and Dent, J. D., 1987, Kinematic properties of mudflows on Mount St. Helens: Journal of Hydraulic Engineering, v. 113, p. 646–660.

Lewis, E. L., and Perkins, R. G., 1986, Ice pumps and their rates: Journal of Geophysical Research, v. 91, no. C10, p. 11756–11762.

Maes, M. A., and Jordaan, I. J., 1986, Arctic environmental design using short data extremal techniques: Proceedings, 5th International Offshore Mechanics and Arctic Engineering Symposium, v. 4, p. 13–19.

Makarov, S. O., 1901, Notes on ice science, *in* Makarov, S.O., ed., Yermak in the Ice: St. Petersburg, 507 p. (in Russian).

Mellor, M., 1974, A review of basic snow mechanics: Grindelwald Symposium on Snow Mechanics, Proceedings, IAHS-AISH Publication 114, p. 152–291.

—— , 1986, Mechanical behavior of sea ice, *in* Untersteiner, N., ed., The Geophysics of Sea Ice: NATO Advanced Science Institutes, Series B—Physics, v. 146, New York, Plenum Press, p. 165–281.

Michel, B., 1978, Ice mechanics: Quebec, Les Presses de L'Universite Laval, 499 p.

OMAE, 1986, 5th International Offshore Mechanics and Arctic Engineering Symposium, Proceedings: New York, American Society of Mechanical Engineers, v. 4, 637 p.

OTC, 1986, Offshore Technology Conference: Dallas, Texas.

Patterson, W.S.B., 1981, The physics of glaciers (2nd edition): Oxford, Pergamon Press, 380 p.

Perla, R., Cheng, T., and McClung, D., 1980, A two parameter model of snow-avalanche motion: Journal of Glaciology, v. 26, no. 94, p. 197–208.

Pilkington, G. R., and Marcellus, R. W., 1981, Methods of determining pipeline trench depths in the Canadian Beaufort Sea: Proceedings, POAC 84, 6th International Conference, v. 2, p. 675–687.

POAC, 1985, Proceedings, 8th International Conference on Port and Ocean Engineering under Arctic Conditions, 2 volumes: Narssarssuaq, Greenland, Danish Hydraulic Institute, 1063 p.

Polartech '86, 1986, International Offshore and Navigation Conference and Exhibition, VTT Symposium 71, 2 volumes: Espoo, Technical Research Centre of Finland, 1168 p.

Ralston, T. D., 1980, Plastic limit analysis of ice sheet loads on conical structures, *in* Tryde, P., ed., Physics and mechanics of ice: New York, Springer-Verlag, p. 289–308.

Rearic, D. M., 1986, Temporal and spatial character of newly formed ice gouges in eastern Harrison Bay, Alaska, 1977–1982: U.S. Geological Survey Open-File Report 86-391, 82 p.

Reed, S. C., 1966, Spread footing foundations on snow: U.S. Army Cold Regions Research and Engineering Laboratory Technical Report 175, 40 p.

Reimnitz, E., Barnes, P. W., and Alpha, T. R., 1973, Bottom features and processes related to drifting ice: U.S. Geological Survey Miscellaneous Field Studies Map MF-532, scale 1:1,000,000.

Rigsby, G. P., 1968, The complexities of the three-dimensional shape of individual crystals of glacier ice: Journal of Glaciology, v. 7, no. 50, p. 233–252.

Robertson, J. D., and Bentley, C. R., 1975, Investigation of polar snow using seismic velocity gradients: Journal of Glaciology, v. 14, no. 70, p. 39–48.

Russell-Head, D. S., and Budd, W. F., 1979, Ice sheet flow properties derived from bore-hole shear measurements combined with ice core studies: Journal of Glaciology, v. 24, no. 90, p. 117–130.

Salm, B., 1966, Contributions to avalanche dynamics: IAHS-AISH Publication 69, p. 199–214.

—— , 1982, Mechanical properties of snow: Reviews of Geophysics and Space Physics, v. 20, p. 1–19.

Sanderson, T.J.O., 1984, Theoretical and measured ice forces on wide structures, *in* Proceedings, 7th International Symposium on Ice, International Association of Hydraulic Research: Hamburg, Hambugische Schiffbau-Verschsanstalt GmbH, v. 4, p. 151–207.

Sunder, S. S., 1986, An integrated constitutive theory for the mechanical behavior of sea ice; Micromechanical interpretation, *in* Murthy, T.K.S., Connor, J. J., and Brebbia, C. A., eds., Ice technology: New York, Springer-Verlag, p. 87–102.

Swinzow, G. K., 1966, Ice cover of an arctic proglacial lake: U.S. Army Cold Regions Research and Engineering Laboratory Research Report 155, 43 p.

Tarr, R. S., and Martin, L., 1914, Alaska Glacier Studies of the National Geographic Society in the Yakutat Bay, Prince William Sound, and lower Copper River regions: Washington, D.C., National Geographic Society, 498 p.

Third IAHR, Ice Forces, 1986, IAHR Working Group on Ice Forces, 3rd state-of-the-art report on ice forces: Proceedings, International Association of Hydraulic Research Ice Symposium 1986: Iowa City, Iowa, Institute of Hydraulic Research, v. 2, p. 283–482.

Timco, G. W., and Frederking, R.M.W., 1984, An investigation of the failure envelope of granular/discontinuous-columnar sea ice: Cold Regions Science and Technology, v. 9, p. 17–27.

—— , 1986, Confined compression tests; Outlining the failure envelope of columnar sea ice: Cold Regions Science and Technology, v. 12, p. 13–28.

Tobiasson, W., 1979, Snow studies associated with the sideways move of DYE-3: Proceedings, 36th Eastern Snow Conference, p. 117–124.

Trunk, F. J., Dent, J., and Lang, T. E., 1987, Computer modeling of large rock slides: Journal of Geotechnical Engineering, v. 112, no. 3, p. 348–360.

Untersteiner, N., ed., 1986, The Geophysics of Sea Ice: NATO Advanced Studies Institutes, series B—Physics, v. 146, New York, Plenum Press, 1196 p.

—— , 1988, Structure and dynamics of the Arctic Ocean ice cover, *in* Grantz, A., Johnson, L., and Sweeney, J. F., eds., The Arctic Ocean region: Boulder, Colorado, Geological Society of America, The Geology of North America, v. L (in press).

Untersteiner, N., and Nye, J. G., 1968, Computations of the possible future behavior of Berendon Glacier, Canada: Journal of Glaciology, v. 7, no. 50, p. 205–213.

Voellmy, A., 1955, Uber die zerstorungskraft von lawinen: Schweizerische Bauzeitung, v. 73, p. 159–165, 212–217, 246–249, 280–285 (English translation, On the destruction force of avalanches, Translation 2, USDA-FS, Alta Avalanche Study Center, Wasatch National Forest, Salt Lake City, Utah, 1964).

Wadhams, P., 1983, The prediction of extreme keel depths from sea ice profiles: Cold Regions Science and Technology, v. 6, no. 3, p. 257–266.

Wheeler, J. D., 1981, Probablistic force calculations for structures in ice-covered seas: Engineering Structures, v. 3, p. 45–51.

Weeks, W. F., and Ackley, S. F., 1986, The growth, structure, and properties of sea ice, *in* Untersteiner, N., ed., The Geophysics of Sea Ice: NATO Advanced Studies Institutes, series B—Physics, v. 146, New York, Plenum Press, p. 9–164.

Weeks, W. F., and Assur, A., 1963, Structural control of the vertical variation of the strength of sea and salt ice, *in* Kingery, W. D., ed., Ice and snow; Properties, processes, and applications: Cambridge, Massachusetts Institute of Technology Press, p. 258–276.

—— , 1968, The mechanical properties of sea ice; Proceedings, Conference on Ice Pressures Against Structures, Laval University, 1966: National Research Council of Canada, Associate Committee on Geotechnical Research Technical Memorandum 92, p. 25–78.

—— , 1972, Fracture of lake and sea ice, *in* Liebowitz, H., ed., Fracture; An advanced treatise; Volume 7, Fracture of nonmetals and composite: New York, Academic Press, p. 879–978.

Weeks, W. F., and Cox, G.F.N., 1984, The mechanical properties of sea ice; A status report: Ocean Science and Engineering, v. 9, no. 2, p. 135–198.

Weeks, W. F., and Gow, A. J., 1978, Preferred crystal orientations in the fast ice along the margins of the Arctic Ocean: Journal of Geophysical Research, v. 83, no. C10, p. 5105–5121.

Weeks, W. F., and Mellor, M., 1983, Mechanical properties of ice in the arctic seas, *in* Dyer, I., and Chryssostomidis, C., eds., Arctic technology and policy: New York, Hemisphere Publishing Company, p. 235–259.

Weeks, W. F., and Weller, G., 1984, Offshore oil in the Alaskan Arctic: Science, v. 225, no. 4660, p. 371–378.

Weeks, W. F., Barnes, P. W., Rearic, D. M., and Reimnitz, E., 1984, Some probalistic aspects of ice gouging in the Alaskan shelf of the Beaufort Sea, *in* Barnes, P. W., Schell, D., and Reimnitz, E., eds., The Alaskan Beaufort Sea; Ecosystems and environment: New York, Academic Press, p. 213–236.

Weeks, W. F., Tucker, W. B., III, and Niedoroda, A. W., 1985, The numerical simulation of ice gouge formation and infilling on the shelf of the Beaufort Sea: Proceedings, 8th International Conference on Port and Ocean Engineering Under Arctic Conditions, v. 1, p. 393–407.

Manuscript Accepted by the Society October 8, 1987

ACKNOWLEDGMENTS

The authors thank Carl Benson, Will Harrison, and Lew Shapiro for their critical reviews of this manuscript.

Geological Society of America
Centennial Special Volume 3
1991

Chapter 18

Regional/areal reconnaissance and investigation of candidate areas/sites

George A. Kiersch
Geological Consultant, Kiersch Associates, Inc., 4750 N. Camino Luz, Tucson, Arizona 85718

INTRODUCTION

The importance of regional/areal geoscience investigations for data critical to selecting candidate areas/sites for engineering works and for related supplies of natural materials is frequently overlooked. Project sponsors have commonly restricted the efforts of the exploration team to only the area near an initially selected candidate site, and sometimes this attitude persists during the planning and design phases (discussed in Chapters 19 and 20, this volume). This limited approach disregards the potential importance of much relevant geological information, because sites rarely conform to a preconceived interpretation. In many cases, an understanding of neotectonic conditions, geomorphic history, and active regional geologic processes can prove more advantageous to an engineering objective than restricted, detailed knowledge of a single site (Fig. 1). A very early example of this occurred during a reconnaissance for the proposed Roundout Valley tunnel alignment for New York Board of Water Supply in 1906. C. P. Berkey recalled (Sanborn, 1950, p. 46), "consultant W. O. Crosby left the party and drove far beyond the line of tunnel; on his return Crosby understood the geology of the site better than anyone could from a mere inspection of the alignment, and it was a good practical lesson in reconnaissance technique."

Several examples of how an inadequate understanding of areal geology subsequently affected project design are given in case histories cited herein and in Chapters 19, 20, and 23, this volume. Moreover, regional/areal geological investigations have been utilized with increasing effectiveness as one basis for the most economic and effective approach to selecting suitable locations for engineering works (Fig. 1). Three specific geologic problems that can affect engineering works and that can be strongly influenced by the regional/areal history and conditions are seismotectonics, subsidence, and foundation defects; these phenomena and others are discussed in Chapter 4 of Jansen (1988).

Specialist-team approach

A specialist-team approach (Arbeitsgemeinschaft) to site evaluation and the environs has been successful in practice (Kiersch, 1964). The team reconstructs the geologic history of an area, determines the physical or chemical causes for particular geologic features and conditions, and predicts the effect of a proposed project on the candidate area/sites. This approach is particularly effective when the individual team members (engineers, geoscientists, economists, planners) develop mutual respect for each other's specialities (Kiersch, 1964).

Although early planning and investigation for any proposed engineering work should begin with an assessment of the regional/areal geologic environs (Fig. 1), the scale of any regional/areal investigation is proportional to the complexity of geology and purpose of the project. The preliminary geological data are synthesized with findings of the other specialist-team members for an interim summation and preliminary decisions (Stage 3, Fig. 1). This first appraisal will invariably disclose some inherent problems that are fundamentally geologic in origin and related to such things as structural features, erosional features, the weathering processes, or surficial deposits (Kiersch, 1962; Fisher and Banks, 1978).

When the team approach is not utilized the project may be burdened by serious errors of judgment, or controversies relative to siting, design, or construction of a project, as demonstrated by cases cited in Chapters 1, 18, 22, 23, and 24, this volume.

Changes with time

The performance histories of engineering works demonstrate that no structure is entirely free from risks of progressive deterioration or destruction by natural forces. Integrity of the engineered works can be influenced by critical geologic elements that may contribute effects ranging from insignificant to destructive during the life of a project. For example, the improved service record for dam projects can be attributed in part to the practice of applying weighting factors for incorporating such geologic elements into decisions on the candidate-preferred site selections and project design (Fig. 1).

Engineering works present three kinds of risks: unknown,

Kiersch, G. A., 1991, Regional/areal reconnaissance and investigation of candidate areas/sites, *in* Kiersch, G. A., ed., The heritage of engineering geology; The first hundred years: Boulder, Colorado, Geological Society of America, Centennial Special Volume 3.

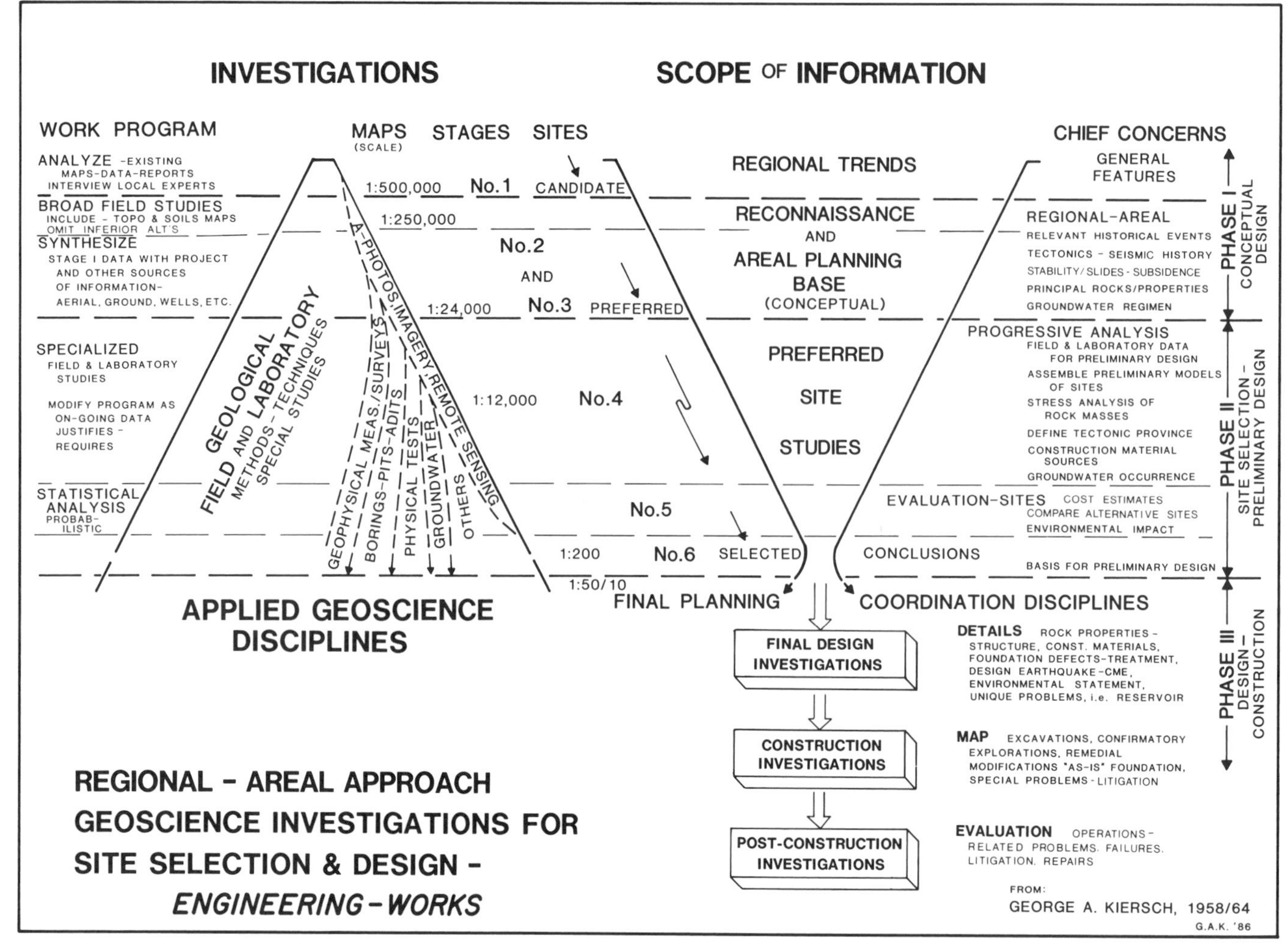

Figure 1. Flow chart for the regional/areal approach to geoscience investigations for site selection and design of engineering works. (From Kiersch, 1958, 1964.)

calculated, and human, as described in Chapter 22, this volume. Because risk involves both uncertainty and potential damage, a systematic evaluation of recognized hazards and their risks over time can provide a measure of the acceptable level or intensity of damage from the risk (see Chapter 23, Fig. 31, this volume). During the past two decades, much progress has been made in recognition of the critical factors that are involved in evaluating potential geologic influences and plotting new paths for a better understanding of what can occur. This trend by the engineering and geological professions has focused on the necessity for an improved knowledge of natural materials and their behavior. Fulfilling this need has required an increased acceptance of applied geosciences as part of the investigative activities for the siting and design of large engineered works and projects. Furthermore, geological specialists increasingly are members of risk analysis teams for large projects (Cooper and Chapman, 1987).

Few natural materials are truly homogeneous in character or uniform in their behavior. The rates at which soils and rocks move toward physical and chemical balances depend on several geologic factors, particularly geologic histories, and a combination of many inherent chracteristics that are affected by the new conditions imposed by the engineered works. Rates can change slightly to dramatically in response to shifts in the geologic environment. Consequently the prediction of changes in conditions or materials ultimately must be based on a detailed understanding of often highly complex geologic systems that are part of the regional setting, such as the cases cited under typical surveys later in this chapter.

The geologist is ideally qualified to correlate the natural processes in the historic and geologic past with the future action of such processes in response to the uses proposed. Every effort should be made to evaluate processes or geologic reactions in terms of their rate—specifically, to distinguish between geologic time and the time span in an engineering or human perspective. Quantitative estimates are needed concerning such features as natural compaction, bearing characteristics, transmissibility rate of ground-water flow, gross stability of earth and rock materials, and the expected yearly rates and magnitudes of erosional processes, weathering, and ground movements. In essence, how will the site react to the proposed engineered works relevant to both natural and man-induced processes? Answering this question commonly requires devising techniques for age-dating the changes or events that are observed and/or predicted (Kiersch, 1964, 1975).

HISTORICAL DEVELOPMENT OF REGIONAL SURVEYS

Pioneer surveys

Railroads across western territory, 1850s to 1870s. The discovery of gold at Sutter's Mill in California in 1848 sparked a frenzied migration across the nearly trackless western territories that was without precedent in this country's early history. So rapid was the settlement of the West Coast, with development of a hub city on San Francisco Bay, that railways were proposed that would cross the entire continent. They would not only serve the population on the coast, but also would further settlement of the broad regions between the Mississippi River and the Pacific Ocean. The following brief summary is largely after Radbruch-Hall (1987).

In 1853 to 1854, U.S. War Department engineers made extensive exploratory surveys to determine the best routes for the railroads. These survey parties, directed by U.S. Army officers, were accompanied by scientists, including geologists. They made detailed studies relevant to their subjects along the various survey routes, and the many volumes of reports are classics on the early exploration of the western United States. Their geological reports dealt mainly with classical aspects and natural resources; little was reported on the geological problems that might be encountered in the construction of the railways. However, the reports dealing with California were an exception, with details and discussions on building stone, railroad grades, water supply, the possible obstruction of roads or railroads by sand-waste areas, and the possibility of irrigation in the desert (U.S. War Department, vol. V, 1856).

Land Grants. The federal government began land grants in 1850 when grants were given states of Illinois, Mississippi, and Alabama to encourage construction of railroads. By 1862, such grants had gone to 12 states (DeFord, 1954, p. 4), after which the government policy changed (Fig. 2). The first direct land grant was made to the Union Pacific and Central Pacific Railroads on July 1, 1862, to build a transcontinental line from the Missouri River to the Pacific Ocean via Nebraska and Wyoming to connect with a line across California, Nevada, and Utah (Fig. 2). A

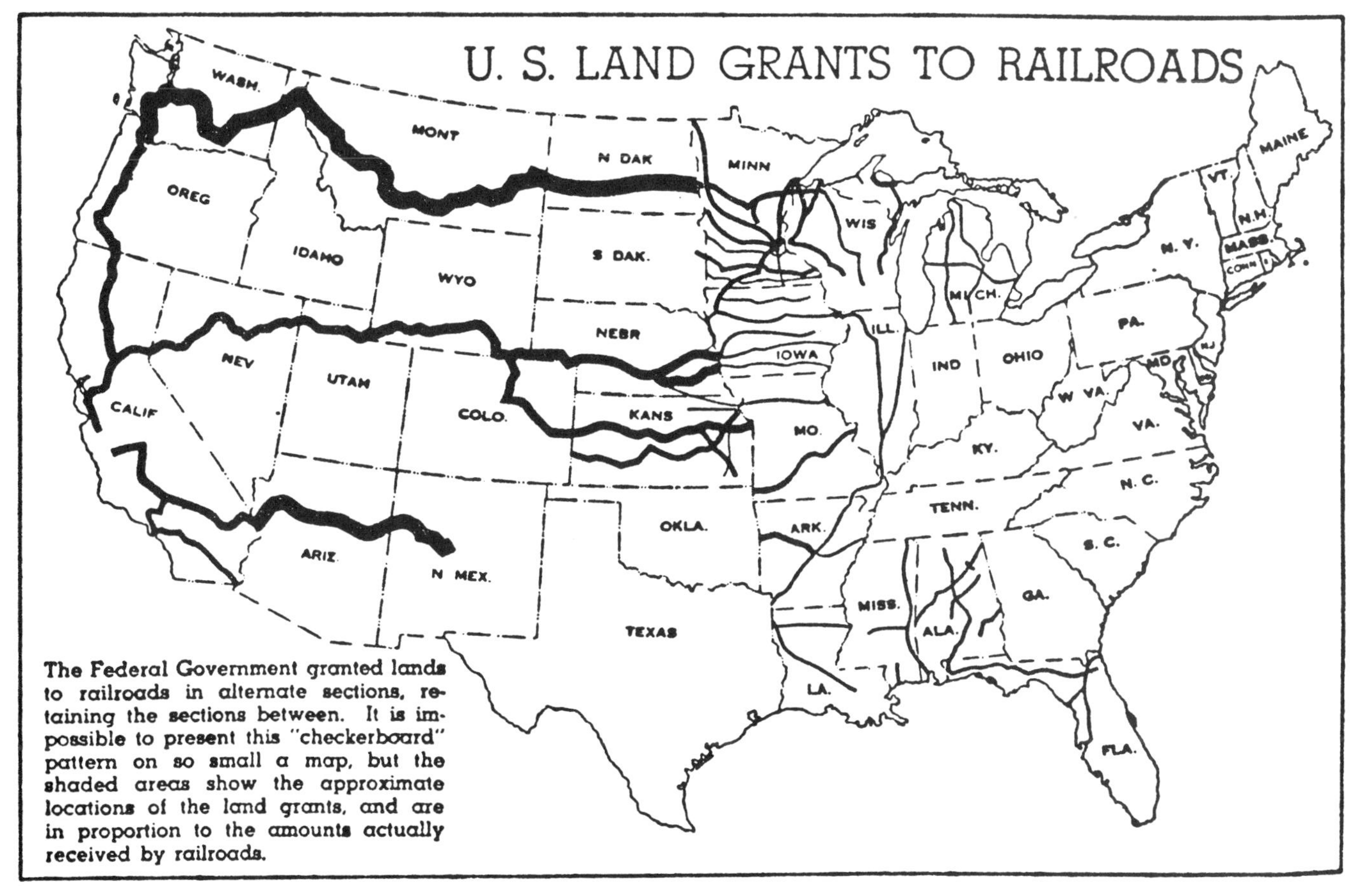

Figure 2. Lands granted by the United States government to aid construction of the railroad system throughout the United States between 1850 and 1871. (From Henry, 1945, map 2.)

second Congressional Act, July 25, 1866, granted to the Central Pacific Railroad a right-of-way and alternate odd sections of land for 20 mi on each side of the railroad from Roseville, California to the Oregon border, and a second act on July 27, 1866, granted to the Southern Pacific Railroad a similar strip from Needles, California, to San Jose, via Coalinga. Another act of 1871 granted alternate odd-section holdings to the Southern Pacific Railroad for a line from the Tehachapi Pass area of California southward to Yuma, Arizona. Land holdings from these four grants that were still retained in the 1940s were geologically mapped for land-use management in the 1950s (see Fig. 12 and discussion later in this chapter). Similar grants were made to other western railroads in 1864, 1866, and 1871, and only lands with coal and iron were excepted and retained by the government; in all, 20 million acres were granted to the Union Pacific Railroad, 17 million acres to the Santa Fe Railroad, 44 million acres to the Northern Pacific Railroad, and 24 million acres to the Central/Southern Pacific Railroads.

Although the land grants served as a mortgage for the pioneer railroad, completion of the lines across the barren "wastelands" of that day immediately enhanced the value of all lands, federal or otherwise, within travelling distance of the railroad. An important aspect to the federal government, however, was the reduced tariff rate for government traffic, a benefit realized on some lines until terminated by a Congressional Act in 1946. For the sparsely settled counties of some western states, the revenue from railroad real estate taxes has been a principal source of income.

Route reconnaissance. The main transcontinental rail line route ultimately selected from Omaha to the Pacific Ocean did not initially include descriptions of the geology. However, a geological reconnaissance along the eastern part of this line was made by a geologist of the Union Pacific Company, David Van Lennep (1867, 1868), but even his reports were concerned mostly with the resource aspects of the geology. The western part of the route, extending from San Francisco over the Sierra Nevada, and eastward to join the Union Pacific north of the Great Salt Lake in Utah was built by the Central Pacific Railroad (see Chapter 23, Fig. 33). The federal government's payment schedule for building the railroad was based on the topography of terrain. For example, payment per mile in mountainous terrain was many times more than flatland or gently rolling hills. Profiles of the early rail routes indicated "rugged high terrain" would be crossed in the gentle areas of "western Nebraska-Wyoming." The most difficult sector of the transcontinental route was the high mountains and rugged terrain across the Sierra Nevada. Although no geologist or engineer made studies specifically related to construction problems that might be encountered over the Sierras, some geologic information was made available and is summarized in Table 1 (from Montague, 1865).

Four expeditionary surveys, 1860s to 1870s. Recognition of the relevance of geology to planning for a continent-wide transportation system developed slowly during the late nineteenth century. Following the railroad surveys, both government and industry became aware of the vast mineral potential of the western United States. This awareness led to the sponsorship of four great geological exploratory surveys of the western United States that preceded the establishment of the U.S. Geological Survey as a government agency in 1879 (Rabbitt, 1979a). The work of these four surveys, as well as other early geological work, are described by Rabbitt (1979b). A brief summary of the four exploratory surveys follows.

By 1867, developing industries in America were making radical demands on the nation's natural resources. Congress reacted and for the first time funded western explorations in which geology would be the principal objective. The Fortieth Parallel survey, the first of the four great surveys, was authorized on March 2, 1867, to explore for the transcontinental railroad under direction of Clarence King and the U.S. Army Engineers. King (1880, p. 4) later remarked that "1867 was the turning point in national geological work when the science took a commanding position in the professional work of the country." After three successful years and further exploration in Colorado and Wyoming, the King survey was placed under the Secretary of the Interior in 1870.

A second survey, concerned with the natural resources of the new state of Nebraska was also authorized on March 2, 1967, under the General Land Office, with Dr. F. V. Hayden in charge. This survey soon increased in scope and became the Geological and Geographic Survey of the Territories under the Department of Interior in 1870.

An expedition in 1869, led by Lieutenant George Wheeler, U.S. Army Engineers, resumed the major exploratory activities of the Fortieth Parallel study. In 1871, this survey explored and mapped the area south of the Central Pacific Railroad in eastern Nevada and Arizona. On May 24, 1869, John Wesley Powell and party began a daring, privately sponsored exploration of the unknown Colorado River Canyonlands in three small boats. In 1870, Powell made a second trip down the Colorado, which was later funded (1872) by Congress and the Department of the Interior; he reported his results to the Smithsonian Institution.

Subsequently (June 10, 1872), Congress authorized geographical surveys west of the 100th Meridian under Lt. George Wheeler and provided further funds for the Powell and Hayden surveys (Rabbitt, 1979a). By 1874, all of these surveys were under the direct control of Congress, which led to the establishment of the U.S. Geological Survey in 1879.

Early geological surveys: 1880s to 1900s

Initial U.S. Geological Survey efforts. The newly organized U.S. Geological Survey in 1879 continued the work of the four expeditionary surveys. Although emphasis was initially on mining aspects of geology and the "classification of public lands," the scope of the agency's geological studies gradually broadened. For example, the drought of 1886 led Congress to authorize a survey to investigate the irrigation of arid regions, and to select sites for reservoirs and hydraulic works and prevention of floods

TABLE 1. SUMMARY OF GEOLOGIC CONDITIONS ANTICIPATED IN TUNNELS OF THE SIERRA NEVADA ROUTE—CENTRAL PACIFIC RAILROAD, 1865

Locality	Length (feet)	Material	Probable time required for construction	Remarks
Boulder Hill	200	Cement and conglomerate	60 days	
Prospect Hill	400	Clay, slate, and cement	100 days	Will require lining
Fort Point	300	Clay, slate, and cement	75 days	Will require lining
Grizzly Hill	600	Conglomerate	150 days	A portion will require lining
Lost Camp Spur	300	Slate and sandstone	100 days	
Red Hill, above Crystal Lake	300	Schist	130 days	
Opposite Jones's Station	200	Granite	100 days	
Summit of Sierras	1,700	Granite	19 months	Will be constructed
Cement Hill	400	Cement	120 days	For double track
Mouth Strong's Canon	180	Granite	90 days	For double track
Coldstream	900	Cement	1 year	For double track
Devil's Grip	175	Cement	80 days	
Total feet	5,655			

Being less than one-third aggregate length of tunneling contemplated by the original surveys.
The above tunnels can all be worked from both ends, and with the exception of the Summit Tunnel, will require no shafting.
After D. Radbruch-Hall, 1987.

(Rabbitt, 1979a, p. 4). In 1890 the Survey published an account of the disastrous earthquake at Charleston, South Carolina (Radbruch-Hall, 1987). The primary investigation was made by Capt. Clarence E. Dutton, geologist (1889), who was then with the U.S. Ordnance Corps, but his report was published by the Geological Survey. The inter-relation of geologic theory and practice, long widely separated, was recognized by Dutton. Although there was no means to determine precisely the magnitude of an earthquake, when investigating the Charleston earthquake, Dutton recognized that surface materials had an influence on the amount of damage sustained, and this observation guided his attempt to determine the location of the epicenter. The large number of sand boils, where quicksand occurred near the surface, was particularly noted: "These craterlets reached greatest development, both in size and number, near Ten-mile Hill, and depended upon the nature of the strata beneath and the depth of water table" (Dutton, 1889). Even though the report emphasized new concepts, no discussion correlated the amount of damage with the type of earth material underlying damaged structures, nor did any conclusion state that areas such as the quicksand tract be avoided in future construction.

In 1888 the application of ground-water geology to irrigation in the arid west was systematically begun with the establishment of the United States Irrigation Survey within the U.S. Geological Survey (Powell, 1890); this group became the separate Reclamation Service in 1907, and the Bureau of Reclamation in 1923 (Chapter 1, this volume).

In the 1890s, geological work was extended into new fields and beyond the national domain. A survey geologist and a hydrographer were detailed to the Nicaraguan Canal Commission in 1897 to study the proposed canal route between the Atlantic and Pacific Oceans. This was one of the first times geological guidance and input was provided for a great regional engineering project in the Americas (Rabbitt, 1979a, p. 19).

In 1894, the application of geologic information to construction problems was discussed in a Geological Survey publication entitled "Common roads of the United States," by N. S. Shaler (1895a). The report is notable for its historical summary, its clear description of the state of road-building at the time, and its recommendations that geologists provide the highway engineers with information about the locations of suitable construction materials. Shaler's report includes sections on block pavements, the effect of geologic structure on grades of roads, and sources of road stone within districts of the country.

In 1895, another U.S. Geological Survey publication described the geology of road materials in Massachusetts (Shaler, 1895b). A report by Merrill (1897) on "Road Materials and Road Building in New York" contained a map showing the distribution of the rocks most useful for road material. This was one of the earliest statewide special-purpose geologic maps for engineering in the United States. Thus, by the end of the nineteenth century, the location of roads and engineering works was being affected by studies of the geologic conditions (see discussions of the Boston Harbor/Charles River Dam, Catskill aqueduct, and associated engineering works in Chapter 1, this volume).

By the early twentieth century, most new settlement throughout the United States was drawing to a close. With urbanization and new construction, the role of applied geology increased proportionally; historical land-use problems in the United States are summarized by Nichols (1982). A short history of the Engineering Geology Branch of the U.S. Geological Survey, which developed in response to this increased role of applied geology, is given by Lee in Chapter 3, this volume.

Southern Pacific lands: 1909 to 1920s. The original grants of 1862 to 1871 to the railroads did not mention mineral rights, and little controversy existed over mineral rights and land values until near the turn of the century when the need for industrial development brought railroad land holdings into the spotlight. As an outgrowth, a far-sighted geological survey and mineral evaluation of their grant lands was undertaken by the Southern Pacific Railroad in 1909 under D. T. Dumble, formerly a professor of geology at the Univeristy of Texas. Efforts were limited to the tools and techniques of the day.

A principal objective of the initial survey was to select all lands for patent that were considered nonmineral, and negotiate the release of the known mineral-bearing lands. This survey was mainly active between 1909 and 1914, although geological work continued sporadically, at reduced intensity, into the 1930s. The survey was responsible for many pioneering firsts in applied geology, among them the introduction of techniques for detailed geological logging of deep borings and oil wells for use by the petroleum industry in 1910s. Many well-known California geologists served on that early survey, including J. T. Taff, S. H. Jester, F. S. Hudson, D. Clark, L. Melhase, W. L. Moody, and C. L. Cunningham.

By 1930, the bulk of the Southern Pacific's remaining lands were patented. Among other things, the early survey delineated the Coalinga region of California as one with excellent potential for petroleum. These lands were subsequently acquired by a rising new company, Standard Oil of California, and the Coalinga area became its principal producing field for about two decades. Several geologists associated with the Southern Pacific survey became the nucleus of Standard's exploration staff; S. H. Jester became chief geologist, serving into the 1940s.

Survey trends

Historically, probably the earliest program for engineering/economic development in the United States, based on the interrelated natural resources, was the Tennessee Valley Authority (TVA), launched in 1933. Although world famous for development of hydroelectric potential, the TVA's activities also included minerals, soil-types, agriculture, timber and vegetation, and industrial potential, as well as wildlife and recreation. Soil, vegetation, and grazing conditions were delimited on aerial photographs; qualities of the three related land classes delineated for agricultural uses were dependent on the geological substratum: (1) suitable without treatment; (2) suitable after cultivation, clearing, and/or soil measures; and (3) sometimes suitable after special treatment according to geology and soil.

Perhaps the earliest large-scale survey in the western United States that focused on a specific region was the compilation of resources in the vicinity of Boulder (Hoover) Dam in 1934 by the U.S. Geological Survey (Hewett and others, 1936). This survey was designed to attract new industry to the low-cost hydropower region around the dam. Soon thereafter, a similar survey program was sponsored by the U.S. Corps of Engineers for the region around Bonneville Dam in the northwest (Hodge, 1935). Since World War II there has been an expansion of the regional survey philosophy, such as the Point-Four mapping programs of the U.S. Geological Survey, to serve undeveloped or growth areas in all parts of the world. In North America during the 1950s, programs with somewhat similar objectives were sponsored by government agencies, industry, and even private organizations of landholders such as the New Mexico–Arizona Cattle Company, Southern Pacific Land Company, Santa Fe Railroad, Kern County Land Company, and several lumber companies.

Another trend in the 1960s is exemplified by the regional survey completed in Manitoba, Canada, and termed "Operation Overthrust" (Huntington Aerosurveys, 1961). This combined effort of several leading mining companies explored large land tracts and located candidate areas for investigations of the mineral resources. The survey of central Alaska in the early 1960s by a team from the Battelle Institute was the first step toward planning a comprehensive regional survey of southeastern Alaska by the Development and Resources Corporation of Alaska. Mineral resources and potential sites for dams and hydroelectric projects that might be sponsored or developed by government agencies or private firms were investigated.

Scope of investigations. Regardless of sponsor or immediate objectives, regional geological surveys are likely to be limited by a number of factors, such as: (1) whether an inventory is comprehensive or restricted to a specific commodity or site consideration; (2) whether the relation between quality of a site/area and its location is dependent on such factors as marketability and transportation; (3) total funds available and availability of a survey team; (4) type of terrain and accessibility for ground-truth studies; (5) adequacy of existing ground control, maps, and remotely sensed coverage; (6) extent of available geologic data, particularly reliable geologic maps; (7) time schedule for performance, survey and report; (8) geologic complexity of the terrane, which affects funding, completion time, and thoroughness and reliability of the investigation; and (9) organization of survey, whether under direct supervision of its sponsor, detached except for adminstration, or a wholly detached contractor.

Survey Personnel. Staffing a large-scale survey calls for judicious thought and planning by its sponsors. High priority must be given to human engineering, a major factor in attaining a smooth and effectively functioning team of specialists. The geoscientists, in addition to being competent and well rounded, must possess a wide interest in the applied aspects of geology. Within the budget constraints, they need to collect all possible data regarding the physical properties and features of the rocks and rock masses that relate to the proposed engineering works or uses.

The experienced geologist of a regional survey might be likened to the industrial inventor: both are scientifically trained and oriented, but each deals with specific problems and applications in a creative manner. Because a survey's timetable does not allow for frequent repeat studies and detailed local investigations, it is imperative that the geoscientist possess a practical bent, a feeling for the quantitative, and the ability to separate geologic facts from inferences and hypotheses (Kiersch, 1958).

Approach philosophy. The most successful method of undertaking a regional survey is the "convergence-of-evidence" plan (Kiersch, 1958). This is illustrated on the right-hand part of Figure 1, although less than optimum conditions or sponsor limitations may result in some phases or steps being omitted.

The approach outlined here is ideally suited for investigating the ground-water regimen, the engineering properties of rock masses or soil deposits, or the principal features or conditions that affect construction sites. Restricted-site or construction-material investigations can also use this approach with only minor variations, which depend on the specific characteristics of the materials or site being sought. Paramount is the ability of the applied geologist to analyze evidence from more than one point of view. Many geologic situations initially require the creative consideration of fragmentary evidence; only later, after exposure by subsurface excavation or after further laboratory testing, might the corroborating evidence become available (Stage 4, Fig. 1; Kiersch, 1964).

SELECTION OF CANDIDATE AREAS/SITES

Selecting a candidate area for a major engineering work, such as a dam, power plant, tunnel, or hazardous waste disposal site, requires balancing many conflicting values and elements that may not be measureable on a common scale, except in the broadest sense—the public interest. Potential sites must be assessed in two distinct phases, assuming the critical engineering factors are satisfied at all candidate sites. Sectors not suitable for the works proposed, and areas of conflicting land use, are excluded, and the remaining potential site areas must be subjected to increasingly more detailed evaluation, as shown on Figure 1.

The investigations are a continually evolving effort, and require parallel studies to evaluate competing sites. A typical example is the site-selection study to identify suitable new power-generation sites in New York State (Kiersch, 1975), discussed later in this chapter. Widespread concern about the safety of nuclear power plants located in seismically active areas led to the acceptance of screening criteria for sites in southern California (Elliott, 1978), including a requirement that all candidate areas/sites have unequivocal age control on the youngest faulted and the oldest unfaulted geologic units in the area.

The initial screening of candidate areas (Stage 1, Fig. 1) consists of excluding areas under consideration for other projects, parks, or population sectors, and known areas of geologic risk, such as areas of potentially high seismic and volcanic activity, or flood- or landslide-prone terrains. This initial consideration is followed by an analysis (Stage 2, Fig. 1) to meet a set of broad feasibility criteria of size, land use, and topography, and to provide a comparison of 27 different attributes or factors relevant to sites, from both environmental/technical and economic aspects (Kiersch, 1975, p. 3–5; Tilford, 1982).

Ultimately, the identification of preferred sites requires an estimation of costs associated with each candidate site, including an evaluation of the trade-offs between environmental and economic considerations.

Evaluation of environmental costs should include water quality, noise, land use, aesthetics, socioeconomics, and the geologic consequences of operation of the planned works (i.e., impact of the proposed project on the geologic environs).

Evaluation of the economic costs should include geological/seismological elements, and the cost of access, transportation of supplies, operation, and transfer of any product to consumption centers (i.e., impact of the geologic environs on the proposed project).

Costs of environmental factors are rated on scales of merit and weighted to reflect relative importance. For example, geology/seismology might have a relative weight of 50 to 60 percent; hydrology 10 percent; land use 5 percent; and population 10 percent. Risk analysis provides a systematic way to attach probability distributions to geologic factors (Cooper and Chapman, 1987). Normally the relative weight of important factors will vary according to the type and design needs of engineering works. Subsequently, individual site data are plotted with the comparative dollar values of environmental and economic elements as axes. Such plots show groupings of preferred sites into those of lowest total cost, least environmental impact, and best combinations of cost and environmental factors (i.e., lowest cost/benefit ratio).

The recommended or preferred site(s) of Stage 6 on Figure 1 (see also Chapter 19, this volume) is based on the extensive on-site and laboratory explorations of Stages 4 and 5. The systematic evaluation may use borings, adits, and geophysical measurements, as well as statistical analysis and a probabilistic assessment. These basic findings are combined with input from other specialists of the exploration team and project staff. Consequently, the site-selection process utilizes criteria and weighting factors that integrate the environmental impacts and ratings of the entire physical complex, including the specialized geoscience factors, the relative costs, and the economic/environmental cost balance.

A comprehensive site-selection study is a methodical process. The preferred site(s) represents the best possible correlation of scientific and engineering knowledge with relevant regional economic and environmental factors. The most successful method of investigation, whether for an unmapped sector or for a comprehensive analysis of several candidate areas to recommend a preferred site, integrates Stages 1 to 3 for the conceptual design of Phase I; and Stages 4 to 6 for the site selection and the preliminary design of Phase II (Fig. 1). Examples of this process, citing five regional/areal projects, completed since 1950, are briefly reviewed in the next section of this chapter. The synthesis and strong geoscience databases compiled by such surveys are best utilized by a team that includes geoscientists, engineers, planners, and economists, as demonstrated by the uniquely integrated San Francisco Bay Area project (Brown and Kockelman, 1983).

Conventional map scales for the geoscience investigations during Stages 1 through 6 are indicated on Figure 1. The designation "regional" normally implies an area of ten to hundreds of square miles around an eventual site; some specialized projects evaluate the geologic conditions within a 200-mi radius of a site. The designation "areal" means at least a 4- to 5-mi radius surrounding a site, while "site-specific" refers to the half-mile radius surrounding a site.

INVESTIGATION OF CANDIDATE AREAS/SITES: PHASE I—CONCEPTUAL DESIGN

Initial stage 1: Indirect methods

The investigation of a candidate area/site begins with a synthesis of existing information to provide a basis for initial comparison. The reconnaissance investigation consists of three separate steps, which compile the framework for base maps, interpretations, and predictions required during Stages 2 thru 6.

Step 1. The initial effort is to compile and assimilate all existing geological, hydrological, and seismological data and related materials. Such data include published and unpublished reports and theses, surveys, and interviews with local geoscience experts; topographic, soil, and geologic maps; aerial-photographs, selected imagery, or remote sensing; and some geophysical and ground-water base data. This compilation provides a reconnaissance map with the principal geologic features and rock units to orient and guide the field staff for Steps 2 and 3. If the region is uncharted or only sketchy data exists, a remote sensing/aerial-geologic data acquisition mission is desirable as part of the Step 1 effort.

Step 2. Additional pre-field investigations include preparation of composite maps that utilize the geology of an individual state, province, or republic at 1:500,000, or ideally, the popular $1^\circ \times 2^\circ$ quadrangle maps or Defense Mapping Agendy maps at 1:250,000; satellite imagery is commonly available at the same scale. The assimilated elements can guide conceptual planning by use of transparent overlay maps for the general geologic features, tectonics, seismology, ground water, and other appropriate elements as warranted by the geologic setting and the needs of the proposed engineering works. Frequently the most critical factor concerns the use of preexisting data; it should not be accepted without an adequate field evaluation for accuracy and for any errors of judgment by the earlier investigators.

Step 3. The findings and predictions relative to both concealed subsurface and historic geologic events are interpreted and synthesized. The preliminary conclusions are correlated with such interpretative databases as paleotectonic maps, environmental or hazard-risk maps, neotectonic maps, and regional stress-pattern plots for relevance to the proposed engineering works.

The base map and data assembled for Stage 2, the ground reconnaissance, may indicate that certain geologic features are important to the conceptual plan, such as buried stream valleys, distinct structural trends, or features or lineaments of the basement rocks (Fig. 1). This second physical screening of the region, based on the broad geologic elements, will likely result in the exclusion of some sites or areas, reducing the number to a realistic group for the proposed engineering works.

Stage 2: Direct methods

The principal Stage 2 effort is a geological field reconnaissance and ground-truth check of the broad regional geologic trends identified by Stage 1 (Fig. 1). This evaluation should emphasize the physical and engineering properties of rock units and soil or surficial deposits, along with any tectonic or structural patterns, and the selected ground-water regimen, noting any significant features or special conditions of importance. The data should be plotted on moderate-scale maps. Routinely, the initial compilation maps of Stage 1 are modified by the supplemental observations and data collected, and the maps are thus upgraded to conform with project standards. Areas with limited or no preexisting geologic data are mapped to provide areal continuity and uniform reliability. Concurrent aerial photographic interpretation and/or satellite imagery studies may be advantageous to evaluate complicated structures or reveal otherwise undetected features and relations. Digital elevation data have been used as an aid for tectonic evaluations (Schowengerdt and Glass, 1983) and other spatial assessments. Similarly, airborne geophysical surveys may provide critical data for evaluating an important but otherwise undetected feature during the conceptual planning of Phase I.

Stage 3: Direct and indirect methods

The interim Stage 3 summation includes conclusions, interpretations, and forecasts or predictions based on the synthesis of Stage 1 and the field reconnaissance of Stage 2. The extrapolation, correlations, and evaluation of the Stage 1 and Stage 2 data are the basis of a Phase I conceptual design report and the selection of several preferred sites (Fig. 1). The potential problems inherent to the studied areas are described and delimited for attention by the specialized investigations of Stage 4.

SELECTION OF PREFERRED SITES: PHASE II—PRELIMINARY DESIGN

Stage 4: Direct methods

Progressive analysis, evaluation, and broad criteria for selecting a preferred site are reviewed in Chapter 19, this volume. Stage 4 continues the field exploration effort by utilizing such detailed mapping techniques and supplemental studies by a team of geological specialists as the areal site conditions may warrant. The findings of the specialists are extrapolated and interpreted or projected for each geological discipline, and the data are correlated using overlay maps for the separate geological elements (Fig. 1). This compilation should focus on the design criteria relative to earthquakes and ground water or other specific hazards, such as rock stresses or subsidence, that must be considered in the preliminary design of Phase II. A wide range of subsurface exploration techniques may be appropriate, from indirect probes, soundings, and geophysical measurements, such as cross-hole, down-hole, or acoustic logging, to direct video closed-circut borehole observations and exposure through cored borings, adits, trenches, shafts, and pits. Investigations at several of the preferred sites may proceed concurrently, with the less favorable site(s) being dropped after appropriate consideration of the data.

This selection process will likely include geologic circumstances and critical evaluations that warrant the logic-tree and/or decision-tree approaches (Figs. 3, 4), in which the relative weight of evidence prevails. The geologic uncertainties of seismic and volcanic risks are commonly evaluated by probabilistic analysis (Coppersmith and Young, 1989; Mabey and Youd, 1989). Detailed maps (1/12,000) with additional basic data are frequently advisable; they may require aerial photo analysis, and airborne or ground geophysical techniques to provide the insight for extending or extrapolating bedrock conditions beneath surficial deposits or thin cappings of thrust plates or layered rocks. Anomalies detected by aerial surveys are refined by gravity, magnetic, or resistivity ground methods. Growing airborne adaptations of radar (SLAR), thermal and reflected infrared sensoring, special-mission imagery, and multispectral scanning offer advanced techniques for gaining detailed data on the physical properties of rock masses, especially in regions of heavy forest cover.

The progressive investigations of Stage 4 should provide a basis for delimiting a seismotectonic province, locating sources for construction materials, and developing an engineering geologic model of each preferred site.

Several suitable geological mapping techniques for plotting a quantitative understanding of the rock units and features, such as Genesis-Lithology-Qualifiers (GLQ), Alpha-Numeric, and the Unified Systems are discussed in Chapter 19, this volume. Any number of special-purpose maps may be appropriate: for example, construction materials, the common natural hazards-risks, ground-water occurrence, structural features, and recency of fault movement. A specialized system of maps for interpreting the geotechnical conditions has been prepared by Hannan (1984), while mapping for dams and reservoirs has been described by A. B. Arnold (1977). The GLQ systems (Keaton, 1984) is an excellent basis for standardization of geologic map symbols or documenting geologic facts relevant to their engineering significance, and it enhances communication between geologists and engineering specialists. The logic of geologic maps with emphasis on the problems, definition, and classification of map units as intended for use in engineering works has been outlined by Varnes (1974).

Stage 5: Direct and indirect methods

The preferred site investigations are progressively refined by synthesizing the Stage 4 compilation of field and laboratory investigative data. Frequently, a further evaluation is made on site-specific characteristics, by an overview and by the application of probability analysis and statistics relative to the potential for failure of a rock mass, slope, or foundation. Other special-function exploration tools and/or specialists may be appropriate according to circumstances, such as a suspected but concealed "problem" feature. Special time-slice plots or maps that depict geologic structures or formational thicknesses at selected earlier times can bring out many geologic facts otherwise unnoticed.

The extensive physical and engineering data collected on the specific area/site by the Stage 5 investigations should be adequate for preparation of the environmental impact statement for Phase II design. The final site(s) evaluations are completed through the site selection and preliminary design of Phase II.

Stage 6: Direct and Indirect methods

The summary report and conclusions of Stage 6 are a guide in the preparation of construction design (Fig. 1). A specific preferred site is selected for Phase III design-construction and the final investigation/planning for the engineering works; the relative weights of geoscience parameters and other factors are important to the decision. The report is organized to serve both project personnel and the sponsor concurrently; the significance of both the surface and subsurface geoscience data and conditions relevant to the site selection are explained and documented.

The general criteria for selection of a designated site incorporate how the proposed works might impact on the surrounding environs. Besides the conventional techniques developed and utilized by practitioners throughout this century, computer-aided programs such as TECHBASE (MINEsoft, 1982) can provide colored geologic maps and cross-sectional plots, and specialty maps/plots of many kinds, including the ground-water regimen, seismic and/or liquefaction potential, structural features, rock facies and inherent engineering properties, and a contour map of weathered/fresh bedrock. Many commercially available programs, such as Auto CAD, Golder Software SURF, and DISSPIA, provide the capability of displaying three-dimensional perspective drawings/models of the basic or special elements needed to enhance the understanding and communication of

Sense of Slip	Maximum Displacement per Event	Computed Maximum Magnitude
Strike Slip (0.4)	2 m (0.3)	7.1 (P=0.12)
	3 m (0.5)	7.3 (P=0.20)
	4 m (0.2)	7.4 (P=0.08)
Reverse (0.6)	1 m (0.4)	7.0 (P=0.24)
	1.5 m (0.4)	7.3 (P=0.24)
	2 m (0.2)	7.5 (P=0.12)

a) Logic Tree for evaluating maximum magnitude

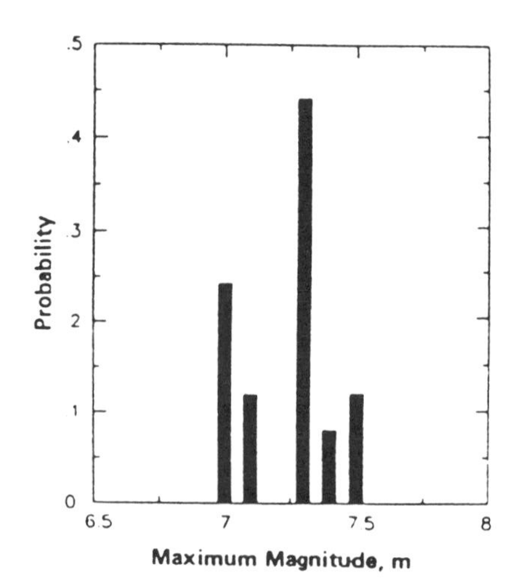

b) Discrete distribution for maximum magnitude

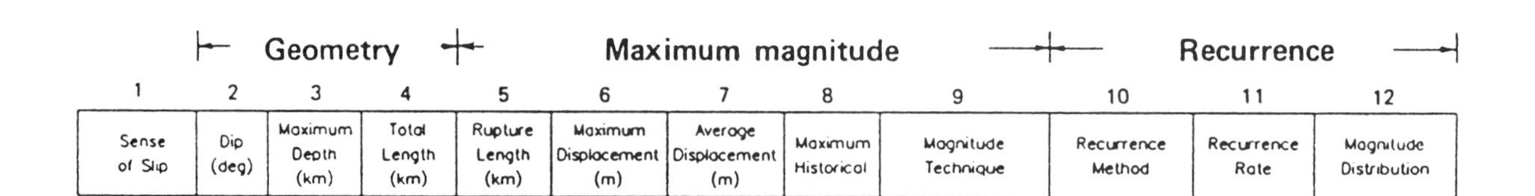

Geometry
Maximum magnitude
Recurrence
1 Sense of Slip
2 Dip (deg)
3 Maximum Depth (km)
4 Total Length (km)
5 Rupture Length (km)
6 Maximum Displacement (m)
7 Average Displacement (m)
8 Maximum Historical
9 Magnitude Technique
10 Recurrence Method
11 Recurrence Rate
12 Magnitude Distribution

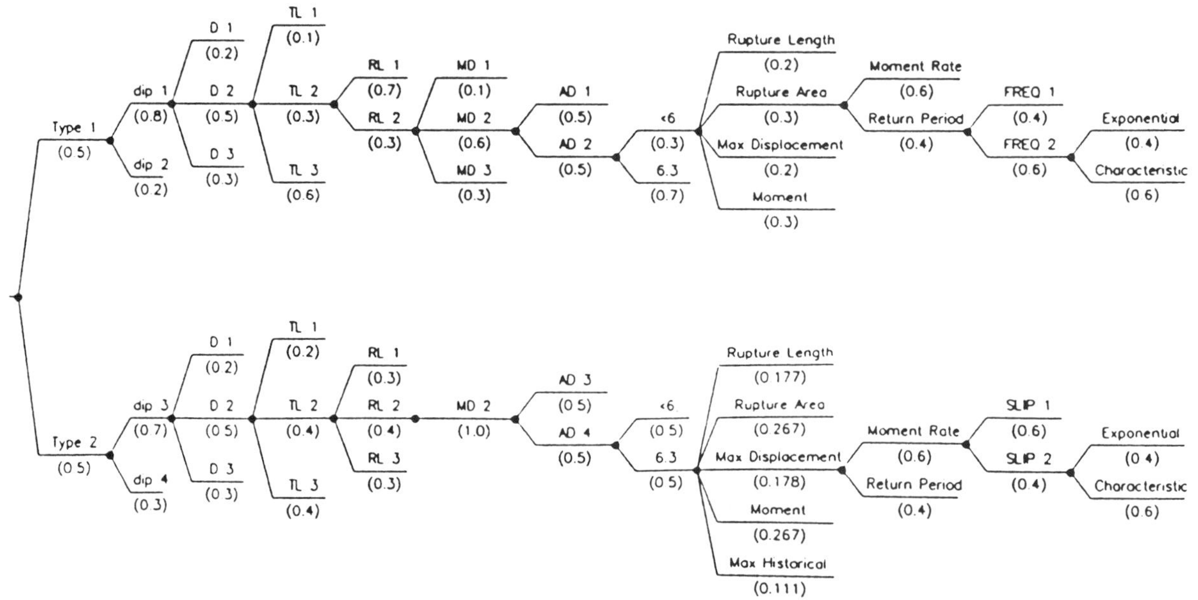

Figure 3. A seismological example of the logic and decision-tree approach for evaluating critical geologic features and/or uncertainties. (From Coppersmith and Young, 1989.)

often complicted relations. An example of specialty maps prepared by a combination of a computer-aided program and deterministic ground surveys is the natural hazards study on surface fault rupture and liquefaction potential areas of Salt Lake County, Utah, by Nelson (1989) (Fig. 5).

SOME REGIONAL SURVEYS FOR ENGINEERING WORKS, 1950s to 1980s

Dams and reservoirs along Rio Grande, United States and Mexico, 1948 to 1952

The regional occurrence of ground water in the Cretaceous limestone terrain and the southward regional flow pattern aided by structural localization was of major concern when selecting preferred sites for international dam projects throughout the Big Bend District of the Rio Grande. The 1948 to 1952 regional study for designated sites, sponsored by the International Boundary and Water Commission, United States and Mexico, is briefly summarized herein after Kiersch and Hughes (1952). The principal geological objectives were to reconnoiter a 300-mi stretch of the river (Fig. 6), map a strip 4 to 12 mi wide athwart the international border, and investigate by cored borings and other means the main geologic features relevant to the potential dam and reservoir sites therein. The regional stratigraphic investigations and geologic mapping database were correlated with ground-water data compiled from cored borings and "pumping-in" tests, water wells, and springs to supply average transmissibility rates of the limestone formations involved with candidate areas/sites.

The information collected throughout the downstream stretch from Reagan Canyon to Del Rio–Villa Acuna (Fig. 6) was critical to the selection of preferred sites. The structural and physiographic features in this sector provided a strong potential for excessive bypass leakage to the southeast, driven by the induced ground water of a reservoir, with the water loss moving into Mexico. After realizing this potential for leakage, the survey was broadened to cover a much wider area throughout this part of Mexico. Besides the significant findings relative to the regional ground-water regimen and localized flow in limestones, other critical factors related to reservoir storage capacity and to long-term reservoir operation at the candidate site/areas were studied.

Geologic setting. The Big Bend District lies within the mountainous Mexical Highlands section of the Basin and Range Province and the westernmost part of the Edwards Plateau section of the Great Plains Province, where structure, both regional and areal, strongly controls the spectacular topography. The northwest-trending, asymmetrical folds of the Mexican Highlands are combined with steep gravity faults; roughly parallel "monoclinal" folds are commonly associated with faults or step-faulted structures. In contrast, the Edwards Plateau sector is characterized by a few shear zones and low, gentle flexures (Fig. 6). The Rio Grande flows in an ovaloid syncline throughout most of this lower stretch, and gentle flexures partially bound the margins.

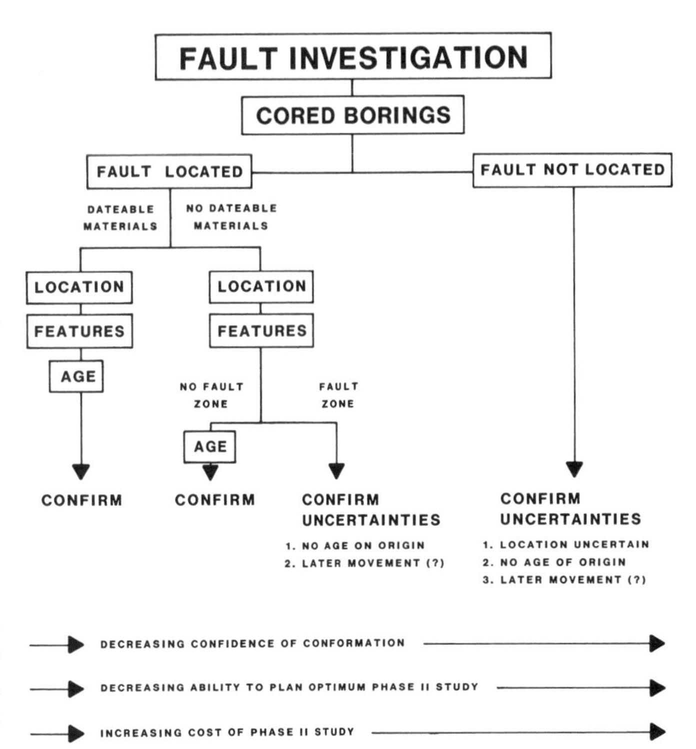

Figure 4. An evaluation of fault characteristics using the decision-tree approach. (From Kiersch, unpublished report, 1978.)

Cretaceous marine limestones are widespread, with near-surface distribution largely controlled by regional and local structures. Generally, older Cretaceous formations are most prevalent upstream. The principal structural features and progressive development of integrated solutioned openings in the limestones are related to historical events and structural development of the district. The interconnected network of openings affords avenues for leakage from potential reservoirs (Fig. 7).

During the mid- to late Tertiary, widespread structural disturbances caused recurring movement along the northwest structures of the Mexican Highlands sector and the main northeasterly trending flexures, shears, and gravity faults of the Edwards Plateau sector. Solution action was apparently accelerated at this time along channelways and throughout a network of newly formed openings. Prolonged erosion since has backfilled the intermontane, down-faulted blocks, and accelerated the solution of limestones. The ancestral Rio Grande was progressively downcutting throughout the district and formed several distinct terrace levels corresponding to solution base levels in the limestones (Fig. 7).

Ground-water regimen. Regionally, the direction of underground flow is normal to the Rio Grande with parallel and oblique flow locally. The principal aquifer is the Edwards Limestone with the younger Georgetown Limestone a local source of water (Fig. 8). Ground-water movement is controlled by both structural and stratigraphic conditions; solution is the dominant factor, with movement via a skeletonized pattern of openings.

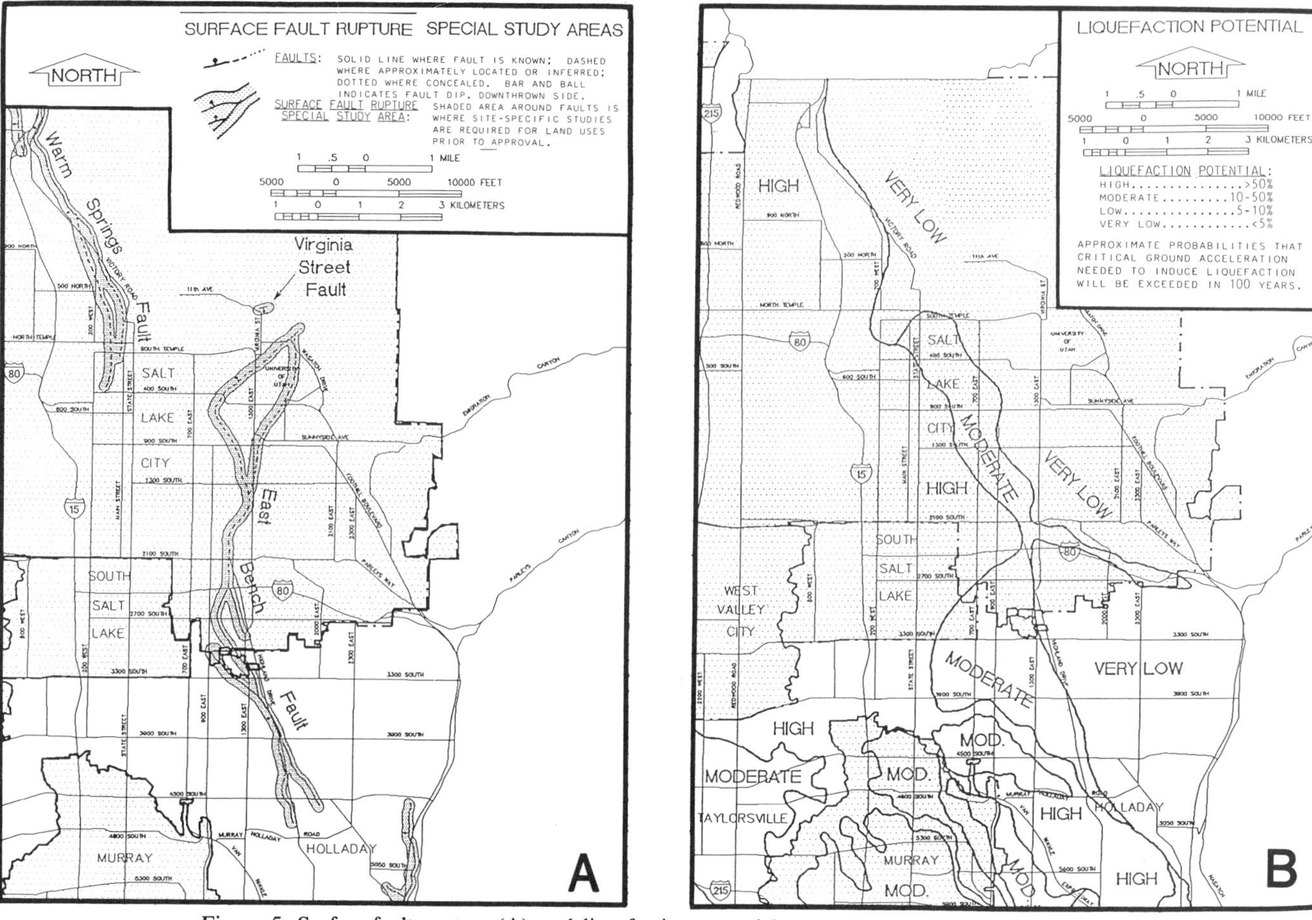

Figure 5. Surface-fault rupture (A) and liquefaction-potential areas (B), Salt Lake County, Utah. (Modified from Nelson, 1989.)

Cored borings demonstrated that solution action, active over a long period, occurs hundreds of feet below the static water level and the Rio Grande and substantiates the theory of arcuate circulation (Rhoades and Sinacori, 1941).

Engineering Works: Reservoirs. The impoundment of a reservoir normally induces a new ground-water regimen that changes the ambient areal gradient and hydrostatic head. Most of the preferred dam sites involved one or more limestone formations (Fig. 8). Districtwide the environmental conditions dictated a limited runoff for impoundment, and a high rate of evapotransportation losses. Consequently, reservoir sites needed to minimize water loss by eliminating leakage to other drainages or "short-circuiting" southeastward across Mexico and by offering the minimum water-surface exposure of deep, narrow canyons to reduce evaporation losses.

Rock units within the lower levels of canyon sections throughout the district show only weak evidence of solution, while rocks of the upper levels are modified by several base-levels of solution. Consequently, high dams would involve extensively solutioned rocks in the abutments or foundation, and the reservoir rim would probably require remedial treatment. A set of generalized reservoir conditions was formulated for the district (Fig. 7) based on permeability factors and geologic parameters of the rock units.

Industrial development: Navajo country (1952 to 1955)

Introduction. The reservations of the Navajo and Hopi Indians, known as the Navajo country, cover 25,000 mi^2 within parts of northern Arizona, southern Utah, and western New Mexico. World renowned for its scenic beauty, this colorful island of semiarid mesas is home to one of the most rapidly increasing, and no longer self-sufficient, population groups in America. At the time of the survey in the 1950s, three persons lived per square mile throughout the Navajo country, while adjacent land with a non-Indian population supported about one person per square mile.

It had been widely accepted for several decades that many of the innumerable problems throughout the Navajo country could be solved by a greater exploitation of the region's natural resources. Such a maximum development program, with its increased employment, was a major premise for undertaking a regional geological survey. Under terms of the Navajo-Hopi Rehabilitation Act of 1950 (U.S. Congress, public law no. 474) the

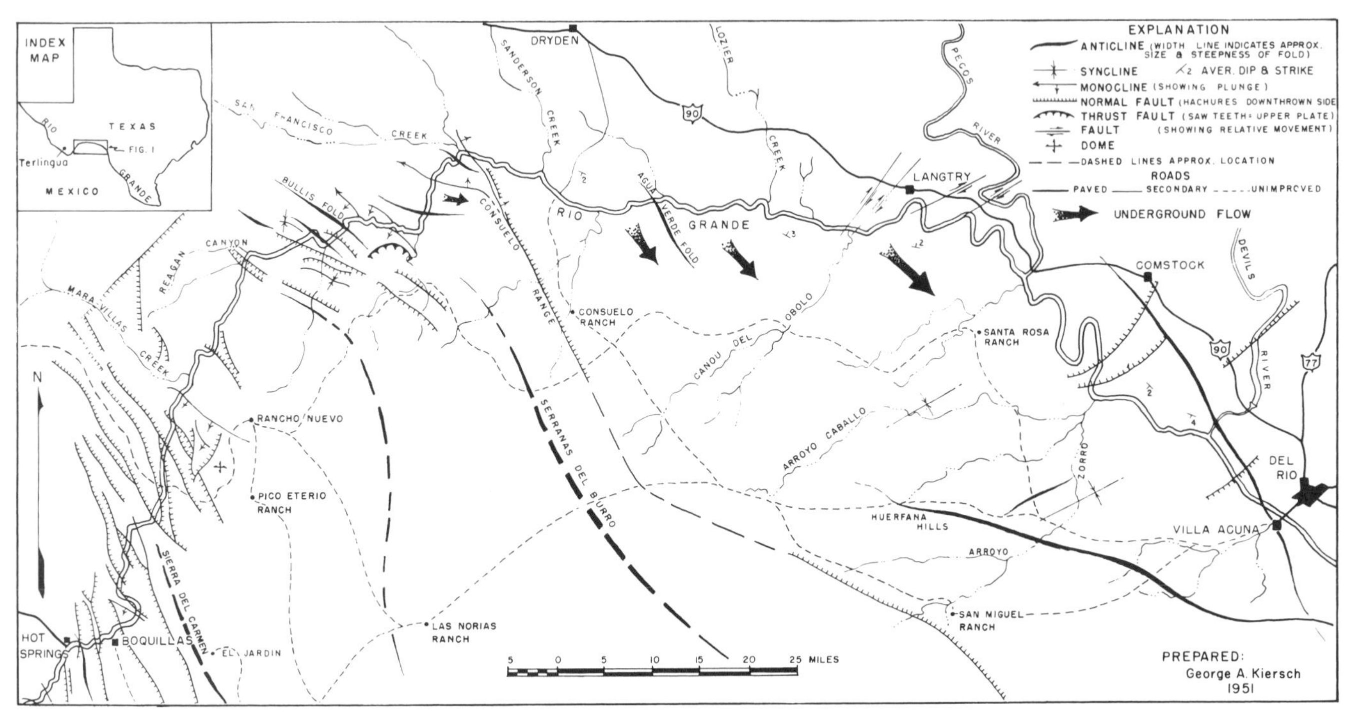

Figure 6. Structural map along the Rio Grande from Boquillas to Del Rio–Villa Acuna, Texas-Mexico; predominant direction of ground-water movement from the proposed reservoir impoundments is indicated. (From Kiersch and Hughes, 1952.)

survey was designed to "make a mineral and physical resources investigation of the Reservations available to promote a self-sufficient economy, a self-reliant community, and moreover provide a basis for diversified economic and engineering endeavors of the environs which together would attain a standard of living comparable with that of other citizens." The comprehensive investigations of the Navajo and Hopi Reservations lying within Arizona and southern Utah are described in three volumes (Kiersch and others, 1955); the survey's goals were to: (1) locate and study sources of construction materials for the accelerated building and modernization of a highway network throughout the reservations; (2) investigate, evaluate, and make recommendations for exploitation of known mineral resources and/or abandoned mines; and (3) inventory mineral resources of current and future value with site-specific investigations and recommendations regarding promising deposits.

The successful development of new industries and exploitation of resources depended on many factors, among them a reasonably inexpensive transportation system. Consequently, development of the highway system (Fig. 9) throughout the Navajo country was critical to enhanced development of the natural resources and construction of the engineering works as planned, which included additional water supplies. Those engineering works, and specific deposits with immediate commercial potential, were investigated and evaluated first, while deposits of long-range or future potential were studied in a generalized manner only for consideration as changing economic conditions might warrant.

Geologic data in construction. Topographic and geologic maps of the Navajo country contain much of the basic data needed for planning any engineering works to be compatible with the environs. The maps, supported by physical data on the representative geologic units, contain a large reservoir of facts of critical importance to the engineer at the planning-design state of a project (Fig. 1). The relation between a general-purpose geologic map and a map of the same area prepared to depict specific engineering properties of the rock units can be seen on Figure 10. The "special-purpose" geologic map delineates the types of construction materials that can be produced from the different geologic formations and their subdivisions. Rock units are subdivided on the basis of physical properties and possible uses; no attention is given to formational names or their geologic ages.

The Navajo country was covered in the 1950s by general-purpose geologic maps prepared by the U.S. Geological Survey and other agencies during development of a network of water wells (Harshbarger and others, 1957). This database was supplemented by resources maps (Kiersch and others, 1955) that designated the location of known and potential construction material deposits, interpretations related to foundation and excavation conditions for specific sites or highway routes, the areal soil-alluvial distributions, and known and potential mineral resources. Although general-purpose areal geologic base maps provide pertinent data that can be used for engineering endeavors, the numerous "special-purpose" maps of localities throughout reservations provided much more essential data for planning and special uses.

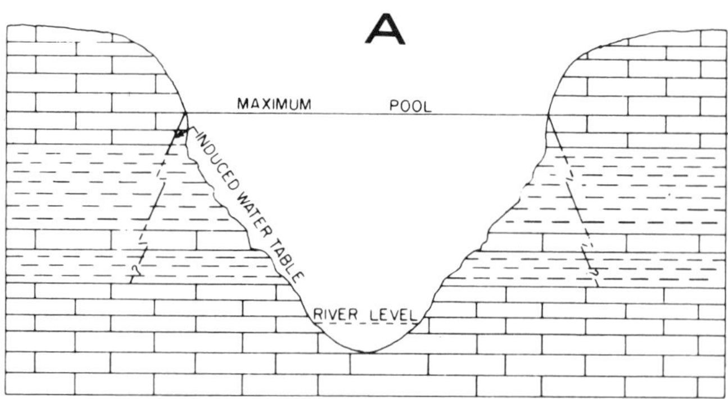
A
MAXIMUM POOL
INDUCED WATER TABLE
RIVER LEVEL
Reservoir located in impermeable rocks. Induced water table has an exceedingly steep gradient — position at depth unknown

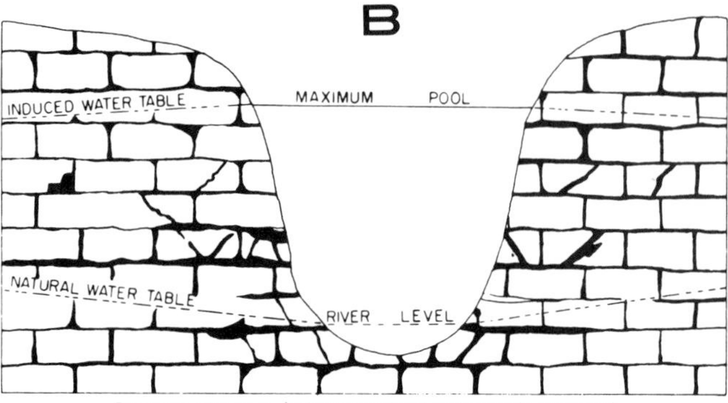
B
INDUCED WATER TABLE
MAXIMUM POOL
NATURAL WATER TABLE
RIVER LEVEL
Reservoir located in faulted, jointed, and subsequently solutioned (permeable) limestone. Reservoir locally reverses the natural water table. Consequently considerable bank storage with "short-circuit" leakage possible.

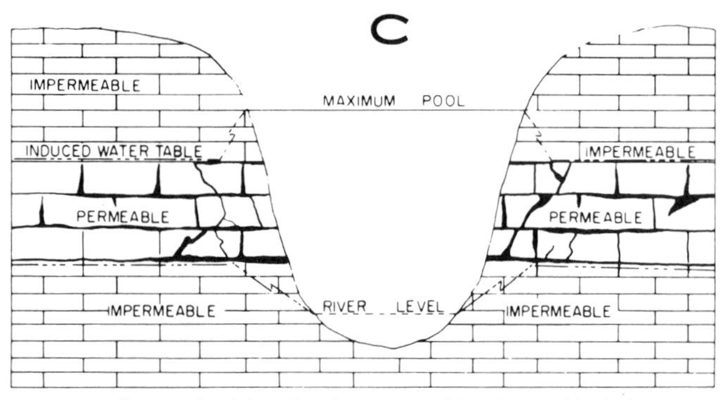
C
IMPERMEABLE
MAXIMUM POOL
INDUCED WATER TABLE
IMPERMEABLE
PERMEABLE
PERMEABLE
IMPERMEABLE
RIVER LEVEL
IMPERMEABLE
Reservoir located in alternating impermeable and permeable strata — none to small amounts of ground water in latter before construction. Perched water table created would soon reach equilibrium if downstream outlet was not available—bank storage recoverable.

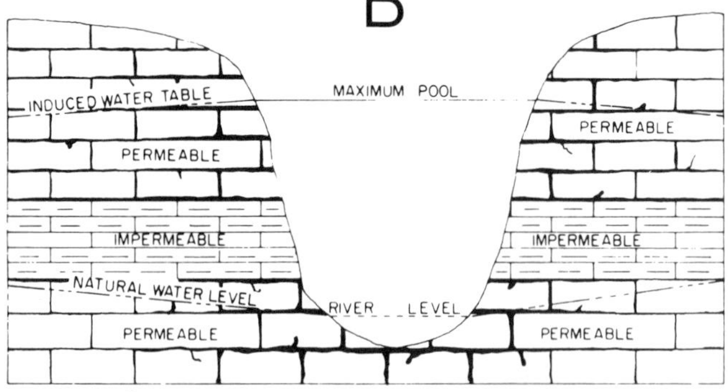
D
INDUCED WATER TABLE
MAXIMUM POOL
PERMEABLE
PERMEABLE
IMPERMEABLE
IMPERMEABLE
NATURAL WATER LEVEL
RIVER LEVEL
PERMEABLE
PERMEABLE
Reservoir located in alternating permeable and impermeable strata Induced (perched) water table created above impermeable beds. Considerable bank storage with "short-circuit" leakage possible

Figure 7. Schematic diagrams of some possible reservoir conditions, Big Bend District. For stratigraphy see Figure 8. A. Located in impermeable rocks; includes some preferred sites in the Boquillas, Walnut-Comanche Peak, Maxon, and Glen Rose Formations of the upper reaches of the Rio Grande. B. Located in faulted, jointed, and subsequently solutioned limestone; includes some preferred sites in the Georgetown Limestone. C. Located in alternating impermeable and weakly permeable strata; includes some preferred sites in the Kiamichi, Edwards, and Walnut-Comanche Peak Formations. D. Located in permeable and moderately permeable strata; includes some preferred sites in the Georgetown, Kiamichi, and Edwards Formations. (From Kiersch and Hughes, 1952)

Natural and manufactured aggregate. Any evaluation of the natural aggregate deposits of the Navajo country must ascertain both their beneficial and harmful qualities relative to their chemical and physical properties. The principal natural deposits, as shown on Figure 9, include terrace or pediment gravels and stream alluvium, beds of weakly consolidated sand or conglomerate ("lag" gravel), and cinders. The properties and characteristics of the Navajo country aggregates usually result from the origin of source material, and any modifications from transport and deposition by water, wind, or gravity (Kiersch and others, 1955, vol. III). Some mineral constituents are known to react with high-alkali cements and cause deterioration of concrete (see Chapter 16, this volume).

Rock formations of the Navajo country are generally "soft," so the natural aggregates are largely composed of soft sandstones and shales, although locally some hard limestone, quartzite, chert, and/or igneous rock fragments may occur. Much of the natural aggregate is neither suitable or durable enough for construction purposes, a factor that increases costs many-fold in parts of region. Representative test data were tabulated for the principal deposits (size analysis, absorption, specific gravity, and abrasion) along with remarks on the available tonnage, accessibility, suitable uses, overburden conditions, and a general appraisal of the deposits (Kiersch and others, 1955, p. 18, vol. III). The geographic grouping of the natural aggregate deposits serves as an inventory for proposed projects in an area. A review of the occurrence of aggregates and their physical and chemical properties is given elsewhere by Rhoades (1950).

Deposits suitable for the production of crushed stone and manufactured aggregate as supplements to the natural sources usually occur in Navajo country as hard beds or as blocky rock. Crushed aggregates are made from selected bodies of sandstone, limestone, and conglomerate; granite, basalt, and agglomerate occurrences; or a few occurrences of natural aggregate that require crushing. Suitability for crushing depends on physical and chemical qualities—which can be critical (Kiersch and others, 1955, vol. III)—mineral composition, and textural and structural features. The product ideally is composed of roughly equidimensional particles, with a minimum of fine sizes. Among such deposits throughout the reservations, the granite mass exposed in a

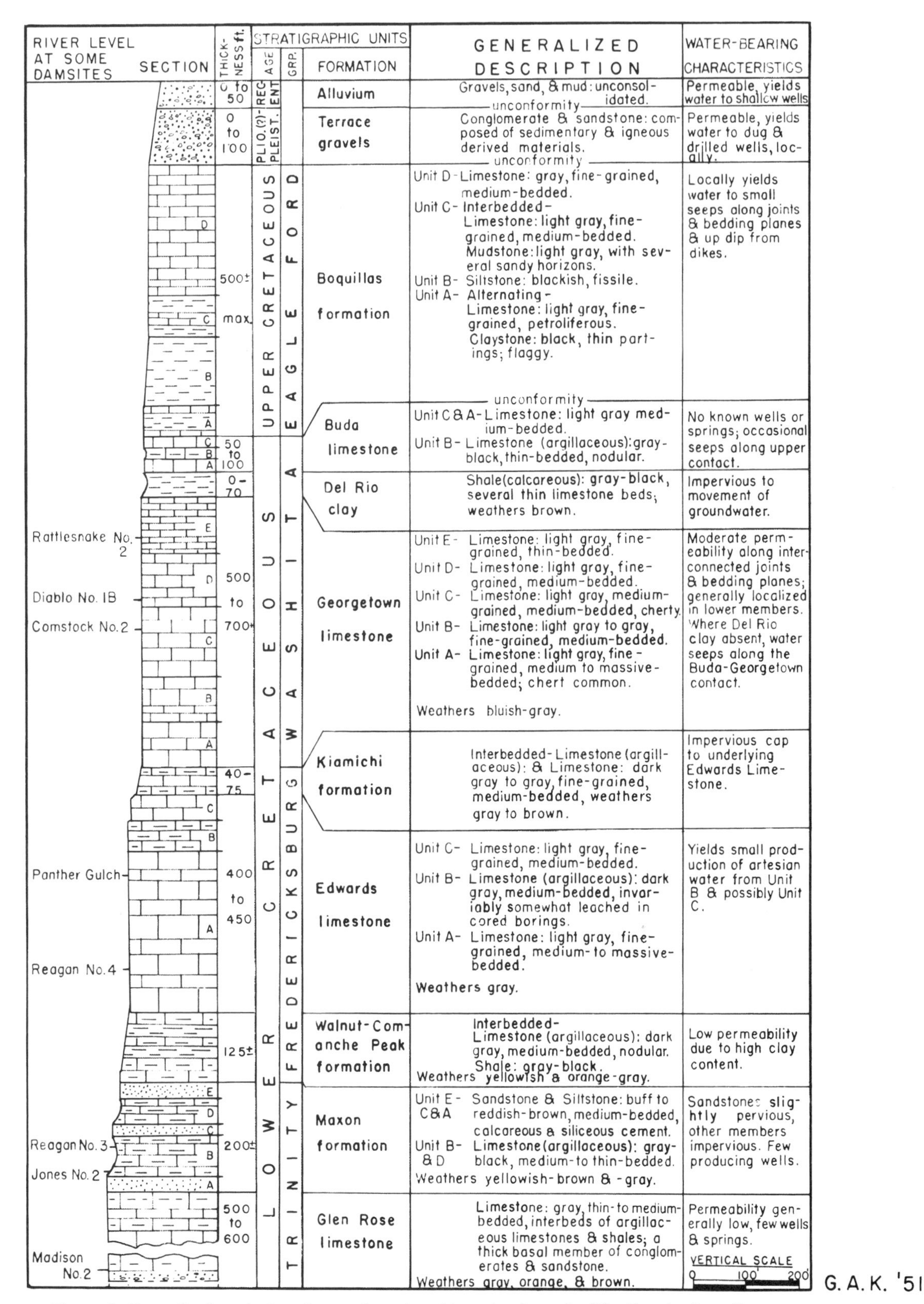

RIVER LEVEL AT SOME DAMSITES
SECTION
THICK-NESS ft.
STRATIGRAPHIC UNITS
AGE
GRP.
FORMATION
GENERALIZED DESCRIPTION
WATER-BEARING CHARACTERISTICS
0 to 50
REC ENT
Alluvium
Gravels, sand, & mud: unconsolidated.
unconformity
Permeable, yields water to shallow wells
0 to 100
PLIO.(?)-PLEIST.
Terrace gravels
Conglomerate & sandstone: composed of sedimentary & igneous derived materials.
unconformity
Permeable, yields water to dug & drilled wells, locally.
UPPER CRETACEOUS
EAGLE FORD
500± max.
Boquillas formation
Unit D - Limestone: gray, fine-grained, medium-bedded.
Unit C - Interbedded - Limestone: light gray, fine-grained, medium-bedded. Mudstone: light gray, with several sandy horizons.
Unit B - Siltstone: blackish, fissile.
Unit A - Alternating - Limestone: light gray, fine-grained, petroliferous. Claystone: black, thin partings; flaggy.
Locally yields water to small seeps along joints & bedding planes & up dip from dikes.
unconformity
WASHITA
50 to 100
Buda limestone
Unit C & A - Limestone: light gray medium-bedded.
Unit B - Limestone (argillaceous): gray-black, thin-bedded, nodular.
No known wells or springs; occasional seeps along upper contact.
0-70
Del Rio clay
Shale (calcareous): gray-black, several thin limestone beds; weathers brown.
Impervious to movement of groundwater.
Rattlesnake No. 2
Diablo No. 1B
Comstock No. 2
500 to 700+
Georgetown limestone
Unit E - Limestone: light gray, fine-grained, thin-bedded.
Unit D - Limestone: light gray, fine-grained, medium-bedded.
Unit C - Limestone: light gray, medium-grained, medium-bedded, cherty.
Unit B - Limestone: light gray to gray, fine-grained, medium-bedded.
Unit A - Limestone: light gray, fine-grained, medium to massive-bedded; chert common.
Weathers bluish-gray.
Moderate permeability along interconnected joints & bedding planes; generally localized in lower members. Where Del Rio clay absent, water seeps along the Buda-Georgetown contact.
LOWER CRETACEOUS
FREDERICKSBURG
40-75
Kiamichi formation
Interbedded - Limestone (argillaceous): & Limestone: dark gray to gray, fine-grained, medium-bedded, weathers gray to brown.
Impervious cap to underlying Edwards Limestone.
Panther Gulch
Reagan No. 4
400 to 450
Edwards limestone
Unit C - Limestone: light gray, fine-grained, medium-bedded.
Unit B - Limestone (argillaceous): dark gray, medium-bedded, invariably somewhat leached in cored borings.
Unit A - Limestone: light gray, fine-grained, medium- to massive-bedded.
Weathers gray.
Yields small production of artesian water from Unit B & possibly Unit C.
125±
Walnut-Comanche Peak formation
Interbedded - Limestone (argillaceous): dark gray, medium-bedded, nodular. Shale: gray-black.
Weathers yellowish & orange-gray.
Low permeability due to high clay content.
TRINITY
Reagan No. 3
Jones No. 2
200±
Maxon formation
Unit E - C & A - Sandstone & Siltstone: buff to reddish-brown, medium-bedded, calcareous & siliceous cement.
Unit B - & D - Limestone (argillaceous): gray-black, medium-to thin-bedded.
Weathers yellowish-brown & -gray.
Sandstones slightly pervious, other members impervious. Few producing wells.
Madison No. 2
500 to 600
Glen Rose limestone
Limestone: gray, thin- to medium-bedded, interbeds of argillaceous limestones & shales; a thick basal member of conglomerates & sandstone.
Weathers gray, orange, & brown.
Permeability generally low, few wells & springs.
VERTICAL SCALE
0
100'
200'
G. A. K. '51

Figure 8. Generalized geologic column of stratigraphic units along the Rio Grande, from Reagan Canyon to Del Rio–Villa Acuna, Texas-Mexico, with their water-bearing characteristics. Some of the potential damsites are plotted with respect to river level and rock units (see Fig. 7). (From Kiersch and Hughes, 1952.)

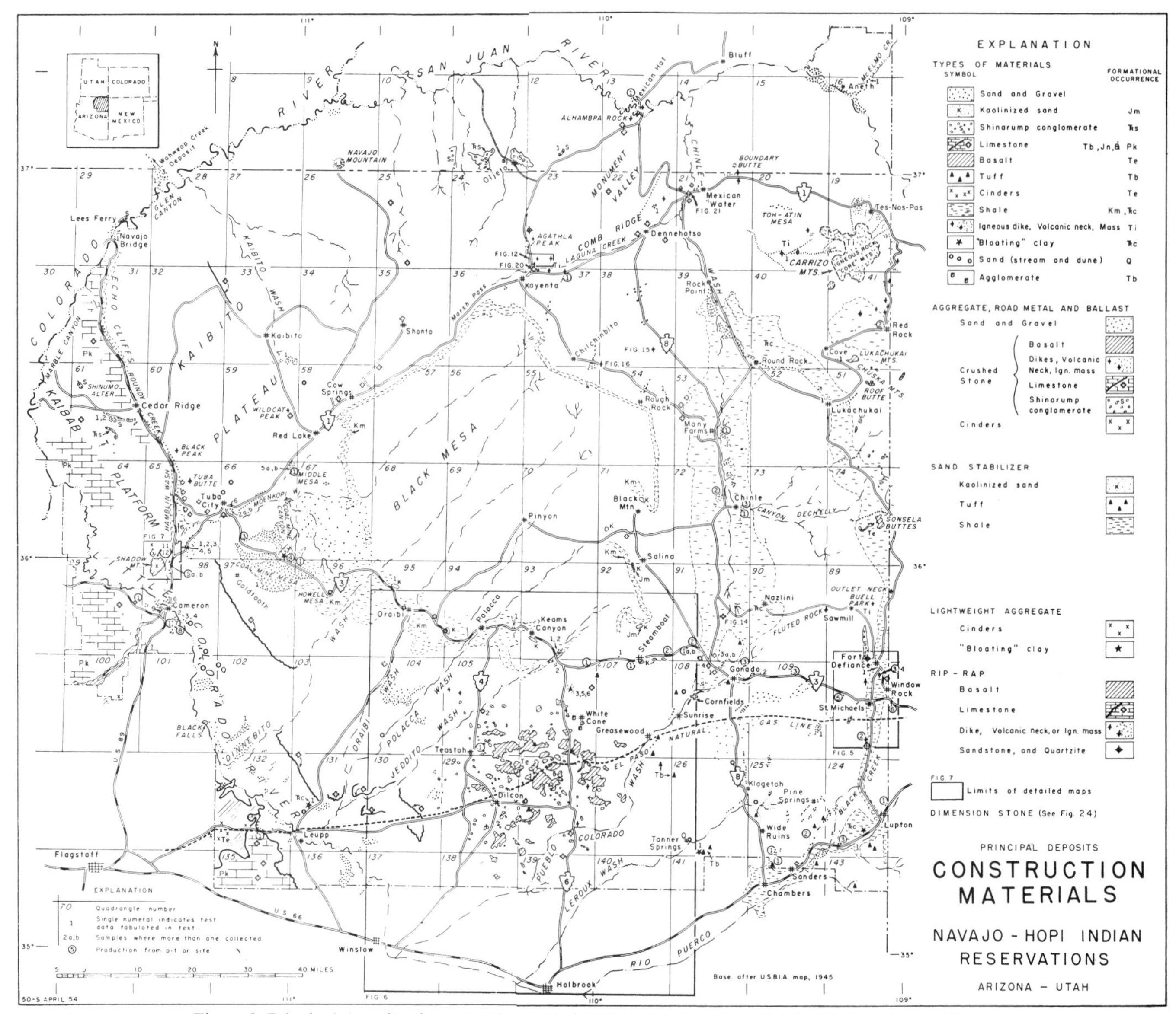

Figure 9. Principal deposits of construction materials throughout the Navajo-Hopi Indian Reservations, Arizona-Utah. (From Kiersch and others, 1955, v. 3.)

canyon near Hunters Point Trading Post, which is the inner core rock of the Defiance Uplift, has been the most widely exploited since discovery in 1953 (Fig. 11).

Construction materials. One of the most important problems faced by the Navajo construction program was the abundance of sand and sandy material over large sections of the reservation either as dune and drift sand or as sandy blankets. As the industrial program developed, stabilizing these sands and sandy soils with bitumen proved too expensive in the remote Navajo country. A successful construction technique was devised that stabilized the "blow" sand and sandy soil by filling the interstices with clayey fines; locally, fine-grained mixtures used as stabilizers were obtained from kaolinized sandstones of the Mesaverde Group rocks, high-grade lenses and irregular-shaped bodies of kaolinized sandstone in the Morrison and Cow Springs Formations, and whitish gray tuff beds in the upper Bidahochi Formation (Fig. 9; Kiersch and others, 1955, vol. II and III).

Another resource of the Navajo country widely used for construction is hard, uniformly dense rock units that can be quarried and cut to desired size and shape for dimension stone, whether as building-cut, rough-cut, flagstone, or ornamental polished stone. Dimension stone in the Navajo country is produced from sandstone, limestone, and some igneous rock (Peirce, 1955); the upper unit of the De Chelly Sandstone offers the best potential for a good commercial stone, with other sources shown on Figure 11.

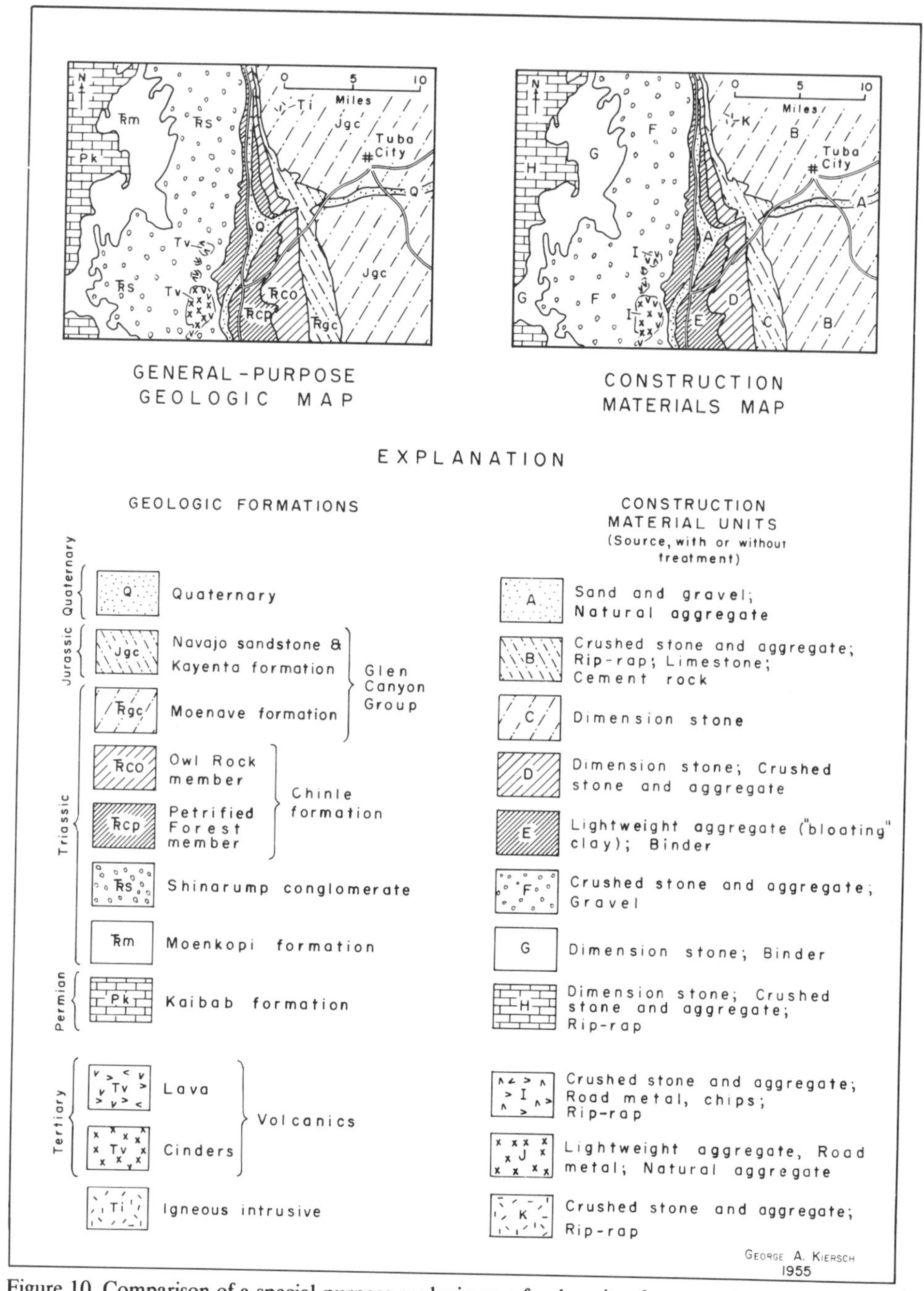

Figure 10. Comparison of a special-purpose geologic map for deposits of construction materials and a general-purpose geologic map, Tuba City–Shadow Mountain area, Arizona. (From Kiersch and others, 1955, v. 3.)

Management of land holdings and related engineering works, 1955 to 1961

The most ambitious private geological mapping project of the 1950s was completed between 1955 and 1961 by the Southern Pacific Corporation (SPCo), and covered its landholdings in California, Nevada, and Utah (Fig. 12). This broad-based, far-sighted survey and related special projects was designed to provide technical guidance and a comprehensive database to manage the company lands for the ensuing 35- to 50-year period, as demands and technology dictated. The population expansion to the western states by the 1950s placed the company landholdings in the center of future industrial developments in California and Nevada. Reacting to this change and opportunity in 1954, the SPCo Board of Directors, at the urging of three prominent engineer-scientist members (E. E. DeGolyer, Henry T. Mudd, and Cleveland E. Dodge), and with the support of president D. J. Russel, authorized a geological survey with exploration and eva-

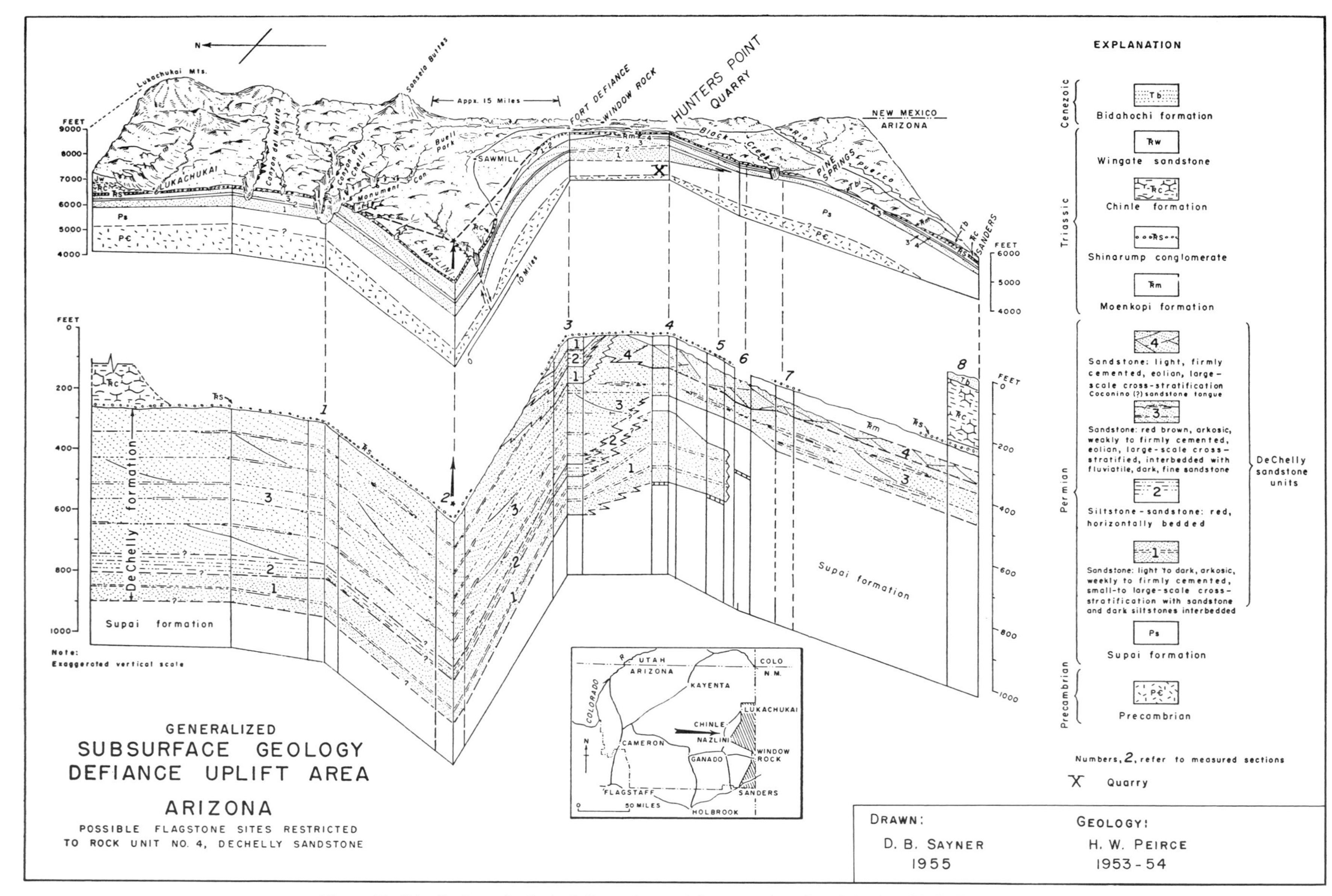

Figure 11. Distribution of principal dimension and flagstone deposits of the De Chelly Sandstone, Defiance Uplift region, Navajo Country, Arizona. (Modified from Peirce, 1955, Fig. 25.)

luation of company lands for guidance in management, industrial and resources development, and related engineering projects.

Survey Approach. The survey, operating at full strength by late 1955, consisted of a regional mapping and related investigations group of about 25 field and office personnel, with a smaller special-projects staff for the follow-up phase. About 22,000 mi^2, the equivalent of 93 15-minute quadrangles, were mapped at a scale of 1/24,000 on standardized two-township base maps. The regional strips, 40 mi wide, athwart the railroad, included the alternate, odd-numbered sections of Southern Pacific's land-grant holdings, about 38 percent of the area investigated (Kiersch, 1958; Fig. 12). This second geological survey of the landholdings utilized all the principles and geological techniques available in the 1950s and followed the steps outlined in Stages 1 through 6 described this chapter with the customary geological and exploration tools. The standardizing approach for field activities, along with the preparation of the maps and interpretation of data, is set forth in a handbook for the project personnel (Kiersch, 1959). The positive results of the SPCo survey and widespread usefulness of the assembled database changed the originally apathetic attitude of many individuals about this process, and they became strong supporters of the comprehensive geological survey concept. Parts of the survey mapping in California were incorporated into the 1959 to 1969 edition of the *Geologic Atlas and Map of California* (Jennings, 1969) and an early edition of the *Geologic Map of Nevada* (Webb and Wilson, 1962). Format of a typical two-township geologic map prepared by the survey is shown on Figure 13.

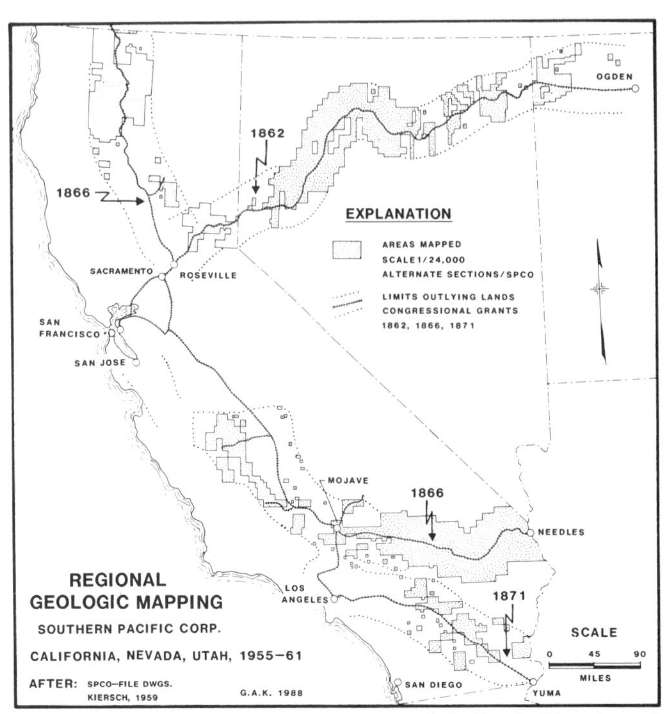

Figure 12. Grant lands and associated areas of California, Nevada, and Utah mapped by the Regional Geological Survey of the Southern Pacific Corporation. Extent of the original Central (Southern) Pacific land grants of 1862 to 1871 is inferred; parts disposed of prior to 1949 are based on a 1909 survey.

Utilization. Large-scale geologic mapping combines with a systematic inventory of known or potential resources to serve a host of uses for construction, maintenance of engineering works and railroads, land management, and to attract new industry and developments. Resources data on minerals, fuels, water, soils, or engineering materials became an asset, whether for maintaining the railroad trackway and network in hazardous and slide-prone terrain, or providing additional freight and/or lease revenue from untapped deposits. Lands designated nonmineral could be managed exclusively for their surface value or ground-water without concern for possible future mineral leasing. Moreover this provided a workable understanding of the geologic features inherent to future utilization, such as for agriculture, grazing, timber, recreation, and commercial plant sites, and even as guidance for litigation and claims filed against SPCo.

The geological database and survey personnel were also utilized to select new sites for major industrial developments and provide geological guidance on problems arising from operation of the railroad system and an associated oil distribution pipeline network, such as the mitigation of slides, earthquake damage, and tunnel failures. For example, the major Alta landslide of April 1958 near Baxter, California, blocked the main west-east rail line and threatened to close U.S. Highway 40; the financial losses to be sustained by SPCo approached $500,000/day. Geological knowledge collected by the area survey from nearby lands provided an understanding of ground-water conditions and physical properties of the tuffaceous Tertiary rock units involved with the unstable slope, trackway collapse, and subsurface conditions downslope from the railroad that affected Highway 40. Fortunately, the principal cause of failure could be temporarily mitigated within two days and the main line returned to continuous traffic, thus avoiding a potentially large financial loss. Long-term stabilization was achieved only after installing an extensive drain system.

Although not directly involved with the design and construction phases of the SPCo causeway across the Great Salt Lake, survey personnel provided background data on potential quarry-rock sites, fill material, and the potential for high-yield water wells near the west end of the construction site (see Chapter 23, Fig. 33, this volume).

Urban hazards and land use, San Francisco Bay area, 1970s to 1980s

Early surveys. Regional geological investigations to initially evaluate the natural hazards and risks in the San Francisco Bay region, which were later expanded into established guidelines for land use and the siting of new engineering works, were initiated by the U.S. Geological Survey in 1947. The initial San Francisco mapping project was organized under Clifford Kaye, the first quadrangle, San Francisco North, was mapped by M. G. Bonilla, Julius G. Schlocker, and Dorothy Radbruch (Schlocker,

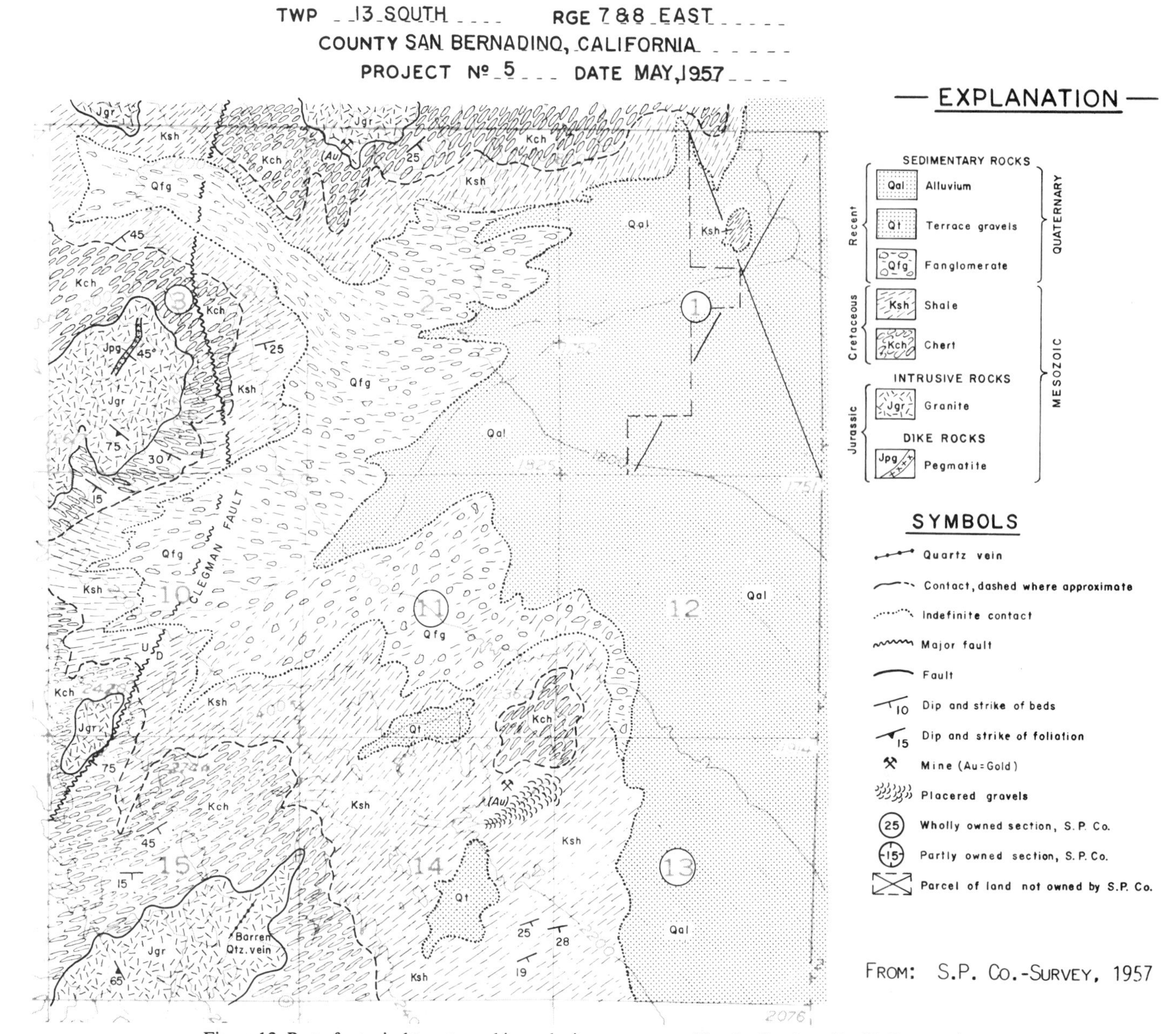

Figure 13. Part of a typical two-township geologic map, prepared by the Southern Pacific Survey, that covered both Southern Pacific Corporation lands and intervening sections at a scale of 1:24,000. Area shown is southeast of San Bernardino, California. (From Southern Pacific Corporation, unpublished report, 1957.)

1974). This was followed by San Francisco South (Bonilla, 1960) and the Oakland West and East Quadrangles by Radbruch-Hall (1957, 1969). These mapping activities extended into the 1960s and included attention to ongoing construction projects, such as the Broadway tunnel—described in Chapter 23, this volume. The areal mapping supplied both surface and subsurface information critical to an improved understanding of causes for some common urban hazards of bay areas and their relative risks throughout the greater region, such as landslide susceptibility (Bonilla, 1960; Brabb and others, 1972) and land use for a housing subdivision (Kachadoorian, 1956).

Earl E. Brabb and others, in the 1960s, initiated a San Francisco Bay area study on a wide range of urban hazards and geologic conditions affecting the growth, population, and cost of construction in progress. By 1970, these investigations were concentrated on a broad "Landslides and Seismic Zonation Study of San Francisco Bay Region" (Brabb, 1979). These and other detailed studies on seismic zonation of the Bay Region were summarized by Borcherdt (1975), wherein is found an excellent set of guidelines for similar surveys. For example, the seismic zonation based on the 1906 earthquake events correlated with the areal distribution of the principal geologic materials (Fig. 14).

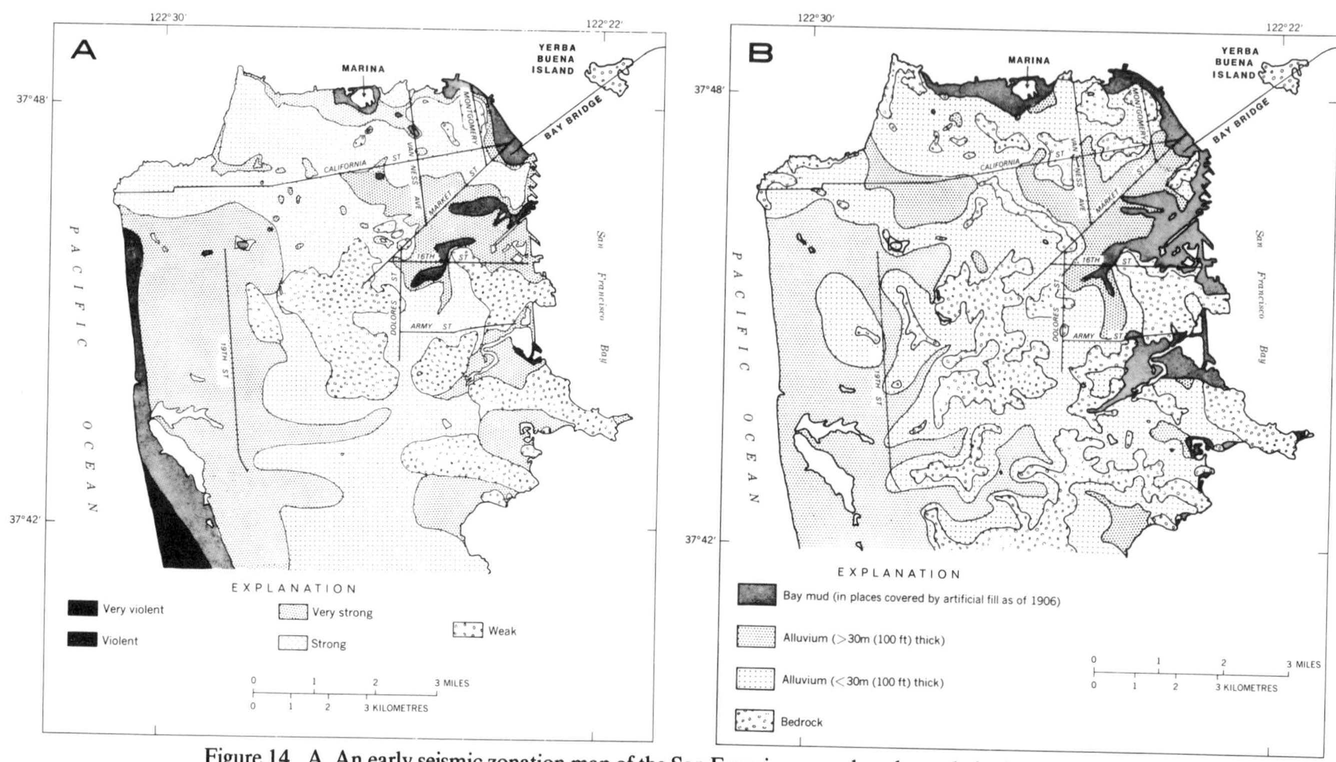

Figure 14. A. An early seismic zonation map of the San Francisco area, based on relative intensities felt throughout the city during the 1906 earthquake. B. Surficial materials map of the same area. Note the correspondence between surficial materials and felt intensities. (From Borcherdt, 1975, Figs. 1, 2.)

Regional hazards and resources planning. During the 1970s, several urban studies projects that evaluated the risk of natural hazards were begun in the United States. One of these, the San Francisco Bay Region Environment and Resources Planning Study (SFBRS), was cooperatively sponsored by the U.S. Geological Survey (USGS) and the U.S. Department of Housing and Urban Development (HUD).

The SFBRS study utilized USGS expertise in geology, geophysics, geochemistry, hydrology, and cartography and HUD expertise in planning. Numerous ways of disseminating the SFBRS information were used to ensure that the findings would reach planners and decision-makers. This project focused on the geological hazards of the environmentally sensitive area surrounding San Francisco and San Pablo Bays. The aim of the Survey's study was to increase the use of geology in solving urban and regional development problems, and to provide guidance and suggest techniques to the scientific, engineering, and planning professions concerned with application of the earth sciences to urban development (Brown and Kockelman, 1983). For example, the faults of the San Francisco Bay region that may cause damaging earthquakes, and a classification of their surface displacement (Fig. 15) can be correlated with areas in the region likely to experience common hazards such as shaking, flooding, liquefaction, landsliding, and related geologic consequences resulting from seismic events (Fig. 16).

The aim of the HUD study was to demonstrate the significance of geologic hazards in urban and regional planning and development (Little, 1975), and to improve urban planning techniques in a real-life situation (Laird and others, 1979). More than 160 maps and reports, many of them planned and written for use by nonscientists (planners, engineers, and decision-makers) were published and disseminated. An overview of the project (Brown and Kockelman, 1983) reminds decisionmakers of the intimate relation between geology and land use and recommends that they demand information based on valid technical or scientific evidence.

The varied provinces of land-use concern and associated natural processes and hazards covered by the Bay area project include the estuaries with marshlands, saltponds, and bay waters (Atwater and others, 1979); the retreating coastal bluffs with erosion and landslides (Atwater and others, 1977; Newman and others, 1978); the lowlands underlain by water-bearing sediments where flooding is periodic (Laird and others, 1979; Helley and others, 1979); the hillsides and uplands with erosion and unstable foundations and slopes (Laird and others, 1979; Nilsen and others, 1979); and the faults and earthquakes that trigger landsliding and liquefaction (lowlands adjoining the bay) as outlined by Figures 15 and 16 and Borcherdt (1975) and Borcherdt and others (1979). This monumental benchmark-type survey and special studies have since served as guides for investigations in other states and regions of America.

The Loma Prieta earthquake of October 17, 1989, and ser-

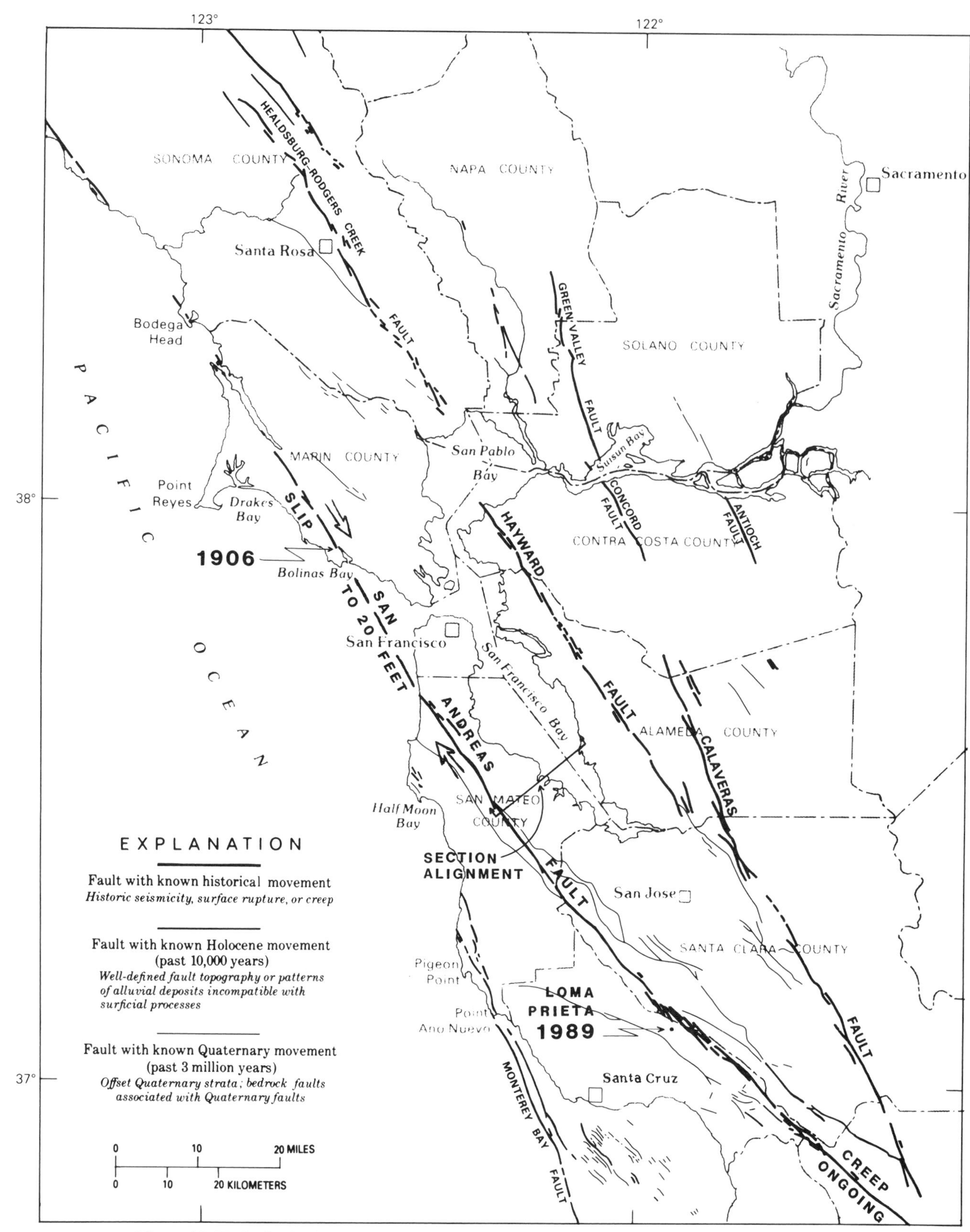

Figure 15. Faults in the San Francisco Bay region that may cause damaging earthquakes, surface displacements, or both; most are members of the San Andreas fault system. Line of section for Figure 16 is indicated. (Modified from Borcherdt, 1975, Fig. 3.)

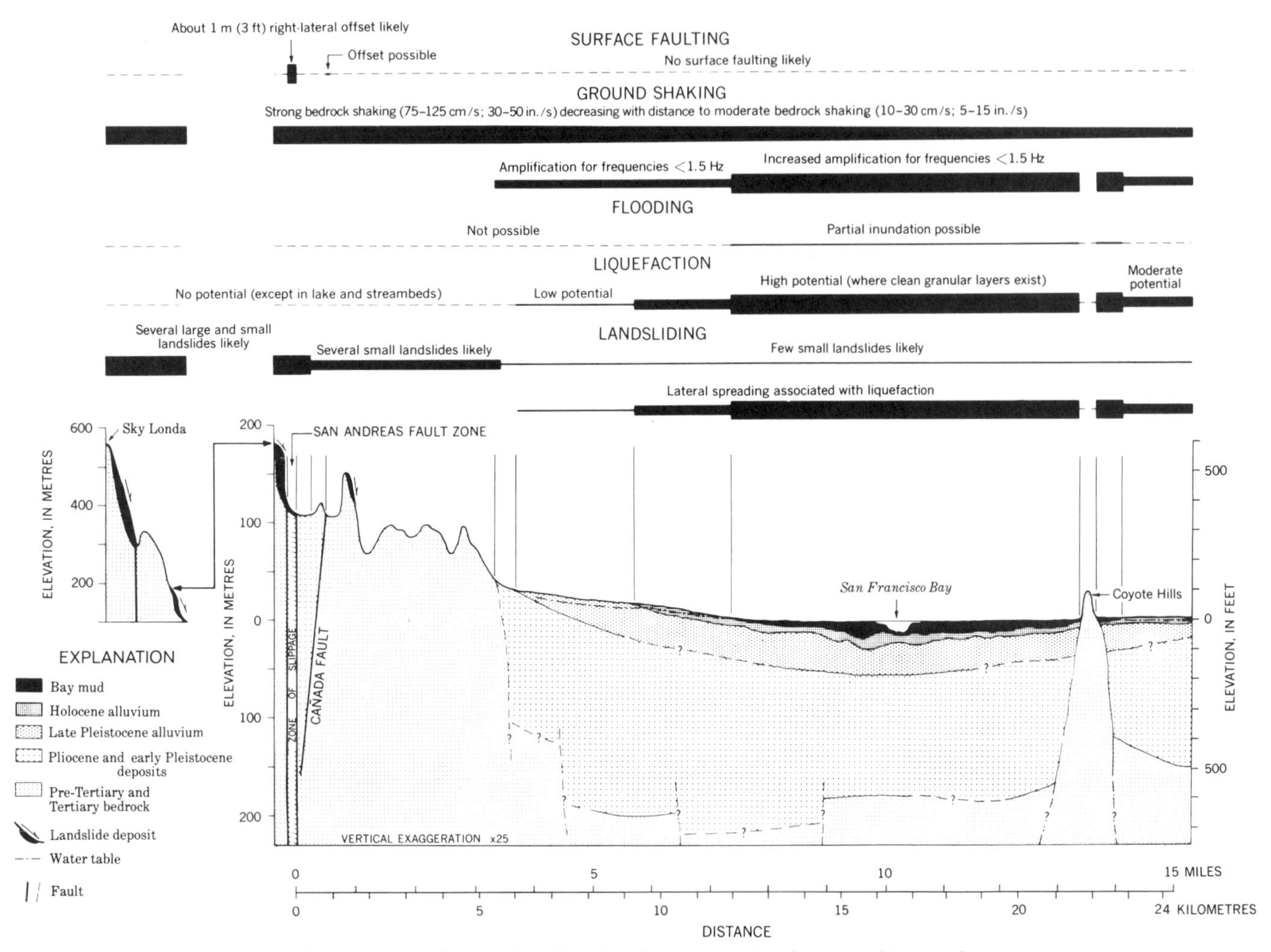

Figure 16. Cross section of lower San Francisco Bay region showing potential areas for common hazards and processes that can be generated by movement along faults of the San Andreas fault system. For section location, see Figure 15. (From Borcherdt, 1975, Fig. 68.)

ies of prominent aftershocks afforded a further opportunity to demonstrate the amplification of ground motion throughout the Bay area relative to the principal geologic materials (Fig. 14). The heavily damaged Marina District was studied by recording the vertical velocities of a magnitude 4.6 aftershock at three seismograph stations founded on bedrock, sand, and fill materials, respectively (Plafker and Galloway, 1989). The potential for damage to surface buildings was clearly demonstrated, as shown on Figure 17.

Power plant sites: statewide New York, 1975 to 1979

A statewide study in New York was designed to identify suitable new power generation sites within a minimum period of time, yet with maximum effectiveness and potential for licensing by the regulatory agencies. The initial study in 1975 was guided by several broad assumptions and constraints, as agreed between New York State Electric & Gas Corporation (NYSEG) and United Engineers & Constructors Inc. (UEC, 1975). In mid-1975 about 2,400 megawatts of base-load generating capacity, either fossil or nuclear, was projected for operation by 1986 to 1988. The New Site Selection Team Survey was authorized to provide the best available sites to meet the projected needs; the entire state of New York was screened, except New York City/Long Island, under the guidance of the geological contractor and a seismological consultant.

The initial selection process, consisting of two phases and five separate stages, was designed to identify at least three potentially licensable sites (Kiersch, 1975). The recommended sites subsequently chosen by the study team were subjected to intensive environmental impact analysis, and technical investigations during 1976 to 1978. Following this, a further analysis by state and federal regulatory agencies led to a final choice being made on the basis of public hearings (UEC, 1976).

The selection process consisted of two distinct phases: an exclusion phase that reduced the area under consideration in two stages, and a selection phase, wherein the large number of potential sites was reduced in three successive stages by detailed comparative evaluations. The details of each stage and the manner in which geological information was utilized at each of five successive evaluation levels are summarized in the project report (UEC, 1976). Guidelines and considerations for each stage of the site evaluation, as well as criteria and methodology, were established, and a standard sequence established that involved formulating criteria, review and concurrence, cataloging data, analyzing criteria relative to data, assessing results for reasonableness, and ascertaining results as they related to the problem.

Methodology. The selection process consisted of definable Stages 1 thru 6 of increasingly detailed analysis, as discussed earlier. The candidate regions were identified on maps at a scale of 1:500,000 utilizing broad criteria such as insufficient cooling water, major parks and restricted land use, regions of recognized geologic uncertainty and possible earthquake hazards, or where no additional electrical capacity was needed. The candidate areas were subsequently identified on maps at a scale of 1:250,000 on the basis of criteria that would influence smaller sectors, such as conflicting land uses, population factors, nearness to major airport installations, and transmission-related costs.

The geological criteria for evaluating and selecting candidate sites from within the candidate areas utilized a set of ratings of 2g, 1g, and 0g, where "g" signifies geological criteria (Fig. 18).

Serious adverse features included (1) faults or zones of tectonic structures and folds, particularly any such features that might conceivably be approaching reactivation; (2) limestone formations, salt, and/or gypsum units in the rocks of a candidate area that might be cavernous with natural openings at the surface or at depth, or might contribute to near-surface slump features; (3) man-made openings, due to mining, quarrying, abandoned construction, and disposal areas; and (4) general overburden conditions, such as active flood-plain deposits or ancient drainages, deep soil and/or weathered bedrock, soft glacial tills and debris, deep glacial lake beds and/or gravelly backfill in ancient valleys, or special features like expansive soils, sinkhole terrain, and areas of active subsidence.

The seismological criteria and methodology for evaluating the candidate areas also consisted of a rating system of the respective geologic sectors outlined on the base maps (Fig. 18). The seismic criteria for evaluating the preferred nuclear sites utilized ratings of 2s, 1s, and 0s; a separate rating system was used for evaluating plant sites utilizing fossil fuels.

Candidate sites were identified by Stage 3 on the basis of the general criteria of earlier stages, but the potential sites were compared on the basis of simplified scoring of a large number of environmental, technical, and economic attributes. The better sites had a large number of favorable characteristics and scored high by this selection process. The original 542 potential sites were reduced to 101 candidate sites when evaluated on the basis of 16 attributes, including the geological criteria. The geological

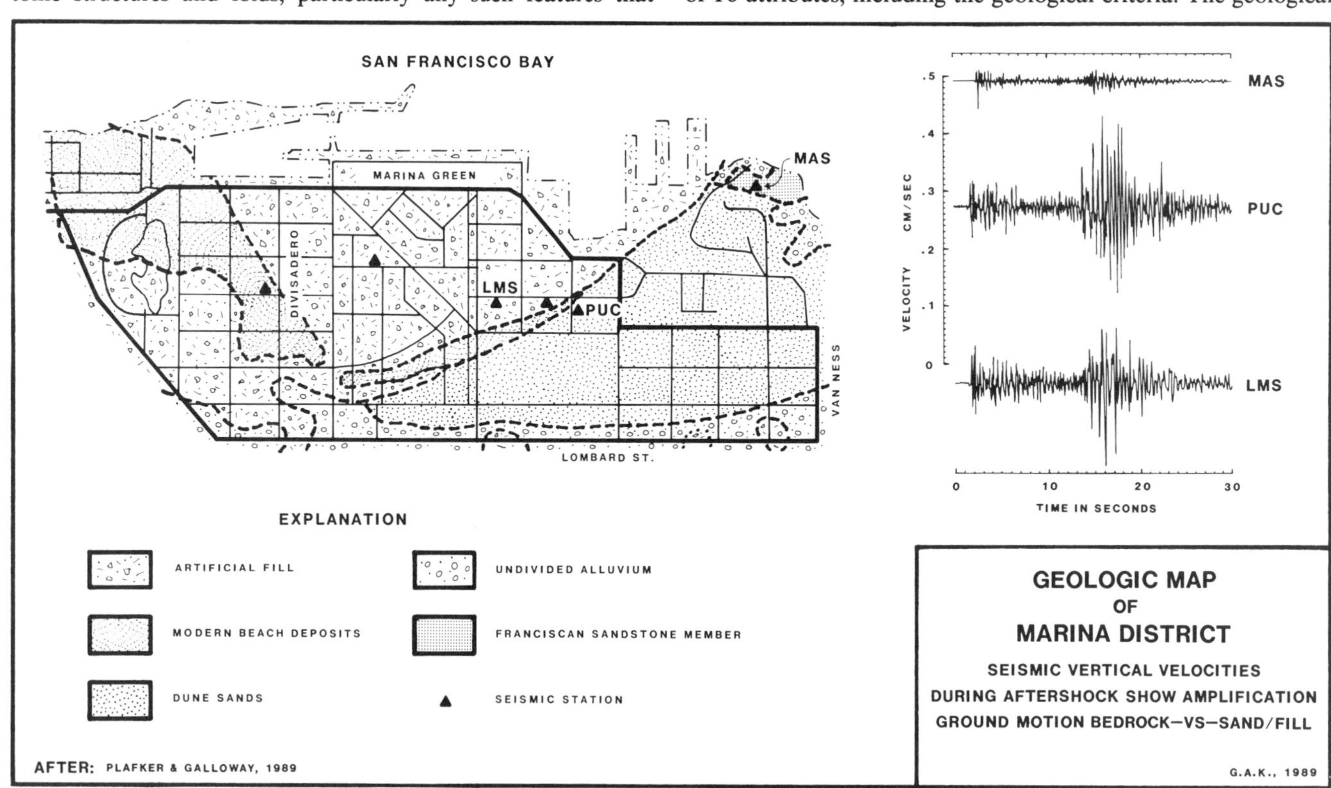

Figure 17. Vertical velocities during a magnitude 4.6 aftershock of the Loma Prieta earthquake of October 17, as recorded on October 21, 1989, at three temporary seismograph stations in the Marina district of San Francisco (cf. Fig. 14). Note comparative amplification of ground motion in damaged (LMS) and undamaged (PUC) areas, and areas of bedrock (MAS). (Modified from Plafker and Galloway, 1989, Fig. 27.)

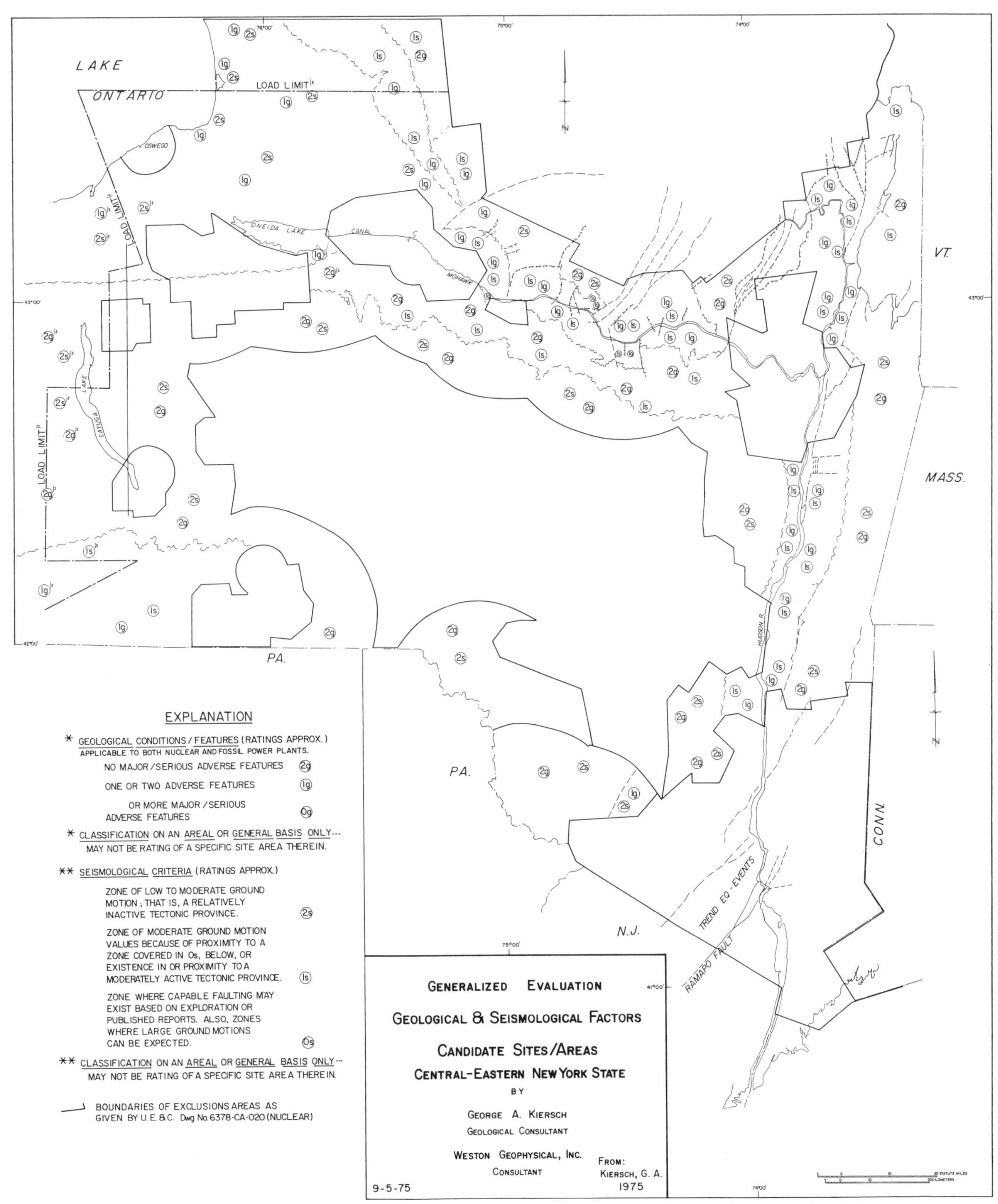

Figure 18. Generalized evaluation of geological and seismological factors to be considered in selection of candidate sites and areas for power plants in central and eastern New York State. The guidelines for geological and seismic ratings are indicated. This evaluation became the basis for subsequent preferred- and selected-site investigations. (From Kiersch, 1975.)

ratings (2g, 1g, 0g) were correlated with the type and depth of overburden materials, type and depth to bedrock, and the reliability of source of data, such as published state maps, county reports, personal observations by the geological contractor's personnel, water-well data, water-resources bulletins, or unpublished materials.

Preferred sites were identified by Stage 4 and reflected all major effects described earlier. They included economic factors (geologic features; construction costs of roads, railroads, piplines, and transmissions; cooling water pumping costs; transmission losses/costs; fuel and fuel-transportation costs) and environmental factors (air quality, meteorology, water quality, aquatic and terrestrial ecology, noise, land use, aesthetics, population, and socioeconomics). The environmental factors were scored on scales of merit and included weight factors, with economic factors expressed in dollars. Each subsequent stage of the geological study defined the selected criteria and the methodology applied for that stage of the project and interaction with the other disciplines of the evaluation team. From this rating sequence, 22 preferred-candidate sites were selected (Fig. 19) and later investigated in the field by the geological reconnaissance and evaluation of Stage 5. Each site was evaluated with respect to location and its geologic setting and features, including depth and character of surficial deposits and bedrock units, ground-water occurrence, potential problems, and the scope of additional investigations anticipated; site reports provided sources of data for documentation.

Evaluation of the designated sites. A geological evaluation of each Preferred Site was made on the basis of the point rating system. Many sites attained a rating of 2, with the remainder a rating of at least 1 (summarized by Kiersch, 1975, part V). Likewise, a seismological rating of each Preferred Site was given for either a nuclear or fossil-fuel plant.

The five most attractive preferred-candidate sites were further evaluated on the ground by the selection team, which looked particularly for any adverse features that might have been missed that would be critical to the design engineer or the environmental impact specialist; initially only one to three days on the ground were budgeted to analyze each candidate site. After several weeks of reviewing the five sites, two recommended sites, both for nuclear power plants, were chosen in early 1976 for feasibility and design-licensing investigations: The Stuyvesant site near Kinderhook, south of Albany, and a second site at New Haven, near Oswego, New York. Other sites were selected for fossil-fuel power-generating plants, but investigations were delayed awaiting the outcome of the proposed nuclear plants.

The extensive background on the regional geology and principal elements of a site needed to acquire a construction license from the National Regulatory Commission are given in Appendix A; the group of required regional and areal maps to accompany the text is given in Appendix B.

The site-specific investigations and follow-up licensing procedures with state and federal agencies moved ahead in 1977 to 1978, and both sites proved licensable. However, both of the projected nuclear plants were dropped by NYSEG in 1979 with the nationwide forecast for a decreased need and cutback in new electrical generation capacity throughout the 1980s.

Low-level radioactive waste disposal

The Low-Level Radioactive Waste Policy Amendment Act of 1985 (Public Law 99-240) reiterates state responsibility for disposal of commercial low-level waste, encourages interstate compacts for disposal on a regional basis, and defines compliance deadlines and payment of surcharges at existing sites as well as surcharge rebates for performance or penalties for noncompliance. Guidance on land disposal of low-level radioactive waste is provided by the Nuclear Regulatory Commission under Title 10 Code of Federal Regulations Part 61 (10 CFR 61); land-disposal facilities must be sited, designed, operated, closed, and controlled thereafter for assurance that any human exposure to radiation is within established limits. A brief review follows on the regional screening of sites, based on an unpublished summary by J. R. Keaton (written communication, 1989).

The minimum technical characteristics for acceptable low-level, land-disposal facilities require that candidate sites be identified by a systematic regional screening process. Some specific criteria for site selection are: (1) capable of being characterized, modeled, analyzed and monitored; (2) projected population growth and development will not affect the facility and its performance objectives, such as exploitation of known resources; (3) location is not within a 100-year flood plain, an area of flooding or frequent ponding, a coastal high-hazard area, or a wetland, and site must be well drained; (4) minimal drainage area exists upstream, so runoff cannot exhume or inundate buried waste; (5) sufficient depth to water table beneath site, so ground water cannot intrude the waste; (6) hydrogeologic unit used for disposal does not discharge ground water to the surface within the site; (7) tectonic processes such as faulting, folding, seismic activity, or volcanism cannot occur with a frequency or extent that affects ability of site to meet objectives; (8) areas must be avoided where surface geologic processes such as mass wasting, erosion, slumping, landsliding, or weathering occur with such frequency and extent as to significantly affect the ability of the site to meet performance objectives.

Cederborg and Tosetti (1982) note that additional criteria may be needed to ensure an adequate consideration of engineering, economic, environmental, and sociological factors as required by the National Environmental Policy Act of 1969; namely, availability of transportation, site capacity, construction costs, presence of endangered species, and land use.

Three types of siting criteria suitable on a regional basis for low-level radioactive waste disposal facilities are given by Golder Associates and Weston (1987). *Exclusionary* criteria set absolute limits that may be applied at the beginning of regional screening, such as the boundaries of state or national parks and monuments. *Avoidance* criteria define undesirable siting conditions, such as sites with upstream drainage areas larger than 1 mi^2, or complicated geologic conditions. *Preference* criteria define desirable sit-

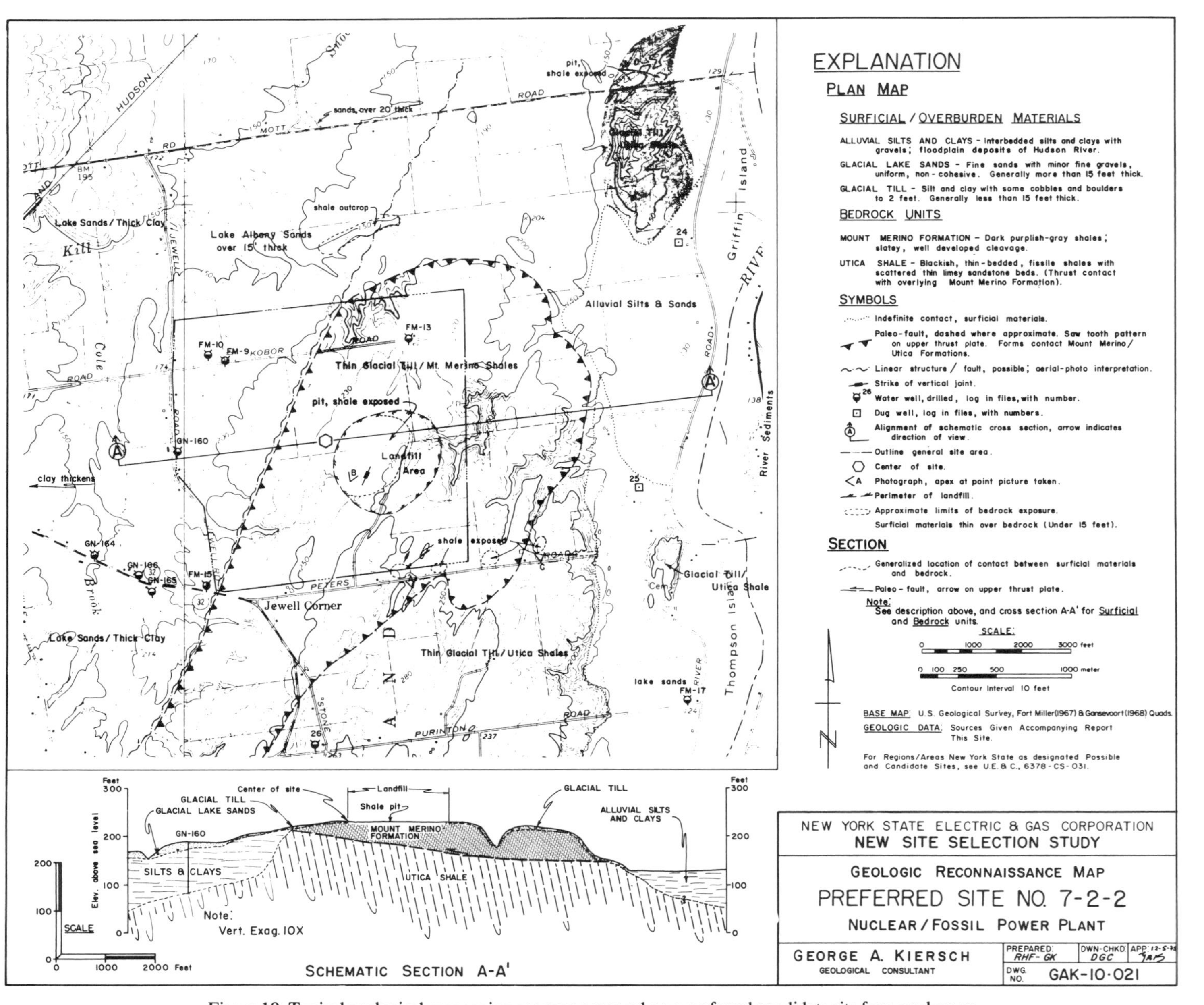

Figure 19. Typical geological reconnaissance map prepared on a preferred candidate site for a nuclear or fossil energy power plant in New York State. (From Kiersch, 1975.)

ing conditions that may be encouraged but not required, such as being within 5 mi of an interstate highway.

Cederborg and Tosetti (1982) describe a computer-aided methodology called SITE SCREEN developed in the 1970s by Bechtel Corporation that largely parallels the site selection/screening process described for New York State power plant sites (Kiersch, 1975). Essential elements for regional screening to identify candidate sites include (1) defining limits of the territory and regions therein to be screened, with each having a merit value based on a computer program equation; (2) distribution of engineering, economic, environmental, and sociological attributes within the region; (3) criteria for site selection, with relative importance using a rating scale; (4) identification of potential sites, in lieu of any stated criteria and importance values; (5) evaluation of each potential site relative to stated criteria; and (6) selection of preferred sites and rationale.

Primary screening of the territory starts with a definition of regions on the basis of natural or man-made barriers such as rivers, interstate highways, major suburbs, and national forests. Not all criteria are relevant for a regional screening; although areas subjected to seismic activity and flooding can be screened, the suitability relative to erosion, ground water, and drainage cannot be determined adequately on a regional basis. Moreover, construction costs will vary from one region to another, and some aspects cannot be assessed until specific sites are identified (Cederborg and Tosetti, 1982). The regions are ranked in descending order by merit value with the aid of a computer program based on an equation; sensitivity analyses assess the regional ranking in relation to changes in weighting factors. A secondary screening is conducted to identify potential disposal facility sites within candidate regions.

This methodology (Cederborg and Tosetti, 1982) provides a systematic framework for regional assessments, based on an integration of geologic information with other important factors in site selection. Graphic computer output serves to document the decisions and judgments of the site-selection team. The principal emphasis in regional geology investigations is on quantification of factors for computer manipulation. As noted by Varnes (1974), if the geologist hesitates in quantifying geologic data, the nongeologic computer operator will be forced to do the quantifying and make the geological interpretations.

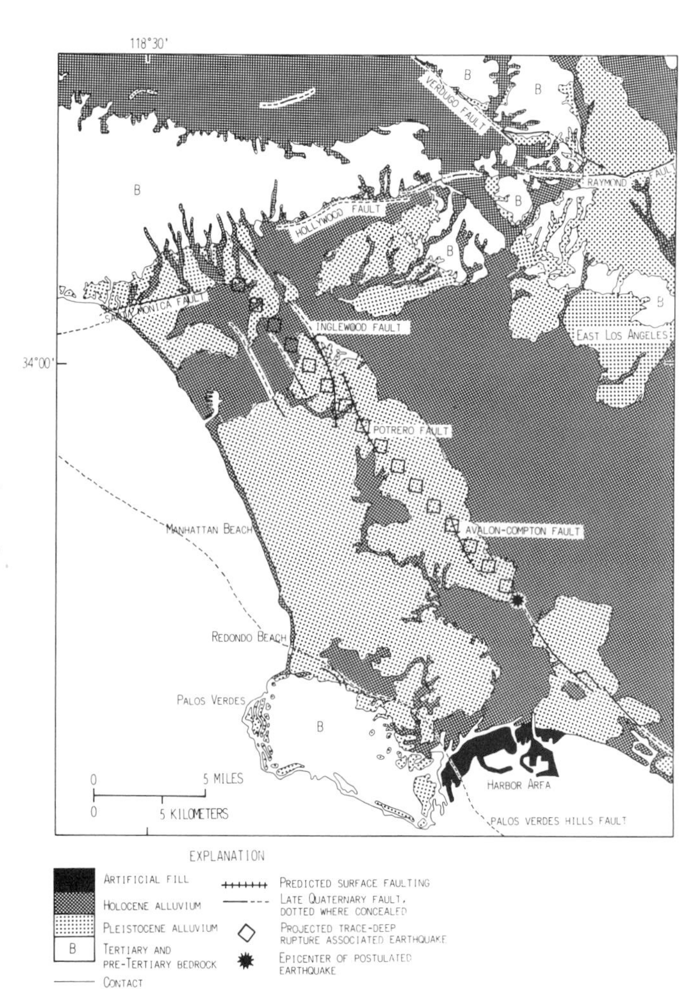

Figure 20. Regional geology of the Los Angeles basin and environs, showing the epicenter of a postulated magnitude 6.5 earthquake, the presumed trace of the associated fault rupture at depth along the Newport-Inglewood zone, and predicted locations of associated surface faulting. (Modified from Ziony and others, 1985, Fig. 210.)

Earthquake damage-response planning scenarios

Potentially destructive earthquakes are inevitable in the Los Angeles region of California, but hazards prediction can provide a basis for reducing damage and loss. The characteristics of principal geologically controlled earthquake hazards of the region—surface faulting, strong shaking, ground failure, and tsunamis—suggest opportunities for their reduction. Two systems of active faults generate earthquakes in the Los Angeles region: northwest-trending, chiefly horizontal-slip faults, such as the San Andreas; and west-trending, chiefly vertical-slip faults, such as those of the Transverse Ranges. Faults in these two systems have produced more than 40 damaging earthquakes since 1800. Ninety-five faults have slipped in late Quaternary time (approximately the past 750,000 years) and are judged capable of generating future moderate to large earthquakes and displacing the ground surface (Ziony and others, 1985). The following text is adapted from a short, unpublished review by J. R. Keaton (written communication, 1989), and focuses attention on yet another use that regional geologic data can serve—forecasting areas of high risk from earthquakes.

A creative use of regional geological information is the earthquake-planning scenarios prepared in California in the early to mid-1980s. Davis and others (1982) predicted geologic effects of a magnitude 8.3 earthquake along the San Andreas fault in southern California. Everenden and Thompson (1985) predicted geologic and seismologic effects of a postulated magnitude 6.5 earthquake along the northern part of the Newport-Inglewood zone in the Los Angeles basin (Fig. 20).

The assessment of Davis and others (1982) was based on their forecast that a strong earthquake on the San Andreas fault in southern California was likely in the near future. Regional information on the distribution of geologic and ground-water conditions, combined with a projected attenuation of strong ground shaking was used to predict seismic intensities and ground failure over much of southern California (Fig. 21). Based on their assessment of likely distribution of the seismic intensities and ground failure, the affected communities and utilities were able to assess potential losses and identify areas where mitigation activities might be effective. Emergency planning groups were then able to evaluate and forecast locations of the most concentrated damage, and hence the greatest need for emergency response.

The Los Angeles basin was the area of a demonstration project described by Ziony and others (1985). Their project was intended to demonstrate some of the predictive methods recently developed primarily by the U.S. Geological Survey for earthquake hazard reduction. The Los Angeles basin and environs were chosen for the demonstration because it is the most densely populated part of southern California and, furthermore, it contains a wide range of geologic conditions. Consequently, the project could consider the likely effects of surface faulting and related deformation, ground shaking, liquefaction, and landsliding as might be expected from a moderate-sized earthquake.

The regional geology and distribution of major rock units throughout the demonstration project area (Fig. 20) were utilized to predict the geologic and seismologic effects of a postulated earthquake; a major factor is the physical contrast between bedrock and alluvium or fill materials, with the extent of damage dependent on the location of the epicenter of the postulated earthquake and of the predicted surface faulting. Ziony and others (1985) demonstrated these relations with a series of maps showing geologic ground conditions, the distribution of predicted Modified Mercalli intensities for the postulated earthquake, predicted mean shear-wave velocities in the Los Angeles basin, predicted relative potential for liquefaction-related ground failure (Fig. 21), and the fifty-percent probability limits of earthquake-induced coherent and disrupted landslides in susceptible materials.

Ziony and others (1985) caution that the maps and evaluations of predicted geologic and seismologic effects of the magnitude 6.5 earthquake on the Newport-Inglewood zone should not be used for site-specific purposes; the geologic information varies in completeness across the demonstration area, and testing of subsurface materials usually is required to reach conclusions about a site. Moreover, as is clear from this case history, the quality and completeness of the regional geologic information are important to its reliability.

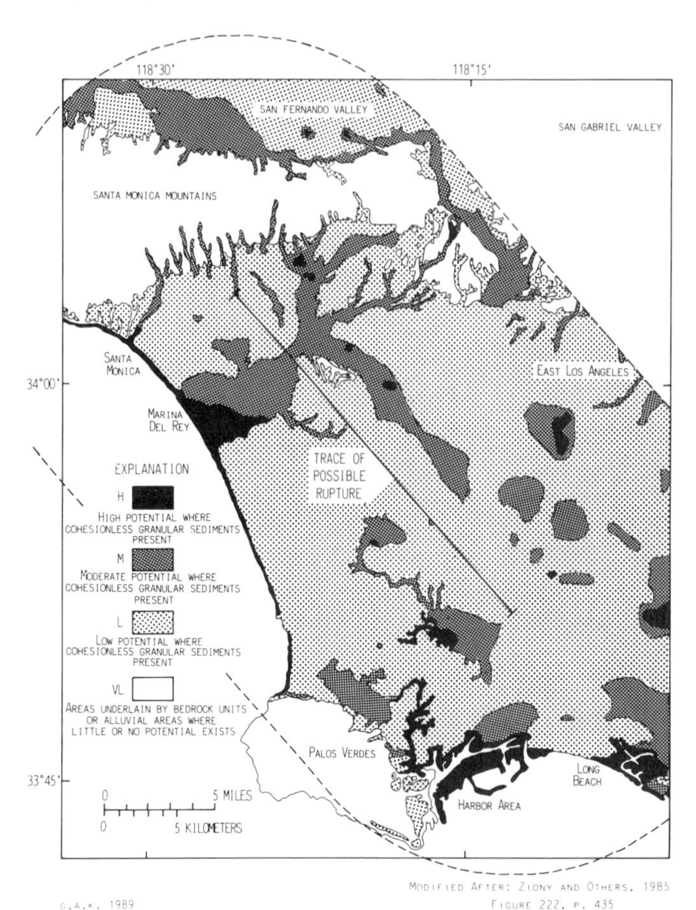

Figure 21. Predicted relative potential for liquefaction-related ground failure from a postulated magnitude 6.5 earthquake on the Newport-Inglewood zone in the Los Angeles Basin. (Modified from Ziony and others, 1985, Fig. 222.)

Underground disposal of high-level radioactive waste (1970s to 1990s)

The United States has the largest nuclear energy program in the free world, currently with 110 operable nuclear reactors and another 10 with construction permits. The safe disposal of high-level waste (HLW) has been under study since 1955; current plans are based on isolation of HLW in a geologic repository until such time that it is either rendered harmless to humans or can be retrieved for further processing. Current efforts, under the Office of Civilian Radioactive Waste Management Program, are directed toward developing a geologic repository in welded ash tuff at the Yucca Mountain site in Nevada (DOE, 1988; and Eriksson, 1989).

The screening process that led to the selection of Yucca Mountain for characterization started with regional assessments and location of candidate areas throughout the nation in 1977. The U.S. Geological Survey, after investigating the isolated area of the Nevada Test Site at Yucca Mountain, proposed (DOE, 1988, p. 11) the general region for further consideration as the HLW repository because (1) throughout southern Nevada, ground water does not discharge into rivers that flow to major bodies of surface water; (2) many of the rocks at the Nevada Test Site have geochemical characteristics that retard migration of radionuclides, (3) paths of ground-water flow between potential sites for a repository and the points of ground-water discharge are long, and (4) the rate of ground-water recharge is very low, and thus the amount of moving ground water is also low, especially in the unsaturated rock.

The nation's first geologic repository for disposal of commercial spent nuclear fuel and HLW from United States defense activities is planned for operation in the year 2003, if the Yucca Mountain site proves feasible and receives approval of the U.S. Nuclear Regulatory Commission (NRC) for a license to construct. Procedures for investigation of the HLW site, including appropriate scientific studies, are set forth in the National Environmental Policy Act and the Nuclear Waste Policy Amendments Acts of 1987, as specified in the Nuclear Waste Policy Act (NWPA) of 1982.

Since 1988 the Department of Energy has been conducting extensive tests at Yucca Mountain, Nevada. These site characterization studies, which involve both areal and regional investigations, will determine how well the site's geologic and hydrologic setting can isolate HLW from the human environment. Geologists, in association with geochemists and engineers, will determine if the rocks beneath Yucca Mountain are capable of isolating radioactive wastes, whether earthquakes or volcanic activity pose a threat to the underground repository, and whether potentially valuable natural resources exist within the site area. A critical factor will be whether the dry rocks of the site can prevent radionuclides from entering the ground-water reservoir that lies hundreds of feet below repository level, or if rainfall in the region or changes in the water table might affect repository operation. Many of the necessary investigations will require drilling of both shallow and deep boreholes to obtain rock or water samples, measuring the inherent rock pressures and physical properties, and monitoring the ground-water properties (Fig. 22).

An important physical feature of site characterization will be two vertical shafts and an exploratory shaft facility to investigate in detail the geologic and hydrologic conditions at Yucca Mountain. Underground test facilities will be constructed at the depth of the planned repository, 1,020 ft below the surface. One shaft will serve as an underground laboratory for detailed scientific studies, while the second will provide ventilation, construction access, and a safety exit. A series of rooms at repository depth will connect the two shafts so scientists can carry out experiments that simulate the environment of an operating repository, and evaluate such effects on the surrounding rock mass.

In addition to the site characterization investigations, environmental, socioeconomic, and transportation studies will be conducted to determine potential impacts of repository construction and operation on the surrounding region. These include such aspects as land use, water and air quality, potential for impact on threatened or endangered species, and the archaeological, cultural, and historical resources. An excellent overview of the potential problems, both technical and administrative, of the high-level waste disposal program in the United States has been prepared by Eriksson (1989). To protect and assure the health and safety of the public over the long term, multiple independent barriers, both natural and engineered, will be used in constructing the depository.

IMPACT OF REGIONAL FEATURES: ENGINEERING WORKS

Many case histories citing the importance of regional geologic features to the planning, design, construction or operation of engineering works are available (Hodge, 1935; Hewett and others, 1936; Kiersch and Hughes, 1952; Kiersch and others, 1955; Kiersch, 1975; Brown and Kockelman, 1983). Brief summaries of some typical cases follow that demonstrate how regional features not only can control or influence the site selection, design, and/or operation of an engineering works or complex, but conversely, can provide advantageous dimensions to the proposed site, route, or geologic reaction, and the changes induced by the works.

Downie slide, Columbia River

The Downie slide, an early Holocene rockslide, is located 44 mi (70 km) north of Revelstoke on the western slope of the Columbia River valley in British Columbia (Fig. 23). This critical geologic feature was not identified during the regional investigative phase for dams on the Columbia River, yet since its discovery in 1956 it has become critical to subsequently planned projects. The following brief review is derived from Piteau and others (1979), Brown and Psutka (1980), and W. I. Gardner (personal communication, 1982). The rockslide has a maximum thickness of 886 ft, a volume of 53×10^9ft^3 of rock and debris, and covers some 3.5 mi^2. The slide mass is bounded by a vertical-head escarpment of more than 410 ft, and lateral boundaries are a prominent, east-west–trending scarp on the south and a linear ridge on the northeast (Fig. 23). The slide toe forms the west bank of the Columbia River channel. The mass is 52 mi downstream from the proposed Downie dam.

The principal rocks involved are high-grade pelitic schists with foliation dips of 20° eastward toward the river channel. The region has undergone three distinct phases of deformation with prominent shears and fold phases; the slide mass broke along the hinge zone of a monoclinal flexural fold that formed a valley wall. The Columbia River fault zone, a major regional feature with a long history of movement, extends from south of Revelstoke to beyond Goldstream River (Fig. 23). Although initially believed inactive, the fault was reported to be active in 1977 with annual movements of an inch or more (W. I. Gardner, personal communication, 1982).

Regional geological investigations for the Revelstoke and Mica Dams and other potential reservoir sites throughout this stretch of the Columbia River did not recognize the existence of the Downie slide. Instead, a major fault in the river channel was believed to account for the marked difference in stratigraphic units on opposite sides. Actually, had systematic areal investigations been undertaken during the initial dam-site studies, the

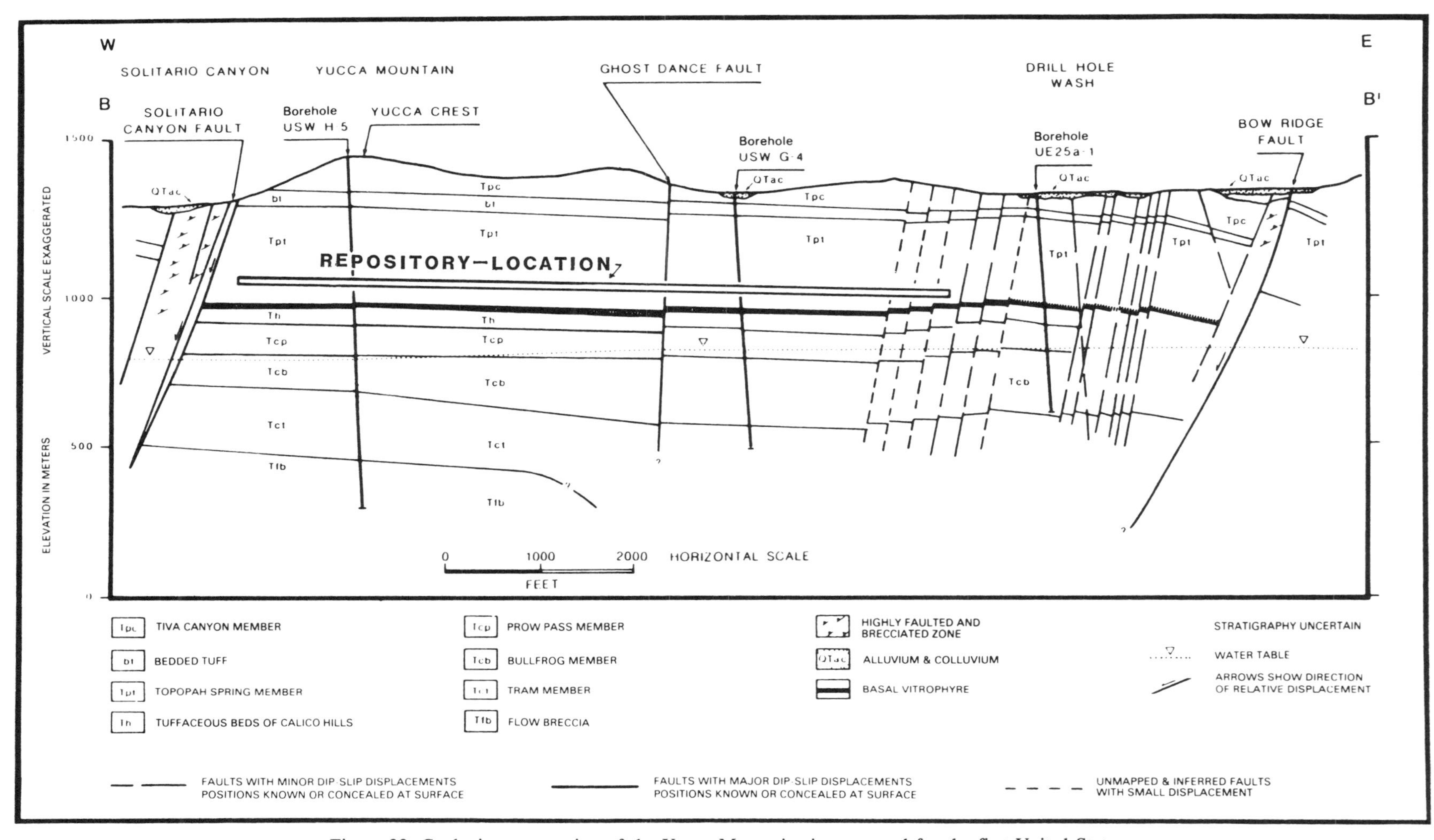

Figure 22. Geologic cross section of the Yucca Mountain site proposed for the first United States underground repository for high-level radioactive waste. (From DOE, 1988, Fig. 2–3.)

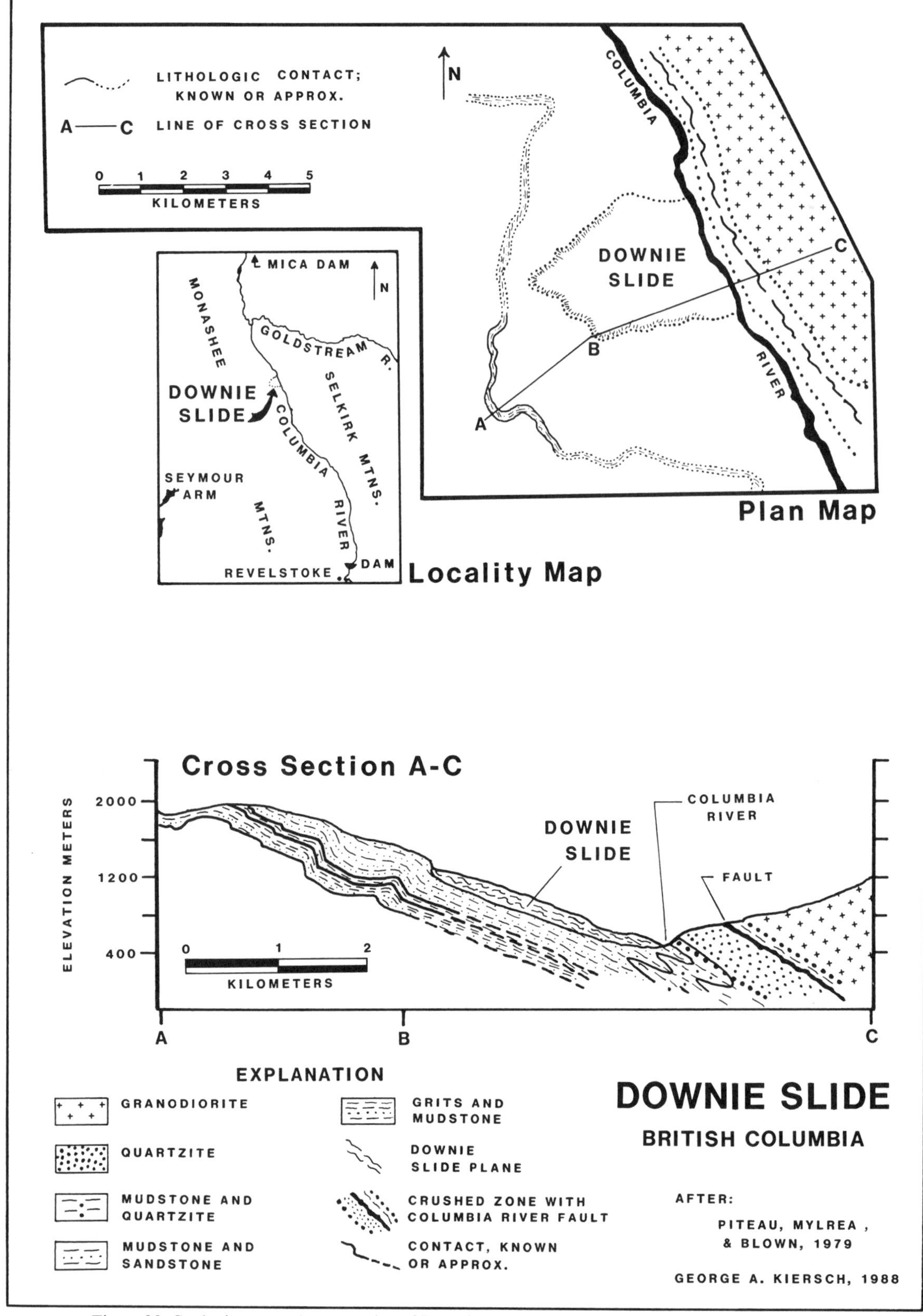

Figure 23. Geologic map and cross section of Downie slide and vicinity. Several phases of deformation/folding, which affected the rock units involved with the Downie slide, are indicated. (From Brown and Psutka, 1980.)

Downie slide area would have been identified; the normal rock sequence of metamorphic units crosses the river channel immediately upstream of the Downie slide mass (W. I. Gardner, personal communication, 1982). Moreover, the regional Columbia River fault zone does not border the toe of the slide mass, but is located east of the river on a terrace opposite the Downie slide (Fig. 23). Since it was recognized, the Downie slide has been the subject of extensive geological investigations. Should the slide mass fail suddenly and dam the river, the Revelstoke project downstream would be affected, as would—even more seriously—the proposed Mica project upstream and the Downie project immediately downstream.

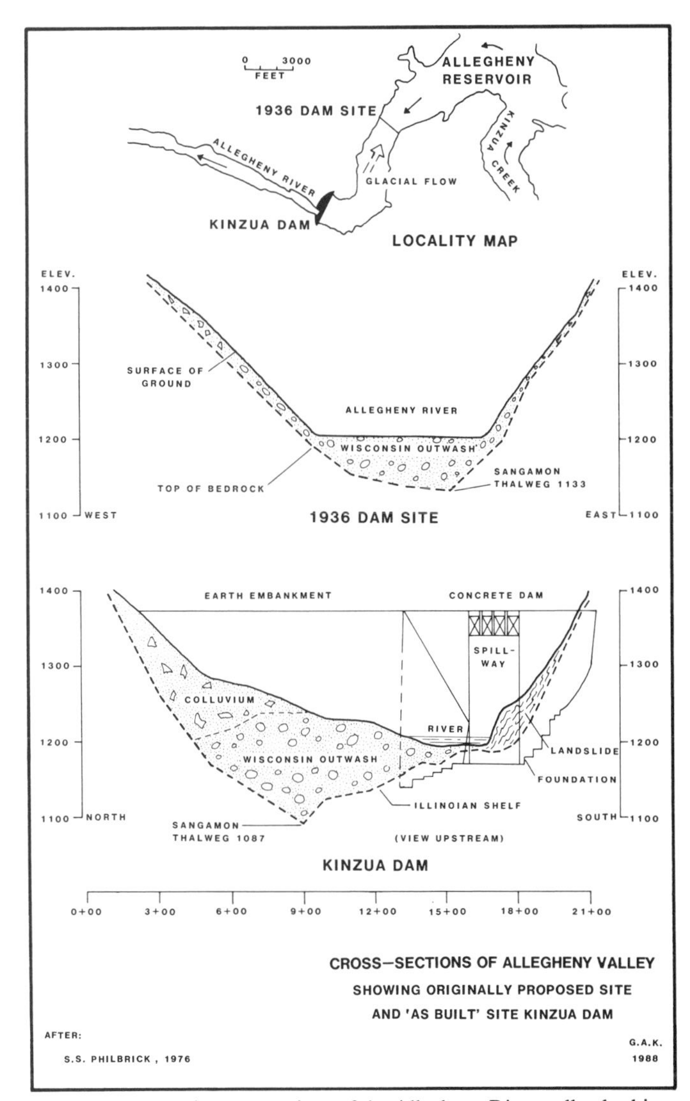

Figure 24. Geologic cross sections of the Allegheny River valley looking upstream, showing the geology at the originally proposed site for the Kinzua Dam in 1936 (above), and the more advantageous and less costly site where the dam was actually built in 1966, located after evaluating the regional bedrock features and ancestral drainage pattern. The location map shows the locations of the two sites and the ancestral and present flow directions of the river. (From Philbrick, 1976.)

Kinzua dam, Pennsylvania

The Kinzua dam on the Allegheny River near Warren, Pennsylvania, is an outstanding example of changes that occurred in both location of a dam and its design during the period of project studies (1936, and 1955 to 1966) due to regional geologic features. Recognition of the ancestral regional drainage and its impact on geologic events that affected the glacial history of the river system has been described by Philbrick (1976).

The "obvious" location for the proposed dam, as explored in 1936, was downstream from the confluence of Kinzua Creek and the Allegheny River, at the upper end of a gorge-like stretch of valley (Fig. 24). Bedrock occurs 75 ft below the streambed, and early designs assumed a solid concrete gravity dam. Delays and World War II suspensions terminated further consideration of the project until 1955, when studies resumed. Fortunately, the revised project included a regional investigation of the complex geology of the Allegheny Valley; attention was focused on surficial features and bedrock relations, so critical to siting a dam in an area changed by several glacial intervals and a complex ancestral drainage. Carll (1880) had noted in reports on oil and gas wells that the bedrock surface, beneath glacial cover, slopes northward—opposite to present topography and flow of the river. He deducted that the preglacial drainage was northward rather than southward as today. Following this concept, geophysical surveys and cored borings explored the river channel downstream of the "obvious" site and located a bedrock high beneath the channel cover near Big Bend. This relic of the bedrock, with a deeper interglacial channel on its margin, was 450 ft wide and occurred 37 ft higher in elevation than the bedrock channel at the 1936 site upstream (Fig. 24).

The 20-year delay in construction of the Kinzua dam, which enabled a systematic investigation of the relevant regional geologic features, became the basis for a major savings in project costs of $4 million in 1957 dollars. Furthermore, the concept for the eventual new and improved site location came from a pioneering geologist's regional observation and early oil wells, the database for a reexamination of the areal features and glacial history, with reevaluation by Philbrick (1976).

Vaiont Reservoir, Italy

The Vaiont disaster is considered a turning-point case by the engineering geology profession, because it dramatically demonstrates the consequences of both inadequate concepts and insufficient regional geological investigations before the dam was designed and built (see chapter 22, this volume). An ancient, large-scale slide plane or decollement (more than 10,000 years old) was not recognized and/or acknowledged during the site exploration phase, nor was conceived of as a potential contributor to massive slope instability (Fig. 25). Yet such features had occurred elsewhere in the region and were a warning of the complex, slide-prone terrain (Fig. 26). Consequently, when the higher-level reservoir water acted on the delicately balanced,

Figure 25. View of Vaiont reservoir site in 1961, before impoundment, from the crest of the dam. The spectacular V-shaped inner gorge is carved in limestones with interbeds of sandstones and clayey units. Note the exposure of the 1960 slide area (right foreground), the location of the Old Slide plane that was reactivated by 1963 sliding (dashed line), and the rugged features of the site. The 1963 slide mass jumped the reservoir area and moved up the opposite wall of the canyon as much as 460 ft (140 m), nearly to the upper road shown in the left center of the photo. Preconstruction investigations of 1959 to 1960 did not recognize either the Old Slide plane or the impact that reservoir water would have on the delicately balanced canyon wall. (Photo by Gerald T. McCarthy, 1961.)

steeply dipping complex of broken limestones, sandstone-shale, and clayey beds that occurred as a deformed valley-wall rock mass in 1961 to 1963, the uplift pressures induced movement. Combined with other inherent adverse geologic conditions, the creep of the valley slopes ultimately became a sudden, large-scale rapid movement, and a catastrophic collapse occurred. A full geological understanding of the pertinent regional features, such as solutioned limestone (Fig. 27), ancient slide planes, and clayey beds, during the planning phase of the project, and how they would be critical to an operating reservoir at the Vaiont site, would have alerted the design engineers to the serious hazards and risks. In all likelihood, remedial measures could have been incorporated in the design and operating phases that would have averted this tragedy. Instead the Vaiont disaster stands as a monument to inadequate preconstruction geological investigations, compounded by errors of geologic judgment in evaluating and interpreting the available field data.

Baldwin Hills dam, Los Angeles

The occurrence of a young regional fault system was a critical factor in the eventual failure of the Baldwin Hills dam (chapter 22, this volume). The faults were affected when the areal rock column was depressurized by the withdrawal of fluids and gas, which contributed to areal subsidence. The subsequent movement by this and other causes along faults traversing the site resulted in a rupture of the reservoir lining, leading to leakage and eventual piping beneath the dam. The water progressively breached the embankment and caused a dramatic failure.

Figure 26. View of a large-scale landslide that occurred downstream from the Vaiont area during Holocene time and blocked the Piave River, creating Lake Croce. The upstream flooding subsequently forced the river to carve a new channel through Belluno (see inset map, Chapter 22, Fig. 29, this volume), with the flow joining the ancestral channel upstream from Venice and continuing to the Adriatic Sea. The Faldo Power Station (left foreground) is an example of utilization of a natural dam for hydroelectric power generation. Premier Benito Mussolini ordered construction of this unit as his first public works project after being elected in the early 1920s. Diversion tunnels driven into the left wall of the canyon feed the penstock and station in the center foreground. (Photo by Kiersch, 1963.)

River locks and reservoirs, east-central United States

Three examples in which regional features influenced the siting and operation of engineering works were described by Fisher and Banks (1978); they concern dams, reservoirs, and river locks built in the greater Mississippi River drainage system by the U.S. Corps of Engineers between 1961 and 1973. Stress-relief structures occur throughout the canyon walls and channels of river valleys in the Allegheny Plateau region (Fig. 28); slope failures occur along glacially scoured trench valleys in clayey rocks of the Missouri River valley and affect dams and impoundments; and alluvial backfill deposits overlying bedrock at sites for locks along the lower Mississippi River valley affect the construction of levees, as occurred at the Port Allen Lock (Saucier, 1969). In this case the geological database obtained from regional mapping provided the guidance for initial planning and siting. Geological predictions were verified by detailed investigations during the design and construction stages.

Pa Mong reservoir, Mekong River

Extensively solutioned limestone terrain is widely considered too hazardous and not acceptable for large water impoundments and reservoirs. Several cases where the areal features and hazardous solutioned-limestone contributed to unacceptable leakage from a reservoir have been reviewed by James (1988). Whether the solution action has been extensive and beds are sufficiently dissolved, with openings and karstic features, is dependent on many factors, such as composition, geologic history, inherent structures, paleohydrology, and climate. The Big Bend reservoir sites on the Rio Grande, discussed earlier, confront the planner-designer with the hazard of excessive reservoir losses, while the high level of water loss at the Pa Mong reservoir site, discussed here, is a project asset.

Quite in contrast to the usual unacceptable high losses from a limestone reservoir is the estimated leakage via the solutioned Rat Buri Limestone at the proposed Pa Mong reservoir (triple the

Figure 27. Typical solution-widened openings in limestone beds exposed on canyon walls upstream from the Vaiont Dam, and within the area of the slide mass (the surface has been washed clean by waves in the reservoir). Karst features are exposed in displaced limestone beds near the old and the reactivated slide planes of 1963. (Photo by Kiersch, 1963.)

size of Lake Mead) on the Mekong River, Thailand-Laos (Gardner and others, 1968). The high water losses would mainly move southward via channelways and a network of openings in the limestones and reappear in other stream valleys (Fig. 29). This redistributed water would be recaptured in reservoirs built on adjacent drainages in the region, partly for this purpose. The 8 to 10 percent water loss through leakage from the Pa Mong reservoir via the network of solutioned openings in limestones thus would be advantageous for the project because the annual flow of the Mekong River is far greater than can be accommodated by a single reservoir at the Pa Mong site. Consequently, this major leakage serves to distribute excess water to other drainages in the region where additional reservoirs can serve the local needs for irrigation and power.

Ancient slides: Bearpaw Shale, Missouri River project

The dark gray to blackish, poorly-bedded Bearpaw Shale is widespread throughout the Missouri River region where it is the bedrock at the sites of many engineering works, such as the Fort Peck dam and power plants discussed in chapter 23, this volume. The clayey shale or claystone contains abundant seams of bentonite up to 2 ft thick, along with thin limestone beds and pyrite horizons. The firm shale weathers readily on exposure, producing a clay soil of high plasticity that extends to depths of 10 to 20 ft; less weathered shale sometimes is not encountered above depths of 50 ft (Fleming and others, 1970; Hamel and Spencer, 1984).

Early reconnaissance surveys in the 1930s for dam and powerhouse sites throughout the Missouri River system failed to identify the occurrence of ancient landslides throughout the Bearpaw Shale terrane, even though available aerial photographs showed potentially unstable slopes and old features that have since been recognized as slides (Fig. 30). Furthermore, the existence and implication of these ancient landslides were not recognized when the Fort Peck Dam and reservoir outlet works were designed. Construction excavations undercut the unstable slopes in 1934, causing progressive failure of colluvium on slopes that were not finally stabilized until 1974 (Hamel and Spencer, 1984). Such unstable masses should be recognized during the systematic regional-areal investigative phases (Fig. 1) and prior to the design and construction phases.

U.S. Air Force Academy site, Colorado

The U.S. Air Force Academy, located about 11 mi north of Colorado Springs, Colorado, includes 28 mi^2 miles of the Great Plains east of the Rampart Range. The area chosen for the Academy and new small community was not heavily populated or well developed, yet it contained major rail and highway routes (Fig. 31). Ideally, the planning phase could proceed without land-use restrictions or problems with prior municipal works, and with complete freedom to anticipate future needs. The best solution for the problems common to municipal planning took into account the topography and the physical nature and geology of the site. Each phase of planning and investigation was done, therefore, in logical sequence prior to the design phase.

Geologic studies of the Air Force site were done in October and November 1954. A descriptive text and geologic map (1:12,000) of the site were prepared by Varnes and Scott (1967) along with statements on water supplies, local aggregate supplies, and availability of building stone provided by the architect-engineer in December 1954. During the following year, many water wells were completed, and U.S. Geological Survey hydrologists performed pumping tests and evaluated the potential yield of both the Dawson Arkose and the deeper Fox Hills Sandstone. The following brief summary of the areal geologic studies for planning of the Academy is derived from Varnes and Scott (1967).

The predominant bedrock within the Academy area is Dawson Arkose of Cretaceous and Paleocene age, which consists of coarse arkosic sandstone and interbedded lenticular siltstone and clay. Several stages of downcutting and alluviation have produced gravel-covered bedrock surfaces at three levels. Remnants of these pediments that trend eastward from the mountain front form narrow fingerlike mesas at two levels and broader valleys at the third and youngest level (Fig. 31).

In planning the Academy, the topographic and geologic environments were considered so that the required facilities would be arranged attractively and efficiently. Topography was a dominant influence on the early planning stages. Geology entered to the degree that distribution of rock types and their structure had influenced the development of erosional and depositional landforms. All the geologic units upon which the Academy facilities are constructed are predominantly granular in texture, except

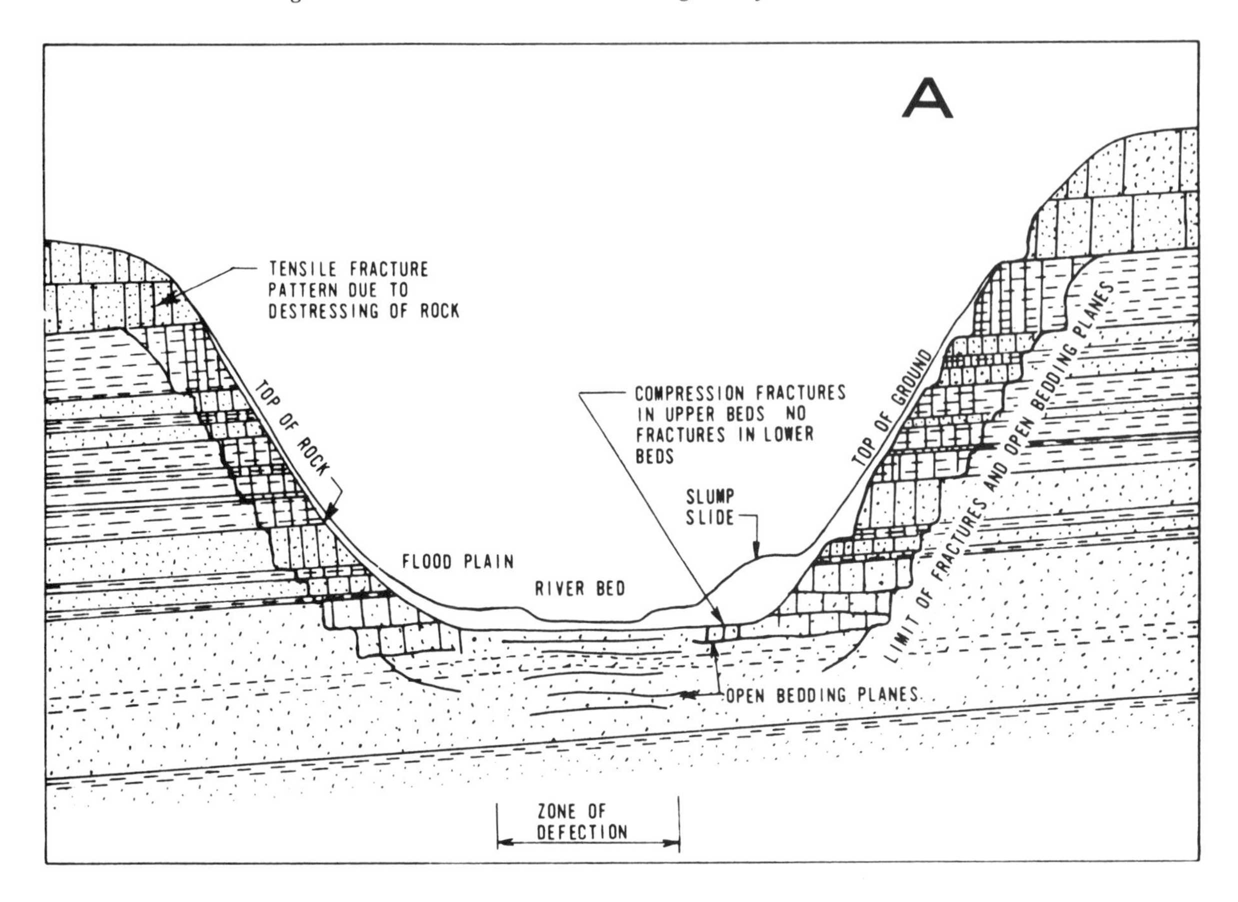

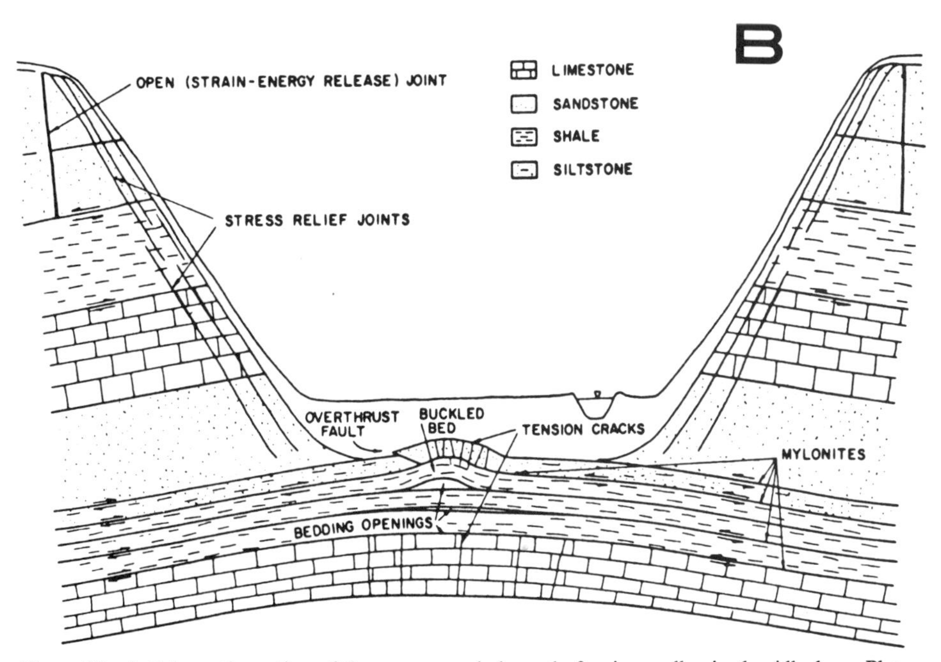

Figure 28. A. Schematic section of the canyon and channel of a river valley in the Allegheny Plateau region. The occurrence and distribution of stress-relief structures so common to the sandstone and limestone beds are shown. B. Schematic section of a river channel carved in a bedded rock sequence, showing the common structural features that are formed in association with the valley down-cutting. (Modified from Fisher and Banks, 1978.)

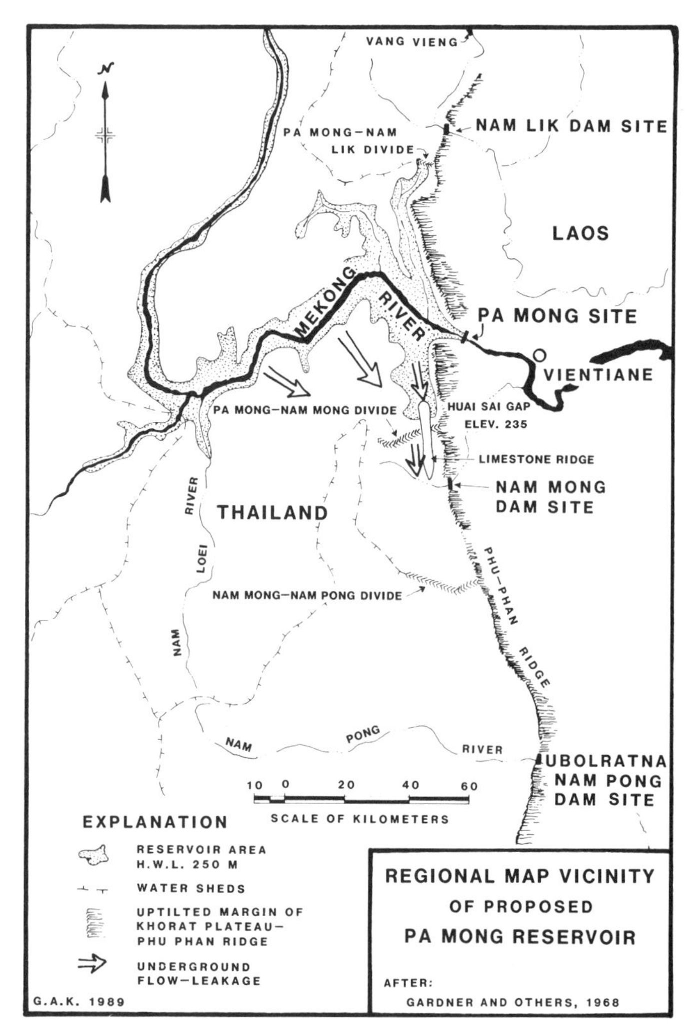

Figure 29. Regional map in the vicinity of the proposed Pa Mong Dam and reservoir, Thailand-Laos. Predicted leakage southward from the reservoir via the karstic limestone terrane would be recaptured in other drainages and reservoirs as indicated. (From Gardner and others, 1968.)

silty and clayey sections in the Dawson Arkose and peaty facies of the Husted alluvium. Hence, no very serious engineering geologic problems were encountered during construction. An airfield was necessarily located in the more nearly level southeastern part of the site (Fig. 31). The large buildings of the cadet and academic areas were placed on mesas toward the north and near the western border. The gravel cover on this mesa also allowed a large amount of grading to be done economically, and the underlying Dawson Arkose furnished a suitable foundation for load-bearing caissons. Valleys in the west-central part of the site were used for two large housing developments, and the mesas between them for community facilities. The whole Academy is served by a peripheral highway and an interior network of roads.

The need to balance cuts and fills—particularly along roads in the academic areas, at dams and reservoirs, and at the airfield—required that thorough exploration and classification of the geologic materials be made so that they could be used with a minimum of hauling and waste. The architect-engineer carried out two extensive test programs: experimental fills at the airfield, and a caisson load test program in the academic area. Prior appropriation of surface-water rights required that irrigation water be derived from underground sources; the two principal aquifers are sandstone beds of the Fox Hills Sandstone and Laramie Formation, which are considered as a single water-bearing unit (L-F aquifer) at depths of 1,493 ft, and the Dawson Arkose at depths of 826 ft.

OVERVIEW

Common factors contributed by regional-areal studies

Significant geological data collected throughout the regions of candidate sites can be utilized to advantage in evaluating conditions at the subsequently selected preferred and specific sites. The relatively high cost of cored borings and subsurface openings or cuts generally restricts their use to near the preferred site. The understanding of the rock units and historical events gained from regional investigations often provides a perspective that enables optimum use of such relatively expensive exploration techniques.

Much significant information and insight can be determined from regional-areal investigations (Kiersch, 1962, 1988) that will strengthen the preferred and site-specific investigations. Suggested actions include:

- *Rock sequence/column.* Subdivide rock masses or bodies on the basis of their physical properties and their responsiveness to engineering needs; extrapolate concealed lithologies; estimate or establish boundaries; forecast expected site conditions/characteristics.
- *Rock properties and stability.* Determine pattern, cause, and age of structural features; rock-soil weathering and alteration; slope stability and downslope movement; susceptibility to solution action, cavitation, and erosion; subsidence activity (shallow and/or deep) and extent.
- *Geologic phenomena.* Determine distribution, cause, rate, and age of areal subsidence, downslope movement, weathering processes, sediment-filling, inherent structure, and seismic events.
- *Underground fluids.* Determine characteristics of alluvial-filled basins, skeletal pattern within fractured rock mass, and ground-water occurrence and controls.
- *Geomorphic history, near-surface conditions.* Evaluate surficial sediments for origin, properties, and correlation with available laboratory test data; determine extent and characteristics of special features, such as caliche horizons, lateritic soils, glacial deposits, organic soils, and muck; determine origin of construction material deposits; evaluate correlation with paleoflood evidence.
- *Tectonic geomorphology.* Determine morphology of escarpments and mountain fronts, indices and landform assemblages to evaluate active tectonic processes, rates of geomorphic processes, response of alluvial rivers to active tectonics, paleo-

Figure 30. Aerial view of slide-prone Bearpaw Shale terrain showing ancient slides encountered in the areas of the Fort Peck dam and powerhouses. These were not recognized during predesign stages and became a source of serious maintenance problems. The reactivated slide no. 4 of September 21, 1934 is in the foreground. (From Hamel and Spencer, 1984, Fig. 5.)

faults, fault-scarp morphology, paleoseismology and slip rates (Wallace, 1986).

• *Surface fluids—flooding.* Evaluate paleoflood indicators relative to the nature, magnitude, risk, and consequences of extraordinary flood phenomena (Baker, 1987).

Geologic factors affecting economic parameters include legal counsel and contractural claims, land management, classification of excavation, and possible impact of some unit-bid construction costs.

Design-related engineering characteristics of natural materials include:

• *Excavation characteristics.* Establish classification of potential excavation materials for subdividing into rock, common (soil) or unclassified; evaluate drilling, blasting, and fragmentation properties; estimate broad costs and powder consumption.

• *Cut-slope characteristics.* Cut-slopes converge rapidly whether on a 1:1 or 1½:1 slope angle, with most yardage in the upper half of the excavation; costs to remove the toe of the slope can be up to ten times the unit cost for other yardage.

• *Construction materials.* Locate adequate quantities of acceptable materials for construction of works; evaluate service record of specific materials, including occurrence of deleterious minerals in aggregates.

Drainage pattern—climatic effects are a critical factor in highway design and maintenance; sediment load and scouring pattern control design of crossings, maintenance of channels, bridge piers, and embankments.

Conclusions

Gone is the attitude of yesteryear when a geologist visited a new region in the manner of "I went, I saw, I conquered." Today the applied geologist's actions involve assembling a systematic

Figure 31. Oblique aerial photograph of the Air Force Academy site prior to construction, looking southward toward Colorado Springs. Most of the large installations are in the belt of east-trending mesas and valleys between the foot of the Rampart Range and Monument Creek. Lehman Mesa was chosen for the cadet academic area. (From Varnes and Scott, 1967, Fig. 2.)

interpretation of a site or region based on a "team" compilation of data from a variety of scientific techniques through the integrated efforts of specialists. Although many ways have been noted to improve the success of a regional geologic reconnaissance survey, it should be remembered there are at least as many common pitfalls.

A regional survey should map and/or investigate the entire area in some degree, although the investigative sponsor may be seriously interested in only parts of the region. The maximum effective geological projection, with selection of the strongest candidate areas, is gained from an integrated interpretation of all relevant data collected by various geoscientific techniques.

APPENDIX A. PRINCIPAL REGIONAL AND AREAL GEOLOGY ELEMENTS REQUIRED BY NUCLEAR REGULATORY COMMISSION PUBLIC SAFETY ANALYSIS REPORTS (PSAR) FOR CONSTRUCTION AND LICENSING OF NUCLEAR POWER PLANT FACILITIES—1970

REGIONAL GEOLOGICAL ELEMENTS

Hazards of significance to site/area

Briefly discuss geologic, seismic, and man-made hazards of significance to the safety of the site. If no hazards are known or inferred within a 200-mi radius of the site, indicate and repeat statement at end of each successive section (below) that deals with regional physiography, surficial geology, bedrock geology, regional tectonics, and history.

Physiography and geomorphology

State the physiographic section and province in which the site is located, and describe the essential physiographic characteristics of that area.

Describe the physiography of surface features of the 200-mi region around the site.

Discuss the geomorphology of the 200-mi region with interpretation of origin or historical development of present landforms and their relationships to underlying geologic materials or structures.

Discuss or deny geologic, seismic, or man-made hazards of safety-significance to the site as reflected in this regional physiography.

Surficial geology

Describe the general characteristics of the surficial deposits occurring throughout the 200-mi region of the site.

Describe briefly the general characteristics of the surficial geology in the 5-mi area of the site.

Discuss or deny geologic, seismic, or man-made hazards of safety-significance to the site as reflected in the regional surficial geology.

Bedrock geology

Describe generally the bedrock geology of the 200-mi region surrounding the site; include lithologies, stratigraphy, and structures of the bedrock and whether these features might logically extend from geologic province to province as outlined on the regional tectonic map.

Further define the relationship between geologic provinces by discussion of regional geologic profiles relative to the areal bedrock geology; utilize areal geologic profiles and stratigraphic sections.

Briefly describe the areal bedrock geology and reference to site geology.

Discuss or deny geologic, seismic, or man-made hazards of safety-significance to the site as reflected in the regional bedrock geology.

Tectonics

Identify and describe the essential tectonic elements of the 200-mi region surrounding the site; include systematic discussion on the faults, folds, domes, basins, intrusives, and any other tectonic elements that can be controlled geographically on a geologic province-to-province basis.

Each tectonic element should be defined as to location, regional extent, nature or style of deformation, and reported or inferred age of deformation. Use of special maps such as gravity or magnetic anomaly maps may aid dicussion.

Discuss or deny geologic, seismic, or man-made hazards of safety-significance to the site that are reflected in the regional tectonic elements.

Geologic history

Discuss regional geologic history from earliest applicable (Precambrian) to Recent, and include what has happened in the region in the way of sedimentation, erosion, intrusive activity, crustal adjustments (epeirogenetic), orogenesis, and faulting, separately for each relevant time period, or group them where little is known. Use available radiometric dates to define ages of such elements as folding, faulting, and orogenesis.

This section should provide a time-framework of the crustal geologic processes in order to show whether bedrock structures, of possible safety significance to the site, are considered old and dead, or young and capable of rejuvenation to the jeopardy of the plant facilities.

Discuss or deny historical geologic implications of safety-significance to the site.

AREAL GEOLOGICAL ELEMENTS

Site physiography

Describe the areal physiography and local landforms, and relationship between regional and site physiography.

Discuss the historical development of the areal landforms with reference to the site topographic map.

Lithologies and stratigraphy

Define the stratigraphic column and describe the physical characteristics, degree of consolidation, thicknesses, and origin of the rock types and overburden.

Define the relationship of the local stratigraphy to that of the region and reference logs of borings, test pits, and trenches.

Areal-site structural geology

Describe the bedrock structure in the vicinity of the site in terms of distribution bedrock outcrops, surface contours, structural features, intrusives, etc.

Define and substantiate any differences between published and site-generated interpretations of geologic structure in area.

Areal-site geologic history

Discuss the local geologic history and relate it to regional geologic history.

Areal-site ground-water conditions

Briefly describe local ground-water occurrence.

Areal-site engineering geology

Describe the geologic conditions underlying all Category I structures such as dams, dikes, and pipelines, under the following headings: Shears, slickenslides, Joints, fractures; Foliation, cleavage, fissility; Folds; Faults; Alteration, weathering; Unrelieved residual stress; Unstable lithologies or soils, Cavernous lithologies, Collapse structures; Effects of man's activities; Dynamic behavior during prior earthquakes; and other.

APPENDIX B. PRINCIPAL ELEMENTS REQUIRED ON REGIONAL AND AREAL GEOLOGY MAPS BY NUCLEAR REGULATORY COMMISSION PUBLIC SAFETY ANALYSIS REPORTS (PSAR) FOR CONSTRUCTION-LICENSING NUCLEAR POWER PLANT FACILITIES–1970s

REGIONAL GEOLOGY–MAPS AND DRAWINGS

Physiographic map 200-mi radius. Utilize physiographic or topographic base with an overlay of physiographic province or section boundaries and their names. Site location (center of map) is defined as "Site."

Surficial geology 200-mi radius. Prepare a generalized map, based on Glacial Map of United States (Canada), United States Glacial or Soils Map. Locate "Site."

Bedrock geology 200-mi radius. Prepare a standard geologic map, with rocks subdivided according to age: sedimentary, metamorphic, plutonic rocks differentiated: regional faults shown. Locate "Site."

Geologic profiles and stratigraphy. Use suitable scales; include a regional geologic profile; an areal (5 to 20 mi) geologic profile, and graphic stratigraphic section(s). Locate "Site."

Regional tectonic map with 200-mi radius. Define geologic provinces; show all faults, folds, domes, basins, and intrusives (youngest defined separately). Locate "Site."

Regional gravity anomalies 200-mi radius or the appropriate region. Show geologic provinces, regional faults, and relevant features as under- or overlay. Locate "Site."

Regional magnetic anomalies. If useful data available, use any appropriate regional radius. Locate "Site."

Special-purpose map(s). One or more regional maps useful to regional geologic or tectonic discussions, such as ERTS, U-2, SPACELAB, areal-sensory, or other photo impressions; correlate with regional oil or coal fields, karst terrane, etc. Locate "Site."

AREAL AND SITE GEOLOGY–MAPS AND DRAWINGS

Topography of local area. Variable scale: 5- to 20-mi radius. Use U.S. Geological Survey topographic base showing site-area topography. Locate "Site."

Site topography. Scale 1 inch = 200, 500, or 1000 ft, depending on the presence of such features as landslides, subsidence, caverns, karst, mines, and oil fields. Map shows landforms in site area; outline principal plant facilities of the "Site."

Site stratigraphic column. Scale as appropriate; show overburden and bedrock stratigraphic columns at the site, based on trenches, test pits and borings.

Bedrock geology. Variable 5- to 20-mi radius. Standard geologic map. Show mines, wells, caverns, etc. Locate "Site."

Surficial geology 5-mi (8 km) radius. Show surficial deposits and classification of units of surrounding area. Located "Site" and plant facilities.

Site structural geology. Scale 1 inch = 200 to 500 ft as appropriate; show bedrock contours, outcrop areas, structural symbols, geologic formation patterns, folds, faults, deformational zones, etc. "Site" plant facilities: Category I structures in heavier lines. May require separate figure to define features of specific engineering or safety significance.

REFERENCES CITED

Arnold, A. A., 1977, Geologic mapping of dam site, *in* Golze, A. R., ed., Handbook of dam engineering: New York, Van Nostrand Reinhold Co., p. 204–207.

Atwater, B., Hedel, C. W., and Helley, E. J., 1977, Late Quaternary depositional history, Holocene sea-level changes, and vertical crustal movement, southern San Francisco Bay, California: U.S. Geological Survey Professional Paper 1014, 15 p.

Atwater, B., and 5 others, 1979, History, landforms, and vegetation of the estuary's tidal marshes, *in* Conomos, T. J., ed., San Francisco Bay; The urbanized estuary: San Francisco, California Academy of Sciences, p. 347–385.

Baker, V. R., 1987, Paleoflood and extraordinary flood events: Journal of Hydrology, v. 96, p. 79–99.

Bonilla, M. G., 1960, Landslides in the San Francisco South Quadrangle, California: U.S. Geological Survey Open-File Report.

Borcherdt, R.D., ed., 1975, Studies for seismic zonation San Francisco Bay region: U.S. Geological Survey Professional Paper 941-A, p. A1–A102.

Borcherdt, R.D., Gibbs, J. F., and Fumal, T. E., 1979, Progress on ground motion predictions for the San Francisco Bay region, California, *in* Brabb, E. E., ed., Progress on seismic zonation in the San Francisco Bay region: U.S. Geological Survey Circular 807, p. 26–36.

Brabb, E. E., ed., 1979, Progress on seismic zonation in San Francisco Bay region: U.S. Geological Survey Circular 807, 91 p.

Brabb, E. E., Pampeyan, E. H., and Bonilla, M. G., 1972, Landslide susceptibility in San Mateo County, California: U.S. Geological Survey Miscellaneous Field Studies Map MF-360, scale 1:62,500.

Brown, R. D., and Kockelman, W. J., 1983, Geologic principles for prudent land use; A decisionmaker's guide for San Francisco Bay region: U.S. Geological Survey Professional Paper 946, 97 p.

Brown, R. L., and Psutka, J. F., 1980, Structural and stratigraphic setting for the Downie slide, Columbia River Valley, British Columbia: Canadian Journal of Earth Sciences, v. 17, no. 6, p. 698–709.

Carll, J. F., 1880, The geology of the oil region of Warren, Venango, Clarion, and Butler Counties: Second Geological Survey of Pennsylvania Report of Progress 1875–79, p. 352–353.

Cederborg, E. A., and Tosetti, R. J., 1982, Site selection for low-level radioactive waste disposal sites: University of Arizona and U.S. Department of Energy Waste Management, v. 2, p. 215–233.

Cooper, D. F., and Chapman, C. B., 1987, Risk analysis for large projects, models, methods, and cases: New York, John Wiley and Sons, 160 p.

Coopersmith, K. J., and Young, R. R., 1989, Addressing geological uncertainties in seismic hazard [abs.], *in* Common geologic hazards; Natural and man-induced: Washington, D.C., 28th International Geological Congress, Abstracts Volume, p. 1-325-326.

Davis, J. F., and 5 others, 1982, Earthquake planning scenario for a magnitude 8.3 earthquake on the San Andreas fault in southern California: California Division of Mines and Geology Special Publication 60, 128 p.

DeFord, P. V., 1954, Southern Pacific outlying lands and railroad rights of way acquired by congressional grant: Southern Pacific Corporation Legal Department, 36 p.

DOE, 1988, Overview, site characterization plan, Yucca Mountain site, Nevada Research and Development Area, Nevada: U.S. Department of Energy Office of Civilian Radioactive Waste Management, 150 p.

Dutton, C. E., 1889, The Charleston earthquake of August 31, 1886: U.S. Geological Survey Annual Report 9, p. 203–528.

Elliott, W. J., 1978, The role of geology in California nuclear power plant siting [abs.], *in* Association of Engineering Geologists Annual Program: Association of Engineering Geologists, p. 12.

Eriksson, L. G., 1989, Underground disposal of high-level radioactive waste in the United States of America; Program review: International Association of Engineering Geologists Bulletin, no. 39, p. 25–51.

Everenden, J. F., and Thompson, J. M., 1985, Predicting seismic intensities, *in* Ziony, J. I., ed., Evaluating earthquake hazards in the Los Angeles region; An earth-science perspective: U.S. Geological Survey Professional Paper 1360, p. 151–202.

Fisher, P. R., and Banks, D. C., 1978, Influence of the regional geologic setting on site geologic features, *in* Dowding, C. H., ed., Site characterization and exploration; Proceedings of a Workshop at Northwest University: American Society of Civil Engineers Geotechnical Division, p. 163–185.

Fleming, R. W., Spencer, G. S., and Banks, D. C., 1970, Emperical study of behavior of clay shale slopes: Livermore, California, U.S. Army Engineering Cratering Group NCG Technical Report 15, vols. 1 and 2.

Gardner, W. I., Kiersch, G. A., Moneymaker, B. C., and Waggoner, E. B., 1968, Report of Board of Geological Consultants on Pa Mong Reservoir, Pa Mong Project, Lower Mekong River Basin, Thailand/Laos: U.S. Bureau of Reclamation/AID, 22 p.

Golder Associates and Weston, R. F., 1987, Site selection handbook: Idaho Falls, EG&G Idaho, Inc., National Low-Level Waste Management Program Report DOE/LLW-64T, various pagination.

Hamel, J. V., and Spencer, G. S., 1984, Powerhouse slope behavior, Fort Peck Dam, Montana, *in* Prakash, S., ed., Proceedings, International Conference on Case Histories in Geotechnical Engineering, St. Louis, Missouri: New York, American Society of Civil Engineers, v. 2, p. 541–551.

Hannan, D. L., 1984, Geotechnical mapping symbols (GEMS): Association of Engineering Geologists Bulletin, v. 21, no. 3, p. 343–344.

Harshbarger, J. W., Repenning, C. A., and Irwin, J. H., 1957, Stratigraphy of the uppermost Triassic and Jurassic rocks of the Navajo country: U.S. Geological Survey Professional Paper 291, 74 p.

Helley, E. J., LaJoie, K. R., Spangle, W. E., and Blair, M. L., 1979, Flatlands deposits of the San Francisco Bay region, California: Their geology and engineering properties, and their importance to comprehensive planning: U.S. Geological Survey Professional Paper 943, 88 p.

Henry, R. S., 1945, The railroad land grant legend in American history texts: Mississippi Valley Review, v. 32, no. 2, p. 171–194.

Hewett, D. F., and 5 others, 1936, Mineral resources of the region around Boulder Dam: U.S. Geological Survey Bulletin 871, 183 p.

Hodge, E. T., 1935, Industrial mineral development potential Bonneville Dam Project, Columbia River: Portland, Oregon, U.S. Army Corps of Engineers, 125 p.

Huntington Aerosurveys, 1961, Operation overthrust, Canada: London/Elstree, Wyoming, Borehamwood, Huntington Aerosurveys, Ltd., no. 21, 8 p.

James, L. B., 1988, Reservoirs, *in* Jansen, R. B., ed., Advanced dam engineering for design, construction, and rehabilitation: New York, Van Nostrand Reinhold, p. 722–729.

Jansen, R. B., ed., 1988, Advanced dam engineering for design, construction, and rehabilitation: New York, Van Nostrand Reinhold, 797 p.

Jennings, C. W., compiler, 1969, Geologic atlas and map of California; Olaf P. Jenkins edition 1958–1969: California Division of Mines and Geology, scale 1:250,000.

Kachadoorian, R., 1956, Engineering geology of the Warford Mesa Subdivision, Orinda, California: U.S. Geological Survey Open-File Report, 13 p., scale 1:2,400.

Keaton, J. R., 1984, Genesis-lithology-qualifier (GLQ) system of engineering geology mapping symbols: Association of Engineering Geologists Bulletin, v. 21, no. 3, p. 355–364.

Kiersch, G. A., 1958, Regional mapping program of Southern Pacific Company: Geological Society of America Bulletin, v. 69, p. 1691.

——, 1959, Handbook for geologists, engineers, and draftsmen: San Francisco, California, Southern Pacific Company Land Department, 174 p.

——, 1962, Regional/areal geologic investigations in highway geology, *in* Scott, L. W., ed., Proceedings, 13th Annual Highway Geology Symposium: Phoenix, Arizona Highway Department, p. 121–160.

——, 1964, Trends in engineering geology in the United States, *in* Rock mechanics and engineering geology, supplement 1: Wien, Springer-Verlag, p. 31–57.

——, 1975, Geological reconnaissance and evaluation of 22 preferred sites, New York Sites; New site selection study: Boston, United Engineers Constructors; Binghamton, and New York State Electric and Gas, 195 p.

——, 1988, Geological considerations, *in* Jansen, ed., Advanced dam engineering for design, construction, and rehabilitation: New York, Van Nostrand Reinhold, p. 106–117.

Kiersch, G. A., and Hughes, P. W., 1952, Structural localization of ground water in limestones; Big Bend District, Texas-Mexico: Economic Geology, v. 47, no. 8, p. 794–806.

Kiersch, G. A., and others, 1955, Mineral resources, Navajo-Hopi Indian Reservations, Arizona-Utah, 3 volumes: Tuscon, University of Arizona Press, v. 1, 72 p.; vol. 2, 98 p.; v. 3, 76 p.

King, C., 1880, First annual report of the United States Geological Survey, 79 p.

Laird, R. T., and 7 others, 1979, Quantitative land-capability analysis; Selected examples from San Francisco Bay region, California: U.S. Geological Survey Professional Paper 945, 115 p.

Little, A. D., Inc., 1975, An evaluation of the San Francisco Bay region environmental and resources planning study: U.S. Department of Housing and Urban Development Office of Policy Development and Research, 93 p.

Mabey, D., and Youd, T. L., 1989, Probability liquefaction severity index map of Utah: Salt Lake City, Utah Geological and Mineral Survey, Special Map, Hazards.

Merrill, G. P., 1897, Road materials and road building in New York: New York State Museum Bulletin, v. 40, no. 17, 134 p.

MINEsoft, 1982, TECHBASE; Engineering software for solving real-world problems: Denver, Colorado MINEsoft, Lts., 8 p.

Montague, S. S., 1865, Reports of the president and chief engineer, upon recent surveys, progress of construction, and estimated revenue of the Central Pacific Railroad of California: 22 p.

Nelson, C. V., 1989, Surface fault rupture and liquefaction potential; Special study areas, Salt Lake County, Utah: Salt Lake County Public Works Planning, 6 maps, various scales.

Newman, E. B., Puradis, A. R., and Brabb, E. E., 1978, Feasibility and cost of using a computer to prepare landslide susceptibility maps of the San Francisco Bay region, California: U.S. Geological Survey Bulletin 1443, 27 p.

Nichols, D. R., 1982, Application of earth sciences to land-use problems in the United States, with emphasis on the role of the U.S. Geological Survey, *in* Whitmore, D. F., and Williams, M. E., eds., Resources for the twenty-first century; Proceedings of an International Centennial Symposium of the U.S. Geological Survey: U.S. Geological Survey Professional Paper 1193, p. 283–291.

Nilsen, T. H., Wright, R. H., Vlastic, T. C., and Spangle, W. W., 1979, Relative slope stability and land-use planning in the San Francisco Bay region, California: U.S. Geological Survey Professional Paper 944, 96 p.

Peirce, H. W., 1955, Dimension stone, *in* Kiersch, G. A., Haff, J. G., and Peirce, H. W., Mineral resources, Navajo-Hopi Reservations, Arizona-Utah: Tucson, University of Arizona Press, v. 3, p. 60–74.

Philbrick, S. S., 1976, Kinzua dam and the glacial foreland, *in* Coates, D. R., ed., Geomorphology and engineering: Stroudsburg, Pennsylvania, Dowden,

Hutchison, and Ross, p. 175–197.
Piteau, D. R., Mylrea, F. H., and Blown, I. G., 1979, Downie slide, Columbia, British Columbia, Canada, *in* Voight, B., ed., Rockslides and avalanches: Amsterdam, Elsevier Publishing Co., v. 1, p. 365–392.
Plafker, G., and Galloway, J. P., 1989, Lessons learned from the Loma Prieta, California, earthquake of October 17, 1989: U.S. Geological Survey Circular 1045, 48 p.
Powell, J. W., 1890, Irrigation survey, first annual report, *in* Tenth Annual Report of the U.S. Geological Survey 1888–89: U.S. Geological Survey, p. 1–65.
Rabbitt, M. C., 1979a, A brief history of the U.S. Geological Survey, 2nd ed.: U.S. Geological Survey, 48 p.
—— , 1979b, Minerals, lands, and geology for the common defence and general welfare; Volume 1, Before 1897: U.S. Geological Survey, 331 p.
Radbruch-Hall, D., 1957, Areal and engineering geology of Oakland West Quadrangle, California: U.S. Geological Survey Miscellaneous Investigations Map I-239, scale 1:24,000.
—— , 1969, Areal and engineering geology of the Oakland East Quadrangle, California: U.S. Geological Survey Geologic Quadrangle Map GQ-769, 15 p., scale 1:24,000.
—— , 1987, The role of engineering-geologic factors in the early settlement and expansion of the conterminous United States: International Association for Engineering Geology Bulletin, no. 35, p. 9–30.
Rhoades, R., 1950, Influence of sedimentation on concrete aggregate, *in* Trask, P. D., ed., Applied sedimentation: New York, John Wiley and Sons, p. 437–463.
Rhoades, R., and Sinacori, M. N., 1941, Pattern of groundwater flow and solution: Journal of Geology, v. 49, no. 8, p. 785–794.
Sanborn, J. F., 1950, Engineering geology in the design and construction of tunnels, *in* Paige, S., ed., Application of geology for engineering practice: Geological Society of America Berkey Volume, p. 46–81.
Saucier, R. T., 1969, Geological investigation of the Mississippi River area, Artonish to Donaldsonville, Louisiana: U.S. Army Waterways Experiment Station Technical Report S-69-4.
Schlocker, J., 1974, Geology of the San Francisco North Quadrangle, California: U.S. Geological Survey Professional Paper 782, 109 p.
Schowengerdt, R. A., and Glass, C. E., 1983, Digitally processed topographic data for regional tectonic evaluations: Geological Society of America Bulletin, v. 94, p. 549–556.
Shaler, N. S., 1895a, Preliminary report on the geology of the common roads of the United States: U.S. Geological Survey 15th Annual Report, p. 259–306.
—— , 1895b, The geology of road-building stones of Massachusetts, with some consideration of similar materials from other parts of the United States: U.S. Geological Survey 16th Annual Report, part 2, p. 277–341.
Tilford, N. R., 1982, Power plant siting: Association of Engineering Geologists Bulletin, v. 19, no. 2, p. 187–196.
UEC, 1975, Report on new site selection study: United Engineers and Constructors, Inc., New York State Electric and Gas Corporation (September, 1975), 15 p.
—— , 1976, Report on new site selection study/recommended sites: United Engineers and Constructors, Inc., New York State Electric and Gas Corporation (April 1976), volume 1, 195 p.
U.S. War Department, 1856, Reports on exploration and surveys to ascertain the most practicable and economical route for a railroad from the Mississippi River to the Pacific Ocean: Made under Secretary of War in 1853–54, 12 v., 33rd Congress, 2nd Session, Executive Document 78.
Van Lennep, D., 1867, Report of David Van Lennep, geologist, *in* Union Pacific Railroad Co., 1868, report of the chief engineer with accompanying reports of division engineers for 1866: Washington, D.C., Philip and Solomons, p. 97–123.
—— , 1868, Report of David Van Lennep, geologist, *in* Union Pacific Railroad, 1868, report of G. M. Dodge, chief engineer, with accompanying reports of chiefs of parties for the year 1867: Washington, D.C., U.S. Government Printing Office.
Varnes, D. J., 1974, The logic of geological maps with references to their interpretation and use for engineering purposes: U.S. Geological Survey Professional Paper 837, 48 p.
Varnes, D. J., and Scott, G. R., 1967, General and engineering geology of the United States Air Force Academy site Colorado: U.S. Geological Survey Professional Paper 551, 93 p.
Wallace, R. E., ed., 1986, Studies in geophysics; Active tectonics: Washington, D.C., National Academy Press, p. 80–94; p. 125–154.
Webb, B., and Wilson, R. V., (compilers), 1962, Progress geologic map of Nevada: Nevada Bureau of Mines, Map 16, scale 1:500,000.
Ziony, J. I., and 10 others, 1985, Predicted geologic and seismologic effects of a postulated magnitude 6.5 earthquake along the northern part of the Newport-Inglewood zone, *in* Ziony, J. I., ed., Evaluating earthquake hazards in the Los Angeles region; An earth-science perspective: U.S. Geological Survey Professional Paper 1360, p. 415–442.

ACKNOWLEDGMENTS

This chapter has benefited from the thoughtful review and technical input of Jeffrey R. Keaton, which included a summary on low-level radioactive wastes and the earthquake scenario of the Los Angeles basin. The author appreciates the assistance and ideas on regional investigations gained throughout four decades of collaboration with colleagues on several large-scale surveys, particularly: Lyle H. Henderson on the Big Bend, Texas, project; Edwin D. McKee, W. D. Keller, H. Wesley Peirce, and Donald B. Sayner on the Navajo Country project; Hal F. Bonham, Warren L. Coonrad, Edward A. Danehy, and William A. Oesterling on the Southern Pacific Corporation survey; and Richard H. Holt and Edward N. Levine of New York State Gas and Electric Corporation Survey. A. R. Palmer kindly offered editorial assistance with a revew that improved the content and flow of the text.

Manuscript Accepted by the Society June 8, 1990

Printed in U.S.A.

Geological Society of America
Centennial Special Volume 3
1991

Chapter 19

Investigation of preferred sites for selection and design

John G. Cabrera
Consultant Engineering Geology, 2936 Valle Vista, Las Cruces, New Mexico 88001
Allen W. Hatheway
Department of Geological Engineering, University of Missouri, Rolla, Missouri 65401

INTRODUCTION

Preferred sites represent those locations that have been identified and subsequently survived a Phase I screening against a set of general locational, design, and environmental requirements, which provide the basic ingredients for functionality. Ordinarily, the candidate sites are identified on a broad basis, using small-scale topographic maps (less than 1:250,000 scale), remote images such as aerial photographs, and aerial reconnaissance; geologic factors are partly downgraded, due to the lack of specific details provided during this second phase of feasibility investigations (Chapter 18, Fig. 1).

Preferred sites, representing Phase II of a project, are those considered worthwhile on the general basis of location and topography, design requirements, environmental impact potential, seismic and groundwater characteristics. By their nature, Phase II efforts are geologically intensive. During Phase II investigations, reconnaissance-level mapping and subsurface exploration, sampling, and testing of the materials can provide enormous returns when planned and conducted by mature and field-experienced applied geologists.

The concept of the alternative site-selection process emerged in the 1930s, as a result of developments in the internal combustion engine, which served as a power plant to drive mass-produced heavy construction equipment. Thus, early, large-scale heavy construction projects, such as dams, canals, and tunnels, were possible. With this capacity, national governments and cities alike became owners of transportation and water supply projects that could have, by nature of their function, several choices for location.

Heavy construction was widespread during the 1930s, with projects like dams of the Tennessee Valley Authority, and Boulder (Hoover), Coulee, and Shasta dams, and dozens of military bases and installations for World War II. The Alaskan Highway and Ledo Road (Burma to China), and such massive and rapid undertakings as hundreds of airfields and port construction projects in the 1940s developed a resilient and daring breed of engineers, geologists, and skilled laborers who improved the role of heavy equipment. No longer was construction capability open to question. Civil engineers could design and build almost any project, with cost usually the overall constraint.

By the 1950s, most-favorable sites were becoming scarce, so that through the 1960s and 1970s, site-specific costs provided a strong impetus for detailed geological investigations of preferred sites (Fig. 1).

Beginning in the 1970s, environmentally associated considerations joined the standard pre-1970 siting factors. Requirements generated in California in the 1960s blossomed into the National Environmental Policy Act (NEPA) of 1969 and the Environment Quality Act of 1970; any project receiving even partial federal funding was subject to disclosure of the potential environmental impacts. Geologic features are the basis for many environmental aspects, and most of the impacts are associated with more rural developments. Today, geological investigations for preferred sites generally continue to follow the guidelines that emerged in 1970s.

History

Participation by geologists in the specific investigation of site locations and foundations for major engineering works in North America began with the efforts of William O. Crosby and James F. Kemp in the 1890s. The contributions of these pioneers to the development of modern engineering geological practice are reviewed in Chapter 1 for such projects as the Boston Harbor, Charles River Dam, Catskill Aqueduct, and New York City water supply systems.

The need for a more detailed and accurate geological site evaluation was dramatically emphasized in 1928 by the tragic failure of St. Francis Dam in California (frontspiece and Chapter 22, this volume). F. L. Ransome (1928) wrote "the plain lesson of the disaster is that engineers, no matter how extensive their experience in the building of dams or how skillful in the design of such structures, cannot safely dispense with the knowledge of the character and structure of the adjacent rocks, such as only an expert and thorough geological examination can pro-

Cabrera, J. G., and Hatheway, A. W., 1991, Investigation of preferred sites for selection and design, *in* Kiersch, G. A., ed., The heritage of engineering geology; The first hundred years: Boulder, Colorado, Geological Society of America, Centennial Special Volume 3.

Figure 1. Weak Pliocene rock of the Ecuadorian coast, folded and closely jointed, and typical of marginally favorable geologic conditions that are encountered at many sites, and which significantly affect construction and operating costs. (Photo by A. W. Hatheway.)

vide." The unknown presence of water-soluble gypsum was a major factor.

Soon after the St. Francis disaster, state and federal agencies, along with private firms, began to employ applied geologists (Chapters 1, 2, and 5, this volume). Throughout the 1930s and 1940s, geoscience began to be applied enthusiastically to major engineering works (Lugeon, 1933).

By the 1950s and 1960s, engineering geologists had learned to extend the methods of traditional surface mapping and underground mine-like mapping to the collection of a wide variety of subsurface data supplied by numerous new geological techniques. All available data were being interpreted to reveal the physical characteristics that influence the integrity, strength, deformability, and permeability of foundation materials. Furthermore, specific geological opinions were routinely expected and invited as a factor in project design requirements. By the 1980s, the capacity of digital computers had grown to unexpected proportions and a general trend to emphasize the physical-property measurements of earth materials emerged. Unfortunately another tendency also emerged among some technicians: the relegation of basic geologic data collection and interpretation to a secondary, less glamorous importance. If such a trend is followed, a real potential for subsequent difficulties exists in Phase III site-specific investigations.

A strong case for the basic economic value of detailed Phase III investigations is the notable savings of both construction time and overall costs at the TVA Chickamauga Dam on the Tennessee River. This dam, located a few miles upstream from Chattanooga, was completed in 1940. Areal-site geology consists of a complex series of thin-bedded limestones, dolomites, and shales, with some interbedded zones of volcanic tuff or swelling clay (TVA, 1949). Beginning as far back as the initial investigation of 1938 (Fox, 1941), TVA permanently maintained two geologists in the area for dam siting efforts. Numerous candidate sites were investigated, upstream and downstream of the final site, and all revealed the common presence of cavernous limestone. Within this framework, a careful geological analysis of the areal dissolution patterns throughout the calcareous rock units, combined with defining the stratigraphic position of the tuff/swelling clay layers, led to the rational Phase II designation of the best and preferred dam site. Greater foundation stability and a minimum of potential permeability of the rocks were the basis for the selection.

Goals and methods

A schematic diagram (Chapter 18, Fig. 1) of the regional-areal stages concerned with the site investigations portrays the development and synthesis of the Phase II site-selection data. All preexisting (Phase I) data are used to evaluate the candidate sites in the following steps.

Investigation of a preferred site is undertaken to determine the areal suitability of the site, particularly in terms of impacts on or from the project within a radius of 8 to 16 km, and to improve background data on geotechnical characteristics of the underlying soil/rock units as foundation and/or construction materials. Consequently, candidate sites and foundation areas for a specific structure are investigated initially for a three-dimensional characterization, by such techniques as core borings, exploratory pits and trenches, occasional inspection adits and shafts, sampling of undisturbed soil, geophysical surveys (surface and borehole), and in-site testing. These data are correlated with surface and underground mapping.

As in the Phase I selection of candidate sites, it is advisable that Phase II evaluations tabulate all of the known and possible geological-geotechnical "defects" related to soil/rock properties and structural features (Fig. 2). This data analysis can lead to the identification and elimination of many possible construction and/or site-operational problems. Some identified problems may require correction by project design (Phase III of Chapter 18, Fig. 1). The area investigated by detailed site-specific studies is normally 1 km (.5 mi) around the site foundation.

It is critical to link the detailed site interpretations with the regional-areal, geologic, geomorphic, and tectonic characteristics already determined by Phase I studies (Fisher and Banks, 1978; Chapter 18, Fig. 1). A preferred-site evaluation normally relies on detailed surface and subsurface observations, combined with a continuing interrelation of the facts supplied by a series of flexible alternate interpretations, as the field data become available for assessment. All observations must be carefully recorded graphically and tabulated (preferably in a computer database) for ready review and comparison. Rapid retrievability of pertinent information is essential to timely and cost-efficient site investigations.

Phase II engineering analyses should entail the utmost degree of data comparison, from all sources. Often, site-specific features will take on new relevance because of regional tectonic history (Fig. 3), and regional data may take on new relevance when reviewed in light of such site-specific features as lithology, stratigraphy, or individual discontinuities.

Surface observations should be used to test and expand the details provided by borings and other subsurface investigations. All data collected in the course of Phase II investigations should be thoroughly evaluated in terms of all of the growing bank of geological/geotechnical interpretations, and should be completely integrated into the construction and operational concepts of the project. As a balancing mechanism, all intuitive opinions should be tested by appropriate in-place observations.

Phase II demands flexibility of thought; when a preponderance of evidence indicates that a working hypothesis could be inaccurate, the conclusion should be reformulated to respond to all of the available evidence.

Even the slightest hint of the existence of any geologic feature that might compromise the bearing capacity or functionality of the foundation material is of great importance to data evaluation. On the other hand, poor exploration techniques may falsely produce negative evidence regarding these features by damaging somewhat fragile rock, later described as an "extremely fractured or blocky zone" (Fig. 4). In some of these cases, the fine material between the rock-core fragments has been washed out by the circulating drilling fluids or air-rotary flow, when in reality the zone is a moderately acceptable breccia with mylonitic gouge. When such zones are encountered, they should be sampled by slow, pressure-balanced drilling, with multiple-tube core barrels. This is a good example of the maxim that the unrecovered material is usually the weakest and corresponds to zones of important geologic defects.

Experienced engineering geologists think in three dimensions when correlating features. All appropriate techniques are utilized, including physical models and perspective-oriented computer graphics, to represent to others the potential variations of observed geologic features.

Approach. Results of regional-areal investigations should be carefully evaluated to determine relevant lithologic, tectonic, and hydrogeologic characteristics of each alternative preferred site. Since 1975, the digital computer has been routinely capable of software programming for statistical analysis, and the emergence of accessible computer graphics (about 1985) has provided ideal accessory tools to weigh the consequences of numerous variables such as discontinuities, irregular-bedrock surfaces, and various geologic constraints.

Figure 2. Typical geologic/geotechnical defects encountered in otherwise suitable lower Paleozoic metavolcaniclastic rock of the Ammonusac Formation on the Vermont–New Hampshire border. Phase II site evaluations should identify key engineering geologic units, along with their typical and unpredictable features, such as the open foliation planes (inclined to the left of this view) and the near-vertical joint set, which together form a rhombehedral intersection of discontinuities. (Photo by A. W. Hatheway.)

Geological evaluation during the selection process should evolve from the general to the particular. It is best to initially obtain a broad perspective of the characteristics of the project area, then to identify and focus on the specific foundation problems. There are many outstanding examples of problems caused by concentrating attention exclusively on the foundation of a structure, with disregard to important regional constraints or influences. A case in point is that of Vaiont Dam, Italy (Kiersch, 1964), in which the arch-dam foundation was almost too care-

Figure 3. Paleo-liquefaction and fissures, here shown to crosscut bedding of the Older Alluvium of Singapore. Features such as these are often direct evidence of ancient tectonic activity that may play a part in project design. (Photo by A. W. Hatheway.)

fully studied, and much less attention was given the disastrous potential for post-reservoir-filling slope failures along the canyon wall. In this now-familiar case, the deformed slope rocks became saturated, and the effect of water pressure in clefts led to rapid slope failure and nearly instantaneous displacement of massive amounts of reservoir water (described in Chapter 22, this volume).

Tectonics and hydrogeology are additional regional concerns. Late discovery of a nearby active fault (Fig. 5) may so increase construction costs that it may be more economical to select an alternate site sufficiently distant from the fault to eliminate the need for special structural reinforcement.

An illustrative case is the proposed (early 1980s) Rio Pita water supply project of the Empresa Municipal de Agua Potable (EMAP), City Water Board of Quito, Ecuador. This otherwise fine, preferred site was undergoing a limited Phase II investigation when it was determined by one of us (AWH) that the consequences of a possible future eruption of the volcano Cotopaxi would govern the feasibility of the project. This mountain, the highest active volcano in the world, would most certainly produce lahars that would totally fill the reservoir within a calculated time period of 90 minutes, following an eruption equal to that of 1878. The dam could be designed to withstand seismic ground motion related to the chain of pre-eruptive earthquakes, and the dam and reservoir could be protected by lahar routing to an adjacent valley (quebrada). Lahar resistance could be designed into the dam so that impounded waters could be released at a rate that would not threaten human life in the watershed below, all based on a carefully devised system of preeruption warnings. The costs associated with either mucking out the reservoir or constructing the lahar-routing embankment were computed to be equivalent to the cost of the original dam. The risks associated with a predicted 75-percent chance of a lahar-producing eruption within the 100-yr lifetime of the project, as evaluated in 1981, were too great for the owners to accept the preferred location.

Criteria. The preferred site must conform to specific topographic, hydrologic, geologic, engineering, and economic requirements. Some of the more common problems are: underseepage or leakage of dams, slope instability, settlement or subsidence, ground support of tunnels, excessive inflow of ground water, ground motion or rupture, channel scour, siltation, contamination by encroaching sea water, and long-term deterioration of mass concrete, rock, or earth as embankment fill (Kiersch, 1955).

Borings should be initiated only after geological mapping of the site area (5 to 8 km radius) has been completed, and should be positioned on the basis of findings derived from that mapping, and from the results of any preliminary geophysical surveys. They should always be located to sample the characteristics of local structure (Meade, 1936, 1937a, b; USCE, 1946; USBR, 1963) and, in dipping strata, should be carried to depths that will permit correlation of the strata encountered in adjacent holes.

When the foundation area is blanketed by superficial deposits, exploratory borings should be carried into bedrock no less than 2 m and, preferably, 3 to 5 m when the presence of large boulders or highly weathered bedrock is suspected. This criterion is especially important in formerly glaciated areas, where there is always a strong possibility of the existence of buried preglacial or postglacial valleys.

Exploration activities should always respect the three-dimensional character of foundation geology, principally in relation to continuous or persistent discontinuities that could span an entire dimension of a constructed facility, thereby profoundly affecting structural stability (Fig. 6) (Hoek and Bray, 1981, Chapter 4). This three-dimensional concept refers not only to spatial but temporal factors. Certain processes (such as under-seepage, uplift pressures, or carbonate dissolution) can affect a foundation throughout the lifetime of the project. Although modern engineered works assume a 50-yr lifetime, we are reminded that numerous ancient Roman structures are today in daily use. In this connection, special attention should always be given to hydrogeologic conditions, by including careful, periodic observation of ground-water levels in monitoring wells as an indication of potential long-term ground-water problems.

Preferred sites are selected to become designated Phase III sites by a system of progressive elimination on the basis of their major geological, environmental, and engineering constraints, in addition to the critical cost/benefit ratio.

In order to determine the relative merits of two or more preferred sites, probabilistic estimates can be made of the potential for occurrence of individual geologic hazards that would affect the operation and maintenance of each site. All other aspects being similar, the site that portrays the minimum operational risk, and associated costs, is likely to be selected.

INVESTIGATION OF PREFERRED SITES

The main purpose of a site investigation for proposed engineering works should be to determine site suitability and geologic and hydrologic impacts of its environs (5 to 8 km radius). These determinations form the basis for preparation of an adequate and economical design, prediction, and accommodation of possible geology-related problems during and after construction, and also to form the quantitative baseline needed to identify and overcome subsequent changes in dynamic conditions or geologically based failures that may occur during construction. Guidelines on exploration techniques and appropriate investigations were established early by the U.S. Army Corps of Engineers (USCE, 1946) and the U.S. Bureau of Reclamation (USBR, 1963, 1988).

Figure 4. Blocky rock of a sequence of flysch-type sandstones and shales of the Pennsylvanian Atoka Formation of northeastern Arkansas. Such rock may have appreciable rock-mass strength for many engineering purposes, yet may represent significant ground-water-flow pathways at sites experiencing contaminant transport. (Photo by A. W. Hatheway.)

Figure 5. Holocene micro-grabens typical of the central Andean valleys. This view is about 4 m wide. (Photo by A. W. Hatheway.)

Objectives

The primary objective of Phase II investigations is to determine if foundation materials are suitable to support the proposed engineered works with respect to their durability, strength, deformability, and permeability. Consideration should be given to the behavior of all materials during and after excavation below the general ground level, and a predictive evaluation made of the tendency for the sides of excavations or cut slopes to ravel, creep, or slump.

Sequence of site investigations

Engineering geologists beginning the detailed investigation of a specific site can make profitable use of all previous data and interpretations assembled during the Phase I regional-areal investigations, the conceptual design stages (Chapter 18, Fig. 1, this volume), and the subsequent site selection process in order to adequately interpret the geologic setting for existing lithologic and structural conditions.

Roberts (1964) described the prevalent post–World War II site investigation sequence as: (1) analysis of previously acquired data with modification or reevaluation when appropriate, (2)

Figure 6. Intersections of continuous (persistent) joints or shear planes can produce significant impact on the structural stability of cuts and underground openings. At this drydock, under construction in lower Paleozoic limestones of the Gaspe Peninsula of Quebec, three intersecting shear planes and the excavated face have produced one failed block (to the left) and one incipient failure (to the right). The block to the right must either be pinned by rock bolts or removed. Such features are difficult to detect during the preferred-siting phase, but are sometimes detectable on the basis of photo-lineaments or geomorphic features. (Photo by A. W. Hatheway.)

investigation of the site area, (3) interpretation of results, and (4) development of recommendations.

Geological mapping of a preferred site should begin with recognition of the geologic and geomorphic processes that have produced the existing natural conditions. Mapping the details should suit the project purposes because, for example, valley-bottom glacial scouring may have created higher benches of more resistant rock that can be incorporated into the project. "Design with and not against nature" is always a good precept (McHarg, 1969).

Nambe Falls Dam, northern New Mexico, is an example of a radical and consequently expensive design modification of the type that may occur when foundation conditions are not fully investigated before construction begins (ENR, 1974a). Unfortunately, investigation of the site began after the initial design, common occurrence in public-sector works. A broad fault zone, nearly 36.5 m wide, was delineated within the left abutment. Consequently, the design was modified and the concrete structures located outside the fault zone (Fig. 7). Fortunately, a savings in the concrete volume was gained by the modification to a thin arch design, but this was offset by the costs of pre-stressing with large vertical flat jacks in the central contraction joint of the arch, and displacement of the thrust block, which formed the left abutment of the arch, both to be moved away from the fault. Although not suitable to support a concrete structure, the fault breccia and gouge was satisfactory foundation for an earth-fill embankment that abuts the massive thrust block.

The principal investigations that should be carried out at a project site are determined by the type and dimensions of the proposed engineered works. The earth materials upon which the structures are to be founded, or through which they traverse, must be well characterized, and the geologic history of the site area understood, as related to potential impacts on the design or operation of the project (Cabrera, 1974).

Engineering soils/unconsolidated deposits. Soil-like deposits introduce many potentially relevant factors, such as: capability of particles to transmit superimposed load without crushing; degrees of expected consolidation; hydrostatic deformation or widening of pore spaces, which enhances the possibility of shearing; potential for under-seepage beneath the structures, or seepage-induced piping (erosion of pores and voids; Fig. 8) of fine particles; susceptibility to erosion (Fig. 9); collapse potential on wetting; and expansion potential of clayey soils.

Parts of the Florida Everglades, previously drained and reclaimed for agriculture, show a high subsidence rate (more than 2.5 cm/yr; Stephens and others, 1984). Some structures founded

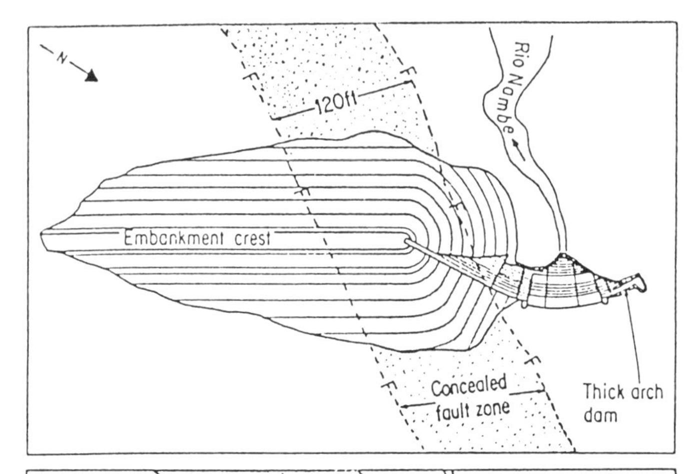

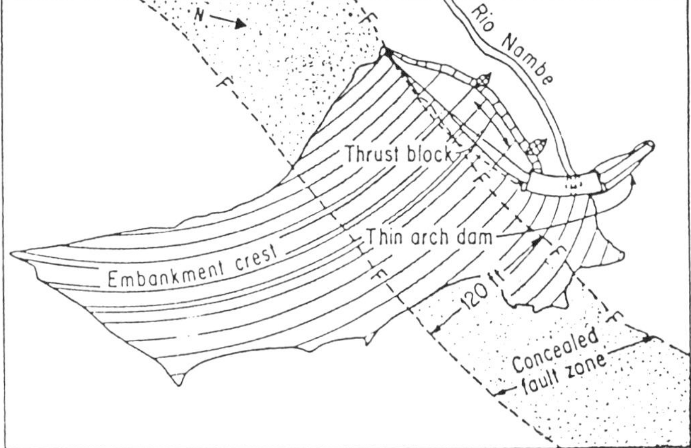

Figure 7. Design modification of Nambe Dam, New Mexico, after discovery of a major fault zone during the preferred-site exploration phase (ENR, 1974a). The fault was discovered after original dam was designed (top); concrete sections were moved out of the fault zone following redesign (bottom).

Figure 8. Regional soil and rock types should be reviewed during the preferred-siting phase for potential geotechnical problems. Pleistocene loess of the Denver basin is known (as elsewhere) as a collapse-prone material. This photograph shows piping developed in a cutbank of loess at spring evaporation ponds, then under construction, as a result of a single high-intensity, 24-hr rainfall event. (Photo by A. W. Hatheway.)

on piles have required periodic replacement of their access steps and entryways. This structural damage is typical of subsidence related to organic soils, primarily caused by pressure-induced expulsion of pore water, resulting in desiccation cracks, consolidation, erosion, and biochemical oxidation.

Severe damage is frequently caused by the opposite effect: soil expansion. Several environmental factors (Mathewson, and others, 1980), all independent of soil properties, are typically responsible, including ground slope, rainfall, surface drainage, and vegetation. Even in the case of light foundations, the effects are site-specific, requiring individual site evaluations.

An interesting problem of this nature occurred in Texas. A church was founded on a stratum of weathered shale 1.8 m thick, containing a high content of expansive clay minerals (Gieseke, 1922). It was built during dry months that preceded a year of abnormally low precipitation. Soon after completion, rainfall more than doubled; this caused the foundation material to swell an average of 10 cm and to break the concrete floor slab. Interior columns and portions of the concrete foundation wall were also lifted and damaged.

If the shale stratum had been tested for swelling potential, a competent engineering geologist could have alerted the design engineer to remove the clayey layer. This was, however, prior to 1922, and a general professional awareness of the nature and hazards of expansive earth materials was not realized until the 1960s.

Rock units. Rock origin, the events of its past geologic history, and physical properties are critical factors in evaluating any rock mass used for engineering purposes (Hatheway and Kiersch, 1982). Evaluation begins with determining the degree of weathering, hardness, and durability, and the frequency and pattern of discontinuities (Fig. 10). Weak and volcaniclastic rocks have a particular tendency to slake when excavated and relieved

Figure 9. Coastal areas of recent uplift are typically blanketed with late Tertiary to Holocene sands, often underconsolidated and uncemented. This exposure, in Newport Beach, California, shows striking erosion rilling throughout the entire 30-cm depth of a poorly sorted, medium-grained sand exposed for only one month at a construction site. (Photo by A. W. Hatheway.)

of in situ stress, or when subjected to alternate wetting and drying, especially if alteration-related clay minerals are present. Subsequent swelling-induced disaggregation of the rock mass is to be expected. Weak rocks, in general, are sedimentary units of Cretaceous or younger age, or those that have unconfined compressive strengths of less than 5 MPa.

The need to carry out observations in situ and to base conclusions on direct observations was demonstrated by Gignoux (1945). Although the granite foundation for a small arch dam was adequately investigated and known to be satisfactory, the design incorporated a conical Howell-Bunger-type discharge valve that directed a powerful jet of water on the granite wall downstream from the dam. The planners assumed the granite to be hard and resistant, as exposed at the dam abutments, and no efforts were made to investigate the face. The initial discharge from the dam revealed the effects of hydrothermal alteration from a nearby intrusion throughout the granite; the bluff was quickly undercut, and a sizable rockfall brought about the need for a heavily reinforced concrete impact wall. This is an excellent example of the need for areal investigations so that construction design can compensate for any adverse effects to the site environs (Chapter 18, Fig. 1).

Factual versus interpreted data. Geological and geotechnical data are of two types, when considered in terms of documentation for engineered construction: factual data and interpreted data. The former category is information collected in an observational or measured mode, but which represents actual conditions, characteristics, or properties, as directly observable or measurable. It is generally agreed that borehole logs, field and laboratory tests, and bedrock and face maps are representative of this category. Interpreted data are an extension of factual data. They provide additional information dealing either with concealed or obscure (e.g., subsurface) data, or the expected results or implications of factual data. Surficial engineering geological maps are usually considered to be in this category.

Mapping of site. A site map should be an early-work product, to be subsequently utilized for locating borings and geophysical traverses. This site map, in the general scale of 1:1,000, should be evaluated along with the findings of the areal–regional and neotectonic mapping to provide for rational locations, inclination, and orientation of exploratory borings. The number of borings will increase in accordance with the complexity of areal geology, the size and critical aspects of the project, and the relative impact of the project on its environment.

In addition to structural and lithologic characteristics, geological maps for engineering purposes show the properties of materials to be used as a foundation or for construction. A single lithologic unit, e.g., a specific granitic phase, may vary in degree of weathering, alteration, and foliation to such an extent that it will be nonuniform for engineering requirements.

Modern engineering geological maps are highly influenced by the late 1940s innovations of Clifford A. Kaye, U.S. Geological Survey, who produced the first engineering-specific map explanations in the U.S. (Hatheway and Feininger, 1988). These geologic maps include two categories of earth material units: rock and surficial alluvial-soil bodies that are generally found on conventional geological maps; and units based on characteristics that indicate durability, strength, deformability, hardness, and permeability.

It is not generally possible to make the early field or laboratory tests quickly enough to support conceptual or feasibility-level judgments. Therefore, opinions on physical characteristics and engineering properties of materials must be formulated on the basis of outcrop examination and mature geological experience and judgment. Specialty-trained engineering geologists and some geological engineers have been taught these techniques as part of academic training since the 1970s. Varnes' (1974) masterful work on the logic of geological maps provides the reader with the necessary rationale.

Figure 10. Geologic history must be factored into preferred-siting explorations. These 4 NX-core specimens of the Cambridge Formation were all recovered from the same site in Cambridge, Massachusetts (left to right): (1) elliptical reflection of high-angle jointing; (2) typical appearance of a failure plane from penecontemporaneous turbidity slumping, now completely healed and shown by its truncation of laminar bedding; (3) a portion of breccia zone representing shearing; and (4) a specimen of totally altered volcanic tuff, now a mass of kaolinitic clay. (Photo by A. W. Hatheway.)

Qualitative evaluation of rock/alluvial-soil properties can be made by careful field observations, including tendency to swell, shrink, or slake; presence of deformable clay minerals; size and interconnection of pores; resistance to hammer impact; spacing and openness of discontinuities; and degree of rock weathering, alteration, and erodability.

Various authors have proposed symbols that describe origin, earth material type, and thickness, along with qualifiers denoting type of deposits. For example, on Figure 11, the map symbol "Amgs" describes alluvial silty gravel and sand. These symbols all derive from the early works of soils engineer Arthur Casagrande, who originated the Unified Soil Classification System (USCS) in 1942 for the U.S. Army Corps of Engineers, one of the simple tools used by the Corps in its massive military construction program of World War II. Casagrande's system and the geotechnical implications that flow from it today are limited; they apply only to normal or underconsolidated, nonvolcanic soils of the mid-latitudes (USBR, 1974).

Engineering geological map symbols are basically of two types: (1) those dealing with rock (the Unified Rock Classification System, or URCS, of Williamson [1984]); and (2) those used to portray the origin and depth of materials that are essentially alluvial soils (the Genesis Lithology Qualifier, or GLQ system). The latter set of symbols was advanced by Galster (1977) and subsequently refined by Keaton (1984).

Mapping must be carried out in the detail necessary to define foundation conditions for any engineered structures of the project. Besides the usual documentation of the extent and location of overburden and top-of-rock, the preferred-site map should record the various types of soil, surficial deposits, and rock, and the principal structural features of the bedrock. Depending on the size of the project area, the mapping scale should be increased in order to better portray features of importance to the project (Geological Society of London, 1972; Varnes, 1974; Radbruch-Hall and others, 1979). It is apparent in this brief summary that engineering geological mapping underwent considerable refinement during the period from 1965 to 1975.

Mapping underground openings, excavations, and trench walls. Walls of underground openings, exploratory pits, and open cuts normally provide the best earth-material exposures at any proposed construction site (Fig. 12). This form of exploration was developed by the U.S. Army Corps of Engineers, beginning in the late 1940s (USCE, 1946; and Kiersch and Treasher, 1955). The principles developed by earlier workers were published as an Engineer Technical Letter by the Corps under the impetus of deputy chief geologist Lloyd B. Underwood (USCE, 1970).

Mapping detail should incorporate observed lithologic and structural features in relation to precisely marked distance intervals, or in the case of a high bluff or wall, to a regular grid of stations and elevations (Proctor, 1971; Hatheway, 1982). Map scale should be selected according to the dimensions of the surface to be mapped. For small adits and trench walls, a scale of 1:50 is suitable; for more extensive areas, 1:100 is generally adequate. In addition to symbolic maps, a photographic record of the exposed rock surfaces can be obtained, and with very little additional work, a series of overlapping stereoscopic photos can be produced at locations with the necessary access (Low, 1957).

Foundation mapping (prior to excavation). Foundation areas of designated sites should be mapped in detail in the scale range of 1:200 to 1:500, depending on the area, before removal of any overburden. Where topography is subdued, a grid of survey lines and points can be installed as a reference network for the geological mapping.

As the foundation area is investigated by core borings, it is generally possible to contour the top of the bedrock surface, whether fresh or mildly weathered, that is suitable for foundation bearing (Keaton, 1984). Whenever considerable thickness of overburden soil and weathered rock are to be excavated above foundation level, isopach maps can be prepared that differentiate between materials that can be excavated directly by scraper and shovel and those that will require ripping or blasting.

Foundations of major projects, investigated by numerous borings, are ideal candidates for preparation of a sequential map series showing geologic conditions expected at specific intervals of excavation depth; such maps have been widely used in ore-deposit modeling since the 1930s (Chapter 20). These stack-unit maps, as adapted to engineering geological use, are extremely useful in making preexcavation decisions relating to the adequacy of foundation materials and their expected depths (Kempton and Cartwright, 1984).

Computer-drawn maps, cross sections, and perspective

views are now routinely available to assist in displaying subsurface data and relations. Computer science developments brought these techniques forward, via computer assisted design (CAD) in the 1980s. By the late 1980s, the same capabilities were appearing on personal computer (PC) systems in the less-than-$10,000 range, including software and the necessary laser printer for publication-quality output. By use of these new technologies, engineering geologists can produce three-dimensional graphics that enhance the data interpretation, a feature not available even from "smart computers" (employing expert systems); such graphics are of great assistance to design engineers.

The final mapping of exposed-foundation areas at scales of 1:100 or even 1:50 constitutes the record of as-built conditions (described in Chapter 20), an essential element of all major engineered works.

Quarry mapping. Two types of maps are commonly utilized for engineered quarry sites: the overburden isopach map, used to show the amount of material that must be removed prior to quarry development, and a suitable-quality rock isopach map, for estimation of available volume, down to the maximum depth of mining.

Subsurface investigation methods

Prior to the late 1930s, labor was cheap, and earth-moving equipment was limited to the bulldozer and power shovel; site investigations were often carried out by a large force of manual labor, just ahead of foundation excavations. Seminoe Dam, part of the U.S. Army Corps of Engineers Project in Wyoming, was being investigated in that manner when two intersecting fault

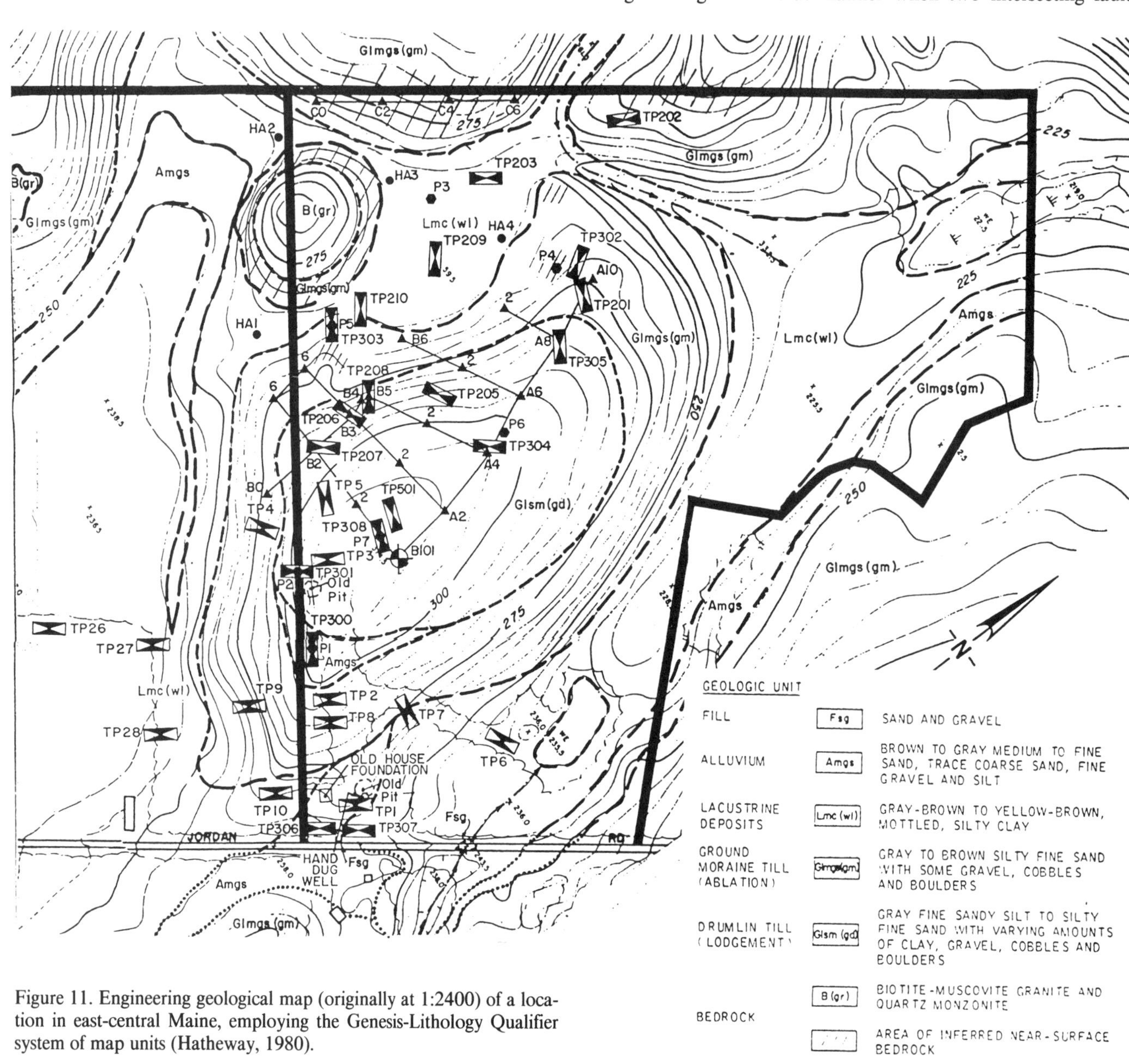

Figure 11. Engineering geological map (originally at 1:2400) of a location in east-central Maine, employing the Genesis-Lithology Qualifier system of map units (Hatheway, 1980).

zones appeared in the heavily jointed foundation granite (ENR, 1939). These two zones of weakness crossed at nearly right angles, near the dam axis, on the left side of the stream valley. One zone was oriented essentially parallel to the left bank of the stream and had an average dip of 70° into the left abutment; the other zone crossed just downstream from the dam axis and was inclined 50° downstream. To avoid a general halt in construction, it was decided to carry out foundation treatment by mining out the weak shear-zone materials from galleries driven below the base of the dam, and replace them with structural concrete.

Today, it would be financially disastrous to encounter serious, unforeseen, geologic defects during construction. Since World War II, site explorations have been allocated proportionately larger sums for investigation to avoid the risk of encountering serious flaws during construction (Fig. 13). Even now, however, it is uncommon for the entire geo-exploration budget to exceed 1.5 percent of total construction costs; it is more likely to be less than 0.5 percent (Fig. 14) (Legget and Hatheway, 1988).

Direct methods. This approach offers the engineering geologist first-hand observation of in-place materials and subsurface relations, in addition to the opportunities for direct sampling.

Exploratory openings such as backhoe trenches and small-diameter bored or drilled shafts are ideal sites for in situ mapping of subsurface materials (USCE, 1946; Kiersch, 1955, 1959; USBR, 1963; Hatheway, 1982; Cregger, 1986). Observational openings have been in use since 1846, when French engineer Alexander Collin called for sampling of foundation clay (Schriever, 1956), but detailed geological mapping began only in the 1930s. Standards were not set until 1970 (USCE, 1970), and the relative obscurity of the methods has meant that relatively few engineering geologists have received training in these methods. The technique gained prominence in the nuclear-power-plant siting studies of 1970 to 1980, mainly as a means to assess potential fault activity and to satisfy the as-constructed foundation (power block) mapping called for in the Final Safety Analysis Report required by the Nuclear Regulatory Commission prior to issuance of a limited operating license.

Small adits (2 by 2 m) are extremely useful for investigation of open and filled stratigraphic or metamorphic contact zones, shear zones, and fault zones. These galleries can be widened locally to permit erection of equipment for plate-bearing tests and direct shear tests. Determination of the elastic modulus of the rock can also be made in the adit walls using large-area flat jacks (Rocha, 1974).

Long-distance (270-m) horizontal exploratory boreholes have recently been perfected in tests conducted at Navajo Dam, Arizona, by the U.S. Bureau of Mines for the U.S. Bureau of Reclamation (Kravits and others, 1987). The tests employed a small-diameter (7.30 cm, outside diameter) boring machine. The technique has not become popular yet, due to the developmental nature of the equipment and associated costs.

Soil-type borings. These are the oldest of subsurface investigations; their origins are lost in antiquity, when they were probably limited to some sort of push probes made to test the resistance of potential foundation ground. Collin expressed the need for such subsurface sampling in 1846 (Schriever, 1956), and Sir Standford Flemming's crude boring-plate tests were conducted along the Montreal-Halifax rail line in 1860. Exploratory rock drilling (only cuttings) was first performed in Switzerland in 1864 at the Mont Cenis tunnel, driven from 1856 to 1872 (Engineering, 1868). Hvorslev's 1949 comprehensive state-of-the-art report, *Subsurface exploration and sampling of soil for engineering purposes,* remains the standard, as amended by the reader's awareness of certain improvements in technologies.

Soil-type borings still remain the best method of obtaining samples of undisturbed or disturbed, unconsolidated materials for

Figure 12. A combination bulldozer and backhoe trench of unusual depth (combined at 17 m) to expose a regional photo-lineament at the site of a proposed nuclear power plant in karstic terrane in northern Puerto Rico. The upper cut, excavated by bulldozer, exposed the lineament, which on a map scale of 1:20,000 could not be located precisely enough for a short excavation. After logging of the cutface, the exposure was deepened by backhoe (5 m) and timber-shored, according to federal regulations for worker safety. Excavations of this type have become common for investigation of the earthquake-generation potential of faults since the late 1960s. (Photo by A. W. Hatheway.)

Figure 13. Contacts in fluvial or eolian depositional environments can be highly variable over short distances. In this view, Holocene volcanic ash horizons of variable physical character are juxtaposed and overlain by fluvial sands and gravels, in Ecuador. (Photo by A. W. Hatheway.)

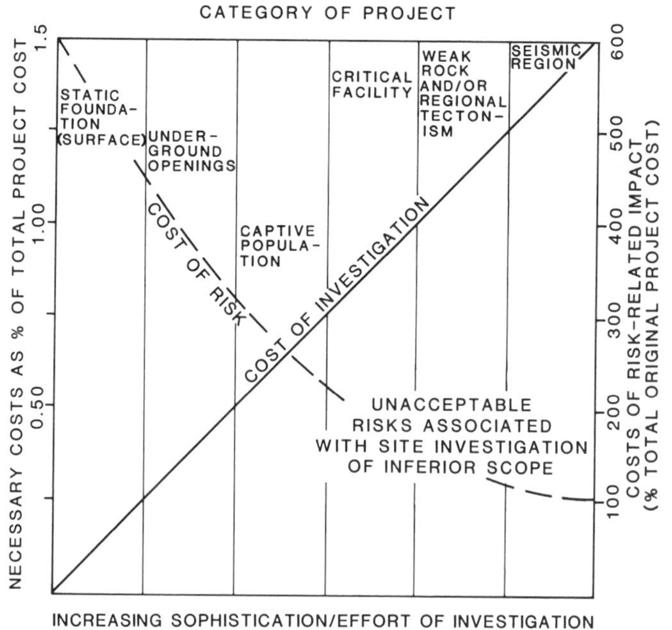

Figure 14. General relations of risk to exploration costs. The preferred-siting phase is the time when the major elements of risk during construction, operation, and maintenance should be defined and given a relevant evaluation. If adequate and competent exploration is undertaken, this effort is characterized by enhanced returns related to the amount of money allocated to site exploration (after typical cost percentages) (Kent, 1981; Legget and Hatheway, 1988).

field classification and laboratory characterization testing (USBR, 1963). The sampling equipment should be carefully selected to insure adequate sample recovery. In situ testing of soils and unconsolidated alluvial units has become a common practice (AASHTO, 1988).

Borings will likely remain the prime means of characterizing soils for design, but in situ testing is gaining in application, because it avoids the costs and disturbance of sampling. Since 1970, use of cone penetration, a Dutch innovation of the 1930s (Sanglerat, 1972), has gained acceptance in North America; the technique is limited to normally consolidated cohesive soils and relatively cohesionless and unconsolidated alluvial deposits. Sites subjected to penetrometer testing should be investigated first by borings and direct observational techniques in order to calibrate the cone and frictional components of penetrometer resistance. Even then, a considerable degree of familiarity with regional materials and direct geotechnical experience with the resistance relations is required for an accurate interpretation of penetrometer signatures.

Careful field investigation should be undertaken before design, in order to determine subsurface soil conditions. Deep excavation in soft to medium-stiff clayey soils was possible at a sewage treatment plant in Milwaukee (Riker and Daller, 1988) on the basis of soil borings, complemented by in-place vane shear tests and laboratory tests of samples recovered from the borings. An accurate characterization of those soil samples made it possible to provide the contractor with information on the minimum earth pressures necessary for design of the earth support system; the minimum tip elevation for the steel support (soldier beam) piles to be used; and the ground-water conditions, so that criteria could be developed regarding the piezometric drawdown levels required to avoid hydrostatic blowout of the excavation.

Core borings provide cuttings or physical cores of subsurface materials (Fig. 15). The diamond drill was originally developed for investigating deep into bedrock, and was represented in the United States by an 1867 patent; rock borings are advanced at the same time a continuous core sample is produced by the annular cutting core barrel, equipped with a diamond-studded coring bit at its lower end. By 1870, a 250-m-deep borehole had been drilled by steam, in search of coal near Pottsville, Pennsyl-

Figure 15. Exploratory drilling associated with preferred-siting investigations often is conducted in remote or marginally accessible terrain. Particularly suited to these conditions, or wherever environmental considerations preclude construction of access roads, are a wide variety of portable rotary drilling machines. Shown is the dismountable and man-portable Neptune machine, manufactured in Spain and capable of reaching 900 m depth. (Photo by A. W. Hatheway.)

vania (Fornwald, 1958). By the 1890s, the rotary-powered drill was well developed, and by the 1930s, this means of exploration was employed widely for siting of engineered works (McKinstry, 1948). Nearly all modern diamond drills are hydraulically pressure-advanced, as circulating air, water, or a mud slurry are forced under pressure through the cutting bit. The pressure of the circulating fluid helps prevent caving of the borehole, as well as removing cuttings and improving the quality and percentage recovery of rock core. The core barrel is provided with an inner tube suspended by a ball-bearing swivel to keep the rock core from being abraded and to protect core lengths from rotational damage at intersecting joint surfaces. Most rock coring now relies on the wireline method, developed by the Longyear Corporation in the late 1940s. This method employs a recoverable core barrel inner tube that is retrievable through the hollow drilling stem and brings each core to the surface in a matter of minutes without the need to "trip" the drill string (USBR, 1987; Fig. 16).

Rotary core drilling by air was a development of the 1950s, stemming from the water-deficient uranium exploration grounds of the Colorado Plateau. This type of drill is especially useful at sites where drilling fluid may harm the penetrated strata. Since the early 1980s, nontoxic chemical foam mixtures have also been available; foams assist in returning cuttings without buildup of borehole annulus plugs, which may tend to force drilling-air pressure against the host ground, producing some form of pneumatically induced rock breakage. The double-walled drill stem has a center opening large enough to permit cuttings as large as cobbles to be lifted, intact, by air flow. When core is not required, drilling and casing advancement are combined so that there is no need to first drill, then to case, and finally, to clean out the casing. Unconsolidated sands, gravels, and boulders are penetrated and sampled at an unprecedented rate. When the top of bedrock is reached, either air-rotary or hydraulic rotary drilling can continue, utilizing a double-wall or triple-wall stem, which serves also as a casing.

Numerous subsidiary exploration techniques have been developed to overcome the recovery problem. Borehole film, first employed by the U.S. Army Corps of Engineers at Folsom Dam, California, in 1949 (Burwell and Nesbitt, 1954), and video cameras are used to view the condition and orientation of discontinuities. These devices transmit images in color or black and white, operate in clear borehole water, and can be obtained with interchangeable side-scan or vertical-downward viewers. Side-scan can be used to measure dip directly in most cases, and strike and dip in bedded or foliated rock of moderate to steep dip (greater than 10° from vertical).

Oriented core differs from ordinary rock core only by the fact that its long axis is scribe-cut with a north-oriented notch. Once recovered, the core is placed on a portable goniometer, and strikes and dips of all planar features can be measured to within degrees of their in situ orientation. This technique costs, on the average, about four to five times that of ordinary diamond-drilled rock core.

The integral sampling method facilitates recovery of spatially oriented rock of all types, and has the added advantage of retaining all of the fabric and fracture relations of jointed, faulted, weathered, or altered rock. Conventional core-drilling equipment supports four sequential recovery operations (Rocha, 1974): (1) a 75-mm-diameter pilot hole is drilled to the top of the horizon to be sampled, and an inner hole one-third its diameter is advanced ahead to the horizon or segment of interest; (2) an azimuth-oriented pipe of roughened steel is then grouted into the smaller hole; (3) a contrasting-color grout is then forced out into the immediately surrounding rock; and (4) the grouted pipe and surrounding rock are overcored by conventional means and recovered for inspection (Figs. 17 and 18). Core that would otherwise be broken into many disoriented segments by its own discontinuities and the drilling process thus is fixed to a central reinforcing spine of steel and can be extracted from the core

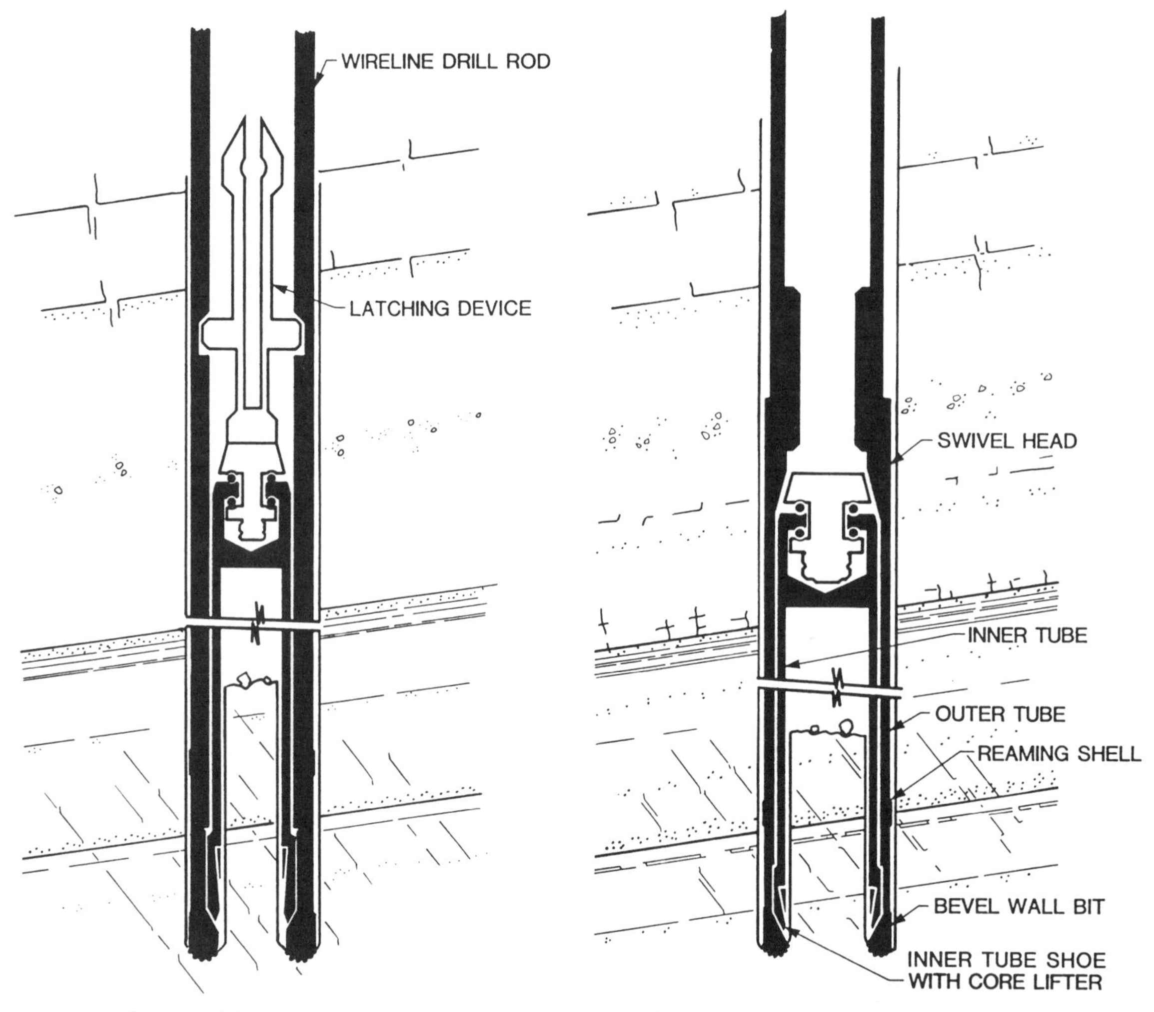

Figure 16. Schematic drawing of a wireline rock-coring barrel. The left-hand view illustrates the latching device that is necessary for recovering the core barrel while leaving the wireline drill rods in place. The right-hand view presents details of the core barrel, as advanced over a segment of rock being annularly cut to produce core. As the core run (nominally 1.5 m) is completed, the barrel and core are raised separately from within the stationary wireline rods (Charles O. Riggs, Sverdrup Corp., St. Louis).

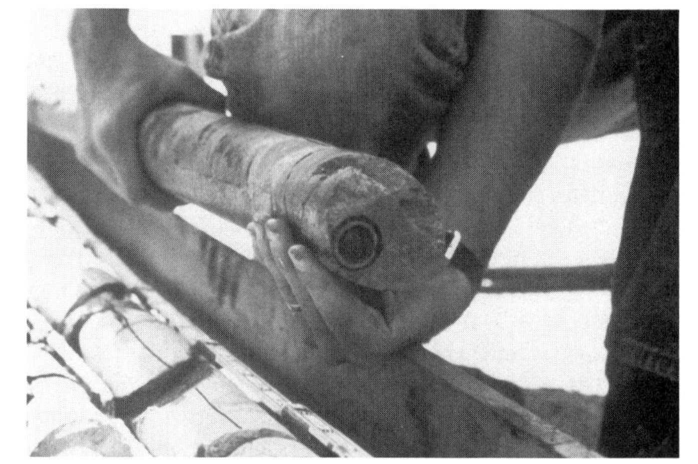

Figure 17. Rock core produced by the integral sampling method, in which all in situ discontinuities are preserved and sealed with grout, and by which rock coring fractures (mechanical breaks) are reduced to nil. (Photo by A. W. Hatheway.)

barrel as a continuous cylinder, without differential rotation of rock fragments, and without loss of significant infilling mineralization or gouge. Cost of core produced by the integrated sampling method is about three times that of oriented rock core.

Site-exploration geologists should always insist that every effort be made to obtain 100 percent rock-core recovery. Usually, most of the core that is lost represents highly fractured or altered rock, soft infillings, thin partings, interbeds, and weakly consolidated layers; all of which are the very features that may jeopardize foundation or project requirements, or that may cause construction problems in general. Core recovery is improved considerably by using a triple-tube core barrel, with a split inner sleeve that protects the core, and a core lifter case sleeve, which extends to the bit face and also reduces abrasion. Costs of obtaining high recovery in bad ground are significant; yet the alternative of bad-ground difficulties during construction far outweighs these expenditures.

The minimum diameter that allows both good core recovery and passage of a borehole video camera is NWT-size (75.7 mm). This diameter also permits measurements with geophysical probes to identify basic stratigraphy through physical, electrical, and radioactive (natural and induced) responses. Quantification includes such properties as bulk density, pore-water content, and resistivity. Consequently, relative porosity, seismic wave velocity, and several suites of geophysical logs can be produced for non-geologically logged boreholes, at a fraction of the cost of coring and logging. Borehole geophysical logs (Fig. 19) are interpreted as to lithologic/stratigraphic content, on the basis of carefully compiled geologic logs of calibration-control borings (ASCE, 1976). The authors understand that Tennessee Valley Authority geologists employ a general 3:1 ratio of geophysically logged to geologically logged boreholes, whenever the geology is simple enough to warrant the inherent cost savings.

Diamond core drilling can furnish excellent, economical information in competent rock and in the hands of competent drillers. It has been common knowledge, since the mid 1960s that drilling contracts let for recovery of high-quality core should be based on hourly drill-rig rates, rather than on the footage-payment method, which tends to make the drill foreman concentrate on rate of progress, rather than on recovering high-quality core. An alternative method to promote good core recovery is a monetary bonus for recovery percentages greater than 85, 90, and 95 percent. Weak rock, including fractured and weathered or altered rock, presents difficult core-recovery challenges, and drilling must be slowed accordingly in such rock. Infillings of shear planes and fault zones are generally washed away by the circulating fluid. Horizontal or subhorizontal discontinuities or bedding planes may allow individual core segments to rotate differentially, abrading ends and displacing the original piece-to-piece orientation. As is always the case, the poorest core recovery represents the rock most important to designers!

If the drilling contract calls for payment based on recovered core length, and there is no on-site geologist (an unforgivable situation), special attention should be given to all intervals

Figure 18. Determination of spatial orientation for discontinuities in oriented core. Core is recovered with a directionally oriented scribe mark that is used to set the rock in its natural position with respect to north. Strikes and dips are then read directly from the goniometer. (Photo by A. W. Hatheway.)

marked "clay seams" by the drillers. In the record of some dam and tunnel site investigations (generally pre-1950s in North America), there have been "erasure" cases in which several meters of voids corresponding to dissolution cavities have gone unreported by the drillers. Driller's revenue for lack of core is then made up by drilling deeper. Unless the observing geologist knows the terrane, and has prior evidence of dissolution cavities from the regional-areal investigations, the "clay seams" could be accepted as present, and the true rock section would not be known until exposed during construction (see cases mentioned in Kiersch, Chapter 1, this volume).

Geological logs provide the standard subsurface information on borehole parameters, such as location coordinates, inclination and orientation of strata, ground elevation, total depth, water levels during drilling, depth and diameter of casing, type and size of bit, as well as much descriptive data on soil, rock, geologic structure, and ground-water conditions (Figs. 20 and 21). Besides a pictorial representation of the core, the log should give graphic information on the percentage of core recovered or the rock-

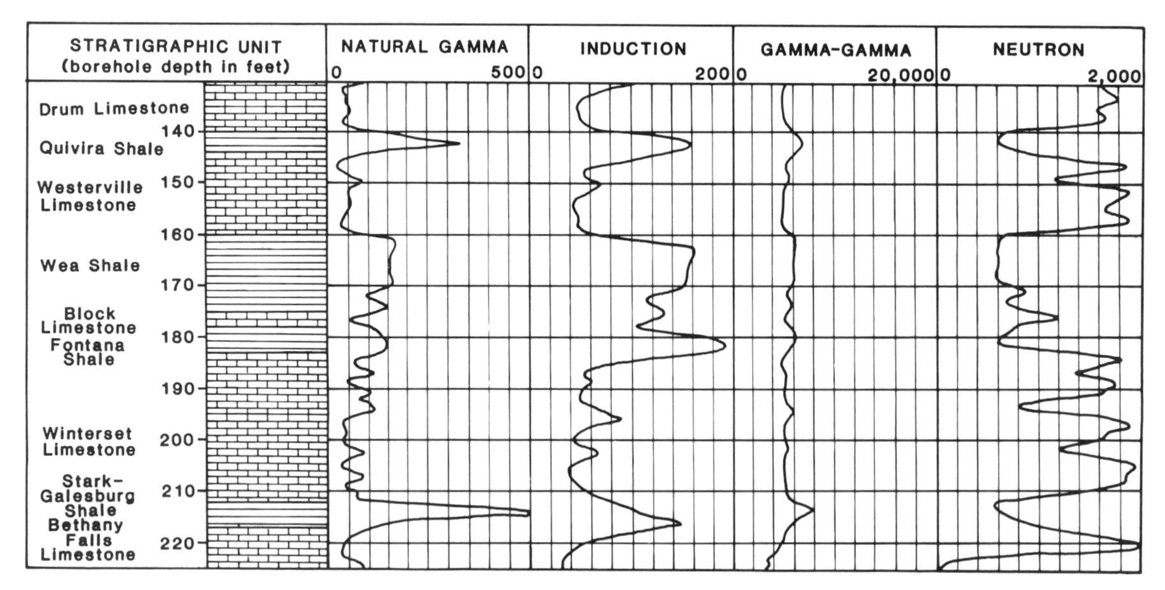

Figure 19. Borehole geophysical logging has gained firm acceptance as a means of extending site exploration through correlation of the various geophysical signatures with calibration borings that have been visually logged against rock core. This illustration is in a Pennsylvanian limestone/shale cyclothem sequence of greater Kansas City (TECHINOS Inc., Miami, Florida).

quality designation (RQD of Deere, 1964), degrees of weathering and/or alteration, discontinuity spacing, hardness, and the specific water losses at the pressure-tested horizons (liters per minute, per meter of hole, at 0.1 MPa pressure). Particularly helpful in later geological analyses are accurate, as-drilled records of such "fugitive data" as drill rates, water return and color changes, speeds of drill and advance of bit, as well as the driller's opinion on drill performance. All of these assist in establishing the characteristics of the rock penetrated.

Simple field tests, performed at the rig or in a field laboratory, can determine porosity, point-load compressive strength, and torsional shear strength (soil and weak rock only), with results plotted in a separate log column. A graphic representation of a boring can include most of the important characteristics required to evaluate rock quality.

Indirect methods. The most reliable of the "direct" exploratory methods permit visual examination of the features of materials investigated, along with recovery of core samples. Indirect geophysical methods, such as seismic refraction and reflection, resistivity, and ground-probing radar are frequently useful and very economical during the initial stages of the investigations (Chapter 18, Fig. 1). Engineering geophysics began with the efforts of Dart Wantland, who left the Colorado School of Mines faculty in 1942 to develop and employ such techniques at Bureau of Reclamation projects, such as Folsom Dam in 1949 (Wantland, 1950).

This approach may delimit or otherwise indicate the occurrence of anomalies representing broad and/or persistent discontinuities requiring direct investigations. Furthermore, the first core borings should be located and completed in a way that will provide the end-anchors and calibration for the indirect "sensing" traverses that are to follow. Combining of coring and geophysical techniques gives a very economical means of interpolating between widely spaced borings, and forms the basis for compilation of geologic profiles.

Indirect methods, such as geophysical measurements, require calibration against direct, known, geological observations. Although the geophysicist sees virtually everything within the subsurface range of penetration, it is without supporting geological observations, and these responses are often inaccurately or inappropriately interpreted. Only direct methods can provide a positive identification and confirmation of the geophysical subsurface anomalies.

During the 1970s, a change occurred in the terminology of geologic profiles and sections. The deeper and smaller-scale vertical representations were termed "sections," and the shallow and/or larger-scale versions, "profiles" (AASHTO, 1988). Furthermore, sections usually are used to portray general details of the site area or region and are regarded as most similar to the conventional cross section. In contrast, profiles are compiled to portray specific, geotechnically related geologic conditions, such as the units of soil, weak or weathered rock, top-of-rock line, ground-water surface, zones or pockets of weak materials, specific discontinuities, and cavities. Boreholes and other explorations are also shown in direct location on the profile or as "offset" in near proximity to the profile line.

The correct approach to any site investigation is stepwise, thereby progressively adding to the database as pertinent facts are disclosed. One of the most useful visual aids is a three-dimensional model showing geophysical and geologic profiles

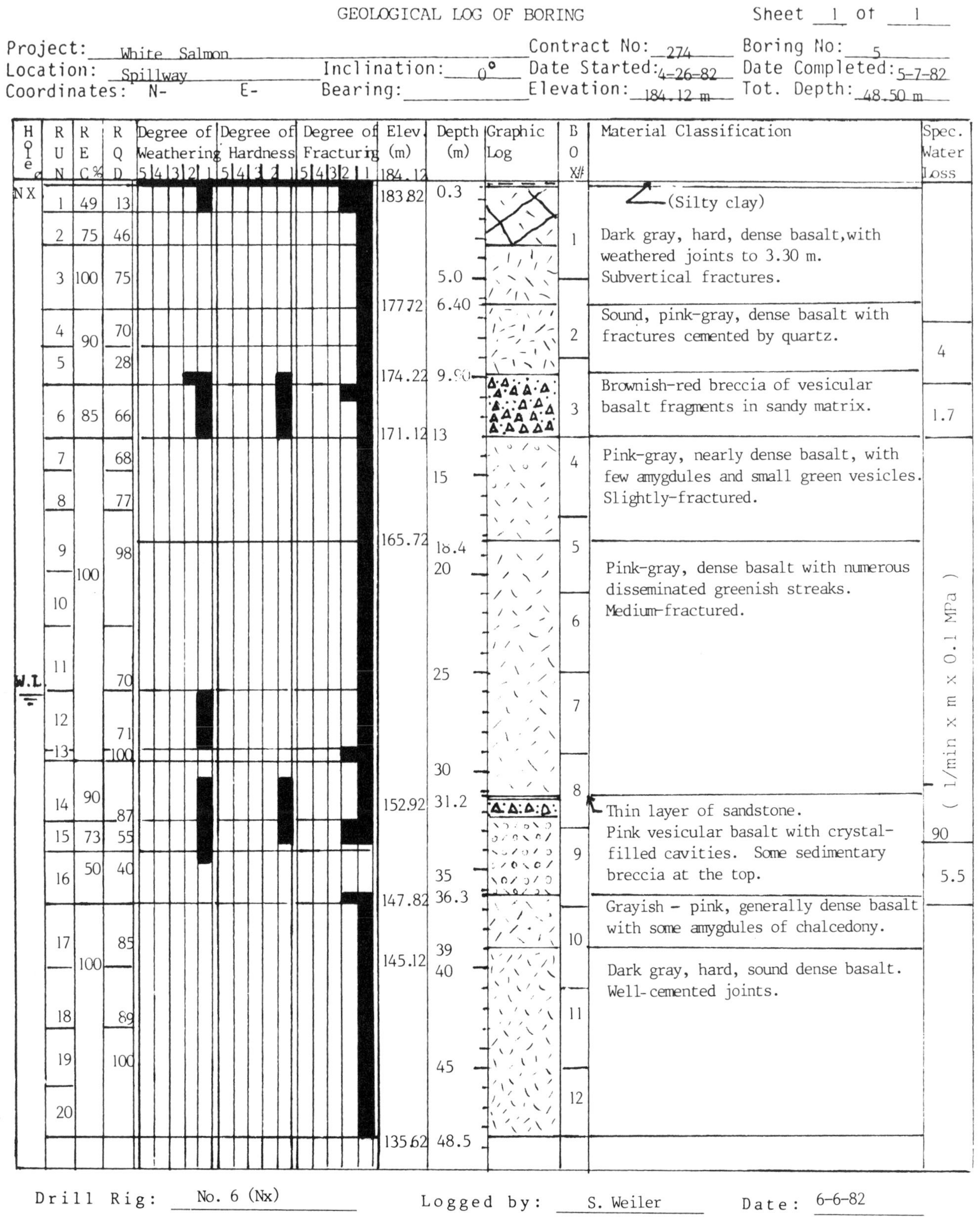

GEOLOGICAL LOG OF BORING

Sheet 1 of 1

Project: White Salmon | Contract No: 274 | Boring No: 5
Location: Spillway | Inclination: 0° | Date Started: 4-26-82 | Date Completed: 5-7-82
Coordinates: N- E- | Bearing: | Elevation: 184.12 m | Tot. Depth: 48.50 m

Hole: NX; W.L.

RUN	REC %	RQD
1	49	13
2	75	46
3	100	75
4	90	70
5		28
6	85	66
7		68
8		77
9	100	98
10		
11		70
12		71
13		100
14	90	87
15	73	55
16	50	40
17		85
18	100	89
19		100
20		

Degree of Weathering 5 4 3 2 1 | Degree of Hardness 5 4 3 2 1 | Degree of Fracturing 5 4 3 2 1

Elev. (m)	Depth (m)	BOX #	Material Classification	Spec. Water Loss
184.12				
183.82	0.3		(Silty clay)	
	5.0	1	Dark gray, hard, dense basalt, with weathered joints to 3.30 m. Subvertical fractures.	
177.72	6.40	2	Sound, pink-gray, dense basalt with fractures cemented by quartz.	4
174.22	9.90	3	Brownish-red breccia of vesicular basalt fragments in sandy matrix.	1.7
171.12	13	4	Pink-gray, nearly dense basalt, with few amygdules and small green vesicles. Slightly-fractured.	
	15			
165.72	18.4	5, 6, 7, 8	Pink-gray, dense basalt with numerous disseminated greenish streaks. Medium-fractured.	
	20			
	25			
	30			
152.92	31.2	8, 9	Thin layer of sandstone. Pink vesicular basalt with crystal-filled cavities. Some sedimentary breccia at the top.	90
	35			5.5
147.82	36.3	10	Grayish – pink, generally dense basalt with some amygdules of chalcedony.	
145.12	39	10, 11, 12	Dark gray, hard, sound dense basalt. Well-cemented joints.	
	40			
	45			
135.62	48.5			

(1/min x m x 0.1 MPa)

Drill Rig: No. 6 (Nx) Logged by: S. Weiler Date: 6-6-82

Figure 20. Typical geological log of a drilled boring that illustrates common data collected: principal engineering properties of rocks, drill performance, physical geologic conditions, and a description of rock units penetrated (after J. G. Cabrera).

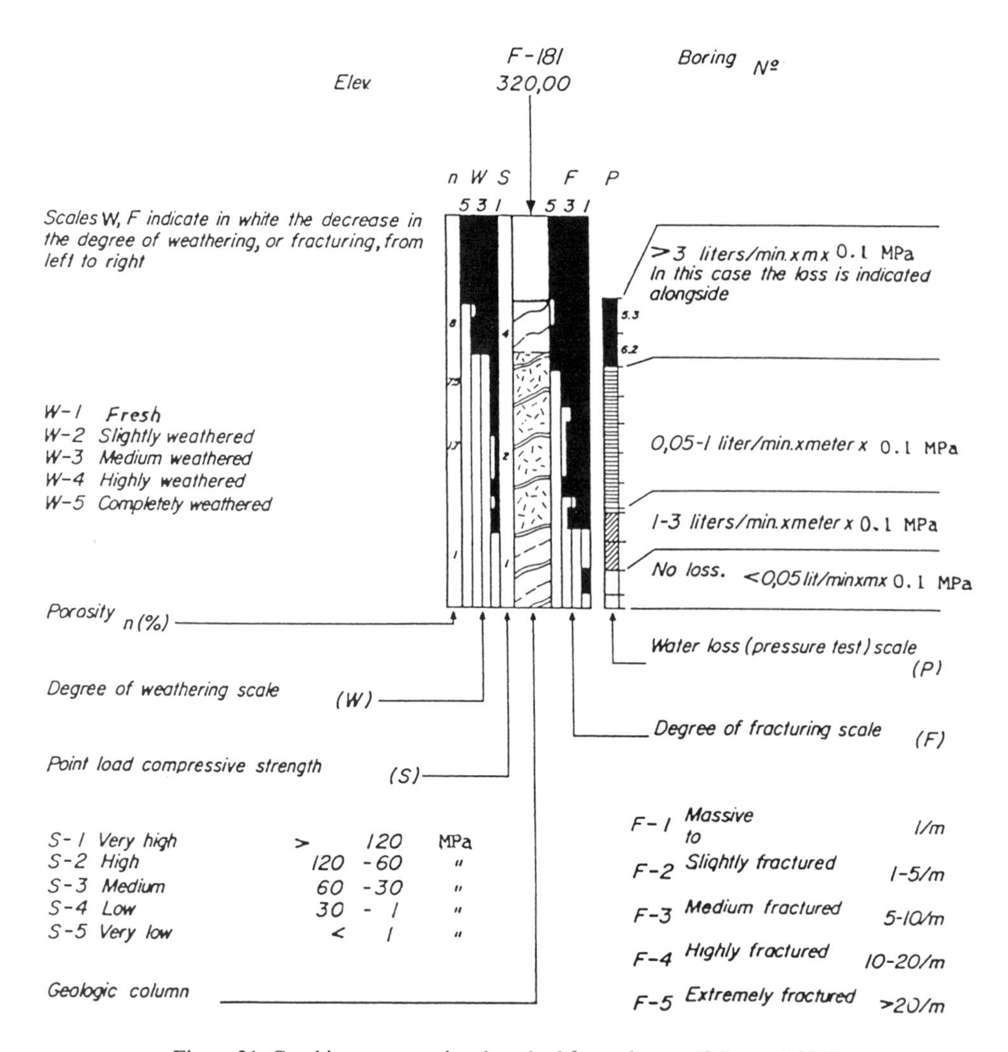

Figure 21. Graphic representational method for rock core (Cabrera, 1974).

with the borings (Chapter 18). Such models are easily constructed of fairly rigid, clear acetate sheets; they are preferably joined at right angles, like cardboard bottle spacers, by slotted intersections. The varied soil and/or surficial units, along with the rock lithology and structural features, are plotted on sheets using erasable transparency pens. The model should be updated with new and additional field data or interpretations and will be useful in the construction stage (discussed in Chapter 20). Such a model can be very useful during progress reviews and meetings of consulting boards.

Ground penetrating radar (GPR) operates on an electromagnetic pulse transmitted from one antenna and received by reflection at a second antenna. The reflected pulse is then digitized to produce a graphic record. GPR was developed in the early 1970s to determine thickness of near-surface ice sheets on Greenland. The technique was extended to geotechnical applications by the late 1970s and to hazardous-waste site characterization in the 1980s. For geotechnical applications, GPR is used as an aid in investigation of large volumes of rock for defects ahead of excavation or construction, at a relatively low cost and in short time frames (USBR, 1987; Geological Society of London, 1988). The general characteristics of GPR surveys are similar to seismic techniques, although depth of penetration today rarely exceeds 10 m. GPR is most useful in clay-rich soils and in the presence of significant anomalies in the dielectric constant of earth material, the physical property that controls the geophysical response. Improvements in GPR equipment and data processing continue to make this technology attractive for near-surface mapping of broad discontinuities such as cavities, solution channels, and brecciated fault zones concealed by overburden.

Signal enhancement refraction seismography became possible in the early 1970s, with the advent of the electronic microchip. The equipment circuitry samples incoming wave signals, digitizes them, and stores each arriving wave form. Repeated signals are added and random vibrations are eliminated when they do not conform to the growing (enhanced) refracted wave record.

Seismic refraction remains an extremely valuable indirect method of investigation for some foundation areas and near-surface alignments, especially tunnel portal areas. The technique was borrowed from the field of petroleum exploration in the late 1940s and, after refinement for engineering works, was widely used on dams and the U.S. Interstate Highway system in the 1950s. Refraction is often supplemented by reflection traverses, both ideally utilized to locate second-order borings for intersection and sampling of potentially problematic areas or zones (Geological Society of London, 1988).

In addition to providing information on depth to bedrock, to accuracies of 60 cm (Legget and Hatheway, 1988), the configuration of the top-of-rock surface, and vertical or horizontal variations in rock properties, well-planned and executed seismic traverses can be used to determine some of the physical properties of surficial deposits and rock. Ease of rock excavation can be estimated on the basis of compression wave (P wave) velocities, which permit differentiation between materials that can be removed by power shovel and rock that requires ripping or blasting (Weaver, 1975) (Fig. 3).

Seismic refraction methods are commonly used to estimate overburden thickness and to furnish positive information on the location of unconfined ground water, due to the tendency for compressional P waves to increase in velocity from 900 to 1,200 m/s in unsaturated earth materials to about 1,500 m/s in saturated earth materials. A special technique, known as fan shooting, can be employed to locate buried valleys filled with alluvium or glacial debris; often sources of ground water in otherwise impervious areas.

In situ shear-wave velocity measurements can be made within and between borings as a means of estimating dynamic response of foundation materials to machine vibrations, hydroelectric generation, and earthquake soil-structure interaction (Dobecki, 1979). Because shear waves are slower than their compressional counterparts, they often can provide better resolution of subsurface conditions than P-wave surveys.

For this purpose, geophones are built for horizontal orientation and are constructed with lower natural vibrational frequencies than those normally used for surface refraction/reflection surveys (USBR, 1987).

Resistivity surveys rely on differential electrical conductivity of adjacent subsurface soil and rock materials. This technology was known in the late 1920s (Heiland, 1946; Wantland, 1950) and became one of the principal geophysical methods for site investigation by the late 1950s. Spurred by hazardous-waste applications since 1985, there has been a significant improvement in the quality of data collection and interpretation, primarily in hydrogeologic and contaminant transport studies. Resistivity surveys can locate and delimit zones of relatively high conductivity, such as clay-rich soils or weak and weathered rock, by the presence of naturally brackish water, or from many forms of chemical contamination. In some instances, unfilled man-sized cavities in carbonate rocks create gaps of high resistivity.

Magnetic and gravity methods have been commonly available since the mid-1970s, but only a few exploration organizations maintain the equipment and staff expertise for this specialized geophysical technique. For Phase II investigations (preferred sites), these techniques are generally incorporated into searches to locate the presence of possible linear anomalies, such as fresh or altered dikes, as well as the larger (more than 15 cm wide) shear zones and faults, and concealed contacts between rock units. Proton magnetometers are now available in small, hand-carried versions, built with microchip circuitry. Both techniques are frequently useful in the final stage of site mapping. Standard gravimetric surveying is significantly less precise than magnetic surveying, but microgravimetric techniques can be used successfully to detect buried valleys, the presence of faults, and even the existence of some subsurface cavities within depths of interest to foundation stability (Geological Society of London, 1988).

Investigation of preferred sites should always avoid adherence to preconceived notions and should constantly strive to improve the on-going three-dimensional model of subsurface conditions (Chapter 18, Fig. 1). Concern should be given specifically to those locations where geomorphic and geophysical data indicate the possible presence of structural features or defects, such as gullies or other incised paleo-erosion features that may relate to subsurface zones of low seismic velocity or anomalous resistivity data.

Testing rock strength. Field and laboratory testing for the various components of rock strength (tensile, shear, compression, cohesion, and friction angle) began in the late 1930s (Obert and others, 1946). Standardization of those tests was begun by the U.S. Bureau of Reclamation and the U.S. Bureau of Mines in the 1950s, followed by the U.S. Army Corps of Engineers Ohio River and Pittsburgh and Missouri River Divisions in Cincinnati and Omaha, respectively. The American Society for Testing and Materials (ASTM) has supported a major effort in standardizing such tests since the 1960s, and the International Society for Rock Mechanics has been the current leader in the standardization effort since the 1970s. It is now accepted practice that Phase II site investigations for major engineering works incorporate the principal tests (Goodman, 1978) to provide the basic ranges of strength and deformability of foundation rock. All sampling and many of the testing conditions are determined by the efforts of engineering geologists to acquire and classify test specimens.

In situ shear tests on block samples determine the influence of irregularities or asperities on shearing resistance along discontinuities being tested (ISRM, 1981). A typical question, yet to be resolved, is whether a particular vector of shear stress acting on an identified discontinuity will force the upper surface of the fracture to "ride up" the dip of the plane, thereby increasing the shearing resistance, or conversely, to encounter resistance from asperities (Fig. 22) along the general dip of the fracture.

Large-scale deformability tests are the most common of in situ tests. In the plate-loading test, the deformation of a rock mass is determined when loaded by a plate that is pressed against the rock by hydraulic jacks. However, deformability and, consequently, the elastic modulus, can more easily be determined using flat jacks. In this process, a line is marked on the wall of the test gallery perpendicular to the stresses to be measured. Reference pins are inserted and cemented on either side of this line, and the

Figure 22. In situ testing of actual project rock by means of jacking against reaction frames excavated and cased into rock at a site in Colombia. The schist bedrock has been isolated by excavation so that the shear ram can be emplaced (right side) and there is room for displacement of the sheared block (left side). (Autopista de Bucaramonga, Colombia; K. Dragert, Kassel, West Germany.)

distance between them is precisely measured. A slot wide enough to fit the flat jack is cut along the line, and stress-relief deformations are recorded. The flat jack is then cemented into the slot and expanded until the distance between the reference pins is restored to the pre-slot distance (Pratt and Voegele, 1984). The elastic modulus can be calculated from the stress required to cancel the observed strains. Use of this test is mainly restricted to weak and soft rocks amenable to slot cutting.

Determination of in situ stress at many project sites slated for underground or deep-valley construction requires knowledge of the state of stress within the host rock. Flat-jack tests and strain-relief overcoring of deformation gages cemented within boreholes, both taking place in exploratory adits, are commonly used to evaluate the existing state of stress. However, even the presence of the adit and the access borehole somewhat affects the distribution of virgin stress (Leeman and Hayes, 1966). Beginning in the late 1970s, hydrofracturing of boreholes emerged as a sometimes promising technique. A packer-isolated borehole segment is fluid-pressurized in stages, bringing about failure and fracture development in the plane of least stress (Haimson, 1981). The pressure required to cause the rock to crack and to maintain the fracture in an open condition can be related to the in situ principal stresses. Principal stress vectors can also be calculated by use of an impression packer to determine the orientation of the resulting fracture. The technology is not completely perfected at the present time, and costs of determining one three-dimensional ellipsoidal stress orientation can easily exceed $15,000 (in 1989).

The U.S. National Research Council has made forthright recommendations (USNC/RM, 1981) to improve data-reduction methods when carrying out in situ stress determinations by hydrofracturing; e.g., consider the presence of nearby discontinuities, time-dependent deformation, and hysteresis in cyclic loading.

Site-specific investigation of tunnel and canal alignments

Investigations for tunnel alignments should include a characterization of tunnel portal rock conditions (degrees of weathering, hardness, and fracturing) and measurement of ground-water levels and pressures by means of piezometers installed in deep borings along the planned tunnel axis. Pilot adits, in which in situ tests can be made to determine rock-strength parameters, are generally included in the investigation of important tunnel projects (Underwood, 1974). The most extensive of such pilot-adit studies to date was performed for the proposed Cumberland Gap Tunnel, Kentucky, by the U.S. Federal Highway Administration, in 1982 to 1984. A study of comparable scope, made in the 1960s, that of the Straight Creek Tunnel in Colorado, is discussed in Chapter 23.

Stability calculation for proposed canal cuts should be based on rock properties from core borings along the alignment. Canal cuts are generally benched for long life; these are largely affected by the upgradient intercepted ground-water "shadow" along one side of the canal.

Planning excavations for foundation levels

Foundation levels are tentatively selected during Phase II, based on the preliminary determinations of the depth and elevation of suitable rock, as shown on geologic profiles and excavation contour maps. This practice was widely accepted by the 1930s; Nickell (1942) gave examples for five dam foundations at which the criterion has improved only on the basis of refined geological and geophysical techniques. As the Phase II civil, structural, and hydrologic requirements are prepared for a first estimate, the selected foundation materials are integrated into a preliminary loading concept, and overall deformation is often predicted through analysis by the finite-element method. This prediction includes the two-dimensional deformations as an effect of foundation geometry, and the orientation of adjacent cuts or abutments (Londe and Tardieu, 1977). Routine use of the finite-element method, by the early 1980s, has permitted the identification and elimination of needless stress concentrations at specific edges and surfaces of foundation blocks (Goodman, 1976, p. 300–368).

The three basic elements to be evaluated for a major construction site are slope/base stability, ground-water conditions, and the excavation characteristics of subsurface materials (Kent, 1981), unless special conditions such as seismicity or impact on the environment are critical. Stability of excavated side slopes can be calculated from conservative assessments of the range of measured cohesion and the angle of shearing resistance, as applied to worst-case interpretations of the attitudes of major observed or expected discontinuities. Since the early 1970s, there has been a general awareness of the danger of continuous or persistent discontinuities, which may traverse the entire cut slope or join in pairs to produce dangerously large tetrahedral wedges (Fig. 23) (Hamel, 1976; Hoek and Bray, 1981). If the foundation pads are separated from an adjacent lower groundlevel by a slope, the most appropriate inclination of the slope and location of benches can also be determined from material properties.

Figure 23. Tetrahedral wedge formed in andesite by intersecting tensional-stress joints developed during the cooling process. Surfaces such as these have comparatively small frictional resistance and are uncemented. This amounts to negligible shear resistance when faces are exposed by excavation. (Photo by A. W. Hatheway.)

Investigation for foundation treatment and drainage

Poorer quality foundation rock, in otherwise superior sites, has been grout-treated since the mid-1930s (Albritton and others, 1984). Injectability and permeability of the foundation host rock is generally determined by a Phase II pilot program of borehole test grouting. Variable grout viscosities (from 1:1 water/cement ratio to 3:1 or a chemical mix) are the main feature of the trials. After the grout sets, core borings are made to determine the effectiveness of the grouting in reduction of fracture permeability, consolidation of the fragmented rock, and/or increase in the strength of the site rocks that are porous and fractured. However, grouting still remains an almost constant source of construction claims, based either on too little or too much grout-take, in variance with the quantities estimated in the contract documents (Albritton and others, 1984).

Evaluating instrumentation requirements

Soil/surficial deposits and rock instrumentation form the basis for the "observational method" of determining geologic and geotechnical conditions, so successfully advanced by Peck (1985). Modern foundation instrumentation began to appear in the late 1950s, with the pioneering work of Wilson (1970), in the form of slope inclination measurements. Other investigations at the Bureau of Reclamation and Army Corps of Engineers developed means of monitoring load accumulations and resulting deformations. Mining engineers with the Bureau of Mines, and the mammoth Pit Slope Manual, of the Canada Centre for Mineral and Energy Technology, made significant contributions. The Canadian effort, a 5-yr project (CANMET, 1976 to 1981) directed by the late Donald F. Coates, consists of 10 separately published chapters and numerous free-standing supplements.

By the mid 1960s, a definite specialty field in instrumentation had emerged, to be followed by general acceptance and full inclusion of an amazing and wide variety of mechanical, optical, pneumatic, hydraulic, and electronic devices, in all major geologically related studies. This stage was completed by 1975, and was followed by miniaturization and great improvements in accuracy and detail of measurement, remote recording (where necessary), and extended instrument life. During this time, the instrumentation industry, still mainly made up of small, independent shops, became very cost competitive, with international dominance still held by Americans, Canadians, French, and Germans. In North America, instrumentation is regarded as a "technical service," and instrumentation specified in contract documents is almost always installed on a low-bid basis by contractors. This situation, of course, is highly uncomfortable to the applied geologists and engineers who are best qualified to specify, locate, install, monitor, and interpret the results of an instrumentation program. An excellent and comprehensive treatise by Dunnicliffe (1988) deals with all aspects of geotechnical instrumentation for monitoring field performance of engineered works (further discussed in Chapter 21).

From the geological standpoint, however, any instrumentation installed without the fullest appreciation of geologic conditions such as contacts, discontinuities, and ground water is doomed from the start as an expensive and likely misleading frivolity. Instrumentation is emplaced solely to measure stresses, deformation, and the occurrence, head, and movement of ground water in soil-alluvial deposits, weak rock, and/or a rock mass. The instruments must be correctly placed to sense conditions along discontinuities, and between separate blocks of the host rock mass. If they are improperly aligned or anchored, instruments cannot possibly deliver meaningful information. Table 1 lists relative merits of displacement measuring instruments that are widely used in evaluating proposed sites for engineering works (Pratt and Voegele, 1984).

INVESTIGATION OF CONSTRUCTION MATERIAL SOURCES

The search for fine and coarse aggregate and earth and rockfill for embankments must be carried out within an economical hauling distance, but not so close that the presence of a borrow area can have an impact on a project through slope instability or sedimentation. Engineering geologists had an early and traditional involvement in such investigations. Guidelines were in place by the mid-1930s, beginning with awareness of the importance of mineral content, followed by the realization that the presence of borrow areas can affect site functionality. In one case (Cabrera, 1963), the client mistakenly allowed the contractor to borrow soil from the upstream slope of a dam abutment. Underlying weathered, pervious quartzite was exposed, and reservoir water seeped through the abutment to emerge downstream as alarming boils carrying fines in suspension. The costly solution was to lower the reservoir level, replace the stripped soil with a compacted earth blanket, and install a system of drains where the water emerged. Quarry sites also should be located far enough from the areas where concrete structures are to be sited to prevent quarry-mining vibrations from being detrimental to fresh concrete undergoing curing.

Investigations for construction rock should determine the suitability of the rock for the intended use and verify the existence of a sufficient volume of uniform quality. If the quarried rock is to be used as concrete aggregate, the recommended tests are petrographic analysis, determination of specific gravity, compressive and point-load strength, Los Angeles abrasion resistance, alkali-silica reactivity, and breakdown due to cyclic wetting and drying (discussed in Chapter 16).

Borings made in a proposed quarry area should be distributed in a manner that will permit calculating the available volume of satisfactory rock. In general, thinly bedded or foliated rocks are not suitable for manufactured aggregate because they crush into flat-shaped particles. Rocks containing high percentages of deleterious minerals, such as free silica (opaline chert), sulfides (pyrite), or sulfates (gypsum), are generally unsuitable because they do not allow formation of strong bonds and may later cause cracking and spalling due to continued alteration after the cement sets.

Sand for fine aggregate and filter zones can be sought along river banks, some riverbed erosion pools, alluvial terraces, colluvial fans, glacial outwash plains, moraines, and eskers (Rhoades, 1950).

Soils for embankment fill should not be excessively plastic, have a high moisture content, or be composed principally of silt, all of which lead to difficult compaction or placement. Silts and clayey silts also have this disadvantage. Sandy clays or clayey sands and gravels remain the usual choice of soil borrow for compacted fills (USBR, 1963; ASCE, 1976).

After initial identification on aerial photographs, potential soil borrow areas are generally investigated by establishing a grid of auger borings, from which samples are extracted at each planned excavation horizon to determine mineralogy, grain-size distribution, specific gravity, moisture content, and Atterberg limits. Determination of density, related moisture contents, permeability, compressibility, and shearing strength of the compacted soil complete the required material characteristics. Most of this related soil-mechanics technology was thoroughly developed by the time Lambe's (1951) still-standard laboratory manual was released.

TABLE 1. PRINCIPAL MEANS OF MEASURING DEFORMATION OF ROCK MASSES*

Instrument Type	Sensitivity	Accuracy (Field)	Range	Reliability	Advantages	Disadvantages
Probe-type extensometers						
Magnetic sensor type	0.1 mm	±0.4 to 0.7 mm	1 to 30 m	Good	One probe can be used for several installations.	Low accuracy; requires hand readings, suitable only for vertical holes.
Sliding micrometer	0.0001 mm	±0.0015 mm	?	Excellent	Accurate, simple, can be used for all hole orientations, relatively insensitive to borehole curvature.	Requires hand readings.
Sonic probe extensometer	0.025 mm	?	1 to 8 m	?	Simple to install and operate. Can be modified for automatic readout; useable in all hole orientations.	New and hence untested; limited to depths of 8 m or less.
Fixed extensometers						
Rod extensometer	0.002 to 0.02 mm	0.002 to 0.1 mm, over 30 m	10 to 150 mm	Good	Accurate; can be used with automatic readout.	Subject to errors caused by rod sag, friction, and borehole curvature.
Variable tension wire extensometers	0.002 to 0.02 mm	0.05 to 0.1 mm, over 10 m	10 to 75 mm	Fair	Allows the placement of more measurement points downhole. Suitable for automatic readout.	Subject to same errors as rod extensometer, but of greater magnitude.
Constant tension wire	0.002 to 0.02 mm	0.02 to 0.1 mm, over 10 m	10 to 75 mm	Fair	Same as variable.	Same as variable, although errors are reduced.
Downhole transducer type	0.001 to 0.02 mm	0.001 to 0.04 mm, over 1 m	10 to 75 mm	Good to fair	Reduces the errors associated with long instrument lengths for measurements at depth.	Some models not removable for repair or recalibration.
Convergence meters						
Portable rod extensometer	0.02 to 0.2 mm	0.08 to 0.2 mm	1 to 8 m	Excellent	Simple, reasonably accurate; can be used for many measurements.	Limited in range and somewhat cumbersome to use.
Tape extensometers	0.02 to 0.2 mm	0.1 to 2 mm	1 to 50 m	Excellent	Extremely portable.	Accuracy requires considerable care in usage.
Portable wire extensometer (Telemac type)	?	0.02 mm, over 6m; 0.05 mm, over 24 m	1 to 50 m	?	Simpler to use and thus less prone to operator errors.	May be subject to instrument errors.
Other instruments						
Pendulums	0.005 mm	?	Variable	Excellent	Simple and accurate; measures two axes of displacement.	May be difficult to install.
TSR gage	0.0003 mm	0.5 percent of measurement	8 mm	?	Accurate, able to measure tunnel deflection prior to mining.	Complicated.
Fixed multipoint borehole inclinometers	?	±0.02 mm, over 3 m	±30°	Good	More accurate than probe inclinometer.	Provides less measurement data than probe inclinometer.

*After Pratt and Voegele, 1984.

TABLE 2. GEOLOGICAL AND GEOTECHNICAL INPUTS TO ENVIRONMENTAL IMPACT STATEMENTS*

Environmental Parameter	Potential Negative Impacts	Causative Activity	Means of Estimation	Prime Indicators	Type of Quantification Usually Possible
Land surface	Erosion	Grading; surface disturbance and disposition of excavated materials	Geologic reconnaissance; interpretation of USDA County Soil Survey Maps and geologic maps	Expected physical properties of each soil or rock unit	According to U.S. S.C.S. Standards
	Cuts	Grading; surface disturbance and disposition of excavated materials	Preliminary slope stability analysis; suggests stable slope angle; hence extent of cut	Expected physical properties of each soil or rock unit	High degree of definition of aeral extent of cut
	Fills	Alteration of land surface by placement of earthworks	Preliminary slope stability analysis; suggests stable slope angle; hence extent of fill	Expected physical properties of each soil or rock unit	High degree of definition of aeral extent of fill
	Slope movements	Inadvertent activation of ancient/dormant unstable slopes	Detection of existing unstable masses	Aerial photographic interpretation	±20 percent of area
Surfical Soil	Denial of agricultural access	Placement of earthwork and structures; determination of extent of cut/fill slopes; delimitation of right-of-way	Preliminary geological and geotechnical findings are incorporated by highway engineers; minor explorations may be required	Overlap of construction features on USDA soil series of agricultural value	According to U.S. S.C.S. Standards
	Quality	Surface disturbance of land, for marshalling, maintenance, and access; subsequent release from Agency jurisdiction as surplus to needs	Planner's stated requirements	Exposure of poor quality soil/rock units	According to U.S. S.C.S. Standards
Surface Water	Water quality	Grading; relocation of drainage courses and waterways	Field reconnaissance Topographic interpretation	Exposure of unwanted soil/rock minerals/ groundwater; uncontrolled erosion and sedimentation	Mainly presence, volume estimations to ±20 to 40 percent by volume
	Water quantity	Grading and diversion	Field reconnaissance Topographic interpretation	Diversion measures	±10 to 40 percent by volume
Ground Water	Water quality	Structures which promote infiltration of roadway runoff within or in near vicinity of right-of-way	Well survey Chemical analyses	Chemical analyses and knowledge of hydrogeologic regime	Estimated conformance/nonconformance with Federal Drinking Water Standards
	Water quantity	Earthworks; cuts that intercept piezometric surface; fills and paved surfaces that obstruct infiltration	Well survey Geologic reconnaissance Geologic interpretation	Water levels Stratigraphy Topography Construction elevations	Estimated gross raising or lowering of water table; estimated "shadow" (area of extent) of this effect

TABLE 2. GEOLOGICAL AND GEOTECHNICAL INPUTS TO ENVIRONMENTAL IMPACT STATEMENTS* (continued)

Environmental Parameter	Potential Negative Impacts	Causative Activity	Means of Estimation	Prime Indicators	Type of Quantification Usually Possible
Air Quality	Fugitive dust	Grading; earthwork placement; open-air blasting	Analysis of nature of soil/rock expected to be encountered	Lithology, grain-size hardness, compressive strength, degree of weathering/alteration, rock structure	General estimate of probable occurrence.
	Blasting and fumes	Open-air blasting	Analysis of nature of soil/rock expected to be encountered	Nature of expected	Specifications can be set for minimal or acceptable human impact; probable duration noted.
Sociologic	Noise	Blasting	Comparison of nature of rock with established empirical relations	Lithology, grain-size hardness, compressive strength, degree of weathering/alteration rock structure	Specifications can be set for minimal or acceptable human impact; probable duration noted.
	Vibration	Blasting	Comparison of nature of rock with established empirical relations	Lithology, grain-size, hardness, compressive strength, degree of weathering/alteration, rock structure	Specifications can be set for minimal or acceptable human impact; probable duration noted.
	Air pressure	Blasting	Comparison of nature of rock with established empirical relations	Lithology, grain-size, hardness, compressive strength, degree of weathering/alteration, rock structure	Specifications can be set for minimal or acceptable human impact; probable duration noted.

*From AASHTO, 1988.

EVALUATION OF AREAS NEAR PLANNED STRUCTURES

Nearly all proposed engineering works will have some potential effect on the site-specific area, as well as the surrounding environs. The early concerns for this began with the soil conservation efforts of the 1930s, dealing with erosion and sedimentation; although major, they were manageable concerns for construction heavy in earth work (Maddock and Borland, 1951). By about 1960, superior-quality construction sites had become noticeably scarce in the developed nations. Competition for land use was common, including for government, commercial, residential, agricultural, mining, and recreational purposes. Sites that did not impact somehow on neighbors were becoming rare. This, coupled with the environmental impact assessment provisions of the National Environmental Policy Act of 1969 (USCEQ, 1970) foreclosed the days when the engineers could cut the trees, move the earth, divert the water, and get on with building. The present list of geological and engineering inputs to environmental impact statements is considerable (Table 2; AASHTO, 1988), and the array of potentially negative construction-related environmental impacts (Table 3; AASHTO, 1988) demands equal input.

Some particular environmental impacts and recommended applied geological references are: reservoir-induced seismicity (Meade, 1983; Kiersch and James, 1988); downstream erosion from recently regulated flows (Dirmeyer, 1950); sediment loss (Maddock and Borland, 1951); instability of slopes created by undercutting or saturation (Schuster, 1979); subsidence of overlying surface areas due to caving and stoping of ground within tunnels or underground openings (Gray, 1988); reservoir sedimentation (Trask, 1950; Kiersch and James, 1988); and subsidence and failure of higher land adjacent to an excavation (Hoek and Bray, 1981; and Chapter 9, this volume).

Investigation of a reservoir area

In most cases, the level of risk associated with reservoir geology is far less than that associated with the foundation of the retention dam. The first order of verification of reservoir integrity is its ability to hold water, and secondly, stability of the slopes

TABLE 3. POTENTIALLY NEGATIVE CONSTRUCTION-RELATED ENVIRONMENTAL IMPACTS*

Environmental Parameter	Potential Negative Impacts	Baseline Desirable or Possible?	Applicable State of Project Planning or Development†						Remarks
			Grading	Underground Blasting (Tunnels)	Surface Blasting	Excavation of Cuts	Embankment Construction	Operation	
Land Surface	Erosion	Estimate/define present susceptibility to erosion; use photography	■			■			Data collected in the course of normal preliminary-level field investigations should suffice
	Cuts	Not indicated			■	■			
	Fills	Not indicated					■		
	Slope movements	Photogeologic interpretation	■			■	■		
Surfical Soil	Denial of agricultural access	Map the extent of existing soil units	■				■		Data collected in the course of normal preliminary-level field investigations should suffice
	Quality	Perform index tests of soil agronomic properties	■			■	■		
Surface Water	Water quality	Chemical analysis prior to construction	■			■	■	■	Observational approach can be applied toward accurate estimation
	Water quantity	Visual estimate of seasonal discharge	■			■	■	■	
Ground Water	Water quality	Chemical analysis prior to construction				■		■	Most difficult impact, of these listed, to quantify; requires careful attention to indicators; in-situ hydrogeologic tests are expensive and sometimes equivocal
	Water quantity	Visual estimate of seasonal well/spring discharge and water level prior to construction				■		■	
Air Quality	Fugitive dust	On-site monitoring at intervals during construction	■						Can be controlled by Contract Specifications
	Blasting and fumes	On-site monitoring at intervals during construction			■				
Sociologic	Noise	On-site monitoring at intervals during construction			■				Can be controlled by Contract Specifications
	Vibration	On-site monitoring at intervals during construction			■				
	Air pressure	On-site monitoring at intervals during construction			■				

*From AASHTO, 1988.
†Use of symbol (■) denotes that the stated parameter applies to the particular stage.

surrounding the reservoir. The lessons of such calamitous failures as at Vaiont Dam (Italy) or the common slides around the margins of Lake Roosevelt (Jones and others, 1961), are classic reminders of the need for investigations and geological mapping.

Slope stability and permeability. Reservoir storage will eventually affect, in some way, the stability of slopes that front the impoundment. Awareness of the reduction of this stability was initially heightened by the failure of the first Fort Peck Dam embankment in 1938 (Middlebrooks, 1942), and furthered by many minor slope failures along reservoir margins. Such slope movements cause a loss of storage volume and may require expensive route relocations, such as at the Dalles Dam, Washington. In most of these cases, foundation areas were adequately studied, but minimal attention was given to the potential problems that could arise within or surrounding the project area (Kiersch and James, 1988).

In an attempt to impound more water, or to accommodate increased estimates of the probable maximum flood, owners of existing reservoirs may decide to heighten the retention dam. This frequently focuses investigation on the abutment slopes for the dam, and only slight attention is given to the conditions of the reservoir rim. Sometimes those peripheral slopes planned for saturation will suffer instability or may be situated in pervious strata that will allow leakage. Occasionally, buried paleochannels (Fig. 24) have gone unnoticed in the cursory survey of the reservoir rim (Mackin, 1941). During the late 1960s, with the emerging quality assurance regulations for nuclear-power-plant siting and permitting, such oversights were significantly decreased for other engineered works.

Seismicity affecting the area. General awareness of hydraulically induced seismicity came about in the late 1950s with the Rocky Flats Arsenal case in Colorado (Evans, 1966), where industrial wastewater injected into deep wells affected the microseismicity of the surrounding area.

Due to the general concern for safety-related aspects of dam maintenance, there has been considerable interest in reservoir-induced seismicity since the 1970s. Meade (1983) undertook a survey of world sites and concluded that if certain impoundment characteristics of the reservoir footprint and storage volume are not present, induced earthquakes of consequence are not likely. Kiersch and James (1988) reported that many of the cases cited since 1970 have not been wholly factual and/or correlative with the geologic environs, so judgment must be applied carefully in assessing the reservoir-induced seismicity reported at a specific site. The actual phenomenom of reservoir-induced seismicity is not common and has been fully substantiated as a serious event at only a few sites worldwide. (See Chapter 2, this volume).

Study of regional and site-area seismicity is a definite Phase II activity (Caggiano, 1979; and chapter 12, this volume). General sequential steps in engineering seismicity evaluations are: evaluation of the tectonic and seismic history of a province (320^+ km radius), region (320^- km radius), and zones within regions (Figs. 25 and 26), and a reconnaissance of the reservoir area to identify and evaluate the principal structural features. The assessment can be aided by previous interpretations of multispectral satellite imagery and aerial photographs, and by establishment of a microseismic network of six to eight recording seismometers within the seismic zone or adjacent zones of the reservoir. The array should be in place for at least two years, and preferably longer, before the preliminary project design. Furthermore, an interpretation and evaluation of the microseismic history of the area should be made both before and after the reservoir impoundment.

Figure 24. Buried paleochannels have been a constant source of concern at sites in formerly glaciated terrane or in near-shore marine sedimentary sequences. In most cases, these are some sort of channel-fill deposit in which the fill is potentially more permeable or is filled with larger-than-expected alluvial clasts (such as boulders that damage drived piles). An example is this sand channel in a sandstone-siltstone-claystone sequence of Cretaceous marine sedimentary rock, east of Denver, Colorado. (Photo by A. W. Hatheway.)

Figure 25. Exploratory trench (U.S. Geological Survey) exposure of the Hayward-Calaveras fault, near Gilroy, California. This technology was largely developed in California in response to age-determination and capability studies of possibly active faults capable of generating earthquakes. (Photo by A. W. Hatheway.)

SELECTION OF DESIGNATED SITE

A well-designed and well-conducted series of Phase II investigations should determine which of the candidate sites offers the most suitable and advantageous siting characteristics. Alternative sites would be prioritized in order of decreasing suitability, except for those with serious adverse features, which would be dropped from further consideration.

Selection criteria

Many engineered works are ultimately sited on the basis of areal and site features and the associated cost-benefit analysis. Important criteria are the relative suitability of particular component sites and how such sites will interrelate with other elements of a project, besides the overall environmental impacts of the proposed locations. As an example, a candidate dam site has excellent foundation rock, yet the location may lead to severe downstream erosional damage. It may therefore be more economical to locate the dam upstream, even though the foundation rock conditions there will require extensive treatment; the long-term cost of counteracting the erosion/sedimentation damage must be weighed against the immediate cost of additional excavation and foundation treatment.

There sometimes is a tendency on the part of an organization to attempt to fit the wrong structure into the wrong place. The principle of natural determinism, whereby naturally existing conditions aid in determining the type of structure best suited for a site, is illustrated by the case of Crystal Dam, on the Gunnison river of Colorado. The basic design was changed from an earthfill embankment to a double-curvature arch dam (ENR, 1974b), not only for economic reasons but to retain the natural state of the picturesque Gunnison Canyon. An earthen embankment would have scarred adjacent areas to obtain sufficient borrow material, and furthermore, natural runoff would have increased the amount of sedimentation in the river during the construction period; also a large spillway cut would have been necessary in the right canyon wall. This feature was replaced by the negligible impact of a free-fall spillway designed into the arch. Access and construction roads were placed in tunnels rather than blasted into the ledges with switchbacks on the canyon faces; and the outlet structure was placed below minimum tailwater level.

The Bureau of Reclamation had planned in another case to build a long, thin, double-curvature, concrete arch dam at the Auburn site, on the North Fork of the American River in central California (Bellport, 1970). Foundation rocks are a complex sequence of foliated amphibolitic schists of metavolcanic origin, and interbedded metasediments that are interlayered with thin zones of weak talc and chlorite schist, and lenses of serpentinite. Both abutments are cut by faults and shear zones, oriented both parallel and transverse to foliation, which dips downstream. Geologic complexity was increased by the presence of nine significant faults, numerous shear zones, and five different lithologies. The Bureau guidelines for the Auburn project called for strengthening the foundation rock to accommodate the design loads of the dam and impounded reservoir. Treatment was generally specified to concentrate on any weak zones by means of concrete-filled cutoff shafts and trenches.

Considering these complex foundation conditions, the attempt to span such a relatively wide valley with a single arch 1,265 m in length and 209 m in height presented a major challenge. During the Auburn studies in 1975, a magnitude 5.7 earthquake occurred about 70 km to the northwest of the site, an event considered by some seismologists (Beck, 1976; Rajendran and Gupta, 1986) to have been induced by the Oroville Dam reservoir. Evidence of reactivation of the Foothills fault system, which has branches passing within a few kilometers of the dam site, were observed near Oroville (CDMG, 1979). An independent seismic-risk study by Woodward-Clyde Consultants (Packer and others, 1978) concluded that the probability of an active fault

Figure 26. Holocene fault exposed in a roadcut along the Pan American Highway, north of Quito, Ecuador. Repeated movement is indicated by a thin (5 mm) veneer of gouge and rudimentary slickensides. Coin is 20 mm in diameter. (Photo by A. W. Hatheway.)

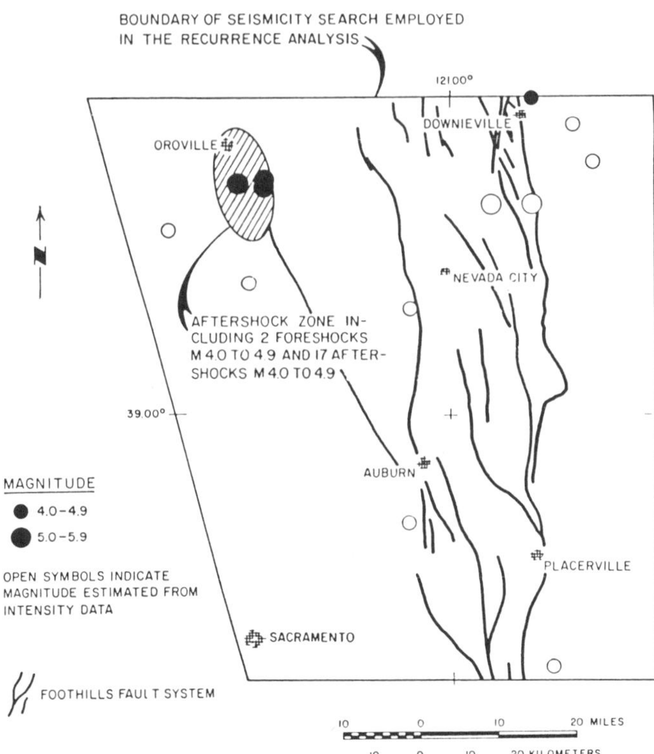

Figure 27. Historic (1900 to 1977) earthquakes of magnitude greater than 4.0 in the site area surrounding the proposed Auburn Dam, east-central California (CDMG, 1979).

traversing the site was low, although six faults occur within 30 km of the site and have been judged capable of producing a maximum credible earthquake of Richter magnitude 6.5 (Fig. 27).

The Auburn site certainly did not lack geological investigations. Before the 1975 earthquake, nearly 8 km of surface trenches, six exploratory adits (over 2 km in extent), and more than 30,000 m of core borings were drilled. After the earthquake, when it was suspected that the Foothills fault system could extend through the dam site, nearly 800,000 m^3 of rock and overburden soil were removed in order to expose and investigate the fault system. An additional 32 km of trenches were excavated and 6,100 m of core borings were completed.

An objective and retrospective analysis of the investigations indicates that several main concerns should have been recognized more fully and at an earlier time in the siting and design process: (1) the thin-arch dam had an unusually high crest-length to height ratio of 6:1; (2) embankment-type dams are normally considered most appropriate for this crest-length ratio in the presence of complex foundation geology; and (3) many faults traversed the foundation, raising the possibility of small movements caused by natural nonseismic strain release or by reservoir-induced seismicity. These should have raised concerns and been resolved prior to initiating an expensive foundation-preparation treatment.

As with other engineered works, it is necessary to consider potential impacts of the project on the surrounding area; conversely, how can geologic and hydrologic conditions of the site area affect each of the component structures planned?

Reports of findings

Phase II engineering geologic reports should be organized and compiled to provide input and advice relating to the preliminary engineering design. All components of the report should begin to develop the substance and elements of the bid documents (USNCTT, 1984), and particularly the geologic and geotechnical conditions of the earth materials expected to be encountered in foundations, open-cut excavations, or underground openings. Descriptions should be in simple, direct terms, understandable by engineers, and in unambiguous and forthright wording, as typical of good contract specifications; descriptions should include the manner in which a material is expected to behave or perform under the imposed conditions of static and dynamic loading and/or unloading, both dry and saturated. The description of materials to be excavated should include size gradation, shape (length-to-width ratios), placement characteristics, and durability under expected secondary seepage. Each rock unit should be clearly differentiated in terms of expected project usage, based on samples from outcrops or cored borings. The nature of ground-water flow should be carefully described, indicating known depths, flow directions, potential points of seepage, and avenues of infiltration, as well as how ground water can possibly affect the excavation, construction, operation, and maintenance activities. Definitive logs (Fig. 20) of borings should record any features or rock characteristics needed for design or construction planning. Contract documents should alert or warn prospective bidders to any potentially hazardous conditions that could develop during construction, such as rockfalls, methane gas concentrations, unstable slopes, and disaggregation of rockfill material.

SPECIALIZED INVESTIGATIONS FOR DESIGN

Frequently, detailed site investigations must incorporate unconventional methods to secure earth-material characterization and property data sufficient to meet all forthcoming design requirements. If the engineering geologist suspects the presence of certain subsurface defects that can initially be determined qualitatively by geophysical methods, those means should be employed to focus on the foundation area where problems exist. For example, a portable gravimeter was used to locate cavities in dissolution-prone marble lenses in mica schist at a site after excavation to grade. A plot of the mass distribution of the gravitational intensity demonstrated that the mass deficiencies approximately corresponded to cavities (Sumner and Burnett, 1974).

During any site investigation, the engineering geologist should compile a handy reference of the possible constraints that can be associated with each geologic unit; for example, although all limestone is not cavernous, it is logical in a carbonate regime to be keenly aware of possible dissolution along bedding planes, joints, and faults. The geologic history of a limestone mass should be studied carefully so as to deduce the possibility of paleo-ground-water flow and levels as a potential for dissolution effects. If the host rock has undergone structural deformation and was subsequently subjected to fluctuations of the ground-water surface, it is likely that some dissolution action has occurred.

An innovative new approach has been the use of seismic tomography to obtain preliminary information on geologic conditions along a proposed tunnel route for a hydroelectric project (Cosma, 1983). Seismic refraction traverses were made parallel and perpendicular to the tunnel alignment, and the depth of rock was incorporated into a computer surface-plotting program, which provided a three-dimensional graphic view of the top-of-rock surface. This perspective plot assisted in placing additional borings and exploratory adits at the most advantageous locations to determine details of suspected problems, character of individual units, and other features.

OUTLOOK

In future years, Phase II geological investigations will take on additional importance in the siting and design process. Although more will be expected and required, there will be a continuous struggle to seek adequate funding for these activities. As in other areas of professional services, the burden will fall on the practitioners. They must be more knowledgeable, more observant, and solve the problems at an ever faster pace.

In providing assistance to applied geologists and engineers, the engineering geophysicist will continue to miniaturize and refine instrumentation, which will peer farther into or across an alluvial-soil and/or rock mass, and thereby deliver better resolution. Most of the electromagnetic spectral variations of geophysics for engineering purposes have been established over the past 20 yr, but continued improvements in efficiency and clarity of record can be expected. Borehole video is already a useful competitor for the more expensive oriented coring, bringing the ability to deduce strike and dip from the borehole traces of discontinuities and bedding/foliation features. This technique, however, is currently ineffective in crystalline rocks that contain sufficient magnetite to have a slight magnetism, because this fouls the compass reading for the spatial orientation of the interpreter; borehole magnetism may be natural or induced by the drilling operation.

Drilling techniques should also improve, although not as dramatically as those of engineering geophysics. We can expect that air rotary drilling, assisted by improved air delivery, and innovations on nontoxic chemical "fluids" should increase future penetration rates as well as core recovery. Oriented core and the integral sampling method will likely find wider use, even at the considerable costs they represent.

Research needs and investigative techniques

Foundation investigations are trending toward in situ determination of inherent properties of soils, unconsolidated deposits, and rock masses (Mitchell, 1978). New techniques for exploration and interpretation were predicted by Underwood (1974) and are evolving rapidly. There is still considerable need to improve drilling methods, with respect to core recovery, cutting speed, and especially, hole alignment. Perhaps future exploration will be accomplished by hydraulic cutting under downhole power, as already demonstrated by the U.S. Bureau of Mines in the case of horizontal drilling (Kravits and others, 1987). There will always be growing pressure to replace more boreholes with geophysical imaging to reduce costs. For the foreseeable future, a practical goal in this arena may be the continued use of uncored boreholes, each logged initially by the crude cuttings and then by geophysical measurements; subsequently each hole can serve as a pathway or source for crosshole electromagnetic and acoustic-sensor imaging (King and others, 1978). However, a limited number of geologically logged, cored borings will always be required to calibrate the host lithology and geologic structure.

Improved methods of determining ground-water conditions in fractured rock and estimating flows into subsurface excavation are urgently needed, both for major engineering works and hazardous-waste site remediation.

Simple methods of investigating geologic conditions ahead of a tunnel face are now emerging. Besides improved methods of directionally controlled horizontal drilling now commercially available, it is anticipated that ground-probing radar (Cook, 1974) and acoustical holography (Price, 1974) will become accepted methods of predictive geological investigation.

The finite-element method mathematical analysis of estimated stress distribution and resulting deformation has been used in evaluating rock masses since the early 1970s, and currently with increasing success, to determine rock-slope stability and estimate the distribution and effect of stresses on foundations and underground openings. The 1983 rock slide that interrupted major transcontinental rail transportation along the Spanish Fork river, near Thistle, Utah, was analyzed by back-calculation using the finite-element method (Leonard and Olsen, 1988); the survey revealed that cleft-water seepage forces were important in bringing about the failure. High precipitation caused hillside saturation to a relatively shallow depth, with sliding initiated from the toe and progressing uphill. This analytical method will be increasingly utilized for the investigation of geologic features as microcomputer storage capabilities grow to meet the large capacities required for the finite-element-method matrix representation and computation.

Techniques of probabilistic analyses have also been refined with advances in computer capabilities. These techniques help applied geologists portray the geologically associated risks that face owners and that will require costly solutions to mitigate or remediate. For example, Wolfe and others (1988) set forth probability estimates of problems for an earthen flood-control dam due to a postulated recurrence of the New Madrid, Missouri, earthquake.

The future

The degree of intensity of a site investigation is controlled chiefly by project requirements and geologic complexities. This intensity is also a function of the owner's degree of acceptance of risk relevant to encountering unexpected construction conditions, or failure during operation, as measured in terms of costs. The cost versus risk comparison of Kent (1981) indicates a general balance point between the cost of investigations and the cost of risks, beyond which an increase in the overall investigations may not be justified. Most geological and geo-related investigations still fall short of reaching this point and more likely do not exceed about 0.5 percent of the total construction costs. Because the permissible risk is related to this balance, a main objective of the preferred investigations is to forecast what can be expected at a site and reduce the risk of unforeseen failure (discussed further in Chapter 22).

REFERENCES CITED

AASHTO, 1988, Hatheway, A. W., ed., Manual on subsurface investigations: Washington, D.C., American Association of State Highway and Transportation Officials, various pagination.

Albritton, J., Jackson, L., and Bangert, R., 1984, Foundation grouting practices at Corps of Engineers dams: Vicksburg, Mississippi, U.S. Army Engineer Waterways Experimental Station Technical Report GL-84-13, 404 p.

ASCE, 1976, Subsurface investigation for design and construction of foundations for buildings: New York, American Society of Civil Engineers Manuals and Reports on Engineering Practice 56, 61 p.

Beck, J. L., 1976, Weight-induced stresses and the recent seismicity of Lake Oroville, California: Bulletin of the Seismological Society of America, v. 66, p. 1121–1131.

Bellport, B. P., 1970, Auburn Dam will be the world's longest concrete arch dam: The Japan Dam Association, p. 343–347.

Burwell, E. B., and Nesbitt, R. H., 1954, The NX borehole camera: Mining Engineering, v. 6, no. 8, p. 805–808.

Cabrera, J. G., 1963, Geology of the reservoir basin in relation to its impermeabilization, *in* Proceedings 2nd Pan-American Conference on Soil Mechanics and Foundation Engineering, Belo Horizonte, Brazil, v. 2, division 3b: Belo Horizonte, Associacao Brasiliera de Mecanica dos Solos, p. 691–694.

——, 1974, The importance of structural analysis and direct methods of foundation investigations for concrete structures of large dams, *in* Proceedings 2nd Congress, International Association of Engineering Geology, Theme 6, v. 2: San Paulo, Brazil, Associacao Brasiliera de Geologia de Engenharia, p. 5.1–5.7.

Caggiano, J. A., Jr., 1979, A three-phase program of investigation for site selection and development, *in* Hatheway, A. W., and McClure, C. R., Jr., eds., Geology in the siting of nuclear power plants: Geological Society of America Reviews in Engineering Geology, v. 4, p. 13–25.

CANMET, 1976–1981, The pit slope manual: Ottawa, Ontario, Canadian Centre for Mineral and Energy Technology, 10 chapters and numerous supplements.

CDMG, 1979, Technical review of the Auburn Dam site: Sacramento, California Division of Mines and Geology Special Publication 54, 17 p.

Cook, J. C., 1974, Status of ground probing radar and some recent experience, *in* Subsurface exploration for underground excavation and heavy construction: New York, American Society of Civil Engineers, p. 175–194.

Cosma, C., 1983, Determination of rock mass quality by the crosshole seismic method: International Association of Engineering Geology Bulletin 26-27, p. 219–225.

Cregger, D. M., 1986, Shaft construction in crystalline rock: Bulletin of the Association of Engineering Geologists, v. 23, no. 3, p. 287–295.

Deere, D. U., 1964, Technical description of rock cores for engineering purposes: Rock Mechanics and Engineering Geology, v. 1, p. 17–22.

Dirmeyer, R. D., 1950, Geology and irrigation engineering, *in* Van Tuyl, F. M., and Kuhn, T. R., eds., Applied geology: Golden, Colorado School of Mines Quarterly, v. 45, no. 1B, p. 123–154.

Dobecki, T. L., 1979, Measurement of in situ dynamic properties in relation to geologic conditions, *in* Hatheway, A. W., and McClure, C. R., Jr., eds., Geology in the siting of nuclear power plants: Geological Society of America Reviews in Engineering Geology, v. 4, p. 201–214.

Dunnicliffe, J., 1988, Geotechnical instrumentation for monitoring field performance: New York, John Wiley and Sons, 577 p.

ENR, 1939, Mining under Seminoe Dam: New York, Engineering News Record, v. 122, no. 15, p. 490–492.

——, 1974a, Fault moves BUREC to prestressed dam design: New York, Engineering News Record, v. 193, no. 15, p. 76–79.

——, 1974b, Crystal Dam earth fill changed to arch due to cramped site: New York, Engineering News Record, v. 194, no. 22, p. 16–17.

Engineering, 1868, Mont Cenis tunnel, Alps: London, Engineering, July 24, 1868, p. 71.

Evans, D. M., 1966, Man-made earthquakes in Denver: Geotimes, v. 10, p. 11–18.

Fisher, P. R., and Banks, D. L., 1978, Influence of the regional geologic setting on site geologic features, *in* Dowding, C. H., ed., Site characterization and exploration: New York, American Society of Civil Engineers, p. 163–184.

Fornwald, W. L., 1958, Recent devevelopments in soil sampling and drilling, *in* Proceedings 9th Conference on Geology as Applied to Highway Engineering: Charlottesville, Virginia, University of Virginia and Virginia Department of Highways, p. 31–38.

Fox, P. P., 1941, Foundation experiences, Tennessee Valley Authority; Foundation exploration and geologic studies at Chickamauga Dam: Transactions, American Socoety of Civil Engineers, no. 106, paper 2113, p. 765–779.

Galster, R. W., 1977, A system of engineering geology mapping symbols: Bulletin of the Association of Engineering Geologists, v. 14, no. 1, p. 39–47.

Geological Society of London, 1972, The preparation of maps and plans in terms of engineering geology: Engineering Group Working Party, Quarterly Journal of Engineering Geology, v. 5, no. 4, p. 293–382.

——, 1988, Engineering geophysics: Engineering Group working Party, Quarterly Journal of Engineering Geology, v. 21, no. 3, p. 207–271.

Gieseke, F. E., 1922, Columns and walls lifted by swelling clays under floor: New York, Engineering News Record, v. 88, no. 5, p. 192–193.

Gignoux, M., 1945, General geological conditions of hydroelectric projects in Portugal: Lisbon, Tecnica, Review of Engineering, no. 158, p. 541–550.

Goodman, R. E., 1976, Methods of geological engineering in discontinuous rocks: St. Paul, Minnesota, West Publishing Company, 472 p.

——, 1978, On field and laboratory methods of rock testing for site studies, *in* Dowding, C. H., eds., Site characterization and exploration: New York, American Society of Civil Engineers, p. 131–151.

Gray, R. E., 1988, Coal mine subsidence and structure, *in* Siriwardane, H. J., ed., Mine induced subsidence: New York, American Society of Civil Engineers Geotechnical Special Publication 19, p. 69–86.

Haimson, B. C., 1981, In situ stress measurements in hydro projects, *in* Kulhawy, F. H., ed., Recent developments in geotechnical engineering for hydro projects: New York, American Society of Civil Engineers, p. 207–222.

Hamel, J. V., 1976, Libby Dam left abutment rock wedge stability, *in* Rock engineering for foundations and slopes; Proceedings, Speciality Conference, Boulder, Colorado: New York, American Society of Civil Engineers, v. 1, p. 361–385.

Hatheway, A. W., 1982, Trench, shaft, and tunnel mapping: Bulletin of the Association of Engineering Geologists, v. 19, no. 2, p. 173–180.

Hatheway, A. W., and Feininger, T., 1988, Memorial to Clifford Alan Kaye, FGSA, 1914–1986: Geological Society of America Memorials, v. 18.

Hatheway, A. W., and Kiersch, G. A., 1982, Engineering properties of rock, *in* Carmichael, R. S., ed., Handbook of physical properties of rock: Boca Raton, Florida, CRG Press, v. 2, p. 289–331.

Heiland, C. A., 1946, Geophysical exploration: New York, Prentice-Hall, 1013 p.

Hoek, E., and Bray, J., 1981, Rock slope engineering, revised 3rd ed.: London, Institute of Mining and Metallurgy, 358 p.

Hvorslev, M. J., 1949, Subsurface exploration and sampling of soil for engineering purposes: Vicksburg, Mississippi, U.S. Army Engineer Waters Experimental Station, 521 p. (reprinted 1962, 1965 by Engineering Foundation, United Engineering Center, New York).

ISRM, 1981, Suggested methods for determining shear strength, *in* Brown, E. T., ed., Rock characterization, testing, and monitoring; International Society for Rock Mechanics suggested methods: Oxford, Pergamon Press, document 1, p. 129–140.

Jones, F. O., Embody, D. R., and Peterson, W. L., 1961, Landslides along Columbia River valley, northeastern Washington: U.S. Geological Survey Professional Paper 367, 93 p.

Keaton, J. R., 1984, Genesis-Lithology-Qualifier (GLQ) system of engineering geology mapping symbols: Bulletin of the Association of Engineering Geologists, v. 21, no. 3, p. 355–364.

Kempton, J. P., and Cartwright, K., 1984, Three dimensional geologic mapping; A basis for hydrogeologic and land-use evaluation: Bulletin of the Association of Engineering Geologists, v. 21, no. 3, p. 317–335.

Kent, M. D., 1981, Site investigation for construction excavation: Bulletin of the Association of Engineering Geologists, v. 28, no. 1, p. 71–76.

Kiersch, G. A., 1955, Engineering geology, historical development, scope, and utilization: Golden, Colorado School of Mines Quarterly, v. 50, no. 3, 123 p.

——, 1959, Handbook for geologists, engineers, and draftsmen: San Francisco, California, Southern Pacific Land Company, 185 p. (Company publication).

——, 1964, The Vaiont reservoir disaster: Civil Engineering, v. 34, no. 3, p. 32–39.

Kiersch, G. A., and James, L. B., 1988, Reservoirs, *in* Jansen, R. B., ed., Advanced dam engineering for design, construction, and rehabilitation: New York, Van Nostrand-Reinhold, p. 722–750.

Kiersch, G. A., and Treasher, R. C., 1955, Investigations, areal and engineering geology—Folsom Dam Project, central California: Economic Geology, v. 50, p. 271–310.

King, M. S., Stauffer, M. R., and Pandit, B. I., 1978, Quality of rock masses by acoustic borehole logging, *in* Annals of the 3rd International Congress, International Association of Engineering, Geology: Madrid, Spain, Asociacion Espanola de Geologia Aplicada a la Ingenieria, v. 1, no. 17, p. 156–164.

Kravits, S. J., Sainato, A., and Finfinger, G. L., 1987, Accurate directional borehole drilling; a case study at the Navajo Dam, New Mexico: U.S. Bureau of Mines Report 9102, 25 p.

Lambe, T. W., 1951, Soil testing for engineers: New York, John Wiley and Sons, 165 p.

Leeman, E. R., and Hayes, D. J., 1966, A technique for determining the complete state of stress in rock using a single borehole, *in* Proceedings 1st Congress, International Society of Rock Mechanics: Lisbon, Portugal, Laboratorio Nacional de Engenharia Civil, v. 2, p. 17–24.

Legget, R. F., and Hatheway, A. W., 1988, Geology and engineering, 3rd ed.: New York, McGraw-Hill, 613 p.

Leonard, B. D., and Olsen, J. M., 1988, A finite element analysis of the Utah "Thistle" failure, *in* Prakash, S., ed., Proceedings 2nd International Conference on Case Histories in Geotechnical Engineering: Rolla, University of Missouri Department of Civil Engineering, v. 1, p. 593–598.

Londe, P., and Tardieu, B., 1977, Practical rock foundation for dams, *in* Fairhurst, C., and Crouch, S. L., eds., Design methods in rock mechanics: New York, American Society of Civil Engineers, p. 115–138.

Low, J. W., 1957, Geologic field methods: New York, Harper and Sons, 489 p.

Lugeon, M., 1933, Barrages et geologie: Lausanne, Switzerland, University of Lausanne, 170 p.

Mackin, J. H., 1941, A geologic interpretation of the failure of Cedar reservoir, Washington: Seattle, University of Washington Engineering Experimental Station, Bulletin 107, 30 p.

Maddock, T., Jr., and Borland, W. M., 1951, Sedimentation studies for the planning of reservoirs by the Bureau of Reclamation, *in* Proceedings 4th Congress on Large Dams: New Delhi, India, International Commission on Large Dams, Question 14, p. 41.

Mathewson, C. C., Dobson, B. M., Dyke, L. D., and Lytton, R. L., 1980, System interaction of expansive soils with light foundations: Bulletin of the Associa-

tion of Engineering Geologists, v. 27, no. 2, p. 55–94.
McHarg, I. L., 1969, Design with nature: New York, The Natural History Press, 197 p.
McKinstry, H. E., 1948, Mining geology: Englewood Cliffs, New Jersey, Prentice-Hall, p. 82–106.
Meade, R. B., 1983, Reservoir-induced macroearthquakes; A reassessment, *in* Howard, T. R., ed., Seismic design of embankments and caverns: New York, American Society of Civil Engineers, p. 23–40.
Meade, W. J., 1936, Engineering geology of dam sites: 2nd Congress on Large Dams, Washington, D.C., Question VI, p. 1–22.
——, 1937a, Geology of dam sites; 1, Dam sites in hard rock: Civil Engineering, v. 7, no. 5, p. 331–334.
——, 1937b, Geology of dam sites; 2, Dam sites in shale and earth: Civil Engineering, v. 7, no. 5, p. 392–395.
Middlebrooks, T. A., 1942, Fort Peck slide: Transactions of the American Society of Civil Engineers, v. 107, p. 723–764.
Mitchell, J. K., 1978, In situ techniques for site characterization, *in* Dowding, C. H., ed., Site characterization and exploration: American Society of Civil Engineers, p. 107–124.
Nickell, F. M., 1942, Development and use of engineering geology: American Association of Petroleum Geologists Bulletin, v. 26, p. 1797–1826.
Obert, L., Windes, S. L., and Duvall, W., 1946, Standardized test for determining the physical properties of mine rocks: U.S. Bureau of Mines Report of Investigations 3891.
Packer, D. R., Cluff, L. S., Schwartz, D. P., Swant, F. H., and Idriss, I. M., 1978, Auburn Dam; A case history of earthquake evaluation for a critical facility: San Francisco, California, Woodward-Clyde Consultant.
Peck, R. B., 1985, The last 60 years, *in* Proceedings 11th International Conference on Soil Mechanics and Foundation Engineering, San Francisco, California; Golden Jubilee Volume: Rotterdam, Balkema, p. 123–133.
Pratt, H. R., and Voegele, M. D., 1984, In situ tests for site characterization, evaluation, and design: Bulletin of the Association of Engineering Geologists, v. 21, no. 1, p. 3–22.
Price, T. O., 1974, Acoustical surveying: Annual Mineralogy Symposium Proceedings, no. 17, p. 187–205.
Proctor, R. J., 1971, Mapping geological conditions in tunnels: Bulletin of the Association of Engineering Geologists, v. 8, no. 1, p. 1–43.
Radbruch-Hall, D., Edwards, K., and Baston, R. M., 1979, Experimental engineering geology maps of the conterminous United States prepared using computers: Bulletin of the Association of Engineering Geologists, v. 19, p. 358–363.
Rajendran, K., and Gupta, H. K., 1986, Was the earthquake sequence of August 1975 in the vicinity of Lake Oroville, California, reservoir induced?: Physics of Earth and Planetary Interiors, v. 44, p. 142–148.
Ransome, F. L., 1928, Geology of the St. Francis Dam site: Economic Geology, v. 23, p. 553–563.
Rhoades, R., 1950, Influence of sedimentation on concrete aggregate, *in* Trask, P. D., ed., Applied sedimentation: New York, John Wiley and Sons, p. 437–463.
Riker, R., and Daller, D., 1988, Design, construction, and performance of a deep excavation in soft clay, *in* Prakash, S., ed., Proceedings 2nd International Conference on Case Histories in Geotechnical Engineering: Rolla, University of Missouri Department of Civil Engineering, v. 2, p. 1263–1269.
Roberts, G. D., 1964, Investigation versus exploration: Bulletin of the Association of Engineering Geologists, v. 1, no. 2, p. 37–42.
Rocha, M., 1974, Present possibilities of studying foundations of concrete dams, *in* Proceedings 3rd Congress of International Society for Rock Mechanics, Denver, Colorado: Washington, D.C., National Academy of Sciences, theme 3, v. 1, p. 879–897.
Sanglerat, G., 1972, The penetrometer and soil exploration: Amsterdam, Elsevier, 464 p.
Schriever, W., 1956, translation of Collin, A., 1846, Landslides in clay: University of Toronto Press, 181 p.
Schuster, R. L., 1979, Reservoir-induced landslides: Bulletin of the International Association of Engineering Geology, no. 20, p. 8–25.
Stephens, J. L., Allen, L. H., Jr., and Chen, E., 1984, Organic soil subsidence, *in* Holzer, T., ed., Man-induced land subsidence: Geological Society of America Reviews in Engineering Geology, v. 7, p. 107–122.
Sumner, J. R., and Burnett, J. A., 1974, Use of precision gravity survey to determine bedrock: Proceedings of the American Society of Civil Engineers, Journal Geotechnical Division, v. 100, no. GT1, p. 53–58.
Trask, P. D., ed., 1950, Applied sedimentation: New York, John Wiley and Sons, 707 p.
TVA, 1949, Geology and foundation treatment; Tennessee Valley Authority projects: Washington, D.C., U.S. Government Printing Office, Tennessee Valley Authority Technical Report 22, 548 p.
Underwood, L. B., 1974, Exploration and geologic prediction for underground works, *in* Subsurface exploration for underground excavation and heavy construction: New York, American Society of Civil Engineers, p. 65–83.
USBR, 1987, Design of small dams, Washington, D.C., U.S. Government Printing Office, U.S. Bureau of Reclamation, 3rd edition, 595 p.
——, 1974, Earth manual: Washington, D.C., U.S. Government Printing Office, U.S. Bureau of Reclamation, 2nd edition, 810 p.
——, 1988, Engineering geology field manual: Washington, D.C., U.S. Government Printing Office, U.S. Bureau of Reclamation, 598 p.
USCE, 1946, Engineering manual for civil works; Chapter 4, Geological investigations: Washington, D.C., U.S. Army Corps of Engineers Office of the Chief of Engineers, various paging.
——, 1970, Geologic mapping of tunnels and shafts by the full periphery method: Washington, D.C., U.S. Army Corps of Engineers Office of the Chief of Engineers, Engineers Technical Letter 1110-1-37, 6 p.
USCEQ, 1970, National Environmental Policy Act of 1969 (PL 91-190, 1 Jan. 1970): Washington, D.C., U.S. Council on Environmental Quality, 1st Annual Report, Appendix A, p. 243–249.
USNC/RM, 1981, Rock mechanics research requirements for resource recovery, construction, and earthquake hazard reduction: Washington, D.C., National Academy Press, U.S. National Committee for Rock Mechanics, 222 p.
USNCTT, 1984, Geotechnical site investigations for underground projects; v. 1, Overview of practice and legal issues, evaluation of cases, conclusions, and recommendations; v. 2, Abstracts of case histories and computer-based data management system: Washington, D.C., National Academy Press, U.S. National Committee for Tunneling Technology, v. 1, 182 p., v. 2, 241 p.
Varnes, D. J., 1974, The logic of geological maps, with reference to their interpretation and use for engineering purposes: U.S. Geological Survey Professional Paper 837, 48 p.
Wantland, D., 1950, Geophysical investigations of the lower Ashland, Folsom, and Nimbus Dam sites, American River Divisions, Central Valley Project, California: U.S. Bureau of Reclamation Geology Report G-109, various paging.
Weaver, J. M., 1975, Geological factors significant in the assessment of rippability: Die Siviele Ingenieur in Suid-Afrika, December, p. 313–316.
Williamson, D. A., 1984, Unified rock classification system: Bulletin of the Association of Engineering Geologists, v. 21, no. 3, p. 345–354.
Wilson, S. D., 1970, Observational data on ground movements related to slope stability: Proceedings of the American Society of Civil Engineers, Journal of Soil Mechanics and Foundation Division, v. 96, no. SM5, p. 1519–1544.
Wolff, T. E., Hempen, G. L., Dirnberger, M. M., and Moore, B. H., 1988, Probabilistic analysis of earthquake-induced pool release, *in* Prakash, S., ed., *in* Proceedings 2nd International Conference on Case Histories in Geotechnical Engineering: Rolla, University of Missouri Department of Civil Engineering, v. 1, p. 787–793.

MANUSCRIPT ACCEPTED BY THE SOCIETY OCTOBER 17, 1989

ACKNOWLEDGMENTS

We are most appreciative to George A. Kiersch for the meaningful reviews and input of information to earlier draft versions of the chapter, and particularly for his patience and personal efforts in assembling the final version of this text.

Boulder (Hoover) Dam site under construction in the early 1930s. Note diversion tunnels in each abutment, and blasting for the keyway of the dam in volcanic rocks. (Photo courtesy of the Heinrich Ries collection, Cornell University.)

Printed in U.S.A.

Geological Society of America
Centennial Special Volume 3
1991

Chapter 20

Construction geology; As-built conditions

Jeffrey R. Keaton
Sergent, Hauskins and Beckwith, Consulting Geotechnical Engineers, 4030 South 500 West, Salt Lake City, Utah 84123

INTRODUCTION

Most of the historic milestones in engineering geology resulted from the construction of important, large-scale projects, as described in Kiersch (this volume, Chapter 1). The construction phase of a project follows final design and precedes operation and maintenance. Careful documentation of geologic conditions exposed in significant excavations and modification of designs to cope with "changed conditions" are routine elements of major construction projects such as dams, tunnels, power plants, and highways. Unless required by a local ordinance, as-built geological mapping on smaller projects, such as industrial developments and residential subdivisions, usually is not requested by a project owner. However, the increasing attention to economic considerations and the contemporary legal atmosphere are contributing to the expanded use of geological documentation during construction on smaller projects. Roberts (1973, p. 145) analyzed more than a thousand project reports and case histories and concluded that "the comparatively small investment required to obtain a foundation 'as-built' report provides a very substantial return." He notes that preparation of foundation "as-built" reports assures: (1) the owner received what was paid for; (2) the design engineer that construction was in accordance with design assumptions, or any modifications necessary were made in a timely and economical manner; and (3) a record is provided to guide any subsequent structural modifications, remedial treatment of failures, or design of nearby new construction.

The initial involvement of geologists on construction projects most likely occurred in response to problems or failures caused by an inability of the engineer's conservatism or "factor of safety" to compensate for the geologic conditions, which at most sites are unique and cannot necessarily be treated as a random variable with some average value. Karl Terzaghi, in the Foreword to his classic 1943 text *Theoretical Soil Mechanics,* noted that failures in earthwork engineering commonly occur due to unanticipated action of water, which "depends much more on minor geological details than does the behavior of the soil. As a consequence the departure from the average expressed by empirical rules such as those which are used in the design of dams on permeable strata is exceptionally important." The complex and variable nature of many aspects of the geology of a construction site leads the geologist to speculate sometimes about critical elements during the design phase of a project. Frequently, this has resulted in (1) designs based on incomplete geological information, or (2) the modification of designs during construction to accommodate unanticipated conditions. The cost of acquiring subsurface geological data is sufficiently great that interpretations based on limited data and considerable experience often are more valuable in design than evaluations based on considerable data but limited field experience (discussed in Kiersch and James, Errors in Judgment, this volume).

Berkey and Sanborn (1923, p. 91), following their work on the extensive Catskill aqueduct for New York City, commented, "In an important case, where all the facts and conditions are not perfectly clear, the geologist should know enough not to give a final opinion without additional exploratory evidence; and the engineer should be wise enough not to treat an opinion as if it was a proven fact and then blame the geologist for mistakes which belong to both." Burwell and Roberts (1950, p. 9) discussed the role of the geologist in the engineering organization and noted that the construction phase of a large engineering project provides a view of the results of the geologist's work to an extent not commonly met in other fields of geology. Yet the geologist will be exceedingly fortunate if the scattered geologic data obtained during the investigation phase are interpreted sufficiently well that no modification of the construction plans or specifications are needed. Among the more important services provided by the geologist are detailed mapping of foundation excavations; logging of drill holes and test pits; providing geologic information and advice for the guidance of the engineer; and preparing documentary reports. Detailed "as-built" records of ground-water, excavation, and foundation conditions that affect construction methods and procedures may be of inestimable value should the contractor, at some later time, file claims based on changed conditions. Kiersch (1955, p. 4) noted that insufficient geological data contribute to increased construction costs, contract change orders, legal suits, construction delays, and possibly to structural failures. Yet he cautioned the geologist not to withhold opinions to avoid possible mistakes because such action forces the engineer in charge, who is responsible for ultimate success or failure of the project, to make geological decisions.

Burwell and Roberts (1950) implied that geologists gener-

Keaton, J. R., 1991, Construction geology; As-built conditions, *in* Kiersch, G. A., ed., The heritage of engineering geology; The first hundred years: Boulder, Colorado, Geological Society of America, Centennial Special Volume 3.

ally represent the owners of large projects and not the contractors. However, the contracts of most large projects are based on performance standards. Consequently, ingeneous and competent applied geologists on the contractor's staff can utilize the favorable geologic conditions and compensate for the unfavorable ones; such recognition and action can be critical to the successful completion of a contract, as well as increase the contractor's profitability. Furthermore, should "changed conditions" actually be encountered during performance of a contract, the contractor's geologist should be capable of documenting the nature and extent of deviations from the geological information described in the bid documents. An understanding of the actual geologic conditions and communication between the owner's geologist and the contractor's geologist can lead to rapid and equitable settlement of construction claims. Alternatively, in the event that a dispute cannot be avoided, both the owner and the contractor ultimately may require geological expertise during litigation, as discussed in Waggoner and Kiersch (this volume).

GEOLOGY DURING CONSTRUCTION

Purpose

Geological principles can be applied and guidance given during construction for pragmatic, engineering, scientific, and/or legal reasons. The pragmatic and engineering justification for the practice of construction geology is to create a safe and economical project that is based on reasonable design recommendations rather than overly conservative ones, as discussed in Cabrera (this volume). This is accomplished by geological investigations and observations during construction to permit verification and/or modification of the geological interpretations and recommendations on which the design is based. Federal agencies, such as the U.S. Army Corps of Engineers and the U.S. Bureau of Reclamation, have developed guidelines for systematic collection of geological data during construction. Additionally, in the event that a geology-related problem develops during operation and maintenance, the as-built geologic map could provide invaluable guidance for a rapid repair. Leighton (1966, p. 177) notes that treatment of unanticipated ground-water problems during hillside grading is "much simpler and less expensive if caught at an early stage of grading."

The scientific and engineering-related rationale for the practice of construction geology is to develop the best possible understanding and most accurate representation of the site-specific geology. This should provide the best possible design for the engineering works, as well as contribute to a better understanding of the areal-regional geologic setting. The quality and extent of new geologic exposures created during construction is unique and frequently unparalleled in nature. Additionally, geological investigations made with new field exposures during construction provides a means for comparing the conditions encountered with those anticipated on the basis of the design investigations. Furthermore, the adequacy of the scope and techniques used in design investigations can be evaluated and strengthened to improve the results and interpretations of future assignments.

The primary legal impetus for owners of sites and project sponsors to pay for geological documentation is the need for: (1) a basis for evaluating potential contractors' claims that cite "changed conditions," and (2) "base-line" data in the event of future litigation or maintenance needs. Similarly, the contractor may utilize geological investigations during construction to compare the anticipated conditions on which the bid was based to those actually encountered.

Scope

The scope of geological investigations conducted during the construction process can range widely based on the scale and nature of the project and the complexity and severity of geologic conditions at the site (Leighton, 1966, p. 177). When a foundation is excavated, geologic features and conditions are sometimes revealed that require design adjustments or changes in the scheduled construction procedures to successfully create safe works (Farina, 1987). Kent (1981, p. 75) recommends that observations be made "continuously during construction to confirm and refine the subsurface geological model for possible design modifications, and to observe compliance with specifications." The construction geologist must observe the geologic conditions exposed in construction excavations, evaluate the compatibility of the design to the foundation geology, assess potential safety hazards related to the geology, and document the geologic features for: (1) general guidance during operation and maintenance; (2) any potential claims of changed conditions by the contractor; and/or (3) eventual changes of foundation conditions or failures with time (Farina, 1987).

Water-retention dams probably represent the class of project that historically demonstrates the most intensive use of geology during construction. The enormous cost of a dam, combined with the potential for loss of life and widespread damage in the event of a failure, demands that attention be paid to every detail, particularly in the foundation materials. Mead (1936, 1937) summarized many pertinent engineering geology issues for dam sites at a time when their importance was being realized. Irwin (1938) mapped the foundation excavation of the Grand Coulee dam in Washington as it was exposed in a series of square panels, each about 15 m by 15 m. A "simplified" geological map of a portion of the foundation of Grand Coulee Dam is presented in Figure 1. Irwin (1938) noted that (p. 1630):

> The final cleaning process involved the removal of all debris and loose fragments, scrubbing with wire brushes and sand blast, if necessary, and a thorough washing with high-pressure streams of air and water. After this final cleaning, the separate blocks of bedrock surface were mapped individually on a scale of 1 inch = 12 feet [1:144]. A mosaic office map was then constructed on a reduced scale of 1 inch = 20 feet [1:240] to emphasize the major geologic and structural features of the rock. The unusual opportunity to study thousands of square feet of freshly cleaned and washed rock surface made possible a detailed analysis of the successive stages in the history of the bedrock.

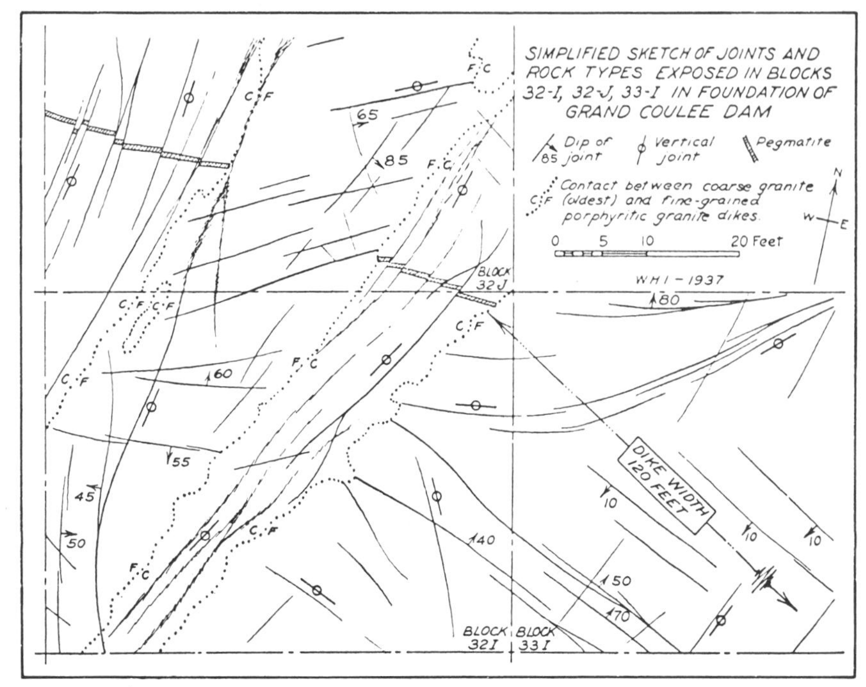

Figure 1. Simplified geologic map of the granitic rocks exposed in a small portion of the foundation of Grand Coulee Dam, Washington. (From Irwin, 1938, Fig. 4.)

Smaller engineering works may require less extensive geological investigation during construction. Rock slopes on highway projects might be mapped at a scale of 1:200 to 1:500, while graded portions of residential subdivision projects might be mapped at scales of 1:600 to 1:2,400, depending on the size of the grading plan. Other even smaller projects might feature intermittent site visits by an engineering geologist without preparation of a comprehensive geologic map.

Instrumentation may be installed early in the construction phase or even during the design phase of projects on which slope problems or ground-water fluctuations are anticipated. Since the 1940s it has been recognized that instrumentation can provide a basis for decisions to modify plans when deformations, stresses, or ground-water levels exceed a threshold established during design (cf., Hunt, 1984, p. 8).

On some projects involving rock slopes, the primary issues might be the spacing and type of rock support, the loading pattern of blast holes to minimize overbreak, or even the final angle of design slope to compensate for variations in bedding or joint orientation. Large grading projects involving weak rock or surficial materials ("soil" in the engineering sense) might involve primary issues such as the depth to uniformly firm material to provide the support for a massive fill slope or the suitability of excavated material for use as compacted fill. Many issues are of combined concern to geologists and engineers, particularly repair of slope failures caused by construction activities.

Techniques

Techniques used by geologists during construction do not differ significantly from those used during feasibility and design investigations. Yet, because of the economics of a major construction operation, whatever field and/or laboratory techniques are utilized by the geologist must yield rapid and usable results. Documentation of direct observations of construction excavations on traditional, appropriate-scale geological maps or logs is still the engineering geologist's primary tool for making interpretations. Construction excavations can include tie-backs, soldier beams, and dewatering wells as well as building footings, utility trenches, sumps, cutoff trenches, shear keys for buttress fills, tunnel headings, grout holes, and blast holes. Meaningful geological information can be developed from examination and interpretation of drilling rates, circulation losses, drill cuttings, and down-hole inspections. The results of routine engineering tests, such as pile-load tests, tie-back pullout tests, and other proof-load tests, should be reviewed and interpreted to aid in understanding variations in the site-specific geologic conditions.

Detailed documentary terrestrial (hand-held) photography is a rapid means of recording conditions at a moment in time. Changes can be documented by reoccupying a bench mark reference point periodically and taking pictures in the same orientation with a high-resolution camera and film. Hunt (1984, p. 259) suggests that "moving the camera a short distance laterally (about 1 m) and taking a subsequent photo provides photo pairs suitable for stereoscopic viewing when enlarged." The writer has achieved good stereoscopic results with 35 mm transparencies taken with lateral spacing of about a shoulder width (less than 50 cm). If suitable topographic control is visible on the photographs to permit creation of stereo models, stereoscopic terrestrial photographs can be used to produce pseudo-topographic or orthotopographic maps, which can be extremely useful for mapping joints and other discontinuities on rock slopes. If prints of photographs can be developed quickly, they could be used as a base for annotations of geologic details such as joint orientations and contacts separating engineering geologic units. Down-hole photography and closed-circuit television cameras, first used in the 1950s (Kiersch and Treasher, 1955, p. 301), provide direct observation of subsurface conditions in small-diameter bore holes. Use of video recording cameras, a common practice today, permits narrated documentation of construction excavations suitable for communication purposes, long-distance transmission, archival storage, and court exhibits.

Instrumentation established early in the construction phase or earlier during the design investigations can provide a nearly continuous input of information, depending on the frequency of readings and data reduction procedures. Typical instruments for detecting and monitoring deformations are simple levelling for survey and resurvey of "benchmark" monuments and settlement plates, electric tiltmeters, strain meters, extensometers, and inclinometers (Hunt, 1984, p. 252). Open-tube, pneumatic, and electric piezometers are used routinely to monitor changes in ground-water levels, which can serve as early warning indicators for increased dewatering requirements.

RELEVANCE OF GEOLOGIC FACTORS

Introduction

Representative case histories have been selected to illustrate the importance and possible impact of geological investigation for construction and the as-built records. The case histories are based on construction of dams, tunnels, a navigation lock, highways, and a service road. Many other fine case histories are available that demonstrate the importance of geologic factors during construction. Numerous classic case histories have yielded significant lessons resulting from postconstruction failures rather than discoveries during construction. Some are reviewed in chapters by James and Kiersch, and Kiersch and James (this volume). The cases described below represent discoveries or lessons learned from changed or unsuspected conditions encountered during construction or revealed during late stages of preconstruction investigation.

One of the early practitioners and pioneers, Frank A. Nickell, geologist for Boulder Dam and formerly chief geologist of the Bureau of Reclamation (1935 to 1941), published "Development and use of engineering geology" in 1942. He used the five largest concrete dams in the world at that time as examples of projects with geologic problems of engineering significance. The details contained on the "simplified" geologic maps of the foundation or abutment excavations are described by Nickell (1942).

Figure 2 is a geologic map of the foundation of Boulder dam, underlain by Tertiary volcanic rocks. Most of the foundation is andesitic tuff breccia, which includes zones of close and widely spaced shearing. The abutments include latite, latite-flow breccia, a biotite latite sill, and basalt dikes. The left abutment (on the Arizona side) also includes relatively minor zones of conglomerate between latite flows. A number of faults cutting the bedrock formations were exposed in the foundation excavation. More attention would be paid today than was considered necessary in the 1930s regarding potential geoseismic aspects of faulting at such a site.

Figure 3 is a geologic map of the foundation of Grand Coulee Dam, which is underlain by intensely jointed, but relatively fresh granitic rocks. The intensity of jointing is shown on Figure 4. The relative freshness of the rock resulted from deep scouring by Pleistocene glaciation, which removed the weathered rock from the site—a sharp contrast to conditions at the Folsom site described in the next section of this chapter. Thus, despite the degree of fracturing of the rock, its freshness resulted in excellent foundation support.

Faulted, jointed, and intruded metamorphic rocks in the southeast abutment of Shasta dam are shown on Figure 5. Four varieties of faults or shear zones are distinguished on this map, as well as two kinds of joint conditions and two degrees of alteration. It is clear from the geologic complexity shown on Figure 5 that different degrees of treatment would be required in different areas of the foundation. Thorough geological mapping and investigation of the foundation during construction can provide timely information needed by the project engineer.

These projects are important, not only with respect to improvements in design and construction methods for dams on such foundations, but also to advancements in the state of knowledge of areal-regional geology and geologic processes.

Weathered rock

Folsom Dam. The Folsom Dam project is on the American River, 32 km northeast of Sacramento in north-central California. The project consists of a concrete gravity section 427 m long and 108 m high with compacted earth-wing dams, another high earth dam 3.5 km to the east, and eight dike sections around the reservoir rim. Design investigations for the project began in 1948, and construction of the tunnels and earth embankments began in 1950. The following discussion of geologic conditions is from Kiersch and Treasher (1955).

The geology of the site area is dominated by igneous and metamorphic rocks of pre-Cretaceous age. The concrete section and most wing dams are founded on Upper Jurassic quartz diorite. The high Mormon Island earth dam and one dike are founded on rocks of the Amador Group, chiefly schists, metasediments, and metavolcanics. Three ages of alluvial deposits associated with the drainage history provided fill material and concrete aggregate. The highest and oldest alluvial deposit, upper Pliocene to Pleistocene in age, indicated a substantial period of exposure during most of Neogene time for extensive weathering of the older bedrock materials. Major northeast- and northwest-trending faults occur in the project area, and quartz veinlets with associated mineralization and alteration are common along the northeast-trending faults. Shear zones and closely spaced fracture zones are common in the exposed foundation (Fig. 6).

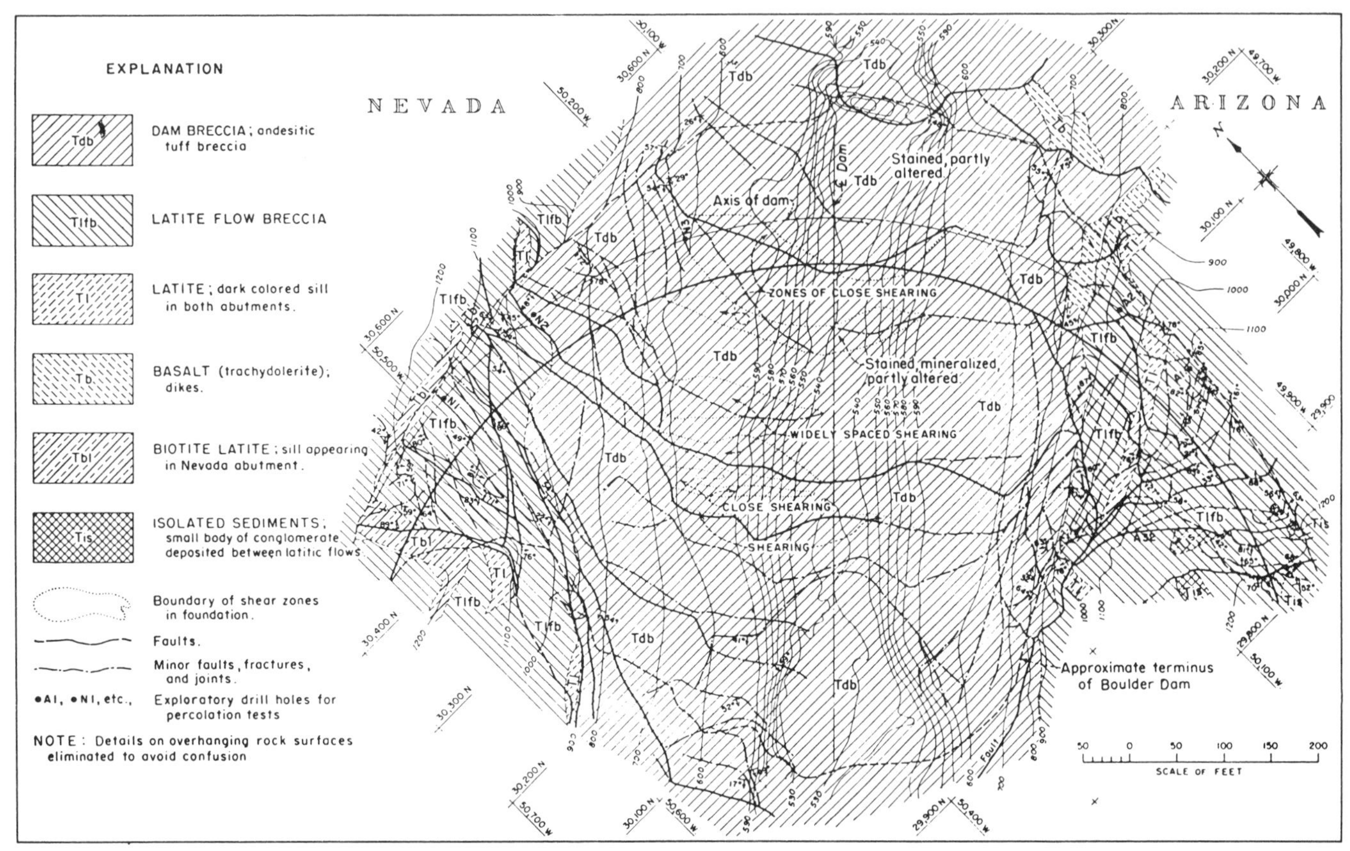

Figure 2. Generalized geologic map of the volcanic rocks exposed in the foundation and abutments of Hoover (Boulder) Dam, Nevada and Arizona. (From Nickell, 1942, Fig. 4.)

An extensive weathering pattern associated with the faults and abundant fracturing of the quartz diorite constituted a serious problem in selecting an adequate foundation. "Outcrops" of quartz diorite frequently proved to be resistant residual boulders underlain to considerable depth by highly weathered rock. In some cases, rock material exposed in pre-contract stripping (Fig. 7) was expected, on the basis of drill core, to represent the main rock mass broken by an occasional weathered joint, fracture, or fault; it often proved, however, to be boulders of fresh rock surrounded by weathered rock of varying degrees.

Excavation of the foundation for the main dam was performed in three contract stages based on the original concept of the weathered rock conditions (Fig. 6). Conditions exposed at each level (Fig. 7) aided in predicting the conditions at subsequent levels and selecting the final foundation elevation (Kiersch and Treasher, 1955, p. 298–300). This preliminary stripping (Stage 1) consisted of three 3-m (10-ft) lifts; after each lift, a grid of wagon drill and jackhammer holes served to guide the planning of the successive lift. The second stripping contract (Stages II and III) consisted of one 3-m (10-ft) lift in parts, and up to 9.1 m (30 ft) in other sectors, according to the drilling data (Fig. 7). The final excavation contract in 1953 involved a sector on the left abutment overlying the projected Channel fault.

During Stages I and II of construction excavation, it became clear that exploration limited to small-diameter core borings was inadequate at this complexly weathered site; the irregular occurrence of weathered rock led to incorrect conclusions. Also, the severely weathered granite from a freshly cored boring was soft and friable, while the core from earlier, pre-1948 exploration had case hardened upon drying, providing misleading information about its in-situ physical characteristics. Consequently, "man-sized" calyx and auger holes were drilled, shafts and adits were driven, and additional NX core borings and percussion holes were drilled. In addition to examining core from the borings, a bore-hole camera lowered into the 7.6-cm-diameter holes provided photographic documentation of their entire lengths, one of the first uses of such equipment (Kiersch and Treasher, 1955, p. 300–301). Detailed petrologic analyses of the quartz diorite permitted the rock mass to be differentiated into four categories; three of weathering and one of fresh, as shown on Table 1 and Figure 8.

A fault zone in quartz diorite was delineated in the excavation for the left abutment of the concrete section during the final excavation contract in 1953. This fault zone was projected to correlate with the Channel fault, known from the initial exploration (1948 to 1950), as shown on Figure 8. This zone contained highly weathered rock and very plastic clay ranging in thickness from a few cm to about 1 m. Some initial NX core borings

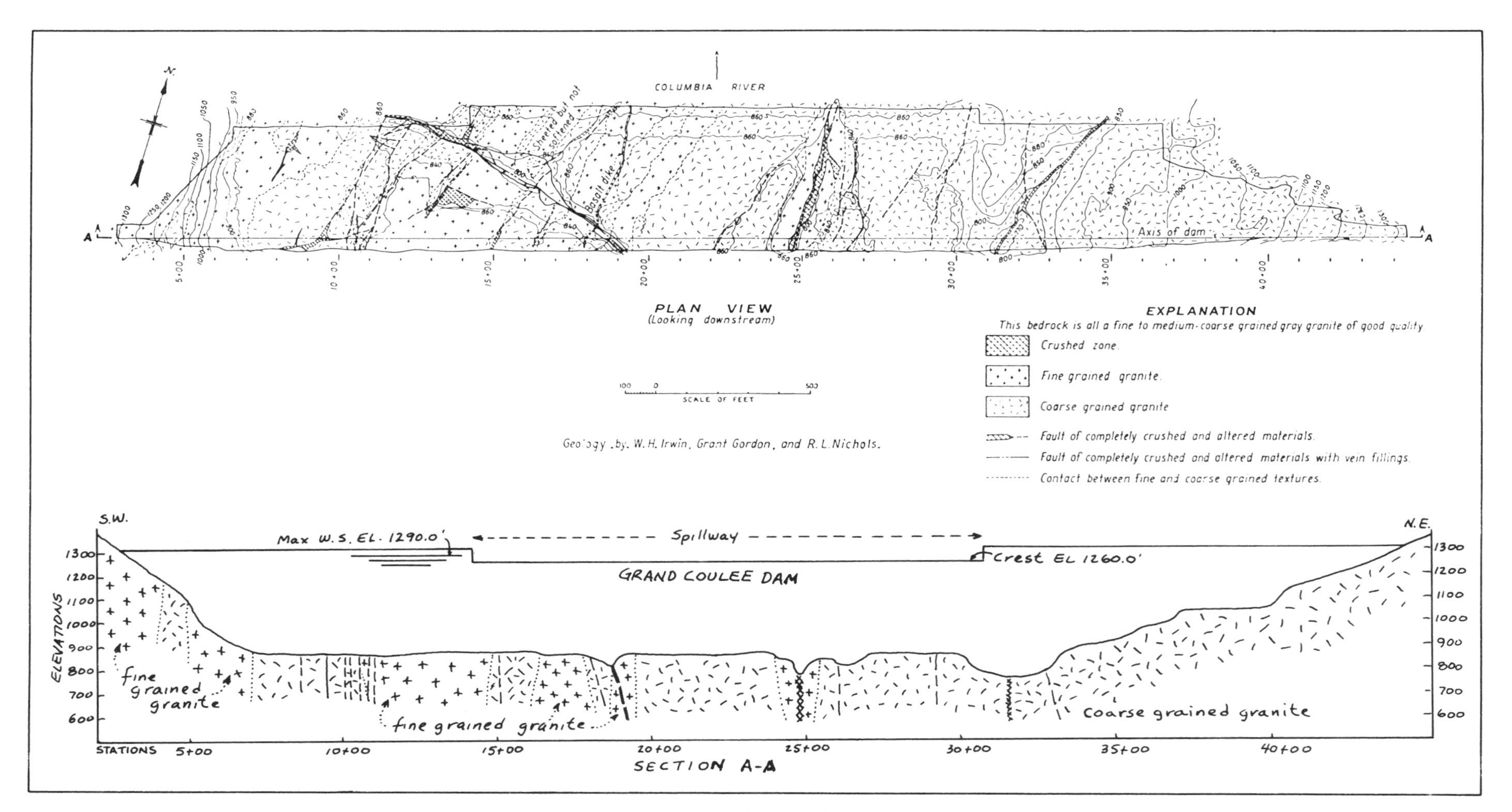

Figure 3. Generalized geologic map and section of the foundation of Grand Coulee Dam, Washington. (From Nickell, 1942, Fig. 6.)

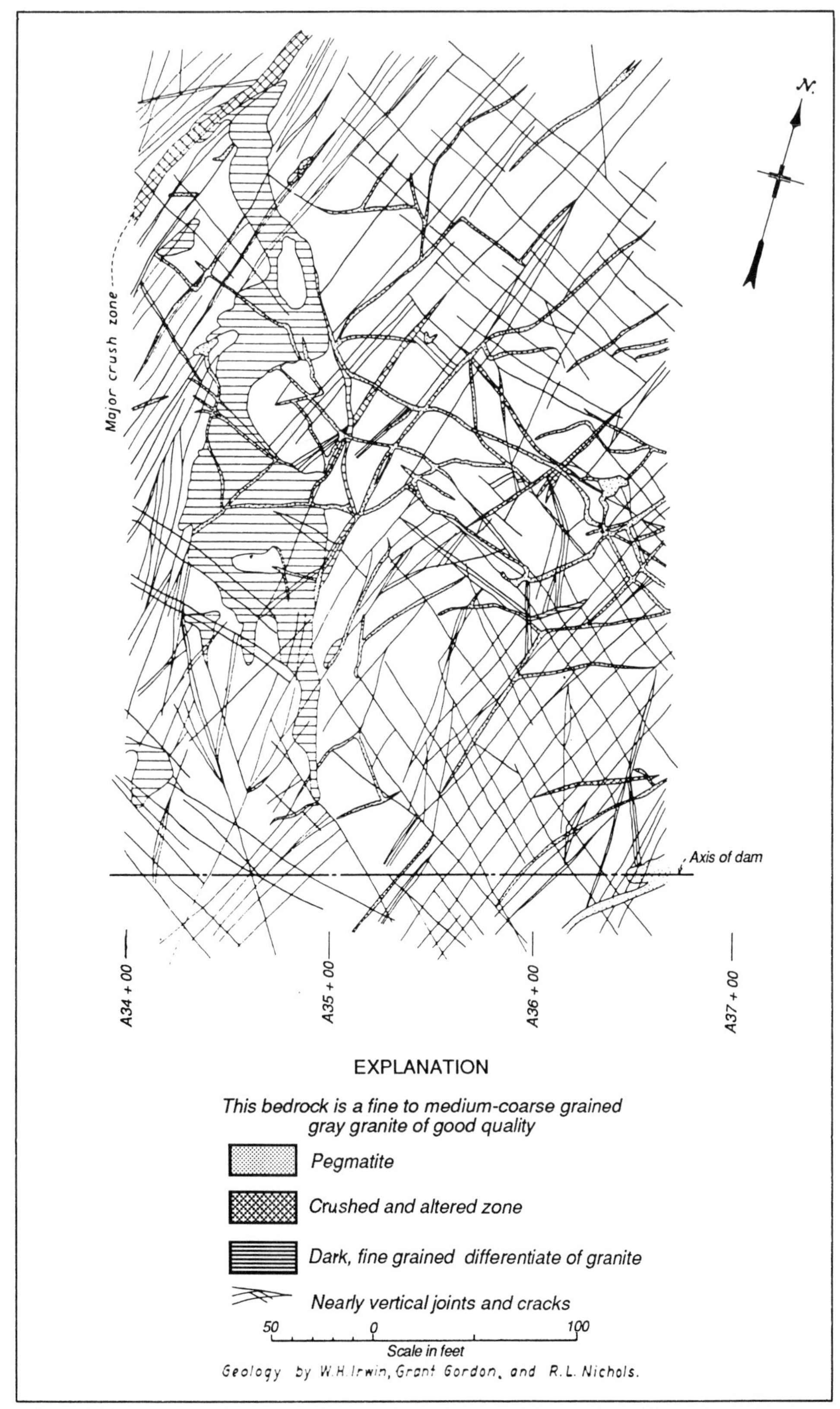

Figure 4. Generalized geologic map of the granitic rocks exposed in a portion of the northeast abutment of Grand Coulee Dam, Washington. (From Nickell, 1942, Fig. 7.)

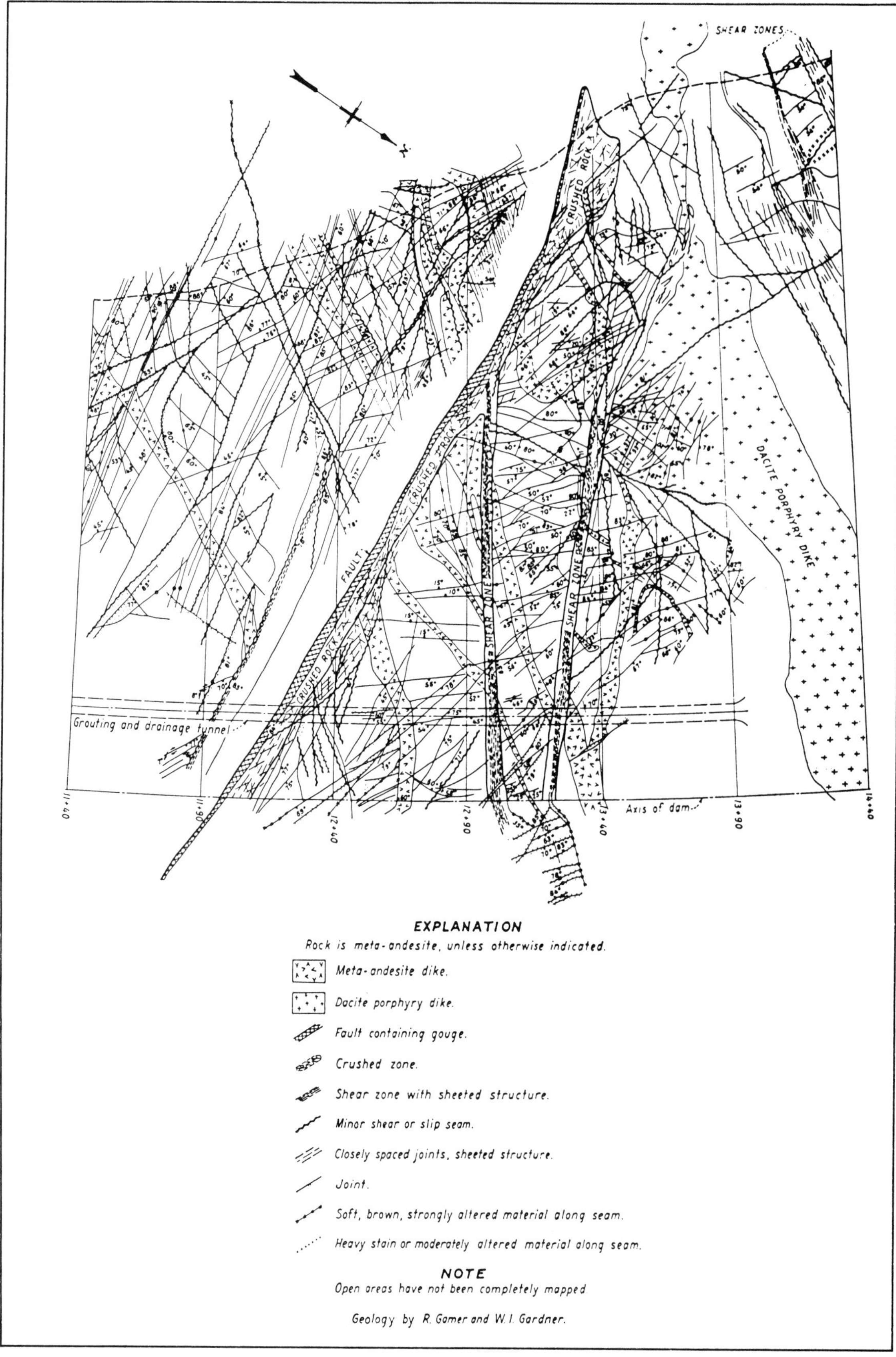

Figure 5. Generalized geologic map of the metamorphic rocks exposed in a portion of the southeast abutment of Shasta Dam, California. (From Nickell, 1942, Fig. 12.)

penetrated the fault zone with no core recovery, and the full character of the fault zone was unanticipated.

Geological investigations conducted during the planning and construction of Folsom Dam were critical to successful completion of the project. Furthermore, geological efforts for construction permitted a further evaluation of the adequacy and effectiveness of the techniques employed during investigations for design. The ability to distinguish and classify the varied degrees of weathering by petrologic means enabled the construction geologists to provide more accurate and reliable predictions of the amount and limits of excavation required to expose foundation rock suitable for support of a rigid concrete structure. Physical characteristics of the three weathered categories and fresh granitic rock related to exploration data and excavation-construction properties are given by Kiersch and Treasher (1955, p. 297–298). This petrologic approach to evaluating complexly weathered rock masses and respective physical-engineering properties at Folsom Dam served as a basis for subsequent research and advancement of weathering classifications (Dearman, 1974, 1976) and for predicting engineering properties of weathered granites (Dearman and others, 1978; Irfan and Dearman, 1978).

Cavernous and solutioned limestone

Rondout Pressure Tunnel. The Rondout Pressure Tunnel, the first of its kind in the United States, is part of the Catskill Aqueduct. Topographic conditions along the aqueduct alignment, located approximately 130 km north-northwest of New York City, required tunnels through mountain ridges that exceeded the hydraulic gradient for gravity flow. An economic analysis of alternatives indicated that tunnels in bedrock under valleys that were below the hydraulic gradient would be less expensive than surface facilities that extended great distances up the valleys or crossed valleys on bridges (Berkey and Sanborn, 1923, p. 2). The hydraulic pressures that would be developed in the tunnels on the Catskill Aqueduct (122 m of head or 1,200 kPa) were substantially greater than those in tunnels built earlier (Berkey and Sanborn, 1923, p. 23). The Catskill pressure tunnels were designed on the basis of the static weight of rock and earth above the tunnels equalling or exceeding the hydraulic pressure, or a minimum of about 45 m of sound rock (Sanborn, 1950, p. 50).

A generalized geologic section along the Catskill Aqueduct is presented on Figure 9. The 6.4-km long, 4.4-m diameter Rondout tunnel was the first pressure tunnel of the Catskill system and the first high-pressure tunnel constructed in the United States (1906). During the investigation for design of the tunnel, twelve separate sedimentary rock formations were identified, each with distinct physical properties and engineering characteristics. The folded, faulted, and deeply eroded rock units were buried by glacial and other surficial deposits. Borings to investigate the alignment across Rondout Valley delimited an unknown 61-m-deep buried valley, several faults with associated zones of crushed rock, and cavernous limestone (Sanborn, 1950, p. 56), as shown on Figure 10.

Cavernous conditions in the Lower Devonian Onondaga and Helderberg Limestones were recognized during design investigations (Fig. 9). Tunnel construction involved large sections in the Helderberg Limestone, which consists of five distinct units: Manlius, Coeymans, New Scotland, Becraft, and Port Ewen (Fig. 10). Berkey and Sanborn (1923, p. 27) note a serious concern of the early 1900s about limestones that has since been disproven:

> If these limestones, which have proven to be soluble when exposed to ordinary meteoric water circulation, should yield in the same way to the tunnel waters, not only would the concrete lining be subject to disintegration, but the supporting rock immediately outside of them might be, in part, removed and the escape channels enlarged, seriously increasing the loss of water from the tunnel. These considerations led to the tunnel being placed below the zone most affected by surface-water circulation, avoiding as far as possible the most susceptible strata. It led also to the rejection of limestone aggregate in the concrete lining.

The limestone was not used in concrete aggregate because of experiences with similar limestone in the Thirlmere Aqueduct in England where severe dissolution of the concrete occurred (Wiggin, 1923, p. 81). Today, a potential limestone aggregate is tested and analyzed for its intended use, rather than being rejected outright on the basis of a descriptive classification only. Limestones are a common concrete aggregate (described in Mather and Mather, this volume).

The possibility of a tunnel-lining failure was anticipated in the cavernous formations. However, it was considered more economical to locate any points of weakness within the concrete-lined tunnel by conducting a pressure test on completion rather than building unnecessary reinforcement into all parts of the tunnel (Berkey and Sanborn, 1923, p. 33). To test the lining, the construction shafts were sealed with concrete plugs, and the end shafts were filled with water and subjected to the design pressure; the tunnel lining ruptured between shafts 3 and 4, the area of known cavernous limestone (Fig. 10). During construction, zones of obvious rock weakness were reinforced with steel shells, and none of these reinforced zones were damaged during the pressure test.

It was later realized that the damage to the concrete lining was due to yielding of the soft clay that surrounded blocks of limestone, although in tunnel exposures the rock had appeared to be intact and sound (Berkey and Sanborn, 1923, p. 33). The tunnel lining throughout the failed section was reinforced with a 2.5-cm-thick steel shell; fractures in the rock caused or enlarged by the failure were grouted under high pressure to minimize water losses.

This description of the recognition of the importance of geology from the Rondout pressure tunnel illustrates only one of many lessons learned by the close association of geologists and engineers during construction of this complex project. The successes achieved were widely recognized and attributed to "proper utilization of the skill of the geologist as an adjunct to good engineering" (Flinn, 1923, p. 68–69). Furthermore, Flinn recog-

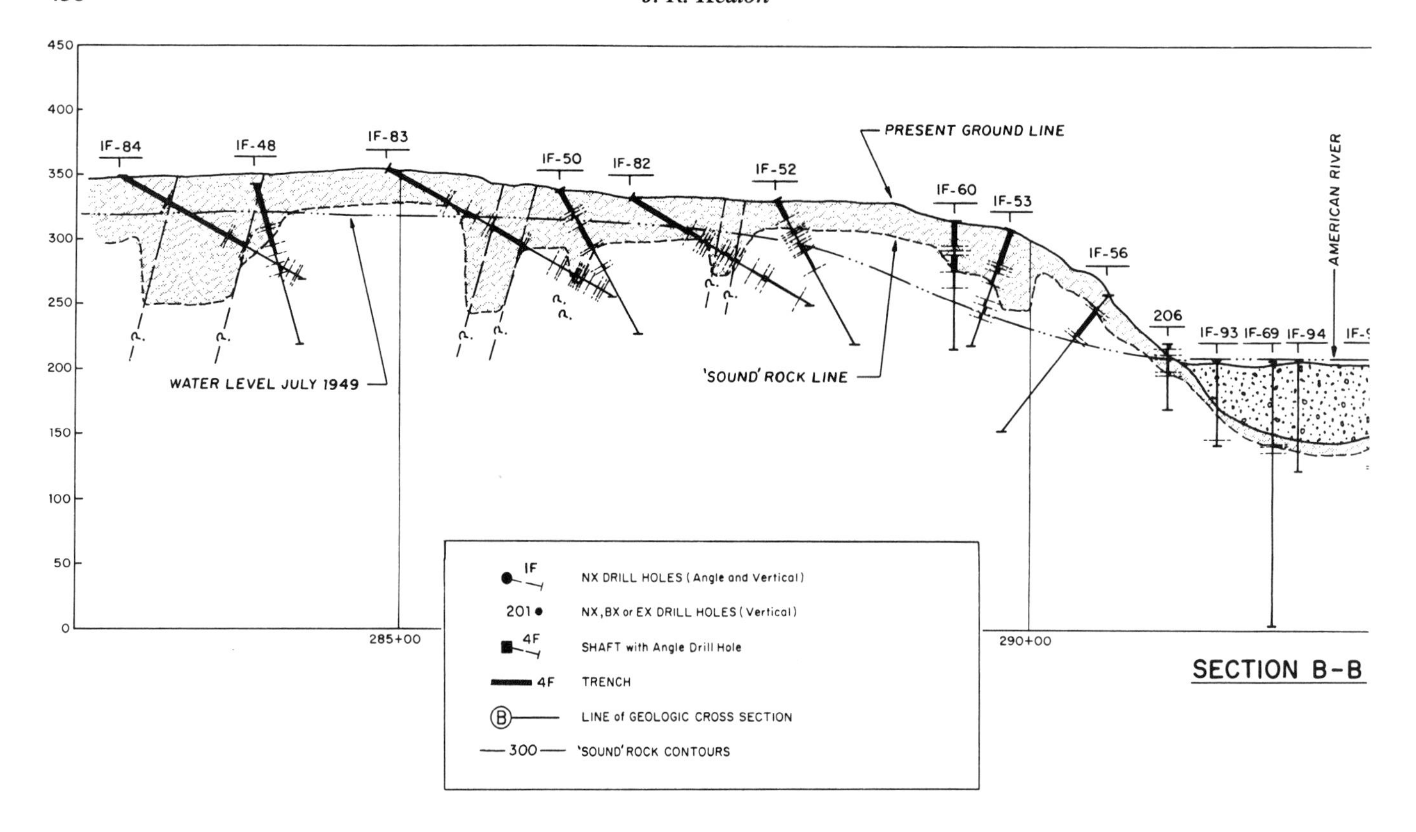

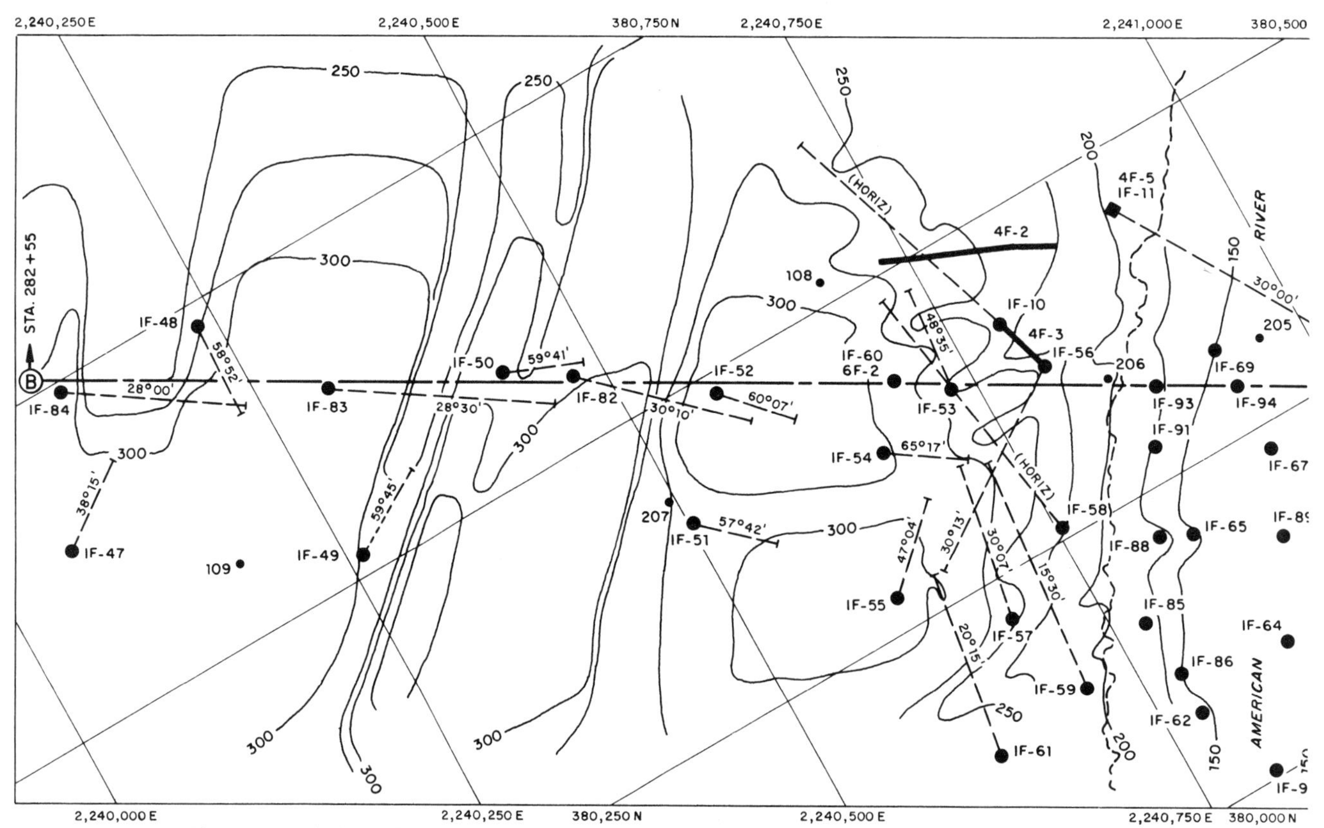

Figure 6. Preliminary forecast of depth to adequately "sound" foundation rock at Folsom Dam site California—as shown on geological cross section and rock contour plot by G. A. Kiersch in 1949; from U.S. Army Corps of Engineers, Folsom Project drawing AM- 1- 20- 289 and 290 of 1950. Compare

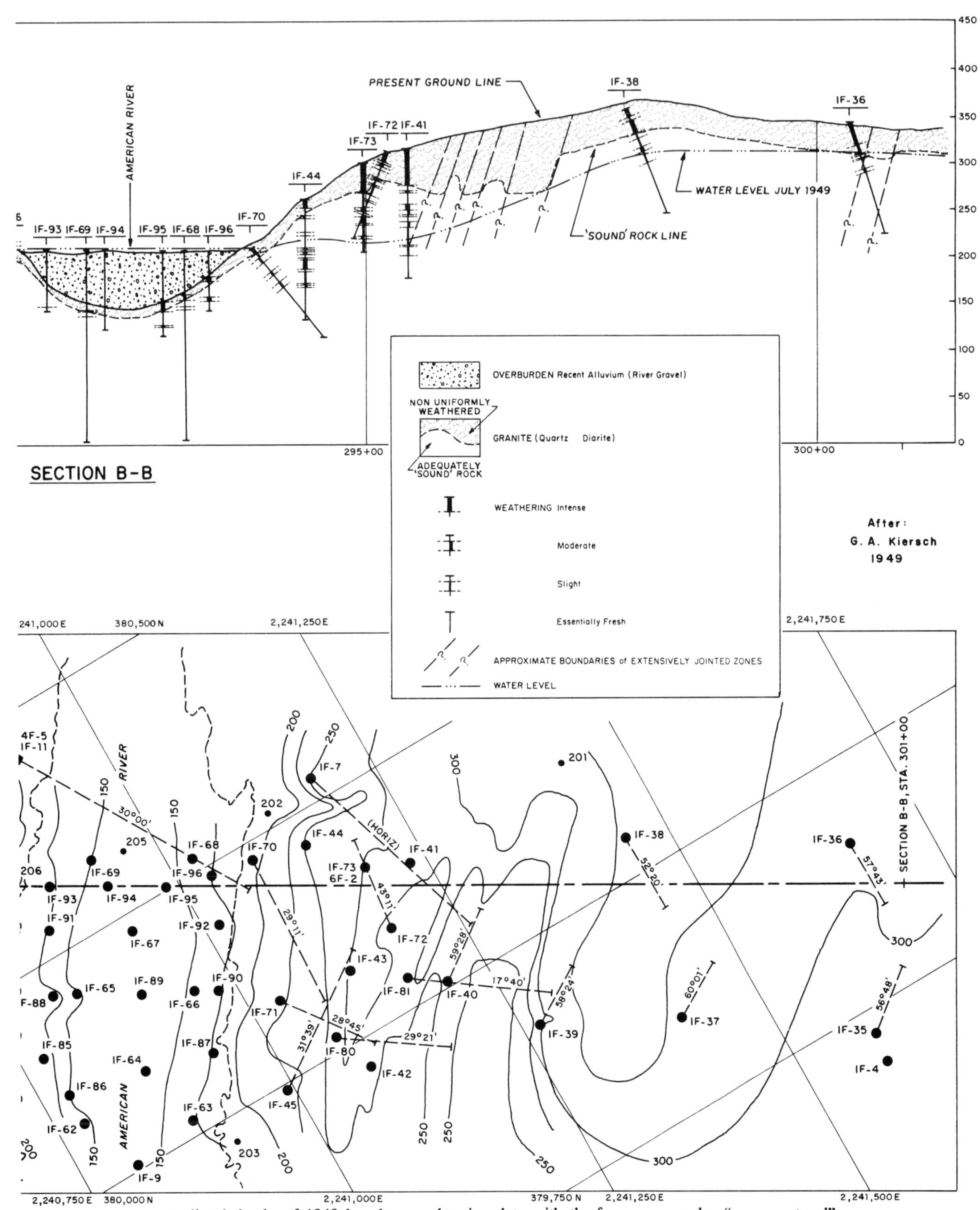

predicted depths of 1949 based on exploration data with the far more complex "as-encountered" conditions shown on stage diagrams of 1954, in Figure 7.

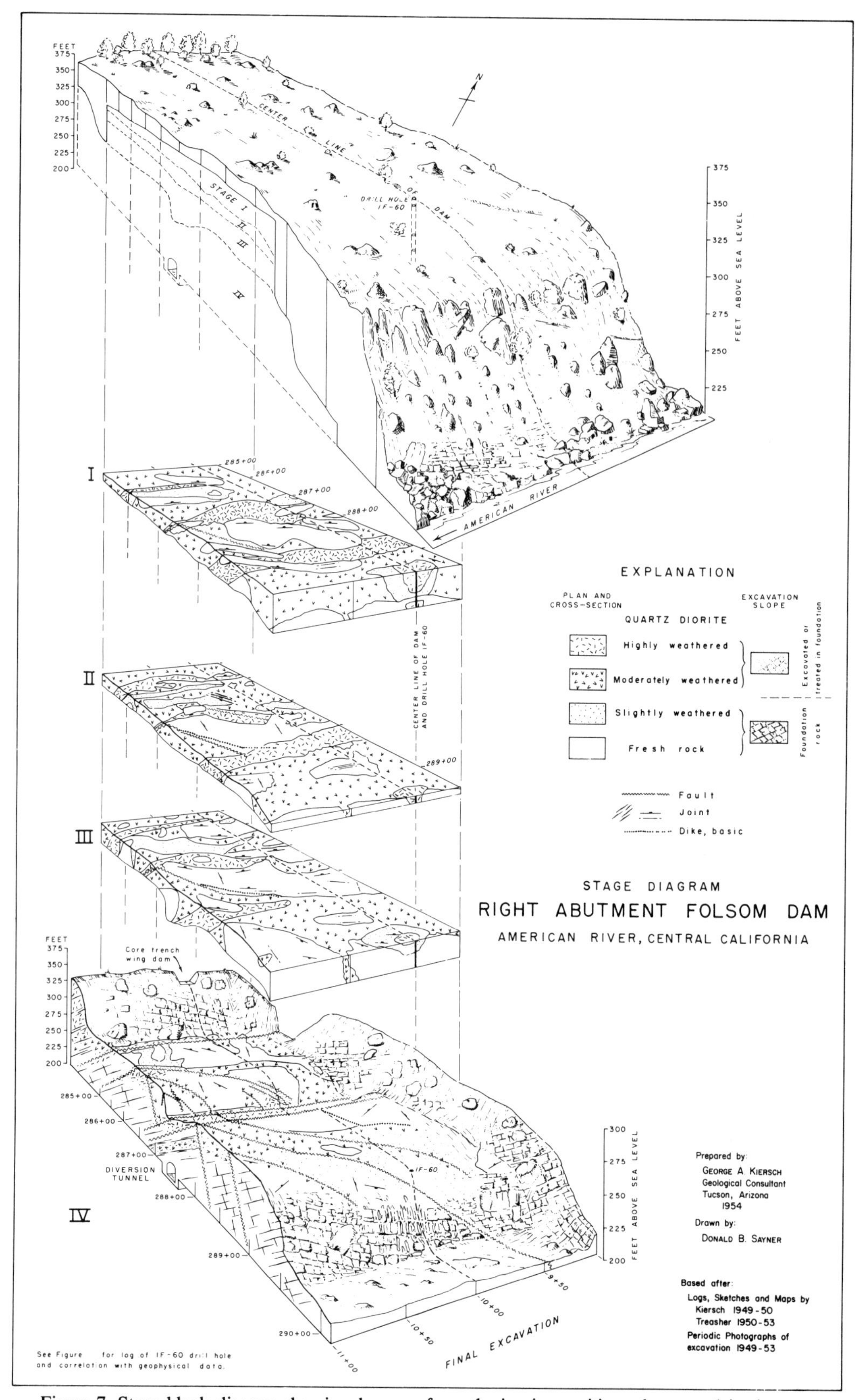

Figure 7. Stage block diagram showing degrees of weathering in granitic rock exposed in the west abutment at Folsom Dam, California. (From Kiersch and Treasher, 1955, Fig. 11.)

nized that geological input is essential for the construction of safe and economical engineering projects, and noted that geologists and engineers should work together from the beginning of a project (early history of cooperation is discussed in Kiersch, Chapter 1, this volume). Moreover, full use should be made of all geological data in the site-selection, design, and construction phases, and all geological data and interpretations should be made available to contractors and project bidders (discussed in Waggoner and Kiersch, this volume).

Kentucky Dam. During construction of the Kentucky Dam on the Tennessee River in southwestern Kentucky, an unusually large, deep, and extensive solution zone was discovered in the foundation at the critical position of a lock required to maintain navigability. The Kentucky Dam has a concrete gravity spillway section that is connected to a navigation lock 33.5 m wide by 182.9 m long. The dam and lock were built from 1937 to 1941; further details are given by Moneymaker and Rhoades (1945, p. 40) and Tennessee Valley Authority (1949).

The rock foundation is Mississippian Fort Payne Formation, which consists of a thick-bedded, fine-grained, dark siliceous limestone with nodular masses and layers of black chert (Burwell and Moneymaker, 1950, p. 35). The limestone is gently folded into a broad, low anticline; the bedding dips from 3° to 5° and is cut by numerous vertical joints.

The Fort Payne limestone is extensively solutioned at the dam site despite its siliceous nature. The solutioning was con-

TABLE 1. PETROLOGIC CHARACTERISTICS OF QUARTZ DIORITE AT THE FOLSOM DAM SITE, CALIFORNIA*

Scale	Essentially Fresh	State of Weathering		
		Slight	Moderate	High
Megascopic	Mottled. Light gray, medium- to coarse-grained. Unaltered, high quality and durable	Essentially as durable and high quality as rock in fresh state. ***Feldspars***—"visibly" fractured, weakly bleached a whitish gray. ***Limonite***—specks scarce. In logging cored borings, lighter colored feldspars used as criteria for separating from fresh rock.	Firmly coherent, somewhat friable, with original texture well preserved. ***Feldspars***—moderately fractured, bleached grayish white. ***Quartz***—very slightly rounded. ***Biotite***—weakly bleached. ***Limonite***—common, as specks coating minerals and along the boundaries of individual grains, cracks. Moisture content somewhat increased.	Loosely coherent, friable, with original texture and structure mostly preserved. ***Feldspars***—highly fractured, bleached white. ***Quartz***—rounded. ***Biotite***—strongly bleached. ***Limonite***—abundant, as specks and coating minerals, along grain boundaries and cracks. Moisture content increased by weight; rock very soft.
Microscopic	Fairly uniform texture; subhedral to euhedral crystals. Some fracturing of grains, but a "tight," welded appearance.	Weakly altered. ***Feldspars***—mildly shattered, "tight" welded appearance, grains interlocked. Minor sericitization and "dusty" clay covering calcic "cores."	Only slightly altered. Grains shattered, particularly feldspars; many "open," others grains interlocked. ***Feldspars***—sericitization and "dusty" clay covering part of calcic "cores." ***Biotite***—partially altered to limonite. ***Quartz***—slightly rounded. Some limonite along ramifying cracks and individual grain boundaries. Some ferrous iron oxidized to ferric iron.	Less altered than expected from iron stained, "rotten" appearance; grains highly shattered, particularly feldspars, more so in uniformly textured rock. ***"Open" fractures***—with limonite. ***Feldspars***—sericitization and "dusty" clay covering calcic "cores." ***Quartz***—partially rounded by solution of silica from grains at points of contact near surface. ***Biotite***—bleached, appreciable alteration to limonite. ***Limonite***—abundant as coating on mineral grains, in ramifying cracks, and along mineral boundaries. Ferrous iron oxidized to ferric oxide.

*From Kiersch and Treasher (1955, p. 288).

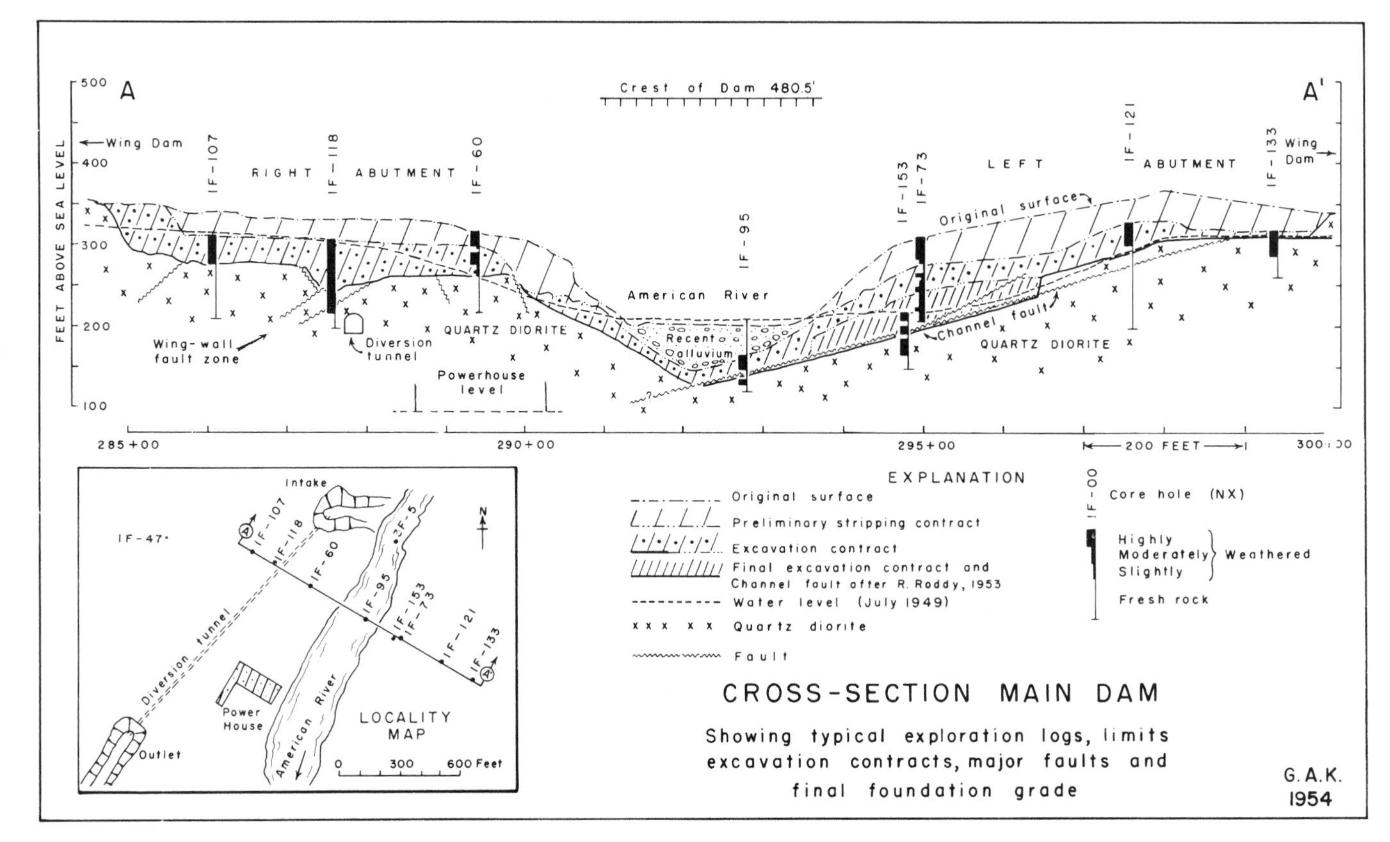

Figure 8. Cross section showing degrees of weathering in granitic rock at Folsom Dam, California. (From Kiersch and Treasher, 1955, Fig. 7.)

trolled by the inherent bedding planes and joints, as well as the composition of the limestone. Expensive, difficult, and time-consuming foundation problems resulted from a series of tabular cavities oriented along the bedding and numerous vertical channels developed along the joints. The vertical channels were generally narrow, so that the concrete dam would have adequate foundation support, but the openings were so deep that seepage became a principal concern. Furthermore, the solution channels were filled with residual clay and chert considered susceptible to erosion and flushing under the anticipated seepage pressures that would be developed by the proposed reservoir.

The particularly large and troublesome solution zone encountered in the foundation of the dam and lock (Fig. 11) coincided with closely spaced joints along the axis of a monocline. Drill data suggested that the zone extended for as much as 600 m, so a minor adjustment of the position of the dam and lock would not have avoided it. In addition to residual clay and chert, the zone was also filled with relict blocks of limestone, as shown in the detailed cross section on Figure 11. Adjacent to the lock site, the solutioned zone was about 21 m wide at a depth of 12 m below the normal level of the Tennessee River. The solution zone remained about 20 m wide to a depth of about 30 m below the river where it bifurcated into two narrower, vertical channels. One of the channels tapered to a termination at a depth of about 70 m. The other channel was much less regular, varying from 1.5 m to 5.5 m wide, and also terminated at a depth of about 70 m.

Moneymaker and Rhoades (1945, p. 44) concluded that the deep solution zone at Kentucky Dam provided additional proof that extensive solution of limestone may occur below the water table, as initially noted at New Croton Dam in 1895 (see Kiersch, Chapter 1, this volume). They found that the chemical purity of limestones becomes an unimportant factor under the field conditions of continuously circulating ground water over a long period of time. Sand grains reworked from nearby Cretaceous deposits had been transported and redeposited in the solution zone. This evidence convinced Moneymaker and Rhoades (1945, p. 43) that inception of the deep solution zone preceded incision of the Tennessee Valley in late Pliocene or early Pleistocene time, and likely in the Miocene or earlier. Moreover, "long periods of geologic time are represented in the development of large caverns," and the solution action in the Kentucky Dam region represents a continuous process of long duration (Moneymaker and Rhoades, 1945, p. 44).

The width and depth of the solution-zone cavity represented a hazard to Kentucky Dam relative to both seepage losses and foundation support. Consequently, seepage control was provided by constructing a "subterranean arch dam" cutoff across the solution zone at the axis of the dam. The cutoff was completed by

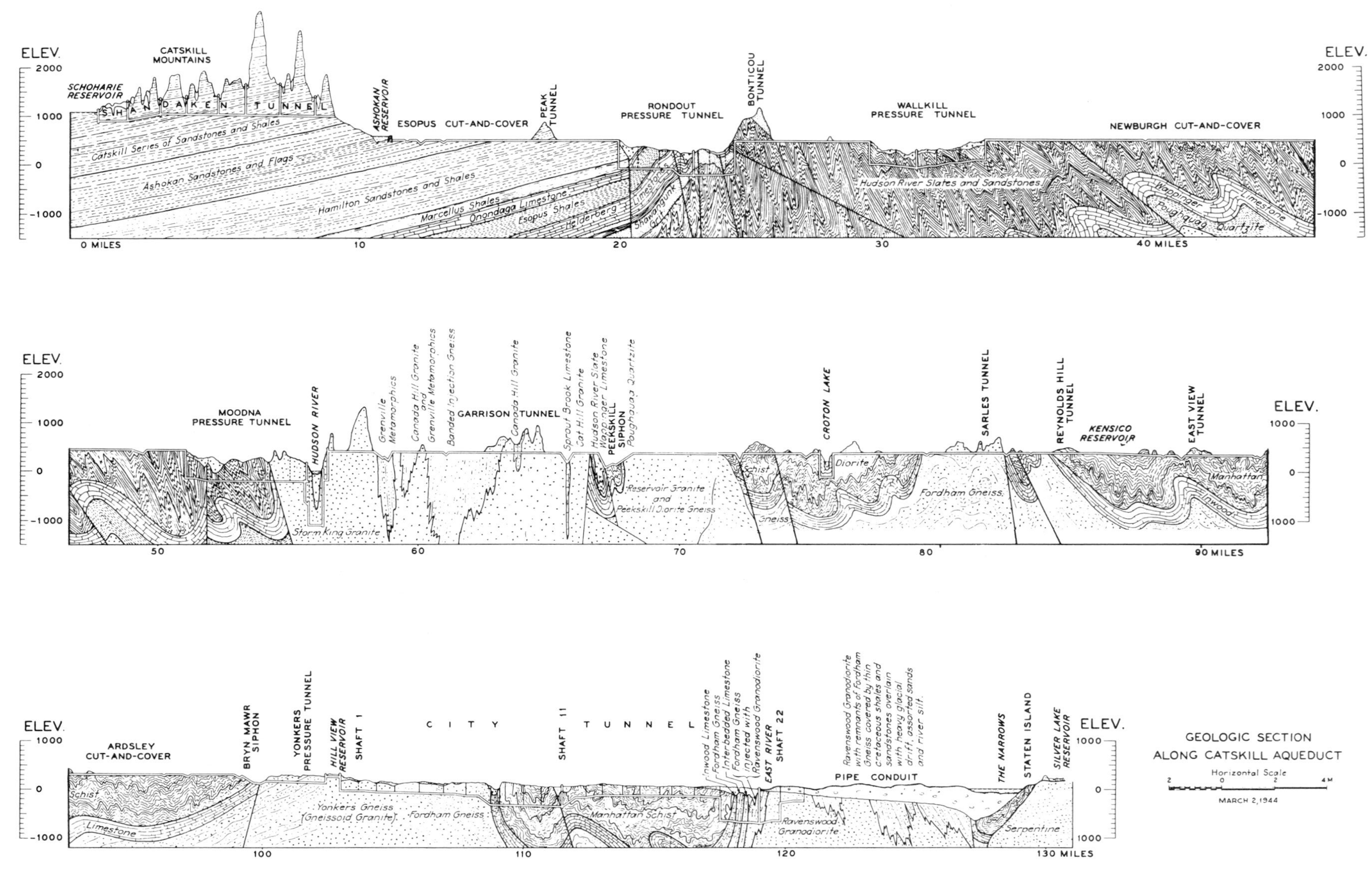

Figure 9. Geologic section along the Catskill Aqueduct, New York. (From Sanborn, 1950, Plate 1.)

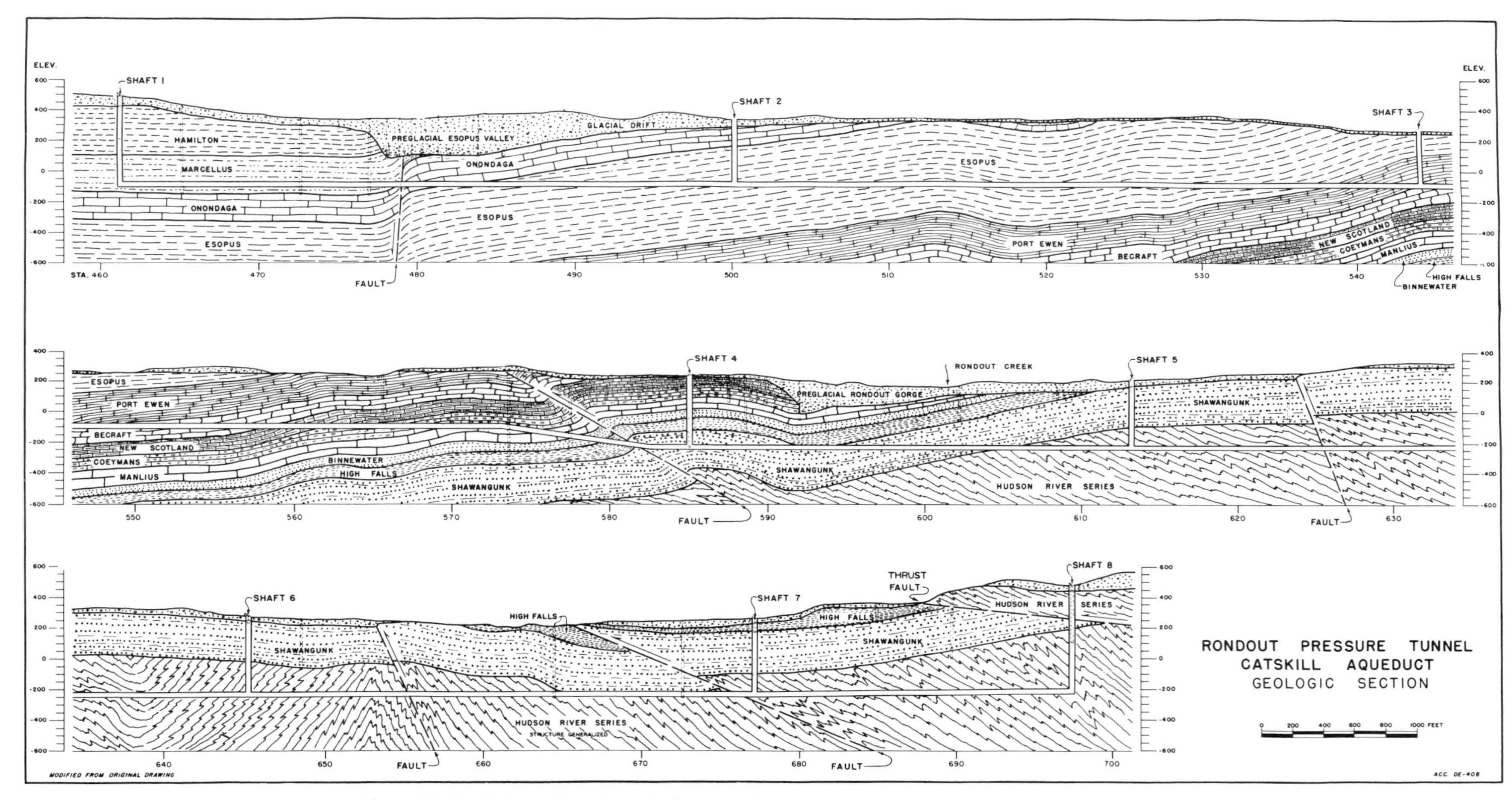

Figure 10. Geologic section along the Rondout Pressure Tunnel part of the Catskill Aqueduct, New York. (From Sanborn, 1950, Plate 2.)

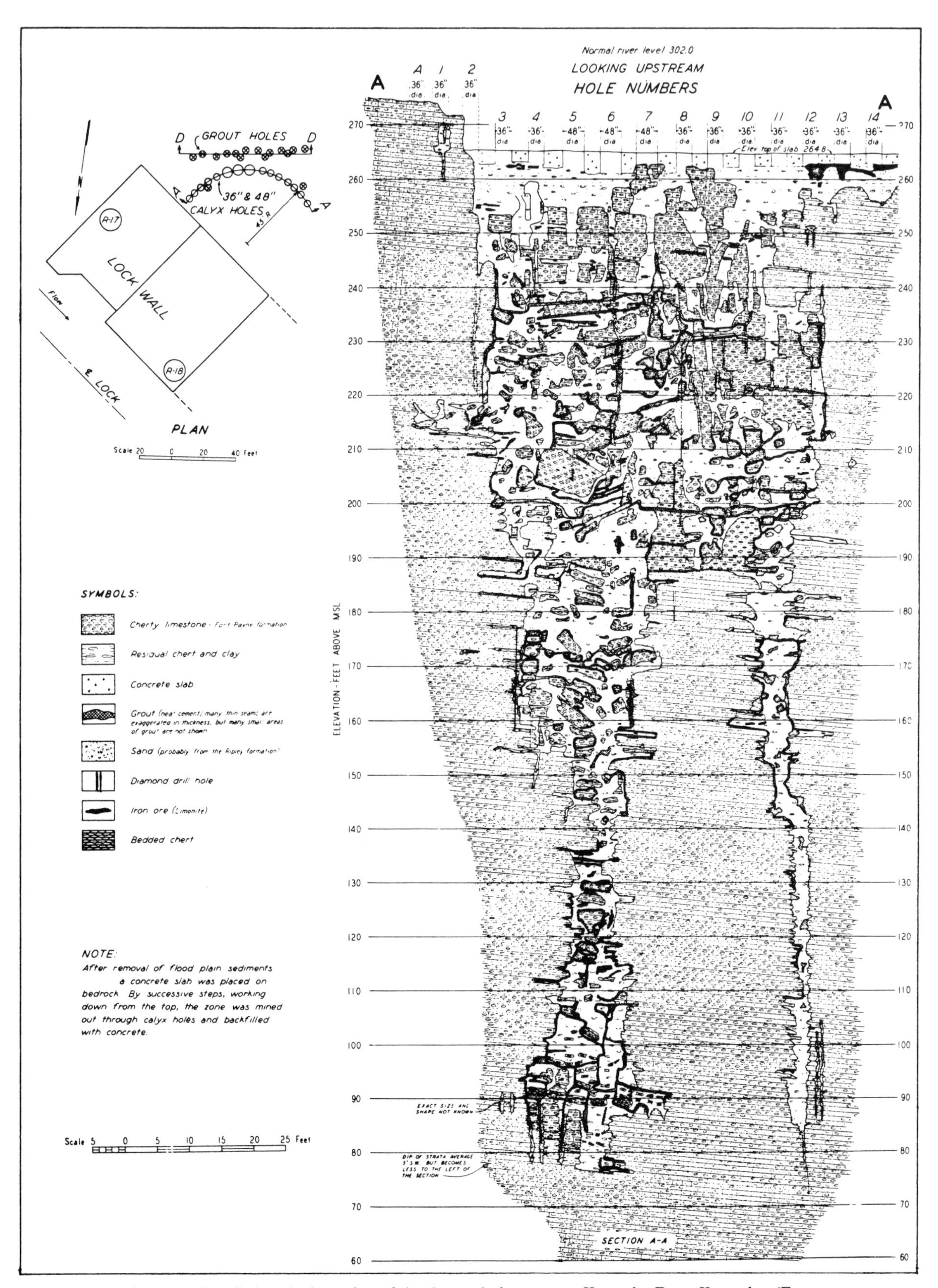

Figure 11. Detailed geologic section of the deep solution zone at Kentucky Dam, Kentucky. (From Burwell and Moneymaker, 1950, Plate 4.)

mining the material filling the solution zone, using 91- and 122-cm-diameter borings as shafts. The cutoff, about 60 m high, was constructed downward from the top by successive excavation of small chambers and backfilling with concrete (Hays, 1943, p. 1403). Improved foundation support was provided by constructing a thick reinforced concrete slab over the channel.

The position and large size of the solution channel required that a thorough investigation provide basic data for a design modification during construction. The thoroughness of the investigation is evident in the detail displayed on Figure 11, the data for which were collected during progressive removal of the channel-filling sediments as the cutoff was constructed. Although the presence of solutioned channels in limestone was recognized prior to construction of Kentucky Dam, the Kentucky project contributed significantly to an improved understanding of the complexity and variability of these important geologic features. The Kentucky project was a very early instance where an extensive deep solution zone was adequately treated in a foundation.

Barkley Lock. The Barkley Lock structure is situated on the lower Cumberland River in western Kentucky approximately 4 km northeast of Kentucky Dam. Exploration for design of the 33.5-m-wide by 243.8-m-long navigation lock started in 1955; construction began two years later and was completed in 1961 (Clark, 1963, p. 20).

The geologic setting at Barkley Lock is similar to that of Kentucky Dam. The bedrock is essentially horizontally bedded siliceous limestone of the Mississippian Fort Payne Formation. Overlying alluvial deposits in the 2.4-km-wide floodplain of the Cumberland River are up to 27 m thick. The project layout and network of exploratory borings are presented on Figure 12; many borings were inclined at 45° to permit an assessment of vertical joints after a 10-m-long void in a vertical boring was found to be a solution-widened joint rather than a significant cavity (Clark, 1963, p. 21).

The geological problems associated with construction of the Barkley project are similar to those related to Kentucky Dam, namely numerous small, solution-widened channels and one large solution zone. However, the method of treating similar fea-

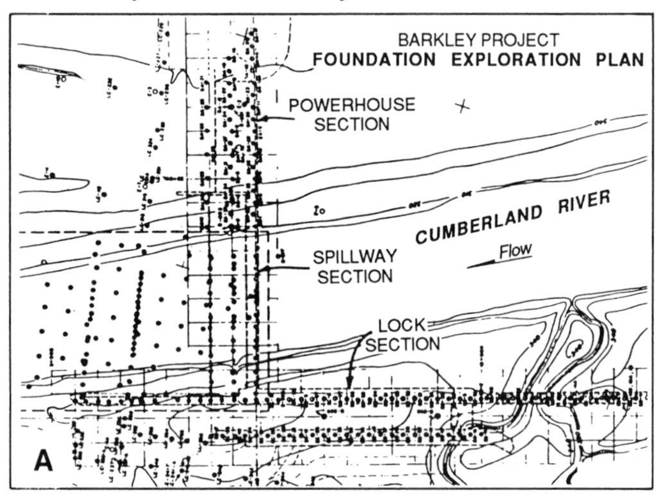

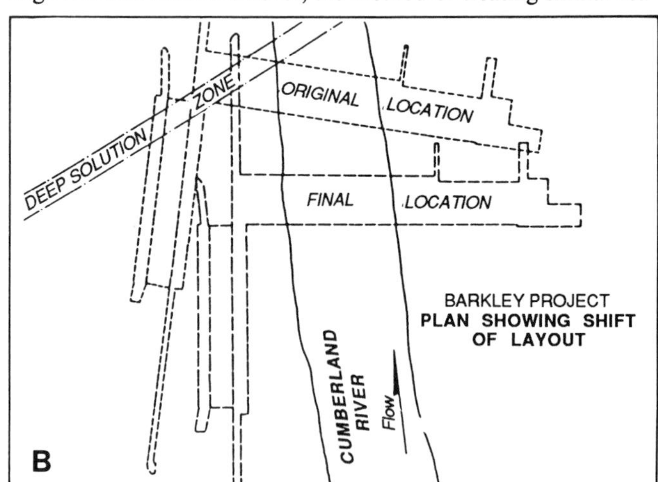

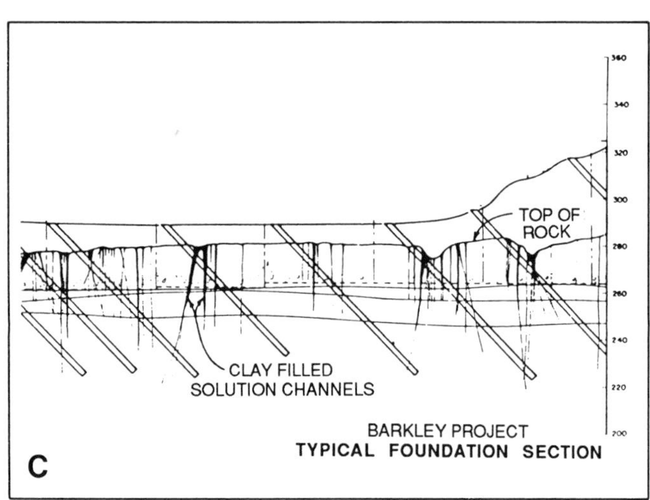

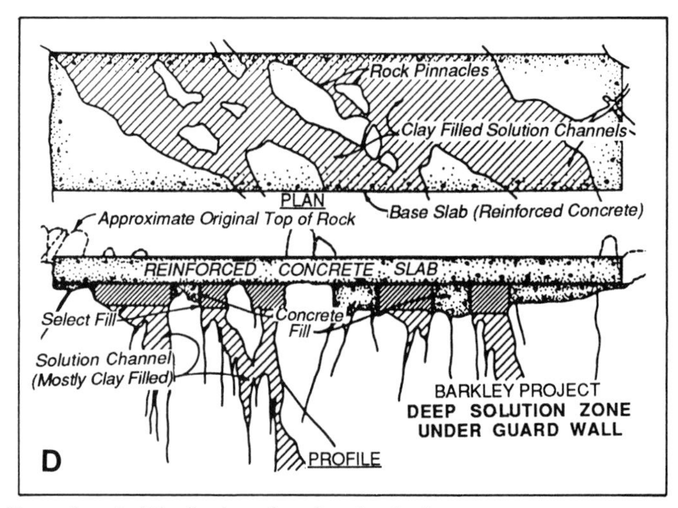

Figure 12. Selected details regarding Barkley lock, Kentucky. A. Distribution of exploration borings. B. Original and final location of spillway and lock. C. Typical foundation section showing top of rock and clay-filled solution channels. D. Plan and profile showing rock pinnacles, clay-filled solution channels, and remedial backfill and concrete slab. (From Clark, 1963.)

tures was different at the Barkley project. Early exploration located a deep solution zone within the downstream lock gate foundation. Unlike the Kentucky Dam remedy of an expensive cutoff structure, the centerline of the Barkley Dam was shifted about 60 m upstream and rotated a few degrees. This shift in alignment of the project structures placed the less critical, downstream guide wall of the lock over the solution zone and the major portion of the main structure over more competent rock.

Subsequent cored borings indicated that the solution zone consisted of numerous channels, which in combination constituted a large, deep clay-filled cavity. However, excavation of the foundation revealed that rock pinnacles surrounded by clay were extensions of bedrock at depth and not relict blocks. Remedial treatment consisted of compacted fill placed on the existing clay to the level of the higher rock pinnacles capped with a 1.5-m-thick, heavily steel-reinforced concrete slab. The slab essentially was supported by the rock pinnacles.

The design and construction of the Barkley project benefitted from lessons learned earlier at Kentucky Dam. For example, had the foundation exploration at the Barkley project not been exceptionally detailed, Clark (1963, p. 25) speculated that serious foundation trouble probably would have been encountered during construction of the lock.

Hades tunnel. The Hades tunnel is located about 125 km southeast of Salt Lake City, Utah, on the south side of the Uinta Mountains; it is part of the Strawberry aqueduct of the Central Utah project. Several types of geologic problems were experienced during construction of the Hades tunnel. Three categories are discussed below.

The straight, 6.74-km-long tunnel was bored with a laser-guided, Robbins model 1011-98 tunnel boring machine (TBM). Boring began at the southwest portal on December 1, 1980, and holed through on November 30, 1981. The completed water conveyance tunnel is 251.5 cm in diameter with a nominal 36.85-cm-thick concrete lining.

The Hades tunnel penetrated 12 Paleozoic formations that dip 5° to 25° to the southwest, as shown on Figure 13 (Stevenson, 1987, Volume 3). Folding of this thick sequence of sedimentary rocks in response to uplift of the Uinta Mountains formed a west-trending syncline, the Uinta basin. The tunnel penetrated the rocks on the north limb of the Uinta basin, which included microcrystalline dolomite, cherty limestone, black shale, thin-bedded siltstone, and calcareous sandstone. The preconstruction geology reports expressed caution and anticipated significant ground-water inflows, overbreak in the tunnel crown, possible gas and oil occurrences, squeezing ground, and numerous fault zones. Some of the limestone formations were cavernous, and one particular solution cavity caused difficulty during construction.

The Lower Member of the Permian Park City Formation, which consists of sandstone, limestone, and siltstone, was encountered at Station 696+71* on December 15, 1980. Solutioned joints, first encountered at Station 696+40 in the Lower Member, had been encountered earlier during tunneling in the Upper Member. Water inflows from the solutioned-joint features ranged from about 0.06 to 2.5 L/s at a relatively constant temperature of 6°C (Stevenson, 1987, Volume 4). Tunneling advanced about 55 m on December 22, 1980, but only 2.1 m on December 23, and 0.9 m on December 28. On December 29, a sudden inflow estimated at 150 L/s temporarily terminated tunneling operations (Stevenson, 1987, Volume 1, p. 52–53).

The large inflow issued from a solution cavity in cherty limestone with interbedded shale in the northwest tunnel wall that extended from Station 690+01 to 690+08 (Fig. 14). A substantial amount of sand and silt was flushed into the tunnel by the discharge from the cavity, which was 10 m long, 3.0 to 3.7 m wide, and 1.2 to 1.8 m high (Fig. 15). After draining, the cavity remained partially filled with blocks of limestone that had collapsed from the roof. Two tube-like openings extended away from the main cavity; one, a horizontal tube extending to the southwest, was the source of the water flow, which diminished to about 19 L/s within 24 hours and flowed at some 12.6 L/s for the duration of tunneling (Stevenson, 1987, Volume 1, p. 54).

Tunneling advanced beyond the cavity after a delay of 3 days while the sand and silt was removed from around the TBM and the water was controlled. A concrete bulkhead was constructed with some 2.4 m inside the cavity and capped with shotcrete. Although similar silt-filled cavities were encountered during tunneling in the Lower Member of the Park City Formation, none produced large and troublesome water inflows (Stevenson, 1987, Volume 1, p. 54). The contractor was prepared for large water flows from possible solution cavities and reacted quickly to each occurrence with a minimum of lost time.

Unstable slopes

Libby Dam. Libby Dam, located in northwest Montana about 17 km east of Libby, was constructed between 1966 and 1973. The concrete gravity dam is 930 m long and 128 m high; construction required relocation of Montana State Highway 37 on the east side of the Kootenai River valley (Hamel, 1976, p. 363). The bedrock at the dam is meta-argillite with subordinate calcareous argillite and quartzite assigned to the Wallace Formation, part of the Precambrian Belt Series (Voight, 1979, p. 285). Relic bedding features dip uniformly at 42° to the southwest; two sets of joints dip uniformly to the northwest at 72° and to the northeast at 65 to 66° (Voight, 1979, Table 1). Slip has occurred along the relict bedding planes in numerous places, and many are bedding-plane faults (Voight, 1979, p. 288).

Topographic contours on the left abutment of Libby Dam exhibit a "chevron" configuration, which is characteristic of bedrock terrain susceptible to wedge rockslide failures. Two large prehistoric wedge-shaped rockslides were identified during the early stages of investigation for design of the dam (Voight, 1979, p. 289). The larger slide was recognized as a prominent "V"-shaped notch in the hillslope about 760 m upstream from the left

*Stations are in feet (1 ft = 0.3048 m); tunneling progressed up-station from the outlet portal at Station 704+11 to the inlet portal at Station 483+02.

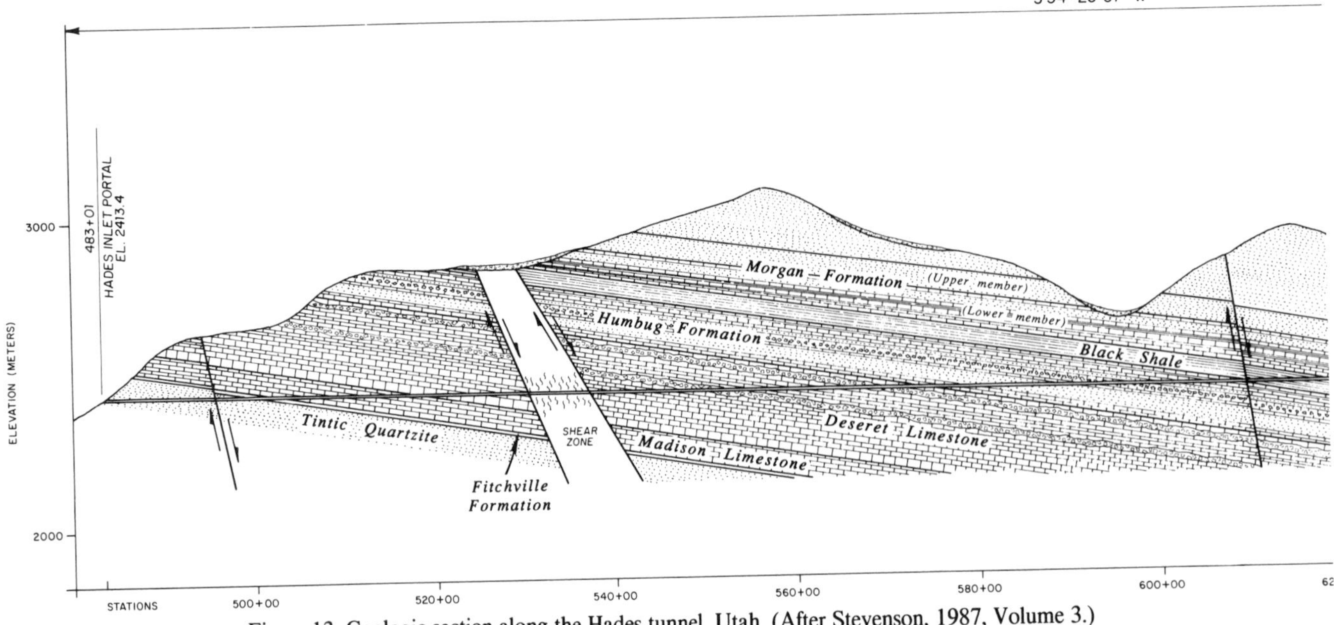

Figure 13. Geologic section along the Hades tunnel, Utah. (After Stevenson, 1987, Volume 3.)

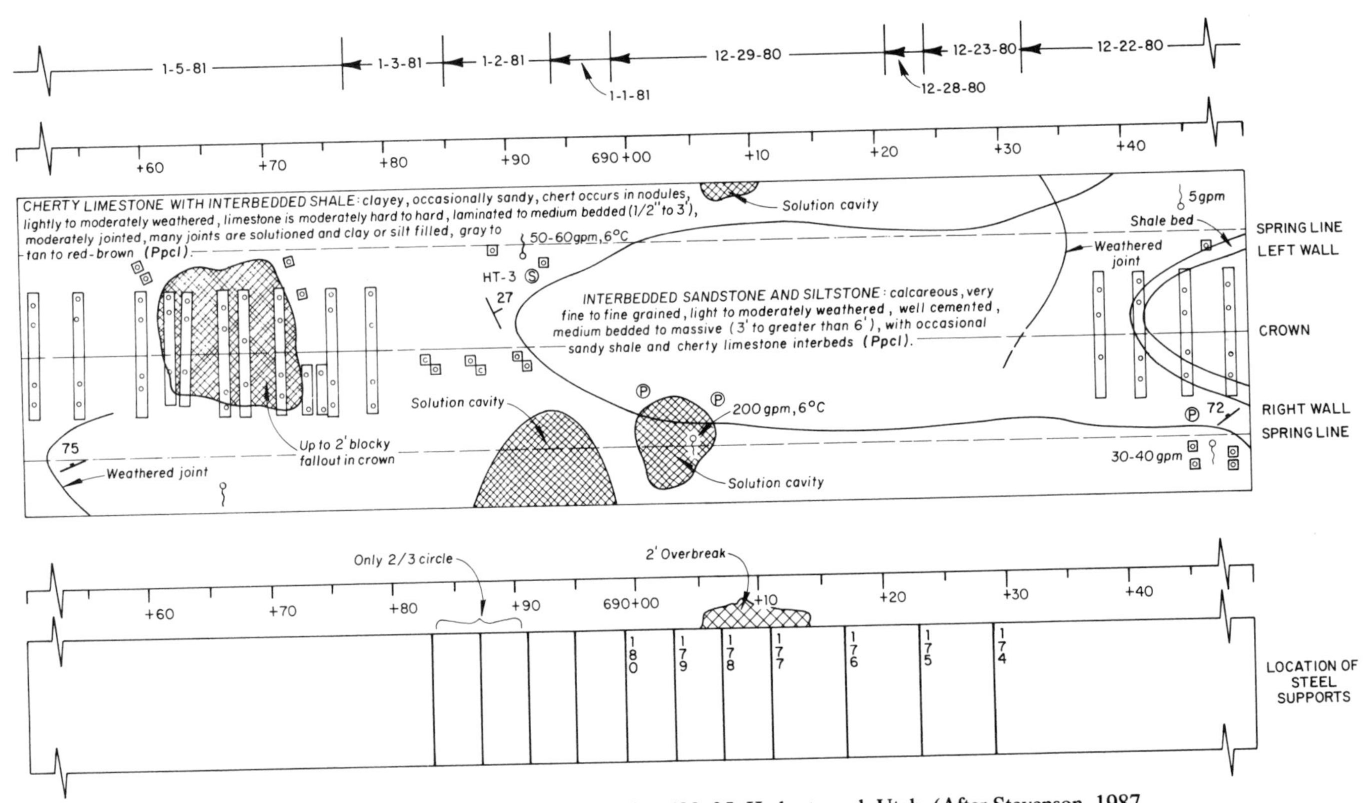

Figure 14. Geologic log in the vicinity of Station 690+05, Hades tunnel, Utah. (After Stevenson, 1987, Volume 3.)

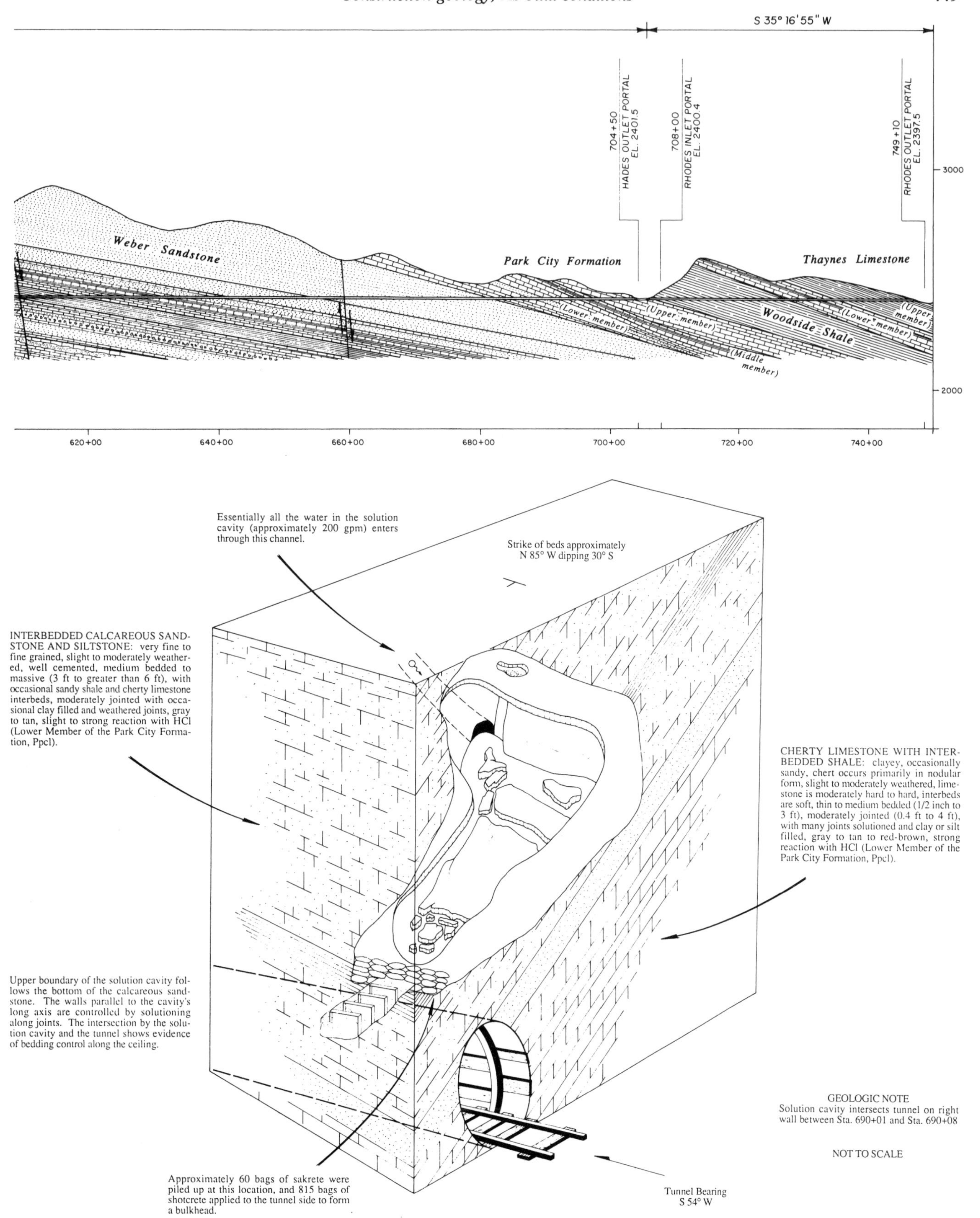

S 35° 16'55" W
704+50 HADES OUTLET PORTAL EL. 2401.5
708+00 RHODES INLET PORTAL EL. 2400.4
749+10 RHODES OUTLET PORTAL EL. 2397.5
3000
2000
Weber Sandstone
Park City Formation
Thaynes Limestone
(Lower member)
(Upper member)
Woodside Shale
(Lower member)
(Upper member)
(Middle member)
620+00
640+00
660+00
680+00
700+00
720+00
740+00
Essentially all the water in the solution cavity (approximately 200 gpm) enters through this channel.
Strike of beds approximately N 85° W dipping 30° S
INTERBEDDED CALCAREOUS SANDSTONE AND SILTSTONE: very fine to fine grained, slight to moderately weathered, well cemented, medium bedded to massive (3 ft to greater than 6 ft), with occasional sandy shale and cherty limestone interbeds, moderately jointed with occasional clay filled and weathered joints, gray to tan, slight to strong reaction with HCl (Lower Member of the Park City Formation, Ppcl).
CHERTY LIMESTONE WITH INTERBEDDED SHALE: clayey, occasionally sandy, chert occurs primarily in nodular form, slight to moderately weathered, limestone is moderately hard to hard, interbeds are soft, thin to medium bedded (1/2 inch to 3 ft), moderately jointed (0.4 ft to 4 ft), with many joints solutioned and clay or silt filled, gray to tan to red-brown, strong reaction with HCl (Lower Member of the Park City Formation, Ppcl).
Upper boundary of the solution cavity follows the bottom of the calcareous sandstone. The walls parallel to the cavity's long axis are controlled by solutioning along joints. The intersection by the solution cavity and the tunnel shows evidence of bedding control along the ceiling.
GEOLOGIC NOTE
Solution cavity intersects tunnel on right wall between Sta. 690+01 and Sta. 690+08
NOT TO SCALE
Approximately 60 bags of sakrete were piled up at this location, and 815 bags of shotcrete applied to the tunnel side to form a bulkhead.
Tunnel Bearing S 54° W

Figure 15. Block diagram of solution cavity in Lower Member of the Park City Formation (Permian) in vicinity of Station 690+05, Hades tunnel, Utah. (After Stevenson, 1987, Volume 3.)

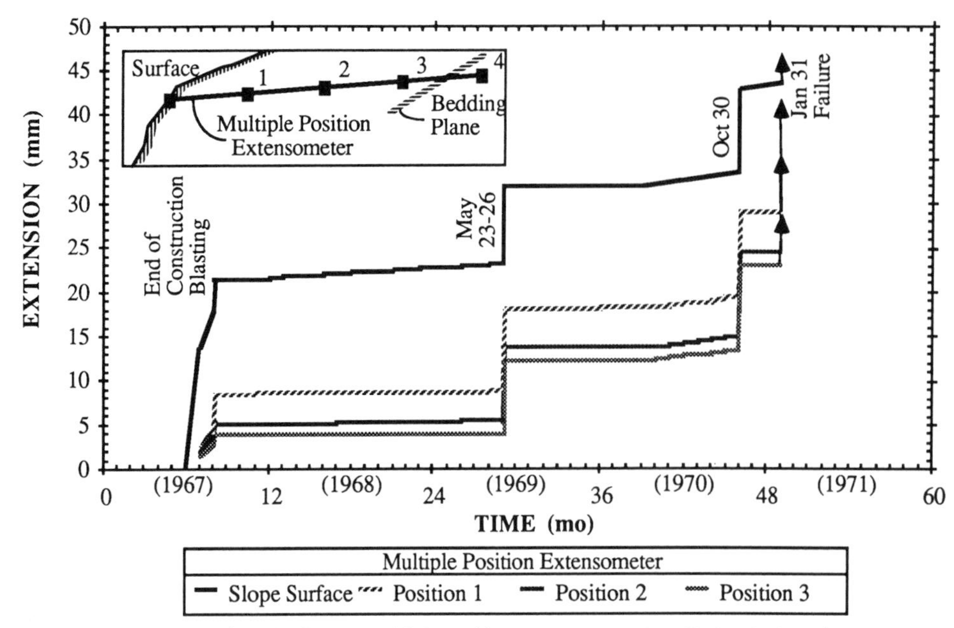

Figure 16. Time-extension plot from a multiple position extensometer installed at the location shown on Figure 17 following the 1967 rockslide. (Modified from Hamel, 1976, Fig. 2.)

abutment of the dam. The debris from this rockslide overlies postglacial alluvium of the Kootenai River on the west side of the valley, and is believed to be Holocene in age. The slide debris was estimated to be 2.4 million m^3 on the basis of the topographic configuration and the interpretation of a seismic survey (Voight, 1979, p. 289). The smaller prehistoric rockslide is located 300 m upstream from the left abutment, with an estimated volume of 0.27 million m^3.

Relocation of Highway 37 at the left abutment began in 1967 with design-cut slopes at inclinations of 0.25:1 using presplit techniques. A 6,100-m^3 wedge rockslide occurred a few minutes after the blast muck was removed from the base of an 18-m-high slope at the left abutment of the dam (Hamel, 1976, p. 366; Voight, 1979, p. 291). The failed wedge was bounded by a bedding plane (N24.5°W142°SW), a downslope joint (N74.5°E/72°NW), and an upslope joint (N81°W/66°NE), which opened as a tension crack during the slide (Hamel, 1976, p. 366). A pocket of soil in the rock updip from the wedge failure suggested that previous movement had occurred (Voight, 1979, p. 291).

The potential for additional wedge rockslides was recognized, and seven multiple-position borehole extensometers were installed in the left abutment slope in 1967 to monitor deformations. Approximately 20 mm of deformation of the slope face occurred adjacent to the 1967 slide during construction of the relocated highway, as documented by the deformation-time plot for one of the extensometers shown on Figure 16.

Following the 1967 failure of the highway slope at the left abutment, an investigation of the potential for future similar slides was conducted, and a potentially unstable rock wedge was identified adjacent to the 1967 failure (Ward and Galster, 1969, in Voight, 1979, p. 293). This wedge was defined by the same bedding plane as the 1967 failure and was being monitored by the extensometer plotted on Figure 16. The intersection of the potential failure planes was exposed in the road cut about 3 m above highway level; the defined wedge of potential failure had an estimated volume of about 50,000 m^3 (Voight, 1979, p. 293). Recommendations for reinforcement of the slope with rock bolts and anchors were made in 1969, but were not followed. As documented by the extensometer (Fig. 16), on January 31, 1971, a wedge rockslide of some 33,000 m^3 failed and caused surface shock waves recorded at the Libby seismograph station located 8 km north of the dam. The initial event recorded at the station had a duration of 17 seconds, and is thought to have been caused by the impact of large blocks on the road (Voight, 1979, p. 293). Seismic tremors caused by rockslides are well-recognized phenomena; for example, the Vaiont, Italy, reservoir slide event sent surface waves recorded in Rome, Vienna, and even Brussels, as discussed in James and Kiersch, this volume.

The wedge failures at Libby Dam are well documented, and the largest was monitored to failure (Fig. 16). The abundance of data regarding these failures permitted back-calculation of the strength parameters on the failure planes by Hamel (1974). After a thorough evaluation of the failures he concluded that, for effective normal stresses of less than about 200 kPa, the strength parameters required to maintain equilibrium (factor of safety = 1.0) for the five failed rock wedges at Libby Dam consisted of an effective cohesion (c') of 0 and an effective friction angle (ϕ') of 30° to 35° (Hamel, 1976, p. 383). Residual strength values for geologic discontinuities determined by in-situ field tests were found to be somewhat greater than those measured in the laboratory; the differences were attributed to "joint asperities, wedges sliding in directions different from tectonic shearing directions, non-planeness of failure surfaces, and gouge particle interaction" (Hamel, 1976, p. 384). Hydrostatic pressures were considered in tension cracks for the 1971 failure but were considered to be zero for the 1967 failure. Laboratory-measured strength parameters were found to be: $c' = 0$ and $\phi' = 25°$ to $30°$. Cable bolts were installed to increase the normal force on bedding planes (Fig. 17), and a buttress fill was placed along the toe of the slope upstream from the left abutment of the dam.

An additional concern on the Libby Dam project was the potential for rockslide-generated waves in Lake Koocanusa. Model studies were conducted to estimate wave heights, runup, and potential overtopping for four wedges near the left abutment (Davidson and Whalin, 1974, in Slingerland and Voight, 1979). A peak velocity of 22 ± 7 m/s was estimated for the largest prehistoric rockslide at Libby Dam, depending on assumptions of friction coefficients and normal-force time histories of moving deformable slide blocks (Voight, 1979, Table V).

The wedge rockslides that occurred during construction at Libby Dam focused attention on this type of slope failure. Fortunately, the geologic conditions contributing to the failure were well understood and provided a sound basis for design of the remedial treatment.

Cochabamba-Villa Tunari highway. A new highway under construction in 1969 was to provide access from Cochabamba, in the Bolivian Andes at an altitude of about 3,000 m, to Villa Tunari, on the Chapare River in the upper Amazon Basin at an altitude of about 300 m (Sowers and Carter, 1979, p. 401). Although the direct distance between the cities is about 75 km, the road distance is approximately 150 km. The highway must cross rugged mountains with steep slopes and peaks rising to 5,000 m and must cross the Andes at an altitude of 4,500 m. Highway construction in such rough terrain is exceptionally difficult; fills cannot be placed on the steep slopes, and excavation is hazardous due to incipient instability (Sowers and Carter, 1979, p. 402). The narrow existing road followed ridge crests and featured numerous switch-back turns and steep grades. The new highway was designed with more modest grades without switch-backs by using tunnels through narrow ridges, and construction of thick fills in deep valleys and deep cuts into the mountainside.

Without warning, a rockslide occurred during construction on July 9, 1969, on the highway about 85 km from Villa Tunari at an altitude of 1,860 m. The slide occurred several months after the last significant rain, and days after the cut had reached a height of 15 m (Sowers and Carter, 1979, p. 409). The major rockslide, comprising some 60,000 m^3, had a nearly triangular configuration of 160 m length along the road, 40 to 50 m high, and 10 to 12 m thick. Seven men were killed, and many pieces of construction equipment were lost in the rapid slide (Sowers and Carter, 1979, p. 410).

The highway alignment is underlain chiefly by interbedded sandstone and shale with some limestone. The rock sequence has been weakly metamorphosed, strongly tilted, and is locally faulted (Sowers and Carter, 1979, p. 404). In the area of the rockslide, well-indurated sandstones are interbedded with shale seams 5 to 15 cm thick at intervals of 1 to 2 m. The bedding dips 30° to 45° to the west and controls the steepness of the natural hillslopes so the west mountain face is the dip slope (Sowers and Carter, 1979, p. 405). The tilted beds have been folded into a series of small-scale, gentle anticlines and synclines, which trend roughly parallel to topographic contours at a wavelength on the order of 30 cm. The dip of the sandstone at road grade in the failure area ranges from 41° to 45°, while 350 m upslope the dip is about 33° (Fig. 18A).

The mountainside was too steep to permit construction of fill without retaining structures. Therefore, the 9-m-wide right-of-way was designed to be on a cut bench, and excavated material was cast downslope. A nearby old road constructed in this manner had performed satisfactorily; however, the dip of the sandstone along the old road was less than 30° (Sowers and Carter, 1979, p. 406). The excavation of the cut bench was made in lifts of about 3 m at a design slope of 0.2:1, and the final 3-m lift was being drilled on July 9 at the time of failure. The cut design and the sequence of the rockslide failure are shown on Figures 18B, C, D, and E.

Post-failure investigations by Sowers and Carter (1979, p. 411–413) documented previous slope failures in the same area. The relative ages of the old slides could be estimated by differences in vegetation growth. Clearly, exposure of the weaker shale seams in the road cut had caused the rockslide. Direct shear testing of large samples of the slide surface indicated friction angles that ranged from 29° to 42° dry and 24° to 25° wet. Testing of samples smaller than the dimension of the 30-cm undulations of the bedding revealed friction angles ranging from 27° to 34° dry.

Minor dampness existed in the shale interbeds months after the last significant rain, and the lack of free water and seepage indicated that hydrostatic pressure had not contributed to the rockslide. During the post-failure investigations, "loud cracking and popping sounds, like small-arms fire," were generated as to the rock in the cut expanded in the morning sunshine; similar sounds occurred late in the day as the rock cooled (Sowers and Carter, 1979, p. 413). Thus, a principal underlying cause of the failure was interpreted to be thermal expansion and contraction, which resulted in a gradual, progressive downslope movement. The progressive movement caused the block above the failure surface to "climb" out of the undulations on the bedding surface. The initial condition of the block seated in the undulations is shown on Figure 19A; as the downslope movement occurred, the condition shown on Figure 19B resulted. Sudden failure occurred as the effective slope angle exceeded the threshold condition where the slope angle equals the angle of friction on the surface (Sowers and Carter, 1979, p. 415).

Corrective measures considered by Sowers and Carter (1979, p. 416–417) consisted of (1) removing the rock above the shale seam, (2) supporting the unstable block with a concrete buttress, (3) increasing normal stresses on the block with tensioned rock bolts, and (4) draining any subsurface water. Ultimately, no corrective measures were implemented due to cost

Figure 17. Photographs of the 1971 wedge failure on the left abutment of Libby Dam as it appeared in February 1988. Surface anchors for tensioned cable bolts are visible. a. From right abutment: looking at areal surroundings of left abutment and wedge. b. From top of dam: looking across wedge on left abutment.

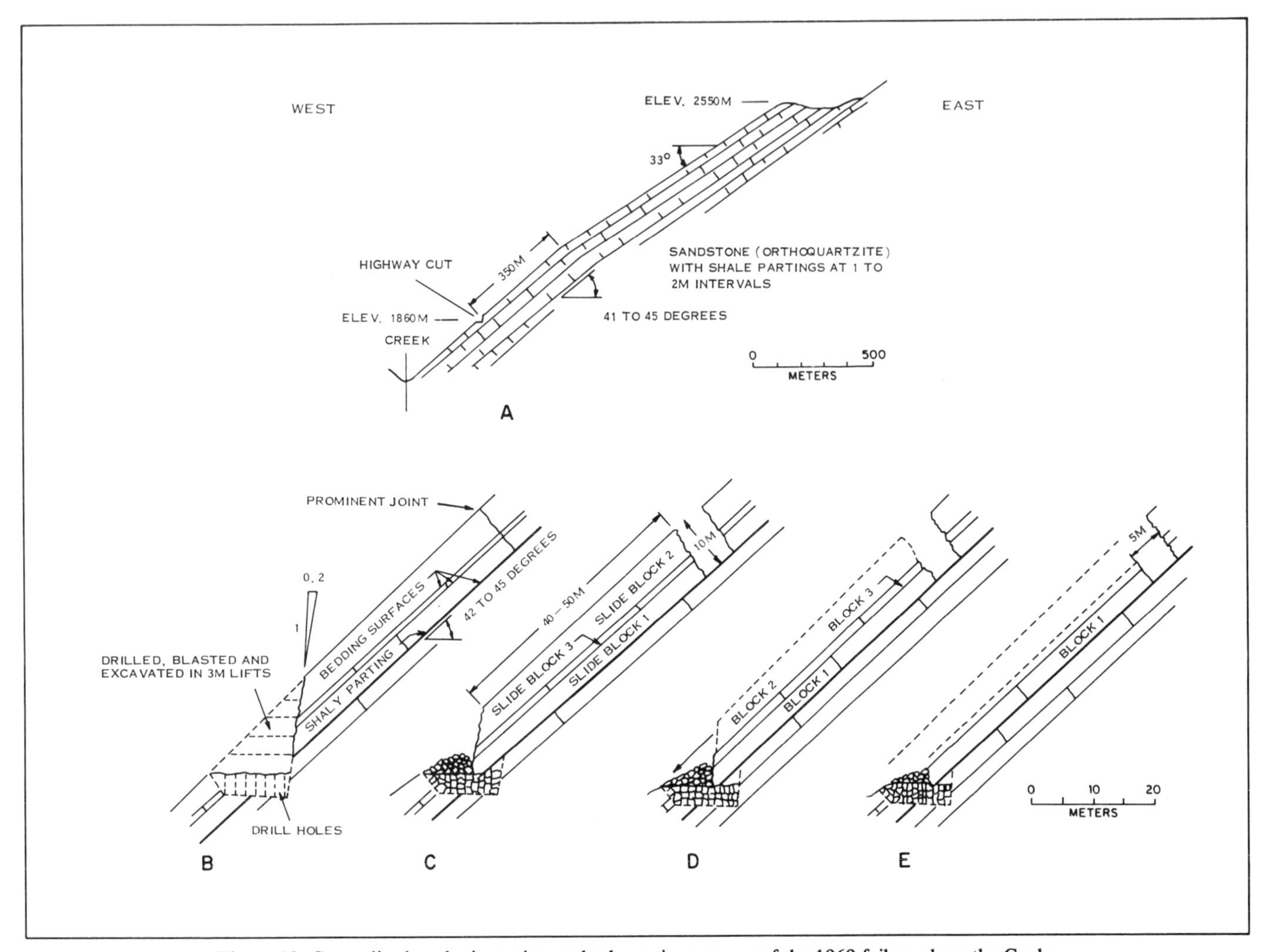

Figure 18. Generalized geologic section and schematic sequence of the 1969 failure along the Cochabamba–Villa Tunari Highway, Bolivia. A. Generalized geologic section at failure location. (Modified from Sowers and Carter, 1979, Figure 1.) B. Schematic section of failure location before sliding; holes for blasting of last lift were being drilled at the time of sliding. C. Initial sliding of Blocks 1, 2, and 3 simultaneously. D. Subsequent sliding of Block 2 within seconds of initial movement. E. Subsequent sliding of Block 3 two days after initial movement. (Modified from Sowers and Carter, 1979, Fig. 4.)

and time factors and, in 1973, a rockslide of some 40,000 m^3 occurred in the same area as the 1969 failure. Additional slides are expected in the future.

The 1969 rockslide during construction of the Cochabamba–Villa Tunari highway was located in a geologic setting that provided an unfavorable exposure (daylighting) of sensitive, steeply dipping bedding planes; past slides were evident, and future slides are predicted. Yet the investigation of this rockslide advanced the understanding of failure along dry bedding planes and, as such, illustrates the usefulness and kinds of input construction geology practices can contribute to applied geosciences for engineering works.

Interstate Highway 80. The opening of the Ohio Turnpike, the initial segment of Interstate Highway 80, was scheduled for October 1, 1955. On September 1, field evidence suggestive of a forthcoming landslide was observed in a section of the turnpike south of Toledo, Ohio. The slide-prone area occurs in a flat lake plain, underlain by sediments of Pleistocene Lake Maumee, of which Lake Erie is the remnant (Supp and others, 1957, p. 58). The lake beds consist of varved silt and clay with very fine sand partings. Reworked glacial till and comparatively minor areas of sandy and gravelly beach ridge, dune, and near-shore deposits are present locally. Bedrock in the area is gray dolomitic limestone of the Upper Silurian Salina Group, which is buried by more than 20 m of surficial deposits.

Supp and others (1957) present a detailed description of the landslide and associated features that occurred after the pavement had been placed and one month before the turnpike was to be opened. The pattern of surface deformation (Fig. 20), such as bulging of the toe, cracking, and sagging of the head, indicated a

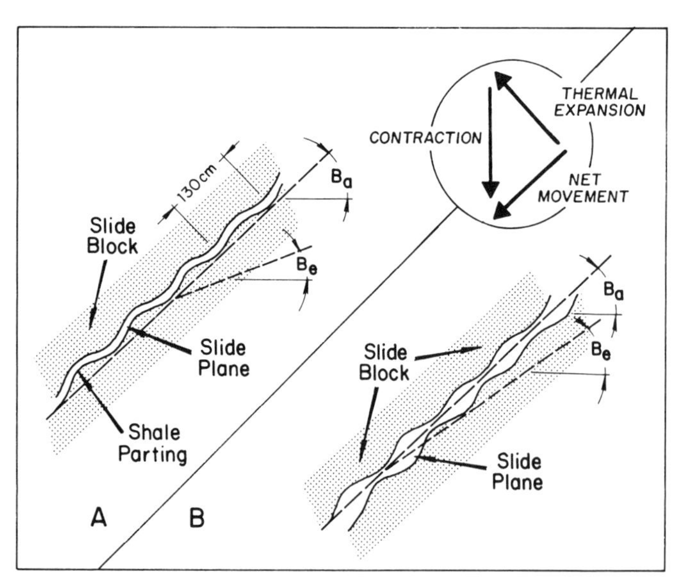

Figure 19. Schematic detail of slide surface with suggested failure mechanism. A. Before sliding the undulations on bottom of the slide block are seated in the undulations on the slide plane; the average slope angle, B_a, is steeper than the friction angle on the plane, θ, but the effective slope angle due to the undulation, B_e, is flatter than θ. B. Sudden failure occurs as the effective slope angle, B_e, equals the friction angle, θ; B_e gradually increases as thermal expansion and contraction causes the slide block to climb out of the undulations. (Modified from Sowers and Carter, 1979, Fig. 11.)

rotational-circular slump movement. Although similar subsurface geologic conditions existed for a distance of some 365 m along the highway, only 125 m of 11-m-high road embankment was affected by the slump. The initial assessment suggested that the movement was triggered by excavation of a drainage ditch at the toe of the road embankment, so emergency remedial measures included backfilling the ditch. If not halted immediately, further movement could cause complete failure of the embankment and inflict serious and costly damage on the completed concrete pavement (Supp and others, 1957, p. 59).

A comprehensive subsurface investigation program was initiated while construction of an earth buttress began at the toe of the failed slope. The concrete pavement was releveled and cracks grouted when survey data confirmed that slide deformation had stopped. Compacted fill buttresses, totalling about 65,000 m^3, were placed along the 365-m-long unstable section. The original 2:1 embankment slope was increased by the buttress to a maximum of 3:1, with a 10:1 segment over the toe of the original slope (Fig. 21).

The subsurface investigation consisted of nearly 300 m of drilling in 23 borings, including about 180 m of continuous samples taken with a Swedish Foil Sampler, the first use of such a sampler in the United States. A discontinuous horizon of thin sand lenses and partings was delimited in the lake sediments making up the foundation of the embankment. Piezometers installed in the exploratory borings (Fig. 21) disclosed an excess pore pressure equivalent to a head of water 4 m above the original ground, or some 38.9 kPa, which dissipated over the subsequent few months. The road embankment apparently was on the verge of failure under the as-built configuration with the excess pore water pressure. Thus, excavation of the surface drainage ditch further reduced the forces providing stability, and the slope failure of September 1 occurred. The excess pore water pressure dissipated slowly in the area of the failure because the sand lenses and partings in the lake deposits were discontinuous; elsewhere, the lenses and partings apparently were more continuous, which permitted the excess pore water pressure to dissipate more rapidly, maintaining stability.

Supp and others (1957, p. 65) concluded that "near-miracles can be achieved by the joint efforts of persons and organizations of different backgrounds and interests when motivated by a common goal." The dedicated efforts, concerted abilities, and cooperation among geologists, engineers, and contractors led to an early solution to the failure. Serious expense, inconvenience to the public, and delay were averted.

Grand Coulee Dam. Grand Coulee Dam is on the Columbia River in north-central Washington. Construction of the massive, 1,310-m long and 168-m high concrete dam began in 1933 and ended in 1942. Charles P. Berkey supervised the detailed geological investigations, which were conducted from the beginning of construction by Irwin, (1938, p. 1628). A "simplified" geologic map of a portion of the dam foundation is presented on Figure 1, while a more generalized geologic map and a representative cross section are presented on Figure 3.

The geologic setting of the dam includes granitic rocks of the Colville batholith, Tertiary shale, Columbia River basalt, and surficial deposits of glacial, lacustrine, and fluvial sand, gravel, silt, and clay (Irwin, 1938, p. 1632–1633). The granitic rocks, which are durable and of high quality, provide excellent support for the dam over its entire foundation and abutments (Berkey, 1935, p. 67). The eastern portion is an equigranular biotite granite, while the western portion is a porphyritic granite with minor syenite and monzonite (Fig. 3). The granitic rocks are jointed, but preparation of the rock for use as a dam foundation required minimal effort (Irwin, 1938, p. 1633). The sound granitic rock conditions at Grand Coulee are in sharp contrast to the variable degrees of weathering at Folsom Dam, described earlier in this chapter. The chief difference in rock weathering is attributed to glacial scour at Grand Coulee removing the Neogene weathered veneer that remained at Folsom.

The Tertiary shale and Columbia River basalt occur outside the immediate dam site and are of no direct concern (Vehrs, 1974, p. 252). In contrast, the surficial deposits (Fig. 22) proved to be of great concern and required specialized treatment to excavate the foundation.

The ancient Columbia River was dammed at least once downstream from the Grand Coulee site by a late Pleistocene glacier (Irwin, 1938, p. 1633). The resulting lake accumulated a thick sequence of varved clays and silts at the dam site. Near the left (west) abutment, the varved sediments grade upward through

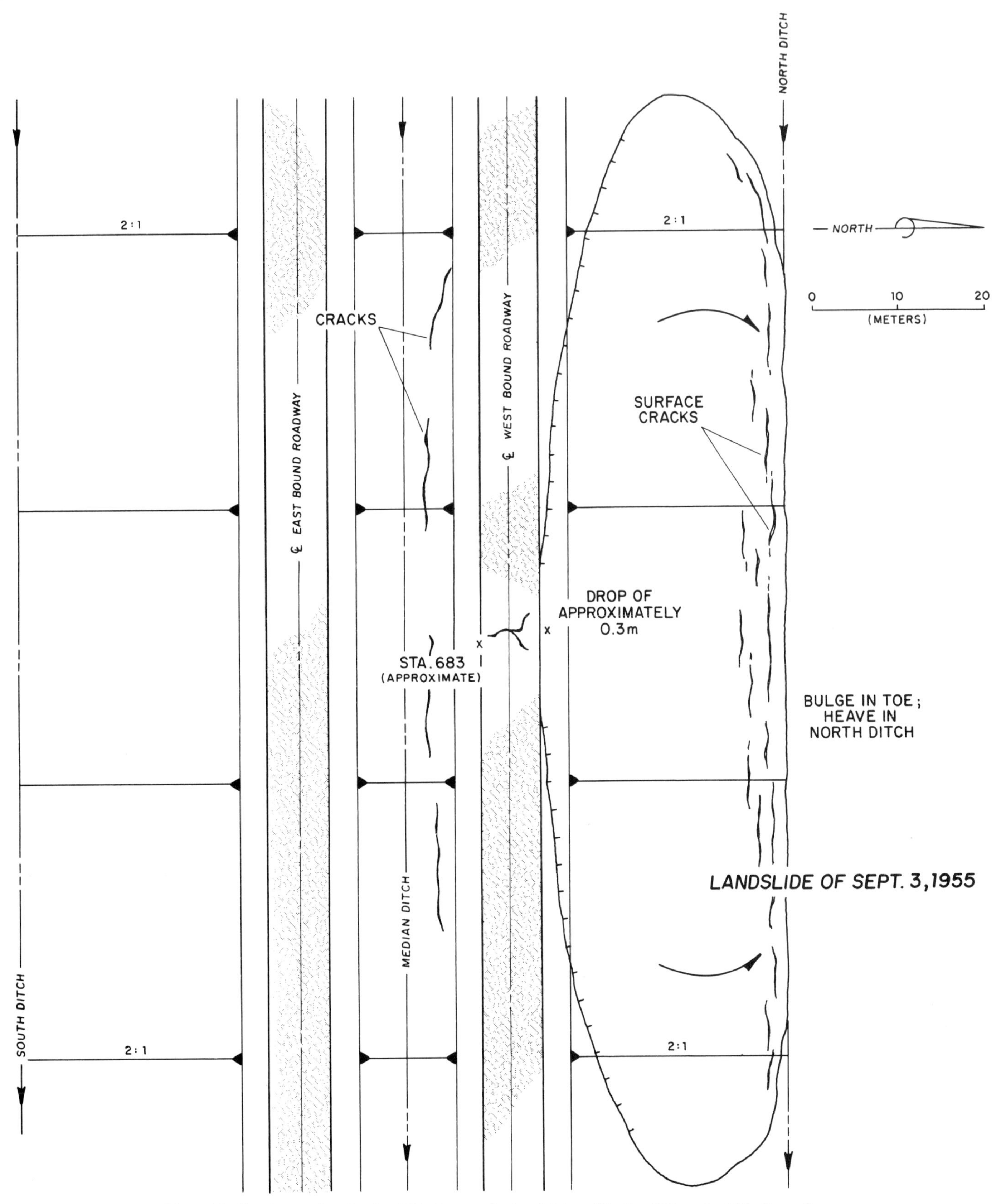

Figure 20. Sketch map showing landslide features in the vicinity of Station 683 on the Ohio Turnpike (Interstate 80) south of Toledo, Ohio. (Based on description by Supp and others, 1957.)

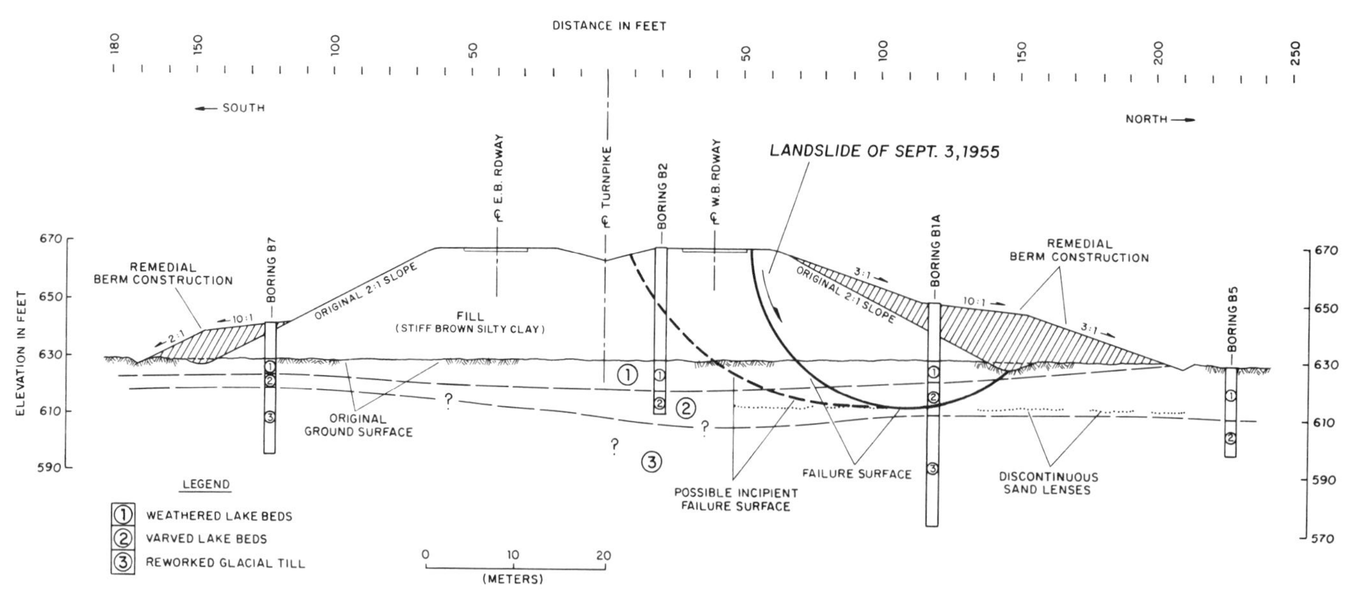

Figure 21. Geologic section showing the assumed failure surface of the 1955 landslide at Station 683 on the Ohio Turnpike. (Modified from Supp and others, 1957, Fig. 1.)

stratified silt and sand and lenses of till into an overlying member of stratified sand, silt, and gravel (Fig. 22). The combined stratigraphic thickness of these sediments in the vicinity of Grand Coulee Dam is 245 m. However, at the dam the excavated thickness of surficial deposits ranged from about 15 m in the center to 90 m near the left abutment (Fig. 22). Distortion of the stratification in the surficial materials is common locally. The distortion has been attributed to (1) the thrust of advancing ice, (2) the slumping of high-level deposits during ice retreat, and (3) landslides caused by undercutting during subsequent river erosion (Irwin, 1938, p. 1635).

Excavation to bedrock required removal of more than 13.4 million m^3 of surficial deposits, which was accomplished by large (1.0 to 3.8 m^3) shovels, trucks, and conveyor belts. The excavation proceeded with little difficulty except in the thicker sections of varved clay and silt, which had a tendency to lose strength when disturbed (Vehrs, 1974, p. 253). Numerous old landslides encountered in the deep excavation near the left (west) abutment were reactivated during construction. The design slope of 1.5:1 was flattened to 2:1, requiring an additional 1.2 million m^3 of excavation. Additionally the toe of the slope was loaded and supported by a rock blanket, and water in the slide mass was drained into a collection tunnel driven in bedrock beneath the slide in order to effectively stabilize the excavated slope (Vehrs, 1974, p. 252; Irwin, 1938, p. 1635).

Much more difficult stability problems were experienced with the surficial materials on the right (east) abutment. Varved clay and silt that backfilled a 30-m-deep ancient channel carved in the bedrock became unstable at the conventional cut slope of 1.5:1. In an attempt to achieve stability, the slope was flattened to 2:1, and a concrete retaining wall was built across the channel feature with a timber crib wall at the top. This was unsuccessful: the clay and silt became mobilized, overran the wall, and filled the excavation. Further attempts to stabilize by flattening excavation slopes to 2.5:1 to 3:1 also failed. Flatter slopes would have required several million cubic meters of added excavation. An innovative and less costly engineering solution to the slope stability problem was achieved at Grand Coulee. The fluidized clay and silt directly upslope from the crib wall across the backfilled channel feature was frozen in place by circulating a refrigerated ammonia-brine solution in 377 evenly spaced tubes, which created an arch dam of frozen earth 32 m long, 6 m wide, and 12 m deep (Leggett and Karrow, 1983, p. 19–27). The refrigeration system had a capacity to produce 72,000 kg of ice per day. The frozen arch dam cost $30,000, but saved much more than that in excavation costs and time (Gordon, 1937, p. 211).

Recognition that frozen soft clay and silt can develop sufficient strength to act as a temporary retaining structure resulted in a viable economic solution to a difficult stability problem at Grand Coulee.

Ground water

San Jacinto tunnel. The San Jacinto tunnel is located in the northern San Jacinto Mountains between the San Jacinto and San Andreas fault zones (Henderson, 1939, p. 316) and is part of the Los Angeles–Colorado River aqueduct. Construction of this 21-km-long structure began in 1935, and serious ground-water problems were encountered almost immediately. A 250-m-deep shaft was constructed at the Potrero heading of the tunnel, approximately 5 km northeast of the intersection of the tunnel and the San Jacinto fault. The tunnel had advanced only 50 m east from the bottom of the shaft when a shear zone was encountered. A short penetration into this fault zone resulted in a sudden

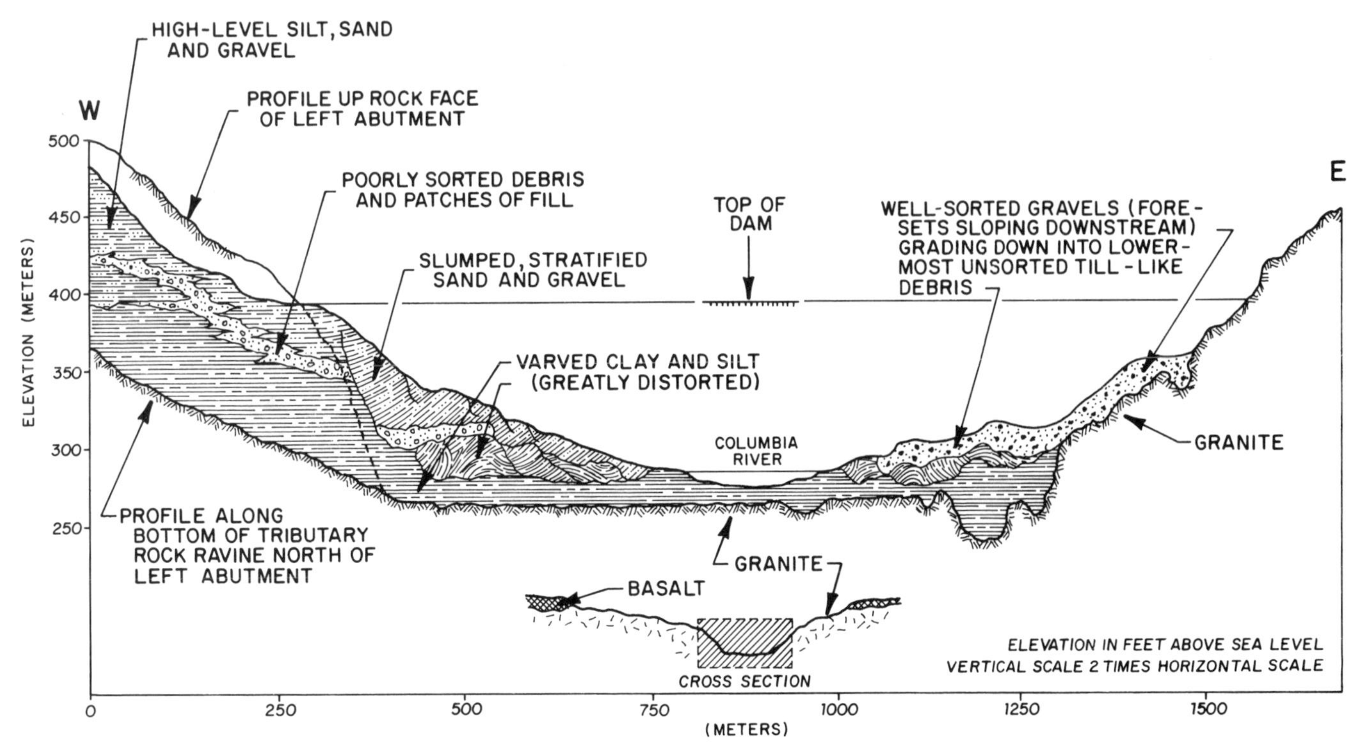

Figure 22. Geologic section showing surficial deposits excavated during construction of Grand Coulee Dam, Washington. (After Irwin, 1938, Fig. 2.)

inflow of ground water, estimated at 475 L/s, which flushed more than 750 m^3 of rock debris into the heading (Thompson, 1966, p. 105). The water discharge from the fault zone soon flooded the tunnel and filled the shaft to within 45 m of the ground surface, indicating a hydrostatic pressure of about 1,990 kPa at the tunnel heading.

Geological investigations in the early 1930s, before construction of the tunnel started, indicated that hard, mostly massive granitic rocks of Cretaceous age, including granodiorite, quartz monzonite, and quartz diorite, would be the predominant type through which the tunnel would be driven, but that at least one long section of older metamorphic rocks (mica schists, quartzites, and crystalline limestones) and possibly several smaller bodies would be encountered. Although some suggestion of faulting was noted, it was believed that the faults would not represent serious obstacles to advancement of the work since no one visualized the tremendous volumes of water that were later found to be associated with the faults (Thompson, 1966, p. 105). A detailed geological evaluation of the tunnel alignment was undertaken in 1935 after the first water-bearing faults were encountered to assess the possible presence of additional faults of a similar nature. Aerial photograph interpretation and field verification of about 1,550 km^2 of area in the northern San Jacinto Mountains revealed 21 northwest-trending faults. The attitudes of the fault traces measured on the ground surface were found to be unusually consistent with depth throughout the area of the "Potrero fault field" (Henderson, 1939, p. 316).

As the tunneling advanced, water consistently occurred in "open interconnected fractures and joints in the hanging (northeast) wall of the northwest-striking faults" (Thompson, 1966, p. 105). The fractures in the hanging wall rock became increasingly tight with increasing distance from the faults. Consequently, tunnel headings approached the faults from the hanging-wall side; this encountered water under pressure some distance from the faults, with flow volumes increasing toward the faults. Normally, by the time the footwall was reached, water pressures had been largely dissipated, and little difficulty was experienced in driving through soft fault gouge (Thompson, 1966, p. 106). Conversely, tunnel headings approaching faults from the footwall side encountered little ground water until the shear zone was penetrated; abruptly, large volumes under high pressure flowed into the tunnels. Sudden surges of water commonly carried large quantities of broken rock, sand, and fault gouge into the tunnel, such as at the Goetz fault zone in 1937. The fault here was penetrated from the footwall side and a heavy surge of ground water flushed about 2,280 m^3 of sand into the tunnel.

The peak ground-water flow from all headings in the San Jacinto tunnel was approximately 2,525 L/s; the maximum inflow from a single station was 1,010 L/s. Hydrostatic pressures commonly ranged from 1,035 to 2,070 kPa, with pressures up to 4,140 kPa being measured in a few locations (Thompson, 1966, p. 106).

The results of engineering geological studies undertaken during the construction of the San Jacinto tunnel not only greatly

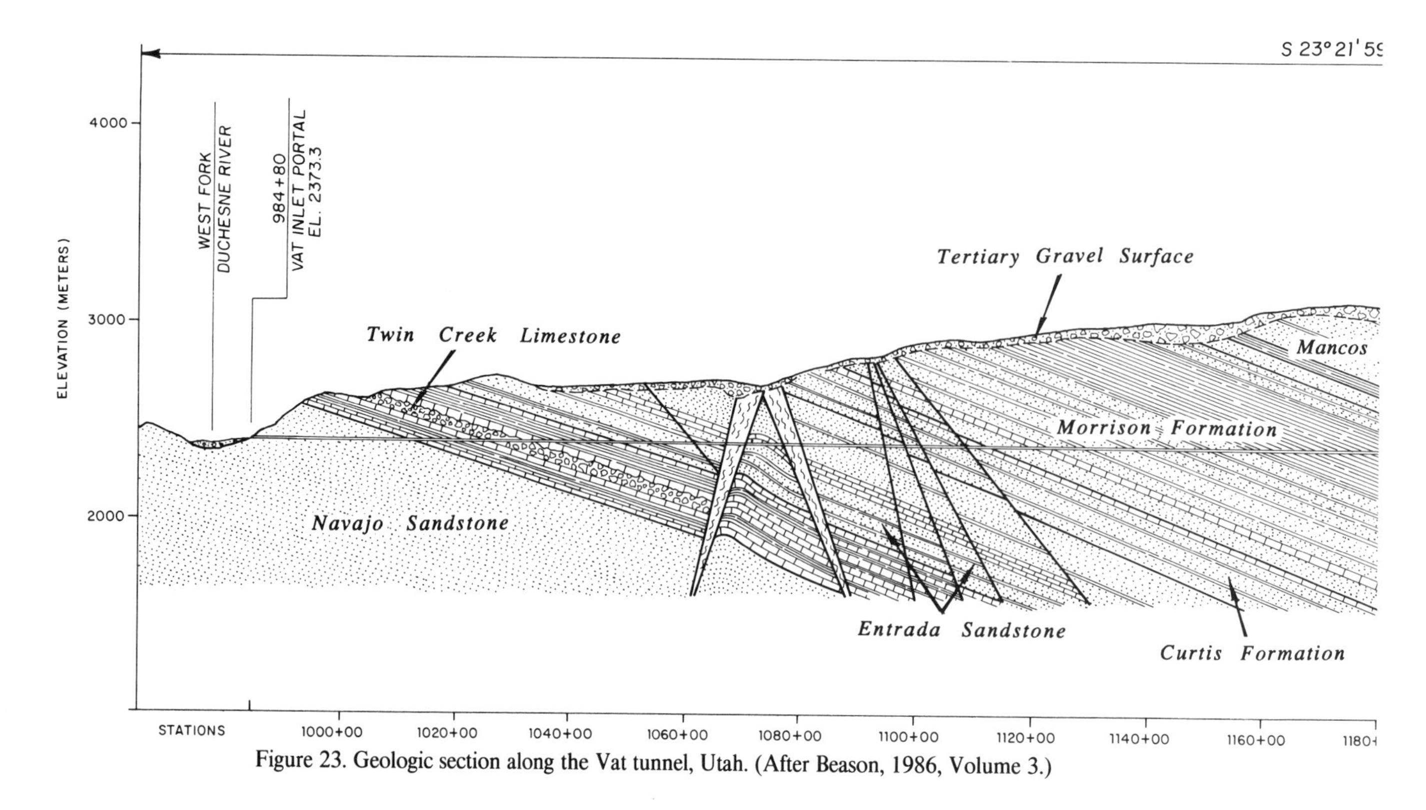

Figure 23. Geologic section along the Vat tunnel, Utah. (After Beason, 1986, Volume 3.)

assisted in the successful completion of the project, but contributed to geological understanding for the practice of tunneling with specific development of innovative approaches to strongly faulted rocks in a semiarid climate. Some of the more outstanding of these innovations, as summarized by Thompson (1966, p. 106–107), are:

1. Pilot holes drilled horizontally ahead of the tunnel face to permit detection of high water pressures or soft ground.

2. Pilot bores (tunnels smaller in section than the main tunnel) driven parallel to and ahead of the main tunnel when difficult ground was known to lie ahead. These pilot bores aided in reducing water pressures by permitting fault zones to be approached and drained from both sides.

3. Minor realignments of the tunnel to approach a fault zone normal to the strike can minimize the distance of fault gouge.

4. Techniques for high-pressure grouting (up to 10.35 MPa) ahead of the driving face to inject neat cement grout into open fissures. This not only reduced the hydraulic conductivity of the broken rock, but also tended to consolidate heavy ground, which reduced tunneling difficulties.

5. The use of steel-reinforced gunite rings, in lieu of the conventional horseshoe sets, in zones of heavy ground where inflows of water continued after the face had been advanced. These ring sets permitted high-pressure consolidation grouting before placing the conventional tunnel lining.

6. Procedures for grouting through concrete or sandbag bulkheads in particularly difficult ground.

Additional discussion of the San Jacinto tunnel as related to a depletion of the normal ground water supply in the vicinity and litigation are given in James and Kiersch (this volume).

Vat tunnel. The Vat tunnel, located approximately 80 km southeast of Salt Lake City, Utah, is part of the Strawberry aqueduct system of the Central Utah project. The northeast portal of the Vat tunnel is located about 8 km southwest of the southwest portal of the Hades tunnel. The geology of the region is dominated by the Uinta basin; consequently, the rocks encountered in the Vat tunnel, being closer to the axis of the basin, are younger than the rocks in the Hades tunnel. As shown on Figure 23, eleven formations of Mesozoic age were penetrated throughout the Vat tunnel and the axis of an asymmetric syncline was crossed in the vicinity of Station 1350* (Beason, 1986).

The straight, 11.75-km-long tunnel was bored with a laser-guided, Robbins model 1011-179 tunnel boring machine (TBM). Boring began at the southwest portal on October 4, 1976, and holed through on July 17, 1981, after considerable difficulty with large inflows of water. The TBM was a fully shielded, hard-rock-type machine 312.4 cm in diameter with 26 cutter heads on a rotating base capable of exerting a maximum of 9.40 MN of thrust at the face. The completed water conveyance tunnel was 251.5 cm in diameter with a nominal 36.45-cm-thick concrete lining (Beason, 1986, Volume 1, p. 50–51).

Considerable difficulty was experienced with the TBM due to its length. Additional difficulty was experienced when water flows at the face exceeded about 3.2 L/s. The muck conveyor

*Stations are in feet (1 ft = 0.3048 m); tunneling progressed up-station from the outlet portal at Station 1371+88 to the inlet portal at Station 985+65.

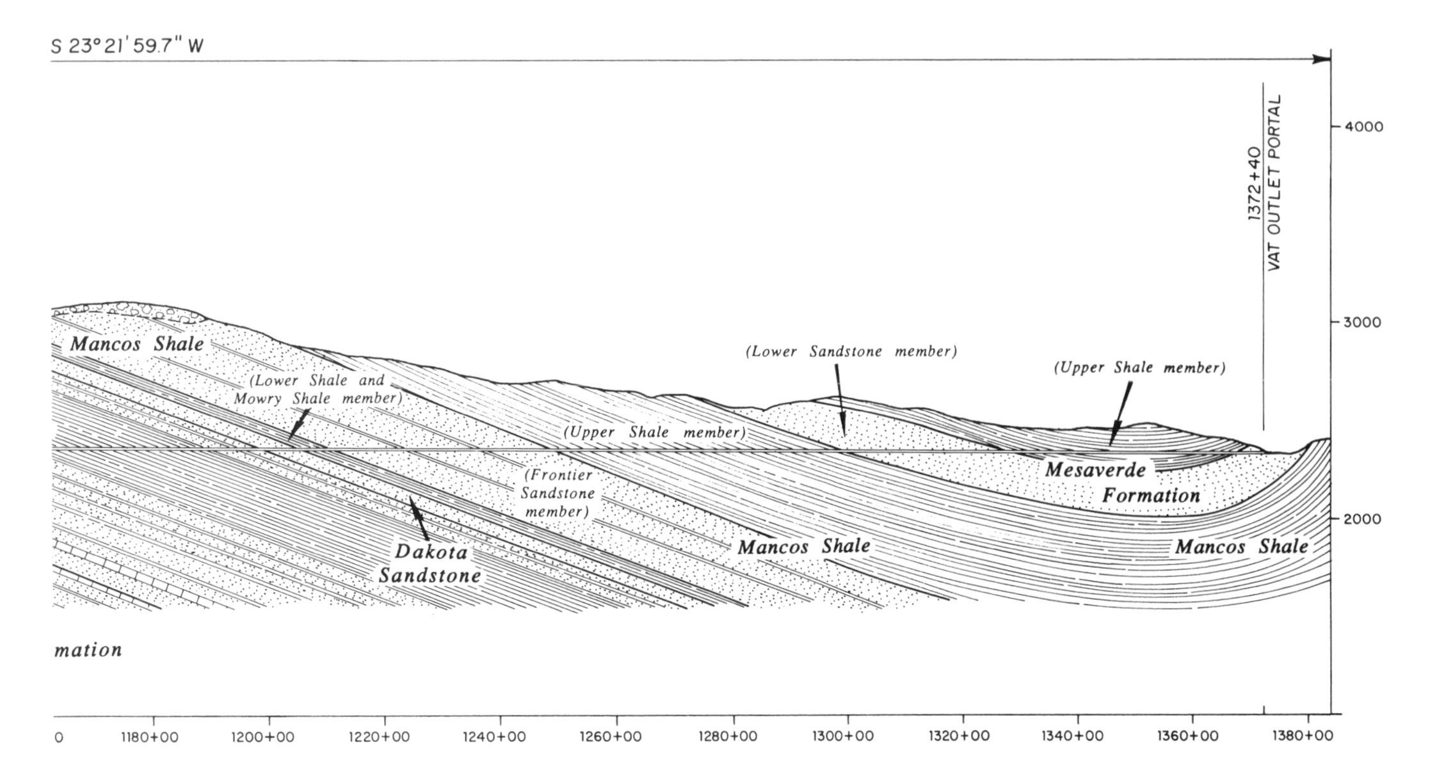

sloped upward at 13° through the throat of the TBM to the trailing gear. Saturated muck would not travel up the incline with the constant water inflow and spilled over the conveyor, filling the space around the TBM and requiring hand mucking (Beason, 1986, Volume 1, p. 73 and 77).

The tunnel is below the regional ground-water surface throughout its length, and water inflows were anticipated, particularly from the limestone and sandstone formations, and from a fault zone in calcareous siltstone of the Jurassic Entrada Formation. A major fault zone was encountered at about Station 1060+38, but no particular difficulty was experienced until the fault zone had been penetrated at Station 1058+50 three days later (Fig. 24). Excessive water inflows occurred in a 10-m segment between Stations 1058+30 and 1057+97; initial flows estimated to be 31.5 L/s increased to about 82 L/s after three weep holes were drilled in the tunnel heading to facilitate drainage of the ground-water reservoir. Maximum hydrostatic pressure measured in the weep holes was 1,654.8 kPa, indicating a ground-water level nearly 170 m above the tunnel (Beason, 1986, Volume 1, p. 74). The fault zone had a low hydraulic conductivity and acted as a ground-water barrier due to recemented gypsiferous fault gouge in many areas. The major water flows issued almost exclusively from a network of open joints apparently connected hydraulically with collapse breccia of the underlying Jurassic Twin Creek Limestone. Ground-water flow and pressure declined with increasing distance past the fault zone, as experienced at the San Jacinto tunnel discussed above.

The excessive water inflow from the faulted calcareous siltstone overwhelmed the drainage system, and water flowed down the invert of the tunnel to the outlet portal. This allowed a large volume of water to come in contact with the relatively dry, but expansive clays in the Jurassic Morrison Formation and the Cretaceous Lower Mancos Shale (Beason, 1986, Volume 1, p. 77). The swelling clay heaved the invert of the tunnel, ponding water, causing some timber lagging to fail, and locally impeding access through the tunnel. No steel supports were damaged by the swelling, and the lining had not been installed at the time of the problem. Some cracking of the concrete lining was noticed as early as 1983 and attributed to continued swelling of the sensitive formations (Beason, 1986, Volume 1, p. 87).

The water inflows encountered between Stations 1058+30 and 1057+97 were sufficiently great to halt the TBM operation; the wet muck was washed off the conveyor and jammed the machine. A bypass drift 1.2 m wide and 2.1 m high was driven by conventional mining along the left (southwest) wall of the tunnel to provide access. A bulkhead was constructed and a concrete plug emplaced. Core borings 30 m or more ahead of the face were used to plan a grout pattern using holes with pipes embedded in the concrete plug (Beason, 1986, Volume 1, p. 80).

Pressure grouting began at Station 1058+20 on February 28, 1979, and continued to Station 1012+97, a distance of more than 1,375 m. Grout reaches ranged from 26.8 to 247.5 m, averaging 91.4 m (Beason, 1986, Volume 1, p. 83). Thick grout mixtures (1:1 to 3:1) were injected initially, but cored borings showed minimal grout penetration due to the relative tightness of the joints. Frequently, grout flow occurred between holes, an indication of interconnecting joints. The TBM was reactivated after completing each grout reach, and tunneling continued until water inflow increased to the point where saturated muck spilled off the conveyor and the TBM became impractical. The last of 15 sepa-

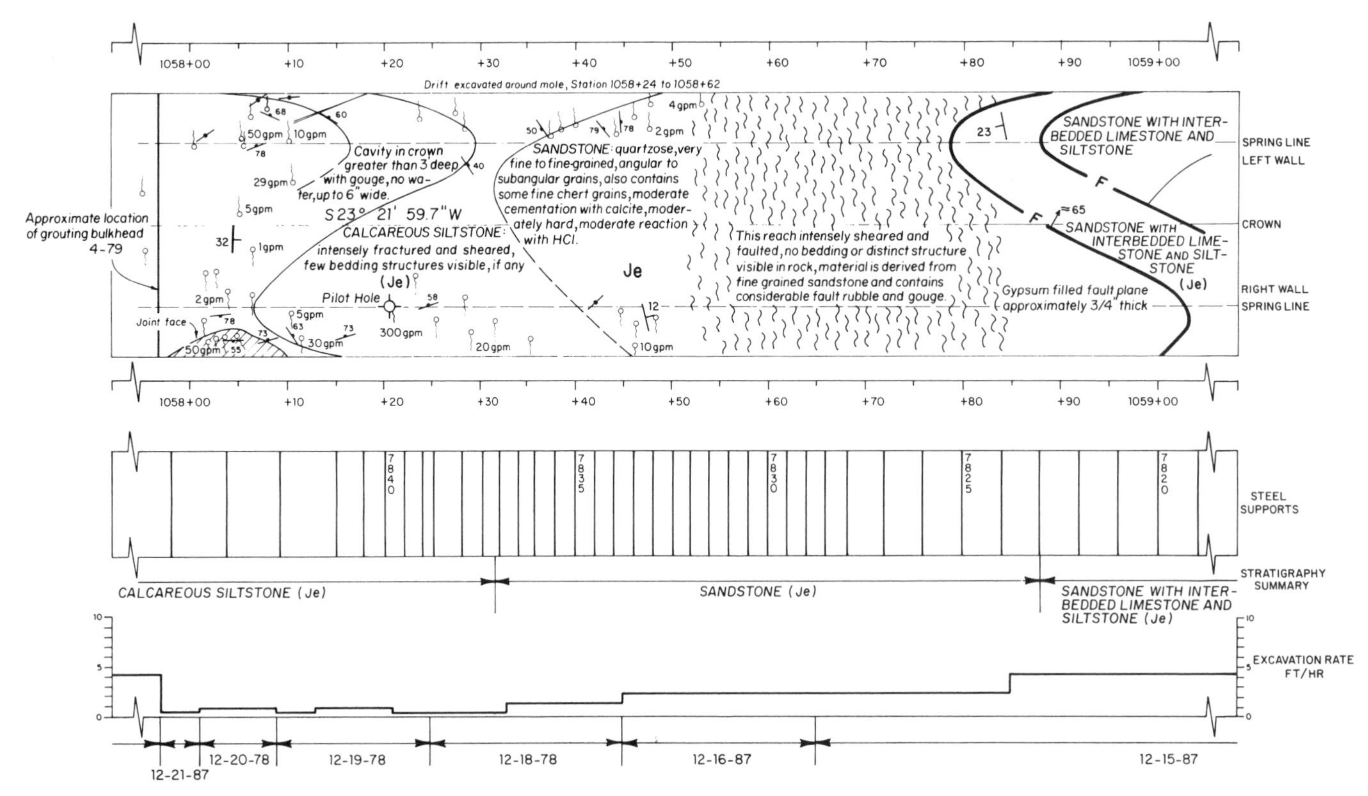

Figure 24. Geologic log in the vicinity of Station 1058+50, Vat tunnel, Utah. (After Beason, 1986, Volume 3.)

rate grout reaches were completed on March 15, 1981, a total grout distance of 1,378.6 m. A total of 55,082 sacks of cement were pumped into 8,068 lineal m of grout hole, an average of 6.83 sacks/m of hole, throughout the zone of excessive water inflow (Beason, 1986, Volume 1, p. 82).

At the beginning of construction, an administrative decision omitted regular periodic geological mapping of the tunnel face in the Vat tunnel because of personnel ceilings. This decision hampered data collection for countering contractor claims; and even though the tunnel was mapped eventually, the full value of the geological data was diminished by delaying the mapping (Beason, 1986, Volume 1, p. 1). Farina (1987) notes that TBM equipment conceals the rock surface near the face, requiring that detailed geologic mapping be done behind the equipment. The Vat tunnel was lined after it was completely excavated, but steel supports and timber lagging were installed behind the TBM in the reaches where the rock would not stand safely. Had a resident engineering geologist been available during construction, valuable and timely geological data would not have been lost. The excavation of the Vat tunnel required nearly 6 years to complete. The 1,280-m-long zone of excessive water inflows with grouting and sporadic TBM tunneling consumed more than 2 years, or an average advance rate of 1.75 m/d in this reach.

Lagoon underpass. The East Service road at Walt Disney World near Orlando, Florida, had to pass underneath a waterway between the Lagoon and Bay Lake. The perimeter of the Magic Kingdom theme park is within 450 m to the northwest, and the resort hotel area is 150 m to the northeast (Fig. 25). The road passes under the lagoon in a cut-and-cover tunnel 61 m long, 11.9 m wide, and 4.4 m high; at the deepest point, the tunnel excavation is approximately 11 m below the ground surface at elevation 17.7 m. Open-cut access ramps slope down from the ground surface and join the tunnel at elevation 20.4 m (Dames and Moore, 1969, p. 2). Site investigations in 1969 disclosed three separate layers of surficial deposits overlying limestone bedrock (Fig. 26). An upper layer of silty sand is 2.7 to 4.3 m thick, the middle layer of clayey sand contains occasional interbeds of clay and silt and totals 6.7 to 9.1 m thick, and the lower layer, the Hawthorn Formation, of silty sand 1.8 to 4.3 m thick. Bedrock is the cavernous Floridian Limestone aquifer, which exhibits artesian pressures of nearly 15 m. Ground water in the upper silty sand is perched on the middle clayey sand and recharged from the lagoon and Bay Lake. Ground water in the Floridian Limestone and Hawthorn Formation is confined by the middle clayey sand.

The principal geological concerns during project design were twofold: (1) effective dewatering during construction with minimal risk of causing imbalanced hydrostatic pressures and a blowout, and (2) control of areal depression of the piezometric surface associated with the Floridian Limestone aquifer to minimize the risk of developing sinkholes throughout the nearby Magic Kingdom theme park and resort hotel areas. Early project planning considered construction of recharge wells with capacities of about 945 L/s in the resort area and 315 L/s in the park area to counteract the areal depression of the piezometric surface, which ranged from a seasonal high elevation of about 29 m above sea level to a construction elevation of about 15 m. Five pumping wells, each 30 m deep and about 90 cm in diameter, with a combined capacity of about 1,580 L/s, and three to four recharge wells about 150 m deep were considered (Dames and Moore, 1969, p. 9). Other provisions for constructing dewatering and permanent underdrainage were also formulated.

An alternative plan to constructing dewatering wells utilized a few large-diameter, relatively deep wells extending into the Floridian Limestone aquifer, combined with a larger number of small-capacity wells tapping the uppermost meter or two of the aquifer. The alternative dewatering plan would reduce the potential for serious sinkhole development at the Magic Kingdom theme park and resort hotel areas; this scheme minimized the amount of drawdown and eliminated the need for costly recharge wells. The concept was verified by pump testing a 25.4-cm-diameter well, cased 14 m to the top of the aquifer, but drilled to 35 m, or 21 m into the limestone. Water levels in surrounding observation wells stabilized after 11 hours at a pumping rate of 25.7 L/s. After 18 hours of pumping, drawdown in the pumped well was 13.7 m, while an observation well 63.7 m away (across the underpass site) showed 49 cm of drawdown, and a well 452 m away in the resort hotel area showed only 5 cm of drawdown (Fig. 27). The specific capacity proved to be relatively constant at about 1.5 L/s/m for pumping rates, which ranged from about 4 to 27 L/s (Dames and Moore, 1970a). The dewatering system was designed on the basis of this pump test data.

The initial construction for the underpass was the installation of eight dewatering wells (Fig. 25), of which one was the pump test well. The wells were initially intended to be 45 m deep with 30.5 m of 25.4-cm-diameter steel casing. However, difficulty was encountered in advancing casing deeper than 18.9 m in four of the wells. Fortunately, testing indicated these wells could produce substantial drawdown in the excavation area without causing water levels to decline below the bottom of the casing. During excavation of the underpass, pumping rates in the eight wells ranged from 15.8 to 37.9 L/s and caused drawdowns that ranged from 6 to 18.3 m (Dames and Moore, 1970b). The drawdown at each of the eight pumping wells varied because of the nonuniform

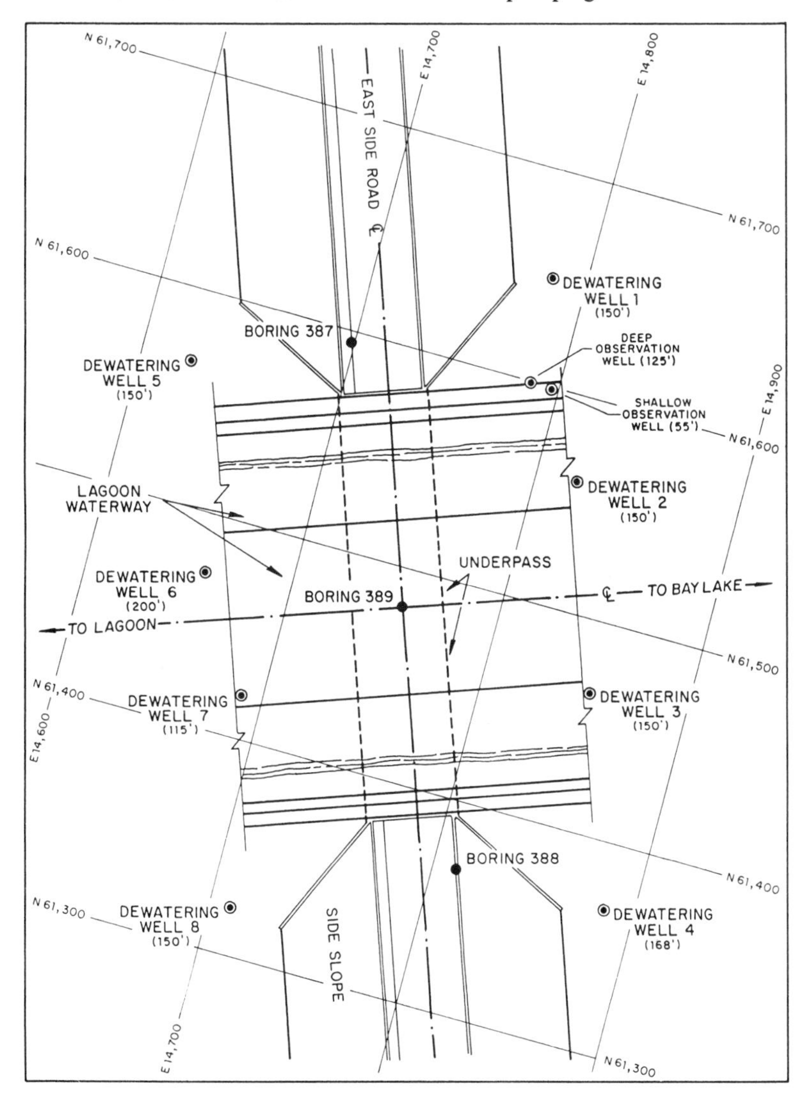

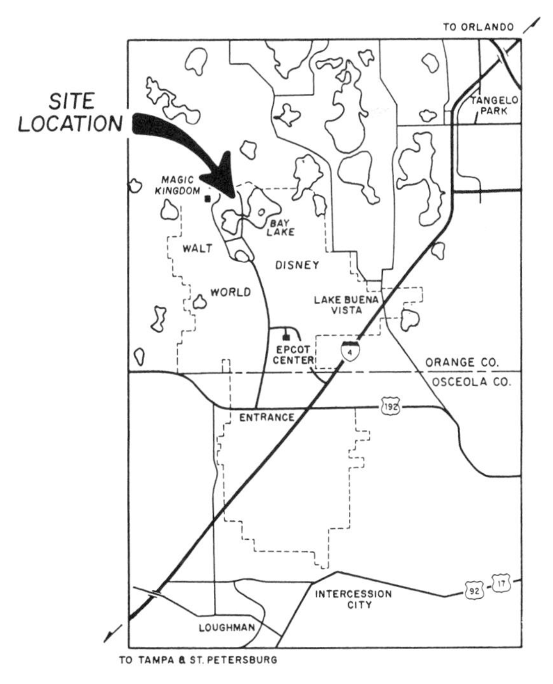

Figure 25. Map showing locations of borings and dewatering wells at the Lagoon underpass, Walt Disney World, Florida. (Modified from several consulting reports by Dames and Moore, 1969 to 1970.)

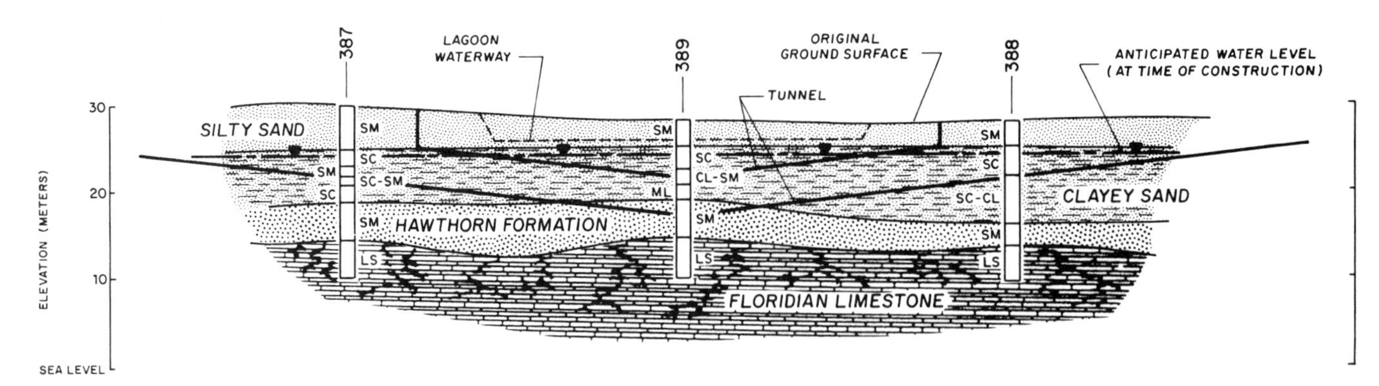

Figure 26. Geologic section along the Lagoon underpass, Walt Disney World, Florida. (After Dames and Moore, 1969, Plate 3.)

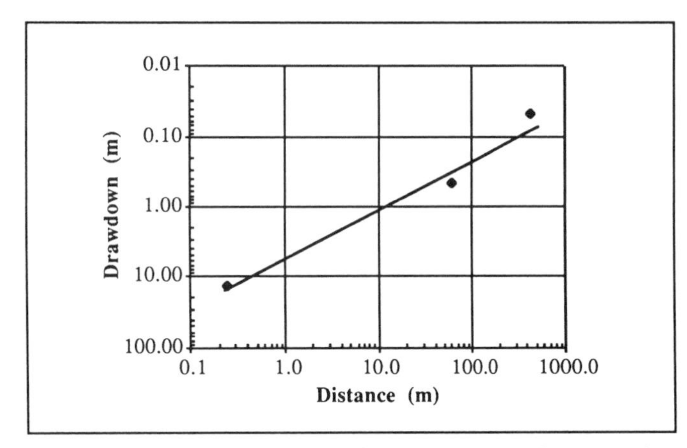

Figure 27. Drawdown-distance plot for pump test at the Lagoon underpass, Walt Disney World, Florida. Pumped well is at the construction site and experienced a drawdown of 13.7 m; observation well at 63 m distance is in contemporary resort hotel area and experienced a drawdown of 0.49 m; observation well at 452 m distance is in the Magic Kingdom theme park area and experienced a drawdown of 0.05 m. Regression equation: $y = 5.79\ x^{-0.72}$ ($r^2 = 0.98$); where y is drawdown in meters, and x is distance from pumped well in meters.

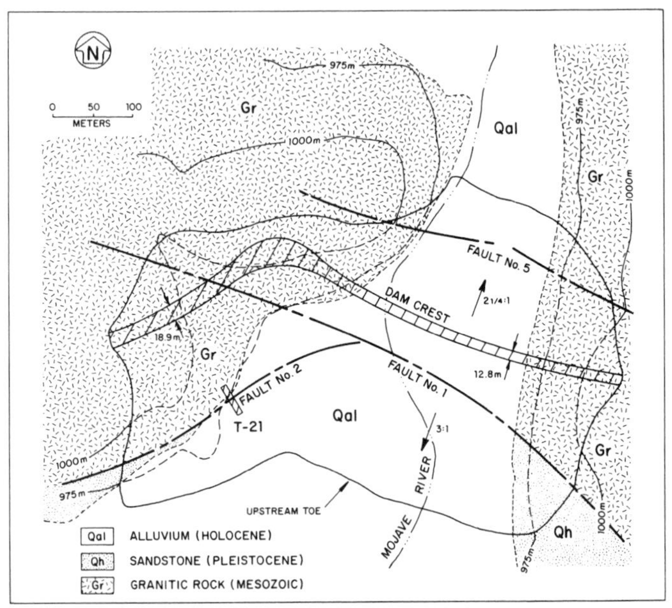

Figure 28. Geologic map showing faults at the Cedar Springs dam site, California. Fault no. 1 separates weakly cemented sandstone of the Harold Formation on the south from hard granite on the north in eastern part of site; fault no. 2 separates sandstone from granite in western part of site. (After Sherard and others, 1974, Fig. 12; and Stankov, 1982, Fig. 2.)

characteristics of the Floridian Limestone aquifer, even over short distances. The eight-well system produced a combined discharge of 252 L/s and created a cone of depression with a 610-m radius. Pumping rates during construction could be "fine tuned" to precisely control water levels (W. J. Attwooll, oral communication, 1987).

Construction of the underpass was successfully completed in 1970, and the pumps were removed from the dewatering wells. An underdrain system was installed in the underpass to collect seepage after construction. Combined flow in the underdrain system was estimated to be 32 L/s (Dames and Moore, 1970c).

The construction of dewatering wells at the lagoon underpass at Walt Disney World provides an excellent example of the application and correlation of ground-water geology and hydrology principles to control and remedy a subsurface process (sinkhole development) and its potential impact on the construction of an engineered facility.

Geoseismic considerations

Cedar Springs Dam. The Cedar Springs Dam is located on the Mojave River in southern California near Cajon Pass and is a major component of the California aqueduct system. The dam was planned to be 92 m high, about 1,000 m long, and impound a reservoir of 160 million m^3 of water. However, the discovery of faults cutting Holocene alluvium in the foundation resulted in a reduction of the dam to the minimum dimensions

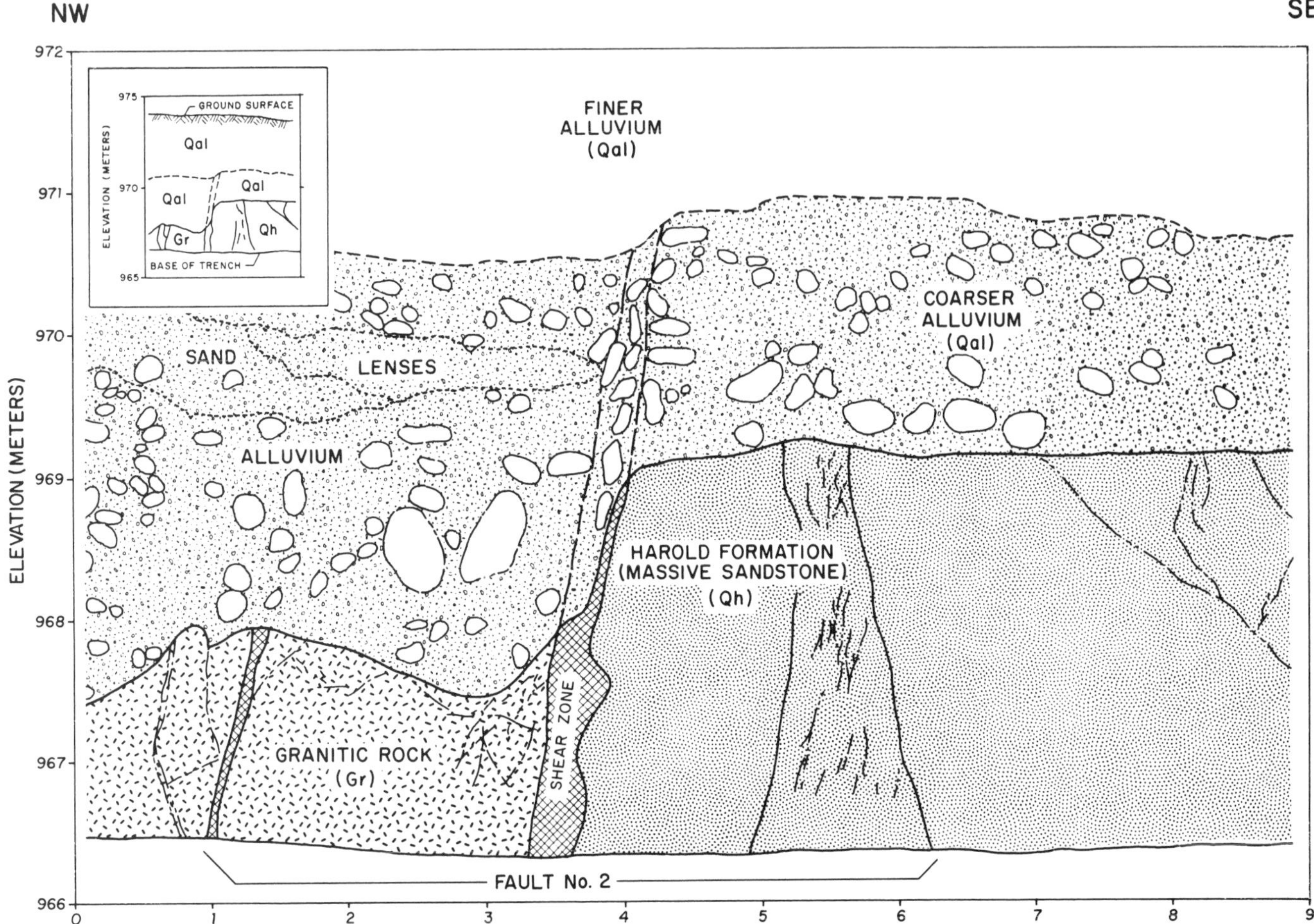

Figure 29. Geologic log of part of excavation T-21-B showing structural and stratigraphic relationships of alluvium and bedrock at Fault no. 2 discovered at the Cedar Springs dam site, California. (After Stankov, 1982, Fig. 5.)

that would satisfy project requirements, an impoundment of 60 million m^3 of water behind a dam 66 m high and 660 m long (Sherard and others, 1974, p. 376–379).

Several faults were located during foundation exploration in 1964 (Fig. 28), and exposed in trenches excavated through the alluvial deposits to the top of bedrock. Structural and stratigraphic relationships of the Holocene alluvium and underlying sandstone and granite bedrock at one of the faults delineated (fault no. 2) are shown on Figure 29. The faults exhibited a general northwest areal trend, as shown on Figure 28. The alluvium/bedrock contact as exposed (Fig. 29) was offset vertically 0.9 to 1.5 m, but the magnitude and direction of possible horizontal displacement could not be determined. The faults clearly cut the gravelly alluvial deposits (Fig. 29) and were assessed to be capable of future ground-surface displacements of 1 to 1.5 m (Stankov, 1982, p. 669). Areally, the dam is located about 8 km north of the main trace of the northwest-trending San Andreas fault, which contributed to the concern regarding the faults at the site (Sherard and others, 1974, p. 376).

Fault no. 1 (Fig. 28) separates weakly cemented Harold Formation sandstone on the south from typical granitic rock on the north in the southeastern part of the foundation, while fault no. 2 separated the two formations in the southwestern part (Stankov, 1982, p. 668). The massive sandstone generally consists of arkosic silty sandstone with minor lenses of claystone and siltstone. The much older granitic rock is quartz monzonite and gneissic granodiorite in which foliation trends N70°W and dips 5° to 15° to the northeast. Minor chlorite schist and marble occur nearby, but outside the foundation area (Stankov, 1982, p. 665).

Lengthy deliberation and attention was given to the design and construction of a major dam at Cedar Springs, a site traversed by faults known to be capable of movement. Subsequently, an independent board of consultants reviewed the earthquake analysis, and a second independent board reviewed the earth-dam design. The owners, California Department of Water Resources, after considering the seismologic and geologic factors and conditions, decided that a conservatively designed, well-built dam would represent a minimum risk to safety (Sherard and others, 1974, p. 376). Besides reducing the scale of the dam, other design modifications implemented were: (1) the axis was shifted so the entire clay core could be founded on the hard granite north of fault no. 1; (2) the design concept was radically changed from an ordinary zoned section with a narrow chimney drain to a special zoned section with thick core, transition, and shell zones; and (3) the crest width of the dam was increased some 40 percent to 12.8 m, and within the segment where fault no. 1 passed under the dam axis, it was increased over 100 percent to 18.9 m (Sherard and others, 1974, p. 379–380). Locally available core material was judged to be inadequate, and high-quality clay was transported 50 km from the bed of a playa lake.

The Cedar Springs Dam was completed in 1971 and has been in continuous successful operation since. This dam project demonstrates that instead of abandoning a site or just ignoring and avoiding geologic hazards and risks, alternative procedures can be implemented to accommodate natural hazards in a responsible manner. At Cedar Springs, modification of the hazard (preventing faults from moving) was technically unfeasible. Therefore, the potential risk was modified by designing the dam more conservatively and shifting the position of the axis, which together provided an acceptable alternative.

OUTLOOK

Construction is the logical extension of design. During construction, interpretations of geologic conditions developed during design are proved or disproved, and modifications to project designs can be made with a minimum of expense to achieve optimum future performance. The neglect of geologic factors during construction can lead to expensive problems during project operation, as well as an inability to rapidly formulate repair methodologies when required, and ultimately to unnecessary and costly litigation. Thus, the application of geological principles during construction should be a matter of high priority in virtually all projects.

The precision and sophistication of geological investigations in general has improved remarkably during the past 50 years, contributing to a better understanding of geologic processes. Construction geology should be viewed as an excellent opportunity to confirm or modify design assumptions to provide optimum project performance. Descriptions in James and Kiersch and Kiersch and James (this volume) are based in part on lessons learned from failures and errors in judgment that could have or should have been avoided during construction by appropriate investigation, interpretation, and application of geological principles and guidance.

Despite the advancements in geological knowledge and the improvement in sophistication of geological investigations, a primary problem seems to remain: the inability of many geologists to communicate with nongeologists, particularly engineers. The engineer's need for quantitative design values requires the geologist to present relevant information in specific terms. Where uncertainties exist, as they do in most instances, the engineer needs the geologist's considered opinion ("best prediction") as to the most likely conditions that will be encountered. Geologists with extensive experience on major construction projects develop an appreciation of the engineer's needs and an ability to express geologic conditions in terms of relevant value to the engineer. As Kiersch (1955, p. 4) emphasizes, the geologist who withholds opinions in order to avoid making possible mistakes forces the engineer to make geological decisions, which, during construction, must be made expeditiously.

The geologist's perspective of "geologic time" is considerably different from the engineer's perspective of "design life." The probabilistic "100-year flood," with specific risk connotations and boundaries of potential damaging effects, is considerably different than the deterministic "probable maximum flood." Engineered works with low to moderate consequences if damaged (i.e., most projects) would be uneconomical if designed deterministically. Geological understanding of relevant hazardous processes must be put into probabilistic magnitude/frequency terms to permit intelligent decisions, not only when evidence of a hazard is revealed for the first time during construction, but likewise during design. Probabilistic risk assessments for potential earthquake damage (Shah, 1984) and flood damage (Costa and Baker, 1981, p. 362–363) are well established, but the application of probabilistic techniques to hazards other than geoseismic and flooding has not yet been accomplished and/or accepted by the profession. In engineering, advances have been made in probabilistic assessments of homogeneous embankment stability (Vanmarcke, 1977a, b), but deterministic approaches are still used for natural slopes. Estimates of a priori probability, based on experience, can provide valuable insight into possible needs for additional subsurface data for site investigations (Attewell and Farmer, 1976, p. 547). Statistical distributions and probability concepts in rock discontinuity analyses (Attewell and Farmer, 1976, p. 355) are applied routinely, but the types of analyses needed for risk assessment for multiple hazards and natural systems are still being developed (Bowles and others, 1986). In essence, risk is probabilistic, but decisions are binary. Thus, the geologist must give clear, straightforward interpretations about obscure, complicated situations.

It has been long recognized that dams are located where topographic conditions provide adequate reservoir storage, not necessarily where geologic conditions provide trouble-free construction (Weaver and others, 1942, p. 1795). Future major projects, dams as well as other types of facilities, are most likely to be located at sites with increasingly difficult geologic conditions; and consequently, these sites will require more detailed geological investigations for project design, more careful geological guidance during construction, and more extensive communication with engineers and related professionals during both the design and construction.

Design and construction of many projects in the United States are governed by provisions of the Uniform Building Code (International Conference of Building Officials, 1988). Section 7015 (a) 3 states:

> Upon completion of the rough grading work and at the final completion of the work the building official may require . . . [a] geologic grading report prepared by the engineering geologist [which includes] a final description of the geology of the site including any new information disclosed during the grading [? and their influence] on recommendations incorporated in the approved grading plan. [The engineering geologist] shall provide approval as to the adequacy of the site for the intended use as affected by geologic factors.

This requirement appears to be positive and worthwhile; however, the words *the building official may require* tend to weaken its impact. Too often, the building official is not qualified to review geological reports, does not have assistance from qualified individuals, and therefore, does not require as-built geological reports. If a report is not required, one usually is not done. This common waiver of the as-built geology requirement soon may be a thing of the past in light of local governments being held liable for damage occurring at sites for which building permits were issued. The "implied warranty of habitability" attached to building permits seems to be forcing agencies that issue such permits to transfer the burden of "proof" of site suitability to the developers and their consultants. During construction of the Catskill water supply system in 1923, Flinn (1923, p. 68) concluded that owners or other persons in control of projects must have "foresight and courage" to obtain adequate geological information and interpretation of relevant factors. Such information is not *expensive,* it is *invaluable.*

As recently as 1976, the Bureau of Reclamation, under Federally imposed personnel constraints, waived the requirement for systematic geological mapping of the Vat tunnel in Utah (described above in the section regarding ground water), a tunnel that was advanced at an average rate of 5.4 m/d and was beset with geological difficulties and resulting contractor claims (Beason, 1986, Volume 1, p. 1–4). Conversely, the nearby Hades tunnel was started before the Vat tunnel had been completed and benefitted from the systematic application of geological guidance during construction. Although beset with geologic problems, such as cavernous limestone, unstable formations, and ground water, the Hades tunnel advanced at an average rate of 18.5 m/d and was part of a larger project completed 450 days ahead of schedule (Stevenson, 1987, Volume 1, p. 2). Factors other than the geology likely contributed to the marked differences in construction of these two tunnels. Nonetheless, selective and astute application of geologic principles and guidance undoubtedly contributed to the success of the Hades tunnel.

Many critical geological lessons must be learned more than once, and some learned by first-hand experience only, rather than vicariously. The importance of geological guidance during construction is emphasized whenever a problem involving unanticipated geologic conditions is experienced. If geological monitoring of the construction activity is conducted systematically, geologic problems can be anticipated before costly delays of litigation are incurred. Advances in the understanding of geologic processes, an expanded development of probabilistic approaches to natural systems, improvements in the ability to communicate geological information to team members, and numerous examples of the growing reliance on mature geological recommendations all indicate a continued increase in the application of geoscience principles to meet the demands of future construction.

REFERENCES CITED

Attewell, P. B., and Farmer, I. W., 1976, Principles of engineering geology: London, Chapman and Hall, 1045 p.

Beason, S. C., 1986, Final construction geology report for Vat tunnel, Strawberry aqueduct, Bonneville Unit, Central Utah project: Duchesne, Utah, Uinta Basin Construction Office, U.S. Bureau of Reclamation Report G-358, 4 volumes.

Berkey, C. P., 1935, Foundation conditions for Grand Coulee and Bonneville projects: Civil Engineering, v. 5, no. 2, p. 67–71.

Berkey, C. P., and Sanborn, J. F., 1923, Engineering geology of the Catskill water supply: American Society of Civil Engineers Transactions, v. 86, p. 1–91.

Bowles, D. S., Anderson, L. R., and Glover, T. F., 1986, Risk assessment for dams: Logan, Utah State University Department of Civil and Environmental Engineering (unpublished), 27 p.

Burwell, E. B., Jr., and Moneymaker, B. C., 1950, Geology in dam construction, *in* Paige, S., ed., Application of geology to engineering practice: Geological Society of America, Berkey Volume, p. 11–43.

Burwell, E. B., Jr., and Roberts, G. D., 1950, The geologist in the engineering organization, *in* Paige, S., ed., Application of geology to engineering practice: Geological Society of America, Berkey Volume, p. 1–9.

Clark, B. E., 1963, Geologic exploration and foundation treatment, Barkley Lock, Cumberland River, Kentucky, *in* Trask, P. D., and Kiersch, G. A., eds., Engineering geology case histories: Geological Society of America, p. 19–25.

Costa, J. E., and Baker, V. R., 1981, Surficial geology, building with the earth: New York, John Wiley and Sons, 498 p.

Dames and Moore, 1969, Report, site investigation, service road underpass, Walt Disney World, near Orlando, Flordia, for WED Enterprises, Inc.: Los Angeles, (unpublished), 15 p.

——, 1970a, Supplement No. 3, report, results of pumping tests, proposed tunnel, Walt Disney World, near Orlando, Florida, for WED Enterprises, Inc.: Los Angeles, (unpublished), 6 p.

——, 1970b, Supplement No. 4, report, installation and operation of relief wells at lagoon underpass site, Walt Disney World, near Orlando, Florida, for WED Enterprises, Inc.: Los Angeles, (unpublished), 7 p.

——, 1970c, Supplement No. 5, water flow from springs, excavation for lagoon underpass, Walt Disney World, near Orlando, Florida, for WED Enterprises, Inc.: Los Angeles, (unpublished), 3 p.

Davidson, D. D., and Whalin, R. W., 1974, Potential landslide generated water waves, Libby dam and Lake Koocanusa, Montana: Vicksburg, U.S. Army Engineers Waterways Experiment Station Technical Report H-74-15, 33 p.

Dearman, W. R., 1974, Weathering classification in the characterization of rock for engineering purposes in British practice: International Association of Engineering Geology Bulletin 9, p. 33–42.

——, 1976, Weathering classification in the characterization of rock—a revision: International Association of Engineering Geology Bulletin 13, p. 123–127.

Dearman, W. R., Baynes, F. J., and Irfan, T. Y., 1978, Engineering grading of weathered granite: Engineering Geology, v. 12, no. 4, p. 345–374.

Farina, R. J., 1987, Geology in construction: Denver, unpublished manuscript, 45 p.

Flinn, A. D., 1923, Discussion, *in* Berkey, C. P., and Sanborn, J. F., eds., Engineering geology of the Catskill water supply: American Society of Civil

Engineers Transactions, v. 86, p. 68–69.
Gordon, G., 1937, Arch dam of ice stops slide: Engineering News Record, v. 118, p. 211.
Hamel, J. V., 1974, Rock strength from failure cases, left bank slope stability study, Libby dam and Lake Koocanusa, Montana: U.S. Army Corps of Engineers, Missouri River Division Technical Report MRD–1–74, 239 p.
—— , 1976, Libby dam left abutment rock wedge stability, *in* Proceedings, Rock engineering for foundations and slopes, Volume 1: New York, American Society of Civil Engineers (University of Colorado, Boulder), p. 361–385.
Hays, J. B., 1943, Deep solution channel, Kentucky, *in* Symposium on unusual cut-off problems—dams of the Tennessee Valley Authority: American Society of Civil Engineers Proceedings, v. 69, no. 9, p. 1400–1416.
Henderson, L. H., 1939, Detailed geological mapping and fault studies of the San Jacinto tunnel line and vicinity: Journal of Geology, v. 47, p. 314–324.
Hunt, R. E., 1984, Geotechnical engineering investigation manual: New York, McGraw-Hill Book Company, 983 p.
International Conference of Building Officials, 1988, Uniform building code: Whittier, California, International Conference of Building Officials, 734 p.
Irfan, T. Y., and Dearman, W. R., 1978, Engineering classification and index properties of a weathered granite: International Association of Engineering Geology Bulletin 17, p. 79–90.
Irwin, W. H., 1938, Geology of the rock foundations of Grand Coulee dam, Washington: Geological Society of America Bulletin, v. 49, p. 1627–1650.
Kent, M. D., 1981, Site investigation for construction excavation: Bulletin of the Association of Engineering Geologists, v. 28, no. 1, p. 71–76.
Kiersch, G. A., 1955, Engineering geology: Golden, Colorado School of Mines Quarterly, v. 50, no. 3, 122 p.
Kiersch, G. A., and Treasher, R. C., 1955, Investigations, areal and engineering geology; Folsom dam project, central California: Economic Geology, v. 50, no. 3, p. 271–310.
Legget, R. F., and Karrow, P. F., 1983, Handbook of geology in civil engineering: New York, McGraw-Hill Book Company, various pagination.
Leighton, F. B., 1966, Landslides and hillside development, *in* Lung, R., and Proctor, R., eds., Engineering Geology in Southern California: Association of Engineering Geologists, Los Angeles Section Special Publication, p. 149–207.
Mead, W. J., 1936, Engineering geology of dam sites: Washington, D.C., Second Congress on Large Dams, Question VI, p. 1–22.
—— , 1937, Geology of dam sites—I. Dam sites in hard rock; II. Dam sites in shale and earth: Civil Engineering, v. 7, no. 5, p. 331–334, and no. 6, p. 392–395.
Moneymaker, B. C., and Rhoades, R., 1945, Deep solution channel in western Kentucky: Geological Society of America Bulletin, v. 56, p. 39–44.
Nickell, F. A., 1942, Development and use of engineering geology: American Association of Petroleum Geologists Bulletin, v. 26, p. 1797–1826.
Roberts, G. D., 1973, Foundation "as-built" reports: Bulletin of the Association of Engineering Geologists, v. 10, p. 145–155.
Sanborn, J. F., 1950, Engineering geology in the design and construction of tunnels, *in* Paige, S., ed., Application of geology to engineering practice: Geological Society of America, Berkey Volume, p. 45–81.
Shah, H. C., ed., 1984, Fundamentals of probabilistic risk assessment: Earthquake Engineering Research Institute Publication 84–06, 155 p.
Sherard, J. L., Cluff, L. S., and Allen, C. R., 1974, Potentially active faults in dam foundations: Geotechnique, v. 24, p. 367–428.
Slingerland, R. L., and Voight, B., 1979, Occurrences, properties, and predictive models of landslide-generated water waves, *in* Voight, B., ed., Rockslides and avalanches, Volume 2; Developments in Geotechnical Engineering, v. 14B: Amsterdam, Elsevier Scientific Publishing Company, p. 317–397.
Sowers, G. F., and Carter, B. R., 1979, Paracti rockslide, Bolivia, *in* Voight, B., ed., Rockslides and avalanches, Volume 2; Developments in Geotechnical Engineering, v. 14B: Amsterdam, Elsevier Scientific Publishing Company, p. 401–417.
Stankov, S., 1982, Cedar Springs dam, San Bernardino County, California—a review of a dam designed against potentially active faults, *in* Fife, D. L., and Minch, J. A., eds., Geology and mineral wealth of the California Transverse Ranges: Santa Ana, California, South Coast Geological Soicety, p. 665–671.
Stevenson, T. K., 1987, Final construction geology report for Hades tunnel, Wolf Creek siphon, and Rhodes tunnel (a portion of the Wolf Creek complex), Strawberry aqueduct, Bonneville unit, Central Utah Project: Duchesne, Utah, Uinta Basin Construction Office, U.S. Bureau of Reclamation Report G–379, 4 Volumes.
Supp, C.W.A., Whikehart, R. E., and Obear, G. H., 1957, Occurrence, investigation, and treatment of an embankment failure on Ohio Turnpike Project No. 1, *in* Trask, P. D., ed., Engineering Geology Case Histories 1: Geological Society of America, p. 57–66.
Tennessee Valley Authority, 1949, Geology and foundation treatment, Tennessee Valley Authority projects: Knoxville, U.S. Tennessee Valley Authority Technical Report 22, 548 p.
Terzaghi, K., 1960, From theory to practice in soil mechanics: New York, John Wiley and Sons, 425 p.
Thompson, T. F., 1966, San Jacinto Tunnel, *in* Lung, R., and Proctor, R., eds., Engineering Geology in Southern California: Association of Engineering Geologists, Los Angeles Section, Special Publication, p. 104–107.
Vanmarcke, E. H., 1977a, Probabilistic modeling of soil profiles: American Society of Civil Engineers, Geotechnical Engineering Division Journal, v. 103, no. GT11, p. 1227–1246.
—— , 1977b, Reliability of earth slopes: American Society of Civil Engineers, Geotechnical Engineering Division Journal, v. 103, no. GT11, p. 1247–1265.
Vehrs, R., 1974, A summary of the engineering geology of Grand Coulee Dam and related features, *in* Voight, B., and Voight, M. A., eds., Rock mechanics; The American northwest: University Park, The Pennsylvania State University College of Earth and Mineral Sciences Experiment Station, p. 252–253.
Voight, B., 1979, Wedge rockslides, Libby Dam and Lake Koocanusa, Montana, *in* Voight, B., ed., Rockslides and avalanches, Volume 2; Developments in Geotechnicial Engineering, v. 14B: Amsterdam, Elsevier Scientific Publishing Company, p. 281–315.
Ward, T. E., and Galster, R. W., 1969, Rock stability report—left abutment and MSH 37 cut area—Libby Dam: August 28 memorandum (unpublished), Seattle District, U.S. Army Corps of Engineers, 4 p.
Weaver, P., Aurin, F. L., Owen, E. W., Markham, E. O., and Ver Wiebe, W. A., 1942, Foreword *to* Development and use of engineering geology: American Association of Petroleum Geologists Bulletin, v. 26, p. 1795–1796.
Wiggin, T. H., 1923, Discussion, *in* Berkey, C. P., and Sanborn, J. F., eds., Engineering geology of the Catskill water supply: American Society of Civil Engineers Transactions, v. 86, p. 79–83.

Manuscript Accepted by the Society November 11, 1988

ACKNOWLEDGMENTS

Reference materials and advice are gratefully acknowledged from William J. Attwooll, William E. Collins, Leo D. Handfelt, George A. Kiersch, Ian Kinnear, William G. Pariseau, Richard J. Proctor, and Fred Thompson. This chapter benefitted from a critical review of the early draft by Raymond H. Rice and Ellis L. Krinitzsky, and the input of George A. Kiersch to subsequent editions. Walt Disney, Inc. kindly granted permission for publication of the description of the Lagoon underpass at Magic Kingdom Theme World. Technical support was generously provided by EarthStore, a former Division of Dames and Moore, Salt Lake City; Scott A. Sanders skillfully prepared the illustrations.

Printed in U.S.A.

Geological Society of America
Centennial Special Volume 3
1991

Chapter 21

Investigations of existing engineered works

Paul R. Fisher
Consulting Engineering Geologists, 436 Cranes Roost Court, Annapolis, Maryland 21401

INTRODUCTION

Investigations in conjunction with operating engineering works and structures are receiving increasing emphasis throughout North America. The most obvious reason for the investigations is observed distress of the works, while less obvious but common causes are decaying infrastructure, advances in state-of-the-art engineering, and increased regulatory activity due to safety concerns. These latter causes frequently lead to modification of the original design.

Decaying infrastructure

In the early 1980s, the need for maintenance and rehabilitation of the nation's engineering-works infrastructure was recognized and widely published as a national problem of great magnitude. Phrases such as "decaying intrastructure" and "crumbling public facilities" were used by the engineering and construction industry publications or by the general news media. The Public Works Improvement Act (Public Law 98-501) constituted a significant step in formulating a national response to the long-term intrastructure needs of our modern society (ENR, 1985; Pender, 1985).

The relative cost of continuing operations, maintenance, and rehabilitation versus the investment in new construction is increasing. One indicator of the shift is seen in the civil works budget of the U.S. Army Corps of Engineers. In fiscal year 1967, construction and operations and maintenance accounted for 79 percent and 16 percent, respectively, of their budget. In fiscal year 1983, construction accounted for 42 percent and operations and maintenance for 40 percent. In fiscal year 1986, the construction budget was $0.8 billion, and the operations and maintenance budget was $1.4 billion!

Advances in state of the art

Advances in state-of-the-art engineering have produced improvements in design criteria in such areas as earthquake engineering, maximum hydrologic events, strength and deformation properties of soil and rock materials, and methods of analyzing the response of structures to forces acting upon them.

Concepts in earthquake engineering have evolved considerably in the period from the mid-1960s to the mid-1980s. Twenty years ago most earthquake design analyses involved the static application of a force equivalent to a small percentage of gravity to the structure being analyzed. Dynamic analyses, using vibratory earthquake motions were rarely attempted. Currently, where any significant earthquake risk is anticipated, complex dynamic analyses, using anticipated earthquake motions, are performed. The development of those motions for input to the analyses is a complex exercise in the analysis of regional and site-specific geologic history and conditions.

Furthermore, advances in the science of meteorology and a greatly increased amount of available meteorologic data have, almost invariably, resulted in increases in the forecasted amounts of maximum precipitation and run-off for water-resource projects. Often these increases result in the need to increase the size of emergency outlets (spillways) for water-retention structures. Another problem that is associated with emergency spillways is excessive erosion. Although it was understood that erosion would occur in unlined emergency spillways and in the unlined getaway channels, rarely was the degree of erosion subjected to a design exercise. Recent occurrences of spillway erosion have underscored the need for geologic evaluation and input to the analyses of the erosion problem, especially where hydrologic analyses have shown the spillway to be inadequate.

Regulatory activity

For decades, federal agencies involved in construction of various types of public works have operated programs of inspection, investigation, analysis, repair, and rehabilitation to assure public safety. Over the last 15 years, federal and state regulatory agencies have required an increasing amount of inspection and investigation of existing state, municipal, and private structures. Three highly visible examples of this increased regulatory activity are seen in the Nuclear Regulatory Agency, the Federal Energy Regulatory Agency, and in the proliferation of state dam safety activities. This increased regulatory activity requires initiation of investigations to assure safety and to assure that older projects meet present-day design criteria.

Fisher, P. R., 1991, Investigations of existing engineered works, *in* Kiersch, G. A., ed., The heritage of engineering geology; The first hundred years: Boulder, Colorado, Geological Society of America, Centennial Special Volume 3.

TYPES OF INVESTIGATIONS

The principal geological and geotechnical investigations of existing structures can be grouped into three logical phases: (1) observation, inspection, instrumentation, and monitoring; (2) surface and subsurface explorations; and (3) analysis and design of remedial work, when necessary.

Monitoring

Nearly all major projects are subjected to a formal program of instrumentation and periodic inspection for the purpose of monitoring performance and assuring project safety as part of the design, construction, and operation sequence. In modern structures, instruments are installed as part of construction. Surprisingly, many structures built before the 1950s were constructed without any instrumentation. Increased awareness of physical safety problems and regulatory pressure has resulted in instrumentation retrofitting programs for most major structures. Table 1 briefly illustrates the types of instruments used for geological and engineering monitoring purposes (USCE, 1971, 1976).

Periodic instrument readings during the project operation are compared with performance predictions made during the project design state (discussed by Keaton and Kiersch, this volume). As instrument readings accumulate, baselines of performance can be developed. The magnitude and suddenness of excursions from those baselines can provide valuable insight and improved understanding of the actual performance. For example, a piezometer located in a foundation filter blanket downstream from an embankment dam cut-off may be showing long-term stable readings. If the hydraulic head starts to rise, this may indicate an increase in conductivity through, under, or around the cut-off.

For an instrumentation program to be effective, the instruments must be read at the correct intervals, and the instrument data must be reduced and analyzed by experienced personnel. In the past this has been a labor-intensive effort and frequently neglected. Recent equipment improvements and data-handling advances tend to reduce this problem.

Periodic inspections are usually performed by a team of design engineers, geologists, and operations personnel who visit the project at predetermined intervals to observe the condition of project features. The interdisciplinary inspection team is allowed a competent inspection of all project aspects. The instrumentation data collected since the last inspection are reviewed and compared with the design predictions and the on-going observations of the inspection team. If a deficiency is observed, the inspection-team report may become the basis for follow-up investigations. Inspections are normally frequent (e.g., yearly) during the first few years of project operation and less frequent thereafter. However, if problems are observed, the inspection frequency will likely increase.

In addition to the periodic inspections, available project personnel are trained to observe project conditions in order to detect and report any anomalous behavior.

TABLE 1. TYPES OF GEOLOGICAL AND ENGINEERING INSTRUMENTATION

Movement Instruments

- Vertical movements
  - Surface monuments
  - Single-point settlement gauges
    - Plates
    - Probes
  - Liquid-filled settlement gauges
  - Multi-point movement devices
  - Extensometers
  - Heave points
- Horizontal movements
  - Surface monuments
  - Horizontal-displacement devices
  - Extensometers
  - Inclinometers
- Seismic measuring devices (strong motion)
  - Strong-motion accelerographs
  - Peak- recording accelerographs
  - Seismoscopes

Pressure-Measuring Devices

- Total pressure
  - Pressure cells
    - Diaphragm
    - Electrical
    - Piezoelectric/resistive
    - Pneumatic/hydraulic
  - Stress meters
  - Flat jacks
- Pore water pressure
  - Open standpipe (open system) piezometers
  - Hydraulic (closed system) piezometers
  - Diaphragm piezometers

Surface and subsurface geological investigations

This exploratory phase of post-construction investigations, including both surface and subsurface explorations, is usually undertaken as a result of one or more of the following circumstances: (1) observed structural distress, such as slumping or slope failure; (2) observed symptom of anomalous behavior, such as anomalous piezometer readings; (3) deficiency noted during a periodic inspection, such as clogged drains; or (4) internally or externally (regulatory) imposed requirement to analyze some aspect of the project according to current criteria, such as seismic design.

The common geologic events and conditions associated with operating engineering works that can be evaluated by surface and subsurface explorations are given in Table 2. A few of the characteristic symptoms of each condition are indicated.

TABLE 2. COMMON GEOLOGICAL EVENTS OR CONDITIONS AND THEIR SYMPTOMS ASSOCIATED WITH OPERATING ENGINEERING WORKS

Event/Condition	Symptom/Indication
Leakage and internal erosion	Variations in piezometer response to changes in reservoir or regional ground-water levels.
	Changes in the behavior of foundation or slope drainage system.
	Changes in volumes of seepage.
	Suspended sediment in seepage.
	Boils, toe of slope.
Surface erosion	Directly observable.
Solutioning and collapse	Rapid changes in ground-water levels.
	Circumferential and radial cracking patterns at potential site.
	Instrumented surface deformations.
	Observed holes from openings stoping to the surface.
Foundation deformation or incipient bearing-capacity failure	Instrumented surface and/or subsurface deformations.
	Visible bulging and/or cracking near the structure base.
Sliding instability	Instrumented surface deformation.
	Change in behavior of foundation drains.
	Change in tiltmeter readings.
	Displacements at the structure/abutment foundation contact.
	Displacements between the structure monoliths.
	Structural cracking.
	Offsets in foundation drains and/or instrument holes.
Slope instability	Instrumented surface and/or subsurface deformations.
	Change in behavior of slope drains.
	Change in inclinometer readings.
	Cracks in or at the top of slope.
	Bulges in or at the bottom of slope.
	Offsets in drain or inclinometer holes.
Instability of underground structures	Cracking in permanent lining.
	Distress in temporary support.
	Changes in behavior of drain holes.
	Loosening and/or detachment of rock blocks.
Susceptibility to earthquake motions	Evidence of historic sand boils.
	Existence of nearby capable fault.
	Landslide scars in vicinity.
	Known seismic history in area.

Analyses and design of remedial works

The procedure for analyzing the geological and engineering data collected in conjunction with an investigation at an existing engineering work is not unlike that for a new project. A geological model is developed for the subsurface below a foundation, or behind a slope. The engineering data is integrated into the geological model. A team of geologists and engineers assesses the model and problem symptoms, and analyzes the problem, its severity, and a preliminary approach to remedial action. This analysis will provide the technical input for the management decision on implementing remedial or rehabilitation work. Other important input to the management decision includes the benefit-cost ratio, environmental impact and the consequences of no action. If the decision is to proceed with the remedial or rehabilitation work, a design effort is initiated.

IMPACT OF GEOLOGIC EVENTS/CONDITIONS ON OPERATING WORKS

Delineation and repair of foundation erosion

An investigation to delineate and repair foundation erosion occurred at Wolf Creek Dam, Kentucky. The types of investigations performed when a major engineering structure exhibits symptoms of foundation erosion are summarized from reports by

Figure 1. Location of Wolf Creek Dam (from Simmons, 1982).

personnel of the Nashville District, U.S. Army Corps of Engineers (Simmons, 1982).

Wolf Creek Dam is located on the Cumberland River in south-central Kentucky. It was constructed from 1938 to 1952 by the U.S. Army Corps of Engineers (Fig. 1). Construction on the dam, a combination of concrete-gravity and earth-fill structures, was interrupted from 1943 to 1947 because of World War II. The concrete-gravity structure has a maximum height of 258 ft and extends 1,796 ft from the left abutment across the stream channel (Fig. 2). The earth-fill structure is a homogeneous embankment of low- to medium-plasticity clay and extends from the gravity section for a distance of 3,940 ft to the right abutment. Its maximum height is 205 ft. The gravity section supports a gated spillway and contains intakes to a six-generator powerhouse. The impoundment, Lake Cumberland, has a volume of 6,089,000 acre ft, making it the largest reservoir east of the Mississippi.

The area of Wolf Creek Dam is underlain by flat-lying Ordovician to Mississippian shales and limestones (Fig. 3). The lower reaches of the river at the dam are underlain by argillaceous limestones of the Leipers and Catheys Formations. The concrete gravity section was founded on limestones of the lower Leipers and the underlying Catheys Formations. The earth fill was founded on overburden. Because of extensive solution activity discovered in the limestone underlying the embankment, a cut-off trench was excavated at a location about three-fifths of the distance from the centerline to the upstream toe. The extensive solutioning in the Leipers Formation was further documented during the excavation of the cutoff trench, which had a minimum bottom width of 10 ft and was carried to a depth considered to be "sound rock." Differential bedrock relief along the trench was as much as 60 ft. Because of the fracture-controlled solutioning orientation, the trench had irregular and steep sides. A single-line grout curtain was constructed from the bottom of the cut-off trench to a depth of about 50 ft.

The reservoir, impounded in 1950, was operated with pool levels fluctuating between 692 and 722 ft in elevation. In August 1967, wet areas were first noted near the downstream toe of the embankment on the right abutment about 2,000 ft from the powerhouse tailrace. Subsequently, muddy water was observed entering the tailrace near the right-bank retaining wall, located downstream from the powerhouse. By January 1968, muddy water was seen issuing from the rock floor of the tailrace near the

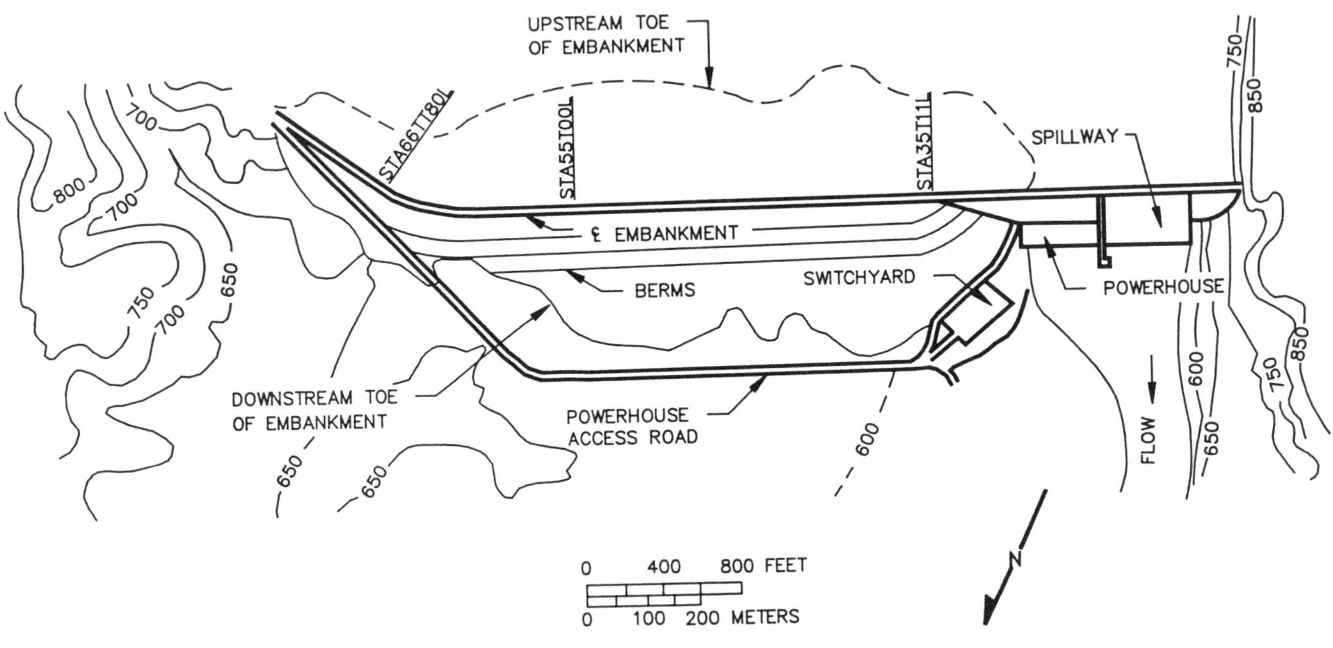

Figure 2. Plan of Wolf Creek Dam (from Simmons, 1982).

retaining wall (Fig. 2). Suddenly, in March 1968, a sinkhole appeared near the downstream toe of the embankment immediately adjacent to the gravity structure. This occurrence triggered an extensive investigation to determine the cause of the muddy flows and the sinkhole opening. During the field explorations of April 1968, a second sinkhole appeared near the first sinkhole at a location where recent subsurface exploration had indicated no disturbance of the overburden or fill materials.

Analyses of the new data from the exploratory efforts confirmed that the limestone solutioning observed in the excavation of the cut-off trench was extensive and tended to follow the regional jointing pattern. Furthermore, water was seeping either through and/or under the cut-off and subsequently moving through the area where the sinkholes appeared and then draining into the tailrace. Piping of the natural materials that filled the solution cavities had progressed, and collapse of the overburden was underway. Internal collapse of embankment materials could be occurring. Moreover, erosion of the natural cavity fillings had allowed dangerous hydraulic pressures to migrate downstream of the centerline of the embankment and to be present near the surface at the downstream toe.

Evaluation of the geologic environs and subsurface data indicated that the most expeditious way to treat the high pressures and gradients near the downstream toe was by grouting. Three grout lines were constructed along the dam centerline for a distance of 200 ft from the gravity section. Another three lines were constructed perpendicular to the centerline curtain at a point 150 ft from the concrete; these ran downstream for a distance of about 250 ft. This produced a t-shaped curtain and isolated the wrap-around section where the sinkholes had occurred. After completion of the first six lines of grout holes, additional grouting was performed along embankment berms in the wrap-around section and at the contact of the gravity structure and the embankment. The two-year grout program included 174,666 ft of overburden drilling, 97,032 ft of rock drilling, and 290,087 ft^3 of solids injected. The high downstream piezometric heads were reduced

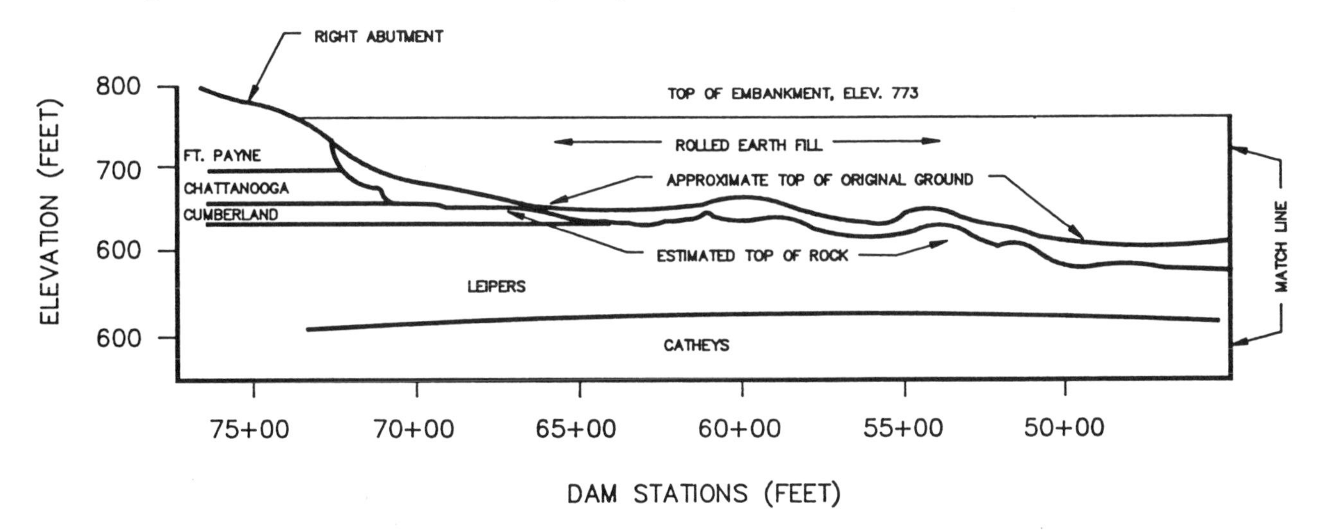

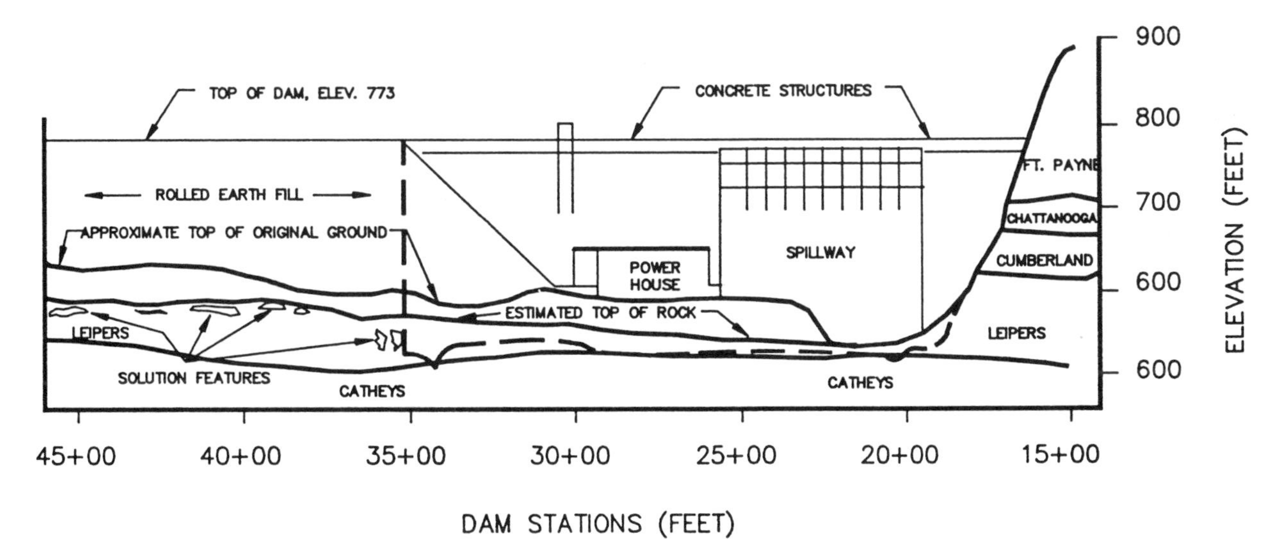

Figure 3. Geologic cross section along axis of Wolf Creek Dam (from AEG, 1977).

and the muddy flows ceased; the emergency grouting probably saved the dam from a major piping failure. However, the grouting had only filled, at best, the voids caused by the erosion of cavity-filling materials. Undoubtedly, the solution network remained partially filled with erodible materials and the unsafe condition could recur. Consequently, a permanent solution was necessary.

Various types of explorations and analyses were carried out from 1972 to 1975 to arrive at a permanent solution. The most favorable solution was construction of a concrete diaphragm wall that extended along the upstream crest of the embankment for a distance of 3,000 ft from the concrete structure, and that penetrated through the solutioned Leipers beds into the Catheys Formation. The subsurface explorations undertaken for the design of the diaphragm were in two phases: (1) investigations to further define the seepage problem and assist in selection of a permanent solution, and (2) explorations to design the diaphragm wall.

A large number of exploratory methods were used during the emergency grouting period and afterward to define the subsurface conditions for the permanent diaphragm wall. For example, more than 325 piezometers, installed beginning in 1968, were read on a regular basis; they confirm that high pressures downstream were reduced by grouting. Subsurface-water-temperature surveys were made on a regular basis, utilizing the open-tube piezometers. The temperature-contour plot map indicated low temperatures around the end of the concrete structure that trended toward the sinkholes and the tailrace. Dye-tracer tests between boreholes located close to sinkholes determined the velocity of subsurface seepage, a maximum of 18 ft/min.

Surface resistivity surveys were unsuccessful in delineating the subsurface solution features. Consequently, precision depth soundings were made in the reservoir; profiles 5 ft apart extended 800 ft upstream from the heel of the dam. Six areas of depressions were delineated by soundings that required further investigation. Free divers investigated all unusual bottom features and the areas of depressions located by the depth soundings. In addition, an underwater television survey was attempted but was unsuccessful due to water turbidity.

Investigations using colored infrared aerial photographs and thermal infrared imagery of the dam site and environs identified only the surface jointing and drainage patterns and provided no information on subsurface seepage.

A test trench was excavated immediately downstream of the embankment, where the top of bedrock was at a depth of less than 5 ft. The rock was traversed by solution channels as much as 1 ft wide, and joint orientations were consistent with the surrounding areas. A thermonic survey was performed by measuring the thermal gradients in 140 piezometers (installed by 1971) that were mostly set at the top of bedrock. The investigators identified major zones of seepage with as much as ten times greater seepage than the main foundation mass. These zones coincided well with the known orientation of solution features in the bedrock. Such analyses and exploration data led to the 1974 decision to construct the concrete diaphragm wall along the upstream edge of the embankment crest.

The design depth extended the wall 5 ft into the Catheys Formation, an elevation believed to be below all solution activity. The maximum depth of the wall, 278 ft, was constructed through more than 100 ft of solutioned rock. A minimum wall thickness of 2 ft and the single-element method of construction proved the most feasible. Detailed geological investigations were required to define wall limits and identify the locations of solution features.

Investigations were undertaken to characterize the physical condition of limestones between boreholes. Acoustical surveying in 21 borings utilized both pulse-echo and through-transmission methods. The pulse-echo technique places the acoustic transmitter and receiver in the same boring, while the through-tranmission technique locates the transmitter and receiver in adjacent borings, 20 to 30 ft apart. The 21-hole survey determined a potential for distinguishing joints, bedding planes, and interbeds in rock but was ineffective in the overburden.

Most of the existing downhole geophysical logging methods were attempted, and correlation between geophysical data and the cored boring logs was very good. Cavities that intersected the borings were detected and displayed on the caliper, gamma-gamma, neutron, 3-D acoustic velocity, and televiewer logs. However, none of the methods used detected cavities away from the boring.

The most reliable method for detecting cavities was exploratory grouting. This program consisted of exploratory drilling and grouting along the alignment of the diaphragm wall. Primary holes were on 25-ft centers and approximately 300 ft in depth; they extended approximately 50 ft into the Catheys Formation. The holes were split-spaced down to 3.2-ft centers and gravity grouted. Grout placed above the top of rock had the same strength as the embankment materials (cement, clay, sand, and water mix), while a mix of cement, sand, and water was used in rock. Overall, 852 grout holes totaling 239,460 ft were drilled along the alignment: 51 holes took more than 1,000 ft^3 of solids with a maximum of 6,971 ft^3 in 1 hole. The total grout injected was 146,461 ft^3. This "brute-force" exploration method had two distinct advantages: (1) the conditions and wall limits were defined beyond reasonable doubt; and (2) the rock was preconditioned before the hazardous construction of deep, fluid-supported excavations was begun in the solutioned rock. In addition, this method saved about $19 million in overall construction cost, and exploration costs were less than $4 million. Since completion in September 1979, the wall has performed satisfactorily.

Stability of gated concrete structure

The investigation to evaluate the foundation of the Lake Superior Regulatory Structure is summarized from reports by personnel of the U.S. Army Waterways Experiment Station (Thornton, 1981).

The Lake Superior Regulatory Structure is a key Great Lakes structure located at the head of St. Mary's Rapids between the cities of Sault Ste. Marie, Michigan, and Sault Ste. Marie, Ontario. The St. Mary's River, the only outlet from Lake Super-

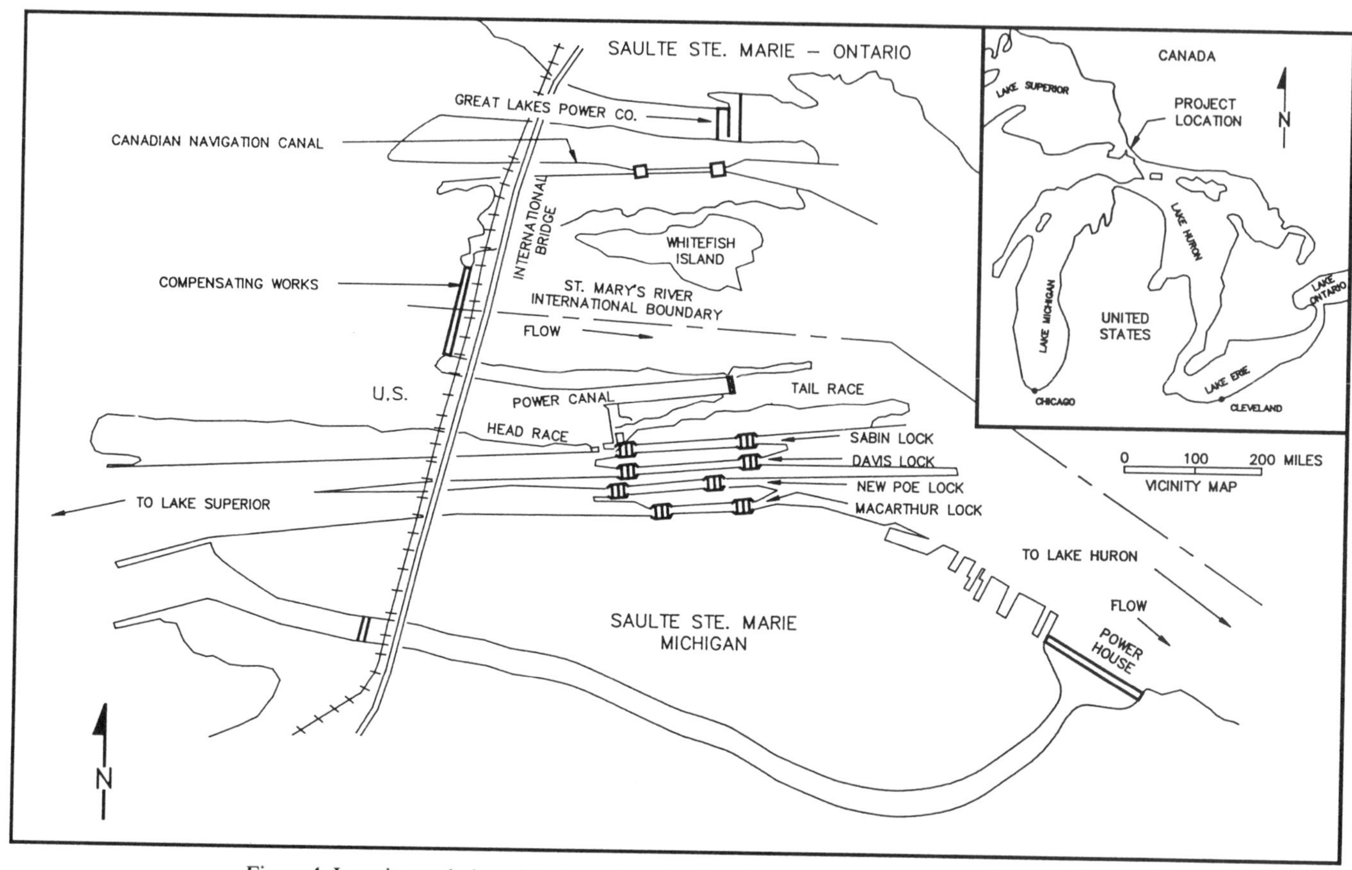

Figure 4. Location and plan of the Lake Superior Regulatory Structure (from Thornton, 1981).

ior, connects Lake Superior with Lake Huron (Fig. 4). There are several structures on the St. Mary's River: power canals adjacent to each bank; 3 ship canals with locks; and the regulatory structure consisting of 16 gates. Eight of the gates are operated by Canada and 8 by the United States.

The gate piers are concrete and are 50 ft long, 8 ft wide, and rise about 20 ft above the river bed. The gates themselves are cast-iron lift gates, which close onto a concrete sill between the piers. The piers are founded on the Cambrian sandstones of the Jacobsville Formation.

The regulatory structure was constructed by private firms between 1913 and 1919. In 1976 the Lake Superior Control Board requested a testing program as a basis for future decisions on the structure. The 8 U.S. structures are under the jurisdiction of the Detroit District, U.S. Army Corps of Engineers, and the U.S. Army Waterways Experiment Station was asked to design and accomplish the requested testing, under guidance of the International Joint Commission for Regulation of the Great Lakes water levels. Although the overall scope of the evaluation program included tests of all steel structures and machinery, as well as sampling, testing, and analyses of the concrete and foundation materials, this discussion deals only with the investigation of the foundation characteristics.

Pre-construction exploration of the foundation area was not performed for the regulatory structure, although investigations were performed for later engineering works constructed in the vicinity. The preliminary stability analyses used a conservative assumption of no cohesion between the piers and the foundation, and a 30° friction angle at the base. This analysis indicated the structure was stable against overturning, but might be inadequate in resistance to sliding.

Thirty investigative core borings were drilled through the concrete into the foundation rock; 7 recovered 6-in core and 23 returned 4-in core. The borings penetrated 31 ft into the sandstones. The total footage included 28 ft of fill and overburden, and 464.25 ft of rock. Nine borings were logged with the Corps of Engineers' borehole camera to determine orientations of detailed geologic structure. Scour surveys indicated that rock had been removed in some areas downstream from the gate apron. Underwater inspections and video films documented the scouring and rock structure.

Although the region was subjected to extensive glacial action during the Pleistocene, no glacial deposits were penetrated in the borings. Locally they were probably removed by erosion of the St. Mary's River. The bedrock of Jacobsville Formation is fine- to medium-grained arkosic sandstone with thin shale interbeds and clay seams. Engineering-wise, the rock was divided into 3 sandstone units: very hard, moderately hard, and shaly.

TABLE 3. TYPES OF LABORATORY TESTING ON ROCK MATERIALS

Characterization Tests
- Wet and dry unit weight.
- Water content.
- Unconfined compressive strength.
- Direct tensil strength.

Engineering Design Tests
- Elastic moduli.
- Poisson's ratio.
- Triaxial strength.
- Direct shear strength.
  - Concrete to rock bond strength (maximum).
  - Concrete on rock, precut (minimum).
  - Intact (maximum and residual).
  - Precut (residual).
  - Clay and shale seams (maximum and residual).
  - Natural joints (maximum and residual).
  - Cross bed (maximum).

Thin shale interbeds and clay seams occur in the 2 sandstone units, and the shaly sandstone contains thin beds of hard sandstone. The sandstones are cemented with secondary quartz mixed with sericite, illite, and iron oxide; the clay from seams has a composition of illite, chlorite, and mixed-layer clay, in part, smectite. The beds dip a few degrees westward, and high-angle jointing is widely spaced. The significant structural geologic features relevant to foundation stability are the thin interbeds of shale and the clay seams.

Representative core specimens were selected from the 3 bedrock units that included the soft clay and shale interbeds, as well as the natural joints; the characterization and engineering design tests performed are summarized in Table 3.

The very hard sandstone unit had an unconfined compressive strength of 14,700 psi; the moderate sandstone, 8,800 psi; and the shaly sandstone, 7,600 psi. The modulus of elasticity and Poisson's ratio ranged from 5,000,000 psi and 0.2 for the very hard sandstone to 2,000,000 psi and 0.34 for the shaly sandstone. The shear strengths ranged from a cohesion of 12 to 44 tsf and friction angles of 57° to 68° for intact rock to a cohesion of 0 and residual friction angles of 23° and 21° for the clay seams and shale interbeds, respectively.

Sliding stability analyses indicated that the critical case would be sliding along a soft seam just below the concrete foundation interface (this condition occurred at Wheeler Locks; Kiersch and James, this volume). Using the low measured residual strengths, the factor of safety is below unity for the winter loading conditions of ice pressure if a continuous soft-shale interbed exists within the top 4 ft of foundation rock. The Waterways Experiment Station analyses recognized that shallow continuous shale-seam interbeds occur in the foundation and recommended remedial-stability measures that consisted of post-tensioned anchors. The recommended tendons were installed, on the Canadian side as well as on the American side, and the scoured areas downstream from the piers and apron were backfilled with rip-rap and tremie concrete.

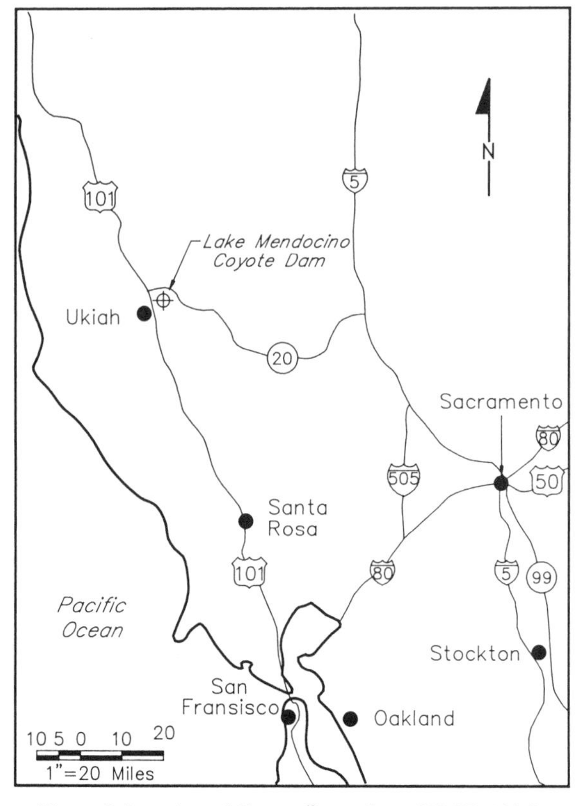

Figure 5. Location of Coyote Dam (from USCE, 1986).

Seismic stability of embankment dam

This evaluation of the Corps of Engineers' Coyote Dam in northern California is based on investigations and reports of the U.S. Army Corps of Engineers, Sacramento District (USCE, 1986).

The Coyote Dam project, completed in 1959, is located on the East Fork of the Russian River, 5 mi north of Ukiah, California, and 110 mi north-northwest of San Francisco. The dam is a zoned compacted-earth embankment with a maximum height of 164 ft and a crest length of 3,525 ft. Control of the 122,500-acre-ft Lake Mendocino reservoir is provided by a conduit 12.5 ft in diameter and an uncontrolled spillway situated in a natural saddle about 3,000 ft east of the left abutment, as shown on the location map and plan of the dam (Figs. 5 and 6).

Because state-of-the-art earthquake engineering has evolved significantly since the construction of Coyote Dam, a re-evaluation of the stability under earthquake loading was undertaken in 1979. The evaluation consisted of: a seismicity study to develop the design earthquake parameters; exploratory borings to sample embankment and foundation materials for static and dy-

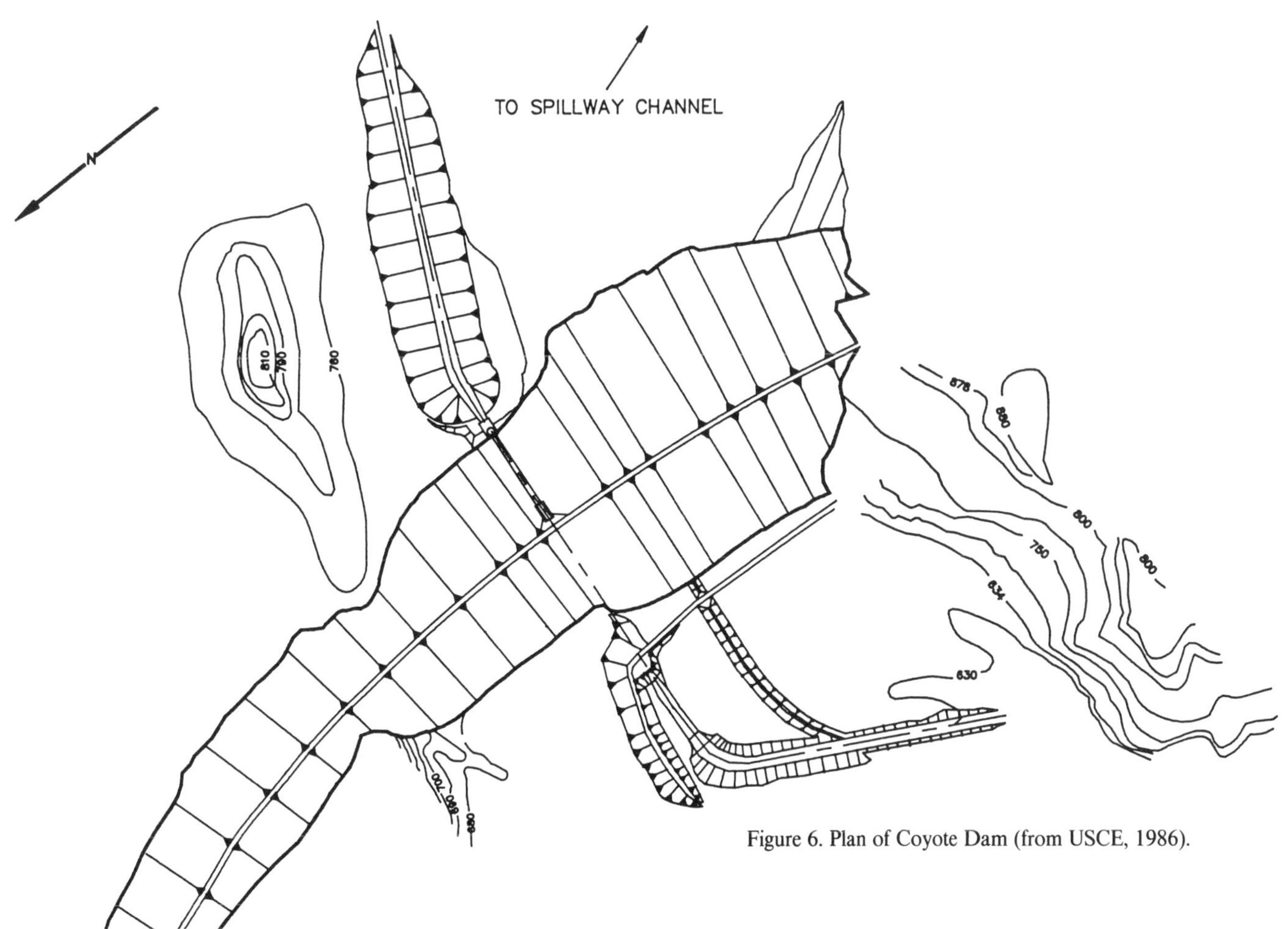

Figure 6. Plan of Coyote Dam (from USCE, 1986).

namic testing; geophysical investigations to provide compression and shear-wave velocities and dynamic elastic properties; and dynamic analyses. Subsequent to the initial evaluation, a supplemental evaluation consisted of test pits and samples from certain alluvial materials, reassessment of the design earthquake, and additional analyses of post-earthquake stability.

Coyote Dam is located in topography that ranges from rounded, gently rolling hills to rugged, steep, mountainous terrain. Drainage patterns reflect the northwest-trending regional structures. The principal rocks are the Franciscan assemblage of coastal marine sediments, mélange, and graywacke; the marine sediments of the Great Valley sequence; and Cenozoic marine sediments, volcanics, and alluvial deposits.

The known capable faults and resultant maximum credible earthquakes are shown on Figure 7; Coyote Dam and Lake are located fewer than 100 mi from 8 capable faults. Of these, 2 were determined to be significant and able to generate the controlling earthquake event. The San Andreas fault system occurs 30 mi to the west and is capable of generating a magnitude 8.3 earthquake; a peak bedrock acceleration at the dam of 0.2 g was assigned to potential San Andreas events. The Maacama fault zone passes within 1 mi of the dam and roughly parallels other regional features, the Healdsburg, Rodgers Creek, and Hayward fault zones (Fig. 7). Two interpretations have been advanced concerning these fault zones: (1) they are separate but en echelon faults, and (2) they make up one large fault zone. After considering available data, the Corps of Engineers adopted the interpretation of separate but en echelon faults (USCE, 1986). A maximum potential magnitude of 7.0 was assigned to the Maacama fault zone with a peak acceleration at the dam of 0.6 g.

The oldest rocks in the area of the dam and lake represent the Upper Jurassic Franciscan-Knoxville Group. These rocks crop out along the reservoir rim, upstream from Coyote Valley, and underlie the lower reservoir and dam at depths below the original subsurface exploration. Near the dam, the Franciscan rocks are overlain by gravelly to fine-grained sediments of the upper Pliocene or early Pleistocene Ukiah beds and younger terrace deposits (reworked Ukiah material), and recent alluvium.

The dam embankment was constructed on the young terrace deposits that overlie the Ukiah beds. During dam construction, portions of the overburden were observed to be slickensided and of low natural strengths. In these areas, the unsuitable materials were removed and the dam embankment was placed directly on the Ukiah beds. In addition, the cut-off trench was excavated to

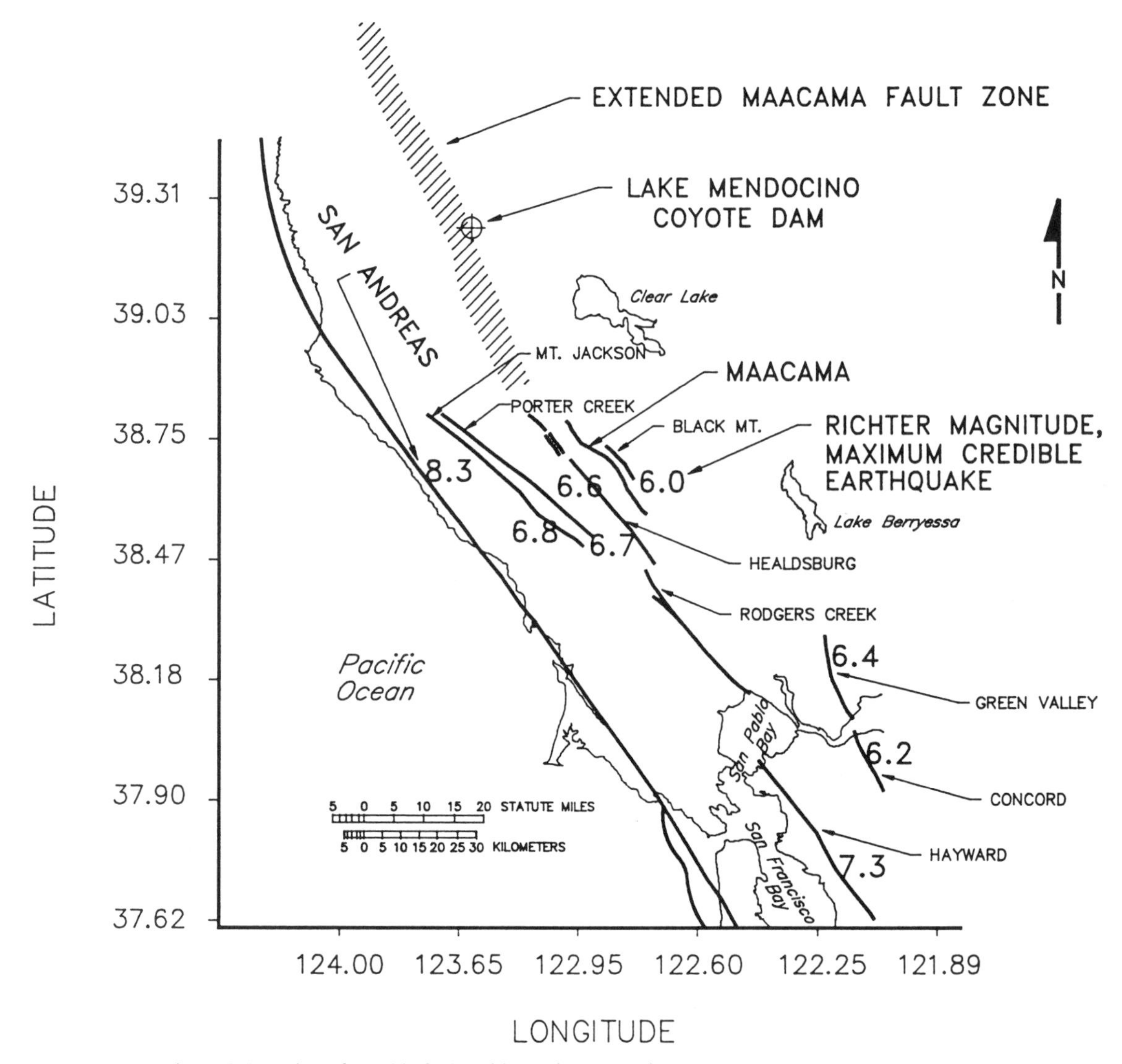

Figure 7. Location of capable faults with maximum credible earthquakes (from USCE, 1986).

the Ukiah beds. These consist of bluish, sandy, and clayey gravels with locally calcareous cement. The gravel sizes are mostly subangular to subrounded with a high percentage of disintegrated sandstone that breaks down to sand sizes; the matrix varies in clay content. The overburden materials are from 6 to 24 ft thick.

Although categorized in the design documents as reworked older alluvium, older terrace deposits, Recent terrace deposits, and Recent alluvium, these classifications denote location and not materials. In general, the deposits are gradational, and it is difficult to delimit each one. Overburden in the valley section, upstream from the cut-off trench, was considered susceptible to liquefaction and was assigned zero shear strength for the stability analyses. However, this was not a controlling situation because the overburden was confined and isolated by the 60-ft-wide cut-off on the downstream side and the shell material in contact with the Ukiah beds on the upstream side. The 4 embankment zones were combined into 2 zones for the seismic evaluation: shell and core.

The explorations for the dynamic analysis of Coyote Dam consisted of 6-in core borings and geophysical studies. Eighteen borings were drilled in groups of 3 at 6 separate locations. They sampled core, shell, and foundation materials at the downstream abutment. The borings were grouped, spaced, and drilled so that in-hole geophysical measurements could be performed. These consisted of 6 sets of uphole/downhole and crosshole tests and 4 surface refraction lines, as shown on Figure 8. Strain- and stress-controlled triaxial shear and resonant column tests were performed on samples of core and shell materials.

The 1986 evaluation of Coyote Dam under earthquake loading consisted of the following 4 steps: (1) static finite-element analyses to evaluate the state of stress within the embankment under steady-state seepage conditions, (2) a dynamic response

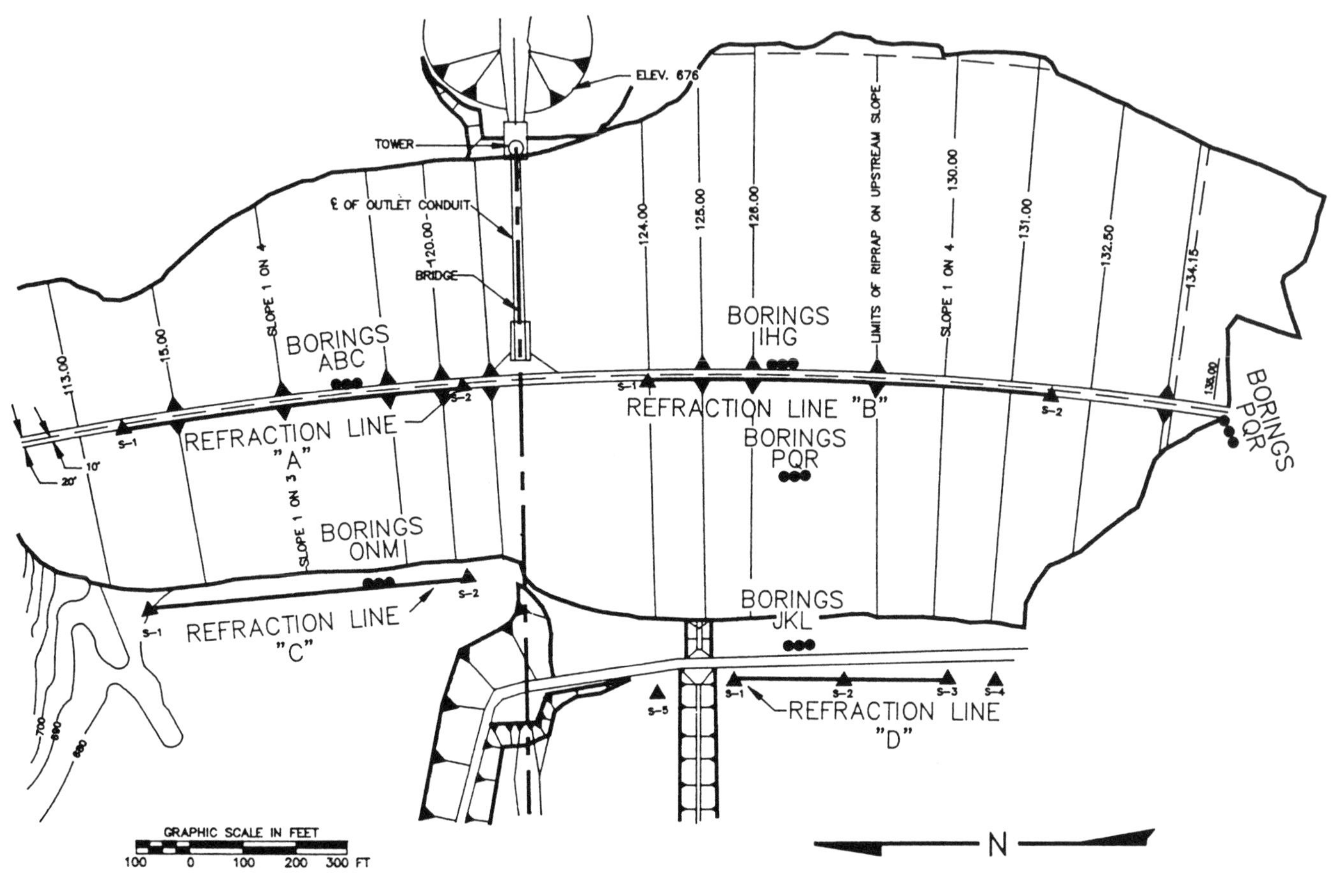

Figure 8. Location of seismic-evaluation exploratory borings and geophysical studies (from USCE, 1986).

evaluation of the dam from strong-motion records obtained during the Willits earthquake of March 1978, (3) dynamic finite-element analysis that evaluate the dynamic stresses induced by the design earthquake, and (4) a post-earthquake stability analysis of the dam.

The comprehensive analyses indicated the dam embankment would remain stable and not experience any significant deformations, even under shaking by the severest earthquakes. Yet the dynamic analysis report exposed 3 areas of concern relevant to the seismic stability of the dam: (1) the design earthquake parameters relative to the San Andreas fault, (2) liquefaction potential of the terrace deposits, and (3) the need for additional post-earthquake stability analyses using wedge methods.

The maximum earthquake parameters were reviewed by two prominent consultants who concluded that the most likely peak acceleration from a San Andreas fault event would be 0.24 g, and a Maacama fault event 0.65 g. These differences did not warrant reanalysis.

The liquefaction potential of the older alluvial terrace deposits was investigated by in situ density tests in 6 test pits, supplemented by laboratory tests on the terrace soils. In addition, triaxial shear tests determined that at low confining pressures soils were dilative, while at higher confining pressures they were slightly contractive and subject to limited liquefaction. The reanalyses of post-earthquake stability, using wedge methods, yielded safety factors in excess of that required by current criteria.

The seismic re-evaluation of the Coyote site concluded that the dam would be safe under loading from the maximum credible earthquake, as constructed.

OUTLOOK

The growing importance of regulatory statutes and their effect on public facilities and the nation's infrastructure has placed a strong emphasis on the rehabilitation and/or design re-evaluation of major engineering works in North America (Jansen, 1988). Of necessity, applied geologists will be members of the team of experts involved with the forthcoming rehabilitation phase of the works.

This re-emphasis is further evidenced by planned changes in policy of one federal agency—the U.S. Bureau of Reclamation—heretofore a major builder of dams, powerplants, and irrigation projects in the western states since 1902. Beginning in 1989, construction of major new projects will be restricted, and the focus will be on maintaining and rehabilitating existing facilities. With the federal government, which controls some 25 percent of all water resources in the western states, the Bureau of Reclamation will serve as a facilitator for water marketing proposals between willing buyers and sellers.

The basic considerations of surveillance activities for operating dams and major engineering works have been reviewed and the critical parameters outlined by Duscha and Jansen (1988). The principal problems that might occur with operating large reservoirs have been reviewed by James and Kiersch (1988), as have the instrumentation techniques for monitoring the performance of embankments and foundations by Carpenter and others (1988).

REFERENCES CITED

AEG, 1977, Geology of the Cumberland River Basin and the Wolf Creek Dam site: Association of Engineering Geologists Bulletin, v. 14, no. 4, p. 245–269.

Carpenter, L. R., Lytle, J. D., Misterek, D. L., Murray, B. C., and Raphael, J. M., 1988, Instrumentation, *in* Jansen, R. B., ed., Advanced dam engineering for design, construction, and rehabilitation: New York, Van Nostrand Reinholdt, p. 751–776.

Duscha, L. A., and Jansen, R. B., 1988, Surveillance, *in* Jansen, R. B., ed., Advanced dam engineering for design, construction, and rehabilitation: New York, Van Nostrand Reinholdt, p. 777–797.

ENR, 1985, Infrastructure bill aired: New York, Engineering News Record, v. 214, no. 18, p. 11.

James, L. B., and Kiersch, G. A., 1988, Reservoirs, *in* Jansen, R. B., ed., Advanced dam engineering for design, construction, and rehabilitation: New York, Van Nostrand Reinholdt, p. 722–750.

Jansen, R. B., ed., 1988, Advanced dam engineering for design, construction, and rehabilitation: New York, Van Nostrand Reinholdt, 797 p.

Pender, M. R., 1985, Coming to grips with the nation's infrastructure: Constructor, v. 67, no. 6, p. 38.

Simmons, M. D., 1982, Remedial treatment explorations, Wolf Creek Dam, Kentucky: Proceedings, Society of Civil Engineers Journal of the Geotechnical Engineering Division, v. 108, no. GT7, p. 966.

Thornton, H. T., Jr., 1981, Evaluation of condition of Lake Superior Regulatory Structure Sault Ste. Marie, Michigan: U.S. Army Engineer Waterways Experiment Station Miscellaneous Paper SL–81–14, 93 p.

USCE, 1971, Instrumentation of earth and rock-fill dams, Part 1 of 2: Washington, D.C., U.S. Army Corps of Engineers Engineer Manual EM 1110–2–1908, (August), 68 p.

—— , 1976, Instrumentation of earth and rock-fill dams, Part 2 of 2: Washington, D.C., U.S. Army Corps of Engineers Engineer Manual EM 1110–2–1908, (Nov.), 87 p.

—— , 1986, Dynamic analysis of Coyote Dam: Sacramento, California, U.S. Army Corps of Engineers, Sacramento District supplemental report, Sacramento, California, 17 p.

MANUSCRIPT ACCEPTED BY THE SOCIETY MARCH 20, 1989

ACKNOWLEDGMENTS

Early drafts of the chapter received input and the benefit of critical reviews by George A. Kiersch, who also assisted in assembling the completed text.

Levee along the Susquehanna River at Sunbury, Pennsylvania, showing the flooding of June 24, 1972. Photo looking north from the now-dismantled Bainbridge Street bridge linking the towns of Sunbury and Shamokin Dam. (Photo from *The Daily Item*, Sunbury, by Perry H. Rahn.)

Printed in U.S.A.

St. Francis Dam. A fault through the right abutment is clearly discernible at the end of the shadow cast by the remnant concrete "tombstone." The fault marks the contact between an erodible conglomerate (dark) and mica schist. A zone of gouge up to 1.2 m in width, which exists at this contact, is described as firm and coherent when dry but becoming soft and plastic when wet (Wiley, 1928). (Photo courtesy of California Division of Water Resources, 1928.)

Geological Society of America
Centennial Special Volume 3
1991

Chapter 22

Failures of engineering works

Laurence B. James
Chief Engineering Geologist (retired), California Department of Water Resources, 120 Grey Canyon Drive, Folsom, California 95630
George A. Kiersch
Geological Consultant, Kiersch Associates, Inc., 4750 North Camino Luz, Tucson, Arizona 85718

INTRODUCTION

During the twentieth century, remarkable changes have transpired in the attitudes of engineers and related specialists toward the importance and relevance of geological input into the success of most engineering works. During the early decades it was difficult to determine whether it was negligence or an inadequacy of geologists that resulted in failure to convey the pertinent data needed by engineers and associated professionals to quantify the geologic setting and features. All too often the geologist was prone to a lofty, scientific style of report or oral presentation that was foreign to the intended user. We (the authors) were introduced to this situation early in our careers. One prominent project engineer caustically informed one of us (LBJ), "Engineers are not interested in a classification of rocks as 'herbiverous or carniverous,' but rather with the physical evaluation of geological conditions and a prediction of the potential effects of areal-site conditions on the safety and performance of project."

In a similar thrust, dam-builder H. F. Bahmeier (personal communication, 1936) counselled the other (GAK) with, "Professor Berkey is my preferred consultant because he is the only geologist I can understand."

Unfortunately, past misunderstandings frequently created schisms and polarized the two disciplines. In extreme instances, some project engineers minimized the importance of adequate geological investigations or scorned the data and recommendations of geologists as of little practical value. Sadly, some of the resulting engineering works ultimately experienced difficulties and/or failure due to the geologic conditions, such as the Baldwin Hills and Vaiont Reservoirs, and the Malpasset and Teton Dams. Fortunately, today a mutual interdisciplinary respect exists for the importance of mature, field-experienced geological input for engineering purposes.

The literature relates many examples of communications failures between geologists and engineers that subsequently resulted in engineering failures. A breakdown in communications between the geologist and design engineer can be as critical as between the field and office engineer, or when a contractor fails to appreciate the meaning of the geologic conditions on a project. The burden for a mutual understanding of the geologic environment and its relevance to project design is on the geologist (Judd, 1967). He or she must use language and explanations that are readily understandable by associates, whether engineers, other scientists, lawyers, economists, or planners. A fine group of project examples, in which geotechnical communications became a factor, was assembled by Judd (1967). Two common illustrations of language difficulties noted by Judd (1967, p. 1) are, "Soil mechanics engineers describe a poorly-graded sand for concrete or filter material and mean all grains have the same size while a geologist generally means poorly-graded represents grains of all sizes." Or the geologist may describe a dam site as "underlain by a granodiorite with frequent intercalations of mylonite that probably formed during Keewatin orogenies"—but this does not explain to the design engineer (Judd, 1967, pp. 1, 2) that "the dam will be founded on hard, quartz-like rock that has numerous fractures filled with clay-like material." Conversely, the engineer must clearly and intelligently explain his requests for field data. How many geologists would understand a request to drill exploratory borings in the foundation at the quarter-points of an arch dam and at angles equal to the thrust lines at these points? Practically, the engineer should provide the geologist with a plan of the dam and show the locations of the quarter-points, and the dip and azimuth of the thrust lines.

Today, both the interrelated education and indoctrination of applied geologists, engineers, and associated professionals has led to a general understanding of the respective disciplines and contributed to eliminating this "gap" between professions. With this broadening of capabilities, several geological- and/or engineering-oriented specialties have emerged, and are discussed in Chapters 1 and 2.

Over the past two centuries, engineering principles and design have been perfected and today approach an exact science. However, the natural sites upon which the engineered works are founded all too often contain defects or are subject to ongoing geologic processes and changes that become hazards which defy quantitative analysis. Therefore, many designs for engineering works involve a significant geological risk—for which the assessment is largely a matter of mature judgment. Today, the

James, L. B., and Kiersch, G. A., 1991, Failures of engineering works, *in* Kiersch, G. A., ed., The heritage of engineering geology; The first hundred years: Boulder, Colorado, Geological Society of America, Centennial Special Volume 3.

designer looks to the applied geologist for assistance in evaluating such risks, whether natural or man induced. The experience and knowledge or database derived from the investigation of past failures of engineered works are critical to any evaluation/assessment of sites proposed for new projects. The following sections include a review of selected case histories concerned with a failure, or a significant geological problem, experienced by major engineered works. Such cases ultimately contributed to an improved geological and/or engineering understanding of risks and underlying causes involving geology, and furthered the increasing acceptance of geological input by all concerned parties. Although necessarily brief, the case discussions focus largely on the lessons learned; sources of detailed information on each are given in the references cited.

The efforts of the many early professionals who were instrumental in the growth and acceptance of geological concepts and input, so important to the success of engineered works, are discussed at length in Chapter 1.

HISTORICAL MILESTONES. DEFINITIONS

The disappointment of failure has affected engineers and their works since ancient times. Today there are vestiges of ancient roads, canals, dams, tunnels, and other works in Asia, the Middle East, Africa, Europe, and the Western Hemisphere, and many provide evidence of failure. Why have these failures occurred in the past, and do they occur in a similar manner today? Are failures inevitable? Have actions been taken by the "engineer" to avert failure in the past? Before attempting to assess the success of efforts designed to thwart the repetition of failures, it is necessary to understand and evaluate the technical and natural systems under which the engineering works operate (FitzSimons, 1986). For example, the Tunnel of Eupalinus on the Island of Samos, Greece, was the first tunnel successfully driven from both portals simultaneously (530 to 526 B.C.); the headings joined beneath Mount Castro (see Chapter 1). The tunnel, about 1,000 m long and more than 2 m high, transported spring water from the Valley of Mytilina to the seaside Polycrates City (Pythagorian). The two headings missed meeting each other by some 2 m horizontally under the mountain. Was this an engineering failure? Indeed not, for under the circumstances of excavation and surveying techniques, the tunnel was a resounding success and a world-class achievement. Today the same endeavor and near miss would be a failure.

The law concerning engineering failures has been established on the basis of three major historical actions (FitzSimons, 1986). Hammurabi, although known as the great Babylonian King, performed his greatest achievement with a written legal code. Relative to construction, his ideas were quite simple: "If a contractor builds a house and it collapses killing the owner, the contractor will be killed. If a son of the owner is killed, then so will be the son of the contractor." From this simple beginning the laws associated with "engineering" failures grew more complex, as did the means of adjudication.

A milestone in English common law occurred in 1782 in the case of Folkes versus Chadd and others. The latter, as trustee of the harbor of Wells on the east coast of England, obtained an injunction against Folkes to remove a protective dike he constructed around low pasture lands, claiming the embankment was causing the silting and backfilling of the harbor. Folkes countered at his trial with testimony by John Smeaton that, in Smeaton's opinion, there was no connection between the dike and the silting action. The court ruled that Smeaton was qualified to offer a judgment of facts and his opinion was proper evidence.

Subsequently, the Napoleonic Code of 1804 promulgated an interesting interaction between the political and economic system: "If the serviceability of a constructed works is lost within 10 years of completion because of foundation failure or poor workmanship, the contractor and architect will be sent to prison."

The impact of Folkes versus Chadd is evident on today's litigation; the cost of proceedings includes not only legal fees and expenses, but also those for technical experts of all kinds. The impact of Hammurabi's or Napoleon's Code on the reduction of failures is questionable. Are there fewer failures in regions under the Napoleonic Code than under English common law? Certainly the human factors are at least as important as the physical and technical factors (FitzSimons, 1986).

Failure

One widely accepted attitude toward failure has been—you will never get anywhere without it! Furthermore, one does not really profit from experience, one merely learns to predict the next mistake.

A definition of "failure" does not necessarily denote a catastrophic event or collapse of an engineered works. A failure generally implies that a facility does not perform as was originally intended by the designer-engineer, contractor, or owner (Cooper, 1986). For example, unexpected foundation conditions may be encountered that allow excessive leakage, resulting in potential piping, or a mechanical system that functions improperly. Verified pre-event data may be the critical component of any investigation into the cause and/or responsibility for this failure. Normally there are many contributing factors, and conflicting opinions are plausible. Frequently, expert witnesses do not come to the same conclusions regarding causes, whether due to the effects of age on a project, ongoing geologic processes or reactions, inadequate inspection and maintenance programs, or unforeseen natural or man-induced events. Moreover, structural failure ranges from catastrophic collapse of an engineered works to the inconsequential bending of a few beams in multistory buildings; the latter actually means certain code-specified "allowable" stresses or strains are numerically exceeded (Meehan, 1984, p. 58).

The varied causes of dam failures and accidents have been analyzed and summarized in Jansen (1983, ch. 2). Interestingly, natural conditions and geologic features are often main causes of failure, particularly regarding foundation defects and piping. A classification of dam failures, whether total (F1) or repairable (F2) has been established by Jansen, as are three separate categories of such accidents that might occur while a project is in operation, during initial filling of a reservoir, or during construction.

Frequently used descriptive terms to denote failure are: catastrophic—many lives lost; heavy—severe damage, some lives lost; severe financial loss—no fatalities; and tolerable loss—some hardship.

Any satisfactory classification of failure must address the needs of each party involved in a project, because each involvement is unique and occurs at different times from inception to demolition. Consequently, failure means or implies quite different circumstances or actions to each party.

Geologists are concerned with changes and/or modifications to physical properties of rock and soil-sediment masses with time, and particularly how the actions may interact to cause rupture, movement, sliding, deformation, deterioration, or saturation and thereby affect engineered works. The potential for geologic hazards to develop during the life operation of a project must be recognized, and their influence on safety taken into account. Geologic changes may be due to tectonic or nontectonic reactions, ongoing geologic processes, or man-induced modifications to the natural conditions. Many works have failed because of the geologic environs. Unfortunately, ingenuity and long-range assessments ("calculated risks") have not been part of the responsible thinking of some mission-oriented geologists.

Designers-engineers are interested in failures as they involve unsatisfactory behavior due primarily to technical and functional shortcomings. The design must be possible to construct in a reasonable time and with a suitable profit, and also be one the owner can afford with an accepted factor of safety (Chesson, 1986, p. 46).

Contractors are primarily concerned with structural failures caused by workmanship, inferior materials, and temporary construction, but must consider relevant costs and potential profits, despite uncertainties of supplies and future costs of materials and labor (Chesson, 1986).

Owners are concerned with many factors outside the technical fields and are strongly influenced by economics, return on investment, and the public image. Failure may be any behavior of the completed works that differs from what is desired by the owner (Roberts, 1959), and may be involved with poor performance by the siting team, designers, contractors, and even operating personnel.

Insurers are concerned with classifications of failures that permit the setting of premiums and making equitable profits, whether insurance is for the designer, contractor, or owner (Chesson, 1986).

Risk

Risk is a problem that faces each person daily; it means exposure to loss or injury. In many ventures, risk, or the probability of loss, can be predetermined with a high degree of accuracy. Relative to engineering, the concept of risk evaluation has a much less precise meaning, and can mean different things to different people.

In essence, terms like "failure," "precision," "factor of safety," "economical solutions," and "risk" are somehow interrelated, but the meanings of these terms depend on one's viewpoint. For simplicity, concepts such as "failure," "factor of safety," and "risk" can be considered from three hypothetical viewpoints: (1) by a geoscientist who is primarily interested in soil, unconsolidated deposits, and rock as structural materials; (2) by an engineering designer who is concerned about the behavior of a structure placed on a given combination of soils and rock deposits and the active geologic processes of the environs; and (3) by the owner of an engineering project who is interested in a financial return on his investment, besides maintaining good public relations and a safe and trouble-free operation of the engineered works.

The three types of risk that must be considered in the planing and construction of engineered works have been defined by Casagrande (1965). Because risks are inherent to any project, they should be recognized and steps taken to deal with them as appropriate in order to have a strong balance between economy and safety.

Unknown risk is the chance of experiencing a failure as the result of some defect or geologic condition that heretofore had not been recognized as hazardous.

Relevant cases. The failure of Wheeler Lock, described in Chapter 23, was due to an "unknown" thin shale seam that extended throughout the foundation near the surface. Likewise, the high rock wall at the Schoellkopf generating station below Niagara Falls, New York, collapsed in 1956 due to tensional stress induced by high-pressure grouting of the wall. Collapse was attributed to unknown risk—grout pressures injected into the canyon walls triggered movement along major open fractures, causing the collapse (described in Chapter 23).

Calculated risk is the possibility of failure due to geologic conditions/features for which no practical solutions are known at the time. However, a satisfactory partial solution is assumed, which would be implemented during the construction. Such risks are deemed justifiable in the absence of any threat of fatalities or serious injuries, and where the probable benefits to a project outweigh the possible losses (discussed further in Chapters 22 and 23).

In recent years there has been an increase in application of the principle of probability (Risk Analysis, this chapter) as part of the decision-making process for such applications as stability of slopes, safety factors relevant to liquefaction and evaluation of potentially liquefiable soils, and risk evaluation of earth dams (Whitman, 1984). Analyses of reliability and risk are probably most valuable during the early stages of a project, for guidance about whether to proceed, and for selecting design criteria.

Relevant case. The route of the California Aqueduct through San Joaquin Valley crossed land that had undergone subsidence

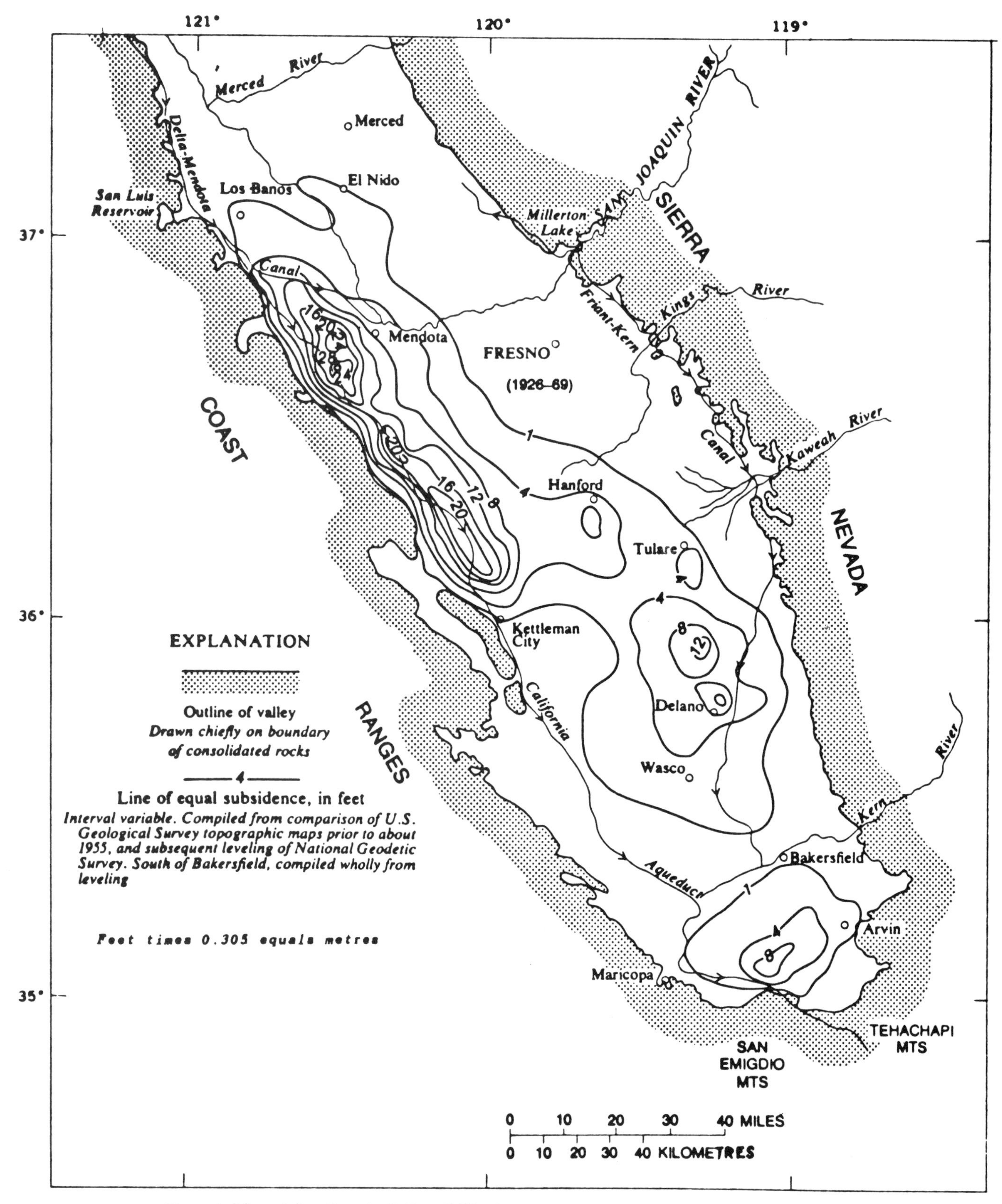

Figure 1. Map of San Joaquin Valley, California, showing the areal extent and amount of land subsidence along the route of the California aqueduct. Subsidence during period 1926–1970 is due to ground-water withdrawal. (Courtesy U.S. Geological Survey, circa 1958.)

(1926 to 1970) of up to 6 m due to lowering of ground-water levels in underlying deep aquifers (Poland and Lofgren, 1984). The canal alignment (Fig. 1) was adopted with full knowledge of the subsidence problem. Its selection was governed by economic factors. Any route circumventing subsidence areas would have entailed at least one additional pump lift or a longer canal through materials more costly to excavate. Either of these alternatives would have resulted in increasing the cost of delivered water sufficiently to jeopardize the economic feasibility of the entire State Water Project. Two kinds of subsidence were involved: that caused by consolidation of deep sediments due to heavy ground-water withdrawals for irrigation, and that resulting from collapse of low-density, near-surface deposits, which were susceptible to hydrocompaction when saturated (Fig. 2). A full discussion of hydrocompaction is given by Holzer (Chapter 10).

With respect to subsidence caused by ground-water withdrawals, it was reasoned that once the project was put into operation the cost of water delivered by the canal would be significantly less than that for the water then being pumped from deep wells. As a consequence, the aqueduct would replace wells as the source of water, thereby eliminating the cause of the problem.

It was determined that approximately 122 km of the alignment overlay unconsolidated deposits that were subject to hydrocompaction. In some areas these materials extended to depths up to 60 m (Fig. 2). Large-scale field experiments demonstrated that even slight leakage from the concrete-lined canal would initiate subsidence, causing a progressive failure of the concrete lining, with ultimate collapse of the canal prism (Fig. 3). Although subsidence of this type and/or scale had not been experienced previously by a major project, it was decided to proceed and build the other elements of the State Water Project, anticipating that an understanding of the geologic hazard was forthcoming and an economical solution possible. Extensive research concluded the only acceptable remedial treatment was to presubside the susceptible sediments (Lucas and James, 1976). An array of ponds along the canal alignment supplemented by injection wells, up to 60 m deep proved effective for this purpose, as shown on Figure 4.

Figure 3. Prototype section of test canal demonstrated beyond question that concrete lining by itself was incapable of preventing hydrocompaction, but by presubsiding susceptible soils, subsidence could be reduced to acceptable limits. The ruptured segment of canal (lower right) had been excavated in virgin soil, whereas the undamaged lining rests on ground that was presubsided by ponding prior to excavation of the canal prism. (Courtesy California Department of Water Resources, 1966.)

Figure 2. Hydrocompactable soils along the California aqueduct alignment in San Joaquin Valley were identified and delimited during an intensive program wherein open-bottom tanks were embedded in suspect soils and the resulting subsidence was measured over a period of months. The vertical rods, which extend through casing to subsurface bench marks embedded at depths ranging from 7.6 to 70 m, were periodically observed to determine subsidence rates at depth and the thickness of susceptible soil. The concentric surface cracks are typical manifestations of hydrocompaction. (Courtesy California Department of Water Resources, 1958.)

Thus, two major calculated risks were taken during early stages of the State Water Project. Both have proven justified; no significant subsidence has been experienced along the aqueduct since water deliveries were commenced in 1968.

Human risk is the chance that failure or damage will result because of a physical modification of the site-specific conditions during construction or operational activities, or from human failings and mismanagement. This might include organization-related events, weak management, technical incompetence, adverse political influence, inadequate office-field communication, imprudent use of advice from interdisciplinary experts, or even a total collapse by internal corruption.

Relevant cases. Administrative policies discouraged the project designers for Teton Dam from visiting the site regularly and entering into some of the critical construction decisions. Consequently, decisions were made that weakened the dam's resistance to piping and led to failure (described below).

A more specific case is the failure of an embankment at

Figure 4. The California aqueduct crosses an extensive region underlain by hydrocompactable soil in the San Joaquin Valley. An array of ponds was constructed to presubside the alignment and stabilize the canal foundation prior to emplacement of the concrete lining for 121 km. At places, particularly where susceptible soils were thick, the ponds were augmented by infiltration wells to accelerate compaction at depth. (California Department of Water Resources, 1966.)

Walter Bouldin Dam that involved fill placed during marginal weather conditions between Thanksgiving and New Year's Day; holidays interrupted the momentum of construction, and quality control was affected when the chief inspector was hospitalized (Sowers, 1986; Leps, 1988).

Risk analysis. The benefits and virtues of probabilistic risk analysis methods have been given a certain degree of legitimacy since a 1977 decree by President Jimmy Carter to the principal dam safety agencies within the United States government. His April 23rd memorandum recommended considering probabilistic risk in evaluating dam sites and construction.

Risk involves both uncertainty and damage and implies that danger can be minimized by increasing safeguards (Shah and McCann, 1982), but never reduced to zero. Practically, risk analysis is a combination of art, judgment, and science—in that order. Furthermore, probabilistic analysis cannot be replaced by geological or engineering judgment and/or mature, field-experienced input.

Peck (1982) has evaluated risk analysis for dams, and points out that both the geologic setting/features and precedent are critical to the design and site specifics, while the usefulness of probabilistic methods is not clear. For example, the stability of embankment materials, complex geologic environment, or possible occurrence of a weak feature(s) in a foundation cannot be adequately evaluated by current statistical methods. Consequently, the most serious inherent circumstance or defect must be assumed, irrespective of the probabilistic forecast.

FAILURES RELATED TO GEOLOGIC ORIGIN

Foundation defects

The principal foundation defects are fault and fracture zones, both open and filled, with the potential for tectonic displacement or effects of areal subsidence; stress-relief joint systems; seepage piping aided by permeable rock units or structural zones; soft erodible materials; and solutioned-porous limestones.

Malpasset Dam. The physical cause of the catastrophic failure in 1959 of the concrete arch dam (Figs. 5 and 6) located

Figure 5. Malpasset Dam. This double-curvature arch dam, 66.5 m high and 222 m long at the crest, failed catastrophically on February 12, 1959. During its first filling, when the reservoir level reached within 30 cm of the spillway crest, the foundation ruptured along two discontinuities in the left abutment, one a previously undetected fault plane. During the failure, the left abutment thrust block (right center) was displaced 2 m. Collapse of the arch occurred almost instantaneously, probably lasting only a few seconds (Londe, 1988). (Courtesy Coyne and Bellier/P. Londe.)

near Frejus, France, has been attributed to the presence of a clay-filled fault plane in the left abutment, and a second ruptured surface aligned parallel to the rock schistosity (Londe, 1988). These physical discontinuities intersected beneath the abutment to form an isolated wedge of rock. The designers directed the arch thrust against this detached wedge-block rather than against the intact rock of the abutment. Consequently, the left end of the dam, supported by only a loose, near-surface wedge of rock with limited dimensions, was dislodged as the thrust forces acted during the initial filling of the reservoir (Fig. 7). As the wedge was displaced upward along the clayey fault plane, the dam was raised in its entirety and rotated downstream, as if the entire monolith were hinged at the right abutment (Fig. 8). The displacement created some open space and subsequently a blowout beneath the left abutment, which led to the breakup and collapse of the concrete dam. The existence of the causative fault beneath the dam was unknown from the sketchy geological investigations (no systematic mapping, borings, etc.). After the disruption and removal of the soil mantle and rock wedge by the flood waters in 1959 (Fig. 9), the fault plane was exposed, and the site-specific geologic features and rock conditions became evident (Fig. 10). The Malpasset dam failure furthermore involved "errors in geologic judgement" (Chapter 23) due to the inadequate feasibility-planning investigations as described in this section and shown on the figures.

Waco Dam slide. The Waco slide of 1959 was a failure from an unknown risk (West, 1962). During the geological inves-

Figure 6. Aerial view of Malpasset Dam and vicinity, after failure, 1959. Blocks of concrete deposited by the sudden release of water that followed collapse of the arch remain in the channel. Traces of the flood, evidence as high as 20 m below the reservoir level, attest to the rapidity of the failure. The surge of water, in part, overtopped the saddle (right center) depositing debris along the alignment of the auto route, which was under construction at the time. (Courtesy Coyne and Bellier/P. Londe.)

Figure 7. View of left abutment of Malpasset Dam taken from remnant of dam in channel (bottom foreground). Remains of the thrust block and wing wall are shown (upper right). The thrust of the arch was directed approximately parallel to the rock schistosity (center left). The intersection of schistosity planes with a fault lies beneath the debris (right center). These two discontinuities undercut the abutment, forming a detached wedge of rock against which the thrust was directed. (Photo by L. B. James, 1964.)

tigations for planning and design of the dam, the potential for weaknesses within the shale of the Pepper Formation was not realized. Specifically, at the time of the field studies, no satisfactory methods had been developed for determining or measuring the potential for a buildup of dangerous pore pressures in Pepper-type shales. The slide experience emphasized the necessity for conducting an investigation of any shale foundation mass in sufficient detail to identify accurately not only the stratigraphic elements and structural features of site, but also the inherent physical properties and geologic reactions of the shale materials to the conditions induced by the engineered works proposed.

Slope collapses/Slides

Slope collapses/slides may be rapid, short-duration, catastrophic events such as rock-soil slides, rockfalls, or waves generated by falls; they can cause local earthquakes. The slow progressive failures of long duration due to rock or soil creep and slides are less hazardous. Damage losses from slope collapses/slides and subsidence alone totalled over $1.5 billion per year in the United States over the past 50 years, irrespective of more precious loss of life. This average loss is much more than the total loss per year from earthquakes, floods, hurricanes, and tornadoes (Kockleman, 1986). The loss of life from slope failures averaged 600 per year worldwide during 1971 to 1974, according to Varnes (1981).

Kansu Province, China. The great earthquake that occurred in Kansu Province on December 6, 1920, created immense slope collapses/slides and resulted in about 100,000 fatalities within an area of 162 by 486 km. Ten large cities and numerous villages within the area were damaged (Close and McCormick, 1922).

The region of damage is underlain by loess deposits that are a partially cemented mixture of clays and fine-grained sand. These fine-grained materials were apparently liquefied by earthquake vibrations, giving rise to the giant slides that caused most of the devastation. For example, a 0.4-km section of road was reportedly displaced about 1.6 km in a few seconds. Earthflows were frequent and described as follows (Close and McCormick, 1922, p. 447): "The earth materials which collapsed bore the appearance of having been shaken loose—clod from clod and grain from grain—and cascaded like water down slopes forming vortices, swirls, and other convolutions and shapes of a torrent."

The Kansu slope collapse awakened the interest of geologists and engineers in the phenomenon of soil liquefaction as a result of seismic events and provided an early stimulus to research on the mechanics of earthquake-related slope collapse/slides.

Solution cavities

McMillan Reservoir, New Mexico. This project was one of the first to alert geologists to the potential for serious leakage from reservoirs flanked or underlain by evaporite beds (Cox, 1967). Since first filled in the 1890s, McMillan Reservoir's problems have been investigated by numerous geologists, hydrologists, and civil engineers, in search of solutions to the reservoir losses.

Leakage when the reservoir is full was determined by measuring the flow of rising water in the river channel downstream from the dam—about 1.3 m^3 per second. Experiments have measured the velocities of underflow using dyes; a transmissivity on the order of 58 million liters per day per meter was calculated. This indicated the leakage probably occurs as turbulent flow through solutioned-open conduits. Dikes were constructed to isolate the gypsum beds that crop out along the left flank of the reservoir; this reduced the leakage, but the storage capacity of the impoundment was substantially reduced.

Figure 8. Malpasset Dam, right abutment, showing the opening that developed during failure. The dam pivoted downstream about the right abutment; the movement opened this wide gap along the contact between the concrete and rock. (Photo by L. B. James, 1964.)

Figure 9. View of left abutment, Malpasset Dam, showing the fault plane along which failure occurred. This fault was not detected prior to the collapse of the dam. A concrete remnant rests upon the fault plane. (Photo by L. B. James, 1964.)

A solution to the dilemma, proposed by the U.S. Bureau of Reclamation, would be to construct Brantley Dam a few kilometers downstream at a site underlain by more massive carbonate rocks. The "Brantley" Reservoir would recapture the leakage that occurs from McMillan Reservoir.

Carbonate terrains and related problems

Canelles Reservoir, Spain. Topographically, Canelles Reservoir in Spain is an excellent site for the 148-m-high concrete-arch dam. Unfortunately, the site is underlain by over 400 m of limestone, the upper part of which is solutioned-karstic and contains interconnected openings (Therond, 1972). One network of interconnected cavities near the reservoir crest level has been explored for more than 800 m. Regionally, the limestone beds dip 35° upstream, which exposes the upper cavernous layers throughout the reservoir floor near the dam. Farther upstream, relatively impermeable marly beds overlie the solutioned limestone, and prevent bank and channel leakage throughout the upper reaches of the reservoir.

Originally, engineers anticipated the reservoir margins could be sealed by constructing a grout cutoff curtain at the dam site extending to a marly unit 25 m thick located within the lower third of the limestone series. Unexpectedly, when reservoir filling commenced in 1958, serious leakage appeared in the vicinity of the underground power station. As filling progressed to a lake level of 75 m, the leakage had increased to 1.6 m^3 per second. Above this reservoir level a pronounced flushing and declogging of the interconnected karstic-openings was initiated, so when the lake level rose to 85 m, leakage reached 8 m^3 per second. At this rate of reservoir loss, no further impoundment of the inflow was possible, although 51 m of freeboard remained behind the dam. Expensive remedial measures were undertaken, including an extension in the length of the grout curtain, to reduce the large losses of reservoir water.

Apparently, project planners had been attracted by the exceptional topographic setting of the dam and reservoir site, and this feature had outweighed the potential for negative geologic factors such as solutioned-karstic limestones. The Canelles project is an exceptional case of insufficient field investigation for the feasibility and design phases of a dam-reservoir project. This error in judgment subsequently resulted in far more costly remedial measures and treatment during the operational phase.

Logan-Martin Dam, Alabama. The dam consists of a concrete gravity-type spillway section about 30 m high in the center of the Coosa River with earth dikes on each flank and abutment that extend to high ground. The foundation consists of a massively bedded limestone-dolomite complex of undetermined thickness. Sinkholes are present in the reservoir, and solution-openings and cavities have been identified and extend to great depths (FPC, 1976) beneath the site.

During construction, a continuous grout curtain was emplaced beneath both the earth and concrete sections of the dam, which extends from 46 to 91 m in depth. However, since operation, reservoir seepage losses have been excessive. Strong boils discharge large volumes of normally clear water into the tail race, and new sinkholes up to one or two meters in diameter appear periodically near the dam. The underseepage is considered potentially hazardous to the safety of the project.

Figure 10. Typical metamorphic rocks of the site area of Malpasset Dam are shown with some of the common crosscutting structural features, right abutment. (Courtesy Coyne and Bellier/P. Londe.)

Attempts to control the leakage by grouting have been hindered by the clay and residual materials throughout the cavities and openings. The residual clayey debris within openings prevents the grout from completely filling and plugging the leakage channels. Consequently, after each grouting attempt, new channelways are opened through the weakly consolidated residues, and leakage subsequently resumes at former rates.

The experience at Martin-Logan Reservoir typifies the problems encountered at several other reservoir projects in the Appalachian region where grouting programs were implemented to curtail leakage through cavities in weathered carbonate rocks, such as at Kentucky, Hales Bar, Douglas, and Great Falls Dams (TVA, 1949).

Lessons from earthquakes

Hebgen Dam, Montana. This dam is a fill-type embankment with a central concrete core that rises 26 m above stream bed (Sherard, 1959). During the magnitude 7.5 West Yellowstone earthquake of August 17, 1959, the dam embankment suffered a settlement of more than a meter (Fig. 11); the core wall was cracked in four places, and the concrete-lined spillway was severely damaged. Of further significance was the pronounced regional tectonic deformation, which tilted the reservoir floor and created water waves that overtopped the dam (Myers and Hamilton, 1964). The ruptured spillway was overtaxed by spring runoff and the channel seriously eroded. Although severely damaged, the dam embankment was not breached.

This experience has focused attention on the dangerous wave effects that a strong earthquake event can create in reservoirs and lakes. According to the geologic potential for a seismic event, this possibility should be considered in planning freeboard allowances and designing spillways, particularly for embankment-type dams. Furthermore, seismic events with tectonic deformation may be of sufficient magnitude to tilt and affect the gradients of such water-distribution facilities as canals, drains, and irrigation ditches, even though they may be centered some distance from a capable fault (largely controlled by areal geologic setting/features).

Niigata, Japan. Foundation liquefaction beneath high-rise buildings in Niigata during the June 16, 1964, earthquake resulted in as much as an 80° tilt from the vertical, but buildings remained intact (Seed and Idriss, 1967). Occupants were able to evacuate the structures by stepping out of the windows and walking along the sides. This 1964 incident was another early stimulant for research in dynamic analysis of foundations, to quantify the assessment of earthquake hazards.

Water supply, San Francisco. San Andreas Dam, owned by the city of San Francisco, lies immediately west of the rupture surface of San Andreas fault, which offset approximately 2 m in this locality during the 1906 earthquake. The dam, of earth-fill construction, withstood the severe shaking with only minor damage. However, the tunnel spillway, which passed through the San Andreas fault, was ruined (Fig. 12A).

Figure 11. Embankment settlement at Hebgen Dam, Montana, resulting from the 1959 West Montana earthquake. Prior to the event, the top of dam embankment (foreground) was level with the top of the concrete core wall, a settlement of some 1.5 m. (Photo by C. J. Cortright, 1959.)

The Pilarcitos pipeline, one of the principal water conduits to San Francisco, also sustained severe damage as described in the following quote from Schussler (1906):

> The earthquake having torn a crack or fault, several miles in length, along and across the upper or 30-inch Pilarcitos pipeline, had either completely destroyed it, tearing and telescoping it in a number of places, or at least had so injured it, that it would be many, many months, to say the least, before it could be put into service again, if at all. At the large Frawley Canyon the Pilarcitos 30-inch pipe was thrown some 60 feet to one side. It was torn in two for over 100 feet and thrown bodily, in two parts and about at right angles to its original line; so that it could not have been maintained there, owing to the evidently great violence of the shock. Other portions of the Pilarcitos 30-inch pipe were destroyed by the earthquake, pulling the pipe apart in many places, while at other places, it was telescoped.

Pampeyan (1986) describes an inspection made in 1983 of part of the outlet works for San Andreas reservoir, which were located within the fault zone (Fig. 12B). His investigation revealed that the west end of the Bald Hill outlet tunnel had been displaced at least 2.5 m right laterally since construction in 1869 to 1870. Rather than being sheared along a single fracture plane, the tunnel had been bent over a distance of at least 23 m, the brick lining articulating to accommodate most of the strain. The system reportedly sustained some damage within the fault zone, but surprisingly "not so as to interfere with the flow of water" (Schussler, 1906, from Pampeyan, 1986). The water system was taken out of operation in 1947. The tunnels located to the west of the brick forebay (Fig. 12B) are also within the fault zone; however, they remain under water and are not accessible. The effect of the earthquake on that part of the outlet facilities was not determined.

San Andreas Reservoir was constructed in 1868 to 1870, before the San Andreas fault was recognized. The damages sustained during the 1906 earthquake alerted builders of water-supply systems to the importance of identifying capable faults and avoiding the construction of facilities upon or through them where possible.

Ground-water related problems

San Jacinto Tunnel, southern California. During the 1935 to 1936 excavation of the Los Angeles–Colorado River aqueduct tunnel beneath San Jacinto Mountain, faults and fracture zones were encountered that conveyed enormous water inflows (e.g., 28,387 Lpm, with 60,560 Lpm at one location; Fig. 13). This unexpected development caused costly construction delays, and required the adoption of novel tunneling methods (Henderson, 1939). One lesson learned was that low annual precipitation and arid terrain are not necessarily indicative of a dry tunnel in a highly faulted terrain (Fig. 14). The original alignment was changed after detailed areal geologic mapping located and evaluated 21 fault zones; the new alignment intersected only 11 zones.

The underground water occurred in open, interconnected fractures and small faults throughout the hanging-wall side of the steeply dipping faults as well as within the gouge and crushed material of the main zone. Control of inflow into the tunnel was possible by approaching a fault zone from the hanging-wall side with a gradually increased flow that largely drained the interconnected fractures and major gouge zone in advance; a sudden rush of water was thereby averted when the main zone was crosscut. If the heading was approached from the foot-wall side, the full impact of the interconnected, water-bearing storage network plus the main zone came as one heavy surge of water when excavation broke into the gouge and crushed-rock zone (Henderson, 1939).

The water drained from the network of faults throughout San Jacinto Mountain during the tunnel construction depleted the normal supply of ground water to springs and wells in the vicinity of the alignment. Investigations and assessment of damage to the local ground-water supplies were complicated because there was no systematic collection of information on the flow of springs and wells up to 20 km away; this resulted in costly litigation for more than a generation (Henderson, W. H., oral communication, 1951). This experience demonstrates why the geological investigation of a tunnel alignment should entail a

Figure 12. Photo: San Andreas Dam of the City of San Francisco experienced some damage from the 1906 earthquake. The tunnel spillway for the dam passed through the San Andreas fault zone and was ruined by movement associated with the earthquake (from Schussler, 1906). Drawing: Outlet works for San Andreas Reservoir located within the fault zone and showing the effects of the 1906 earthquake on two of the forebay structures, one of concrete, the other brick. Location of associated cracking and movement are plotted on both the areal (A) and site plan (B) maps. (From Schussler, 1906; and Pampeyan, 1986.)

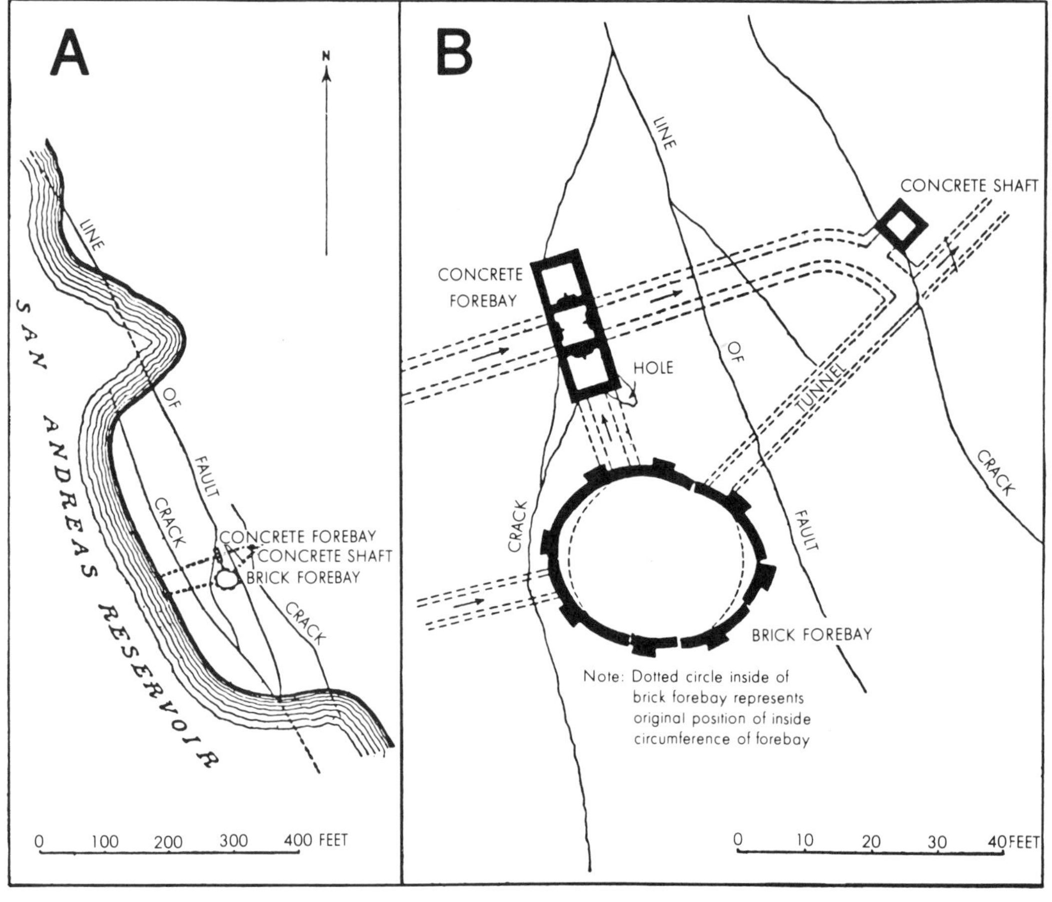

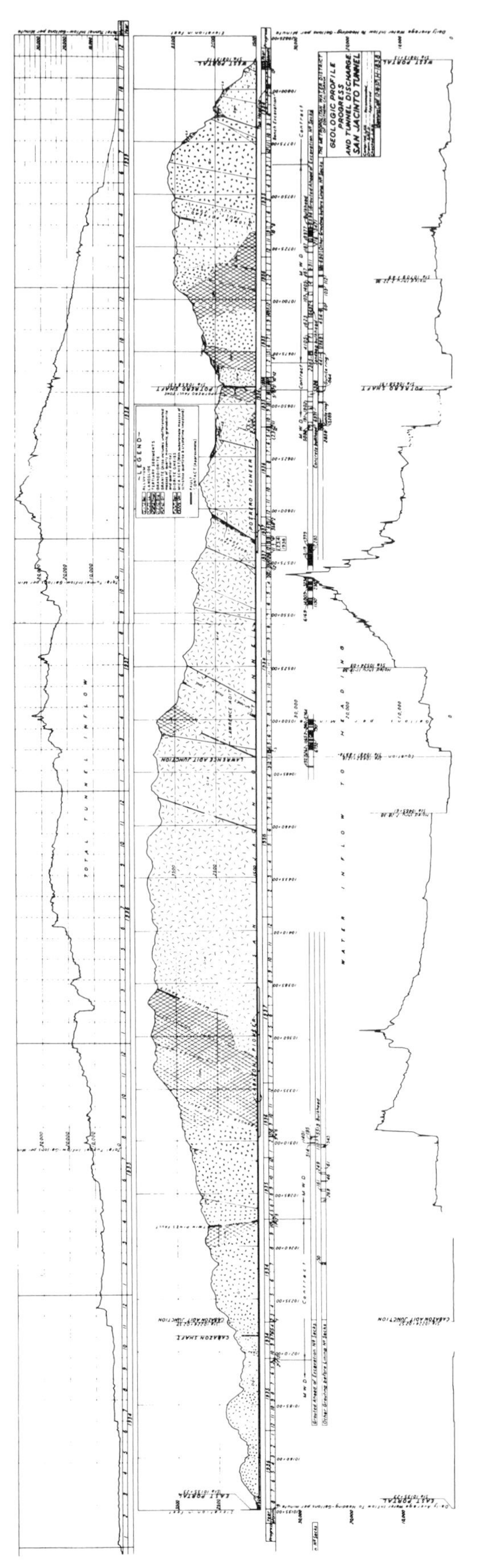

Figure 13. Geologic conditions along the San Jacinto Tunnel of the Colorado River aqueduct in southern California, driven 1934 to 1939. The abundance of strong fault zones throughout the alignment and the measured quantities of ground-water inflow associated with each feature are shown and correlated with the total tunnel inflow and construction progress (from Thompson, 1954).

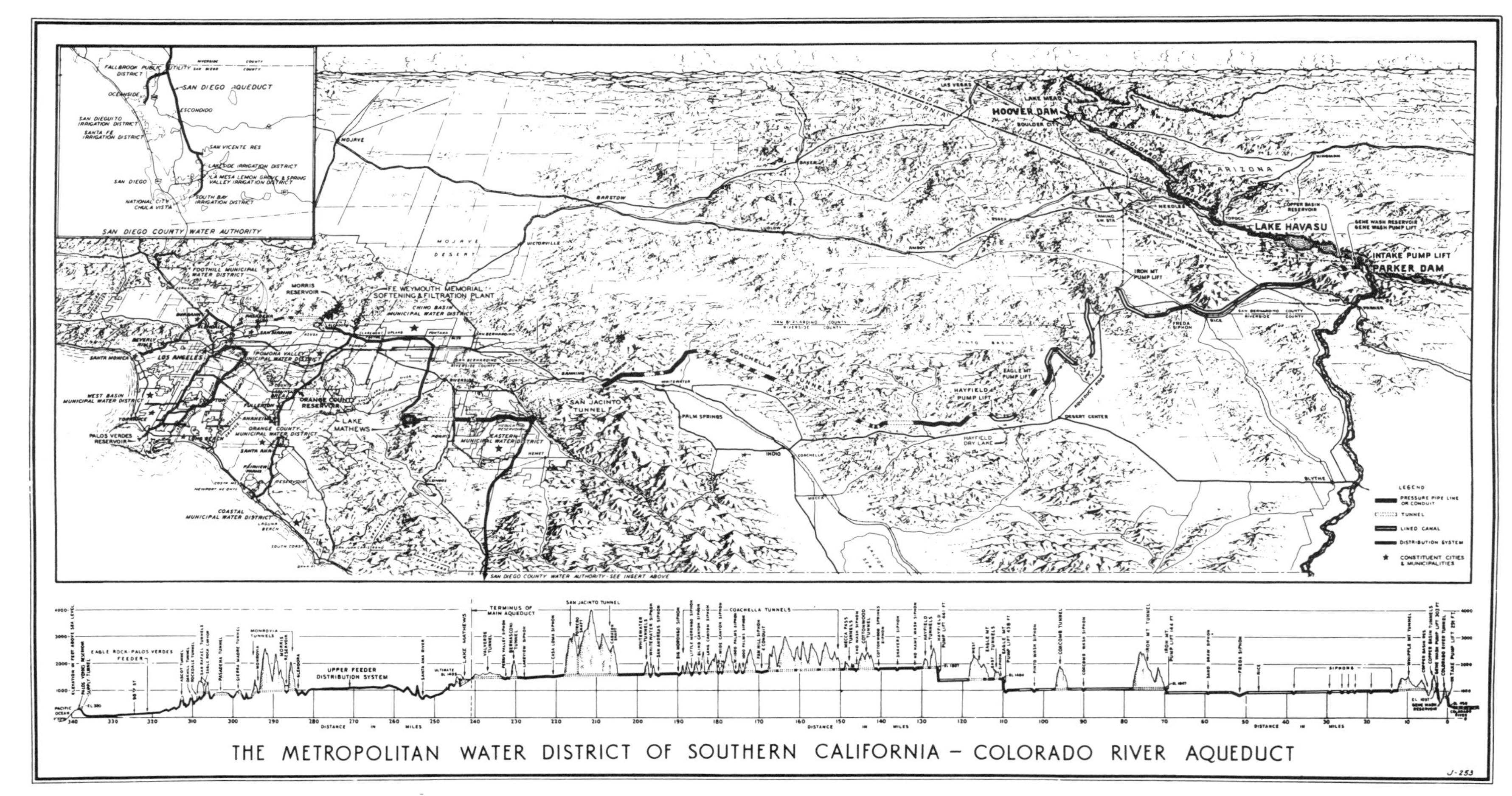

Figure 14. The Colorado River aqueduct system of the Metropolitan Water District of southern California. The plan map shows the main features and location of San Jacinto Tunnel (Fig. 13) and its regional geographic setting (from Thompson, 1954).

detailed study of springs and wells in the vicinity, if costly litigation is to be avoided. Before a major tunnel is constructed, periodic measurements of spring flow and the ground-water levels throughout the area should be made over a sufficiently long period to establish the regional base level from which any subsequent effects of a tunnel on the ground-water resources can be monitored and calculated. The flows encountered along the San Jacinto tunnel both locally and in total are plotted on the structural/geologic section (Fig. 13).

Tecolote Tunnel, Santa Barbara, California. High inflows of unpredicted hot water were encountered during construction of this tunnel, which delayed progress, requiring a renegotiation of the prime contract and adoption of more costly construction methods (USBR, 1959; Trefzger, 1966). Heading temperatures reached 47°C, and outflows at portal were over 1 m in depth (observed LBJ), with total inflow of 3,600 to 9,100 gpm maximum (Fig. 15). A recent inspection of the tunnel observed leakage of about 0.06 m^3 per second of hot (47°C), sulfate-bearing water that persists through the lining (Trefzger, personal communication, 1988). This experience further emphasizes the importance of a detailed, preconstruction investigation of all springs and water wells in the vicinity of a proposed alignment. A systematic record of measurements, both flow rates and water temperatures, may portend an abnormally wet or hot project. Consequently, a decision may be made to steepen the tunnel gradient or to redesign the ventilation system in order to improve working conditions. As learned from the San Jacinto Tunnel experience, a preconstruction inventory is critical to assessing any postconstruction claims for damages to regional ground-water resources.

FAILURES RELATED TO HUMAN ACTIVITIES

A common problem resulting from inadequate geological consideration during design or construction is the failure of excavations or slopes due to undercutting at the base, overloading of the upper parts, or by neglecting to provide suitable drainage. Other failures may be due to land subsidence resulting from withdrawal of ground water, petroleum, or geothermal fluids; ground-water pollution caused by indiscriminate disposal of industrial waste fluids or from sea-water intrusion exacerbated by pumping from carelessly located water-well fields; and damages wrought by reservoir-induced seismicity. Some examples follow that were strongly influenced by site alterations performed during construction, man's activities remote from the specific site, and human failings such as project mismanagement, inadequate supervision, technical errors, political intervention, and corruption.

Poor management practice and an ineffectively organized project team have been accused of contributing to more than one major project failure (Malpasset, Vaiont, and Teton Dams). Occasionally, these situations appear in most organizations in greater or less severity. The more common problems of this nature at the management level are:

- The responsibility for the project is neither centralized nor clearly defined. This situation occurs most frequently in large organizations where the home office, regional headquarters, and field operations are all involved in building the project. The result is inefficient use of available specialized talents, unhealthy intra-organizational rivalry, and most importantly, breakdown in communications between the designers and the field office performing the actual construction.
- Management yields to pressure, either proprietary or political, and accelerates the project dangerously.
- Management fails to seek or heed advice of consultants, as in the St. Francis and Teton cases discussed below.
- Insufficient funds are appropriated for site investigation, as in the Malpasset and Vaiont cases, discussed elsewhere in this chapter.
- Responsibility is centered in a strong-willed, unyielding project management that shortcuts investigations and pushes for project completion relentlessly, as in the St. Francis case discussed below.

Slope failures caused by excavations

Culebra Cut, Panama Canal. A historic case of miscalculating the effects of excavation on slope stability, overall cost, and scheduling of a major engineering project occurred at the Culebra Cut of the Panama Canal (MacDonald, 1915). The original estimate called for excavation of 17 million m^3 of slide material. Ultimately, 114 million m^3 had to be removed (Fig. 16). This resulted in not only an enormous cost overrun and delay in completion of the canal of nearly two years, but also in the temporary closing of the canal on seven occasions after it was initially opened in 1914 (Cameron, 1972).

This experience, early in the twentieth century, quickly and dramatically convinced many civil engineers of the critical importance of the geologic environs and rock features to constructing major works. Furthermore, the Panama Canal experiences were an early milestone toward developing specialists capable of interpreting geologic conditions and their potential effects on proposed engineering works, such as deep, open cuts.

Big Rock Mesa slide, Malibu Coast, Los Angeles. A large, creeping mass of rocks and finer sediments at Big Rock Mesa, was activated late in the summer of 1983 and threatened to dump 49 ha that overlooked the Pacific Coast Highway into the ocean (Schuster and Fleming, 1986). Progressive failure resulted in condemnation of 13 houses by March 1984 and threatened more than 300 additional homes valued at $400,000 to $1,000,000 each. Throughout 1984 a plethora of lawsuits related to the large-scale mass movement and slope failure was filed by property owners against Los Angeles County and a number of geotechnical consultants. The aggregate of legal claims against the county was reportedly in excess of $500 million (Los Angeles Times, March 3, 1984).

Hillside instability in southern California has resulted in many costly slope failures, such as at Portuguese Bend (described this chapter) and the Pacific Palisades (Fig. 17). Grading ordi-

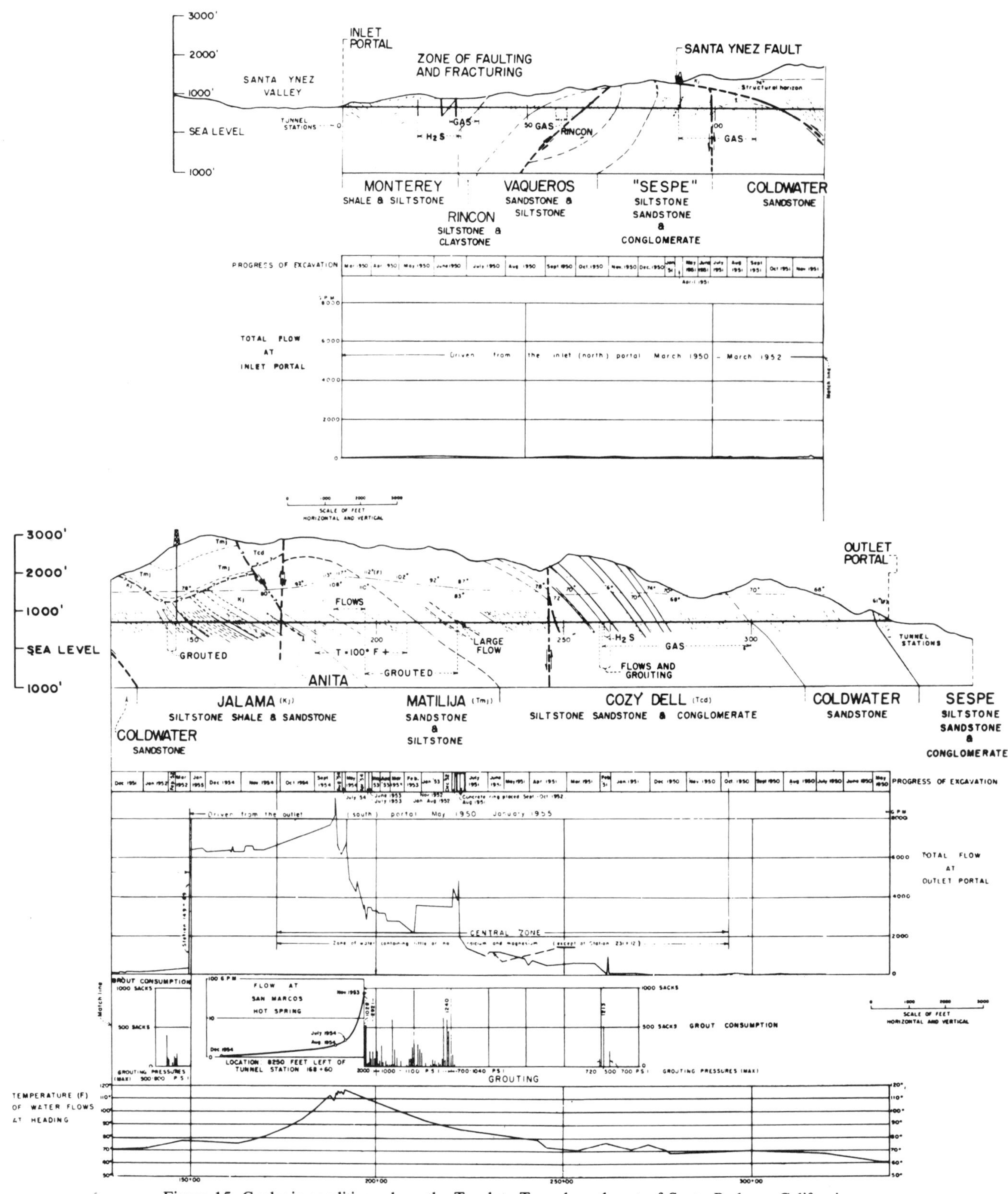

Figure 15. Geologic conditions along the Tecolote Tunnel northwest of Santa Barbara, California, driven 1950 to 1955. The Santa Ynez fault divides the tunnel into two major structural units as shown throughout the Inlet or North portion (above) and Outlet or South portion (below). The amounts and temperature of ground-water inflow encountered were moderate within the older rock units of Inlet portion. However, the quantities and temperatures increased greatly within the Outlet portion of young, fractured rock units. (Courtesy USBR/R. E. Trefzger, 1966.)

Figure 16. Culebra Cut–Panama Canal near Contractor's Hill on February 7, 1913. A post-excavation, active-slope collapse of soft Cucaracha beds put a total of 2,190,000 m^3 of material in motion. This slide is typical of the slope failures that delayed completion of the canal. (Courtesy David J. Varnes/U.S. Geological Survey.)

nances initiated in 1952 (Scullin, 1966) require geological and engineering investigations prior to issuance of a grading permit in all areas of hillside slopes throughout the Los Angeles region. The engineering geologists of southern California provided an important input and guidance in the technical and political processes involved to establish these ordinances. Unscrupulous builders have been known to attempt to influence the findings set forth in geological reports since enactment of the ordinances, and engineering geology consultants have been accused wrongly, on occasion, of an overly conservative assessment of the slopes (too treacherous for construction). Landslides and hillside development constraints are reviewed in detail by Leighton (1966) for the active terrain and environs of southern California.

Land subsidence

Picacho Dam, Arizona. The Picacho Dam, near Coolidge, Arizona, is one of the few cases where damages were sustained by an earthfill dam and embankment system as a result of ground-water withdrawal (James, 1988). Drawdown of water wells in the surrounding area lowered the ground-water levels and contributed to regional subsidence of the land surface, which caused the dam to likewise subside and crack (Fig. 18). With time, the dam embankment cracked and required expensive repairs (mid-1950s), but the dam was not breached. The reservoir occupies a natural surface depression enclosed by a circular earthen embankment with the dam on the south end. Subsidence activity is apparently due to several subsurface geologic factors, such as evaporite beds and/or an old sink-like feature, in addition to the fluid withdrawal.

Because of the ever-increasing development of ground-water resources throughout much of the world, it is incumbent on applied geoscientists to assess the potential for land subsidence when performing site-specific investigations for proposed large-scale projects or local works. An example of such a case with ultimate breaching of an earthen dam occurred in the Lerma Valley of Mexico, as shown on Figure 19. Progressive subsidence facilitated piping beneath the dam; eventually, it was fully breached, similar to the Baldwin Hills case described this chapter.

At Picacho Dam, the legal responsibility for inducing subsidence activity and causing damage to the dam was not contended or established. Interestingly, legal barriers may prohibit any restriction on pumping subsurface fluids, even though such doctrines are in conflict with other rights to claim damages (Amandes, 1984).

Damage from land subsidence caused by withdrawal of underground fluids has been costly in many parts of North America (Chapter 10), particularly in Texas, California, and Arizona.

Figure 17. Slope failure, the Flagg's restaurant slump-type slide of 1958, Pacific Palisades, near Malibu, Los Angeles. The slope failure covered U.S. Coastal Highway 101. Photo looks southeastward. (Courtesy California Division Highways/A. B. Cleaves.)

Damages in Texas are largely due to reactivation of faults and submergence, while in Arizona differential subsidence and fissuring are the main causes. The legal approaches to abating subsidence vary in the different stages; the landmark Friendswood Development Co. versus Smith-Southwest Industries, Inc. (1978) decision in Texas, conservation measures in California water districts, and minor provisions of the 1980 Arizona Ground Water Management Act are examples. The effort to control subsidence and related damage is in its infancy. Effective long-range planning is needed to arrive at corrective measures that equitably distribute risks and costs (Carpenter and Bradley, 1984).

The effects of subsidence on engineering works, the natural terrain, and society include the legal implications, which are reviewed by Teutsch (1979). Attention was first raised in 1925 when subsidence of 1 m accompanied development of the Goose Creek oil field in Texas (Chapter 10, Fig. 9) between 1918 and 1925 (Pratt and Johnson, 1926). This was the first recorded occurrence of subsidence caused by sediment compaction, and it triggered a unique bench mark decision in a court of law. Even more critical, however, the subsidence submerged low-lying parts of the oil field to below sea level, and the state sought title to the oil under the Texas submerged land law. Subsequently, the court ruled the subsidence was an act of man, and not a natural process, so all mineral rights were retained by the lease. This decision provided the incentive to determine the geologic causes of subsidence at Goose Creek and other fields (Chapter 10).

Ground-water pollution

The protection of ground-water resources from pollution caused by improper disposal of toxic wastes, both chemical and radioactive residues (man-induced), and from sea-water intrusion (natural) has become an important environmental issue, and such issues involve an increasing number of applied geologists. The magnitude of the ongoing problem is exemplified by separate experiences at the Aerojet General Corporation and McClellan Air Force Base, both located in Sacramento County, California. Both cases demonstrate convincingly that a well planned geological investigation of any proposed waste-disposal site is far less expensive than the cost of rehabilitating a polluted subsurface environment, with the inevitable litigation and damage awards. Moreover, the cases emphasize the need for a working knowledge of the principles and practical application of ground-water geology and hydrology as an important prerequisite for practice as a professional engineering geologist.

Aerojet General Corporation. The Aerojet General facilities are located on a 3,440-hectare tract situated about 24 km east of downtown Sacramento. The American River, a major source of water supply, flows westward less than 1.6 km north of the site (Fig. 20).

In 1979, law suits were filed against Aerojet General by the state of California and the federal government after discovery that disposal of hazardous wastes—including rocket propellants, herbicides, organic solvents, inorganic compounds, and sewage—had invaded the subsoils and sediments, surface water, and ground water, both on and off the company property. Aerojet General and governmental agencies have since detected over 100 contaminants (Consent Decree, U.S. District Court, Eastern District California, 1986) at over 250 locations.

Aerojet General began operations in 1951. During the early years they disposed of hazardous wastes by burial, open burning, discharge into unlined ponds, and injection into deep wells.

Aerojet General is now engaged in a massive effort to protect drinking-water supplies and prevent endangerment to public health and the environs. An intensive investigation is underway to locate and track the paths followed by the pollutants and to develop procedures for controlling the problem. Up to fall of 1988, Aerojet General had spent approximately $50 million to investigate the extent of the ground-water contamination and to control and treat ground water that would otherwise move off the site (EPA, 1988). Over 650 monitoring wells have been drilled and a system of extraction and treatment facilities installed. Aerojet General has agreed to reimburse the EPA and the state of California $3.4 million for civil and monetary claims and to pay up to $625,000 per year for the cost to the government for monitoring the activities. The remedial investigation involves data collection and analysis, including mapping, geohydrological studies, sampling, exploratory drilling, air monitoring, and com-

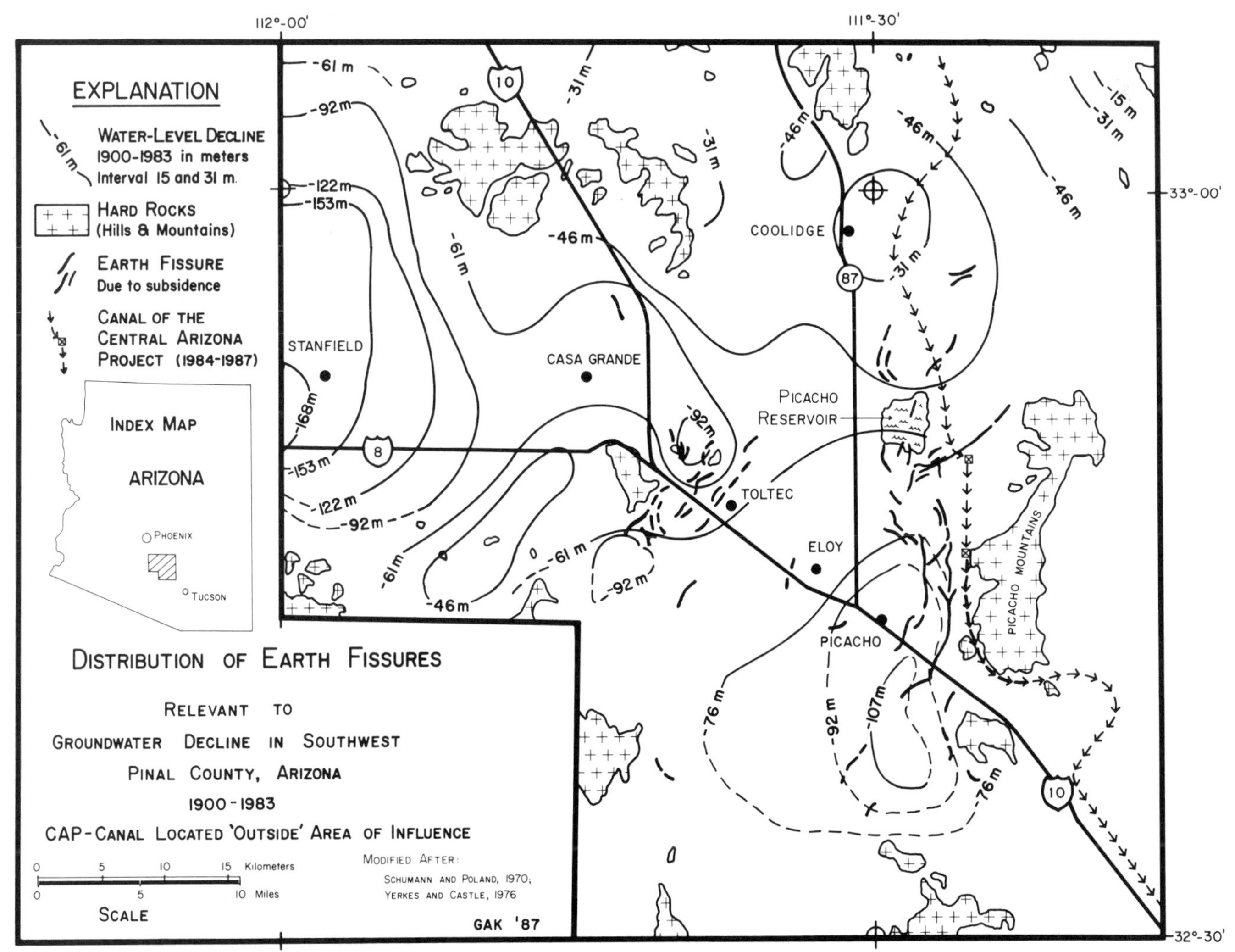

Figure 18. Plan map of area in the vicinity of Picacho Dam, south-central Arizona, showing the distribution of earth fissures and associated subsidence features. Decline of the ground-water level, 1900 to 1983, affected the stability of the dam and reservoir beginning in the early 1950s.

Figure 19A. Lerma Dam, located west of Mexico City in Lerma Valley. The dam failed due to subsurface action along an earth fissure beneath the embankment. (Courtesy T. Holzer, 1977.) B. The surface opening created by the piping-like action along an earth fissure beneath Lerma Dam in the background (see Figure 19A). The failure is somewhat like the cause and sequence of Baldwin Hills Reservoir, California. (Courtesy T. Holzer, 1977.)

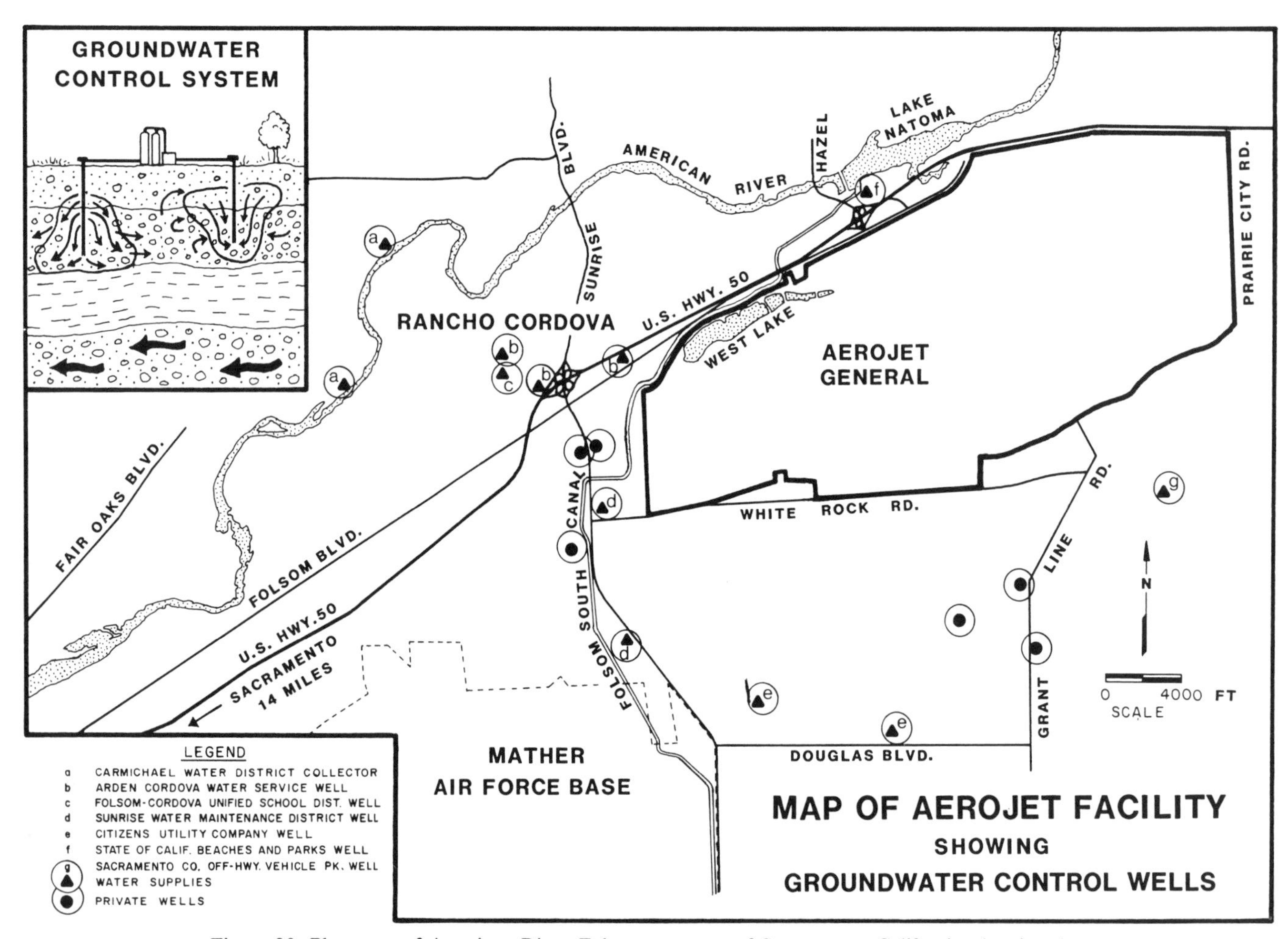

Figure 20. Plan map of American River–Folsom area east of Sacramento, California, showing the location of Aerojet General facilities and the sector affected by ground-water contamination.

puter modeling. The data will be analyzed in order to develop comprehensive clean-up alternatives and provide protection of the public health and the environment. Alternatives will then be evaluated on the basis of the degree of protection provided, technical feasibility, reliability, and costs. On completion of these studies, the appropriate cleanup remedy will be undertaken. This may be accomplished through new and separate legal agreements between governmental agencies and Aerojet General, or by governmental funding and enforcement action.

McClellan Air Force Base. Another typical case of large-scale pollution due to seepage of toxic wastes disposed in open pits has occurred at McClellan Air Force Base, Sacramento County, California. Pollutants buried in the 1950s have encroached on privately owned wells adjacent to the facility. An extensive investigation has involved detailed geological and hydrological studies. Residents affected filed claims against the federal government of one billion dollars for damage to the ground-water resources (Sacramento Union, April 24, 1986).

Sea-water intrusion, Los Angeles. Beginning in 1940s, costly water works and well fields beneath the coastal plain had to be abandoned, due to the intrusion of sea water induced by ground-water overdraft. This experience and the impact on regional resources stimulated the participation of engineering geologists in the planning of ground-water resources development and protection in coastal areas. This further led to the location and operation of a network of recharge wells to restrict sea-water encroachment and, with time, drive out the saline waters and cleanse the ground-water resources (Zielbauer, 1966).

Historically, sea-water intrusion was recognized initially in Los Angeles County, California, in 1912, at Redondo Beach, in 1944 along the coast of Orange County, and in 1951 at Port Hueneme and Point Mugu of Ventura County and the San Luis Rey Canyon area of San Diego County. Subsequently in all cases, the intrusion fronts spread along the coastal margins and steadily advanced inland (CDWR, 1957). Today, successful sea-water barrier projects are formed by operating well-injection fields that protect and recharge the ground-water basins (Fig. 21).

FAILURES RELATED TO A COMBINATION OF GEOLOGIC AND HUMAN ACTIVITIES

At times, geology-related failures are initiated unintentionally when humans undertake activities that disturb the balance of natural forces. Often such accidents are triggered where condi-

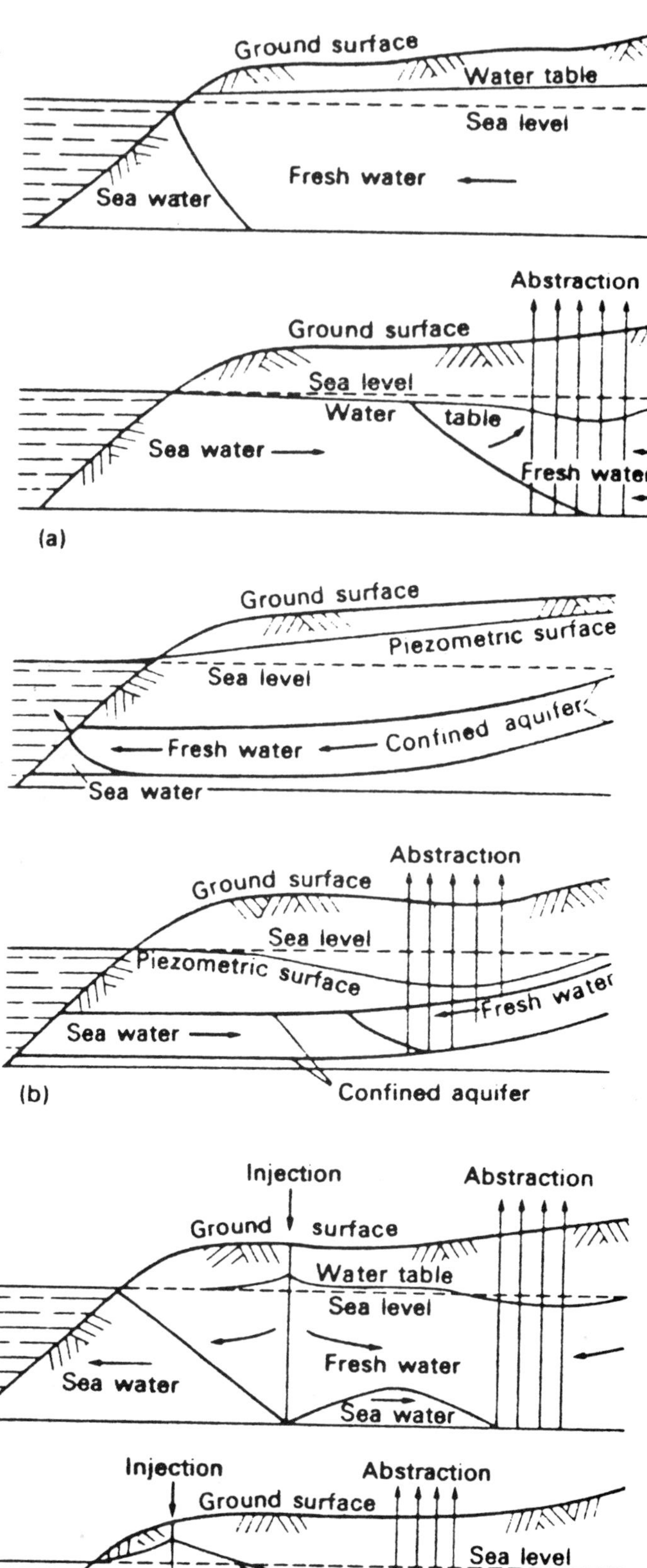

Figure 21. Stage diagram of sea-water intrusion, southern California, showing initial intrusion from overpumping and remedial techniques. The relation between fresh ground water and sea water in a coastal aquifer is shown as: (a) an unconfined aquifer under natural conditions and subject to saline intrusion; (b) a confined aquifer in continuity with the sea under natural conditions and subject to saline intrusion; and (c) the control of saline intrusion with a fresh-water ridge in an unconfined and confined aquifer. (Courtesy California Department of Water Resources/Downing, 1978.)

Figure 22A. Aerial view of Portuguese Bend area looking north on September 3, 1947, prior to reactivation of Pleistocene and historic slides. The shoreline is in the foreground, with several levels of prominent terraces and the major erosional features of the landslide complex. (Courtesy Spence Aerial Surveys, and Ehlig, 1987.)

tions in the natural state are already precarious and man's influence simply hastens an event that probably would have occurred eventually without intervention. For example, seismic activity induced by filling of reservoirs occurs in regions where existing faults have been stressed previously to the verge of displacement. Landslides activated by overloading or undercutting slopes frequently happen where the original stability was questionable. Thus, many of the accidents attributed to man's activities are in a strict sense the result of imposition of conditions upon an existing hazardous setting. The identification of potential hazards and assessment of the effects of proposed projects on them is the responsibility of the applied geoscientist working with planners, designers, and engineering teams.

Selected cases are briefly described in which failure or significant geology-related problems resulted from the impact of construction activities on an unfavorable geologic environment or from the effects of site defects aggravated by management-related deficiencies.

Induced landslides, subsidence, fault movement

Portuguese Bend slide, Los Angeles. One hundred thirty-four residential houses had to be abandoned by 1984, and several roads and utilities were damaged by the Portuguese Bend slides that were initiated on a small scale in 1929, with subsequent recurrences (Merriam, 1960; and Reiter, 1984). The area of slope failure and creep occupies about 109 ha on the slopes of the Palos Verdes Hills, situated west of Los Angeles, overlooking the Pacific Ocean (Figs. 22A, B, and 23).

The slides since 1929 are the most recent activity within a complex, unstable slope mass of several ancient landslides activated sporadically over the past million years (Merriam, 1960) and shown on Figure 24. A recent episode of serious movements at Portuguese Bend commended during the summer of 1956 and involved about 60 million tons of rock and soil material. Property owners affected were awarded $9,500,000 in property damages on the grounds that road construction and maintenance in the area by Los Angeles County were largely responsible (error in judgment) for reactivating the 1956 movements throughout the steep ocean-front slope (Reiter, 1984). Total horizontal displacements occurring throughout the Portuguese Bend and adjoining Abalone Cove landslides from 1956 to 1986 are shown on Figure 25.

The heavy damages sustained in 1956 prompted extensive geological exploration of the Portuguese Bend slide with borings,

Figure 22B. Aerial view of Portuguese Bend area looking north on September 26, 1956, one month after the Portuguese Bend landslide complex was reactivated (correlate with Figure 22A). Note Crenshaw Boulevard under construction on right side of photo. Slide movement was first observed in road fill at right center of photo; the white scar on the hill northeast of the fill is the landslide caused by road excavation. (Courtesy Whitehouse Aerial Surveys, and Ehlig, 1987.)

tests, and maps. Ten separate consulting firms prepared reports on the causes of slope failures and recommended corrective measures. In addition to a reactivation of the slides by road building activities, a rise of ground-water levels throughout the area (both natural and man-induced) contributed to the instability of the slopes.

An early attempt was made to prevent further slope movement by installing precast concrete pins in 1957 (Reiter, 1984). These extended through the slide material and 3 m into the underlying bedrock. Each pin measured 1.2 m in diameter and 6 m in length. Initially this procedure apparently decreased the velocity of the slide by 50 percent. However, after about five months, movement accelerated and soon approached the average rate of 1956; the concrete pins had failed, toppled, or been bypassed.

At present, a carefully planned project is underway to stabilize the slide over a period of several years (Ehlig, 1982, 1987). Phase 1 (now completed) entailed the installation and operation of eight dewatering wells and movement of 436 m^3 of earth to improve surface drainage and increase slope stability. Culverts were also installed to convey water from the head of the slide to the ocean. These efforts almost stopped movement at the head of the slide and reduced movement within the eastern 24 ha to less than 15 cm per year. The velocity of the remainder of the slide has been reduced to less than half of its former rate of 12 m per year. Phase 2 includes road relocations, improvement of drainage and stability by regrading the seaward part of the slide, and the operation of five additional dewatering wells, all scheduled for completion by early 1988. By that time, expenditures on the project are expected to total about 1 million dollars. Future phase 3 plans include construction of a revetment to protect the toe of the slide from wave erosion, further improvement of surface drainage, and installation of a sewer to service residents in the slide area. Cost estimates for this third phase await completion of the revetment design for preventing wave erosion. Phase 1 and 2 efforts are expected to reduce costs for maintaining roads and utilities in the area by about $250,000 per year (Ehlig, 1987).

The intensive investigation of the Portuguese Bend landslide has had positive results for applied geologists, possibly as great as any other major landslide event. The mechanics of the slide have been studied in meticulous detail and are probably as well understood as any comparable slide in the world. Several important precedents have been established and are described in the follow-

Figure 23. Aerial view of Portuguese Bend area on December 17, 1986, when it was near completion of Phase I grading and remedial treatment. Note the culvert installed, from the two canyons on either side of the slide, which extends to the ocean. (Courtesy Ace Aerial Photography, and Ehlig, 1987.)

ing slightly edited excerpt from a letter by P. F. Ehlig to the authors (August 21, 1987):

The Abalone Cove landslide is a westward continuation of the prehistoric Holocene Protuguese Bend landslide reactivated in 1978 (Ehlig and Bean, 1982). The first precedent was the legal decision that the County of Los Angeles must compensate property owners for having caused the slide; a decision which led to engineering geology requirements by both the County and City of Los Angeles. After the Abalone Cove landslide started moving in 1978, citizens within Abalone Cove established the Abalone Cove Homeowners Protective Association (ACHPA) and promptly raised funds to start dewatering activities (not a precedent). In the meantime, the City of Rancho Palos Verdes sought and got enabling legislation which led to the establishment of the Abalone Cove Landslide Abatement District (ACLAD), the first such district in California. Since it was organized, ACLAD has operated very smoothly, and its policies have stopped the slide at a relatively small cost to the benefiting property owners. The Klondike Canyon Geologic Hazard Abatement District, adjoining the Portuguese Bend landslide, is the second such district in California. It was formed to stop the Klondike Canyon landslide. The third precedent came in 1984 when the City of Rancho Palos Verdes established the City Redevelopment Agency, covering the two-square-mile area of the ancient Portuguese Bend landslide complex. The agency was organized to stop the Portuguese Bend landslide and take mitigating actions as needed to improve the stability of the area so that at some future time the existing building moratorium could be lifted. The City obtained a grant of $2,000,000 from the State (tideland oil revenues) to help stabilize the Portuguese Bend landslide. About 100 acres of the slide has nearly stopped moving (1987). Present maximum rate of movement is less than 1 foot per year.

The other 165 acres have slowed greatly during 1986–87 and I expect all of the slide landward of Palos Verdes Drive South to be nearly stable when Phase 2 grading is completed. However, complete stabiliza-

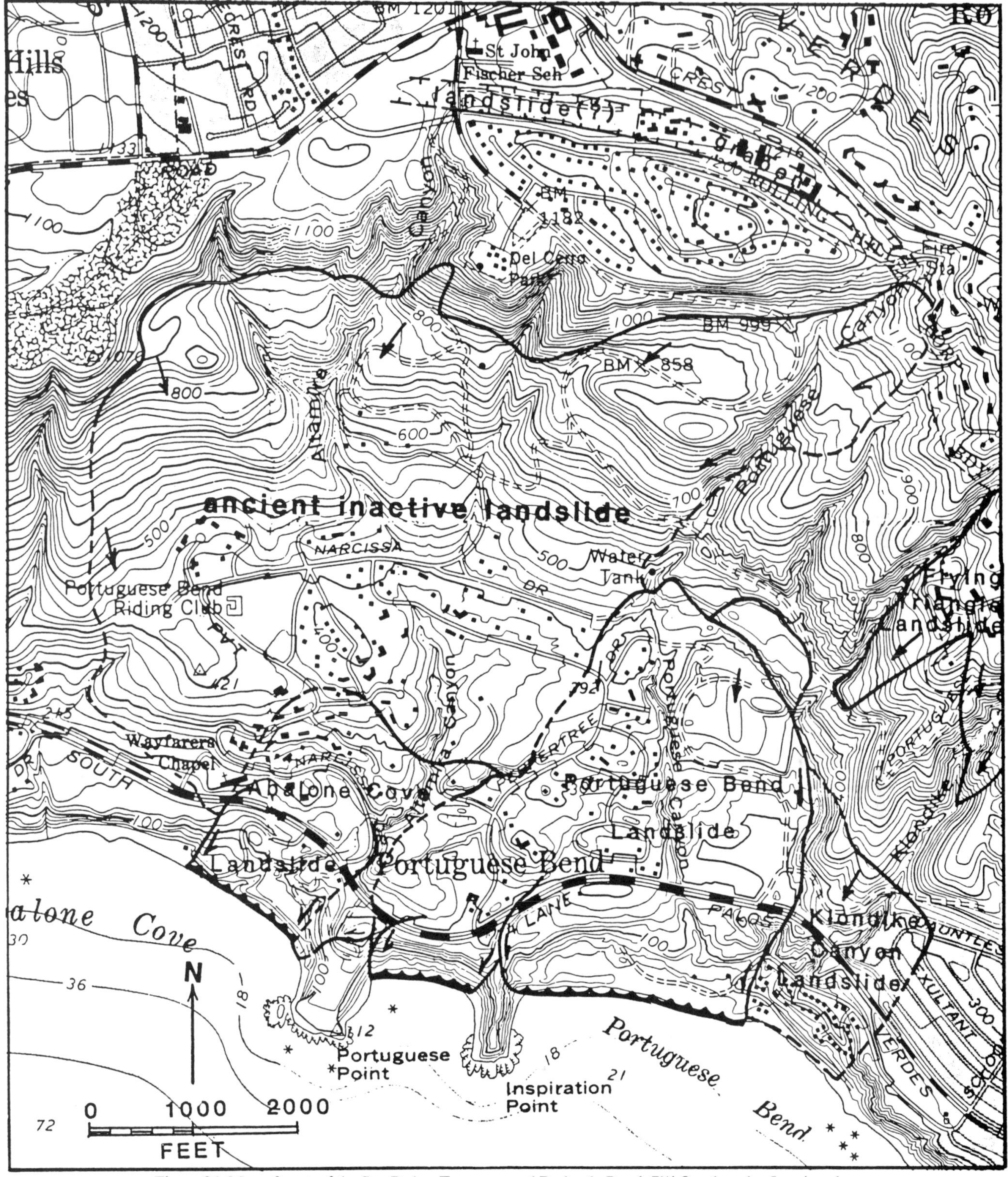

Figure 24. Map of parts of the San Pedro, Torrance, and Redondo Beach 7½′ Quadrangles, Los Angeles County, California, showing locations of the Portuguese Bend, Abalone Cove, and Flying Triangle landslides and the adjoining ancient inactive landslide area. The active Flying Triangle landslide (right center) is not connected to the others. Slide movement has greatly modified the topography in the Portuguese Bend landslide area from that shown on this early base map. (From Ehlig, 1987.)

tion of the coastal area will probably not be achieved until a revetment is constructed to prevent wave erosion (Phase 3).

Yet another apparent precedent is the terms of the settlement from the Abalone Cove landslide litigation (Horan and others vs, County of Los Angeles and others, 1987). The lawsuit specifically requested money to mitigate the slide, and a technical panel was established that made recommendations for improving the stability of the area. The case was settled in July 1987 without going to trial. Although full details were not released, it is known that some $5,000,000 will be available to improve stability of the long-time slide area; the priorities and specifications for all work will be established by a technical panel.

The obvious lesson, reemphasized by the Portuguese Bend experience, is that any investigation of large-scale slope instability is usually complex and requires the participation of field-experienced geologists and engineers acting as a team. Stabilization of large-scale, ancient slide areas that are reactivated can be both difficult and expensive. Areal geological investigations as part of the site-selection process for planning construction work is obviously less costly and beneficial than a postconstruction sequence of events, litigation, and property losses, as demonstrated at Portuguese Bend.

Baldwin Hills Reservoir, Los Angeles. The failure of Baldwin Hills Reservoir resulted in five fatalities and property damages estimated at $15,000,000 (Fig. 26). The rolled-fill earth dam, 47 m high, was constructed near the head of a ravine carved from loose to moderately consolidated, young marine sediments. The completed project had a storage capacity of 1,106,001 m^3 and a water surface area of 7.92 ha. The following account is after CDWR (1964), James (1968), Leps (1972), and Leonard (1987).

The site is cut by minor faults associated with the seismically active Newport-Inglewood uplift, source of the catastrophic Long Beach earthquake of 1933; the region has been affected by moderate-scale warping and uplift during the late Pleistocene. Periodic levelings conducted since 1917 throughout the project

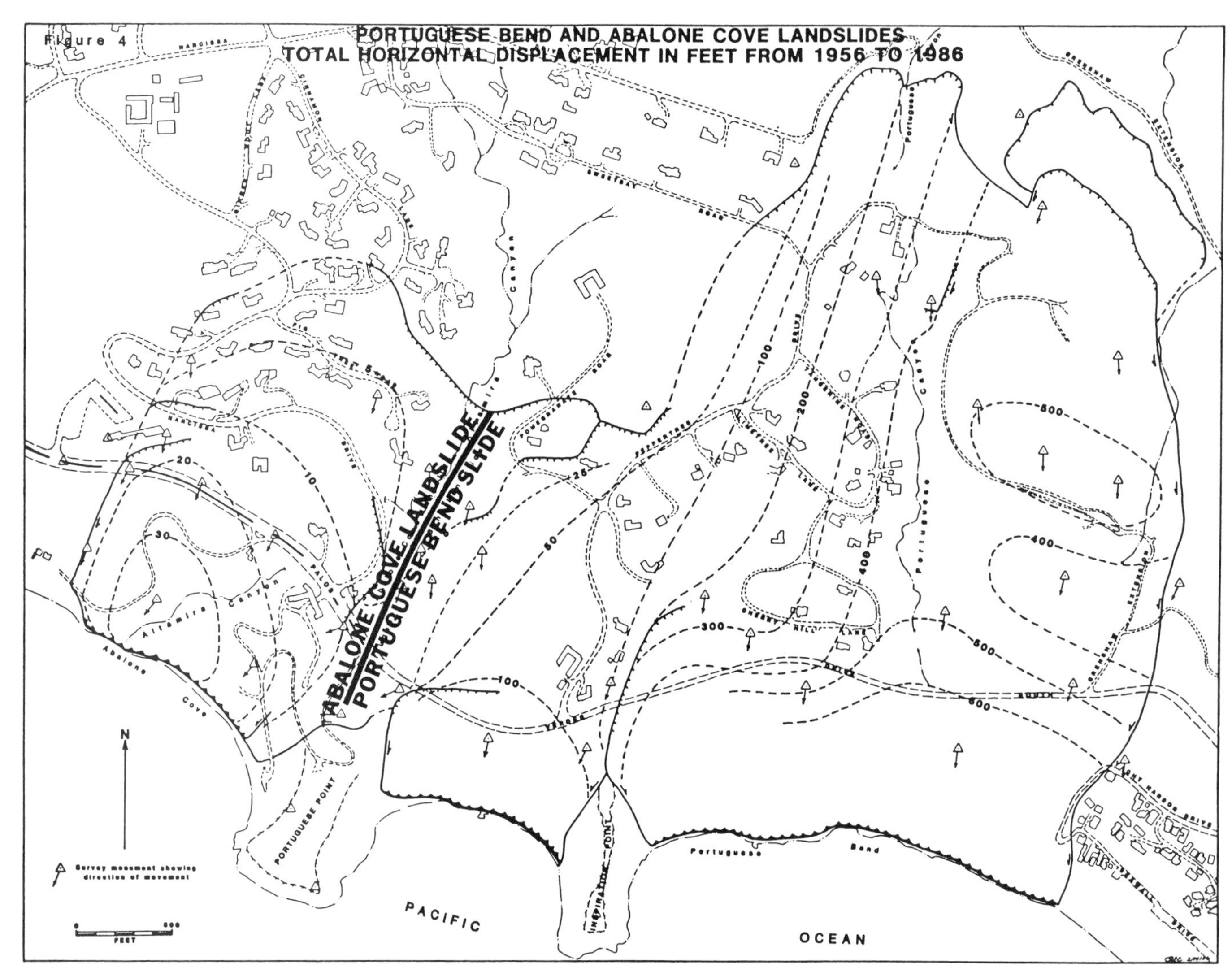

Figure 25. Plan map of Portuguese Bend landslide area. The limits of the Abalone Cove and Portuguese Bend slides are shown, along with the total horizontal displacement (feet) and direction of movement of each feature during the period 1956 to 1986. (From Ehlig, 1987).

Figure 26. A post-failure aerial view of the Baldwin Hills Reservoir, California. The breach through the dam embankment (lower left) follows a fault disclosed during construction. Failure was attributed to subsidence caused by production from the nearby Inglewood oil field that flanks the reservoir on the south and west. The earth fissure that developed along the fault affected the subsurface stability and provided a pathway for piping and eventual collapse. (Courtesy California Department of Water Resources, 1964.)

area disclosed that the slopes, including those at the reservoir location, were subsiding. The maximum subsidence of about 2.7 m occurred 800 m west of the site. Earth cracks up to 760 m in length were detected immediately adjoining the reservoir on the southeast. The subsidence and consequent cracks are mainly attributed to withdrawal of fluids and gas throughout the producing Inglewood oil field, which adjoins the Baldwin Hills Reservoir on the west and south, in an area traversed by tectonic features.

The disaster is specifically attributed to an earth crack, which opened along a minor fault initially mapped by geologists during construction of the project in 1948. This structural feature split the reservoir floor and the dam; displacement due to subsidence created a narrow passageway through the erodable foundation sediments that was rapidly enlarged by the escaping water (Fig. 27A and B). Consequently, in the final stages of failure, Baldwin Hills reservoir emptied through a spectacular breach in the dam and ultimately gullied the foundation itself (Fig. 27C).

The experience at Baldwin Hills Reservoir served as a reminder that investigations for engineering works must extend beyond the foundation and site for the proposed structures to the surrounding area and region (Fig. 28). Furthermore, feasibility and design investigations must include a consideration of the effects of both current and future human activities.

Vaiont Reservoir disaster. One of the worst dam disasters in history occurred in 1963 at the Vaiont Dam, Italy. This tragedy, caused by a tremendous rock/soil slide into the reservoir, resulting in a massive downstream flood, is a benchmark case of both inadequate regional/areal geological investigations and a misconception of the site-specific geologic features relevant to the engineered works. Geologically, the case involved judgmental errors, in part due to insufficient field data needed for adequate evaluation of suspected hazards of the area and site (Fig. 29). This brief summary on the geologic environs, causes of sliding, and engineering implications of the Vaiont disaster is largely after papers by Kiersch (1964, 1965), who investigated immediately after the catastrophe, supplemented by the 1978 to 1982 reevaluation of field and laboratory data by Hendron and Patton (1985).

The findings of these investigators further emphasized the necessity for mature, field-experienced geological input into the decision-making process for major dam-reservoir projects. The massive slope collapse occurred along a preexisting old slide plane and/or low-angle fault zone. This structural zone was not recognized, according to sketchy exploration data available to the designers in 1960 to 1961.

Several critical areal structural features and geologic parameters caused the slope failure at Vaiont:

1. Occurrence of clay interbeds, seams, and lenses throughout the Cretaceous Malm beds and widespread association of clays with the zone of failure (Figs. 30 and 31).

2. Occurrence of solution-widened openings in some limestone beds of the slide mass and the underlying limestone units that crop out upslope as "karstic terrain" (Fig. 31).

3. Ground-water levels induced by reservoir water increased the buoyancy effect/uplift pressures within the slide mass and contributed to instability. The clay interbeds and claystone enhanced the potential for local uplift pressures and contributed to instability; a multilayer artesian system was created.

4. The suspected old preexisting slide plane and associated conditions were not adequately investigated in 1960 to 1961; the significance of low- or no-core recovery was misunderstood, and the subsequent geological decisions did not reflect a serious concern. For example, the adits of 1961 crossed a strongly sheared zone, which postslide exploration determined to represent previous slide planes.

5. A misunderstanding of the geologic environs influenced yet another critical decision: that stabilization of the slide mass by drainage was not practical. Actually, in-depth field studies would have indicated that the heavy surface runoff from the slopes of

Figure 27. The progressive collapse of Baldwin Hills Reservoir December 14, 1963, in three stages. (Courtesy Los Angeles Herald-Examiner; From L. B. James, 1968.) A. Hole in main dam due to piping at 2:20 p.m. on December 14. B. Hole in main dam greatly enlarged at 3:30 p.m. C. Collapse of hole and breaching of dam at 3:38 p.m.

Mt. Toc largely disappeared upslope of the old slide plane in "karstic terrain" (Fig. 31). Consequently the interconnected fractures and openings within the limestones would have responded to drainage adits and galleries for controlling the ground-water level throughout the slope above lake level, and likely the reservoir bank could have been stabilized and disaster averted.

The lack of mature geological guidance during the planning-design phase of Vaiont project was an issue raised during the litigation proceedings of 1969 to establish responsibility for the disaster. The Italian Courts accepted the testimony of Nino A. Biadene, Director of Construction for Owner (SADE), that inadequate and misleading geological information was a major factor in causing the tragedy.

Vaiont, like Malpasset in 1959 (described this chapter), reflects the frequent European attitude of that date: wherever possible, restrict geological exploration to a minimum for a dam-reservoir project, in spite of on-site geologic circumstances. Furthermore, Vaiont demonstrates the long-standing attitude of some engineers to accept the opinion of a well-known professional or individual as adequate in lieu of the "team approach" accompanied by overlapping data as practiced widely and discussed in Chapters 18 thru 21 (this volume) and in Kiersch (1988).

Foundation defects combined with management deficiencies

St. Francis Dam, California. The erodible foundation upon which the St. Francis Dam rested (Fig. 32) was unsuitable for a concrete gravity structure (Wiley, 1928; Ransome, 1928). Had an adequate program of site exploration and laboratory tests been conducted, the fatal defects would have been realized during planning (Frontispiece of this section). Project management was highly centralized and accustomed to rendering unilateral decisions, and unyielding with respect to meeting construction sched-

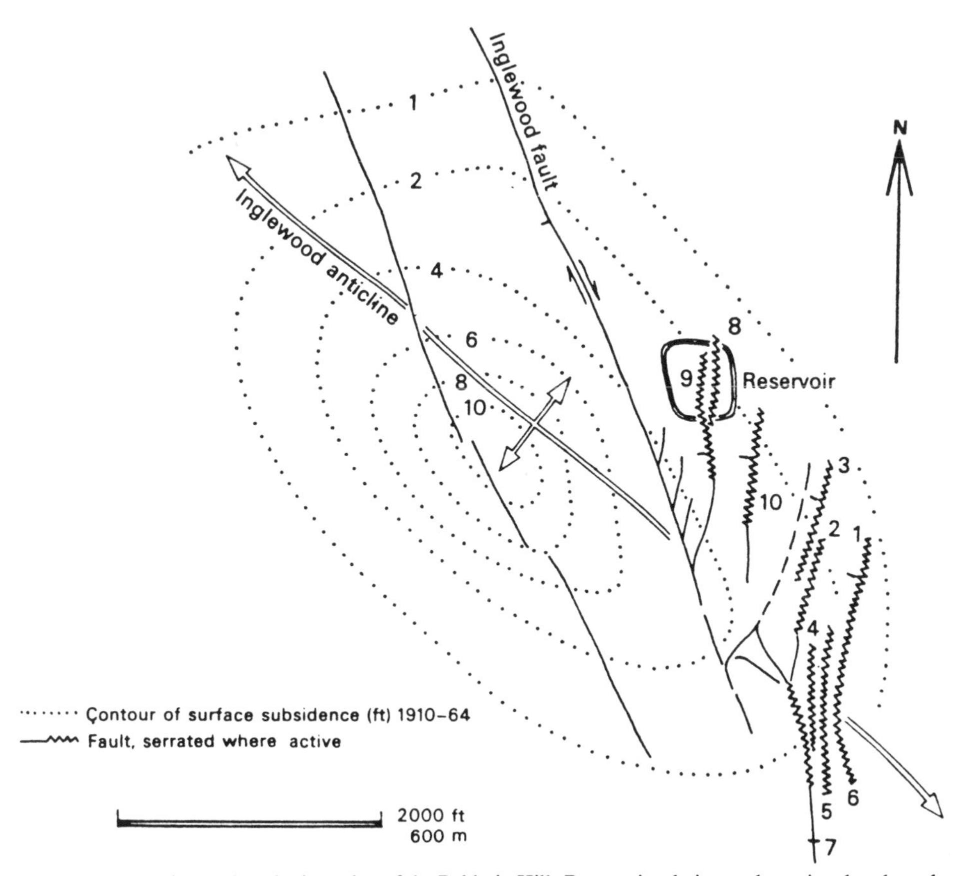

Figure 28. Location and geologic setting of the Baldwin Hills Reservoir relative to the regional and areal structural features: Inglewood anticline; Inglewood strike-slip fault; and the pattern of branching tensional faults activated by recent movement due to subsidence of the region. Two of the activated faults traverse the reservoir floor and contributed to piping and failure of the dam (De Freitas, 1978).

ules. Furthermore, the chief engineer for the project pursued geology as a hobby (Outland, 1977). A decade before the application for land use of the site was made to the federal government, the chief engineer had excavated shafts and adits into the red conglomerate (Fig. 33) to determine the physical characteristics of the rock spur that formed the right abutment. The results convinced him that the rock spur would provide a satisfactory abutment—if the need arose (Outland, 1977, p. 34). Selection of the St. Francis site was further aggravated and influenced by political efforts to speed completion of the Owens Valley–to–Los Angeles Aqueduct. The dam was completed May 4, 1926. Obviously, less than adequate attention was given to investigating and evaluating the site. For example, by March 5, 1928, with the first full reservoir, some seepage was noted through the conglomerate beds of the right abutment. Within days, the leekage softened the conglomerate foundation, and failure occurred on March 12, almost instantaneously.

As a consequence of this 1928 catastrophe, the State of California established the office of Supervision of Safety of Dams in 1929 within the State Division of Water Resources. The agency was assigned responsibility for approval of all proposals to construct dams of significant height and impoundment and also for the periodic monitoring of dams for safety. Today the division's staff of 62 includes four full-time engineering geologists, with jurisdiction over more than 1,100 nonfederal dams in California.

Teton Dam, Idaho. Another case of a foundation failure that was aggravated by organizational deficiencies is the Teton Dam (Fig. 34A, B, C, D). The June 5, 1976, breaching of the earthfill dam has been widely attributed to the absence of adequate remedial treatment of a defective foundation. On-site geological staff had accurately identified and mapped the foundation defects, but disastrously, the significance of the inherent problem—open-jointed foundation rock—was not adequately recognized at the organizational headquarters level in Denver. Postfailure inquiries suggest management-related deficiencies that included inadequate communication between headquarters and field-office staffs, less than desirable coordination between the design and construction staffs, and the nonexistence of an independent, technical review group or board. Reportedly, some of the headquarters decision-making personnel involved never visited the project site during construction; and furthermore, only a few

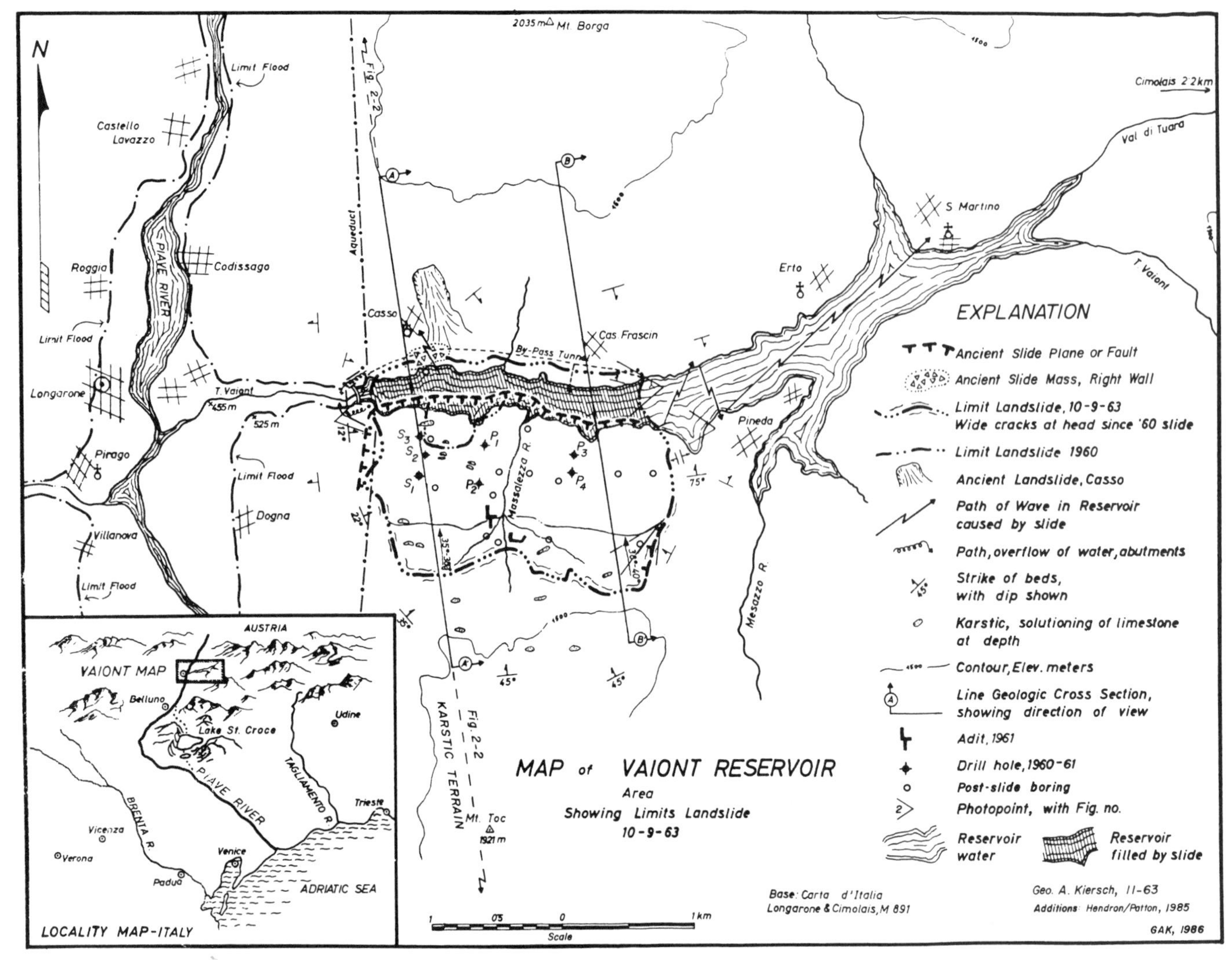

Figure 29. Areal geologic map of the Vaiont Reservoir, Italy, showing limits of the landslide, principal features, and location of some pre- and post-collapse subsurface exploration. (From Kiersch, 1964, 1988.)

observed the foundation when it was sufficiently exposed to evaluate the potential difficulties. At Teton, there was considerable dependence placed on field reports by mail, and much of the focus of those reports was on contract administration, rather than the site conditions and geologic aspects (Chadwick, 1976). Fortunately, such internal deficiencies with this dam-building agency (Bureau of Reclamation) have been resolved as a result of the Teton experience (Leonard, 1987).

Technically, the Teton incidence focused attention on the necessity for incorporating backup defensive construction measures to eliminate any possibility of piping with time, whenever embankment dams are constructed on open-jointed or cavernous foundation rocks (Fig. 35) (Chadwick, 1976). Furthermore, Teton should remind all applied geologists dealing with engineering works to be on the alert for deficiencies in project management, particularly any oversights that could adversely affect the acceptance or application of geological findings. When necessary, the applied geologist should respond to management in accordance with the principles of ethical professional behavior (see Chapter 24).

CONCLUSION AND OUTLOOK

Over the past 100 years, the safety and quality of engineering works have been enhanced through continued improvement in the application of geology to the planning, design, and construction stages. During the early years, few geologists possessed significant familiarity with the engineering fundamentals, construction practices, and economics involved with real-life projects. The early geologists tended to regard these factors as unrelated to their professional role. As a consequence, geologists before the 1930s often worked independently of engineers and produced

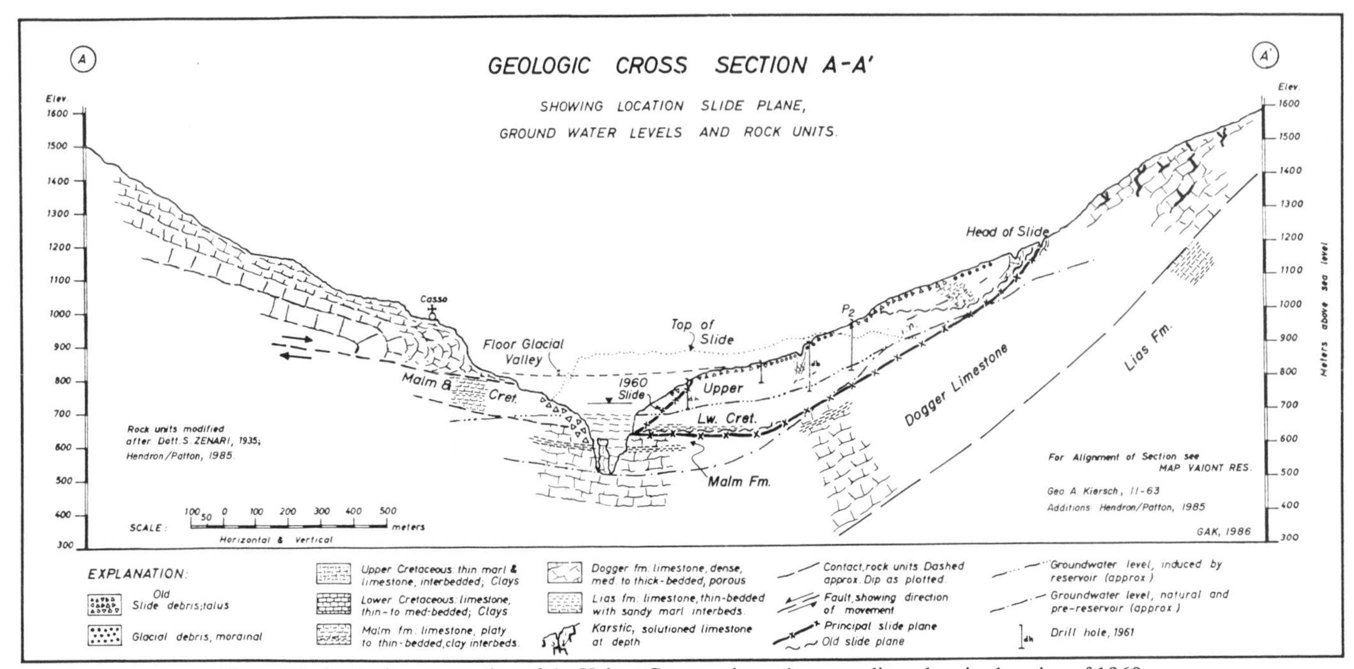

Figure 30. Geologic cross section of the Vaiont Canyon site and surroundings showing location of 1960 and 1963 slide planes, ground-water levels, and rock units. (From Kiersch, 1964, 1988.)

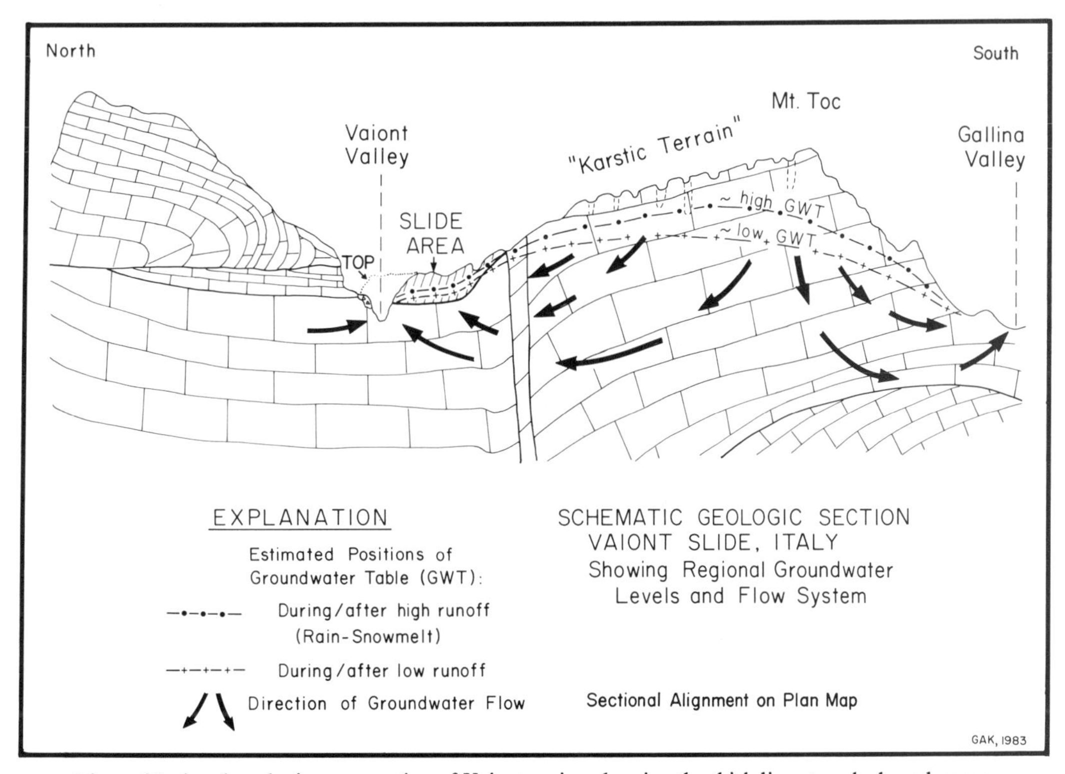

Figure 31. Areal geologic cross section of Vaiont region showing the thick limestone beds and occurrence of associated solution features. (From Kiersch, 1988.)

Figure 32. St. Francis Dam, California, looking upstream during the initial filling of the reservoir; failure occurred at 11:58 p.m. on March 12, 1928. The west or right abutment of the dam was founded upon a reddish conglomerate (Fig. 33); when dry the material was of decidedly inferior strength, and when wet the conglomerate became so soft that most rock characteristics were lost. (From files, California Department of Water Resources, 1928.)

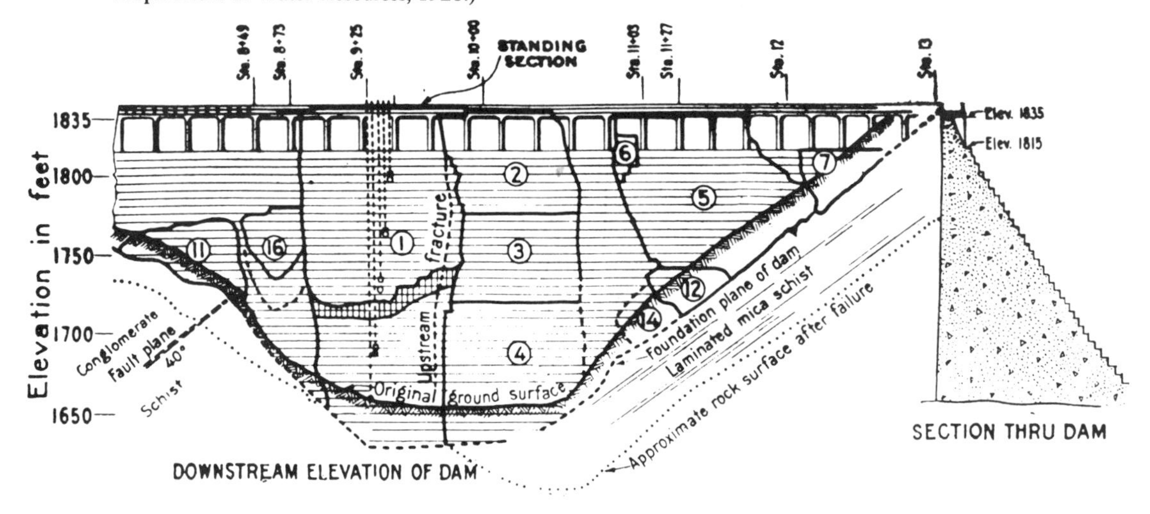

Figure 33. Geologic cross section of St. Francis Dam site, showing location of rock units and the weak reddish conglomerate, structural features of the foundation, and post-collapse outline of the canyon wall. Blocks of the concrete dam identified downstream after failure are shown by number and original position on section. (After Wiley, 1928.)

Figure 34. A. Teton Dam, Idaho, showing leakage through the open foundation along joints of right abutment groin, where gulley is being rapidly eroded on downstream face at 10:35 a.m. The dam crest breached about one hour and ten minutes after this photo was taken. (Courtesy Chadwick and others, 1976). B. Photo of dam face and leakage at 10:45 a.m. on June 5, 1976. C. Photo of dam face and major breach in progress after collapse of dam crest at 12:10 p.m. D. Photo of dam after failure and full reservoir bypass in progress, early afternoon. (Courtesy Department of Earth Sciences, Notre Dame University, 1976.)

academic-oriented reports that contained information either irrelevant or too technical for the engineer's needs. Because such reports were disregarded as impractical by the practicing engineer and staff, this unfortunate noncommunicative situation contributed to the failures of works—sometimes catastrophic. It is likely such early failures could have been avoided had the critical geologic facts on a specific site been effectively and understandably provided to the project managers and designers.

Early in the twentieth century, the need to utilize applied geology for engineering works became apparent to both geologists and engineers, and at the same time was recognized to be critical to the success of a major project by the principal federal agencies engaged in construction. The development of personnel with the training and experience required to fill this gap was undertaken with assistance from academic and professional organizations. The caliber and scope of modern engineering geology—applied geoscience for engineering works—has developed as a result of these long-time efforts.

The mature engineering geologist—an applied geoscientist with a general knowledge of engineering fundamentals and construction practices—is responsible for understanding the broad needs of a project builder, and for effectively communicating his relevant geological findings and ideas. Unfortunately, geologists occasionally serve individuals who are reluctant to work with the data assembled unless it is accurate to four decimal places, a difficult and unrealistic attitude. To overcome such roadblocks, a good personality and an ability to communicate convincingly are major assets for a successful practice.

Besides qualifications and field experience on projects, an understanding of the causes for geology-related failures of engineering works remains basic to a well-rounded, engineering geology professional. Any investigation for a proposed engineering works requires the detection and assessment of the site-related rock and soil defects and a comparison of them with the natural conditions known to have caused or influenced failures at similar engineering projects. Certainly, development of the ability to assess the site conditions critically and profit from past experiences is one of the applied geologist's best assurances for eliminating costly mistakes.

Although avoidance of physical failure of an engineering structure is the principal concern, today's engineering geologists are becoming increasingly involved with improving the economics of a project, particularly elimination of financial failures and cost overruns. In some agencies, geologists have participated directly in cost estimating. After all, who is better prepared to estimate the cost of a tunnel through a complex terrain, someone with little or no geologic background or the engineering geologist with a knowledge of construction practices and cost-estimating procedures?

Moreover, the engineering geologist's participation in management roles is increasing, and a trend has been established. This has resulted in geologists becoming more deeply involved in making decisions concerning project financing, personnel, and political ramifications.

Figure 35. Teton Dam and Reservoir, Idaho: One of several cavities detected in rhyolite tuff during post-failure exploration of the reservoir. One fissure uncovered during construction excavation for the right abutment key trench extended some 30 m below the trench invert and was explored for about 30 m both upstream and downstream of the dam axis. The opening in the dam foundation was described as fairly consistently 1.2 m wide, pinching and turning upstream; it was blocked downstream by a rock "the size of a pickup truck." During construction of the dam this fissure was backfilled with concrete, but possibly some others were not located. (Photo by L. B. James, 1976.)

The risk of human failure is an important consideration inherent in these added responsibilities. Case histories that demonstrate or relate to geological responsibilities in such categories as failures due to geologic origin, human activities, or a combination of geologic and human circumstances have been briefly reviewed. The failures of engineering works cited involve foundation defects, natural and man-induced slope collapse earthquake activity, ground water, land subsidence, induced landslides, and management deficiencies.

REFERENCES CITED

Amandes, C. B., 1984, Controlling land surface subsidence; A proposal for a market-based regulatory scheme: University of California at Los Angeles Law Review, v. 31, no. 6, p. 1208–1246.

Cameron, I., 1972, The impossible dream; The building of the Panama Canal: New York, William Morrow and Company, Inc., 284 p.

Carpenter, M. C., and Bradley, M. D., 1984, Legal perspectives on subsidence caused by ground water withdrawal in Texas, California, and Arizona, USA: Third International Symposium on Land Subsidence, Internatonal Association Hydrological Science Publication 151, p. 817–826.

Casagrande, A., 1965, Role of the calculated risk in earthwork and foundation engineering: American Society of Civil Engineers Journal of Soil Mechanics and Foundation Engineering, p. 1–38.

CDWR, 1957, Sea water intrusion in California: California Department of Water Resources Bulletin 63B, Appendix B, 141 p.

——, 1964, Investigation of failure, Baldwin Hills Reservoir, Los Angeles, California: California Department of Water Resources, 64 p.

Chadwick, W. L., chairman, 1976, Report of panel to review causes of Teton Dam failure in 1976: Washington, D.C., Superintendent of Documents, U.S. Government Printing Office.

Chesson, E., 1986, Failure classifications, *in* Cooper, K. L., ed., Forensic engineering: New York, American Society Civil Engineers, p. 46–61.

Close, U., and McCormick, E., 1922, Where the mountains walked: National Geographic, v. 41, no. 5, p. 445–464.

Consent Decree, 1986, U.S. District Court, Eastern District of California, United States of America vs. Aerojet General Corporation and Cordova Chemical Company: Documents of June 1986 submitted Courts, Sacramento, California, 137 p.

Cooper, K. L., ed., 1986, Forensic engineering: New York, American Society Civil Engineers, 95 p.

Cox, E. R., 1967, Geology and hydrology between Lake McMillian and Carlsbad Springs, Eddy County, New Mexico: U.S. Geological Survey Water Supply Paper 1828, 48 p.

De Freitas, H. H., 1978, Geologic hazards, *in* Knill, J. L., ed., Industrial geology: Oxford University Press, 300 p.

Downing, R. A., 1978, Groundwater, *in* Knill, J. L., ed., Industrial geology: Oxford University Press, 244 p.

Ehlig, K. A., and Bean, R. T., 1982, Dewatering of the Abalone Cover landslide, City of Rancho Palos Verde, Los Angeles County, California: Association of Engineering Geologists, Southern California Section, p. 67–74.

Ehlig, P. L., 1982, Mechanics of the Abalone Cove landslide including the role of groundwater in landslide stability and a model for development of large landslides in Palos Verdes Hills: Association of Engineering Geologists, Southern California Section, p. 57–66.

——, 1987, The Portuguese Bend landslide stabilization project: American Association of Petroleum Geologists Field Guide, Los Angeles, California meetings, p. 2-17–3-24.

EPA, 1988, Aerojet General superfund site, Sacramento, California, fact sheet: San Francisco, California, U.S. Environmental Protection Agency, Region IX, 8 p. (September).

FPC, 1976, Investigation of failure of Walter Bouldin Dam and safety of other dams, Alabama Power Company: Washington, D.C., Federal Power Commission, Bureau of Power, p. 2-79–2-93.

FitzSimons, N., 1986, An historical perspective of failures of civil engineering works, *in* Cooper, K. L., ed., Forensic engineering: New York, American Society of Civil Engineers, p. 38–45.

Henderson, L. H., 1939, Detailed geological mapping and fault studies of the San Jacinto Tunnel line and vicinity: Journal of Geology, v. 47, no. 3, p. 314–324.

Hendron, A. J., and Patton, F. D., 1985, The Vaiont slide; A geotechnical analysis based on new geological observations of the failure surface: Washington, D.C., U.S. Army Corps of Engineers Technical Report GL-85-5, v. 1, 104 p. v. 2, 47 p.

James, L. B., 1968, Failure of Baldwin Hills Reservoir, Los Angeles, California, *in* Kiersch, G. A., ed., General case histories on dams, tunnels, highways, and underground construction: Geological Society of America Engineering Geology Case Histories 6, p. 1–11.

——, 1988, Subsidence, *in* Jansen, R. B., ed., Advanced dam engineering for design, construction, and rehabilitation: New York, Van Nostrand Reinhold Co., p. 124–133.

Jansen, R. B., ed., 1983, Safety of existing dams, evaluation and improvement: Washington, D.C., National Academy Press, 338 p.

Judd, W. R., 1967, Geotechnical communication problems; Alex L. du Toit Memorial Lectures Number 10: The Geological Society South Africa, Annexure to v. 70, 45 p.

Kiersch, G. A., 1964, Vaiont Reservoir disaster: Civil Engineering, v. 34, no. 3, p. 32–39.

——, 1965, The Vaiont Reservoir disaster, *in* Landslides and Subsidence, Geologic Hazards Conference Proceedings: The Resource Agency, State of California, p. 136–145.

——, 1988, Vaiont Reservoir disaster, *in* Jansen, R. B., ed., Advanced dam engineering for design, construction, and rehabilitation: New York, Van Nostrand Reinhold Co., p. 41–53 and 106–117.

Kockelman, W. J., 1986, Some techniques for reducing landslide hazards: Association of Engineering Geologists Bulletin, v. 23, no. 1, p. 29–52.

Leighton, F. B., 1966, Landslides and hillside development, *in* Lund, S. P., and Proctor, R. J., eds., Engineering geology in southern California: Association of Engineering Geologists Special Publication, p. 149–193.

Leonard, G. A., ed., 1987, Dam failures, Proceedings of Workshop at Purdue University, West Lafayette, August 6–8, 1985: Amsterdam, Elsevier Publishing Co., Engineering Geology, v. 24, no. 1–4, 612 p.

Leps, T. M., 1972, Analysis of failure of Baldwin Hills Reservoir, *in* Proceedings, Specialty Conference on Performance of Earth and Earth Supported Structures, Purdue University: American Society of Civil Engineers, p. 507–550.

——, 1988, The Walter Bouldin Dam failure, *in* Jansen, R. B., ed., Advanced dam engineering for design, construction, and rehabilitation: New York, Van Nostrand Publishing Co., p. 53–59.

Londe, P., 1988, Malpasset Dam, *in* Leonard, G. A., ed., Dam failures, Proceedings of Workshop at Purdue University, West Lafayette, Indiana, August 6–8, 1985: Amsterdam, Elsevier Publishing Co., Engineering Geology, v. 24, no. 1–4, p. 295–329.

Lucas, C. V., and James, L. B., 1976, Land subsidence and the California State Water Project: International Association Hydrological Sciences Publication 121, p. 533–543.

MacDonald, C. W., 1915, Some engineering problems of the Panama Canal in their relation to geology and topography: U.S. Bureau of Mines Bulletin 86, 88 p.

Meehan, R. L., 1984, The atom and the fault; Experts, earthquakes, and nuclear power: Cambridge, Massachusetts Institute Technology Press, 157 p.

Merriam, R., 1960, Portuguese Bend landslide, Palos Verdes Hills, California: Journal of Geology, v. 68, no. 2, p. 140–153.

Meyers, W. B., and Hamilton, W., 1964, Deformation accompanying the Hebgen Lake earthquake of August 17, 1959: U.S. Geological Survey, Professional Paper 435–I, 98 p.

Outland, C. F., 1977, Man-made disaster: Glendale, California, A. H. Clarke Company, 249 p.

Pampeyan, E. H., 1986, Effects of the 1906 earthquake on the Bald Hill outlet system, San Mateo County, California: Association of Engineering Geologists Bulletin, v. 23, p. 197–208.

Peck, R. W., 1982, Risk analysis for dams: Proceedings, Dam Safety Research Coordination Conference, Interagency Committee on Dam Safety, Research

Subcommittee, Denver, 11 p.
Poland, J. F., and Lofgren, B. E., 1984, San Joaquin Valley, California, *in* Poland, J. F., ed., Guidebook to studies of land subsidence due to ground-water withdrawal: UNESCO, ISBN 92-3-102213-X, p. 263–277.
Pratt, W. E., and Johnson, D. W., 1926, Local subsidence of the Goose Creek oil field: Journal of Geology, v. 34, no. 7, p. 556–590.
Ransome, F. L., 1928, Geology of the St. Francis Dam site: Economic Geology, v. 23, no. 5, p. 558–563.
Reiter, M., 1984, The Palos Verdes Peninsula: Dubuque, Iowa, Kendal/Hunt Publishing Co., p. 26–33.
Roberts, D. V., 1959, The evaluation of risk in foundation engineering: Los Angeles, Dames and Moore, Engineering Bulletin 9, p. 5–8.
Schumann, H. H., and Poland, J. F., 1970, Land subsidence, earth fissures, and ground-water withdrawal in south-central Arizona, U.S.A., *in* Land Subsidence: Tokyo, International Association of Scientific Hydrology Publication 88, v. 1, p. 295–302.
Schussler, H., 1906, Water supply of San Francisco, before, during, after earthquake April 18, 1906: City of San Francisco (unpublished). Also *in* Second report Governor's earthquake council, September, 1974: California Department of Water Resources, 86 p.
Schuster, R. L., and Fleming, R. W., 1986, Economic losses and fatalities due to landslides: Association of Engineering Geologists Bulletin, v. 23, no. 1, p. 15.
Scullin, C. M., 1966, History, development, and administration of excavation and grading codes, *in* Lund, R., and Proctor, R. J., eds., Engineering geology in southern California: Association Engineering Geologists Special Publication, p. 227–236.
Seed, H. B., and Idriss, I. M., 1967, Analysis of soil liquefaction in the Nigata earthquake: American Society of Civil Engineers Journal of Soil Mechanics and Foundation Division, v. 93, no. SM3, p. 83–108.
Shah, H. C., and McCann, M. W., 1982, Risk analysis—it may not be hazardous to your judgement: Denver, Colorado, Proceedings Dam Safety Research Coordination Conference, Interagency Committee on Dam Safety Research Subcommittee, pt. 2, p. 12–19.
Sherard, J. L., 1959, What the earthquake did to Hebgen Dam: Engineering News Record, v. 163, no. 11, p. 26.
Sowers, G. F., 1986, Failure investigation for forensic engineering, *in* Cooper, K. L., ed., Forensic engineering: New York, American Society Civil Engineers, p. 7–21.
Teutsch, J., 1979, Controls and remedies for ground water-caused land subsidence: Houston Law Review, v. 16, p. 283–331.
Therond, R., 1972, Recherche sur l'etancheite des lacs de barrage en pays karstique [Thesis D. Eng.]: l'universite Scientifique et Medicale de Grenoble, 9 Jun 1972, 444 p.
Thompson, T. F., 1954, The San Jacinto Tunnel on Colorado River aqueduct, Metropolitan Water District of Southern California: Geological Society of America Los Angeles Annual Meeting, November, 1954 Engineering Geology packet, 7 p.
Trefzger, R. E., 1966, The Tecolote tunnel, *in* Lund, R., and Proctor, R. J., eds., Engineering geology in southern California: Association Engineering Geologists Special Publication, p. 108–113.
TVA, 1949, Geology and foundation treatment, Tennessee Valley Authority, projects: U.S. Tennessee Valley Authority Technical Report 22, 548 p.
USBR, 1959, Tecolote Tunnel; Technical record of design and construction: Denver, U.S. Bureau of Reclamation, p. 109.
Varnes, D. J., 1981, The principles and practices of landslide hazard zonation: International Association Engineering Geology Bulletin, no. 23, p. 13–14.
West, R. P., 1962, Waco Dam slide; Its cause and correction: Engineering News Record, v. 169, no. 5, p. 34–36.
Whitman, R. V., 1984, Evaluating calculated risk in geotechnical engineering: American Society of Civil Engineers Journal of Geotechnical Engineering, v. 110, no. 2, p. 145–188.
Yerkes, R. F., and Castle, R. O., 1976, Seismicity and faulting attributable to fluid withdrawal: Engineering Geology, v. 10, nos. 2-4, 151 p.
Wiley, A. J., 1928, Causes leading to the failure of the St. Francis Dam: Sacramento California State Printing Office, 73 p.
Zielbauer, E. J., 1966, Sea water intrusion and the barrier projects, *in* Lund, R., and Proctor, R. J., eds., Engineering geology of southern California: Association Engineering Geologists Special Publication, p. 265–269.

Manuscript Accepted by the Society March 20, 1989

Printed in U.S.A.

Geological Society of America
Centennial Special Volume 3
1991

Chapter 23

Errors of geologic judgment and the impact on engineering works

George A. Kiersch
Geological Consultant, Kiersch Associates, Inc., 4750 Camino Luz, Tucson, Arizona 85718
Laurence B. James
Chief Engineering Geologist (retired), California Department of Water Resources, 120 Grey Canyon Drive, Folson, California 95630

INTRODUCTION

Many of the more common causes of physical failures or serious remedial problems of engineering works are related to the construction or operation and maintenance phases of a project and involve either controversies or errors of geologic judgment. Mistakes and misunderstandings that impact on a project, including an incorrect design, may be due to any of several causes, such as a misinterpretation of available geological facts, inadequate factual background data, or misunderstanding, misjudgment, or poor forecasting of the geologic changes that occur with time and/or operation. Frequently the errors of judgment committed are related to the cost/benefit and/or calculated-risk evaluations of a project, as discussed later in this chapter.

Many faulty interpretations of areal and site-specific geologic conditions have been traced to a lack of input by mature, field-experienced geologists. Even more frequently the errors are a function of insufficient funds for adequate investigations. Management is often under pressure to expedite the investigative stage of the project; in a few instances, site assessment has been attempted by inexperienced individuals or non-geologists, as in the case of the St. Francis dam failure (Chapter 22, this volume). On rare occasions, accusations have claimed incompetence or intentional misinterpretation of the geologic environs for self-serving purposes. However, the authors are unaware of any such instance of fraudulent practice by a professional geologist.

Judgment plays a particularly important role in the planning stage of most major engineering works; for example, in comparing alternative sites for structures or determining routes for aqueducts or highways. Errors of judgment during planning may lead to design problems or result in a failure to achieve the maximum potential benefits from the project. Commonly, funds for exploration are restricted during the planning period; often, ample funds become available only after the site-selection process is completed and economic feasibility has been established. Consequently, the engineering geologist may be called upon for important predictions based on scattered data, thus relying on experience and technical skills, as well as judgment.

An example is the early investigations for the multibillion dollar California State Water Project. Nearly 100 alternative routes or variations of routes were considered for the 400-mile-long California aqueduct. Some of these routes traversed subsidable soils (described in Chapter 22, this volume); some crossed active faults or potential gas-bearing structures; other routes avoided such hazards but were longer and required added pump lifts or costly excavation. In addition to problems created by the physical hazards, political pressure was exerted to speed these preliminary studies so that a statewide project plan could be presented to the California electorate for approval. Perhaps the most critical decision addressed was whether to cross the Tehachapi Mountains in a deep tunnel or near the surface. One tunnel route that was seriously considered would have penetrated the San Andreas and other faults at tunnel depth. In contrast, the surface crossing required a pump lift of unprecedented height and size, which would be subject to high and accelerating costs for energy.

After deliberation, and based almost entirely on geologic factors, a near-surface route crossing all major faults on the surface was selected and subsequently constructed. The alignment of the California aqueduct through the Tehachapi Range was chosen and authorized based on limited funding for investigative geological exploration and a heavy reliance on geological judgment; generous funding for site-specific exploration only became available during the later stages of construction.

During the construction stage of most large projects, geological investigations are generally directed to site specifics and concentrate mainly on foundations and on detailed, as-constructed mapping (discussed in Chapter 20, this volume). Good judgment dictates that, wherever possible, each successive drill hole, as well as each stage of exploration, should be guided by all preexisting data. However, occasionally the entire exploration program for a project is completed prior to onset of construction. This policy should be discouraged because unforeseen problems may be encountered. Ample funds should be reserved to pay for late-stage

Kiersch, G. A., and James, L. B., 1991, Errors of geologic judgment and the impact on engineering works, *in* Kiersch, G. A., ed., The heritage of engineering geology; The first hundred years: Boulder, Colorado, Geological Society of America, Centennial Special Volume 3.

exploration that may be required to justify changes in design or construction.

Karl Terzaghi, sometimes known as the father of modern foundation engineering, introduced the observational method now followed by builders of most dams (Peck, 1973). Using this method, a reservoir is slowly and only partially filled during construction of a dam. During this initial filling or field-experiment phase, subsurface conditions that otherwise could only be deduced or assumed on the basis of the site exploration data may be evaluated. Following this observational approach, it is not necessary to design for the worst possible geologic conditions. Instead, design may be based on the most probable conditions, although the plans may incorporate in advance provisions for any serious problems believed possible but not probable. This procedure imposes fewer constraints on the engineering geologist, and judgment can be weighted toward less conservative possibilities. This enhances the feasibility of engineering works while assuring that safety remains the foremost consideration.

Modern dams seldom, if ever, fail because of incorrect or inadequate numerical analyses. Dams fail most frequently because inadequate judgment was brought to bear on problems related to the foundation or to the interface between the embankment and the foundation. Geological investigations for dams have frequently focused on details of foundation or design while giving inadequate attention to the reservoir or to possible adverse effects of natural or human activities in the vicinity. Serious problems, and even some catastrophic failures discussed in this chapter, have resulted from a short-sighted concentration on mathematical analysis, computer studies, and modeling at the expense of field observations and mature judgment. Dam failures can be decreased significantly if the underlying causes of potential failure can be addressed. Peck (1980) feels that 90 percent of recent dam failures occurred because of avoidable oversights, either due to a lack of communication among the designer and constructors, or an overly optimistic interpretation of geologic conditions (discussed in Chapter 22, this volume). Realistically, the best means of identifying and coping with such problems is through the coordinated efforts of researchers, designers, and management.

According to Peck (1980), much responsibility for judgment now resides where the rewards of professional recognition and advancement are greatest: in the design office, where analysis and aesthetics are often separated from reality; or in the research institutions, with their efforts to idealize the properties of real materials for analysis, and to solve stress distribution and deformation problems using these idealized materials.

Contractual arrangements, possibilities of claims for extras, and delays are all common problems that must be recognized and treated satisfactorily, irrespective of project specifications, or else failures will continue (Peck, 1980). Moreover, if the myth persists that only such calculated answers as mathematical analysis, including finite elements and computer language, constitute engineering or applied science, practitioners will lack the incentive or the opportunity to apply the best-judgment approach to crucial problems that cannot be solved by standard calculations. There is overwhelming evidence that in many geological cases, some of which are cited in this chapter, mature judgment is the logical and best approach.

PROBLEMS OF JUDGMENT INVOLVING INDUSTRIAL SITING

Some errors of judgment may have a reverse impact. Cases exist where evaluation of the geologic environs of a major engineering work was incorrect and caused the project to be unjustly terminated. One such case involved the location and permit to continue construction of a nuclear power plant on the south coast of Puerto Rico in 1970–1972. The seismo-tectonic potential for recurring movement along the Great South Puerto Rico (GSPR) fault zone near the proposed site (Fig. 1) became a controversial factor in evaluating the site for a construction license. After lengthy investigations the owner's geoscience consultants (PRWRA, 1971–1972) had determined that the zone was within the allowable design risk for the plant proposed. A subsequent, brief, routine review study by an alternate firm, less experienced with the Puerto Rico geologic setting, hastily labeled the major fault zone as active and urged the owner to move from the site to the less favorable north coastal sector of the island. This forecast, with implications of risk on the on-going project, caused the plant owner to stop pursuing a construction license for the south coast site (Fig. 1) and seek an alternate site on the less favorable north coast of the island (see Chapter 19, Fig. 12, this volume). After many frustrations, a large expenditure of funds, and no satisfactory licensed site, the utility terminated the nuclear plant project.

In retrospect, the on-going areal and regional geological and seismological studies by U.S. Geological Survey personnel in the 1960s and 1970s (Glover, 1971; McCann, 1985; Hays, 1985) had documented the Great South Puerto Rico fault zone as an old inactive zone, with little or no movement in the past 20 m.y. (McCann, 1985). Abandonment of the south coast site in 1972 was based on a hasty, inaccurate evaluation and error of geological judgment. As the owner's original site was suitable for the power plant, tens of millions of dollars were wasted by seeking an alternative, and ultimately, no power plant was built at all.

Errors in geological judgment can also have a devastating effect on an industry that is already in place. This is illustrated by the controversy surrounding the geologic setting and possible implications of the Verona fault relative to the Vallecito test reactor (GETR) near Pleasanton, California (Fig. 2).

Vallecito, a small facility of General Electric Company, was the first U.S.-licensed commercial reactor in 1959; the neutron-radiography facility produced one-half of the free world's medical radio-isotopes. In 1977, the U.S. Geological Survey reviewed the geology and seismology of the reactor area, as was required for renewal of the operating license.

Extensive trenches and exploration in the vicinity of the GETR revealed at least three northwest-trending, thrust-like faults bracketing the reactor site (Fig. 2). The fault zones exposed

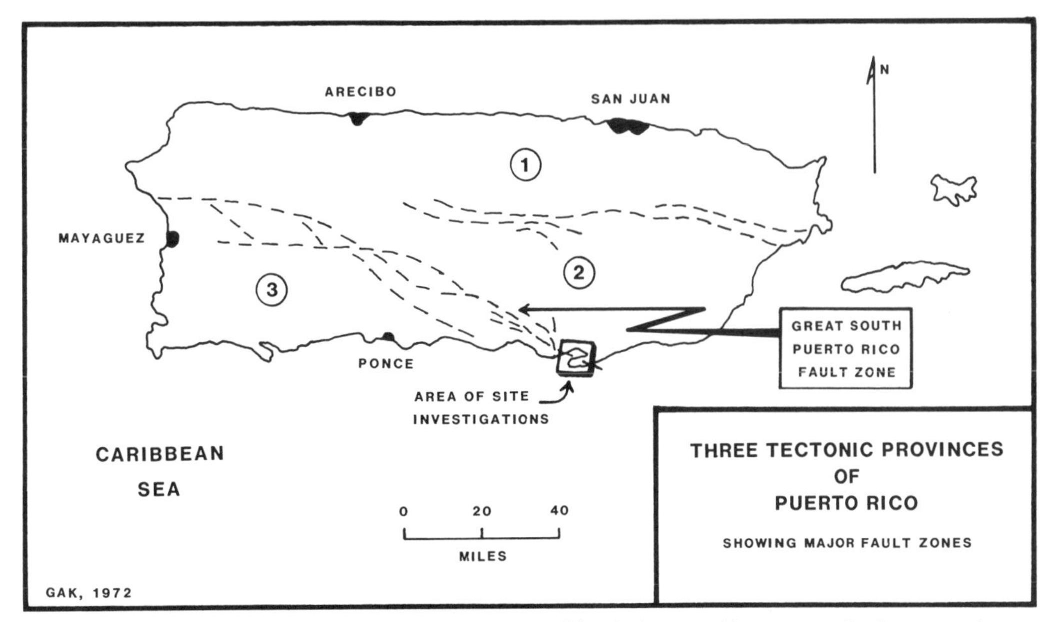

Figure 1. Location of the regional Great South Puerto Rico fault zone with respect to the three tectonic provinces of Puerto Rico. The area investigated for a plant site east of Ponce is shown. (From: PRWRA, 1971–1972.)

in the trenches are similar to each other in minor structural features and age of displacement, and none intersects the GETR foundation (Fig. 2). All of the rupture planes displace Plio-Pleistocene Livermore Gravels and younger colluvium and paleosols. The attitudes of the rupture planes and their relation to local topographic features, together with other geologic evidence, suggested two plausible origins: tectonic thrust movement (Herd, 1977), or large-scale landslides (Rice and others, 1979). The mapped length of Herd's Verona fault is less than 6 miles. This new mapping placed the potentially active Verona fault close to the reactor (Herd, 1977), whereas a 1958 map, which was the basis for the construction license, only showed a fault about 2,800 ft from the reactor.

After many technical investigations by U.S. Geological Survey and the owner's consultants, Earth Science Associates, including hearings over a 5-year period, the U.S. Nuclear Regulatory Commission ruled in 1982 that the zone was not a hazard to operation of the commercial reactor (discussed in Chapter 10, this volume). The arguments for landsliding were accepted by a majority of investigators. However, the 5-year shutdown due to the controversy over the interpretation of geological issues destroyed a profitable business which could not be revived by General Electric; there was another serious impact on the U.S. nuclear industry (Meehan, 1984, p. 37) when the unit was abandoned.

A prominent error of judgment was central to the eventual siting and design of the Golden Gate Bridge over the entrance to San Francisco Bay in the 1930s. Fortunately, the misinterpretation of a fault location by Professor Bailey Willis (Stanford) and his claim that the quality of the serpentine foundation rock at the South Pier was unsafe were overridden. The site was pronounced safe and adequate by Professor A. C. Lawson (Berkeley) after a lengthy investigation and costly hearings (described in Chapter 1, this volume). The bridge opened in 1937, and no stability problems have been experienced over the intervening years despite several strong earthquakes.

JUDGMENTS INVOLVING SITE CONDITIONS

Misjudgments, inadequate site evaluation and/or engineering control

The following brief summaries are examples of the effect that geological misinterpretations and/or errors in understanding site conditions can have on existing engineering works.

Schoellkopf power plant, Niagara Gorge, New York

Massive, closely timed rockfalls destroyed the Schoellkopf power plant stations 3C and 3B and damaged 3A on June 7, 1956 (Fig. 3). The stations were founded in the Grimsby Sandstone at the base of the Niagara gorge wall (Fig. 4). The rockfalls were a series of progressive, large-scale slab failures that started 200 ft south of station C and worked northward across stations C and B (Figs. 3 and 5). The sequence of events that led to the actual cliff failure at 5:32 p.m., for which a technical explanation was never released, is given below.

Failure sequence. The sequence of events began with increased seepage behind stations C and B in the early morning of June 7 (Bernstein, 1956). Modest, jet-like leaks appeared high up on the cliff by 9:30 a.m., and construction drains in the cliff near penstocks 20 and 21 (Fig. 5) were leaking 6 cfs and 9 cfs, respec-

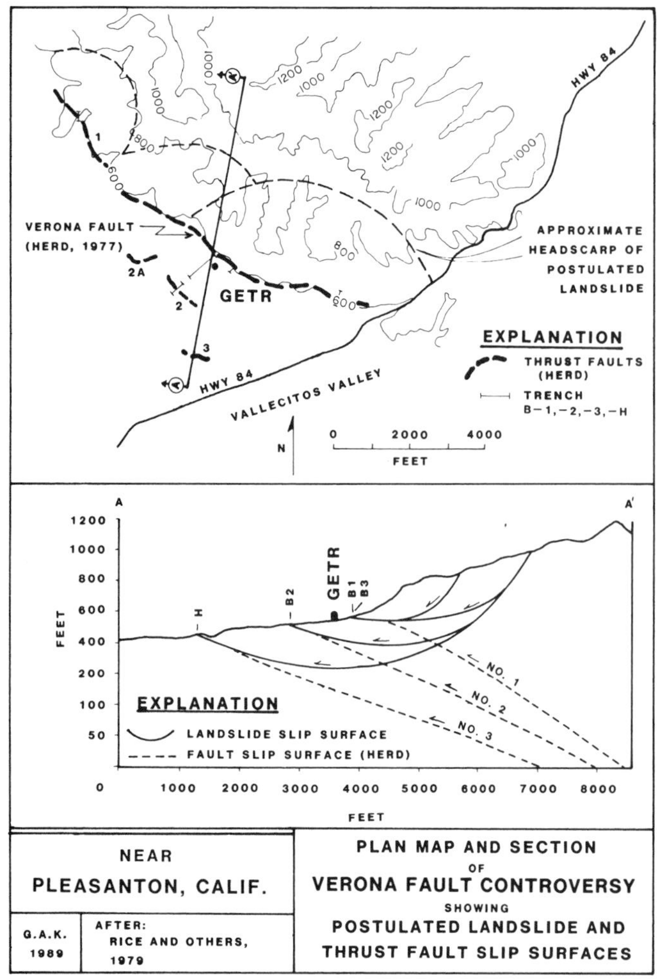

Figure 2. The Verona fault controversy relative to the Vallecito, General Electric nuclear reactor facility. The plan map shows the distribution of postulated thrust faults and headscarps of landslides; the section shows an interpretation of landslide slip surfaces and fault slip surfaces in the vicinity of the GETR facility. (After: Rice and others, 1979; and Earth Science Associates/R. C. Harding.)

tively; several leaks of 4 to 6 cfs were also issuing near the floor. At 3:16 p.m., the jet-like leak about 90 ft below the top of the cliff dramatically increased, followed soon by a second breakout close by (Fig. 5). At about 5:00 p.m., the floor of station A buckled with a pop-up; this had been preceded by two earlier pop-ups near turbine no. 1—about noon and at 3:40 p.m. Although a few glass window panes had been cracking in station A all day, this increased sharply at 5:15 p.m. when the major, jet-like leaks high on the cliff south of station C were at a rate of about 100 cfs and pieces of cliff-rock (4 to 12 in) had begun to break out and fly off. The south wall of station C then started to break away and pieces fell into river (Fig. 5).

The lower part of a cliff-slab of Lockport Dolomite located behind station C failed first, leaving the upper part unsupported; this caused a further cliff failure that extended to the top of a main relief-joint feature (Fig. 5A). The joint planes bounding the failed blocks (about 60 ft long) were oxidized and weathered (J. R. Dunn, personal communication, 1988), indicating they had been exposed for a long time. At the south end of station C, many thin rock slivers fell from the cliff as separate pieces. Dunn (personal communication, 1988) calculated that a column of water as much as 200 ft high acted on the cliff face. This force both rotated the blocks that spalled off the canyon wall and rotated the floor of station C as much as 23 ft toward the gorge (Fig. 5). One block fell toward the south, while a block alongside on the north side fell westward toward the river. This mixed motion of the blocks torn from the cliff face strongly infers hydrostatic pressure as the main driving force. In the Rochester Formation, about 15 ft below the Lockport Dolomite, or 125 ft from the top of the cliff (Fig. 4), vertical relief-joint planes were about 10 ft apart where the thick slab-mass broke away and failed above an impervious shale unit. A small amount of water continued to flow from the newly exposed joint face at 7:05 p.m. 1½ hr after the rock fall. However, the amount of water that flowed from the cliff above the penstock levels did not compare with the amount of water observed flowing before the failure at 5:32 p.m.

The initial rock slab that failed and fell onto station C was about 10 ft thick at the top and 65 ft thick at the base; the overall length extended from a point 200 ft south of station C northward past this station and part of station B. The base of this failure was within shales of the upper Rochester Formation (Fig. 4).

Causes of collapse. Based on information from various sources, three major factors contributed to the cliff failure: (1) removal of rock at the base of the cliff; (2) weak shale zones within some rock units and particularly in the "Clinton"/Rochester Formations at the base of the failed rock slabs; and (3) the widespread occurrence of open vertical joints and fractures, some of which were induced by blasting while others were rebound-relief structures (Fig. 4). The latter are located both parallel and normal to the cliff face. This network of structurally controlled, interconnected openings throughout the gorge wall aided and permitted high joint-water pressures to build up when charged by leakage from the unlined canal basin and penstocks on top of and within the cliff. Strong joint planes extend downward for at least 130 ft from the top of the cliff into the Rochester Formation in the vicinity of the penstock openings.

Grouting operations by the Pennsylvania Drilling Company for the Niagara Mohawk Company, on-going for several weeks prior to the collapse, were designed to reduce leakage losses from the unlined penstocks (Calkin and Wilkinson, 1982; Fig. 4). Actually, the high-pressure injection fluids became the triggering force for the cliff failure. Grout injection penetrated to the Neagha Shale level. J. R. Dunn (personal communication, 1988) noted 16 grout holes behind the terminal buildings (Fig. 4) that were 35 to 115 ft from the edge of the cliff and inclined 75 to 80 degrees westward. Most of the grout holes were filled to refusal, but a few seemed bottomless and inferred the occurence of an extensive interconnected network of joint planes throughout part of the cliff. The failure of entire rock slabs in the cliff face was apparently due to the injections of grout slurry blocking past drainage

Figure 3. Schoellkopf power plant, looking across Niagara Gorge from Canada. A. Collapse began at 5:32 p.m. on June 7, 1956. The first serious failure was a large block of the cliff face (center) above station 3B, which fell to the right onto station 3C; water jets south of station 3C are shown on the cliff face. B. Collapse of cliff in progress, shortly after the first failure shown on A. Note the very large block of the cliff that is failing (center) and that will demolish station 3B and part of 3C. Station 3A, on the left, is largely intact and was returned to service. C. After the collapse of stations 3B and 3C. The area of cliff-block failures and the full destruction of both buildings on June 7 are shown. Water is draining from the penstocks and lowering the induced ground-water level throughout the margin of the cliff. D. Oblique aerial view of the failed cliff face above destroyed stations 3B and 3C on June 7. The size of one cliff block is evident, but many broke apart on impact. Note the forebay building, station 3C, and the canal in the upper corner. One jet-like leak south of station 3C is marked by a small flow from the cliff face. Joint-fracture control of block failure is clearly evident throughout the cliff face. (A, B: Niagara Falls Heritage Foundation, George Siebel, 1987; C, D: T. A. Wilkinson, USCE, 1987).

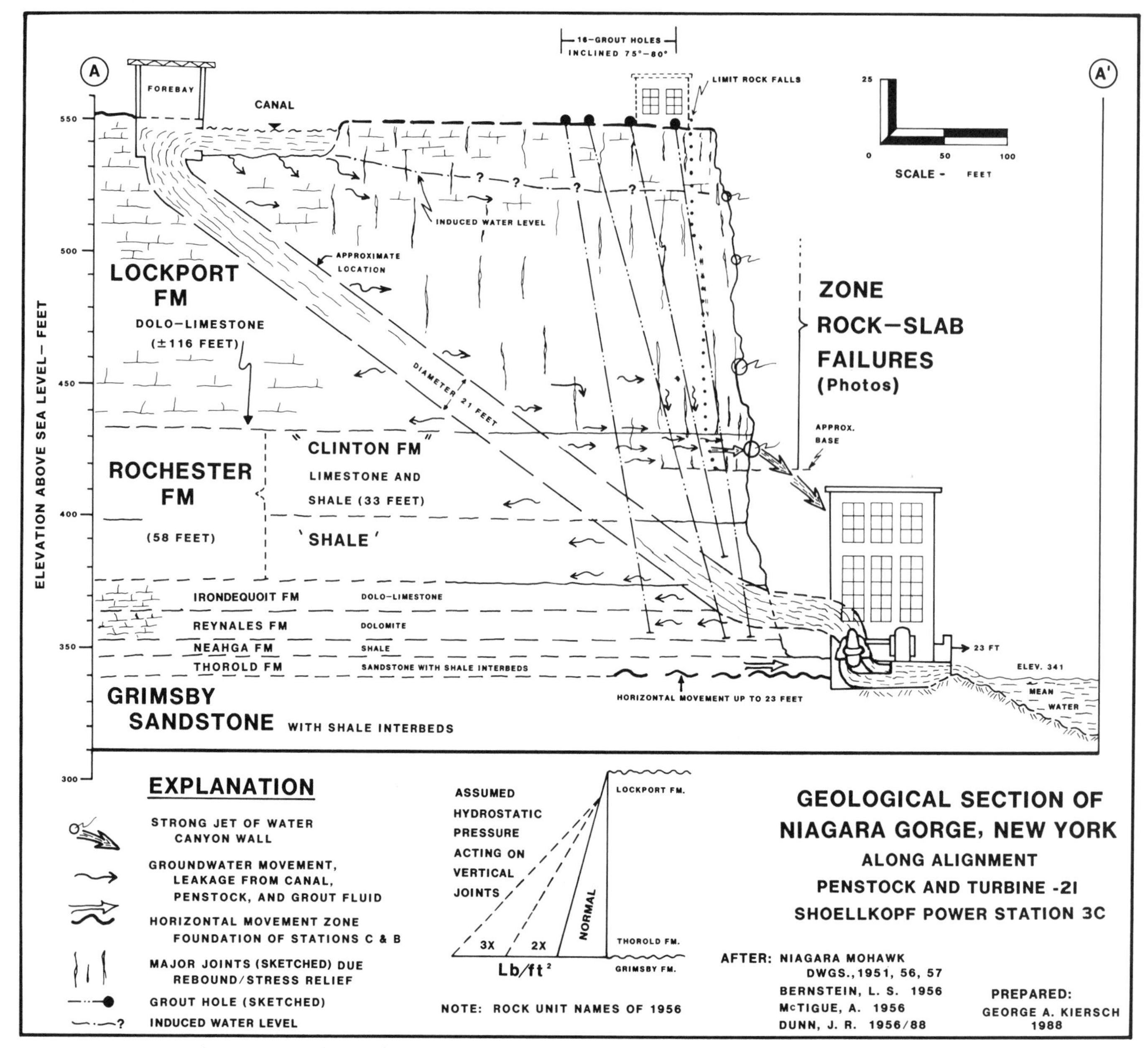

Figure 4. Geological section of the Niagara Gorge along the alignment of the penstock and turbine number 21 of station 3C, relative to the intake canal and forebay. Inherent structural features of the rock units are sketched, as are likely paths of leakage from the penstocks that contributed to the cliff collapse and destruction of the power plants as shown on Figures 3A through 3C. (After notes by J. R. Dunn 1956, personal communication, 1988; L. S. Bernstein, 1956; and A. McTigue, 1956.)

channels that extended throughout the interconnected joints and openings within the cliff (Bernstein, 1956). The grouting created water pressures as great as three times the normal head throughout the cliff. Although the grout injection was intended to seal off excessive leakage from the unlined penstocks, the high fluid pressures instead accelerated the last phase of joint widening, changed the natural balance of stress conditions throughout the gorge wall, and induced progressive rock slabbing.

For several weeks before June 7, small leaks were evident along the cliff face; however, the volume of perceptible leakage increased dramatically on the day of collapse. Dunn (personal communication, 1988) concluded that a column of water due to the combined penstock leakage and induced grouting fluids acted within much of the cliff, and the hydrostatic forces moved large, joint-bound blocks from the unconfined cliff face. Rotation of the individual rock blocks, as they spalled off the cliff, was recorded in a movie sequence of the June 7 failure by Virgilio (1986; Fig. 3). Lateral pressures near the base of the cliff, due to about 200 ft of hydrostatic head, may have approached 13,000 psf (Dunn, personal communication, 1988) and are reflected in the fact that

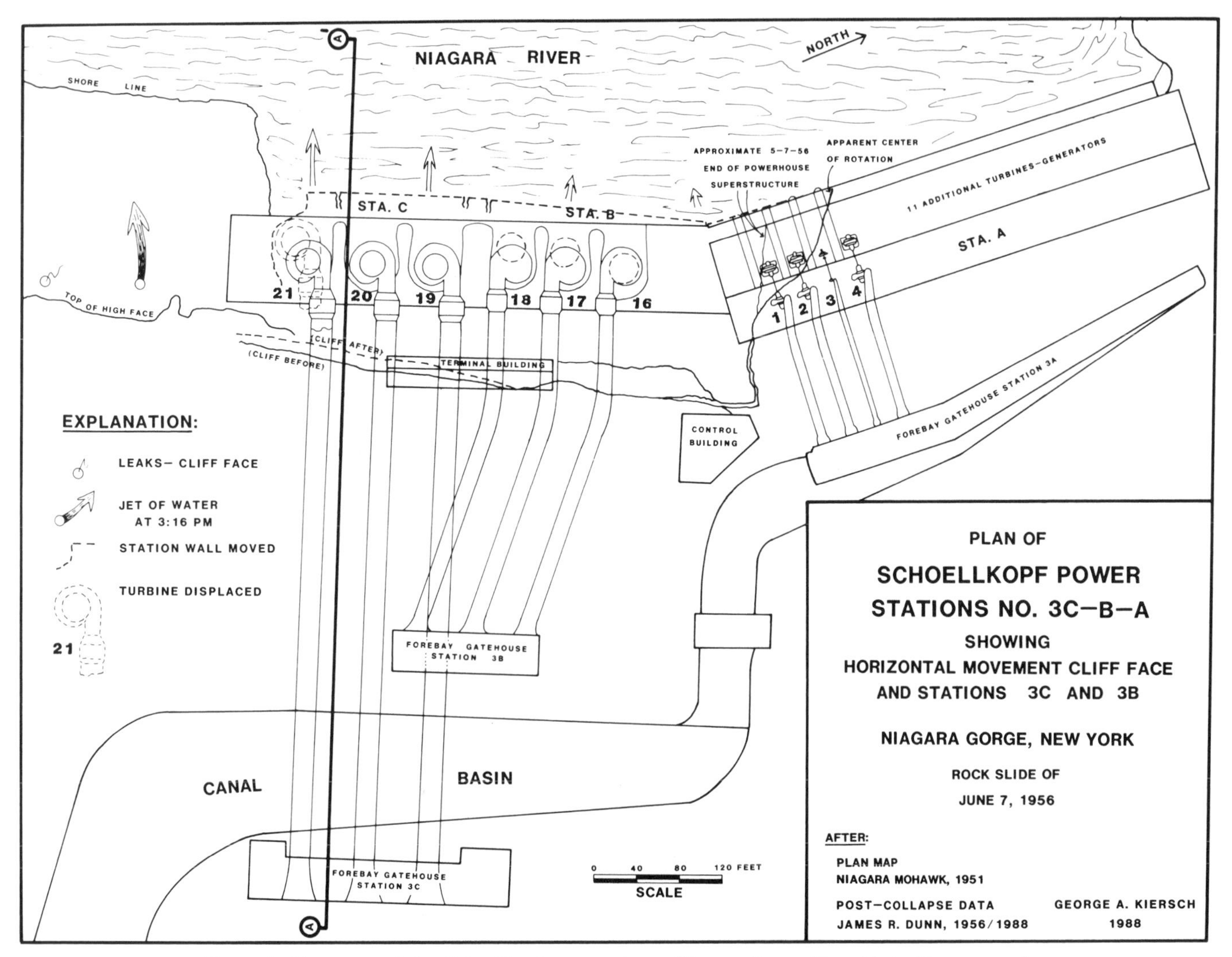

Figure 5. Plan map of the Schoellkopf power plants, Niagara Gorge, New York. The areas of deformation, tilting, and lateral movement that affected the turbines of stations 3C and 3B, and the cliff face are shown. (After notes by J. R. Dunn 1956, personal communication, 1988; and Niagara Mohawk file drawing.)

part of the station C foundation was moved 23 ft toward the river gorge (Figs. 4 and 5).

Field observations and the progressive increase of leakage during June 7 strongly infer a widening of the joint planes, which continued until the large jet-like streams began to flow from two separate locations on the cliff south of station C. These flat streams of water occurred at locations 200 ft and 30 ft south of station C and about 100 ft below the top of the cliff (Fig. 5A). The thin, muddy streams, about 8 ft wide and 1 in deep, were under sufficient pressure to push Louis S. Bernstein and his engineer-companion Draper into the Niagara River while investigating the leak. Draper was swept away and drowned (Bernstein, 1956).

The predicted reaction of the cliff-rock column to the injection of grout was clearly an error of geological judgment. Instead of attaining stability, massive rock-slab failures occurred, along with substantial horizontal movements of the cliff and foundation, and heaving of the power station rock floors that destroyed power stations C and B (Fig. 5). Power station A was repaired and operated into the 1960s when it was replaced by the generating capacity of New York Power Authority facilities nearby.

Vakhsh-Yavan tunnel portal, USSR

The first stage of the Baipaza hydroelectric power station located in the spectacular canyon of the Vakhsh river south of Dushanbe (Fig. 6) in the Tajik Republic, USSR, was completed in 1968. The 55 m-high dam was "built" by a direct quarry blast of rock using explosives. Thirteen coyote-blast holes were driven into the steeply dipping Cretaceous limestone and sandstone beds of Karatau ridge that formed the right wall of the river canyon (White, 1968; IGG, 1984; inset map, Fig. 6).

The first-stage dam created by the blast initially allowed substantial leakage through the crushed blast-rock. However, following the natural silting action of one major flood, the blast-fill dam was largely sealed. Sealing was completed by loading with fine-grained crushed rock and embankment materials. This raised the water level in the reservoir to the level of the diversion tunnel constructed to transport water from the reservoir directly to Yavan valley, about 8 km away, for irrigation.

The second stage of the project began in the early 1980s and involved raising the original blast-fill dam height and applying further treatment to reduce leakage through the rock-fill structure. Construction of the 600,000-Kw Baipaza hydroelectric power plant at the dam required relocation of the intake structure for the Vakhsh-Yavan tunnel to a new portal downstream of the dam. During the planning and construction for the new tunnel portal, an error of geological and engineering judgment was made in predicting the effects of blasting within the Karatau limestone ridge for excavation of the portal area (Fig. 7).

Adits driven in the steeply dipping (60°) limestone and sandstone beds of Karatau ridge were reportedly loaded with 19,000 tons of explosives to break ground around the new portal (IGC, 1984). Effect of this single blast far exceeded the engineer's calculations, creating an excessive volume of broken surface and underground rock. Even more unexpected was the large-scale V-shaped opening 60 to 80 ft deep across the upper ridge wall (Fig. 7). The massive rock breakage was not anticipated, although the site possessed strong physical elements that are known to influence blasting patterns; the inclined attitude of the free-face slope with weak bedding planes, the contrasting physical properties of the bedded rocks, and the prominent fracture pattern and numerous faults throughout the rock ridge (FIg. 7). The deep opening formed across the ridge likely correlates with a preexisting fracture zone.

Cedar reservoir, Washington

A serious blowout in 1918 through morainal deposits in the right bank of this reservoir (Fig. 8) is ascribed to the almost complete disregard for geologic conditions by the builders (Mackin, 1941). Operated and owned by the city of Seattle, the reservoir is situated approximately 40 mi southeast of the city. The impounding concrete structure, known as Masonry dam, was completed in October 1914 with a crest length of 795 ft and a height of 217 ft above foundation rock. Selection of the site was

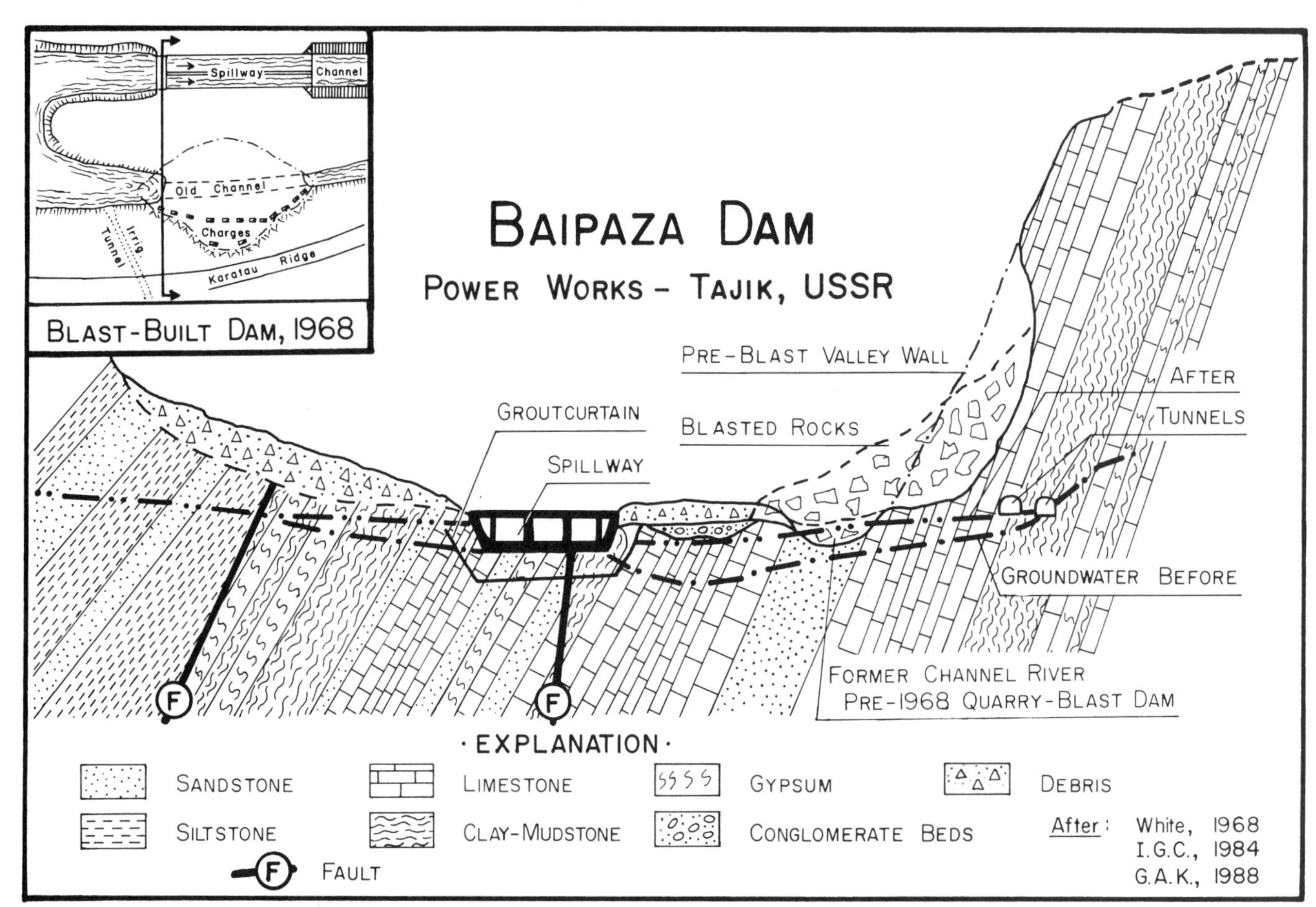

Figure 6. Geologic cross section of the Baipaza dam and hydroelectric project on Vakhsh River, USSR, showing the area of blast-rock moved into place by the explosion of 1984 for a tunnel portal. Inset shows the design changes from the original blast-rock dam in 1968 (Sources: IGS, 1984; White, 1968.)

based on the assessment of an adequate foundation and appropriate topography that allowed a short penstock alignment. An evaluation of geologic conditions throughout the reservoir area and the vicinity of the Cedar delta morainal deposits was largely neglected.

The site, except where the dam is on bedrock in Cedar Ridge gorge, is underlain by lacustrine sediments and flanked by glacial outwash gravel from the ancestral Puget Sound glacier (Fig. 8). During low-water stages, the reservoir pool is underlain by fine-grained lacustrine clay, silt, and sand of low permeability. However, these deposits are capped by coarse outwash gravel, with which the pool comes in contact when the reservoir is partially filled.

Prior to the 1918 blowout, leakage downstream from the dam had been measured and was seen to increase as the reservoir level was raised. During the initial filling in 1914, when the depth of water below the spillway crest reached approximately 80 ft, seepage loss was about 30 million gal/day (Henry, 1915; Mackin, 1941). In 1915 the pool was raised 14 ft above the previous high level, which caused a marked increase in discharge from springs downstream: the town of Moncton was flooded and destroyed.

Attempts were made to seal off the leaks, and the reservoir basin was partially filled several times between 1916 and 1918. To test the sealing in 1918 the reservoir level was permitted to rise; leakage increased markedly, and the Boxley failure occurred on December 23, 1918. The blowout was located 6,000 ft northeast of the dam and washed out an estimated 800,000 to 2 million yd^3 of detritus within less than 2 hr, and perhaps as little as 20 min. The bank failure created a crater known as Boxley

Figure 7. Vakhsh-Yavan tunnel portal, USSR. Disastrous effects of the single blast to initiate excavation at the portal of the relocated water-distribution tunnel downstream from the Baipaza dam. (Photo by G. A. Kiersch, 1984.)

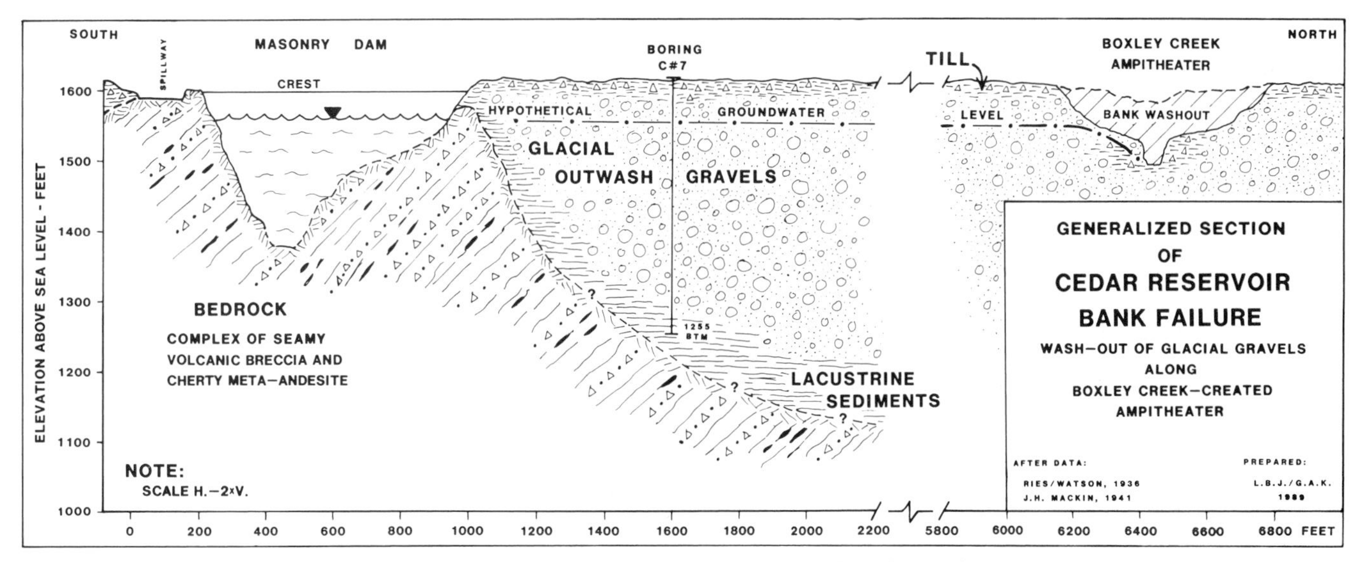

Figure 8. Cedar reservoir, Washington. Generalized geologic section of dam site and the northern reservoir rim. The bank washout creating Boxley Creek ampitheater apparently resulted from an oversteeping of the phreatic line near Boxley Creek due to the bouldery-clay veneer in the gravels; the glacial materials eventually liquefied. (after data of Mackin, 1941, p. 89, 22–24; Ries and Watson, 1936, p. 47)

Figure 9. View of the 1963 slide mass looking west at the upstream/eastern end, with the toe on the right; the remainder of the reservoir water is at the bottom. The character of intact limestones with sandstones and interbedded clayey units and seams is further indication of the speed of movement, with "arching" of beds due to movement up the right canyon wall. All of the slide mass is above the former reservoir level. For areal relations of the slide to the canyon-reservoir, see Chapter 18, Figure 27. (Photo by G. A. Kiersch, 1963.)

amphitheater (Fig. 8). Peak discharge was estimated to have been between 3,000 and 20,000 ft/sec. The town of Edgwick, tracks of the Milwaukee railroad, saw mills, and other properties were destroyed. Litigation resulted in a judgment of $337,000 against the city of Seattle.

Prior to construction of the dam, University of Washington geologists K. K. Landes and W. A. Roberts had reported to the city that open-textured gravel was exposed along the walls of Cedar Gorge upstream from the dam site. They recognized the danger of serious leakage and recommended further investigation of the potential problem, as had a board of engineers in 1912 (Mackin, 1941). The recommendations of Landes and Roberts were disregarded, however, and construction proceeded. Had the recommended geological investigation of the reservoir area been made, the designers would have been alerted to the threat of piping action through the morainal reservoir banks. If this hazard had been realized by the designers and builders it is doubtful a high dam would have been constructed at this location.

Important lessons from the Cedar Gorge incident include: (1) evaluating a site based on borings and foundation test data only, without understanding the regional geologic features, can lead to dangerous consequences; and (2) the geomorphologic history of a proposed site may be critical to the successful construction and/or operation of an engineered work. Since the 1918 blowout, Cedar reservoir has remained operational, but with the pool kept below the bottom of the Boxley channel.

Today, any proposed reservoirs involved with permeable deposits should consider the installation of piezometer arrays to monitor the phreatic line near likely seepage outlets. The first filling should be controlled in increments, with analysis by flow nets (Cedergren, 1989) when appropriate.

Vaiont reservoir, Italy

Four specific geologic conditions and defects of this site, and potential results, were either undetected or erroneously interpreted by the project's geotechnical staff and geological consultants during the feasibility, design, and construction phases of 1956–1960, and the operation phase of the Vaiont project in 1960–1963. As described in Chapter 22 (this volume), they included:

- Clay interbeds throughout the reservoir wall rocks and widespread clays associated with old failure zones (Chapter 22, Fig. 30) (Fig. 9).
- Solution-widened openings in limestones of the reservoir walls and the "karstic terrain" upslope (Chapter 22, Fig. 31) that influenced runoff and bank storage.
- The reservoir-induced ground-water level extended more than 2,000 ft into the bank slope and old slide mass, aided by the flattish configuration of folded beds (Chapter 22, Fig. 30).
- An undetected old fault and slide zone throughout the left reservoir canyon wall (Chapter 18, Fig. 27), which was reacti-

vated by the reservoir impoundment (Fig. 10); the fault was exposed along the eastern edge of the main slide area (Fig. 11).

Furthermore, the geotechnical advisors concluded that there was no evidence for preexisting sliding of the left canyon wall, in spite of broken rock and "tectonic" features observed in exploratory borings and adits; they pronounced the slope stable. Muller (1961) likened any movement of the left slope to that of a glacier with zero velocity at the base, increasing upward. This erroneous interpretation affected any hopes of realistically controlling the subsequent large-scale movements of 1963, and avoiding the tragedy.

Lengthy investigations from 1964 to 1969 exposed these and other errors of geological judgment relative to the guidance provided the designers, and moreover, exposed the inadequacy of the exploration and evaluation, which did not identify the old slide and past failure zones. The reservoir impoundment-bank storage affected these zones (Fig. 10), which contributed to the major slope failure and loss of 2,034 lives. An Italian Board of Inquiry blamed a group of eight engineers, officials, and one geologist for the disaster. The year-long trial of 1968–1969 to establish "responsibility" ultimately convicted three engineers, while two others, including the geologist, were dead at verdict time. Moreover, the Italian Courts accepted testimony of one convicted official, N. A. Biadene, director of construction, that "inadequate and misleading geological information was a major factor in causing the tragedy" (ENR, 1969).

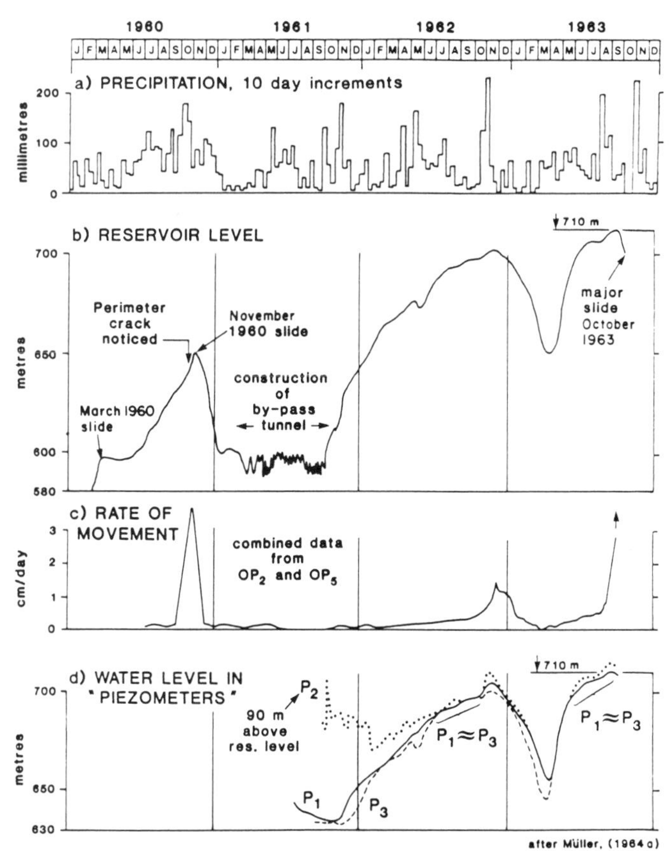

Figure 10. Comparison of water levels throughout the Vaiont slide mass based on readings of two piezometers, P1 and P3; their locations on the left canyon slope are shown on Figure 29 of Chapter 22. The error of judgement at Vaiont is demonstrated by the correlation between the reservoir level and the rate of movement throughout the eventual slide mass from 1961 to 1963. The 'permeable' reservoir wall rocks are due to inherent properties and secondary openings. This unrecognized "network" permitted lateral saturation of the canyon slope, thereby increasing uplift pressures, which culminated in the slide of 1963 (after Muller, 1964.)

Portuguese Bend slide, California

The experience at Portuguese Bend, California (Chapter 22) demonstrates how an insufficient geological investigation prior to large-scale land development can result in serious property damage and costly litigation at a later stage. The slope consisted of several old, coalescing slides that displayed evidence of sporadic movement over several million years. In view of this evidence, the land development project should have been prohibited pending completion of a coordinated geological and soils-engineering study to establish zoning restrictions and construction controls.

A comprehensive preconstruction report by knowledgable geologists and engineers undoubtedly would have advised against indiscriminate road building or undercutting the toe of the latent slide; furthermore, a tean investigation would have recommended installation of surface-water diversion facilities and subsurface drains throughout the potentially unstable slopes.

PREDICTIONS BASED ON EXTRAPOLATION

Engineers rely on geologists for a prediction of conditions to be encountered during construction of tunnels and deep excavations. The geologist's response usually includes a geologic section based on surface observations supplemented by information from exploratory drilling and/or geophysical surveys. How reliable is this customary procedure? What are its limitations? What are the factors that determine its accuracy? These questions were considered by independent investigators during the planning and construction stages for the Harold D. Roberts and Straight Creek projects, two major tunnels located in Colorado (Wahlstrom, 1964; Robinson and Lee, 1974). In each case the surface area above the tunnel alignment was mapped by experienced geologists and supplemented by geophysical surveys and limited core drilling. During construction, the as-encountered geology was mapped, and a log described the size and spacing of supports, excavation progress, groundwater inflow, and other relevant observations. When completed, preconstruction predictions were compared with the as-encountered conditions.

Straight Creek tunnel, Colorado

Prior to construction, the U.S. Geological Survey, in cooperation with the Colorado Department of Highways, initiated a geological investigation in the vicinity of the proposed tunnel alignment (Fig. 12). A major objective was to evaluate whether

Figure 11. General view of the eastern part of the main 1963 slide plane. Note the eastern margin of the slide (center); it coincided with a large-scale, ancient slide plane or ancestral décollement feature that deformed the left canyon wall more than 10,000 yr ago. Clays associated with the failure surface (whitish) and step-like failure blocks are clearly evident. (Photo by G. A. Kiersch, 1963.)

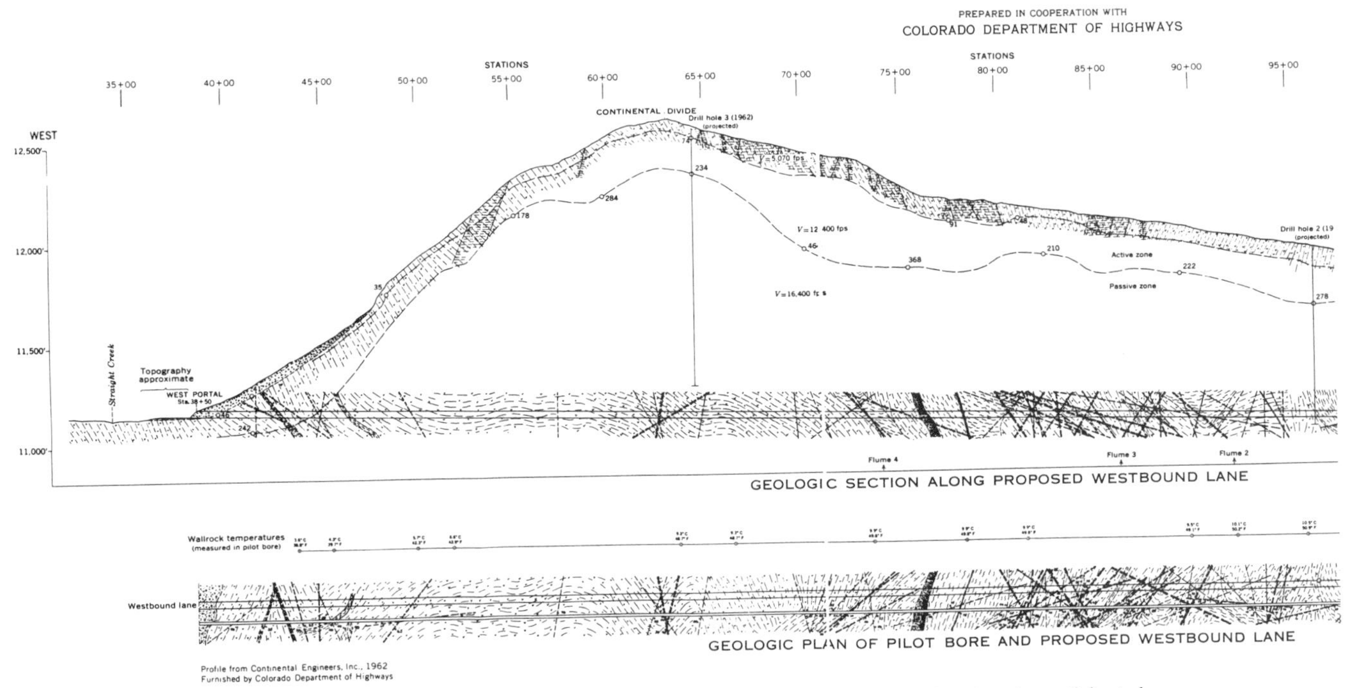

Figure 12. Geologic cross section along the Straight Creek tunnel, Colorado, showing the anticipated conditions along the full-scale westbound lane based on a correlation with the as-encountered features along the pilot bore completed during the design phase. (Source: Robinson and Lee, 1974.)

surface geology could be projected to tunnel depth with sufficient accuracy and detail to assist the planning, design, and construction of the tunnel and provide a database for future tunnels in similar geologic settings (Robinson and Lee, 1974).

The following briefly describes the outcome, when the areal map data along the tunnel alignment is correlated with the as-encountered conditions in a pilot bore 13 ft in diameter, driven before the excavation of the main tunnel. The serious problems encountered in driving the full-size twin tunnels have been described by McGlothlin (1973) and Hopper and others (1972).

The Straight Creek tunnel is located within an area of very adverse conditions. Many consider the tunnel to be the toughest ever built; portals are located at about 11,000 ft elevation, making it one of the highest major tunnels in the world. The tunnel penetrates the continental divide near Loveland Pass about 60 mi west of Denver and, besides shortening travel distances, avoids the hazardous avalanche areas along the old highway. The two-lane vehicular tunnel is 8,900 ft long; each bore is approximately 64.7 ft high by 56.2 ft wide through the most difficult stretch (McGlothlin, 1973); maximum cover is about 1,400 ft.

Bedrock throughout the tunnel consists of Precambrian granite and metamorphic rocks of sedimentary origin, which occur as inclusions in the granite. Locally, the bedrock is cut by Tertiary diorite dikes, with common faults, shears, joints, and altered zones. The Loveland fault, a major tectonic feature, is penetrated by the tunnel (Robinson and Lee, 1974).

The investigative program was designed to compare the percentages of different rock types; categories of fractures, faults, shear zones, and joints; along with their spacing and attitudes as

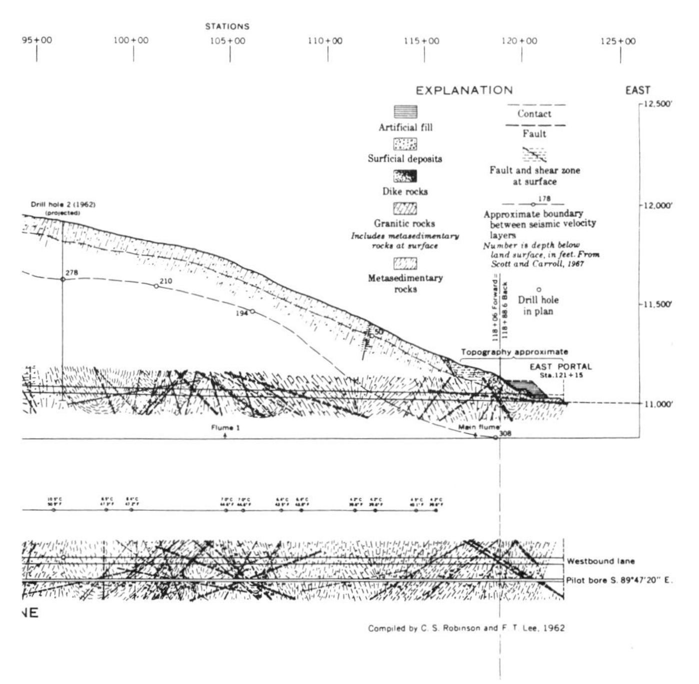

mapped on the surface with the percentages of these features exposed in the pilot tunnel. Surface mapping, at a scale of 1:2,000, covered a strip 6 mi wide athwart the alignment, with particular focus on faults, joints, rock lineation, fracture trends/attitudes, and spacing. Surface data was limited because bedrock only cropped out over 4 percent of the area mapped. The map data was augmented by geophysical survey measurements and a few cored drill holes.

Extrapolation of pilot bore data. Data were gathered during construction of the bore by underground mapping and geophysical surveys to compare with the surface information; in addition, a record on rock temperature and water inflow was kept as the heading advanced. The pilot bore was mapped every 8-hr working shift.

The statistics obtained from geologic mapping of the surface exposures and those from the pilot tunnel are compared in Table 1 and are in surprisingly close agreement. The greatest divergences are in the requirements for the number and spacing of tunnel supports, lagging and blocking, and feeler holes. However, the support requirements are to some extent a judgmental decision and subject to influence by intangible factors, such as the personal attitude of an underground foreman and/or the contractor. Consequently, personal judgment, bid prices, and geologic factors (rock load, etc.) may all contribute to the selection of the support. The correlations for support items obtained during this study are about as close as can be expected, however, considering the complexity of the geologic setting, depth of tunnel cover, and the fact that bedrock is obscured by overburden in 96 percent of the area along the alignment.

Similarly, the requirements for feeler holes and grouting ahead of a tunnel heading may be influenced by intangible factors in addition to the as-encountered conditions. Consequently, it is not surprising that the requirements for these items also exceeded the estimates (Table 1). Rock temperatures ranged between 4.5°C (39.6°F) and 10.5°C (50.9°F); natural gas was neither expected nor encountered.

Water inflows during tunnel construction fluctuated seasonally; heaviest inflow during the spring melt was more than 7 times that of the winter and summer seasons. Consequently, it was not possible to make meaningful predictions of flows to be encountered as the heading advanced.

The results of the U.S. Geological Survey and Colorado Highway Department geological study of the surface area were made available to the contractors, prior to preparation of bids for the construction. Cost of the pilot bore was $400,000 less than the engineer's estimate.

The pilot tunnel study by Robinson and Lee (1974) concluded that geology can statistically predict the kinds and percentages of the different rock conditions at depth and that engineering requirements equated with predicted geologic conditions can provide a sound basis for estimating probable cost of construction. The Straight Creek tunnel was constructed in a bedrock area with a relatively small number of geologic variables, which partly accounts for its success. However, a similar approach can be equally successful when applied to the projection of geologic conditions to depth in any environment, if the structural framework of the areal geology is properly analyzed and thoroughly understood.

JUDGMENTS RELATED TO FOUNDATION MATERIALS AND FEATURES

Errors of judgment and misinterpretation of subsurface data often occur in assessing the physical properties and characteristics of rock masses. One of the most common circumstances involves the misinterpretation of cored boring data for the design of tunnels and dams, demonstrated by three cases that follow.

Broadway tunnel, San Francisco

The Broadway tunnel, constructed by Morrison-Knudsen Company, Inc., was the culmination of 85 years of endeavor by advocates of a tunnel under historic Russian Hill that began with proposals in 1865. After many attempts, voters approved a bond issue in 1946, and construction began May 10, 1950. The tunnel consists of twin bores, 28.5 ft wide and 35 ft apart, which provide a traffic artery from the metropolitan district to the northwestern part of San Francisco and the Golden Gate bridge.

Exploration. In planning for the project, the city of San Francisco engaged a consulting engineer/geologist in 1944 to direct and supervise the drilling of cored borings and to provide an interpretation and evaluation of the geological data obtained along the proposed alignment. The subsurface investigations were critical to understanding the rock and soil conditions to be encountered along the urbanized and built-over tunnel alignment (Cooney, 1952). The feasibility study by the consultant reported many small faults, some breccia and shear zones, and a general deformation of the Franciscan rock mass. However, the full geological implications and meaning of the boring data, both physical and "fugitive," were not realized and/or not well interpreted (Forbes, 1945). The evaluation of the low core recovery (less than 50 percent) erroneously placed too much emphasis on core losses due to mechanical or drilling causes, and insufficient attention was paid to the highly fractured conditions of the recovered core as an indication of rock in place. Tunneling quickly revealed a more extensively fractured and less healed rock mass with weathering more widespread than forecast. Unfortunately the consultant had not investigated the surface outcrops near the tunnel to establish the occurrence and characteristics of the sandstone and shale units involved with the alignment (Marliave, 1951), a serious oversight in any geological judgment of a site. The alignment largely coincided with a topographic saddle between Nob and Russian Hills, which usually indicates faulting in Franciscan rock, while the eastern portal area of Russian Hill was on an old, back-filled stream channel with no bedrock for the planned tunneling.

As the tunnel progressed, geologic conditions continued to be very different from those expected by the contractor from bid documents, and a new geological investigation was made for the contractor in early 1951 by Chester Marliave. By the summer of 1951 the city of San Francisco was also concerned and made two separate studies: one by consultant John P. Buwalda (1951), and the other by the renowned consultant Karl Terzaghi of Boston. In addition, the as-encountered geologic conditions throughout the tunnel were mapped by U.S. Geological Survey personnel. The consultants and the USGS investigators concluded that the geo-

TABLE 1. STRAIGHT CREEK TUNNEL PILOT BORE*

Geologic Measurements		Predicted	Findings
Percent of tunnel length:			
Rock type:			
Granite		75.0	75.4
Metasedimentary rock		25	23.8
Diorite dikes		±	0.8
Fracture density:			
<0.1 ft to 0.5 ft		40.1	38.7
0.5 ft to 1 ft		49.3	42.6
>1 ft		10.6	18.7
Faults and shear zones		51	49
Faults	Strike	N20 to 50°E	N20 to 45°E
	Dip	75°NW/SE	40 to 75°SE
Joints	Strike	180°	180°
	Dip	45 to 90°	8 to 90°
	Average dip	60°	45°
Foliation	Strike	N to N 30°E	N 10 to 60°E
	Dip	60 to 90°SE/NW	10 to 50°SE
Engineering Measurements and Practices		**Predicted**	**Findings**
Rock loads		6,970 psf	6,300 psf
Ground water:			
Maximum initial flow from any section		1,000 gpm	750 gpm
Maximum flow from portal		500 gpm	800 gpm
Flow at portal 2 weeks after completion		300 gpm	130 gpm
Set spacing (% of length):			
1 ft		1.6	2
2 ft		23	22
3 ft		40	30
5 ft		35	20.9
Invert struts		1.4	8
Total number of sets		2,691	2,059
Total number of invert struts		113	210
Lagging and blocking:			
100 to 67 percent		1,731 ft	1,456 ft
66 to 34 percent		3,659 ft	2,447 ft
33 to 0 percent		2,660 ft	4,447 ft
Feeler holes		2,905 ft	9,816 ft
Grout		403 ft	0

*Source: Robinson and Lee, 1974.

logic setting of the tunnel site had not been fully evaluated and correlated with the cored-boring data for purposes of either design or bidding.

Changed conditions. The Broadway tunnel project became the center of a long public controversy between the contractor and the city during 1951, after tunneling was well advanced and the actual rock conditions were realized. The contractor contended the original consulting report of 1945 was misleading: the substantially changed conditions that were encountered confirmed that errors of judgment had resulted in a less than adequate and safe tunnel-support design.

The tunnel was largely driven in distorted, strongly jointed and fault-sheared Franciscan Formation rocks, with slippage along bedding planes that ranged from those in soft and clayey weathered sandstone and shale units to some hard gray sandstones with clay interbeds (Fig. 13). The principal contentions of the contractor, regarding misrepresentation by the city's consultant in 1945 and the owner, were that the descriptions of rock conditions stated or implied: (1) no bedding planes in the sandstone unit, no faults parallel to the tunnel alignment, and all faults were healed so that the rock mass was generally intact; (2) no slickenside features or swelling were noted in the cores (thus implied for mass), and there was no indication of swelling ground; (3) no air-slaking materials were known from the cores, and the shales were of limited extent; and (4) no ground-water inflow was to be expected.

The original consultant's report in 1945 stated that the core recovery (less than 50 percent) was the lowest in his experience on many projects around San Francisco and attributed this to both natural fracturing and to mechanical grinding, blocking, and overruns by the driller. The rock conditions described in 1945 influenced the contractor to bid with the expectation of using a full-face mining method throughout much of the alignment. Yet the experience at the east portal of the north bore soon changed the outlook; an old stream channel, since back-filled, existed throughout the east portal sector, and the forecast hard rock did not exist, which necessitated extending the use of Type-A steel supports.

The contractor was soon convinced that a top-heading method or any full-time tunneling method, as planned, was impractical. The rock mass consisted of thick-bedded sandstones, thin-bedded sandstones with interbeds of soft shale, and thin-bedded sandstone and shale with widespread clay gouge and seams. The rock mass varied from hard to soft, and the highly weathered, sheared, and fractured rocks air-slaked on exposure. Large slabs could be air-spaded and chipped off with little effort; blasting was controlled and used sparingly.

Due to slow progress and excessive settling the contractor changed the top-heading mining method after driving 178 ft from the east portal of the north bore (Cooney, 1952). The plumb-post method was adopted, with two header foot-block drifts 9 ft wide on each side at the base of the tunnel section (Fig. 13). The remaining 60 percent of the face excavation was removed with a breast-board machine. A design controversy arose based on an

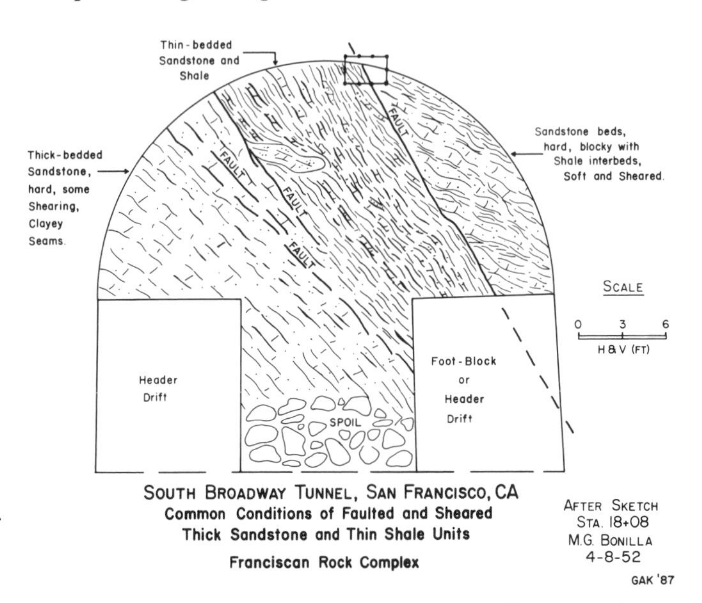

Figure 13. Broadway tunnel, San Francisco. Geologic cross section at station 18+08 of the southbound lane. The common conditions encountered in the Franciscan rock complex are represented, along with a plot of faulted and sheared rock units and mining methods. (after field notes of M. G. Bonilla, 1952; U.S. Geological Survey Library.)

assessment of the as-encountered rock conditions and inherent strength of the Franciscan rock mass. The contractor originally recommended that the stronger Type-B (a full-circle) tunnel support be substituted for Type-A (a three-quarter support) in several sectors of the twin tunnels. The city of San Francisco considered this proposed change in design and decided the $1.5 million modification would be a waste of taxpayers' money (Brooks, 1952; Forbes, 1951).

Overview. The contractor's basis for contending that the weaker Type-A support was inadequate in some sectors of the tunnel involved the width of the excavated bores and the extent to which the highly fractured rock conditions and areas of soft and weathered rock material impaired the natural arch action of the rock mass (Cooney, 1952). The Broadway tunnel experience emphasizes the importance of accurate interpretations of drill core data, and particularly the ability to distinguish geologic defects inherent in the undisturbed rock mass from mechanical flaws created by drilling operations. Such ability is largely a matter of good judgment, and best available to a project by the on-site participation during drilling operations of an experienced applied geologist. All too often the least qualified personnel are assigned this task, and in this case the consultant was weak on interpreting data from the borings.

Golder dam, Arizona

This privately built reservoir project is an unusual example of an earthfill structure that failed. The puddled clay foundation cut-off was an inadequate design and did not reduce underseep-

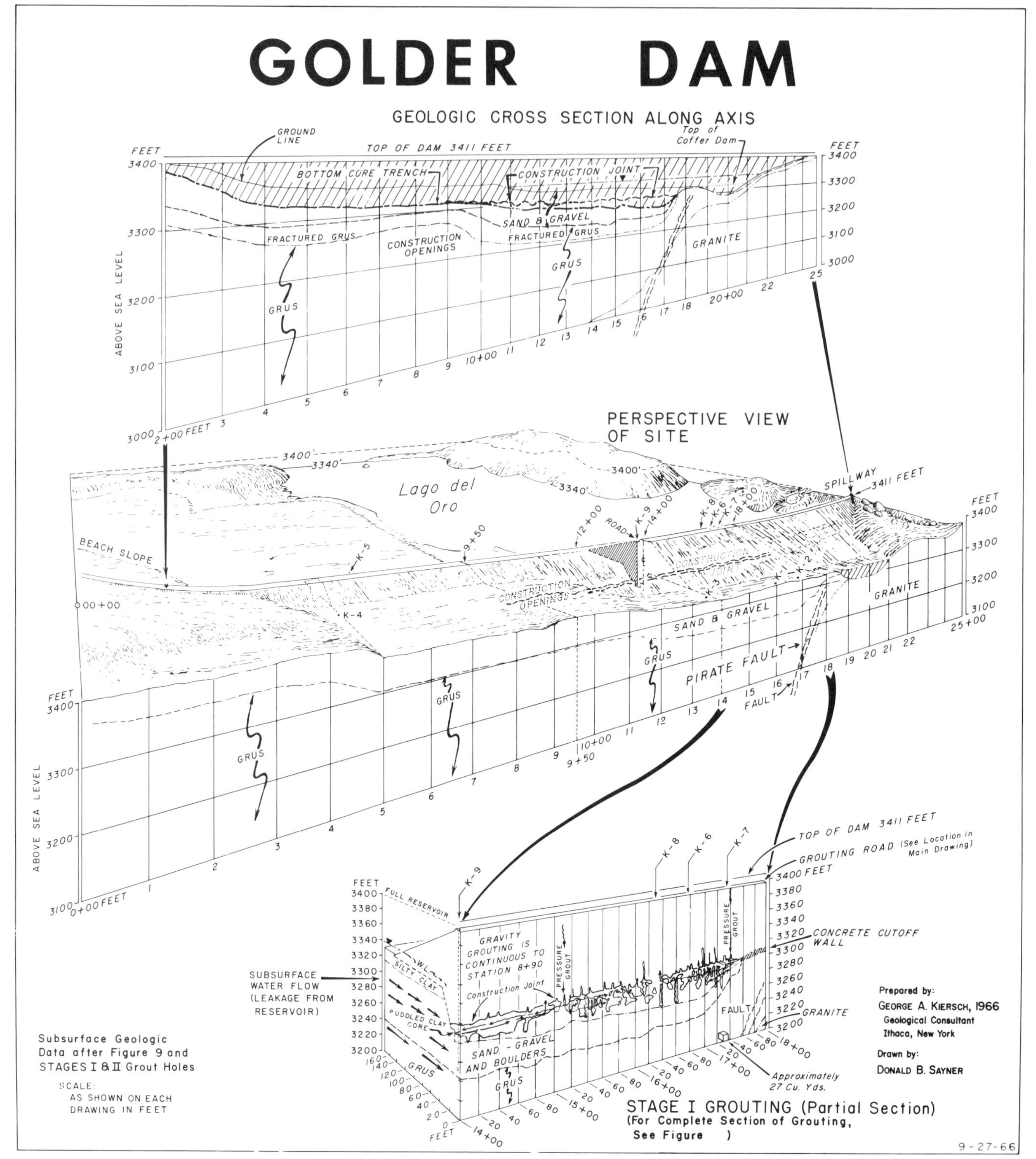

Figure 14. Golder dam and reservoir area. Perspective view with subsurface geologic section along the dam axis that shows distribution of granite, grus, and alluvial deposits. The eastern sector of the stage 1 grouting by gravity to seal the construction joint within the dam is shown in a separate three-dimensional drawing. (From: Kiersch, 1968a.)

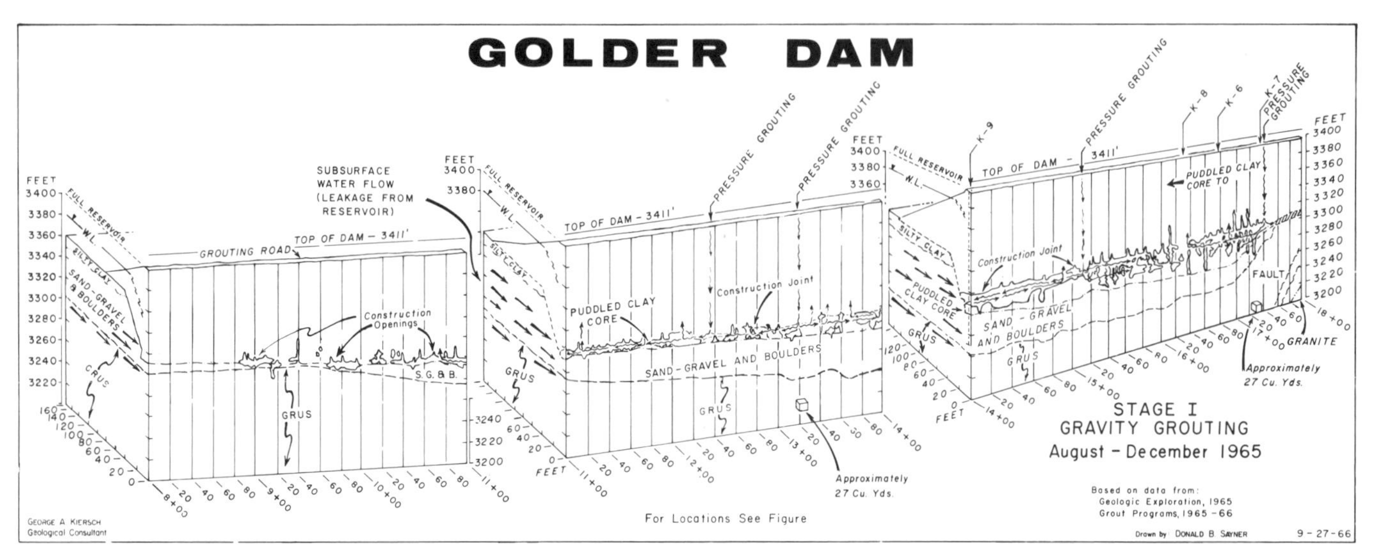

Figure 15. Golder dam. Three-dimensional view of the dam showing the location and distribution of two avenues of leakage: the construction-joint opening within the impervious core; and the section of permeable alluvial deposits and uppermost grus beneath the puddled clay core. (From: Kiersch, 1968a.)

age. In addition, a construction opening developed within the dam, between the main embankment core and the deeper cut-off, through which major leakage occurred on the first filling. Constructed by the Rail-N-Ranch Corporation of Arizona in 1963–1964, the 130-ft-high dam with a 280-acre reservoir was planned as the nucleus for a new commercial and residential community in southern Arizona. The following review of the project and litigation is derived from reports by Kiersch (1966, 1968a).

Golder dam is in the Canada del Oro along the northern edge of the Santa Catalina Mountains, 18 mi north of Tucson. The eastern part of the dam site and reservoir is underlain by granitic rocks and weathered detritus: grus. Alluvial deposits of sand and gravel, with a thick section of grus at depth, occur beneath the western sector (Fig. 14). Throughout the design and construction phases, insufficient attention was given to the importance of the sand and gravel formation, with its highly permeable interbeds and lenses, in which the major underseepage flows occurred. The gravels and a seasonal fluctuation of the water table combined to cause development of an opening within the earthfill at the top of the puddled core section (construction joint).

Golder dam was completed in February 1964, but the reservoir remained dry, due to a low ground-water level, until summer storms on July 26th, 1964. With this first reservoir impoundment, leakage was observed at the downstream toe of the dam 6 days later. The leakage increased and by March 1965 approached 5,000 gpm with a partially filled reservoir.

Investigation of leakage. During 1964–1965 a number of remedial measures were attempted, such as a clay blanket for the reservoir floor, but all proved ineffective. A systematic geological investigation was subsequently undertaken in 1965–1967 (Kiersch, 1968a) to determine the causes of the leakage. This included cored borings, pumping-in and pumping-out water tests, and daily inflow and outflow measurements of the reservoir and the downstream leakage. These investigations determined that the dam was inadequately designed and built because the impervious core cut-off stopped 5 to 40 ft above the "bedrock" and was founded on permeable sands, gravels, and boulders (mistaken for bedrock) for a distance of 800 to 1,000 ft; and because the upper 10-ft of "bedrock" was highly permeable grus, and additional bank storage losses occurred throughout the sand and gravel formation along the western margin of the reservoir. The Pirate fault zone (Fig. 14) within the channel section had been assumed by some geologists to be the main avenue of water escape from the reservoir; yet several borings across the fault zone showed no losses when tested by the pumping-in pressure test. Actually, the fault zone in the granite is tight and intruded by stringers of dark greenish, basaltic rock. A construction joint (maximum 1 ft in width) developed after completion (1964–1965) within the core section of the dam between the lower puddled core section and the overlying main core (Fig. 14). This extended for 850 ft along the dam site. Thus, the dam experienced two avenues of leakage: the construction-joint opening and the underlying alluvial deposits and uppermost grus. Leakage reached 11,000 gpm in April 1966 when the reservoir was nearly filled (Figs. 14 and 15).

Remedial treatment. Repairs to the dam consisted of an initial stage of primary gravity holes (6¾-in rotary) on 20-ft centers, with secondary holes on 10-ft centers and a few on 5-ft centers, to fill and grout the construction-joint openings, and a second stage of pressure grouting primary holes on 10-ft centers and secondary holes on 5-ft centers to seal the 5 to 40 ft of permeable sands and gravels and the underlying upper 10 ft of the permeable grus (Fig. 15).

The gravity grouting was completed by December 1965; major winter storms largely filled the reservoir during January 1966, which field tested the adequacy of the gravity grouting program. Although some seepage occurred at the downstream toe in early January, characteristics were different than previously; this was underseepage instead of the through-seepage sealed off by the gravity grouting. The pressure grouting phase to seal the underseepage began in March 1966 and continued until August when 500 ft of the thickest section of permeable gravels had been sealed using a chemical grout. Lack of finances and legal entanglements stopped the pressure grouting with as much as 500 ft of permeable gravels remaining. Subsequent hydrologic studies documented that the pressure grouting was effective and reduced underseepage losses.

Litigation. Claims filed by the owner regarding the inadequate design and construction of Golder dam centered on responsibility for design of the foundation cut-off that allowed excessive underseepage, and the design and construction of the impervious core that caused an open joint to develop, causing major leakage through the core.

Lengthy litigation terminated in March 1969: the constructor (Tanner Brothers), the engineer-design firm (Hoffman-Miller Engineers), and the consulting geologist (S. F. Turner) and geophysicist (Paul Menera) accepted responsibility for committing errors of technical judgment and for incorrectly interpreting the geologic facts and core-boring data. The owner, Rain-N-Ranch Corporation, was awarded a sizeable settlement, sufficient to fully repair the faulty dam through grouting techniques and construction of a downstream berm and dewatering well system.

Unfortunately, the owner did not choose to adequately repair the underseepage. After several legal decisions, the state of Arizona breached Golder dam in 1982 to remove any dangers of failure due to thru-leakage or piping. In 1983 the downstream, residents of Canyon del Oro requested that the dam be rebuilt to provide flood protection. By 1986, plans and field investigations were underway to rebuild the breached dam (ATC, 1988); an out-of-court settlement was reached with residents in 1989.

Silent Valley dam, Ireland

A project with some similarities to Golder dam is Silent Valley dam, built by the city of Belfast. During the geological investigations, large granite boulders were encountered in borings at a depth of 50 ft and mistakenly interpreted to be bedrock (Walters, 1971). Actually the depth to bedrock was 150 to 200 ft below the surface, as demonstrated by subsequent investigations and borings.

Similar experiences at many other projects have convinced geologists that exploratory borings must confirm the bedrock positions and penetrate at least 10 ft into rock in place. Forecasting the depth to bedrock and elevation of the upper contact frequently requires a general understanding of the geologic environs, history of events, and rock units, besides cored boring data and frequent geophysical profiles between the drill holes.

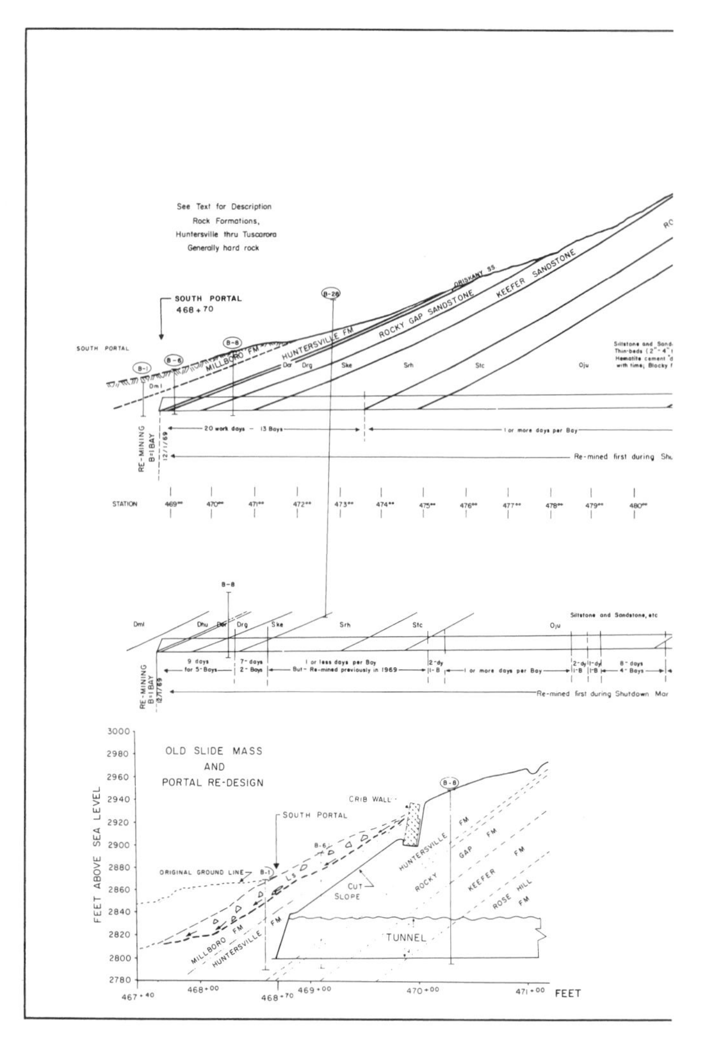

INADEQUATE GEOLOGICAL DATA BASES

Errors in design of major engineering works are frequently due to an inadequate investigation and database for the site selected. The four cases that follow, a major interstate highway tunnel and three dams built in very different geologic environs, illustrate how the inadequate database resulted not only in errors in design and expensive construction changes, but also in long periods of maintenance and remedial treatment of previously unrecognized adverse features, such as solutioned limestone.

Big Walker Mountain tunnel, Virginia

The Big Walker Mountain tunnel on Interstate Route 77 in Wythe and Bland Counties, Virginia, was one of the first twin-bore tunneling projects (1961–1970) of the interstate highway system. In the 1960s, the tunnel project, with a planned cost of $23,000,000, ranked as the most expensive 0.8 mi of highway in the Virginia highway system (Cooper, 1968).

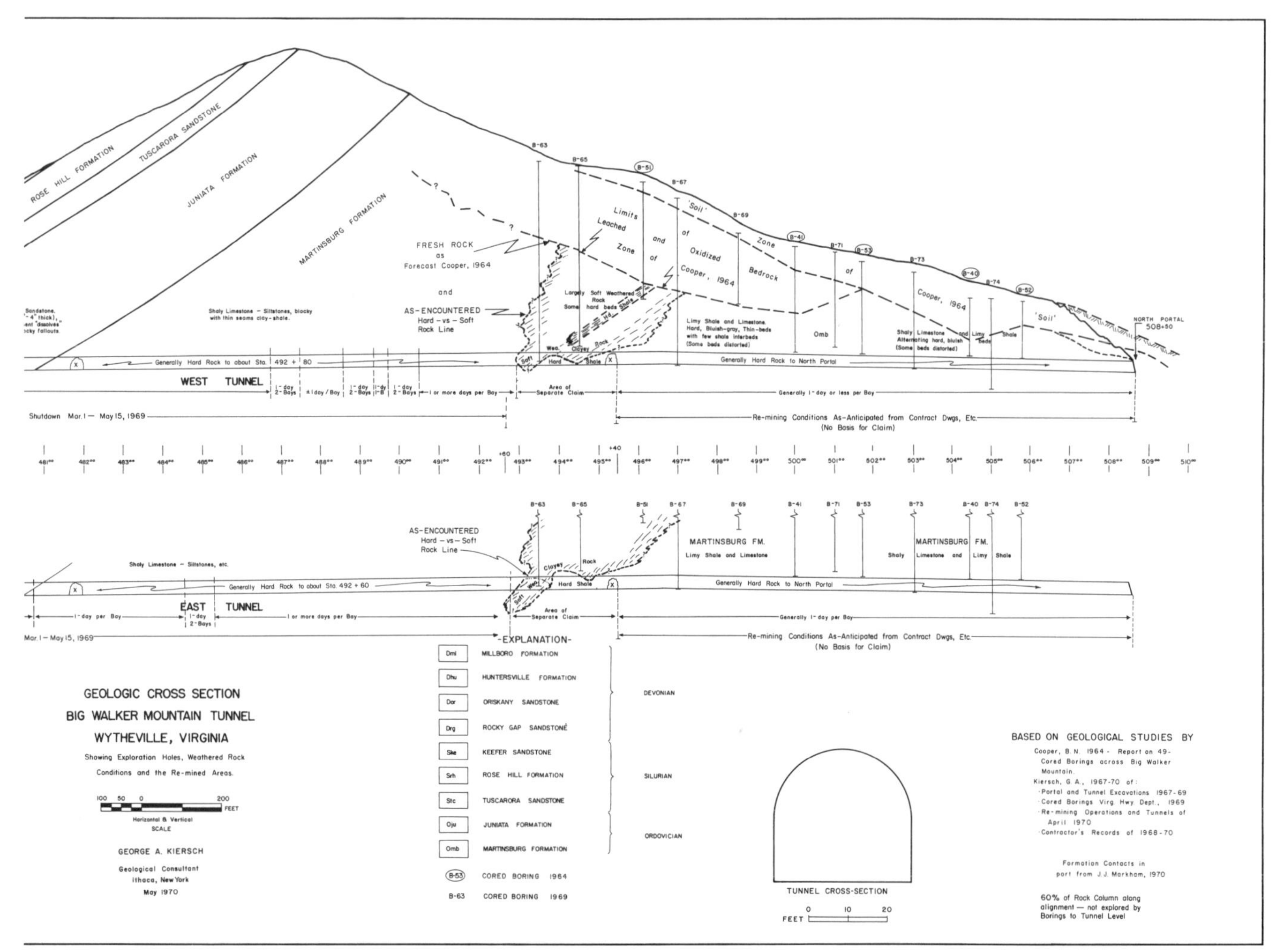

Figure 16. Big Walker tunnel, Virginia. Geologic cross section of both the west- and east-lane bores showing rock units, location of weathered Martinsburg rocks at tunnel grade, distribution of exploratory borings, and the slide mass encountered at the south portal. (Sources: Kiersch, 1968b, 1969; Markham, 1970.)

Walker Mountain is in many respects a typical Appalachian ridge. The rock formations of the ridge form the northwest slope of a major synclinorium. Walker Mountain is composed of a Paleozoic succession at least 14,500 ft thick (Fig. 16). The Tuscarora Formation (85 ft) forms the crest, and the Rose Hill Formation (110 ft) forms the main dipslope on the southeast side. The thick Martinsburg Formation (1,600 ft) was penetrated, except for the basal 150 to 200 ft at the north portal, as were the Juianta sandstones and shales (320 ft); the remaining six younger formations were penetrated around the south portal. Exploration by cored borings for the feasibility and alignment of the tunnel (Cooper, 1964) did not investigate the uppermost 400 ft of the Martingsburg beds, which constituted about one-half of the stratigraphy of the tunnel alignment at grade level (Fig. 16).

Exploration and tunneling. Difficulties for the contractor began with the initial Langenfelder-Raymond construction at the south portal based on the subsurface geological investigations in 1964. Two serious errors of judgment are incorporated in the initial report to the owner, the state of Virginia, by Cooper (1964): the geological exploration of the south portal area failed to recognize an old slide feature (Kiersch, 1968b); and the rock-mass conditions to be expected at tunnel grade beneath Big Walker Mountain were inadequately evaluated (Kiersch, 1969). Both resulted in changed conditions for mining and design and caused lengthy delays. Most critical was the lack of knowledge of the depths to which the uppermost Martinsburg Formation, of alternating limestone and shale units, had been decalcified and weathered to a soft, clayey residual mass (Fig. 16). Extensive areas were weathered to depths below the selected design grade of the twin tunnels; when encountered, the weak, clayey masses collapsed in the tunnel face and required redesign of the mining/tunneling method (Fig. 17) to header drifts and plumb posts. Moreover, this condition required extensive sections of the tunnels to be reinforced with steel supports (Fig. 16).

Figure 17. Big Walker tunnel. West-lane bore, full face at station 493+00 on June 19, 1969. After encountering soft, weathered shale–clayey beds the mining method was redesigned to header drifts and plum posts. (Source: Langenfelder-Raymond, personal communication, 1969.)

If the decalcified and weathered limestone, with its soft masses at depth, had been delimited before the tunnel grade was selected, most of the as-encountered "changed conditions" could have been avoided. During the shutdown caused by these conditions in 1968–1969, a supplemental cored-boring exploration program of the alignment was completed by the state of Virginia. This included deep borings, probes ahead of the tunnel faces, and geophysical tests and surveys. The boring data indicated that a simple lowering of the tunnel grade by about 15 ft (Fig. 16) would have placed the twin bores wholly within good quality limestone, shale, and sandstone rocks and beneath the lower limits of the weathered rock units.

The unscheduled changes in mining/tunneling methods, and redesign, resulted in months of down time and financial losses to the contractor. Furthermore, this delay triggered other geologic actions throughout the excavated sections of the tunnels (Fig. 18): a destressing of retained strain energy occurred in the rock zone around the tunnel opening during the months before the final concreting and full support were placed. This rebound effect caused extensive fracturing and spalling within the zone of loosened rock and required excessive re-mining (120 days) before the tunnel could be lined. The mined-out space between the intact rock and the tunnel supports required 24 percent more concrete backfill than is normal.

Changed conditions. The initial south portal excavations for the east tunnel encountered changed conditions where an old landslide zone and thick debris were discovered. Subsequently, major unanticipated problems began at station 492+88 when soft and clayey altered rocks collapsed the face. Tunneling was stopped while the geologic conditions were reevaluated; as commonly occurs, major differences of opinion were expressed about the same factual data by the state of Virginia and by the contrac-

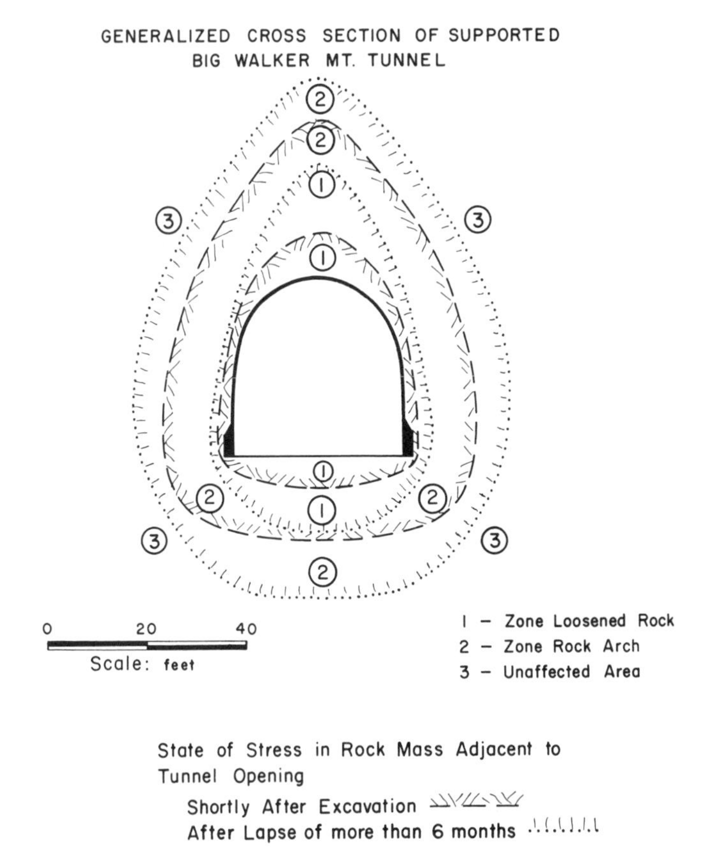

Figure 18. State of stress in the rock mass adjacent to the Big Walker tunnel opening, shortly after excavation and then after the lapse of more than six months. (Source: Kiersch, 1970.)

tor's representatives. Some of these differences and beliefs revolved around the pre-bid documents (Cooper, 1964), which were proved incorrect by the as-encountered geologic conditions during construction. The problems included inadequate presentation of the distribution of subsurface rock conditions, as to weathered versus fresh rock, on the drawings; and omission from drill logs of mention of excessive moisture in the clayey overburden at the north portal. Other problems with presentation of data and with errors throughout the initial geological report misled the contractor. For example, "No additional information is needed to present a full picture of the basic conditions [implies geologic] that will be encountered in building the tunnel through Big Walker Mountain" and "the State of Virginia has investigated rock conditions in sufficient detail to design the complete tunnel" (Cooper, 1964, p. 9).

Later the state of Virginia agreed that changed geologic conditions were encountered and authorized extensive exploration and physical testing to be undertaken. This included in situ pressure tests by the contractor to determine the engineering properties of the soft versus fresh rock exposed in tunnel. Menard pressure-cell tests were particularly meaningful in determining the strength characteristics and value of stresses active in the weak soft-rock pillars between bores.

Geologic causes for excessive re-mining included a combination of inherent features of the rock units and several geologic changes induced by tunneling that impacted Big Walker Mountain (Kiersch, 1970). The southward dip of beds (Fig. 16), which allowed the penetration of air and caused hydration of the inclined seams and blast-fractured beds, contributed to the increased rate of rock deterioration. The longer the deterioration-prone beds were exposed, the greater the depth of air penetration. Because of construction delays, this phenomenon was responsible for excessive re-mining costs for removal of broken rock that progressively deteriorated around the opening (Fig. 18). The destressing phenomenon contributed to the large overrun of concrete required when pouring the anchor curbs for the lining; this averaged 48 percent above the anticipated volume in the contract and bid.

The unexpected delays more than doubled the length of time the wall rock was exposed to deterioration processes in both tunnels; some of the delay reflected the slow rate of tunneling dictated by changes in mining method through the soft ground (Fig. 17). Overall, the changed conditions caused tunneling operations to extend 10 months beyond the originally anticipated schedule: 3½ months shut down for reevaluation, plus months of slow mining afterwards.

Although several years of geological exploration and design preceded construction, the conditions encountered were not anticipated. Lack of an adequate initial geological base led to submission of and payment for five major claims by the contractor for added expenses incurred due to changes in character of the work.

Lotschberg tunnel, Switzerland

A deeply incised glacial valley filled with morainal debris was intersected during construction (1906–1911) of the 9.25-mi Lotschberg tunnel. This unexpected feature was responsible for catastrophic flooding of the tunnel heading. Preconstruction investigation of the surface had concluded the tunnel alignment would penetrate only granite, gneiss, or limestone. However, in the section beneath Kander Valley, an ancestral channel was cross-cut, and some 9,200 yd^3 of saturated sand and gravel burst into the heading; 25 workmen were killed (Anderson and Trigg, 1976). No detailed geologic investigation had been undertaken prior to commencing construction.

Following the accident, the tunnel was realigned to pass beneath the gorge at a point where granite was exposed in the bed of the channel. The realignment necessitated the excavation and construction of an additional one-half mile to the originally planned tunnel length. This historic tragedy has been a reminder that the bedrock profile must be established whenever a proposed tunnel may intersect an ancient back-filled channel in the bedrock, particularly in glaciated regions. A similar experience al-

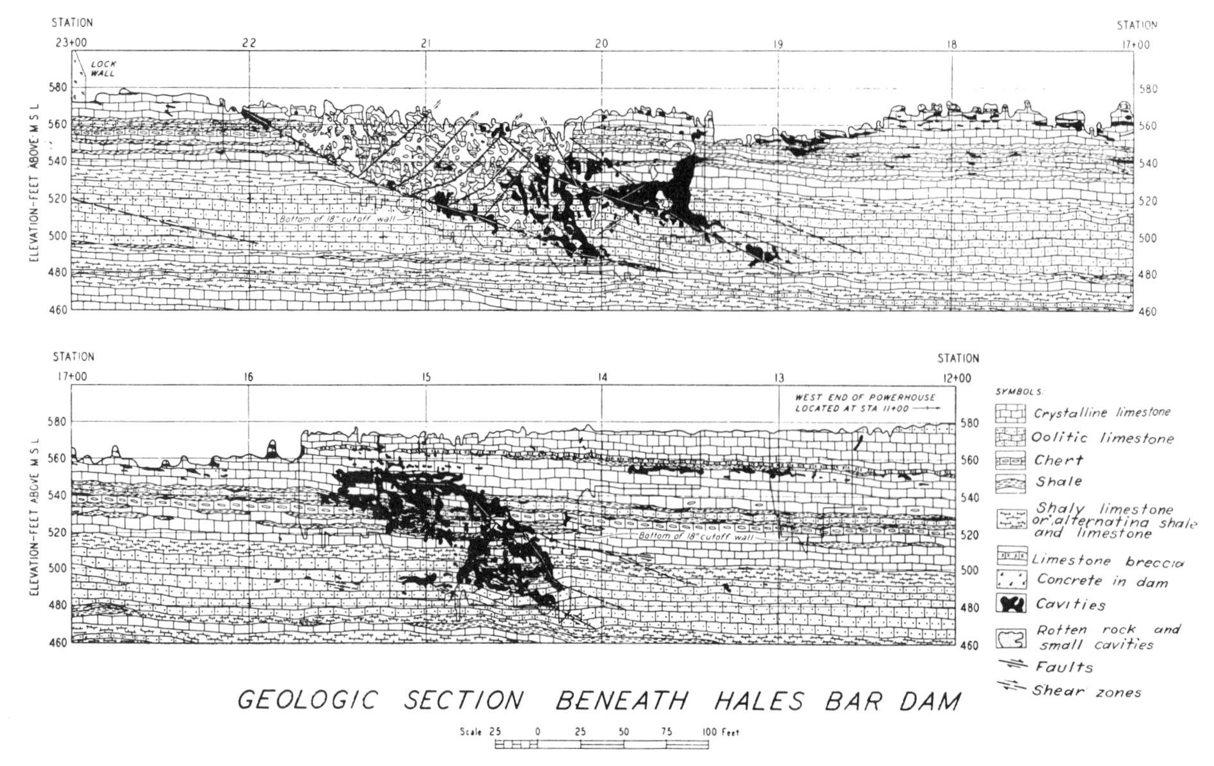

Figure 19. Hales Bar dam. Geologic cross section along the dam axis, showing the distribution of cavernous limestone units and associated structural features beneath the dam. (Source: Schmidt, 1943.)

most occurred during construction of the St. Gotthardt tunnel in the 1870s (see Chapter 1).

Hales Bar dam, Tennessee

This dam site has been described as the most drilled and most cavernous site in the world (Frink, 1946; Moneymaker, 1968). More than 2,000 drill holes disclosed 8,411 cavities with vertical dimensions ranging from 0.1 to 34 ft, the deepest lying 148 ft below the surface (Fig. 19). The Hales Bar experience was one of the earliest to demonstrate that carbonate rocks may contain extensive, partially filled cavities that extend more than 100 ft below the water table and river channels (described at Croton dam, Chapter 1).

This project, undertaken by private interests, included an earth-embankment dam with a concrete core wall 745 ft long, a navigation lock 60 ft wide and 267 ft long, and a power plant. Construction commenced in 1905, and severe leakage developed shortly after the completion in 1913. Several attempts to seal off the leakage by grouting were unsuccessful. The Tennessee Valley Authority (TVA) took over the project in 1939, at which time leakage had accelerated to 1,700 cfs and the stability of parts of the dam had become questionable. After an extensive investigation (Frink, 1946), TVA undertook two grouting programs to seal off the leakage, but they reduced the reservoir losses only temporarily. Particularly troublesome were clay, silt, and sand that filled or partially filled solution cavities in the foundation of the dam. These fillings obstructed grout penetration and prevented the cavities from being entirely sealed. Following each grouting operation, the inevitable seepage through these residual deposits eroded new channels; this progressive increase in leakage ultimately reached unacceptable levels. A subsequent review of the project conditions and economics resulted in abandoning the project for a replacement dam downstream; Hales Bar dam was demolished in 1968.

Under the circumstances of extensive solution action and subsequent cavitation throughout a proposed site, abandonment of the location for an alternate site with better quality rocks may be suitable, as at Lone Pine, discussed below. However, not all reservoir sites situated in limestone or dolomite rock terrain should be avoided. Sites prudently selected on the basis of adequate explordation and an understanding of the solutioned rock (Kiersch and Hughes, 1952) can usually be made watertight by well-planned foundation preparation and rim tightening.

According to Frink (1946, p. 567), the Hales Bar dam site "was selected almost wholly on the basis of physiographic character of the gorge and geology played a very small part in the choice. No effort was made, as far as known, to determine the character or condition of the rock before construction commenced."

Lone Pine reservoir, Arizona

Failure of Lone Pine reservoir to hold water illustrates a complete disregard of geologic conditions. The 101-ft earth- and rockfill dam on Show Low Creek, a tributary of the Little Colorado River, is located 7 mi northwest of Show Low, Arizona (Fig. 20, inset map). No geological advice was used in selecting the site or designing and constructing the dam (Kiersch, 1958), which was a part of a Public Works Administration (PWA) program, dedicated in March, 1936.

In April, 1936, when the reservoir stood 60 ft deep at the dam and extended 2 mi upstream (Fig. 20), noticeable leaks began to develop. Both the PWA and state of Arizona engineers became concerned for the first time about the possible influence of the geologic setting. A geological study was made by Parry Reiche (1936), who concluded, "geologic conditions are so adverse that a remedy for the leakage is not economical." This recommendation was disregarded by the engineers in charge and local irrigation officials because they were convinced the sinkholes identified could be plugged, as indicated by an excerpt from the Snowflake Herald newpaper: "A sinkhole running one second-foot was discovered and sealed with clay soon after three earlier sinkholes leaking 2, 3, and 4 second-feet were discovered and treated." The water level receded from the high mark of 74 ft of April 7, 1936, to the 43-ft mark on May 8, 1936; thus 5,700 acre-ft were lost in 31 days.

During the late 1930s and mid-1940s, the U.S. Bureau of Reclamation studied the feasibility of its Snowflake project, and Nickell's (1939) conclusions were similar to those of Reiche (1936), as were those of Lasson (1947) who recommended that the reservoir site be abandoned as a possible storage basin. In 1953, the Arizona Game and Fish Commission proposed integrating the Lone Pine reservoir into their statewide system of lakes and reservoirs. A geological study of the Lone Pine reservoir by Kiersch (1953) again concluded that the reservoir site did not appear feasible for storage. Instead, a new structure, located nearby at a favorable upstream site, was considered more promising and economical (inset map, Fig. 20).

Geologic setting. The Show Low–Snowflake area is located a few miles north of the Mogollon Rim. Show Low Creek and its tributaries have dissected this flattish upland area, and the youthful valleys are steep walled and a few hundred feet deep.

Most of the reservoir basin is underlain by sandstones, sandy limestones, and limestones, gently dipping northward, which lithologically resemble the Coconino and Kaibab Formations (Fig. 20). Where limestone crops out, the joints and bedding planes are widened by solution action. The limestone is relatively thick immediately upstream from the dam (Figs. 20 and 21) where wide openings, numerous sinkholes, and an interconnected system of caverns have developed. The widespread overlying Pliocene Bidahochi Formation consists of a lower basalt, which forms the upper two-thirds of the dam abutments, and thick fluviatile gravel, sand, and tuffaceous beds. An extensive late Quaternary basalt caps the section. The composition and lithology of the Coconino and Supai Formations at depth are important to the reservoir operation and are discussed elsewhere (Kiersch, 1958). Some of the moderate-scale folds, monoclines, and domal uplifts that interrupt the regional homoclinal structure influence conditions at Lone Pine directly (Fig. 20, vicinity map).

Sinkholes of the reservoir basin are associated with the minor folding and faulting that fractured all rocks and facilitated the solution of limestone and limy sandstone beds. The sinkholes that are located near the fold axis west of the reservoir occur in areas thinly blanketed by gravel; they were reopened when water pressure of the nearly full reservoir breached the gravel cover (Fig. 20). Reiche (1936) described the sinkholes as ranging from 2 to 30 ft in diameter.

Reservoir leakage and failure was due to the secondary permeability of limestone, the gentle northward dip of the pre-Tertiary beds, and the inherent porosity of the sandstone. If the lower reservoir rocks had been impervious, then serious leakage would have occurred through the open joints of the overlying lower basalt (and underlying ash bed), which forms the dam abutments, and the open texture of the Bidahochi sands and gravels above.

Overview. Kiersch (1958) deduced that the hydrostatic pressure exerted by a full reservoir (90 ft) would probably rupture and flush the sealing material in a patched opening, unless it was strongly bonded. Undoubtedly the silt that naturally seals the ponding basin (Fig. 20) would be flushed by the hydrostatic head of a near-capacity reservoir. Because all reservoir rocks are permeable to varying degrees, parts of the impoundment area could not be diked off to reduce leakage, and any attempts at repair would be uneconomical.

Satisfactory reservoir sites in acceptable rocks occur upstream from Lone Pine (Fig. 20, inset map) where a major share of the design capacity of Lone Pine reservoir could be impounded at a reasonable cost. Had a geological study been a part of the site selection and planning phases of Lone Pine dam, this conclusion would have been apparent; the total loss of project funds, besides the personal losses of local inhabitants, could have been avoided.

Malpasset dam, France

Preconstruction geological investigations of this site were limited to inspection of the site by a local professor of geology, plus eight holes 65 to 100 ft in depth to evaluate grout-take and information on the foundation excavation (for details, see Chapter 22). The site was not explored by systematic cored-borings for guidance in selecting the alignment or design. The partial geological investigations were based on the initial supposition that a gravity dam would be constructed. It is not clear whether the geologist involved was advised and subsequently participated in the design change to an arch dam (Goguel, 1963).

This design change contributed to the foundation failure described in Chapter 22. One obvious lesson in judgment learned from this tragedy was, "The feasibility study for an arch dam must henceforth include detailed subsurface investigations and guidance by an applied geologist who is familiar with arch dam design as well as the areal and regional geological context; geological input should be accompanied by in-situ and laboratory tests" (Post and Bonazzi, 1987, p. 350). Experience at Malpasset also emphasizes the importance of changes induced with time and

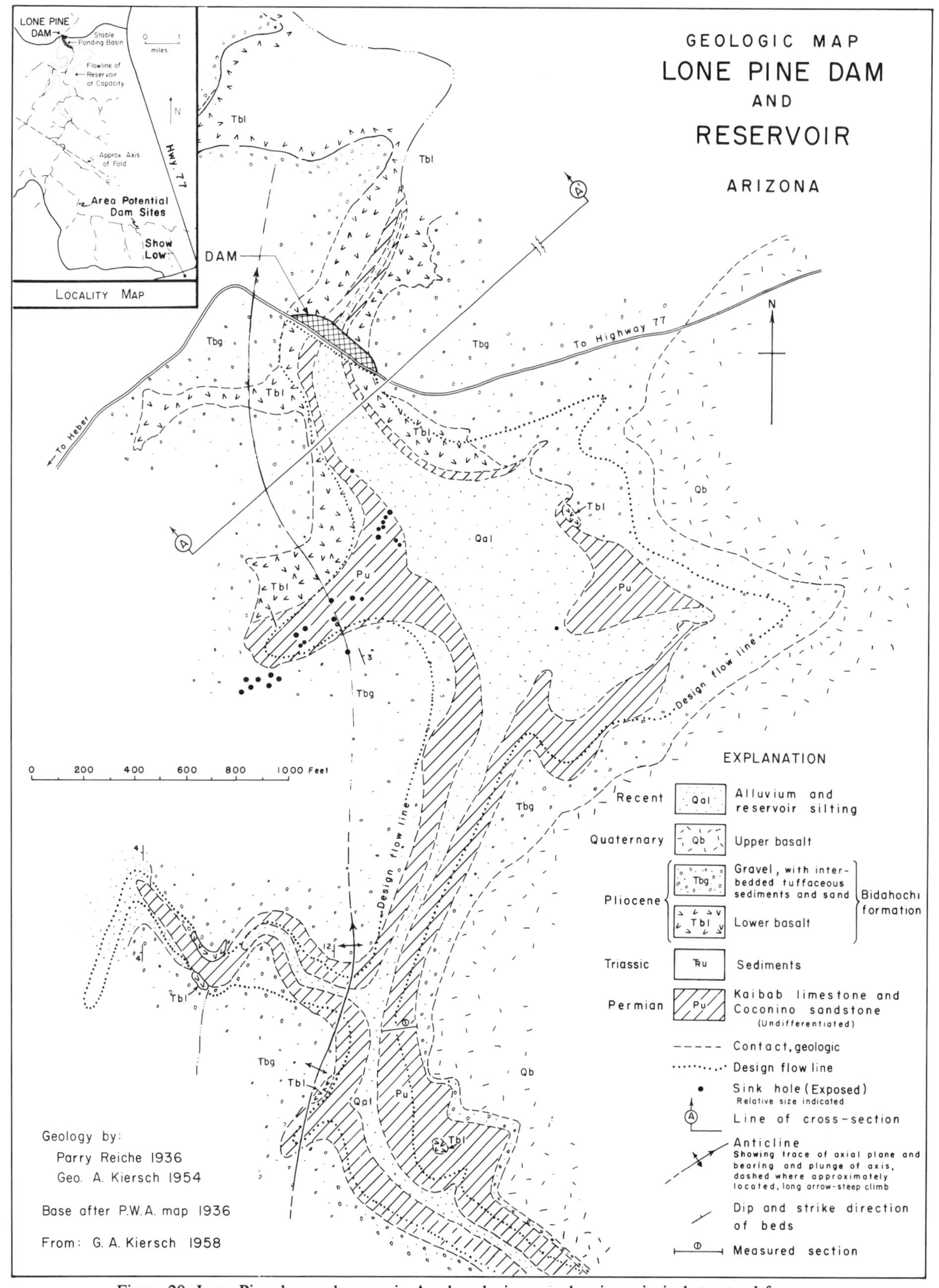

Figure 20. Lone Pine dam and reservoir. Areal geologic map, showing principal structural features, distribution of the solutioned limestone unit (Pu), and location of some sinkholes in the reservoir floor. (Source: Kiersch, 1958.)

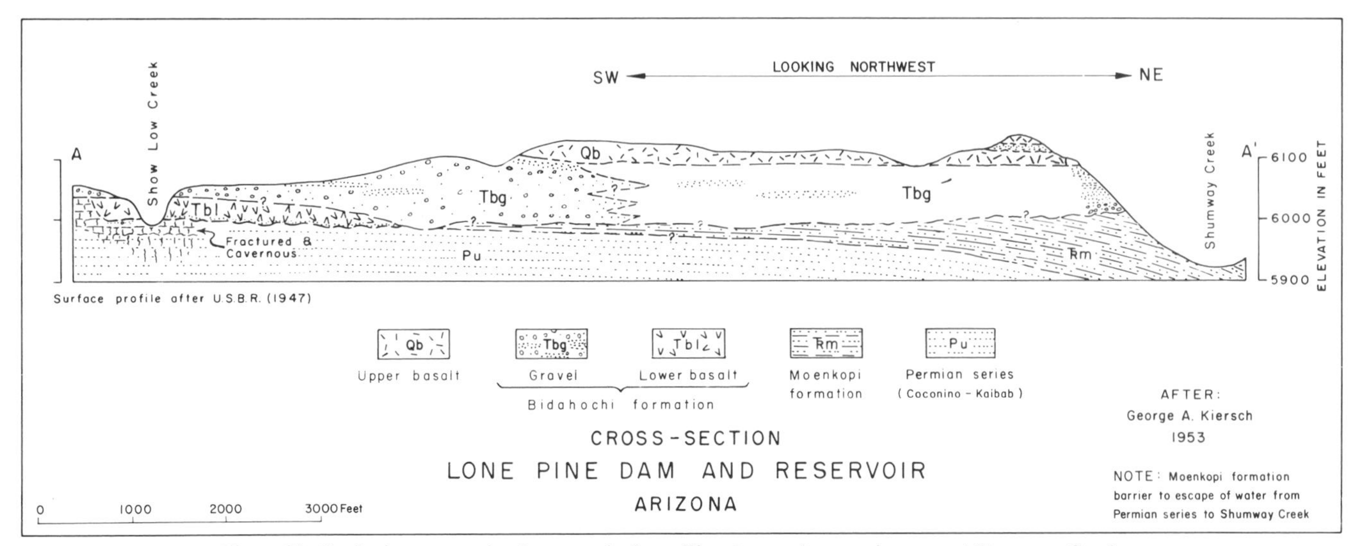

Figure 21. Geologic cross section between the Lone Pine dam and reservoir area and Shumway Creek to the northeast. Locations are shown of the fractured and cavernous limestone, basalt, and younger gravel units that would transmit leakage if a reservoir were impounded. (Source: Kiersch, 1958.)

operation of a reservoir dam. For example, the altered and highly fissured gneissic rocks throughout the left abutment were greatly weakened when saturated by the reservoir (Habib, 1964, 1987) and exhibited a variation of permeability as a function of effective stress.

CHANGES WITH TIME AND OPERATION OF ENGINEERING WORKS

The impact of the geologic environs on an operating engineering work, with changes induced over time due to on-going geologic reactions, is frequently misjudged or not recognized. Three well-known cases demonstrate the changes induced to a dam foundation, reactivation of ancient landslides within a foundation, and the collapse of a major highway bridge by undermining scour of a pier during high water due to errors of design and construction within the channel.

East Branch dam, Pennsylvania

This well-constructed, 170-ft-high, earth-embankment dam represents one of the few cases where failure occurred several years after the dam and reservoir were placed in operation (Bertram, 1974). The sudden appearance of 300 l/sec of muddy leakage from the downstream rock-fill toe drain brought immediate response. The reservoir was lowered, and a series of borings drilled through the dam located and delimited an irregular cavity (10 by 20 ft) in the clayey silt core above the bedrock near the right abutment (Fig. 22A). The engineering investigation concluded that differential settlement with cracking had occurred after a delay of more than four years because the cavity was directly adjacent to a horizontal bench or discontinuity on the rock abutment slope that had been used as a construction road. The cavity was filled by sand-cement grouting, and the leakage was brought under control.

The explanation for the development of the cavity, however, seems to reflect an error of judgment. Philbrick (1986), who was the geologist in charge of subsurface exploration and design, including grouting, has an alternative explanation. The right side of the dam rested on a cutoff trench into bedrock with twin grout curtains spaced 5 ft apart extending into the unweathered rock (Fig. 22A). The valley wall was highly fractured by relief-joints and weathering action induced when the valley was eroded at an earlier time to about 40 ft deeper than its present floor (phenomenon discussed in Chapter 18). This exposed the walls to prolonged weathering and rebound fracturing to a far greater depth and more closely spaced than the other valleys of the region. During construction, grout slurry was injected to refusal in deep holes to produce a continuous impervious curtain that extended from the base of the cutoff trench to the bottom of the grout curtain (Fig. 22A). Both up- and downstream from the grout curtain, the bedrock was criss-crossed by open fractures. The full reservoir head, measured from the bottom of the pipe-cavity to the reservoir surface, existed in the bedrock fractures at the upstream side of the cutoff trench. The top of the grout curtain lay in the cutoff trench below the top of the rock and the top of the curtains where the full reservoir pressure was acting, yet the width of the impervious dam section there was minimal. Downstream from the grout curtain, the water pressure active throughout the network of fractures was from tailwater, yet the head differential across the top of the grout curtains was from the nearly full reservoir. Little wonder the embankment piped and a cavity formed, according to Philbrick (1986). The leakage, related to piping action within the dam, is due to simple overtopping of the grout curtain at the base of the earthfill. This kind of action is known to occur over time and further explains the

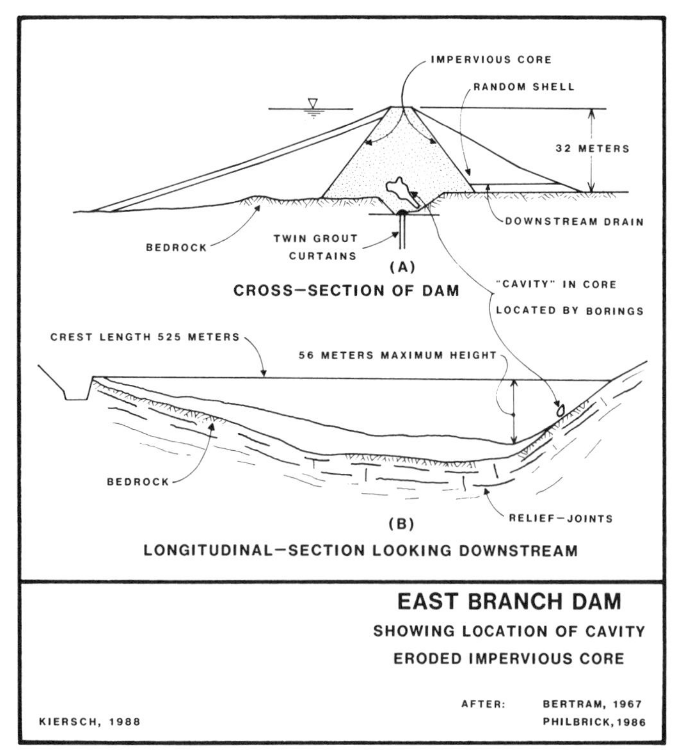

Figure 22. East Branch dam. A. Cross-section view of the impervious core. B. Longitudinal section of the dam showing the location of the cavity eroded in the core. (After: Bertram, 1974, Fig. 10; Philbrick, 1986.)

intervening years of no leakage after the reservoir was in operation.

Slides at Fort Peck dam, Montana

Fort Peck dam is located on the Missouri River approximately 70 mi south of the Canadian border. Shale and sandstone formations of Cretaceous age underlie and flank the stretch of the river that is occupied by the dam (Fig. 23). The Bearpaw Shale at the dam site is partly consolidated, subject to air slaking, and contains zones of bentonite ranging from a fraction to several inches in thickness. These inherent defects and seams have contributed to many of the ancient slides so evident along the banks of the river and were a factor in a serious sliding involving the dam in 1938. The bentonite seams also contributed to more recent slides on the slopes adjacent to the power plant and to deformation of the concrete-lined spillway.

The dam slide. This slide occurred on September 22, 1938, with only 20 ft of embankment remaining to be placed; the immense slide mass resulted in a remarkable displacement on the upstream slope of the dam. Slide debris filled the 500-ft-wide intake channel located 800 ft upstream from the toe of the dam, and sections of the dam shell were moved intact more than 1,000 ft. Although the damage was costly, no injuries were incurred. The postslide investigation was unprecedented in extent and detail; the latest exploration and soil-mechanics techniques were utilized, and several highly regarded civil engineers and geologists were retained as a board of consultants. The board's findings were summarized in a detailed report (Middlebrooks, 1942) along with the comments of seven prominent engineers and geologists. The following are highlights from reports on the project by geologist Lloyd B. Underwood of the U.S. Corps of Engineers (summary of March 15, 1985 to G. A. Kiersch).

Failure of the embankment commenced gradually in the dam foundation near the right abutment and within a partially disintegrated zone of Bearpaw Shale. Deformation was observed in progress about three hours before the actual slide, and may have occurred earlier but was undetected. Postfailure exploration confirmed that masses of the weathered shale foundation were displaced, based on prefailure drill hole data (Gilboy, *in* Middlebrooks, 1942); the hydraulic-fill embankment performed well during the sliding action; and the weak shale foundation failed and was responsible for the sliding.

The prefailure exploration had disclosed a zone of blocky shale approximately 30 ft thick in the foundation at station 15+00 (Fig. 23). Although a stability analysis on soil samples prior to the slide had not indicated this zone to be critical, the postfailure exploration of this weathered shale, identified weak bentonite seams in the foundation. Furthermore, observations afforded by postfailure borings disclosed substantial hydrostatic pressure acting within the foundation both below the slide and in the abutment. For example, the piezometric head at one hole exceeded the maximum height of the fill by 60 ft, while other holes displayed pressures greater than 100 ft above the lake level. These excessive pressures were attributed to the load imposed by some 200 ft of fill emplaced on the trapped water within the fractured shale and the bentonite seams. Such an artesian effect had not been observed during the site exploration phase. The high pore pressure is believed to account in large part for the initial failure, and for the speed and great displacement of the dam's toe, which made up the slide mass (Middlebrooks, 1942; Ryan *in* Middlebrooks, 1942). Additionally, unequal settlement of the fill, due to a sudden change in foundation slope and in the thickness of the weak shale near the point of failure (station 15+00, Fig. 23), was likely a contributing factor (Feld *in* Middlebrooks, 1942).

A. Casagrande (1974), a member of the board of inquiry, has more recently offered the following opinion on the causes. The initial reservoir filling imposed a considerable load increase on the shale underlying the dam and reservoir; when this stress was transferred to the bentonite seams it resulted in an increase in pore pressure with a consequent reduction in effective stress, which led to slippage along the seams. This caused widespread liquefaction of sand members within the foundation, and the massive slide followed. Casagrande (1974) noted that preslide borings had disclosed sand in the foundation in the area of the failure; whereas holes drilled after the slide showed no sand. All indications are that the sand had liquefied and flowed from its original location, thus reducing the strength of the foundation to the point of failure (Casagrande, 1974).

Although the soils and weathered rock mantle were tested extensively prior to construction, the investigations failed to fully identify the physical characteristics of the foundation rock, and furthermore did not predict that excessive pore pressure could develop due to the superimposed weight of the dam. After analyzing the failure and type of foundation rock, it was concluded that undisturbed samples of sufficient size should be obtained to permit determination of the representative characteristics of the bedrock (Middlebrooks, 1942). Moreover, the investigation of sites should include a systematic development of the pertinent areal geologic history in order to better evaluate the site. This background information should provide guidance to the sampling and testing program if a failure should occur (Fahlquist, and Crosby *in* Middlebrooks, 1942). The engineer-reviewers recognized that too much dependence can be given to laboratory soils test data, and cautioned for restraint (Gerig *in* Middlebrooks, 1942).

Dam builders profited from the intensive postfailure investigation of the Fort Peck slide. Concepts and precautions in dam design today, such as the effects of pore pressure, liquefaction, the treacherous characteristics of bentonite seams—not the least of the problems associated with embankments that are emplaced by hydraulic-fill methods—were pioneered by this experience.

Although the immediate tendency is to criticize the builder's judgment in failing to recognize the geologic imperfections at the Fort Peck site, recall that in 1938 the engineering and geological professions had acquired little experience with such problems. This is clear when one realizes that the postfailure investigation was the most thorough study of an earth-dam slide during construction that had been undertaken to that time (Justing *in* Middlebrooks, 1942).

The damaged section of the dam was reconstructed with compacted earthfill designed on the basis of criteria developed during both the pre- and postfailure investigations. A dynamic analysis by the U.S. Corps of Engineers showed that the factor of safety is greater than two for much of the dam and the embankment is safe under all earthquake loadings considered possible (Marcuson and Krinitzsky, 1976).

The powerplant slides. Since the Fort Peck, Montana, project construction in 1934, there have been many investigations of the slope adjacent to the powerplant by the U.S. Corps of Engineers and others. Several reports have been released, but most are unpublished. The following short review is based on observations and findings of several of these studies summarized by Hamel and Spencer (1984).

The powerplant and switchyard are located close to the dam, near the base of the right abutment. The slope, which extends to a maximum height of about 200 ft above the outlet portal, consists of Bearpaw Shale with seams of bentonite. Preconstruction maps and aerial photographs show evidence of ancient landslides, which indicate that the natural slopes at the site have long been close to the critical angle for failure. Therefore, it is not surprising that excavation for construction of the powerhouse created stability problems. In fact, during early stages of construction, a direct relation became apparent between excavation near the toe of the slope and ensuing landslides.

Excavation for the outlet works involved maximum cuts of 50 ft deep (average of 35 ft) at the toe of the original slope. Several slides occurred (Fig. 24), which were systematically mapped, instrumented, and monitored. All involved translational blocks with a basal failure surface along a bentonite bed and termination at a fault plane (Fig. 25), and are believed to reflect reactivations of ancient landslide masses. Fortunately, none of the slides affected operation of the project.

Analysis diagrams showed that the relation between height

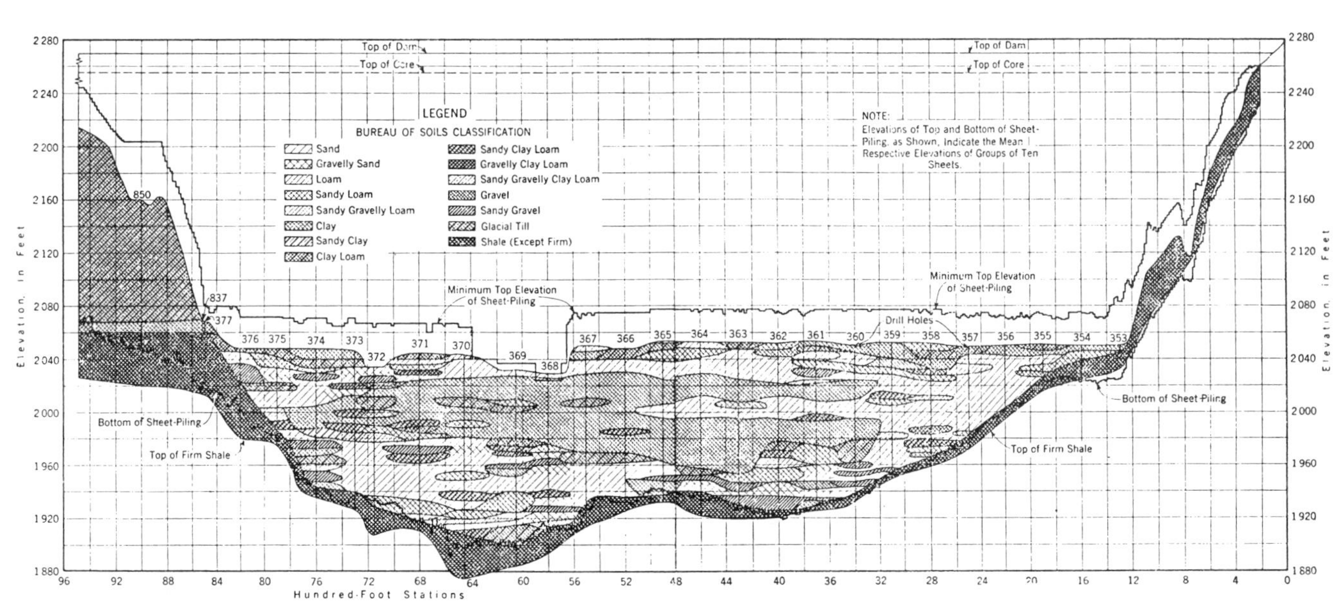

Figure 23. Fort Peck dam. Soils and geologic cross section along the dam axis, showing the distribution of young sediments and soils within the channel carved in the Bearpaw Shale. (Source: Middlebrooks, 1942.)

Figure 24. Fort Peck power plant. Typical slide-prone terrain encountered in the area of the dam and generating facilities on the right abutment of the site showing the Bearpaw Shale and reactivated slide no. 4 of October 2, 1934; the location of the power plant is shown on Figure 18 of Chapter 28. (from Hamel and Spencer, 1984, Figure 7; USCE, Omaha, 1989).

and inclination for stable and unstable slopes required an overall inclination of 1 on 6 to assure stability. Consequently, in 1974, 1.6 million yd^3 of the shale beds were removed to flatten the slope and achieve stability (Hamel and Spencer, 1984).

Evidence or implications of ancient landslides were not always recognized by field investigators in the early 1930s. At the powerhouse site, excavation at the toe of a previously unrecognized sensitive slope caused progressive failure and required costly corrective measures. In the 1930s, design decisions were relying chiefly on the results of near-surface soils tests, and the importance of regional and areal features to a site-specific geologic assessment were yet to be accepted. Builders of most modern engineering works now recognize the benefit of the team approach (see Chapter 18).

The dam spillway. The Fort Peck dam spillway is constructed on clayey Bearpaw Shale with bentonite seams and numerous shallow faults. It is situated in a natural saddle on the reservoir rim about 3 mi east of the dam, and consists of a partially lined approach channel, a gate control structure, a lined concrete channel, and an unlined channel. The trapezoid-shaped, concrete-lined discharge channel is about 5,030 ft long, and varies in width from 800 ft at the upstream end to 130 ft downstream.

Observations along the concrete-lined discharge channel covering a period of 42 yr have disclosed progressive deformation attributed to rebound related to construction excavation (summary by L. B. Underwood, March 15, 1985, to G. A. Kiersch). Maximum movement has occurred in a reach about 150 ft long; there the concrete floor is elevated about 2 ft, and the walls about 3 ft. The deformation, though readily apparent, has not affected operation of the spillway, and rebound continues but has decreased with time.

The interpretation of observation from 20 slope indicators emplaced around the spillway since 1953 indicates that the most serious deformations and movements occur along bentonite seams and intersecting faults. Stability of the spillway has been improved by flattening the side slopes at critical locations and installing horizontal drains to dewater joints and faults in the clay shale. Monitoring of the structure continues.

Schoharie Creek bridge, New York

A 300-ft length of the Schoharie Creek bridge fell into the raging Schoharie Creek on April 5, 1987, claiming ten lives (Fig. 26). The 25-ft-deep floodwaters were twice the normal stream depth (ENR, 1987a). The bridge failure, caused by an error of construction judgment, was influenced by the geologic setting.

The bridge was designed by conventional 1950 standards and consisted of five H-shaped piers, three on the creek bed and two on the abutments (Fig. 27). The center span between piers 2

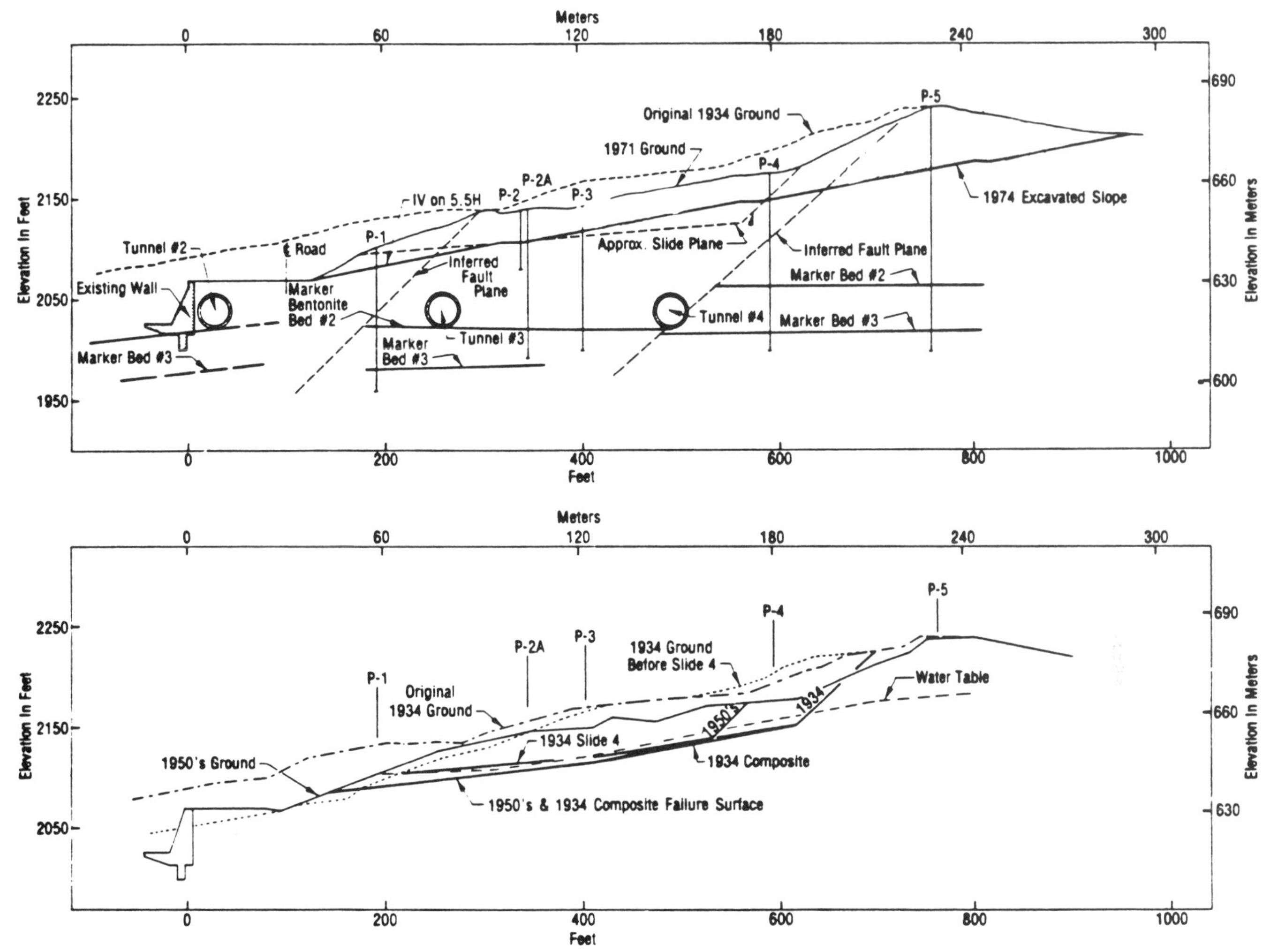

Figure 25. Slides near the Fort Peck power plant. Upper: Detailed section showing the former surface profiles, subsurface marker beds, and exploration borings used to interpret the locations of faults in order to design the stabilization in 1974 by excavating the slope. Lower: section showing profiles from 1934 and the 1950s, relative to pre- and postslides and their respective failure surfaces, which were closely associated with the ground-water table. (Source: Hamel and Spencer, 1984.)

and 3 collapsed first, due to deep foundation scour beneath pier 3, followed after 10 and 90 minutes by the spans between piers 3 and 4, and between pier 4 and the east abutment. Pier 4 tilted heavily upstream (ENR, 1987b; Fig. 26B).

Reportedly, an engineering consultant in 1981 had recommended placing an extra 800 yd^3 of riprap around the two submerged footings, but the recommendation was not followed; New York State engineers have expressed uncertainty as to whether this measure would have prevented the collapse of the bridge (MRCE, 1987). Design for the bridge included a provision for narrowing the stream channel at the site from 600 to 540 ft by extending landfill into the channel, a modification that shortened the length and cost of the bridge. This action increased the velocity and erosive power of the stream during the flood, a factor in the failure (MRCE, 1987).

Geologic setting. The Schoharie Creek bridge is near the upstream limit of the flood plain formed around the juncture of Schoharie Creek with the Mohawk River, which at this point is part of the New York State barge canal system. Bedrock consists of shale and limestone units of the Canajoharie Formation beneath an overburden of glacial and alluvial deposits; the bedrock dips to the southwest. The shales consist of black, uniformly fine-grained material with bands of silty calcareous shale and interbeds of limestone (Fig. 27). Overburden materials are associated with Quaternary or Holocene glacial events. The massive unit directly involved with the failure consists of younger ice-contact-stratified drift that ranges from a till of debris flows to layered sediments. Distribution of other glacial deposits, varved clay and outwash silt, and the Recent alluvium are shown on Figure 27. The basal till overlying bedrock acts as an aquitard,

Figure 26. Schoharie Creek bridge. A. Before collapse. Four-span bridge over Schoharie Creek on New York State Thruway I-90, completed in 1955. (Source: New York Times/L. B. James, 1988.) B. Collapse of Schoharie Creek bridge on April 5, 1987. Pier 3 is in the center; pier 4 is on the right. Note the correlation of collapsed piers with the geologic cross section and scouring shown in Figure 27. (Source: New York Times/L. B. James, 1988.)

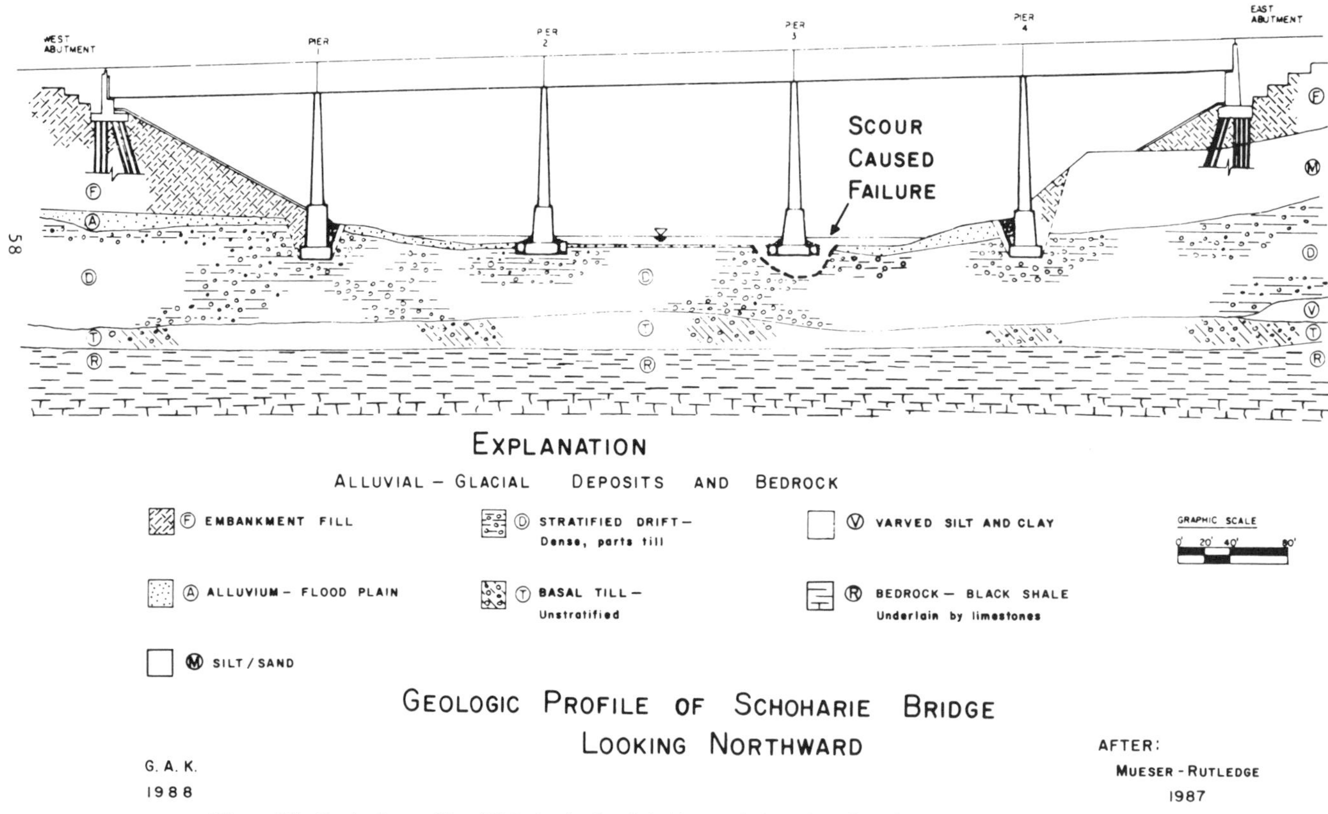

Figure 27. Geologic profile of Schoharie Creek bridge and the subsurface foundation of alluvial and glacial deposits overlying the shale bedrock. Deep scour of pier 3 caused the collapse. (Modified Mueser-Rutlege, *in* MRCE, 1987.)

separating the hydrostatic pressure of the stream from piezometric levels in the underlying shale.

Schoharie Creek experienced several major historical floods between 1900 and the one in 1987 that collapsed the bridge. The flood of March 21, 1901, discharged 50,000 cfs just downstream at Ft. Hunter. Floods of March 1936 and September 1938 reached stages of 58,000 and 55,000 cfs, while the flow in Schoharie Creek was 76,500 cfs on October 16, 1955, one year after the bridge was opened to traffic. The disastrous April 5, 1987, event constituted a 50-yr flood, while the October 16, 1955, event was a 100-yr flood. Stream velocities approaching 12.5 fps occurred with flows of 55,000 cfs at the site (MRCE, 1987). On April 5 the recorded flow in Schoharie Creek increased in 24 hr from 4,000 cfs to a peak of 64,900 cfs; the flood wave reached approximately 15 fps at pier 3 because of the constricted flood plain created by the thruway embankment on the west bank. The change in velocity resulted in high turbulence at the bridge; and flows in excess of 50,000 cfs continued for 9 hr.

Conclusions. Causes for the collapse of the Schoharie Creek bridge were determined after an extensive geological and hydrological investigation—including soil-rock borings—of the glacial deposits underlying the foundations of the two interior piers, the concrete in pier 3 (Fig. 27), and a review of the structural design and maintenance of the bridge (MRCE, 1987; RCI, 1987). The principal causes were:

1. The Schoharie Creek bridge collapsed due to extensive undermining scour of the pier 3 footing when the flood flow peaked at 15 fps. Undermining extended at least 25 ft upstream, and 30 or more feet in depth.
2. Although the largest flood on record occurred in 1955 (76,000 cfs), the effects of the 1987 flood were more damaging because the subsequent construction on the west bank had narrowed the channel.
3. Although the piers of the bridge were constructed on a strong foundation, an error was made in design of the plinths on top of the footings with regard to resisting the effects of soil pressure.
4. Riprap of adequate size and essential to protection of the footings was missing. Over time, the initial riprap had shifted downstream from pier 3 (Fig. 27).
5. Errors in construction judgment to protect the pier foundations were worsened by errors in design of the subsequent repairs, which made the bridge susceptible to a catastrophic failure.

PROBLEMS RELATED TO ECONOMICS AND COST BENEFITS

Construction costs for many projects may be substantially reduced if an adequate geological investigation is undertaken. For example, the cost of correcting unsuspected subsurface conditions usually far outweighs the cost of an adequate exploration program; also, when funds for geological investigations are imprudently restricted, the geologist becomes more conservative in his assessment of a site, and the engineer in turn incorporates a greater factor of safety in his design. The project geologist must therefore lobby convincingly for funds sufficient to conduct an adequate investigation. Examples of the consequences of insufficient funding for geological investigation are given below.

Sweasey dam, California

The Franciscan Group is renowned for the problems encountered during construction of engineering works, such as the Broadway tunnel in San Francisco, discussed earlier in this chapter. Sweasey dam (Fig. 28), located on the Mad River, is no exception (Brown and James, 1960). The dam, owned by the city of Eureka, was completed in 1938 for storage and diversion of a city water supply. The designed dam was 75 ft high and originally stored 3,000 acre-ft of water; however, siltation has greatly reduced the storage capacity. Sweasey is the only arch-type dam of consequence located on Franciscan Group rocks.

The local rock units are largely sandstones and shales that dip 50 to 60° downstream and strike across the river. The shales are typically hard and clayey with crushed zones locally. Beds exposed in both abutments indicate an offset of about 400 ft has occurred along a fault that traverses the river channel. Competent sandstone beds in the right abutment form a prominent cliff; whereas the softer, thinner beds of the left abutment are weaker, distorted, crushed clay-shale strata (Fig. 29).

During original exploration of the site by borings, core recovery was low and generally negligible in the soft shales and gouge of the left abutment. Consequently, further exploration was done by means of a tunnel under the river, and by shafts excavated into the left abutment to determine whether satisfactory foundation rock occurred at depth and in the channel section.

Although the findings were discouraging, excavation for an arch dam abutment proceeded. As the cut became deeper, the left abutment hill began to slide (Fig. 29). When excavation reached an estimated 90,000 yd^3, the site preparation was stopped and more exploratory shafts and drifts were driven. This exploration data confirmed that the subsurface rock conditions were not adequate for construction of a full masonry arch-type dam at the site. To compensate, the dam was redesigned as a composite section; this employed an arch dam across the channel section, which abutted a massive concrete thrust block at the left end. An earthfill dam section then connected the thrust block with the deeply excavated left abutment. Because of the additional costs for the deep left-abutment excavation and the changes in dam design, it was necessary to reduce the height of the dam 60 ft in order to stay within the authorized funds.

We must conclude that the original exploration was incorrectly interpreted. Planning a concrete masonry arch dam at the site was an error of judgment. The Sweasey project would normally have been redesigned or abandoned during the early stages if the site conditions had been understood. The outcome had a substantial adverse effect on the ratio of costs to benefits for the completed project.

Figure 28. Sweasey dam. This arch-type dam, founded on Franciscan Group rocks, encountered unforeseen conditions that required a change in the design; the main dam was lowered 60 ft in order to meet budget constraints (photo by L. B. James, 1960).

Cuber reservoir, Majorca

The Cuber reservoir is located in the mountains east of Palma, the capitol city of Majorca, for which it provides a water supply. The reservoir is situated in terrain that consists of interlayered limestone and consolidated noncarbonate rocks. The dam occupies a narrow canyon, which opens upstream into an expansive valley, the reservoir site. The foundation and dam abutments are massive and impermeable; topography and areal geology seem ideal and undoubtedly influenced the project location.

Unfortunately, the areal geologic conditions, including the reservoir site, proved to be very unfavorable. The diagrammatic section through the dam and reservoir (Fig. 30) was sketched in 1973 by one of the authors (LBJ) during a tour conducted by the International Congress for Large Dams. The sketch is based on descriptive material given by the Spanish engineer guides. Although generalized, it illustrates a basic problem associated with a reservoir, as well as a solution. The reservoir site is located within a synclinal structure that consists of near-surface impermeable beds underlain by permeable strata. The dam and most of the reservoir rests on the impermeable rocks, while the upper reservoir sector is underlain by permeable limestone beds with a network of solution conduits. In the upper reservoir floor the permeable limestone beds are exposed at an elevation significantly lower than the spillway crest. The limestone stratum dips downstream; water can pass beneath the watertight layer and issue from outcrops in the canyon downstream from the dam (Fig. 30), a major by-pass and loss of the water stored.

Engineers realized the error of judgment in selecting the site during the first filling of reservoir; the water level could not be raised above the elevation of limestone outcrops upstream. Above this level, the limestone beds acted like an inverted siphon, conducting water beneath the reservoir floor and the dam, and resulting in discharge in the canyon downstream.

The by-pass leakage problem was solved by constructing a dike (Fig. 30) to isolate the permeable limestone outcrops from the remainder of the reservoir area. Although this remedial design was effective, it substantially reduced the total storage capacity of the reservoir, and furthermore, was expensive. Cost of the dike construction exceeded that of the dam, and active storage was confined to the area of impermeable beds.

Areal geologic mapping of the reservoir site and vicinity during the planning stage, combined with a modest-size core-drilling program and downhole pressure tests, would undoubtedly have revealed the potential problems. Such a database is essential before estimating the risks, costs, and benefits to be realized by a project, and furthermore reduces costly errors in planning.

Figure 29. Sweasey dam, left abutment. (Source: W. A. Brown, circa 1938).

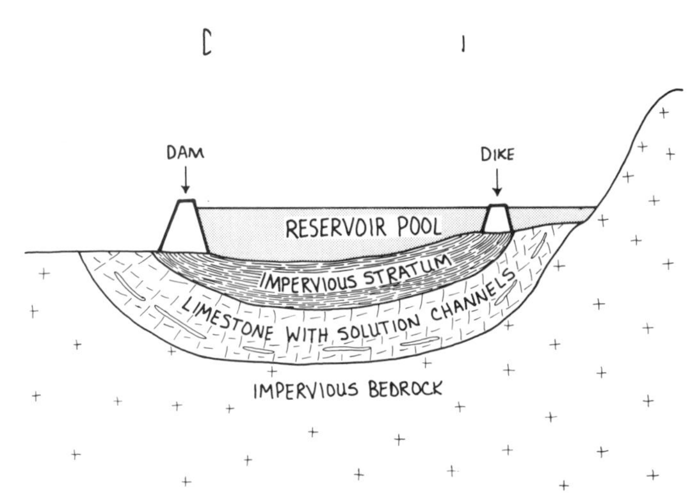

Figure 30. Cuber reservoir. Diagrammatic geologic sketch along the axis of the reservoir, showing how the thick limestone beds with inherent solutioned channels and openings acted as a siphon beneath the reservoir. (from notes by L. B. James, 1973.)

Site 19 dam, New York

This 65-ft-high, earthfill, high-hazard flood-protection dam is located in Cattaraugus County in western New York State. It contains 95,000 yd^3 of embankment, and was constructed in 1971 and rehabilitated in 1981.

During a routine dam safety check, unaccountable seepage was detected at the left toe drain, which prompted a detailed, geological and engineering investigation. This entailed both drilling and sampling of embankment fill and foundation materials as well as measurements of the permeability of the foundation and trench drains. Zones of high permeability were revealed near the embankment-foundation contact. Because the situation was judged potentially hazardous, the dam was breached in 1980 to remove the threat of failure. The breaching allowed further geological investigation of the foundation conditions, because no "as-constructed" geologic map or report had ever been prepared.

Breaching entailed removal of the left section of dam and exposure of the abutment and much of the drainage system. Exploration holes drilled during the earlier safety investigation were uncovered and closely analyzed, some of these holes displayed radiating cracks that were attributed to hydraulic fracturing associated with the drilling or permeability tests. They could be traced only a few centimeters into the fill and were not important (Kirkaldie and Thomas, 1984). The embankment that was exposed was tightly compacted and apparently unchanged over the nine years the dam was in use. Significant leakage had not occurred through the fill.

However, previously unrecognized landslide material was exposed in the left abutment excavation. Shale and siltstone units were observed to overlie younger deposits of glacial till, lacustrine deposits, alluvial sand, and rounded gravel. The Pleistocene(?) slide extended 50 ft upstream and 90 ft downstream from the dam axis; the failure zone at the base averaged 5 to 6 in thick. Although the slide was not detected during the preconstruction investigations, the project geology report had recommended a cutoff trench into firm bedrock. Apparently, the boulders and rock fragments encountered in the slide material were interpreted as firm bedrock by construction personnel. No geological guidance or input was involved with the cutoff construction. In order to rehabilitate the dam, the cutoff trench had to be deepened to bedrock on the left abutment, and 65,000 yd^3 of new material were excavated. Expressed in terms of 1967 dollars, the costs for constructing the dam and its subsequent breaching and rehabilitation were, respectively, $167,000 and $78,000 (Kirkaldie and Thomas, 1984; Agnew, 1985). Failure to recognize the ancient landslide in the abutment resulted in substantial economic loss.

PROBLEMS RELATED TO CALCULATED RISK

Concepts and limitations

The past decade has experienced rapidly growing interest in probabilistic reasoning and an increase in its application to geological uncertainties and geotechnical engineering practice. Unfortunately, probability still remains a mystery to many scientists and engineers, in part because of a language barrier and in part for lack of examples that demonstrate how the methodology can be used in the decision-making process. Whitman (1984) has provided some general illustrations regarding the geological applications of probability and calculated risks. These appliations: (1) separate systematic and random errors when evaluating uncertainty relative to slopes; (2) establish safety factors when analyzing liquefaction; (3) optimize the design of an embankment, although uncertainty exists concerning stability; (4) set a risk evaluation for an industrial facility built over potentially liquefiable soils; and (5) provide a risk evaluation for embankments and earth dams founded on weak rocks. Even though a precise quantification of probability of failure is not possible, the systematic formulation of an analysis aids greatly in understanding the major sources of risk; this can lead to cost-effective remedial measures. Analyses of reliability and risk are potentially most valuable during the early stages of a project as a guide to deciding whether or not to proceed and, if so, to establishing design criteria.

Casagrande (1965) went to considerable length to explain what he meant by the words "calculated risk." The phrase implies the process of recognizing and dealing with risks, in two steps: (1) using imperfect knowledge, guided by judgment and experience, to estimate the probable ranges for all pertinent quantities that enter into the solution of the problem; and (2) taking into consideration economic factors and the magnitude of losses that would result from failure, to make a decision on an appropriate margin of safety or degree of risk.

Responsible management of hazards and risks begins with recognition of how the hazard affects an area or site, as shown on Figure 31. Failure to recognize hazards in the legal atmosphere of the United States may constitute negligence, and probably invites liability for damage in the event a hazard occurs with a damaging intensity. Similarly, ignoring hazards after earlier recognition probably constitutes negligence (Keaton, 1988). Evaluation of both hazards and risks includes quantifying the probabilities as to degree and a specific exposure duration or design life.

Probability means different things to different people and is a large and diverse discipline. Among the applications of interest to applied geologists and geotechnical engineers are: (1) an optimized search, exploration, and testing procedure in which probability theory is used to formulate a network of borings that minimizes missing a significant weak zone within a rock-soil mass; (2) the optimization of design to reduce uncertainty concerning loads and behavior; (3) risk evaluation, which implies basic concepts and procedures such as logic and decision trees (discussed in Chapter 18), to study a structure or geologic setting with many components or modes of failure; and (4) a use of reliability theory as a factor relative to the safety of geologic components or subsurface conditions.

The degree of geologic risk changes with the type of engineering works, because the foundation characteristics are provided by nature (Nickell, 1966). Thus, the engineering geologist must reach many conclusions that exercise his opinion and a considerable element of judgment. Furthermore, judgment risks of one generation may seem unimportant in the next generation; judgment of risks between two individuals also may vary depending on their personal backgrounds.

Geologic hazards frequently cannot be calculated with certainty. Such an example was the successive 1,000-year floods during construction at Kariba dam, Central Africa, an inevitable risk that resulted from a paucity of data in a sparsely inhabited region. Moreover, failure of a dam in an uninhabited region may result mostly in monetary loss and inconvenience for those dependent on the project, and is a less important risk than a dam in populated areas.

Similarly, the best of engineering and human evaluation for the railroad causeway across the Great Salt Lake (described below) did not prevent partial failure during construction, but it emphasized that investigation of the lake bed foundation by cored borings only was inadequate. Calculations, as well as judgments, may only be the best estimates based on probability (Nickell,

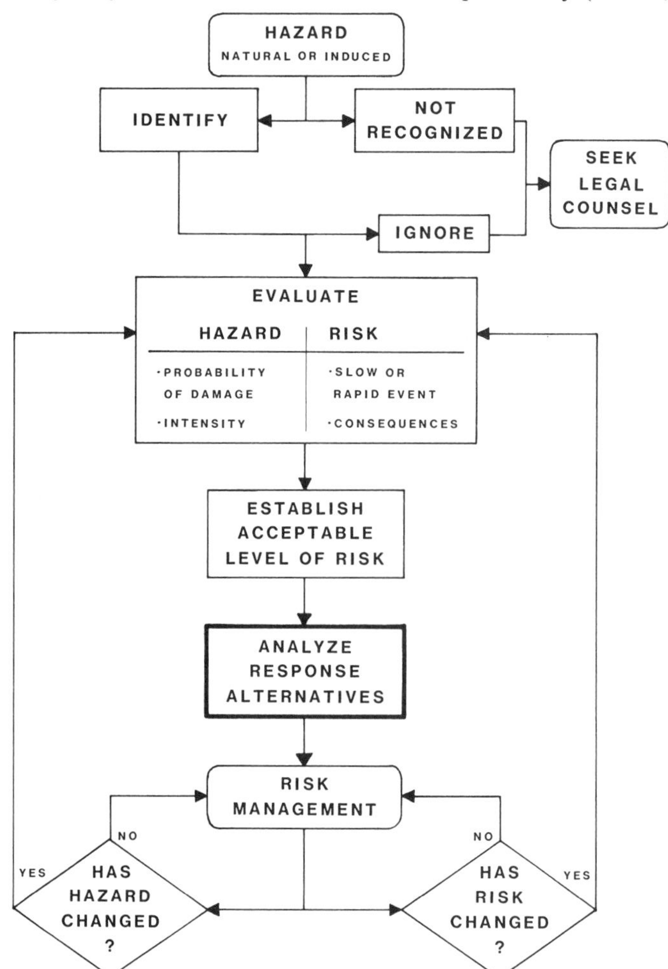

Figure 31. A model for the evaluation of natural or man-induced hazards and the management of associated risk. The hazard-response alternatives can vary from doing nothing to avoidance, to modifying the hazard by an appropriate procedure, according to the costs. (Source modified after: Keaton, 1988.)

Figure 32. Pillsbury dam, looking from the left abutment. This dam, constructed in Franciscan-type rocks in northwestern California encountered serious foundation defects in the left abutment that necessitated a redesign and bend in the axis to reach a stable foundation. (Source: L. B. James, circa 1960.)

1966). Methods to assure wider knowledge of risks and the consequences are a challenge for the next decade.

Any definition of a geologic risk must include five separate but related factors: (1) The nature hazard, H; (2) vulnerability, V; (3) specific risk, R_s); (4) element of risk, E; and (5) total risk, R_t. The relation has been expressed by UNESCO as: $R_t = (E)(R_s) = E(HV)$.

Although the engineer may assume a calculated risk when designing and building engineering works in complex geologic environs, this uncertainty can be reduced by negotiating a fair contract with a notion of limited liability. Moreover, this is quite different from boldness without due caution. Simply stated, risk = uncertainty × damage.

Pillsbury (Scott) dam, California

The Pillsbury dam is a concrete gravity structure, 138 ft in height and 815 ft long, constructed in 1921 on the South Fork of the Eel River, in northern California. The foundation consists of Franciscan-type rocks that contain irregular bodies of serpentine. During excavation for the foundation, it was revealed that the rock "outcrop" intended for the left abutment of the dam was actually a "floater," and during the wet winter it slid into the excavation. The situation was rectified by bending the dam axis and extending it to abut against a massive outcrop located farther downstream (Fig. 32).

The costly experience, somewhat like the outcome at Malpasset dam, emphasizes the importance of ascertaining that the abutments proposed for a concrete dam are intact and not disconnected from the underlying rock mass.

None of the records available indicate that any geological exploration was undertaken at the Pillsbury dam site prior to construction of this dam. Obviously the earlier engineers and constructors assumed a greater calculated risk than is common today.

Railroad causeway, Great Salt Lake, Utah

An outstanding example of a calculated risk is the design and construction of a Southern Pacific railroad embankment across the Great Salt Lake to replace the 12-mi-long timber trestle that was built at the beginning of this century (inset, Fig. 33). The new fill, which was built from 1956 to 1959, started from the existing old fill sections and then ran parallel to the trestle, located 1,500 ft to the north, to ensure that construction of the fill would not endanger the trestle (Casagrande, 1965).

Preconstruction laboratory strength tests on undisturbed samples of the soft and sensitive silty clay and Glauber's salt units, which underlie the lake to a great depth (Fig. 33) and the foundation for the causeway, indicated the design would involve great uncertainties. The Glauber's salt layer, which would underlie the fill for many miles, with its upper surface at a depth of 20 to 30 ft

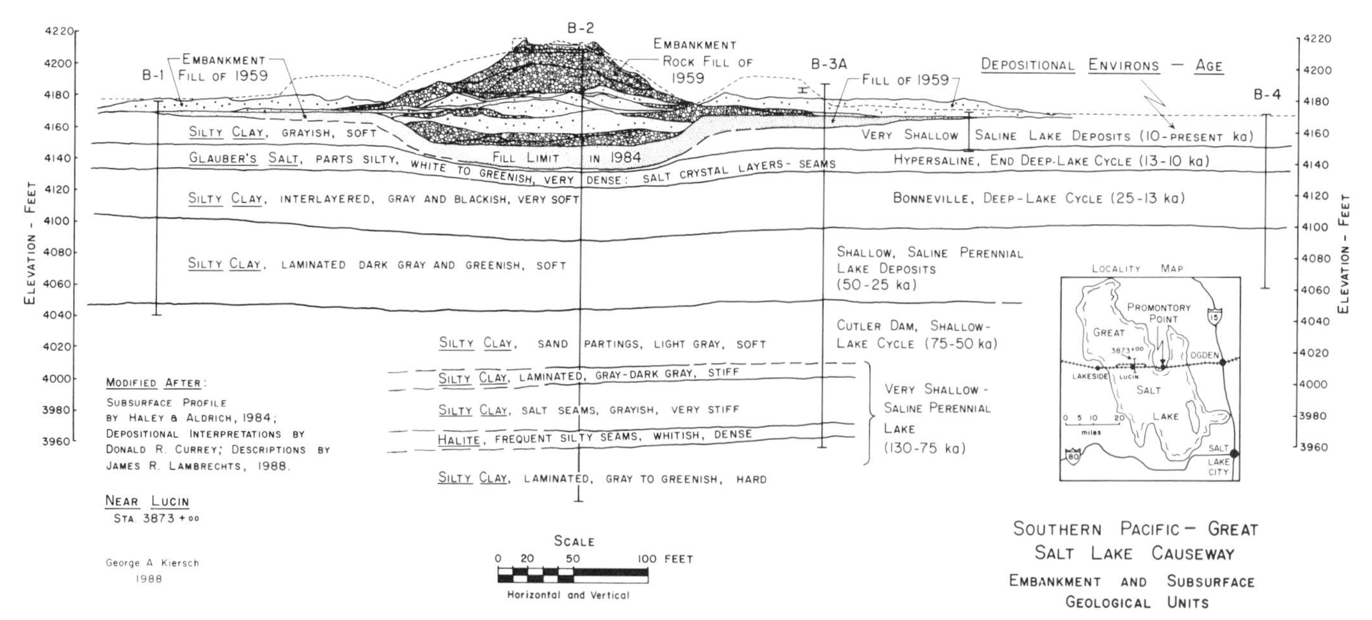

Figure 33. Southern Pacific Railroad, Great Salt Lake causeway. Embankment construction and the reaction of subsurface geologic units, particularly the near-surface bed of Glauber's salt. The section shows exploration holes and composition of the lake beds to depths of more than 200 ft at Lucin, Utah. (After data by: Haley and Aldrich Co., personal communication, 1984; D. R. Currey and J. R. Lambrechts, personal communications, 1988).

below lake bottom, was found to vary greatly in thickness and consequently in strength; this seriously complicated the design and construction of the fill (Fig. 33). Several design stages for the cross section on soft clay (no salt layer) are given by Casagrande (1965).

To achieve an economical design it became necessary to build full-scale test fills and induce failures, which constituted a field test and developed data on the in situ strength of the foundation units. Particularly impressive were the failures of fill founded on the Glauber's salt layer (Fig. 33). For practical purposes the salt had to carry the entire lateral thrust of the fill, and when the salt buckled, the fill sank into the soft clay with extraordinary speed. Even the most pessimistic initial assumptions of the consultants did not prepare them for the very low in situ strength of the soft clay, determined from analysis of the test section failures and other fill sections. Although the consulting board first recommended construction of full-scale test sections for design data, they soon learned that a test fill could be built only by mobilizing most of the expensive equipment needed for construction of the entire embankment. Consequently, the "as-built" causeway became the "test section," which was closely monitored and modified according to the "as-encountered" foundation units. This resulted in successful completion of the project one year ahead of schedule (Casagrande, 1965). Success would not have been achieved however, if the consultants had known before construction that the "as-is" strength of the Glabuer's salt and clay beds would control the stability of the embankment. Moreover, Southern Pacific probably would not have authorized the project. Based on the "as-built" knowledge of the foundation units, the conventional factor of safety required would have forced a design fill costing far in excess of the 1955 estimate and $50 million limit established by the Board of Directors of Southern Pacific Corporation. The initial misinterpretation of subsurface units and their inherent strengths allowed the project to get underway; however, the adoption of field test data and evaluation of the risk as construction progressed resulted in its successful completion.

A fill built on normally consolidated clay has its lowest factor of safety against foundation failure during construction or immediately after its completion. Subsequently, progress of consolidation steadily decreases the possibility of failure. Since the summer of 1959, when this new railroad crossing was put in operation, the rate of settlement has gradually decreased in a consistent pattern, reflecting a steadily increasing strength of the clay.

The Great Salt Lake causeway project is a good example of what Karl Terzaghi liked to call the "observational approach," i.e., the continuous evaluation of observations and new information for redesigning as needed while construction is in progress. It also illustrates another statement that Terzaghi made many times: "A design is not completed until the construction is successfully completed" (Casagrande, 1965).

The calculated risks involved in this project would not be complete without mentioning another risk. The "sand and gravel" used for the main body of the underwater fill was largely a silty sand. The question of its stability under dynamic stresses was of serious concern to the consultants. What one can see throughout the entire railroad embankment across Salt Lake is only a very small portion of the total volume of the causeway fill.

This ratio is a monument to the concept of a calculated risk, which could have defeated those individuals responsible for the design and ultimate construction.

Wheeler Lock, Alabama

Dramatic failure of the foundation beneath the Wheeler Lock wall, built in 1936, occurred during construction of the adjacent second lock on June 2, 1961 (Fig. 34). Foundation movement and failure involved a previously unrecognized and unsuspected small-scale clay seam at the base of a shale interbed within the limestone foundation (West, 1962). After three months of intensive geological investigations, the "unknown" and almost imperceptible clay seam within the thin shale (0.5 ft) was located and delineated within the Fort Payne Formation limestone foundation of the old Wheeler Lock wall (Fig. 35). Two workers lost their lives by drowning when movement broke the lock and flooded the adjacent excavation; severe financial losses were incurred by the wall collapse and blockage of river traffic. The incident was due to an apparent misunderstanding of the subsurface geological conditions beneath the old lock that led to an error of judgment by designers of the new lock; yet the possibility of such unanticipated developments is ever-present if geological studies do not exclude the possibility of weak zones or seams (Terzaghi, 1929). Such possibilities of the influence of minor geological features on the foundation of large concrete works were reviewed by Burwell (1948, 1951).

Micro-seams associated with interbeds of such wide areal extent had not been encountered or forecast prior to the 1961 Wheeler Lock incident. The feature escaped detection at the time of design and construction of the first lock in 1936 and again in the 1961–1962 exploration for the second lock. Excavation for the construction of a second Wheeler Lock parallel to and landward of the old structure exposed the Fort Payne Formation to a design depth 5 ft below the foundation level of the landwall of the existing Wheeler Lock (Fig. 35). This excavation resulted in the old lock being founded on a 5-ft slab of limestone resting on the shale interbed, and the toe of the old landwall being at the free-face margin of the new construction (Fig. 36). Without warning the old lock wall moved laterally up to 50 ft, into and across the excavation for the river wall of the new lock, fracturing the limestone floor of the new lock (Fig. 35). The old lock was wrecked and had to be rebuilt, and the Tennessee River could no longer pass barge traffic. This confronted the TVA with two major issues.

Geological investigations revealed the limestone bed overlying the thin shale interbed was not continuous to the riverbank (S. (S. S. Philbrick, written communication, 1987), which likely con-

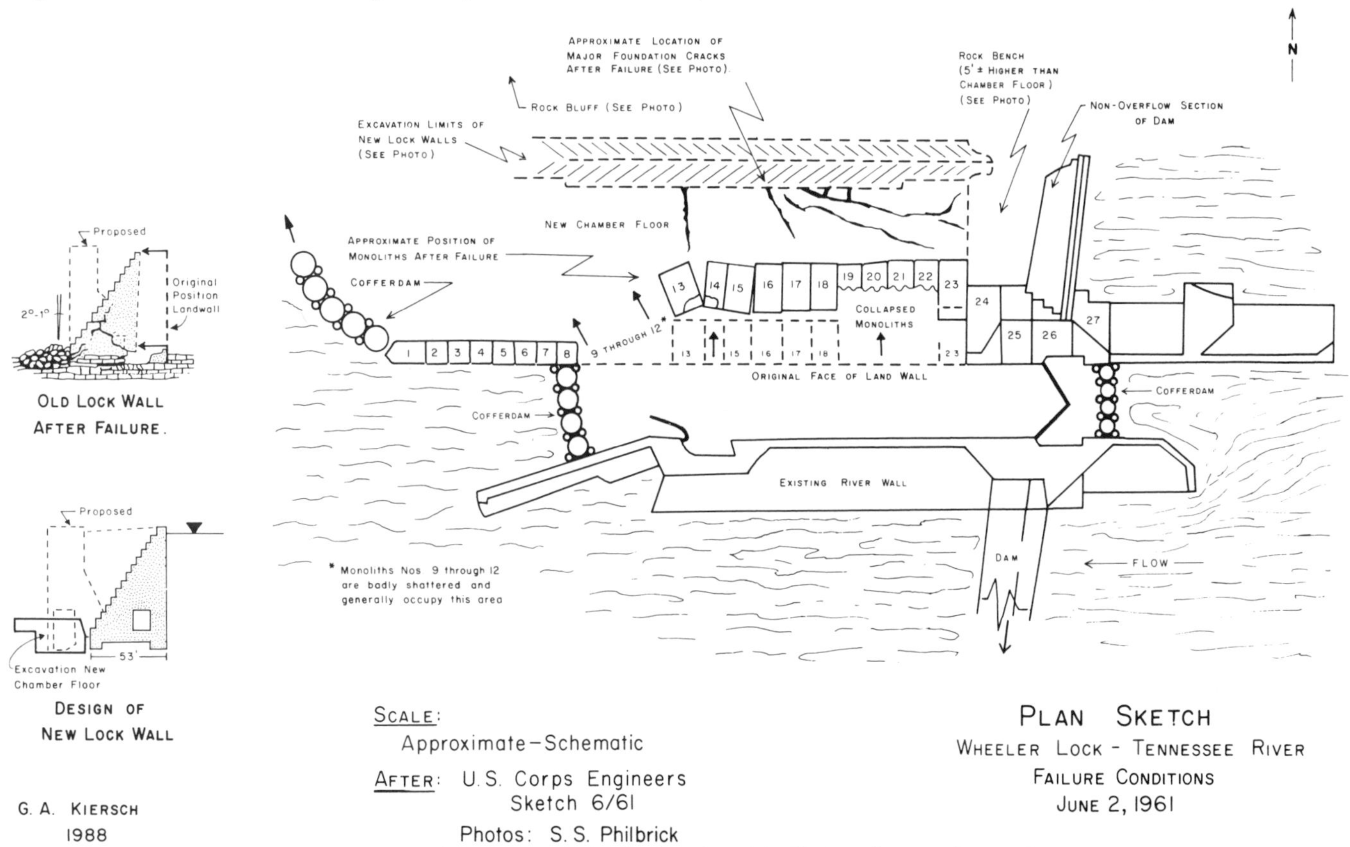

Figure 34. Plan sketch of the Wheeler locks showing the original lock and a second one under construction in 1963 when the failure occurred. Note the original versus collapsed position of monoliths that composed the land wall (Figs. 35 and 36), cross sections of the design for the old lock after failure, and the new lock wall. (After: USCE, and S. S. Philbrick, photos, 1961).

Figure 35. Wheeler Lock after failure, looking across the new lock excavation, which shows exposures of Fort Payne limestones (foreground). The displaced and shattered monoliths, numbers 9 thru 12 and 19 and 20, are shown in the center along with the relatively intact but moved numbers 13 thru 18 to the left (see Fig. 36). (Photo by S. S. Philbrick, 1961.)

tributed to oversight of the dangerous condition in the 1930s. Moreover, the excavation and foundation trench of 1961 cut and removed some of the limestone mass (slab) that was within the influence area of the old lock foundation, thereby reducing some of its shearing resistance (Fig. 36).

Geological investigations and borings in the 1930s for the first lock indicated that the foundation consists of a fine- to coarse-grained, clayey to crystalline limestone with a thin shale layer, averaging 0.5 ft in thickness, slightly below the foundation levels. The limestone beds and shale partings are nearly horizontal; a vertical joint system divides the rock into large blocks.

N-diameter (3-in) borings investigated foundation conditions for the new lock prior to the failure, and a 36-in calyx hole revealed the presence and width of the shale band; yet, as in previous borings, the seam of clay 1/16 to 3/8 in thick at the base was not detected (West, 1962).

The consultants' report of geologist Nickell and engineers Warren and Wilbur concluded that exploration by small-diameter borings and the inspection hole overlooked the weak clay seam at the base of the dark shale interbed (0.5 ft) due to the strong similarity in color and texture.

To avoid overlooking an unexpected geologic feature or hazard, one widely accepted practice is to excavate the proposed foundation during the design phase of a project so it becomes a large-scale test pit. Subsequent detailed mapping data can then aid in evaluating the foundation prior to final design and construction (discussed in Chapter 20, this volume). Obviously, any thin, weak rock zone or seam, such as a shale, that occurs near the foundation level of a rock mass should be removed. Moreover, suspected soft zones should be carefully tested by indentations for relative hardness across the bedding or foliation.

CONCLUSIONS

Technical incompetence of geologists was not a critical factor in most of the errors of judgment described in this chapter. Many of the serious problems are attributable to project managers (usually engineers) who had little or no background in geology. Some project managers failed to seek the assistance of qualified geologists or disregarded advice that was offered (cf. Cedar reservoir). Management's lack of appreciation of the importance of geological investigation in many instances contributed to costly and even fatal mistakes.

Some errors made at early projects are excusable in view of the primitive technology available. For example, in Spain, the failure of Puente dam in the early 1800s was due to a blow-out of deep channel gravels and occurred before the mechanics of piping were understood (see Chapter 1). In California, the construction of Pillsbury dam on the Eel River was undertaken before the treacherous characteristics of the Franciscan Formation were realized. However, inexcusable judgment has been involved to some extent in most of the serious problems experienced during the 20th century. It has resulted from combinations of the following deficiences, for which only a few of many possible examples have been cited.

Mistakes by management were responsible for most catastrophic failures. For example, the city of Los Angeles, against the advice of its engineering staff and with no geological investigation, had the height of St. Francis dam raised above the level originally planned. It then ordered that the first filling of the reservoir be accomplished rapidly to save water, thereby violating a cardinal engineering rule (Reisner, 1986). Another error was made by Los Angeles at Baldwin Hills reservoir where the city failed to realize the threat posed by the encroaching bowl of subsidence that ultimately destroyed the dam (see Chapter 22). Technical errors and controversies among geologists have accounted for some costly decisions, such as that involved with the site of the Vallecitos test reactor where an old slide plane was incorrectly reidentified as a dangerous, capable fault, thereby terminating the operation of an important facility. In another instance, an erroneous assessment of a site on the south coast of Puerto Rico relative to historical seismicity led to abandonment of a promising nuclear power project.

Poor communication between headquarters and project office and between geologists and their engineer counterparts has been responsible for design errors and faulty construction. For example, the designers of Teton dam (see Chapter 22) were located in Denver, and seldom visited the dam site; coordination with the field was poor, and the designers were not fully aware of the findings and significance of the geological exploration.

Insufficient geological investigation has contributed to many problems, and particularly to serious failures. In most instances, the cost of an adequate exploration would have been far less than the costs incurred by the problem (cf. Aerojet General, in Chapter 22, and Golder dam in this chapter).

Where projects are located in complex geologic settings, it is difficult to estimate how much exploration may be required. Consequently, funds budgeted for the investigations are frequently overspent, and the design or construction stages are undertaken before foundation conditions are adequately understood (Sweasey dam, this chapter).

Technical competence is a product of training, experience, and motivation. Obtaining and maintaining a competent staff is a management responsibility. Good management recognizes the benefits derived from on-the-job training and opportunities for educational advancement.

Undoubtedly, management's lack of appreciation for the importance of geological investigations has contributed significantly to many faulty decisions (cf. Malpasset and Cedar dams). The managers responsible for the major decisions were intelligent and experienced engineers. Why did they fail to seek or respect the advice of geologists? Obviously, they underrated the importance of a geological investigation and its findings. Why? Poor geologist-engineer relations can usually be identified as a contributing factor; management was likely ignorant of the geologists' capabilities, and the geologists had failed to gain the confidence of the engineers. Many engineers who have had little geological indoctrination are confused by geological reasoning and terminology, are prone to discredit geological reports, and can regard them to be of little practical value. Likewise, geologists have neglected to present their findings in language understandable to the intended readers, and explain the engineering significance of a geologic feature or setting. This dilemma is changing. Training in the basic engineering principles is now included in most applied geosciences curricula, and efforts by interprofessional societies have helped to promote interdisciplinary respect.

Figure 36. Wheeler Lock after failure, looking downstream along the toe of the displaced land wall, showing excavation for the new lock. Note disturbed and distorted limestone beds at the toe (mid-distance), the prominent shear fracture in the floor of the excavation at right-center, and the shattered monoliths, number 19 thru 23, on the left. (Photo by S. S. Philbrick, 1961.)

The California Department of Water Resources addressed this problem in the 1950s and 1960s by offering courses in geology for engineers and by encouraging geologists to enroll in extension courses in civil engineering. Interest in geology was stimulated among administrators by field trips or helicopter visits to remote sites where geology-related problems were explained in nontechnical language, e.g., following the trace of the San Andreas fault zone and explaining the significance of the fault to the location and design of proposed hydraulic structures. In addition, experienced applied geologists were invited to speak to local sections of the America Society of Civil Engineers, and visits were reciprocated by engineering speakers at the local sections of geological groups. These efforts helped establish an engineer-geologist team relation, and increased management's interest in geology. This awareness among professionals resulted in a simplified procedure for funding geological programs. Moreover, geologists have been assigned the added responsibility of signing off on plans for structures that involved their participation, and have been invited to participate in making decisions that involve geological judgment.

Errors of geological judgment can be attributed to both engineers and geologists. Risks can never be eliminated completely; however, they have diminished in the past two decades largely because of technological advances, better geologist-engineer relations, and recognition that mature, field-experienced geological judgment is an important element of any investigative program. Developing such insight will benefit from the granting of equal prestige and responsibility for solutions of judgment, even when that judgment is not expressed in numerical form.

REFERENCES CITED

Agnew, A. F., 1985, Comment on dissecting a dam: Association of Engineering Geologists Bulletin, v. 22, no. 4, p. 458.

Anderson, J.G.C., and Trigg, C. F., 1976, Case histories in engineering geology: London, Elek Sciences, p. 67–68.

ATC, 1988, Design report for reconstruction of Golder dam: Denver, ATC, Inc., for Rail-N-Ranch Corp, Tucson, Arizona, 315 p. (private publication).

Bernstein, L. S., 1956, Report on Niagara Gorge rock slide on June 7, 1956: Niagara Mohawk Power Corp., 5 p. (private publication).

Bertram, G. E., 1974, East Branch dam, Pennsylvania, *in* Hirschfeld, R. C., and Poulos, S. J., eds., Embankment-dam engineering, Casagrande volume: New York, John Wiley and Sons, p. 280–281.

Brooks, T. A., 1952, Tunnel experts consider adequacy of lining design; Western Construction, v. 52, no. 2, p. 90–91.

Brown, W. A., and James, L. B., 1960, California's Franciscan Group as a foundation for dams and appurtenant structures: Second Annual Meeting of the California Association of Engineering Geologists (unpublished).

Bruhne, G., 1965, Anhydrite and gypsum problems in engineering geology: Association of Engineering Geologists Bulletin, v. 2, no. 1, p. 26–38.

Burwell, E. B., 1948, Foundation engineering for large concrete dams: Third Congress on Large Dams, International Committee on Large Dams, Stockholm, Session C-6: Stockholm, Sweden.

—— , 1951, Influence of minor geologic structures on dam construction: Geological Society of America Bulletin, v. 62, p. 1427–1428.

Buwalda, J. P., 1951, Geological report on Broadway tunnel, San Francisco: Consultants report to City Engineer, San Francisco (unpublished).

Calkin, P. E., and Wilkinson, T. A., 1982, Glacial and engineering geology aspects of the Niagara falls and Gorge, *in* Buehler, E. J., and Calkin, P. E., eds., Geology of the northern Appalachian Basin, western New York; Guidebook, New York State Geological Association 54th Annual Meeting: Buffalo: State University of New York, p. 247–284.

Casagrande, A., 1965, Role of the "calculated risk" in earthwork and foundation engineering: Journal of Soil Mechanics and Foundation Division, v. 91, no. 4, p. 1–40.

—— , 1974, Presentation on Fort Peck dam failure: Vicksburg, Mississippi, U.S. Army Corps of Engineers Waterways Experiment Station (from written communication by L. B. Underwood, March 15, 1985).

Cedergren, H. R., 1989, Seepage, drainage, and flow nets, 3rd edition: New York, John Wiley and Sons, 440 p.

Cooney, J. E., 1952, Compilation-engineering geology of Broadway tunnel, San Francisco: Consultants report to Morrison-Knudsen Co., 18 p. (unpublished).

Cooper, B. N., 1964, Summary report on core drilling across Big Walker Mountain on Interstate Route 77, Wythe and Bland Counties, Virginia: Consultant's report to Singstad and Kehart, Engineers, New York City, 10 p. (unpublished).

—— , 1968, Geology at Big Walker Mountain tunnel on Interstate Route 77, Wythe and Bland Counties, Virginia, *in* Erwin, R. B., ed., Proceedings of the 19th Annual Highway Geology Symposium: West Virginia Geology and Economic Survey Circular 10, p. 83–90.

ENR, 1969, Vaiont trial defendent blames geologist: Engineering News Record, v. 182, no. 9, p. 13.

—— , 1987a, Scouring suspected in collapse of Schoharie bridge: Engineering News Record, v. 218, no. 15, p. 11.

—— , 1987b, Piers still focus of Schoharie bridge collapse probe: Engineering News Record, v. 218, no. 16, p. 20.

Forbes, H., 1945 Report of diamond drill exploration, Broadway between Mason and Larkin Streets: Report to City Engineer, San Francisco (unpublished).

—— , 1951, Interpretation of ground conditions Broadway tunnel, San Francisco, February 8, 1951 (unpublished).

Frink, J. W., 1946, The foundation of Hales Bar dam: Economic Geology, v. 41, no. 6, p. 567–597.

Glover, L., III, 1971, Geology of the Coamo area, Puerto Rico, and its relation to the volcanic arc-trench association: U.S. Geological Survey Professional Paper 636, 102 p.

Goguel, J., 1963, The geological reports: London, International Water Power, v. 15, no. 2.

Habib, P., 1964, Physical properties of gneissic foundation rocks, Malpasset dam, France: Report to Coyne et Bellier Engineers, Paris, 65 p. (private publication).

—— , 1987, The Malpasset dam failure, *in* Leonard, G. A., ed., Dam failures: Engineering Geology, v. 24, no. 1-4, p. 331–338.

Hamel, J. V., and Spencer, G. S., 1984, Powerhouse slope behavior, Fort Peck dam, Montana, *in* Parakash, S., ed., Proceedings, International Conference on Case Histories in Geotectnical Engineering, St. Louis, Missouri: American Society of Civil Engineers, v. 2, p. 541–551.

Hays, W., 1985, Fundamentals of geology and regional geology for solving earthquake engineering problems in the Puerto Rico area: U.S. Geological Survey Open-File Report 85-731, p. 22–52.

Henry, D. C., 1915, Blanket versus cut-off wall for scaling bank at Cedar River dam: Engineering News Record, v. 74, p. 609 and 1053.

Herd, D. G., 1977, Geologic map of the Los Positas, Greenville, and Verona faults, eastern Alameda County, California: U.S. Geological Survey Open-File Report 77-689, 25 p.

Hopper, R. C., Lang, T. A., and Mathews, A. A., 1972, Construction of Straight Creek tunnel, Colorado, *in* Lane, K. S., and Garfield, L. A., eds., Proceedings 1st North American Rapid Excavation and Tunneling Conference, Chicago, Illinois: New York, Society of Mining and Metallurgical Engineers, v. 1, p. 501–538.

IGC, 1984, Baipaza hydroelectric power station, *in* Guidebook Excursion 30, Engineering geological problems of central Asia: Moscow, USSR, 27th International Geological Congress, p. 20–23.

Keaton, J. R., 1988, A probabilistic model for hazards related to sedimentation processes on alluvial fans in Davis County, Utah [Ph.D. thesis]: College

Station, Texas A&M University, 441 p.

Kiersch, G. A., 1953, Reservoir conditions, Lone Pine dam, Arizona: Consultant's report to Arizona Game and Fish Commission, Phoenix, 6 p. (unpublished).

——, 1958, Geologic causes for failure of Lone Pine reservoir, east-central Arizona: Economic Geology, v. 53, no. 7, p. 854–866.

——, 1966, Testimony, Superior Court, State of Arizona, County of Pima, State of Arizona vs. Rail-N-Ranch Corp, no. 94745, March 28, 1966, Judge Alice Truman, 92 p. (unpublished).

——, 1968a, Engineering geology of Golder dam project near Oracle, Arizona: Consultant's report to Rail-N-Ranch Corp., December 23, 102 p., Appendices on exploration data, tests, hydrology, grout data (private publication).

——, 1968b, Report on geological conditions, Crib Wall and South Portal area, Big Walker Mountain tunnel project, Route I-77, Virginia: Consultant's report to Langenfelder-Raymond, Wytheville, Virginia, February 7, 8 p. (unpublished).

——, 1969, Changed geologic conditions, Big Walker project beginning station 492+88, east tunnel: Consultant's report to Langenfelder and Sons, Baltimore, Maryland, March 18, 5 p. (unpublished).

——, 1970, Geologic causes for the excessive remining requirements throughout parts of Big Walker tunnel; Consultant's report to Langenfeld-Raymond, Wytheville, Virginia, June 23, 13 p. (unpublished).

Kiersch, G. A., and Hughes, P. W., 1952, Structural localization of groundwater in limestones; "Big Bend District" Texas–Mexico: Economic Geology, v. 47, no. 8, p. 794–806.

Kirkaldie, L., and Thomas, L. E., 1984, Dissecting a dam: Association of Engineering Geologists Bulletin, v. 21, no. 4, p. 509–514.

Lasson, G., 1947, Description of geology, Lone Pine reservoir, *in* Snowflake Project, Arizona: Boulder City, Nevada, U.S. Bureau of Reclamation Project Planning Report 3-8b 2-1, p. 94–98 (mimeographed).

Livingston, C. W., 1961, The natural arch, the fracture pattern, and the sequence of failure in massive rocks surrounding an underground opening, 4th Symposium on Rock Mechanics: University Park, Pennsylvania State University Mineral Industries Bulletin 76, p. 197–204.

Mackin, J. H., 1941, A geological interpretation of the failure of the Cedar reservoir, Washington: Seattle, University of Washington Engineering Experiment Station Bulletin 107, 30 p.

Marcuson, W., and Krinitzsky, E., 1976, Dynamic stability of Fort Peck dam: Vicksburg, Mississippi, U.S. Army Waterways Experiment Station Technical Report S-76-1.

Markham, J. J., 1970, The engineering geology of the Big Walker Mountain tunnel, Wytheville, Virginia [M.S. thesis]: Ithaca, New York, Cornell University Graduate School, 83 p.

Marliave, C., 1951, Geological report on Broadway tunnel, San Francisco: Consultant's report to Morrison-Knudsen Co., February 19, 1951, 8 p. (unpublished).

McCann, W. R., 1985, On the earthquake hazards of Puerto Rico and the Virgin Islands: Bulletin of the Seismological Society of America, v. 75, p. 251.

McGlothlin, W. K., 1973, Straight Creek tunnel: Mines Magazine, v. 63, no. 3, p. 4–13.

McTigue, A. C., 1956, Preliminary report on rock slide, Niagara Mohawk Hydroelectric Plant, Niagara Falls, New York, on June 7, 1956: Niagara Mohawk Power Corp, 10 p. (private publication).

Meehan, R. L., 1984, The atom and the fault: Cambridge, Massachusetts Institute of Technology Press, 161 p.

Middlebrooks, T. A., 1942, Fort Peck slide: Transactions of the American Society of Civil Engineers, v. 107, p. 723–764.

Moneymaker, B. C., 1968, Reservoir leakage in limestone terrain: Association of Engineering Geologists Bulletin, v. 6, p. 3–30.

MRCE, 1987, Collapse of the Thruway bridge at Schoharie Creek: Wiss, Janney, Elstner Associates, and Mueser-Rutledge Consulting Engineers, New York, for New York State Thruway Authority, 220 p.

Muller, L., 1961, Die felsgleitung im bereich Toc: Talsperrve 15, Baugeologischer Bericht to Societa Adriatica di Elettricita (SADE) (unpublished).

——, 1964, The rock slide in Vaiont valley: Rock Mechanics and Engineering Geology, v. 2, p. 148–212.

Nickell, F. A., 1939, Description of geology, Lone Pine reservoir, *in* Geologic considerations of dam sites on Little Colorado and tributaries, Northern Arizona: Boulder City, Nevada, U.S. Bureau of Reclamation mimeographed report, p. 21–22.

——, 1966, Discussions on role of the "calculated risk" in earthwork engineering: American Society of Civil Engineers Journal of Soil Mechanics and Foundation Division, v. 92, no. 1, p. 185–194.

Peck, R. B., 1973, Influence of nontechnical factors on the quality of embankment dams, *in* Hirschfield, R. C., and Poulos, S. J., eds., Embankment dam engineering, Casagrande volume: New York, John Wiley and Sons, p. 201–208.

—— 1980, Where has all the judgment gone?: Canadian Geotechnical Journal, v. 17, p. 584–590.

Philbrick, S. S., 1986, Response to receiving the Distinguished Practice Award: Geological Society of America Engineering Geology Division Newsletter, v. 21, no. 2-3, p. 5–6.

Post, G., and Bonazzi, D., 1987, Latest thinking on the Malpasset incident, *in* Leonard, G. A., ed., Dam failures; Proceedings of a workshop at Purdue University, August 6-8, 1985: Engineering Geology, v. 24, no. 1-4, p. 339–353.

PRWRA, 1971–72, Aguirre nuclear power plant, unit 1, PSAR: Puerto Rico Water Resources Authority, San Juan, Puerto Rico, Geological Amendment 11, 11 p., Amendment 17, 44 p.

RCI, 1987, Hydraulic, erosion, and channel stability analysis of the Schoharie Creek bridge failure, New York: Fort Collins, Colorado, Resource Consultants, Inc., 350 p.

Reiche, P., 1936, Memorandum report on Lone Pine reservoir: Albuquerque, New Mexico, U.S. Department of Agriculture Soil Conservation Service, 5 p.

Reisner, M., 1986, Cadillac, Desert: New York, Penguin Books, p. 101 and 103.

Rice, S., Stephens, E., and Real, C., 1979, Geologic evaluation of the General Electric test reactor site, Vallecites, Alameda County, California: California Division of Mines and Geology Special Publication 56, 19 p.

Ries, H., and Watson, T. L., 1936, Engineering Geology: New York, John Wiley & Sons, 5th edition, 750 p.

Robinson, C. S., and Lee, F. T., 1974, Engineering geologic, hydrologic, and rock mechanics investigations of the Straight Creek tunnel and pilot bore, Colorado; Comparisons of predictions and findings: U.S. Geological Survey Professional Paper 815-I, p. 123–132.

Schmidt, L. A., 1943, Flowing water in underground channels, Hales Bar dam, Tennessee: Proceedings of the American Society of Civil Engineers, v. 69, no. 9, p. 1417–1446.

Szechy, K., 1967, The art of tunnelling: Budapest, Akademiai Kiado, p. 141–147 and 230–233.

Terzaghi, K., 1929, Effect of minor geologic details on the safety of dams, *in* Geology and engineering for dams and reservoirs: American Institute of Mining and Metallurgical Technical Publication 215, p. 31–46.

Virgilio, B. J., 1986, Movie on "Rockfall failure, Schoellkopf plant on June 7, 1956": Niagara Falls, New York, Schoellkopf Geological Museum, 17 minutes.

Wahlstrom, E. E., 1964, The validity of geologic projections; A case history: Economic Geology, v. 59, no. 3, p. 465–474.

Walters, R.C.S., 1971, Dam geology: London, Butterworths, 470 p.

West, R. P., 1962, Clay seam wrecks Wheeler Locks: Engineering News Record, v. 168, no. 1, p. 19.

White, S., 1968, Building a dam with a bang: New Scientist, November 14, 1968, p. 358–359.

Whitman, R. V., 1984, Evaluating calculated risk in geotechnical engineering: Journal of Geotechnical Engineers, v. 110, no. 2, p. 145–168.

Manuscript Accepted by the Society May 25, 1990

ACKNOWLEDGMENTS

Preparation of this chapter has benefited from the constructive comments and input of many colleagues and individuals, and particularly the editorial advice of DNAG editor A. R. Palmer. The authors appreciate this help and gratefully acknowledge indebtedness to the following: Richard C. Harding of Earth Sciences Associates provided report data on the Verona fault controversy; B. J. Virgilio, curator of the Schoellkopf Geological Museum, Niagara Falls, kindly showed one author (GAK) the ten-minute movie that spans the June 7th cliff failure and provided access to the file reports of consultants A. C. McTigue and L. S. Bernstein on the effects of collapse; J. R. Dunn, geological consultant to the Schoellkopf power plant owner, provided unpublished data, maps, and three written communications on the collapse, and reviewed the text; Thomas A. Wilkinson, United States Corps of Engineers (USCE) Buffalo, made a sequence of documentary photographs on collapse of the Schoellkopf power plant available and short descriptions were offered by Owen L. White, Toronto, and Brian Greene, Pittsburgh.

Factual data from on-site observations and a meaningful review and interpretation of the geologic conditions of two projects, Wheeler locks and East Branch dam, were provided by S. S. Philbrick. Fitzhugh T. Lee, U.S. Geological Survey, supplied reports and counsel on the Straight Creek tunnel investigations for predicting conditions at depth. Langenfelder-Raymond, Baltimore, authorized use of file data on Big Walker tunnel and the progress records compiled by the staff of A. A. Mathews, Inc. and R. E. Huera. Lloyd B. Underwood, consultant and retired from the USCE, prepared a summary overview with analysis on the numerous reports and findings surrounding the Fort Peck dam, powerhouse, and spillway slides along with supporting published data; J. V. Hamel arranged for his co-author, G. S. Spencer, USCE, to supply photos of early slides in the Bearpaw Shale near Fort Peck. J. R. Dunn of Dunn Geosciences, and James P. Gould, partner in Mueser Rutledge engineers, provided comments and published data on the Schoharie Creek bridge failure, and Resource Consultants, Inc., provided a hydraulic and channel stability analysis. Donald R. Currey, University of Utah, authorized use of unpublished research data on the Great Salt Lake sediments relevant to the Southern Pacific Company causeway project and strength of foundation materials.

Contractors Hill, Panama Canal, looking southward on November 14, 1910. The rock slide is typical of the slope failures that occurred after initial excavation for the canal, and before the cut slope was investigated and redesigned for stabilization. (Photo courtesy of the U.S. Geological Survey and D. J. Varnes.)

Printed in U.S.A.

Geological Society of America
Centennial Special Volume 3
1991

Chapter 24

The geologist and legal responsibilities

Eugene B. Waggoner
Consulting Engineering Geologist, 336 Seawind Drive, Vallejo, California 94590
George A. Kiersch
Geological Consultant, Kiersch Associates, Inc., 4750 North Camino Luz, Tucson, Arizona 85718

INTRODUCTION

Throughout the nineteenth century and the early part of the twentieth, the involvement of geologists in legal matters as part of their normal professional services was rare. A geologist's principal endeavors prior to the early 1900s were mainly restricted to the more classical and academic studies of the Earth's features and resources. An exception, and one of the earliest recorded cases of geological litigation and the as-encountered site conditions, involved excavation to enlarge the Erie Canal Locks at Lockport, New York, in 1839. James Hall of the New York Geological Survey was asked to evaluate and classify a "Slate Rock and Shale" sequence; the engineer's contract quoted a unit price for "solid rock" and a lower price for "Slate Rock and Shale" (described in Chapter 1, this volume). Terms of the 1839 contract made a clear distinction between the rock types impossible.

Even the early applied geologists (Chapter 1, this volume) were mainly involved with collecting and describing the geologic setting of a proposed project and providing the relevant information on areal and site maps with accompanying texts that described the general geologic conditions to be expected. Only on rare occasions would the highly respected early applied geologists, such as W. O. Crosby, James Kemp, Heinrich Ries, Charles P. Berkey, or Warren J. Mead, be invited to serve in litigation, and never as part of a large-scale claim or changed-conditions argument so common since the 1960s. Rather, those early applications of geology for legal purposes were usually restricted to a single geologic entity, such as the two detective problems of a geological flavor in the early 1900s solved by Professor Berkey:

"The Ward Steamship Company carried a shipment of goods from Italy, but the goods were stolen enroute and rocks were substituted. By identifying the petrographic details of rock, he was able to demonstrate the theft occurred in Naples, Italy. On a second occasion, a shipment of rubber from the Amazon was transported by the New York Central railroad to its destination in the midwest and on arrival a portion was found to consist of "rocks." Actually the material was concrete made from Cow Bay Sand, so the theft had occurred in Brooklyn. The concrete came from a nearby dock undergoing repairs" (Sanborn, 1950, p. 40).

Successful detective work by applied geologists continues today, although on a much smaller scale than the typical claims and associated litigation. For example, Donald B. Coates (personal communication, November, 1987) described such a case in 1987 that involved $7.3 million dollars. The victim died due to a rock accident in Watkins Glen, New York, and the state was sued for negligence (dangerous rock slopes). Coates determined the rock that struck the victim was thrown and not a rockfall, because the petrographic details were not those of the nearby rock outcrop; therefore, New York State was not responsible.

Similarly, John S. Rapp (1987) reported on several criminal cases in California during the period from 1958 through 1983 that were solved with the assistance of geological input and documentation by members of the California Division of Mines and Geology. Three of the cases demonstrate yet other ways of utilizing geological input and its implications.

In one case, sand and rock pebbles found on the floor of an accused murderer's auto matched the gravelly materials used by the accused to bury an earlier, known murder victim. The burial was near an oil well in Colusa County, California, where aggregate fill material from Kern County (320 mi away) had been transported to the well site. Small rock clasts in the auto of the suspect proved to be the well-site aggregate, which established the character and past activities of the accused beyond any reasonable doubt.

In another case, a prominent 1967 kidnapping in Los Angeles involved holding a 10-year-old boy captive in a quarry—the source of industrial diatomaceous earth for the region. The auto of the suspected kidnapper was later found abandoned, and the search for evidence turned up a mixture of soil and diatomaceous earth from the floor of the car. A geological investigation of the regional resources and locations established that the earth mixture came from the quarry where the victim had been held. The suspect was subsequently convicted of the kidnapping. An extensive review of forensic geology and criminal investigations has been published by Murray and Tedrow (1975).

Waggoner, E. B., and Kiersch, G. A., 1991, The geologist and legal responsibilities, *in* Kiersch, G. A., ed., The heritage of engineering geology; The first hundred years: Boulder, Colorado, Geological Society of America, Centennial Special Volume 3.

Geology began to play an important part in litigation for engineering works when geologists were brought into the planning, design, and construction of projects during the 1930s and 1940s. As the projects became larger and more complex, geological information became more and more critical for site selection and assistance in designing engineering works. Consequently, with the expansion of activities, contractors found themselves more frequently encountering unexpected geologic conditions at compled sites. As a result, geologists were called upon more frequently to assist contractors in preparation of unit-bid costs or to testify, at a later stage, in settling or adjudication of construction-related claims. Because of this, the U.S. Bureau of Reclamation created a formal engineering geologic staff by 1944 in anticipation that the staff of geologists assigned to serve the design and construction branches would be effective in both improving designs and reducing potential construction claims.

Early history

The first construction claims of importance that required expert professional geological input occurred in the early 1930s and 1940s. They mostly involved governmental or quasi-governmental agencies such as projects of the Bureau of Reclamation, Corps of Engineers, Tennessee Valley Authority, state highway and water resources departments, regional irrigation districts, and large private power companies. The majority of these early geology-related claims involved differing or unexpected site conditions encountered in the construction of such major works as dams, tunnels, canals, highways, and power plants. This concept of modification and change was recognized by early practitioners like Karl Terzaghi who stated that no foundation design is completed until the construction has been finished. In retrospect, very few major engineering works in the United States were completed without some claim being filed on changed or differing as-encountered site conditions. Most of these early claims for differing site conditions were in the millions of dollars, and a few in those days were in the tens of millions. An outgrowth of these claims was the need to mitigate future problems, as well as for designers to better understand the influence of geology on their project designs; this created a further demand for capable applied geologists, and this group expanded rapidly. Similarly, the impact of geology on project construction in two areas—unit-bid preparation for contracts, and claim litigation for changed conditions—was particularly rapid. Unfortunately, very few cases that are concerned with details of geologic causes and/or results of differing site-condition claims are published, particularly because many were settled out of court. For cases taken to court, the circumstances are released in the court records and frequently in brief engineering magazine articles or news items.

The list of projects on which there were claims filed relating to geologic conditions is very long; it includes such widely known projects as Hoover Dam, Grand Coulee Dam, St. Lawrence Seaway, California Aqueduct, Washington, D.C., metro transit, and the Eisenhower Tunnel, Colorado. In addition to being involved in the legal aspects of design and construction, geologists were also brought into litigation and/or investigations resulting from failures of project structures affecting the public, where those failures involved geologic materials, conditions, or phenomena (e.g., dam failures such as the St. Francis, Malpasset, Vaiont, Baldwin Hills, and Teton Dams, or large landslides such as Malibu and Portuguese Bend, all of which are described in Chapters 22 and 23 of this volume, and in Jansen, 1988).

THE GEOLOGIST'S RESPONSIBILITY

The brief summary above indicates the extent, impact, and growing importance of the involvement of geologists in litigation. The how, where, and what of their legal activities must be considered in principle, and are discussed in this section.

When acting in a professional capacity, the primary legal responsibility of geologists is to their employer/client, whether they serve as a member of a government agency, a private company, or as independent consultants. Once they agree to accept money for their professional service and provide a geological opinion as part of that professional service, they have accepted a legal responsibility. As professionals, geologists may also feel a responsibility to their profession; however, such responsibility is secondary and rarely of a legal nature and usually concerns their general social and business behavior. It also is reflected in their willingness to serve their peers and associates in a beneficial manner (Kiersch and Cleaves, 1969).

Geologists also have a responsibility to protect the health, safety, and welfare of the public as a part of their professional service, besides being ethical and law abiding. One ramification of this concern lies in the geologist's evaluation of what position a prospective client or employer may ask him to support when giving testimony in court, before special commissions, or in public print. The geologist must separate unacceptable advocacy from sincere, documented beliefs, and should not publicly express or agree to opinions or conclusions that are not believed correct. Geologists must be willing to inform an employer/client of their opinions and what they can or cannot ethically do or say. If conflicts arise, the geologist should be willing to withdraw from service to the client. This matter is so important that numerous papers have been written on the subject (Hammon, 1968; Kiersch, 1969a, 1977; Dunn, 1974; and Tank, 1983). The Association of Soil and Foundation Engineering has prepared a set of 13 recommended practices for design professionals engaged for expert testimony. Publication of the recommendations constitutes a profession-wide response to the common problems created by "hired gun" witnesses (ASFE, 1988). The set of guidelines has the endorsement of several national engineering societies.

LIABILITY AND MALPRACTICE

Since World War II the greatly expanded role of applied geologists in construction, combined with environmental concerns, has brought increased responsibility by applied geologists for the safety and welfare of society. Under these circumstances, the probability of litigation is constantly increasing. Today geologists are far more likely than in past decades to be involved in some aspect of litigation, either as members of a team, as individuals, or serving a client that is a plaintiff or defendant. Moreover, even when applied geologists perform their services in a wholly conscientious and competent manner, they may on occasion face personal litigation and be potentially liable for real or apparent malpractice.

Liability is the legal responsibility for any loss or damage that occurs as a result of one's actions, performance, or statements. Malpractice is improper, negligent, or unethical conduct or practice that results in damage or injury. Consequently, a claim of negligence leading to economic or physical damage or to injury is basically a charge of malpractice. Although there is no public law that specifically defines the limits, lawsuits and court decisions imply a legal responsibility to practice according to the current "state of the art" or "standard of practice." This means that an expert witness must be informed as to the latest technical, theoretical, and analytical aspects of the profession and investigative methodologies and practice. Even when their practice is as up to date as possible, however, geologists can in some circumstances make an incorrect judgment or interpretation and ultimately be found liable for negligence.

As a practical matter, the probability of a geologist being sued for negligence or malpractice as an individual varies enormously depending on the nature of the work. No case is known where geologists working as teachers or research scientists or as mining or petroleum geologists employed by companies or by governmental agencies have been sued by their employers. In these circumstances, any problems of incompetence or malpractice are likely to be detected and compensated for by such mechanisms as education, closer supervision, or severance. Geological malpractice, however, may result in the employer of a geologist being sued, in which case the geologist could conceivably be sued by the employer. More likely, however, the geologist involved will help defend the employer, who takes responsibility for the geologist's work. The personal liability of the geologist may be minimal in such cases, unless the professional works with a consulting firm. In the latter case, a litigator may attempt to go within the firm to the individual who prepared the drawings or reports, as well as to the supervisor, the senior partner, or an owner. The employee may also be named in the suit if the harmed party thinks a "deep pocket" can be tapped, such as insurance coverage.

As a group, individuals serving as geological consultants, or those in private practice associated with geological-geotechnical engineering consulting firms, are the geologists most vulnerable to charges of malpractice or negligence. In accepting a client, a geological consultant firm assumes responsibility for the quality of their work. The client has the expectation—and the right—to receive high-caliber professional geological work that utilizes the relevant state-of-the-art or standard-of-practice concepts and techniques. Practically, geologists or their firms must be able to legally defend their practice on this basis. In the event that errors occur, the consultant, individual or firm, is legally vulnerable if it can be demonstrated that applicable state-of-the-art or standard-of-practice concepts or techniques were not employed.

In practice, flawless professional work may be nearly impossible, as anyone knows who has had to defend such a position on the witness stand. Eliminating errors such as misconceptions due to typographical mistakes is expected. Preventing major potential mistakes in geological work requires a realistic and humble assessment of the nature of the geologic and test data and its interpretation. Two long-standing rules are that humbleness minimizes error and ego, or the molding of data to fit preconceptions, causes errors. Geologists must exercise a fine sense of balance and practicality in evaluating the degree of certainty of the geological conclusions made from the data available. One safe rule is that when conclusions are not obvious, one should conclude there is either insufficient information or only a qualified opinion can be given (Litigation, 1977).

When information is insufficient to reach a meaningful conclusion, frequently due to inadequate time and/or money being available, it is the responsibility of the geologist to so state, regardless of whether to an employer, client, or themselves. As an example, in a mining project an employer may request that a block of proven ore of a specific grade be delineated. Realistically, project funding is insufficient, and only ore of a probable or possible category can be ascertained within the budget. Any such limitations should be clearly stated in writing so the employer is not misled. Furthermore, any such limitations should always be explained in project reports even though both the geologist and employer are fully aware of them. At some future date the report may be used by someone not familiar with the original budget and scope limitations; furthermore, the report might be used out of its original context. Finally, and apart from the general statements on limitations in stating conclusions, geologists should exercise a conservative sensitivity relative to the degree of certainty, and this awareness should be expressed in each conclusion. Unfortunately, test data and conclusions are frequently quoted out of context in litigation and mislead, instead of support, the meaning expressed by the full report.

The geological practitioner should be aware of any state or county requirements for professional certification, registration, or licensing when working in a new region. Geologists not fulfilling such legal requirements may not only be breaking laws but, in the event of any litigation that involves possible malpractice, may find their work defined as unqualified or unacceptable.

In large firms, geologists frequently work under a principal or group head who is registered, and in such cases, that person and/or the firm is responsible for the staff members. The same is true of a public agency that exempts its employees from being registered; the public agency is usually held responsible for the quality and accuracy of the geological work performed under its jurisdiction. In private practice, where a license is not required and work is alleged to result in loss or damage, the individual practitioner is commonly held to be primarily responsible. In some cases, a client—whether engineer, owner, or sponsor—may, by contract, specifically hold harmless, indemnify, or agree to defend a geologist (or other professional) who provides services. However, it is becoming increasingly evident that attorneys can, in searching for the "deepest pocket," file "shotgun" suits and include anyone even remotely connected with a damage or loss situation. Since geologists are somehow involved with nearly everything built on or within the earth, or out of earth materials, their degree of liability is growing. In most cases, it is the consulting geologists involved with engineering, construction, and environmental matters who have the greatest liability risk; this is largely because public health, safety, and welfare are so deeply involved with engineering works.

Geologists practicing in the fields of applied geosciences relevant to the environment, hazard-risk mitigation, pollution, hazardous waste, and general engineering works are the practitioners most likely to require registration or licenses and are at the greatest risk of liability. Many public agencies do not require their geologists to be registered/licensed, nor are those in academic and research work. Most geologists operating in the mineral or resources-related disciplines with petroleum, mining, and engineering corporations are usually legally exempt from any registration or licensing requirements.

There is a difference of legal opinion as to who is responsible whenever geological problems arise. When geologists are licensed and work within the jurisdiction of registration as consultants, they are legally responsible for the quality of their work. If the geologists are unlicensed in a state where registration is required, they normally must work for someone who has a license and will approve their work; the licensee can be held responsible. Another common alternative is to seek a reciprocity privilege to practice in a state other than the licensee's.

The practice of applied geoscience by the typical, science-oriented research organization, whether a mining, petroleum, or water-resources corporation, or similar engineering groups, rarely leads to problems of individual professional liability; if cases arise, the employer is generally held liable for any errors that cause damage to an outside partner. Moreover, even though the responsibility for any error or negligence may be the geologist's, no cases are known where a geologist working for a private firm has been sued by his employer.

A new option, mini-trials, to resolve disputes arising during construction, is being employed in some cases today. The relatively short proceedings emphasize an attitude of arbitration, as demonstrated by the settlement of claims involving the Tennessee Tombigbee Waterway (Henry, 1988).

RESPONSIBILITIES OF GEOLOGIST-EMPLOYER-CLIENT-OWNER-ENGINEER-CONTRACTOR

Introduction

The foregoing comments have dealt principally with how geologists fit generally into the question of who is held responsible for the work product and to whom practicing geologists are responsible.

In the case of failures or claims related to geologic conditions, parties other than geologists may be involved, such as the owners, employers, clients, engineers, contractors, and others. Much depends on what actually occurred or went wrong. Was there a construction-related problem, the failure of specific geologic materials to perform properly, a catastrophic failure of an engineered-structure or a natural slope, or an incorrect geological evaluation of some source of material or product? Lawyers tend to place these liability situations in general categories such as: (1) differing or changed site conditions, (2) failure to disclose information, (3) misrepresentation of geologic conditions, (4) errors and omissions of critical facts, and (5) failure to perform as predicted/intended. Although there are other categories, these will serve to illustrate the point that a geologist's responsibilities may range from being the individual's alone, to being shared by a group of entities.

As a further example of the growth of responsibility of engineering geologists, the first pre-bid geological reports for contractors in the United States were done in the 1950s (Kiersch, 1957). Today, they are common practice for most major projects. The pre-bid geological reports are used to assist the contractors in understanding the geologic data provided with bid documents. Better understanding allows the contractor to prepare his construction plan and select equipment with more confidence and thus become a more competitive bidder. The practice of contractors hiring applied geoscientists for pre-bid studies has led to geologists having more involvement in the entire construction process. Engineering geologists are called for counsel on geologic problems that show up during construction and often to map the "as-built" geologic conditions encountered in the construction. This procedure makes it imperative that today's applied geologist, in addition to being a well-informed geoscientist, also understands basic construction procedures and equipment. This background is particularly important relevant to design and the unit costs and construction methods.

Because the geologic conditions of a project site are rarely, if ever, fully determined by the preconstruction site investigations, no matter how thorough, it is not unusual for a contractor to encounter unexpected geologic conditions in some part of the project. When this occurs, the contractor may file a "differing site conditions" claim. Encountering unsuspected geologic conditions and filing such a claim may place the owner/engineers' geologist in opposition to the contractor's geological consultant in the settlement of the "differing site conditions" claim. In some cases, "disputes settlement" clauses have been included in the bid contracts, providing a board of experts (usually three persons) who hear and rule on disputes as the project proceeds. This has brought about the opportunity for differing geologists to present their opinions and interpretations in a more reasonable manner than is possible in a courtroom. How the geologist meshes his responsibility with that of others is demonstrated by the following brief cases.

Differing or changed site conditions

This situation, in which geologic information is supplied to contractors for use in preparing unit-cost bids and planning for construction, is probably the most common circumstance to result in litigation involving applied geologists. The amount and quality of this information relates to how much the owner/client is willing to spend on collecting geologic data and specific information the designer and engineers need. If the owner budgets too little for an adequate exploration program, the geologist will have insufficient baseline data with which to make interpretations and predictions; in this circumstance the potential for changed-conditions claims will be large. Furthermore, an inadequately substantiated analysis for the design engineer may lead to project designs not wholly adapted to the actual conditions at a site. Consequently, it is the geologist's responsibility, with the help of the designers, to explain to the owner the need for a complete and thorough investigation and testing program. They should also explain the inherent risks of an inadequate investigation. If the owner still refuses to provide funds for an appropriate level of investigation, the liabilities should pass to the owner. If the designer agrees to accept less than what should be provided, then part of the responsibility lies with the design engineers. If the geologist submits a report based on less information than recommended and believed necessary, this circumstance should be made clear. Moreover, what sorts of different conclusions might be reached if more complete baseline data were available is an appropriate comment in the report.

Failure to disclose

This condition occurs when one of the parties involved with a project has either intentionally or accidentally failed to provide some relevant data from project files that could affect the design, construction, or performance of engineering works. Generally this failure is due to an oversight, but cases are known where it was the planned decision of a designer or an owner. The primary responsibility of the geologist is to provide the owner/client and the designer with all available geologic data, along with one or more likely interpretations, irrespective of the client's beliefs. At this point, it becomes the responsibility of the designer or owner/employer to make sure the full range of baseline data is made available to the contractor.

Misrepresentation

Misleading data and/or misrepresentation can be deliberate or unintentional where geology is concerned. The geologist is usually responsible to oversee and make sure misrepresentation does not happen. Misrepresentation can range from ambiguous, inaccurate, or incomplete borehole logs to erroneous or unsupported geological opinions or interpretations. All geologists should constantly keep in mind how their data and opinions are to be used and be responsible to see that they are as complete and logical as possible. Where they cannot be certain of their opinions, they are responsible to let the users of the data know the probability of their being right. They are responsible to see that those using the data understand correctly what they are being told. Furthermore, geologists are responsible to see that users of their data are not led astray by strange, ambiguous, or unclear words and statements.

Errors or omissions

Errors and omissions are those mistakes that are made unintentionally and usually unwittingly or carelessly. They can be incorrect calculations, misprints, misclassifications, or wrong interpretations. Such things will never be completely eliminated, but through constant review, geologists and professional peers can greatly reduce errors or omissions. The responsibility for errors and omissions can rest only on those professionals who make them.

Failure of works to perform as predicted, designed, or intended

This type of failure happens when several kinds of events occur: for example, a slope is predicted to be stable yet fails, a building foundation settles or swells even though declared stable, a tunnel roof collapses after supports were declared unnecessary, or a quarry depletes all the satisfactory rock when the user relied on a geological report that assured more reserves of rock than needed. It is usually difficult to place responsibility on any one person, but on occasion a single person is clearly at fault. The failure may be due to the negligence of the geologist interpreting the facts, the designer misusing the geologic data, or the contractor doing poor-quality work and/or not following specifications. One can be sure, however, that if litigation results from such errors or omissions, and the responsibility is not clear, everyone having anything to do with developing the geologic data or using it in design or construction of the project will be brought into the case. Court action will determine responsibility and make proportionate assignments of liability.

TYPICAL LITIGATION OF CONSTRUCTION

Several examples follow of project litigation or settlement arbitration related to the five geological categories. They are not all cited with specific references because in most cases the pertinent claims or litigation points are not in print, but come from the author's personal acquaintance with the projects.

A few of the hundreds of claims litigated or arbitrated are representative of the points and categories of liability relevant to geologic conditions. Two worthwhile references on many such claims are the U.S. National Committee on Tunneling Technology report "Geotechnical Site Investigations for Underground Projects" (USNCTT, 1984) and "Legal Geology Cases" (Tank, 1983). The first report describes 87 case histories, of which many illustrate the impact of geology on claims litigation or arbitration; the second report is more general and includes details on litigation of all aspects concerned with geologic conditions and features. The following projects demonstrate one or more of the liability categories discussed.

Tecolote Tunnel

This tunnel project is located in the coastal range north of Santa Barbara, California. The U.S. Bureau of Reclamation performed a considerable amount of geological investigations to select the alignment and grade for design purposes (USBR, 1959; Trefzgar, 1966). However, the Bureau decided not to include all this data and geologic background as part of the bid document, so the selected contractor was not fully aware of the geologic setting. The tunnel-bore encountered major problems of excessive hot water inflow that in some zones reached temperatures up to 118°C (USBR, 1959). Inflows of methane gas were also encountered that resulted in several explosions and injuries to crews (see chapter 22, Fig. 15, this volume). The Bureau's decision not to disclose geological data that would have prepared the contractor for some of these problems eventually resulted in a legal decision against the Bureau. The Bureau was required to pay the costs and damages of the first tunnel contractor with the project uncompleted, and subsequently was required to engage a new contractor to finish the last mile of the six-mile tunnel. The second contract was at a price equal to the original bid for the entire six miles (further discussed in Chapter 22, this volume).

The St. Lawrence Seaway

This huge project required the construction of new channels, dams, bridges, locks, powerhouses, and other structures. The principal excavations were in a very dense glacial till (see discussion by Coates, Chapter 15, this volume). Geological investigations and findings completed by the governmental agencies, together with the resulting data, were included in the project documents. The specifications classified excavation areas as either "common" or "rock," with different unit prices. The borehole logs of the till available to the contractor showed geologic descriptions of the drill cuttings as predominantly sand, silt, clay, and some gravel. From this it appeared that all of the excavations, except bedrock under the till, would be "common," low-priced, excavation; this evaluation was based on samples of drill cuttings of till.

The till was actually dense and tough, and had to be blasted before it could be efficiently excavated (Cleaves, 1963). Without blasting, even the heaviest bulldozers, rippers, and shovels were torn up in an unsuccessful attempt to excavate the cemented till. What had not shown up in the borehole logs was that, although the cuttings were sand, silt, clay, and some gravel, the till in place was a dense, 140 to 150 pcf, rock-like material, with a high compressive strength. The till was actually tougher than the underlying Paleozoic bedrock. In examining the logs during the litigation it became evident that the original field logs had been condensed and simplified several times in the process of preparing specifications. In this compilation process, much important and significant geologic data was eliminated. These data would have told the contractor that what he thought would be "common" excavation might be "rock" excavation. The owner ultimately paid a large claim for misrepresenting the properties of the till.

Allegheny Dam

This dam in Pennsylvania required placing a large quantity of rock rip-rap on the upstream face. Specifications stated the rock had to be "durable" and suitable for the purpose intended (Fuquay, 1967). The desired amounts of 6 in to 12 in rock and over 12 in rock were also stated. The specifications also provided the location, and delimited the boundary, of a quarry site acceptable to the engineer and owner. If the contractor elected to use a different quarry site, he would be required to perform a group of designated tests on the new site rock and submit the results for the engineer's approval. Interestingly, the project engineers had not performed all of the tests to be required on the designated quarry-site rock. The project engineers had drilled five core holes on or near the designated quarry site and provided the boring logs to bidders. The boring logs described the cores and carried statements indicating "top of durable rock" and "bottom of durable rock." Based on these statements, the contractor calculated the amount of "durable" rock in the quarry deposit and estimated that the amount of waste would be about 35 percent. File documents later released indicated the project engineers had made similar estimates. Contrary to the two estimates, the "unacceptable" rock actually quarried from within the zone bounded by "top" and "bottom of durable rock" yielded rock with waste of more than 70 percent of the quarried material. Consequently, the contractor had to go outside the designated quarry limits for a major quantity of the rock required and was delayed in completing the project. Furthermore, quarry operations proved that much of the rock classified on the logs as "durable" was not of acceptable quality for rip-rap.

Apparently, the site geologist relied mainly on a general judgment of the rock mass and an examination of case-hardened outcrops and much less on the visual appearance of the cores to assess the durability. Had the owner run the suite of tests that the contractor would have been required to perform to demonstrate suitability of rock from a new site, or had the owner made a pilot field blast of the rock, the concept of rock durability within the designated site rock would have been more accurate. This changed-conditions case was carried through all levels of hearings to the top court and ended with a favorable decision and award to the contractor.

Interstate highway, Utah

Construction materials for this highway pavement baserock were specified to be taken from a quarry on the north side of the proposed highway alignment. The bidders were shown a previous quarry located on the south side of the highway that had been used for earlier road construction in the area. Bidders were also shown the preliminary stripping of the proposed quarry on the north side, which had very little overburden, and the waste appeared minimal. However, before the project was completed, site rock meeting the specified quality was depleted and the depth of overburden above acceptable rock increased to 30 ft or more. During the litigation it was revealed that the owners had failed to provide bidders with several drill hole logs and file correspondence that indicated the top of some acceptable rock was covered by overburden and thus was deeper by 30 ft in one area to be quarried; in addition, the quarry contained much unacceptable rock. From bid data the contractor had expected there would be only a few feet of overburden. The failure to disclose all boring logs and the true conditions of the proposed quarry, as known by the owner, resulted in a settlement in favor of the contractor (Jack B. Parsons Construction versus Utah Department of Transportation: Supreme Court of Utah, April 1, 1986).

Pipeline, Las Vegas, Nevada

A large-diameter pipeline for water supply was constructed from Lake Meade to the city of Las Vegas by the U.S. Bureau of Reclamation. The excavation required an open cut about 20 ft wide and 15 to 20 ft deep across recent alluvial outwash deposits and unconsolidated fan sediments. The Bureau of Reclamation had bored exploration holes at regular intervals along the proposed route, sampled the materials to be excavated, and established the water-table conditions. The project bidding and construction occurred several years after the borings were made; in the meantime the city was growing, and some of the open desert along the alignment had become residential subdivisions. A considerable proportion of the trench excavation encountered wet and unstable soil and alluvial conditions; heavy wall supports, dewatering, and even soldier piles were required to hold the trench open until the pipe could be installed. What had not been foreseen by the geologists and engineers was the impact of lawn and shrub irrigation in the new residential areas; the local groundwater level was raised by an amount equivalent to the effect of several additional inches of rain per year. This higher level of ground water and resulting trench instability was settled as as true differing (changed) site condition.

Sewer tunnel, Chicago

The contractor who used a boring machine on the sewer tunnel excavation encountered a large number of boulders that damaged his equipment and slowed his operation. Project specifications classified the material to be excavated as "glacial outwash" and logs of 28 boreholes were provided. None of the logs showed any boulders within the outwash material, and the contractor assumed there would be none. There was nothing anywhere in the specifications to even suggest boulders might occur in the materials to be excavated. Clearly, nothing was done illegally, and no information was withheld or misstated by the geologists or engineers involved. However, had a thoughtful geological statement been included that regardless of the logs of the 28 borings, one should expect to encounter occasional boulders in any glacial till, a million dollars or more could have been saved in unnecessary operating costs and maintenance.

Tunnel, Boston Transit

The impact of geologic conditions on the line and grade of the Boston Metro Transit Authority's red line extension tunnels was critical. Initially the tunnel grade was established as parallel to the surface, with rail level at about 30 ft, except at the stations. The grade was subsequently relocated as shown on Figure 1, because geological exploration for the original alignment-grade established that the Porter Square Station excavation would be in both rock and the overlying till deposits, and the tunnels north of the station would require a mixed-face construction and be within 15 ft of surface buildings. The new alignment grade placed tunnels and stations mainly in the argillite bedrock— a good to excellent tunneling rock (Keville and Sutcliffe, 1983).

This alignment-grade was chosen after the weighted consideration of geologic conditions, construction needs, real estate restrictions, shaft siting, ventilation egress, and train operation and traffic. The sequence of glacial and postglacial deposits along the red line tunnel alignments (Meyer, 1983) is shown on Figure 1 and consists of:

Surficial deposits–of silty sand with varying proportions of gravel, clay and organic materials;

Marine clay–with varying proportions of zones or lenses composed of silt and fine sand with an irregular thickness of 10 to 20 ft;

Glacial till–generally consists of dense, silty, fine- to coarse-clayey sand with varying amounts of gravel, cobbles, and boulders. The irregular thickness of 50 to 85 ft reflects the ancient profile of the argillite erosional contact/surface;

Argillite–is an indurated alternating sequence of claystone, siltstone, and shale that is locally faulted and deeply weathered, and in gradational contact with the overlying till.

Construction of the Harvard Square to Porter Square extension, two parallel tunnels, encountered locally unstable conditions due to ground-water conditions inherent to glacial till deposits with zones and lenses of pervious materials. Although both areal and alignment borings made up the design geologic logs that accompanied the bid-contract text, the contractor misinterpreted the reasons for the ground-water occurrence in the till and believed it was associated with horizontally interconnected pervious zones. On this assumption the contractor expected to dewater the alignment with large-diameter wells spaced on 100-ft centers; however, on pumping, the water table was not lowered as expected in wells located south of station 195+00 (Meyer, 1983).

Because as little as 0.5 gpm water inflow would cause the till to become unstable, three alternative methods for coping with the

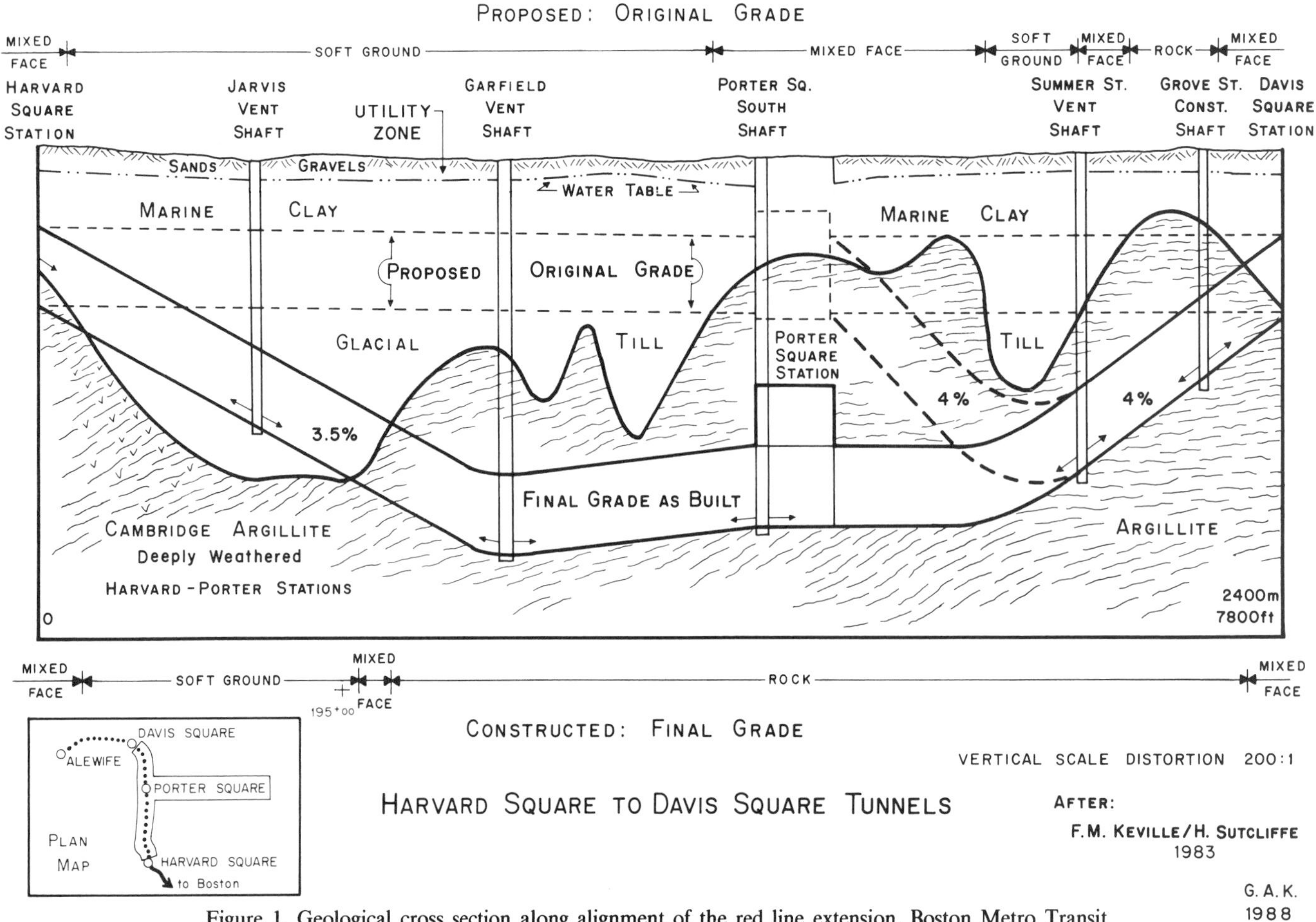

Figure 1. Geological cross section along alignment of the red line extension, Boston Metro Transit Authority, from Harvard Square to Davis Square. The originally proposed tunnel grade required several tunneling changes of a mixed face, soft ground, and rock. The final grade minimized soft-ground tunneling and is mainly in Cambridge Argillite, an excellent rock for a tunnel.

face instability were considered: (1) grouting prior to excavation, (2) excavation under compressed air, and (3) installation of an extensive dewatering system. Options 1 and 2, grouting and compressed air, were discarded due to costs and/or ineffectiveness in clayey till deposits. A reinvestigation of the geologic conditions and preconstruction exploration data proved the till more complex than initially considered. The hard glacial till (soft-ground to some) was composed of many thin, water-filled mini-lenses, each one surrounded by impervious silt and rock flour; essentially no interconnection existed between the water-bearing lenses of sand and gravel and the overall watertight mass. Consequently, the costly option 3 was selected for dewatering. Three hundred units of a well-point system with four-phase ejectors were installed on 20-ft centers on either side of each tunnel. Although the ejector systems tapped the pervious zones, they only produced a maximum of 60 gpm; this dewatered most of the small lenses and locally regained ground stability for the tunnels. Also, inflow from the water-bearing till at the large Harvard Square Station (Fig. 1) excavation never exceeded this total rate.

A postconstruction claim for ground settlement was considered an "unforeseen condition" by the contractor, but the factual and informative pre-construction geologic logs with detailed borehole data, and subsequent as-encountered mapping of the tunnels was critical in settling such claims. With this database, post-construction litigation awards for claims relevant to the geologic conditions were minimal; most claims were for surface interference or third-party public issues that slowed progress on the joint-venture tunneling contractors, Morrison-Knudsen, White, and Margentime (Meyers, 1983).

The high-quality geologic logs and tunnel mapping provided an excellent basis on which to forecast conditions to be encountered along the 24-ft, open-face tunnels. The manager of a tunnel project recently remarked that "contractors should listen to their own geologists" (H. Sutcliffe, personal communication, 1988). For example, the contractor continued driving a tunnel with full face rounds after his geologist forecast problems and advised on a change to multiple-drifts through an area with thin overlying bedrock. Soon thereafter, the back of the outbound tunnel broke, with severe stoping and within two feet of a predicted problem zone located in a "ravine sector" between the Porter and Garfield shafts (Fig. 1; Meyers, 1983, p. 947).

How the glacial till deposits would perform or behave under construction conditions was a matter of misinterpretation by some involved with the Boston Metro project. The tunnel engineer's work and the contractor's preparation of a contract bid would have been significantly aided by a clearer understanding of the detailed geologic features and water-bearing properties inherent to the till deposits. Yet in spite of this "error in judgment" (discussed in Kiersch and James, this volume) involving tunneling conditions, the contract that began on October 25, 1979, was completed in 1982 for about $8 million less than the original engineer's estimate of 1978 (H. Sutcliffe, personal communication, 1988) and about $2 million below the estimate, following all post-construction settlements.

Residential buildings: Land

Probably the most common and costly area of alleged liability for applied geologists and geotechnical engineers is related to residential construction. Tens of millions of dollars in damages are alleged each year by home and building owners who say their structures were damaged by swelling or subsiding foundations, slope failures, wet basements, or fault movements. Most of these cases are generally quite similar; damages to residences occur years after the structures are built. All too frequently the question of who is truly at fault is circumvented; settlement is based on who has the most effective lawyer or the deepest pocket. Consequently every applied geologist who accepts work involving residences or similar structures should be aware of the potential for being held responsible for such damages, regardless of how thorough the work. Furthermore, all geological reports and opinions on residential construction should be properly qualified as to the relative potential for and nature of possible foundation or slope failures.

Case histories. Summaries of several prominent court actions relevant to landslides and slope instability are given in Tank (1983); supporting readings on surficial processes include: (1) *Massei versus Lettunich* (reading 22-1 in Tank, 1983, p. 489–493), which involves action for damages to land that was graded and filled under supervision of a developer in Watsonville, California. The actions were tried on theories of negligence and deceit, with a significant role by expert witness testimony; (2) *Albers versus County of Los Angeles* (reading 22-3 in Tank, 1983, p. 498–514), which involves the major landslide area of Portuguese Bend in the Palos Verdes Hills. Los Angeles County road construction in 1956 triggered and reactivated old and prehistoric slides identified in 1946; numerous attempts have been made to stabilize the slide mass (see James and Kiersch, this volume); and (3) *Finley versus Teeter Stone* (reading 22-4 in Tank, 1983, p. 515–525), which involves the liability of a nearby quarry operator whose subsurface dewatering activities below 80 ft led to subsidence and opening or flushing of sinkhole features in the underlying limestone beds. The subsequent development of large, pipe-like voids or sink features at irregular intervals over the adjacent farm land became a serious danger to lives and property.

Landslide insurance and dwelling foundation—Pfeiffer case. This precedent-setting case illustrates how geologic facts, events, and terminology can be central to the outcome of a controversy between an insured party and the insurance company. Gravity sliding damaged the dwelling of the insured, and geological testimony determined the causes, possibility of recurrence, and the advisability of rebuilding on the site. In addition, such testimony was judicious regarding whether there was negligence on the part of the builder or the insurance company because neither realized the geologic setting was hazardous. Geologic facts revealed that causes were visible, and that the gravity sliding was foreseeable and could not be passed over by the insurance company as "an act of God" (Kiersch, 1969b).

The Pfeiffer case (Appendix A) involved a decision concerning landslide insurance and the payment of damage to the insured. The sole issue was "whether an insurance policy that insures a dwelling against the perils of landslide includes the foundation underlying the house as well as the house structure itself." This crucial point had not been clarified by a court decision prior to the Pfeiffer case in 1960. Insurance companies had customarily accepted no responsibility for the subsurface foundation part of the dwelling, and awarded damages for repair of the building only.

The Pfeiffer dwelling is situated on the slope of a northwest-trending ridge of the Berkeley Hills at Orinda, California (Fig. 2). The area is underlain by rocks of the Orinda Formation, which consists of alternating beds of siltstone, soft, fine-grained sandstone and conglomerate, clay shale, and clays.

At the site the underlying rock sequence dips as much as 45°, more or less parallel to the natural slopes of 20° to 40°. The soft, fractured, and saturated Orinda beds are overlain by as much as 10 ft of colluvium. Details of the gravity sliding movement, as determined by surface and subsurface investigations and borings, are described elsewhere by Kiersch (1969b).

The history of sliding. Sliding began soon after Robert J. Pfeiffer purchased the newly constructed home in December 1957. An insurance policy written by the General Insurance Corporation insured the dwelling against hazards of fire, and an attached endorsement provided protection against natural events such as landslides.

Major gravity sliding began in the soil and bedrock upslope from the dwelling on April 3, 1958; the building was partially distorted and twisted, and the subsurface foundation moved. Within two days the house was unsafe for occupancy, and the family moved out. The sliding continued intermittently for several months; damage to the dwelling and building structures increased until the garage section was demolished, the house structure was further damaged, and the foundation was weakened and became unstable due to being fractured, deformed, and displaced.

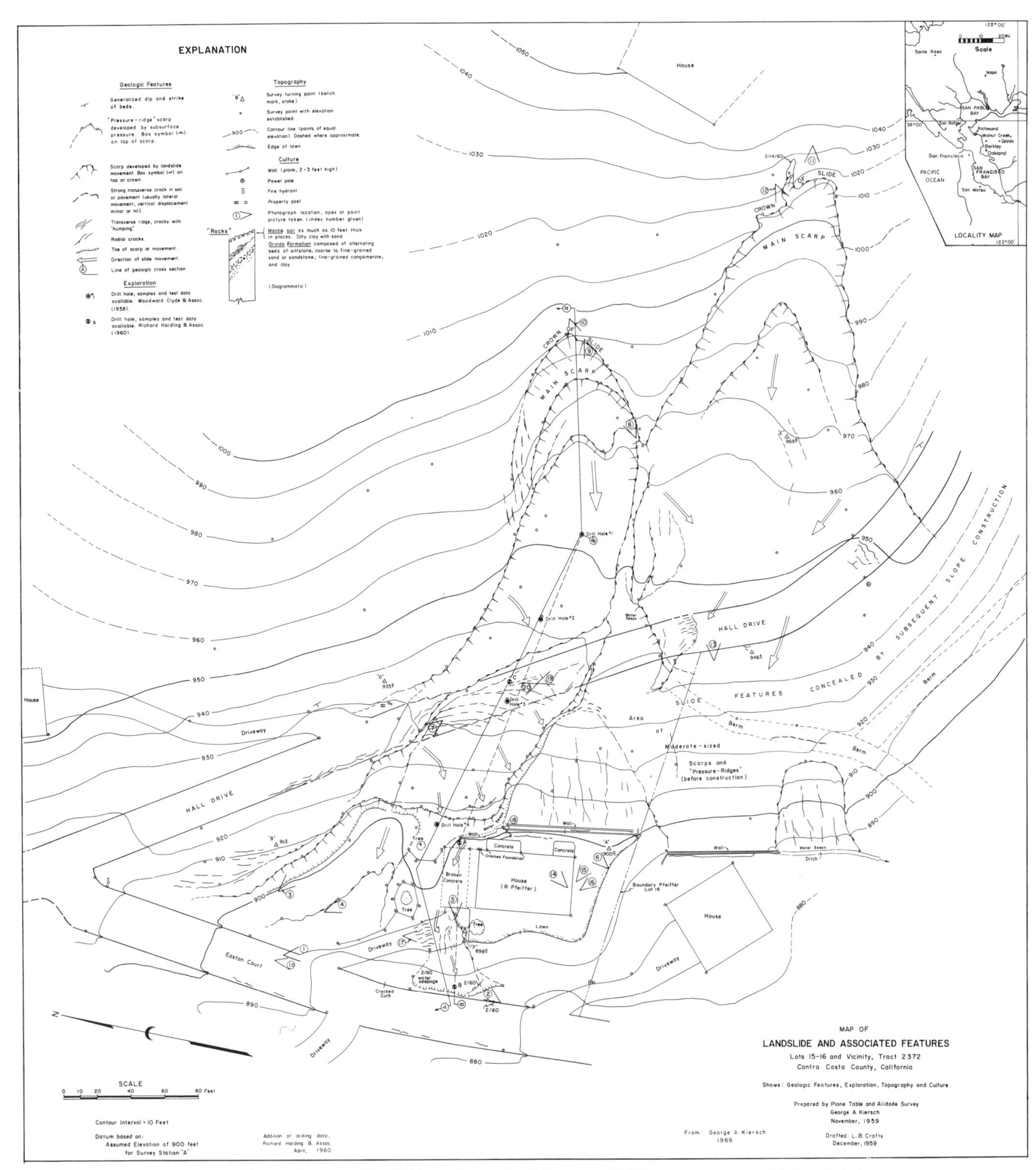

Figure 2. Geologic map of landslide and associated features, Pfeiffer property and vicinity, Orinda, California. (Source: Kiersch, 1969b).

Geologic causes. The multiple landslides shown in Figure 2 were due to a series of geologic events, both ancient and recent, that combined to cause sliding in April 1958. Ten contributing causes were responsible, among them: the slide-prone characteristics of the weathered Orinda rocks and soil, an old, active landslide within the 1958 slide mass, heavy inflow of surface runoff from storms and a high ground-water level, construction activities for Hall Drive and the Pfeiffer dwelling that affected the toe of the natural slope, and the additional surface water that infiltrated the hillside slope from a broken water main along Hall Drive, caused by earlier sliding (Fig. 3).

Litigation. Pfeiffer filed litigation against the General Insurance Corporation because they refused to accept responsibility for the total damage to the dwelling. Rather, they agreed to repair of the house structure only, at an estimated cost of up to $8,000. General Insurance adamantly refused to acknowledge any responsibility for repair of the subsurface and the damaged foundation of the housing structure, part of an active landslide mass. Furthermore, after repairing the damaged housing structure, they would cancel the insurance policy. This action and attitude was contrary to Pfeiffer's interpretation of the all-physical-loss coverage, with endorsements that encompassed landslides. Consequently, the Pfeiffer suit sought to recover the maximum amount set forth in the policy for landslide damage to their dwelling. The Pfeiffers were awarded $31,000 in a court judgment of August 1960 for repair of their dwelling, which included return to the as-built conditions of the house and foundation prior to the landslide damage.

The importance of geologic conditions and a competent explanation of the interpretation for the court, as relevant to insurance coverage and landslides, is further demonstrated by several other precedent-making cases briefly summarized by Kiersch (1969b); all cases were decided uniformly in favor of homeowners on all major issues. For those readers who may wish a layman's explanation of certain legal terminology, the Appendix in Tank (1983) is suggested.

Relevance of Pfeiffer to other cases. Other cases directly preceded the Pfeiffer opinion and lent the authority of precedent to that case, which was mainly the attempt by the insurance company to cancel a policy. One of these concerned the landslide in the Portuguese Bend area of southern California described in Chapter 22, this volume). Fire-insurance policies were in force covering two family structures located in that area when a massive and continuing land movement, which affected a large number of properties, commenced in 1956. With full knowledge of the ongoing land movement, the insurance company canceled two policies. Could an insurance company legally do this? The Harman (Supra) court held that liability, where a continuing loss had already commenced, could not be terminated or avoided. The insurance company was required to continue covering the loss, until the loss had occurred in its entirety, or the cause of the loss had ceased. Once the event has occurred, the insuree cannot alter the policy, and the contractual obligation must be carried out.

The Harman case, like Pfeiffer, involved a policy with an "all physical loss" provision. A court will always endeavor to resolve doubt and ambiguities in the interpretation and construction of policies in favor of the insured homeowner (Appendix A: Harman versus American Casualty Company of Reading, Pennsylvania, 1958, p. 614), and prevent insurance companies from taking unfair advantage of the insured.

The Pfeiffer case was followed in this progression of landslide insurance by the case of *Hughes versus Potomac Insurance Company* (1962). Again, for the same type of landslide, the interpretation of the term "dwelling" in the insurance policy was an "all physical loss" endorsement, and the outcome was similar to the Pfeiffer award.

THE GEOLOGIST AND CONTRACTUAL RISK

When determining professional responsibility or liability, the questions of who is at risk, when were they at risk, and for what, can be very fuzzy. The kind of risk under discussion may result in a large financial loss, a damaged reputation, job security, and sometimes health or life. The degree of risk is a constantly changing kaleidoscope. Attorneys, insurors, professionals, manufacturers, and the public all have their opinions about what constitutes liability and who should be made to pay. Basic law interpretations are constantly being changed by precedent-making court decisions or new laws; hence, lengthy and costly court hearings are constantly being conducted to decide on liability issues. Because the final product of geological services is subjectively observed from the investigation and analysis of earth materials, crustal structures, and the earth's behavior, the geologist's work products can rarely be certain. Consequently, geologists can make incorrect prediction, and therefore are constantly at risk in this litigious world. The risk is generally contractual, and the contract can be either written or oral, as long as there is a meeting of the minds of those involved and a common understanding of what is agreed upon. This risk is in contrast to "strict liability," which is described as follows (K. H. Harbinger, written communication, 1987): "Strict liability is a legal concept which has been widely adopted in the United States. In essence, if a producer has control of the means, methods, and components of a manufactured product, and a defect in that product causes injury to a user, the producer is strictly liable, even if the injury is not the result of negligence in the manufacture of that product." There have been some efforts to construe professionals and their services as equivalent to producers or manufacturers and manufactured goods. Although to date the work product of the professional has not been held by the doctrine of strict liability, this could happen in the future.

Although contractual risk is inherent to geologists in private practice, those who must deal with such contractual circumstances can at least seek ways to mitigate the consequences of the risk (Kiersch, 1977; James and Kiersch, this volume). The beginning of any risk lies in what one does for the client or

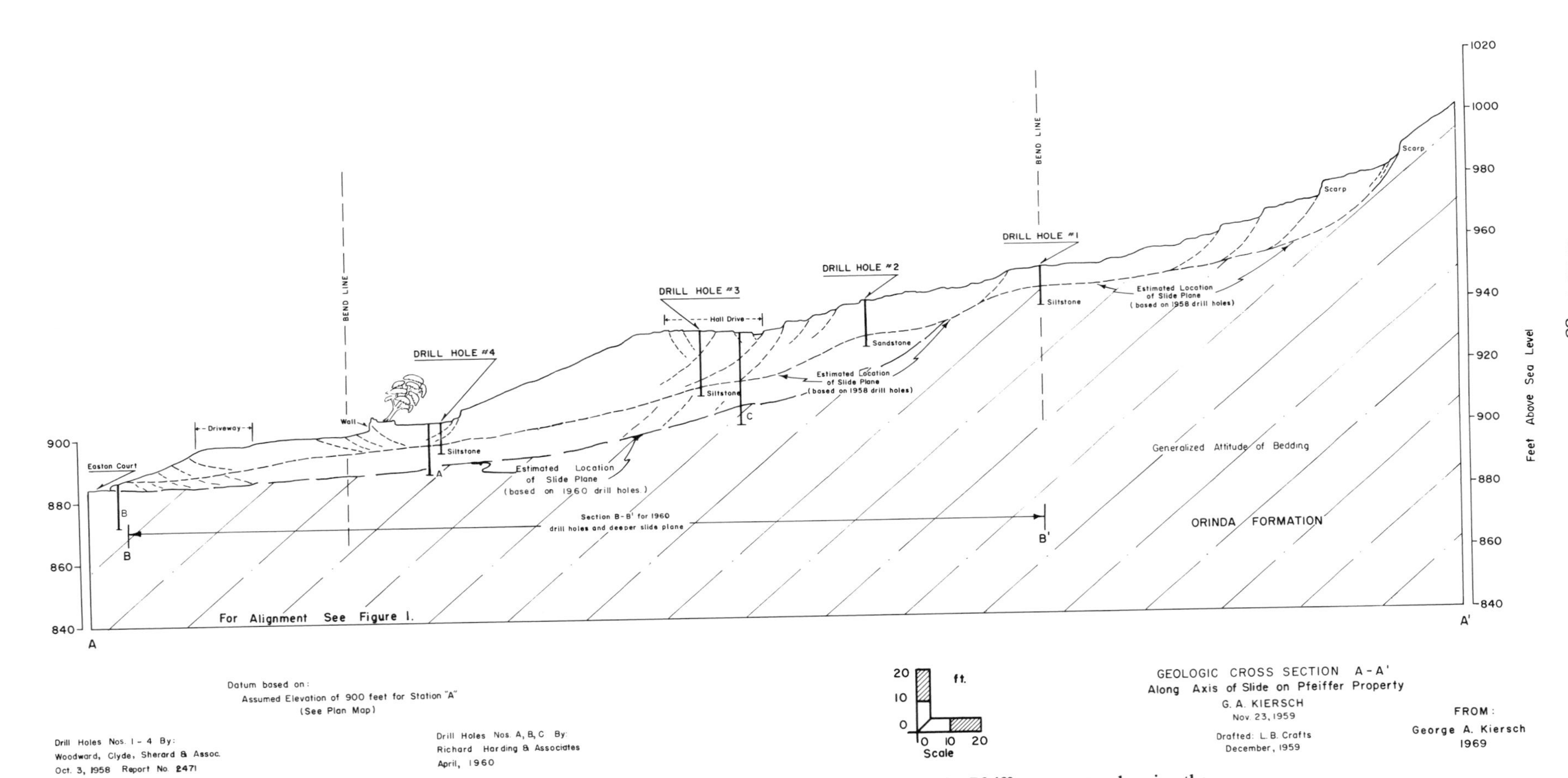

Figure 3. Geologic cross section along axis of slide (A-A′) through the Pfeiffer property, showing the progressive changes of major slide plane with time. For alignment, see Figure 2. (Source: Kiersch, 1969b).

employer. For example, it is important to state precisely what you will do and, in some cases, not do. Clarify that you are responsible only for your own actions or statements. Learn how and when to use "hold harmless" and "limited liability" clauses (Meurer, 1987). Watch your semantics and be sure you know how the courts define certain words, such as inspection and supervision, which carry different degrees of responsibility. Do not use words that state or imply a guarantee, a warranty, or a certification. Provide your client with things "as they are" and do not hesitate to qualify statements as to their degree of probability. Moreover, remember that public services rendered as a good citizen may involve public health, safety, and welfare and carry a real risk of liability. Do such services carefully and with a reduction of risk in mind.

In making information available to prospective contractors and bidders, it is important to distinguish between specific facts and data gathered and those interpretations, interpolations, and inferences made as a result of professional evaluations and studies. If properly stated, it is possible to make the contractor and his staff responsible for interpretation of the data and for application of their conclusions to the construction process. However, almost invariably, it is not possible to make the contractor responsible for significant changes in factual conditions encountered, which would materially affect his operations (Moore, 1977).

How a geologist can reduce contractual risk

The applied geologist can reduce the potential for contractual risk to the project team of designer-engineer-contractor-owner by providing the most factual and meaningful geological explanation possible, relevant to such inherent conditions and features as:

- The character, properties, and type of rock units and alluvial deposits (overburden) to be encountered. Clarify contractual definitions.
- The approximate location and shape of the foundation surface. Estimate volume and type of rock and/or overburden to be excavated, along with the thickness of units, degree of weathering, and water-table location.
- Geologic structures. Evaluate the attitude and structural characteristics of bedded deposits and their effect on the excavation, faults and fractures with associated weathering, likelihood of requiring foundation treatment and its impact on excavation, and joints and rebound structures and their potential effect on or control of foundation shape.
- The ground water conditions. Determine whether dewatering or drainage will be required, or whether grouting will become necessary.
- The potential for slides and unstable slopes in the proposed excavations. Evaluate these relative to slope adjustment, rock strata, rock falls, artificial fill, and swelling or decomposition of natural materials. Recommend controls to minimize potential for slope failure.

Disclaimer clauses

Many owners attempt to avoid liability by adding clauses to their contracts that shift risk to the contractor. These "disclaimer" or "exculpatory" clauses are generally carefully examined by a court before being deemed enforceable.

A wide array of cases have held that the disclaimer clause will not relieve the owner from differing site conditions and delays or other conditions for which the owner is contractually responsible. Similarly, courts may not enforce a disclaimer if the result would be inequitable. For example, *Depot Construction Corporation versus State,* NYS 246, 2nd 527 (1964), illustrates one type of claim that involved a disclaimer clause (Kiersch, 1969a, p. 3). The contractor was required to bid a rock excavation project with the following disclaimer:

> Holes drilled on the site are shown on the plan drawing. Test hole data are not guaranteed by the state in any respect, nor represented by it as being worthy of reliance. The state makes them available as information in its possession without intent or attempt to induce the bidders to rely thereon; they should make their own independent evaluation.

The court decided the disclaimer clause on test hole data was confusing, illogical, contradictory, and even deceptive, and furthermore was an attempt by the state to escape prime responsibility for efficient data on the subsurface of the site under the contract, as well as other aspects (summarized by Kiersch, 1969a, p. 3).

A recent Pennsylvania decision, regarding the case of *Coatsville Contractors and Engineers, Inc. versus Borough of Ridley Park,* 506A 2nd 862 (Pa. 1986), demonstrates how a court balances disclaimer language. The case involved excavation at Ridley Park Lake of a silt-filled lake basin that was dry on the pre-bid inspection, but on the date work began, the lake area was flooded. A disclaimer by the owner of any responsibility for flood-changed conditions was overruled by the court, as discussed by Loulakis (1986).

CONCLUSIONS—OUTLOOK

Geologists have always had a moral responsibility to perform their work carefully and make the conclusions available to the public, both as a service to science and as knowledge to help people understand the environment better and thus improve the quality of their lives. More recently, with the extensive construction of engineering works, man has become increasingly involved with legal responsibilities as well (Moore, 1977; Waggoner, 1981a).

These responsibilities involve the liability for damage to others that results from the performance of services and may ultimately cause monetary penalties, loss of reputation, and/or loss of employment. The degree of legal responsibility for a geologist varies considerably, depending on the discipline or area of his employment (i.e., academic, government service, or private business). The greatest risk of liability lies in the field of private practice, particularly in design and construction of major engineering works (Kiersch, 1969a; Waggoner, 1981c). Although there are other areas of liability, the most common and costly are:

Residential structures and foundation stability.

Slope stability, whether landslides, mud debris flows, or rockfalls (Leighton, 1966; Ehlig, 1986).

Construction claims due to differing site conditions or materials sources (Moore, 1977; Waggoner, 1981b; Halligan and others, 1987; Ruttinger, 1986).

Catastrophic events that cause structure failures due to earthquakes, floods, landslides, eruptions, or subsidence (Kiersch, 1964, 1969a; Tank, 1983).

Environmental, hazardous wastes, and ground-water pollution (Coates, 1976).

Accretion, reliction, erosion, and submergence (Tank, 1983).

Since geologists are usually members of a professional team responsible for the design and construction of a project, they can expect to be included in liability suits or construction claims regardless of their personal actions. This catch-all approach is a common practice of the plaintiff with a "shot gun" suit designed to assure catching someone. Frequently, geologists will be used by plaintiffs to prove that someone else's geological reports, maps, or statements have damaged them (Waggoner and Bachner, 1976). This places the expert witness geologist in the position of being a "hired gun." Since contractual liability is almost unavoidable in some aspects of practice, applied geologists should seek to mitigate the effects of the liability. Some measures lie in preparation of written contracts, others rely on habits of practice. The basic rules of operation should include: know what you need to do; plan how to do it; and know who you are providing information to, and how the recipient plans to use it (ISRM, 1975; Waggoner, 1981a). Above all, make sure your client or associate understands the information you provide. Likewise, be sure you know current state of the art and standards of practice, as well as the importance of semantics of key report and contractual words, their legal connotations, and particularly the meaning of such words as guarantee, warranty, certify, and assure. Finally, consider the degree of certainty of your statements and predictions and let your employer, client, or associate know what the degree of probability is (Ruttinger, 1986). Furthermore, say it in writing, as well as orally (UTRC, 1987).

It is widely recognized that insurors play an important role in the matters of legal responsibility of geologists, but their role is paid-for protection regardless of the geologist's actual responsibility. Thus, it becomes more a matter of business than a directive of how to perform one's services. That story is important but belongs in a business-practice discussion, not in a review of how applied geology is relevant to litigation or how its acceptance has progressed in the past century.

APPENDIX A—CASES CITED

Harman versus American Casualty Company of Reading, Pennsylvania, C. D. Cal. S. D., 155 F. Supp. 612(1957).
——, 1958, *in* Federal Supplement, v. 155, p. 612–615. West Publishing Co., St. Paul, Minnesota.
Hughes versus Potomac Insurance Company, 18 Cal. Rptr. 650, 199 C.A. 2d 239 (1962).
—— , 1962, *in* Nankervis, Wm., Jr., Reports of cases determined in The District Court of Appeals, state of California: Appendix California Supplement, v. 199 2nd series, p. 239–254. Bancroft-Whitney Co., San Francisco, California.
Pfeiffer versus General Insurance Corporation, N.D. Cal. S.D., 185 F. Supp. 605 (1960).
—— , 1960/1961, *in* Federal Supplement, v. 185, p. 605–609. West Publishing Co., St. Paul, Minnesota.
Snapp versus State Farm Fire and Casualty Company, 24 Cal. Rptr. 44, 206 C.A. 2d 827, (1962).
—— , 1962, *in* Nankervis, Wm. Jr., Reports of cases determined in the District Court of Appeals, state of California: Appendix California Supplement, v. 206 2d series, p. 827–834, Bancroft-Whitney Co., San Francisco.
Zimmerman versus Continental Life Insurance Company, 99 Cal. App. 723, 279, p. 464 (1929).
—— , 1929, *in* Pacific Reporter, v. 279, p. 464–466. West Publishing Co., St. Paul, Minnesota.

REFERENCES CITED

ASFE, 1988, Recommended practices for design professionals engaged as experts in the resolution of construction industry disputes: Silver Springs, Maryland, American Society of Foundation Engineers, xx p.
Cleaves, A. B., 1963, Engineering characteristics of based till, *in* Trask, P. D., and Kiersch, G. A., eds., Engineering geology case histories, no. 4: Geological Society of America, p. 51–57.
Coates, D. R., 1976, Geomorphology in legal affairs of the Binghamton, New York, metropolitan area, *in* Coates, D. R., ed., Urban geomorphology: Geological Society of America Special Paepr 174, p. 111, 147.
Dunn, J. R., 1974, The professional geologist as an expert on the witness stand: American Institute of Professional Geologists Guide Services, 7 p.
Ehlig, P. L., 1986, Landslides of Palos Verde Peninsula, *in* Ehlig, P. L., ed., California: Geological Society of America Cordilleran Section Guidebook and Volume, Los Angeles, March 25–28, Trip No. 16, p. 195–225.
Fuquay, G. A., 1967, Foundation cutoff wall for Allegheny reservoir dam: American Society of Civil Engineers Journal of Soil Mechanics and Foundation Division, v. 93, no. 3, p. 37–59.
Halligan, D. W., Weston, H. T., and Randolf, T. H., 1987, Managing unforeseen site conditions: American Society of Civil Engineers Journal of Construction Engineering and Management, v. 113, no. 2, p. 273–287.
Hammon, S., 1968, The expert: Journal of the American Bar Association, v. 54, no. 583, p. 7.
Henry, J. F., 1988, ADR and construction disputes; the mini-trial, CF: Journal of Performance of Constructed Facilities, v. 2, no. 1, p. 13–17.
ISRM, 1975, Commission report on recommendations on site investigation techniques: International Society of Rock Mechanics, 56 p.
Jansen, R. D., ed., 1988, Advanced dam engineering for design, construction, and rehabilitations: New York, Van Nostrand Reinholt, 797 p.

Keville, F. M., and Sutcliffe, H., 1983, Metropolitan Boston Transit Authority red line extension northwest; The influence of geology on alignment and grade, *in* Sutcliffe, H., and Wilson, J. W., eds., Proceedings Rapid Excavation and Tunneling Conference, Chicago, Illinois: New York, American Institute of Mining and Metallurgical Engineers, v. 2, p. 62–69.

Kiersch, G. A., 1957, Engineering geology; An aid in estimating contract bids for heavy construction: Geological Society of America Bulletin, v. 68, p. 1831–1832.

—— , 1964, Vaiont reservoir disaster: Civil Engineering, v. 34, no. 3, p. 32–39.

—— , 1969a, The geologist and legal cases; Responsibility, preparation, and the expert witness, *in* Kiersch, G. A., and Cleaves, A. B., eds., Legal aspects of geology in engineering practices: Geological Society of America Engineering Geology Case Histories 7, p. 106.

—— , 1969b, Pfeiffer versus General Insurance Corporation; Landslide damage to insured dwelling, Orinda, California, and relevant cases, *in* Kiersch, G. A., and Cleaves, A. B., eds., Legal aspects of geology in engineering practices: Geological Society of America Engineering Geology Case Histories 7, p. 81–85.

—— , 1977, Geologist and legal cases; Responsibility, preparation, and the expert witness, *in* LeRoy, L., ed., Subsurface geology: Golden, Colorado School of Mines Publication, p. 747–752.

Kiersch, G. A., and Cleaves, A. B., eds., 1969, Legal aspects of geology in engineering practice: Geological Society of America Engineering Geology Case Histories 7, 93 p.

Leighton, F., 1966, Landslide and hillside development, *in* Lund, L., and Proctor, R. J., eds., Engineering geology in southern California: Glendale, California, Association of Engineering Geologists Special Publication, p. 149–193.

Litigation, 1977, Preparing and examining witnesses: American Bar Association Journal of the Section of Litigation, v. 3, no. 2, p. 7.

Loulakis, M. C., 1986, Disclaimers of liability: Civil Engineering, v. 56, no. 10, p. 32.

Meurer, C. D., Jr., 1987, Lets stick it to the designers: K&H Communications Harbinger, April, p. 2.

Meyer, D. F., 1983, Metropolitan Boston Transit Authority red line project, Harvard Square to Porter Square tunnels, Cambridge, Massachusetts, *in* Sutcliffe, H., and Wilson, J. W., eds., Proceedings Rapid Excavation and Tunneling Conference, Chicago, Illinois: New York, American Institute of Mining and Metallurgical Engineers, v. 2, p. 941–951.

Moore, W. W., 1977, Construction claims, *in* LeRoy, L., ed., Subsurface geology: Golden, Colorado School of Mines Publication, p. 742–746.

Murray, R. C., and Tedrow, C. F., 1975, Forensic geology; Earth Sciences and criminal investigations: Newark, New Jersey, Rutgers University Press, 217 p.

Rapp, J. S., 1987, Forensic geology and a Colusa County murder: California Division of Mines and Geology California Geology, v. 40, no. 7, p. 147–153.

Ruttinger, G. D., 1986, The differing site conditions clause; What are "Contract Indications?": National Contract Management Journal, v. 20, p. 67–76.

Sanborn, J. F., 1950, Engineering geology in the design and construction of tunnels, *in* Paige, S., ed., Applications of geology in engineering practice: Geological Society of America, Berkey Volume, p. 45–81.

Tank, R. W., 1983, Legal aspects of geology: New York, Plenum Press, 580 p.

Trefzger, R. E., 1966, The Tecolote tunnel, *in* Lund, R., and Proctor, R. J., eds., Engineering geology in southern California: Glendale, California, Association of Engineering Geologists Special Publication, p. 108–113.

USBR, 1959, Tecolote tunnel; Technical record of design and construction: U.S. Bureau of Reclamation, 109 p.

USNCTT, 1984, Geotechnical site investigations for underground projects; v. 1, Overview of practice and legal issues, evaluation of cases, conclusions, and recommendations; v. 2, Abstracts of case histories and computer-based data management system: Washington, D.C., National Academy Press, U.S. National Committee for Tunneling Technology, v. 1, 182 p., v. 2, 241 p.

UTRC, 1987, Avoiding and resolving disputes in underground construction; Successful practices and guidelines: American Society of Civil Engineers—American Institute of Mining Engineers, Underground Technology Research Council, 60 p.

Waggoner, E. B., 1981a, The expert witness: Bulletin of the Association of Engineering Geologists, v. 28, no. 1, p. 29–38.

—— , 1981b, Construction claims; Spurious and justified: Bulletin of the Association of Engineering Geologists, v. 28, no. 2, p. 147–150.

—— , 1981c, Engineering geology for earth dam foundations: American Society of Civil Engineers preprint, p. 81–174.

Waggoner, E. B., and Bachner, J. P., 1976, What to do when the suit is served: Civil Engineering, v. 46, no. 4, p. 63–64.

Manuscript Accepted by the Society October 18, 1989

Printed in U.S.A.

Geological Society of America
Centennial Special Volume 3
1991

Chapter 25

Forensic geoscience for engineering works; Litigation, hearings, and testimony

James R. Dunn
Dunn Geoscience Corporation, 12 Metro Park Road, Albany, New York 12205

INTRODUCTION

Forensic geoscience, as paraphrased from a definition of "forensic," is the component of geological sciences that belongs to or is used suitably in courts of judicature and/or public discussion and debate. Murray and Tedrow (1974) in their text, *Forensic Geology,* describe only those areas of forensic geology that primarily relate to criminal investigations. In the ensuing decade and a half, however, the field has expanded and changed radically. This review defines and describes the subject in a much broader sense, including such forums as public discussion and debate. The principal examples cited are the applied geosciences component of engineering works.

Clearly, forensic geoscience, like consulting geology, is not a branch of geological sciences but rather a category of geological application. Yet it is important to separate forensic geoscience from the other areas of applied geoscience because the orientation and methodology of forensic geoscientists is normally quite different from that of the main body of practicing geoscientists. Furthermore, because much of forensic geological practice is focused on influencing public decisions of a judicial, regulatory, or legislative nature, forensic geoscience today owes its existence to government policies as expressed through laws, regulations, the judiciary, and public works.

In the late 1800s and early 1900s, the greatest volume of the litigation component of forensic geoscience was concerned with mining and water supplies, along with occasional criminal cases. However, by the 1930s, more forensic geological work was gradually being applied to engineering works, primarily for determining responsibility for engineering errors, unforeseen adverse conditions, and the determination of mineral values in eminent domain cases when land was taken for engineering projects. By the 1960s, the nature of forensic geological work had changed radically because of governmental activities related to the environmental movement. Although the application of forensic mineralogy to criminal cases and of forensic geoscience to mining and mineral resources, hydrology, and engineering works constitutes a substantial part of the field today, the majority of forensic work is related to environmental laws. Some 70,000 cases are before the courts in asbestos litigation alone (Anonymous, 1988).

A major difference between forensic geoscience and other categories of geological practice is that the forensic geoscience orientation is totally outward, and influencing the lay public is its justification or *raison d'etre.* The largest part of forensic work, serving as an expert witness in judicial matters, has two other major differences: first, the requirement to adhere to legal procedures, and second, the almost total lack of geological speculation with respect to the latter constraint. The great intuitive leaps that led to such major advances as the theory of evolution, or concepts of granitization and plate tectonics, are not appropriate for forensic work. Probably many of the well-known geologists of the past would be uncomfortable serving in forensic work; actually their intuitive strengths would be considered weaknesses in the public arena of litigation.

The purposes of this chapter are: to show how the nature of forensic geoscience has evolved; to show its place in the field of applied geosciences; and to review the main directions and procedures that are typical of forensic work. Finally, there is speculation about the outlook for forensic geoscience.

THE CHANGING SCOPE OF FORENSIC GEOSCIENCE

Whereas professional geologists have always been expected to document their determinations as though they might have to defend their reports in court or in other public tribunals, actual litigation pertaining to the quality of geological work has been relatively rare. Few practicing geologists presented testimony in court cases before the 1960s. Forensic geoscience was largely limited to three areas: (1) relatively few appearances before legislative bodies; (2) litigation regarding the taking of mineral property for such public works as highways or dams; and (3) litigation involving engineering works, mines, or mineral resources.

The forensic work with the highest visibility and in which

Dunn, J. R., 1991, Forensic geoscience for engineering works; Litigation, hearings, and testimony, *in* Kiersch, G. A., ed., The heritage of engineering geology; The first hundred years: Boulder, Colorado, Geological Society of America, Centennial Special Volume 3.

the most geologists have participated has been litigation. Litigation and forensic work in general have evolved from the late 1800s to the present through three different periods: most early forensic work was resource oriented; the middle period was mostly engineering oriented; and the current period is mostly environment oriented. These three periods reflect changes in national economic priorities and evolving technology. To some extent, geologists may have had some impact on national priorities, but to a much larger degree, geology—and forensic geoscience in particular—has changed in response to the external pressures of changes in national interests.

The early period: Resource oriented (began 1870s)

In the mid- to late 1800s, both the economy and the area of the nation were expanding rapidly, with the federal government strongly encouraging development. For example, Congress granted vast tracts of land to four major railroads—the Central Pacific, Southern Pacific, Santa Fe, and Northern Pacific—as incentives to construct rail links to the Pacific Coast in the 1860s to 1870s. The strips were 64 km wide astraddle the trackway with the alternate sections in each township. Many other laws were passed to help develop western lands, development that included their mineral wealth. (See Ch. 18, this volume).

The Mining Act of 1872 (see Peele, 1966, p. 24–06), including the famous Law of Apex, controlled mineral staking and mining developments. Its major concern was to ". . . promote the development of the mining resources of the United States" by giving prospectors firm rights to minerals found on public lands provided they conformed to certain rules for claim staking, filing, and performing assessment work; bona fide claims could be patented with time, and the land owned in fee. The law achieved its purpose very well because the rate of discovery of great new deposits of gold, silver, copper, lead, and zinc was extraordinary. The feverish rate of discovery, development, and exploitation in vast and sparsely settled regions provoked few or no contemplative thoughts about the long-term impacts of mining and development on the natural environment.

The Mining Law of 1872 (Peele, 1966) provided that "all valuable mineral deposits" under lands belonging to the United States shall be free and open to exploration and purchase. Conflicts frequently arose in attempts to answer two questions about a "valuable mineral deposit": (1) is it locatable? and (2) is it valuable? In addition, as part of the law, a miner had the right to follow ore from the apex of ore exposed on his claim within the projected parallel end boundaries down-dip under adjacent properties or claims, the so-called apex rule. All of these rules were subject to litigation, and geologists were frequently involved as expert witnesses.

Perhaps the most celebrated cases involving geologists were those, starting about 1903, that pertained to the application of the apex rule to mining at Butte, Montana. Mining began at Butte Hill in the late 1860s as gold placers in Missoula and Dublin gulches, but by about 1885 there were 25 companies mining copper because the ore changed character at depths. Although some conflicts were inevitable, they were minor until after 1900 when litigation about the apex rule involved interpretation of the geology of the Butte ore bodies. Despite injunctions, mining companies deviously mined high-grade copper ore under other company properties; in some cases two companies were mining an ore body simultaneously. Gathering geological information about violations of injunctions involved surreptitious surveying of mine workings at a time when miners were hurling rocks and homemade hand grenades at each other underground and, in other cases, were fouling ventilation air with smoke from burning tires or with quicklime (Sales, 1964). Practicing applied geology for forensic purposes has rarely been so hazardous as during the Butte litigation.

In court, the geological witnesses on both sides were impressive. They included Clarence King, founder of the U.S. Geological Survey; Rossiter W. Raymond, Secretary of the American Institute of Mining Engineers; Nathaniel Shaler, Professor of Geology at Harvard; David Brunton, an outstanding mining engineer; W. S. Keyes, a Freiberg graduate in charge of the first smelter at Argenta, Montana; and Horace B. Winchell, geologist for Anaconda Company. The geologists disputed the supposed continuity of veins, disagreed on whether certain faults existed, and generally seemed to find little basic geology about which to agree. In addition, some of the geologists had an unfortunate ability to support concepts they were paid to uphold, regardless of the facts. Such experts often acted "not as scientific witnesses but as hired advocates" (Rickard, 1932, p. 364).

During this early period, forensic geoscience developed for use in criminal cases. Celebrated cases involved determination of the origin of sands and dirt particles on the clothing of suspects. According to Murray and Tedrow (1974), the use of mineralogy in criminal cases started in Europe in the 1890s. Its use was pioneered in the early 1900s in United States by C. P. Berkey (see Ch. 24, this volume) and in the 1920s by Edward Oscar Heinrich, a chemist, sometimes known as the "wizard of Berkeley," who used mineralogy of earth materials in several celebrated cases.

The intermediate period: Engineering oriented (began 1930s)

By the end of the 1920s the free-wheeling days of early mineral exploration started to change. The era of large-scale construction and major government-sponsored engineering works—dams, water systems, railroads, and highways—became of greater importance, and a proportionately greater need for litigation evolved. Engineering geology now made a significant contribution to planning and construction of major engineering works (as reviewed in Ch. 1, this volume), and gradually achieved increasing stature. The scope and nature of forensic geology underwent further changes because of the increased case load of litigation relevant to engineering works. This was particularly true in categories of (1) changed geologic conditions encountered, or

(2) insufficient base data made available to contractors for bid purposes.

Lands affected by the construction of dams or highways were often underlain by mineral deposits, and conflicts arose about the value of minerals in the ground. In fact, disagreement about value by more than an order of magnitude was common. At first, the major accepted method for establishing value in much of the eastern United States was the comparable sales technique, in which value of mineral deposits was demonstrated by finding sales of similar mineral properties and comparing value. The comparable sales approach was borrowed from real-estate appraisers who used the method as the primary means of evaluating residential real estate.

An example of conflicts in the land-taking process in eminent domain proceedings was the litigation involving Champion Brick Company (Supp, 1969) in which the Maryland State Roads Commission condemned 22 ha of brick clay reserves for the purpose of building a highway interchange. Forensic geological work involved an analysis of the local geology, drilling, sampling, laboratory testing, and determination of the reserves. The type of analysis described by Supp is typical of forensic work in such cases. In this situation, the conflict was between the mineral-resource orientation of early forensic work and the engineering orientation of the middle period of forensic geoscience. This kind of geological expertise is still commonly requested of field-experienced geologists, but is a decreasing fraction of the total volume of forensic geoscience work.

Throughout the eastern United States, this important area of forensic expertise for the applied geologist has experienced changes in recent years due in part to a technical publication describing capitalized income techniques (Paschall, 1984). These techniques of mineral appraisal have a long history in the western states where they were used in eminent domain cases as well as for a basis of taxation. Based on his experience in the West, Paschall challenged the comparable sales approach, pointing out that very rarely are mineral properties comparable, and furthermore, sales of mineral properties are rare. Actually the mineral producers themselves usually use other methods of property appraisal. As a consequence, the comparable sales approach for appraisal has given way in most states to the various capitalized income methods of evaluation. No longer can the forensic geologist restrict investigations to the quantity, quality, and mineability determinations; for most typical cases the assessment also involves a determination of the value of the mineral deposits.

A common type of litigation involving engineering geologists has been in conflicts over "changed geologic conditions" of engineering projects. The bid documents that are prepared by government agencies or the private owners of projects usually describe the character of subsurface soil-rock conditions and the environs. Such data as the quantity, quality, and distribution of sand and gravel for a bid preparation are usually based on mapping, borings, and other tests as well as geophysical measurements and laboratory analyses. Frequently, the conditions encountered are not as described, and cost overages result for design or construction. Litigation may be undertaken to achieve a financial settlement. Several such examples are described in Kiersch and Cleaves (1969). In addition, Waggoner (1981) cites some additional cases involving misinterpretation of geologic conditions that led to problems with the production of natural aggregates for highway construction. In addition, many cases during this period involved ground-water geologists in disputes centered largely in the arid western United States. Some examples are described by Mann (1969) and Tank (1983), and the subject is reviewed in Chapter 24 (this volume).

Foose (1969) describes a typical example of conflict between mineral development and the environment. Several quarries in the vicinity of Hershey, Pennsylvania (Fig. 1), required dewatering. This caused a drawdown of the local water table that resulted in the development of sinkholes. Hershey Chocolate Company started a recharge program to retard sinkhole development, which caused nearby Annville Stone Company to pump more water to keep their operation dry. Annville Stone in 1950 sought an injunction to prevent further ground-water recharge. Forensic geoscience work involved an analysis of the water-table elevations and directions of water movement in the Hershey Valley, along with a correlation of sinkhole formation and mine dewatering.

The current period: Environment oriented (Began 1960s)

Starting in the early 1960s, a major socioeconomic upheaval—the environmental movement—began, radically altering the nature of professional practice for major segments of the geosciences, particularly mineral exploitation and ground-water geology. The change of public orientation, and the inevitable legislation and regulations required, increased participation by geoscience professionals. Forensic geoscience practice became a larger component of the professional geologist's work.

The shifting national context of forensic geoscience and the government's influence on that context is well illustrated by Coates (1976). He described certain laws, starting with Flood Control Act of 1936 through the National Environmental Policy Act (NEPA) of 1969, which influenced the nature of litigation involving flood control of the Susquehanna and Chenango Rivers and the construction of Interstate Highways 81 and 88. All four of the cases described resulted from governmental policies; the earlier cases in the engineering-oriented category, and the last case a result of conflict between the engineering orientation and the environmentally oriented laws of the "environmental revolution." On a microscale, the litigation described by Coates for the Binghamton, New York, area is typical of the national transition in thought from a construction-engineering orientation to an environment-preservation orientation when engineering works are involved.

Because of NEPA, preservation-oriented citizens in the Schenectady-Albany area were encouraged to challenge the wisdom of government policies that were responsible for the interstate highway system. The citizens' suit that started in 1973 was

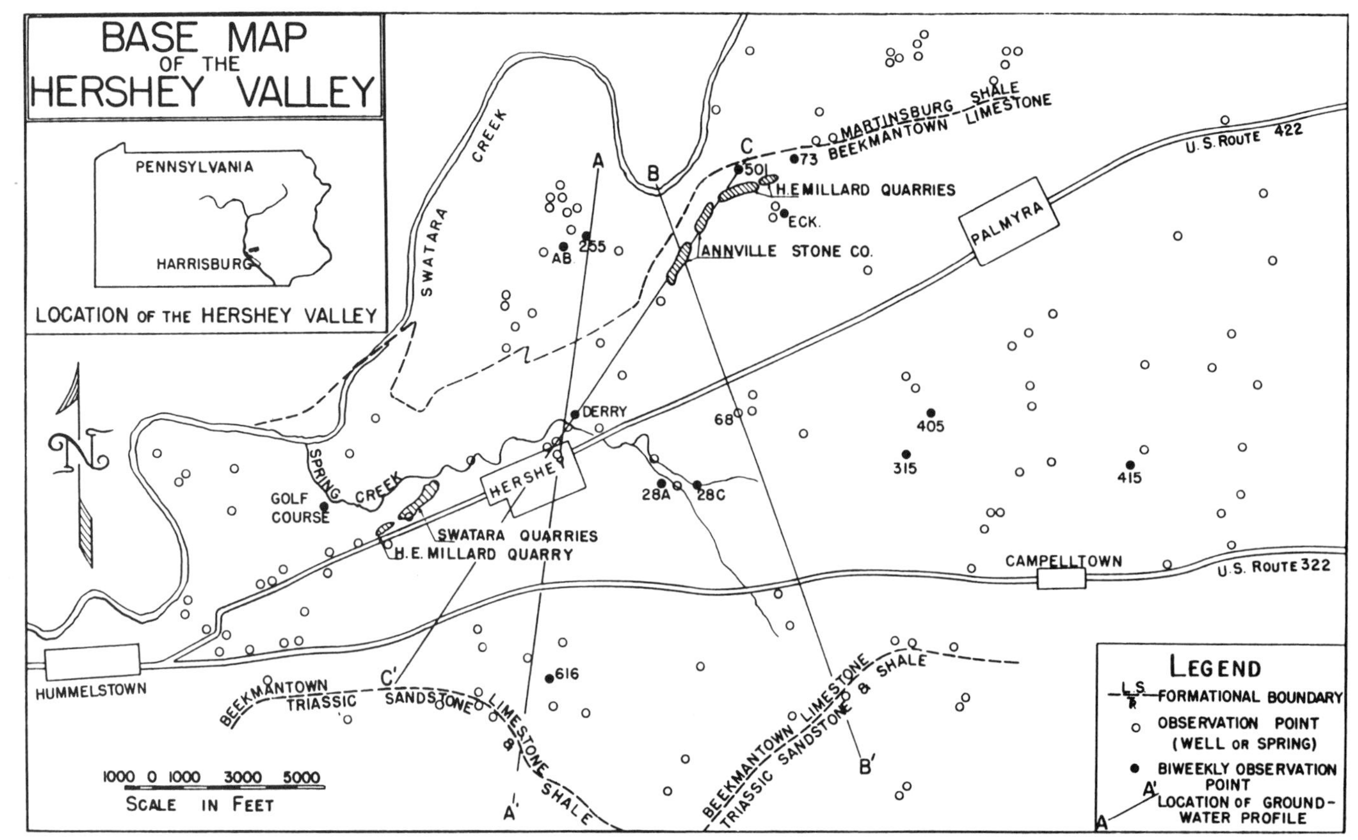

Figure 1. Base map of the Hershey Valley (from Foose, 1969).

successful in delaying the construction of a section of I-88 along the Chenango River east of Binghamton until 1987. It is not at all clear that the environment was improved as a result of the I-88 suit. According to Coates (1976) (also Ch. 15, this volume), excessive erosion was one result of halting construction, and the delays were costly to the taxpayers. Like so much environmental legislation, NEPA's motivation of improving the environment is difficult to contest. Some environmental legislation, however, may be counterproductive.

NEPA legislation, Title 1, Sec. 102, states "... (2) all agencies of the Federal Government shall - ... (C) include in every recommendation or report or proposals for legislation and other major Federal actions significantly affecting the quality of the human environment, a detailed statement by the responsible official on - (i) the environmental impact of the proposed action, and (ii) any adverse environmental effects which cannot be avoided should the proposal be implemented, ..."

The environmental impact statement (EIS) is now used throughout the United States at the federal, state, county, and town levels to evaluate actions influencing the environment. The typical EIS is multidisciplinary and may require the input of archaeologists, biologists, meterologists, sociologists, and engineers, among others. Analyses by engineering geologists are needed because of the adverse impacts of improper construction, such as dams in the wrong place (James, 1968), construction of buildings in unstable areas (Woods, 1964; Crawford and Eden, 1964), and excessive ground-water withdrawal during mining (Foose, 1969). EIS requirements generate large amounts of work for forensic geoscientists.

Coinciding with the environmental movement that began in the 1960s—and to some extent because of it—the exploration, development, and exploitation of mineral, energy, land, and water resources grew less important and even became anathema to the American public. Geological work in these areas decreased, and segments of the profession found that survival depended on developing skills related to the new laws and regulations. Although forensic work in resource and engineering-oriented geoscience continued, working on problems related to improvement and/or preservation of the environment became the order of the day for many geologists. The change, so far as professional geology is concerned, caused some firms that were devoted to engineering, applied geosciences, and resource exploration and development to become largely environmental-oriented. Many firms that failed to adjust to the changes in public priorities fell on bad times.

Probably the philosophical orientation most influential for the expansion of forensic geology was the concept that most of the nation's environmental ills were caused by industry. This firm belief easily led to a view of industry as environmentally guilty until it could prove itself innocent. In addition, the idea evolved that one molecule or "one particle of a carcinogen can kill you," and in response, technology became ever more efficient at detecting such molecules or particles. Any chemical, so long as it was among the 575,000 manufactured or used by industry, became suspect, and science showed how to detect them if they were contaminants. Curiously, science and technology developed the tools that could be used against them by segments of the public who were suspicious of science and technology. For example, highly sophisticated radiation-measuring equipment was used to detect the traces of radioactive compounds released by the Three Mile Island nuclear power plant in Pennsylvania during its accident. This information was then used by anti-nuclear-power advocates to advance their views against nuclear technology (U.S. House of Representatives, 1983).

In response to public and media perceptions of the needs of the environment, many new laws and regulations were written. Because the public perceptions were often erroneous, and the laws were hurriedly written, one result was conflict. Forensically oriented geologists found themselves—or placed themselves—in position to take advantage of this confusion. In addition, an increasing number of lawyers were willing to take almost any case, and some lawyers encouraged environmental conflicts. The amount of litigation, public hearings, or other types of testimony increased radically.

The following are additional examples of government activities and perspectives that have greatly influenced forensic geology. (1) The Delaney Amendment to the Food, Drug, and Cosmetic Act (1982) states that "no additive shall be deemed to be safe if it is found to induce cancer when ingested by man or animal," (setting the stage for concerns about traces of contaminants in ground water). (2) The governmental redefinition of asbestos includes those brittle amphiboles that have an asbestiform counterpart, and treats all forms of asbestos the same, even though they vary considerably in the amount of health hazard. The Federal Environmental Protection Agency (EPA) did not state specifically that asbestos products in schools could in many cases merely be covered or otherwise neutralized, rather than removed. This was finally mentioned in the 1985 "purple book" [EPA, 1985], but only after a public mind-set made less costly remedies unlikely. (3) As previously noted, there has been a philosophical perspective held by many in government that one particle or molecule of a carcinogen can kill. (4) The licensing for construction of nuclear power plants by the Atomic Energy Commission following such Federal laws as the 1978 Title 10 (Reactor Site Criteria) includes hearings and litigation as a normal part of geology-related activities for nuclear plants, and particularly, documentation about the "age of last movement on pertinent faults" and the seismic-tectonics of a region, area, or site. (5) The ubiquitous requirement for an Environmental Impact Statement for any action that could adversely affect the environment requires geoscience input in most cases and virtually always leads to hearings and often to litigation.

Geologists have found themselves in the middle of many major controversies since the 1960s. Four examples follow. (1) In the Reserve Mining case in Minnesota (Reserve Mining Company, 1974 to 1975), it was contended that Reserve Mining was polluting Lake Superior with carcinogenic amphibole (brittle cummingtonite-grunerite). Although the carcinogenicity was never demonstrated, Reserve Mining was dumping wastes into the lake and was ordered by the court to desist. (2) The Love Canal toxic waste dump problem at Niagara Falls (EPA, 1982), was a ground-water problem to a considerable degree, and ground-water geologists guided the drilling of over 2,000 wells, as well as water sampling, analysis, and interpretation. Over 200 homes were torn down and a school abandoned because of fear of toxic wastes (Fig. 2). (3) Scores of suits sought to determine who, if anyone, should pay the billions of dollars needed to remove asbestos insulation products from schools. Identifying the manufacturers of asbestos-bearing products is largely a mineralogic problem. (4) The entire licensing process for atomic power plants "developed into an unsettled atmosphere of legalistic uncertainty surrounding the efforts of conscientious scientists and engineers . . ." (McClure and Hatheway, 1979).

Such situations also gave forensic geoscientists insights into the nature of the socioeconomic environment in which they operated. The Reserve Mining case was tried first before a judge with a strong environmental orientation. His perspectives were sharply criticized by appeals courts, and his decisions were reversed. In all four situations cited above, public hysteria in response to media reporting caused court costs to rise astronomically. The political orientation of public officials, the perspectives of the information media, and the emotional state of the public can have enormous influence on the nature and need for forensic geoscience.

Some governmental activities do not necessarily lead to litigation but still have a considerable impact on what is, broadly, forensic geology. One major example is governmental funding to clean up toxic wastes—the 1980 Superfund Legislation and its 1986 reauthorization total $10.1 billion in available funds. Because cleaning up toxic wastes usually involves ground-water–related problems, ground-water geologists and engineering geologists who feel comfortable with forensic work are heavily involved.

While the analysis and recommendations that lead to remedial environmental actions are not, strictly speaking, in the area of forensic geoscience, in many or perhaps most such cases the geologists' work leads to hearings, to litigation, or to assigning responsibility for cleanup costs. Consequently, forensic geoscientists find that much of their work may ultimately come under the microscope of public scrutiny, often embodied as a cross-examining attorney. Furthermore, they likely will spend increasingly more time talking to lawyers than to corporate officers, top governmental employees, or other scientists.

Coinciding with, and to some extent because of the environmental revolution, the United States has become increasingly litigious because (1) people have become aware of their rights; (2) large financial settlements have encouraged litigation; (3) many lawyers seem willing to take even very weak cases; (4) some lawyers encourage litigation; and (5) some laws

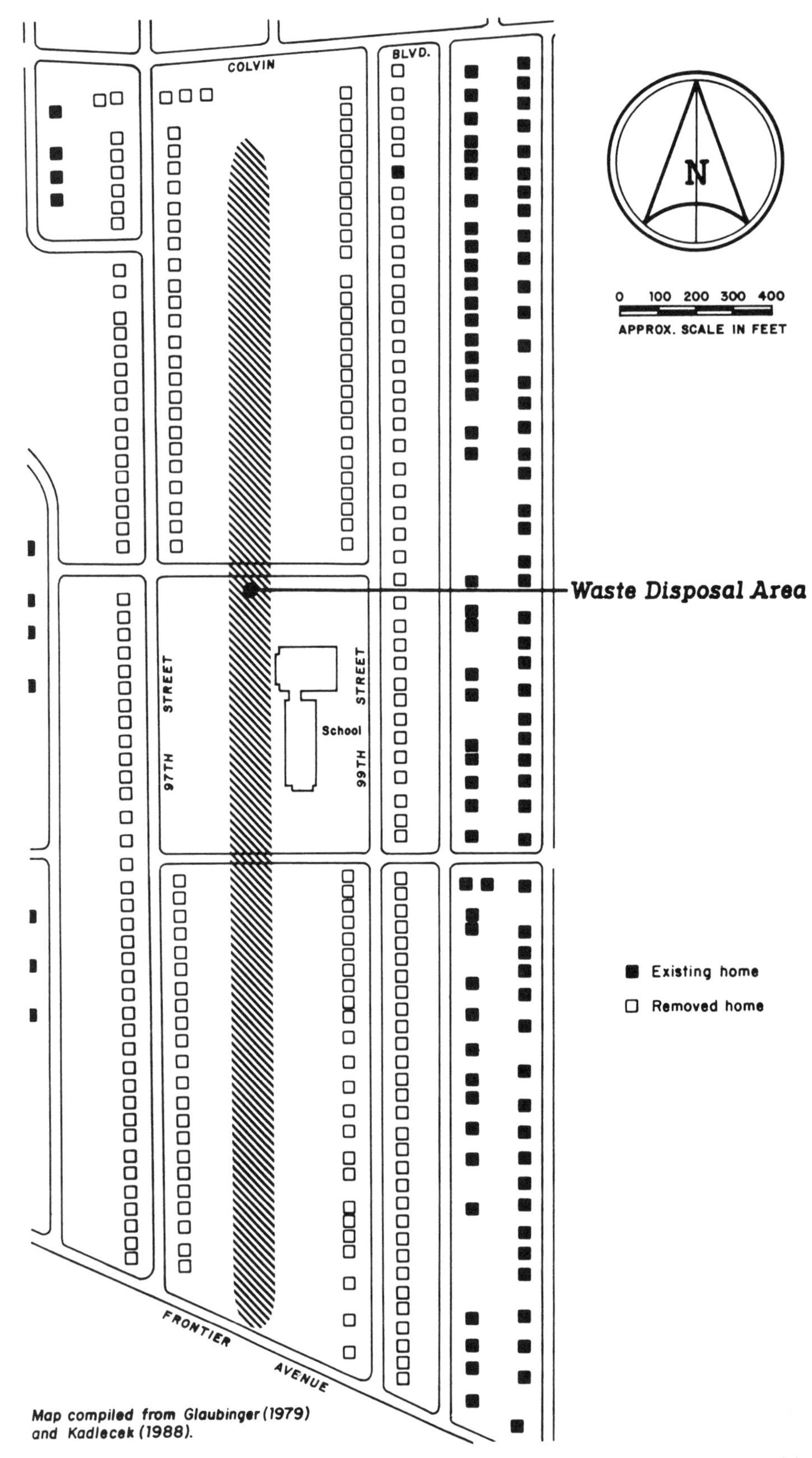

Figure 2. Love Canal waste disposal area, Niagara Falls, New York. Chemical wastes, municipal garbage, and fly ash are in about equal proportion. Map compiled from Glaubinger and others (1979) and Kedlecek (1988).

encourage litigation, such as the 1986 California Safe Drinking Water and Toxic Enforcement Act, which pays those who bring civil actions against polluters a fraction of any fines levied.

FORENSIC GEOSCIENCE PROCEDURES

The practice of most forensic geology is quite different in character from other areas of applied geosciences. A very high order of measurement precision and structure is absolutely essential because geologists may have to defend every research step, procedure, and conclusion to a cross-examining attorney. Written quality assurance and quality control procedures (or protocols) are often mandatory. Detailed procedures (or protocols) for drill-site selection, sample handling (from retrieval through analysis to sample disposal), and finally, interpretation are usually required. Geologists may have to answer such questions as these: Who collected the samples (whether they be core, water, or outcrop)? Were they ever out of that person's sight? What was the condition of the samples on arrival at the laboratory? Can you prove there was no tampering and are the chain-of-custody forms complete and accurate? Questions of this type must be anticipated for every stage of an investigation. Frequently more time is spent demonstrating and defending the validity and accuracy of work than is spent explaining or demonstrating geologic or hydrologic conditions and data.

Procedures used in forensic work should, to whatever extent possible, be "tried-and-true." The legal basis for this is from Frye versus United States in 1923 " . . ., the thing from which the deduction is made must be sufficiently established to have gained general acceptance in the particular field in which it belongs" (Black, 1988, p. 1508). Innovative research techniques, while not ruled out, must be so thoroughly rationalized in scientific terms that their use is discouraged. Preferably, procedures should follow governmental recommendations by such agencies as the EPA or the quasi-governmental American Society of State Highway Officials. Procedures described by the American Society for Testing and Materials (ASTM), by the American Institute of Professional Geologists (AIPG), or the Association of Engineering Geologists (AEG) may be applicable. Methods described in such standard references as the *Mining Engineers Handbook* (Cummins and Given, 1973) or similar works may be used. In-house procedures, if written in sufficient detail and thoroughly justified, may be used. To the greatest degree, procedures should be standard, acceptable, and well known to the profession. In addition, descriptive classifications should be standard, preferably those described in standard references. Geological intuition, speculation, and basic research are generally not applicable to forensic work, or at least, they should rarely be a visible part of a final product. However, understanding the relevance of fundamental processes and/or phenomena of geoscience may be a critical factor in the geologist's testimony and opinions. Finally, it is axiomatic that if speculation is possible among cooperating forensic geologists on the same side, insufficient work has been done.

Chain-of-custody procedures in which an individual or organization has specific responsibility for handling, storage, or disposal of samples at all times are usually mandatory, and the use of standard forms is encouraged. Some typical forms are shown in Figures 3 and 4.

FORENSIC GEOSCIENCE COSTS

Needless to say, data gathering in forensic geoscience is expensive, often several times as costly as it would be if litigation or public debate were not potentially involved. In much applied geological work, a 90 or 95 percent probability may be sufficient, and frequently all that is justifiable or cost effective. But when conclusions may be under public scrutiny, in particular the scrutiny of a hostile attorney whose primary objective is to "win," such probabilities may be insufficient. Costs may be astonishingly high for determining the quantity of a contaminant as it approaches zero, or when a required degree of certainty approaches 100 percent. Proving a negative or determining and quantifying a miniscule quantity of anything is very expensive, as is absolute perfection. We are faced with the irrationality that the closer a defendant is to innocence, the more expensive it is to prove. It is as though it costs far more to dispute a ticket for speeding than an accusation of murder.

Asbestos contamination

An example of such high costs is the determination of very small quantities of amphibole in rock products. The June 20, 1986, Federal Register contained the Occupational Safety and Health Administration's (OSHA) new regulations for asbestos (U.S. Department of Labor . . ., 1986). In that regulation, the previous governmental definition of asbestos, which included brittle particles of the amphiboles actinolite, tremolite, and anthophyllite (ATA), was finally corrected after nearly 20 years of debate. However, OSHA's 1986 asbestos regulation said that those brittle amphiboles (though now omitted from the asbestos classification) would still be regulated as though they were asbestos. Further, according to the U.S. Department of Labor (1987, p. 946), any product containing more than 0.1 percent of a carcinogenic substance must be labeled as such. A gneissic rock containing over 0.1 percent actinolite and used as crushed stone, thus, must be advertised as potentially carcinogenic. For a rock containing abundant amphibole, the mineralogical problem is trivial, although the problem for the stone producer certainly may not be. But when the quantity of amphibole approaches 0.1 percent, the cost of determining and demonstrating the quantity of such a mineral in all future products can be very high. In one producer's experience, over $400,000 was spent for drilling and mineralogical work to determine the amount of ATA for a 100-acre crushed stone property—and that did not include the cost of on-going quality control.

The above regulation does not appear to have been applied to students studying hand specimens, sawing rock slabs, or crushing rock for analyses, but there is no apparent reason why it

Client Name:	Contact:
Project No.:	Laboratory Contact:
Site Location:	Lab Identification:
	Date Report Required:
Sampler:	

Sample Identification	Date	Time	Sample Matrix	Collection Vessel	Lowering Device	Sample Container	Preserv.	Filter: Pore, Type	Comp or Grab	Analysis Required	Comment

	Name	Affiliation	Date	Time	
Relinquished by:					Received by Laboratory:
Received by:					Samples Intact & Properly Preserved:
Relinquished by:					
Received by:					

Figure 3. Chain of custody form. This type of form is used to assure proper handling of rock, water, asbestos-bearing insulation, or other samples of potential forensic value.

should not be. Further, there is no apparent reason why students exposed to brittle ATA in quantities larger than 0.1 percent should not sue a university for exposing them to a regulated carcinogen. The cost to universities of such litigation could be enormous, needless to say. Forensic geoscience would seem to benefit from the regulation, but the education system would not.

The determination of quantities of amphibole in a single sample can be expensive. For example, limestones are certainly not considered to be amphibole-rich rocks by geologists; in fact, the observation of amphiboles in limestones is rare. However, because amphiboles are ubiquitous in detrital accumulations from air or water, most limestones do contain some amphiboles. Making insoluble residue and heavy liquid concentrations of limestone from a quarry will almost certainly produce a concentrate in which some amphibole can be observed under the petrographic or electron microscope. This is where the difference between normal geological endeavors and forensic geology becomes very apparent. In typical geological work, if any amphibole were observed, it would be noted and it would usually be sufficient to state that the quantity was trivial. In forensic geoscience practice, because of the common public emotionality and the nature of geo-related questions that could be raised, it becomes mandatory to quantify the amphibole content and also relate the quantity to the deposit as a whole and to daily quarry production. The cost of quantification and correlation with regulations on such minerals is likely to be several orders of magnitude greater than a single qualitative observation. In addition, the cost of determining innocence falls on the accused, while the plaintiff's responsibility often ceases with the accusation.

Chemical contamination

The situation described for asbestos and ATA particles applies as well to chemical contaminants. The cost of quantification as quantities approach zero can be astronomical. Further, the likelihood that such costs will increase seems assured, not only because of the Delaney Amendment to the 1982 Food, Drug, and Cosmetic Act, but because California's 1986 Safe Drinking Water and Toxic Enforcement Act, in effect, makes any toxic chemical in a detectable amount a matter of concern. This con-

ASBESTOS PRODUCT ANALYTICAL SHEET

DATE ______________________________ SAMPLE NO. ______________________

PETROGRAPHER ______________________ CLIENT SAMPLE NO. ________________

Firing loss, percent ______; Acid loss, percent ______.

Constituent	Vol., %	Comments	Reviewer Check
chrysotile			
amosite			
anthophyllite			
crocidolite			
tremolite-act.			
quartz			
perlite			
black rock wool*			
white rock wool*			
slag			
glass wool			
fiber glass			
vermiculite			
muscovite			
clay			
carbonate			
(or hydroxide)			
cellulose			
other organic			
gypsum			
diatoms			
portland cement			
other			

Constituent summary:

____

____

Petrographic Reviewer: ______________________ Comments: ______________________

Signature

*May be slag wool.

Figure 4. Asbestos Product Analytical Sheet. This form is used to minimize error in analyses of asbestos-bearing products that are used to determine their manufacturers.

cern can be translated into a civil penalty not to exceed $2,500 per day, 25 percent of which could be awarded to whomever brought the civil action in the public interest. Because California is a bellwether state, there is a high probability that this sort of law will spread to other states. The cost of quantification can be high, partly because drilling and sampling is expensive. Also, the cost of analysis is high because an analysis of a single sample for the full range of regulated organic chemicals can cost up to $4,500. In addition, documenting the movement of chemicals in the subsurface environment, particularly of minute traces, can be a complex and costly hydrogeologic problem; finally, the ever-present need for duplicate and confirming data can add orders of magnitude to the cost of supporting data. Clearly, defining such a situation in terms suitable for forensic work is very expensive. Obviously, collecting appropriate data and describing its significance requires the services of mature and experienced forensic geoscientists.

Erroneous or conflicting geological data can also cause problems by extending the time and cost of the forensic work involved in litigation. For example, a single inconsistent or incorrect result on one water sample can cast doubt on an entire data base used in the litigation. Such conflicting results could be caused by laboratory error or contamination during collection or shipment. Because discrepancies can be a nightmare for a forensic geoscientist or for the testing laboratory, duplicate sampling, spiked samples, and the use of standard samples is mandatory.

The philosophical justification for the group of scientific and technical disciplines directed toward environmental improvement would seem to be that a type of "environmental perfection" can be achieved. It is not at all certain, however, that the enormous expenditures of money and energy are justifiable in terms of environmental results.

SOME CURRENT FORENSIC CATEGORIES

There are several different surroundings in which the forensic geologist may have to present his findings. While the environments may differ legally, emotionally, and sociologically, the common thread is that the geologist must be expert in his field of testimony and thoroughly prepared to answer questions in depth. Although all testimony must be based on fact, the geoscientist will virtually always be asked to express his opinion as to the interpretation of the data. Experience and maturity are usually a requirement to make the conclusions of the geoscientist credible.

Litigation

Much has been written about the expert witness in litigation, and the periodical, *The Expert and the Law* (The National Forensic Center, since 1981), is a publication devoted to the subject. Guidelines for conduct during litigation have been written for geologists, and actually, similar guidelines have been written for engineers and many others who find themselves in court as witnesses. The guidelines are useful, but your attorney will generally tell you what you need to know in the way of procedures, presentation, and decorum. The bottom line usually is: listen to your attorney.

Relationship between the geoscientist and client. Although the geoscientist is representing one side in litigation, he should feel that in most cases he can represent either side in all matters of fact. Thus, in such things as the taking of mineral reserves in eminent domain proceedings, the quantity, quality, and value should be the same regardless of the side represented. However, it is basic that a witness should not do the other side's "homework," particularly where there are matters of interpretation.

The geoscientist should always have a clear picture of whom he is working for and who is billed. In most cases, the client is the lawyer, although billing may go directly to the lawyer's client. Ethically, the fees of witnesses cannot be contingent upon the outcome of a case, although a lawyer's fees frequently are. The expert witness's fees should be the same for all clients, and the witness's fee schedule and some estimate of cost should be made in advance. In addition, the client should be informed on a regular basis about the expenditures on his behalf.

Generally, communications are with the lawyer, and the forensic geoscientist is his witness. Suggestions directly from the lawyer's client are acceptable only if the lawyer agrees.

Ethically, the forensic geoscientist gives impartial, loyal, honest, professional service, with truth absolutely mandatory, not only for ethical reasons but because this is the law. The forensic geoscientist should realize, however, that even an honest and conscientious geoscientist may be considered by some to be an advocate of a position. In the event that a proceeding appears dishonest or you are asked to testify to things that you know to be untrue, disassociate yourself from the endeavor. Serving an unethical lawyer can be a professional disaster for a geologist (Kiersch and Cleaves, 1969; Dunn and Kiersch 1974).

As in all client relationships, discussion with others about the client's business is not ethical.

Gathering and presenting data. The general principles for gathering data have been previously described. It bears repeating that a detailed record of what was done, by whom, when, and why should be kept. The geologist should use relevant sampling procedures and sample-size reduction such as found in Cummins and Given (1973, p. 5–72). The witness should personally do a significant amount of the work, and necessarily delegated work, such as chemical analyses, gas chromatography, age-dating, engineering tests, or soundness tests for mineral aggregate should be under the witness's direction. The geoscientsit should have personal experience with the personnel or companies doing the tests. Even so, it is usually necessary to run duplicates or spiked samples in order to check the accuracy of the work.

A basic rule is that interpolation of information is usually acceptable; extrapolation is not.

Whatever the geoscientist does, the investigations and results should be reviewed and explained to the lawyer so that he understands the purpose and the methodologies used. Furthermore, the lawyer may have some useful suggestions about documentation and related matters that will improve the geological relevance to the case.

Preparation for trial. As preparation before a trial, a geological expert witness usually reviews with the attorney the scope and main body of testimony to be covered. The specific points to be brought out about a witness's experience or opinions are summarized, and the answers to expected questions are given as a dry run. Detailed, rehearsal-like review can be particularly important when the testimony is technically complex and therefore likely to be difficult for a judge or jury to understand. The attorney will usually outline and write out his sequence of principal questions designed to lead the judge and jury slowly and clearly through the technical complexities. Questions leading to vague answers or rambling by the witness are discouraged. This pre-appearance preparation by geologist and attorney is a critical factor for a strong court appearance. Most importantly, the geologist should refuse to appear in court if the attorney is not fully aware of his findings and of the thrust of his testimony. An uninformed lawyer in court can be a disaster for the geological expert/witness (Kiersch and Cleaves, 1969).

Pretrial discovery/examination. The purpose of the pretrial discovery (deposition) is to streamline court procedures by giving the litigants the opportunity to know the nature of opposing witnesses and their fundamental conclusions. Usually opposing lawyers, the witness and sometimes his counterpart, and a court reporter are present at an Examination Before Trial (EBT). The witness is sworn in, and his testimony becomes part of the court record. Although the atmosphere is less formal, the same rules of evidence prevail as in court.

There is, however, no direct examination as in court. The witness for one side is presented and the opposing attorney is free to ask him any questions relevant to the case. Generally, the witness's background is thoroughly explored first; then the attorney probes to determine the witness's position on various issues critical to the case. The attorney on the witness's side is free to object to what he considers improper questioning.

The preparation for the EBT is basically the same, so far as technical aspects are concerned, as for court appearances. It is axiomatic that a lawyer does not want to be surprised by his own witness; hence, the lawyer on the witness's side will try to thoroughly understand all of the witness's perspectives and findings likely to be given in testimony.

The trial. The courtroom environment is highly structured and has as its basic purpose the eliciting of relevant information from witnesses so that decisions can be made on controversial issues. A case may be tried before either a judge, or a judge and jury, whereas some technical cases are before a panel of three experts serving as judge and jury. In the former case, the witness talks to the judge, in the latter case, his audience is the jury. Each side presents its case through its witnesses in direct examination, and the opposing attorney has the right of cross-examination of each witness. A witness is bound by his EBT testimony: changes can be made later only for very strong reasons. Cross-examination varies from none to questions requiring several days.

There are many lists of admonitions for the expert witness, such as Kiersch and Cleaves (1969), Dunn and Kiersch (1975), and Waggoner (1981). The witness's lawyer will probably repeat many of them and will probably have admonitions of his own that are relevant to the particular situation. Similar admonitions are available for engineers, architects, or other professionals who get involved with forensic work. Anyone anticipating being an expert witness might find an article by Lance (1985) of considerable interest, because it describes a laywer's view toward expert witnesses along with suggestions for cross-examination techniques.

Legislative testimony

A very important area of forensic geology is testimony before legislative committees. Too often, testimony before legislative groups has been by politically motivated people rather than by technically competent professionals. Usually the busy professional would much rather avoid potential conflicts in favor of doing his job. This position is often reinforced when expenses for testifying are likely to be out of pocket.

The testimony may be in a casual manner between the expert and legislative staff members. Although the surroundings may be informal and a court reporter is usually not present, such an appearance can be very important, because the average legislator is often heavily influenced by his staff. Good preparation is important, and providing a written summary of the testimony can be very helpful to your presentation.

Similarly, testimony may be presented in a more formal manner to committees of legislators. Such testimony may be read, or the expert may speak from notes. In this case a reporter records everything that transpires, and the witness has a microphone for the presentation. Because many legislators are lawyers and few have difficulty with words, the witness should expect questions from legislators pertaining to the testimony. Some legislators may be competent technically and consequently may ask sophisticated questions, so be prepared. Generally, providing testimony before legislators is more relaxed than in a courtroom, but on highly partisan, emotional issues the witness may receive some sharp and demanding questions. If the witness is testifying on matters of fact, questions are usually not a problem. However, if the witness draws sweeping conclusions of a philosophical nature, extensive questioning should be expected.

Hearings before local boards or committees

Frequently, forensic geologists testify before city/town or country boards, planning and zoning boards, or similar local entities. These are political units at the grassroots level, and the hearing body, usually acting in the capacity of a judge, is sensitive to local opinion. Consequently, such meetings are rarely as tightly structured as courts. The normal courtroom decorum may be absent or lax, and judicial rules of evidence are often waived.

The geological witness, however, must still gather his data

and reach his conclusions in a professional manner not unlike that required for a court appearance. This is especially true because a common outgrowth of such hearings is litigation; therefore, statements by the witness at the hearing can be used in later court testimony. At most hearings, lawyers from opposing sides—if appointed—will be present.

Because such hearings lack normal courtroom constraints, and because many environmental issues are emotional, the hearing atmosphere can be tense. A description of the hearing environment from a mineral producer's perspective is given by Dunn and Hart (1984). Such hearings tend to attract those with the strongest opinions, and if several hearings are held, the audience may ultimately consist of only those with the strongest opinions. Consequently, an extreme polarization of positions is the frequent outcome. The political entity charged with judging the proceedings then has the problem of gauging the extent to which the testimony and mood of the audience at the hearings actually represents average public opinion. Issue decisions made under such circumstances are often overturned by later court rulings.

The forensic geoscientist must be prepared for the pressures exerted by an emotion-filled setting, particularly if he is supporting an unpopular position. Most important, the geologist should maintain composure, keeping in mind that the comments he offers may ultimately be used in a trial.

Administrative hearings

Forensic geoscientists may participate in administrative hearings sponsored by federal, state, or county regulatory agencies in which the hearing officer acts in a judicial capacity. Commonly the hearing officer will have only limited legal training. Although the atmosphere in such hearings is usually more casual than in a courtroom, the expert witness should prepare as much as for litigation in court. The expert witness is usually sworn in, and the proceedings are recorded. If conflict develops and litigation results, the witness's testimony at the hearing could be used in the court case. Ideally, even though research and preparation should be as thorough as for a court case, cost considerations may at times dictate otherwise. In such an event, the forensic geoscientist should be acutely aware of the facts and able to distinguish them from inferences and speculation; most importantly he should make certain that the hearing officer is cognizant of the distinctions in degree of certainty.

OVERVIEW

The benefits of forensic geoscience

Coates' (1976) statement that geomorphologists "can, should, and must be involved in the decision-making process" is equally true for engineering, ground-water, mining and other categories served by geoscientists. Yet for many years geologists were seldom involved in the public forum, largely because most geoscientists, neither by inclination nor training, could be comfortable in that environment. The input of more experienced geoscientists who were knowledgeable about the history of the Earth and its atmosphere, location and management of mineral resources, and knowledge of what is normal in water and rocks could have been helpful in supplying background data for many laws and regulations. The need for the input of competent professional geoscientists in litigation, whether concerned with engineering works or environmental issues, should increase for three basic reasons: (1) because the public seems to expect and accept ever more protection and control from government; (2) because of the extreme complexity of engineering and environmental disciplines; and (3) because of the increasingly litigious nature of our society. Moreover, the need for and importance of geological input is directly proportional to the quantity of proposed and passed legislation pertaining to environmental matters.

Who should be forensic geologists?

Obviously the practice of forensic geology is not for everyone. Geoscientists who are inexperienced or who cannot handle the highly charged atmosphere should avoid the field. If a geoscientist feels uncertain about a forensic situation, he should not participate. Geoscientists who measure their worth largely on the basis of the quality and number of contributions to basic research or on the basis of measurable economic or engineering activity may feel frustrated in serving the needs of forensic practice. Forensic work only rarely contributes to scientific knowledge or to the world's wealth. Consequently, many geologists, for these or other reasons, may feel that forensic geology is not an appealing endeavor.

Outlook for forensic geoscience

Understanding the nature and evolution of forensic geoscience allows some speculation about its future. The current nature of forensic geology was probably not predictable in the 1800s, nor could it have been clearly predicted as late as the 1960s. In fact, like many trends, its nature was not particularly clear to many who were in the midst of the revolution.

Forensic geoscience has two characteristics that partly define its nature: (1) its character and magnitude are always in response to national priorities as reflected by public attitudes, policies, laws, and regulations; and (2) it has evolved to such a status that it appears "recession-proof"—having a life of its own, that is, so far, independent of the economy. Further, we should realize that forensic geological work of the type practiced in the United States is largely an American phenomenon and is virtually absent in less developed nations. A geologist who had practiced for a considerable length of time in Africa, Asia, or South America, on returning to this country would find himself almost totally out of touch with a large segment of the geological work currently being done in the United States. In addition, environmental forensic geoscience in the European Common Market nations is a shadow of what it is in the United States, because their environmental regulations tend to be more related to the magnitude of the dangers involved. For example, asbestos regulations are far more lenient, and its removal from buildings is far less common. In addition, Europeans are also more sophisticated about radioactivity, making atomic power plants easier to licence and build; irradiated food is accepted and common in Europe.

Forensic geoscience has one other critical characteristic: it does not create wealth. In fact, quite the opposite, it is a major consumer of wealth: the magnitude of that consumption is often inversely proportional to the quantity of any measurable environmental or economic benefits. This offers one clue to the future of forensic geology as now practiced. A major financial disruption, such as a 1930s-type depression, could cause a national reordering of priorities, and many environmental programs could find themselves competing with an expanding poor population for funds. Many environmental regulations could simply go unenforced as national priorities change. In that case, of course, the amount of forensic geoscience would greatly decrease.

Four other developments could cause a slow change in the nature and amount of forensic geological activity. First, shifting the burden of proof to the plaintiffs in environmental cases would likely reduce the amount of environmental litigation. Second, making plaintiffs liable for damages if a lawsuit is found to be capricious would have a sobering influence. Third, higher quality public education in the sciences could reduce some of the current environmental hysteria that appeals to a scientifically naive public. (For example, if public misconceptions about chemicals or radioactivity were lessened, the amount of environmental litigation could decrease.) Fourth, legislators and government regulatory agencies could make use of risk-assessment analyses in which the risk of death or disease from various activities or influences is evaluated (Ames and others, 1987; Wilson and Crouch, 1987; Slovic, 1987; and Russell and Gruber, 1987). The implication of risk assessment analyses will be discussed at some length, because of its possible influence on the future of forensic geoscience.

Risk assessment. An internal memorandum in the EPA, quoted by Ray (1988, p. 2), stated, "Our priorities, [in regulating carcinogens] appear [to be] more closely aligned with public opinion than with our estimated risks." The greater use of risk-assessment analyses by the EPA and other regulatory agencies could help funding and legislative concern to be directed more toward solving high-risk problems than low-risk problems (i.e., there would be a cost-benefit relationship). For example, the asbestos-abatement industry market has been forecast as high as $100 billion (Krizan, 1988), and this does not consider the cost of 70,000 abestos-related lawsuits previously noted. However, Lave (1987, p. 292) suggests that because of the low risk of mesothelioma or lung cancer, of five per million lifetimes for the average exposure to asbestos in buildings, concentrating on reducing much higher risks (such as not wearing seat belts, smoking, or some personal habits of consumption) would produce far more health benefits for the money spent. Smoking, for example, is responsible for 307,000 deaths per million deaths; motor vehicles (U.S.) are responsible for 16,000 per million; and cycling, 750 deaths per million (Commons, 1985). Topping all of these causes of death appears to be diet—mostly too many calories—which accounted for two-thirds of the 2.1 million deaths in the United States in 1987 (Touferis, 1988).

To a very large extent, environmental regulations have erred on the "conservative" side so far as human safety is concerned. Even though this seems logical, regulations that reflect the philosophy that one molecule or particle of a carcinogen can kill can result in potentially insupportable environmental cost. The excessive high costs of compliance with environmental laws based on emotionalism rather than scientific evidence inevitably lead to economic inefficiency.

The high cost of society's trying to create a "perfect environment" has another aspect: history suggests that economic costs that lead to a decrease in the national material wealth will also ultimately lead to environmental costs. For example, only since about 1900, because of their wealth and high technology, have the industrial nations of the world been able to expand forests, increase wild game, clean their waters, and enjoy greater health and longevity. This is in sharp contrast with the situation in poor nations. Yet economic problems could reverse these environmental benefits of industrialization. Huge sums of money spent to solve minor environmental problems could, thus, ultimately contribute to a deteriorating environment.

Currently, it appears that applied risk assessment may have a good chance of having a major impact on the future of environmental regulations, laws, and expenditures. For example, courses in risk-assessment analysis are now taught at many colleges, and risk assessment is being used to a greater degree by governmental agencies. But the general public is not now well informed. A more knowledgeable public could cause an evolution in the nature of forensic geoscience, with the trend being toward more cost-effective analyses and remedies and, perhaps, toward a reduction in the amount of emotion that influences public decisions in the environmental area.

It should be noted, however, that the pressures against the expanded use of risk-assessment analyses are also strong, for example: (1) the political pressures of some environmentalists who are able to skillfully use (or misuse) modern communications media; (2) the political expediency of writing legislation based on emotional issues of the moment; (3) the driving needs for legislators to legislate and regulators to regulate; and (4) the apparent increasingly litigious nature of our society.

Legislation. Finally, forensic geoscience may be influenced by changes that may restrict the qualifications of expert witnesses, as a result of Maryland's Senate bill NO. 559 effective July 1, 1986 (Anonymous, 1986). The law states that, "an attesting expert must devote a minimum of 80 percent of his or her annual professional time to actual practice in the area of expertise." The purpose is to eliminate the too-often-used "hired-gun" type of witness in medical cases: commonly almost all of such an expert's time may be devoted to being a witness, and the rationale for the law is that they may lose contact with changes in practice. The law is controversial in that some of the most competent medical doctors and those most sought could be prevented from testifying. Were such a law to be applied to all experts and were such a law to be passed in many states, the problem for geological or engineering firms that are now largely environmentally oriented (and hence forensically oriented) could be severe.

The outlook for forensic geoscience practice will be governed by a balance (or imbalance) of the above factors—and, probably, other factors not visualized. The current strong trend toward more laws and more regulations has long been with us. It appears that the trend will continue. As a consequence, the field of forensic geoscience should continue to flourish—at least for a few years.

REFERENCES CITED

Ames, B. N., Magaw, R., and Gold, L. S., 1987, Ranking possible carcinogenic hazards: Science, v. 236, p. 271–280.

Anonymous, 1986, Limitations on expert testimony set by new Maryland law: The Expert and the Law, v. 6, no. 2, p. 1–2.

——, 1988, Internal rifts closing asbestos claims group: Engineering News Record, June 30, p. 8.

Black, B., 1988, Evolving legal standards for the admissibility of scientific evidence: Science, v. 239, p. 1508–1520.

California Safe Drinking Water and Toxic Enforcement Act, 1986, *in* California Health and Safety Code Division 20, Miscellaneous Health and Safety Provisions, Chapter 6.6, CHSC 2524.5, et seq.

Coates, D.R., 1976, Geomorphology in legal affairs of the Binghamton, New York, metropolitan area, *in* Coates, D. R., ed., Urban geomorphology: Geological Society of America Special Paper 174, p. 111–148.

Crawford, C. B., and Eden, W. J., 1964, Nocolet landslide of November 1955, Quebec, Canada, *in* Trask, P. D., and Kiersch, G. A., eds., Engineering geology case histories no. 4: Geological Society of America Engineering Geology Case Histories Numbers 1–5, p. 229–234.

Commins, B. T., 1985, The significance of asbestos and other mineral fibres in environmental ambient air: Maidenhead, Commins Associates, 1 p.

Cummins, A. B., and Given, I. A., eds., 1973, Mining engineering handbook, v. 1: American Institute of Mining, Metallurgical, and Petroleum Engineers Society of Mining Engineers, 1185 p.

Dunn, J. R., and Hart, M. J., 1984, Public hearings and the mine permitting process; Mineral aggregates: Washington, D.C., National Research Council Transportation Research Board Transportation Research Record 989, p. 7–9.

Dunn, J. R., and Kiersch, G. A., 1974, The professional geologist as an expert witness, *in* Suggested practices and guides: American Institute of Professional Geologists, 19 p.

EPA, 1982, Environmental monitoring at Love Canal, v. 1: Washington, D.C., U.S. Environmental Protection Agency Office of Research and Development, 290 p.

——, 1985, Guidance for controlling asbestos containing materials in buildings: Washington, D.C., U.S. Environmental Protection Agency Office of Pesticides and Toxic Substances, 118 p.

Foose, R. M., 1969, Mine dewatering and recharge in carbonate rocks near Hershey, Pennsylvania, *in* Kiersch, G. A., and Cleaves, A. B., eds., Legal aspects of geology in engineering practice: Geological Society of America Engineering Geology Case Histories 7, p. 45–60.

Glaubinger, R. S., and others, 1979, Love Canal aftermath: Chemical Engineering, v. 86, no. 23, p. 86–92.

James, L. B., 1968, Failure of the Baldwin Hills Reservoir, Los Angeles, California, *in* Kiersch, G. A., ed., Engineering geology case histories no. 6: Geological Society of America Engineering Geology Case Histories 6–10, p. 1–12.

Kedlecek, M., 1988, Love Canal, 10 years later: The Conservationist, November/December 1988, p. 40–43.

Kiersch, G. A., and Cleaves, A. B., eds., 1969, Legal aspects of geology in engineering practice: Geological Society of America Engineering Geology Case Histories 7, 112 p.

Krizan, W. G., 1988, Asbestos: Hazard and hysteria: Engineering News Record, June 2, 1988, p. 20–24.

Lance, M., 1985, The law and Herald Fahringer: The Expert and the Law, v. 5, no. 3, p. 2–5.

Lave, L. B., 1987, Health and safety risk analyses: Information for better decisions: Science, v. 236, p. 291–300.

Mann, J. F., Jr., 1969, Groundwater management in the Raymond Basin, California, *in* Kiersch, G. A., and Cleaves, A. B., eds., Legal aspects of geology in engineering practice: Geological Society of America Engineering Geology Case Histories 7, p. 61–74.

McClure, C. R., Jr., and Hatheway, A. W., 1979, An overview of nuclear power plant siting and licensing, *in* Hatheway, A. W., and McClure, C. R., Jr., eds., Geology in the siting of nuclear power plants: Geological Society of America Reviews in Engineering Geology, v. 4, p. 3–12.

Murray, R. C., and Tedrow, J.C.F., 1974, Forensic geology, earth science, and criminal investigation: New Brunswick, New Jersey, Rutgers University Press, 217 p.

Paschall, R., 1984, The appraisal of mineral-producing properties: American Society of Appraisers, v. 29, no. 2, p. 56–61.

Peele, R., 1966, Mining engineers handbook, 3rd ed., v. 2: New York, John Wiley and Sons, (includes Mining Act of 1872), 1001 p.

Ray, D. L., 1988, Who speaks for science?: Imprimus, v. 17, no. 8, p. 1–7.

Reserve Mining Company and others v. U.S.A. and others, 1974–1975, U.S. Court of Appeals for the 8th Circuit, St. Louis, Missouri, no. 74-1291 Civil, memoranda, and orders.

Rickard, T. A., 1932, A history of American mining: New York, McGraw-Hill Book Company, American Institute of Mining Engineering Series, 419 p.

Russell, M., and Gruber, M., 1987, Risk assessment in environmental policy making: Science, v. 236, p. 236–290.

Sales, R. H., 1964, Underground warfare at Butte: Caldwell, Idaho, The Carton Printers, Ltd., 77 p.

Slovic, P., 1987, Perception of risk: Science, v. 236, p. 280–285.

Supp, C.W.A., 1969, Maryland State Roads Commission versus Champion Brick Company; Condemnation of clay-bearing lands, *in* Kiersch, G. A., and Cleaves, A. B., eds., Engineering geology case histories, no. 7: Geological Society of America Engineering Case Histories 6–10, p. 95–112.

Tank, R. W., 1983, Legal aspects of geology: New York, Plenum Press, 583 p.

Toufexis, A., 1988, The food you eat may kill you: Time, August 8, p. 66.

U.S. Department of Labor, Occupational Safety and Health Administration, 1986, Occupational exposure to asbestos, tremolite, anthophyllite, and actinolite, final rules, Part II 29 CFR, parts 1910 and 1926: Federal Register, v. 51, no. 119, 178 p.

U.S. Department of Labor, 1987, Code of federal regulations, Title 29: Washington, D.C., U.S. Government Printing Office, 8 vols., 4,725 p.

U.S. House of Representatives, 1983, Current status of the Three Mile Island nuclear generating station, units 1 and 2: Committee on Interior and Insular Affairs Oversight Hearing, serial 98–10, 211 p.

Waggoner, E. B., 1981, The expert witness: Association of Engineering Geologists Bulletin v. 28, no. 2, p. 147–150.

Wilson, R., and Crouch, E.A.C., 1987, Risk assessment and comparisons; An introduction: Science, v. 236, p. 267–270.

Woods, H. D., 1964, Causes of the Sear's Point landslide, Sonoma County, California, *in* Trask, P. D., ed., Engineering geology case histories, no. 2: Geological Society of America Engineering Geology Case Histories 1–5, p. 107–109.

MANUSCRIPT ACCEPTED BY THE SOCIETY MARCH 20, 1989

ACKNOWLEDGMENTS

This chapter has benefited from the thoughtful review and comments of Donald B. Coates and the input of background materials and comments on earlier drafts by George A. Kiersch.

Printed in U.S.A.

Index

[Italic page numbers indicate major references]

Typeset by WESType Publishing Services, Inc., Boulder, Colorado
Printed in U.S.A. by Malloy Lithographing, Inc., Ann Arbor, Michigan